Kent and Riegel's
HANDBOOK OF INDUSTRIAL CHEMISTRY AND BIOTECHNOLOGY

ELEVENTH EDITION

Kent and Riegel's
HANDBOOK OF INDUSTRIAL CHEMISTRY AND BIOTECHNOLOGY

Volume I

ELEVENTH EDITION

Edited by
James A. Kent, Ph.D.

 Springer

James A. Kent
Professor of Chemical Engineering and Dean of Engineering
Kentjamesa@aol.com

ISBN: 978-0-387-27842-1   e-ISBN: 978-0-387-27843-8

Library of Congress Control Number: 2005938809

Cover illustration: Abigail Kent

Printed on acid-free paper

9 8 7 6 5 4 3 2 1

springer.com

To my Wife
ANITA

Preface

The central aim of this book is to present an up-to-date account of the science and engineering and industrial practice which underlie major areas of the chemical process industry. It attempts to do so in the context of priorities and concerns which characterize the still early days of the new millennium and, perhaps more important, it provides various tools for dealing with those factors through, for example, an extensive discussion of green engineering and chemistry and related topics. The heart of the book is contained in twenty eight chapters covering various areas of the chemical process industry. It is to be noted that the products and processes associated with a particular area are discussed in the context of the corresponding chapter rather than in the isolated manner characteristic of an encyclopedia.

This work, Kent and Riegel's Handbook of Industrial Chemistry and Biotechnology, is an outgrowth of the well known Riegel's Handbook of Industrial Chemistry, the last edition of which, the tenth, was published in 2003. It follows the essential arrangement of earlier versions, i.e., several chapters devoted to general or "infrastructure" topics, with most of the book being given over to the various areas of the chemical process industry. However, this version introduces a wealth of new, timely, and very useful "infrastructure" material, and greatly enhances the process industry content. (The latter is most noticeable in this book by increased emphasis on biotechnology, although all of the chapters have been reviewed and updated as necessary by their respective authors.) In keeping with past practice, all of the new chapters have been written by individuals having demonstrated expertise in their respective fields. All told, the work may in many respects be regarded as a sourcebook for practice in the chemical process industries.

Concerning the infrastructure or contextual material mentioned above, the Handbook contains three new chapters which lie in the area often referred to as "green chemistry". The first and most comprehensive of these is titled Green Engineering: Integration of Green Chemistry, Pollution Prevention and Risk Based Considerations. It provides an excellent guide for applying the methods of green chemistry and engineering to process and product development activities, whether for new products and processes, or for upgrading older ones. Written by a team of experts in the field, the chapter can be of enormous help to all practicing chemists and chemical engineers, as well as to students studying in either discipline. Another new chapter, Industrial Catalysis; A Practical Guide, is a valuable adjunct to the "Green" chapter since catalysis is an important aid in the practice of Green Chemistry. The third new chapter in what might be termed the "green" group is Environmental Chemical Determinations.

Succinctly put, *green chemistry,* also termed *sustainable* chemistry, is described by that chapter's authors, as "the use of chemistry to reduce pollution at the source, through the design of chemical products and processes that reduce or eliminate the use or generation of unwanted or hazardous substances." *Green engineering* is defined as "the design, commercialization, and

use of processes and products that are feasible and economical, yet at the same time minimize
1) generation of pollution at the source, and 2) risk to human health and the environment." Risk
assessment methods used in pollution prevention can help quantify the degree of impact for
individual chemicals and thus is a valuable tool for intelligent design of products and processes
by focusing on the most beneficial methods to minimize risk.

Even a superficial look at the literature on green chemistry shows that catalysis is regarded
as a very important tool. After all, if in the idealized case one can produce desired product B
from A, with no unwanted side reactions or by-products, by choosing appropriate reaction conditions and a suitable catalyst, one will have done a great deal to promote efficiency and prevent pollution. Therefore, another of the new chapters, Industrial Catalysis, a Practical Guide,
is of special relevance. Finally, this particular portion of the new material is rounded off with
the chapter Environmental Chemical Determinations, which discusses the many complex factors involved in detecting, tracking, and measuring chemical species which have found their
way into the environment.

Additional chapters in the grouping broadly referred to as infrastructure include the new
Recent History of the Chemical Industry: 1973 to the Millennium: and an update of the chapter titled Economic Aspects of the Chemical Industry, in which some of the material extends
information provided in the former. Rounding out the infrastructure group are yet another new
chapter Nanotechnology: Principles and Applications, together with the earlier ones which
cover such diverse and fundamental topics as process safety, emergency preparedness, and
applied statistical methods.

Biotechnology first appeared in the Riegel's Handbook some time ago as a chapter titled
Industrial Fermentation. It has since been updated several times and more recently was joined
by a chapter on Industrial Cell Culture. For this Handbook, the biotechnology content, rather
than being updated, has undergone a major reorganization, including revision of content and
emphasis. The former fermentation chapter has become two which are titled, respectively,
Industrial Biotechnology: Discovery to Delivery, and Industrial Enzymes and Biocatalysis.
This revision was accomplished by two teams from a major biotech company and thus reflects
that background. It is informative to interpose at this point a statement (edited) by the authors
of the first of these two chapters. They describe it thus: *"The chapter uses an approach to integrate gene discovery, functional genomics, molecular evolution and design, metabolic pathway
engineering, and production processes including formulation of delivery systems. The chapter
walks the reader through biomolecule discovery, development and delivery, by starting from
screening millions of natural and designed gene variants in the mountains of DNA sequences
available today. Also included are several state-of-the-art examples of purposeful modifications of cellular metabolism, and descriptions of unit operations and unit processes which link
the upstream and downstream technologies to manufacture biochemicals, enzymes, peptides,
and other products on an industrial scale. "Commercializing new bioproducts is a complex,
time consuming process, and therefore an integrated biotechnology approach is necessary. It
is the authors' hope that the chapter will help readers learn how to design and produce biotechnology products rapidly and successfully."* Revision of the cell culture chapter was accomplished by a team from another biotech company. The new title, Industrial Production of
Therapeutic Proteins: Cell Lines, Cell Culture and Purification, reflects its new content and
orientation. Also, as might be expected by persons knowledgeable in the field, the chapter
Animal and Vegetable Oils, Fats and Waxes is rich in related biotechnical content, as is effectively described in the chapter's early pages.

Finally, addressing an area of great interest in connection with world energy needs, we have
added a chapter in a related area, Biomass Conversion. Written by a team whose primary work
lies in that area, it provides comprehensive coverage of the subject from biomass structure and
composition to thermochemical and biological routes for conversion to energy and a host of

chemicals and products including liquid transportation fuels. This chapter defines the opportunity for using sustainable sources of biomass as feedstock for new refineries that will produce fermentable sugars and chemical intermediates from which much needed forms of fuels can be made.

As mentioned earlier, the crux of the Handbook comprises twenty eight chapters which are devoted to various areas of the chemical process industry. This information, together with supporting "infrastructure" material described above, viz., process safety, emergency preparedness, statistical methods, green engineering and chemistry, provides *in toto* many sophisticated and useful tools to aid in the design of new products and processes and for study and evaluation of older ones. The handbook should prove useful also to individuals who possess a background in chemistry or chemical engineering and work in related areas such as regulatory agencies and environmental organizations. Among other benefits, it will help ensure that the work of such individuals reflects knowledge of relevant contemporary science and engineering and industry practices. Reflecting new realities in the world energy situation, this edition also includes a chapter titled The Nuclear Industry.

Individuals who have responsibilities in the chemical process industries are usually engaged, consciously or otherwise, in continually reviewing their operations to ensure that they are safe, efficient, and in compliance with current environmental regulations. They are also, or should be, anticipating future needs. It is hoped that the information contained herein will provide the wherewithal by which chemists, chemical engineers, and others who have a peripheral interest in the process industries, for whatever reason, can ensure that they have touched every base, dotted every *i*, and crossed every *t* in their quest to make the processes and products for which they are responsible as environmentally sound, safe, and efficient as possible.

Because of the scope of the book and the large number of products and processes it covers, some redundancy is inevitable. For example, more than one chapter includes discussions of gasification and hydrogen production. However, there are significant differences in emphasis in the various discussions. Thus, rather than distract readers by referring them to information in locations other than the one of their primary interest, such topics have been left intact in the context in which they are discussed.

As in all the earlier versions of this work for which I have been privileged to serve as designer and editor, I am happy to acknowledge again the unselfish and enthusiastic manner in which the contributing authors have shared their knowledge and insights so that many others may learn and still others may benefit. The picture of a bit of knowledge, acting like a stone tossed into a quiet pond, spreading the result of the impact ever more widely, is, I think, apt. There is a saying that knowledge is power, and the authors who have contributed their knowledge and expertise to this work are pleased to have had the opportunity to empower others. All have been unstinting in their efforts to make their contributions as complete and informative as possible, within the space available, and I am indeed humbled and honored to have had a part in bringing it about. Needless to state, errors of omission and shortcomings in organization are mine.

Grateful acknowledgement is made to the publishing houses and technical/scientific organizations for permission to reproduce copyrighted illustrations, tables, and other materials, and to the many industrial concerns which contributed drawings, photographs, and text material. And finally, I wish to express my thanks to Springer editor, Dr. Kenneth Howell, for his many helpful suggestions and support along the way, and for leveling several bumps on the road to publication.

James A. Kent
Morgantown, West Virginia USA

Contents
Volume I

Volume II

1

Recent History of the Chemical Industry* 1973 to the Millenium: The New Facts of World Chemicals Since 1973

Fred Aftalion

I. OVERCAPACITIES AND THE SEARCH FOR REMEDIES

The first oil shock that occurred at the end of 1973 with the Yom Kippur war served to pinpoint the crisis which world chemicals were already undergoing.

The chemical industry's soaring development after the war was due to the extraordinary burst of innovations occurring between 1935 and 1955 and coinciding with an explosion of world demand in a variety of sectors served by chemicals. Production units multi-

*This chapter consists of two chapters taken from a book by Dr. Fred Aftalion. *A History of the International Chemical Industry,* Second Edition, translation by Otto Theodor Benfy, Copyright © the Chemical Heritage Foundation, Philadelphia, PA (2001). This material is reprinted by permission of the copyright owner and Fred Aftalion. All rights reserved. The book traces the development of the Industry from its earliest days, describing the activities of the pioneers of chemical science and the entrepreneurs who built on their work to create the chemical industry as we know it. Space limitations permit the inclusion of only Chapter 6. "World Chemicals Since 1973," and Chapter 7, "The Period of the 1990s." Noteworthy changes that have occurred in the industry since 2000 are mentioned in the following chapter, "Economic Aspects of the Chemical Industry."

plied in Europe as well as in the United States and Japan.

Two other factors contributed to this rapid growth. The use of oil as a substitute for coal provided the chemical industry with abundant, cheap raw material that was easy to transport. With interest rates lagging behind the rate of monetary erosion over a number of years, industry leaders were tempted to carry out investments that they would not have made had currencies remained stable and interest rates higher. The fear of these leaders that competition would get the better of them if they slowed down their investments, the race for market shares advocated by a number of consultant firms like the *Boston Consulting Group*, the belief—quite widespread among world chemicals leaders—that they had to keep building new units to keep up with forecast needs, all had a share in building up production overcapacities which were already becoming apparent before 1973 in certain sectors of heavy chemicals (petrochemicals, synthetic fibers, thermoplastics, and fertilizers).

The establishment of an OPEC cartel that led to a rise in the price of a barrel of crude oil from $3 to $12, then the 1979 Iranian

1

Revolution which made it soar to $40, and finally the publication of the gloomy forecasts of the Club of Rome experts which mistakenly saw oil shortages ahead when, in fact, these had been artificially engineered by the Cartel members—all these facts upset chemical leaders in industrialized countries. And yet some of them still continued to invest in new plants during the stock-building lulls that occurred in 1974 and 1979 through consumers' speculating on new price rises.

This only made the necessary adjustments much harder when they had to be carried out at the beginning of the 1980s. Companies were to suffer greatly from an error of judgment, building new plants at great expense at the same time that economic growth rates tumbled from over 10 percent to a mere 2 to 3 percent. Caught between the increasing cost of their hydrocarbon raw materials and the ever-lower prices they had to use to sell their products in markets where offer exceeded demand, leading chemical companies in industrialized countries were forced to go through agonizing reappraisals.

This led them to act in a number of different directions. First and foremost, they had to lower their operating costs by cutting down on excess personnel and taking the measures needed to increase the productivity of each company. At the same time, they had to reduce, in a concerted way if possible, the overcapacities affecting the hardest-hit sectors. Finally, it seemed advisable to redirect production into areas that were less sensitive to economic change. This meant increasing the share of specialties in relation to commodities in overall turnover.

A new generation of leaders was called upon to carry out the socially painful and politically delicate job of rationalizing and restructuring the chemical industry through layoffs and plant closures. These same leaders were also given the more exalting, but just as difficult, task of defining the redeployment strategy that needed to be followed and of determining on a case-by-case basis the sectors that should be abandoned and those that, on the contrary, had to be invested in force.

By 1973, it was obvious that the chemical industry had reached a degree of maturity to the extent that all the companies involved in that area in industrialized countries were long established and that no discovery likely to affect its development had been made over the last two decades. While new areas of research like composite materials and biotechnologies had emerged, no immediate fallout was expected for a number of years. Thus failing any rapid internal growth brought about by major scientific breakthrough, the strategy of leaders anxious to refocus or diversify their portfolio of activities very often consisted of a kind of Monopoly game, as a range of production was shifted from one enterprise to another without anything new being created.

THE RESTRUCTURING OF SECTORS IN DISTRESS

Priority action was required in petrochemicals, in the large thermoplastics, in fertilizers, and in synthetic fibers where the most serious investment mistakes had been made. The hardest cases were those of *petrochemicals* and *thermoplastics*. For one thing, a steam cracker cannot technically operate under 60 percent of its capacity. For another, the products that emerge are linked to one another in almost invariable proportions. Finally, a polymerization unit cannot have its pace slowed down without this affecting the upstream monomer unit to the same extent.

In addition to such rigidities, there was the need to reduce not only the quantities produced but also the number of production units. The problem then arose of sharing the sacrifices among the different producers within an economic area.

The problem was most easily solved in Japan because of the discipline which MITI managed to establish within the country's petrochemical industry. Making the most of a new law that allowed competing producers to act in concert, a cartel was set up with the object of cutting down ethylene production. Four groups of petrochemical producers were formed within which the necessary arbitrations took place. This led Sumitomo to close its Niihama units, Mitsubishi a number of its

Mizushima plants, and Showa Denko two of its Ohita installations.

At the same time, producers reached agreements on cutting down competing PVC and polyolefin sales networks, while MITI authorized the import of naphtha through an organization consisting of Japan's petrochemical producers. Its price served as a marker for naphtha produced in Japan.

In Europe, of course, it was difficult to show such disregard for market laws. The views of the European Economic Community Commission in Brussels had to be taken into account, and they upheld the principle of free competition as set down in article 85 of the Rome Treaty. Moreover, in Western Europe there were a number of petrochemical industries that operated according to the rules of private capitalism while there were others, as in France, Italy, Austria, Norway, and Finland, that were state-controlled and more concerned about retaining market share than ensuring profitability.

Despite such obstacles, unilateral decisions were taken and bilateral arrangements carried out among firms, leading to some measure of production rationalization. Between 1980 and 1984, twenty-five ethylene and eight polyethylene units were scrapped in Western Europe while ethylene oxide capacities were reduced by 10 percent.

The 1983 agreement between ENI and Montedison put some order in Italy's chemical industry, as ENI took over the PVC and polyethylene operations of Montedison. Previously in France, Rhône-Poulenc had sold its petrochemicals division and its thermoplastics to the Elf Aquitaine group. At the same time, steam crackers were being shut down in Feyzin and Lavera, and a vinyl chloride unit in Jarrie. The association between BP Chimie and Atochem in polypropylene and the exchange of Atochem's Chocques unit for ICI's Rozenburg polyethylene unit were other instances of rationalization.

The Brussels Commission also gave its approval to three large-scale operations: the ICI and BP Chemicals exchange of polyethylene/PVC, the vertical integration of vinyl chloride involving AKZO and Shell Chemicals, and the recent Enichem and ICI association, which produced European Vinyls Corporation and was intended to lead to major capacity cuts in PVC.

In West Germany, rationalization measures were less spectacular because the heads of Germany's leading chemical companies had not waited for the crisis to delineate their respective fields of operation and to establish close links with international oil companies, either through long-term supply contracts or through parity associations.

A number of American companies became involved in restructuring. Union Carbide sold its Antwerp site to BP Chemicals; Monsanto, its Seal Sands acrylonitrile unit to BASF; Esso, its Stenungsund steam cracker to Statoil; while Hercules joined up with Montedison to set up the Himont company, which accounted for 20 percent of the world polypropylene market.

In the *United States* the petrochemical industry set its house in order along purely capitalistic lines. Each company involved acted alone for fear of infringing antitrust legislation and the main concern was to restore profitability. Unlike Du Pont, which acquired Conoco, other chemical companies tried to get rid of their petrochemicals. Hercules sold its DMT units to Petrofina, and subsequently its 40 percent stake in Himont to Montedison, while Monsanto was shedding its Texas City petrochemical site.

Major divestments took place, particularly in the major thermoplastics, which were taken over by individual entrepreneurs who bought up the units the chemical giants wished to get rid of. As Hoechst, Union Carbide, Du Pont, Monsanto, ICI, and USS Chemicals withdrew from a number of the major oilbased intermediates as well as from polystyrene, polyethylene, and PVC, a number of large, hitherto unknown companies emerged: Huntsman Chemical, El Paso Products, Aristech, Vista Chemical, Sterling Chemicals, and Cain Chemical.

At the same time, oil companies were integrating downstream petrochemicals and polymers. Such was the case of Occidental Petroleum, which through its chemical subsidiary Hooker (later Oxychem) bought up

Tenneco and Diamond Shamrock's PVC in 1986, becoming the largest American producer in this area. Likewise, BP Chemicals fully acquired its subsidiary Sohio. The long-standing petrochemical divisions of the large oil groups returned to profits in 1986 after some painful tidying up but no agonizing reappraisal, helped along by falling oil prices and dollar rates.

Most of them had cut down on operating costs and diversified to the point at which they were able to face up to the economic ups and downs without too much apprehension. Productivity improvements and a better utilization of existing capacities because of higher demand put Exxon, Mobil, and Texaco on the way to prosperity in petrochemicals in 1986.

Standard Oil of California added the petrochemicals of Gulf Oil, purchased in 1984, to its subsidiary Chevron Chemical. Other United States petrochemical producers took advantage of special circumstances. Amoco was served by a strong terephthalic (TPA) base and its good performance in polypropylene; Arco, by its Lyondell subsidiary in Channelview, Texas, and by its development of the Oxirane process through which propylene oxide could be produced by direct oxidation with styrene as a coproduct. The process also led to MTBE (methyl tertiary-butyl ether), the antiknock agent used as a substitute for tetraethyl lead.

Even Phillips Petroleum, badly affected by Boone Pickens' takeover attempt, managed to make substantial profits from its petrochemicals because of drastic restructuring. New prospects were also opening up for the United States chemicals industry as needs grew for butene and hexene comonomers used to produce linear low-density polyethylene (LLDPE), also as consumption of higher olefins to prepare detergent alcohols increased and as demand for MTBE used as a gasoline additive soared.

The problem of overcapacities in *chemical fibers* in each economic region was both easier to overcome because of the small number of producers and more complicated because of outside factors. In *Europe*, producers suf-

fered heavy losses from 1973 onward. For one thing, the Europeans were not particularly suited to manufacture chemical fibers at satisfactory cost, a fact that was proved by growing imports from Southeast Asia. For another, the capacity increases decided upon did not tally with any comparable increase in demand in the foreseeable future.

In view of such imbalance, one might have thought that a number of producers would withdraw from the market. But this did not happen because some of them had to heed government instructions to maintain employment. Also textiles accounted for only a share of the business of the companies involved and could be kept up through the profits generated in other areas. From 1978 to 1985 two agreements were implemented with the blessing of the European Economic Community Commission. The first aimed for a linear reduction of existing capacities; the second and more important one allowed each producer to specialize in those areas where it held the best cards, giving up what amounted to marginal productions.

Thus Courtaulds withdrew from polyester and from Nylon to concentrate on its acrylics and cellulose fibers; ICI focused on Nylon and Bayer on acrylics; Rhône-Poulenc withdrew from acrylics but revamped its Nylon and polyester units well-integrated in upstream intermediates; Montedison decided in favor of polyester and acrylics; AKZO focused on polyesters and on aramide fibers while keeping up its profitable rayon sector. Such efforts, which aimed to reduce European chemical fiber capacities by 900,000 tons and to increase productivity through specialization, undoubtedly corrected the situation.

Nonetheless, European producers are still faced with two kinds of competition: first imports of synthetic fibers from Turkey, Taiwan, South Korea, and Mexico, against which it is hopeless to expect that the multifibers agreements —which contravene GATT rules—will constitute a permanent obstacle; and second, imports of natural fibers such as cotton, for which prices have fallen spectacularly in recent times.

The Japanese solution to chemical fiber overcapacities naturally involved MITI which pushed through a 17% cut in existing polyester, Nylon filament, and acrylic fiber capacities between 1978 and 1982. These were linear cuts, however, and did not restrict the range of synthetic fibers developed by each producer, contrary to the specializations that marked the second stage of Europe's approach.

The United States was faced with an additional problem because its market remained wide open to textile imports from developing countries. These imports constituted an indirect threat to American producers of chemical fibers. Their first reaction was to reduce their bases in Europe. Du Pont closed its acrylic units in Holland in 1978 and in Northern Ireland in 1980; the following year it ceased production of polyester thread in its Uentrop unit in Germany. Monsanto did likewise in 1979, shutting down its Nylon units in Luxembourg and Scotland and selling its acrylic fiber installations in Germany and Ireland to Montedison.

In the United States itself, capacity cuts were not so substantial and the 1983 upturn boosted utilization of remaining units to 80 percent of their capacities. Major American producers such as Du Pont, Celanese, and Monsanto returned to satisfactory profit margins. Other companies for which fibers were not an essential sector withdrew from this area. Chevron Chemical, for instance, shut down its Puerto Rico Nylon and polypropylene fiber units between 1980 and 1982 as well as the polypropylene fiber unit in Maryland.

The *fertilizer* market was in no better shape than the petrochemicals and chemical fibers markets, for world producers had largely allowed supply to exceed demand.

The situation in this area was further complicated by the unequal distribution worldwide of the raw materials required to produce fertilizers and the special attention which governments bestowed on agriculture. Such attention had led to a surfeit of production units and their increasing control by governments, either directly through taking a stake in the companies concerned, or indirectly through establishing ceiling prices for home sales or export subsidies. The emergence of new producers in Eastern countries and in developing areas increased the share of state-controlled companies in world production from 30 to 64 percent for ammonia, from 40 to 65 percent for potash, and from 10 to 46 percent for phosphoric acid between 1967 and 1986.

In Western Europe, nitrate fertilizer producers had deemed it expedient to set up a cartel arrangement for exporters called Nitrex. But the collapse of demand in countries outside its area had prevented it from functioning properly, sparking a fight for market shares even within the community.

As a country like Morocco switched from its long-established role as phosphate exporter to downstream ammonium phosphate and superphosphate integration, traditional fertilizer producers were forced to reappraise their strategy and take severe rationalization measures.

Japan which had none of the required raw materials and, accordingly, had high production costs, began, as early as the 1970s, gradually to cut down capacities along the lines jointly agreed upon by the authorities and the five main Japanese producers of nitrate and phosphate fertilizers.

In *Europe*, the pressure of events disrupted the whole market as the number of producers was drastically reduced. Because of market proximity, production both from Eastern Europe of nitrates and from Africa of superphosphates were becoming dangerously competitive. Supply conditions for natural gas varied according to each country's policies. France, for instance, agreed to pay extra for Algeria's gas, while Holland's Groningen gas, which Dutch ammonia producers were getting at a very favorable price, was linked to the price of petroleum products. On the other hand, a number of Scandinavian state-controlled companies like Norsk Hydro and Kemira, were pushing ahead with ambitious fertilizer programs, taking advantage of their interests in North Sea oil or of the conditions under which they were being supplied with oil and gas from the Soviet Union.

Between mergers and acquisitions, the structure of the fertilizer industry in Western Europe was spectacularly pared down. A few giants emerged to dominate the market. In France, there was CdF Chimie, later to be known as ORKEM, which had just taken a 70 percent stake in Air Liquide's subsidiary La Grande Paroisse, and Cofaz, which was taken over by Norsk Hydro; in Western Germany, there was BASF and Ruhr Stickstoff; in Britain, ICI and Norsk Hydro, which bought up Fisons; in Italy, ANIC and Montedison's subsidiary Fertimont; in Holland, DSM's UKF and Norsk Hydro NSM; in Finland, Kemira, which took over both Britain's Lindsay and its Kesteven facilities.

But the scene has not yet become sufficiently clear, since the competing companies do not all enjoy, within the community, the same raw materials supply conditions, and Europe is still open to imports from other countries that do not apply the rules of market economy.

In the *United States*, the situation was in many ways different. With its large sulfur, natural gas, phosphate, and even potash resources, America's fertilizer industry rested on a sound base. It was an exporter of minerals and fertilizers, and did not have to worry to the same extent as Europe's industry about competing imports from Socialist countries. But reserves of sulfur extracted by the Frasch process have been depleted in Louisiana and Texas, and President Ronald Reagan's "payment in kind" (PIK) farm-acreage cuts reduced the fertilizer requirement of American farmers. These farmers are also much in debt and are having trouble selling their products on saturated markets.

Consequently, very little money has been sunk into extracting phosphate rock in Florida or in increasing nitrogen fertilizer capacities, for a new ammonia and urea unit can cost as much as $250 to $500 million to build in the U.S., depending on the state of the existing infrastructure.

With such dim market prospects, it is understandable that W. R. Grace has decided to shut down its Trinidad ammonia unit, or that a company as large as International Mineral Chemicals has tried to diversify through purchase of Mallinckrodt and has put half its fertilizer assets up for sale.

THE NATIONALIZATION OF FRANCE'S CHEMICAL INDUSTRY

When a left-wing government came to power in 1981, France's chemical industry was in dire straits judging from the losses of the major groups: CdF Chimie was losing 1,200 million francs; Péchiney Ugine Kuhlmann 800 million francs; Rhône-Poulenc 330 million francs; Chloé Chimie 370 million francs; Atochimie 130 million francs; and EMC 100 million francs. Admittedly world chemicals were in poor shape. But while French leaders were posting losses amounting to 7 to 10 percent of their turnover, Hoechst and BASF were still making consolidated profits that year of 426 million DM and 1,290 million DM, respectively, even though they had noticeably slumped.

There were many reasons, some of them old, for the difficulties of France's chemical industry as illustrated by losses of 7 billion francs in seven years—4 billion francs in 1981 alone. Caught between increasingly heavy charges and price controls on the home market, France's chemical entrepreneurs never managed after the war to achieve sufficiently profitable margins. They ran up high debts to make up for their lack of funds, building up ever heavier financial costs.

A further disadvantage of France's chemical industry was its scattered production sites, originally due to the need during the two World Wars to keep plants far from the battlefields. For both social and political reasons, it was inconceivable in France to have a site like BASF's Ludwigshafen where 52,000 people are concentrated on six square kilometers with three thermal power plants and countless production sites. The first concentrations which President Georges Pompidou sought to carry out had not changed things much, neither had they cut down increased operating costs. Indeed, the leaders of merged companies had not cared at the time to close sites down and reduce personnel, two moves that might have improved the performance of the new groups.

Although the state spent considerable sums for chemical research, particularly through CNRS and the universities, the fallout for industry was scarce because of the persistent lack of communication between industry and those doing research.

The research and development sectors of the companies themselves made few breakthroughs, so that the chemical industry had to rely for a large part on foreign technologies, a fact that left little room for maneuver.

In addition to the difficulties inherent in their environment, France's companies also suffered the effects of bad management decisions in specific areas. Rhône-Poulenc had been badly prepared for the chemical fibers slump and had sunk too much money in heavy chemicals. These did not fit in with the group's original calling, as its leaders demonstrated when they withdrew, at the height of the crisis, from petrochemicals and the base thermoplastics, concentrating on specialties. The purchase of GESA from PUK in 1978, of Sopag the following year from the Gardinier brothers, and the sale of Lautier were hardly fortunate decisions for a group that could draw no advantage from getting further into fertilizers and that could have diversified to good purpose on perfumes through Lautier.

PCUK had never managed to strengthen Francolor's international base to good purpose and had finally sold it to ICI. Also, it wasted a lot of money in belatedly trying to develop a PVC chain. In 1981, PCUK was negotiating with Occidental Petroleum the sale of its chemical division, which had long since ceased to be of interest to the group's leaders.

At no time since it was set up was CdF Chimie master of its destiny, subject as it was to political pressures rather than economic rationality. Constantly in the red despite a number of worthwhile activities, it received the final blow when the untoward decision was taken in 1978 to build, on borrowed money, a one-billion-franc petrochemical site in Dunkirk in the framework of Société Copenor set up in joint venture with the Emirate of Qatar.

Elf Aquitaine had established under Sanofi a small conglomerate with profit-making subsidiaries involved in pharmaceuticals and perfumes. But Atochem, set up on a joint basis by Total and Elf, was a loss-making concern, as was Chloé Chimie, a cast-off of Rhône-Poulenc, which retained only 19.50 percent of its capital, while Elf and Total each acquired a 40.25 percent stake in the new chemical entity.

EMC was more a mining than a chemical company. It focused on potash, having restricted its diversification to the purchase of the animal food company Sanders and to a subsidiary in Tessenderloo, Belgium.

It was in this environment that the nationalization measures decided upon by the new Socialist government took place. The state took control of 40 percent in value of production of commodity chemicals and 70 percent of petrochemicals in France, an event that had no precedent in the free world's industrial countries.

Société L'Air Liquide, which figured as one of the companies to be nationalized on the initial Socialist list, escaped this fate, no doubt because the disadvantages of taking over this star multinational had been pointed out to the President of the Republic by one of his brothers, who was adviser to the group. On the other hand, Roussel-Uclaf, which had never needed state funds, found the government partly in control of its capital in addition to the main shareholder Hoechst.

Short of the extreme solutions advocated by some Socialists in favor of a single French chemical entity, the nationalized part was cut up along the lines announced by the Ministry of Industry on November 8, 1982. The restructuring signaled the death of PCUK as an industrial enterprise. Its various sectors were shared out among the other state-controlled groups. Most favored was Rhône-Poulenc, which received the agrochemicals and pharmaceuticals sectors with Sedagri and Pharmuka as well as the Wattrelos and La Madeleine sites in the north of France, together with a plant in Rieme, Belgium. At the same time, its fluorine division was boosted. The lion's share went to Elf Aquitaine with what amounted to two-thirds of PCUK's turnover, including, in particular, the halogen and peroxide products.

Complex negotiations with Total (Compagnie Française des Pétroles) ended with the group's withdrawing from Atochimie and Chloé Chimie, after which Elf Aquitaine set up its Atochem subsidiary to encompass all its chemical activities. After a long and brilliant independent career, Rousselot was split between Atochem and Sanofi.

Already sorely tried, CdF Chimie came out the worst from the restructuring. It inherited the Oxo alcohols and organic acids of the Harnes unit and had to call upon Esso Chimie to ensure their survival; it also got an ABS unit that was too small, which it exchanged with Borg Warner for a 30 percent share in their European subsidiary company—the Villers Saint-Paul site, which could become profitable only with the help of the industries to be set up there; the polyester resins division of the Chauny unit, and the downstream activities of the Stratinor subsidiary, both open to stiff competition. Among the lot there were some profitable sectors, however, such as Norsolor's acrylics, well integrated on the Carling site, and Société Lorilleux, a small ink multinational of PCUK's. But CdF Chimie was left to manage the difficult fertilizer sector swollen by Rhône-Poulenc's and EMC's divestments (GESA and APC), as well as a petrochemical branch set off balance by the unfinished Dunkirk site. As for EMC, all it got from PCUK was the historic site of Loos, which nevertheless served to boost its chlorine and potash divisions.

This enormous restructuring job, no doubt, did produce chemical groups with sounder bases and a more promising future. But the financial cost to the country was considerable, for not only were the shareholders refunded with public money to compensate for nationalization, but the companies that were now state-controlled had to be bailed out: their losses in 1982 were even higher than those registered the previous year. Just as high was the social cost. Manpower cuts which the former company leaders had been loath to carry out had become not only absolutely

necessary but also easier to implement by a leftwing administration.

RESTRUCTURING IN ITALY AND SPAIN

As was to be expected, the path to overcapacities aided by state subsidies had brought *Italy's chemical industry* to the edge of the precipice. In 1981, SIR and Liquichima, on the brink of bankruptcy, had been taken over by ENI, the state-controlled oil group whose own chemical subsidiary ANIC was also losing considerable sums of money. Montedison had been able to show balanced books only once in ten years, in 1979. Its debts had soared to $2 billion in 1984.

The rather belated restructuring measures consisted, in their first stage, in the sale of the state's 17 percent share in Montedison to private interests. Then Italy's petrochemicals and plastics companies were shared out between Montedison and ENI's chemical subsidiary Enichem.

These two groups then set out to concentrate their efforts on polyesters and acrylics in the fibers area. At the same time, Montedison gave up control of SNIA Viscosa, specializing in polyamides, to Bombrini-Parodi-Delfino (BPD). The restructuring, carried out together with manpower cuts and unit shutdowns, made it possible for Montedison in 1985 and Enichem in 1986 to post operating profits after long years in the red. Enichem received a further boost from association with ICI in PVC and with BP and Hoechst in polyethylene, for it had emerged from the restructuring in a less favorable position than Montedison because it was still saddled with commodity chemicals.

Montedison, now 45 percent owned by the Ferruzzi sugar group, reinforced its strategic sectors by purchasing Allied-Signal's fluorine polymers through its stake in Ausimont, by fully acquiring the Farmitalia and Carlo Erba pharmaceutical subsidiaries, and by buying from Hercules its 50 percent share in Himont, the joint subsidiary set up in 1983 in polypropylene.

The two Italian giants were still very much in debt, a fact that could lead to further divestments. But their leaders could nevertheless contemplate the future with some equanimity. Their heavy chemical sectors were finally merged under Enimont in 1988.

The *Spanish chemical industry* was also faced with considerable difficulties. Short of innovations, it had developed through foreign technologies and had lived a sheltered life behind customs barriers and import licenses not conducive to cost cuts. Neither Spain's petrochemicals industry, which was in the hands of the Enpetrol state group and the private company CEPSA, nor the main national companies Explosivos de Rio Tinto (ERT) and Cros, were in a position to face without transition the pressure of competition felt when Spain joined the Common Market. This was particularly true of ERT, which had missed bankruptcy by a hair, and Cros, which had remained in the red for a long time. Neither would be able to avoid severe restructuring.

Their total merger project failed through lack of financial means, and it was Kuwait in the end which, through the Kuwait Investment Office, took a 47 percent share in ERT and 24 percent in Cros in 1987 and promised to provide the necessary cash for the two groups to form a joint fertilizer subsidiary.

ARAB COUNTRIES GAIN A FOOTHOLD

As soon as OPEC was set up, Middle Eastern countries had sought to find ways to invest their oil revenues in downstream industries. *Kuwait*'s approach was, preferably, to acquire shares in existing companies. It thus bought up Gulf Oil's interests in Europe, took a share in Germany's Hoechst, and injected considerable capital into ERT and Cros in Spain.

Qatar had chosen to associate with CdF Chimie to set up a petrochemical base in the Emirate and to build the Dunkirk site through Copenor.

Saudi Arabia's policy has been to develop a national petrochemical industry that would sell its products worldwide. More than Qatar and Kuwait, it had abundant supplies of ethane and methane extracted from gases that were being flared. The ethane separation capacities of its refineries alone accounted for a potential of 3.5 million tons a year of ethylene.

Sabic, the body in charge of the project, had cleverly involved itself with major international groups such as Mobil, Exxon, Shell, and Mitsubishi. Production would then be easier to place in Europe, North America, and Southeast Asia without wounding national feelings. The first giant methanol unit came on stream in 1983, while the other Saudi productions located in Al Jubail and Yanbu have gradually begun supplying low- and high-density polyethylene, ethylene glycol, ethanol, dichloroethane, vinyl chloride (monomer and PVC), and styrene as the relevant units came on stream.

Since 1970, Saudi Arabian Fertilizer has been producing urea and melamine in Dammam, in association with Sabic; the two companies have scheduled construction of a 1,500-tons-a-day ammonia unit in Al Jubail.

Because of the obviously low cost of the principal local methane and ethane raw materials, and because the fixed costs of the installations are high with regard to variable costs, European petrochemical producers were afraid that Saudi Arabia with its low home consumption, would flood outside markets with its ethylene derivatives and methanol at cut prices. So far, however, Saudi exports have not shaken up the market because they have been carefully channeled through the distribution networks of Sabic's international partners.

Taking a different course than Algeria with its liquefied natural gas, the Gulf States have thus upgraded their natural resources and already account for 10 percent, 5 percent, and 4 percent of world production of methanol, ethylene, and polyethylene respectively.

THE AMERICAN CHEMICAL INDUSTRY CAUGHT OFF BALANCE

The difficulties resulting from world overcapacities were enhanced in the United States by the behavior of financial circles and the reaction to this behavior of the U.S. chemical industry leaders. America's chemical giants had reached their advanced stage of development because of the long patience of their shareholders and the acumen of their leaders based on thirty years of product and process innovation. Just like their German and Swiss counterparts, U.S. chemical industry leaders had upheld the notion of long-term interest over the more immediate concern of the various types of shareholders.

The shock waves sent out by the two oil crises, which had not spared the United States, the growing influence of financial analysts on the behavior of shares quoted on the Stock Exchange, and the arrival at the head of the large industrial groups of graduates from glamorous business schools trained more in finance than in technology gave the scene a new twist. Shareholders were more interested in the instant profits they could draw from breaking up a group than with the added value that could be patiently built up through its development.

Drawn along by their own convictions or under pressure from bankers and "raiders," U.S. chemical leaders were constantly redeploying their activities. The *leveraged buyout* (LBO) system had already been applied by the leaders of FMC's American Viscose division when they sought to buy, with the help of the banking world in the early 1970s, the Avtex rayon and polyester producer, which thereby became a successful company. Despite the risk to buyers in borrowing from financial organizations as much as 90 percent of the amounts needed for the purchase, the system was eagerly seized upon by individuals wishing to set up their own business and taking advantage of the disenchanted mood of potential sellers. This is how *Huntsman* became the world's leading producer of styrene and polystyrene after buying up the relevant sectors from companies like Shell and Hoechst, which wanted to pull out of them.

Likewise, it is because Du Pont, having spent $7.4 billion to acquire Conoco, sought to reduce its debts by selling part of Conoco's chemicals and also because Monsanto, ICI, and PPG were withdrawing from petrochemicals, that firms like *Sterling Chemicals, Vista Chemical*, and *Cain Chemical* have emerged since 1984. Cain Chemicals was itself to be taken over by Oxychem (Occidental Petroleum)

in 1988. Various acquisitions made at the right moment turned Vista within three years into one of the leading PVC and detergent alcohol producers in the United States. Through purchases made in its behalf by Sterling Chemicals, Cain Chemicals became a major petrochemical company with assets worth $1 billion in 1987, including ethylene, ethylene oxide, glycol, and polyethylene units, all strategically located in the Gulf of Mexico area. A further newcomer on the American scene was *Aristech*, which emerged through the takeover by its management of the heavy chemicals division of USX (U.S. Steel).

All these companies were acquired under very favorable conditions, as more often than not they were sold by the large groups at 25 percent of their replacement value. Contrary to assumed notions, individual entrepreneurs were thus able to acquire installations which until then only the most powerful groups could afford to run. These groups gave up whole sections of their traditional chemicals to redeploy in specialties for which they had no particular disposition and, at times, in areas even further removed from their original areas of competence. Thus *Diamond Shamrock* gave up its chemicals to Occidental Petroleum at the worst possible time, to devote itself exclusively to the energy sector, which in fact failed to live up to expectations.

One of the most powerful of America's chemical companies, *Allied Chemical*, became a high-technology conglomerate under the leadership of Edward L. Hennessy, Jr., who was formerly with United Technology. After acquiring Bendix and Signal, it took on the name of *Allied-Signal* and is now focusing on electronics and space, having entrusted a large part of its chemicals to the portfolio subsidiary Henley, which will sell them to the highest bidder. As for *Monsanto*, it shed a number of fibers, plastics, and petrochemical units both in Europe and in the United States and decided to hinge its further development on biotechnologies, a new area for the group. It bought up in particular the aspartame producer *Searle* for $2.6 billion.

At the same time as these changes were being wrought by the protagonist themselves, other major changes were taking place under outside pressure. Wily businessmen acting as "raiders," with the help of financial concerns that issued high-risk and high-interest "junk bonds" to finance a large share of the targeted acquisitions, set their sights on large companies quoted on the Stock Exchange: they acted in the belief that the company's parts would be worth more sold separately than as a whole.

The raiders' takeover bids had instant attraction for shareholders, and their criticism of the way the firms they were after were being managed was often not without truth. But it stood to reason that once the raiders had bought the company, they would break it up to reduce financial charges and to refund the money borrowed for the raid. The more interesting assets were often the first to be sold off, for they found ready buyers. To counter the raiders, the managers of the targeted firms were likely to raise the ante. But this only aggravated the financial problem, and the group's dismantling was unavoidable.

The instant advantage which both shareholders and raiders drew from these operations was obvious. But their consequence was, sooner or later, to destabilize the enterprises concerned, when these did not disappear altogether. The most spectacular case was *Union Carbide*, coveted in 1985 by the real estate developer S. Hayman, who had already taken over GAF Corporation.

To fight off the raid, Union Carbide had to borrow $3 billion. To reduce such an unbearable debt, the group's management was forced to sell its best sectors (batteries, consumer products, engineering plastics, agrochemicals) and even its headquarters in Danbury, Connecticut. This was how one of the best chemical concerns in the United States, with sales amounting to $10 billion, was left with only three areas of business after divesting to the tune of $5.3 billion. Even these areas—industrial gases, petrochemicals and plastics, and graphite electrodes—were faced with stiff competition. And with debts that still remain three times as high as the industry's average, Union Carbide is in no position to invest in the short term in anything likely to push it back to its former major rank in chemicals.

Other U.S. companies involved in chemicals were also the victims of raiders in 1985. To fight off C. Icahn, *UniRoyal* was taken over by its management and was forced to sell off its chemicals to Avery, which in turn placed them on the block, before accepting a leveraged buyout by the management. *Phillips Petroleum* had to buy back its shares from C. Icahn and B. Pickens and was forced to sell $2 billion worth of assets to refund part of its debt. And what about *Gulf Oil*, which sold itself to Standard Oil of California to escape the clutches of Boone Pickens, or *Stauffer Chemical*, which changed hands three times within a single year from Cheeseborough Pond to Unilever and finally to ICI, when it was broken up among ICI, AKZO, and Rhône-Poulenc?

Attracted to the U.S. market, European investors had also joined the raiders' ranks. This is how the Britain-based Hanson Trust managed to acquire *SCM*. This was a company that had just completed its restructuring; but after Sylvachem was sold off by the new owners, it retained only chemical production of titanium dioxide.

Anglo-French tycoon J. Goldsmith, unable to take control of Goodyear, nevertheless made substantial profits from his raid on the company. Goodyear was left with the sole alternative of withdrawing from all the sectors except chemicals in which it had diversified outside of tires.

In a number of cases, transactions led to an agreement between the heads of companies that had stock options and were eager to make a profit, and the potential buyers. This was how *Celanese*, an able and well-diversified company that had the means to retain its independence and competitiveness with regard to any major company, was acquired by Hoechst following a transaction that was satisfactory both to the German buyer and to the shareholders of the American group, at least for the time being.

The fear that their company might be the target of an "unfriendly" takeover bid induced the boards of directors of some of the well-managed chemical companies to guard against such attacks either through deceptively appealing offers—"poison pills"—or through purchase of their own shares. This was certainly not the best for industrial firms to make use of their funds.

COPING WITH SAFETY AND ENVIRONMENTAL PROBLEMS

Handling chemicals has never been without danger, if only because of the unstable and harmful nature of a number of substances when they are placed in certain conditions of temperature, pressure, or concentration.

Chemists have always been haunted by the risks of explosion. The explosion which occurred on September 3, 1864, in the Heleneborg laboratory near Stockholm, where Alfred Nobel was handling nitroglycerin, caused the death of five persons, including Emile Nobel, his younger brother. The ammonia synthesis unit set up by BASF within the Oppau plant was totally destroyed in 1921 by an explosion causing the death of over 600 people. In 1946, the French cargo ship *Le Grand Camp*, carrying 2,500 tons of ammonia nitrate, exploded in Texas City, killing 512 people. Other disasters, such as that of Flixborough in England, which took place through rupture of a Nypro caprolactam pipe within the plant in 1974, or again the one caused in a holiday camp in Los Alfraques in Spain when a tankwagon carrying propylene exploded in 1978, are reminders of the explosive nature of certain chemical products and of the need to handle them strictly according to the prescribed security rules.

A number of chemicals, fortunately a limited number, become dangerous either when they are used wrongly, or when they are accidentally set free. *Thalidomide*, put on the market in 1957 by the German company Chemie Gruenenthal, was indeed a powerful sedative. But it took three years to perceive that when prescribed to pregnant women, it dramatically crippled the newborn children. The synthetic intermediate for insecticides, *methyl isocyanate*, which Union Carbide has used for years without incident in its West Virginia Institute plant, caused over 2,000 deaths when it escaped in 1984 from a storage tank in Union Carbide's Bhopal plant in India.

Other products act insidiously, so that it is harder to establish their effects on human and animal health and more generally on the environment. Indeed, progress in understanding the safe dosage of minute quantities of impurities has enabled governments to fix with greater care the maximum allowed content of *vinyl chloride monomer, formaldehyde,* and *benzene* beyond which these products could become dangerous for workers to handle.

Lessons have been drawn from accidents caused by faulty handling of certain substances. Through the work carried on by Alfred Nobel, we know how to stabilize nitroglycerin in the form of dynamite, and since 1946 methods have been devised to avoid the spontaneous explosion of ammonium nitrate. Ammonia units with capacities of 1,500 tons a day have been operating for decades without incident.

Because of the painful thalidomide episode, long and costly tests are now carried out to study the possible secondary effects of pharmaceutically active substances. A great number of drugs that today save many lives would not have been available had they needed to go through the long periods of tests that are now required by legislation.

Likewise, in industrial countries, increasingly stringent regulations limit noxious vapor discharge from chemical plants, which are required to treat their effluents effectively. The transport of dangerous substances is also closely monitored by the authorities. Such precautions stem not only from the publicity which the media now gives to any catastrophe worldwide, but also from the public's instinctive distrust of chemistry, which it still regards as a mysterious science.

But just as an air crash does not mean the end of commerical aviation, neither does the damage caused by improper use of certain substances mean the end of the chemical industry. The image of chemicals is tarnished, however. Citizens who deliberately risk their own death, when they are not actually killing others, because of speeding on the roads or because they are addicted to alcohol, tobacco, or drugs, are less and less inclined, for all that, to accept accidental security breaches when these are not caused by themselves.

Politicians in our parliamentary democracies who wish to please public opinion feel the urge to take into account demands that are more emotional than scientific, and advocate restrictions even when these go against the best interests of the citizens. The *Three Mile Island* nuclear power plant accident in the United States which resulted in no fatalities, the more recent *Chernobyl* explosion which, as of 1988 had directly caused two deaths, have, with no good reason, prevented any resumption of the U.S. nuclear program and have aroused fears in European countries in people least likely to give way to mass hysteria.

The *Seveso* leak, which occurred in Italy on July 10, 1976, in the trichlorophenol unit belonging to Hoffmann-La Roche's subsidiary Givaudan, did have an impact on the immediate environment and a number of people were temporarily affected by the dioxin vapors. But the accident caused no lasting harm. It was the publicity which the media gave to it that forced Hoffmann-La Roche to close down the unit, turning Seveso into a dead city.

The litigation over residues left in the ground by Occidental Petroleum's affiliate Hooker, in *Love Canal*, in the state of New York, led to the evacuation of all the area's residents, beginning in 1978. But no clear explanation has yet been given of the ailments some of the inhabitants have been complaining about.

The lack of universally accepted scientific explanations for certain phenomena has often meant that the precautionary measures taken by one country do not necessarily apply in another. Where sweeteners are concerned, for instance, some governments have banned *saccharin* and other governments allow its use. The same is true of *cyclamates* and *aspartame*.

DDT was banned as an insecticide as early as 1974 by most industrial nations. But it is still widely used in many developing countries. The risks of *eutrophication* are perceived differently by governments, so that legislation applying to products for the production of detergents, like *alkylbenzene sulfonate, tripolyphosphate*, or *nitriloacetic acid* (NTA) differs from country to country.

The agreement which a number of nations reached in 1987 to ban the use of *chlorofluorocarbons* in aerosols is so far the only instance of harmonized legislation, even though no one has so far managed to prove scientifically that the chlorofluorocarbons really destroy the atmosphere's ozone layer.

Thus while it is understandable that authorities must be careful to soothe the fears of a public that is insufficiently informed of the dangers that threaten it, it must also be aware of the economic and social costs of refusing to accept the risks inherent in any human activity, and also conscious of the uncertainties surrounding the rules and regulations taken to satisfy its demands.

Some companies are turning the necessity of cleaning up the environment into new opportunities to improve their profitability. Thus Du Pont has found a useful application as a building material for the calcium sulfate that was piling up as a by-product in one of its Texas plants.

SCIENTIFIC AND TECHNOLOGICAL BREAKTHROUGHS

Short of fundamental discoveries over the past fifteen years, the chemical industry has gone forward by systematically developing its store of knowledge in processes and products.

Process Improvement

Higher crude oil prices had revived studies in the use of coal as a chemical feedstock. But while the Fischer-Tropsch synthesis was still used in South Africa by Sasol, the only other industrial gasification unit was the one Eastman Kodak brought on stream in Kingsport, Tennessee, in 1983, to produce *coalbased acetic anhydride*. The coal came from the Appalachian mountains and was cheap enough relative to oil prices at the time to warrant such an installation, and the plant is now to be expanded.

Together with these studies on synthetic gas, some progress has been achieved in the use of a group of alumino-silicates, the *zeolites*, as selective catalysts to boost certain reactions. Half the world production of p-xylene and a quarter of the production of ethylbenzene, an intermediate required to prepare styrene, are carried out using the zeolite-based ZSM-5 catalysts developed by Mobil Oil, which played a pioneer role in this area.

Applications of the olefin *metathesis* reversible chemical reaction, discovered by Phillips Petroleum in the 1960s, were also developed in the subsequent years. By this reaction, Arco produces propylene from ethylene and butene-2; Hercules prepares its plastic, Metton, from dicyclopentadiene; and Shell synthesizes its C_{12}-C_{14} SHOP (Shell Higher Olefin Process) alcohols used for detergents.

The application of *electrochemistry in organic synthesis* had already served to bring on stream in the United States in 1965 Monsanto's first industrial adiponitrile process from acrylonitrile. This was followed in 1977 by a similar installation in Seal Sands, England, which was later bought up by BASF.

The former *Reppe chemistry*, still practiced in Germany by BASF and in the United States by GAF, also led to new developments as demand for certain intermediates such as the 1,4-butanediol increased. This diol, now also obtained from maleic anhydride, is used to produce PBT polyesters through reaction with terephthalic acid and leads to other major derivates (tetrahydrofuran, butyrolactone, N-vinylpyrrolidone).

New synthetic processes for the preparation of established products were also industrially developed: in Japan the manufacture of methyl methacrylate from C_4 olefins, by Sumitomo and Nippon Shokubai; in France, the simultaneous production of hydroquinone and pyrocatechin through hydrogen peroxide oxidation of phenol by Rhône-Poulenc; in the United States the production of propylene oxide through direct oxidation of propylene operating jointly with styrene production, developed by Ralph Landau and used in the Oxirane subsidiary with Arco, which the latter fully took over in 1980; in Germany and Switzerland, the synthesis of vitamin A from terpenes, used by BASF and Hoffmann-La Roche.

Processes apparently well established were still further improved, such as the *electrolysis of sodium chloride*, dating back to the last century: diaphragm and then membrane cells were substituted for mercury cells, which were a possible source of pollution.

Important progress was also made in *chemical engineering*, such as use of rotary compressors in ammonia synthesis or ICI's fermentation reactors in Billingham to produce the Pruteen protein from methanol reactors, having no mobile parts.

Product Development

Although research was not as fruitful after 1960, new materials put on the market in the 1970s were the outcome of research in high polymers essentially conducted within industry.

It was through such research that ICI's PEEK (polyether ether ketone), one of the first high-performance aromatic polymers, was put on sale, as well as Du Pont's aramide fibers Nomex and Kevlar, more resistant than steel in like volume.

To the range of engineering plastics were added polyethylene and polybutylene terephthalates (PET and PBT), as well as General Electric's polyethers, the PPO (polyphenylene oxide) produced through polymerization of 2,6-xylenol and the Noryl plastic produced by blending PPO with polystyrene. Other special polymers, derived like the polycarbonates from bisphenol A, were added to this range: polyarylates, polysulfones, polyetherimides.

A major step forward was taken in the area of base thermoplastics with the application of Union Carbide's Unipol process. Variations of this were subsequently offered by other low-density polyethylene (LDPE) producers such as Dow and CdF Chimie (now ORKEM).

Under a process that consisted in copolymerizing in the existing highpressure installation ethylene with 5 to 10 percent of an α-*olefin* (butene-1, hexene-1), a stronger linear low-density polyethylene (LLDPE) was produced with a higher melting point than LDPE. Thinner films could thus be produced that were just as strong but required less material.

The new polymers opened up an unexpected market for producers of C_4, C_6 and C_8 α-olefins like Shell, Ethyl, and Chevron. Their higher linear α-olefins were also used either for polyalphaolefins (PAO) intended for synthetic lubricants or to prepare detergent alcohols.

While no great new plastic has emerged over the last fifteen years, researchers in major chemical companies did their utmost to improve both the features and the performance of known polymers.

As we have just seen, they improved LLDPE by adding comonomers in the carbon chain. But also through additives they managed to render polymers more resistant to fire, to oxidation, and to alteration through ultraviolet rays.

This slowly gave rise to a new industry that consisted in supplying polymer producers and plastic processors, not only pigments and charges, but also antioxidants, light stabilizers, and fireproofing agents. Added in small doses to the polymer, they added to its value by extending its life span. Such an activity, in which the Swiss firm Ciba-Geigy plays a noteworthy role, was boosted by the spectacular development of polypropylene, a particularly sensitive polymer that has to be stabilized with appropriate additives.

Another way of improving the performance of polymers consisted in blending them either with other polymers, or with inert materials such as glass fibers, carbon fibers, or various mineral fillers. Thus were produced a series of *alloys and composite* materials. Glass fiber-reinforced polyester has long been in common use. But the possibility of introducing carbon fiber obtained through pyrolysis of polyacrylonitrile (PAN) fibers already developed in aeronautics, opened up fresh prospects, particularly in the area of sports articles. The need, in turn, to link organic polymers and mineral fillers led to coupling agents such as the silanes which Union Carbide and Dynamit Nobel have put on the market.

This is how, little by little, spurred on by the demands of the processing industries which are also under pressure from major clients like the automobile industry, a number of companies have brought a large number of improvements to plastics. While not very spectacular, these improvements have appreciably added value to existing materials.

More generally, the requirements of many downstream industrial sectors have hastened the development of derivatives that otherwise might have remained laboratory curiosities. Discoveries of new molecules have been particularly inspired by the needs of plant protection. This was because agriculture, before it became a crisis sector, offered worldwide markets for crop protection agents, and also because product approval was easier to obtain, and therefore less costly, than in the case of pharmaceuticals. The success of glyphosate, which Monsanto put on the market in 1971 under the trade name Round Up, has made it the world's leading selective herbicide, for it can be used throughout the year and becomes harmless when absorbed into the ground. A new range of synthetic pyrethroids, developed in the United Kingdom by Elliott of the National Research and Development Corporation, (NRDC), a government agency, was marketed from 1972 onward under the trademarks of Permethrin, Cypermethrin, and Decis. These wide-spectrum insecticides owe their success to the fact that they are exceptionally active in small doses and are not toxic to humans. With increasingly strict legislation and stiff competition among pesticide producers at a time of slumping agricultural markets, the golden days could well be over for crop protection products, so that the years ahead are likely to be more favorable for restructuring than for new discoveries.

Over the last fifteen years, the *pharmaceutical sector* also made great demands on the ingenuity of chemists. But from the time of the thalidomide drama, the testing times required by health authorities have increased, to the point that since 1980 ten to twelve years are needed instead of the three to four previously required to bring a drug on the market from the time of its discovery. Research and development costs, accordingly, have grown fourfold over the last ten years, dangerously reducing the number of new specialties provided for patients each year. Because of such delays, a patent protecting a new substance may be left with but a few years of validity when final approval is granted to the laboratory that made the discovery.

Such difficulties have apparently not affected the zeal of researchers. Nor have they diminished the sums devoted each year to research and development, which on the contrary have been constantly on the increase. This is because any major discovery may have worldwide portent. And in most developed countries there is a system of refunding to patients the cost of ethical drugs, so that a new active principle may provide the laboratory that has exclusive rights over it with a considerable source of profits even if such refunds are coupled with tight price controls.

And while it is also true that thirty pharmaceutical companies alone account for 60 percent of worldwide ethical drug sales, the sums of money invested in research do not always get their full return. Thus it is that a small company like Janssen's laboratory, Janssen Pharmaceuticals, in Belgium, which was acquired in 1979 by Johnson and Johnson and which has among its discoveries diphenoxylate (1963) and loperamide (1975), has proved more innovative over the last fifteen years than the Rhône-Poulenc group, which has produced no major new molecule during the same time, although it devotes far more money to its research.

Indeed, success depends at least as much on chance, the ability of researchers, and the strategy of management in that area as on the sums expended. Valium and Librium, which have been providing Hoffmann-La Roche with its largest profits since the end of the 1960s, were the outcome of Leo Sternbach's acumen. Instead of merely modifying the meprobamate molecule as management had requested, he began studying the sedative properties of benzodiazepines used as dyestuff intermediates and on which he had worked for twenty years previously at Cracow University.

One of the most prolific inventors of the 1960s was most certainly Sir James Black, a Nobel laureate in 1988. While working for ICI, he discovered the first β-blocking agent Propranolol in the early 1960s. He also discovered Cimetidine, sold under the trade name of Tagamet as an anti-ulcer agent by SmithKline & French from 1974 onward, and which has become the world's largest-selling specialty. After working successively for ICI, SmithKline & French, and for Wellcome in Britain, Sir James now has his own business, and he is convinced that small competent teams are, by nature, more innovative than the large armies of researchers which many of the big companies have set up.

Likewise, the successful ventures of Merck Sharp & Dohme cannot be dissociated from the work of its president, Roy Vagelos. This biochemist, a latecomer to research, supervised the whole process of work to bring Mevacor, the new cholesterol miracle drug, onto the market. It has just been approved by the U.S. Food and Drug Administration. Mevacor was but the crowning touch to Merck's scientific tradition with its long series of discoveries: α-methyldopa against hypertension, indomethacin and sulindac to fight arthritis, and cefoxitine, an antibiotic.

At a time when pharmaceutical research is becoming increasingly costly and the likelihood of a great discovery remains hazardous, success will come to laboratories which not only sink large sums of money into research but also rely on teams where competence does not necessarily rhyme with size, and whose management has reached a sufficient level of scientific maturity.

THE CRAZE FOR BIOTECHNOLOGY

The catalytic action of living organisms, or rather of the proteins they contain, had received the beginnings of an explanation with the experiments of Payen and Persoz on malt amylase separation in 1833 and with J. J. Berzelius's catalyst theory in 1835. In 1897 Eduard Büchner demonstrated that a yeast extract could turn sucrose into ethyl alcohol. Fermentation took place without the presence of living organisms through enzymes. In this case zymase was the catalyst.

Ethyl alcohol, already known to alchemists, was used by industry towards the middle of the last century when continuous distillation in columns was devised by Ireland's Aeneas Coffey in 1830 and when it became exempt from excise duties on alcohol if methanol was added to it.

After alcohol, *lactic acid* was the second product obtained industrially from sugar fermentation, starting in 1880. The levo-isomer is still made this way to the tune of 20,000 tons a year.

In 1890, the Japanese chemist Jokichi Takamine had introduced a fermentation process in the United States by which an enzyme blend was produced. This takadiastase catalyzed starch and protein hydrolysis. Some years later in 1913, Boidin and Effront discovered the "*bacillus subtilis*" that produced an α-*amylase* stable under heat. This enzyme was used to desize cloth and later in the sugar fermentation process.

During World War I, Chaim Weizmann had succeeded in producing for the British Admiralty acetone and butanol on a large scale through anaerobic fermentation of starch. The Germans were then producing as much as 1,000 tons a month of glycerin from sugar. These war productions proved no longer competitive in peacetime. But *citric acid*, which Pfizer began producing in 1923 from sucrose, is still biochemically made today from *Aspergillus niger*, which Currie advocated in 1917.

The discovery of *penicillin* and its industrial development during World War II have led the pharmaceuticals industry increasingly to resort to *biosynthesis* for the preparation of its active principles. Through rigorous selection of the microorganisms extracted from the soil or from various molds, the cost of an antibiotic like penicillin has been brought down to $30 per kilo, compared with $25,000 per gram initially—an impossible target if the exclusively synthetic process had been used. Moreover, it became possible to extend the range of antibiotics that could be used. The antianemia *vitamin B$_{12}$* and most of the amino acids were prepared in the same way through culture of microorganisms in selected environments containing precursors.

In the case of *steroids*, biosynthesis permitted reactions that could not be achieved through direct synthesis. In 1952, this was how Upjohn researchers in the United States managed to introduce on carbon atom 11 of the steroid nucleus, a hydroxyl group –OH, using the *Rhizopus arrhizus* fungus, making the switch from the pregnancy hormone progesterone to cortisone and its derivatives.

Microorganisms are also capable of separating optical isomers. In the case of sodium glutamate, where it is necessary to start from levo-glutamic acid to obtain the desired flavor, and where synthesis produces only a racemic blend, it was a particular yeast called *Micrococcus glutamicus* that led to the required isomer through carbohydrate fermentation.

Considering that sodium glutamate, like other amino acids, is contained in soy sauce, which is a traditional Japanese food, it is not surprising that Japan should have become interested very early in this type of fermentation. Firms like Ajinomoto and Kyowa Hakko dominate the world market for amino acids and particularly for *glutamic acid* and *l-lysine*. It is also through enzymes that the resolution of *dl-methionine* into its optical isomers is achieved since its laboratory synthesis yields the racemic form.

Heat-stable amylases are frequently used in both the United States and Japan to produce *syrups with a high fructose content* from corn starch.

Single-cell proteins such as ICI's Pruteen were produced through culturing microorganisms on a bed of organic material.

Interest in biosynthesis grew still further with the discovery in 1953 of the structure of DNA, then in the 1960s of the genetic code of proteins. It then became possible to clone microbe or plant cells, through *genetic engineering*, by recombination of fragments of genetic material from different species. Thus, towards the end of the 1970s, the biotechnology firm Genentech succeeded in isolating the human insulin gene and to insert it into the DNA of the *Escherichia coli* bacteria: through reproduction, these bacteria produced the *first human insulin*, which Eli Lilly and Company has been marketing since 1982.

The *human growth hormone* (HGH), which can only be extracted in minute quantities from the pituitary glands, can now be isolated in larger quantities through genetic engineering.

Monoclonal antibodies (mabs), which replicate the antibodies in the organism with the added advantage of being "immortal," were discovered in 1975 by scientists working at the Cambridge Medical Research Council in the United Kingdom. They serve more particularly as reactive agents for medical diagnostic purposes.

Through *plant genetics*, it has also been possible to render plants resistant to chemical agents (Calgene, Monsanto) as well as to improve crop yields (Pfizer) with new seeds.

With the prospects which *biogenetics* was opening up for medicine and agriculture, a number of private laboratories sprang up in the United States between 1971 and 1978— *Genentech, Cetus, Genetic Institute, Biogen, Amgen*, and *Agrigenetics* to mention but the principal ones. These laboratories managed to finance their work with the help of venture capital, research contracts with the major chemical firms like Du Pont, Monsanto, Eastman Kodak, W. R. Grace, or shares purchased on the stock exchange.

Vast sums of money have been spent over the last ten years but with small tangible results, prompting the definition of biogenetics as a business likely to bring in a small fortune as long as a large one is invested! Thus far the only commercial fallout of biogenetic research involved human insulin (Eli Lilly), the human growth hormone HGH (Genentech, KabiVitrum), the hepatitis B vaccine (Merck, Smith, Klein-RIT), interferon (Boehringer, Ingelheim), the amylase enzyme (Novo), a number of veterinary vaccines (AKZO Pharma), and monoclonal antibodies for diagnostic reactive agents. Hopes raised by interferon and interleukin-2 as cancer cures have not materialized, but the tissue plasmogen activator (TPA) as a blood clot dissolver in heart attacks was approved by the U.S. Food and Drug Administration (USFDA).

Plant genetic research is encountering opposition from the U.S. Department of Agriculture and the Environmental Protection Agency. Pressured by environmentalists, the U.S. administration is loath to approve developments which could affect the environment in unknown ways. In addition to these administrative obstacles, there is uncertainty over patent rights, for there are no legal precedents. Finally, the biocompanies recently set up will need to associate with large pharmaceutical groups to develop and market the products born of their research.

Generally speaking, although *biotechnology* has acquired credibility in many areas, its development is being slowed by scientific, economic and administrative obstacles. First and foremost, proteins are complex substances that cannot be handled as easily as the simple molecules involved in traditional organic syntheses.

It is true that Japan's Ajinomoto and Kyowa Hakko, in particular, have become masters of the art of producing amino acids. Likewise, enzymes have remained the specialty of Novo (now Novo Nordisk) in Denmark, Gist Brocades in Holland, and Bayer's subsidiary Miles in the United States, which together account for 60 percent of the world needs in the area.

Even when they are technologically sound, however, bioproducts may turn out to be economically uncompetitive. The profitability of l-lysine from one year to the next, for instance, depends on soy market prices. In the same way, the single cell proteins which BP produced in 1963 in Lavera from a petroleum base, using a process developed by France's Champagnat, never managed to compete with soy cakes for animal food. ICI has also just been forced to close down its 50,000-ton Pruteen unit in Billingham.

At current crude oil prices, the production of ethanol from biomass is not profitable, either. Whether produced from beets, sugar cane, or corn, it can become competitive only if it is subsidized. And these subsidies would only be forthcoming for political reasons: to please their farmer voters, the French, Brazilian, and United States governments would adopt such a policy to absorb excess agricultural products. From cereals, corn in particular, starch is produced and hydrolyzed to form glucose which ferments to ethanol.

Powerful groups like American Corn Products and France's *Roquette Frères* produce starchy matters in this way. The former is also the leading producer of *isoglucose* (a blend of glucose and fructose) in the United States, while the latter is the largest producer of *sorbitol*. Starch can, therefore, compete directly with saccharose both for foodstuffs and for industrial uses as a fermentation or enzyme-reaction base.

This gives rise to a permanent conflict in Europe between the starch manufacturers on the one hand and the sugar and beet refiners on the other, a conflict that the EEC Commission with its *Common Agricultural Policy* of quotas and subsidies has been unable to settle. The only point of agreement between the two parties is the price which they demand for their production from downstream Community industries, a price that is far higher than world rates.

Spurred on by the Italian sugar group Ferruzzi-Eridiana, Montedison's and now Enimont's main shareholder and an associate of France's Béghin-Say sugar group, there is a campaign under way to introduce ethanol into gasoline. Farmers, of course, support the move because incorporating 7 percent of ethanol in gasoline would mean for a country like France the use of two million tons of sugar or four million tons of cereals. But ethanol happens to be in competition with methanol and the new MTBE antiknock agent as a gasoline additive. More important, a tax rebate would be needed at current gasoline prices to induce the oil industry to incorporate ethanol in prime rate gasoline. So the "farm" lobby can receive satisfaction only at the expense of the taxpayer, whether American, Brazilian, or European.

The rules that have always governed the use of ethanol, government policy favoring one agricultural raw material over another, the new constraints that limit the marketing of genetically engineered products—all these factors serve to remind those interested in the development of biotechnology how narrow is their room for maneuvering.

THE FINE CHEMICALS APPROACH

In their search for products that could provide better margins than those achieved from commodity chemicals, the industry had hit upon *fine chemicals*. These typically involved derivatives from organic synthesis, obtained in multipurpose units and sold in relatively small quantities at high prices.

The German and Swiss *dye manufacturers* (Hoechst, BASF and Bayer, as well as Ciba-Geigy and Sandoz) were in the most favorable position to develop such advanced chemicals. They had a long tradition behind them of multiple-stage syntheses involving intermediate derivatives that could also serve to prepare pharmaceutically active principles or pesticides. Starting from a number of major raw materials and working according to the chemical-tree concept, these producers can work down the line to well-defined molecules which they use in their own downstream production or sell as synthetic intermediates to outside clients.

In Europe, the giant ICI group, which had retained a strong position in dyes, also became involved in this kind of chemicals.

France, with PCUK having closed down in 1980 its Société des Matières Colorantes in Mulhouse and then having sold Société Francolor to ICI, had restricted its ambitions in this area. It retained only a few products of Rhône- Poulenc and of its 51 percent subsidiary *Société Anonyme pour l'Industrie Chimique* (SAIC), located in Saint-Fons and in Mulhouse-Dornach, respectively.

As was to be expected, the U.S. chemical leaders, Du Pont, Allied Chemical, American Cyanamid, GAF, and Tenneco Chemicals, had all withdrawn between 1976 and 1979 from the dyes sector. Only three medium-sized companies were still active in this area: *Crompton & Knowles, American Color*, and *Atlantic Chemical*.

Yet at the end of World War II, America's dye production had been the leading one worldwide. For over thirty years it had enjoyed high customs tariffs protection through the American Selling Price clause. But dyes were produced by giant companies used to large-scale continuous productions. Their engineers were not trained to run month-long syntheses campaigns involving many stages. Moreover, American marketing executives were little attracted to the German methods for "motivating" their clients. There was also the fact that during the 1960s, U.S. dye manufacturers had come to rely on imported intermediates. With rising prices and the textile slump, they found themselves caught between rising purchase costs and falling selling prices. Finally, unlike their European counterparts, U.S. manufacturers had never given international scope to their dye business. It remained restricted to the home market.

For all these reasons and also because they were not tied down like the Germans by any prestigious tradition, they unhesitatingly gave up dyes, losing at the same time the know-how needed to succeed in fine chemicals.

With more modest means, other firms were more successful. They either developed their own "chemical tree," or put to good use the know-how acquired through development of certain processes.

Ethyl became a bromine and derivatives specialist and an expert in orthoalkylation (orthoalkyl phenols and anilines). Its acquisition of Dow's bromine activities has given Ethyl a leading role in this field. *DSM* developed its fine chemicals from the benzoic acid produced during manufacture of synthetic phenol by toluene oxidation. *Atochem* took advantage of the sulfur resources of its parent company Elf Aquitaine to build up successfully a thioorganic chemicals industry (thioglycol, mercaptans, DMSO). Its position will be further strengthened by the takeover of Pennwalt. *PPG* in the United States and *Société Nationale des Poudres et Explosifs (SNPE)* in France are producing a wide range of phosgene-based derivatives to be used in the most varied manner (carbonates, chloroformates). More than any other company, *Lonza* has extended its range of fine chemicals (diketenes, HCN derivatives, pyrazoles,

pyrimidines). *Reilly Tar* has become a world leader in pyridine and derivatives. *Dottikon* in Switzerland and *Kema Nobel* in Sweden have put to use their nitration experience to extend their range of nitrated intermediates. Among others, *Rhône-Poulenc* and *Montedison* are involved in organic fluorine derivatives while *Hüls'* fine chemicals division has specialized in alkylation, hydrochlorination and catalytic hydrogenation.

Thus a number of firms with special know-how in a family of products or in processes that were not among the biggest have succeeded in taking a more than honorable place as suppliers of fine chemical derivatives alongside the organic synthesis specialists originating from the dye business.

THE ATTRACTION OF SPECIALTY CHEMICALS

Besides fine chemicals sold according to specifications but accounting for only a small part of the sales of major companies, *specialty chemicals* held attractions for companies wishing to diversify. These chemicals involved substances or mixtures whose composition mattered less than the function for which they were intended: the test of success lay in performance. Thus old family businesses or more recent companies born of a leader's entrepreneurial spirit had been successful in performance products, whether these were paints, inks, or glues; or in specialties, cosmetics, detergent, or electronics industries.

Indeed, not much capital is needed to manufacture specialty chemicals compared with what is required for commodity chemicals. The development of new products is both quicker and less costly than it would be to find new processes for large-volume products or to bring to the market an original active principle for an ethical drug.

This largely explains why specialty chemicals managed to remain until the early 1970s products for medium-sized private companies. In the long run, however, the internationalization of trade, the size of advertising budgets for consumer products, and the necessary adaptation to new technologies requiring highly qualified personnel all called for funds that were not always available to family businesses. Many small owners were forced to sell out, and their need coincided with the attraction they held for large chemical groups trying to diversify away from heavy chemicals. They hoped to find in specialty chemicals the profit margins which their traditional branch of chemicals no longer supplied.

Barring a few exceptions such as *Gulf Oil* or *Diamond Shamrock*, which withdrew from downstream chemicals, all the major companies, both in Europe and in the United States, decided to make specialty chemicals a priority in their development strategy. In truth, some of them had not waited for the energy crisis for them to take a firm foothold in the specialty market.

In the United States, *Du Pont* and *PPG* had a long-established reputation in industrial and consumer paints. *W. R. Grace* since buying Dewey & Almy, and *Rohm & Haas* because of its age-old tradition in acrylics, drew substantial profits from their specialties. This was also true of *American Cyanamid* (additives for plastics, cosmetics) and of *Monsanto* (products for rubber, special polymers). Since its withdrawal from the tire business, *BF Goodrich*, aside from its PVC lines, is concentrating now on specialties.

In Europe, ICI had already acquired a large paints sector (Duco, Dulux). The three major German leaders—*Bayer, BASF*, and *Hoechst*—had not yet made great inroads into the specialties market, but the Swiss *Ciba-Geigy* could be said to be particularly well established in certain areas like additives for polymers, in which it was a world leader. *Rhône-Poulenc* had assembled some of its activities within a "chemical specialties" division. But on the whole, they could be said to be offshoots of fine chemicals rather than actual specialties, with the exception of the performance products brought out by subsidiaries such as Orogil, SFOS, Soprosoie, and Vulnax. Orogil is now fully owned by Chevron, however, and Vulnax has been acquired by AKZO. Failing to develop through

internal growth, *AKZO* had very early developed its specialties by buying up companies involved in peroxides, paints, oleochemicals, and now rubber additives.

To increase their specialty sectors as fast as possible, the leaders of large companies found it more expedient to do so through acquisitions. The prices paid for the most interesting purchases can be considered high because, very often, they amounted to fifteen to twenty times the profits. But the financial sacrifices made by the buyers seemed worthwhile, for they gained a foothold in the market without the long preliminary work that would otherwise have been needed.

There were, of course, many companies that were sufficiently important or prosperous to escape being bought up. Even then their independence was often at stake. Thus *Nestlé* took a share in the cosmetics group *l'Oréal*; and in the United States, the raider Perelman managed to buy *Revlon*.

Considering that the grass always looks greener on the other side of the fence, for many leaders of the chemical giants diversification into new areas might seem more attractive than mere concentration in well-known sectors; and it was in this sense that specialty chemicals seemed a good proposition. In 1983, *Olin* began to get involved in electronic chemicals by buying up 64 percent of *Philip Hunt Chemicals*, and took a firm foothold in the sector through successive acquisitions. Other groups became interested in enhanced oil recovery and exploration, for the future of oil seemed assured at the time. In both cases, however, the electronics and oil exploration slowdown did not confirm established forecasts. The investments made in these areas have yet to prove their profitability.

Moreover, many firms were unable to contribute anything except capital to the development of sectors far removed from their traditional areas of business. They became discouraged and ended by selling out, not without suffering heavy losses. *Hercules* was seen to back out of its water treatment sector and *Rhône-Poulenc* from its very recently acquired media business.

Even when the businesses acquired are not too different, trouble can arise through disagreement between the new owners and the former boss of the purchased firm over how to manage it. The former tries to impose his own personnel and procedures, while the latter, used to making his own decisions, is unable to fit into a large unwieldy concern. As a large part of the worth of an acquisition in specialty chemicals lies in the competence of the personnel involved, some purchasers have understood that it is to their benefit to leave the day-to-day running of the business to those who have already shown their worth and to centralize only those activities related to the financing of new investments. This was how *Witco* proceeded in the United States, most likely because the father of the current president had founded and built up the business to the point of making it one of America's leading specialty concerns.

ICI followed the same policy when it bought *Beatrice Chemicals* for $750 million in 1985. But in this case, it was important to delegate power, because Beatrice Chemicals consisted of ten distinct companies established in eighteen countries and involved in different businesses (composite materials, vinyl resins for paints, leather auxiliaries). Keeping in mind that cultural differences may produce problems that are not always easy to solve, the strong involvement of the big chemical groups in the specialties area over the past years had drastically changed the structures of the sector.

THE PAINT INDUSTRY

Few industries have been as affected by the restructurings of the past ten years as the paint industry. The extension of markets worldwide owing to the multiplication abroad of client factories of this industry, the technological revolutions brought about by the introduction of electrophoresis, of water-based lacquers, and of powder coatings had the twin effect of pushing the chemical leaders to expand worldwide in this area and to lead those paint companies that were still independent to sell out for want of the funds needed to develop

their research base. *ICI*, which was strongly established only in Britain and in the Commonwealth, became the world's leading paint producer with 750 million liters after buying *Valentine* in France and, especially, the *Glidden* division of the U.S.-based SCM for $580 million in 1986.

PPG has been pushed back to second world position with 450 million liters. But with its 100 percent stake in France's *Corona* and its controlling share in Italy's IVI and in Germany's Wülfing, the U.S.-based PPG has maintained a comfortable technological lead in the application of cataphoresis in automobile bodies, accounting for 60 percent of the world market in this specialty.

Through its costly $1 billion purchase of America's *Inmont, BASF* has become the world's third-ranking paint producer, leaving behind its German Rival *Hoechst*, which was too busy bailing out its British subsidiary Berger Paints to get a foothold in the U.S. market.

AKZO, which holds an honorable place among the leaders, has not been able to penetrate the United States market, either. Most of its recent acquisitions (Blundell, Permoglaze, Sandtex, Levis) were European.

Other companies with comparable 250-million-liter paint capacities are Japan's *Nippon Paint* and *Kansai Paint*, as well as America's *Du Pont*. These three firms, however, have restricted their ambitions to filling the needs of their home markets. With a broader international base, Courtaulds' subsidiary, *International Paints*, ranks among the top ten, although it is mainly involved in the very special sector of marine paints.

Ranking fifth in the world with its 300-million-liter capacity, *Sherwin Williams* is the only large paint company that has retained its independence. It remains focused on the United States, essentially in the decoration market.

Although France is the world's third largest market for paints after the United States and Germany, none of its national manufacturers has thought of striking out beyond its frontiers. Indeed, most of the French companies involved in the sector, with the exception of

Blancomme and *IPA*, have been taken over by foreign groups when they were not merged within statecontrolled entities. *Astral Celluco* was one of the first to sell out to AKZO, *Corona* was taken over by PPG, *Celomer* by International Paints, *Bichon* and *Lefranc Bourgeois* by Sweden's Becker, *Valentine, Julien, Galliacolor* by ICI, *Ripolin Georget Freitag* became part of the *CdF Chimie* group as did *Duco* which has just been sold to *Casco Nobel*, while *La Seigneurie* was taken over by *Elf Aquitaine*.

In 1988 CdF Chimie, later known as ORKEM, took over full ownership of AVI, a profitable company specializing in decorative paints. Another subsidiary of ORKEM was *Lorilleux*, an ink manufacturer merged in early 1988 with *Coates Brothers* to become the third largest group in its field after Dainippon Ink Company (DIC) of Japan and Germany's BASF.

It must be pointed out that all the international groups involved in paints and inks on a worldwide basis produce, in addition, most of the resins and binding agents needed for their formulations. Only the solvents and pigments are likely to be partly brought in from outside sources. The restructuring of the paint industry has, accordingly, been to the advantage of the new groups. On the one hand, it reduced the number of producers and extended the range of products these producers were putting on the market, and on the other, it supplied a captive market for their resins which, until then they had mainly sold to outside customers.

SURFACE-ACTIVE AGENTS

Used for their good performance, more often than not in formulations, surface-active agents can be classified as specialties even though the quantities consumed in certain cases might connect them with commodities. The structure of the major part of the detergents industry has remained rather stable over the last few years despite some frontier adjustments. The *washing powder* sector, where advertising costs are considerable, is dominated by a small number of substantial soapmakers who came into business as far back as the nineteenth century: the American companies *Procter & Gamble* and *Colgate*, the Anglo-Dutch group *Unilever*, and Germany's *Henkel*. They are all, in various degrees, involved in the major world markets.

Then there are the Japanese companies *Kao Soap* and *Lion Oil*, which remain confined to their own home territory and to a few Southeast Asian countries. Behind these giants, a number of firms catering to their home markets stand out, such as *Purex* in the United States or *Benckiser* in Germany.

As in paints, France is curiously absent from the area. Since Germany's Henkel recently took over the detergents division of the *Lesieur-Cotelle* group and its trademarks Mir, La Croix, and Persavon, after buying up the Savon de Marseille soap flakes of *Union Générale de Savonnerie* (UGS), the French market is now 94 percent supplied by the big international soapmakers. The few remaining national firms such as *Chimiotechnic* merely sell their products through the supermarkets.

While the sector now seems to be structurally stabilized, washing powder components are fast changing to take into account the new rules and technologies laid down both by governments and consumers. For the companies which supply the soapmakers, these new rules and regulations are having major consequences throughout the world. Just as the requirement of biodegradability had doomed the use of *branched-chain alkylbenzenes* in industrialized countries in the 1960s and caused the shutdown of a large number of dodecylbenzene sulfonate units, so the new rules established by some governments against *tripolyphosphates* in Europe and elsewhere to ward off eutrophication are likely to wipe out the several-hundred-thousand-ton markets of producers like Rhône-Poulenc, Benckiser, Knapsack, or Montedison.

Replacing TPP by new formulations based on *polyacrylic acid* and *maleic anhydride* would, on the other hand, greatly boost companies like Atochem and BASF, which are very much involved in acrylic chemicals. Likewise the use in Europe of washing machines at temperatures that do not exceed 50° to 60°C, like the ones now used in the United States, should have immediate consequences for the formulation of washing powders. *Perborates*, used extensively in Europe as bleaching agents ever since Henkel invented Persil in 1907, are not very efficient at such low temperatures. Activators such as ethylenediamine tetraacetic acid (EDTA), produced by Warwick in England, are needed to hasten decomposition. *Enzymes* which had been very popular in the 1960s in the United States and Europe, then had disappeared in 1971 because they were considered harmful to the skin, have been reintroduced in washing powder formulations because they do help remove certain stains.

The use of *liquid detergents* is more widespread in the United States, where they account for 20 percent of the market, than in Europe, where their share does not yet exceed 8 percent on an average. This has consequences on the consumption of nonionic derivatives.

Different habits as well as different regulations have therefore led to frequent changes in the chemicals supplied to soapmakers. Few industries have changed as much as the detergent industry since the end of World War II as it shifted from soap to synthetic detergents, from branched alkyl benzenes to linear alkyl benzenes, from anionic to nonionic. TPP and enzyme regulations were changed; preference was given at times to perborates, at others to chlorine-based products such as bleaching agents.

To develop surface-active agents for industrial use did not require the same financing as was needed for washing powder consumer products. Therefore, producers of all sizes could become involved. Some of these producers were chemical giants who had gone into the business because they had the available raw materials or the right markets. Indeed, surface-active agents use a number of major raw materials to which suppliers attempt to add downstream value.

In Europe, for historical reasons, large chemical groups have become involved in this area. Thus the dye manufacturers had very early added to the range of products sold to the textile and leather industries, wetting agents, softeners, and dye auxiliaries. *BASF*—a pioneer in synthetic auxiliaries with its Nekal, patented in 1917—*Hoechst, Bayer*, and *ICI* were in fact interested at the same time in the markets which surface-active agents opened for their ethylene oxide, higher alcohols, sulfonating agents productions, and in the fact that they help provide better services for their traditional textile clients.

Hüls, the subsidiary of the German holding company VEBA, had no dye tradition. But it nevertheless acquired the Dutch surface-active unit Servo to ensure captive use for at least part of its ethylene oxide and alkyl benzene production. *BP* followed a similar line when it took over the Belgian company Tensia, selling back some of its product lines to ICI. Already involved in surface-active products through its Lissapol for many years, *ICI* has expanded in this sector by buying *Atlas Powder* and its special range of Tweens and Spans. *Shell*'s interest in surface agents went back to the development of its Teepol. It completed its range with ethoxylates, the "neodols" which used both its higher alcohols and its ethylene oxide. *Montedison* was also involved in surface-active agents through its stake in *Mira Lanza*.

In France, however, there was no vertical integration between the great national chemical industry and the surfactant sector. Producers of the latter had to find the necessary feedstock—whether ethylene oxide, alkylphenols, fatty acids or higher alcohols—from rival companies, while for instance a medium-sized company like Berol Chemie in Sweden, recently acquired by Nobel Industries, had its own source of ethylene oxide, amines and nonylphenol in Stenungsund to feed its surfactants division.

In the *United States*, vertical integration was not as thorough as in Europe. Although ethylene oxide producers like *Union Carbide, Dow,* or *Texaco* also had their range of ethoxylates, it was mostly specialized firms that produced the surface agents for industrial uses. The same was true in Japan, although a number of producers such as *Nippon Oil & Fats* for fatty acids, *Kao Corporation* and

Lion Corporation for fatty alcohols and amines had direct access to their main raw materials.

Thus in addition to the large chemical and petrochemicals companies that had chosen downstream integration, there were a number of important surfactant producers that, in varying degrees, were integrated upstream. The most striking example of this, besides the three Japanese companies just mentioned, is Germany's *Henkel*. Its natural fatty alcohol production exceeds 170,000 tons capacity, and besides fatty acids, it produces its own range of carboxymethylcellulose-based thickeners. Recently Henkel even associated in this area with Hercules within a company called *Aqualon*, now fully owned by Hercules, and acquired from Quantum Chemicals in the United States their fatty acids subsidiary *Emery Industries*.

A number of surfactant specialists have also chosen the market approach. Because they are not tied down by their own produced raw materials, they can use those that are the most suitable for the type of surfactant they wish to offer their clients.

An independent producer like the U.S.-based *Stepan* is in a position to provide a complete range of anionic, cationic, and nonionic agents because it has flexible units in four areas of the United States as well as one in southeastern France in Voreppe.

Witco is in the same position, but its own policy has been to develop through acquisitions rather than through internal growth, buying Humko Chemical and Onyx Chemical.

Right from the start *GAF* acquired, from IG Farben, experience in surfactants still of use today. This activity sector, however, was sold to Rhône- Poulenc in 1989.

With a market lacking the uniformity of the United States market, the European producers serve in greater numbers clients with standards and habits varying from country to country. The Tenneco group's *Albright & Wilson* has had to cover France, Italy, and Spain with its *Marchon* subsidiaries. Germany's *Hoechst, Henkel*, and *Schering,* which bought up *Rewo*, also have a number of subsidiaries abroad that produce their surfactants. Hüls's

subsidiary *Servo* has only the single production unit in Delden, Holland. But because of the high concentration of its products, Servo manages to carry out three quarters of its sales abroad.

While the range of products offered by these companies is very wide, some of them, nevertheless, focus on specific sectors. Thus the cationic technology acquired in the United States from Armour by *AKZO* and from Ashland by *Schering* has given both these companies a dominant position in the market of textile softeners both in Europe and in the United States.

Companies like *Rhône-Poulenc, Berol* and Witco are, for their part, interested in the pesticide formulation market. Fatty amines are in the hands of such European firms as *AKZO, Kenobel* (Nobel Industries) and *CECA* (Atochem).

Other European companies, such as *ICI* through Atlas and Tensia, *Th. Goldschmidt, Rewo*, and *Servo*, have particularly targeted the lucrative area of beauty care. In the United States, *Miranol* has been very successful with the amphoterics (imidazolines, betaines) for baby shampoos, an activity acquired by Rhône-Poulenc in 1989.

America's *Du Pont* and *3M* and Japan's *Sanyo* pay particular attention to the development of fluorine-based surfactants. *Air Products* with its acetylene derivatives Surfynol and *W. R. Grace* with its sarcosinates (Hampshire Chemicals) have also focused on well-defined segments of the business. With world demand exceeding two million tons, the market of surfactants for industry is of a nature to attract a large number of operators, raw material suppliers, processors of these raw materials into anionic, nonionic, and cationic derivatives, or downstream industries that use surfactants in various formulations.

FLAVORS, FRAGRANCES, AND BEAUTY PRODUCTS

The sector of flavorings, perfumes, and beauty products has also had its share of restructuring and technological changes over the past ten years.

Although many of the raw materials needed in this area still come in the form of essential oils from natural sources like jasmine from Grasse, roses from Bulgaria, ylang-ylang from Madagascar, oak moss from Yugoslavia, an increasingly significant role is now being played by semisynthetic or fully synthetic products.

Thus *terpenes* (α-pinene, β-pinene) can be produced from natural turpentine, as is traditionally done by rosin producers such as *Hercules, Glidden,* or *Union Camp* in the United States, or on a smaller scale in France by *Société des Dérivés Résiniques et Terpéniques, DRT*. BASF and Hoffmann-La Roche, however, have demonstrated that starting from acetylene or isobutylene, terpene chemicals can be synthetically reproduced. Both companies are able to produce both their vitamins and perfume bases in this way.

Likewise, *vanillin* is now largely produced synthetically. The world leader in this area is *Rhône-Poulenc*, which has a unit in Saint-Fons to which was added a unit bought from Monsanto in 1986 on the West Coast of the United States. In the latter plant, vanillin is still extracted from paper pulp liquor. *Menthol* from plantations in Brazil and China is also produced by synthesis since *Haarmann & Reimer*, bought by Bayer in 1954, managed to carry out industrially the resolution of racemic menthol, thus isolating the levoisomer. *Anethole*, synthesized by Hercules from pine oil, is two to three times cheaper than when it is extracted from star anise.

Instead of identically reproducing natural products, chemists have also succeeded in making cheaper substitutes with similar features. Thus nitrated musks and later macrocyclic musks have become substitutes for more rare natural musk. Major chemical companies became interested in the firms that specialized in perfume chemicals. But their involvement in this area was not always successful, for their business views did not necessarily apply to this new activity.

While Bayer's association with Haarmann & Reimer proved successful, it took Hercules several years, from 1973 on, to understand properly how its *Polak Frutal Works* (PFW)

had to be managed. Today it is autonomous and prosperous. In contrast, Rhône-Poulenc ended by selling *Lautier* to *Florasynth* in 1981.

When *Tenneco* bought Albright & Wilson in England, it did not see the point of keeping its *Bush Boake Allen* (BBA) aroma chemicals division. BBA, itself the outcome of a merger of several family businesses, was finally sold to the U.S.-based Sylvachem in 1982. Sylvachem[1] already owned *George Lueders*, an essential oils concern Monsanto sold failing proper management. It would seem, therefore, that among the major chemical companies, only Bayer, Hercules and, more recently, BASF, which bought *Fritzsche Dodge & Alcott* in the United States in 1980, have achieved their downstream breakthrough in the flavor and perfume sector.

On the other hand, the pharmaceuticals group Hoffmann-La Roche, which purchased *Givaudan* in 1963, then *Roure Bertrand Dupont* a little later, has managed to rank third in the world in this difficult area. But the leader is undoubtedly *International Fragrances & Flavors* (IFF), an American company that accounts for 10 percent of the world market. Set up in 1929 by a Dutch immigrant, A. L. van Ameringen, IFF acquired its current form in 1958 and, pushed along by the creative invention of its perfumers and the quality of its compositions, has never ceased growing.

Close on the heels of IFF is the *Unilever* group, which developed in the field through acquisitions. After consolidating in 1983 its three perfume and flavor subsidiaries—PPL, Food Industries, and Bertrand Frères—to form *PPF International*, the group acquired a foothold in the U.S. market in 1984 with *Norda*. Three years later it merged PPF with Holland's *Naarden*, which was on the decline. Called *Quest International* (Unilever) the new company accounts for over 7 percent of the world market in its area.

Amongst the world leaders, the only privately owned company, the Swiss-based *Firmenich*, ranks fourth. It has retained its independence both because it was held

---

[1]Sylvachem belongs to the Union Camp group.

together by the heirs of the founding family and because it produces quality products based on strong research. A number of smaller companies that do not belong to any multinationals are highly competitive. They include Japan's *Takasago*, which began in 1920, America's *Florasynth*, which took over *Lautier*, and Britain's *Pauls Flavours & Fragrances*, which has just established a hold in the United States market by purchasing *Felton International*.

France, which had in its favor the age-old reputation of Grasse and the world image of its perfumes linked to its haute-couture prestige, is nevertheless absent from the fray of large suppliers in this area, even though it has some Grasse-based companies like *Mane* and *Robertet* and despite the efforts made by the Elf Aquitaine group which has assembled, around Sanofi, firms like *Méro et Boyveau*, *Tombarel* and *Chiris*.

Flavors account for a substantial share of the sales of these firms: 30 percent for IFF, 40 percent for Givaudan, 50 percent for Unilever and 100 percent for Sanofi-Méro. They are increasingly being used in foodstuffs since the fashion of fruity yogurts and instant desserts began between 1965 and 1970. The internationalization of food habits and the growing industrialization of the food sector have contributed to the development of demand for flavors and to the gradual substitution of natural substances by synthetic products that are less costly to produce and more active in small doses. Just as the perfume industry composes fragrances for its clients, subtle blends of flavors are now devised for the large food companies. Demanding customers, together with stringent regulations and sophisticated technologies, all combine to build up research costs. This explains the restructuring that has taken place in the sector as family businesses have been taken over by powerful international chemical, pharmaceutical, or food-industry groups, leaving only a few independents willing and able to make the necessary research efforts.

Although it still clings to a long tradition, the *world of perfumes* has also changed both in its structures and in its technologies. The highest volume comsumption derives from products of the soapmakers. *Procter & Gamble* prepares its own compositions, but its competitors mostly rely on the laboratories of their suppliers for fragrance preparation.

With a few rare exceptions, such as *Guerlain*, *Chanel*, and *Patou*, the great names as well as the small perfumemakers do likewise. One of the world's largest-selling perfumes, "Anais-Anais" by *Cacharel* (l'Oréal), is prepared by Firmenich, while *Roure-Bertrand-Dupont* has signed two other recent successful perfumes, *Dior's* "Poison" and *Saint-Laurent*'s "Opium." Launched in 1921, *Chanel No. 5* was the first perfume to carry a synthetic aldehyde note and is still one of the ten world best-sellers. But the market has now moved to floral and oriental fragrances. Perfumes for men with stronger notes have developed spectacularly and now account for 25 percent of alcoholic perfumery. In addition, the aerosol format has boosted sales of toilet waters and deodorants.

The most varied distribution systems have been developed, ranging from door-to-door sales, which *Avon* started, to sales by mail, a specialty of *Yves Rocher*'s, to sales in large department stores, to sales in selected areas such as perfume shops and pharmacies. Few "nonessentials" have become so indispensable. If they cannot be dispensed with, it is through the efforts of the industry, which relies upstream on the suppliers of both contents and containers, who adapted to all requirements, and downstream on efficient marketing networks. It can also devote to advertising the money that it need not spend on research conducted on the industry's behalf by the chemists.

Although it originated in France, the perfume industry is now mostly in the hands of foreign firms. While *Parfums Dior* and *Givenchy* (belonging to the Moët-Hennessy-Louis Vuitton group), as well as *Guerlain, Lanvin, Nina Ricci*, and *Patou* are still under French control, *Cardin Parfums* belongs to American Cyanamid, *Orlane* to Norton Simon, *Chanel* to the Swiss Pamerco group. *Rochas* was owned by Hoechst, which has now sold it, and *Parfums Saint Laurent* is now controlled by Italy's Carlo de Benedetti, who bought it from the U.S.-based Squibb. As for the L'Oréal group, which had taken over the *Lancôme, Jacques Fath, Guy Laroche, Ted Lapidus, Cacharel,* and *Courrèges* perfume brands, it has been within the Nestlé orbit since 1973, although it was arranged that until 1993 it would be managed by those representing the interests of the founding family. Not surprisingly, the same great names recur in the area of *beauty products,* including, besides perfumes, *hair care products* and *cosmetics.* Each firm, indeed, wishes to complete its range by acquiring complementary businesses.

The cosmetics industry was born in the United States with the three great "ladies," *Elizabeth Arden, Harriet Hubbard-Ayer,* and *Helena Rubinstein.* In their wake are now *Estée Lauder,* the giant *Avon, Max Factor,* and *Revlon,* founded by Charles Revson. A number of these firms did not survive their founders. Elizabeth Arden was first bought by Eli Lilly and now belongs to the United States *Fabergé* groups; Helena Rubinstein has disappeared after being taken over by *Colgate Palmolive*; following ten years of poor management and uncontrolled diversification, particularly in pharmaceuticals, Revlon has been grabbed by Pantry Pride, a chain store group belonging to the raider Perelman; Max Factor now belongs to the *Norton Simon* group.

While all this was taking place on the American scene, two groups, one Japan's *Shiseido,* and the other France's *L'Oréal,* were climbing to the rank of leading world producers, raised there by dint of good management and competent research and marketing skills. Although Shiseido was unsuccessful in its bid in the United States to take over the famous *Giorgio* of Beverly Hills, which was acquired by Avon, and the skin care company *Charles of the Ritz,* which Yves Saint-Laurent had sold back to Revlon, it nevertheless ranks second in the world after *L'Oréal* and has very strong positions throughout Asia.

L'Oréal's founder, Eugene Schueller, graduated as a chemist from Institut de Chimie de Paris. He resigned from his job at the Sorbonne to produce a "harmless" hair dye called l'Auréole. The trade name l'Oréal was

adopted the following year. A skillful businessman and a true pioneer of ad campaigns, Schueller bought the Monsavon soap factory in 1928 and, before the war, brought on the market the O'Cap hair lotion, then Ambre Solaire. When he died in 1957, his successors managed to develop the business both through internal growth and an efficient research base and through a series of acquisitions.

In 1961, Monsavon was sold to Procter & Gamble, and L'Oréal purchased the *Cadoricin* firm, which extended its range of hair products, to which were added *Garnier* and *Roja.* Then Lancôme was purchased, introducing high-class products. This was followed by the purchase of other perfumemakers. Tempted by the pharmaceuticals market, the company bought *Synthélabo* in 1973, consisting of four medium-sized laboratories. It is still too early to say whether the money sunk into the sector since then will bring in returns as large as those of the perfumes and skin-care business. Mixing the two has not always been successful.

In the United States the marriages between Pfizer and Coty, Colgate and U.S. Vitamin, Eli Lilly and Elizabeth Arden, Squibb and Charles of the Ritz, Avon and Mallinckrodt, Revlon and Armour Pharmaceuticals all ended in divorce. There was, of course, Bristol-Myers' successful venture with Clairol, and American Cyanamid with its Shulton subsidiary. But these exceptions only confirm the general rule of failure.

In France, while Sanofi can draw satisfaction from its association with Yves Rocher, which enjoys great management freedom, the sector comprising Roger & Gallet, Stendhal, and Charles Jourdan has not yet lived up to the parent company's expectations. Only the British seem to have succeeded in combining such different businesses, possibly because from the start the skincare activities were intimately associated with pharmaceuticals within large groups like Beecham, Glaxo, and BDH.

THE CHEMISTRY OF ADDITIVES

Used in small doses to improve the products in which they are incorporated, additives are to be regarded as specialties with well-stated functions even if, in many cases, they are well-defined chemical entities sold according to specifications. Because of this ambivalence, chemical companies have approached the sector of additives sometimes through the markets they serve, sometimes through the chemicals from which they derive, even at times from both ends.

Additives for Plastics

Additives for plastics have experienced the double approach. The opening up of the markets leading to uniform production of plastics gave worldwide scope to some additive producers. Tasks were shared since polymer producers did not consider it useful to prepare the additives they needed, while additive producers were, as far as possible, careful to avoid competing with their clients in the area of base thermoplastics.

It is true that a major polyolefin producer like *Hoechst* sells its own range of antioxidants and its subsidiary *Riedel de Haen* produces ultraviolet ray absorbers. Likewise, the world ABS leader *Borg Warner*, now acquired by General Electric, has been marketing, since it took over *Weston*, a series of organic phosphites for the stabilization of high polymers. In Japan, *Sumitomo Chemical* is a supplier of large-volume plastics as well as of a rather complete range of stabilizers.

These are exceptions, however. The world's largest additives producer for plastics, *Ciba-Geigy*, remains, for its part, at the sole service of its downstream customers and tries not to appear as a competitor. This is also the position of other additive suppliers like *American Cyanamid, Ferro, Witco, UniRoyal Chemical* in the United States, *AKZO, SFOS* (Rhône-Poulenc) in Europe, *Adeka Argus* and *Dai-ichi Kogyo Seiyaku* (DKS) in Japan.

Ciba-Geigy owes its leading position to a number of factors: long perseverance in the specialty, an efficient research base through which the universally used *Irganox* antioxidants were developed, application services adapted to all the polymers requiring stabilization, worldwide production units estab-lished within large comsumption areas (Europe, America and Japan). Even where Ciba-Geigy did not invent a product but took a license on it as with *HALS* (hindered-amine light stabilizers), licensed from Japan's *Sankyo*, it developed it to the point of acquiring world supremacy in the area. Ciba-Geigy's success in this activity is all the more remarkable as it has no upstream integration on raw materials used in the synthesis of phenol antioxidants, of phosphites, of thioesters, of substituted benzophenones, of benzotriazoles, or of HALS. But this apparent weakness is fully compensated by the dominant position Ciba-Geigy has acquired in the different types of additives for plastics in its range, either through internal growth by its research, or through license acquisitions (Sankyo), or through purchase of relevant companies (Chimosa in Italy), or again through complementary activities (range of Goodrich's Goodrite antioxidants).

The other producers of plastics additives trail far behind Ciba-Geigy in variety of range or in market coverage. *UniRoyal Chemical* produces antioxidants and blowing agents and has production units in the United States, Latin America, Italy, and Taiwan; but its recent restructuring has cut short its development. *American Cyanamid* which pioneered a number of additives (substituted benzophenones, 2246) sold its European business to Ciba-Geigy in 1982 and now operates only in the American market. *AKZO's* range is restricted to antistatic agents, PVC stabilizers, and peroxide catalysts, which it acquired through *Armak, Interstab*, and *Noury van der Lande*. *Borg Warner* is mainly focused on phosphites, which it produces solely in the United States; Elf Aquitaine's subsidiary *M & T* is focused on organotins; *Witco*, through its purchase of *Argus Chemical* and *Humko Products*, is involved in heat stabilizers, antistatic agents, and lubricants. *Ferro*, which also produces master batches, has developed specialties such as fireproofing agents and stabilizers for PVC and has recently joined forces with Italy's Enichem to produce and market new lines of polymer additives in the United States.

Other companies came to additives through the chemical tree, such as *Société Française d'Organo-Synthèse* (SFOS), a subsidiary of Rhône-Poulenc, which by isobutylating phenols produces a whole range of phenolic antioxidants as well as special phosphites, or *Ethyl*, which approached the Irganox family of antioxidants and bromine fireproofing agents through its orthoalkylation technology and its access to bromine. Similarly, because the U.S.-based *Olin* was an important hydrazine manufacturer, it became interested in blowing agents like azodicarbonamide and bought up National Polychemicals, which also provided it with a range of phosphites. The blowing agent line has since been sold to UniRoyal Chemicals.

But the interest which major firms like ICI, Bayer, or even Hoechst still have in the sector is restricted by the small number of additives they supply to plastic producers. Under the circumstances, it is more than likely that Ciba-Geigy's lead in the variety of products offered, in research, in customer service, or in geographic coverage will be hard for competitors to catch up with. Indeed, their narrow approach to the market would hardly warrant the heavy investment to fulfill any high ambitions they might have in the area.

But favored by their access to certain raw materials or by their specialization in a very specific range, such competitors can, at least, be assured of a degree of prosperity inasmuch as the standards required for optimum use of plastics are closely related to incorporation in the high polymer of effective additives at a reasonable cost.

Rubber Additives

The specialists in *rubber additives* are distinctly different from the specialists in additives for plastics, even though the same products are sometimes used in both industries: blowing agents (azodicarboamide), phenol antioxidants (BHT, 2246), phosphites (tris-nonylphenyl phosphite). In the first place, additives for elastomers, unlike those which might come into contact with foodstuffs, do not require official approval, which makes it easier to put them on the market. In the second place, most of them are well-known products sold to specifications by a number of producers. The development of new products protected by patents is rarer than in the case of additives for plastics. In fact, while consumption of plastics has been constantly increasing, stagnant demand for both natural and synthetic rubber has not warranted any significant recent research efforts by suppliers of this industry.

If one considers that the automobile sector accounts for 75 percent of rubber consumption in developed countries, it stands to reason that the longlife radial frame and smaller diameter tires of modern vehicles should require smaller amounts of rubber for the same number of cars produced; in the United States alone, rubber consumption has fallen from 3.2 million tons in 1977 to a little over 2.6 million tons in 1989. This implies a consumption of some 150,000 tons of organic additives.

Faced with such a situation, producers of additives for rubber have either restructured or else rationalized production. In rarer cases, others have offered new products with higher added value than the conventional additives. In the United States, *American Cyanamid* in 1982 halted production in Bound Brook, New Jersey, of its accelerators; *Goodyear* terminated its substituted *p*-phenylenediamine production in Houston in 1984. In 1985, *Allied-Signal* took over *UOP*'s antiozonant unit, while in 1986 *UniRoyal Chemical* became part of *Avery, Inc.*, before becoming the object of a leveraged management buyout in 1989. In Europe, Rhône-Poulenc and ICI merged their rubber divisions within a subsidiary called *Vulnax* and then finally sold it to AKZO in 1987. Atochem, meanwhile, was taking a minority stake in *Manufacture Landaise de Produits Chimiques*, henceforward leaving France and Britain with no significant producer with international clout.

In this changed environment, three major additives manufacturers emerged: *Monsanto* with its plants in the United States, Canada, Britain and Belgium; *Bayer*, which owns two sites in Europe and produces antiozonants in

Pittsburgh through Mobay; and *UniRoyal Chemical*, which has production units in Naugatuck, Connecticut, and Geismar, Louisiana, as well as in Canada, Brazil, and Italy.

UniRoyal Chemical was separated in 1966 from U.S. Rubber, which had provided it with a captive market. But two other tire manufacturers had retained their traditional activities in additives. They were *Goodrich*, which produced only in the United States, and *Goodyear*, which also operates in Europe in its antioxidant units in Le Havre, France. Both these giants sell part of their production through a rubber blend specialist, R. T. Vanderbilt. Goodrich, however, has recently withdrawn from the tire business in order to concentrate on its chemical activities, so that only Goodyear enjoys today the advantage of a captive outlet for the rubber chemicals it produces.

Although Monsanto can rely on only two of its own raw materials, tertbutylamine and *p*-nitrochlorobenzene, for its range of additives, it is regarded as an efficient producer and a pioneer in antiozonants based on *p*-phenylenediamine and prevulcanization inhibitors. It has one of the most complete ranges of additives for rubber and the most modern units to manufacture them.

Because of its long experience in organic synthesis intermediates, *Bayer* is possibly better integrated upstream than Monsanto. Its range of products is just as large, but its production units are essentially restricted to Leverkusen and Antwerp. With the exception of *AKZO*, which, through its purchase of *Vulnax*, seems to want to improve its range of additives and its geographic coverage, no other major European chemical group has gone beyond a small range of special products.

Like their competitors in Europe and in America, the Japanese producers have focused their attention on accelerators (vulcanization activators and agents) and on anti-aging agents (antiozonants, stabilizers). Their automobile exports provide a market for tires that their counterparts in other countries cannot claim to the same extent.

Japan's additives production, however, is too scattered among a large number of producers to be truly profitable. With the exception of two principal companies in the area, *Sumitomo Chemicals* and *Mitsubishi-Monsanto*, firms like *Ouchi, Shinko, Kawaguchi*, and *Seiko Chemical*, which were the first to get into the business in 1930, do not have the required size to be competitive on international markets.

Additives for Lubricants

Additives for lubricants are also greatly dependent on the automobile industry, which alone uses some 60 percent of the lubricants produced worldwide, whether lube oils for engines (gasoline and diesel oil) or for gear boxes. Since the oil-price rise in 1973, lube oil consumption has been affected by a number of factors: smaller vehicles and therefore smaller engines, falling automobile production, larger intervals between oil changes, implying a higher additives dosage to extend oil efficiency. To these various changes should be added increasing use of diesel fuel in Europe because of favorable taxation. The generalized use of multigrade oils and the introduction of unleaded gasoline, and consequently of catalytic exhaust pipes, should lead to enhanced engine oils. In the circumstances, world consumption of additives for lubricants is likely to remain at around two million tons a year over the next few years, with higher additive doses compensated by extended lube oil efficiency and smaller casing size.

With the exception of *Lubrizol*, the world leader in this area, and *Ethyl*, which came to lube oil additives by buying Edwin Cooper off Burmah Oil in 1968, the main suppliers with extensive ranges of additives are the international oil companies *Exxon, Chevron, Amoco*, and *Shell*. The business was a natural extension of their lube oil production, which serves as a captive market.

All these oil companies market their additives as a package, the efficiency of which has been extensively tested. Most of the ingredients in the package are produced by the companies themselves: *detergents* (sulfonates, phenates, naphthenates), *dispersants* (succin-imides, polybutene, succinates), *antiwear agents* (zinc dithiophosphates, chlorinated paraffins, sulfur and phosphate hydrocarbons), *anticorrosion agents* (substituted amines, succinic acid derivatives, nitrites). On the other hand, the antioxidants are often supplied separately, as are pour-point depressants (polymethacrylates) and additives to improve the viscosity index of multigrade oils (polymethacrylates, olefin copolymers).

Originally called Graphite Oil Products, *Lubrizol* was founded in 1928 near Cleveland with a capital of $25,000 by six associates. Their close ties with the Case Institute of Technology gave the concern a strong technical orientation. In this manner, *Lubrizol* played a pioneer role in developing lube oil additives and is still today a world leader in its area, with fourteen plants installed worldwide, including four in the United States, and testing sites in Wickliffe (Ohio), Hazlewood (Britain), and Atsugi (Japan). In its attempts at diversification, the company recently became interested in biotechnology with the purchase of *Agrigenetics* in 1985 and of a stake in *Genentech*. But it is too early to state whether this choice will bring the same long-term satisfactions as the company's traditional business.

Exxon came to chemical additives for lubricants by producing its Paraflow range of freezing-point depressants as early as 1930 in Bayonne, New Jersey. Through the agreements signed in 1937 with IG Farben, Standard Oil of New Jersey (later to become Esso and then Exxon) acquired the thickeners and additives based on polyisobutylene that improve the oil viscosity index. In the 1960s, Exxon further enlarged its range of lubricant additives and in 1979 set up the Paramins special division, which marketed a series of olefin copolymers (OCP) based on the chemistry of the group's ethylene-propylene elastomers. The object was to compete with the polymethacrylates (PMA) in improving multigrade oils (VI improvers). Based in Houston, Texas, and involved in all world markets, Paramins has become Lubrizol's most dangerous rival.

Chevron approached the oil additives market in the 1930s by supplying metal naphthenates to its parent company Standard Oil of California. Some of these additives were marketed under the trademark Oronite from 1948 onwards. Chevron kept its main research center in Richmond, California, even when, in the 1950s, it spread to international markets through subsidiaries set up with local partners: *Orobis* with BP and *Orogil* with Progil in Europe; *Karonite* in Japan, *AMSA* in Mexico. In 1986, BP bought Chevron's 50 percent share in Orobis, and more recently Rhône-Poulenc sold to Chevron its 50 percent share in Orogil.

The interest shown by *Ethyl Corporation* in Edwin Cooper stems from its desire to diversify into the oil sector as unleaded gasoline begins to threaten the future of tetraethyl lead. But in a business in which it is a newcomer, Ethyl still has much to learn before attaining the efficiency and international coverage of its three main rivals. The same is true of *Amoco* and of *Texaco Chemicals*, although they are endowed with a significant captive market through their parent companies, Standard Oil of Indiana and the group made up of Texaco, Caltex, and Getty Oil respectively.

For its part, the *Royal Dutch Shell* group came to additives after the Second World War in the United States with a range of alkaline sulfonates. Subsequently it enlarged its range with new additives (detergents, dispersants, VI improvers) and fuller geographic coverage through production centers located in Berre, France; in Stanlow, England; and in Marietta, Ohio, and in Martinez, California, in the United States. More recently, a common subsidiary with Lubrizol was set up in Brazil.

Besides these large companies, which offer a range of additives as extensive as possible, if only to recoup research expenses and the high cost of tests required to obtain approval of the "packages," there are a number of chemical companies that have also established a foothold in the market of lube oil specialties. Their reason for doing so was that they had acquired know-how in the chemical sector leading to the products marketed.

Rohm & Haas in Philadelphia developed additives to lower the freezing points of oils and to improve their viscosity index through work carried out as early as 1934 by the chemist Herman Bruson on the properties of polymethacrylates (PMA) produced from higher methacrylates. Other companies, such as *Röhm* in Darmstadt and Melle-Bezons (whose Persan unit in France was bought from Rhône-Poulenc by *Société Française d'Organo-Synthèse* [SFOS] in 1978), also supplied PMA for such applications. Through the chemistry of phenol isobutylation, *Ethyl* and *SFOS* took a foothold in the phenol antioxidant market of oil companies, while *Ciba-Geigy* is developing a significant program in this area. But the need to be thoroughly acquainted with the lube oil business and to be well introduced in the world oil circles narrows the scope of chemical firms that have only a small range of additives to offer and precludes their taking a significant place in such a specialized market.

Food Additives

Because their nature, their uses, and their origins are extremely varied, *food additives* are supplied by a large number of different firms. In what is a fragmented industry, some producers stand out more because of the major place they occupy on the market than because of their range of additives.

In the United States, there are only two producers of *citric acid* (Pfizer and Bayer's subsidiary, Miles) and of *vitamin C* (Pfizer and Hoffmann-La Roche) and a single producer of *saccharin* (PMC, which bought Maumee from Sherwin Williams), *sorbates* (Monsanto), and *carrageenates* (FMC since it acquired Marine Colloids).

Because of the very strict rules that in industrialized nations govern additives used in human food, it has become very expensive to introduce new products. In some countries even some of the older derivatives that used to be considered nontoxic have been questioned. This is the case with saccharin, discovered by Ira Remsen in 1879 and used without drawbacks since then. Because of such limitations, few new producers have ventured into the area over the past few years except through purchase of existing companies that already had approved additives. The giants in the business are generally satisfied with being dominant in certain market sectors through their special technologies (fermentation, extraction, synthesis). The problems of excessively high sugar consumption, however, have induced a number of researchers to look for low-calorie substitutes for sucrose other than saccharin. Accordingly, new *synthetic sweeteners* have been discovered: *cyclamate* (sodium cyclohexylsulfamate), synthesized in 1937 and put on the market by Abbott in 1950; *aspartame*, isolated in 1965, produced by reaction of aspartic acid with phenylalanine methyl ester, and developed by Searle, which was susequently purchased by Monsanto); and Hoechst's *acesulfame K*. Despite lack of coordination in this area among the different national legislatures, these synthetic sweeteners, with their low calorie content and a sweetening power that is fifty to two hundred times as great as that of sugar, should sooner or later take root on international markets.

The use of *gelling* and *thickening* agents in foodstuffs goes back to earliest times. In the last few years, progress has been made in the extraction and purification of plant-based hydrocolloids used for the purpose. In addition, the polysaccharide *xanthane*, produced through fermentation, has been developed over the past twenty years to take its place among the water-soluble gums supplied to the food industry. At the same time, a semisynthetic gum, *carboxymethyl cellulose* (CMC), used in a number of industrial applications, was allowed in its purified form, in human foodstuffs.

The U.S.-based *Hercules*, which started by producing precisely this CMC of which it is the world's leading producer, has gradually extended its range of products by purchasing companies. It is now involved in *pectin*, extracted in Denmark and in the United States from lemon peel; and *guar*, prepared in Italy from a bush that grows in India and Pakistan; *carob* developed in Spain; and *carrageenates* extracted from algae growing along the Atlantic and Pacific coastlines. With the exception of pectin, these various gums recently became the business of *Aqualon*, a common subsidiary of Hercules and *Henkel*, its German partner, already involved in the guar and CMC market. In 1989 Hercules became the sole owner of Aqualon.

Hercules never did succeed in developing *Xanthane* through its association with the British company Tate & Lyle in 1979. This gum has remained a specialty of *Rhône-Poulenc* which produces it in Melle, France, and of *Kelco*, a San Diego, California, subsidiary of Merck that also owns *Alginates Industries*.

Other groups have likewise specialized in particular sectors. The Stein Hall subsidiary of Celanese, taken over by the British-based *RTZ Chemicals*, now part of Rhône-Poulenc, has focused on guar, while *Marine Colloids*, a subsidiary of FMC, and *Satia*, of the Sanofi Elf Bio Industries group, specialized in carrageenates. There are a great number of industrial applications for gum, and thus gum producers are usually drawn to the food industry because of their know-how in gum. It is seldom that they have chosen to manufacture thickeners because of their experience in foodstuffs.

The same is true of *antioxidants* like *BHT* (butylhydroxytoluene). Although it is used in purified form in human and animal food, its more common use is as a stabilizer for polymers and lubricants. Only *BHA* (butylhydroxyanisole), α-*tocopherol* (vitamin E), *TBHQ* (tertiary-butylhydroquinone) and *propylgallate*, which are marketed by *Eastman Kodak*, can be considered as purely food antioxidants for the two reasons that they are not toxic and that they are high priced. In fact, Eastman Kodak is the only chemical leader to produce an extended range of food additives: mono- and diglycerides and vitamins.

Producers of *acidulants* came to food applications through chemistry or biochemistry. *Malic acid* is produced by Denka, now owned by Mobay, in the United States and by Croda in England. Like *fumaric acid*, it is a derivative of maleic anhydride production. The major acidulant is *citric acid*, which is also used as a stabilizer. It is a fermentation product that is produced by a few traditional specialists—in Europe by La Citrique Belge, which was bought by Hoffmann-La Roche; in Britain by Sturges, taken over by RTZ Chemicals and now owned by Rhône-Poulenc; and in the United States by Pfizer and Miles.

Phosphoric acid, used in fizzy drinks, is produced in its food quality only by a small number of firms such as FMC and Stauffer, an activity taken over by Rhône-Poulenc in the United States, and by Prayon in Belgium.

Food conservation generally requires the use of chemical additives, although the problem can be solved at times through temperature control (pasteurization or sterilization through heating, freezing, or control of water content [dehydration]). Chemical additives act by working on the metabolism of the microorganisms responsible for food deterioration. More often than not they involve organic acids and their salts, *propionic acid, potassium sorbate, sodium and calcium propionates*, and *sodium benzoate*, traditionally used to preserve cheese, jam, cakes, and fatty materials.

Here again a few large companies such as Monsanto for sorbates and Pfizer for propionates have acquired a leading place on the markets. On the whole, the food additives sector is less open to restructuring and rationalization because it is made up of enterprises that are fundamentally different in size, technologies and in objectives pursued.

PHOTOCHEMICALS

Since the early 1980s, the major photographic companies have made efforts to bring changes to their basic technologies, which had long remained unchanged. The U.S.-based *Eastman Kodak* became interested in reprography, setting up its own range of photocopying machines. It also became involved in electronics and video to counter competition from new Japanese equipment (such as Sony's Mavica filmless cine-camera). The other photographic giants like Bayer's subsidiary

Agfa-Gevaert and *Fuji Photo Film* have also invested heavily in new areas, the former in magnetic tape and reprography and the latter in photo disks. *Polaroid*, whose founder Edwin Land remained to the day of his retirement an advocate of specialization, is also starting to put a range of videocassettes on the market. These changes, however, are essentially intended for the amateur and mainly concern camera manufacturers. Overall, the sensitive surfaces market should receive no shakeup from these new ventures, for there is still a high demand in a number of areas where photography remains irreplaceable (press and publishing, scientific research, industrial applications, radiographic control devices). The industry's structure reflects this stability. It is not likely to be upset in the immediate future because of the power acquired by the few large multinationals, which vie with one another on international markets and give any newcomer little chance of success.

Unable to compete with Kodak on the American market, *GAF* withdrew from the film industry in 1982. Previously, the first European merger had taken place between Belgium's Gevaert and Germany's AGFA, producing Agfa-Gevaert. Its early years were hard ones, and it is now fully owned by Bayer. Italy's *Ferrania* was taken over by America's *3M*, while Ciba-Geigy was bringing together Britain's *Ilford*, France's *Lumière* and Switzerland's *Telko* within the Ilford group based in Britain.

Following these restructurings, which, in many cases, took place some time ago, the photographic film industry is now dominated by three giants: *Eastman Kodak*, with units in Rochester, New York, in the United States, in Châlons, France, and in Hemel Hempstead, England; *Agfa-Gevaert*, which produces its photochemicals in Antwerp, Belgium, and Vaihingen, West Germany; and *Fuji Photo Film*, which produces in Japan and has recently set up a film unit in Holland. *Ilford* and *Polaroid*, which went through difficult periods of readaptation; *3M*, which is involved in other areas besides photography; and *Konishiroku* (*Konika*) in Japan, which

bought Fotomat in the United States, cannot be regarded as dangerous rivals to the Big Three.

The three major companies follow different policies in matters of raw materials. Fuji Photo film, which has no links with the chemical industry, buys 80 percent of its supplies outside, while Eastman Kodak and Agfa-Gevaert supply half their needs through their own production. They all produce their most "sensitive" organic derivatives, which are kept secret since they form the basis of emulsion quality.

Although polyester film, introduced by Du Pont under the trade name Mylar in the 1960s, has been added to the traditional supports like paper and cellulose acetate, the principle of photographic film preparation has remained unchanged since "daguerreotype" was developed. The sensitive surface always contains a silver halide crystal emulsion with a *gelatin* binder. Despite all the efforts to replace them, silver salts remain the basis of these emulsions, and film manufacturers still require gelatin, which they consume at the rate of 20,000 tons a year. The suppliers are few, and they are carefully selected. The world leader in this area is *Rousselot*, now a subsidiary of Sanofi Elf Aquitaine, with four units in Europe and one in the United States.

Reducing agents such as *hydroquinone, metol* (*p*-methylaminophenol) and *p-phenylenediamine* are generally purchased from outside producers. Eastman Kodak produces its own hydroquinone, however. The other producers get their supplies from Rhône-Poulenc or from Japanese firms like Sumitomo Chemicals or Mitsui Petrochemicals.

Color photography, now fully perfected, requires a *developer* like N,N'-diethylphenylenediamine which reacts with silver salts. The oxidized derivative obtained reacts with a coupling agent made up of groups ($-CH=$) or ($-CH_2-$) to produce the desired color.

Formulations for sensitive surface emulsions also include *accelerators* (alkaline carbonates, borax), *stabilizers* (sodium bromide, benzotriazoles), *conservation agents* (sodium sulfite), *hardeners*, which improve gelatin behavior (chloromucic acid, substituted 2, 4-dichlorotriazines).

The great variety of products used by film manufacturers, their stringent quality standards, and their secretiveness, which prevents them from subcontracting their most advanced formulations, are all factors that keep photochemical producers apart from the rest of the chemical industry and keep newcomers out of their sector.

THE ALLIANCE OF CHEMICALS AND ELECTRONICS

Chemical products used today in electronics seem, at first glance, to be a very ordinary kind. They are different from those generally offered, however, by reason of the extraordinary degree of purity which their producers must achieve in order to satisfy the stringent requirements of the electronics industry. The maximum dose of impurities tolerated in monocrystalline silicon amounts to one part in 10^{13}.

Polycrystalline silicon, produced from silane (SiH_4) or trichlorosilane ($SiHCl_3$) forms the upstream part of the *semiconductor sector*. Monocrystalline silicon is extracted from polycrystalline silicon and sliced into wafers 25 microns thick and 8 to 10 centimeters in diameter.

Hoechst's subsidiary Wacker is the world's leading polycrystalline silicon producer, with a capacity exceeding 2,000 tons. The overcapacities that began affecting the electronics industry in the early 1980s forced Monsanto, one of the largest wafer producers, to slow down its silicon production units in 1984. It has since sold this business to Germany's Hüls. Rhône-Poulenc, which had ambitions in the area but lacked the right technology, has withdrawn from the business.

There are enough suppliers of this type of silicon, including, for instance, Dow-Corning, Dynamit Nobel, Shin-Etsu, Tokuyama Soda, Motorola, and Texas Instruments. A possible substitute for the silicon used to produce wafers is *gallium arsenide*, in which Rhône-Poulenc, ICI, and Shinetsu are already involved.

This situation shows how closely suppliers of electronic chemicals need to monitor the very rapid developments taking place in the area; otherwise, their productions run the risk of becoming obsolete before the full payoff.

Photosensitive products are also used for the production of wafers. These *photoresists* polymerize through X-ray treatment. They are called positive or negative according to whether or not they are soluble in solvents when exposed to light. The miniaturization of printed circuits tends to give a boost to positive resins. Germany's Hoechst has pioneered in such photosensitive resins. They are also supplied by Eastman Kodak, Olin Hunt, Ciba-Geigy, E. Merck and Tokyo Ohka Kogyo.

A great number of chemical firms have set up special divisions to manufacture products for the electronics industry, essentially through acquisitions. For example, *Du Pont* bought *Berg Electronics* in 1972, and a little later *Olin* purchased *Philip A. Hunt*. Some companies, such as Du Pont, Olin, and Ciba-Geigy, have chosen an "integrated systems" approach in this area by providing as wide a range as possible of products and services for the electronics industry. Others have elected to remain strictly within the special areas in which they excel through long experience or proper chemical integration. Thus it was the work carried out before the war with AEG that led *BASF* to make its range of magnetic tapes and gave it the supremacy in chromates which it shares with Du Pont. *Hoechst* came to silicon through Wacker and to gases through Messer Griesheim, and now provides, besides high-purity special gases, a range of photosensitive polymers. Rhône-Poulenc became involved in printed circuits through its polyimide resins and Ciba-Geigy through its epoxy resins.

Most of the companies already producing diethylene glycol terephthalate polymers have launched into the applications of polyester film to video and data processing, Hoechst through its Kalle subsidiary, ICI, Rhône-Poulenc, Du Pont, Japan's Toray, Teijin, and Toyobo, the latter in association with Rhône-Poulenc in Nippon Magphane.

Although Rhône-Poulenc has given up direct upstream development after fruitless association with Dysan in magnetic supports and Siltec in silicon, it still believes it can use its know-how in rare earths to develop their electronics applications. Today, Rhône-Poulenc is the indisputable leader in rare earths, accounting for 40 percent of the world market. At its units in La Rochelle, France, and Freeport, Texas, it is capable of extracting from lanthanide sands the fourteen elements they contain. Over the last few years, *samarium*, for instance, has become essential for microelectronics to the same degree that *europium* and *yttrium oxides* already are for color television.

Whether they approach electronics directly, or through chemicals, or both, chemical companies involved in this business can hope to reap the fruits of their efforts in this area, providing, however, that the sector is spared the technological and economic jolts it has suffered over the past ten years.

CATALYSTS

Ever since England's Humphry Davy observed in the early 1800s that water was formed when hydrogen and oxygen react in the presence of a red-hot platinum wire, the phenomenon which Berzelius was to call catalysis has intrigued chemists. The uses of catalysts in industry were first consciously demonstrated by Peregrine Phillips in 1832 when he used platinum to oxidize sulfur dioxide (SO_2) to form sulfur trioxide (SO_3) and by Frédéric Kuhlmann in 1837, when he produced nitric acid from ammonia.

Early in the twentieth century, Germany's Wilhelm Ostwald, France's Paul Sabatier, and America's Irving Langmuir had advanced a step in interpreting the phenomenon of catalysis by showing that it was characterized by an acceleration of the rate of reactions and that it was conditioned by the state of the catalyst's surface. From then on, chemical technology made striking progress through use of catalysts. Between 1905 and 1920, and more particularly in Germany, there was a spurt of new industrial-scale processes, for example,

Fischer-Tropsch synthesis and BASF's use of vanadium oxide to produce sulfuric acid.

It is no exaggeration to say that without catalysts Germany would have been in no condition to pursue its war effort until November 1918. Likewise, if Houdry had not developed in the early days of World War II its "catalytic cracking" process, the United States would have found it very hard to provide its bombers with light fuel. It was also through catalytic reforming that the United States managed to obtain from petroleum the toluene needed to produce TNT between 1941 and 1945.

Since then, catalysts have played an essential role, particularly in the production of ethylene oxide from ethylene (Shell, Scientific Design), in the synthesis of hydrogen cyanide and acrylonitrile through ammoxidation (oxidation in the presence of ammonia), of formaldehyde (from oxidation of methanol), and, of course, in the polymerization reactions to produce plastics, elastomers, and synthetic fibers. It is not surprising, under the circumstances, that a catalyst industry should have developed after World War II through internal growth or through acquisitions. The very diversity of catalysts and of their uses has necessarily led to a fragmented sector.

Some oil companies became involved in the production of catalysts because they needed them in their own refineries. *Mobil* has developed the ZSM 5 catalyst based on zeolite following studies which began as early as 1936 on catalytic cracking; *Shell* has used its own technology to develop the sales of its catalysts for hydrogenation cracking. Other companies became involved in catalysts because of their precious metals business. *Johnson Matthey, Engelhard*, and *Degussa* applied their know-how in platinum metals to industrial catalysts. Chemical firms, for their part, approached the area in different ways. *ICI* made the most of its acquired know-how, particularly in methanol and ammonia, by associating with *Nalco* to form *Katalco*, a catalyst supplier; *American Cyanamid* has set up a subsidiary in Holland with Ketjen; *Rhône-Poulenc* has formed *Procatalyse* in joint venture with Institut Français du Pétrole.

In other cases, the involvement in catalysts has been through acquisitions. *W. R. Grace* bought *Davison Chemical* in 1953, and in 1984 Union Carbide purchased *Katalistics International*, BV. One of the three leading United States companies in cracking catalysts, together with Engelhard and Davison, is *Harshaw-Filtrol*, which is the result of the merger of subsidiaries of Gulf Oil and Kaiser Aluminum & Chemical that have specialized in the area.

The developers of new processes have found it at times more expedient to set up their own separate entities to supply the catalysts they were advocating. Allied-Signal's subsidiary *UOP* did so for its platforming; *Houdry* for its catalytic cracking; *Ralph Landau* for the silver catalyst used for direct ethylene oxidation, which was marketed by *Halcon SD* and subsequently taken over by Denka, then by Bayer; and *Phillips Chemical* for its polyolefin catalysts, sold through its subsidiary, Catalyst Resources.

Through inert supports, a number of firms have succeeded in creating a niche in catalysts—for instance, *Crosfield*, a subsidiary of Unilever in England and a silica producer; or the German *Südchemie* group, which specializes in hydrogenation and polymerization catalysts; or again *Condea*, which produces in West Germany alumina of high purity. The sector also includes a few firms which are only involved in a very special sector. Denmark's *Haldor Topsoe* makes catalysts for the synthesis of ammonia and methanol; and *Lithium Company of America*, an FMC subsidiary, produces lithium, while *Du Pont* makes boron derivatives.

Linked to the oil industry, to petrochemicals, and to the large commodity chemicals, the catalyst industry can hardly escape the economic ups and downs affecting these three large sectors. Its clients are understandably both demanding and prudent, for the catalytic system is basic to the good running of production units. This explains why it is an area of business that is so difficult to penetrate and run profitably. Its structure should therefore remain rather stable even with the development of catalytic exhaust systems. Introduction

in the United States and in Europe of unleaded gasoline and the use of bimetallic systems for catalytic reforming should open up new markets for platinum and rhodium.

RETROSPECT AND PROSPECT

The economic slump that started in 1973 when OPEC pushed up crude oil prices challenged what were until then regarded as indisputable truths.

First came the realization that just as no tree can climb as high as the sky, so *no growth can be guaranteed to be continuous*. Suddenly investments made at a time of high inflation and low interest rates turned out to be disastrous as demand slowed down simultaneously with monetary erosion.

The scale effect, which until then was assumed to be cost-saving, showed its weaknesses as the giant steam crackers proved more expensive to run at low capacity than smaller units already written off and working at full capacity.

The notion that production costs could be improved by grabbing a greater share of the market turned out to be fatal as the gain in sales was wiped out by severe price erosion.

Likewise the assumption that the fruits of research would be propor-tional to the funds devoted to the sector was totally invalidated, for never had the world's chemical industry spent so much money in research and development to so little avail. At the same time, the venture capital poured into biotechnology companies has yet to bring in the returns expected. The managers of chemical plants, wary of world petrochemical and heavy chemical overcapacities, believed they would find in a switch to specialties at least partial compensation for the losses incurred through traditional productions. Although they were not all disappointed in their hopes, some of them found that results obtained fell short of expectations, for until then specialties had been the special field of firms that had acquired experience in what were specific and as yet uncrowded sectors.

Manufactured by too many producers, some specialties were becoming commonplace. For a manufacturer, there are only two kinds of products, those that make money and those that do not. The profits that can be made on a sale are closely related to the number of producers on the market and to the day-to-day relationship between supply and demand. Whether a product is deemed a "commodity" or a "specialty," it is all the more profitable for its being offered by a smaller number of producers for a demand that remains unchanged. In this context, there are some pharmaceutically active materials protected by patents and some secret formulations that are genuine profit centers for their producers. Likewise, should a base product become scarce on the market because of an accident on a petrochemical site or because of sudden high demand, prices soar and the fortunate producer can turn out the product to maximum capacity and profit.

Over the last few years, the high cost of installations and of the money needed to finance them was not conducive to the building of new plants on any large scale in industrialized countries. But as demand trends have been moving upwards lately, *petrochemicals* have at long last *returned to profitability*.

The specialties rush of chemical leaders is, on the contrary, more likely than not to produce a surfeit of products, at least insofar as some specialties are concerned. These will shed their "added value" and consequently lose their attraction for the too numerous industry leaders that had decided to follow that path.

Other disappointments are likely to come from the organizational and managerial differences between a purchaser and the specialties firm acquired. The many divestments that have often followed upon hasty acquisitions show how difficult it is to force on an entrepreneurial company the management methods of a large multinational.

One of the paradoxes of the last few years has precisely been that specialties suitable for medium-sized firms capable of being flexible in their approach to daily matters should have fallen into the hands of chemical giants with necessarily heavier structures, while in the United States, for instance, through the "leveraged buyout" procedure, a few strong-minded individuals have succeeded in taking over large petrochemical and thermoplastic production units considered until then as the rightful field of the industry's greats.

It is not certain whether the errors of the past will be repeated in the future. The thirst for power could indeed lead some company heads to overinvest, especially if they have public funds at their disposal. They would then recreate the overcapacities that have been so harmful to fertilizer, petrochemicals, synthetic fibers, and plastic producers over the last few years. It is also likely that specialties will continue to attract industry leaders anxious to develop fresh prospects.

Let us hope that all the decision makers will bear in mind that capital funds, whether provided involuntarily by the taxpayer or willingly by the shareholder, are a rare resource that must be judiciously allocated and that success in all things comes from mastery acquired through long patience. In this respect, Germany's chemical industry, which has shown continuity from the time it was established in the last century to the present under the guidance of professionals, is a tried and tested model, showing profits even in the most adverse circumstances.

Drawing inspiration from this example for long years, the United States chemical industry, under the pressure of financial analysts and raiders, has in recent years undergone many upheavals. While they provided new opportunities for the fortunate few, they changed the environment and made people forget that to operate efficiently any industry must set its sights on the long term.

For reasons that were more political than financial, France's and Italy's chemical industries have also undergone too frequent changes over the past twenty years—in their structures, their strategies, and their management teams—to have had a chance of getting through the economic slump unscathed. It is only very recently that they have returned to profits by recovering a measure of stability.

Worldwide, in 1987 through 1989 the industry, whether in specialties or in basic chemicals, has certainly had its most prosperous years ever. The chemical industry, on the whole, does not, however, enjoy a very favorable image in the eyes of the public. The harmful spillovers caused by untoward accidents are given wider publicity by the media than the benefits the industry provides. In consequence, administrations that were anxious to soothe the more or less justified fears of their citizens have brought out a spate of regulations often more restraining and therefore more costly than is really necessary.

Since one cannot work simultaneously toward a thing and its opposite, no great spate of discoveries useful to humanity should be expected at a time when everything is being done to make it difficult to bring new products onto the market. For a long time the chemical industry was left free to apply its own safety standards and could devote most of its time to the development of new products. In the last few years, it has had to submit to increasingly costly and prolific rules and regulations that require its attention and delay the development of innovations that could save human lives or at least improve our living conditions. Some balance will have to be found between safety requirements and the wider interest of the public.

As in all history, the story of chemicals recalls past events and makes an attempt to explain them. But it can neither create them nor prevent them from recurring. While such history, therefore, teaches us the essential facts that have taken place within two richly endowed centuries, it does not tell us which major facts will form the threads of the next years. It is this unknown factor which makes up the spice of our professional life. We can at least hope that if we conform to reason, to ethics, and to scientific and economic laws for all that is within our scope, each of us will have served this wonderful science that is chemistry to the best of our capacities and in the interests of the greatest number of people.

II. THE PERIOD OF THE 1990s

THE CHEMICAL INDUSTRY UNDER PRESSURE FROM PUBLIC OPINION AND REGULATORY AUTHORITIES

Prior to the mid-1980s the chemical industry experienced some upheavals of which the thalidomide tragedy of the early 1960s and the Bhopal catastrophe in 1984 are two major examples. However, the lessons learned from such sad events led to a spectacular improvement during the 1990s in the safety record of the industry in the Western world.

At the same time much progress has been made in the abatement of pollution in the air and the treatment of effluents from chemical operations in North America, Western Europe, and Japan. Between 1978 and 1988 the content of SO_2 in the atmosphere was reduced by 30 percent in the United States. Similarly, constant improvements in the way chemicals are manufactured have reduced the amount of by-products resulting from chemical operations and, therefore, of the quantities of effluents to be treated.

Paradoxically, as these improvements were brought about, the chemical industry in the Western world has become the preferred target of environmentalists, and through the influence of the media, its image has been deteriorating in the eyes of the public at large. The time is indeed long gone when a firm like Du Pont could print as a motto on its letterhead "better things for better living through chemistry."

Public opinion was just one area in which the chemical industry of the developed nations suffered setbacks. The industry had to face a more tangible threat in the form of increased pressure from regulatory authorities. For the United States alone, Edgar Woolard, then chairman of Du Pont, cited a figure of $585 billion for the financial burden incurred by industry in 1993 as a result of federal regulations, and he predicted the figure to reach $660 billion in 2000 ("In Praise of Regulation Reform," *Chemistry & Industry*, 5 June 1995). At the same time the Environmental Protection Agency itself projected that by the end of the 1990s the United

States would spend $160 billion per year on pollution control. In 1996 Ben Lieberman, an environmental research associate with the Competitive Enterprise Institute, estimated that in the United States the cost of the phase-out of chlorofluorocarbons (CFCs) in accordance with the 1987 Montreal Protocol on Substances That Deplete the Ozone Layer could reach $100 billion over the next ten years. Indeed chemical manufacturers had to develop eco-friendly substitutes such as hydrochlorofluorocarbon (HCFC) and hydro-fluorocarbon (HFC), which are more costly to make, and hundreds of millions of pieces of air-conditioning and refrigeration equipment using CFCs had to be discarded.

In fact the chemical industry has been affected in many different ways by the flurry of regulatory edicts in the 1990s. For example the banning of CFCs, together with new restrictions on the use of chlorinated solvents, has forced chemical producers to steer their product mix of chlorinated hydrocarbons away from precursors of these two categories of chemicals. Similar disruptions in the "chemical tree" of derivatives have been caused by the phasing out of tetraethyl lead as an antiknocking additive for gasoline and by its replacement with methyl tertiary butyl ether, whose fate is now held in balance by the authorities in California.

The fear of dioxins, which was born from an accident occurring on 10 July 1976 at the unit in Seveso, Italy, of Givaudan, a subsidiary of the Swiss Hoffmann-Laroche, has also changed the way many chemical operations are conducted. That accident, although it caused no human fatalities, damaged the environment by releasing in the atmosphere some 500 grams of a very noxious chlorinated impurity of the family of the dioxins. In order to avoid a release of even the most minute quantities of this particular dioxin again, hexachlorophene, a very useful germicide is no longer manufactured. In addition, pulp and paper mills in North America have been asked to drastically curtail the use of chlorine and chlorates for bleaching the pulp, and industrial and municipal waste incinerators are being submitted to stricter air-pollution con-

trols, which make it necessary to install high-efficiency scrubbers.

The more stringent regulations enacted by Western governments have led, in turn, to a delocalization of chemical activities to places where such rules are less strictly enforced. Nations like India and China have thus become world leaders in the production of some fine chemical intermediates and dyestuffs. Similarly in Mexico many maquiladoras owe their success, in part, to this process of delocalization.

International trade in chemicals has been affected in many other ways by the vagaries of national legislations. For instance, meat treated with hormones is considered safe as food for Americans but not for the citizens of the European Union. Bovine somatropin (BST), a hormone that increases milk production in cows, has long been cleared by the U.S. Food and Drug Administration, yet it is banned by health authorities in Brussels. Products derived from biotechnology have been particularly prone to these inconsistencies. Hoechst spent seven years and 80 million DM to obtain permission to operate its artificial insulin unit in Frankfurt. Meanwhile one of Eli Lilly's plants located in Strasbourg, France, was regularly exporting an identical product to Germany.

While the use of genetically modified organisms in pharmaceuticals has finally been accepted because of their beneficial effects on human health, environmentalists, particularly in Western Europe, strongly object to the application of such organisms to agriculture. Transgenic crops, which offer increased resistance to herbicides and reduce the need for synthetic insecticides, have received the blessing of public authorities in the United States, Canada, and Argentina and have been planted with enthusiasm by farmers in these countries over the last ten years. Even though no harmful effect resulting from their use has ever been detected during that period, Greenpeace and other environmental organizations, with public backing, were able to convince European officials that the planting of transgenic crops should be severely restricted until more is known about how they react with the environment. The European Union's sudden conversion to the "precautionary principle" will have important consequences. International trade of transgenic food and feed based on genetically modified corn or soy will be submitted to stricter regulations. Further development of disease- and pest-resistant seeds may take more time to benefit the farmers and, in turn, the final consumers. As these restrictions are being implemented, chemical companies specializing in life sciences will have to consider a drastic restructuring of their portfolio of products.

THE STATUS OF SCIENCE AND TECHNOLOGY

At the end of the 1990s the chemical industry in its main activities had reached a stage of maturation with respect to innovation. Several factors can help explain this situation. First, the pace at which new discoveries were made between the 1930s and the 1960s was not sustainable for the same reason that new elements of the periodic table were not easily found once most of them had already been described. Second, the part of the turnover that chemical firms were able to devote to their research and development budgets became smaller as more of the available funds were used to cover increasing environmental expenditures. Third, the management of companies whose shares were quoted on the stock market had to pay closer attention to the wishes of stockholders and financial analysts, who were often more interested in short-term accomplishments than in ambitious R&D programs, which are necessarily lengthy, costly, and risky.

However, while no major breakthrough was made during the last ten years of the century in their traditional fields of endeavor, chemical corporations continued to improve the performance of their products by devising new methods for their manufacture. A case in point is supplied by the development of the metallocene catalysts. First described by Walter Kaminsky of the University of Hamburg in the 1980s and pioneered in the field of polyolefins by Exxon and Dow, these

organometallic initiators are still more expensive than the conventional Ziegler-type catalysts. They have nevertheless already gained wide acceptance in the field of polymer production because they make possible, owing to their special configuration, the production of a second generation of polyethylene and polypropylene plastics with improved characteristics. The substitution of butane for benzene in the production of maleic anhydride, which began in the 1980s, has at the same time lowered the cost of manufacture of this intermediate and done away with benzene as an objectionable raw material. In the period covered in this chapter, only biotechnology has offered the chemical industry new opportunities for spectacular developments in applications related to both pharmaceuticals and agrochemicals.

In the 1980s, as has been discussed (see p. 342), only a few bioengineered products were developed: human insulin and human growth hormone, both by Genentech, came on the scene, followed by the antithrombotic tissue plasminogen activator (Genentech, 1987) and the red corpuscle producer erythropoietin (Amgen, 1989). About the same time (1985) Abbott introduced a diagnostic test that could detect the AIDS virus in human blood collected for transfusions. By the early 1990s, however, through alliances with pharmaceutical laboratories, all these bioengineered products were already in commercial use with annual sales passing $100 million, and one third of the research projects of the major pharmaceutical companies were based on biotechnology. It is significant that Genentech, the biotechnology company that Roche now controls with ownership of 59 percent of its shares, has diversified its product range to include oncologic drugs for the treatment of lymphoma and breast cancer as well as cardiovascular products.

The technologies that have thus transformed drug discovery have also been applied to agriculture in such a way that new transgenic plant varieties were produced with characteristics that could not be easily obtained through cross-pollination. Monsanto, which had relied until the end of the 1980s on agrochemicals obtained synthetically, started selling seeds, which yielded transgenic crops engineered in such a way that they either offered outstanding resistance to herbicides or generated insecticides in the form of *Bacillus thuringiensis* toxins. By 1999, 33 percent of America's corn and 55 percent of its cotton crop as well as 99 percent of Argentina's soybeans came from such genetically modified varieties. Contrary to what happened to the pharmaceutical applications, the use of biotechnology in agriculture has been opposed by such pressure groups as Greenpeace and by public-sector agricultural institutions, so it may take some time before genetically modified crops are universally accepted.

In the 1990s the pharmaceutical industry proved to be innovative not only through the use of biotechnology but also through the discovery of new drugs by other methods. It applied the technique of chiral chemistry to isolate from a racemic blend the optically active molecule that is the one desired as a drug. Combinatorial chemistry was another tool the industry began to use: in one stroke thousands of small molecules could be made for screening as drugs rather than having to synthesize the molecules one at a time. It made use of improved drug-delivery systems that could bring new life to older products or maximize the number of potential drug formulations likely to accelerate the path from preclinical trials to final approval.

It even met with luck through serendipity in the field of "lifestyle drugs." Sildenafil, for example, which is being offered by Pfizer under the brand name Viagra as a treatment for male impotence, was initially developed as an antianginal drug before its property of improving male sexual performance was detected. Upjohn first marketed minoxidil (Rogaine) as an antihypertensive before it came to be recommended as a hair growth stimulant for the treatment of male baldness.

THE NEW LANDSCAPE

The 1990s were characterized by two main trends in the world economy. First was a move toward globalization made possible by the further development of free trade between nations; this led to a more competitive environment and made it necessary for corporations to streamline their operations and increase their productivity. Second was an expansion of stock markets, with more attention being paid to the financial performance of companies by the pension funds holding their shares and more generally by the various stakeholders.

In order to cope with the changing conditions, the management of many chemical companies of the Western world concluded that they had to operate along new lines. They gave priority to the concept of being the leaders in a few selected fields, and in order to obtain quick results, they came to favor external growth by acquisition at the expense of internal growth by innovation. This in turn led to a flurry of mergers, joint ventures, and divestitures that radically modified the landscape of the industry. Some companies already involved in pharmaceuticals and agrochemicals decided to concentrate on their life science activities exclusively. Oil companies that had previously diversified into fine chemicals and specialties left these fields and limited their ambitions to being strong in petrochemicals. Conversely, various chemical groups well established in basic chemicals tried to divest product lines considered to be too cyclical in favor of specialties. Meanwhile some individual entrepreneurs and financial buyers became interested in the very commodities that chemical giants were divesting.

At the same time these various acquisitions and divestitures were taking place, a minority of more traditional companies decided to retain the various areas in which they had been operating; they decided to try to grow internally by promoting products related to their "chemical tree" or, less often, by developing new molecules through their own research. As a result of these different attitudes, while the products made by the chemical industry remained generally the same over the decade, the ownership of the plants in which these products were made changed hands rapidly, and some well-known names in the industry disappeared, with new ones springing up in their place.

THE TRIALS AND TRIBULATIONS OF THE PHARMACEUTICAL INDUSTRY

For historical reasons the activities of the pharmaceutical industry in the Western world had been operated either by "stand-alone" pharmaceutical laboratories or by divisions or subsidiaries of diversified chemical groups. To the first category belonged such well-known firms as American Home Products, Bristol-Myers Squibb, Eli Lilly, Merck, Abbott, Upjohn, and Pfizer in the United States; Burroughs-Wellcome, Glaxo, and Beecham in the United Kingdom; Roussel-Uclaf and Servier in France; E. Merck, Schering, and Boehringer, Ingelheim in Germany; and Hoffmann-La Roche (now Roche) in Switzerland. The second category included the life science operations of American Cyanamid, Hoechst, Bayer, Rhône-Poulenc, Elf Aquitaine, ICI, Ciba-Geigy, and Sandoz. However, independently of their origin, all companies involved in pharmaceuticals had to face similar challenges in the period considered in this chapter:

- A growing population of older people with more ailments to be treated;
- Higher R&D expenditures requiring returns between $300 million and $600 million for each approved active pharmaceutical ingredient;
- Patents due to expire for blockbuster drugs, which would invite competition from producers of generic drugs; and
- Higher marketing costs at a time when Social Security institutions and health maintenance organizations insisted on lower selling prices for the drugs being offered.

As if this were not enough, the pharmaceutical industry was also confronted, as were other industries, with the new concepts of globalization and "shareholder value."

In order to increase their geographical coverage, most firms on both sides of the Atlantic resorted to mergers and acquisitions. Thus,

beginning in the late 1980s, Squibb merged with Bristol-Myers, SmithKline & French merged with Beecham, Rhône-Poulenc acquired Rorer, and Bayer took over Miles and Cutter Labs. Other transfers of ownership followed in 1994 and 1995: Roche bought Syntex, the Swedish Pharmacia merged with Upjohn, Hoechst acquired Marion Merrell from Dow and completed its control of Roussel-Uclaf to become HMR, and Glaxo and Burroughs-Wellcome formed a single entity. This was also the time when the pharmaceutical, crop protection, and nutrition operations of Ciba-Geigy and Sandoz were combined into a new company called Novartis (1995) and when similar operations belonging until then to American Cyanamid went to American Home Products (1996).

As the process of globalization proceeded, many diversified chemical groups started to pay more attention to the lucrative market of pharmaceuticals at the expense of their traditional chemical lines and to get interested in the promising field of biotechnology. Already in 1985 Monsanto had bought Searle before entering the field of genetically modified seeds. Du Pont also got involved on a limited scale in pharmaceuticals through a joint venture with Merck and moved into seeds. ICI, after the threat of a takeover by the British conglomerate Hanson Trust, split its life science and specialty chemicals operations from the rest of its portfolio by giving birth to Zeneca (1993).

Eastman Kodak decided in 1994 to part with Sterling Drug: the ethical drug division went to Sanofi (Elf Aquitaine), and over-the-counter drugs went to SmithKline Beecham. Dow also divested its pharmaceutical business by selling Marion Merrell to Hoechst, which later was to announce its intention to become a life science company (1997). After some hesitation Rhône- Poulenc followed the same path as Hoechst, and in 1999, having proceeded with the separation of their chemical activities from their core life science business, both firms put together, under the aegis of a new company to be called Aventis, their pharmaceutical divisions and subsidiaries as well as the former crop protection operations of Rhône-Poulenc and of AgrEvo, jointly owned until then by Hoechst and Schering.

These various moves, made under the pressure of financial analysts and with the purpose of enlarging the pipeline of active pharmaceutical ingredients close to approval, caused drastic changes in the structure of the chemical industry. Some of these changes were also a result of the outsourcing by pharmaceutical laboratories of most of their upstream chemical production, which gave a strong impetus to companies specializing in organic synthesis.

Within the pharmaceutical industry itself, the wish of each company's management to see its firm reach what is considered a "critical size" remained a constant feature as the century came to an end and the new millennium began. It led to new national and transnational mergers. In France, Sanofi (Elf Aquitaine) merged with Synthélabo (L'Oréal) and Laboratoires Pièrre Fabre with BioMérieux. Elsewhere the Swedish Astra and the British Zeneca combined their operations; Pharmacia-Upjohn took control of Monsanto; and two giant firms, Glaxo Wellcome and SmithKline Beecham, announced their intention to combine their activities, which would lead to a group with a turnover of $25 billion, exceeded only by Pfizer after its acquisition of Warner Lambert.

However, serious problems remain after such mergers take place. Shortterm savings do not necessarily produce long-term growth. Furthermore, the executives and their subordinates of the merged companies may not get along with each other, the best researchers sometimes leave, and most of the time size is not the corollary of creativity. For all these reasons some "contrarians" in the industry have decided to pursue a different course at least for the time being.

Roche, with its capital still controlled by the founder's family, has opted for internal growth. Although still a market leader in vitamins, the company is concentrating on its own research for pharmaceuticals. It is developing simultaneously a line of diagnostics after acquiring Boehringer, Mannheim and has gained access to biotechnology through its controlling interest in Genentech. In order to finance these activities, Roche has at the same time announced the spin-off of its flavors and fragrances business known as Givaudan. Among other contrarians, mention should also be made of Bayer, which with a pharmaceutical operation amounting to more than $5 billion in sales, has until now run the business under its existing structure. Other chemical giants like Solvay with Solvay Pharma, BASF with Knoll, and Akzo Nobel with Organon are following a similar strategy. It is possible, however, that because of the relatively small size of their pharmaceutical businesses they may decide to do otherwise in the future.

Stand-alone pharmaceutical laboratories have also in many cases preferred internal growth to mergers and acquisitions and have prospered by doing so. Such is the case particularly in the United States with Johnson & Johnson, Eli Lilly, Schering Plough, and Merck and Company. In addition biotechnology companies, which have developed products until now commercialized by well-established laboratories, are reaching a size (as with firms like Amgen, Chiron, Genentech, and Genzyme in the United States) that will allow them to consider acquiring their own pharmaceutical companies in the nottoo-distant future.

This survey of the Western world should not make us forget that Japan also has a thriving pharmaceutical industry. Three firms in particular—Takeda, Sankyo, and Yamanouchi Pharmaceutical Company—are of international repute. Takeda, the largest, is also the oldest, having been founded by the Takeda family in 1781. It has a joint venture with Abbott in the United States. Out of twenty-five blockbuster drugs currently available in the United States, six were discovered in Japan. But Japan remains weak in biotechnological developments, which are the source of much Western drug innovation.

THE RESHUFFLING OF CHEMICAL ASSETS

Many large groups, once deprived of their life sciences activities, were left with sizable chemical activities that had to be dealt with.

One widely used solution was the creation of new chemical entities through various spin-offs. Thus, between 1993 and 1999, Kodak gave birth to Eastman Chemical Company, American Cyanamid to Cytec Industries, Ciba-Geigy to Ciba Specialty Chemicals, Sandoz to Clariant, Monsanto to Solutia, and Rhône-Poulenc to Rhodia. Some more complex cases had to be solved differently. Hoechst, for example, proceeded in successive steps: in 1994 it set up a fiftyfifty joint venture with Bayer, called Dystar, for the purpose of managing their respective dyestuffs businesses. In 1997 it transferred its specialty chemicals operations to Clariant in return for a 45 percent stake in the enlarged company; finally, in 1999, it passed its industrial chemicals assets on to Celanese—the U.S. corporation it had acquired twelve years earlier—and passed Herberts, its coatings company, on to Du Pont.

ICI also was faced with difficult problems after the creation of Zeneca, as it had an impressive range of commodity chemicals to dispose of and wanted to acquire activities that would be more lucrative and less cyclical than the ones it was left with. In 1997, having changed its top management, ICI was able to acquire from Unilever, which wanted to focus on consumer products, three profitable businesses: National Starch (adhesives), Quest International (fragrances), and Unichema (oleochemicals), now called Uniquema. While taking advantage of this opportunity to re-enter the specialty chemicals markets, ICI began to divest its bulk chemicals and polymer operations: polyester fibers and films went to Du Pont; fertilizers and ammonia to Terra Industries; explosives to AECI and Orica; autopaints to PPG Industries; polyurethane, aromatics, and titanium dioxide (TiO_2) to Huntsman; and acrylics, fluorochemicals, Crossfield silicas, and chloralkali units to a privately held financial company called Ineos. At the end of 2000, with this ambitious program of divestitures completed, ICI had become an efficient specialties and paints company, and the three letters *I*, *C*, and *I*, which originally stood for Imperial Chemical Industries, are the only reminder of

Sir James Whyte Black, one of the recipients of the 1988 Nobel Prize in physiology or medicine. While at ICI, his pioneering work in analytical pharmacology led to the discovery of β-adrenoceptor and histamine type 2 antagonists. Courtesy Nobel Foundation.

its past glory. Other groups did not even retain their former names (for example, American Cyanamid, Sandoz, Hoechst, and Rhône-Poulenc) and therefore will soon survive only as entities in the memory of old-timers. Many other changes were to occur as the reshaping of the chemical industry continued.

THE IMPOSSIBLE MARRIAGE OF PHARMACEUTICALS WITH AGROCHEMICALS

Placing all life science activities under one roof appeared to be a logical decision, and it at first had the blessings of financial analysts. However, it proved unwise. Indeed, no sooner had Novartis, AstraZeneca, Pharmacia-Upjohn-Monsanto, and Aventis been created than their managements were made aware of some important facts:

- There was very little synergy between the pharmaceutical and the cropprotection parts of the business.

- The margins generated in the human health sector were far higher than the ones obtained in selling products to farmers.
- Pesticide sales are cyclical because they depend on the weather and on commodity prices.
- The backlash begun in Europe against genetically modified foods has decreased—at least temporarily—the expectations for high-tech seeds.

As a result all the above-mentioned companies have decided to divest their agrochemical operations through either mergers, spin-offs, or straight sales, and new entities devoted entirely to crop protection are being born: Syngenta from the merging of the agrochemical divisions of Novartis and AstraZeneca; Aventis CropScience, the merger of the crop protection activities of Rhône-Poulenc and of Hoechst Schering, formerly known as AgrEvo, once the problem raised by the participation of Schering in AgrEvo had been solved; and Monsanto, to be left essentially as a separate unit by Pharmacia-Upjohn shortly after having been acquired by them.

While these divestitures are taking place, large chemical groups less spoiled by the high margins of the pharmaceutical industry are reinforcing their position in agrochemicals. Such seems to be the case with BASF, which in March 2000 acquired the pesticides line of American Cyanamid from American Home Products, and of Bayer, which has added strength to its range of fungicide products by buying the Flint product line of Novartis. Other groups like Du Pont that are very active in seeds and Dow Chemical, which purchased Eli Lilly's remaining share in DowElanco in 1997, have also kept faith in the future of agrochemicals, a market worth $30 billion worldwide. Even smaller firms can find the agrochemical business particularly rewarding, provided they focus on "niche" products as FMC and Uniroyal Chemical (now part of Crompton Corporation) have done. In any case the trend seems to be for life science companies to focus on pharmaceuticals and for large chemical groups with relatively small operations in human health to negotiate

a withdrawal and concentrate on agricultural chemicals, an activity closer to their traditional practice. The announcements at the end of 2000 that Knoll, part of BASF, will be sold to Abbott and that Bayer and Du Pont are considering a separation of their pharmaceuticals from the rest of their businesses provide further proof for this line of thinking.

THE FATE OF THE DYESTUFFS SECTOR

Still in 1992 there were six major producers of dyestuffs in Europe—BASF, Bayer, Hoechst, Ciba-Geigy, Sandoz, and ICI—and an American manufacturer of smaller size, Crompton & Knowles. However, the pressure of small competitors in India and China has since led to major changes in the industry. Indeed, with their low overhead, favorable labor costs, and lack of consideration for environmental issues, these Far East firms began to offer intermediates for dyestuffs, then finished dyes, at very attractive prices. The established producers of the Western world reacted initially by using these cheaper intermediates in their own production, but they soon realized that more drastic moves were necessary. In the case of ICI the textile dyes were transferred to Zeneca Specialties, which in turn made toll-manufacturing arrangements with BASF before being taken over by venture capitalists (Cinven and Invest Corp) in 1999 to form Avecia. As has already been mentioned, Hoechst and Bayer combined their textile dyes in a fifty-fifty joint venture called Dystar (1994), which BASF later joined. The dyestuffs lines of Ciba-Geigy and Sandoz went respectively to Ciba Specialty Chemicals and Clariant after the creation of Novartis in 1995. Crompton & Knowles sold its textile dye business in 2000 to a British manufacturer, Yorkshire Chemicals, now called Yorkshire Group.

These various changes of ownership were accompanied by plant closures and the transfer of production to such places as Brazil and the Asia-Pacific area, which offered cheaper labor costs and less severe regulatory constraints. Smaller producers of textile dyes in Europe also lowered their ambitions, with

Holliday Chemical Holdings, now part of Yule Catto, closing their historical site of Huddersfield in the United Kingdom and Yorkshire Chemicals discontinuing production at its unit in Tertre, Belgium. Conversely, large firms like Ciba Specialty Chemicals and Clariant, while delocalizing their commodity dyes to more propitious regions, were able to link their dye expertise with their pigment technologies and to develop more sophisticated products for use in such applications as plastics, paints, and inks.

CONSOLIDATION IN THE FIELD OF SPECIALTY CHEMICALS

In their pursuit of "shareholder value" chemical companies have generally favored developing specialty chemicals over manufacturing large-volume commodities, which are considered to be less profitable and too cyclical. However, for historical reasons, the balance between these two categories of products varied from one group to another. This was particularly evident in the case of the newly born companies that resulted from the split of life science operations. Thus Ciba Specialty Chemicals was from the beginning a company focused on specialties and performance chemicals. It became even more focused with the acquisition in 1998 of Allied Colloids, a British company specializing in flocculants, and with the sale of its epoxy resins, a line of polymers in which it had been a pioneer, to Morgan Grenfell Private Equity. Clariant, with the product ranges inherited from Sandoz and more recently from Hoechst, belonged in the same league. It further enhanced its position by acquiring the former British Tar Products (now BTP) with its fine chemicals arm Archimica, a leading manufacturer of active molecules for the pharmaceutical industry, and PCR, a high-tech product specialist. The new ICI under the guidance of Brendan O'Neill had been re-created into a leading specialty chemicals corporation.

Other newly born companies were not so lucky, and their image remained blurred in the eyes of financial analysts because their product mix still included a significant proportion of commodities. Eastman Chemical Company, for example, still relied heavily on polyethylene terephthalate plastics and cellulose acetate in the field of polymers and on the acetyl chain and oxoalcohols as far as high-volume chemicals were concerned. Its management, by acquiring Peboc in the United Kingdom from Solvay Duphar as well as two other fine chemicals units in the United States, had made plans to enter the sector of organic synthesis for pharmaceuticals, only to give it up a few years later. Instead, Eastman Chemical Company decided to increase its presence in the resin market after it acquired Lawter International, an ink resin manufacturer; in 2000, Eastman also acquired the rosin esters and hydrocarbon resin lines of Hercules.

Rhodia, because of earlier mergers, is also regarded as a conglomerate rather than a purely specialty chemical firm. Having placed its toluene diisocyanate unit outside its perimeter of activity and sold its chloralkali business to a U.S. investor, La Roche, it still retains such commodity chemicals as phenol, phosphoric acid, and sodium tripolyphosphate (STPP) and maintains a presence in such unrelated fields as rare earths, styrene butadiene rubber latex, and cellulose acetate tow. Its core businesses are also very diversified; they include nylon fibers, polyamide engineering plastics, biopolymers (guar and xanthan gum), diphenols and derivatives, silica, silicones, and surfactants. In the future Rhodia intends to develop its activities in organic synthesis for pharmaceuticals, and it has recently acquired ChiRex, a U.S. company that complements the units it already operates in the United Kingdom and in France (ICMD [Industrie Chimique Mulhouse Dornach]). Rhodia recently made a successful bid for the old British firm Albright & Wilson, which gives it a leading position in the field of phosphoric derivatives, while the surfactant line will be resold to Huntsman.

Solutia, a firm created through a spin-off by Monsanto in 1997, also looks more like a conglomerate than like a specialty chemicals company because it retains large operations in nylon and acrylic fibers as well as in upstream commodities. It has entered its phosphorus chemicals operation into a joint venture with FMC called Astaris. Another joint venture, this time with Akzo Nobel, runs its former rubber chemical business under the name of Flexsys. Solutia has acquired, from Akzo Nobel, CPFilms and has purchased from Deutsche Morgan Grenfell the line of Vianova resins that belonged initially to Hoechst. The prospects for the lucrative polyvinylbutyral film for safety glass (Saflex) look particularly promising.

Another newly born corporation, Cytec Industries, having sold its acrylic fibers to Sterling Chemicals, looks more entrenched in the field of specialties than either Rhodia or Solutia. As a spin-off of American Cyanamid, Cytec has gained a leading position in the production of acrylamide derived from acrylonitrile and is a major producer of flocculants destined for water treatment. Recently, however, it sold its line of paper chemicals to Bayer. Among the new entities that appeared after the restructuring of the life sciences operations of several large groups, the former specialty chemical business of ICI, renamed Zeneca Specialties and then Avecia after it was acquired by Cinven and Investcorp, should be mentioned.

Aside from the above-mentioned changes from outside causes, consolidating this sector of the chemical industry has also been to a great extent the work of long-established firms already well positioned in the field of specialties and performance chemicals. In the United States, for instance, Rohm & Haas through its acquisition of Morton International in mid-1999 has greatly enhanced its status as a leading producer of chemical specialties—a status it had earned through its past presence in such lines as plastics additives, biocides, agrochemicals, and electronic chemicals. With the purchase in 1999 of Lea Ronal and of the photo-resist business of Mitsubishi, Rohm & Haas has further increased its presence in the electronic materials market, and its decision to divest the salt production activities of Morton International can only strengthen its position as a supplier of specialty chemicals.

Another American firm, Great Lakes Chemical Company—whose most profitable product, tetraethyl lead, made by its subsidiary Octel was being threatened by regulatory authorities—has also decided to concentrate its efforts on specialties and performance chemicals. Great Lakes, after deciding to focus on plastics additives, acquired successively Société Française d'Organo-Synthèse (SFOS), a Rhône-Poulenc subsidiary; LOWI, an independent German firm; and the antioxidant and ultraviolet absorber lines of Enichem in Italy. Finally, in 1998, Great Lakes demerged Octel and became a company in which over 80 percent of its sales was devoted to specialties. Ethyl, Great Lakes' main competitor in bromine chemistry, took a different path. In 1994 it set up two autonomous companies: Albermarle Chemical Company, which took over the polymer and fine chemicals businesses, and Ethyl, which was to specialize in petroleum additives.

Olin, one of the few conglomerates left on the American scene, clarified its structure in 1999 by regrouping under a new company called Arch Chemicals all its fine chemicals operations, while retaining the metal, ammunition, and chloralkali operations under the existing organization. A year later Arch Chemicals took over the British company Hickson International, which was involved in wood-treating chemicals, coatings, and fine chemicals. The trend toward further consolidation has led FMC to plan the separation of its chemical divisions, which represent almost 50 percent of its turnover from its machinery and engineering operations.

Other large groups, however, have chosen to maintain their specialties in their traditional structure. General Electric, for example, generates over $6 billion worth of plastics and chemical sales, including sales of acrylonitrile butadiene styrene resins and plastics additives acquired from Weston Chemical (a former Borg Warner subsidiary) and now managed by GE Plastics. Similarly Riedel de Haen, the fine chemicals German company taken over from Hoechst by Allied-Signal is run as part of that conglomerate whose turnover of specialty chemicals exceeds $1 billion. Several other American firms belong to the category of companies with at least $1 billion worth of sales in specialties and performance chemicals, including W. R. Grace, Lubrizol, Crompton, Hercules, and B. F. Goodrich. The fate of these companies is worth considering.

W. R. Grace has shrunk considerably from its earlier days as a conglomerate under Peter Grace, and it has to be considered as a specialty chemicals company from now on. Lubrizol is still a lube additives specialist with some growing activity outside its core business. But Crompton is the outcome of two mergers that have greatly expanded the portfolio of products it now manages.

In 1996 Crompton & Knowles, then a manufacturer of dyestuffs and polymer-processing equipment, took over Uniroyal Chemical, a company resulting from a leveraged management buyout after the tire business, which had gone to Michelin, was split off. Three years later Crompton & Knowles and Witco were merged through an exchange of shares, Witco having become vulnerable as a result of unsuccessful restructuring. The new entity, now called Crompton, no longer includes the oleochemicals of Witco, which went to Th. Goldschmidt (SKW Trostberg), or the textile dyes of Crompton & Knowles, sold to Yorkshire

Chemicals. With a turnover of around $3.3 billion it is nevertheless a well-diversified specialty chemicals company, with leading positions in ethylene propylene rubber, plastics additives, and rubber chemicals as well as crop protection products and silanes.

At the same time that Crompton appeared as a strong contender with its wide range of specialties, two well-known American firms were approaching the end of their existence as specialty chemicals manufacturers. B. F. Goodrich, originally a tire maker, had diversified successfully into chemicals. However, at the end of 2000, it had become a major aerospace industry supplier with only $1.2 billion worth of specialty chemicals sales left. This business unit has now been sold to an investor group led by AEA Investors Inc., a closely held business, and B. F. Goodrich has divested the last of its former operations. The other firm, Hercules, was originally a prominent and innovative producer of chemicals and polymers with leading positions in several niche markets. When Thomas Gossage, from Monsanto, took over as chairman in 1991, Hercules had already sold its dimethylterephthalate (DMT) business and the 50 percent share it owned in Himont, a polypropylene joint venture with Montedison. From 1991 to 1996 the company went through a restructuring that combined the sale of various parts of

Crompton's silane unit, Termoli, Italy. Courtesy Crompton Corporation.

its portfolio of activities (aerospace, flavors and fragrances, polypropylene packaging films, electronics, printing materials, and so forth) with an ambitious share repurchase program. This program raised the market value of Hercules stock to a peak of $65 per share in 1995, a year before the chairman retired.

By that time, however, the company was left with only a few businesses, which although quite profitable, offered little growth potential. To give the company a boost, the new chairman considered entering the field of water treatment. In 1997 Hercules made an offer for Allied Colloids, but it was outbid by Ciba Specialty Chemicals. A year later, however, Hercules, by paying a high premium, was able to acquire BetzDearborn, a water-treatment specialist formed when Betz Laboratories bought the Dearborn units of W. R. Grace. The deal boosted Hercules' revenues from $1.9 billion to $3.5 billion, but it also increased its debt substantially at a time when competition in the watertreatment market had become much more acute owing to the consolidation of the two entities Calgon and Nalco Chemical Company under the aegis of the French group Suez Lyonnaise des Eaux. The management of Hercules, hoping to improve the company's debt ratio, decided to part with other assets: their food gum operation is now a joint venture with Lehman Brothers Merchant Banking partners as majority shareholders; the resins operation, for which they held a leading position, has been sold to Eastman Chemical; and FiberVisions, the world's largest producer of thermally bonded polypropylene fibers, is for sale. As was to be expected, the valuation of Hercules stock has been severely downgraded, reaching a low of $14 per share. This decline has in turn attracted the attention of the well-known raider Samuel Heyman, the chairman of International Specialty Products, which is the new name for the former GAF Corporation. With Thomas Gossage now back in the driver's seat, Hercules may not have any other choice but to be sold and disappear as a going concern after a long and often brilliant existence that began in 1912.

This story is typical of what happens to an otherwise healthy company when it is managed for too long under the pressure of short-term financial considerations. The fate of Laporte Chemicals, a British company founded in 1888 by the chemist Bernard Laporte to produce hydrogen peroxide (H_2O_2) for use as a bleach, was not any better than that of Hercules. Yet by 1995 Laporte had grown into a uniquely large specialty chemicals corporation with a turnover of £1 billion. It had quit the phthalic anhydride and TiO_2 sectors in the 1980s and in 1992 dissolved its joint venture with Solvay, Interox making H_2O_2. Having made several acquisitions in the fields of organic peroxides, adhesives, and process chemicals, Laporte was operating through sixteen strategic business units. Between 1995 and 1998, however, the management of the company started to divest some businesses that were underperforming, the remaining portfolio being focused on fine organics, peroxide initiators, construction chemicals, electronic materials, pigments, and additives. This reshuffling of activities culminated in 1999 with the acquisition of Inspec, a British firm that resulted from the leveraged management buyout seven years before of several chemical operations and in particular of parts of British Petroleum's specialty chemicals and of Shell's fine chemicals divisions. No sooner had this last acquisition been completed than the new management in charge of Laporte, with a view to reducing the outstanding debt, decided to part with those units that were not directly involved with fine organics and performance chemicals. As a result the units that represented half of the total turnover were sold at the beginning of 2000 to Kohlberg Kravis Roberts & Co., a private equity investor. Finally, having reached a perfect size for being taken over, Laporte, or rather what was left of it, was acquired a few months later by Degussa AG and disappeared from the British scene as an independent chemical corporation after 112 years of operation.

Degussa AG is the new name of a German group that is slated to become one of the largest fine chemicals and specialty chemicals corporations in the world. But to achieve this status, further restructuring is needed. Indeed Degussa itself, which celebrated its one hundred twenty-fifth anniversary in 1998, was by then a member of VEBA AG, a large German energy concern that owned 100 percent of Chemische Werke Hüls. The next step was for Degussa and Hüls to be merged through an exchange of shares that gave VEBA a stake of 62.4 percent in the new company. Meanwhile, another energy company, VIAG AG, established in Bavaria, had taken full control of SKW Trostberg, to whom it was supplying power. In 1997 VIAG gained a controlling interest in a family business, Th. Goldschmidt, which it merged with SKW Trostberg two years later. VIAG thus owned 63.7 percent of the merged entity. Then VEBA and VIAG announced their intention of combining their chemical assets and gave birth at the end of 2000 to what is now called Degussa AG with initial sales of 14 billion euros per year. In order to streamline its operations into a true specialty chemicals group, Degussa AG will need to dispose of several commodity activities (e.g., phenol, fertilizers, salt and metallic chemicals, dimethyl terephthalate, C-based alcohols, and plasticizers) and to sell its pharmaceutical subsidiary Asta Medica as well as its automotive catalyst sector. It will then be left with leading positions worldwide in such diversified fields as amino acids, carbon black, precipitated silicas, organosilanes, specialty polymers, flavors (Mero & Boyveau) oleochemicals (Th. Goldschmidt), and hydrogen peroxide.

As the consolidation of the specialty chemicals industry continued, it affected many firms in which the founder's family still held substantial portions of the capital, even though some of them kept on thriving. One example was Union Chimique Belge, whose management had been wise enough in the late 1980s to focus on three profitable lines: the pharmaceutical sector, with a blockbuster antiallergic drug called Zyrtec; the chemical sector, with ultraviolet curing and powder-coating technologies; and the film sector, with a leading position worldwide in oriented polypropylene and cellophane films. The German company

Wacker, in which Hoechst had been a partner from the start, also managed to retain its independence, although it was still active in some commodity-type products (acetic acid, vinyl monomers and polymers, and silicon carbide), it chose to emphasize lines that showed greater profitability, such as silicones, hyper pure silicon for semiconductors, and specialty and fine chemicals. In France the well-run starch and derivatives producer Roquette Frères, after a short flirtation with Rhône-Poulenc, was able to recover its autonomy.

Other firms were less fortunate; they lost their independence because the founder's successors either could not agree on a management plan or had to sell their shares to pay inheritance taxes. Thus, Compagnie Française des Produits Industriels, which belonged to the Hess family, was sold in 1996 to Fernz Nufarm, a firm from New Zealand specializing in agrochemicals. As has been discussed, Th. Goldschmidt of Germany was taken over by VIAG. Another German firm, Raschig, which was founded in Ludwigshafen in 1891 and which had remained under the control of its family owners since then, was integrated in the mid-1990s into the PMC Group of Philip Kamins, a California-based entrepreneur with interests in plastics and specialty chemicals. Even a powerful company like the venerable Henkel had to reconsider its position. In 1999 it set apart under the name of Cognis its specialty chemicals activities, including production of an important range of oleochemicals. Henkel, still controlled by the heirs of Fritz Henkel, has made clear its intention to leave this field in order to better compete in the consumer products area with such large groups as Unilever and Procter & Gamble.

The specialty chemicals industry is estimated to have an annual turnover of $200 billion worldwide, with production scattered among a great variety of suppliers. Consolidation of that industry was to be expected owing to the pressure of shareholders and to management eager to rationalize their portfolio of products. Some corporations like Ciba Specialty Chemicals, Clariant, ICI, Rohm & Haas, Crompton, and Great Lakes have succeeded in being considered as specialty chemicals companies in their own right. Others, such as Rhodia and Degussa AG, still have a product mix that is much too diversified to reach that status. They will have to restructure their portfolio further if they want to improve their share value. Still others, because their management was led by short-term objectives or made unwise divestitures and acquisitions, have been penalized by the stock markets and will no longer survive as independent entities: Laporte and Hercules are among these sad cases. Aside from bowing to financial considerations, the consolidation of the industry also had at its root the necessity of reducing the number of participants once their customers became less numerous and acquired a worldwide presence. For instance, with only three major tire manufacturers left—Michelin, Goodyear, and Bridgestone—the number of the main rubber chemicals manufacturers also had to shrink: Solutia and Akzo Nobel merged into Flexsys, facing Uniroyal Chemical (Crompton) and Bayer as their major competitors.

THE CASE OF FINE CHEMICALS

Fine chemicals, sold in relatively small volumes and at rather high prices, are obtained through various organic chemical reactions. Their production does not require heavy investments as they are made in multipurpose units equipped with glass-lined and stainless-steel reactors. For this reason this sector of the chemical industry has attracted skilled entrepreneurs who know how to surround themselves with teams of good chemists capable of conducting multistep syntheses leading to the formation of complex molecules used by pharmaceutical laboratories and agrochemical firms. Over the years, however, these customers of the fine chemicals industry realized that they should make their active ingredients themselves and so created their own chemical departments to carry out at least the final steps of the synthesis of the products they had developed.

During the 1990s a drastic change took place in that many pharmaceutical companies saw their R&D and marketing expenditures rise considerably, so they decided to outsource the synthesis of their proprietary drugs. At the same time many patents protecting well-known ethical drugs expired, and the same drugs came to be offered on the market as generics by competitors. As a result the chemical subsidiaries of major laboratories were offered for sale, and the toll-manufacturing activities of well-established fine chemicals firms increased spectacularly.

Thus, in 1997, Glaxo Wellcome handed over its Greenville site in the United States to Catalytica and its Annan plant in Scotland to ChiRex in exchange for supply contracts running for a five-year period. Similar deals were made when in 1997 Warner Lambert sold Sipsy Chimie Fine, the chemical operation of its French subsidiary Laboratoire Jouveinal, to PPG Industries and a year later when the British firm BTP bought the French firm Hexachimie from Bristol-Myers Squibb. Prior to its merger with Sandoz, Ciba-Geigy had, for its part, divested its French subsidiary La Quinoléine, which was acquired by Organo-Synthése. Sanofi sold its operation in Spain, Moehs SA, to PMC.

While these divestitures were taking place, traditional fine chemicals manufacturers were able to enjoy new contracts for the synthesis of active ingredients from companies involved in crop protection products and pharmaceuticals. Many European companies were the beneficiaries of the new trend. In Switzerland the leader was Lonza, which became independent in 1999 when its mother company, Alu Suisse, merged with the Canadian aluminum manufacturer Alcan. Other Swiss firms—such as Siegfried AG, EMS-Dottikon, Cilag, and Orgamol, Röhner, now part of Dynamit Nobel—took advantage of these developments. Clariant entered the field on a big scale in 2000 with its acquisition of BTP.

In France as well, some companies that had been traditionally involved in fine chemicals through their specific expertise engaged in similar activities. The state-owned Société Nationale des Poudres et Explosifs decided to make use of its skills in phosgene and nitration chemistry by buying Isochem from the

Wirth family and making inroads in the field of polypeptides. Rhodia, which had inherited ICMD from its former links with Rhône-Poulenc, acquired ChiRex at the end of 2000. Another firm, Produits Chimiques Auxilliaires de Synthèse (PCAS), which belonged to Dynaction, a mediumsized French conglomerate, also expanded by taking over Pharmacie Centrale de France's chemical subsidiary and a plant in Limay belonging to the German Schwartz Pharma.

In Holland, under the leadership of Simon de Brée, DSM chose to diversify into fine chemicals first by acquiring the Austrian Chemie Linz from ÖMW, then in 1999 by taking over Gist-Brocades, a Dutch supplier to the pharmaceutical industry. Finally, with the purchase of Catalytica's pharmaceutical subsidiary in 2000, DSM has become a leader in this field and now generates a turnover of $1.3 billion.

Conversely, the United Kingdom has seen many of its fine chemicals companies disappear from the map: Hickson International is now a part of Arch Chemicals, Courtaulds Chemicals has been taken over by Akzo Nobel, and the fine organics and performance chemical lines of Laporte were recently acquired by Degussa AG. Peboc got sold by Duphar to Eastman Chemical Company, which now wants to dispose of it. A similar trend has occurred in Germany where Riedel de Haen was acquired in 1995 by Allied Signal, Raschig by PMC, and Boehringer Mannheim by Roche.

Italy, long the source of expertise in multistep synthesis built up over the years by individual entrepreneurs, did not fare any better. Francis in Milan, after having been owned by Laporte, is now controlled by Degussa AG; Profarmaco was acquired by Nobel Chemical, which in turn is owned by Cambrex; OPOS went successively to Hoechst Marion Roussel and then to Holliday Chemical Holdings, itself a part of Yule Catto.

The United States was not as strongly endowed with fine chemicals expertise as were many European countries since the skills of the organic chemist were less cultivated in America than those of the chemical engineer and only a few American companies have been successful in this field in the past. But mention should be made here of Cambrex, a company that began with the acquisition of the former Baker Castor Oil Company and of Nepera Chemicals, a manufacturer of pyridine and derivatives. In 1993 Cambrex decided to become a global supplier of products to the life sciences industry. It made two major moves to accomplish this: in 1994 it acquired from Akzo Nobel the Nobel pharmaceutical chemistry business, with its lucrative subsidiary in Italy, Profarmaco, a producer of generics; and in 1997 it purchased BioWhittaker, a supplier of human cells and cell cultures, which gave Cambrex a promising future in the field of bioscience. Thus the sector of fine chemicals has been the scene of many divestitures and acquisitions, and since its consolidation is far from complete, it still offers opportunities to those firms that either enjoy some lucrative niche markets or that reach a size that allows them to compete successfully with the established leaders.

THE FURTHER CONCENTRATION OF THE INDUSTRIAL GAS BUSINESS

At the end of the 1990s there were four major producers of industrial gases in the world: L'Air Liquide, British Oxygen Company, Praxair, and Air Products. In addition to these leaders the German firms Messer Griesheim, the Swedish AGA, and the Japanese Nippon Sanso played an important role in their respective markets. If we consider that Praxair was the former Linde Division of Union Carbide Corporation, these companies were well-established firms, the youngest among them, Air Products, having celebrated its fiftieth anniversary in 1990. Since it is not economical to transport liquefied gases long distances, all of these producers operated on a worldwide basis, with plants scattered throughout the five continents.

British Oxygen had invested more heavily than its competitors in the former Commonwealth and in the Asia-Pacific area, while AGA had invested more heavily in the

Nordic countries and in Latin America. Air Products was the only company with a sizable part of its total turnover devoted to commodity and specialty chemicals, which represented over 30 percent of its yearly sales of $5 billion. L'Air Liquide, although initially involved in chemicals, had sold its fertilizer subsidiary, Société de la Grande Paroisse, to Elf Atochem and its pharmaceutical arm, Lipha, to E. Merck, in Darmstadt, Germany. Although the competition for industrial gases was fierce between these companies, new applications were developed in the fields of electronics, semiconductors, health, oil refining, and food processing, and all the firms with standardized plants and similar technologies were enjoying a thriving business.

At the end of 1999, however, British Oxygen, which was overinvested in Southeast Asia, approached Praxair with an offer to merge the two groups. Dating back to 1992, Praxair was a spin-off of Union Carbide, made necessary when in 1985 in order to fight a takeover bid from Sam Heyman, the majority owner of GAF, Union Carbide decided to buy back its shares and as a result considerably increased its outstanding debt. The proposed merger of Praxair and British Oxygen would have had a considerable impact on the industrial gas market. Feeling threatened, Air Liquide and Air Products made the unusual move of a joint bid for British Oxygen, but antitrust considerations led the parties to give up their plan after a long period of negotiation.

Meanwhile less ambitious schemes have materialized. Indeed, in 2000, Linde took over AGA and Messer Griesheim, which was owned initially by Hoechst, now part of Aventis (66.6 percent), and the Messer family (33.3 percent) cut a deal whereby two financial associates, Allianz and Goldman Sachs, will buy from Aventis the shares formerly held by Hoechst. The new landscape offered by these various moves may still change somewhat as the companies just merged will for financial or antitrust reasons have to dispose of assets that should be of great interest to the majors.

THE CHANGING TIES BETWEEN THE OIL AND CHEMICAL INDUSTRIES

Beginning with the World War II effort, oil companies in the United States became involved in petrochemicals and polymers and contributed to major innovations in those fields. The process of downstream integration was then seen by them as a way to upgrade their feedstocks and to develop new activities. Over the years oil majors like Exxon and Shell, either through a frenzy of acquisitions or through internal growth, became important producers not only of petrochemicals, synthetic rubbers, resins, and plastics but also of specialties and even fine chemicals. In the 1990s low crude-oil prices and increased competition on a worldwide scale led the management of the oil companies to reconsider their position, with the idea of restricting their chemical operations to base commodities produced at the site of their main refineries. In order to achieve this goal, a vast program of divestitures was begun either through straight sales or spin-offs or through joint ventures run at arm's length from the companies' owners.

Thus in 1994 Texaco sold its chemical operations to Huntsman. ARCO Chemical Company, after floating its Texas-based petrochemical activity under the name of Lyondell in 1989, sold to that new entity nine years later the remaining operations of ARCO Chemical. The joint-venture route was used by Shell when in 1989 it bought the stake Montedison held in Montell, their polypropylene subsidiary, before merging it with the poyolefin operation of BASF, now called Basell. Similarly, Occidental Petroleum, which with Oxychem still holds strong positions in chloralkali products and specialties inherited from the former Hooker Chemical, is trying to unload some of its petrochemical operations. To this end it has become a shareholder of Equistar Chemicals, now the largest U.S. producer of ethylene, which is as a result a joint venture between Lyondell (41 percent), Millennium (29.5 percent), and Oxychem (29.5 percent). Millennium Chemicals was itself born in 1996 from a demerger of Quantum Chemicals, formerly owned by

Hanson Trust, and like Oxychem it wishes to find a buyer for its shares in Equistar.

The consolidation of the oil industry, which started at the end of 1998 with the merger of BP-Amoco and was followed in 1999 by BP-Amoco's acquisition of ARCO and then by the still larger Exxon-Mobil merger, can only further dilute the part of petrochemicals in the product mix of these new entities. The announced takeover of Texaco by Chevron will have the same effect. The chemicals of Chevron and Phillips Petroleum have already been combined under a new entity, Chevron Phillips Chemical Company. Oil companies everywhere have followed a similar trend.

Thus in 1999 the Spanish Repsol, by acquiring YPF in Argentina, and the Norwegian Norsk Hydro, by taking over Saga Petroleum in Norway, have both contributed to a decrease in the share of chemical sales in their consolidated turnover. Norsk Hydro, still strong in fertilizers, no longer considers its ethylene, vinylchloride, PVC, and chlorine operations as core businesses. ENI in Italy is also prepared to part with some of Enichem's activities: it disposed of its acrylic fibers and sold at the end of 2000 the polyurethane business of Enichem to Dow Chemical while trying to take over Union Carbide's share of Polimeri Europa, following Dow's acquisition of Union Carbide and the resulting antitrust rulings. Similarly, the Nesté Chemicals subsidiary of the Finnish Nesté Oil has been divested and acquired by a Nordic investment firm, Industri Kapital, which merged it with Dyno, a recent acquisition specializing in explosives and fine chemicals.

The only important exception to this policy of unloading chemical activities not strictly related to the immediate downstream production of refinery operations appears to be the French oil company TOTAL. Historically, TOTAL, the former Compagnie Française des Pétroles, had only a limited presence in petrochemicals and polymers through aromatics produced at the Gonfreville refinery and minority participations in alkylbenzene, polyolefins, and butyl rubber joint ventures in France. The situation changed in 1990 when ORKEM, a company born from the restruc-

turing of the state-owned French chemical industry, was split and its assets transferred to TOTAL for the specialties and to Elf Aquitaine's chemical arm Atochem for the petrochemicals and polymers. Through a deal completed in June 1999, TOTAL made a public exchange offer for Petrofina of Belgium. This was followed a few months later by the merger of the newly created TotalFina with Elf Aquitaine. That way a new group, TotalFinaElf, was born, with an impressive chemical arm, Ato-Fina, having global sales of $16 billion a year.

Ranking as the fifth largest chemical company in the world, after BASF, Bayer, Du Pont, and Dow, once its merger with Union Carbide is completed, AtoFina is a very diversified producer. Its product mix includes fertilizers and a variety of agrochemicals, four of the major thermoplastics (polyethylene, polypropylene, polystyrene, and PVC), acrylics and polymethyl-methacrylates, a chloralkali chain, oxygenates (H_2O_2 and organic peroxides), hydrazine and derivatives, and fluorinated products. While commodity petrochemicals and plastics represent less than 20 percent of its business, AtoFina has inherited from TOTAL a lucrative portfolio of specialties in such varied fields as resins (Cray Valley), adhesives (Bostik), radiation curing coatings (Sartomer), paints (Kalon), and rubber articles (Hutchinson). Originating from various restructuring steps and acquisitions over the last ten years, AtoFina, because of its size and diversity, does not resemble any of the chemical operations of the major oil companies.

Its future within TotalFinaElf will depend on whether the management of the group feels that its chemical activities should be closer to the cracker or can be run as a full-fledged chemical operation. In any case some streamlining cannot be avoided. In fact, prior to the merger, TOTAL had already sold its Ink Division (Coates Lorilleux) to Sun Chemical, a subsidiary of the Japanese Dainippon Ink & Chemicals. More recently a department specializing in metal treatment has gone to Henkel, whereas unsuccessful attempts have been made to dispose of the agrochemicals of

ElfAtochem. At this stage anyway and for historical reasons the approach of TotalFinaElf can be considered unconventional when we compare it with that of the oil companies. With improving crude oil prices and better capacity utilization in their refinery units, the oil companies are concentrating their major reinvestments on what they consider to be their core business and limiting their production of chemicals to the base commodities derived from the olefins of the steam cracker or other feedstocks.

THE ROLE OF ENTREPRENEURS AND PRIVATE EQUITY FUNDS

The various mergers and divestitures described in this chapter have provided opportunities for entrepreneurs and venture capitalists to make acquisitions in fields that were no longer of interest to major chemical firms. Already in the 1980s a similar situation had given birth to new chemical entities launched by such daring entrepreneurs as Gordon Cain and Jon Huntsman. Cain Chemical, the result of a highly leveraged management buyout, was for a while a large petrochemical concern before being taken over in 1988 by Oxychem.

Huntsman Chemical, however, has proved longer lasting. Its first steps in the chemical business started with the acquisition of the polystyrene activities of Shell, followed three years later by those of Hoechst. By 1999, through external growth, Huntsman Chemical ranked as the largest privately held chemical company. That year its revenues reached $8 billion as a result of the acquisition of ICI's polyurethane, titanium dioxide (TiO_2), aromatics, and petrochemical operations by Huntsman ICI Holdings (70 percent Huntsman Chemicals and 30 percent ICI). ICI's Wilton, Teesside cracker has in particular given Huntsman a strong position in olefins.

With the cash obtained from the sale of its styrene business to the Canadian NOVA Chemicals, Huntsman could afford to proceed with more acquisitions, and in 2000 it took over the surfactants operations of Albright & Wilson, a British firm that is now part of Rhodia.

Another American entrepreneur, D. George Harris, who had managed the SCM chemical conglomerate before it was acquired and dismantled by Hanson Trust, had also shown an interest in basic commodities being divested by the majors. During the 1990s Harris was able to set up operations in such fields as soda ash, phosphate, and boron, both in the United States and in Europe. In 1998, facing a difficult period, Harris managed to sell these operations to IMC Global, an American firm active in agribusiness.

Recently, new opportunities were offered to private equity institutions and to ambitious managers who can use the funds that venture capitalists provide to revamp companies in need of restructuring and make substantial money when selling them a few years later. In Europe the PVC field has offered such an opportunity because it suffers from intense competition and is submitted to the pressure of such environmentalists as Greenpeace. As a result several PVC producers have been anxious to sell their assets or at least to participate in the restructuring of the sector. They have found buyers among the private equity institutions. With a consortium of financiers, including Advent International and Candover, George Harris acquired in December 2000 Vestolit, the vinyl business of Degussa-Hüls. For its part Advent has also bought Vinnolit and Ventron from Wacker and Celanese, respectively. European Vinyls Corporation (EVC), the troubled joint venture formed by Enichem and ICI, may well be the next to go to private capital for the consolidation to be further advanced.

The fibers sector is another area in which European producers are suffering because of the presence in Asia of strong competitors, a situation still aggravated for these producers by the gradual elimination of the protective textile quotas set up under the Multi Fiber Agreement. Here again the situation is attracting the attention of financial investors as well as that of private textile firms in countries enjoying the advantage of low labor costs. As an example CVC Capital Partners, in January 2000, purchased a 64 percent stake in Acordis, the new entity resulting from the takeover of Courtauld's fiber business by Akzo Nobel and its subsequent float on the stock market. Du Pont, in its effort to unload part of its activities in polyester fibers, set up between 1999 and 2000 joint ventures with two foreign firms: Alpek, a subsidiary owned entirely by the Mexican firm ALFA, the largest private petrochemical concern in Latin America, and Haci Ömer Sabanci Holding, an important financial and industrial conglomerate in Turkey.

The increasing role played by private equity capital can be explained by the important sums of money they can collect from pension funds and other institutions. It may seem strange that they should direct investments toward chemical operations that are being spun off by the big players. However, some of the assets being divested are offered at attractive prices, and the managers who are put in charge as well as their backers can hope to restore rapidly some profitability to the operations through a thorough program of cost cutting and restructuring. In any event these investors are generally inclined to take a short-term view of the business, and with the exception of entrepreneurs like Huntsman a quick profit is their main objective.

THE EVERLASTING PRESENCE OF CONTRARIANS

In the past ten years many chemical companies have changed their product mix and even their names and have taken entirely new directions in order to cope with a new environment and in the hope of pleasing financial analysts. During that same period, however, some large chemical corporations, while prepared to streamline their operations and take advantage of new opportunities, have maintained their faith in the traditional activities that had been the key to their success. These contrarians have taken a long-term view of the business they are in, in the hope that their approach will in the end meet with the approval of the stock market because internal growth, which takes time to deliver, is an essential part of any industrial undertaking.

The German chemical giant BASF ranks among the top contrarians. Its management still believes in the *verbund* concept of linking products from the raw material down to the most elaborate derivatives. It is a concept particularly suited to a large site like Ludwigshafen where vertical integration can be easily practiced, with products flowing from one unit to another. BASF has also maintained its connection to oil as a raw material, which started in 1952 through Rheinische Olefin Werke in Wesseling, BASF's joint venture with Shell, and was amplified in 1969 when the Wintershall refinery was acquired. The fact that BASF remains diversified in various lines of organic and inorganic chemicals does not prevent it from restructuring its traditional operations when the need arises. Joint ventures were thus entered into at the end of the 1990s, in polyolefins with Shell (Basell), in PVC with Solvay (Solvin), and in dyestuffs with Bayer and Hoechst (Dystar). BASF finally decided to part with its pharmaceutical subsidiary Knoll, sold at the end of 2000 to Abbott, and it is negotiating with Kali und Salz its withdrawal from some fertilizer operations. With a turnover around $30 billion per year in group sales, BASF retains a leading position in the world chemical industry.

Dow Chemical has followed a similar path. Over the last five years it has reinforced its position in Europe in basic chemicals and commodity polymers by the acquisition and reconstruction of the Buna Sow Leuna Olefinverbund complex in eastern Germany, which has been fully owned by Dow since June 2000. At the same time the olefin crackers in Terneuzen, Holland, have been greatly expanded, while methylene diisocyanate (MDI) capacities for use in polyurethanes are being increased in Stade, Germany, and in Estarreja, Portugal. Above all, Dow has seen at the beginning of 2000 its acquisition of Union Carbide approved by both the Federal Trade Commission in Washington and the European Commission in Brussels. This merger, leading to yearly sales of $28.4 billion, puts the new Dow Chemical in the same league with BASF and Du Pont. The approval of the merger implies that Dow sell its ethylene amine business, which will go to Huntsman Chemical, and its ethanol amine operations, which will be acquired by Ineos, a financial institution. It also commits the new entity to part with the 50 percent stake Union Carbide held in Polimeri Europa, a joint venture with Enichem to manufacture polyethylene. At the same time Dow will further strengthen its position in polyurethanes by acquiring the polyurethane line of Enichem, which includes TDI, MDI, and polyols.

Like BASF, Dow is a firm believer in vertical integration. It has always tried to connect its thermoplastic and elastomer productions to upstream steam crackers while making its chloralkali operations less dependent on PVC than other producers by developing various lines of chlorine derivatives on site. Even before BASF, Dow left the field of pharmaceuticals with the sale of Marion Merrell to Hoechst and has increased its stake in crop protection chemicals through the acquisition of Eli Lilly's interests in DowElanco. The announced takeover of Rohm & Haas fungicide and insecticide lines will further enhance Dow's presence in this sector.

Among the contrarians, Bayer has had for historical reasons a different approach to the problem of remaining a well-rounded producer of chemicals. It has always been a leader in such specialty chemicals as plastics and rubber additives and supplies master batches through its subsidiary Rhein Chemie. Its polymers are mainly centered on synthetic elastomers and specialty plastics like polyurethanes and polycarbonates that Bayer itself develops. These are not subjected to the cutthroat competition that commodity thermoplastics are. Bayer has remained involved in aroma chemicals with a well-known subsidiary, Haarmann & Reimer. In pharmaceuticals the company occupies an honorable rank, although in the future its management may want to separate that branch from the rest of its activities in preparation for further alliances. Like BASF, Bayer considers its agrochemical line as a core business. However, unlike BASF, it is less interested in integrating upstream production to oil raw materials and has therefore decided to sell to its partner BP-Amoco its 50 percent stake in their EC Erdölchemie joint venture. When necessary, Bayer is also prepared to divest non-core activities as it did in mid-1999, with 70 percent of Agfa-Gevaert or with its Dralon fiber operation sold to the Fraver Group in Italy. Taking these divestitures into account, Bayer's yearly turnover should approach that of the new Dow Chemical–Union Carbide merger.

Another chemical giant, Du Pont, while still broadly based, has been less inclined than its above-mentioned rivals to remain focused on its traditional lines of activity and has changed its course of action several times over the last ten years. In 1981 Du Pont purchased Conoco in order to protect itself from oil shocks; twenty years later its management was no longer interested in maintaining that business, which was then spun off. Du Pont also seemed eager at one time to enter the pharmaceutical sector and had made an alliance with Merck in that field. In 2000 Du Pont tried to acquire the former Romainville research center of Roussel-Uclaf in France, which belonged by then to Aventis. After failing in that attempt, Du Pont became somewhat disenchanted with its pharmaceutical business, which had yearly sales of only $1.6 billion out of a total turnover of $27 billion; it is predicted that they will eventually give up that business. Du Pont remains determined, however, to become a leader in seeds after its acquisition in 1999 of Pioneer Hi-Bred. In spite of some temporary setbacks in Europe for genetically modified organisms, this branch of life science should provide the company with a bright future. Like other Western synthetic fiber manufacturers, Du Pont has been trying to unload some of its assets in that area, which has suffered from the impact of Asian producers. And as discussed earlier, it has to that effect made joint ventures in polyesters with such private groups as ALFA in Mexico and Haci Ömer Sebanci in Turkey.

Akzo Nobel, with yearly sales about half as large as those of the four majors, belongs nevertheless to the league of contrarians in that it occupies leading positions in many traditional lines of the chemical industry. Under a five-year plan initiated in the second half of the 1990s, the company, which resulted then from the merger of the Swedish Nobel Industries with the Dutch AKZO, undertook an ambitious restructuring program. The merger had given Akzo Nobel the number-one position worldwide in coatings and a valuable stake in surfactants and in pulp and paper chemicals. From the former AKZO the company had inherited a lucrative pharmaceutical line with the Organon prescription drugs and with Diosynth, a producer of raw materials for pharmaceuticals, which was reinforced by the Buckhaven unit of Courtaulds in the United Kingdom. The animal health care business Intervet International is also profitable. The restructuring of the group involved the creation of joint ventures for rubber chemicals with Monsanto (Flexsys) and for PVC with Shell (Rovin), acquired in late 1999 by the Japanese Shin-Etsu. The soda-ash operation was sold early in 1998 to Brunner, Mond. The problems of the fibers division were treated first by a merger with the British Courtaulds, followed by a spin-off of the fibers operations of the joint entities, sold in 2000 as Acordis to its new owners, which included—aside from a minority stake held by the management—the financial institution CVC Capital Partners and Akzo Nobel itself.

Another contrarian, DSM in the Netherlands, had been a state-owned company before it became privatized, a process that began in 1989 and was completed in 1996. From its past it had inherited positions in fertilizers, industrial chemicals, and such intermediates as melamine and caprolactam as well as polyolefins, with access to basic olefins through its own crackers in Geleen, Netherlands. In 1997 DSM acquired the polyethylene and polypropylene operations of Hüls (VEBA) with the Gelsenkirchen site. The company had also diversified into elastomers, having purchased in the United States the Copolymer Rubber and Chemical Corporation, which contributed to DSM's expansion into the fields of ethylene propylene, styrene butadiene, and nitrile rubbers. DSM is also a supplier of industrial resins and engineering plastics products. It had made inroads in the fields of fine chemicals in 1985 when it bought Andeno, a producer of synthetic intermediates for the pharmaceutical industry. By the end of the 1990s this initial acquisition had been followed by that of Chemie Linz; of Gist Brocades, a biotechnology firm; and in 2000 of Catalytica, thus giving DSM a yearly turnover of $1.3 billion in life science operations out of a total yearly turnover of around $7 billion. The surge in such activities will probably force DSM to slow down its heavy investments in olefins and petrochemicals, giving the company a better balance between commodities, performance, materials, and fine chemicals.

The last contrarian worth mentioning, with a turnover at the level of that of DSM, is Solvay, a company that dates back to 1863 and is still involved in most of its traditional lines. It is organized around four main sectors—chemicals, plastics, processing, and pharmaceuticals—with no intention of spinning off any of them. Solvay remains a leader in such commodities as soda ash, chloralkali chemicals, and oxygenates (H_2O_2 and peroxides). For its lines of polymers, it set up joint ventures with BASF (PVC) and Petrofina (highdensity polyethylene) in order to improve its access to raw materials and acquire new technologies. Solvay also has considerable experience in plastics processing, an activity that represents more than one third of its total sales. In the health sector the group—after selling its crop protection line to Uniroyal Chemical (now Crompton) and the U.K. fine chemicals producer, Peboc, to Eastman Chemical—has decided to concentrate all its efforts on Solvay Pharma, its pharmaceutical unit that already accounts for 22 percent of its earnings and specializes in such areas as gastroenterology, psychiatry, and gynecology.

It takes courage these days on the part of the management of the contrarians to go on thinking of the chemical activities in which they operate as made up of several parts to be kept under one roof. Indeed many investors have come to feel, as do financial analysts, that a split-up of the activities of large and diversified chemical groups would unlock more value in the shares they hold. This makes the temptation to dismantle existing organizations sometimes hard to resist.

THE CASE OF JAPAN

In the ten years discussed in this chapter, the economy of Japan languished, and the Japanese chemical industry, the third largest in the world, could not avoid the consequences of that situation. It also suffered from several factors specific to Japan. First, the largest chemical companies belonged to the "Keiretsu" system, which had replaced the "Zaibatsus" of pre–World War II, and featured a complex network of cross-shareholdings dominated by banks. Second, these companies remained under the influence of the Ministry of International Trade and Industry (MITI), which proved to be a hindrance when the time came to restructure the chemical sector in the early 1990s. Third, internationally, with the exception of such firms as Dainippon Ink & Chemicals and Shin-Etsu, Japanese chemical groups had not developed a strong presence outside Asia and depended heavily on indirect exports through their domestic customers or on licensing for their activities overseas. Fourth, Japanese society, used to lifetime employment, was not prepared for the social upheavals inherent in any serious restructuring. For all these reasons the changes required from the chemical industry by the new conditions of world trade have been slow in coming and were only partially accomplished by the end of the 1990s.

As an example, the necessary merger of Mitsui Petrochemicals and Mitsui Toatsu Chemicals became effective only in 1997 and was long delayed because of the reluctance of each senior management group of these two members of the Mitsui group to merge. Similarly difficult to achieve has been the consolidation of Mitsubishi Chemicals and Mitsubishi Petrochemical, which finally took place in October 1999 and led to the birth of the eighth largest chemical producer in the world, with a yearly turnover equivalent to $15.7 billion in fiscal year 1999–2000. The proposed merger of Sumitomo Chemical with Mitsui Chemicals should create an even larger group when it is implemented.

These three majors of the Japanese chemical industry can be considered as contrarians in their own right since they maintain a strong presence in all facets of their industry. Contrary to their American or European counterparts, they are not submitted to the pressure of investors eager to obtain the best value for their shares. While this could be an advantage in that it allows the management to undertake long-term projects, it may lead to some complacency at a time when cost cutting and restructuring are the necessary requirements for companies to survive in a fiercely competitive environment.

THE CHEMICAL INDUSTRY AT THE BEGINNING OF THE THIRD MILLENNIUM

The general trends described at the end of the previous chapter have been amply confirmed over the last ten years. Regulatory burdens have multiplied, especially environmental ones, causing higher costs to the industry and yielding either no benefits or negative consequences. Globalization has proceeded at a faster pace, and as products have matured and technologies have become more readily available, competition between chemical firms has been fiercer on the international scene. The expansion of stock markets, with the increased interest paid to profitability by financial analysts, pension funds, managers, and individual investors in the Western world, has given prominence to the concept of "share value." The management of chemical companies on both sides of the Atlantic could not remain indifferent to these trends and reacted in several ways to the heavier pressure they implied. In an effort to meet regulatory expenditures and at the same time remain competitive while pleasing their shareholders, Western chemical groups had to resort to severe cost-cutting measures.

Except in such fields as biotechnology R&D budgets were made smaller, and as a result organic growth was somewhat sacrificed, with fewer innovative products coming to the market. Preference was given instead to growth by acquisition, which offered companies an immediate way of achieving a better rank and a higher turnover in some selected activities, if not always better profits. More generally, short-term considerations took precedence over long-term ones, and chemical groups went into a frenzy of asset transfers through mergers, acquisitions, divestitures, and joint ventures. Since 1994 the total value of merger and acquisition deals across the chemical industry in the United States and in Europe has reached a figure of $200 billion, and there is no end in sight to this restructuring fever, greatly facilitated by the ample availability of funds. These moves provided an opportunity for new players to enter the game.

Individual entrepreneurs not accountable to outside shareholders tried their luck by acquiring at rebate prices businesses that larger companies did not want to keep anymore. Private equity institutions bought pieces of business that were for sale with the hope of combining and rationalizing them before returning them to the market three to five years later through an initial public offering. Financial services groups played their role either by financing some of the deals or by acting as brokers between the parties. While profitable to some individuals or banking institutions, these transfers of assets did not always meet the expectations of those who had engineered them or for that matter of the new shareholders.

Indeed the creation of pure play companies focusing on selected fields in which they come to command leading positions cannot achieve the required results if the remaining competitors keep fighting each other in order to maintain or even to improve their market share in such a way that the benefits of the consolidation are lost through the erosion of selling prices. When the number of suppliers is reduced, there is often a corresponding reduction in the number of customers, and the pressure on prices remains unabated. Furthermore, when acquisitions are purchased at too dear a price, they affect the purchaser's balance sheet by denting the bottom line and lowering the operating profit. Aside from these facts that can be assessed quantitatively, there is the less visible aspect of mergers and acquisitions, which has to do with the "morale" of the employees involved in the deals.

Sometimes at all levels valuable people do not like the new "environment" in which they would have to operate, so they decide to leave, offering their talents to competition and thereby weakening the newly born organization. Some chemical groups acting as contrarians have opted for a more stable course in the way they shape their future. Although they are prepared to take advantage of opportunities in order to enlarge or streamline their activities, they keep their diversified operations under one roof, making use of existing synergies and managing their assets with a medium- to long-term objective. It is to be hoped that these contrarians can remain successful because the chemical industry is too vital to our well-being to be submitted entirely to the whims of short-term financial considerations.

2

Economic Aspects of the Chemical Industry

Joseph V. Koleske*

Within the formal disciplines of science at traditional universities, through the years, chemistry has grown to have a unique status because of its close correspondence with an industry and with a branch of engineering—the chemical industry and chemical engineering. There is no biology industry, but aspects of biology have closely related disciplines such as fish raising and other aquaculture, animal cloning and other facets of agriculture, ethical drugs of pharmaceutical manufacture, genomics, water quality and conservation, and the like. Although there is no physics industry, there are power generation, electricity, computers, optics, magnetic media, and electronics that exist as industries. However, in the case of chemistry, there is a named industry. This unusual correspondence no doubt came about because in the chemical industry one makes things from raw materials—chemicals—and the science, manufacture, and use of chemicals grew up

together during the past century or so. In addition, the chemical industry is global in nature.

Since there is a chemical industry that serves a major portion of all industrialized economies, providing in the end synthetic drugs, polymers and plastics, fertilizers, textiles, building materials, paints and coatings, colorants and pigments, elastomers, and so on, there is also a subject, "chemical economics," and it is this subject, the economics of the chemical industry, that is the concern of this chapter. Of course, the chemical industry does not exist alone, rather it interacts with many aspects of the global economy.

DEFINITION OF THE CHEMICAL INDUSTRY

Early in the twentieth century, the chemical industry was considered to have two parts: the discovery, synthesis, and manufacture of inorganic and organic chemicals. Later, and until about 1997, the Standard Industrial Classification (SIC) of the U.S. Bureau of the

*Consultant, 1513 Brentwood Rd., Charleston, WV 25314.

Census defined "Chemical and Allied Products" as comprising three general classes of products: (1) basic inorganic chemicals such as acids, alkalis, and salts and basic organic chemicals; (2) chemicals to be used in further manufacture such as synthetic fibers, plastic materials, dry colors, pigments; and (3) finished chemical products to be used for ultimate consumer consumption as architectural paints, cosmetics, drugs, and soaps or to be used as materials or supplies in other industries such as industrial paints and coatings, adhesives, fertilizers, and explosives.[1] The SIC system was a series of four-digit number codes that attempted to classify all business by product and service type for the purpose of collection, tabulation, and analyses of data. It used a mixture of market-based and production-based categories.

In 1997, the SIC classification was replaced by the "North American Industry Classification System" (NAICS).[2] The system is a major revision based on six-digit numerical codes, and it allows for new or relatively new industries to be included in what is termed "Chemical Manufacturing." It also reorganizes all categories on a production/process-oriented basis. Further, NAICS establishes a common numerical code among Canada, Mexico, and the United States that is compatible with the two-digit level of the United Nations' "International Standard Industrial Classification of All Economic Activities" (ISIC).

The NAICS code for "Chemical Manufacturing" is "325" and there are 49 subclassifications with four- to six-digit codes. The four-digit codes, which are a description of the manufacturing segments included in chemical manufacturing, the value of shipments, and the number of employees in the manufacturing segment are listed in Table 2.1.[2] Each of these four-digit segments may have five-digit subclasses associated with them, and the five-digit subclasses in turn may have six-digit subclasses associated with them. This hierarchy is exemplified for Manufacturing Segment 3251, which is titled "Basic Chemical Manufacturing," and one of its sub components, Code 32519, in Table 2.2.

While it may seem that Code 325199, "All Other Basic Organic Chemical Manufacturing," is too general in nature for its size, one needs to consider that by delving into the makeup of this component, about 150 individual compounds or groups of compounds are found. These contain a diverse group of chemicals including manufacturing of acetic acid and anhydride, calcium citrate, cream of tartar, ethylene glycol ethers,

TABLE 2.1   Chemical Manufacturing, NAICS Code 325, and Its Four-Digit Area Components. Shipment Value and Employees are from 1997 U.S. Economic Census[2]

NAICS Code[a]	Description of Area	Shipments Value ($1000)	Percentage of Total	Employees
325	Chemical manufacturing	419,617,444	100	884,321
3251	Basic chemical manufacturing	115,134,992	27.44	202,486
3252	Resin, synthetic rubber, artificial and synthetic fibers, and filament manufacturing	63,639,476	15.17	114,792
3253	Pesticide, fertilizer, and other agricultural chemical manufacturing	24,266,513	5.78	37,206
3254	Pharmaceutical and medicine manufacturing	93,298,847	22.23	203,026
3255	Paint, coating, and adhesive manufacturing	26,594,550	6.34	75,100
3256	Soap, cleaning compound, and toilet preparation manufacturing	57,507,318	13.7	126,895
3259	Other chemical product manufacturing	39,175,748	9.34	124,816

[a]Codes 3257 and 3258 were not used.

TABLE 2.2   Basic Chemical Manufacturing, NAICS Code 3251, and its Five-Digit Components and Other Basic Organic Chemical Manufacturing, Code 32519, and Its Six-Digit Components. Shipment Value and Employees are from 1997 U.S. Economic Census[2]

NAICS Code[a]	Description of Area	Shipments Value ($1000)	Percentage of Total	Employees
NAICS Code 3251 and Its Components				
3251	Basic chemical manufacturing	115,134,992	100.00	202,486
32511	Petrochemical manufacturing	20,534,750	17.84	10,943
32512	Industrial gas manufacturing	5,231,468	4.54	12,492
32513	Dye and pigment manufacturing	6,427,357	5.58	17,289
32518	Other basic inorganic chemical manufacturing	20,716,361	17.99	60,056
32519	Other basic organic chemical manufacturing	62,225,056	54.05	101,706
NAICS Code 32519 and Its Components				
32519	Other basic organic chemical manufacturing	62,225,056	100.00	101,706
325191	Gum and wood chemical manufacturing	815,201	1.31	2,267
325192	Cyclic crude and intermediate manufacturing	6,571,093	10.56	8,183
325193	Ethyl alcohol manufacturing	1,287,273	2.07	1,890
325199	All other basic organic chemical manufacturing	53,551,489	86.06	89,366

[a]Codes 325194 through 325198 were not used.

ethylene oxide, solid organic fuel propellants, hexyl and isopropyl alcohols, perfume materials, peroxides, silicone, sodium alginate, sugar substitutes, tear gas, synthetic vanillin, vinyl acetate, and so on. All these compounds have the code number 325199. Compounds with the code number 325211, "Plastic Material and Resin Manufacturing" are about 80 in number and may be exemplified by acrylic and methacrylic polymers; cellulose derivatives such as acetates, nitrates, xanthates, and the like; phenolics, polyesters, polyolefins, polystyrene, poly(vinyl halide)s, polyurethanes, and, again, and so on.

The new NAICS has broadened the definition of the chemical industry, and it now is more encompassing than in the past. The broadening is reasonable, and it improves on the goals of collecting and tabulating data so that it is available for study and analysis. One might say the data could be timelier, but collecting, amassing, and breaking down the information so it is understandable is a difficult, time consuming task that is dependent on many people. The Internet is a major factor in making the data available to the general public almost as rapidly as it is compiled. The NAICS system is recognized and accepted by the North American countries, and the system appears to be global in nature by being at least partially in line with the United Nations' classifications. Those interested in markets and market areas and in their size including their relation to other markets will find the U.S. Census Bureau's web site pages well worth visiting.

THE PLACE OF THE CHEMICAL INDUSTRY IN THE ECONOMY

Because the chemical industry is a major sector of any advanced national economy, a forecast of trends in the chemical industry must fall within certain general guidelines that are established by the national economy. A forecast for the chemical industry in the United States must be within the general boundaries set for the overall societal, financial, environmental, governmental, and economic forecasts for the country. However, such forecasts should be carefully considered for they may or may not accurately predict the future.

It had been said that it was clear for many years that certain demographic and societal issues would have a dominant effect on the U.S. economy of the 1990s. In the previous edition of this Handbook, it was pointed out that there was an expectation that from the late 1980s through the year 2000 there would be a decline in the growth of the work force in the United States. This was predicated on the number of women within the usual childbearing age group of 18–35 and by family-size decisions that were made in the 1960s. Shortages of chemists, chemical engineers, and other scientists were predicted for the 1990s. Supposedly, such predictions can be made from census data that was obtained in the prior two or three decades. There is a direct relation between the growth of the workforce and the growth of Gross Domestic Product (GDP). Although this was the prediction, it is not what happened.

During the 1990s and through the start of the twenty-first century, due to events put in motion during the late 1980s, the United States and many world economies experienced unprecedented growth. During this time unemployment decreased and reached very low percentages on an absolute and a historical level. This factor was coupled with significant productivity increases throughout the economy. The productivity increases resulted from a better-trained workforce, from new tools such as computers and allied software, and from just plain harder, more conscientious working during regular and overtime hours. An important factor during this period of growth was that the productivity increases were obtained without inflation raising its ugly head. Company mergers and the spinning off or the selling off of business segments to stockholders or to allied businesses played an important role through these years. These actions resulted in new stand-alone businesses that were operated by new owners and managers when the units were spun off. The mergers or unit-sales resulted in a restructuring or downsizing

as duplicated efforts were eliminated. Productivity increased because of these actions, and many workers were displaced. But, the man power hungry economy quickly absorbed for the most part these displaced workers. The hunger for manpower was partially, but importantly, related to the electronic, computer, telecommunications, and related industries that provided many jobs in previously non-existent sectors. Chemicals were used in various ways in these new growth areas—as, for example, wire coatings, solder masks, conformal coatings, optical fiber coatings and marking materials, magnetic tape coatings, and so on.

Mergers and acquisitions certainly played a role in shaping today's chemical manufacturing industry. Included among the notable mergers are the Pfizer Inc. merger with Warner Lambert Company. The Dow Chemical Company acquisition of Union Carbide Corporation, Exxon merging with Mobil, and many others in the United States. Larger companies acquire smaller companies to expand business through new or expanded opportunities, diversify, reduce research and development expenditures, improve negotiations with suppliers and customers, and improve operating efficiency. For example, in the water industry, Aqua America, Inc., the largest United States-based, publicly traded water company, acquired 29 small companies in 2004 in line with their growth target of 25 to 30 acquisitions per year. In 2005, diversified 3M Company acquired CUNO, Inc., a water filtration products company, to capitalize on the global need for water purification, a market that is growing at more than 8 percent per year. Great Lakes Chemical Corp. and Crompton Corp. merged to form a new company known as Chemtura Corp., a company that is focused on the future's specialty chemical needs. In the paint and coating segment of the chemical business, major changes have taken place through mergers and acquisitions. In 1990, the five largest producers had 37 percent of the market and the ten largest producers had 52 percent. By 2003, the top five had 51 percent and the top ten had 74 percent of the market.[3] Included in the acquisitions are Akzo Nobel's purchases of Courtalds and Ferro's powder coating business; Dow acquired Celanese's Taxas acrylic monomers plant and output will be used for superabsorbants and paint emulsions; Sherwin William's acquisitions of Duron, Krylon, Pratt & Lambert, Paint Sundry Brands (a manufacturer of high quality paint brushes), and Thompson Miniwax; and Valspar's purchase of Lilly Inc. and Samuel Cabot Inc.

In Great Britain, Glaxxo Holdings PLC first merged with Wellcome to form Glaxo Wellcome, this combination then merged with Smith Kline to form GlaxoSmithKline PLC. Malvern Instruments Ltd. In the United Kingdom and Perkin Elmer of the United States agreed to a collaborative sales agreement that will offer customers material characterization instruments of both companies: rheology, thermal analysis, and rheometers. Degussa AG acquired Cytec Industries holdings in CYRO Industries to consolidate its position as a leading global supplier of methyl methacrylates. In The Netherlands, Arnhem, Akzo Nobel's Coatings business, acquired Swiss Lack, Switzerland's leading paint company. Larger, improved efficiency companies resulted. One result of such national and international mergers and acquisitions is a shrinking of *Chemical and Engineering News'* top 100 companies to the top 75 companies.[4]

The electronics/computer industry grew rapidly during the past decade or so, and new company names appeared during this time period. Its growth was spurred by the productivity increase even as it was a participant in causing the increased productivity. Computers began to be used to control processes and training personnel with the skills to run such computer-operated processes was high on many companies' lists of important projects. A decade or so ago, computers were available in companies on a limited basis. Today, there is a computer on essentially every desk and portable computers to carry out work during trips, and the like. The "dot com" companies started their appearance through the Internet, and they grew rapidly. Later, when business turned down, many of these companies disappeared—

they merged with or were purchased by other companies. However, overall prosperity reigned during this time period, and, as it did, the chemical manufacturing industry, which was allied with a broad variety of these industries, also prospered.

At the turn of the twenty-first century, it was becoming apparent that the economy was at a high point and could be expanding too rapidly. Inflation was still low, and there was even talk that deflation might come into play. They latter did not happen. Price-to-earnings ratios were very high for many companies, and it did not appear that future growth would expand sufficiently to accommodate such high price-to-earnings ratios and large additions to the work force. The national economy, which certainly includes chemical manufacturing, entered a recession in March 2000. However, the economy grew in the first three months of 2001 indicating that the economic recovery could be beginning. The improvement was led by new automobile purchases and increased government spending. The fourth quarter of 2001 was small and considered by some as flat, but it built on the preceding quarter and in early 2002 there was a belief that the economic recovery has begun.

The events of September 11, 2001, changed many aspects of our lives with chemical manufacturing included. The terrorist attacks rocked many markets on a short-term basis, but before long the markets stabilized, but did not really grow in the recent past. Overall, in the first half of 2002 the world economy remained in a recession. However, because of the constant threat of terrorism, national corporate spending will increase as military, security, and other government expenses increase and transportation costs and its allied security measures come into play. Chemical manufacturing of basic chemical, polymer, and pharmaceuticals are expected to increase. Yet, there are no expected productivity increases as was seen in the 1990s associated with the increased spending.[5] In early 2002, the Chairman of the Federal Reserve predicted that the recovery was apparent, but would be a mild recovery.[6]

As can be seen from the previous discussion, economic forecasts are subject to all of the uncertainties and unpredictabilities of national, international, and societal events. With this in mind, at the present time the forecast for the ensuing part of this decade is for improved growth in national chemical manufacturing with growth and the profit picture beginning an upturn in the second or third quarter of 2002. (Note that in the 1980s, it was not predicted that the 1990s would show strong growth, yet strong growth did take place.) At present the concerns with global terrorism, low interest rates, high but constant productivity, oil and gas prices, and other factors are components of a mixture that will dictate the future. None of these factors will remain constant. Rather, they will change individually at times and with some factors in concert at other times. These variations along with the size of the workforce and its attitude will dictate the future for chemical manufacturing and the global economy.

Against this brief discussion of the general demographic, societal, and economic factors that govern forecasting economic prospects, a general picture of the economy of the United States can be given by the GDP and chemical and allied products portion of GDP as described in Table 2.3.

This reasoning is a way to highlight the sensitivity and place chemical manufacturing has in the national economy, which is becoming more and more entangled with the countries of the North American Free Trade Agreement

TABLE 2.3   U.S. Economy and Chemical Manufacturing[7]

Year	U.S. Gross Domestic Product (GDP) Current Dollars, Billions	Chemicals and Allied Products Portion of GDP, Current Dollars, Billions
1987	4,742.5	83.8
1990	5,803.2	109.9
1995	7,400.5	150.8
1996	7,813.2	153.6
1997	8,318.4	164.8
1998	8,781.5	164.8
1999	9,268.6	175.1
2000	9,872.9	191.1

(NAFTA) and with the global economy. Thus the chemical manufacturing industry is worldwide and interconnected in many ways. These factors play important roles in the importance of imported raw materials such as petroleum products and the cost of labor. Businesses or parts of businesses can be transported across the southern U.S. border to take advantage of more favorable labor costs. Through this, successful partnerships have been forged and welded together between border countries. As mentioned earlier, other partnerships are developing through purchase of assets in other countries by the United States and by other countries in the United States. Today, the United States is entrenched in the age of a global economy and all it ramifications.

The United States imports and exports a wide variety of raw materials and chemical products. Major U.S.-based chemical companies have manufacturing and sales facilities abroad and a large number or foreign-based companies have similar facilities in the United States. The U.S. economy is dependent on the balance of trade, that is, on the difference between the dollar value of exports and imports. A negative trade balance means that dollars spent abroad to import goods and services exceed the value of goods and services exported. In effect such an imbalance increases the cost of goods and services purchased in the United States and results in a net inflationary effect. To a large extent during the 1980s, this potential inflationary effect was offset by foreign investment in the United States. In the 1990s and through the early years of the twenty-first century, foreign investment in the United States has increased, productivity has increased without major wage increases, and interest rates were managed with the net result that inflation remained low.

In foreign trade, the chemical industry of the United States has consistently performed in an outstanding manner. While the overall balance of trade has been negative, the chemical industry has been one of the truly strong sectors in the economy of the United States, Table 2.4. Year after year, the trade balance of chemicals has been positive and thus has had

TABLE 2.4   U.S. Balance of Trade[8]

Year	Total Trade Balance[a] (Billions of Dollars)			Chemical Trade[b] (Billions of Dollars)		
	Export	Import	Balance	Export	Import	Balance
1987	250.2	409.8	−159.6	26[c]	16	+10
1990	389.3	498.3	−109.0	39[c]	22	+17
1995	575.8	749.6	−173.8	32.18	20.59	11.59
1996	612.1	803.3	−191.2	31.4	21.81	9.59
1997	678.4	876.5	−198.1	34.6	23.5	11.10
1998	670.4	917.1	−246.7	33.32	23.38	9.94
1999	684.6	1,030.0	−345.4	34.09	23.82	10.27
2000	772.2	1,224.4	−452.2	38.42	27.12	11.30

[a]International Trade Accounts (ITA); [b]Chemicals-Fertilizer, -Organic, -Inorganic, and -Other; [c]Amounts for 1987 and 1990 are taken from Bailey and Koleske.[9]

a positive impact on the national economy. When the total world export market for chemicals is considered, that is the sum of all the chemicals exported by all the world's national economies, the U.S. chemical manufacturing industry has held a significant market share, about 15 percent, for the past three decades.

The export and import values for chemical segments described in Table 2.4, chemical-fertilizer, chemical-organic, chemical-inorganic, and chemical-other, are detailed in Table 2.5. The magnitude of the individual items varies from year to year, but overall, the balance is favorable and these four segments of chemical manufacturing are usually positive values. It should be pointed out that various items (plastic materials, pharmaceuticals, etc.) that make up chemical manufacturing have been excluded, but this was done without bias. The four items used in Tables 2.4 and 2.5 are directly related to what has been traditionally known as the "chemical industry." The less favorable Total Trade Balance of the United

TABLE 2.5   Chemical Export/Import Segments for Year 2000

Chemical Segment	Exports (Billions of Dollars)	Imports (Billions of Dollars)	Balance (Billions of Dollars)
Fertilizer	4.098	3.388	+0.710
Inorganic	4.180	4.414	−0.234
Organic	16.505	13.779	+2.726
Other	13.636	5.525	+8.111

States is principally due to imports of manufactured good and petroleum products.

To support the U.S. chemical manufacturing economy (Code 325) in 1997 (see Table 2.1), there was a workforce of more than 884,000 of which about one-fourth were employed in basic chemical manufacturing (Code 3251) and about one fourth were employed in pharmaceutical and medicine manufacturing (Code 3254). The next largest area of employment, about 14 percent of the workforce, was the soap, cleaning compound, and toilet preparation manufacturing component (Code 3256), which was closely followed by the polymer manufacturing area (Code 3252) at 13 percent. In such comparisons, one might argue that the paint, coating, and adhesive manufacturing component (Code 3255), with about 8.5 percent of the employment figure, should be included with polymer manufacturing. The remainder of the workforce is employed in the agricultural chemical and other chemical manufacturing components. The value of the chemical manufacturing business produced by this workforce was $419,617,444,000 in 1997. To maintain market share and grow this huge business, the companies in chemical manufacturing invest to various degrees in research and development efforts, which are carried out by scientists within the organizations. The percentage of sales varies with the particular component, and the pharmaceutical firms will spend much more than say a fertilizer manufacturer. The average for many chemical companies varies from about 3 to 5 percent of sales. In the 1999–97 period, about seven billion dollars were spend annually on research and development by the chemical industry.

CHARACTERISTICS OF THE CHEMICAL INDUSTRY

Investment Trends

The U.S. chemical industry is the world's largest and it accounts for about one fourth of global chemical production. The industry, which is a part of the non-durable goods manufacturing industry, is a high capital investment business. Capital spending by the chemical and allied products industry in the United States has been a sizable percentage of that spent for all manufacturing. In 1999, non-durable goods manufacturers spent about $80 billion on capital goods. This was a decrease of about 7 percent from that of 1998, which was approximately $85 billion. Most of this decline can be attributed to decreased spending by the basic chemical industry. In 1996 and 1997, capital expenditures in the chemical industry were about $15.5 and $16.4, respectively. During the late 1980s and 1990s, a significant portion of these capital expenditures was made for pollution control and other environmentally related efforts.

Much of the capital investment in the chemical industry is spent for facilities to produce major chemicals in enormous quantities. The huge volume of the chemicals produced and consumed is reflected in the size of plants being built to achieve the required economies of scale, which in turn allow for competitive pricing. The fact that such economies are achieved is seen in the relatively modest increases in chemical producers' price indices relative to the inflation levels in the general economy. The competitive nature of the chemical business also plays a role in this matter of price. (Economy of scale refers to the relative cost of building a larger plant; a rule of thumb is that the relative cost of building a smaller or larger plant is the ratio of the productivities of the two plants being considered raised to the 0.6 power. In other words, the unit cost of producing a chemical markedly decreases as the size of the plant producing it is increased, providing the plant can be operated near capacity.)

Along with these very large plants and the associated enormous investments, most of the chemical industry is characterized by high investment versus low labor components in the cost of manufacture. The National Industrial Conference Board statistics list the chemical industry as having one of the highest capital investments per production worker. The investment per worker in a base petrochemicals olefins plant may be in the neighborhood a half-million dollars. A profitable chemical specialties manufacturer may have capital investments as low as 10 percent of such values per employee. Of course, sales per employee are also important and large. From Table 2.1, it can be seen that annual sales per employee in the overall chemical manufacturing area (Code 325) are about $475,000. Such ratios vary with the market segment and depend on the labor intensity needed within the segment. For example, the number is about $569,000 for basic chemical manufacturing (Code 3251), $652,000 for agriculture chemicals (Code 3253), $459,541 pharmaceuticals (Code 3254), $354,000 coating chemicals (Code 3255), and so on. Note that number for pharmaceutical sales per employee is quite close to the overall chemical manufacturing sales per employee. The average hourly pay of production workers in the chemical industry was $18.15 in 2000.

Commercial Development and Competition Factors

During the earlier period of the chemical industry's development, chemical companies were generally production oriented, wherein they would exploit a process to produce a chemical and then sell it into rapidly expanding markets. The investments and plant sizes required for participation were a small fraction of that required to participate today. Raw materials were often purchased to produce chemical intermediates for sale. Small-sized units operating in small manufacturing facilities do not present the obvious problems of environmental pollution, a factor about which everyone has become more aware in the past two or three decades. A new investment in chemical production facilities today must include a sizable proportion of the total outlay for pollution abatement and control of environmental intrusion. The chemical industry spends about $5 billion annually on pollution abatement.

As the chemical industry has grown, there has been a strong tendency toward both forward and backward integration. Petroleum producers have found opportunities based on their raw materials—natural gas, condensates, and oil—to move into chemical refining. Chemical companies, on the other hand, have moved to assure their access to low-cost raw materials through contract purchases and hedging contracts. Similarly, producers of basic plastic materials have forward integrated to produce compounder materials and fabricated products such as consumer items, fibers, and films. At the same time, fabricators have installed equipment to handle and formulate or compound the basic plastic materials and thus provide a ready, constant supply at the lowest possible cost. With the global economy in place and relative ease of moving around the world, large investments are now made in far-off countries such as Malaysia and Saudi Arabia, for example, to be near raw material supplies and to meet large market needs.

With ever-larger investment costs and increasing cross-industry competition, markedly greater sophistication has been required of marketing analysis coupled with cost analysis when selections of investment opportunities are made. The enormity of investment capital required in today's marketplace to successfully participate does not permit multiple approaches for the private investor. Consequently, a high degree of market orientation tends to predominate in the chemical industry along with increasingly targeted and pinpointed research and development programs. In 2000, the industry spent about $31 billion on such research and development efforts.[10]

A major trend in industrial chemistry has been an emphasis on improved processes for the production of major chemicals such as ethylene, propylene, vinyl chloride, styrene, alkylene oxides, methanol, terephthalates, and so on. The necessity for higher efficiency, lower cost processes has been accentuated by the relatively slow growth rates of major industrial chemicals over the past two decades or so. The fertilizer portion of the agricultural chemicals market as described in Table 2.6 is an example of the slow growth.

How well has the chemical industry developed? At the beginning of the twenty-first century the United States accounted for 27 percent of the world's chemical production, making it the world's largest chemical producer.[10] In 2000, chemical shipments reached $460 billion, and, at this level, it provides about 1.2 percent of the national GDP and almost 12 percent of the manufacturing GDP. As such, it is the largest factor in the manufacturing segment of the economy. The chemical industry continues to grow, and it attained an all-time high in profits by netting $44 billion. Globally, chemicals are almost a $1.5 trillion dollar business.

With its large size, the chemical industry is a large user of energy, and it consumed about 7 percent of all domestic energy and about 25 percent of all energy used in manufacturing. In 1985, the industry used 3,567 trillion Btu, in

TABLE 2.6   Annual Production of Inorganic Chemicals Used in the Fertilizer Industry (Note Break in Years between 1996 and 1993)[8]

Chemical	Production Amount (Billions of Pounds)		
	1997	1996	1993
Ammonia	34.68	35.85	34.39
Ammonium Nitrate	17.21	17.00	16.56
Ammonium Sulfate	5.42	5.32	4.87
Nitric Acid	18.87	18.41	16.51
Phosphoric Acid	26.32	26.42	23.03
Sulfuric Acid	96.04	95.54	79.68
Super Phosphates and others	20.86	21.09	17.6
Urea	15.33	17.10	16.66

1991 usage increased to 5,051 trillion Btu, and in 1994 it had increased still further to 5,328 trillion Btu.[11] The energy is used to supply heat and power for plant operations and as a raw material for petrochemicals, plastics, and fibers production. Feedstocks represent a little less than half of the total usage, a number that varies from year to year.

Thus, the chemical industry is a key component in the U.S. economy. It converts raw materials such as gas, oil, condensates, water, metals, and minerals into more than 70,000 products that are used in a variety of ways. In some fashion, this industry impacts the daily lives of everyone. Industrial customers for chemicals are many, but some of the major ones are apparel, plastic and rubber products, petroleum refining, textiles, pulp and paper, primary metal, and the like.

Information technology and E-commerce have become increasingly important assets to the chemical industry. Spending on information technology reached $10.2 billion in 2000 and this represented a 75 percent increase over such expenditures of 1990.[10] Selling via the Internet or E-commerce resulted in sales of $7.2 billion in 2000. Projections indicate that this type of business will grow rapidly and are expected to reach $150 billion by 2006. This means about one third of shipments will be via E-commerce transactions in five years.

Technological Orientation

The chemical industry is a high technology industry, albeit now is more marketing oriented and competitive than in its earlier period of development. Chemists and materials scientists held about 92,000 jobs in 2000. Over half of these are employed in manufacturing companies and most of these companies are in the chemical manufacturing industry, that is in firms that produce synthetic materials, plastics, drugs, soaps and cleaners, paints, industrial organic and inorganic chemicals, and other chemicals.[12(a)] Other chemists and chemical engineers are found in various government Departments and Agencies, in teaching, and in research, development, and testing firms. The latter

firms are becoming more and more a growth area in the chemical industry. A bachelor's degree in chemistry or a related discipline is the minimum education requirement for these technical positions. To work and grow in research positions, a Ph.D. is required. There will be strong demand for those people who have a masters or Ph.D. in the future with job growth concentrated in the pharmaceutical companies and in research, development, and testing services firms.

The contemporary scientist or engineer engaged in research and development in the chemical industry is a highly trained individual who is a part of a high-investment occupation. Since about the mid 1950s, much of chemistry has become increasingly an instrumental science, and the instruments routinely used by investigators are highly sophisticated, reliable, and costly. In the laboratory, a scientist has available mass, infrared, visible, and ultraviolet spectrometers; various chromatographs; physical and chemical property determination devices such as those used for molecular structure, size, and conformation determinations; and others used for reaction kinetic studies. Pilot plants and many production facilities are highly instrumented and automated. The basic scientist doing research, laboratory workers, pilot plant and process development chemical engineers, and plant production workers require at a minimum access to excellent computer facilities. All engineers and scientists require computers to analyze the massive amounts of data that are generated and to aid in the design of manufacturing processes and equipment.

Employment of chemists and chemical engineers is expected to grow at about a 10–20 percent rate between 2000 and 2010.[12(a)] Predictions indicate that job growth will be concentrated in drug manufacturing and in research, development, and testing services companies. Demands will be for new and better pharmaceuticals, personal care products, and specialty chemicals designed to solve specific problems or applications. Demand will be high for personnel who have a Ph.D. degree and the opportunities will be in biotechnology and pharmaceutical firms. An aging, better informed population will want products that treat aging skin, that are milder on the body, new and innovative drugs, reliable medical devices, and so on. The population in general will be interested in chemical processes that are more benign in nature to produce all types of products and thus in an industry that is more friendly to the environment.

In the year 2000, the median salary of chemists was $50,080.[12(a)] The lowest 10 percent earned less than $29,620, and the highest 10 percent earned more than $88,030. The middle 50 percent earned between $37,480 and $68,240. It is interesting to point out that the median annual salary of chemists employed in the Federal Government was $65,950 or about 30 percent higher than the overall median. In 2001, chemists in non-supervisory, supervisory, and managerial positions in the Federal Government averaged $70,435.[12(a)] As is the usual case, chemical engineer salaries were higher than those of chemists by about 10–25 percent.[12b] Median experienced and starting salaries for the various degrees can be found in Table 2.7.

In 2003, median starting salaries for industrial chemists were $32,000 (B.S.), $44,500 (M.S.), and $63,000 (Ph.D.).[12c] The Ph.D. starting salary actually dropped from $67,000 in 2002 due to Ph.D.s' finding academic

employment. In 2003, 35.3 percent of new Ph.D. graduates went into academic positions compared with 20.5 percent in 2002. A 2005 salary survey of chemical industry professionals with a Bachelor's degree in chemical engineering with 22 years of experience in the chemical industry (pharmaceuticals, organic chemicals, construction, and consulting) indicated that the average salary was $85,234.[12d] the 1205 nationwide respondents were male, had an average age of 47, and rated their overall job satisfaction as "satisfied." During 2004, about 85 percent of the participants received a raise and about 65 percent also received a cash bonus. Health insurance costs increased with many companies requiring employees to pay a larger portion of the insurance costs. The satisfaction portion of the survey indicated that challenging work was the most important factor (44 percent of responses) and other factors such a salary and benefits (17.5 percent), job security (13.5 percent), advancement opportunities (7.4 percent), recognition (12.6 percent), and other (5.1 percent) were secondary in nature.

Historical

How did the chemical manufacturing industry get its beginning? To get this answer, we need to go back to the latter part of the eighteenth century.[13] The availability of alkali or soda ash (sodium carbonate) for the growing manufacture of glass, soap and textiles in France was becoming a major concern. At that time, the chemical was obtained from plant materials, principally from wood ashes that were leached with hot water to obtain potash and from marine plants such as *barilla*, which grows mainly along the Spanish Mediterranean coast and in the Canary Islands. Other plant sources existed. The main exporter of soda ash was Great Britain with whom France was at odds and there was concern about the chemical's availability. In 1783, the French Academy of Sciences was offered a handsome prize by Louis XVI to develop a simple process for "decomposing" sea salt on a large scale and securing alkali from it. Eight years later Nicholas Leblanc, a 49-year-old French

TABLE 2.7 Chemist[12(a)] and Chemical Engineer[12(b)] Salaries in 2000

Degree	Year 2000	
	Overall Median Salary	Inexperienced Median Starting Salary
Chemists		
Bachelor	$55,000	$33,500
Master	$65,000	$44,100
Ph.D.	$82,200	$64,500
Chemical Engineers		
Bachelor	a	$51,073
Master	a	$57,221
Ph.D.	a	$75,521

aThe median annual salary for all chemical engineers was $65,960. The salary of the middle 50% ranged between $53,440 and $80,840. The salary of the lowest 10% was less than $45,200 and of the highest 10% was greater than $93,430.

physician, devised a scheme to commercially obtain soda ash from sea salt. The process became known as the Leblanc process, and this process is considered the basis for development of the first chemical industry. For almost a century, this process was the most important method known for producing chemicals. Basically, Leblanc's process involved reacting sodium chloride with sulfuric acid to produce sodium sulfate. The product was then reacted with calcium carbonate and carbon to form a "black ash" that contained sodium carbonate and other compounds. The "black ash" was extracted with water followed by an evaporation process to obtain soda ash.

Leblanc's process had many disadvantages; it was complicated, was dirty and polluting, and was materials and fuel inefficient. This set other scientists to working on development of a new process. In about 1872, Ernst Solvay developed what became known as the Solvay process, and this resulted in establishment of the French firm Solvay & Cie. By 1890 the Solvay process dominated the world's alkali production.[14] Leblanc's process was obsolete.

Obsolescence and Dependence on Research

The high technology level that characterizes the chemical industry, and which is reflected in heavy research and development investments, generally concerns discovery and development of new products as well as improvements in the manufacture of known products. New product discovery and development may be typified by a new pharmaceutical product for a specific disease, by a stealth aircraft and all its special polymer and composite needs, by development of a new non-polluting technology for a known process, a uniform molecular weight polymer designed and made by nanotechnology, and so on. Improvements in the manufacture of known products might be typified by producing a modified form of a pharmaceutical that is easier to dissolve, by a new or modified higher efficiency catalyst for a known process, by toughening a brittle plastic

material, by improving the strength of a composite, and so on. The development of a new, lower cost process for a commercial product can permit development of a profitable opportunity or it can spell disaster for a company with existing investment in a plant made obsolete by the competitor's new process as in the preceding soda ash example.

Major reductions in manufacturing cost can be achieved, for example, by reducing the number of reaction steps required in a process, by changing to a lower cost or more available raw material, or by eliminating by-products and co-products, costly separation, and environmental intrusions. The ability of a process scheme to contain or avoid a pollutant can be a deciding factor in continuance of a manufacturing operation. At times new regulations, such as the Clean Air Act (CAA), or shortages can spawn new ideas and technologies if the people involved are astute and react positively to the new, developing environment. The brief discussion of Leblanc's process being replaced by Solvay's process for soda ash is an example of how economic consequences can change if a competitor finds a process better than the one being practiced. The following detailed examples will make the matter even more clear.

Acetic Acid. Acetic acid production in the United States has increased by large numbers in the last half century, since the monomer has many uses such as to make polymers for chewing gum, to use as a comonomer in industrial and trade coatings and paint, and so on. In the 1930s, a three-step synthesis process from ethylene through acid hydrolysis to ethanol followed by catalytic dehydrogenation of acetaldehyde and then a direct liquid-phase oxidation to acetic acid and acetic anhydride as co-products was used to produce acetic acid

$$CH_2{=}CH_2 \xrightarrow{H_2SO_4/H_2O} C_2H_5OH$$

$$C_2H_5OH \xrightarrow{Cu/Cr} CH_3CHO$$

$$CH_3CHO \xrightarrow{[O], Co} CH_3COOH + (CH_3CO)_2{=}O$$

Then, in the 1940s, a major process change was introduced. In this new process, butane was directly oxidized to acetic acid and co-products such as methylethylketone.

$$C_4H_{10} \xrightarrow{\text{[O]}} CH_3COOH + CH_3COC_2H_5 + \text{others}$$

The novel synthesis required fewer process steps, and this resulted in lower costs and investment. In 1969, another advance was announced—the synthesis of acetic acid from methanol and carbon monoxide with essentially no by-products or co-products.[15,16]

$$CH_3OH + CO \xrightarrow{\text{I/Rh}} CH_3COOH$$

The use of readily available raw materials and absence of co-products reduces production costs and investment needed for distillation and other separation systems. Such simplification results in a very attractive process in an industry where the principally accepted measure of business quality is return-on-investment.

Acetic Anhydride. Acetic anhydride is required as a process intermediate in acetylations. To obtain acetic anhydride from acetic acid, acetic acid is first pyrolyzed to ketene, which then reacts with recovered acetic acid to yield the anhydride.

$$CH_3COOH \xrightarrow{\text{Heat}} CH_2=C=O$$
$$CH_2=C=O + CH_3COOH \longrightarrow (CH_3CO)_2-O$$

In 1980, the Tennessee Eastman unit of Eastman Kodak announced that it would begin construction of a facility to make acetic anhydride from coal, which was readily available at reasonable cost.[17,18] This decision reflected a changing of the raw materials base of much of the chemical industry due to such factors as the rising cost of natural gas and petroleum and the large coal reserves of the United States.

In the new Eastman process, synthesis gas (carbon monoxide and hydrogen) is made from coal. Then, from the generated synthesis gas, methanol was prepared. (Prior to this time, methanol had been made from methane, i.e., natural gas.)

$$CO + 2H_2 \longrightarrow CH_3OH$$

Methanol was next reacted with acetic acid to form methyl acetate.

$$CH_3OH + CH_3COOH \longrightarrow CH_3COOCH_3 + H_2O$$

Acetic anhydride was then obtained by the catalytic carbonylation of methyl acetate with carbon monoxide.[16]

$$CH_3COOCH_3 + CO \longrightarrow CH_3CO-O-OCCH_3$$

There are two major points that make this process attractive. First, the raw material base of synthesis gas is coal. The second point is avoidance of the energy-consuming manufacture of ketene by pyrolyzing acetic acid.

Vinyl Chloride. The increase in the production of vinyl chloride, which is the principal monomer for poly(vinyl chloride) plastics and various vinyl copolymers that are used in vinyl flooring, shower curtains, car-seat upholstery, house siding, pipe, beverage can coatings, and so on, is an even more spectacular example. This polymer is used in multibillion pound quantities. It is an interesting sidelight to point out that the polymer has poor thermal stability, and its huge penetration into the marketplace is attributable to the development of highly efficient thermal stabilizers.

During the early monomer development in the 1930s, vinyl chloride was produced by means of a catalytic addition of hydrogen chloride to acetylene.[19]

$$CH\equiv CH + HCl \xrightarrow{\text{HgCl}} CH_2=CHCl$$

Later, what was called a "balanced" process was introduced. In this process, chlorine was added to ethylene and ethylene dichloride was produced.

$$CH_2=CH_2 + Cl_2 \xrightarrow{\text{[O]/Cu}} CH_2ClCH_2Cl$$

The ethylene dichloride was then cracked to vinyl chloride and hydrochloric acid with the hydrochloric acid recycled to produce vinyl chloride from ethylene as shown above.

$$CH_2ClCH_2Cl \xrightarrow{\text{Heat}} CH_2=CHCl + HCl$$

At this point in time, vinyl chloride was being produced from chlorine, acetylene, and ethylene. After these processes, a catalytic oxychlorination has been developed in which vinyl chloride is produced from ethylene and hydrogen chloride in the presence of oxygen.[20,21]

$$CH_2{=}CH_2 + HCl \xrightarrow{\text{[O]/Cu}} CH_2{=}CHCl + H_2O$$

If desired, the hydrochloric acid can be obtained via cracking of ethylene dichloride. The oxychlorination process freed vinyl chloride production from the economics of a more costly raw material, acetylene. Deliberate acetylene manufacture is energy intensive and relatively expensive. By-product acetylene from gas cracking is less expensive, but it has not been available in sufficient supply for the large, approximately billion-pound-per-year plus, vinyl chloride production units.

During the long development and commercialization of poly(vinyl chloride) into one of the major plastic materials, several basic processes of making the polymer evolved. In all of these processes, vinyl chloride was handled as a liquid under pressure. Other than the relative ease with which the monomer could be free radically polymerized, vinyl chloride was regarded as an innocuous, relatively inert chemical. During the 1960s, the monomer sold for five or six cents a pound. Because of the low cost, it was uneconomical to recover and compress the monomer for recycle during stripping and drying operations at the end of the process. The monomer was often vented into the atmosphere.

Then, in the 1970s, a number of poly(vinyl chloride) producers were completely surprised when it was found that long-term (20-year) exposure to vinyl chloride could cause rare forms of tumors.[22] After the discovery that vinyl chloride was a carcinogen, venting was not permissible. Containment and recovery of the monomer was mandatory. As a result, some older processes and manufacturing facilities could not be economically modified to incorporate containment, and as a result such operations were discontinued. This case is but one example of the impact that necessary and regulated environmental controls can have on manufacturing processes and operations.

Coatings Technology. Environmental regulation also had a major impact on the coatings industry. Before, around 1970, almost all (between 90 and 95%) industrial coatings were applied at low-solids (about 10–20%) contents from solvents. Many trade or house paints were also solvent based, since aqueous latex technology did not yet have its dominant position. Solvents were inexpensive and they did an excellent job of dissolving the high molecular weight polymers needed to obtain good performance characteristics. The high molecular weights used necessitated the large quantities of solvent to be used—about 4–9 lb of solvent were venting to the atmosphere for each pound of final coating film.

Then in the early 1970s, two factors affected the coatings industry. One of these was cartel oil pricing—both unexpected and quickly imposed. This was a factor that increased solvent cost and potentially its availability in needed quantities. At almost the same time, Government regulations requiring less solvent usage (the CAA) were imposed on the industry. Solvent cost was not a major problem—just raise prices, but many coatings are not inelastic commodities and there still was the threat of non-availability. The availability of oil to manufacture solvents was a totally different matter, and anyone who suffered through the gasoline shortages of this time knows well what effect oil availability can have on an economy. Also, government regulations were not just a temporary measure. Many of the regulations were difficult to meet and could not be met with the technology in hand.

Here a large industry (NAICS Code 3255) that represented about 6 percent of all chemical manufacturing was being asked to change the way they had been doing business for many, many years. Some companies responded well, but others thought solvents were too important to literally be taken out of such a large industry. Many companies innovated and came up with radiation—curable coatings, powder coatings, high-solids coatings, two-package coatings, and others. The new technologies did not take over the marketplace overnight, but with time, each found a niche and in so doing took away a portion of the original market. Today one does not find

$$C_6H_5{-}C_2H_5 \xrightarrow{\text{[O]/V}} C_6H_5{-}CH(CH_3)OOH$$

$$C_6H_5{-}CH(CH_3)OOH + CH_3CH = CH_2 \longrightarrow CH_3CH{-}CH_2 + C_6H_5{-}CH(CH_3)OH$$
$$\underset{O}{\diagdown\diagup}$$

low-solids, solvent-borne coatings in the market to any great extent. One does find companies that have changed the nature of their business favorably by innovating and changing, that have lost market share by taking a "wait and see" attitude, and that are new on the scene and growing. Thus, increased raw material pricing, shortages—real or created, and regulation can have a positive effect on the overall chemical economic picture.

Propylene Oxide. Propylene oxide is another basic chemical used in manufacturing intermediates for urethane foams (cushioning and insulation), coatings, brake fluids, hydraulic fluids, quenchants, and many other end uses.[23] The classic industrial synthesis of this chemical has been the reaction of chlorine with propylene to produce the chlorohydrin followed by dehydrochlorination with caustic to produce the alkylene oxide, propylene oxide, plus salt.

$$CH_3CH = CH_2 + Cl_2 + H_2O \longrightarrow CH_3CH(OH)CH_2Cl$$

$$CH_3CH(OH)CH_2Cl \xrightarrow{\text{Caustic}} CH_3CH{-}CH_2 + Salt$$
$$\underset{O}{\diagdown\diagup}$$

In this reaction sequence, both the chlorine and the caustic used to effect the synthesis are discarded as a valueless salt by-product.

A more economic process has been commercialized. In one version, the hydroperoxide is produced by catalytic air-oxidation of a hydrocarbon such as ethylbenzene (see top of page). Reaction of this hydroperoxide with propylene yields propylene oxide as a co-product.

This direct peroxidation scheme can be carried out with other agents to give different co-products such as *t*-butanol or benzoic acid.[24,25]

When the economics of the direct peroxidation system are balanced, a significant cost reduction in the preparation of propylene oxide is achieved by eliminating the co-product, salt, which is of low to nil value and thus it presents a disposal problem coupled with all of the related environmental ramifications. Note that in the previous reaction scheme the weight of salt is almost the same as the weight of the epoxide produced and thus great quantities of salt would be produced. In addition, the process can be designed to produce a co-product that can be used or sold as a chemical intermediate. For example, in the case of using isobutene as the starting hydrocarbon, the by-product is *t*-butanol, which can then be converted to methyl *t*-butyl ether, which is a gasoline additive.

Ramifications. If a company is in the business of making and selling products such as those used in the above examples, that is, acetic acid, vinyl chloride, and propylene oxide, as well as other chemicals and if it has plans to stay in business and to expand its facilities and workforce thereby serving growing markets, it must have at least economically competitive processes. Today, this means being competitive with not only any new process developed in the United States, but also with any new process technology developed in the global economy. Another need for market maintenance and growth is a management system that is forward-looking, keenly aware of potential competitive threats, and ready to change directions when needed. The same management team must maintain a highly skilled research, development, and engineering staff to provide the new products and processes as well as product and process innovations and thereby create and maintain the pipeline to future sales, growth, and profitability as existing

TABLE 2.8 Sales for the Top 10 Global Chemical Companies in 2000[26]

Company	2000 Chemical Sales (Billions of Dollars)	Rank in 1999
BASF	30.8	1
DuPont	28.4	2
Dow Chemical[a]	23.0	4
Exxon-Mobil	21.5	5
Bayer	19.3	3
Total Fina Elf	19.2	11
Degussa	15.6	9
Shell	15.2	7
ICI	11.8	6
British Petroleum	11.3	10

[a]Sales numbers were before the merger with Union Carbide Corp. Merger is expected to raise Dow Chemical to the No. 1 position.

products or processes become obsolete. Anticipation of market needs must be recognized by both the scientific and management components of a successful firm. In the year 2000 the top ten global companies in sales are listed in Table 2.8.[26]

The companies spending the most for research and development are given in chronological order in Table 2.9.[27]

Not noted above is the profound effect that environmental concerns have on new products and processes. It was mentioned earlier that the chemical manufacturing industry spends about five billion dollars annually on pollution and environmental control. Thus, the entire staff of an organization must be aware of these costs and have concerns for the environment when new products and processes are created.

THE FUTURE

What will this huge manufacturing giant called the chemical manufacturing industry look like five or ten years from now? That changes will be made is certain, and an ability to predict and anticipate those changes and to guide the industry or segments to certain changes will certainly bode well for the economic health of any particular company. It may be recalled that earlier in this chapter, it was mentioned that more and more sales of chemicals would be done via E-commerce on the Internet. If the projections are correct, there will be a compounded annual growth rate of 66 percent in chemical E-commerce—from $7.2 billion in 2000 to $150 billion in 2006. Such sales will be

TABLE 2.9 Research and Development Spending by Chemical and Pharmaceutical Companies[27]

	Spending in Millions of Dollars					Spending as % of Sales
	2000	1999	1998	1997	1996	2000
Chemical Companies						
Dupont	1,776	1,617	1,308	1,116	1,032	6.3
Dow Chemical	892	845	807	785	761	3.9
Rohm and Haas	259	236	207	200	187	3.8
Union Carbide	152	154	143	157	159	2.3
Eastman Chemical	149	187	185	191	184	2.8
Air Products and Chemicals	124	123	112	114	114	2.3
International Flavors	123	104	98	94	94	7.7
Pharmaceutical Companies						
Pfizer	4,435	4,036	3,305	2,536	2,166	15.0
Pharmacia	2,753	2,815	2,176	2,144	1.936	15.2
Merck	2,344	2,068	1,821	1,684	1,487	5.8
Ely Lilly	2,019	1,784	1,739	1,382	1,190	18.6
Bristol-Myers Squibb	1,939	1,843	1,577	1,385	1,276	10.6
American Home Product's	1,688	1,740	1,655	1,558	1,429	12.7

carried out through buyers and sellers of chemicals as they develop agreements on purchases; cost estimates including the carrier method—land, water, truck, train; custom matters such as documentation, regulatory fees and taxes; insurance; warehousing matters; and so on. This will have an economic effect on the industry.

The coating and paint industry has undergone huge changes in the past two decades, but there probably will be further change as the new technological areas are sorted out. In paint or trade sales, it would appear that aqueous latexes will dominate the industry for the foreseeable future. It would take a major breakthrough to dislodge them from their place. The industrial-coating industry and their customers must continue to sort through a number of new technologies—conventional solvent-based, high-solids, powder, radiation, water-borne, two-package, and others—in the future. In the past some technologies have been given emphasis at coating companies because they fit existing technology, but they are not necessarily the final winners. Powder coating has captured more than 10 percent of the industrial coating market, and it will probably grow in the future but perhaps not as rapidly as it did in the past decade. Radiation curing appears to be poised for rapid growth. It is on an upsweeping growth curve, and if new products, new end uses, and interest through meeting attendance is any measure, this technology will have a significant portion of the market five–ten years from now.

The opening of trade with China has resulted in several chemical company expansions in that country and nearby countries. In 2005, the two national chemical giants, Dow Chemical Company and Dupont Company have large-scale plans for investment in China. Dow is planning a large coal-to-methanol-to-olefins complex that includes chlor-alkali facilities that will take five to ten years for completion. Dupont has recently doubled its investment in China and plans to spend another $600 million BY 2010, a portion of which may go to build a titanium dioxide plant. Dupont employs about 5000 people on Mainland China and recently built a Research and Development center near Shanghai. In 2005, Dupont will also build a laboratory in Japan to facilitate worldwide technical approvals for automotive coatings and to support home-country assembly operations. The laboratory will be established by Dupont and operated in cooperation with Shinto Paint Company.

In the pharmaceutical area, there are many new compounds that will reach the marketplace in the near future. One in particular is insulin for diabetics that can be inhaled or taken orally rather than by injection. The former is in final testing stages and could be on the market in the very near future. The oral type many take longer. Either of these will be "blockbuster" drugs and will mean major sales, growth, and profit for the winning company or companies. The revolution that has taken place in automated drug synthesis and screening will continue and improve. Workers entering the industry or relocating will be expected to have knowledge of parallel and recombinational synthesis methods,[28] or they will be rapidly trained within the company in the area. Such screening methods and facilities can screen 200,000 drug candidates in a single day.[29] This technology is so useful that it will also creep into other areas of chemical manufacturing that can benefit by screening large numbers of candidate materials as, for example, the paint, coating, and adhesive area.

Interrelated with the pharmaceutical industry is the biotechnology area. Biotechnology products or products derived from biotechnology processes, are expected to account for 30 percent of the total chemicals market by 2010. Large national and international chemical companies are getting started and have a position in the biotechnology area. Examples of such companies are The Dow Chemical Company, DuPont Company, and Monsanto Company in the United States and Bayer AG, BASF AG, Alusuisse Lonza Group AG, and Degussa AG in Europe. Biotechnology offers both cost-effective and environmentally friendly technology and products. The technology will produce proteins and vitamins for animal feed; genetically modified vegetative plants that will resist drought, insects, and cold; new enzyme controlled processes for production of specific chemicals, new fibers for textiles that are derived from renewable raw materials and are biodegradable. Such agro-growth aspects will result in significant losses in other agrochemicals areas such as herbicides and insecticides as resistant, modified plants are developed.

Research, development, and testing will be carried out more and more at independent facilities according to Federal Government reports, and this function will be a significant growth area. This is related to the high cost of specialized investigative tools, which can be shared by a number of companies. Also important is the ability to temporarily hire highly skilled personnel to carry out testing and developmental efforts when they are needed rather than have them as permanent members of a firm's staff. The independent agencies are the ones who will be able to set up the combinational programs for some of the chemical manufacturing segments so set-up costs, in effect, are shared by a number of companies. These programs are expensive to develop and maintain. Many smaller firms would not be able to afford the technology unless some centralized, independent source was available to allow them to share the cost rather than the whole cost burden.

Composites have not been previously mentioned, but they form an important area that is sizable and that will grow in the future. These graphite- or glass-reinforced materials are useful in many markets that need strong, shaped articles including the aircraft/ aerospace, automotive, recreation, general industrial, and similar markets. In addition to strength, composites often offer weight savings plus the ability to rapidly produce complex-shaped, small to large articles.

Energy sources continue to be highly important. Oil prices have recently skyrocketed making gas-to-liquid technology by means of syngas technology take on a new importance.[30,31] At today's oil prices, gas-to-liquid is becoming quite attractive. Renewable energy sources such as hydropower, biomass, wind, and solar photovoltaics are also receiv-ing a great deal more attention. Although such energy sources only accounted for about 8 percent of energy consumption early in the 21st century,[32] the field is rapidly growing. Major oil companies are investing in wind power generation to ensure their future growth. For example, Royal Dutch/Shell Transport is concert with The Netherlands power company Nuon N.V. recently announced that 36 three-megawatt wind turbines located about six to eleven miles (10–18 km) off the Dutch coast will be built and should be in commercial operation in 2006.[33] Wind and solar renewable energy have a combined compound annual growth rate of 30 percent on a global basis.[32] The same article indicated that $20.3 billion or about 16 to 17 percent of total world investment in power generation equipment was invested in developing renewable energy in 2003. In a general sense, renewable energy is becoming more affordable, and as fossil fuels become increasingly scarce and high priced, such sources, as well as replacement ethanol from grain, will take on more and more importance and will aid in providing a cleaner environment. The high price of oil also makes development of oil sands in Canada and elsewhere viable energy sources. Some plants are now on stream and production will be markedly increased by 2010 to 2012 as huge capital investments are made by companies such as Canadian Natural Resources, EnCana, and Suncor.

Nanotechnology has great promise for the chemical industry. This is an emerging technology whose aim is to place atoms and molecules in particular arrays, a technique termed "positional assembly," and to have this done repetitively through "self-replication."[34] This sounds more like science fiction than chemistry. But, nanotechnology already produces significant sales, and the sales are predicted to grow from a base of $200 million in 2002 to $25 billion by 2012.[35] This is an astounding 62 percent compounded growth rate for 10 years.

What is nanotechnology? Imagine having a machine that can go forward, backward, right, and left as well as up and down at various angles. Now imagine that this machine is very, very small—in fact, so small that it is approaching atomic dimensions and it is measured in terms of nanometers. A nanometer is 1×10^{-9} meters, and we certainly cannot see things this small. However, regardless of these difficulties, we want to build and control this machine—that is, have it do whatever we want it to do, which is to place atoms and molecules in particular arrays so we end up with a desired product. Still not satisfied, we want the machine to do this over and over and over again.

One might ask why we want to do this. We all know that everything is made up of atoms. Chemicals are everywhere and in everything. The difference in the carbon in coal and in diamonds is the way the atoms of carbon are arranged. Arrange the atoms properly and a worthless pile of carbon becomes a precious diamond. If we were able to arrange the atoms in air, dirt, and water into a desired configuration, it would be possible to make, for example, carrots, beets, potatoes, and so on. At present, the transformations are made by nature, using a gene system to combine the ingredients in the proper way. Properly align the atoms of a material that is to be used as a filler or reinforcing material, and one can envision super strong composite materials resulting. If one were to rearrange the atoms of sand and in so doing add a few trace elements, the end result could be a computer chip.[36] The goals of nanotechnology are to:

- Arrange every or almost every atom in a desired structure in its proper place,
- Make effectively any structure that can be atomically specified and that does not violate laws of chemistry and physics, and

- Have manufacturing costs that are basically energy and raw material costs.

The other concepts associated with nanotechnology are those described above, positional assembly and self-replication. Nanotechnology is currently being used for light-emitting polymer films, in computer applications, electrically conductive adhesives, and other areas.

These are but only a few of the potential areas for chemical manufacturing during the first decade of the twenty-first century. Most probably, many of the products and processes that will be in place in 2012 or so cannot even be imagined today.

New emerging markets for nano-based products include nanomemory products used in mobile communications and computers wherein it is expected the market will grow to $8.6 billion by 2007 and $65.7 billion by 2011; nanosensors for medical, homeland security, and aerospace applications wherein the market is expected to be $446 million in 2007 and should grow to $5.6 billion in 2011; nanoengineered display technology for roll-up displays using plastic electronics, other platforms, and carbon nanotubes for large-size, high-definition television sets should grow to $1.6 billion in 2007 to about $7.5 billion by 2011.[37] A very recent article[38] indicated the global market for nanophotonic devices is to reach $9.33 billion by 2009; photonics is the technology of generating and controlling light and using it to conduct information. All in all, one finds nanotechnology being mentioned, discussed, and applied in application after application, technical, scientific meetings are being organized and held, and nanotechnology along with its architecture certainly will grow in importance in a wide variety of uses.[39]

REFERENCES

1. *Chemical Economics Handbook*, Stanford Research Institute, Menlo Park, CA.
2. "Frequently Asked Question on NAICS, SIC, and Business to Business Market," Internet, http://www.naics.com/faq.htm. (Jan. 11, 2002).
3. Dykes, D.R., "Acquisitions Reshape the Face of the Paint Industry, "*Paint Coat Ind,* **21**(5), 32 (2005).
4. Anon, "Top 100 Shrinks to 75," *Chem. Eng. News,* **77**, 18 (May 3, 1999).
5. Smith, G.A., "A Kinder, Gentler, Less Productive Capitalism," in *Fearless Forecast*, p. 5, Prudential Financial, Jan. 2002.

6. "Greenspan Sees Increasing Signs of U.S. Recovery," Internet, http://www.foxnews.com/0, 3566,46664,00.html (Feb. 27, 2002).

7. "Gross Domestic Product by Industry in Current Dollars," U.S. Census Bureau, 2001.

8. U.S. Department of Commerce, Bureau of Economic Analysis, "U.S. International Trade in Goods and Services," BEA News Releases for various years.

9. Bailey, Jr., F.E., and Koleske, J.V., Chapter 1, "Economic Aspects of the Chemical Industry," in *Riegel's Handbook of Industrial Chemistry* (9th ed.), James A. Kent (Ed.), Van Nostrand Reinhold, New York, 1992.

10. "Industrial Profile," U.S. Office of Industrial Technology, http://www.oit.doe.gov/chemicals/profile.shtml.

11. "Chemical-Energy Use," U.S. Office of Industrial Technology, http://www.eia.doe.gov/emeu/mecs/iba/chemicals/page2.html.

12. (a) "Chemists and Material Scientists," http://stats. bis.gov/oco/ocos049.htm.

    (b) "Chemical Engineers, Earnings," U.S. Department of Labor, http://stats.bls.gov/oco/ocos029.htm.

    (c) Kasper-Wolfe, J., "Mixed News for Grads," *Today's Chemist at Work,*" **13**(6), 17 (2004).

    (d) Anon, "2005 Salary Survey: Field of Greens" http://www.chemical processing.com/articles/ 2005/431.html.

13. Kiefer, D.M., "It Was All About Alkali," *Today's Chemist*, **11**(1), 45 (Jan. 2002).

14. Kiefer, D.M., "Soda Ash, Solvay Style," *Today's Chemist*, **11**(2), 87 (Feb. 2002).

15. Belgian Patent 713,296 issued to Monsanto Co.

16. Parshall, G.W., "Organometallic Chemistry in Homogeneous Catalysis," *Science*, **208**, 1221 (1980).

17. Anon, *Chem. Week*, **126**(3), 40 (1980).

18. Anon, *Chem. Eng. News*, **58**, 6 (1980).

19. Koleske, J.V., and Wartman, L.H., *Poly(vinyl chloride)*, Gordon and Breach, New York, 112 pp, 1969.

20. British Patent 1,2027,277 to PPG Industries and British Patent 1,016094 to Toyo Soda.

21. Fedor, W.S., *Chem. Eng. News*, **43**, 80 (Sept. 12, 1965).

22. Anon, *Chem. Week*, **125**(6), 22 (1979).

23. Bailey, Jr., F.E., and Koleske, J.V., *Alkylene Oxides and Their Polymers*, 261 pp. Marcel Dekker, New York, 1991.

24. French Patent 1,460,575 to Halcon Corp.

25. Anon, *Chem. Week*, **127**, 58, No. 1, 6 (1980).

26. "2001 Chemical Industry Review," *Chem. Eng. News*, **79**(52), 13 (Dec. 24, 2001).

27. Anon, "Facts and Figures for Chemical R&D," *Chem. Eng. News* (Oct. 29, 2001), Internet, http://Pubs.ACS.org/CEN.

28. Czarnik, A.W., and DeWitt, S.H., "A Practical Guide to Combinational Chemistry," *Am. Chem. Soc.*, 360 (1997).

29. Drews, J., *Science*, **287**, 1960 (2000).

30 Brown, D., "GTL on Verge of Coming of Age?," *Explorer,* http://www.aap.org/explorer/2003/09sep/ gastoliquid.cfm.

31 United States Patent 6,534,551 issued to Air Products and Chemicals, Inc. (2003).

32. Anon, Dow Jones Newswires, "Vestas Gets Order for 36 Wind Turbines for Dutch Project," (May 31, 2005).

33. Anon, "Renewable Energy Set to Take Off," *Thomas Industrial Marketing Trends,* **4**(20), (June 8, 2004).

34. Drexler, K.E., *Nanosystems: Molecular Machinery, Manufacturing, and Computation*, Wiley, New York, 1992.

35. Kramer, K.L., "Nanotechnology Set for Exponential Growth," *Adhs. Seal. Ind.*, **9**(2), 6 (Mar. 2002).

36. Merkle, R.C., Internet Website http://www.zyvex. com/nano/.

37. Nolan, R., "New NanoMarkets Report Predicts $10.8-Billion Nanoelectronics Market by 2007," *NanoMarkets Res Rep,* http://nanomarkets.net/press-release10-27-04.html.

38. Rajan, M., "Global Market for Nanophotonic Devices to Reach $9.33 Billion by 2009," Business Communications Company, Inc., Norwalk, CT, Report RGB-314, http://www.bccresearch.com/ editors/RGB-314.html.

39. McGuire, N.K., "The Architecture of the Very Small," *Today's Chem. Work,* **12**(11) 30 (2003).

3

Safety Considerations in the Chemical Process Industries

Stanley M. Englund*

INTRODUCTION

There is an increased emphasis on chemical process safety as a result of highly publicized accidents. Public awareness of these accidents has provided a driving force for industry to improve its safety record. There has been an increasing amount of government regulation.

The chemical industry is one of the safest industries, but its safety record in the eyes of the public has suffered. Perhaps this is because sometimes when there is an accident in a chemical plant it is spectacular and receives a great deal of attention. The public often associates the chemical industry with environmental and safety problems, which results in a negative image of the industry.

Some of the important changes in this chapter since the ninth edition was published involve the area of governmental regulations. These are discussed in the section titled "Regulations." A great deal of information on regulations is available on the Internet and will be discussed briefly in this chapter.

The Internet also provides considerable information on incidents, good industry practice, and design guidelines. The best practices in industry are briefly discussed in this chapter. Details are readily available from resources listed in the references section at the end of the chapter.

Hazards from combustion and runaway reactions play a leading role in many chemical process accidents. Knowledge of these reactions is essential for the control of process hazards. Much of the damage and loss of life in chemical accidents results from a sudden release of material at high pressures, which may or may not result in fire; so it is important to understand how sudden pressure releases can occur. They can be due, for example, to ruptured high pressure tanks, runaway reactions, flammable vapor clouds, or pressure developed from external fire. Fires can cause severe damage from thermal radiation. Chemical releases from fires and pressure releases can form toxic clouds that can be dangerous to people over large areas. Static electricity often is a hidden cause in accidents. It is important to understand the

*The Dow Chemical Co., Midland, MI.

reactive nature of the chemicals involved in a chemical facility.

Chemical process safety involves both the technical and the management aspects of the chemical industry, and this chapter addresses both. It is not enough to be aware of how to predict the effect of process hazards and how to design systems to reduce the risks of these hazards. It also is important to consider how chemical process safety can be managed. Technical people at all levels have a degree of management responsibility, and can contribute to the overall management of safe plants.

Loss of containment due to mechanical failure or misoperation is a major cause of chemical process accidents. The publication *One Hundred Largest Losses: A Thirty Year Review of Property Damage Losses in the Hydrocarbon-Chemical Industry*[1] cites loss of containment as the leading cause of property loss in the chemical process industries.

INHERENTLY SAFER PLANTS*

The design of chemical plants to be more nearly "inherently safe" has received a great deal of attention in recent years. This is due in part to the worldwide attention to safety issues in the chemical industry brought on by the gas release at the Union Carbide plant in Bhopal, India, in December 1984. This and the fairly frequent occurrence of other chemical plant incidents has raised the issue of chemical plant safety to a very high level of visibility and concern. The major factors that should be considered in the planning, design, and operation of chemical plants are described below. The reader is referred to the list of recommended supplementary reading at the end of the chapter for more detailed information.

The term, "inherent," means "belonging by nature, or the essential character of something."

An "inherently safe" plant is safe by its nature and by the way it is constituted.

No facilities can be completely "inherently safe," but they can be made "inherently safer" by careful examination of all aspects of plant design and management. It is possible to identify "inherently unsafe" equipment and management practices and try to avoid them. It has been found that the basic or root cause of most chemical process accidents can be traced to some failure of a management system. Human error plays an important role in many process accidents and is often closely related to problems with management systems. Major efforts are under way to address the potential problems of human error. By being aware of modern advances in design and management techniques, and putting these techniques into practice, it is possible to have facilities that are safer than ever.

The term "intrinsic" has a meaning similar to that of "inherent," but the common usage of "intrinsic" in the chemical industry usually means a protection technique related to electricity. Intrinsic safety is based on the restriction of electrical energy to a level below which sparking or heating effects cannot ignite an explosive atmosphere (Lees,[2] p. 513).

Responsibility for Safety in Design and Operation

It is very important that responsibility for the safe design and operation of a plant be clearly defined in the early stages. This means that competent and experienced people should be made responsible and held accountable for decisions made from the start of plant design on through plant start-up and operation.

Review of Design Alternatives

Hazards should be considered and eliminated in the process development stage where possible. This would include considerations of alternative processes, reduction or elimination of hazardous chemicals, site selection, etc. By the time the process is developed, the process designers already have major constraints imposed on them.

---

*Portions of this section have been reproduced from "Opportunities in the Design of Inherently Safer Chemical Plants," by Stanley M. Englund in Advances in Chemical Engineering Vol. 15, edited by Bodansky, O. and Latner, A., copyright (1990). Elsevier Science (USA), with permission from the publisher.[31]

Hazards should also be identified and removed or reduced early in the design. Adding protective equipment at the end of the design or after the plant is operating can be expensive and not entirely satisfactory.

Allowing time in the early stages of design for critical reviews and evaluation of alternatives would involve studies such as an early hazard and operability (HAZOP) study, using flowsheets, before final design begins.[4] Fault tree analysis," quantitative risk assessment (QRA), checklists, audits, and other review and checking techniques can also be very helpful. These techniques are extensively discussed in the technical literature and will not be discussed in detail here.

Emergency Planning

Emergency planning is primarily for the protection of plant personnel and people in nearby areas and the environment that could be affected by plant problems. It should be considered early in the design and should be coordinated with the existing site emergency plan.

Emergency planning includes tornado and storm shelters, flood protection, earthquakes, proximity to public areas, and safe exit routes. It also includes planning for the effect that an emergency in the "new process" would have on other plants, and the effect that an emergency in another plant would have on the new process. The effects of potential spills on waterways and aquifers should be considered.

Emergency response planning is discussed fully in Chapter 4.

Placement of Process and Storage Areas

The Bhopal plant of Union Carbide was built originally 1.5 miles from the nearest housing (see below under "Case Histories"). Over time, a residential area grew up next to the plant. This demonstrates the need to prevent hazardous plants from being located close to residential areas and to prevent residential areas from being established near such plants.[4] If possible, the cost of a plant should include an adequate buffer zone unless other means can be provided to ensure that the public will not build adjacent to the plant. The nature and size of this buffer zone depend on many factors, including the amount and type of chemicals stored and used.

Storage of Hazardous Materials

The best way to minimize releases of a hazardous or flammable material is to have less of it around. In the Flixborough disaster[2] on June 1, 1974, the process involved the oxidation of cyclohexane to cyclohexanone by air (with added nitrogen) in the presence of a catalyst (see below under "Case Histories"). The cyclohexanone was converted to caprolactam, which is the basic raw material for Nylon 6. The reaction product from the final reactor contained approximately 94 percent unreacted cyclohexane at 155°C and over 20 psig. The holdup in the reactors was about 240,000 lb, of which about 80,000 lb escaped. It is estimated that about 20,000–60,000 lb actually was involved in the explosion. The resulting large unconfined vapor cloud explosion (or explosions—there may have been two) and fire killed 28 people and injured 36 at the plant and many more in the surrounding area, demolished a large chemical plant, and damaged 1,821 houses and 167 shops. The very large amount of flammable liquid well above its boiling and flash points contributed greatly to the extreme severity of the disaster (Lees,[2] p. 863).

The results of the Flixborough investigation make it clear that the large inventory of flammable material in the process plant contributed to the scale of the disaster. It is concluded that "limitations of inventory (of flammable materials) should be taken as specific design objectives in major hazard installations." It should be noted that reduction of inventory may require more frequent and smaller shipments, and this would entail more chances for errors in connecting and reconnecting. These possibly "negative" benefits should also be analyzed. Quantitative risk analysis of storage facilities has revealed solutions that may run counter to intuition.[5] For example, contrary to popular opinion, reducing inventories in tanks of hazardous materials does little to reduce risk in situations where most of the exposure arises from the number and extent of valves, nozzles, and lines connecting the tank. Removing tanks from the service altogether, on the other hand, helps.

A large tank may offer greater safety than several small tanks of the same aggregate capacity because there are fewer associated nozzles and lines. Also, a large tank is inherently more robust, or can economically be made more robust by deliberate overdesign, than is a small tank and therefore is more resistant to external damage. On the other hand, if the larger tank has larger connecting lines, the relative risk may be greater if release rates through the larger lines increase the risk more than the tank's inherently greater strength reduces it. In the transportation of hazardous materials, the benefits of head shields and shelf couplers have been shown through quantitative risk assessment. Maintaining tank car integrity in a derailment is often the most important line of defense in the transportation of hazardous materials.

Liquefied Gas Storage

Usually, leaks of liquefied gases are much less serious if such gases are stored refrigerated at low temperatures and pressures than if they are stored at ambient temperatures under pressure. A leak of a volatile liquid held at atmospheric temperature and pressure results in only a relatively slow evaporation of the liquid. Escape of a refrigerated liquefied gas at atmospheric pressure gives some initial flashoff, and then it evaporates at a rate that is relatively slow but faster than the first case, depending on weather conditions.

Loss of containment of a liquefied gas under pressure and at atmospheric temperature, however, causes immediate flashing of a large proportion of the gas, followed by slower evaporation of the residue. This is usually a more serious case than if a refrigerated tank is used. The hazard from a gas under pressure is normally much less in terms of the amount of material stored, but the physical

energy released if a confined explosion occurs at high pressure is large.

Use of Open Structures

There are many examples of serious fires and explosions that probably resulted in part from handling moderate to large quantities of flammable or combustible liquids and liquefied flammable gases inside enclosed structures. If a sufficient quantity of flammable mixture should ignite inside an ordinary chemical processing building, it is highly probable an explosion will occur that will seriously damage the building. For this reason, processing equipment is often installed in a structure without walls, usually called an "open structure." This permits effective ventilation by normal wind currents and aids the dispersion of any vapors that do escape. If ignition of gas occurs within the structure, the absence of walls minimizes the pressure developed from the combustion and the probability of flying shrapnel from a shattered structure.[6]

Substantial explosion damage will be done to a building by combustion of a surprisingly small quantity of a flammable gas–air mixture. If there is an explosion of a flammable gas–air mixture in a building where the flammable gas mixture occupies a space equal to only 1 or 2 percent of the building volume, the building may be seriously damaged if it does not have adequate explosion venting. This is because most buildings will suffer substantial structural damage from an internal pressure appreciably less than 1 psi (0.07 bar). Thus, a building does not need to be "full" or even close to "full" of a flammable mixture for a building explosion to occur that can cause considerable damage.

In 1950, a serious hydrocarbon explosion occurred in an enclosed chemical processing unit of The Dow Chemical Company. It was instrumental in causing Dow to establish a policy of using open structures for chemical processes that use substantial quantities of flammable liquids and liquefied flammable gases, and combustible liquids above their flash points.

Need to Understand Reactive Chemicals Systems

The main business of most chemical companies is to manufacture products through the control of reactive chemicals. The reactivity that makes chemicals useful can also make them hazardous. Therefore, it is essential that a process designer understand the nature of the reactive chemicals involved in his or her process.

Usually reactions are carried out without mishaps, but sometimes chemical reactions get out of control for many reasons, including:

1. Wrong raw material used
2. Operating conditions changed
3. Time delays unanticipated
4. Equipment failure
5. Reactants contaminated
6. Materials of construction wrong
7. Misoperation
8. External fire

Such mishaps can be worse if the chemistry is not fully understood. A chemical plant can be "inherently safer" only if knowledge of the reactive chemicals systems is carefully used in its design.

Reactive Hazard Evaluations. Reactive hazard evaluations should be made on all new processes, and on all existing processes on a periodic basis. There is no substitute for experience, good judgment, and good data in evaluating potential hazards. Reviews should include:

• Process chemistry:
  reactions
  potential pressure buildup
  intermediate streams
  side reactions
  heat of reaction
• Reactive chemicals test data for evidence of:
  flammability characteristics
  shock sensitivity
  exotherms
  other evidence of instability
• Planned operation of process, especially:
  upsets
  delays

critical instruments and controls
mode of failure
redundancy
worst credible case scenarios

Worst-Case Thinking. At every point in the operation, the process designer should conceive of the *worst* possible combination of circumstances that could *realistically* exist, such as:

air leakage
deadheaded pumps
instrument failure
loss of agitation
loss of compressed air
loss of cooling water
plugged lines
power failure
raw material impurities
wrong combination or amount of reactants
wrong valve position

An engineering evaluation should then be made of the worst-case consequences, with the goal that the plant will be safe even if the worst case occurs. A HAZOP study could be used to help accomplish "worst-case thinking." When the process designers know what the worst-case conditions are, they should:

1. Try to avoid worst-case conditions.
2. Be sure adequate redundancy exists.
3. Identify and implement lines of defense:
   preventive measures;
   corrective measures;
   sometimes, as a last resort, containment
or, possibly, abandoning the process if the hazard is unacceptable

It is important to note that the worst case should be something that is realistic, not something that is conceivable but which may be unreasonable.

Reactive Chemicals Testing. Much reactive chemical information involves thermal stability and the determination of:

• the temperature at which an exothermic reaction starts
• the rate of reaction as a function of temperature
• heat generated per unit of material

The information can be obtained by a variety of types of laboratory tests and by thermodynamic calculations, although reliable kinetic data cannot usually be obtained solely by calculations.

Losses from Dust Explosions

Most organic solids, most metals, and some combustible inorganic salts can form explosive dust clouds. In order to have a dust explosion, it is necessary to satisfy certain conditions:

- suitably sized dust particles
- sufficient source of ignition energy
- dust concentration within explosive limits
- explosible dust
- oxidizer must be present

If an explosive dust in air that meets the above criteria occurs in a process, an explosion should be considered as inevitable. The process designer should take into account the possibility of dust explosions and design accordingly.

In dust explosions the combustion process is very rapid. The flame speed is high compared with that in gas deflagrations. Detonations normally do not occur in dust explosions in industrial plants.

The sequence of events in a serious industrial dust explosion is often as follows:

1. A primary explosion occurs in part of a plant, causing an air disturbance.
2. The air disturbance disperses dust and causes a secondary explosion, which is often more destructive than the primary explosion.

If the occurrence of a flammable (explosive) dust is inevitable in a particular process, several design alternatives or combinations of alternatives are available:

- containment (maximum pressure of a dust explosion is usually below 120–150 psig)
- explosion venting to a safe place
- inerting (most organic dusts are non-flammable in atmospheres containing less than about 10% oxygen)
- suppression

A fundamental solution to the dust explosion problem is to use a wet process so that dust suspensions do not occur. However, the process must be wet enough to be effective. Some dusts with a high moisture content can still be ignited.

Dust concentrations in major equipment may be designed to be below the lower flammable limit, but this often cannot be depended on in actual operation. Dust concentrations cannot be safely designed to be above an upper flammable limit because such a limit is ill-defined.[2]

For a large number of flammable dusts, the lower explosion limit lies between 0.02 and 0.06 kg/m^3. The upper explosion limit is in the range of 2–6 kg/m^3, but this number is of limited importance.

A small amount of flammable gas or vapor mixed in with a flammable dust can cause an explosive mixture to be formed even if both are at concentrations below the explosive range by themselves. These mixtures are called "hybrid" mixtures. The ignition energy to ignite a hybrid mixture is often less than that required for the flammable dust by itself.

Venting is only suitable if there is a safe discharge for the material vented. Whenever an explosion relief venting device is activated, it may be expected that a tongue of flame containing some unburned dust will first be ejected. The unburned dust will be ignited as it flows out of the vent and can produce a large fireball that will extend outward, upward, and downward from the vent. It is essential for protection of personnel that venting is to an open place not used by people. If a duct must be used, the explosion pressure in the enclosure will be increased considerably. Therefore, particular attention must be paid to the design of the enclosure in which the explosion could take place.

The NFPA 68 guide issued in 1998[7] has nomographs, which can be used to select relief areas required for combustible dusts when test data on the dusts are available. The nomographs in NFPA 68 are considered by many to be the preferred way to design dust explosion relief devices.

Relief venting to reduce dust explosion pressure requires the equipment to be protected to have a certain minimum strength. If the enclosure strength is too low, the enclosure will be damaged or destroyed before the explosion relief device can function. NFPA 68[7] states that the strength of the enclosure should exceed the vent relief pressure by at least 0.35 psi. For industrial equipment such as dryers and baghouses, it is often desirable to have considerably more strength built into the structure to reduce the size of the vent area required. Also, the supporting structure for the enclosure must be strong enough to withstand any reaction forces developed as a result of operation of the vent.

Inerting is a very good preventive measure against dust explosions. The maximum oxygen concentration at which dust explosions are "just not possible" cannot be predicted accurately, as it depends on the nature of the combustible material; testing is usually required. It has been found that in an atmosphere of 10 percent oxygen and 90 percent nitrogen, most combustible organic dusts are no longer explosive. To allow a safety margin, it is good industrial practice to maintain oxygen concentrations below 8 percent. For metal dusts, the allowable oxygen content is about 4 percent.[7]

Inerting leads to the possibility of asphyxiation by operating personnel if they were exposed to the inert gas. Strict precautions must be taken to prevent exposure of personnel to inerting atmospheres.

Explosion suppression systems are designed to prevent the creation of unacceptably high pressure by explosions within enclosures that are not designed to withstand the maximum explosion pressure.[8] They can protect process plants against damage and also protect operating personnel in the area. Explosion suppression systems restrict and confine the flames in a very early stage of the explosion. Suppression systems require more maintenance than do relief venting devices. Explosion suppression systems are made by only a few manufacturers and are quite expensive. This may be the reason why this type of safeguard has not been as widely used in industry as one might expect, although its effectiveness has been proved by much practical experience.

Explosion suppression is a proven technology and should be considered as a candidate for explosion protection. The NFPA has published a standard reference on explosion-suppression protection.[9] Manufacturers should be consulted on design, installation, and maintenance.

Even with explosion suppression, it is common for the explosion pressure to reach one atmosphere before it is suppressed. The added pressure surge from the injection of the suppressing agent must also be considered. Therefore, sufficient mechanical strength is always required for enclosures protected by explosion suppression.

Substitution of Less Hazardous Materials

It may be possible to substitute a less hazardous material for a hazardous product. For example, bleaching powder can be used in swimming pools instead of chlorine.[4] Benzoyl peroxide, an initiator used in polymerization reactions, is available as a paste in water, which makes it much less shock-sensitive than the dry form. Other substitutions that have been used to make transportation, storage, and processing safer include:

1. Shipping ethylene dibromide instead of bromine.
2. Shipping ethyl benzene instead of ethylene.
3. Storing and shipping chlorinated hydrocarbons instead of chlorine.
4. Storing and shipping methanol instead of liquefied methane.
5. Replacing flammable refrigerants by halogenated hydrocarbons that are known to have acceptable environmental effects.
6. Storing and shipping carbon tetrachloride instead of anhydrous hydrochloric acid. The CCl_4 is burned with supplemental fuel to make HCl on demand at the user's site.
7. Using magnesium hydroxide slurry to control pH instead of concentrated sodium hydroxide solutions, which are corrosive to humans and relatively hazardous to handle.
8. Using pellets of flammable solids instead of finely divided solids to reduce dust explosion problems.

The use of substitutes may appear to be more costly. The added safety provided by substitutes may make their use worthwhile and can in some cases actually lower the true cost of the project when the overall impact on the process, surrounding areas, and shipping is considered.

Substitutes should be employed only if it is known that overall risk will be reduced. Inadequately tested processes may introduce unrecognized health, safety, and environmental problems.

Catastrophic Failure of Engineering Materials[10–12]

Uniform corrosion of metals can usually be predicted from laboratory tests or experience. Corrosion allowances, which will require thicker metal, can be called for in the design of equipment when uniform corrosion rates are expected.

The most important materials failure to avoid in the design of metal equipment is *sudden catastrophic failure*. This occurs when the material fractures under impulse instead of bending. Catastrophic failure can cause complete destruction of piping or equipment, and can result in explosions, huge spills, and consequent fires. Causes of some of the more common types of catastrophic failures are:

- low-temperature brittleness
- stress corrosion cracking
- hydrogen embrittlement
- high penetration rates involving pitting and corrosion
- fatigue failure
- creep
- mechanical shock
- thermal shock
- high rates of temperature change in brittle materials
- zinc embrittlement of stainless steel
- caustic embrittlement
- nitrate stress corrosion

Redundant Instrumentation and Control Systems[13]

Computer-controlled chemical plants have become the rule rather than the exception. As a result, it is possible to measure more variables and get more process information than ever, and chemical plants can be made "inherently safer" than ever before. However, it must be kept in mind that instruments and control components *will* fail. It is not a question of *if* they will fail, but *when* they will fail, and what the consequences will be. Therefore, the question of redundancy must be thoroughly considered. *The system must be designed so that when failure occurs, the plant is still safe.*

Redundant measurement means obtaining the same process information with two like measurements or two measurements using different principles. Redundant measurements can be *calculated* or *inferred* measurements. Two like measurements would be two pressure transmitters, two temperature measurements, two level measurements, and so on. An example of inferred measurement would be using a pressure measurement and vapor pressure tables to check an actual temperature measurement.

A continuous analog signal that is continuously monitored by a digital computer is generally preferable to a single point or single switch, such as a high level switch or high pressure switch. A continuous analog measurement can give valuable information about what the value is *now* and can be used to compute values or compare with other measurements. Analog measurements may make it possible to predict future values from known trends. Analog inputs may be visual, and one can see what the set point is and what the actual value is. The software security system should determine who changes set points, and should not be easy to defeat.

A single point (digital) signal only determines whether switch contacts are open or not. It can indicate that something has happened, but not that it is going to happen. It cannot provide information to anticipate a problem that may be building up or a history about why the problem happened. Single point signals are easy to defeat. Some single point measurements are necessary, such as fire eyes, backup high level switches, and so on.

As a rule, it is best to avoid:

- both pressure transmitters on the same tap
- both temperature measuring devices in the same well
- both level transmitters on the same tap or equalizing line
- any two measurements installed so that the same problem can cause a loss of both measurements

It is a good idea to use devices that use different principles to measure the same variable, if possible.

An alarm should sound if any time redundant inputs disagree. In many cases the operating personnel will have to decide what to do. In some cases the computer control system will have to decide by itself what to do if redundant inputs disagree.

The more hazardous the process, the more it is necessary to use multiple sensors for flow, temperature, pressure, and other variables.

Since it must be assumed that all measuring devices will fail, they should fail to an alarm state. If a device fails to a nonalarm condition, there can be serious problems. If a device fails to an alarm condition, but there is really not an alarm condition, it is also serious, but generally not as serious as if it fails to a nonalarm condition, which can provide a false sense of security.

Usually it is assumed that two devices measuring the same thing will not fail independently at the same time. If this is not acceptable, more than two devices may be used. If this is assumed, one can consider the effects of different levels of redundancy:

Number of Inputs	Consequence
One	Failure provides no information on whether there is an alarm condition or not.
Two	Failure of one device shows that there is a disagreement, but without more information, it cannot be determined whether there is an alarm condition or not. More information is needed; the operator could "vote" if there is time.
Three	Failure of one device leaves two that work; there should be no ambiguity on whether there is an alarm condition or not.

Pressure Relief Systems

The design of relief systems involves, in general, the following steps:

1. Generate scenario. What could reasonably happen that could cause high pressures? This could be fire, runaway reactions, phase changes, generation of gases or vapors, leaks from high pressure sources, and so on.
2. Calculate the duty requirements—the pounds per hour of material that has to be vented, and its physical condition (temperature, pressure, ratio of vapor to liquid, physical properties). This is a rather involved calculational procedure.
3. Calculate the relief area required based on the duty, inlet and outlet piping, and downstream equipment. This is also a rather involved calculational procedure.
4. Choose the relief device to be specified from vendor information.

A group of chemical companies joined together in 1976 to investigate emergency relief systems. This later resulted in the formation of The Design Institute for Emergency Relief Systems (DIERS), a consortium of 29 companies under the auspices of the AIChE. DIERS was funded with $1.6 million to test existing methods for emergency relief system design and to "fill in the gaps" in technology in this area, especially in the design of emergency relief systems to handle runaway reactions.[14] DIERS completed contract work and disbanded in 1984.

Huff was the first to publish details of a comprehensive two-phase flow computational method for sizing emergency relief devices, which, with refinements, has been in use for over a decade.[15–18] The most significant theoretical and experimental finding of the DIERS program was the ease with which two-phase vapor–liquid flow can occur during an emergency relief situation. *The occurrence of two-phase flow during runaway reaction relief almost always requires a larger relief system than does single-phase flow vapor venting.* The required area for two-phase flow venting can be from two to much more than two times larger to provide adequate relief than if vapor-only venting occurs.[15] Failure to recognize this can result in drastically under-sized relief systems that will not provide the intended protection.

Two-phase vapor–liquid flow of the type that can affect relief system design occurs as a result of vaporization and gas generation during a runaway reaction or in many liquid systems subjected to fire (especially tanks that are nearly full). Boiling can take place throughout the entire volume of liquid, not just at the surface. Trapped bubbles, retarded by viscosity and the nature of the fluid, reduce the effective density of the fluid and cause the liquid surface to be raised. When it reaches the height of the relief device, two-phase flow results. Fauske and Leung[19] described test equipment that can be used to help determine the design of pressure relief systems for runaway reactions that often result in two- or three-phase flow.

Safe and Rapid Isolation of Piping Systems and Equipment

It should be possible to easily isolate fluids in equipment and piping when potentially dangerous situations occur. This can be done using emergency block valves (EBVs). An EBV is a manually or remotely actuated protective device that should be used to provide manual or remote shutoff of uncontrolled gas or liquid flow releases. EBVs can be used to isolate a vessel or other equipment, or an entire unit operation. Manual valves are often used on piping at block limits where it is unlikely that there would be a hazard to personnel if an accident occurs. Remotely controlled EBVs are recommended on tanks and on piping in areas where it may be hazardous for personnel in the case of an accident, or where a quick response may be necessary.

EBVs used on tanks should be as close as possible to the tank flange and not in the piping away from the tank. In cases where EBVs may be exposed to fire, the valve and valve operating mechanism must be fire tested.

In one design case involving large quantities of highly flammable materials, a HAZOP showed that suitably located EBVs were one of the most important features that could be installed to reduce the possibility of loss of containment and serious fire and explosions.

Piping, Gaskets, and Valves

Piping.[20] All-welded pipe and flanges should be used in "inherently safer" chemical plants. Since flanges are a potential source of leaks, as few flanges as possible should be used. One, of course, has to be realistic: if it is necessary to clean out pipes, flanges must be provided at appropriate places to make it possible. Also, enough flanges must be provided to make maintenance and installation of new equipment reasonably easy.

Threaded piping should be avoided for toxic and flammable materials. It is very difficult to make threaded fittings leakproof, especially with alloys such as stainless steel. Where threaded piping is necessary, use schedule 80 pipe as a minimum. Pipe nipples should never be less than schedule 80.

Pipe support design should be given special attention. It may be desirable to increase pipe diameter to provide more pipe strength and rigidity and make it possible to have greater distance between supports. Normally, in chemical plants, it is not desirable to use piping less than $\frac{1}{2}$ in. in diameter and preferably not less than 1 in. in diameter, even if the flow requirements permit a smaller pipe, except for special cases. Pipe smaller than $\frac{1}{2}$ in. has insufficient strength and rigidity to be supported at reasonable intervals. Tubing, properly supported, should normally be used for anything smaller than $\frac{1}{2}$ in. Tubing is not as fragile as pipe in small sizes. It can be bent so that the number of fittings required is reduced. If it is necessary to use smaller pipe, or small tubing, special provisions should be made for its support and mechanical protection. Also, consideration should be given to using schedule 80 or schedule 160 pipe if small pipe is required to provide extra mechanical strength, even if the fluid pressure does not require it.

Gaskets.[21] Gaskets are among the weakest elements of most chemical plants. Blown-out or leaky gaskets have been implicated in many serious incidents. A leak at a flange can have a torch effect if it is ignited. A fire of this type was considered as a possible cause of the Flixborough disaster.[2] (See below under "Case Histories.")

Modern technology makes it possible to greatly reduce the incidence of gasket failure by the use of spiral wound gaskets. These are sold by several manufacturers, including Flexitallic, Parker Spirotallic, Garlock, and Lamons. A spiral wound gasket is considered the safest gasket type available because of the metal inner and outer rings which contain the filler material at the ID and OD. This sturdy construction of a spiral wound gasket does not permit blowout of the gasket material, which is the potential for gaskets made of sheet material.

Bolting with spiral wound gaskets is very important. *Use of plain carbon steel bolts (such as A307 Grade B) with spiral wound gaskets is not permitted.* They are not strong enough. High strength alloy bolts such as A193-B7 (contains Cr and Mo) with A194 heavy hex nuts should be used. To properly seal spiral wound gaskets, it is necessary to tighten the bolts to specified torque limits, which are generally higher than those of conventional gaskets. Compared with conventional gaskets, spiral wound gaskets require better flange finish, heavier flanges, and better flange alignment.

Valves. It is desirable and inherently safer to use fire-tested valves whenever it is necessary to isolate flammable or combustible fluids in a pipeline or tank or other type of equipment. Fire-tested valves were formerly called firesafe valves. They were not really firesafe, and the name has been changed to fire-tested valves. Fire-tested valves should be considered for handling most fluids that are highly flammable, highly toxic, or highly corrosive and that cannot be permitted to escape into the environment.

Fire-tested valves should be used to isolate reactors, storage vessels, and pipelines. They can be used wherever EBVs are required. With the popularity of automated plants, quarter-turn valves are very popular and are used in most installations. The only common quarter-turn valves that are fire-tested are ball valves and high performance butterfly valves. For special purposes, there are other special fire-tested valves.

Plastic Pipe and Plastic-Lined Pipe. Plastic-lined pipe is excellent for many uses, such as highly corrosive applications, or where sticking is a problem, and where ease of cleaning is a factor. It is often the cheapest alternative. However, if there is a fire, there may be "instant holes" at each flange because the plastic will melt away, leaving a gap. Therefore, plastic-lined pipe should not ordinarily be used for flammable materials that must be contained in case of a fire. An exception to this is a firesafe plastic-lined pipe system such as that made by the Resistoflex Corp., which provides a metal ring between each flange that will make plastic-lined pipe firesafe. The pipe will probably have to be replaced after a fire, but the contents of the pipe will be contained during a fire.

In general, all types of solid plastic or glass-reinforced plastic pipe should be avoided, if possible, for use with flammable liquids. Compared with metal, plastic piping:

- will melt and burn easier
- is more fragile and can be easily mechanically damaged
- is harder to support adequately

Plastic pipe should be used with appropriate judgment.

Avoidance of Inherently Unsafe Equipment

Some equipment items can be regarded as "inherently unsafe" for use in flammable or toxic service and should be avoided if possible. These items include a variety of devices.

Glass and transparent devices, such as sight glasses, bull's-eyes, sightports, rotameters, and glass and transparent plastic piping and fittings, may be hazardous. Glass devices are sensitive to heat and shock. Transparent plastic devices may be resistant to shock, but are not resistant to high temperatures. If they fail in hazardous service, severe property damage and personnel injury can result. The guidelines to follow are:

1. If broken, would they release flammable material?
2. If broken, would they expose personnel to toxic or corrosive materials?

Flexible or expansion joints in piping should be eliminated wherever possible. Flexible joints and expansion joints are any corrugated or flexible transition devices designed to minimize or isolate the effects of:

- thermal expansion
- vibration
- differential settling
- pumping surges
- wear
- load stresses
- other unusual conditions

The need for flexible joints sometimes can be eliminated by proper design so that solid pipe will be able to handle misalignment, thermal expansion, and so on, by bending slightly. In many cases electronic load cells can be used to accurately weigh large reactors and process tanks that may have pipes attached to them, with no flexible or expansion joints. This is done by cantilevering the pipes to the equipment using sufficient runs of straight horizontal unsupported pipe with 90°C elbows to take up movements and vibration without interfering significantly with the operation of the load cells.

Flexible joints should not be used as a correction for piping errors or to correct misalignment.

Pumps for Hazardous Service

A wide variety of excellent pumps is available in the chemical industry. It is sometimes a problem to choose the best from the large number available. This discussion will be limited to centrifugal pumps. Assuming that one has sized the pump, decided on a centrifugal pump, and chosen a suitable list of vendors, the main choices involve (1) metallurgy, (2) seal-less pumps versus conventional centrifugal pumps, and (3) the type of seals for conventional pumps.

Metallurgy.[22] Cast iron should not be used for flammable or hazardous service. The minimum metallurgy for centrifugal pumps for hazardous or flammable materials is cast ductile iron, type ASTM A 395, having an ultimate tensile strength of about 60,000 psi.

Seal-less Pumps.[23] The most common maintenance problem with centrifugal pumps is with the seals. Seals can be essentially eliminated with seal-less pumps. Seal-less pumps are manufactured in two basic types: canned motor and magnetic drive. Magnetic drive pumps have thicker "cans," which hold in the process fluid and the clearances between the internal rotor and "can" are greater compared with canned motor pumps. Seal-less pumps are becoming very popular and are widely used in the chemical industry. Mechanical seal problems account for most of the pump repairs in a chemical plant, with bearing failures a distant second. The absence of an external motor (on canned pumps) and a seal is appealing to those experienced with mechanical seal pumps. However, it cannot be assumed that just because there is no seal, seal-less pumps are always safer than pumps with seals, even with the advanced technology now available in seal-less pumps. Seal-less pumps must be used with considerable caution when handling hazardous or flammable liquids. A mistreated seal-less pump can rupture with potentially serious results. The "can" can fail if valves on both sides of the pump are closed, and the fluid in the pump expands because of heating up from a cold condition with the pump off, or if the pump is started up under these conditions. If the pump is run dry for even a short time, the bearings can be ruined. Seal-less pumps, especially canned motor pumps, produce a significant amount of heat because nearly all the electrical energy lost in the system is absorbed by the fluid being pumped. *If this heat cannot be properly dissipated, the fluid will heat up with possibly severe consequences.* Considerable care must be used in installing a seal-less pump to be sure that misoperations cannot occur.

Properly installed and maintained, seal-less pumps, especially magnetic drive pumps, offer an economical and safe way to minimize hazards and leaks of hazardous liquids.

TECHNICAL MANAGEMENT OF CHEMICAL PROCESS SAFETY

Although understanding and using sound technology is important in designing and operating safe process plants, technology is not enough. As the chemical process industries have developed more sophisticated ways to improve process safety, safety management systems have been introduced to make process safety engineering activities more effective. The following is a brief summary of a recent approach to the technical management of chemical process safety.[24]

Twelve process safety elements have been identified as important in the context of plant design, construction, start-up, operation, maintenance, modification, and decommissioning. This does not include personnel safety, transportation issues, chronic releases to the environment, or community response, which are separate and important issues. Process safety management must deal with each element. Even the best companies, with the best safety records, have room for improvement.

Accountability. Accountability begins with a clear, explicit, and reasonably specific statement of a company's expectations, objectives, and goals. Example: "Process safety audits must be conducted based on the relative risk involved."

Process Knowledge and Documentation. Capturing process knowledge is a foundation upon which many aspects of a process safety program are built. This is sometimes referred to as "company memory." Preserving this knowledge and making it available within a company are important for process safety for a number of reasons, including:

- keeping a record of design conditions and materials of construction to help ensure that operations remain faithful to the original intent

- providing a basis for understanding how and why the process should be operated a certain way
- making it possible to pass information from older to younger workers
- providing a baseline for process changes
- recording causes of incidents and accidents and corrective action for future guidance

Process Safety Reviews for Capital Projects. The need for process safety reviews of capital projects, whether new or revised facilities, has been a recognized feature of engineering organizations. These reviews must be comprehensive and systematic. It is best to do reviews early to avoid costly modifications later. Not only do process hazards need to be identified and addressed by the design, but broader issues should be considered, such as:

1. Is the company prepared to accept the hazards and risks of this project?
2. Is the location appropriate?

Process Risk Management. Process risk management involves the systematic identification, evaluation, and control of potential losses that may arise in existing operating facilities from events such as fires, explosions, toxic releases, runaway reactions, or natural disasters. If risks are not identified, they cannot be considered. Whether resulting losses are measured in terms of direct costs, impacts on employees and/or the public, property or environmental damage, lost business, or various liabilities, the possibility of experiencing such losses is considered a risk. When risks have been identified, it is possible to plan for their reduction, and it can be determined whether the remaining, or residual, risk is acceptable. Risk can never be entirely eliminated. It is the purpose of this component of risk management to manage the risks that remain after implementation of risk controls.

Management of Change. Changes to process facilities are necessary for many reasons. Change includes all modifications to equipment, procedures, and organizations that may affect process safety. Normal operation of a process should be within defined safe limits; operation outside these limits should require review and approval. Replacement personnel should be trained before moves of people are made. There should be no difference between things to consider for changes intended to be permanent and changes that have a limited life, that is, are "temporary" in nature. A hazard will proceed to an incident whether the change is permanent or temporary. In any operation, situations arise that were not foreseen. To ensure that these deviations from normal practice do not create unacceptable risks, it is important to have a variance procedure incorporated into the management system.

Process and Equipment Integrity. Equipment used to handle or process hazardous materials should control the risk of releases and other accidents. Management systems should ensure that equipment is designed properly and constructed and installed according to the design. Components of the process and equipment integrity element include: reliability engineering, materials of construction, inspection procedures, installation procedures, preventive maintenance, and maintenance procedures. It is very important that a program of tests and inspections be available to detect impending failures and mitigate their potential before they can develop into more serious failures.

Incident Investigation. Incidents can be defined broadly as unplanned events with undesirable consequences. Incident investigation is the management process by which the underlying causes of incidents are uncovered and steps are taken to prevent similar incidents. Almost always, it has been found that process safety incidents are a result of some failure of the management system, which, had it functioned properly, could have prevented the incident. Incident investigations should identify all underlying causes and management system failures. It is not enough to discover the immediate cause, such as operator error, but the investigation should go deeper and find out why the operator made the error and possibly discover training and equipment deficiencies. The incident investigation should report facts and conclusions and make recommendations.

Training and Performance. Training is an essential part of any process safety management program. Proper training of personnel is an absolute requirement for keeping complex process equipment and machinery operating safely. Good training programs tell the student not just what to do, but also how to do it and why it is important. Teaching should be a high-priority task. There should be specific criteria for instructor selection. It is not enough to know the subject, although that is important; the teacher should have good presentation skills. The effectiveness of training should be evaluated regularly to determine the effectiveness of the instructors and the performance of the students.

Human Factors (Ergonomics). Human factors, or ergonomics, refers to human characteristics that must be considered in designing technical systems and equipment so that they can be used safely and efficiently. Anything that increases the difficulty of an operator's job may result in errors if shortcuts are taken. Such shortcuts may work under normal conditions, but are unacceptable in upset or abnormal conditions. The process safety management system should address the human–hardware interface, in both automated and manual processes. Safety reviews and HAZOP studies should include examination of the human–hardware interface. Even in a nearly perfectly designed situation, operators will make occasional errors. Consideration should be given to involving specialized experts in human error assessment, especially in processes involving high potential risks and extensive operator control.

Company Standards, Codes, and Regulations. Company standards, codes, and regulations are intended to communicate minimum acceptable safe practice and to ensure that all operating locations within the company share a common approach to process safety. All U.S. chemical plants are also subject to federal government regulations, and in some states there are specific environmental and toxic laws that apply. (See the section "Regulations" below.) Plants in many countries outside the United States have specific laws that apply to those countries. Many companies also have internal standards and guidelines to ensure consistency in decision-making by design engineers and plant personnel. There should be clear documentation so that it is known which codes, standards, and guidelines are to be followed.

Audits and Corrective Actions. Audits are methodical, independent, and usually periodic examinations of local installations, procedures, and practices. Audits help ensure compliance with a sound process safety program and that risks are being properly managed. The frequency of audits should be determined by the hazards of the facility. Corrective actions are the steps taken by a company in response to the recognition of a process safety deficiency, either through an audit or by other means. It is important that an audit team have expertise in facility operations, safety disciplines such as fire protection, and management systems. Corrective action includes the process of addressing identified deficiencies, weaknesses, or vulnerabilities. It also includes processes for corrective action planning and follow-up.

Enhancement of Process Safety Knowledge. Organizations with strong process safety programs can contribute to advancing the latest process safety technology. They should share nonproprietary results of safety research and support process safety-oriented research and development programs of professional and trade associations and colleges. Improved process safety knowledge can produce a competitive advantage by improved yields, better quality, increased productivity, and less downtime. Consideration should be given to encouraging technical staff participation in professional and trade association programs. The Center for Chemical Process Safety of the American Institute of Chemical Engineers, which is supported by many large chemical and chemical-related companies, is an example of a technical group that sponsors conferences, training programs, and publications in the area of chemical process safety.

PROCESS SAFETY MANAGEMENT SYSTEMS

The public, customers, employees, and local and federal government regulatory agencies all demand that companies take the actions needed to reduce the possibility of the release of hazardous materials. As the chemical process industries have developed better ways to improve process safety, safety management systems have been introduced to help implement improvements in process safety engineering activities. Management systems for chemical process safety are comprehensive sets of policies, procedures, and practices designed to ensure that barriers to major incidents are in place, in use, and effective.[24,38] The following are some definitions that are useful in discussions on managing process safety.

1. *Accident.* An unplanned event with undesirable consequences.
2. *Consequences.* Damage from a scenario; for example, the number of people exposed to a chemical cloud.
3. *Event.* The most elementary action in an accident; for example, an operator response or action.
4. *Hazard.* The result of combining a scenario and a consequence; for example, a chemical cloud is formed, and people are exposed. Also, the inherent potential of a material or activity to harm people, property, or the environment.
5. *Incident.* An occurrence that may be either an accident or a near miss.
6. *Near miss.* Extraordinary event that could have resulted in an accident but did not.
7. *Probability.* The likelihood of the occurrence of an event or a measure of degree of belief, the values of which range from 0 to 1.

8. *Quantitative risk analysis (QRA).* The systematic development of numerical estimates of the expected frequency and/or consequence of potential accidents associated with a facility or an operation. Often used interchangeably with "quantitative risk assessment."

9. *Risk.* The result of combining scenario, consequence, and likelihood; for example, the likelihood (probability) of a cloud being formed and people being exposed, given the process specifics. Or, a measure of potential economic loss or human injury in terms of the probability of the loss or injury occurring and the magnitude of the loss or injury if it occurs. Or, a combination of the expected frequency (events/year) and/or consequence (effects/event) of a single accident or a group of accidents.

10. *Risk assessment.* The systematic evaluation of the risk associated with potential accidents at complex facilities or operations. Or, the process by which the results of a risk analysis are used to make decisions, either through relative ranking of risk reduction strategies or through comparison with risk targets. The terms "risk analysis" and "risk assessment" are often used interchangeably in the literature.

11. *Risk management.* A part of chemical process safety management. Risk analysis (often used interchangeably with risk assessment) is a part of risk management.

12. *Scenario.* The end result of a series of events; for example, the release of a chemical cloud.

Some Tools for Evaluating Risks and Hazards

Dow Fire and Explosion Index. The Dow Fire and Explosion Index (F&EI), developed by The Dow Chemical Company, is an objective evaluation of the potential of a facility for a fire, an explosion, or a reactive chemical accident. Its purpose is to quantify damage from incidents, identify equipment that could contribute to an incident, and suggest ways to mitigate the incident; it also is a way to communicate to management the quantitative hazard potential. It is intended for facilities handling flammable, combustible, or reactive materials whether stored, handled, or processed. The goal of the F&EI evaluation is to become aware of the loss potential and to identify ways to reduce the potential severity in a cost-effective manner. It does not address frequency (risk) except in a general way. The number is useful mainly for comparisons and for calculations of damage radius, maximum probable property damage, and business interruption loss, and to establish frequency of reviews. The method of carrying out an F&EI evaluation is available to the public from the American Institute of Chemical Engineers, 345 E. 47th St, New York, NY 10017.

Failure Modes and Effects Analysis (FMEA). FMEA is a systematic, tabular method for evaluating the causes and effects of component failures. It represents a "bottom-up" approach, in contrast with a fault tree, where the approach is "top-down." In large part, HAZOP is a well-developed form of FMEA.[2]

Fault Tree. A fault tree is a logical model that graphically portrays the combinations of failures that can lead to a particular main failure or accident of interest. A fault tree starts with a top event, which is usually a hazard of some kind. The possibility of the hazard must be foreseen before the fault tree can be constructed. A fault tree helps reveal the possible causes of the hazard, some of which may not have been foreseen.[2]

Safety Audit. A safety audit is a method of reviewing the actual construction and operation of a facility. Often, safety audits are conducted by a small interdisciplinary team. At least some of the members of the team are not connected with the plant. The audit may be carried out before startup and also is repeated later at intervals of, typically, one to five years.

Chemical Exposure Index. The Chemical Exposure Index is a technique for estimating the relative toxic hazards of chemicals, developed by The Dow Chemical Company. It provides for the relative ranking of toxic chemicals in a given facility, including factors relating to toxicity, quantity volatilized, distance to an area of concern, and physical properties. A description of the method can be found in *Guidelines for Safe Storage and Handling of High Toxic Hazard Materials*, Center for Chemical Process Safety, American Institute of Chemical Engineers.[50]

EVALUATION OF HAZARDS AND RISKS

HAZOP[2,50-53]

HAZOP stands for "Hazard and Operability Studies," a set of formal hazard identification and elimination procedures designed to identify hazards to people, processes, plants, and the environment. The techniques aim to stimulate the imagination of designers and operations people in a systematic way so they can identify potential hazards. In effect, HAZOP studies assume that there will be mistakes, and provide a systematic search for these mistakes. In some European countries, HAZOP studies are mandatory and attended by observers from regulatory authorities to ensure that the studies are carried out correctly. The examination of accidents[52] during 1988 at a large U.S. chemical company revealed that the accidents could be classified as follows:

Spills: 52 percent
Emissions: 30 percent
Fires: 18 percent

Of the fires, about 50 percent occurred during construction, 25 percent were due to pump seal failure, and the remaining 25 percent resulted from engineering and operational oversights that a HAZOP study possibly could have prevented.

Of the emissions, 37 percent were due to piping failure, with lined pipe being the largest contributor. Operational and procedural issues accounted for 53 percent of the remainder.

Of the spills, 11 percent were due to equipment failures. Piping failures (especially lined pipe and gaskets) accounted for 30 percent, and 56 percent were caused by various types of operational errors, noncompliance with procedures, or nonexistent procedures.

Material handling was a factor in many spills and emissions.

The most frequent type of operational error was a valve being left in an improper position, either open or closed.

HAZOP studies probably could have reduced the number and seriousness of the problems experienced.

Some investigations have shown that a HAZOP study will result in recommendations that are 40 percent safety-related and 60 percent operability-related. Thus, HAZOP is far more than a safety tool; a good HAZOP study also results in improved operability of the process or plant, which can mean greater profitability.

The HAZOP technique can be used to identify human error potential. From a practical point of view, human error and its consequences can occur at all levels of a management structure as well as in the operation of a particular plant or process. Carried out correctly, Technica[54] states that a HAZOP study will identify at least 70–75 percent of potential operational and safety problems associated with a particular design process, including human error.

The HAZOP technique also can be used for the evaluation of procedures. Procedures may be regarded as a "system" designed to "program" an operator to carry out a sequence of correct actions. Deviations from intent are developed, with the emphasis on "operator action deviation" rather than "physical property deviation." It is the procedure, not the hardware, that is the object of study, but hardware modifications may be recommended to cover potential problems identified from procedure deviations.

Operating Discipline

HAZOP can be an important part of establishing the operating discipline of a plant, which can be defined as the total information required to understand and operate the facility.

By recording recommendations and actions in detail on all parts of the process, the rationale behind the way the process is designed and intended to operate and key details of the process will be available, which are rarely on record from other sources. This can be especially helpful when plant changes are made, and when new plants based on the same design are built.

Risk Analysis and Assessment

Risk is defined (see above) as the combination of expected frequency (events/year) and consequence (effects/event) of accidents that could occur as a result of an activity. Risk analysis evaluates the likelihood and consequence of an incident. Risk assessment is an extension of risk analysis that includes making judgments on the acceptability of the risk. It may be qualitative or quantitative, or it can range from simple, "broad brush" screening studies to detailed risk analyses studying large numbers of incidents and using highly sophisticated frequency and consequence models. Throughout the published literature, the terms "risk analysis" and "risk assessment" are used interchangeably.

Quantitative Risk Analysis (QRA)

Quantitative Risk Analysis (QRA) models the events, incidents, consequences, and risks, and produces numerical estimates of some or all of the frequencies, probabilities, consequences, or risks.[38,55] QRA can be done at a preliminary level or a detailed level, and in all cases may or may not quantify all events, incidents, consequences, or risks.[56] QRA is the art and science of developing and understanding numerical estimates of the risk associated with a facility or an operation. It uses highly sophisticated but approximate tools for acquiring risk understanding.

QRA can be used to investigate many types of risks associated with chemical process facilities, such as the risk of economic losses or the risk of exposure of members of the public to toxic vapors. In health and safety applications, the use of QRA can be classified into two categories:

1. Estimating the long-term risk to workers or the public from chronic exposure to potentially harmful substances or activities.
2. Estimating the risk to workers or the public from episodic events involving a one-time exposure, which may be acute, to potentially harmful substances or activities.

QRA is fundamentally different from many other chemical engineering activities (e.g., chemistry, heat transfer), whose basic property data are capable of being theoretically and empirically determined and often established experimentally. Some of the basic "property data" used to calculate risk estimates are probabilistic variables with no fixed values, and some of the key elements of risk must be established by using these probabilistic variables. QRA is an approach for estimating the risk of chemical operations by using the probabilistic approach; it is a fundamentally different approach from those used in many other engineering activities because interpreting the results of QRA requires an increased sensitivity to uncertainties that arise primarily from the probabilistic character of the data.

COMBUSTION HAZARDS

Introduction

The enchanting flame has held a special mystery and charm the world over for thousands of years. According to Greek myth, Prometheus the Titan stole fire from the heavens and gave it to mortals—an act for which he was swiftly punished. Early people made use of it anyway. Soon the ancients came to regard fire as one of the basic elements of the world. It has since become the familiar sign of the hearth and a mark of youth and blood, as well as the object of intense curiosity and scientific investigation.

Suitably restrained, fire is of great benefit; unchecked or uncontrolled, it can cause immense damage. We respond to it with a powerful fascination coupled with an inbred respect and fear. A good servant but a bad master is Thoreau's "most tolerable third party."[25]

Fire[26]

Fire or combustion is normally the result of fuel and oxygen coming together in suitable proportions and with a source of heat. The consumption of a material by a fire is a chemical reaction in which the heated substance combines with oxygen. Heat, light, smoke, and products of combustion are generated. The net production of heat by a fire involves both heat-producing and heat-absorbing reactions, with more heat being produced than is absorbed. Energy in the form of heat is required:

1. To produce vapors and gases by vaporization or decomposition of solids and liquids. Actual combustion usually involves gases or vapors intimately mixed with oxygen molecules.
2. To energize the molecules of oxygen and flammable vapors into combining with one another and so initiating a chemical reaction.

The amount of energy required to cause combustion varies greatly. Hydrogen and carbon disulfide can be ignited by tiny sparks, or simply may be ignited by static generated as the gases or vapors discharge from pipes into air. Other materials, such as methylene chloride, require such large amounts of energy to be ignited that they sometimes are considered nonflammable. Fire also can result from the combining of such oxidizers as chlorine and various hydrocarbon vapors; oxygen is not required for a fire to take place.

There are exceptions to the general rule that a solid must vaporize or decompose to combine with oxygen; some finely divided materials such as aluminum powder and iron power can burn, and it is generally accepted that they do not vaporize appreciably before burning.

Products of Combustion. Heat, light, smoke, and asphyxiating toxic gases are produced by fire. In a hot, well-ventilated fire, combustion usually is nearly complete.

Almost all the carbon is converted to carbon dioxide, and all the hydrogen to steam, and oxides of various other elements such as sulfur and nitrogen are produced.

This is not the case in most fires, where some of the intermediate products, formed when large complex molecules are broken up, persist. Examples are hydrogen cyanide from wool and silk; acrolein from vegetable oils; acetic acid from timber or paper; and carbon or carbon monoxide from the incomplete combustion of carbonaceous materials. As the fire develops and becomes hotter, many of these intermediates, which are often toxic, are destroyed (e.g., hydrogen cyanide is decomposed at a significant rate at 538°C).

Small airborne particles of partially burnt carbonaceous materials form smoke, which is often thickened by steam, when there is only partial combustion of fuel.

Solids. Ordinarily, combustible solids do not combine directly with oxygen when they burn. They give off vapor and gaseous decomposition products when they are heated, and it is the vapors or gases that actually burn in the characteristic form of flames. Thus, before a solid can be ignited, it usually must be heated sufficiently for it to give off flammable concentrations of vapors. Glowing, which is combustion in the solid state, is characteristic of materials in the final stages of a fire's decay when flammable gases have been burned away, or when the production of gases and vapors has been suppressed.

Solids with larger surface areas, in relation to their volume, burn more readily than those that are more compact when exposed to heat and oxygen in the air. Common materials such as textiles in the form of fibers or fabrics, foamed rubber, foamed plastics, thin sheets of plastic, paper, corrugated cardboard, combustible dusts, dry grass and twigs, and wood shavings are examples of materials with large surface areas in relation to their volume. In a well-established fire, materials with relatively small surface areas, such as chunks of coal or logs, burn readily.

Combustion is self-propagating; burning materials produce heat which causes more of the solid to produce flammable vapors until either the fuel or oxygen is exhausted, or until the fire is extinguished in some other way.

Dusts. Most combustible solids can produce combustible dusts. Combustible dusts are particularly hazardous; they have a very high surface area to volume ratio. When finely divided as powders or dusts, solids burn quite differently from the original material in the bulk. Dust and fiber deposits can spread fire across a room or along a ledge or roof beam very quickly. Accumulations of dust can smoulder slowly for long periods, giving little indication that combustion has started until the fire suddenly flares up, possibly when no one suspects a problem.

Many combustible dusts produced by industrial processes are explosible when they are suspended as a cloud in air. Even a spark may be sufficient to ignite them. After ignition, flame spreads rapidly through the dust cloud as successive layers are heated to ignition temperature. The hot gases expand and produce pressure waves that travel ahead of the flame. Any dust lying on surfaces in the path of the pressure waves will be thrown into the air and could cause a secondary explosion more violent and extensive than the first.

Liquids. A vapor has to be produced at the surface of a liquid before it will burn. Many common liquids give off a flammable concentration of vapor in air without being heated, sometimes at well below room temperature. Gasoline, for example, gives off ignitable vapors above about −40°C, depending on the blend. The vapors are easily ignited by a small spark or flame. Other liquids, such as fuel oil and kerosene, need to be heated until sufficient vapor is produced.

Many liquids can be formed into mists that will burn at temperatures where the vapor pressure is insufficient to produce a flammable mixture of the vapor and air.

For any flammable vapor there are maximum and minimum concentrations of the vapor in air beyond which it cannot burn. When the mixture of vapor in air is too weak, there is insufficient fuel for burning; when the mixture is too strong, there is insufficient oxygen for burning.

If the density of a flammable vapor is greater than that of air, as is normally the case, flammable concentrations may collect at low levels, such as at floor level or in basements, and can travel a considerable distance to a source of ignition, from which flames will then flash back.

Gases. Flammable gases usually are very easily ignited if mixed with air. Flammable gases often are stored under pressure, in some cases as a liquid. Even small leaks of a liquefied flammable gas form relatively large quantities of gas, which is ready for combustion.

The Fire Triangle. The well-known "fire triangle" (see Fig. 3.1) is used to represent the three conditions necessary for a fire:

1. Fuel
2. Oxidizer: oxygen or other gaseous oxidizer such as chlorine; or liquid oxidizer such as bromine; or solid oxidizer such as sodium bromate
3. Energy, usually in the form of heat

If one of the conditions in the fire triangle is missing, fire does not occur; if one is removed, fire is extinguished. Usually a fire occurs when a source of heat contacts a combustible material in air, and then the heat is supplied by the combustion process itself.

The fire triangle indicates how fires may be fought or prevented:

1. Cut off or remove the fuel.
2. Remove the heat or energy—usually by putting water on the fire.

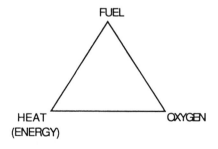

Fig. 3.1. The fire triangle.

3. Remove the supply of oxygen—usually by foam or inert gas.

Flammability

No single factor defines flammability, but some relevant parameters include:

1. Flash point—often considered the main index of flammability; low flash points usually mean increased flammability.
2. Flammability limits; wide limits mean increased flammability.
3. Autoignition temperature; low temperature means increased flammability.
4. Ignition energy; low ignition energy means increased flammability.
5. Burning velocity; high velocity means increased flammability.

A combustion process is an exothermic reaction initiated by a source of ignition that produces more energy than it consumes. The speed at which the reaction proceeds through the mixture of reactants depends on the concentration of the flammable gas or vapor. This speed is lower at higher ("rich") as well as a lower ("lean") concentrations of the flammable gas than at the stoichiometric mixture. There are lower and upper limits beyond which the reaction cannot propagate through the gas mixture on its own. Some definitions follow:

1. *Flammability limits.* The range of flammable vapor–air or gas–air mixtures between the upper and lower flammable limits. Flammability limits are usually expressed in volume percent. Flammability limits are affected by pressure, temperature, direction of flame propagation, oxygen content, type of inerts, and other factors. The precise values depend on the test method.
2. *Upper flammability limit.* The maximum concentration of vapor or gas in air above which propagation of flame does not occur on contact with a source of ignition.
3. *Lower flammability limit.* The minimum concentration of vapor or gas in air or

oxygen below which propagation of flame does not occur with a source of ignition.

The concentrations at the lower and upper flammability limits are roughly 50 percent and 200–400 percent, respectively, of the stoichiometric mixture. The maximum flammability usually (not always) occurs at the stoichiometric mixture for combustion.[2,27]

4. *Flammable limits for mixtures of flammable gases and vapors.* For mixtures of several flammable gases and vapors, the flammable limits can be estimated by application of Le Chatelier's equation, if the flammable limits of the components are known:[2]

$$L = \frac{1}{\sum_{i=1}^{n} \dfrac{y_i}{L_i}} \qquad U = \frac{1}{\sum_{i=1}^{n} \dfrac{y_i}{U_i}}$$

where

L = lower flammability limit of the fuel mixture, vol. %,
L_i = lower flammability limit of fuel component i, vol. %,
U = upper flammability limit of the fuel mixture, vol. %,
U_i = upper flammability limit of fuel component i, vol. %,
y_i = concentration of fuel component i, vol. %.

This equation is empirical and is not universally applicable, but is useful and a reasonable approximation when actual mixture data are not available.

It is possible for a mixture to be flammable even though the concentration of each constituent is less than its lower limit.

5. *Methods of measurement.* Flammability limits are determined by measuring the volume percent of a flammable gas in an oxidizing gas that will form a flammable mixture, thus identifying the lower and upper flammable limits as well as the critical oxygen concentration (the minimum oxidizer concentration that can be used to support combustion).

6. *Uniformity of lower limits on a mass basis.* Concentrations of vapors and gases usually are reported in volume percent. As molecular weight increases, the lower limit usually decreases. On a mass basis, the lower limits for hydrocarbons are fairly uniform at about 45 mg/L at 0°C and 1 atm. Many alcohols and oxygen-containing compounds have higher values; for example, on a mass basis, ethyl alcohol in air has a lower limit of 70 mg/L.[27]

7. *Effect of temperature on flammable limits.* The higher the temperature at the moment of ignition, the more easily the combustion reaction will propagate. Therefore, the reference temperature (initial temperature) of the flammable mixture must be stated when flammable limits are quoted. There are not a lot of data for flammable limits under different conditions of initial temperature. The behavior of a particular mixture under different conditions of initial temperature usually must be determined by tests.

8. *Burning in atmospheres enriched with oxygen.* The flammability of a substance depends strongly on the partial pressure of oxygen in the atmosphere. Increasing oxygen content affects the lower flammability limit only slightly, but it has a large effect on the upper flammability limit. Increasing oxygen content has a marked effect on the ignition temperature (reduces it) and the burning velocity (increases it).

   At the lower explosive limits of gas–air mixtures, there is already an excess of oxygen for the combustion process. Replacing nitrogen by additional oxygen will influence this limit very little.[8]

9. *Burning in chlorine.* Chemically, oxygen is not the only oxidizing agent, though it is the most widely recognized and has been studied the most. Halogens are examples of oxidants that can react exothermically with conventional fuels and show combustion behavior. The applicability of flammability limits applies to substances that burn in chlorine. Chlorination reactions have many similarities to oxidation reactions. They tend not to be limited to thermodynamic equilibrium and often go to complete chlorination. The reactions are often highly exothermic. Chlorine, like oxygen, forms flammable mixtures with organic compounds. As an example: a chlorine–iron fire occurred in a chlorine pipeline, causing a chlorine gas release. Chlorine had liquefied in the lines because of the very cold weather, and the low spot was steam-traced. Steam had been taken from the wrong steam line, using 400 psig steam instead of 30 psig steam. The 400 psig steam was hot enough to initiate the reaction. This serves as a reminder that steel and chlorine can react. The allowable temperature for safe use depends upon the state of subdivision of the iron.

10. *Burning in other oxidizable atmospheres.* Flames can propagate in mixtures of oxide of nitrogen and other oxidizable substances. For example, Bodurtha[27] reports that the flammability limits for butane in nitric oxide are 7.5 percent (lower) and 12.5 percent (upper).

11. *Flame quenching.* Flame propagation is suppressed if the flammable mixture is held in a narrow space. There is a minimum diameter for apparatus used for determination of flammability limits. Below this diameter the flammable range measurements are narrower and inaccurate.

   If the space is sufficiently narrow, flame propagation is suppressed completely. The largest diameter at which flame propagation is suppressed is known as the quenching diameter. For an aperture of slotlike cross section there is critical slot width.

   The term "quenching distance" sometimes is used as a general term covering both quenching diameter and critical slot width, and sometimes it means only the latter.

   There is a maximum safe gap measured experimentally that will prevent the transmission of an explosion occurring within a container to a flammable mixture outside the container. These data refer to a stationary flame. If the gas flow is in the direction of the flame propagation, a smaller gap is needed to quench the flame. If the gas flow is in the opposite direction, a larger gap will provide quenching. If the gas velocity is high enough, the flame can stabilize at the constriction and cause local overheating. These quenching effects are important in the design of flame arrestors.

12. *Heterogeneous mixtures.*[28]   In industry, heterogeneous (poorly mixed) gas phase mixtures can lead to fires that normally would be totally unexpected. It is important to recognize that heterogeneous mixtures can ignite at concentrations that normally would be nonflammable if the mixture were homogeneous. For example, one liter of methane can form a flammable mixture with air at the top of a 100-L container although the mixture only would contain 1.0 percent methane by volume if complete mixing occurred at room temperature, and the mixture would not be flammable. This is an important concept because "layering" can occur with any combustible gas or vapor in both stationary and flowing mixtures.

Heterogeneous gas phase mixtures can lead to unexpected fires if a relatively small amount of flammable gas is placed in contact with a large amount of air without adequate mixing, even though the average concentration of flammable gas in the mixture is below the flammable limit. *Heterogeneous mixtures are always formed at least for a short time when two gases or vapors are first brought together.*

13. *Effect of pressure.* Flammability is affected by initial pressure. Normal variations in atmospheric pressure do not have any appreciable effect on flammability limits.

A decrease in pressure below atmospheric usually narrows the flammable range. When the pressure is reduced low enough, a flame or an explosion can no longer be propagated throughout the mixture.

An increase in pressure above atmospheric usually (not always) widens the flammability range, especially the upper limit.

14. *Explosions in the absence of air.* Gases with positive heats of formation can be decomposed explosively in the absence of air. Ethylene reacts explosively at elevated pressure, and acetylene reacts explosively at atmospheric pressure in large-diameter piping. Heats of formation for these materials are +52.3 and +227 kJ/g/mol, respectively. Explosion prevention can be practiced by mixing decomposable gases with more stable diluents. For example, acetylene can be made nonexplosive at a pressure of 100 atm by including 14.5 percent water vapor and 8 percent butane.

Ethylene oxide vapor will decompose explosively in the absence of oxygen or air under certain conditions when exposed to common sources of ignition if heated to high enough temperatures. One way to prevent the decomposition reaction is to use methane gas to blanket the ethylene oxide liquid. It has also been found that liquid ethylene oxide will undergo a deflagration in the absence of oxygen with a very rapid pressure increase if ignited at a temperature and pressure above a certain level. Fortunately, the conditions required for propagation of the decomposition of liquid phase ethylene oxide are outside the current normal handling and processing ranges for the pure liquid. Propagation has not been observed below 80°C at from 14 to 100 atm pressure.[29] Ethylene oxide also can undergo explosive condensation when catalyzed by a small amount of caustic.[30]

Inert Gases

The addition of inert gases to a mixture of flammable gases and air affects flammability limits. Carbon dioxide causes a greater narrowing of the flammable range than does nitrogen. Water vapor is an acceptable inert gas if the temperature is high enough to exclude much of the oxygen, which requires a temperature of 90–95°C. Because water vapor and carbon dioxide have a higher heat capacity than nitrogen, they are somewhat more effective as inerting agents than nitrogen. Some halogen-containing compounds also can be used for inerting materials at relatively low concentrations. An example of this is the use of Freon-12 (CCl_2F_2). Caution must be used with halohydrocarbons because of the possibility of the halocarbons themselves burning, especially at high pressures. Environmental considerations are making the use of halogenated hydrocarbons for inerting increasingly undesirable. Materials are being developed that are considered environmentally acceptable. Figures 3.2 and 3.3[28] show flammability envelopes for methane and n-hexane for various

air–inert mixtures at 25°C and 1 atm. All flammable envelopes are similar to Figs. 3.2 and 3.3 except in minor detail. The lower limit is virtually insensitive to added inerts. The upper limit, however, decreases linearly with added inert until the critical concentration of inert is reached beyond which no compositions are flammable. In these graphs, C_{st} means the stoichiometric composition.

The limits of flammability are dictated by the ability of a system to propagate a flame front. Propagation does not occur until the flame front reaches about 1200–1400 K. Since the typical terminal temperature for hydrocarbons at stoichiometric conditions is about 2300 K, it can be seen that having only one-half the fuel or oxidizer present will produce about one-half the flame temperature, which is too low to propagate flame.

A useful rule to remember is that the lower flammable limits of most flammable vapors

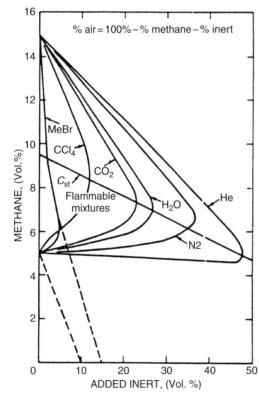

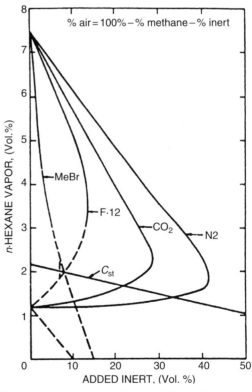

Fig. 3.2. Limits of flammability of various methane–inert gas–air mixtures at 25°C and atmospheric pressure. (*courtesy Bureau of Mines*)

Fig. 3.3. Limits of flammability of various n-hexane-inert gas mixtures at 25°C and atmospheric pressure. (*courtesy Bureau of Mines*)

are close to one-half the stoichiometric composition, which can be calculated. Another easy rule to remember is that about 10 percent oxygen or less in air (assuming the rest is mostly nitrogen) will not support combustion of most flammable hydrocarbon vapors.

The flammability limits of hydrocarbon-type fuels in oxygen and inert gas atmospheres are a function of the inert gas and any fuel or oxygen in excess of that required by the stoichiometry of the combustion process. In systems where fuel content is fixed, inert material having a high heat capacity will be more effective at flame suppression than inert material having a low heat capacity.

Many of the flammable limits reported in the literature are somewhat too narrow, and certain gas compositions regarded as being nonflammable are in fact flammable when given the proper set of circumstances. *In other words, take data on flammability limits from the literature with a grain of salt.* It is best not to design closely on the basis of most available data on flammability limits.

The use of inert gases can cause some serious hazards that must be recognized if inerts are to be used effectively and safely. Considerations in the use of inert gases include:

1. *An inert atmosphere can kill if a person breathes it.* Precautions should be taken to ensure that personnel cannot be exposed to the breathing of inert atmospheres.
2. *Some products need at least a small amount of oxygen to be stored safely.* This includes styrene and some other vinyl monomers, which must have some oxygen in them to make the usual polymerization inhibitor for styrene (*t*-butyl catechol, or TBC), effective. If pure nitrogen, for example, is used to blanket styrene, the inhibitor will become ineffective. TBC customarily is added to styrene monomer to prevent polymer formation and oxidation degradation during shipment and subsequent storage; it functions as an antioxidant and prevents polymerization by reacting with oxidation products (free radicals in the

monomer). If sufficient oxygen is present, polymerization is effectively prevented (at ambient temperatures); but in the absence of oxygen, polymerization will proceed at essentially the same rate as if no inhibitor were present. The styrene may polymerize and can undergo an uncontrolled exothermic reaction, which may generate high temperatures and pressures that can be very hazardous. The inhibitor level of styrene must be maintained above a minimum concentration at all times. The minimum concentration of TBC in styrene for storage is about 4–5 ppm.
3. *To be effective, inert atmospheres must be maintained within certain composition limits.* This requires the proper instrumentation and regular attention to the system.
4. *Inerting systems can be quite expensive and difficult to operate successfully.* Before the use of inert systems, alternatives should be explored, such as the use of nonflammable materials or operating well outside, preferably below, the flammability range.

Mists and Foams

If the temperature of a liquid is below its flash point, flammable concentrations of vapor cannot exist, but conditions still can exist for flammability if mists or foams are formed. A suspension of finely divided drops of a flammable liquid in air has many of the characteristics of a flammable gas–air mixture and can burn or explode. A mist may be produced by condensation of a saturated vapor or by mechanical atomization. Normally, the diameter of drops in a condensed mist is less than 0.01 mm, whereas in a mechanical spray it usually is greater than 0.1 mm.

The commonly accepted fallacy that liquids at temperatures below their flash points cannot give rise to flammable mixtures in air has led to numerous accidents. Flash points are measured under stagnant conditions in carefully controlled laboratory experiments, but in the real world one works with a wide variety of dynamic conditions that can produce mists and foams.

Flammable mist–vapor–air mixtures may occur as the foam on a flammable liquid collapses.[28] Thus, when ignited, many foams can propagate flame. An additional hazard can arise from the production of foams by oxygen-enriched air at reduce pressures. Air confined over a liquid can become oxygen enriched as pressure is reduced because oxygen is more soluble than nitrogen in most liquids. Thus, the presence of foams on combustible liquids is a potential explosion hazard.

The lower flammability limit for fine mists (<0.01 mm diameter) of hydrocarbons below their flash point, plus accompanying vapor, is about 48 g of mist/m^3 of air at 0°C and 1 atm. Mist can occur in agitated vessels under some conditions, especially when an agitator blade is at or near the liquid–vapor interface in the vessel.

Work on condensed oil mists (drop diameter mostly less than 0.01 mm) has demonstrated that they have flammability characteristics similar to those the mixture would have if it were wholly in the vapor phase at the higher temperature necessary for vaporization. The flammability characteristics are affected by drop size. For larger drop sizes (above 0.01 mm) the lower limit of flammability decreases as drop diameter increases. For mists, the amount of inert gas needed to suppress flammability is about the same as that needed to suppress an equivalent vapor–air mixture of the same material if it were vaporized at a somewhat higher temperature.

A useful rule is that mists of flammable or combustible liquids in air can burn or explode at temperatures below their flash points.

Ignition

Flammable gases and vapors can be ignited by many sources. In the design and operation of processes, it is best not to base fire and explosion safety on the presumption that ignition sources have been excluded. Bodurtha[27] reported that of 318 natural gas fires and explosions, the sources of ignition of 28 percent were unknown. All reasonable measures should be taken to eliminate possible sources of ignition in areas in which flammable materials are handled.

Autoignition. If the temperature of a flammable gas–air mixture is raised in a uniformly heated apparatus, it eventually reaches a value at which combustion occurs in the bulk gas. This temperature is defined as the spontaneous ignition temperature (SIT) or autoignition temperature (AIT). The gas–air mixture that has the lowest ignition temperature is called by various names, such as the minimum AIT, the minimum spontaneous ignition temperature, and the self-ignition temperature.[27] Usually the AIT reported in the literature is the minimum AIT.

The AIT of a substance depends on many factors, such as:

- ignition delay
- energy of ignition source
- pressure
- flow effects
- surfaces
- concentration of vapors
- volume of container
- oxygen content
- catalytic materials
- flow conditions

Thus, a specific AIT applies only to the experimental conditions employed in its determination. Usually the values quoted are obtained in clean laboratory equipment.

The AIT of a substance may be reduced below ideal laboratory conditions by as much as 100–200°C for surfaces that are insulated with certain types of insulation, or are contaminated by dust.

Mixtures that are fuel-rich or fuel-lean ignite at higher temperatures than do those of intermediate compositions. Also, in a homologous series of organic compounds, the AIT decreases with increasing molecular weight, as shown in Fig. 3.4.

Ignition Delay. Ignition of a flammable mixture raised to or above the temperature at which spontaneous combustion occurs is not instantaneous; the time delay between the

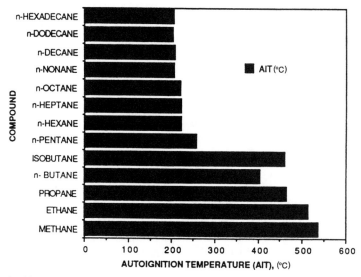

Fig. 3.4. Autoignition temperatures of paraffin hydrocarbons at 1 atm. (*Data from Bodurtha 1980.*[27])

moment of exposure to high temperature and visible combustion is called the ignition delay. This time delay decreases as the ignition temperature increases. The time delay may be as little as a fraction of a second at higher temperatures, or several minutes close to the AIT.

Environmental Effects. It has been found that the AIT becomes lower with increasing vessel size in the range of 35 ml to 12 L. An increase in pressure usually decreases AITs, and a decrease in pressure raises AITs. Usually oxygen enrichment of the air tends to decrease the minimum AIT, and a decrease in oxygen content increases the minimum AIT. Low-temperature oxidation can result in "cool flames," which may grow into ignition.

Catalytic Materials. Ignition may occur where the temperature is less than the minimum AIT. Catalytic materials, such as metal oxides, can promote oxidation on their surfaces, leading to a high local temperature and subsequent ignition. There is a recorded reactive chemical case[31] in which a solvent at 80°C was being pressurized with a gas phase consisting of a high oxygen concentration. The solvent has a flash point in oxygen of greater than 130°C and normally is considered not to be a flammability hazard. There was an

ignition, causing the vessel to rupture its main gasket with major damage to the facility. It was found that a mist had been formed in the vessel by the agitator, and that the source of ignition probably was a trace of palladium catalyst remaining from a previous run.

From this incident, several important lessons can be learned:

1. Ignition of a flammable mixture can result from totally unexpected contamination by trace amounts of catalysts if the oxidizer and fuel are present.
2. Mists of oxidizable liquids may form that can burn or explode at temperatures outside the "normal" flammable range.
3. It can be dangerous to perform experiments with pure oxygen, or air enriched with oxygen, especially under pressure and at elevated temperatures, when oxidizable materials are present.
4. The real criterion regarding flammable mixtures in air should be whether a flammable atmosphere can exist under the given process conditions, rather than whether a flammable liquid is at a temperature below its flash point.

Cleaning Up Spills of Flammable or Combustible Liquids. It is customary to clean up small spills of many liquid materials with sand or other noncombustible absorbent material. Some absorbing agents, such as untreated clays and micas, will cause an exothermic reaction with some liquids, especially monomers, which might ignite the liquid if it is flammable or combustible. Before any material is provided to be used to soak up spills of oxidizable material, tests should be made to determine if the material can cause fires with potential spills.

Ignition Caused by Insulation. Ignition of combustible materials that have been absorbed into commonly used insulating materials is possible at temperatures lower than the AIT for nonabsorbed material. All oxidizable materials oxidize to some extent in air at ambient temperatures, usually at a very low rate. When an absorbent material is absorbed into insulation, it is "spread" over a large area, increasing its access to oxygen. Because the absorbent is an insulator, heat from oxidation is retained rather than dissipated, and the temperature will rise if the heat is produced faster than it can be dissipated. The rate of oxidation increases as the material temperature increases, which produces more heat, compounding the hazard. If the temperature rises enough, the material will ignite ("spontaneous combustion"). This is similar to the classic oily rag and wet haystack phenomenon, which has caused many fires in homes and on farms. In the wet haystack phenomenon, fermentation by microorganisms will create heat. Some air is necessary; too much air will remove too much heat to allow the combustion temperature to be reached. For equipment operating above about 200°C containing combustible liquids with high boiling points, insulation should be impervious to the material handled. To date, only a closed cell foamed glass provides the required degree of protection where oxidizable liquid materials are used above 200°C. Insulation based on glass fiber, silicate, or alumina materials is known to cause hazardous situations and should not be used in this service.

Laboratory tests and actual fires show that Dowtherm A® (a heat-transfer fluid consisting of a eutectic mixture of biphenyl oxide and

TABLE 3.1 Reduction in AITs Caused by Liquids Soaking in Glass Fiber Insulation

Material	Normal AIT (°C)	Ignition Temperature in Glass Fiber Insulation (°C)
Dowtherm A®	621	260–290
Stearic acid	395	260–290

biphenol) can be ignited if it is soaked in glass fiber insulation and in contact with air at temperatures considerably below the normal AIT. This is also true for stearic acid soaked in glass fiber insulation. Table 3.1 shows the reduction in AIT of Dowtherm A® and of stearic acid soaked in glass fiber insulation.

Ignition of this type generally occurs only with materials having a high boiling point. Usually materials with low boiling points will vaporize and cannot remain soaked in hot insulation. There are exceptions. For example, ethylene oxide has a fairly low boiling point, but if it leaks into insulation, a polymer can be formed that has a high boiling point and can autoignite insulation at low temperatures.

Ignition Caused by Impact. Solids and liquids can be ignited by impact. Impact tests are made by having a weight fall freely through a known distance and impacting the sample. Impact can occur, for example, if containers are accidentally dropped. The interpretation of the data from impact tests can be difficult.

Ignition Caused by Compression of Liquids. Liquids can be ignited by sudden compression. This can happen when there is water hammer caused by the pressure surge from quick-acting valves and from the compression in liquid pumps. Sudden compression can occur with liquids, for example if a tank car is bumped rapidly and the liquid goes to one end very quickly, possibly trapping some vapor bubbles that compress and create local hot spots that can cause ignition.

Ignition Caused by Rubbing Friction. Solids can be ignited by frictional sources when rubbed against each other or against another

CHEMICAL PROCESS INDUSTRY SAFETY CONSIDERATIONS 111

material. The frictional heat produced may be enough to ignite other materials, such as lubricants, that are nearby. A common example of this occurs when bearings run hot, causing oil or grease to vaporize and possibly ignite.

Ignition Caused by Glancing Blows. Friction can cause ignition in other ways. Sparks may occur when two hard materials come in contact with each other in a glancing blow (the blows must be glancing to produce friction sparks). These kinds of sparks are not directly related to frictional impact. Hand and mechanical tools are the most likely sources of friction sparks that occur outside of equipment. The need for nonsparking tools is somewhat controversial; Bodurtha[27] states that it is extremely unlikely that anyone would be using tools in a flammable atmosphere, and it is usually more prudent to control the atmosphere than the tools. Sparkproof tools are not really sparkproof in all situations.

Ignition Caused by Static Electricity. Static electricity is a potential source of ignition wherever there is a flammable mixture of dusts or gases (see next section).

Ignition Caused by Compression of Gases. If a gas is compressed rapidly, its temperature will increase. Autoignition may occur if the temperature of the gas becomes high enough (this is more or less the principle of diesel engines).

An advancing piston of high-pressure gas can compress and heat trapped gas ahead of it. For a perfect gas, the temperature rise due to adiabatic compression is given by

$$\frac{T_2}{T_1} = \left(\frac{P_2}{P_1}\right)^{(k-1)/k}$$

where T_1 and T_2 are the initial and final gas absolute temperatures, P_1 and P_2 are the initial and final absolute pressures, and k is the ratio of heat capacity at constant pressure to the heat capacity at constant volume. For air and many other diatomic gases, $k = 1.4$. Many hydrocarbons have k values of between 1.1 and 1.2. The value of k is a function of temperature and pressure.

Energy Levels for Ignition. If a flammable gas mixture is to be ignited by a local source of ignition, there is a minimum volume of mixture required to cause a continuing flame throughout the mixture.

For example, to ignite a methane–air mixture in a cold container, a hot patch of 18 mm² at 1000–1100°C is required in order to heat enough volume of gas to produce a continuing flame,[2] even though the auto-oxidation temperature for methane is 540°C. Ignition of a flammable gas–air mixture by electrical discharge can occur only if the electrical discharge is of sufficient energy.

Minimum Ignition Energy. There is a minimum ignition energy, which usually occurs near the stoichiometric mixture. The minimum ignition energy for some representative substances in air is shown in Fig. 3.5.[2] The energy required to cause ignition frequently is reported in millijoules (mJ). One joule is 0.24 calorie, so 1 mJ is 0.00024 calorie, which is a very small amount of energy.

A person typically has capacitance of 200 picofarads (pF), and if charged to 15 kilovolts (kV) could initiate a discharge of 22.5 mJ. This is enough to ignite many flammable mixtures. The energy in ordinary spark plugs is 20–30 mJ.

The hazard of an explosion should be minimized by avoiding flammable gas–air or dust–air mixtures in a plant. It is bad practice to rely solely on elimination of sources of ignition, as it is nearly impossible to ensure this.

Effect of Oxygen-Enriched Atmospheres. The minimum spark energy to cause ignition varies greatly with the amount of oxygen in oxygen-enriched air. Stull[30] showed that with a composition of 10 percent methane in air, about 0.5 mJ of spark energy is required to initiate a reaction at the lower flammable limit. If the air is enriched with oxygen, the minimum spark energy decreases. If the flammable material is combined with 100 percent oxygen, the spark energy required is only about 1 percent of the required energy in air at 21 percent oxygen! This demonstrates the extremely small amount of energy required to

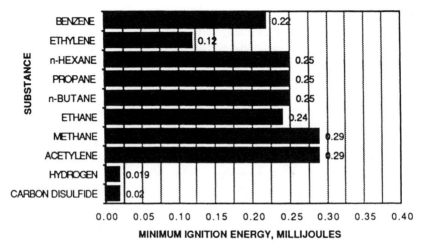

Fig. 3.5. Minimum ignition energy for selected substances. (*Less 1980.*[2])

TABLE 3.2   Comparison of Initiation Energies of Some Common Substances in Air and Pure Oxygen[32]

Flammable Substance	Relative Amount of Energy Required to Initiate Combusion (mJ)	
	In Air	*In Pure Oxygen*
Methane	0.3	0.003
Hydrogen	0.019	0.0012
Acetone	1.15	0.0024
Diethyl ether	0.2	0.0013

initiate the reaction, as well as the additional ease with which oxygen-enriched atmospheres are initiated. Table 3.2 compares initiation energies of some common substances in air and in pure oxygen.

Effect of Pressure. An increase in pressure decreases the amount of energy required to cause ignition. In a mixture of propane, oxygen, and nitrogen, doubling the pressure decreases the minimum energy required to cause ignition by a factor of about 5.

If no other data are available for determination of hazards, minimum ignition energies at ambient temperatures and pressures should be considered as approximately:

- 0.1 mJ for vapors in air
- 1.0 mJ for mists in air
- 10.0 mJ for dusts in air

STATIC ELECTRICITY

Introduction

Many apparently mysterious fires and explosions have eventually been traced to static. In spite of the large amount of information about static electricity, it remains a complex phenomenon not often understood and appreciated. Static electricity is a potential source of ignition whenever there is a flammable mixture of gas or dust.

When two different or similar materials are in contact, electrons can move from one material across the boundary and associate with the other. If the two materials in contact are good conductors of electricity and are separated, the excess electrons in one material will return to the other before final contact is broken. But if one or both of the materials are insulators, this flow will be impeded. If the separation is done rapidly enough, some excess electrons will be trapped in one of the materials. Then both materials are "charged." Electric charges can build up on a nonconducting surface until the dielectric strength is exceeded and a spark occurs. The residual charge could ignite flammable mixtures.

The two materials or phases in initial contact may be:

- a single liquid dispersed into drops
- two solids
- two immiscible liquids

- a solid and a liquid
- a solid and a vapor or gas
- a liquid and a vapor or gas

The important thing to keep in mind is that whenever there is contact and separation of phases, a charge may develop that could be disastrous. Three conditions must be met before an explosion caused by static electricity can take place:

1. An explosive mixture must be present.
2. An electric field must have been produced due to the electrostatic charge that had been generated and accumulated in a liquid or solid.
3. An electric field must be large enough to cause a spark of sufficient energy to ignite the mixture.

In designing preventive measures, all three factors should be controlled.

Static electricity is essentially a phenomenon of low current but high voltage and high resistance to current flow. A low-conductivity liquid flowing through a pipeline can generate a charge at a rate of 10^{-9}–10^{-6} ampere (A). A powder coming out of a grinding mill can carry a charge at a rate of 10^{-8}–10^{-4} A. At a charging rate of 10^{-6} A, the potential of a container insulated from earth can rise at a rate of 1000 V/s and a voltage of 10,000 volts or higher can readily be obtained in this way.

Several electrostatic voltages and energies commonly encountered are typified by the following examples:

1. A person walking on dry carpet or sliding across an automobile seat can generate up to 5000 V in dry weather. An individual having a capacitance of 100 pF, a reasonable figure, could generate a spark energy of 1.25 mJ. This is far more than is needed to ignite some flammable vapor–air mixtures.
2. A person can accumulate dangerous charges up to about 20,000 V when humidity is low.
3. A truck or an automobile traveling over pavement in dry weather can generate up to about 10,000 V.

4. Nonconductive belts running over pulleys generate up to 30,000 V. The voltage generated by a conveyor belt can be as high as 10^6 V; the system can in effect act as a Van der Graaf generator.
5. The energy in the spark from an ordinary spark plug is 20–30 mJ.

The capacitance and the energy for ignition of people and of common objects are important. The capacitance of a human being is sufficient to ignite various flammable gas mixtures at commonly attained static voltages.

Hazard Determinants

Capacitance. The capacitance of an object is the ratio of the charge of the object to its potential. The capacitance gets larger as the object gets larger. With a given charge, the voltage gets higher as the capacity of the object gets smaller. For a sphere, capacitance is given by

$$C = Q(10^{-3})/V$$

The energy stored in a capacitor is[27]

$$W = 0.5CV^2(10-3) = 500Q^2/C$$

where

C = capacitance, pF (1 pF = picofarad = 10^{-12} farad),
Q = charge, microcoulombs (1 coulomb = 1 A/s = charge on 6.2×10^{18} electrons),
V = voltage in kilovolts,
W = energy, millijoules (mJ).

This energy may be released as a spark when the voltage gets high enough. The minimum sparking potential for charged electrodes is about 350 V and occurs at a spacing of 0.01 mm. Sparks from an equally charged nonconductor are less energetic and may contain only part of the stored energy. These comparatively weak sparks are not likely to ignite dust clouds but can ignite flammable gases.

The energy that can be stored by capacitance of an object can be compared with the minimum ignition energies of flammable gas–air mixtures and of dust–air mixtures

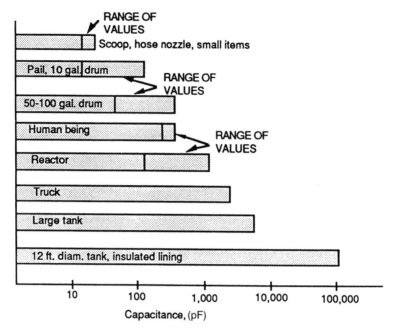

Fig. 3.6. Some typical values of electrical capacitance. (*Data from Eichel 1967.*[33])

to determine the probability that a spark discharge may have sufficient energy to cause ignition. If the charged object is a poor conductor, the calculation of energy available to produce a spark may not be possible because the charge often is not uniformly distributed, and the resistance to flow of current is high. Figure 3.6 shows some typical values of electrical capacitance.[33]

Relaxation Time. When a liquid is flowing in closed metal pipes, static electricity is not a hazard. When the liquid enters a tank, it may become a hazard. Charges caused by liquid separation during pumping, flow, filtration, and other effects such as splashing and agitation can accumulate on the surface of the liquid in the tank and cause sparking between the liquid surface and the tank or conducting objects in the tank. The charge thus generated can be dissipated by relaxation or via discharge through a spark or corona discharge. The relaxation time is the time required for 63 percent of the charge to leak away from a charged liquid through a grounded conductive container. The half-time value is the time required for the free charge to decay to one-half

of its initial value. The half-time is related to the relaxation time by the relationship

$$T_h = T_r \times 0.693$$

where

T_h = half-time,
T_r = relaxation time.

Relaxation times vary from small fractions of a second up to minutes and even hours for some highly purified hydrocarbons that have very low conductivity.

It is important to recognize that a large charge can accumulate in the liquid even in a grounded container. In fact, it was reported that the majority of accidents attributed to static electricity in the petroleum industry have been with liquid in grounded containers.[2]

Relaxation time can be calculated as follows:

$$T_r = E(E_0/k)$$

where

T_r = the relaxation time, in seconds; the time for 63 percent of the charge to leak away,

E = relative dielectric constant, dimensionless,

E_0 = absolute dielectric constant in a vacuum, $= 8.85 \times 10^{-14}$ to less than 1×10^{-18},

K = liquid conductivity, Siemens per centimeter (S/cm).

Siemens (S) are also called mhos

Example: Benzene in a large tank could have a specific conductivity as low as 1×10^{-18} mho/cm and as high as 7.6×10^{-8} S/cm. The corresponding relaxation times for the two conductivities can be calculated as follows. Pure benzene has a dielectric constant of 2.5 to less than 1×10^{-18}. Using the above equation:

(1) $T_r = (8.85 \times 10^{-14})\,(2.5)/(7.6 \times 10^{-8})$
        $= 2.91 \times 10^{-6}$ s
(2) $T_r = (8.85 \times 10^{-14})\,(2.5)/(1 \times 10^{-18})$
        $= 2.21 \times 10^{5}$ s (this is in excess
                    of 60 hr)

Benzene typically contains some water and has a higher conductivity than in the above example and has a much lower relaxation time.

The purity of a liquid has a great effect on its relaxation time, and thus its static hazard potential. In actual practice, relaxation times of a few seconds to an hour are encountered, depending on the purity and dryness of the liquid. This emphasizes the dangers of open sampling of tank contents soon after filling. If it is likely that the liquid being used has a low conductivity, it is important that enough time elapses between activities that can produce a static charge, such as loading a tank, and any activity that could cause a spark, such as sampling from the top of the tank.

In case (2) in the above example, a conductivity of 10^{-18} S/cm is so low that there may be little charge separation and little charge formation, and there may be no hazard even though the calculated relaxation time is extremely long. Materials with a half-time value of less than 0.012 s have been reported not to cause a hazard. A useful rule to remember is that the concept of relaxation is very important because it is possible for liquid in a tank to retain an electric charge for a long time

if the liquid is a poor conductor, even if the tank is grounded. The specific conductivity, and therefore the relaxation time, is greatly affected by impurities. For example, the specific conductivity of benzene can vary from as long as 1×10^{-18} to about 7.6×10^{-8} S/cm, depending on its purity. It can vary significantly with the amount of water or other materials dissolved in the benzene.[27]

Resistivity. The extent of charge separation is dependent on the resistivity of the liquid. Some materials have a sufficiently high conductivity to render them safe in terms of static buildup. If the resistivity is low, charge separation is easy, but so is charge recombination through the liquid. If the resistivity is high, there may be appreciable charge separation without immediate recombination, leading to a high charge. If the resistivity is extremely high, there may be no charge separation, and there is no buildup of a charge. If the conductivity of a liquid falls in the hazardous range, it is possible to modify it by the use of a very small amount of an additive. Additives usually are a combination of a polyvalent metal salt of an acid such as carboxylic or sulfonic acid and a suitable electrolyte. Additives of this type can impart a conductivity of 10^{-8} S/m (Siemens per meter) in a 0.1 percent solution in benzene.[2]

A useful rule to remember is that when the resistivity of a liquid exceeds 10^{15} ohm centimeters (Ω·cm), or is less than 10^{10} Ω·cm, static generation or accumulation is negligible. Between these limits, the net generation of charges increases with the *maximum charge generation at* 10^{3} Ω·cm. Styrene, for example, a commonly used monomer, has a resistivity of 4×10^{13} at 20°C,[34] and therefore is capable of building up a potentially hazardous charge.

Static Charge Development. Static electrification of solids can occur in various ways. Different operations will produce the percentages of the theoretical maximum charge density shown in Fig. 3.7.[33]

It should be noted that pure gases do not generate significant static electricity in transmission through pipes and ducts. Gases contaminated with rust particles or liquid

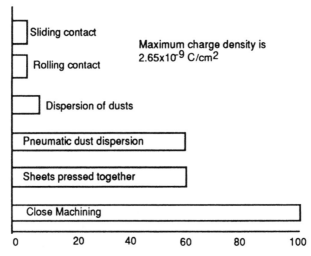

Fig. 3.7. Percentage of maximum theoretical charge produced by various operations. (*Data from Eichel 1967.*[33])

droplets produce static, but this is not a problem in a closed, grounded piping system. *If these gases impinge on an ungrounded, conductive object, dangerous charges can accumulate on that object.* Wet steam, which contains water droplets, can develop charges. *If the water droplets contact an ungrounded conductor, that object can develop a static charge.*

Flammable gases may ignite when discharged to air during thunderstorms, even without a direct lightning hit. Dry hydrogen and occasionally other gases may ignite when they are discharged to air in normal weather. This may be so because the electric field developed by the ejected gases can develop a corona discharge which can cause ignition. The minimum ignition energy of hydrogen is only 0.02 mJ. A toroidal ring developed by the National Aeronautics and Space Administration (NASA) is reported to prevent unwanted discharge and subsequent ignition of a vent-stack outlet.[27]

Humidification. The conductivity of electrical nonconductors, such as plastics, paper, and concrete, depends on their moisture content. Relatively high moisture in these materials increases conductivity and therefore increases dissipation of static electricity. With relative humidity of 60–70 percent or higher, a microscopic film of moisture covers surfaces, making them more conductive.

Humidification can and often should be practiced to reduce the hazard of static electricity, but should not be relied on entirely to remove all possibility of static discharge.

In winter, cold air brought into a building and heated to normal room temperature is extremely dry, often less than 5–10 percent relative humidity. When processing solid materials that can develop a static charge, this air should be humidified to reduce static hazards as well as improve the comfort of personnel.

Filling Liquid Containers. A fire during top loading of a flammable liquid into a tank constitutes a serious problem if there could be a flammable mixture in the vapor space. Static electricity can be generated by splashing if the liquid is top-loaded, so it is normal practice to fill with a dip pipe positioned so the tip of the dip pipe is near the bottom of the tank. This may not be sufficient to prevent static charge buildup, as a charge may be generated in the bottom of the tank before the pipe tip is fully submerged, and it is possible for the liquid to acquire a charge before it reaches the tank.

Product filters using cotton, paper, felt, or plastic elements are prolific generators of static electricity. It is considered that at least

30 s is necessary to dissipate this charge, although with dry nonconductive liquids, it may require as long as 500 s.

Loading a less volatile liquid into a tank where there was previously a more volatile liquid is particularly hazardous because the more volatile liquid may form a flammable mixture, and the less flammable material is often a poor conductor and will not readily dissipate static charge. This type of loading accounts for 70–80 percent of severe losses at terminals.[2] This appears to occur most often when the compartments are one-fourth to one-third full, and when the temperature is close to $-1°C$.

Inerting the tank while it is being filled will reduce the possibility of ignition by static electricity and is highly recommended when it is possible and practical. However, this is not always practical. In any case, if inerts are to be used, they must be added carefully, as the following example illustrates. Two firemen were fatally injured when an explosion occurred as they were attempting to use portable CO_2 fire extinguishers to inert a tank truck. The source of ignition was believed to be a spark from the horn of the extinguisher to the latch on the tank truck. It was found that the voltage on the horn increased as the carbon dioxide "snow" passed down the horn to the outlet side.

Grounding and bonding lines, although very important, will not immediately dissipate the charge on the surface of a nonconducting liquid in a tank. A relaxation time for charge to be a dissipated should be allowed after filling or other operations to permit static charge on the liquid surface to dissipate to the dip pipe or tank shell. *The minimum time is 1 min, but longer periods are advisable with some liquids that have extremely low conductivity. Bottom loading may reduce the static electricity hazard but does not eliminate it.*[35]

EXPLOSIONS

Development of Pressure

Exothermic reactions can led to high temperatures and in the case of large fires to large loss of property and severe damage from radiant energy. However, in many plant accidents it is the sudden generation of pressure that leads to severe damage, injury, and deaths. Hence, it can be stated that "pressure blows up plants, not temperature." Of course, temperature and pressure are closely related, but it is the pressure effect that is of concern in this section.

The word "deflagration" can be defined in several ways. One definition is "a reaction that propagates to the unreacted material at a speed less than the speed of sound in the unreacted substance."[27] Another definition of deflagration is from Latin meaning "to burn down, or to burn rapidly with intense heat and sparks given off."[30] A deflagration may be an explosion, but not all deflagrations are explosions (a violently burning fire may be a deflagration, but that is not an explosion). On the other hand, not all explosions are deflagrations (a steam boiler may explode, but that is not a deflagration).

An explosion is a sudden and violent release of energy. Usually it is the result, not the cause, of a sudden release of gas under high pressure. The presence of a gas is not necessary for an explosion. An explosion may occur from a physical or mechanical change, as in the explosion of a steam boiler, or from a chemical reaction. The explosion of a flammable mixture in a process vessel may be either a deflagration or a detonation, which differ fundamentally. Both can be very destructive. Detonations are particularly destructive, but are unlikely to occur in vessels.

A detonation is a reaction that propagates to unreacted material at a speed greater than the speed of sound in the unreacted material; it is accompanied by a shock wave and extremely high pressures for a very short time. It is debatable whether the flammable range is the same as the detonable range. Detonation limits normally are reported to be within the flammable limits, but the view is widely held that separate detonation limits do not exist.

Unconfined vapor clouds can both deflagrate and detonate, with a deflagration being much more likely. A detonation is more destructive, but a deflagration also can produce a damaging pressure wave. A deflagration can undergo transition to a detonation in a pipeline, but this is most likely in vessels.

If a flammable mixture may be present in process equipment, precautions should be taken to eliminate ignition sources. However, it is prudent to assume that, despite these efforts, a source of ignition will at some time occur.

Deflagration

The conditions for a deflagration to occur are that the gas mixture is within the flammable range and that there is a source of ignition or that the mixture is heated to its AIT.

For the burning of hydrocarbon–air mixtures:

$$\frac{P_{2MAX}}{P_1} = \frac{N_2 T_2}{N_1 T_1} = \frac{M_1 T_2}{M_2 T_1}$$

where

T = absolute temperature,
M = molecular weight of gas mixture,
N = number of moles in gas mixture,
P = absolute pressure,
1,2 = initial and final states,
2MAX = final state maximum value.

The maximum pressure rise for a deflagration of flammable mixtures is approximately as follows for initial absolute pressures of 1–40 bar, for initial temperatures of 0–300°C, and for relatively small volumes of a few cubic meters:

$\frac{P_2}{P_1}$ = approximately 8 for hydrocarbon–air mixtures

$\frac{P_2}{P_1}$ = approximately 16 for hydrocarbon–oxygen mixtures

For conventionally designed pressure vessels:

$$\frac{P_b}{P_1} = \text{approximately } 4-5$$

where

P_b = vessel bursting pressure,
P_1 = normal design pressure,
P_2 = pressure caused by deflagration.

Therefore, in the absence of explosion relief, the deflagration explosion of a hydrocarbon–air mixture is easily capable of bursting a vessel if it is operating near its design pressure when the deflagration takes place. For reactions operating at or near atmospheric pressure, such as many drying and solids processing operations, it may be practical to construct facilities that will withstand the maximum explosion pressure of most dust–air and flammable gas–air mixtures.

Detonations

Detonation of a gas–air mixture may occur by direct initiation of detonation by a powerful ignition source or by transition from deflagration. This transition occurs in pipelines but is most unlikely in vessels. Two useful rules are:

1. Almost any gas mixture that is flammable is detonable if initiated with a sufficiently energetic source.
2. Detonation of a gas–air mixture is possible in pipelines but is unlikely in vessels.

Bartknecht[8] states that the range of detonability is narrower than the range of flammability. For example, the range of detonability of hydrogen in air is 18–59 vol. percent, compared with the flammability of 4–75 vol. percent. With flammable gases in air, if the length-to-diameter ratio of a pipe or vessel is more than about 10 : 1, and the pipe diameter is above a critical diameter, 12–25 mm, a detonation is possible.

Detonation Pressure. In the case of the burning of a flammable mixture of gases in a pipe with one end closed, a series of pressure waves traveling at the speed of sound moves through the unburned gas. Later waves traveling through the unburned gas, which has been heated by compression from the earlier waves, speed up because of the higher temperature and overtake the first wave, and a shock wave develops. Flame follows the shock wave and catches up with it, forming a detonation wave. A stable detonation wave may develop, which moves with supersonic speed relative to the unburned mixture, and peak incident (side-on) pressures are of the order of *30 times* the initial absolute pressure.

Reflected Pressure. Reflected pressure increases the pressure on a rigid surface if the

TABLE 3.3   Overpressure from Detonations[36]

	Pressure (MPa)	Pressure (lb/in.²)
Incident overpressure	3.5	510
Maximum reflected pressure (wave strikes surface head-on)	28	4100
Load the structure feels (due to acceleration)	56	8100

(MPa means pressure in megapascals.)

shock wave impinges on the surface at an angle to the direction of the propagation of the wave. The maximum ratio of reflected pressure to incident (side-on) pressure when a strong shock wave strikes a flat surface head-on is 8 : 1. Furthermore, acceleration from a suddenly applied force of the detonation wave can double the load that a structure "feels." Table 5.3 shows overpressure that can be expected from typical detonations.[36]

Thus, the stable detonation wave may cause enormously high pressures at closed ends of pipes, bends, and tees, where the greatest destruction from a gaseous detonation may occur.

Geometry. The following are some factors to consider when detonation is possible:

1. Large length-to-diameter ratios promote the development of detonations; vessels should be designed with the lowest length-to-diameter ratio practicable if a detonation is possible.
2. Equipment such as tanks (not including pipelines) designed to withstand 3.5 MPa (about 500 psig) usually will be adequate to contain a detonation, with other safeguards, for flammable gases in air at atmospheric pressure.
3. Dished heads survive detonations better than do flat heads because of the more unfavorable incidence of flat heads.
4. If turns in a process line are necessary, two 45° bends or a long sweep elbow will greatly reduce reflected pressure compared with a single 90° elbow.
5. Restrictions such as orifices in pipelines may intensify a detonation by promoting

pressure piling, which results when there are interconnected spaces such that the pressure rise in one space causes a pressure rise in a connected space. The enhanced pressure in the latter then becomes the starting pressure for a further explosion.

6. Detonation may be extinguished when it enters a wider pipe from a smaller one, but the detonation may be regenerated somewhere along the longer pipe.
7. Flame arresters, if properly designed, can arrest detonations.

Explosion Violence

The *rate of pressure rise* is a measure of the violence of an explosion. The maximum rate of pressure rise for confined explosions is greatly affected by the volume of the vessel, with the influence of vessel volume on the rate of pressure rise being given by the following equation:

$$(dp/dt)_{max}(V^{1/3}) = \text{a constant} = K_G$$

where

$$(dp/dt)_{max} = \text{maximum rate of pressure rise, bar/s,}$$
$$V = \text{vessel volume, m}^3,$$
$$K_G = \text{a specific material constant, (bar)(m)(s)}^{-1}.$$

This is the *cubic law*, which states that for a given flammable gas, the product of the maximum pressure rise and the cube root of the vessel volume is a specific material constant, K_G.

The cubic law allows the prediction of the course of an explosion of a flammable gas or vapor in a large vessel, based on laboratory tests. It is valid only for the following conditions:

• the same optimum concentration of the gas–air mixture
• same shape of reaction vessel
• the same degree of turbulence
• the same ignition source

Thus, to characterize an explosion, it is not enough to quote the maximum rate of pressure rise: the volume, vessel geometry,

TABLE 3.4 K_G Values of Gases, Spark-Ignited with Zero Turbulence, Ignition Energy ~10 J, $P_{max} = 7.4$ bar[8]

Flammable Gas	K_G (bar)(m)/s
Methane	55
Propane	75
Hydrogen	550

(From Bartknecht, W., *Explosions Course, Prevention, Protection*, p. 108, copyright Springer-Verlag, Berlin, 1981, by Permission.)

turbulence, and ignition energy must also be stated. Table 3.4 lists the K_G values for some common flammable gases measured under laboratory conditions.

It can be seen that the violence of an explosion with propane is about 1.5 times higher than one with methane, and one with hydrogen is about 10 times higher than one with methane. The explosive behavior of propane is representative of many flammable organic vapors in air. Some important relationships among pressure, temperature, turbulence, and vessel shape are discussed below.

1. *Explosion pressure is primarily the result of temperature reached during combustion, not a change in moles.* With complete combustion of propane in air there is a negligible change in moles of gas:

$$\begin{array}{c} \{\ldots.\text{air}.\ldots\} \\ C_3H_8 + 5O_2 + 18.8\,N_2 \\ = 3CO_2 + 4H_2O + 18.8\,N_2 \end{array}$$

Number of moles at start = 24.8.
Number of moles after complete combustion = 25.8.

Therefore, explosion pressure usually develops principally from an increase in temperature, not an increases in gas moles, during the combustion process of many materials.

Peak explosion pressure at constant volume occurs near the stoichiometric concentration in air. If only a small part of the total volume of a container is filled by an explosive gas–air mixture at atmospheric pressure, and the remainder of the vessel contains air, an explosion of this mixture can create enough pressure to severely damage containers that are designed to withstand only slight pressure—such as buildings and low-pressure storage tanks.

2. *Initial pressure affects maximum explosion pressure and rate of pressure rise.* If the initial pressure is increased above atmospheric pressure, there will be a proportional increase in the maximum explosion pressure and in the rate of pressure rise. Reducing the initial pressure will cause a corresponding decrease in maximum explosion pressure until finally an explosion reaction can no longer be propagated through the gas mixture.

3. *Initial temperature affects maximum explosion pressure and rate of pressure rise.* The maximum explosion pressure decreases when the starting temperature increases at the same starting pressure because of the lower density and thus smaller mass of material within a confined volume at higher temperatures. The maximum rate of pressure rise, $(dp/dt)_{max}$, increases as the initial temperature rises because the burning velocity increases with an increase in initial temperature.

4. *Initial turbulence increases the rate of pressure rise.* Initial turbulence greatly increases the rates of explosion-pressure rise.[8,27] It has been found that with pentane and methane mixtures in air, $(dp/dt)_{max}$ can be five to nine times more with high initial turbulence than with no turbulence. The maximum explosion pressure is raised by about 20 percent. The course of explosions of flammable gases with a low normal speed of combustion, such as methane, is influenced by turbulence to a much higher degree than is the course of explosions with a high speed of combustion, such as hydrogen. Test data usually are obtained in equipment with a high degree of turbulence.

5. *Effect of vessel shape and increased initial pressure.* The maximum explosion pressure in confined vessels is not significantly affected by the volume or shape of the vessel in confined explosions for vessels that approximate the "cubic shape," that is, with a ratio of diameter to length (or vice versa) of about 1 : 1 to 1 : 1.5. In closed elongated vessels with central ignition, spherical ignition of the flame front will cause the flame to proceed swiftly in an axial direction. In the process, it compresses

the unburned gases ahead of it, causing the violence of the explosion to increase, and pressure oscillations may occur.

BOILING LIQUID EXPANDING VAPOR EXPLOSIONS (BLEVES)

Among the most damaging of accidents is a Boiling Liquid Expanding Vapor Explosion (BLEVE, pronounced BLEV-ee). This occurs when a pressure vessel containing liquid is heated so that the metal loses strength and ruptures. Typically, this happens when the vessel failure results from overheating upon exposure to fire. The failure usually is in the metal contacting the vapor phase; the metal in this area heats to a higher temperature because there is no liquid heat sink to keep the metal temperature from rising rapidly, as there is where metal contacts a liquid phase. A BLEVE can occur with both flammable materials and nonflammable materials, such as water. In all cases the initial explosion may generate a blast wave and missiles. If the material is flammable, it may cause a fire or may form a vapor cloud that then gives rise to a secondary explosion and fireball. Kletz states that BLEVEs can cause as many casualties as can unconfined vapor cloud explosions.[2]

The best known type of BLEVE involves liquefied petroleum gas (LPG). Once a fire impinges on the shell above the liquid level, the vessel usually fails within 10–20 min. In the case of a BLEVE involving a flammable material, the major consequences are, in order of decreasing importance:

- Thermal radiation from the resultant fireball
- Fragments produced when the vessel fails
- Blast wave produced by the expanding vapor/liquid

For example, a BLEVE of a propane sphere with a diameter of 50 ft, holding about 630,000 gal, could cause damage as far away as 13,600 ft, and radiation damage and fragmentation damage would each extend to about 3,000 ft.

In a fire, a tank containing liquid is most vulnerable in the shell at the vapor space because very little heat can be absorbed by the vapor, and the metal in the vapor space can heat up

rapidly to a temperature where it will weaken rapidly. The metal contacting the liquid phase will heat up much less rapidly because the liquid can absorb significant amounts of heat, keeping the shell temperature down in that area for a significant amount of time. Thus, there is a dilemma: a partly full vessel may BLEVE sooner than will a full vessel, but a full vessel will have more fuel for the resulting fireball and fire than will a partly empty vessel.

Significant equipment and building damage from radiation is possible from a BLEVE. Wooden structures may be ignited if the radiant heat density at the structure's location exceeds the threshold value for ignition of wood. Severe damage from fragmentation can be expected in the area where 50 percent or more of the fragments may fall, or typically about 300 ft from the vessel.

A BLEVE can lead to shock waves, projectiles, and thermal radiation. The effects of a shock wave and projectiles were dealt with earlier; by far the most serious consequence of a BLEVE is the radiation received from the fireball. The following calculational procedure is used to determine thermal impact (details are available in CPQRA[38]):

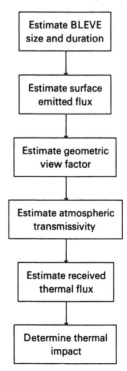

DAMAGE ESTIMATES[37]

Damage estimates deal with the consequences of explosions and thermal radiation to both people and property. Physical models for explosions and thermal radiation generate a variety of incident outcomes: shock wave overpressure estimates, fragment velocities, and radiant flux. These models rely on the general principle that severity of outcome is a function of distance from the source of release. In addition to estimating the damage resulting from an explosion, it is also necessary to estimate how the consequences of these incident outcomes depend on the object of the study. To assess effects on human beings, damage estimates may be expressed as deaths or injuries. If physical property is the object, the damage estimates may be expressed as monetary losses.

Explosion Consequences

A principal parameter characterizing an explosion is the overpressure. Explosion effect modeling generally is based on TNT explosions to calculate the overpressure as a function of distance, Although the effect of a TNT explosion differs from that of a physical or a chemical explosion (particularly in the near-field), the TNT model is the most popular because a large data base exists for TNT explosions. Several kinds of energy may be released in an explosion; three basic types are: (1) physical energy, (2) chemical energy, and (3) nuclear energy. Nuclear energy is not considered here. Physical energy may take such forms as pressure energy in gases, strain energy in metals, or electrical energy. Chemical energy derives from a chemical reaction. Examples of explosions involving chemical energy are runaway exothermic reactions, including decomposition and polymerization.

Table 3.5 summarizes the effects of explosion overpressure on structures. With respect to human casualties, heavy building damage usually is equated to a fatal effect, as the people inside the buildings probably would be crushed. People outside of buildings or structures are susceptible to direct blast injury (blast overpressure) and indirect blast injury (missiles or whole body translation).

Relatively high blast overpressures (>15 psig) are necessary to produce a human fatality from a direct blast. Instead, the major threat is produced by missiles or by whole body translation. Fatalities arising from whole body translation are mainly due to head injury from decelerative impact. Injury to people due to fragments usually results from either penetration by small fragments or blunt trauma from large fragments. TNO[39] suggested that projectiles with a kinetic energy of 100 J can cause fatalities. Table 3.6 shows damage to people (physiological damage) as a function of overpressure.

Radiation Consequences

The effect of thermal radiation on people and objects is determined by one of two approaches:

1. Simple tabulations based on experimental results
2. Theoretical models based on the physiology of the skin burn response

Data on time to pain threshold[40] are summarized in Table 3.7. For comparison, solar radiation intensity on a clear, hot summer day is about 320 Btu/hr ft^2 (1 kW/m^2). Other criteria for thermal radiation damage are shown in Table 3.8.[38]

The effect of thermal radiation on structures depends on whether they are combustible or not, and the nature and duration of the exposure. Thus, wooden materials will fail because of combustion, whereas steel will fail because of thermal lowering of the yield stress.

Unconfined Vapor Cloud Explosions (UVCE)

When a large amount of volatile material is released rapidly to the atmosphere, a vapor cloud forms and disperses. If the cloud is ignited before it is diluted below its lower flammability limit, an uncontrolled vapor cloud explosion will occur. This is one of the most serious hazards in the process industries. Both shock waves and thermal radiation will result from the explosion, with the shock waves usually the more important damage producers. UVCEs usually are modeled by

TABLE 3.5   Effect of Explosion Overpressure on Structures

Pressure (psi)	Damage
0.02	Annoying noise (137 dB if of low, 10–15 Hz frequency)
0.03	Breaking of large glass windows under strain
0.04	Loud noise (143 dB), sonic boom, glass failure
0.10	Breakage of small glass windows under strain
0.15	Typical pressure for glass breakage
0.30	"Safe distance" (probability 0.95 of no serious damage below this value); projectile limits; some damage to house ceilings; 10% window glass broken
0.40	Limited minor structural damage
0.5–1.0	Large and small windows usually shattered; occasional damage to window frames
0.70	Minor damage to house structures
1.00	Partial demolition of houses; houses made uninhabitable
1–2.00	Corrugated asbestos shattered; corrugated steel or aluminum panels, fastenings fail, followed by buckling; wood panel fastenings of standard housing fail; panels blown in
1.30	Steel frames of clad buildings slightly distorted
2.00	Partial collapse of walls and roofs of houses
2–3.00	Concrete or cinder blocks shattered if not reinforced
2.30	Lower limit of serious structural damage
2.50	50% destruction of brickwork of houses
3.00	Heavy machines (300 lb), industrial buildings suffered little damage; steel frame buildings distorted and pulled away from foundation
3–4.00	Frameless, self-framing steel panel building demolished; rupture of oil storage tanks
4.00	Cladding of light industrial buildings ruptured
5.00	Wooden utility poles snapped
5–7.00	Nearly complete destruction of houses
7.00	Loaded railcars overturned
7–8.00	Brick panels, 8–12 in. thick, not reinforced, fail by shearing or flexure
9.00	Loaded train boxcars completely demolished
10.00	Probable total destruction of buildings; heavy machine tools (7,000 lb) moved and badly damaged; very heavy machine tools (12,000 lb) survive
300.00	Limit of crater lip.

(Copyright 1989 by the American Institute of Chemical Engineers, reproduced by permission of the Center for Chemical Process Safety of AIChE.[38])

TABLE 3.6   Physiological Damage as a Result of Overpressure

Effect	Peak Overpressure (psi)
Knock down	1.0
Ear drum damage	5.0
Lung damage	15
Threshold for fatalities	35
50% fatalities	50
99% fatalities	65

TABLE 3.7   Time to Pain Threshold for Varying Levels of Radiation[40]

Radiation Intensity (Btu/hr/ft²)	Radiation Intensity (kW/m²)	Time to Pain Threshold (s)
500	1.74	60
740	2.33	40
920	2.90	30
1500	4.73	16
2200	6.94	9
3000	9.46	6
3700	11.67	4
6300	19.87	2

(Courtesy American Petroleum Institute.)

using the TNT model.[38] The energy of the blast wave generally is only a small fraction of the energy available from the combustion of all the material that constitutes the cloud; the ratio of the actual energy released to that available frequently is referred to as the "explosion efficiency." Therefore, the TNT weight equivalent of a UVCE includes an explosion efficiency term, which typically is an empirical factor ranging from 1 percent to 10 percent. The explosion effects of a TNT charge are well documented.

TABLE 3.8   Effects of Thermal Radiation

Radiation Intensity (kW/m^2)	Observed Effect
37.5	Sufficient to cause damage to process equipment.
25.0	Minimum energy required to ignite wood at indefinitely long exposures.
12.5	Minimum energy required for piloted ignition of wood, melting of plastic tubing.
9.5	Pain threshold reached after 6 seconds; second-degree burns after 20 seconds.
4.0	Sufficient to cause pain to personnel if unable to reach cover within 20 seconds; however, blistering of the skin (second degree burns) is likely; 0% lethality.
1.6	Will cause no discomfort for long exposure.

(Copyright American Institute of Chemical Engineers, reproduced by permission of the Center for Chemical Process Safety of AIChE.[38])

Physical Explosions

A physical explosion usually results from the production of large volumes of gases by non-chemical means. The gases necessary for a physical explosion may be those already existing, such as compressed nitrogen released suddenly from a ruptured cylinder, or steam released explosively from a crack in a steam drum.

The following are some settings and situations in which physical explosions have been known to take place:

- steam boilers
- hydraulic overfill of tanks or pipes with external applied pressure (as in pressure testing)
- compressed air tanks
- deadheaded pumps
- thermal expansion of tanks or pipes
- liquid cryogenic fluids on water (such as liquid methane on water)
- water suddenly mixed with sulfuric acid (also may cause a chemical explosion)

- BLEVE with superheated liquid (flammable or nonflammable) (see next section)
- explosion of grinding wheel at too high a speed
- liquid water in molten $MgCl_2$ solution at high temperatures
- implosions due to vacuum
- overpressured refrigerant systems
- molten metals exploding violently on contact with water
- some molten metals exploding when mixed with each other
- the mixing of two immiscible liquids whose boiling points are not widely separated

Steam boilers are commonly used in power plants and industries of all kinds. They generally are taken for granted now, but in the second half of the nineteenth century boilers blew up with alarming regularity. Records indicate that from 1870 to 1910 there were at least 10,000 boiler explosions in the United States and adjacent areas of Canada and Mexico; that is, more than one recorded explosion every 36 hours! By 1910, the rate had jumped to between 1,300 and 1,400 per year. On October 8, 1894, in the Henry Clay Mine in Shamokin, Pennsylvania, 27 boilers disintegrated almost simultaneously! Mainly because of the incorporation of the ASME Boiler Code into laws, boiler explosions have decreased dramatically.[41]

When a pressurized vessel ruptures, the resulting stored energy is released. This energy can cause a shock wave and accelerate vessel fragments. If the contents are flammable, ignition of the released gas could produce additional effects. There is a maximum amount of energy in a bursting vessel that can be released, and it is released in the following proportions:[36]

Type of Failure	Distribution of Energy When Vessel Ruptures		
	Strain Energy	Kinetic Energy of Fragments	Shock Wave Energy
Brittle failure	<10%	~20%	up to 80%
Plug ejection	small	up to 60–80%	20–40%

The relative distribution of these energy components will change over the course of the explosion, but most of the energy is carried by the shock wave with the remainder going to fragment kinetic energy. To estimate the damage resulting from the shock wave from a physical explosion, the TNT model is used widely. To determine the TNT equivalent of a physical explosion, the total energy in the system must be estimated. For a physical explosion, if the expansion occurs isothermally, and ideal gas laws apply, then the TNT equivalent of the explosion can be calculated. This energy then can be used to estimate overpressure at any distance from the explosion. The analogy of the explosion of a container of pressurized gas to a point source explosion of TNT is not appropriate in the near-field. Prugh[42] suggests a correction method using a virtual distance R_v from an explosion center.

In addition to shock wave effects, a major hazard of a ruptured gas-filled vessel is from projectiles. To estimate damage from projectiles, both the initial velocity and the range are required. A simplified method for calculating the initial velocity uses the following equation:[43]

$$u = 2.05(PD^3/W)^{0.5}$$

where

u = initial velocity, ft/s,
P = rupture pressure, psig,
D = fragment diameter, in,
W = weight of fragments, lbs.

Clancey[44] gives the following values for initial velocity for the majority of fragments from a TNT explosion:

- thin case: 8,000 ft/s
- medium case: 6,000 ft/s
- thick case: 4,000 ft/s

Once the initial velocity has been determined, the maximum range of the fragment, ignoring air resistance, can be estimated from

$$R_{max} = \frac{u^2}{g}$$

where R_{max} is the maximum range of fragments and g is the acceleration of gravity.

If the above values for typical velocity are substituted into the above equation, a maximum range of 5×10^5 ft is possible. Therefore, it is clearly necessary to include air resistance. To include air resistance, a value of C_D, the drag coefficient, must be estimated. The drag coefficient ranges from 0.48 for a sphere to 2 for flow perpendicular to a flat strip, and for most fragments ranges from 1.5 to 2.0.

If one knows the air density, drag coefficient, exposed area of the fragment, mass of the fragment, and the initial velocity, the maximum range R can be calculated with the aid of Fig. 3.8.[45] Although this technique gives the maximum range, most fragments do not travel the maximum distance but fall at distances between 0.3 and 0.8 of the maximum.

The energy required to impart an initial velocity of u to a fragment is

$$E = \tfrac{1}{2}mu^2$$

where

m is the mass of the fragment (lb),
u the initial velocity (ft/s).

Example. A high pressure vessel containing air at 600 bar has ruptured, leading to 15 fragments of approximately equal mass (85 lb), one of which was found as far as 400 ft from the vessel. This fragment has a drag coefficient of 1.5 and an exposed area of 3 ft². Assuming that 20 percent of the explosion energy went to energy of the fragments, estimate the energy of the explosion in weight equivalent TNT. The air density is 0.081 lb/ft³.

Procedure. Assuming that the fragment found at 400 ft is at the maximum range for the fragments, the scaled fragment range R_s can be calculated:

$$R_s = \frac{r_0 C_D A_D R}{M}$$
$$= \frac{0.081 \times 1.5 \times 3 \times 400}{85}$$
$$= 1.7$$

From Fig 3.8. we obtain a scaled force (F_s) of approximately 5. The initial velocity of the

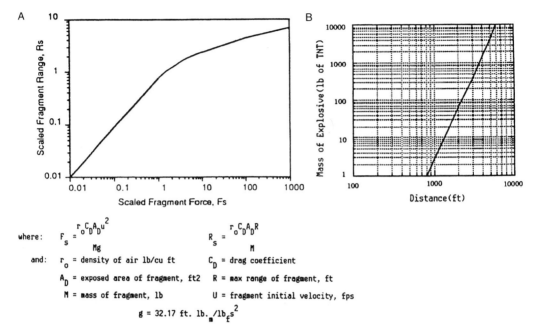

where:

$$F_s = \frac{r_o C_D A_D u^2}{Mg} \qquad R_s = \frac{r_o C_D A_D R}{M}$$

and: r_o = density of air lb/cu ft C_D = drag coefficient

A_D = exposed area of fragment, ft2 R = max range of fragment, ft

M = mass of fragment, lb U = fragment initial velocity, fps

$$g = 32.17 \text{ ft. lb.}_m/\text{lb}_f\text{s}^2$$

Fig. 3.8. (a) Scaled fragment range vs. scaled force. (*Baker et al.* 1983.[45])
(b) Maximum horizontal range of blast fragments. (*Clancy.*[44])

fragment then can be calculated as

$$u = \sqrt{\frac{MgF_s}{r_0 C_D A_D}}$$

$$= \sqrt{\frac{85 \times 32.17 \times 5}{0.081 \times 1.5 \times 3}}$$

$$= 194 \text{ ft/s}$$

The energy required to give the fragment this initial velocity is

$$E = \tfrac{1}{2}(85)(194)^2$$

$$= 1.6 \times 106 \text{ lb ft}^2/\text{s}^2$$

$$= 64 \text{ BTU}$$

Since there were 15 fragments, the total energy of the explosion that went into fragment kinetic energy is 15×635.8 BTU = 9537 BTU. If only 20 percent of the explosion energy went into fragment kinetic energy, then the total explosion energy is 47,680 BTU, which is the equivalent of 23.8 lb of TNT. Using the method of Clancy[44], 2.4 lb TNT can provide a maximum range of 950 ft for projectiles (Fig. 3.8 (b)).

MECHANICAL HEAT

Mechanical motion in fluids becomes kinetic energy and may become heat in devices with rotating parts. Mechanical heat input from rotating agitators, pump impellers, and other mechanical equipment must be taken into account in the design of process equipment, particularly in systems containing reactive chemicals. This section will provide some guidelines for the analysis of individual cases involving pumps and agitated tanks.[46–48]

Some useful rules are as follows.

1. A deadheaded pump is a pump operating full of liquid and with inlet and outlet valves closed.
2. Almost all deadheaded centrifugal pumps with motors of three horsepower or larger are headed for trouble if left deadheaded. (Depending on the horsepower, a few minutes may be too long.)
3. The heat input from the rotating impeller in a deadheaded centrifugal pump is always a large value relative to the heat sink of the fluid and the pump.

4. It is not necessary for there to be a chemical reaction in a pump for an explosion to take place. Deadheaded pumps containing only water or brine have blown up.

5. An agitator or a circulating pump left on in a vessel of a reactive chemical may heat up the contents enough to cause a runaway reaction.

6. All centrifugal pumps with motors larger than 3 hp should be protected in some way to prevent deadheading. A temperature alarm in the casing is a minimum form of protection. A better way may be to have the high-temperature alarm wired to the process control computer, to both alarm and shut off the pump. Other systems are available and may be used; they may include (but are not limited to) a relief valve on the pump, a minimum flow valve, and a flow orifice in the recirculating line. A relief valve on a pump relieving back to the pump inlet may not eliminate the problem of heat buildup in a deadheaded pump and usually should be avoided unless other protective measures are used such as a high-temperature device.

7. An ammeter on the pump motor usually is not a reliable means of detecting deadheaded conditions. The low power factors often experienced with pump motors, and the nature of pump curves, often make it difficult to distinguish between normal running and deadheaded conditions using an ammeter.

8. For mechanical heat equivalent, the following are recommended: (a) For pumps, use 50 percent of the connected motor horsepower for centrifugal pumps that are deadheaded, unless better information is available. (b) For agitators, use 100 percent of the vendor rated shaft input horsepower for the input shaft (total power less drive and bearing inefficiencies) for the actual material in the vessel.

VACUUM[49]

Ask any chemical engineers who have had some plant experience what they know about vacuum, and they probably will smile and tell a tale about some piece of equipment that tried to turn itself inside out. Usually no one was hurt, and often there is no massive leakage—but not always!

The design for the internal pressure condition of vessels usually is straightforward and well understood. Under vacuum conditions, equipment is subject to external pressure from the atmosphere; and the design for external pressures is more difficult than that for internal pressures. The devious ways in which external pressure can be applied often may be overlooked.

The following are some obvious causes of vacuum collapse:

- liquid withdrawal by pump or gravity draining
- removal of gas or vapor by withdrawing with a blower, fan, or jet
- siphoning of liquids.

Less obvious causes include:

- condensation of vapor
- cooling of hot gas
- combination of cooling and condensation of a mixture of gas and condensable vapor.

Sometimes obscure causes of vacuum collapse include:

1. Absorption of a gas in a liquid; for example, ammonia in water, carbon dioxide in water, hydrogen chloride in water.

2. Reaction of two or more gases to make a liquid or solid; for example, ammonia plus hydrogen bromide to form ammonium bromide.

3. Reaction of a gas and a solid to form a solid; for example, corrosion in a tank, air plus Fe or FeO forming Fe_2O_3 in the presence of water.

4. Reaction of a gas and a liquid to give a liquid; for example, chlorination, hydrogenation, ethylation.

5. Sudden dropping of finely divided solids in a silo, creating a momentary vacuum that can suck in the sides of the silo.

6. Flame arrestors plugging; for example:
   (a) In styrene service, vapor may condense in flame arrestors, and the liquid formed is low in inhibitor; the liquid may polymerize and plug off

the arrestor. Possible solutions: clean the arrestor frequently or use a PVRV (pressure-vacuum-relief valve).
(b) Liquid service in cold weather: vapor may condense in a flame arrestor and the liquid formed may freeze and plug the arrestor. Possible solution: heat and insulate the arrestor to prevent condensation.
7. Maintenance and testing. It is not a good idea to apply vacuum on a vessel during maintenance or testing without full knowledge of the external pressure rating unless a suitable vacuum relief device is in place and operable.

Protective Measures for Equipment

If equipment may be subject to vacuum, consideration should be given to designing the equipment for full vacuum. This may eliminate the need for complicated devices such as vacuum relief valves and instruments; if they are used, designing the equipment for full vacuum will prevent collapse of the vessel if the instruments or relief valves fail or plug.

A disadvantage of this approach is that it usually is expensive. However, when the total cost of a suitably instrumented vessel not designed for vacuum is compared with the cost of a vessel designed for vacuum but without the extra equipment, the difference may be small or negligible, and the vessel designed for vacuum will be inherently safer. If a vessel is designed for vacuum, precautions should be taken to ensure that internal or external corrosion will not destroy the integrity of the vessel.

REGULATIONS*

Regulations are a major consideration in the design and operation of chemical facilities. This section provides a description of the sig-

_____
*This section was prepared with the help of William Carmody, Midland, Michigan. Carmody has had more than 30 years experience in chemical and manufacturing operations for The Dow Chemical Company, Midland, Michigan and six years in Safety and Loss Consulting for Midland Engineering Limited, Midland, Michigan. He has developed entire PSM programs and has conducted many Process Hazard Analyses.

nificant process requirements. Details of the regulations are available on the Internet or from government agencies, such as the U.S. Department of Labor, or from publications such as those produced by the Thompson Publishing Group and by Primatech, Inc.

Abbreviations used in Government Regulations information:

CFR	Code of Federal Regulations
EPA	Environmental Protection Agency
EPCRA	Emergency Planning and Community Right To Know Act
HAZWHOPER	Hazardous Waste Operations & Emergency Response
HHC	Highly Hazardous Chemicals
MSDS	Material Safety Data Sheet
NIOSH	National Institute for Occupational Safety and Health
OSHA	Occupational Safety and Health Administrates
PHA	Process Hazard Analysis
PPA	Pollution Prevention Act
PSM	Process Safety Management
RCRA	Resource Conservation and Recovery Act
RMP	Risk Management Plans
SARA	Superfund Amendments and Reauthorization Act
TRI	Toxics Release Inventory

Process Safety Management

On February 24, 1992, the U.S. Department of Labor, Occupational Safety and Health Administration (OSHA) promulgated a final rule, 29 CFR Part 1910.119, "Process Safety Management of Highly Hazardous Chemicals."

OSHA administrates regulations whose objectives are primarily involved with protecting workers. This can be regarded as "inside the fence line." This is a safety issue and is addressed in this section. The rule requires employers to effectively manage the process hazards associated with chemical processes to which the rule applies. OSHA is responsible for the Process Safety Management (PSM) program that is used to prevent or minimize the consequences of catastrophic releases of toxic, reactive, flammable, or explosive chemicals. Standard Number CFR 1910.119 contains requirements for preventing or minimizing the consequences of catastrophic releases of toxic, reactive, flammable, or explosive chemicals. It establishes procedures for PSM that will protect employees by preventing or minimizing the consequences of chemical accidents involving highly hazardous chemicals. The requirements in this standard are intended to eliminate or mitigate the consequences of such releases.

PSM applies to a process involving a chemical at or above the specified threshold quantities listed in 1910.119, Appendix A, and also listed in Table 3.9. The requirements of the rule are also applicable to processes that involve a flammable liquid or gas on-site, in one location, in a quantity of 10,000 lb or more, except for hydrocarbon fuels used solely for workplace consumption as a fuel, or flammable liquids stored in atmospheric pressure tanks.

Process means any activity involving a highly hazardous chemical including any use, storage, manufacturing, handling, or the on-site movement of such chemicals, or combination of these activities. For purposes of this definition, any group of vessels that are interconnected and separate vessels which are located such that a highly hazardous chemical could be involved in potential release shall be considered a single process.

The PSM elements required by 29 CFR Part 1910.119 are briefly described in the following.

Employee Participation. Employers must develop a written plan of action for how they will implement employee participation requirements. Employers must consult with employees, affected contractors, and their representatives on the conduct and development of process hazard analyses and on other elements of the standard. They must have access to information developed from the standard, including process hazard analyses.

Process Safety Information. Employers must compile considerable documented process safety information on the hazards of chemicals used in a covered process as well as information on the process technology and equipment before conducting the process hazard analyses required by the standard.

Process Hazard Analysis (PHA). Employers must perform an analysis to identify, evaluate, and control hazards on processes covered by this standard. The process hazard analysis shall be appropriate to the complexity of the process and shall identify, evaluate, and control the hazards involved in the process. The OSHA standard specifies a number of issues that the analysis must address, as well as requirements for who must conduct the analysis, how often it must be performed, and response to its findings. Methodologies that are appropriate include:

- what-if
- checklists
- what-if/checklist
- Hazard and Operability Study (HAZOP)
- Failure Mode and Effects Analysis (FMEA)
- fault tree analysis

The selection of a PHA methodology or technique will be influenced by many factors including the amount of existing knowledge about the process. All PHA methodologies are subject to certain limitations. The team conducting the PHA needs to understand the methodology that is going to be used. A PHA team can vary in size from two people to a number of people with varied operational and technical backgrounds. Some team members may only be a part of the team for a limited time. The team leader needs to be fully knowledgeable in the proper implementation of the PHA methodology that is to be used and should be impartial in the evaluation. The other full- or part-time team members need to provide the team with expertise in areas such as process technology, process design, operating procedures, and practices.

TABLE 3.9 List of Highly Hazardous Chemicals, Toxics, and Reactive Chemicals (Mandatory)

Standard Number: 1910.119 Appendix A (on the Internet)

This is a listing of toxic and reactive highly hazardous chemicals that present a potential for a catastrophic event at or above the threshold quantity.

Chemical Name	CAS[a]	Threshold Quantity[b]
Acetaldehyde	75–07–0	2,500
Acrolein (2-Popenal)	107–02–8	150
Acrylyl chloride	814–68–6	250
Allyl chloride	107–05–1	1,000
Allylamine	107–11–9	1,000
Alkylaluminum	Varies	5,000
Ammonia, anhydrous	7664–41–7	10,000
Ammonia solutions (greater than 44% ammonia by weight)	7664–41–7	15,000
Ammonium perchlorate	7790–98–9	7,500
Ammonium permanganate	7787–36–2	7,500
Arsine (also called arsenic hydride)	7784–42–1	100
Bis(chloromethyl) ether	542–88–1	100
Boron trichloride	10294–34–5	2,500
Boron trifluoride	7637–07–2	250
Bromine	7726–95–6	1,500
Bromine chloride	13863–41–7	1,500
Bromine pentafluoride	7789–30–2	2,500
Bromine trifluoride	7787–71–5	15,000
3-Bromopropyne (also called propargyl bromide)	106–96–7	100
Butyl hydroperoxide (tertiary)	75–91–2	5,000
Butyl perbenzoate (tertiary)	614–45–9	7,500
Carbonyl chloride (see Phosgene)	75–44–5	100
Carbonyl fluoride	353–50–4	2,500
Cellulose Nitrate (concentration greater than 12.6% nitrogen)	9004–70–0	2,500
Chlorine	7782–50–5	1,500
Chlorine dioxide	10049–04–4	1,000
Chlorine pentrafluoride	13637–63–3	1,000
Chlorine trifluoride	7790–91–2	1,000
Chlorodiethylaluminum (also called diethylaluminum chloride)	96–10–6	5,000
1-chloro-2, 4-dinitrobenzene	97–00–7	5,000
Chloromethyl methyl ether	107–30–2	500
Chloropicrin	76–06–2	500
Chloropicrin and methyl Bromide mixture	None	1,500
Chloropicrin and methyl Chloride mixture	None	1,500
Commune hydroperoxide	80–15–9	5,000
Cyanogen	460–19–5	2,500
Cyanogen chloride	506–77–4	500
Cyanuric fluoride	675–14–9	100
Diacetyl peroxide (concentration greater than 70%)	110–22–5	5,000
Diazomethane	334–88–3	500
Dibenzoyl peroxide	94–36–0	7,500
Diborane	19287–45–7	100
Dibutyl peroxide (tertiary)	110–05–04	5,000
Dichloro acetylene	7572–29–4	250
Dichlorosilane	4109–96–0	2,500
Diethylzinc	557–20–0	10,000
Diisopropyl peroxydicarbonate	105–64–6	7,500
Dilauroyl peroxide	105–74–8	7,500
Dimethyldichlorosilane	75–78–5	1,000

TABLE 3.9 continued

Chemical Name	CAS[a]	Threshold Quantity[b]
Dimethylhydrazine, 1,1-	57–14–7	1,000
Dimethylamine, anhydrous	124–40–3	2,500
2,4-dinitroaniline	97–02–9	5,000
Ethyl methyl ketone peroxide (also methyl ethyl ketone peroxide; concentration greater than 60%)	1338–23–4	5,000
Ethyl nitrite	109–95–5	5,000
Ethylamine	75–04–7	7,500
Ethylene fluorohydrin	371–62–0	100
Ethylene oxide	75–21–8	5,000
Ethyleneimine	151–56–4	1,000
Fluorine	7782–41–4	1,000
Formaldehyde (formalin)	50–00–0	1,000
Furan	110–00–9	500
Hexafluoroacetone	684–16–2	5,000
Hydrochloric acid, anhydrous	7647–01–0	5,000
Hydrofluoric acid, anhydrous	7664–39–3	1,000
Hydrogen bromide	10035–10–6	5,000
Hydrogen chloride	7647–01–0	5,000
Hydrogen cyanide, anhydrous	74–90–8	1,000
Hydrogen fluoride	7664–39–3	1,000
Hydrogen peroxide (52% by weight or greater)	7722–84–1	7,500
Hydrogen selenide	7783–07–5	150
Hydrogen sulfide	7783–06–4	1,500
Hydroxylamine	7803–49–8	2,500
Iron, pentacarbonyl	13463–40–6	250
Isopropylamine	75–31–0	5,000
Ketene	463–51–4	100
Methacrylaldehyde	78–85–3	1,000
Methacryloyl chloride	920–46–7	150
Methacryloyloxyethyl isocyanate	30674–80–7	100
Methyl acrylonitrile	126–98–7	250
Methylamine, anhydrous	74–89–5	1,000
Methyl bromide	74–83–9	2,500
Methyl chloride	74–87–3	15,000
Methyl chloroformate	79–22–1	500
Methyl ethyl ketone peroxide (concentration greater than 60%)	1338–23–4	5,000
Methyl fluoroacetate	453–18–9	100
Methyl fluorosulfate	421–20–5	100
Methyl hydrazine	60–34–4	100
Methyl iodide	74–88–4	7,500
Methyl isocyanate	624–83–9	250
Methyl mercaptan	74–93–1	5,000
Methyl vinyl ketone	79–84–4	100
Methyltrichlorosilane	75–79–6	500
Nickel carbonly (nickel tetracarbonyl)	13463–39–3	150
Nitric acid (94.5% by weight or greater)	7697–37–2	500
Nitric oxide	10102–43–9	250
Nitroaniline (para)		
Nitroaniline	100–01–6	5,000
Nitromethane	75–52–5	2,500
Nitrogen dioxide	10102–44–0	250
Nitrogen oxides (NO; NO(2); N204; N203)	10102–44–0	250
Nitrogen tetroxide (also called nitrogen peroxide)	10544–72–6	250
Nitrogen trifluoride	7783–54–2	5,000
Nitrogen trioxide	10544–73–7	250

TABLE 3.9   continued

Chemical Name	CAS[a]	Threshold Quantity[b]
Oleum (65–80% by weight; also called fuming sulfuric acid)	8014–94–7	1,000
Osmium tetroxide	20816–12–0	100
Oxygen difluoride (fluorine monoxide)	7783–41–7	100
Ozone	10028–15–6	100
Pentaborane	19624–22–7	100
Peracetic acid (concentration greater 60% acetic acid; also called peroxyacetic acid)	79–21–0	1,000
Perchloric acid (concentration greater than 60% by weight)	7601–90–3	5,000
Perchloromethyl mercaptan	594–42–3	150
Perchloryl fluoride	7616–94–6	5,000
Peroxyacetic acid (concentration greater than 60% acetic acid; also called peracetic acid)	79–21–0	1,000
Phosgene (also called carbonyl chloride)	75–44–5	100
Phosphine (Hydrogen phosphide)	7803–51–2	100
Phosphorus oxychloride (also called phosphoryl chloride)	10025–87–3	1,000
Phosphorus trichloride	7719–12–2	1,000
Phosphoryl chloride (also called phosphorus oxychloride)	10025–87–3	1,000
Propargyl bromide	106–96–7	100
Propyl nitrate	627–3–4	2,500
Sarin	107–44–8	100
Selenium hexafluoride	7783–79–1	1,000
Stibine (antimony hydride)	7803–52–3	500
Sulfur dioxide (liquid)	7446–09–5	1,000
Sulfur pentafluoride	5714–22–7	250
Sulfur tetrafluoride	7783–60–0	250
Sulfur trioxide (also called sulfuric anhydride)	7446–11–9	1,000
Sulfuric anhydride (also called sulfur trioxide)	7446–11–9	1,000
Tellurium hexafluoride	7783–80–4	250
Tetrafluoroethylene	116–14–3	5,000
Tetrafluorohydrazine	10036–47–2	5,000
Tetramethyl lead	75–74–1	1,000
Thionyl chloride	7719–09–7	250
Trichloro (chloromethyl) silane	1558–25–4	100
Trichloro (dichlorophenyl) silane	27137–85–5	2,500
Trichlorosilane	10025–78–2	5,000
Trifluorochloroethylene	79–38–9	10,000
Trimethyoxysilane	2487–90–3	1,500

[a]Chemical abstract service number.
[b]Threshold quality in pounds (amount necessary to be covered by this standard).

Operating Procedures. Employers must develop and implement written operating instructions for safely conducting activities involved in each covered process consistent with the process safety information. The written procedures must address steps for each operating phase, operating limits, safety and health considerations, and safety systems and their functions. Included must be normal operation, startup, shutdown, emergency operations, and other operating parameters.

Training. The proposal requires training for employees involved in covered processes. Initial training requires all employees currently involved in each process, and all employees newly assigned, be trained in an overview of the process and its operating procedures.

Refresher training shall be provided at least every three years, and more often if necessary, to each employee involved in the process. After training, employees must ascertain that workers have received and understood the training.

Contractors. Employers must inform contract employees prior to the initiation of the contractor's work of the known potential fire, explosion, or toxic release hazards related to the contractor's work and the process. Contract employees and host employers must ensure that contract workers are trained in the work practices necessary to perform their jobs safely and are informed of any applicable safety rules of the facility work and the process.

Pre-Startup Safety Review. Employers must perform a pre-startup safety review for new facilities and for modified facilities when the modification is significant enough to require a change in the process safety information. The safety review shall confirm that prior to the introduction of highly hazardous chemicals to a process:

1. Construction and equipment is in accordance with design specifications.
2. Safety, operating, maintenance, and emergency procedures are in place and are adequate.
3. For new facilities, a process hazard analysis has been performed and recommendations have been resolved or implemented before startup.
4. Modified facilities meet the requirements contained in management of change.

Mechanical Integrity. Employers must ensure the initial and on-going integrity of process equipment by determining that the equipment is designed, installed, and maintained properly. The standard requires testing and inspection of equipment, quality assurance checks of equipment, spare parts and maintenance materials, and correction of deficiencies. The following process equipment is targeted in this proposal: pressure vessels and storage tanks; piping systems (including valves); relief and vent

systems and devices; emergency shutdown systems; controls, and pumps.

Hot Work Permit. Employers must have a hot work program in place and issue a permit for all hot work operations conducted on or near a covered process.

Management of Change. Employers must establish and implement written procedures to manage changes (except for "replacements in kind") to process chemicals, technology, equipment, and procedures; and, changes to facilities that affect a covered process. Employees involved in operating a process and maintenance and contract employees whose tasks will be affected by a change in the process shall be informed of, and trained in, the change prior to startup of the process or affected part of the process. The procedures shall ensure that the necessary time period for the change and authorization requirements for the proposed change are addressed.

Incident Investigation. Employers must investigate each incident that resulted in, or could reasonably have resulted in a catastrophic release of highly hazardous chemical in the workplace. An incident investigation shall be initiated as promptly as possible, but not later than 48 hr following the incident. A report shall be prepared at the conclusion of the investigation.

Although not stressed by the regulations, the objective of the incident investigation should be the development and implementation of recommendations to ensure the incident is not repeated. This objective should apply not only to the process involved, but also to all similar situations having the same potential. In major incidents, the Chemical Safety Board's investigation of reports serves as a vehicle to communicate to a much broader audience than the organizations that had the incident.

Emergency Planning and Response. Employers must establish and implement an emergency action plan for the entire plant in accordance with the provisions of OSHA's emergency action plan to meet the minimum requirements for emergency planning. This is the only element of PSM that must be carried out beyond the boundaries of a covered process.

Compliance Audits. Employers must certify that they have evaluated compliance with the provisions of this section at least every three years to verify that procedures and practices developed under the standard are adequate and are being followed. The compliance audit shall be conducted by at least one person knowledgeable in the process. The employer shall determine and document an appropriate response to each of the findings of the compliance audit, and document that deficiencies have been addressed.

Trade Secrets. Employers must make all information necessary to comply with the requirements of this section available to those persons responsible for compiling the process safety information, developing process hazard analyses, developing the operating procedures, those involved in incident investigations, emergency planning, and response and compliance audits without regard to possible trade secret status of such information. Nothing in this paragraph shall preclude the employer from requiring the persons to whom the information is made available to enter into confidentiality agreements not to disclose the information.

The above elements outline the programs required by PSM. These programs are performance-type standards. They spell out programs and choices and are not limited to specific details. These elements have served to organize and guide the process safety programs of all who are covered by it. They have served to bring direction to training and publications involving process safety. The AIChE's Center for Chemical Process Safety has publications and training programs to support most of these elements.

Risk Management Plans (RMPs)

The EPA is charged primarily with the responsibility to protect the public and the environment. One could regard this as "outside the fence line." RMPs are required by the Environmental Protection Agency (EPA). Since protecting the public and the environment is mainly an environmental issue rather than a safety issue, this subject will be covered only briefly in this section.

Congress enacted Section 112(r) of the 1990 Clean Air Act (CAA) to address the threat of catastrophic releases of chemicals that might cause immediate deaths or injuries in communities. It requires owners and operators of covered facilities to submit RMPs to the EPA. The final RMP rule was published in 40 CFR 68 in the Federal Register on June 20, 1996. RMPs must summarize the potential threat of sudden, large releases of certain dangerous chemicals and facilities' plans to prevent such releases and mitigate any damage.

Operators of facilities that are subject to the EPA's RMP must perform offsite consequence analyses to determine whether accidental releases from their processes could put nearby populations at risk. In performing a consequence analysis it is assumed that all or part of a hazardous substance escapes from a process at a given facility. It is then estimated how far downwind hazardous gas concentrations may extend.

Facilities that must prepare and submit RMPs must estimate the offsite consequences of accidental releases. This can be done using tables (such as those provided in CAA 112(r) Offsite Consequence Analysis) or a computerized model. There are a number of commercially available computer models. Submitters are expected to choose a tool that is appropriate for their facility.

The owners and operators of stationary sources producing, processing, handling, or storing of extremely hazardous substances have a general duty to identify hazards that may result from an accidental release This includes agents that may or may not be identified by any government agency which may cause death, injury, or property damage. In other words, just because a substance is not listed is not an excuse to fail to consider its hazards.

This section with its emphasis on Process Safety does not cover the considerable other safety, design, and operating requirements of other chemical-related regulations. Many of these requirements also include national codes as guidelines or as adopted regulations. Examples of these are in the American Society of Mechanical Engineers (ASME) *2001 Boiler Pressure Vessel Code*, the National Fire Protection Association (NFPA) which covers a wide range of fire safety issues and the American Petroleum Institute (API) *Recommended Practice 520, Sizing, Selection, and Installation of Pressure Relieving Devices in Refineries.*

An extremely hazardous substance is any agent that may or may not be listed by any government agency which, as the result of short-term exposures associated with releases to the air, cause death, injury, or property damage due to its toxicity, reactivity, flammability, volatility, or corrosivity.

Toxics Release Inventory

Two statutes, the Emergency Planning and Community Right-to-Know Act (EPCRA) and section 6607 of the Pollution Prevention Act (PPA), mandate that a publicly accessible toxic chemical database be developed and maintained by the U.S. EPA. This database, known as the Toxics Release Inventory (TRI), contains information concerning waste management activities and the release of toxic chemicals by facilities that manufacture, process, or otherwise use these materials. The TRI of 1999 is a publicly available database containing information on toxic chemical releases and other waste management activities that are reported annually by manufacturing facilities and facilities in certain other industry sectors, as well as by federal facilities. The TRI program is now under the EPA's Office of Environmental Information. This inventory was established under the EPCRA of 1986 which was enacted to promote emergency planning, to minimize the effects of chemical accidents, and to provide the public. As of November 2001, there were 667 toxic chemicals and chemical compounds on the list.

HAZWOPER

The Hazardous Waste Operations and Emergency Response Standard (HAZWOPER), 29 CFR Part 1910.120, applies to five distinct groups of employers and their employees. This includes any employees who are exposed or potentially exposed to hazardous substances—including hazardous waste—and who are engaged in one of the following operations as specified by 1910.120:

- Clean-up operations
- Corrective actions
- Voluntary clean-up operations
- Operations involving hazardous wastes
- Emergency response operations for releases of, or substantial threats of release of, hazardous substances regardless of the location of the hazard.

In addition, with the passage of the Pollution Prevention Act (PPA) in 1991, facilities must report other waste management amounts including the quantities of TRI chemicals recycled, combusted for energy recovery, and treated on- and offsite.

More Information

For more information on Regulations, the books, magazine articles, and Internet references in the Reference section can be very helpful. Following the requirements of the many aspects of Regulations can be quite complicated and involve a lot of detail. There is a considerable amount of good assistance available which help can make the subject manageable.

THE PRINCIPAL REASON FOR MOST CHEMICAL PROCESS ACCIDENTS

Ask any group of people experienced in chemical plant operations what causes most chemical process accidents, and you will get a variety of answers including: operator error, equipment failure, poor design, act of God, and bad luck. However, in the opinion of representatives of many of the large chemical and oil companies in the United States, these answers are generally incorrect. The Center for Chemical Process Safety, an organization sponsored by the American Institute of Chemical Engineers, includes representatives of many of the largest chemical and oil companies in the United States, and states that "It is an axiom that process safety incidents are the result of *management system failure.*" Invariably, some aspect of a process safety management system can be found that, had it functioned properly, could have prevented an incident (or reduced the seriousness of it). "It is a rare situation where an 'Act of God' or other uncontrollable event is the sole cause of an incident. Much more common is the situation where an incident is the result of multiple causes, including management system failures. Therefore, it is more appropriate to presume that management system failures underlie every incident so that we may act to uncover such failures and then modify the appropriate management systems, rather than presume that if an 'Act of God' appears to be the immediate cause, investigation should cease because there is nothing that can be done to prevent such future incidents."[24]

For example, consider a case where a small amount of hazardous material is spilled while a sample is being taken from a process line. It is not enough to look into the situation and conclude that this is an example of an operator error where procedures were not followed, and then simply to recommend that the employee be instructed to follow procedures in the future. Further investigation may reveal deficiencies in the training system or in the equipment. Still more investigation may reveal deficiencies in the management system that plans resources for training or that provides for proper equipment for sampling. It then may be appropriate to change the management system to prevent repetition of the incident.

Levels of Causes

There are several levels of causes of accidents, usually (1) the immediate cause, (2) contributing causes to the accident or to the severity of the accident, and (3) the "root cause." The root cause is what really caused the accident, and when this is determined, it may be possibly to *prevent* future similar accidents. With the 20–20 hindsight that is available after an accident, the root cause usually can be found. The purpose of the discussion in the next section is to illustrate how knowledge about the root causes of some important accidents can help to keep them from happening again. It will be noted that the root cause is rarely the fault of one person, but instead is the result of a management system that does not function properly.

Following are brief analyses of several case histories that have been of landmark importance in the industrial world, and that have affected the chemical industry all over the world.

CASE HISTORIES

Flixborough, England 1974[2]

On June 1, 1974, an accident occurred in the Nypro plant in Flixborough, England, in a process where cyclohexane was oxidized to cyclohexanone for the manufacture of caprolactam, the basic raw material for the production of Nylon 6. The process consisted of six reactors in series at 155°C and 8.8 bar (130 psig) containing a total of 120 tons of cyclohexane and a small amount of cyclohexanone. The final reactor in the process contained 94 percent cyclohexane. There was a massive leak followed by a large unconfined vapor cloud explosion and fire that killed 26 people, injured 36 people, destroyed 1,821 houses, and damaged 167 shops. It was estimated that 30 tons of cyclohexane was involved in the explosion. The accident occurred on Saturday; on a working day, casualties would have been much higher.

The accident happened when the plant had to replace one of six reactors and rushed to refit the plant to bypass the disabled reactor. Scaffolding was jerry-rigged to support a 20-in. pipe connecting reactor four with reactor six, which violated industry and the manufacturer's recommendations. The reactor that failed showed stress crack corrosion. The only drawings for the repair were in chalk on the machine shop floor. Both ends of the 20-in.

pipe had expansion joints where they attached to the reactors. The pipe was supported on scaffolding-type supports and was offset with a "dog-leg" to fit the reactors, which were at different levels to promote gravity flow. The safety reviews, if any, were insufficient.

Immediate Cause. A pipe replacing a failed reactor failed, releasing large quantities of hot cyclohexane forming a vapor cloud that ignited.

Contributing Causes to the Accident and the Severity of the Accident:

1. The reactor failed without an adequate check on why (metallurgical failure).
2. The pipe was connected without an adequate check on its strength and on inadequate supports.
3. Expansion joints (bellows) were used on each end of pipe in a "dog-leg" without adequate support, contrary to the recommendations of the manufacturer.
4. There was a large inventory of hot cyclohexane under pressure.
5. The accident occurred during startup.
6. The control room was not built with adequate strength, and was poorly sited.
7. The previous works engineer had left and had not been replaced. According to the Flixborough Report, "There was no mechanical engineer on site of sufficient qualification, status or authority to deal with complex and novel engineering problems and insist on necessary measures being taken."
8. The plant did not have a sufficient complement of experienced people, and individuals tended to be overworked and liable to error.

Root Cause. Management systems deficiencies resulted in:

1. A lack of experienced and qualified people
2. Inadequate procedures involving plant modifications
3. Regulations on pressure vessels that dealt mainly with steam and air and did not adequately address hazardous materials

4. A process with a very large amount of hot hydrocarbons under pressure and well above its flash point installed in an area that could expose many people to a severe hazard

This accident resulted in significant changes in England and the rest of the world in the manner in which chemical process safety is managed by industry and government. One of the conclusions reached as a result of this accident, which has had a wide effect in the chemical industry, is that "limitations of inventory (or flammable materials) should be taken as specific design objectives in major hazard installations."

The use of expansion joints (bellows, in this case) which were improperly installed may have been a principal reason for the accident. This provides additional reasons not to use expansion joints (except in special exceptional circumstances).

Bhopal, 1985 (C&EN Feb. 11, 1985; Technica 1989[54])

On December 3 and 4, 1985, a chemical release causing a massive toxic gas cloud occurred at the Union Carbide India, Ltd, plant in Bhopal, India. (Union Carbide is now a part of The Dow Chemical Company.) The process involved used methyl isocyanate (MIC), an extremely toxic chemical, to make Sevin, a pesticide. According to various authoritative reports, about 1,700–2,700 (possibly more) people were killed, 50,000 people were affected seriously, and 1,000,000 people were affected in some way. The final settlement may involve billions of dollars. It was one of the worst industrial accidents in history. The accident occurred when about 120–240 gallons of water were allowed to contaminate an MIC storage tank. The MIC hydrolyzed, causing heat and pressure, which in turn caused the tank rupture disk to burst.

Equipment designed to handle an MIC release included a recirculating caustic soda scrubber tower and a flare system designed for 10,000 lb/hr, which would be moderate flows from process vents. It was not designed to handle runaway reactions from storage. The design was based on the assumption that full cooling would be provided by the refrigeration system. The actual release was estimated to be 27,000 lb over 2 hr, with the tank at 43°C. At the time of the release the refrigeration had been turned off. The flare tower was shut down for repairs. A system of pressurized sprinklers that was supposed to form a water curtain over the escaping gases was deficient, in that water pressure was too low for water to reach the height of the escaping gas.

There have been conflicting stories of how the water got into the tank, including operator error, contamination, and sabotage.

Immediate Cause. The immediate cause was hydrolysis of MIC due to water contamination. The exact source of the water has not been determined.

Contributing Causes.

1. Flare tower was shut down for repair.
2. Scrubber was inadequate to handle a large release.
3. Chilling system was turned off. (It also was too small.)
4. MIC tank was not equipped with adequate instrumentation.
5. Operating personnel lacked knowledge and training.
6. The inventory of MIC was large.
7. There was a lack of automatic devices and warning systems; it has been reported that safety systems had to be turned on manually.
8. When the plant was built, over 20 years before the accident, there were very few people near it. At the time of the accident, a shanty town had grown up near the plant with a density of 100 people per acre, greatly increasing the potential exposure of people to toxic releases. There was no emergency action plan to notify neighbors of the potential for toxic releases or of what to do if there was a release, nor was there a functioning alarm system.

Root Cause. The root cause of the accident appears to be a management system that did not adequately respond to the potential hazards of MIC. There was probably a greater inventory of MIC than was needed. The main process expertise was in the United States. Local management does not appear to have understood the process or the consequences of changes made. This includes plant design, maintenance and operations, backup systems, and community responsibility. (Union Carbide has provided legal arguments alleging that sabotage caused the release; there appears to be enough blame to go around for all those involved in any way in the plant, including government units.

This accident has become widely known. It is an objective of many chemical process safety programs and government actions to "avoid another Bhopal"—that is, to avoid a severe release of toxic chemicals (usually referring to toxic chemicals in the air). Almost every chemical company in the world has been affected by this incident in one way or another, in the design and operation of chemical plants, in community action programs, and in the activities of such organizations as the American Institute of Chemical Engineers, the Chemical Manufacturers Association, and many governmental units.

Phillips Explosion, 1989[57]

On October 23, 1989, at approximately 1300, an explosion and fire ripped through the Phillips 66 Company's Houston Chemical Complex in Pasadena, Texas. At the site, 23 workers were killed, and more than 130 were injured. Property damage was nearly $750 million. Business interruption cost is not available but is probably a very large figure.

The release occurred during maintenance operations on a polyethylene reactor. Two of the six workers on the maintenance crews in the immediate vicinity of the reactor leg where the release occurred were killed, together with 21 other employees of the facility. Debris from the plant was found six miles from the explosion site. Structural steel beams were twisted like pretzels by the extreme heat. Two polyethylene production plants covering an area of 16 acres were completely destroyed.

The Phillips complex produces high-density polyethylene, which is used to make milk bottles and other containers. Prior to the accident, the facility produced approximately 1.5 billion pounds of the material per year. It employed 905 company employees and approximately 600 daily contract employees. The contract employees were engaged primarily in regular maintenance activities and new plant construction.

The accident resulted from a release of extremely flammable process gases that occurred during regular maintenance operations on one of the plant's polyethylene reactors. It is estimated that within 90–120 s more than 85,000 lb of flammable gases were released through an open valve. A huge flammable vapor cloud was formed that came into contact with an ignition source and exploded with the energy of 4800 lb of TNT. The initial explosion was equivalent to an earthquake with a magnitude of 3.5 on the Richter scale. A second explosion occurred 10–15 min later when two isobutane tanks exploded. Each explosion damaged other units, creating a chain reaction of explosions. One witness reported hearing ten separate explosions over a 2-hr period.

In the process used by Phillips at this site to produce high-density polyethylene, ethylene gas is dissolved in isobutane and, with various other chemicals added, is reacted in long pipes under elevated pressure and temperature. The dissolved ethylene reacts with itself to form polyethylene particles that gradually come to rest in settling legs, where they are eventually removed through valves at the bottom. At the top of the legs there is a single ball valve (DEMCO® brand) where the legs join with other reactor pipes. The DEMCO valve is kept open during production so that the polyethylene particles can settle into the leg. A typical piping settling leg arrangement is shown in Fig. 3.9.

In the Phillips reactor, the plastic material frequently clogged the settling legs. When this happened, the DEMCO valve for the blocked leg was closed, the leg disassembled, and the block removed. During this particular maintenance process, the reactor settling leg was disassembled and the block of polymer removed. While this maintenance process was going on, the reaction continued, and the product settled in the legs that remained in place. If the DEMCO valve were to open during a cleaning-out operation, there would be nothing to prevent the escape of the gas to the atmosphere.

After the explosion it was found that the DEMCO valve was open at the time of the release. The leg to be cleaned had been prepared by a Phillips operator. The air hoses that operated the DEMCO valve were improperly connected in a reversed position such that a closed DEMCO valve would be opened when the actuator was in the closed position. In addition, the following unsafe conditions existed:

1. The DEMCO valve did not have its lock-out device in place.
2. The hoses supplied to the valve actuator mechanism could be connected at any time even though the Phillips operating procedure stipulated that the hoses should never be connected during maintenance.
3. The air hoses connecting the open and closed sides of the valve were identical, thus allowing the hoses to be cross-connected and permitting the valve to be opened when the operator intended to close it.
4. The air supply valves for the actuator mechanism air hoses were in the open position so that air would flow and cause the actuator to rotate the DEMCO valve when the hoses were connected.
5. The DEMCO valve was capable of being physically locked in the open position as well as in the closed position. The valve lockout system was inadequate to prevent someone from inadvertently opening the DEMCO valve during a maintenance procedure.

Established Phillips corporate safety procedures and standard industry practice require backup protection in the form of a double valve or blind flange insert whenever a process

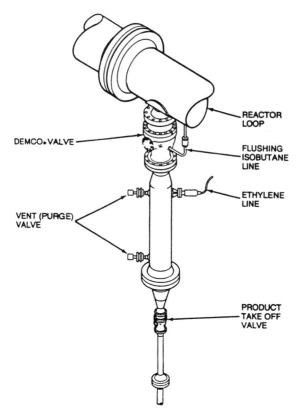

Fig. 3.9. Typical piping settling leg arrangement.

or chemical line in hydrocarbon service is opened. According to OSHA, Phillips had implemented a special procedure for this maintenance operation that did not incorporate the required backup. Consequently, none was used on October 23.

The consequences of the accident were exacerbated by the lack of a water system dedicated to fire fighting, and by deficiencies in the shared system. When the process water system was extensively damaged by the explosion, the plant's water supply for fighting fires was also disrupted. The water pressure was inadequate for fire fighting. The force of the explosion ruptured waterlines and adjacent vessels containing flammable and combustible materials. The ruptured water lines could not be isolated to restore water pressure because the valves to do so were engulfed in flames. Of the three backup diesel pumps, one had been taken out of service and was unavailable, and another soon ran

out of fuel. It was necessary to lay hoses to remote sites—settling ponds, a cooling tower, a water treatment plant, and so on. Electric cables supplying power to regular service pumps were damaged by fire, and those pumps were rendered inoperable. Even so, the fire was brought under control within 10 hr.

In the months preceding the explosion, according to testimony, there had been several small fires, and the alarm had sounded as many as four or five times a day. There had been a fatality at the same plant doing a similar operation about three months before this incident. Some of the employees in the area where the release occurred may not have heard the siren because of the ambient noise level, and may not have known of the impending disaster. Employees in the immediate area of the release began running as soon as they realized the gas was escaping.

The large number of fatalities was due in part to the inadequate separation between buildings in the complex. The site layout and the proximity of normally high-occupancy structures, such as the control and finishing building, to large-capacity reactors and hydrocarbon vessels contributed to the severity of the event.

The distances between process equipment were in violation of accepted engineering practices and did not allow personnel to leave the polyethylene plants safely during the initial vapor release; nor was there sufficient separation between reactors and the control room to carry out emergency shutdown procedures. The control room, in fact, was destroyed by the initial explosion. Of the 22 victims' bodies recovered at the scene, all were located within 250 ft of the vapor release point.

OSHA's investigation revealed that a number of company audits had identified unsafe conditions but largely had been ignored. Thus, a citation for willful violations of the OSHA "general duty" clause was issued to Phillips with proposed penalties of $5,660,000. In addition, proposed penalties of $6,200 were issued for other serious violations. A citation for willful violations with proposed penalties of $724,000 was issued to Fish Engineering and Construction, a Phillips maintenance contractor. Other financial penalties have been proposed. In the investigation it became apparent that Fish had become accustomed to tolerating safety and health violations at the site by its personnel and Phillips personnel, as well as participating in those violations by knowing about them and not taking direct positive action to protect its employees.

Since 1972, OSHA has conducted 92 inspections in the Dallas region at various Phillips locations; 24 were in response to a fatality or a serious accident. OSHA determined that Phillips had not acted upon reports by its own safety personnel and outside consultants who had pointed out unsafe conditions. OSHA also had conducted 44 inspections of the Fish Company, seven of them in response to a fatality or a serious accident.

One of the major findings by OSHA was that Phillips had not conducted a process hazard analysis or equivalent (such as HAZOP) in its polyethylene plants.

Immediate Cause. There was a release of flammable process gases during the unplugging of Number 4 Reactor Leg on Reactor 6 while undergoing a regular maintenance procedure by contractor employees. The unconfined flammable vapor cloud was ignited and exploded with devastating results.

The immediate cause of the leak was that a process valve was opened by mistake while the line was open. The valve was open to the atmosphere without a second line of defense such as another valve or a blind flange.

Contributing Causes to the Accident and the Severity of the Accident.

1. Procedures to require backup protection in the form of a double valve or a blind flange insert were not used. The lockout system was inadequate.
2. Air hoses were improperly connected in the reversed position.
3. The air hoses for the open and closed side of the valve were identical, allowing the hoses to be cross-connected.
4. The DEMCO valve actuator mechanism did not have its lockout device in place.
5. There was not a water system dedicated to fire fighting, and there were deficiencies in the system shared with the process.
6. The site layout and proximity of high-occupancy structures contributed to the severity.
7. There was inadequate separation of buildings within the complex. Especially, there was inadequate spacing between the reactors and the control room.

Root Causes. The root causes of the accident and its extreme severity appear to be failures of the management system, as shown by the following:[57]

1. According to OSHA, Phillips had not conducted a process hazard analysis or equivalent (such as HAZOP) in its polyethylene plants.

2. It was found by OSHA that the contractor, Fish Engineering, had a history of serious and willful violations of safety standard, which Phillips had not acted upon. The same contractor also had been involved in a fatal accident at the same facility three months earlier.
3. A report by OSHA stated that Phillips had not acted upon reports issued previously by the company's own safety personnel and outside consultants. Phillips had numerous citations from OSHA since July 1972. OSHA discovered internal Phillips documents that called for corrective action but which were largely ignored.
4. Safe operating procedures were not required for opening lines in hazardous service.
5. An effective safety permit system was not enforced with Phillips or contractor employees, especially line opening and hot work permits.
6. Buildings containing personnel were not separated from process units in accordance with accepted engineering principles, or designed with enough resistance to fire and explosion.
7. The fire protection system was not maintained in a state of readiness:
   (a) One of the three diesel-powered water pumps had been taken out of service.
   (b) Another of the three diesel-powered water pumps was not fully fueled, and it ran out of fuel during the fire fighting.
   (c) Electric cables supplying power to regular service fire pumps were not located underground and were exposed to blast and fire damage.

SUMMARY

As the tragic case histories unfold, the significance of the process safety consideration presented in this chapter becomes chillingly apparent, and the necessity for inherently safe process design is revealed. The case histories also reveal significant flaws in management systems, which tends to be true for most chemical process accidents. Even an ideally safe process can be transformed into one with a high potential for disaster if a valid system is not in place to ensure that the inherently safe process design retains its integrity.

Because there is always risk when equipment, instrumentation, and human activity are involved, there is no method of making a plant completely safe. However, facilities can be made "inherently safer" by careful examination of all aspects of design and management, using modern techniques that are now available. If we are to improve our process safety performance and our public image, "inherently safe" process design coupled with "inherently safe" process management is imperative.

In addition to the information presented in this chapter and in the publications it has cited, references 58–72 are recommended as appropriate source material.

REFERENCES

1. M & M Protection Consultants, *One Hundred Largest Losses: A Thirty Year Review of Property Damage in the Hydrocarbon-Chemical Industry*, 12th ed., 222 S. Riverside Plaza, Chicago, IL 60606, 1989.
2. Lees, F. P., *Loss Prevention in the Process Industries*, London, Butterworths, 1980.
3. Englund, S. M., "Opportunities in the Design of Inherently Safer Chemical Plants," in *Advances in Chemical Engineering*, J. Wei et al. (Eds.), Academic Press, New York, 1990.
4. Kletz, T, *Cheaper, Safer Plants or Wealth and Safety at Work*, Institution of Chemical Engineers, Rugby, England, 1985.
5. Schaller, L. C., Du Pont, Wilmington, DE 19898, *Plant Operations Progress*, No. 1 (Jan. 1990).
6. Howard, W., Consultant, Monsanto Chemical Company, Personal Communication, 1981.
7. *NFPA 68*, National Fire Protection Association, Batterymarch Park, Quincy, MA, 1998.
8. Bartknecht, W., *Explosions Course Prevention Protection*, p. 108, Springer-Verlag, Berlin, 1981.
9. *NFPA 69* National Fire Protection Association, Batterymarch Park, Quincy, MA, 1986.

10. Liening, G., "Prolonging Service Life of Metals and Alloys in Chemical Processing Plants," *Chem. Proc.*, 22 (Sept. 1986).

11. Liening, G., The Dow Chemical Co., Midland, MI, personal communication (Sept. 5, 1986).

12. *Perry's Chemical Engineers' Handbook*, 6th ed., pp. 23–48, McGraw-Hill, New York, 1984.

13. Grinwis, D., Process Systems Associate; Larsen, Paul, Associate Instrument Engineering Consultant; and Schrock, Luther, Process Consultant, The Dow Chemical Co., Midland, MI and Freeport, TX, personal-communication (Oct. 1986).

14. Fisher, H., *Chem. Eng. Prog.*, 3 (Aug. 1985).

15. Huff, J. E., "A General Approach to the Sizing of Emergency Pressure Relief Systems," *Reprints of Int. Symp. on Loss Prev. and Safety Promotion in the Process Ind.*, Heidelberg, Germany, 1977, p. IV 223, DECHEMA, Frankfurt (1977).

16. Huff, J. E., *Institute of Chemical Engineers Symposium Series, No. 85*, p. 109 (1984).

17. Huff, J. E., "Emergency Venting Requirements for Gassy Reactions from Closed System Tests," *Plant/Operations Progress*, 3(1), 50–59 (Jan. 1984).

18. Huff, J. E., "The Role of Pressure Relief in Reactive Chemical Safety," International Symposium on Preventing Major Chemical Accidents, Washington, DC, sponsored by the Center for Chemical Process Safety of the American Institute of Chemical Engineers, The United States Environmental Protection Agency, and the World Bank, (Feb. 3, 1987).

19. Fauske, Hans K., and Leung, J., "New Experimental Technique for Characterizing Runaway Chemical Reactions," *Chem. Eng. Progr.*, 39 (Aug. 1985).

20. Jackson, B. L., Piping Specialist, The Dow Chemical Co., Midland, MI, personal communication (Sept. 17, 1986).

21. Alexander, S., and King, R., Materials Engineering Dept., Michigan Div., The Dow Chemical Co., Midland MI, personal communication (Feb. 14, 1988).

22. Cromie, J., Engineering Associate, The Dow Chemical Co., Midland, Michigan, personal communication (1986).

23. Reynolds, J. A., Union Carbide Corp., "Canned Motor and Magnetic Drive Pumps," *Chem. Proc.*, 71–75 (Nov. 1989).

24. *Technical Management of Chemical Process Safety*, Center for Chemical Process Safety, American Institute of Chemical Engineers, 345 E. 47th St., New York, NY, 1989.

25. Cloud, M. J., "Fire, the Most Tolerable Third party," *Mich. Nat. Res.*, 18 (May–June, 1990).

26. *Fire Safety Data*, Fire Protection Association, 140 Aldersgate St., London EC1A 4HX, 1988.

27. Bodurtha, F. T., Engineering Dept., Du Pont, *Industrial Explosion Prevention and Protection*, McGraw-Hill, New York, 1980.

28. Zabetakis, M. G., "Flammability Characteristics of Combustible Gases and Vapors," Bulletin 627, U.S. Dept. of the Interior, Bureau of Mines, Washington, DC (1965).

29. Cawse, J. N., Pesetsky, B., and Vyn, W. T., "The Liquid Phase Decomposition of Ethylene Oxide," Union Carbide Corporation, Technical Center, South Charleston, WV 25303 (no date available).

30. Stull, D. R., *Fundamentals of Fire and Explosion*, p. 50, Corporate Safety and Loss Dept., The Dow Chemical Co., Midland, MI, American Institute of Chemical Engineers, New York, 1976.

31. Kohlbrand, "Case History of a Deflagration Involving an Organic Solvent/Oxygen System below Its Flash Point," 24th Annual Loss Prevention Symposium, Sponsored by the American Institute of Chemical Engineers, San Diego, CA (Aug. 19–22, 1990).

32. Mackenzie, J., "Hydrogen Peroxide without Accidents," *Chem. Eng.*, 84ff (June 1990).

33. Eichel, F. G., "Electrostatics," *Chem. Eng.*, 154–167 (Mar. 13, 1967).

34. Boundy, R. H., and Boyer, R. F., *Styrene*, American Chemical Society Monograph Series, p. 63, Reinhold Publishing Co., New York, 1952.

35. *NFPA 77*, National Fire Protection Association, Batterymarch Park, Quincy, MA, 1986.

36. Bodurtha, F. T., "Industrial Explosion Control Course," Center for Professional Advancement, Chicago, IL (Sept. 14–16, 1987).

37. Webley, P., Director, Massachusetts Institute of Technology Practice School, Midland Station, Midland, MI, personal communication (May 1990).

38. *CPQRA (Chemical Process Quantitative Risk Analysis)*, tables 2.12, 2.13, p. 161, 165, Center for Chemical Process Safety of the American Institute of Chemical Engineers, 1989.

39. TNO, "Methods for the Calculation of the Physical Effects of the Escape of Dangerous Materials: Liquids and Gases" ("The Yellow Book"), Apeldoorn, The Netherlands, 1979.

40. API RP 521, 2nd ed., American Petroleum Institute, Washington, DC, 1982.

41. Walters, S., "The Beginnings," *Mech. Eng.*, 4, 38–46 (1984).

42. Prugh, R. W., "Quantitative Evaluation of BLEVE Hazards," AICHE Loss Prevention Symposium, Paper No. 74e, AICHE Spring National Meeting, New Orleans, LA (Mar. 6–10, 1988).

43. Moore, C. V., "The Design of Barricades for Hazardous Pressure Systems," *Nucl. Eng. Des.*, 5, 1550–1566 (1967).

44. Clancey, V. J., "Diagnostic Features of Explosion Damage," Sixth Int. Meeting of Forensic Sciences, Edinburgh, 1972.
45. Baker, W. E., Cox, P. A., Westine, P. S., Kulesz, J. J., and Strehlow, R. A., *Explosion Hazards and Evaluation*, Elsevier, New York, 1983.
46. Ludwig, E. E., *Applied Process Design for Chemical and Petrochemical Plants*, 2nd ed., Vol. 1, Gulf Publishing, Houston, TX, 1977.
47. Brasie, W. C., Michigan Division, Process Engineering, The Dow Chemical Co., Midland, MI, personal communication (Mar. 9, 1983).
48. Brasie, W. C., Michigan Division, Process Engineering, The Dow Chemical Co., Midland, MI, personal communication (Oct. 6, 1982).
49. Allen, W. T., Process Engineering, The Dow Chemical Co., Midland, MI, personal communication (May 1988).
50. AIChE. *Guidelines for Safe Storage and Handling of High Toxic Hazard Materials*, Center for Chemical Process Safety, American Institute of Chemical Engineers, 345 E. 47th St., New York, NY, 1988.
51. Knowlton, R. E., *Hazard and Operability Studies*, Chemetics International Co., Ltd., Vancouver, BC, Canada V6J, 1C7 (Feb. 1989).
52. NUS Corp., HAZOP Study Team Training Manual. *Predictive Hazard Identification Techniques for Dow Corning Facilities*, Gaithersburg, MD (July 1989).
53. Technica Consulting Scientists and Engineers, London, England, 1990.
54. Technica, Inc., "HAZOP Leaders Course," 1989, Columbus, OH, Nov. 6–10; course leaders David Slater and Frederick Dyke.
55. Arendt, J. S., Lorenzo, A. F., and Lorenzo, D. K., "Evaluating Process Safety in the Chemical Industry," *A Manager's Guide to Quantitative Risk Assessment*, Chemical Manufacturers Association, DC, 1989.
56. Burk, Art., Principal Safety Consultant. Du Pont, Newark, DE, personal communication (Feb. 20, 1990).
57. OSHA (Occupational Safety and Health Administration), U.S. Department of Labor, *The Phillips Company Houston Chemical Complex Explosion and Fire* (Apr. 1990).
58. Bartknecht, W., *Dust Explosions Course, Prevention, Protection*. Springer-Verlag, Berlin, 1989.
59. Beveridge, H. J. R., and Jones, C. G., "Shock Effects of a Bursting-Disk in a Relief Manifold," Institution of Chemical Engineers Symposium, Series No. 85.
60. Burgess, D., and Zabetakis, M. G., "Fire and Explosion Hazards Associated with Liquefied Natural Gas," U.S. Bureau of Mines, Report RI 6099 (1962).
61. Burk, A., Principal Safety Consultant. Du Pont, Newark. DE, presentation on "Process Hazards Analysis and Quantitative Risk Assessment" (July 20, 1989).
62. "CEFIC Views on the Quantitative Assessment of Risks From Installations in the Chemical Industry," European Council of Chemical Manufacturers' Federations, Bruxelles (Apr. 1986).
63. *Condensed Chemical Dictionary*, Van Nostrand Reinhold Co., New York, 1983.
64. *Guidelines for Vapor Release Mitigation*, Center for Process Safety, American Institute of Chemical Engineers, 345 E. 47th St., New York, NY (Sept. 1987).
65. Hymes, I., "The Physiological and Pathological Effects of Thermal Radiation," UKAEA Safety and Reliability Directorate, Report SRD R275, Culcheth, UK.
66. Klein, H. H., "Analysis of DIERs Venting Tests: Validation of a Tool for Sizing Emergency Relief Systems for Runaway Chemical Reactions," *Plant Op. Progr.*, **5**(1), 1–10 (Jan. 1986).
67. Nazario, F. N., Exxon Research and Engineering Co., *Chem. Eng.*, pp. 102–109 (Aug. 15, 1988).
68. Pieterson, C. M., and Huaerta, S. C., "Analysis of the LPG Incident in San Juan Ixhatepec, Mexico City, 19 Nov. 1984," TNO Report B4-0222, P.O. Box 342 7300 AH, Apeldoorn, The Netherlands (1985).
69. Roberts, A. F., "Thermal Radiation Hazards from Releases of LPG from Pressurized Storage," *Fire Safety J.*, **4**(3), 197 (1981).
70. Townsend, D. I., and Tou, J. C., "Thermal Hazard Evaluation by an Accelerating Rate Calorimeter," *Thermochimica Acta*, **37**, 30 (1980).
71. Welty, J. R., *Engineering Heat Transfer*, SI Version. John Wiley & Sons, New York, 1978.
72. Wensley, J. H., "Improved Alarm Management through Use of Fault Tolerant Digital Systems," Instrument Society of America, International Conference and Exhibit, Houston, TX (Oct. 13–16, 1986).

ADDITIONAL READING REFERENCES

American Petroleum Institute, *API Recommended Practice 520, Sizing, Selection, and Installation of Pressure Relieving Devices in Refineries, Part I, Sizing and Selection*, American Petroleum Institute, 1995–2000.
American Society of Mechanical Engineers, *2001 Boiler Pressure Vessel Code*, ASME International, 22 Law Drive, Fairfield, NJ 07007-2900, USA, 2001.

Bartknecht, W., *Explosions Course: Prevention, Protection*, Springer-Verlag, 1993.

Bretherick, L., *Handbook of Reactive Chemical Hazards*, 5th ed., Butterworths, London, 1995.

Crowl, D., and Bollinger, R., *Inherently Safer Chemical Processes: A Life Cycle Approach* (Center for Chemical Process Safety (CCPS)). Am. Inst of Chemical, Engineers, 1997.

DIERS (Design Institute for Emergency Relief Systems), American Institute of Chemical Engineers, 3 Park Ave, New York, N.Y., 10016-5991, U.S.A, http://www.diers.net/

Englund, S. M., "Design and Operate Plants for Inherent Safety," *Chemical Engineering Progress*, 85–91 (Part 1) (Mar. 1991), and 79–86 (Part 2) (May 1991).

Englund, S. M., "Process and Design Options for Inherently Safer Plants," in *Prevention and Control of Accidental Releases of Hazardous Gases*, Van Nostrand, 1993.

Englund, S. M., "Chemical Process Safety," in *Perry's Chemical Engineers' Handbook*, 7th ed., D. W. Green (Ed.), published in McGraw-Hill, New York, 1997.

Hendershot, D., *Smaller Is Safer—Simplifying Chemical Plant Safety*, Safe Workplace, National Council on Compensation Insurance, 750 Park of Commerce Drive, Boca Raton, FL 33487, 2000. Hendershot is a senior technical fellow in the Process Hazard Assessment Department of the Rohm and Haas Company, Bristol, PA, and has written extensively on Process Safety.

Hendershot, D., "Chemistry—the Key to Inherently Safer Manufacturing Processes," Presented before the Division of Environmental Chemistry, American Chemical Society, Washington, D.C., August 21, 1994.

Kletz, T., *What Went Wrong?: Case Histories of Process Plant Disasters*, Gulf Publishing Company, Houston, TX, May, 1998. Kletz is well known for his many publications and for bringing the term, "Inherently Safer Plants," into popular usage.

Kletz, T., Process Plants: *A Handbook of Inherently Safer Design*, Taylor and Francis, Philadelphia, PA, 1998.

Lees, F., *Loss Prevention in the Process Industries: Hazard Identification, Assessment, and Control*, Butterworths, London, 1996.

Loss Prevention Committee, Safety, Health Division, AIChE, *Proceedings of the 29th Annual Loss Prevention Symposium—(Serial)*, December, 1995

Publications by National Fire Protection Association (NFPA), 1 Batterymarch Park, Quincy, MA 02269. For a more complete list, see http://www.nfpa.org/Codes/CodesAndStandards.asp

NFPA 30 *Flammable and Combustible Liquids Code* (2000).

NFPA 69 *Standard on Explosion Prevention Systems* (1997).

NFPA 68 *Guide for Venting of Deflagrations* (1998).

NFPA 325 *Guide* to *Fire Hazard Properties of Flammable Liquids*

PHA Software, "PHAWorks," PSMSource (reference tool for OSHA's 1910.119) Primatech Inc., 50 Northwoods Blvd., Columbus, Ohio 43235, http://www.primatech.com

Smith, K. E., and Whittle, D. K., "Six Steps to Effectively Update and Revalidate PHAs," *Chem. Eng. Progr* 70–77 (Jan. 2001).

Thompson Publishing Group, *Chemical Process Safety Report*, 1725 K St.. NW, Suite 700, Washington, D.C. 2006.

Internet References and WEB pages

American Institute of Chemical Engineers, Center for Chemical Process Safety. http://www.aiche.org/ccps/

American Society of Mechanical Engineers, 2001 Boiler Pressure Vessel Code. ASME International, 22 Law Drive, Fairfield, NJ 07007-2900, USA Phone: 1-800-843-2763 or 1-973-882-1167 http://www.asme.org/

CCPS (Center for Chemical Process Safety), American Institute of Chemical Engineers 3 Park Ave, New York, NY, 10016-5991, U.S.A. ccps@aiche.org

Chemical Safety Board (Incident Reports) http://www.acusafe.com/Incidents/frame-incident.htm

Manufacturers Chemical Association: http://es.epa.gov/techinfo/facts/cma/cma.html

Manufacturers Chemical Association (Responsible Care): http://es.epa.gov/techinfo/facts/cma/cmacommo.html

OSHA Regulations & Compliance Links http://www.osha.gov/comp-links.html

OSHA Regulations (Standards—29 CFR) http://www.osha-slc.gov_OshStd_toc/OSHA_Std_toc.html

RMP Regulations http://www.epa.gov/swercepp/acc-pre.html

Publications by CCPS (Center for Chemical Process Safety), American Institute of Chemical Engineers, 3 Park Ave, New York, N.Y., 10016-5991. This is not a complete list. For a complete list, see on the Internet, http://www.aiche.org/pubcat/seadtl.asp?Act=C&Category=Sect4&Min=30

Guidelines for Chemical Process Quantitative Risk Analysis, 2nd ed., 2000.

Guidelines for Engineering Design for Process Safety, 1993.

Guidelines for Auditing Process Safety Management Systems, 1993.

Guidelines for Chemical Reactivity Evaluation and Application to Process Design, 1995.

Guidelines for Consequence Analysis of Chemical Releases, 1999.

Guidelines for Evaluating the Characteristics of Vapor Cloud Explosions, Flash Fires, and BLEVEs, 1994.

Guidelines for Consequence Analysis of Chemical Releases.
Guidelines for Hazard Evaluation Procedures, 2nd ed. with Worked Examples, 1992.
Guidelines for Implementing Process Safety Management Systems, 1994.
Guidelines for Investigating Chemical Process Incidents, 1992.
Guidelines for Safe Storage and Handling of Reactive Materials, 1995.
Guidelines for Technical Management of Chemical Process Safety, 1989.

4

Managing an Emergency Preparedness Program

Thaddeus H. Spencer* and James W. Bowman**

INTRODUCTION

Prevention, Prediction, and Preparation

The preceding chapter explored many technical aspects of chemical process safety and some safety management systems that form the foundation of a comprehensive emergency preparedness program. Clearly, the first step in preparing for emergencies is to identify and mitigate the conditions that might cause them. This process starts early in the design phase of a chemical facility, and continues throughout its life. The objective is to *prevent* emergencies by eliminating hazards wherever possible.

Although hazard elimination is the goal, experience has taught us that guaranteed, failure-free designs and devices have so far eluded human kind, despite astonishing advances in knowledge and technology. Even

the most "inherently safe" chemical facility must prepare to control potentially hazardous events that are caused by human or mechanical failure, or by natural forces such as storms or earthquakes.

The process of careful, structured analysis and evaluation used to eliminate hazards during design and construction will also allow a chemical facility to accurately *predict* unplanned events that may create emergencies, and to effectively *prepare* to manage them should they occur. A comprehensive emergency preparedness program has all of these elements: *prevention, prediction*, and *preparation*.

The fundamental need to predict and prepare for a failure of some kind is familiar to everyone. Fortunately, most of the failures that we encounter create little more than inconveniences in our lives. Others have much more serious potential. Such a failure can trigger an emergency, a term that Webster defines as "an unforeseen combination of circumstances or the resulting state that calls for immediate action." If the immediate action is

*Senior Consultant, Safety and Environmental Management Services, E. I. du Pont de Nemours and Company.
**Senior Consultant, retired, Safety and Fire Protection Engineering, E. I. DuPont de Nemours & Company.

ineffective, the emergency will escalate to a full-blown crisis.

Certainly most if not all of us in our personal experiences have had many opportunities to reconfirm the wisdom of the admonition, "Plan for the best, but prepare for the worst." As a result, we prepare ourselves for human and mechanical failure in a variety of ways, some so simple and familiar that we scarcely are aware that we are managing a personal emergency preparedness program.

To varying degrees, each of us has assessed our personal vulnerability to a specific emergency and the potential consequences to family and property. In some cases the required immediate action may be minor, but elsewhere the same combination of circumstances will demand significant resources to avoid a crisis of major proportions. The loss of household electric power in an urban condominium may only turn out the light, but on a farm, where electricity pumps the well water, milks the cows, refrigerates the produce, and irrigates the land, loss of power can bring disaster to the unprepared.

In an industrial environment, the consequences of human or mechanical failure can be far greater, even threatening the lives of employees and neighbors. Therefore, emergency prevention and preparedness efforts must have high priority, receiving continuous attention from every employee, including those at every level of management and supervision.

In the chemical industry, emergency preparedness programs have long been recognized as vital elements in protecting people, property, and the environment from harm. Few chemical facilities are without an emergency response plan. Still, when the alarms sound and the emergency is real, the response often does not proceed as planned. Too often, preparation for the unlikely event has been inadequate. As a result, many chemical facility managers are taking a fresh and critical look at their existing emergency preparedness programs in cooperation with their communities, and placing a still higher priority on being truly prepared for emergencies. This is consistent with the aggressive approach to other aspects of safety management that has made the chemical industry one of the safest industries.

Although the commitment to *prevent* the events that lead to injuries and emergencies of all kinds remains the first priority in safety management, we must "prepare for the worst." In this chapter we will explore how chemical facilities and their neighbors can better *predict* and *prepare* for unplanned events that threaten lives, property, and the environment—events that call for immediate and coordinated action.

Need for Emergency Preparedness Programs

The need for more effective management of emergency preparedness programs by chemical facilities and their host communities became painfully clear to the world in the 1980s—a decade marred by tragic events linked to the manufacture and distribution of chemical products. These events revealed serious deficiencies, not only in training people to react effectively during an emergency, but also in managing the systems employed to identify, evaluate, and mitigate hazards that may cause an emergency.

The chemical industry has vigorously responded to these problems in a variety of ways. In some cases well before the crisis of the 1980s, individual chemical units took aggressive action to increase the reliability of their operations, to communicate these actions to their neighbors, and to involve the communities in the process of emergency response planning. This proactive interaction at the local level proved fruitful in promoting the active partnership needed to solve mutual problems and to respond to public concerns. In 1986, these individual initiatives were institutionalized by the industry in the Chemical Manufacturers Association's (now the American Chemistry Council) "Community Awareness/Emergency Response (CAER)" program, which has in turn become a major element of the more recent "Responsible Care" program.

The proven success of these activities provides a powerful reason for giving high priority to open and cooperative management of an emergency preparedness program. By implementing such a program, a chemical facility builds important bridges to its neighbors and fosters positive perceptions. Involving the host community provides an opportunity to demonstrate a sincere commitment to the protection of people and the environment, and a route to the mutually profitable solution of real problems through meaningful two-way communication with the public.

Much of what had been voluntary prior to 1986 became law in the United States with the enactment of the Superfund Amendments and Reauthorization Act (SARA). The Emergency Planning and Community Right-to-Know portions of the Act (known as Title III, or EPCRA) place specific organizational, planning, communication, and training responsibilities on the public and private sectors, as do the accidental release provisions of the Clean Air Act amendments enacted in 1990. Other laws, such as OSHA's 29 CFR 1910.114, Process Safety Management of Highly Hazardous Chemicals, also require emergency response plans. In addition, various state and local regulations must be considered, understood, and reflected in a complete emergency preparedness program.

These laws and regulations, which will be covered later in this chapter, should be viewed as describing only the minimum requirements. Beyond the legal and self-protective reasons for chemical facilities to improve their management of emergency response programs lie other important driving forces. One of these is the moral and ethical responsibility to employees and the public to work toward elimination of events destructive to the quality of life. In practice, full recognition of this responsibility requires actions in prevention, prediction, and preparation that go beyond the letter of the law.

The financial benefits of good emergency preparedness program management are important as well. Emergencies are always expensive. Uncontrolled emergencies can become financially devastating crises. Nervous communities, fearful of the dangers dimly perceived through industrial fences, can and have put some chemical operations out of business. Creating and maintaining a comprehensive emergency preparedness program does not come free. However, the investment is proving to be money well spent by most chemical facilities.

PREVENTING AND PREDICTING EMERGENCIES: GETTING STARTED

The objective of a comprehensive emergency preparedness program is the protection of people, property, and the environment from unplanned hazardous events. Organizations handling hazardous materials recognize that the process of creating an effective program starts with the identification, evaluation, reduction, and control of hazards (commonly called risk management), and proceeds through the preparation, drilling, and maintenance of plans and procedures designed to contain an emergency situation should one occur.

To assure an effective emergency preparedness program, chemical facilities need to make periodic, formal assessments of their vulnerability to and preparedness for emergencies. Managers must be involved in developing and monitoring key indicators that will help in assessing an organization's ability to prevent or deal with an emergency. Awareness of any program weaknesses revealed by examining these indicators leads to corrective action to ensure that the potential for incidents decreases, and emergency preparedness improves. Sample checklists including some of these key indicators are shown in Fig. 4.1.

Central to all emergency preparedness programs is a written emergency response plan (ERP). An ERP obviously is a key element of emergency preparedness, yet it is only one procedural part of a systematic process that includes the following sequential steps:

- identify and evaluate hazards
- mitigate hazards wherever possible
- identify and evaluate remaining hazards
- identify and evaluate resources
- develop emergency procedures and ERP
- train facility personnel
- communicate plans to the public

EMERGENCY PREVENTION	EMERGENCY RESPONSE PLANNING
1. Frequent management presence in operating areas.	1. Emergency manual that documents the areas or processes with emergency potential and describes the emergency response plan.
2. Proper storage and identification of hazardous materials.	2. Regular emergency response plan training, and drills each quarter.
3. Formal and systematic inspections of key equipment, safety devices, and safety interlocks.	3. Liaison with the community for response coordination.
4. Audits for compliance with safety rules and procedures.	4. Emergency response management organization with documented functional tasks and assigned personnel.
5. Periodic critical reviews of existing operating and maintenance procedures.	5. Systems to recognize and report an emergency in a timely manner.
6. Training programs updated to meet current needs.	6. Documentation of hazardous chemicals with potential to leave the site, and evacuation plan for affected areas.
7. Control systems for maintenance hot work and changes in process and equipment.	7. Adequate emergency response equipment.
8. Systematic process hazards reviews during design, start-up, and routine operations.	8. Proper method to account for personnel during and after an emergency.
9. Formal reviews all of events that could or did cause serious process incidents.	9. Procedures to review and modify the emergency plan following drills or actual emergencies.

Fig. 4.1. Emergency preparedness checklists.

- integrate with community ERPs
- conduct and critique drills
- review and revise ERPs
- do it again

The creation and maintenance of an ERP requires the allocation of valuable resources, as well as strong leadership from a manager. No one else in the organization has the authority to commit the resources required. Only the power of the manager's office can overcome the organizational inertia that is often encountered.

Many people do not like to "prepare for the worst." They may harbor sincere doubts about the value of planning for events that they consider unlikely to occur. Some people are complacent after years free of serious accidents, and honestly believe that "It can't happen here." Others may be concerned that an open discussion of potentially disastrous events will needlessly upset employees and neighbors. These and other "start-up" problems could make trouble for a manager initiating (or reviving) an emergency preparedness program.

Managers need to find ways to stimulate employee interest and enthusiasm in the planning process. This is best done through the involvement of employees. One strategy that minimizes problems is the early involvement of those employees who serve as emergency responders in their communities. Some of them have witnessed the tragic consequences of failing to heed early danger signs and being unprepared for events. Thus, they can be willing and able catalysts in the emergency response planning process.

Open communication with all employees early in the process also is important. Sharing objectives and encouraging contributions stimulates thoughtful discussion, leads to more complete hazard recognition and mitigation, and ensures a greater chance of effective action when the emergency alarms do sound. Armed with facts, employees acting as informal ambassadors to the community can reduce the potential for public alarm over a facility's preparations to deal with serious but low-probability events.

HAZARD IDENTIFICATION AND MITIGATION

As shown in Fig. 4.2, emergency preparedness begins with the identification and mitigation of hazards. When properly done, the systematic analysis and evaluation of chemical process hazards stimulate actions that eliminate the potential for many emergency situations and pinpoint the situations that remain. The direction of these analytical and corrective efforts is

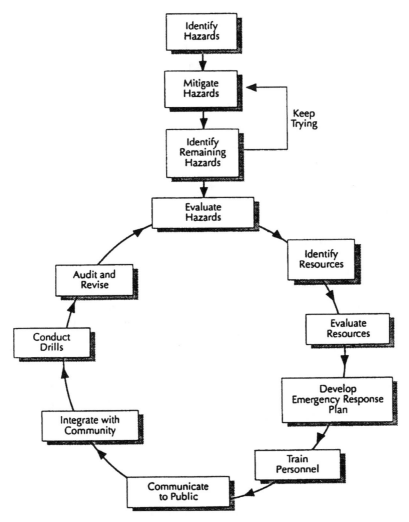

Fig. 4.2. Elements of a comprehensive emergency preparedness program.

an integral part of managing a comprehensive emergency preparedness program.

Process Safety Management Team

Hazard identification and mitigation are the responsibility of a facility's line organization—the same people who are responsible for all other organizational performance parameters. However, a line organization often needs support in identifying and mitigating complex chemical process hazards, support that can be effectively provided through the formation of a standing team functioning under the manager's direction. This team, called a Process Safety Management Team (PSMT), includes representatives of each unit of a facility from various levels of the organization. They meet on a regular basis (usually monthly) and report frequently to the manager on the status of their activities.

Identifying Hazards: PSR Teams

Supported by the PSMT, the line organization develops a structured approach for performing process hazard analyses on a repetitive basis. One effective way to do this is to organize knowledgeable facility personnel into ad hoc Process Safety Review (PSR) teams. These teams, supplemented by outside specialists as required, are responsible for studying all processes, identifying all potential hazards, and recommending appropriate corrective or control measures.

The makeup of a PSR team is critical to the success of this process. The effectiveness of the team depends on the skills, knowledge, and cooperative effort of its members, and the leadership ability of its chairperson. Each member must be familiar with the process being studied, and have at least a working knowledge of the basic engineering principles and chemistry involved. The team should include supervisors from operations and maintenance and a technical support person. Knowledgeable process operators and maintenance personnel would be valuable team members as well. Others who could contribute to the team's work include design engineers, specialists in electrical and instrument

systems, safety engineers, reaction kinetics consultants, and equipment vendors.

The selection and training of PSR team members must be carefully monitored. Most managers make this the responsibility of the standing PSMT, which also may assist a PSR team in choosing the most appropriate hazards evaluation method for a specific process.

Review Methods

There are many structured methods for carrying out effective reviews of process hazards. The four most commonly used methods are:

- what if/checklist
- failure mode and effect analysis
- hazard and operability study (HAZOP)
- fault tree analysis

What If/Checklist. The most frequently used method of process hazard review, the what if/checklist, is effective in reviews of relatively uncomplicated processes from raw materials to final product. The team formulates and answers "What if?" questions at each handling or processing step to evaluate the effects of component failures or procedural errors. They use a checklist to ensure that all important subjects are addressed. This method should be used as the first step in all process hazard reviews.

Failure Mode and Effect Analysis. When the team studies a specific item of equipment, such as a reaction vessel, they often use the failure mode and effect analysis method. Its semi-quantitative approach assists in prioritizing hazards.

Hazard and Operability Study (HAZOP). The HAZOP procedure systematically questions every part of a process to discover how deviations from the intention of the design can occur, and to determine if the consequences of such deviations are hazardous.

Fault Tree Analysis. Fault tree analysis, the most complex of the commonly used methods, is employed to determine the possible causes of a preselected undesired event. Through the use of logic diagrams and failure rate data, the team can make a quantitative evaluation of the frequency of the undesired event.

For additional information on such methods, refer to Chapter 3.

Recommendations and Reports

Regardless of the method used, the PSR team's most important responsibility is to alert management to serious hazards that may have been overlooked or given inadequate attention. To fulfill this responsibility, a PSR team must take the following steps:

1. Identify the hazards that could cause explosion, fire, release of toxic materials, serious injury, or inappropriate exposure to chemicals.
2. Evaluate the magnitude of the hazards for the areas of probable involvement; the consequences of an event in terms of injuries, environmental harm, and property damage; and, qualitatively or quantitatively, the probability of the hazards' occurrence.
3. Develop practical recommendations to eliminate or control the hazards identified.

The PSR team reviews in detail up-to-date reference material such as:

- architectural drawings
- equipment layout drawings
- process schematics
- instrument diagrams
- chemical and physical characteristics of process materials
- equipment design specifications
- operating procedures
- process conditions
- emergency shutdown procedures

When a facility has more than one chemical process, the PSMT develops a priority order for reviews and recommends a review frequency to the facility manager. The frequency usually ranges from once every two to three years for high-hazard-class processes such as explosives manufacture or acetylene purification, to once every five to seven years for low-hazard-class processes such as alcohol

purification, steam generation, and operations involving combustible materials.

Changes not anticipated in the original design of equipment often pose serious problems. Some examples of such changes are:

- introduction of different raw materials
- changes in temperatures, pressures, speeds
- deterioration of equipment

At the completion of each PSR, the team prepares a written report that defines needs, makes recommendations to remedy problems, and recommends priorities for the correction of deficiencies. Following review and acceptance of the report, the facility manager assigns responsibility for corrective action to the appropriate operating personnel. The PSMT then assists the manager in monitoring the status of the recommendations from all PSRs.

Mitigating Hazards: Release Detection and Mitigation

The release of flammable or toxic chemicals from uncontrolled pressure relief vents or as a result of equipment failure may present a serious threat to employees or neighbors who fail to guard against exposure. Every chemical facility must address this potential problem and prepare to protect people from these hazards by working toward reducing the potential hazard and its consequences. Typically, the line organization and PSR teams are responsible for hazard mitigation. They conduct a hazard study that includes the following activities:

1. Detailed appraisal of the potential for a accidental release of toxic gas or vapor.
2. Evaluation of instruments and other methods for detecting such leaks.
3. Provisions for rapid alerting of threatened personnel, and for communicating with emergency responders.
4. Identification of buildings in which people might be trapped by such a release.
5. Assessment of the capability of buildings or rooms to prevent the infiltration of gas or vapor.

6. Evaluation of plans for building evacuations, including the provision and maintenance of appropriate personal protective equipment.

Early *detection* of a chemical leak is necessary in order to limit its effect on people and the environment. The most fundamental method for detecting a chemical release is the systematic patrol of operating units by personnel trained to recognize potentially hazardous vapors using odor or visual observations. The frequency and scope of the patrols will vary with the nature of the process equipment and materials; however, every facility should have documented patrol procedures in place.

When particularly hazardous vapors are present, the patrol procedures should include special provisions to protect the patrollers from the fumes. For example, in facilities producing or consuming hydrogen sulfide, operators may carry emergency respiratory equipment to escape any unexpected fumes that they may encounter, and patrol in pairs or individually, under constant surveillance.

Many chemical facilities supplement operator patrols with an instrumented detection system. Such a system may be a necessary resource when a hazards-study concludes that the system will substantially increase the available escape or emergency response time, or where:

1. The harmful substance is odorless or deadens the sense of smell at hazardous concentrations.
2. The harmful substance is toxic at concentrations undetectable by smell.
3. Large numbers of people may be exposed quickly.
4. Ventilation systems might draw toxic fumes into a building before other means of detection could trigger protective action.

At the core of an instrumented leak detection system is a gas detector. There are many kinds of detectors on the market with varying degrees of sensitivity and selectivity. All require careful regular testing and maintenance.

An engineering study of many site-specific factors is required before one makes a choice and designs an appropriately instrumented system. Some generally accepted guidelines are:

1. Early warning of a leak is enhanced if the sensors can be placed near the process equipment subject to leakage.
2. Air movement characteristics are critical in achieving reliable detection.
3. Most detectors respond to several gases or fumes, so the possible presence of all airborne substances affecting the detection system must be considered.
4. When it is necessary to monitor work areas or the intakes to ventilation systems, a highly sensitive system is desirable.

Most detection systems are designed not only to report the presence of hazardous fumes through instrument readouts but to sound an alarm and automatically initiate corrective or protective action. In an office or shop, for example, the system can be designed to shut down all ventilating fans and close exterior air inlet dampers.

When process safety reviews have identified chemical releases as potential sources of facility emergencies, the organization must provide the training and materials needed to ensure a prompt and appropriate reaction to *mitigate* the hazards. Some countermeasures that are effective in limiting the spread of a hazardous material spill or release should be included in the design of chemical process equipment, and described in emergency response procedures. The following discussion of release mitigation is largely excerpted from *Guidelines for Vapor Release Mitigation*, prepared by R. W. Prugh and R. W. Johnson for the Center for Chemical Process Safety of the AIChE (copyright 1988 by the American Institute of Chemical Engineers, reproduced by permission of the Center for Chemical Process Safety of the AIChE).

"Water, steam, and air curtains and water sprays are primarily effective in dispersing and/or diluting vapors with air to reduce the severity of effects of a hazardous vapor release. In some cases, vapors can be partially 'knocked down' or absorbed after release."

"Ignition source control and deliberate ignition are also possible vapor release countermeasures." "For areas around processes handling flammable vapors, ignition source control is practiced to reduce the probability of vapor ignition if a leak occurs." "Administrative controls are exercised on plants where flammable materials are processed. Such controls may include hot work permits, restricted smoking areas, not allowing lighters or matches on the site, and electrical grounding and bonding procedures."

"Deliberate ignition is a countermeasure against spills of highly toxic materials which are also flammable, such as hydrogen sulfide, hydrogen cyanide, and methyl mercaptan. Igniting nontoxic flammable materials such as hydrocarbons may present hazards outweighing possible advantages." In any case, deliberate ignition must be carefully planned and executed so that the resulting fire is truly controllable.

"Practical methods for combating vapor from liquid leaks are dilution, neutralization, or covering. All three reduce the vaporization rate of the pool. Water dilution is effective for spills of water-miscible or water-soluble material. Spraying water into the spill reduces the vapor pressure by reducing the concentration of the liquid." "For acidic spills, limestone or soda ash is often used" to react with the spilled liquid to produce a less volatile salt or ester. "A foam cover can be effective in reducing vaporization from spills," and "dense liquids can be covered with lighter immiscible nonreactive liquids" to accomplish the same thing.

PREPARING FOR EMERGENCIES: IDENTIFYING AND EVALUATING RESOURCES

The process of hazard identification, evaluation, and mitigation, when sustained over a period of time and coupled with other good safety management practices, can prevent most process-related emergencies. However, the potential for various low-probability process-related events will remain. Should one occur, there must be resources available to promptly bring the event under control. These resources include designated personnel, plans, systems, and facilities that are needed for effective action and communication. The facility line organization is responsible for providing and maintaining these resources.

Personnel

Emergency Preparedness Team. As with the PSMT and the PSR team, the efforts of the line will benefit from the support of a standing Emergency Preparedness Team functioning under the direction of the facility manager. This team, which includes representatives of each unit of the facility, leads the development and maintenance of an ERP for the facility, and monitors specific procedures and training for handling emergency situations of all kinds. It should meet on a regular basis (usually monthly), and report frequently to the manager on the status of its activities.

Emergency Management Organization. Emergencies demand rapid, well-coordinated decisions, communications, and action to bring them under control as swiftly as possible. There must be a formal Emergency Management Organization (EMO) whose purpose is to achieve this objective. The structure of the organization is not critical as long as it is capable of rapid assembly, carries out its assigned responsibilities, and meets the needs of the facility. A generic EMO is shown in Fig. 4.3; the EMO is described below, in the section on "Developing an Emergency Response Plan."

Fire Brigades. Most chemical facilities of any size have established fire brigades as a key resource in their emergency preparedness programs. By virtue of its training and familiarity with the equipment and the physical layout of a site, the brigade can be a major factor in *preventing* incidents as well as in minimizing injuries and property damage due to fire, explosion, or other causes.

The size and the structure of a fire brigade vary, depending upon the hazards present, the size of the facility, and the level of internal

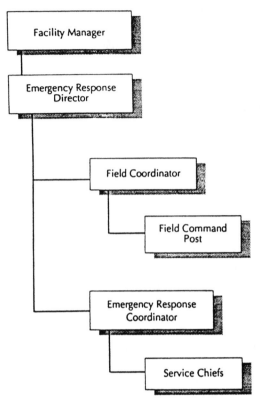

Fig. 4.3 Generic emergency management organization.

fire-fighting capability desired; and large, multi-process facilities usually have several unit or area brigades. Each brigade has a designated captain and an alternate. The training of brigade members must be commensurate with the duties and the functions that they are expected to perform, and thus depends on the fire prevention plan option selected by site management. (Fire prevention plans are discussed later in this section.)

It is best to have a written statement establishing the brigade and its duties, specifying its size and organizational structure, and outlining the type, amount, and frequency of training provided. In the United States, OSHA has promulgated minimum standards for fire brigades, which may be found in 29 CFR 1910.156.

Plans

Work Unit Plans. In a large facility, each work unit has emergency procedures for its own area of operations, and these work unit plans form the foundation of the facility ERP. Thus, the facility's emergency response can be no more effective than the recognition and response capabilities of the employees in the facility work units.

The primary objectives of these unit plans are (1) to control and contain any emergency condition within the unit, and (2) to provide protection for unit personnel and equipment from events originating outside the unit. Unit plans specify who has authority to take emergency action, and how escalation to a full facility emergency occurs. Subjects that must be covered include:

- unit emergency command and coordination
- communications within the unit
- communications with other units
- emergency assessment
- unit evacuation and personnel accounting
- emergency shutdowns
- communications with the facility management
- criteria and procedures for securing resources from outside the unit

Unit plans clearly assign primary responsibility for initial emergency assessment and reaction to the lead unit employee on site at the time of the event. Among the factors to be considered by that individual in making the initial assessment are:

- previous experience with similar situations
- how long the situation has existed
- what might occur "If..."
- properties and hazards of materials involved
- other complicating situations

Using the information gained from the assessment, unit personnel take whatever immediate action is required to protect people and property on and off the site.

Work unit plans should contain guidelines for assuring uniformity in the assessment of and reaction to unusual events. An effective approach is to develop an emergency classification system that includes criteria for classification and guidance for the appropriate response actions for each level of emergency that is defined.

For example, an emergency judged to be controllable within the unit, with no evacuation outside the unit and no impact in the community, is a Level One Emergency. Small releases of hazardous materials usually fall into this category. Appropriate response items to consider would include:

- protection of unit personnel
- shutdown or isolation of affected equipment
- notification on- and off-site
- containment and cleanup

Similarly, other events are categorized by their potential or actual severity. Preplanned response checklists are prepared for each one. An incident that requires response by facility personnel outside the affected unit but has no impact on the community is a Level Two Emergency. A Level Two Emergency requires activation of the facility ERP. If some community impact becomes probable, the emergency is a Level Three Emergency. That level of emergency triggers additional response requirements, including interaction with the community.

The unit plan must be formal and in writing. Because the emergency procedures of each individual operating area or work unit form the foundation of the facility ERP, they should be reviewed by the Emergency Preparedness Team for completeness and consistency across the site before they are included in the facility ERP.

Fire Prevention Plan. One of the most basic resources that all chemical facilities must have is a fire prevention plan. To be fully effective, this plan must be in writing, and it must be reviewed with all employees on an established schedule.

A comprehensive fire prevention plan includes the following subjects:

1. Major facility fire hazards
2. Storage and handling practices for combustible and flammable materials
3. Identification of potential ignition sources

4. Procedures to maintain systems and equipment installed to control ignition sources
5. Names and titles of those responsible for the system and equipment maintenance and the control of fuel source hazards
6. Procedures to minimize accumulations of flammable and combustible waste materials
7. Plans for communicating to all employees the fire hazards and their specific responsibilities in the event of fire

Fire response options open to facility managers range from a plan to evacuate all employees without attempting fire fighting on any scale, to full involvement of an established fire brigade in fighting advanced-stage structural fires. The fire prevention plan must specify which option applies to the facility. The option selected determines the type and extent of education and training required, as well as the type of equipment needed on the site.

Fire prevention and response information may be covered in other written facility documents (such as job descriptions, rules, or procedures) and communicated to employees on the same schedule as the fire prevention plan.

Evacuation Plans. Many emergency events require the evacuation of at least some facility personnel, if only for precautionary reasons. This presents other needs, which must be addressed in the evacuation plan. Personnel who are directed to evacuate their normal work stations need visible wind direction indicators to help them move away from a dangerous gas cloud drifting downwind of the release point. If the building is enveloped in the cloud, however, they will need personal protective equipment in order to evacuate the building safely. When there are processes that must be rapidly shut down and/or isolated as evacuation proceeds, the plan must include written procedures and appropriate means to ensure incident-free action.

Headcount Plan. The evacuation plan also must establish gathering points for all personnel in order to identify those who are missing. The procedure for accounting for personnel, called the headcount, must work effectively and rapidly in the first minutes of the emergency to determine if anyone is missing or known to be injured. Communications to the fire brigades or other first responders must be quick and accurate so that search and rescue operations and medical aid can be successfully implemented.

Each work unit in a facility must have its own plan for headcounts as a part of its emergency preparedness program. This plan must be in writing, and it must be practiced frequently enough to ensure good execution in a time of real emergency. It should include:

- a designated assembly point for each employee
- an alternate assembly point, should the primary location be inaccessible
- a procedure for counting at each assembly point
- a designated unit headcount coordinator and backup
- primary and alternate phone numbers where coordinator may be reached
- phone numbers for reaching the personnel chief, who coordinates and summarizes the site-wide accounting

Headcount procedures must include guidelines for accounting for visitors to the facility at the time of the emergency. These visitors may include truck drivers, vendors, and contractors. Some guidelines are:

1. All visitors must sign in and out, preferably at one designated gate.
2. Visitors will have a designated host employee responsible for their safety whenever possible.
3. Everyone temporarily on the facility must report to the assembly point of the unit he or she is in at the time of the emergency.
4. Unit procedures must include reporting these "extra" people by name to headcount headquarters.
5. The personnel chief of the EMO must have access to the log at the visitors' gate.

Alarm Systems

Another essential resource that every chemical facility must have is an effective alarm system, which initiates action by endangered personnel and emergency responders.

A satisfactory facility emergency alarm system must meet three additional requirements. It must, at a minimum, immediately alert *all* the people on the site. (A single signaling device should be adequate for small sites; however, large facilities may require many devices placed to reach all occupied areas of the site.) The system must function even when the facility has a general power failure. Alarm activation controls should be located so that an emergency condition is unlikely to prevent access to them.

On most older chemical facilities, the basic component of the emergency warning system is the fire alarm. It is essential that there be different signals to distinguish between an actual fire and other threatening events, as the action taken in response to a toxic gas release, for example, may be quite different from that taken in response to a fire. On multi-process sites, the signals also should identify the work unit involved.

Depending upon the number of people at a facility and their familiarity with response and evacuation procedures, it may be necessary to supplement signals with verbal instructions delivered by public address equipment, radios, or automated telephone systems. For reliability at any time of day, a system using verbal instructions must be located in a regularly manned job station such as a control room or gate house.

Any alarm system used must communicate clearly the nature of the emergency event and its severity. In a Level One Emergency some people may have to be evacuated in order to ensure their safety, but by and large the emergency can be contained within the unit. In that case, information usually may be given to unaffected employees through normal telephone communications.

In the event of a Level Two or a Level Three Emergency, each facility must have an easily recognizable signal that clearly communicates that a major incident has occurred. Some facilities have alarm boxes that will automatically sound this signal when pulled. Others have boxes that require manual operation. Still others rely on a telephone message to trigger the signal.

In some facilities, the "major" emergency alarm is a steady blast on the facility steam whistle. Others use sirens or special bells. Whatever is chosen, the alarm must reach everyone on the site, triggering activation of the facility ERP and the initiation of a head-count procedure.

It is essential that every employee on a chemical facility knows how, where, and when to turn in an appropriate and effective emergency alarm. The specifics will vary considerably, but the procedure for turning in an alarm is similar in all cases.

Employees first need to know how to report a fire. On most facilities, this is done by using a fire alarm box or by telephoning a central station such as a guard house from which the fire alarm can be sounded. Chemical facilities need to have well-developed training and drill programs in place to be sure that response procedures and equipment are maintained.

When fire alarms are supplemented by verbal instructions, the facility personnel who are contacted need to know the location and the nature of the fire, the action under way to control it, and the actions required (if any) by those not involved. There are many commercially available communications systems that deliver this information efficiently, even on large sites.

Facilities for Protection and Communication

Safe Havens. When accidental releases of a toxic gas occur or threaten to occur, the immediate protection of on-site personnel is of paramount importance. One resource available for this purpose is a building or room that by the nature of its construction and its heating and ventilating characteristics can prevent the infiltration of intolerable concentrations of the toxic substance. The best location for a safe haven is determined by an engineering study. That location then is documented in the work unit and facility ERPs.

Safe havens may be rare on some sites; however, many buildings can provide personnel with temporary protection until the incident has been analyzed and a decision made on the need for evacuation. To qualify as a designated temporary safe haven, a building must be reasonably well sealed against air infiltration, with adjustable ventilation systems that can reduce or close off exhaust vents and outside makeup air. Emergency procedures should state how long a building can be considered a safe haven under specified exposure conditions.

Main Emergency Control Center. Using a list of safe havens developed by the work units, the Emergency Preparedness Team helps management select one building or room to be used as a control center in case of a major emergency. This Main Emergency Control Center (MECC) should be as remote from potential hazards as possible in order to serve as a reasonably safe haven to those involved in directing, coordinating, and communicating activities for the duration of an emergency.

MECCs need to have basic information readily available and maintained in an up-to-date, standby condition. Such a collection would include:

- copies of the facility ERP
- facility maps and diagrams
- process material isolation points
- fire control maps and diagrams
- maps of the surrounding area
- aerial photos of the facility and the surrounding area
- names, addresses, phone numbers for:
  —all facility employees
  —off-site company people to be notified
  —groups and organizations who may be notified
  —community officials who may be notified
- Material Safety Data Sheets (MSDSs) for facility materials
- MSDSs for the materials of neighboring chemical facilities

- copies of mutual aid agreements
- highlights of the facility's history, products, and performance

The MECC should have the following basic equipment:

- adequate telephones and lines
- unlisted telephone and/or a hotline
- two-way radios
- fume path projector
- FAX machine
- regular radio with tape recorder
- regular television with recorder and play-back video cassette recorders
- chart pads and stands
- battery-powered lighting
- personal protective equipment
- hand-held tape recorders

Alternate Control Centers. Recognizing the unpredictable nature of a developing emergency, it is advisable to select at least one alternate on-site main control center. Both the MECC and the alternate on-site center should have backup power supplies.

The presence of substantial quantities of explosive or toxic materials on-site may justify establishing yet a third location off-site, in case of a complete evacuation of the facility. This could be a fixed location in the nearby community, or a mobile unit such as a truck or van properly equipped for managing an emergency. Some of the supplies and equipment for the alternate off-site main control center may have to be packed in a readily transported kit rather than being on standby at the alternate off-site location.

Media Headquarters. In an emergency, it is essential to maintain regular contact with the public and the media. A separate location for communications will allow that contact to continue without interfering with the operations of the MECC or overloading its communications system. There, the public affairs chief and the facility manager can receive media representatives and provide periodic updates on the emergency situation. The media headquarters should be equipped with several direct outside telephones and the facil-

ities to make the reporters comfortable for the duration of the emergency. Permanently mounted facility plot plans or aerial photographs are helpful resources, as are handouts describing in general terms the facility and its products.

DEVELOPING AN ERP

Following the identification and evaluation of the resources available to the facility, the Emergency Preparedness Team, under the direction of the manager, develops the written ERP for the entire facility. The Emergency Preparedness Team is responsible for ensuring that the written facility ERP informs all employees of their roles in an emergency and the hazards to which they may be exposed.

An ERP must be tailored to a specific facility, reflecting its unique conditions and individual needs. Among the variables that affect the details of a plan are:

- materials used, produced, or stored
- nature of the operations
- available employee skills
- geographic location
- proximity to other facilities
- available emergency resources
- mutual aid agreements

No two ERPs will be exactly alike, nor is there one best outline. The plan of a production facility will differ from that of a warehouse. However, every plan must include:

- initial alarm procedures
- emergency escape procedures and routes
- emergency operating and shutdown procedures for critical operations
- accounting procedures for all on-site personnel
- rescue and medical duties for specified personnel
- procedures for communicating the emergency inside and outside the organization
- structure, duties, and resources of the EMO, including the names or titles of people with detailed knowledge of the plan and its assigned duties
- reference material

Plan Design

The ERP is a working plan, applicable to any event with emergency potential occurring at any time and at any location on the facility. It should be written concisely, with diagrams and checklists used wherever possible, so that it may be effectively used for guidance during an actual emergency. To accomplish this, it is helpful to divide the ERP into several major sections, which are in turn divided by subject.

For example, an ERP may be broken down into three sections: the Emergency Management Organization, Action Plans, and Reference Material.

The first section documents the structure, duties, resources, and communications systems for the facility EMO, and the conditions that will trigger the EMO's response. It includes plans for site-wide notification and response.

The second section includes detailed action plans for each particular type of emergency. This section includes summaries or outlines of the emergency procedures developed by each work unit of a facility, and may contain considerable process-oriented information.

The third section consists of reference material on plan philosophy, training, plan maintenance, drills, and similar supportive data that usually are not needed at the time of an emergency. Members of the EMO and others with assigned response duties should be able to quickly locate and refer to the appropriate action guidelines, to determine that key functions are being performed.

Using this approach, the outline of an ERP for a multiprocess chemical facility would look like this:

Section I
  A. Table of Contents
  B. EMO Structure and Tasks
  C. Notification: On-Site
  D. Accounting for Personnel
  E. Headcount Center
  F. EMO Center
  G. Emergency Scene
  H. Notification: Off-site
Section II
  A. Table of Contents

  B. Work Unit 1
    1. Emergency types
    2. Action checklists
    3. Responsibilities
    4. Communications
    5. Shutdown procedures
  C. Work Unit 2 (same as above)
Section III
  A. Table of Contents
  B. Plan Philosophy and Objectives
  C. Training Programs
  D. Drills and Tests
  E. Return to Normal
  F. Appendices
    1. Special situations (e.g. bomb threats)
    2. Hazardous material data
    3. (Others as required)

EMO Structure

Perhaps the most essential element of an ERP is the establishment of the organization that will manage an emergency response once the event has developed beyond a Level One Emergency. The EMO must be capable of rapid assembly, and must be able to carry out all of its assigned responsibilities.

The EMO is composed of the facility manager, the emergency response coordinator, the field coordinator, and various service chiefs. The number of service chiefs and their duties will vary. There need not be a separate individual in charge of each service, and small facilities may combine two or more services under one individual, as long as each function is performed.

The facility manager, in cooperation with the Emergency Preparedness Team, assigns people, by their title or function at the facility, to the EMO, with designated alternates for each EMO function. The duties and responsibilities of each EMO function are defined and documented in the ERP. Because there must be a functional EMO in place regardless of when a facility emergency occurs, designated shift personnel take specific EMO responsibilities until the primary members can reach the site.

All the positions established for the EMO should have written position descriptions, which may be placed in the third section of the ERP as an appendix to the plan. In the first section of the ERP, the EMO summary need only include a concise checklist for each position, with diagrams showing relationships and major communication flow paths.

Facility Manager. In the EMO, the facility manager has the overall responsibility for protecting people on-site, facility property, the environment, and the public during and after an emergency. With the assistance of the public affairs and communications chiefs, the manager usually serves as the spokesperson for the facility and the company, communicating with representatives of the media and other concerned audiences.

Emergency Response Coordinator. The designated emergency response coordinator, who may also serve as the chairperson of the Emergency Preparedness Team, directs all emergency control activities from the MECC. All other service chiefs on the EMO report to the coordinator. Using information from the emergency scene and from the service chiefs, the coordinator makes the key decisions on what should be done, and coordinates activities on and off the site. The coordinator reports to the manager, who should be available nearby for overall guidance and counsel.

Field Coordinator. The job of the field coordinator is to correct the emergency situation as rapidly as possible with minimum risk to those in or near the affected area. He or she establishes a field command post as close to the scene of the emergency as can be safely done. The post often is in or near a radio-equipped building or vehicle, thus permitting rapid establishment of communications between the command post and the emergency response coordinator.

Ideally, the field coordinator is familiar with the operations and materials involved in the emergency; so she or he often is the highest-ranking supervisor of the affected area who is available at the time of the event. That supervisor may remain as field coordinator for the duration of the emergency, or may be relieved by another designated member of management.

In the latter case, the area supervisor may become a service chief with greater hands-on involvement, with the field coordinator concentrating on marshaling required resources and maintaining effective communications. The choice depends to a great extent on the size and complexity of the facility, and the resources available for the EMO.

Public Affairs Chief. Working in close coordination with the facility manager, the public affairs chief releases appropriate information to the news media, regulatory personnel, government officials, and other public groups and individuals. No information is to be made public by anyone other than the facility manager without specific direction from the public affairs chief. The objective is to provide full and accurate statements in a timely fashion, so that public attention is focused on facts and useful information rather than on rumors and speculation.

The public affairs chief also establishes and monitors the media headquarters, which is isolated from the EMCC to avoid interference with operations there. The public affairs chief's function includes assisting the manager in the preparation of formal statements and background information to be distributed to reporters. It also may include arranging with local radio and television stations to make periodic announcements during an emergency so that the public and the employees not on the site are properly informed. The public affairs chief also arranges to monitor and perhaps record the broadcasts of local radio and television stations in order to determine what additional statements or clarifications to the public may be required.

Communications Chief. The communications chief establishes and maintains communications capability with appropriate people on and off the site. This chief must be familiar with the various communicating systems available, including telephones, public address systems, two-way radios, and messengers. The job includes recommending and coordinating revisions or additions to the communicating systems during an emergency, and assisting the facility manager and the public affairs chief with any communications as needed.

Engineering Chief. The engineering chief's primary responsibility is to maintain electrical power for vital services. These services include on-the-scene portable lighting, continuous fire pump operation, and a steady supply of nitrogen, steam, and process cooling water. Another key duty is to assemble repair groups capable of isolating damaged sections of pipelines, electrical lines, and other necessary equipment in order to contain problems and maintain or restore operations outside the affected area. Because these repair groups must be drawn largely from site personnel such as electricians, welders, riggers, and pipefitters, the position of engineering chief should be filled by someone from the facility maintenance organization.

Emergency/Fire Chief. The emergency/fire chief is responsible for fire-fighting and fume control activities. The best person for this job has good knowledge of and access to the fire-fighting, rescue, and fume control resources available on and off the site. On a small facility this may be the captain of the fire brigade. The job includes direction of designated facility personnel, and coordination with any outside forces brought in to bring the fire or fume condition under control. The environmental chief may assist in determining optimum courses of action based upon actual or threatened adverse effects on air and water emissions from the facility.

Medical Chief. The medical chief ensures that the proper medical care is provided to people on the facility who have been injured or exposed to toxic materials. When facilities have medical professionals on the site, one of them carries out this function. This chief's responsibilities include establishing field stations to treat affected personnel, and, in cooperation with the transportation chief, providing transportation for injured people to other medical facilities.

The medical chief also participates in discussions with community officials and appropriate facility personnel regarding actual or potential medical problems for people outside the site boundaries. Representatives from nearby local hospitals, ambulance services, fire departments, police forces, and emergency management groups also may be involved in these discussions. The medical chief communicates any action or contingency plans developed in the discussions to the emergency response coordinator and the public affairs chief.

Environmental Chief. The environmental chief oversees all activities designed to minimize adverse effects upon the quality of air and water as a result of an emergency. The function includes coordinating air and water quality monitoring on and off the site during and after the emergency, and assisting the emergency/fire chief in selecting the optimum approach to abating a fire or fume condition. The environmental chief also provides assistance in projecting the path and concentration of a fume release, using computer modeling (if available) or maps with plastic overlays, and developing an effective plan of action.

Personnel Chief. The personnel chief is responsible for accounting for all personnel on the facility at the time of the emergency. Unit emergency programs must include training on how this is accomplished. Unit supervisors initiate the process with headcounts at designated rally points. The success of rescue and medical activities depends on how quickly and thoroughly this information is obtained and reported to the personnel chief.

The personnel chief also coordinates communication with relatives of injured or deceased employees, and makes certain that this is completed before any names are released publicly. This requires close coordination with the public affairs chief and the manager.

The personnel chief need not be located at the MECC, but it is imperative that the personnel chief be in close touch with other members of the EMO. Telephones in the headcount center should have answering and message recording equipment to capture any unit reports arriving before the headquarters is staffed.

Security Chief. The security chief makes sure that entry to and egress from the facility are properly controlled. This involves securing gates; limiting entry to authorized personnel; registering all who pass through gates; meeting visitors, including representatives of the media, and escorting them to proper locations; and controlling all traffic on the site. The function also includes communicating with local police so that access to the facility is maintained, and crowd control procedures are enforced. Usually, the individual serving as security/chief at the time of the emergency also is responsible for initiating procedures to summon key facility personnel and urgently needed outside agencies.

Transportation Chief. The transportation chief coordinates and controls all transportation on the facility. This includes directing the assembly of available vehicles and crews, and identifying needs beyond site capabilities such as cranes, trackmobiles, and bulldozers that must be obtained from outside organizations. The function also includes providing suitable transportation for facility employees who monitor the effects of emergencies beyond site boundaries, or interact with community officials at an off-site location.

TRAINING PERSONNEL

Having a written facility ERP that is supported by established work unit emergency plans and procedures it is an important part of the manager's job to ensure that unplanned events will be promptly controlled with minimum risk to people, property, and the environment. There is much important work to be done, however, before the ERP is anything more than a paper resource. Facility personnel must be trained to use the ERP effectively. There must be frequent drills to test the plan and the people against the standards established by management as well as those established by law.

Within the facility, there are three groups of people who require training. First, there are members of the fire brigade, who must be trained to fight fires at the level specified by facility management. Next are the employees who have been assigned active roles in controlling emergencies of all types. This group includes members of the EMO and their alternates, plus designated support people such as headcount coordinators, guards, and emergency repair personnel. Finally, everyone else on the facility must have a basic understanding of the ERP, and must know how to respond when specific alarms sound.

The facility manager is responsible for ensuring that the appropriate training and retraining are done in a timely and effective manner. An employee should be trained when he or she is hired, at least annually thereafter, and when the employee's work area changes or the plan is revised. Most managers make this primarily a line organization function. They expect facility supervisors to use all available resources and means of education to accomplish the tasks. Some specialized training assistance, however, must be provided, particularly for the members of a fire brigade.

Fire Brigade Training

Training programs for fire brigades have two major objectives. One is to inform the brigade members of new hazards at their facility and innovations in fire-fighting techniques and equipment. The other objective is to provide hands-on training for developing skills in emergency operations and using equipment, including:

- portable fire extinguishers
- hoses and accessories
- portable lighting
- forcible entry tools
- ladders
- salvage equipment
- first-aid supplies
- replacement parts
- personal protective equipment
- transportation equipment

A comprehensive training program for fire brigades must include classroom and hands-on training.

Outside resources can provide valuable assistance in the education and training of brigade members. Local fire departments and state fire schools are usually enthusiastic partners in such efforts. Often they are the key to securing adequate resources at a reasonable cost. In industrialized communities, mutual aid agreements may include cooperative training provisions with other chemical facilities, which provide opportunities for even more effective use of available resources. All mutual aid agreements should require cross-training in special hazards at the other facilities.

EMO Training

Training for personnel assigned to the facility EMO can be led by the emergency response coordinator, who meets periodically with each member of the EMO to review and refine position descriptions and the associated functional checklists. An example of such a checklist is shown in Fig. 4.4. The meetings may be followed by limited drills involving only the people and responsibilities included in the individual EMO function. Some facility managers assign specific emergency response training duties to each member of the EMO.

For example, the engineering chief organizes, equips, and trains the repair groups who will be called on to physically stabilize a situation at the time of an emergency. The emergency/fire chief could be given the responsibility for maintaining a trained force of fire and fume fighters, which includes competent leadership on all shifts.

Employee Training

There are many ways to be sure that all employees understand the ERP. The Emergency Preparedness Team can assist in the assessment of existing training programs, alerting the organization to training weaknesses and suggesting or providing creative ways to overcome those deficiencies. Unit supervisors must periodically review the ERP in scheduled group safety meetings or with individuals. Key plan elements can be reproduced on wallet-size cards, desktop displays, or telephone stick-ons. Individual or group discussions of how to react in given situations can detect weaknesses in procedures, training, or understanding.

Training programs for the three groups of employees discussed here should be documented, reviewed regularly, and included in a reference section of the facility ERP. But no matter how comprehensive the training

1. Personnel Chief and aides report to headcount room at the MECC.

2. Replay audio counts already received.

3. Record work unit head counts as they are received.

4. Obtain copy of visitors' log and employee with area head counts.

5. Monitor attempts to locate missing personnel.

6. Report to Emergency Response Coordinator as personnel are confirmed as missing.

7. Issue final report to Emergency Response Coordinator after all work unit head counts are received.

8. Determine from Emergency Response Coordinator names of any injured people.

9. Coordinate communications with families of those missing or injured.

Fig. 4.4. Personnel chief functional checklist.

programs may be, their effectiveness is unknown until a drill of the ERP is conducted and its systems and procedures are tested.

Facility Drills

To evaluate the effectiveness and completeness of an ERP, a facility must conduct periodic announced and unannounced Level Two and Level Three emergency drills. These are in addition to the more frequent Level One unit drills that are held to ensure that the more limited response procedures of a work unit are complete and well understood. Major internal emergency drills should be held at least four times each year, and scheduled to involve each working shift at least once a year.

To gain the maximum benefit from the drills, assigned observers should witness all aspects of response activity, and gather soon after to participate in a verbal evaluation of actions taken. These observers should include members of the Emergency Preparedness Team. This process is enhanced by capturing on-scene action on videotape or in still photographs. Critiquing can be extremely valuable in identifying necessary plan changes, training needs, and resource deficiencies of all kinds. The process of critiquing must be controlled and managed. The leader must be sure to identify the strengths as well as the weaknesses that were revealed in the drill. The objective is to stimulate actions for positive change, not to assign blame.

Drills should be held on weekends and at night occasionally to test segments of the organization that work at times when all of the specialized resources of the facility are not immediately available. Initially, limited scale drills can test segments of the ERP, such as manning the MECC with the shift personnel that are available and carrying out a headcount without involving daytime employees.

As the proficiency of the organization increases, drill scenarios can become more complex. Complicating factors approaching worst real-life conditions should be introduced periodically, including:

- telephone switchboard overload
- absence of key EMO members

- arrival of major TV network anchorperson
- simulated mass casualties
- two-way radio failure
- evacuation of primary and backup MECCs
- major community impact

The Emergency Preparedness Team, with approval of the manager, designs the drill scenarios, monitors the organization's performance, provides leadership for critiquing, and recommends corrective actions. The team also develops and monitors a plan for involving the community in the important task of integrating a facility's ERP into the public emergency preparedness programs of the region.

INVOLVING THE COMMUNITY

Of major concern to the chemical industry is the public perception that facility managers have little concern for the welfare of their neighbors. Managers themselves have contributed to such false impressions by failing to interact with their communities in a consistent and meaningful way.

The public clearly wants to know more about the risks presented by a chemical facility. Increasingly, the public wants to help decide which risks are acceptable and which are not.

Recognizing that industrial facilities exist only with the consent of their host communities, most industrial organizations are assigning a high priority to building stronger bridges with their neighbors. The chemical industry in particular is finding that the involvement of the community in the process of emergency preparedness planning presents an excellent opportunity for constructive two-way communications. This has proved particularly productive if these communications are part of an aggressive and continuous risk communications process.

Communications

Each chemical facility needs to create opportunities for sharing information with its surrounding community; it should not wait for these opportunities to occur. Facility

1. Open communications up and down the line organization.

2. Scheduled two-way communications with all employees at least twice per month.

3. Newsletter for employees, pensioners, and key community leaders.

4. Regular meetings with local media reporters and editors.

5. Facility open house at least once every three years.

6. MSDS available to appropriate organizations in the community.

7. Meetings with political leaders and activists at their locations and at the facility.

8. Leadership and participation in local emergency planning committee.

9. Leadership in community projects such as wildlife protection, public land-use designation, etc.

10. Participation in school programs for children.

Fig. 4.5. Risk communication checklist.

managers in particular should be active in the process, and should monitor the performance of their organizations. Some items that should be on a manager's checklist for risk communications are shown in Fig. 4.5.

The process should start with employee communications. In its employees, a facility has an important, informal communication link with its community that is often ignored. Employees deserve to know at least as much about their facility as their neighbors, and they deserve to know it first. They should be aware through communications (if not through actual involvement) of the facility's entire emergency preparedness program, from prevention to preparation. The employees also should be familiar with the products made and their end uses. They should understand the potential hazards of the processes and materials with which they work, and how to protect themselves and the public from those hazards.

Other communication channels should be developed and regularly used. Scheduled meetings with representatives of community emergency service groups are useful for exchanging information and objectives concerning emergency preparedness, for promoting the sharing of resources, for gaining familiarity with one another's physical facilities and people, for identifying problems, and for recommending action for their solution.

In the 1980s, some chemical facilities and their communities formalized this approach and broadened participation in their meetings to include public officials and representatives of regulatory agencies and the media. Operating as Hazardous Material Advisory Councils (HMACs) and meeting regularly with established leadership and agendas, these groups quickly became key resources in organizing a community's efforts to better understand the potential for hazardous material incidents and to protect against them. Typically the responsibilities of an HMAC include coordinating a regional risk assessment, assisting the development of a community response plan specific to hazardous materials, and assisting with educational programs for various segments of the public.

The Responsible Care program of the American Chemistry Council embraces HMACs as one good way to communicate relevant and useful information that responds to public concerns for safety, health, and the environment. However, managers are finding many other ways to interact with the public to achieve a fuller measure of community awareness and involvement in affairs of mutual interest. Among the many options from which a manager may choose are the following:

- hosting facility tours featuring emergency prevention and mitigation procedures
- speaking at community meetings (service clubs, schools, governing bodies, etc.)
- sending newsletters to selected neighbors
- preparing informative brochures or newspaper inserts
- appearing on local TV or radio

An open and sincere comprehensive risk communication process led by the facility manager creates a better-informed public that is able to understand real risks (vs. perceptions) and is likely to respond effectively in case of an actual emergency.

Integrating Plans

Most communities have long had written ERPs designed for natural events such as floods and windstorms; some communities have had written plans dealing with emergencies created by people, such as bomb threats and civil disturbances; but, until recently, few had specific plans for responding to emergencies involving hazardous materials. As a result, the consequences of accidental chemical releases have been in many cases tragically magnified by the undisciplined reactions of people near the release source. It has been reported, for example, that when the alarms sounded at Bhopal, residents of the nearby homes ran toward the plant rather than taking action to protect themselves from the enveloping fumes.

It is not enough to train the personnel of a chemical facility to implement an ERP effectively. Appropriate people in the community, especially near neighbors, need to understand the elements of an ERP that are designed to protect them and the role they play in making the plan work. There must be a continuous effort to integrate the facility ERP into community emergency planning at local and district levels. Drills involving external resources that test all the plan elements against the standards mutually established with the community and those imposed by laws and regulations are necessary to ensure successful implementation of the plan.

Off-Site Warning

Designing an effective off-site warning system presents some major challenges. Despite excellent ongoing communications between a chemical facility and its neighbors, there is no positive way to ensure that the general public will respond quickly and appropriately to a warning alarm of any kind. Thus, it is essential that the selected warning system be developed with the close cooperation of the community. Even then, it is difficult to predict such factors as the inclination of people to be warned and the degree of public confidence in the validity of an alarm.

No off-site warning system will assure complete coverage of the intended audience. Best results are achieved by combining two or more systems for sequential alerting—the first to trigger preplanned immediate action by the public at greatest risk, followed by other communications that provide further information and guidance to a larger audience. Some of the systems most commonly used are:

- facility fixed-sound sources, such as sirens and whistles
- mobile alerting by police or fire personnel, either from vehicles with loudspeakers or door-to-door
- fixed public address systems in the community or in the facility
- automated telephone calling
- alert radios energized by a special signal to produce a warning tone followed by broadcast messages
- strobe lights in situations where the noise level is a problem
- local radio stations and the emergency broadcast system
- local TV stations

More sophisticated and less commonly used warning systems include helicopters equipped with loudspeakers, modified cable TV installations, and computer networks between a chemical facility and community emergency response groups.

Local Emergency Plans

Existing plans for a coordinated response to emergencies in a community vary greatly in content and organization, but the plans have two common objectives. They are to:

- define authority and responsibilities of various emergency service participants
- describe the interaction between those participants, government, and industry

In creating their plans, most communities draw on the Integrated Emergency Management System (IEMS) developed by the Federal Emergency Management Agency (FEMA).

A local plan has many of the same elements as a chemical facility ERP. It includes:

- an emergency management organization, with designated functional responsibilities
- the location of the emergency operating center and its resources
- guidelines for classifying emergencies
- activation and declaration checklists
- communications requirements and available systems
- evacuation and sheltering plans
- methods for securing added resources
- descriptions of local hazards

Most local plans are written to be nonspecific as to the cause of the emergency, with various appendices describing the details of response to specific events. These appendices are based upon the results of risk assessments made by the community with the cooperation of industry.

One such appendix should relate to emergencies caused by fixed facility or transportation incidents in which hazardous materials are involved. Chemical facilities must provide substantial support to the community in preparing this portion of the local emergency plan, and provide resources and training leadership that are not available elsewhere in the community. Where a Hazardous Material Advisory Council exists, there is an effective forum for doing this. In any case, a chemical facility manager should seek ways to help the community prepare for and recover from incidents of this nature.

Local Emergency Planning Committees

An important contribution to community and industry cooperation in emergency preparedness was the passage in 1986 of the Superfund Amendments and Reauthorization Act (SARA), which contained an emergency planning and community right-to-know provision. Title III, or EPCRA, as this portion of SARA is commonly called, is intended to encourage and support hazardous materials emergency planning efforts at the state and local level, and to provide citizens and local agencies with appropriate information concerning potential hazards in their communities. The major portions of Title III require:

- a statewide organization for planning emergency action and receiving hazardous chemical information
- notification to the community of emergency releases of chemicals
- reports of hazardous chemical inventories and copies of MSDSs to be furnished to the community
- an annual inventory of hazardous chemical releases to the environment

Drills and Critiques

The optimum frequency of major drills involving personnel outside a chemical facility is dependent upon a number of variables:

- location of the facility
- dependence upon community emergency agencies
- size and complexity of the facility
- site and off-site risk assessments
- population patterns

An important element of emergency preparedness is the establishment of an appropriate major drill frequency in cooperation with off-site agencies. A reasonable goal is to hold one such drill each year. The scenario might include an on-site, internally generated hazardous material emergency one year and a transportation emergency somewhere in the adjacent community the following year.

For facilities and communities just beginning to test their plans, desktop or simulated drills are effective for identifying procedural problems that need to be corrected before they proceed to full-scale drills. In these simulations, staffing of the appropriate emergency center would occur, but the emergency service groups would not actually mobilize at the scene of the incident.

As people gain confidence in the completeness and the effectiveness of the emergency response plan, it becomes important to measure the performance of all who are involved. Monitors record and later report on all aspects of response actions, including:

- elapsed times before critical actions occur
- actions and coordination of responding groups
- actions of uninvolved personnel
- alarm and communication effectiveness
- emergency control center management
- control at the emergency scene
- accounting for personnel
- medical aid for simulated casualties
- off-site notifications
- handling media representatives—real or simulated

Following each drill there must be an organized critique that provides the information needed to strengthen the plan and/or its implementation. All the people actively involved should be represented at the critique, and a written report of conclusions and recommendations should be widely distributed. It then is the responsibility of the facility Emergency Preparedness Team and the local emergency planning committee to coordinate and assist in solving any problems identified—a process that begins emergency preparedness activities again: identifying hazards; evaluating and strengthening resources; modifying the emergency plan; training people; communicating and integrating plans; and testing them once again.

LAWS, REGULATIONS, AND SUPPORT

Laws

A number of legal requirements must be incorporated in a facility's ERP. Emergency prevention, preparedness, and response planning are regulated at the federal, state, and, occasionally, local levels. At the federal level, these laws include

- Clean Air Act (CAA)
- Clean Water Act (CWA)
- Comprehensive Environmental Response, Compensation, and Liability Act (CERCLA)

- Emergency Planning and Community Right-to-Know Act (EPCRA, or SARA Title III)
- Energy Reorganization Act (Nuclear Regulatory Commission)
- Hazardous Materials Transportation Act (HMTA)
- Occupational Safety and Health Act (OSHA)
- Resource Conservation and Recovery Act (RCRA)
- Toxic Substances Control Act (TSCA)
- U.S. Coast Guard requirements
- Environmental Protection Agency (EPA) regulations

A list of these laws and their *Code of Federal Regulations* (CFR) citations appears in the bibliography at the end of the chapter. States may have their own laws and regulations that also govern emergency response planning. References to these laws may be found in the Bureau of National Affairs (BNA) *Environment Reporter*.

These laws and their regulations are enforced by all levels of governmental agencies. A knowing or willful violation has serious implications for both companies and individuals, who may be held civilly or criminally liable for noncompliance. The penalties can be severe, ranging from daily-assessed fines to imprisonment. Thus, it is prudent to know the regulations that apply to the facility, and to ensure that the regulatory requirements are met.

The regulations regarding emergency planning and response are comprehensive, covering every aspect from prevention to reporting. RCRA's regulations cover the entire process, from planning to training to formal reports on the facility's response to an emergency involving hazardous waste. Other regulations focus on a specific aspect of emergency response, or part of the facility's operations. For example, the TSCA requires that spills or releases that contaminate the environment be reported orally and in writing within a certain time frame. All but one of the laws put the burden of planning on the facility. EPCRA, the exception, requires state and local agencies to prepare an ERP for the community. Facilities that meet criteria specified in EPCRA regulations have to assist in the development of the plan; however, they are not responsible for creating it.

The OSHA Process Safety Management regulation (29CFR 1910.119) and the EPA's Risk Management Plan regulation require significant attention to emergency planning and response. Inevitably, some of the regulations and requirements overlap. Most chemical facilities are subject to more than one law, and could be expected to prepare separate plans for specific parts of their facility. RCRA's Contingency Plan, for example, must be developed and maintained apart from other emergency response plans. The key to managing all the requirements and satisfying the regulations in an efficient, coordinated manner is first to understand the requirements and how they apply to the facility, and then to look for the common denominators among the requirements. The finished product, or master plan, will satisfy all the common denominators that apply, and will avoid duplication of effort. It also can be used as the basis for plans that must be maintained separately or that have requirements in addition to the common denominators.

Meeting the Requirements

Regulations governing emergency response planning can be broken down into four general categories:

- preparation
- plans
- reports and other communications
- drills and evaluations

Figure 4.6 shows a matrix-type summary of the major federal laws and their requirements for emergency planning and response. Such a matrix is very helpful in determining what the requirements are and how they apply to a facility. It could be further tailored to cover only the requirements that apply to a specific site or operation.

Prevention and Preparation

Some regulations require that a facility conduct a risk assessment and/or other preparatory activities. The RCRA calls it a preparedness and prevention plan. A facility subject to the RCRA must determine how structures, processes, and operations can be changed in order to minimize the possibility of an emergency involving hazardous waste. The facility also has to determine the communications and alarm systems that will be used in the event of such an emergency. The CWA includes prevention in its requirements for the Spill Prevention, Containment, and Countermeasure (SPCC) Plan. The 1990 amendments to the CAA added an accident prevention plan for extremely hazardous substances.

Plans

At the very least, a facility is required to develop a plan describing how it will respond to an incident that threatens human health and/or the environment. Generally, the plan includes notification, evacuation, protection of employees, and control of the incident. This emergency response plan usually must be in writing. For example, the OSHA requires a minimum of three plans: emergency response, emergency action, and fire prevention. The CAA requires that the state implementation plan have an emergency air pollution episode plan.

Communications

There are two aspects to emergency communications: the actual equipment used to communicate information about the incident and the types of communications or information-sharing required. The RCRA has specific requirements for the types of emergency communication equipment (alarm systems, phone or radio communications) that must be present. Under the EPCRA, facilities must provide information about their operations and substances used or stored on site when the Local Emergency Planning Committee (LEPC) or State Emergency Response Commission (SERC) requests it. If the facility uses or stores extremely hazardous substances (EHSs) in reportable quantities, it must appoint a representative to the LEPC. Several laws require that a copy of the ERP be made available to employees and

	RCRA	OSHA	DOT/HMTA	EPCRA	CERCLA
PREVENTION AND PLANNING	Preparedness & prevention plan Contingency Plan (CP) in writing: Emergency coordinator, Evacuation, Access, Equipment, Communications	Emergency Response Plan : For entire site; For employees who respond to uncontrolled re-leases of hazardous substances, including hazardous wastes Emergency Action Plan: Evacuation for employees in case of incidental chemical release; How to report an incident Fire Prevention Plan	Incident preven-tion and response	Local Emergency Planning Committee prepares emergency response plan for community State Emergency Planning Commission (SERC) oversees.	National Contingency Plan (NCP) Facility plan for response and cleanup of oil or hazardous substance must meet NCP Standards
REPORTS	Incidents in transit Transporter must: Notify National Response Center (NRC) at once; Submit written report within 15 days to DOT; Coordinate with DOT. Hazardous waste emer-gency on site: Immediately-NRC Follow-up, in writing to EPA RA within 15 days UST releases	Process hazards review	Report six specific hazardous material incidents, at once to DOT/NRC; written follow-up at once to NRC Spill of RQ into navigable waters	Releases of extremely hazardous substances over reportable quantity (RQ) EHS stored on site in quantities $\geq$ RQ	Release of RQ or 1 pound of hazardous sub-stance-to NRC, immediately
INFORMATION	Maintain copy of CP at site Provide copy to local emergency response organization Establish alarm and communication systems for emergency notification.	MSDS to employees and emergency response organization	Emergency response information available during transportation and at facilities where hazardous materials are loaded or stored.	As required by LEPC and SERC Designate representative to LEPC. Coordinate internal plans or make them available to LEPC.	
TRAINING, DRILLS, AND EVALUATION	Emergency response must be documented and records retained Initial and annual review Evacuation drill	Initial and annual refresher training for employees involved in emergency response; varies with roles-- All employees trained in Emergency Action Plan; initially and with every change to plan. Training in MSDS information			

Fig. 4.6. Emergency prevention, planning, and response.

representatives of government agencies dur-ing working hours. The OSHA requires that facilities provide material safety data sheets (MSDS) for all hazardous substances present on the site.

Reports

Reports are another important communica-tions aspect of the ERP. Most laws insist on prompt notification of the proper agencies immediately after an incident occurs. These

CWA	CAA	TSCA	NUREG 054	USCG	EPA
Spill Prevention Control and Countermeasure (SPCC) Plan Shows how facility will: prevent, respond, follow-up to oil spills in harmful quantities. Must be in writing	State Iplementation Plan (SIP) must include an Emergency Air Pollution Episode plan Accident Prevention Program for EHS States may require prevention and emergency response plan		Emergency response plan, including: Emergency planning zones; Prevention, mitigation, and limitation of core damage and consequences of release. Subject to annual review		Risk management plan
Release of RQ of oil or hazardous substance to NRC, immediately.	Announce uncontrolled releases of pollutants over certain set levels Releases ≥ RQ to NRC	Emergency incidents of environmental contamination At once to EPA, Written follow-up within 10 days	Notify state and local officials: Change in condition Protective action recommendations Notify community within 15 minutes. 100% notification within 15 miles.	All spills into navigable waters of oil, hazardous substances, ≥ RQ, at once, to US Coast Guard or to NRC	Worst case scenarios
Keep copy of SPCC at site available to EPA RA during normal working hours. Submit SPCC amendments to EPA RA and state water pollution control agency					Emergency response plan
Train : SPCC responders Employees who operate, maintain equipment			Annual training On-site Off-site Annual graded emergency response exercises		Training required for all employees in emergency response plan

Fig. 4.6. (Continued)

reporting requirements can be complex, particularly in view of the fact that many laws have their own lists of hazardous substances and reportable quantities. The CERCLA requires that releases of a reportable quantity of what it defines as a hazardous substance must be reported immediately to the National Response Center (NRC). The EPCRA requires facilities that store and/or release reportable quantities of substances on its EHS list to report that information to the LEPC and the SERC. EPA's Risk Management Plan requires prior disclosure of possible "worst case" incident scenarios.

Written follow-up reports are often a requirement. The HMTA has identified six specific hazardous material incidents that must be reported immediately and again in writing. The RCRA gives managers 15 days after a hazardous waste emergency to submit a written report to the EPA.

Training, Drills, Audits, and Evaluations

An emergency plan is relatively useless unless the employees affected by it are trained in its use. The RCRA, OSHA, HMTA, CWA, and the Energy Reorganization Act require annual and refresher training. In addition, the facilities must keep records of the training, and must make them available to the appropriate agency when they are requested. Some laws go so far as to require practice drills. Nuclear power plants must conduct on- and off-site training, and go through annual graded emergency response exercises. The plan and the response executed according to the plan then are evaluated so that the plan can be improved.

Sources of Assistance

Seeing all the requirements together can be overwhelming. Fortunately, there are agencies, associations, and programs that can assist in the preparation of a comprehensive emergency prevention and response plan.

The federal government and the agencies responsible for the laws that govern emergency response planning provide 800-number hotlines and manuals that describe various aspects of emergency prevention, planning, and response.

The volunteer or professional emergency responders in the community have valuable practical experience that can be put to work in developing the facility ERP. Working with them also establishes a forum for communications and understanding with the community.

The American Chemistry Council (ACC) Community Awareness and Emergency Response (CAER) program provides comprehensive guidelines for the development and implementation of an ERP. The CAER program has been expanded to include all aspects of the chemical industry in an initiative called Responsible Care.

Other services of the ACC include CHEMTREC, a 24-hr emergency response service for people who respond to emergencies involving chemicals; CHEMNET, a mutual aid agreement between chemical producers and emergency response contractors; and workshops and videotape training programs for first responders and other emergency response personnel.

Additional sources of assistance and information include other professional associations, such as the American Institute of Chemical Engineers, and publications, seminars, workshops, and videotapes offered by educational organizations. Considerable information is available on the Internet through web sites such as www.fema.gov which is maintained by the Federal Emergency Management Agency.

A bibliography; a list of laws, regulations, and standards; and a compilation of suggested reading material follow.

SELECT BIBLIOGRAPHY

American Chemistry Council (Formerly Chemical Manufacturers Association) *CAER: The Next Phase Program Handbook*, Chemical Manufacturers Association (now American Chemistry Council), Washington, DC, with assistance from HMM Associates, Concord, MA, 1989.

American Chemistry Council (Formerly Chemical Manufacturers Association) *Community Emergency Response Exercise Handbook*, Chemical Manufacturers Association (now American Chemistry Council), Washington, DC, 1986.

American Chemistry Council (Formerly Chemical Manufacturers Association) *Emergency Warning Systems Guidebook*, Chemical Manufacturers Association (American Chemistry Council), Washington, DC, 1987.

American Chemistry Council (Formerly Chemical Manufacturers Association) *Site Emergency Response Planning Handbook*, Chemical Manufacturers Association (now American Chemistry Council), Washington, DC, 1986.

The Conservation Foundation, *Risk Communication: Proceedings of the National Conference on Risk Communication*, The Conservation Foundation, Washington, DC, 1986.

Covello, V. T., Sandman, P. M., and Slovic, P., *Risk Communication, Risk Statistics, and Risk Comparisons*, Chemical Manufacturers Association (American Chemistry Council), Washington, DC, 1988.

Federal Emergency Management Agency, *FEMA Handbook*, Federal Emergency Management Agency, Washington, DC, 1985.

National Response Team, *Hazardous Materials Emergency Planning Guide (NRT 1)*, National Response Team of the National Oil and Hazardous Substances Contingency Plan, Washington, DC, 1987.

National Response Team, *Criteria for Review of Hazardous Materials Emergency Plans*, National Response Team of the National Oil and Hazardous Substances Contingency Plan. Washington, DC, 1988.

Occupational Safety and Health Administration, *How to Prepare for Workplace Emergencies, OSHA 3088 (revised)*, U.S. Department of Labor, Washington, DC, 1988.

Prugh, R. W., and Johnson, R. W., *Guidelines for Vapor Release Mitigation*, Center for Chemical Process Safety, The American Institute of Chemical Engineers, New York, 1988.

SARA Title III (Superfund Amendment and Reauthorization Act, Emergency Planning and Community Right-to-Know)

4OCFR 355.3, Section 302(c), Emergency planning notification Section 303(d), Appointment of emergency coordinator, provision of information; Section 304, Emergency release notification requirements.

EPA 40 CFR Part 68, Risk Management Plan.

U.S. Coast Guard 33CFR 126.9 Reporting requirements for discharge of petroleum products or dangerous liquid commodities into navigable waters of USA 33CFR 153.023 Reporting requirements for discharge of reportable quantity (RQ) of oil or hazardous substance into navigable waters.

Standards

ANSI/National Fire Protection Association Standards
  # 72 National Fire Alarm Code
  # 110 Emergency Power and Standby Systems
  # 600 Private Fire Brigades
  # 1561 Emergency Services Incident Management System

SUGGESTED READING

Regulations

Bureau of National Affairs, *Environment Reporter*, Bureau of National Affairs, Inc., Washington, DC.

ENSR Consulting and Engineering, *Air Quality Handbook: A Guide to Permitting and Compliance under the Clean Air Act and Air Toxics Programs, 10th ed.*, ENSR Consulting and Engineering, Acton, MA, 1988.

ERT, Inc., and Sidley & Austin, *Superfund Handbook, 2nd ed.*, ERT, Inc. and Chicago, IL: Sidley & Austin, Concord, MA, 1987.

Lowry, G. G., and Robert, C., *Lowry's Handbook of Right-to-Know and Emergency Planning*, Lewis Publishers, Inc., Chelsea, MI, 1988.

Office of Solid Waste, U.S. Environmental Protection Agency, *Solving the Hazardous Waste Problem: EPA's RCRA Program (EPA/530-SW-86-037)*, U.S. Environmental Protection Agency, Washington, DC, 1986.

Prevention and Planning

Abrams, M. J., and Lewis, J., "Preplanning, the Key Emergency Response," *Papers Presented at the Spring 1987 National Meeting of the American Institute of Chemical Engineers*, American Institute of Chemical Engineers, New York, 1987.

Adler, V., Sorenson, J. H., and Rogers, G. O., "Chemical and Nuclear Emergencies: Interchanging Lessons Learned from Planning and Accident Experience," *Proceedings of a Joint NEA/CEC Workshop on Technical Aspects of Emergency Planning, Brussels, Belgium (CONF-89906137-1)*, U.S. Department of Energy, Washington, DC, 1989.

U.S. Environmental Protection Agency, *Guide to Exercises in Chemical Emergency Preparedness Programs*, U.S. Environmental Protection Agency, Washington, DC, 1988.

LAWS, REGULATIONS, AND STANDARDS

Laws and Regulations

CAA (Clean Air Act)
   4OCFR 51 SIP Emergency air pollution episode plan
CAA Reauthorization Section 304, Prevention of sudden, accidental releases
CERCLA (Comprehensive Environmental Response, Compensation, and Liability Act)
   4OCFR 300-306 Section 103(a) Spill reporting requirements
   4OCFR 355 Emergency planning
CWA (Clean Water Act)
   4OCFR 112.3-7 Spill prevention, control, and countermeasure plan (SPCC) requirements, modifications, preparation, and implementation Energy Reorganization Act (was Atomic Energy Act)
   1OCFR 50.47, also Appendix E NRC (Nuclear Regulatory Commission) Standards for on-site and off-site emergency plans
HMTA (Hazardous Materials Transportation Act)
   49CFR Parts 171-177 DOT requirements for providing information and advice on meeting emergencies; FEMA requirements for evaluation of training programs for incident prevention and response
OSHA (Occupational Safety and Health Act)
   29CFR 1910.1200 Hazard communication standard
   29CFR 1910-210 and Appendices, Hazardous waste operations and emergency response
   29CFR 1910.35-38 Requirements for evacuation route and exit posting, emergency lighting, accessibility of exits, and emergency action plan
   29CFR 1910.156 Fire brigades
   29CFR 1910.157-164 Fire extinguishing and detection systems
   29CFR 1910.165 Employee alarm systems
   29CFR 1910.119 Process Safety Management of Highly Hazardous Chemicals
RCRA (Resource Conservation and Recovery Act)
   4OCFR 263.30,31 Transporter responsibilities in hazardous waste transportation incidents
   4OCFR 264.30-37 and 4OCFR 265.30-37 Preparedness and prevention
   4OCFR 264.50-56 and 4OCFR 265.50-56 Contingency Plan

American Nuclear Society. *Emergency Response—Planning, Technologies, and Implementation; Proceedings of the ANS Topical Meeting (CONF-880913. UC-610)*, Savannah River Laboratory, E. I. du Pont de Nemours and Company, Aiken, SC, 1988.

Bell, D. W., and Burns, C. C., "Offsite Emergency Plan Development and Maintenance Considerations," *Papers Presented at the Spring 1987 National Meeting of the American Institute of Chemical Engineers*, American Institute of Chemical Engineers, New York, 1987.

Davis, D. S., et al., *Prevention Reference Manual: Overviews on Preventing and Controlling Accidental Releases of Selected Toxic Chemicals (EPA/600/8-88-074)*, U.S. Environmental Protection Agency, Air and Energy Engineering Research Lab., Research Triangle Park, NC, 1988.

Dickerson, M. H., "Emergency Planning, Response, and Assessment: A Concept for a Center of Excellence," *Proceedings from an International Seminar on Nuclear War International Cooperation: The Alternatives (CONF-8608149-1)*, U.S. Department of Energy, Washington, DC, 1986.

Federal Emergency Management Agency, *Hazardous Materials Exercise Evaluation Methodology (HM-EEM) and Manual*, Federal Emergency Management Agency, Washington, DC, 1989.

Fingleton, D. J., Tanzman, E. Z., and Bertram, K. M., "Development of a Model Emergency Response Plan for Catastrophic Releases of Toxic Gases," *Proceedings of the Air Pollution Control Association Annual Meeting (CONF-860606-19)*, U.S. Department of Energy, Washington, DC, 1986.

Gudiksen, P. et al., *Emergency Response Planning for Potential Accidental Liquid Chlorine Releases*, U.S. Department of Energy, Washington, DC, 1986.

International Association of Fire Chiefs, *Fire Service Emergency Management Handbook*, Federal Emergency Management Agency, Washington, DC, 1985.

Jones, E., "Contingency Planning and Emergency Response in Construction Activities: Training the Construction Worker," *Proceedings of an Oak Ridge Model Conference on Waste Problems (CONF-871075-5)*, U.S. Department of Energy, Washington, DC, 1987.

Kalnins, R. V., "Emergency Preparedness and Response," *Symposium on the Characterization of Thermodynamic and Transport Properties of Polymer Systems*, American Institute of Chemical Engineers, New York, 1986.

Krimsky, S., and Plough, A., *Environmental Hazards—Communicating Risks as a Social Process*, Auburn House Publishing Company, Medford, MA, 1988.

Michael, E. J. et al., "Emergency Planning Considerations for Specialty Chemical Plants," *American Institute of Chemical Engineers Summer National Meeting*, American Institute of Chemical Engineers, New York, 1986.

Morentz, J. W., and Griffith, D., "Using Computers for Chemical Emergency Planning and Response," *Papers Presented at the Spring 1987 National Meeting of the American Institute of Chemical Engineers*, American Institute of Chemical Engineers, New York, 1987.

National Response Team, *Criteria for Review of Hazardous Materials Emergency Plans (NRT-1A)*, National Response Team of the National Oil and Hazardous Substances Contingency Plan, Washington, DC, 1988.

National Response Team, *HAZMAT Planning Guide (WH-562A)*, National Response Team, Washington, DC, 1987.

National Response Team, *Technical Guidance for Hazards Analysis: Emergency Planning for Extremely Hazardous Substances (Supplement to Hazardous Materials Emergency Planning Guide NRT 1)*, National Response Team, Washington, DC, 1987.

Philley, J. O., "Emergency Preparedness Training Tips," *American Institute of Chemical Engineers National Meeting—Summer '87*, American Institute of Chemical Engineers, New York, 1987.

U.S. Environmental Protection Agency, *Bibliography on Chemical Emergency Preparedness and Prevention*, U.S. Environmental Protection Agency, Washington, DC, 1986.

U.S. Environmental Protection Agency, *It's Not Over in October: A Guide for Local Emergency Planning Committees*, U.S. Environmental Protection Agency, Washington, DC, 1988.

U.S. Environmental Protection Agency, *Seven Cardinal Rules of Risk Communication (EPA 87-020)*, U.S. Environmental Protection Agency, Washington, DC, 1988.

U.S. Environmental Protection Agency, "Why Accidents Occur: Insights from the Accidental Release Information Program," *Chemical Accident Prevention Bulletin (OSWER-89-008.1)*, Series 8, No. 1 (July 1989), U.S. Environmental Protection Agency, Washington, DC, 1989.

Waldo, A. B. (Ed.), *The Community Right-to-Know Handbook*, Thompson Publishing Group, Washington, DC, 1986.

Transportation

Office of Hazardous Materials Transportation, U.S. Department of Transportation, *Emergency Response Guidebook*. Available through the U.S. Government Printing Office (GPO) U.S. Government Bookstores. Updated every three years.

5

Applied Statistical Methods and the Chemical Industry

Stephen Vardeman* and Robert Kasprzyk**

INTRODUCTION

The discipline of statistics is the study of effective methods of data collection, data summarization, and (data based, quantitative) inference making in a framework that explicitly recognizes the reality of nonnegligible variation in real-world processes and measurements.

The ultimate goal of the field is to provide tools for extracting the maximum amount of useful information about a noisy physical process from a given investment of data collection and analysis resources. It is clear that such a goal is relevant to the practice of industrial chemistry. The primary purposes of this chapter are to indicate in concrete terms the nature of some existing methods of applied statistics that are particularly appropriate to industrial chemistry, and to provide an entry into the statistical literature for those readers who find in the discussion here reasons to

believe that statistical tools can help them be effective in their work.

This chapter will begin with some simple ideas of modern descriptive statistics, including numerical and graphical data summarization tools, and the notions of fitting equations to data and using theoretical distributions. Next, some tools for routine industrial process monitoring and capability assessment, concentrating primarily on the notion of control charting, will be presented. This will be followed by a more extensive discussion of common statistical data collection strategies and data analysis methods for multifactor experimental situations met in both laboratory and production environments. This section will touch on ideas of partitioning observed variation in a system response to various sources thought to influence the response, factorial and fractional factorial experimental designs, sequential experimental strategy, screening experiments, and response surface fitting and representation. Next come brief discussions of two types of special statistical tools associated specifically with chemical applications, namely, mixture techniques and nonlinear mechanistic

*Iowa State University, Departments of Statistics and of Industrial Engineering and Manufacturing Systems Engineering.
**Dow Chemical Company.

model building. A short exposition of chemical industry implications of relationships between modern business process improvement programs and the discipline of statistics follows. The chapter concludes with a reference section listing sources for further reading.

SIMPLE TOOLS OF DESCRIPTIVE STATISTICS

There are a variety of data summarization or description methods whose purpose is to make evident the main features of a data set. (Their use, of course, may be independent of whether or not the data collection process actually employed was in any sense a "good" one.) To illustrate some of the simplest of these methods, consider the data listed in Table 5.1. These numbers represent aluminum impurity contents (in ppm) of 26 bihourly samples of recycled PET plastic recovered at a Rutgers University recycling pilot plant.

A simple plot of aluminum content against time order, often called a *run chart*, is a natural place to begin looking for any story carried by a data set. Figure 5.1 shows such a plot for the data of Table 5.1, and in this case reveals only one potentially interesting feature of the data. That is, there is perhaps a weak hint of a downward trend in the aluminum contents that might well have been of interest to the original researchers. (If indeed the possible slight decline in aluminum contents is more than "random scatter," knowledge of its physical origin, whether in actual composition of recycled material or in the measurement process, presumably would have been helpful to the effective running of the recycling facility. We will save a discussion of tools for rationally deciding whether there is more than random scatter in a plot like Fig. 5.1 until the next section.)

The run chart is a simple, explicitly dynamic tool of descriptive statistics. In those cases where one decides that there is in fact

TABLE 5.1   Twenty-Six Consecutive Aluminum Contents (ppm)[a]

291,	222,	125,	79,	145,	119,	244,	118,	182,	63,	30,	140,	101
102,	87,	183,	60,	191,	119,	511,	120,	172,	70,	30,	90,	115

[a]Based on data in Albin.[1]

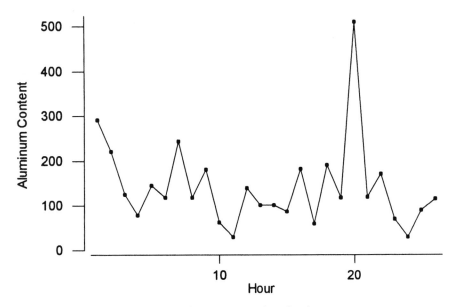

Fig. 5.1. A run chart for 26 consecutive aluminum contents.

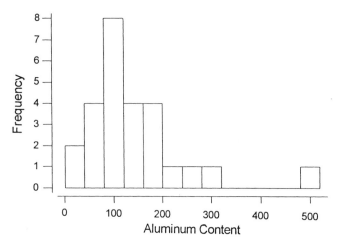

Fig. 5.2. A histogram for 26 aluminum contents.

little information in the time order corresponding to a data set, there are a variety of simple, essentially static, statistical tools that can
be used in describing the pattern of variation
in a data set. Figures 5.2–5.5 show graphical
representations of the data of Table 5.1 in,
respectively, *histogram, stem and leaf plot,
dot plot*, and *box plot* forms.

The histogram/bar chart idea of Fig. 5.2 is
likely familiar to most readers, being readily
available, for example, through the use of
commercial spreadsheet software. It shows
how data are spread out or distributed across
the range of values represented, tall bars
indicating high frequency or density of data
in the interval covered by the base of the bar.
Figure 5.2 shows the measured aluminum
contents to be somewhat asymmetrically distributed (statistical jargon is that the distribution is "skewed right"), with a "central"
value perhaps somewhere in the vicinity of
120 ppm.

Histograms are commonly and effectively
used for final data presentation, but as
working data analysis tools they suffer from
several limitations. In one direction, their
appearance is fairly sensitive to the data
grouping done to make them, and it is usually
not possible to recover from a histogram the
exact data values used to produce it, should
one wish to try other groupings. In another
direction, histograms are somewhat unwieldy,

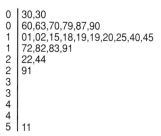

Fig. 5.3. A stem and leaf plot for 26 aluminum
contents.

for example, not being particularly suitable
to the comparison of, say, 10 or 12 data sets
on a single page. The graphical devices of
Figs 5.3–5.5 are less common than the
histogram, but address some of these
shortcomings.

The stem and leaf diagram of Fig. 5.3 and
the dot plot of Fig. 5.4 carry shape information about the distribution of aluminum contents in a manner very similar to the
histogram of Fig. 5.2. But the stem and leaf
and dot diagrams do so without losing the
exact identities of the individual data points.
The box plot of Fig. 5.5 represents the "middle half" of the data with a box divided at
the 50th percentile (or in statistical jargon,
the median) of the data, and then uses so-
called whiskers to indicate how far the most
extreme data points are from the middle half
of the data.

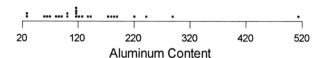

Fig. 5.4. A dot plot for 26 aluminum contents.

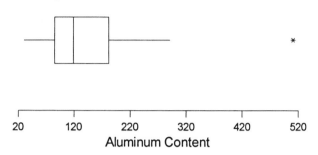

Fig. 5.5. A box plot for aluminum contents.

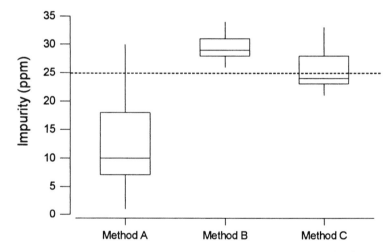

Fig. 5.6. Side-by-side box plots for three laboratory test methods.

Box plots preserve much of the shape information available from the other displays (e.g., portraying lack of symmetry through differing sizes of box "halves" and/or whisker lengths), but do so in a way that is conducive to simultaneous representation and comparison of many data sets on a single graphic, through the placement of box plots side by side. Figure 5.6 illustrates this point with a graphical comparison of three laboratory test methods to a standard.

A total of 90 samples of a stock solution known to contain 25 ppm of an impurity were analyzed by a single lab team using three different test methods (30 of the samples being allocated to each of the three methods), and the box plots in Fig. 5.6 portray the measured impurity levels for the different methods. The figure shows quite effectively that Method A is neither precise nor accurate, Method B is quite precise but not accurate, and Method C is somewhat less precise than B but is accurate. This kind of knowledge can form the basis of an informed choice of method.

Figures 5.2–5.6 give only a hint of the spectrum of tools of statistical graphics that are potentially helpful in data analysis for industrial chemistry. For more details and much additional reading on the subject of modern statistical graphics, the reader is referred to the book by Chambers et al.[2] listed in the references section.

Complementary to graphical data summaries are *numerical summarizations*. For the simple case of data collected under a single set of conditions, the most commonly used measures deal with the location/center of the data set and the variability/spread of the data. The *(arithmetic) mean* and the *median* are the most popular measures of location, and the *variance* and its square root, the *standard deviation*, are the most widely used measures of internal variability in a data set.

For n data values $y_1, y_2, \ldots, y_n$ the median is

$$\tilde{y} = \text{the "middle" or } \frac{n+1}{2} \text{th ordered data value} \tag{5-1}$$

and the mean is

$$\bar{y} = \frac{1}{n} \sum_{i=1}^{n} y_i \tag{5-2}$$

The reader is invited to check that upon ordering the $n = 26$ values in Table 5.1, the 13th smallest value is 119 and the 14th smallest value is also 119, so that the only sensible interpretation of (5-1) for the aluminum content data is that

$$\tilde{y} = \text{the 13.5th ordered data value}$$

$$= \frac{119 + 119}{2} = 119 \text{ ppm}$$

On the other hand, from (5-2) the mean of the aluminum contents is

$$\bar{y} = \frac{1}{26}(291 + 222 + 125 + \cdots + 30 + 90 + 115)$$

$$\approx 142.7 \text{ ppm}$$

The median and mean are clearly different measures of location/center. The former is in the middle of the data in the sense that about half of the data are larger and about half are smaller. The latter is a kind of "center of mass," and for asymmetrical data sets like that of Table 5.1 is usually pulled from the median in the direction of any "skew" present, that is, is pulled in the direction of "extreme" values.

The variance of n data values $y_1, y_2, \ldots, y_n$ is essentially a mean squared deviation of the data points from their mean. In precise terms, the variance is

$$s^2 = \frac{1}{n-1} \sum_{i=1}^{n} (y_i - \bar{y})^2 \tag{5-3}$$

and the so-called standard deviation is

$$s = \sqrt{s^2} = \sqrt{\frac{1}{n-1} \sum_{i=1}^{n} (y_i - \bar{y})^2} \tag{5-4}$$

For the example of the aluminum contents, it is elementary to verify that

$$s^2 \approx \frac{1}{26-1}[(291 - 142.7)^2 + (222 - 142.7)^2$$

$$+ \cdots + (115 - 142.7)^2]$$

$$\approx 9{,}644 \text{ (ppm)}^2$$

so that

$$s = \sqrt{s^2} \approx 98.2 \text{ ppm}$$

An appropriate interpretation of s is not completely obvious at this point, but it does turn out to measure the spread of a data set, and to be extremely useful in drawing quantitative inferences from data. (In many, but not all, circumstances met in practice, the *range* or largest value in a data set minus the smallest value is on the order of four to six times s.) The variance and standard deviation are time-honored and fundamental quantifications of the variation present in a single group of measurements and, by implication, the data-generating process that produced them.

When data are collected under several different sets of conditions, and those conditions can be expressed in quantitative terms, effective data summarization often takes the form of *fitting an approximate equation* to the data. As the basis of a simple example of this, consider the data in Table 5.2. The variable x, hydrocarbon liquid hourly space velocity, specifies the conditions under which information on the response variable y, a measure of isobutylene conversion, was obtained in a study involving the direct hydration of olefins.

For purposes of economy of expression, and perhaps some cautious interpolation between values of x not included in the original data

TABLE 5.2   Seven Liquid Hourly Space Velocity/Mole % Conversion Data Pairs[a]

Liquid Hourly Space Velocity, x	Mole % Isobutylene Conversion, y
1	23.0, 24.5
2	28.0
4	30.9, 32.0, 33.6
6	20.0

[a]Based on a graph in Odioso et al.[3]

set, one might well like to fit a simple equation involving some parameters b, say,

$$y \approx f(x|\underset{\sim}{b}) \tag{5-5}$$

to the data of Table 5.2. The simplest possible form for the function $f(x|\underset{\sim}{b})$ that accords with the "up then back down again" nature of the conversion values y in Table 5.2 is the quadratic form

$$f(x|\underset{\sim}{b}) = b_0 + b_1 x + b_2 x^2 \tag{5-6}$$

and a convenient method of fitting such an equation (that is linear in the parameters b) is *the method of least squares*. That is, to fit a parabola through a plot of the seven (x, y) pairs specified in Table 5.2, it is convenient to choose b_0, b_1, and b_2 to minimize the sum of squared differences between the observed conversion values y and the corresponding fitted values y on the parabola. In symbols, the least squares fitting of the approximate relationship specified by (5-5) and (5-6) to the data of Table 5.2 proceeds by minimization of

$$\sum_{i=1}^{7} [y_i - (b_0 + b_1 x_i + b_2 x_i^2)]^2$$

over choices of the coefficients b. As it turns out, use of standard statistical "regression analysis" software shows that the fitting process for this example produces the approximate relationship

$$y \approx 13.64 + 11.41x - 1.72x^2$$

and Fig. 5.7 shows the fitted (summarizing) parabola sketched on the same set of axes used to plot the seven data points of Table 5.2.

The least squares fitting of approximate functional relationships to data with even multidimensional explanatory variable x typically goes under the (unfortunately obscure) name of *multiple regression* analysis, and is given an introductory treatment in most engineering statistics textbooks, including, for example, the ones by Devore,[4] Vardeman and Jobe,[5] and Vardeman[6] listed in the references. A lucid and rather complete treatment of the subject can also be found in the book by Neter et al.[7]

A final notion that we wish to treat in this section on descriptive statistics is that of representing a distribution of responses and/or the mechanism that produced them (under a single set of physical conditions) by a *theoretical distribution*. That is, there are a number of convenient theoretical distributional shapes, and it is often possible to achieve great economy of expression and thought by seeing in a graphical representation such as Figs 5.2–5.5 the possibility of henceforth describing the phenomenon portrayed via some one of those theoretical distributions. Here we will concentrate on only the most commonly used theoretical distribution, the so-called *Gaussian* or *normal* distribution.

Figure 5.8 is a graph of the function of x

$$g(x) = \frac{1}{\sqrt{2\pi\sigma^2}} \exp\left(-\frac{(x-\mu)^2}{2\sigma^2}\right) \tag{5-7}$$

where $g(x)$ specifies the archetypical "bell-shaped curve" centered at the number μ, with spread controlled by the number σ (and is in fact usually called the Gaussian probability density with mean μ and standard deviation σ).

Figure 5.8 can be thought of as a kind of idealized histogram. Just as fractional areas enclosed by particular bars of a histogram correspond to fractions of a data set with values in the intervals represented by those bars, areas under the curve specified in (5-7) above particular intervals might be thought of as corresponding to fractions of potential data points having values in those intervals. (It is possible to show that the total area under the curve represented in Fig. 5.8, namely, $\int_{-\infty}^{\infty} g(x)\, dx$, is 1.) Simple tabular methods presented in every elementary statistics book

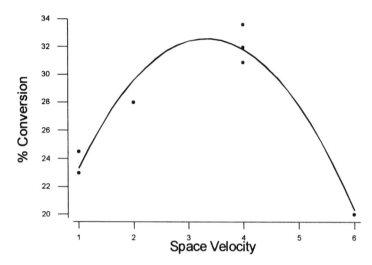

Fig. 5.7. A scatter plot of seven space velocity/mole % conversion data pairs and a fitted parabola.

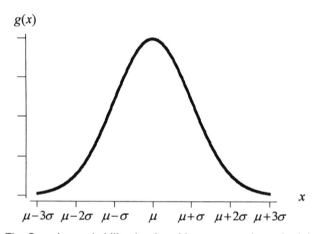

Fig. 5.8. The Gaussian probability density with mean μ and standard deviation σ.

TABLE 5.3   Twenty-Six Logarithms of Aluminum Contents

5.67, 5.40,   4.83,   4.37,   4.98,   4.78,   5.50,   4.77,   5.20, 4.14,   3.40,   4.94, 4.62
4.62, 4.47,   5.21,   4.09,   5.25,   4.78,   6.24,   4.79,   5.15, 4.25,   3.40,   4.50, 4.74

avoid the need to regularly use numerical integration in evaluating such areas. These methods can, for example, be used to show that roughly 68 percent of a Gaussian distribution lies between $\mu - \sigma$ and $\mu + \sigma$, roughly 95 percent lies between $\mu - 2\sigma$ and $\mu + 2\sigma$, and roughly 99.7 percent lies between $\mu - 3\sigma$ and $\mu + 3\sigma$. Part of the convenience provided when one can treat a data-generating process as approximately Gaussian is that, given only a theoretical mean μ and theoretical standard deviation σ, predictions of fractions of future data values likely to fall in intervals of interest are thus easy to obtain.

At this point let us return to the aluminum content data of Table 5.1. The skewed shape that is evident in all of Figs 5.2–5.5 makes a Gaussian distribution inappropriate as a theoretical model for (raw) aluminum content of such PET samples. But as is often the case with right skewed data, considering the *logarithms* of the original measurement creates a scale where a normal distribution is more plausible as a representation of the phenomenon under

```
3 | .40,.40
3 |
4 | .09,.14,.25,.37,.47
4 | .50,.62,.62,.74,.77,.78,.79,.79,.83,.94,.98
5 | .15,.20,.21,.25,.40
5 | .50,.67
6 | .24
```

Fig. 5.9. A stem and leaf plot for the logarithms of 26 aluminum contents.

study. Thus, Table 5.3 contains the natural logs of the values in Table 5.1, and the corresponding stem and leaf plot in Fig. 5.9 shows the transformed data to be much more symmetrically distributed than the original data. The possibility opened up by this kind of transformation idea is one of using statistical methods based on the normal distribution to reach conclusions about ln$y$ and then simply exponentiating to derive conclusions about the original response y itself. The applicability of statistical methods developed for normal distributions is thereby significantly broadened.

In addition to providing convenient conceptual summarizations of the nature of response distributions, theoretical distributions such as the normal distribution form the mathematical underpinnings of methods of formal quantitative *statistical inference*. It is outside our purposes in this chapter to provide a complete introduction to such methods, but thorough and readable accounts are available in engineering statistics books such as those of Devore[4] and Vardeman and Jobe.[5] Here, we will simply say that, working with a Gaussian description of a response, it is possible to quantify in various ways how much information is carried by data sets of various sizes. For instance, if a normal distribution describes a response variable y, then in a certain well-defined sense, based on $n = 26$ observations producing a mean $\bar{y}$ and a standard deviation s, the interval with end points

$$\bar{y} - 2.060s \sqrt{1 + \frac{1}{26}}$$

and   (5-8)

$$\bar{y} + 2.060s \sqrt{1 + \frac{1}{26}}$$

has a 95 percent chance of predicting the value of an additional observation. For instance, applying formula (5-8) to the log values in Table 5.3, the conclusion is that the interval from 3.45 to 6.10 ln(ppm) has a 95 percent chance of bracketing an additional log aluminum content produced (under the physical conditions of the original study) at the recycling plant. Exponentiating, the corresponding statement about raw aluminum content is that the interval from 31 to 446 ppm has a 95 percent chance of bracketing an additional aluminum content. Methods of statistical inference like that represented in (5-8) are called *prediction interval* methods. The book by Hahn and Meeker[8] provides a thorough discussion of such methods, based not only on the Gaussian distribution but on other theoretical distributional shapes as well.

TOOLS OF ROUTINE INDUSTRIAL PROCESS MONITORING AND CAPABILITY ASSESSMENT

Probably the two most basic generic industrial problems commonly approached using statistical methods are those of (1) monitoring and maintaining the stability/consistency of a process and (2) assessing the capability of a stable process. This section provides a brief introduction to the use of tools of "control" charting in these enterprises.

Working at Bell Labs during the 1920s and 1930s, Walter Shewhart developed the notion of routinely plotting data from an industrial process in a form that allows one to separate observed variability in a response into two kinds of variation. The first is that variation which appears to be inherent, unavoidable, short-term, baseline, and characteristic of the process (at least as currently configured). This variation Shewhart called *random* or *common cause variation*. The second kind of variability is that variation which appears to be avoidable, long-term, and/or due to sources outside of those seen as legitimately impacting process behavior. This variation he called *assignable* or *special cause variation*.

Shewhart reasoned that by plotting summary statistics from periodically collected data sets against time order of collection, one would be able to see interpretable trends or other evidence of assignable variation on the plots, and could intervene to eliminate the physical causes of that variation. The intention was to thereby make process output stable or consistent to within the inherent limits of process precision. As a means of differentiating plotted values that should signal the need for intervention from those that carry no special message of process distress, he suggested drawing so-called control limits on the plots. (The word "control" is something of a misnomer, at least as compared to common modern engineering usage of the word in referring to the active, moment-by-moment steering or regulation of processes. The nonstandard and more passive terminology "monitoring limits" would actually be far more descriptive of the purpose of Shewhart's limits.) These limits were to separate plausible values of the plotted statistic from implausible values when in fact the process was operating optimally, subject only to causes of variation that were part of standard conditions.

By far the most famous implementations of Shewhart's basic logic come where the plotted statistic is either the mean, the range, or, less frequently, the standard deviation. Such charts are commonly known by the names *x-bar charts, R charts*, and *s charts*, respectively. As a basis of discussion of Shewhart charts, consider the data given in Table 5.4. These

values represent melt index measurements of specimens of extrusion grade polyethylene, taken four per shift in a plastics plant.

Figure 5.10 shows plots of the individual melt indices, means, ranges, and standard deviations from Table 5.4 against shift number. The last three of these are the beginnings of so-called Shewhart $\bar{x}$, R, and s control charts.

What remain to be added to the plots in Fig. 5.10 are appropriate control limits. In order to indicate the kind of thinking that stands behind control limits for Shewhart charts, let us concentrate on the issue of limits for the plot of means. The fact is that mathematical theory suggests how the behavior of *means $\bar{y}$* ought to be related to the distribution of *individual* melt indices y, provided the data-generating process is stable, that is, subject only to random causes. If individual responses y can be described as normal with some mean μ and standard deviation σ, mathematical theory suggests that averages of n such values will behave as if a different normal distribution were generating them, one with a mean $\mu_{\bar{y}}$ that is numerically equal to μ and with a standard deviation $\sigma_{\bar{y}}$ that is numerically equal to $\sigma/\sqrt{n}$. Figure 5.11 illustrates this theoretical relationship between the behavior of individuals and the behavior of means.

The relevance of Fig. 5.11 to the problem of setting control chart limits on means is that if one is furnished with a description of the typical pattern of variation in y, sensible expectations for variation in $\bar{y}$ follow from simple normal distribution calculations. So Shewhart reasoned that since about 99.7 percent (most) of a Gaussian distribution is within three standard deviations of the center of the distribution, means found to be farther than three theoretical standard deviations (of $\bar{y}$) from the theoretical mean (of $\bar{y}$) could be safely attributed to other than chance causes. Hence, furnished with standard values for μ and σ (describing individual observations), sensible control limits for $\bar{y}$ become

TABLE 5.4 Measured Melt Indices for Ten Groups of Four Specimens[a]

Shift	Melt Index	$\bar{y}$	R	s
1	218, 224, 220, 231	223.25	13	5.74
2	228, 236, 247, 234	236.25	19	7.93
3	280, 228, 228, 221	239.25	59	27.37
4	210, 249, 241, 246	236.50	39	17.97
5	243, 240, 230, 230	235.75	13	6.75
6	225, 250, 258, 244	244.25	33	14.06
7	240, 238, 240, 243	240.25	5	2.06
8	244, 248, 265, 234	247.75	31	12.92
9	238, 233, 252, 243	241.50	19	8.10
10	228, 238, 220, 230	229.00	18	7.39

[a]Based on data from page 207 of Wadsworth, Stephens, and Godfrey.[9]

Upper Control Limit (UCL) for $\bar{y} = \mu_{\bar{y}} + 3\sigma_{\bar{y}}$

$$= \mu + 3\frac{\sigma}{\sqrt{n}}$$

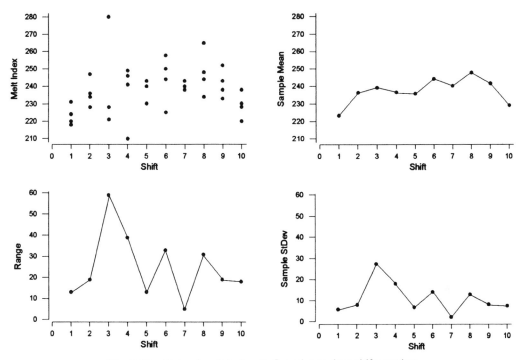

Fig. 5.10. Plots of melt index, $\bar{y}$, R, and s against shift number.

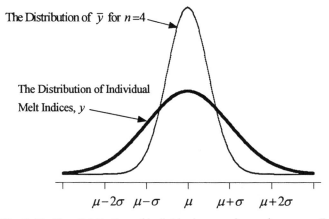

Fig. 5.11. The distribution of individuals, y, and sample means, $\bar{y}$.

and

Lower Control Limit (LCL) for $\bar{y} = \mu_{\bar{y}} - 3\sigma_{\bar{y}}$

$$= \mu - 3\frac{\sigma}{\sqrt{n}}$$

(5-9)

Returning to the context of our example represented by the data of Table 5.4, Wadsworth et al.[9] state that the target value

for melt index in the original application was in fact 235. So if standard process behavior is "on target" behavior, the value $\mu = 235$ seems appropriate for use in (5-9). No parallel value for σ was provided by the authors. Common practice in such situations is to use the data in hand (the data of Table 5.4) to produce a plausible value for σ to use in (5-9). There are many possible ways to produce such a value, but to understand the general logic behind the standard ones, it is important to understand what σ is supposed to measure. The variable σ is intended as a theoretical measure of baseline, short-term, common cause variation. As such, the safest way to try to approximate it is to somehow use only measures of variation *within* the groups of four values in Table 5.4 *not* influenced by variation *between* groups. (Measures of variation derived from considering all the data simultaneously, e.g., would reflect variation between shifts as well as the shorter-term variation within shifts.) In fact, the most commonly used ways of obtaining from the data in hand a value of σ for use in (5-9) are based on the averages of the (within-group) ranges or standard deviations. For example, the 10 values of R given in Table 5.4 have a mean

$$\bar{R} = \frac{1}{10}(13+19+59+ \cdots +19+18) = 24.9$$

and some standard mathematical theory suggests that because the basic group size here is $n = 4$, an appropriate multiple of $\bar{R}$ for use in estimating σ is

$$\frac{\bar{R}}{2.059} \approx 12.1 \qquad (5\text{-}10)$$

(The divisor above is a tabled factor commonly called d_2, which increases with n.)

Finally, substituting 235 for μ and 12.1 for σ in (5-9) produces numerical control limits for $\bar{y}$:

$$LCL = 235 - 3\frac{(12.1)}{\sqrt{4}} = 216.9$$

and

$$UCL = 235 + 3\frac{(12.1)}{\sqrt{4}} = 253.1$$

Comparison of the $\bar{y}$ values in Table 5.4 to these limits reveals no "out of control" means, that is, no evidence in the means of assignable process variation. Figures 5.12 and 5.13 show control charts for all of $\bar{y}$, R, and s, where control limits for the last two quantities have been derived using standard calculations not shown here.

The R and s charts in Figs 5.12 and 5.13 are related representations (only one is typically made in practice) of the shift-to-shift behavior of melt index *consistency*. It is seen that on both charts, the shift #3 point plots above the upper control limit. The strong suggestion thus is that melt index consistency was detectably worse on that shift than on the others, so that from this point of view the process was in fact *not stable* over the time period represented in Table 5.4. In practice, physical investigation and hopefully correction of the origin of the

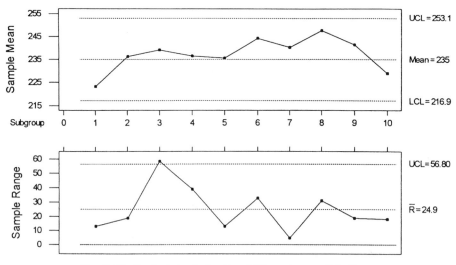

Fig. 5.12. Control charts for $\bar{y}$ and R based on melt indices.

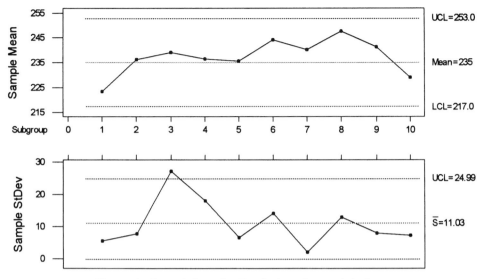

Fig. 5.13. Control charts for $\bar{y}$ and s based on melt indices.

instability typically would follow, as well as some reconsideration of our earlier assessment of 12.1 as a plausible figure to represent the inherent short-term variability of melt index. (If shift #3 could be treated as a special case, explainable as an unfortunate but correctable situation that was not expected to reoccur, there might be reason to revise $\bar{R}$ downward by deletion of shift #3 from the calculation, and thereby to reduce one's view of the size of baseline process variability. Notice that, in general, such a downward revision of $\bar{R}$ might well also have the effect of causing one to need to rethink his or her assessment of the constancy of the melt index *mean*.)

There is a variation on the basic "$\bar{x}$ and R chart" idea that we wish to illustrate here next, because of its frequent use in chemical industry applications. That is the making of a so-called x and MR chart pair. The motivation for this modification of the ideas outlined thus far in this section is that in many chemical process monitoring contexts the natural "group size" is $n = 1$. A mean of $n = 1$ observation(s) is simply that observation itself, and the limits of (5-9) make perfectly good sense for the case of $n = 1$. That is, the analog of an $\bar{x}$ chart for $n = 1$ cases is clear, *at least if one has an externally provided value for σ*. But what, if anything, to do for an

$n = 1$ counterpart of the R chart and how to develop an analog of (5-10) in cases where σ is not a priori known are perhaps not so obvious. Table 5.5 contains data representing moisture contents in 0.01 percent of bihourly samples of a polymer, and the question at hand is what besides simply the bihourly y values might be plotted in the style of a Shewhart control chart for such data.

The final column of Table 5.5 gives 19 so-called moving ranges of pairs of successive moisture contents. It is often argued that although these MR values are actually affected not only by variation within a 2-hr production period but by some variation between these periods as well, they come as close to representing purely short-term variation as any measure available from $n = 1$ data. Accordingly, as a kind of $n = 1$ analog of an R chart, moving ranges are often charted in addition to individual values y. Further, the average moving range is used to estimate σ in cases where information on the inherent variability of individuals is a priori lacking, according to the formula

$$\text{estimated } \sigma = \frac{\overline{MR}}{1.128}$$

where $\overline{MR}$ is the mean of the moving ranges (and plays the role of $\bar{R}$ in (5-10)), and 1.128

is the $n = 2$ version of the factor d_2 alluded to immediately below (5-10).

In the case of the data of Table 5.5,

$$\overline{MR} = \frac{1}{19}(16 + 4 + 5 + \cdots + 16 + 18) \approx 8.2$$

so that a (possibly somewhat inflated due to between period variation) data-based estimate

TABLE 5.5 Moisture Contents for 20 Polymer Samples[a]

Sample	Moisture, y	Moving Range, MR
1	36	—
2	20	16
3	16	4
4	21	5
5	32	11
6	34	2
7	32	2
8	34	2
9	23	11
10	25	2
11	12	13
12	31	19
13	25	6
14	31	6
15	34	3
16	38	4
17	26	12
18	29	3
19	45	16
20	27	18

[a]Based on data from page 190 of Burr.[10]

of within-period variability σ for use, for example in limits (5-9), is

$$\frac{8.2}{1.128} \approx 7.2$$

Figure 5.14 shows both an x (individuals) chart and an MR (moving range) chart based on these calculations. As no standard value of moisture content was provided in Burr's text,[10] the value $\overline{y} = 28.55$ was used as a substitute for μ in (5-9). The MR chart limits are based on standard $n = 2$ (because ranges of "groups" of two observations are being plotted) R chart control limit formulas. Figure 5.14 shows no evidence of assignable variation in the moisture contents.

Statistical research in the last decade has cast serious doubt on the wisdom of adding the MR chart to the x chart in $n = 1$ situations. The price paid for the addition in terms of "false alarm rate" is not really repaid with an important increase in the ability to detect process change. For a more complete discussion of this issue see Section 4.4 of Vardeman and Jobe.[14]

The use of Shewart control charts is admirably documented in a number of statistical quality control books including those by Vardeman and Jobe,[14] Wadsworth et al.,[9] Duncan,[11] Burr,[10] Grant and Leavenworth,[12] and Ott et al.[13] Our purpose here is not to

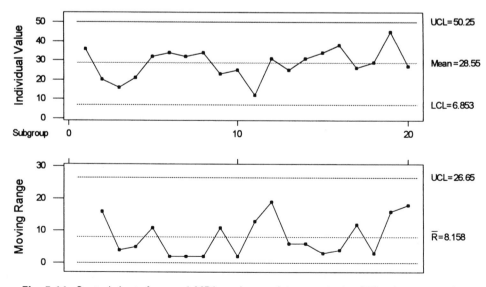

Fig. 5.14. Control charts for y and MR based on moisture contents of 20 polymer samples.

provide all details necessary for their use, but only to give the reader an introduction to the overall function that they serve. It should be said, however, that in recent years other statistical process monitoring tools such as the so-called CUmulative SUM (CUSUM) schemes and Exponentially Weighted Moving Average (EWMA) schemes have been developed as competing methodologies, and can in some circumstances be practically more effective than the original Shewhart charts. Indeed, many computerized controllers for real-time chemical process monitoring and adjustment now employ some form of CUSUM or EWMA logic. For more on these topics, including their integration with model-based process controllers, the reader is referred to Sections 4.1 and 4.2 of Vardeman and Jobe[14] and Vander Wiel et al.[15]

Shewhart's basic conceptualization of common and special cause variation not only leads to control charts as quantitative, rational tools to guide one in knowing when (and when not!) to intervene in an industrial process to correct potential ills, but it also provides a framework for considering the question of what is the best/most consistent performance one can hope for from a particular version of a process. That is, it provides a framework for discussing process capability assessment.

If $\hat{\sigma}$ is some (standard deviation type) estimate of the baseline variation inherent in an industrial process (obtained, e.g., from a calculation such as (5-10) or from data taken from the process after eliminating all physical sources of assignable variation), it essentially specifies what is possible in terms of consistency of process output. There are, however, several common ways of using such an estimate to produce related measures of process capability.

For one thing, remembering again the fact that an interval from $\mu - 3\sigma$ to $\mu + 3\sigma$ (i.e., of length 6σ) will bracket about 99.7 percent of a normal distribution, the figure 6σ is sometimes stated as "the process capability." This usage would say that in the context of the polyethylene melt index example of Table 5.4 the $\hat{\sigma} = 12.1$ figure from (5-10) implies a melt index process capability

of approximately $6 \cdot (12.1) \approx 72.6$. If properly monitored, the process appears capable of producing almost all individual melt indices in a 73-point range.

Where there are stated specifications for individual measurements y, σ is sometimes turned into a kind of index comparing it to the difference in upper and lower engineering specifications. For example, one such *process capability index* is

$$C_p = \frac{USL - LSL}{6\sigma}$$

where $USL - LSL$ is the difference in specifications. Fairly obviously, the larger the value of C_p, the more comfortably (properly targeted) process output values will fit in an interval from LSL to USL.

Another process capability measure that is frequently used in the industrial chemistry sector is

$$C_{pk} = \text{minimum} \left\{ C_{pu} = \frac{USL - \mu}{3\sigma}, \right.$$
$$\left. C_{pl} = \frac{\mu - LSL}{3\sigma} \right\}$$

where μ is an overall process average for an in-control/stable/predictable process, and σ is as before. This measure is clearly similar to C_p, but it takes into account the placement of the process mean in a way that is ignored by C_p. A large value of C_{pk} indicates that not only is the process short-term variation small enough for the process output values to potentially fit comfortably between LSL and USL, but that the process is currently so targeted that the potential is being realized.

STATISTICAL METHODS AND INDUSTRIAL EXPERIMENTATION

One of the most important areas of opportunity for the new application of statistical methods in the chemical industry in the twenty-first century is that of increasing the effectiveness of industrial experimentation. That is, it is one thing to bring an existing industrial process to stability (a state of "statistical" control), but it is quite another to determine how to make fundamental changes in that process that will improve its basic behavior. This second activity almost always involves some form of experimentation, whether it be in the laboratory or in a plant. As we indicated in the introduction, efficient methods and strategies of such data collection (and corresponding analysis) are a central concern of applied statistics. In this section, we hope to give the reader some insight into the kinds of statistical tools that are available for use in chemical industry experimentation.

We will here take as our meaning of the term "experimentation" the observation of a (typically noisy) physical process under more than one condition, with the broad goal of understanding and then using knowledge of how the process reacts to the changes in conditions. In most industrial contexts, the "conditions" under which the process is observed can be specified in terms of the settings or so-called levels chosen for several potentially important process or environmental variables, the so-called factors in the experiment. In some cases, the hope is to identify those (often largely unregulated) factors and combinations of factors that seem to most influence an observed response variable, as a means of targeting them for attention intended to keep them constant or otherwise to eliminate their influence, and thereby to improve the consistency of the response. In other situations the hope is to discover patterns in how one or more critical responses depend on the levels of (often tightly controlled) factors, in order to provide a road map for the advantageous guiding of process behavior (e.g., to an increased mean reaction yield) through enlightened changing of those levels.

This section is organized into two subsections. In the first, we will illustrate the notion of variance component estimation through an example of a nested or hierarchical data collection scheme. In the second, we will discuss some general considerations in the planning of experiments to detail the pattern of influence of factors on responses, consider so-called factorial and fractional factorial experimental designs, illustrate response surface fitting and interpretation tools and the data requirements they imply, and, in the process, discuss the integration of a number of statistical tools in a sequential learning strategy.

Identifying Major Contributors to Process Variation

A statistical methodology that is particularly relevant where experimentation is meant to identify important unregulated sources of variation in a response is that of variance component estimation, based on so-called ANalysis Of VAriance (ANOVA) calculations and random effects models. As an example of what is possible, consider the data of Table 5.6 Shown here are copper content measurements for some bronze castings. Two copper content determinations were made on each of two physical specimens cut from each of 11 different castings.

The data of Table 5.6 were *by design* collected to have a "tree type" or so-called hierarchical/nested structure. Figure 5.15 shows a diagram of a generic hierarchical structure for *balanced* cases like the present one, where there are equal numbers of branches leaving all nodes at a given level (there are equal numbers of determinations for each specimen and equal numbers of specimens for each casting).

An important goal in most hierarchical studies is determining the size of the contributions to response variation provided by the different factors, that is, the different levels of the tree structure. (In the present context, the issue is how variation between castings compares to variation between specimens within a casting, and how they both compare to variation between determinations for a given specimen. If the overall variability observed were considered excessive, such analysis could then help guide efforts at variation reduction by identifying the largest contributors to observed variability.) The structure portrayed in Fig. 5.15 turns out to enable an appealing statistical analysis, providing help in that quantification.

If one lets

y_{ijk} = the copper content from the kth determination of the jth specimen from casting i

$\bar{y}_{ij\cdot} = \frac{1}{2}\sum_{k} y_{ijk}$ = the mean copper content determination from the jth specimen from casting i

$\bar{y}_{i\cdot\cdot} = \frac{1}{2}\sum_{j} \bar{y}_{ij\cdot}$ = the mean copper content determination from the ith casting

and

$\bar{y}_{\cdots} = \frac{1}{11}\sum_{i} y_{i\cdot\cdot}$ = the overall mean copper determination

it is possible to essentially break down the variance of all 44 copper contents (treated as a single group) into interpretable pieces, identifiable as variation between $\bar{y}_{i\cdot\cdot}$s (casting means), variation between $\bar{y}_{ij\cdot}$s (specimen means) within castings, and variation between $\bar{y}_{ijk}$s (individual measurements) within a specimen. That is, it is an algebraic identity that for 44 numbers y_{ijk} with the same structure as those in Table 5.6

$$(44-1)s^2 = \sum_{i,j,k}(y_{ijk} - \bar{y}_{\cdots})^2$$

$$= \sum_{i,j,k}(\bar{y}_{i\cdot\cdot} - \bar{y}_{\cdots})^2 + \sum_{i,j,k}(\bar{y}_{ij\cdot} - \bar{y}_{i\cdot\cdot})^2$$

$$+ \sum_{i,j,k}(y_{ijk} - \bar{y}_{ij\cdot})^2 \qquad (5\text{-}11)$$

The sums indicated in (5-11) are over all data points; so, for example, the first summand on the right is obtained for the copper content data by summing each $(\bar{y}_{i\cdot\cdot} - \bar{y}_{\cdots})^2$ a total of $2 \cdot 2 = 4$ times, one for each determination on a given casting. With the obvious meaning for the $\bar{y}$s and the substitution of the total number of data values for 44, the identity in (5-11) applies to any balanced hierarchical data structure. It is a so-called ANOVA identity, providing an intuitively appealing partitioning of the overall observed variability in the data, an *analyzing of the (observed) variation.*

Some tedious arithmetic "by hand," or use of nearly any commercially available statistical package that includes an ANOVA program, shows that for the copper content data

TABLE 5.6   Forty-four Copper Content Measurements from 11 Bronze Castings[a]

Casting	Specimen	Determination	Copper Content, y, (%)
1	1	1	85.54
1	1	2	85.56
1	2	1	85.51
1	2	2	85.54
2	1	1	85.54
2	1	2	85.60
2	2	1	85.25
2	2	2	85.25
3	1	1	85.72
3	1	2	85.77
3	2	1	84.94
3	2	2	84.95
4	1	1	85.48
4	1	2	85.50
4	2	1	84.98
4	2	2	85.02
5	1	1	85.54
5	1	2	85.57
5	2	1	85.84
5	2	2	85.84
6	1	1	85.72
6	1	2	85.86
6	2	1	85.81
6	2	2	85.91
7	1	1	85.72
7	1	2	85.76
7	2	1	85.81
7	2	2	85.84
8	1	1	86.12
8	1	2	86.12
8	2	1	86.12
8	2	2	86.20
9	1	1	85.47
9	1	2	85.49
9	2	1	85.75
9	2	2	85.77
10	1	1	84.98
10	1	2	85.10
10	2	1	85.90
10	2	2	85.90
11	1	1	85.12
11	1	2	85.17
11	2	1	85.18
11	2	2	85.24

[a]Based on data taken from Wernimont.[16]

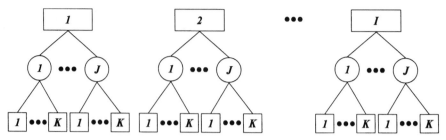

Fig. 5.15. A balanced hierarchical data structure.

of Table 5.6 the numerical version of (5-11) is approximately

$$5.1385 = 3.2031 + 1.9003 + 0.0351 \quad (5\text{-}12)$$

Although we will not provide any details here, the reader is alerted to the fact that it is common practice to present the elements of an identity such as (5-12) in a tabular form called an "ANOVA table." The use for the elements of (5-12) that we wish to illustrate here is their role in estimating casting, specimen, and determination "variance components."

That is, if one models an observed copper determination as the sum of a *random* casting-dependent *effect* whose distribution is described by a variance σ_c^s, a *random* specimen-dependent *effect* whose distribution is described by a variance σ_s^2, and a *random* determination-dependent *effect* whose distribution is described by a variance σ_d^2, the elements of (5-12) lead to estimates of the *variance components* σ_c^2, σ_s^2, and σ_d^2 in the model. Note that in such a random effects model of the data-generating process, copper measurements from the same casting share the same casting effect, and copper measurements from the same specimen share the both same casting and the same specimen effects. The individual σ^2 values are conceptually the variances that would be seen in copper contents if only the corresponding sources of variation were present. The sum of the σ^2 values is conceptually the variance that would be seen in copper contents if single determinations were made on a number of different castings.

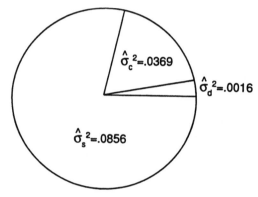

Fig. 5.16. Three estimated variance components for copper contents.

Standard statistical methodology for estimation of the variance components (which we will not detail here but can, e.g., be found in Section 5.5 of Vardeman and Jobe[14] or Chapter 11 of Hicks and Turner[17]) produces

$$\sigma_d^2 = \frac{0.0351}{11 \cdot 2 \cdot (2-1)} \approx 0.0016 \, (\%)^2$$

as an estimate of σ_d^2,

$$\hat{\sigma}_s^2 = \frac{1}{2}\left(\frac{1.9003}{11 \cdot (2-1)} - 0.0016\right) \approx 0.0856 \,(\%)^2$$

as an estimate of σ_s^2, and

$$\hat{\sigma}_c^2 = \frac{1}{2 \cdot 2}\left(\frac{3.2031}{(11-1)} - \frac{1.9003}{11 \cdot (2-1)}\right)$$

$$\approx 0.0369 \,(\%)^2$$

as an estimate of σ_c^2. Figure 5.16 is a pie chart representation of these three estimated variance components as fractions of their sum (the variance predicted if single determinations were made on single specimens from each casting), and graphically identifies inhomogeneity between specimens cut from a single casting as the biggest contributor to observed variation.

On the standard deviation scale the estimates translate to $\hat{\sigma}_d \approx 0.04\%$, $\hat{\sigma}_s \approx 0.29\%$, $\hat{\sigma}_c \approx 0.19\%$. So, for example, the data of Table 5.6 indicate that even if castings and specimens were all exactly alike, it would still be reasonable to expect measured copper contents to vary according to a standard deviation of about 0.04 percent, presumably due to unavoidable measurement error.

Variance component estimation methodology is not limited to balanced hierarchical experiments, but they do provide an important and straightforward context in which to introduce the technology. More detailed information on the case discussed here and extensions to other kinds of data structures can be found in books by Vardeman,[6] Neter et al.,[7] Mason, Gunst, and Hess,[18] and Hicks and Turner.[17]

Discovering and Exploiting Patterns of Factor Influence on Responses

Having discussed statistical methodology particularly appropriate to studies whose primary purpose is simply to identify factors with the largest influence on a response, we will now consider methods aimed more directly at detailed experimental quantification of the pattern of factor influence on one or more responses. As an example, we will use a "sanitized" account of some statistical aspects of a highly successful and economically important process improvement project. (Data presented here are not the original data, but resemble them in structure. Naturally, details of the project not central to our expository purposes and those of a proprietary nature will be suppressed.) A more complete version of this case study appears as Chapter 11 of Vardeman.[6]

The process monitoring, capability assessment, and variance source identification ideas discussed thus far are almost logical prerequisites for industrial experimentation to detail the nature of dependence of response variables on factors of interest. When an industrial process has been made to operate in a stable manner, its intrinsic variability reduced to the extent practically possible, and that baseline performance quantified and understood, the prospects of success are greatly enhanced for subsequent efforts to understand the effects of potential fundamental process changes.

Preliminary work by various groups left a project team with a batch production process behaving in a stable but unsatisfactory fashion. Obvious sources of variation (both in the process itself and "upstream") had been identified and, to the degree practically possible, eliminated. The result was a process with an average output *purity* of 88 percent and an associated purity standard deviation of around 5 percent, and an average *yield* of 43 percent and an associated yield standard deviation of around 5 percent as well. The project team was charged with finding ways to increase the purity and yield means to, respectively, 95 percent and 59 percent while it is hoped, also further reducing the standard deviations. To accomplish this, the team recognized the need for an improved understanding of how various process variables under their control influenced purity (which we will call y_1) and yield (which we will call y_2). Experimentation to provide this was authorized, and, in particular, attention was focused on four factors consisting of three reactant concentrations and the process run time. We will call the Reactant A mole ratio x_1, the Reactant B mole ratio x_2, the Reactant C mole ratio x_3, and the run time (in hours) x_4.

The choice of experimental factors (what to vary in data collection) is a nontrivial matter of fundamental importance that is best handled by people with firsthand process knowledge. There are a number of popular techniques and tools (such as so-called cause and effect diagrams, discussed for instance in Section 2.1 of Vardeman and Jobe[14]) for helping groups brainstorm and reach a consensus on such matters. Further, in cases where a priori knowledge of a process is scarce, relatively small preliminary screening experiments can help reduce a large list of potential factors to a smaller list apparently worthy of more detailed study. (The fractional factorial plans that will be illustrated shortly often are recommended for this purpose.)

Once a particular set of experimental factors has been identified, questions about exactly how they should be varied must be answered. To begin with, there is the choice of levels for the factors, the matter of how much the experimental factors should be varied. Particular experimental circumstances usually dictate how this is addressed. Widely spaced (substantially different) levels will in general lead to bigger changes in responses, and therefore clearer indications of how the responses depend upon the experimental factors, than will closely spaced (marginally different) levels. But they may do so at the expense of potentially creating unacceptable or even disastrous process conditions or output. Thus, what may be an acceptable strategy in a laboratory study might be completely unacceptable in a production environment and vice versa.

Given a set or range of levels for each of the individual experimental factors, there is still the question of exactly what combinations of levels actually will be used to produce experimental data. For example, in the process improvement study, standard process operating conditions were $x_1 = 1.5$, $x_2 = 1.15$, $x_3 = 1.75$, and $x_4 = 3.5$, and the project team decided on the ranges

$$1.0 \leq x_1 \leq 2.5, 1.0 \leq x_2 \leq 1.8, 1.0 \leq x_3 \leq 2.5,$$

and (5-13)

$$2.0 \leq x_4 \leq 5.0$$

as defining the initial limits of experimentation. But the question remained as to exactly what sets of mole ratios and corresponding run times were appropriate for data collection.

A natural (but largely discredited) strategy of data collection is the one-variable-at-a-time experimental strategy of picking some base of experimental operations (such as standard operating conditions) and varying the level of only one of the factors away from that base at a time. The problem with such a strategy is that sometimes two or more factors act on responses jointly, doing things in concert that neither will do alone. For example, in the

process improvement study, it might well have been that an increase in either x_1 or x_2 alone would have affected yield very little, whereas a simultaneous increase in both would have caused an important increase. Modern strategies of industrial experimentation are conceived with such possibilities in mind, and attempt to spread out observations in a way that gives one some ability to identify the nature of the response structure no matter how simple or complicated it turns out to be.

There are several issues to consider when planning the combinations of levels to include in an experiment. We have already said that it is important to "vary several factors simultaneously." It also is important to provide for some replication of at least a combination or two in the experiment, as a means of getting a handle on the size of the experimental error or baseline variation that one is facing. The replication both verifies the reproducibility of values obtained in the study and identifies the limits of that reproducibility. Also, one must balance the urge to "cover the waterfront" with a wide variety of combinations of factor levels against resource constraints and a very real law of diminishing practical returns as one goes beyond what is really needed in the way of data to characterize response behavior. In addition, the fact that real-world learning is almost always of a sequential rather than a "one shot" nature suggests that it is in general wise to spend only part of an experimental budget on early study phases, leaving resources adequate to follow up directions suggested by what is learned in those stages.

It is obvious that a minimum of two different levels of an experimental factor must appear in a set of experimental combinations if any information is to be gained on the effects of that factor. So one logical place to begin thinking about a candidate design for an industrial experiment is with the set of all possible combinations of two levels of each of the experimental factors. If there are p experimental factors, statistical jargon for such an arrangement is to call it a (complete) $2 \times 2 \times 2 \times \cdots \times 2$ or 2^p factorial plan. For example, in the process improvement situation, an experiment consisting of the running of all 16 possible combinations of

$$x_1 = 1.0 \quad \text{or} \quad x_1 = 2.5$$
$$x_2 = 1.0 \quad \text{or} \quad x_2 = 1.8$$
$$x_3 = 1.0 \quad \text{or} \quad x_3 = 2.5$$

and

$$x_4 = 2.0 \quad \text{or} \quad x_4 = 5.0$$

would be called a complete $2 \times 2 \times 2 \times 2$ or 2^4 factorial experiment. Notice that in geometric terms, the (x_1, x_2, x_3, x_4) points making up this 2^4 structure amount to the 16 "corners" in four-dimensional space of the initial experimental region defined in (5-13).

A complete factorial experimental plan is just that, in some sense "complete." It provides enough information to allow one to assess (for the particular levels used) not only individual but also joint or interaction effects of the factors on the response or responses. But when in fact (unbeknownst to the investigator) a system under study is a relatively simple one, principally driven by only a few individual or low-order joint effects of the factors, fewer data actually are needed to characterize those effects adequately. So what is often done in modern practice is initially to run only a carefully chosen part of a full 2^p factorial, a so-called fractional factorial plan, and to decide based on the initial data whether data from the rest of the full factorial appear to be needed in order adequately to characterize and understand response behavior. We will not discuss here the details of how so-called 2^{p-q} fractional factorials are intelligently chosen, but there is accessible reading material on the subject in books by Box, Hunter, and Hunter,[19] and by Vardeman and Jobe.[5]

In the process improvement study, what was actually done in the first stage of data collection was to gather information from one-half of a full 2^4 factorial (a 2^{4-1} fractional factorial) augmented by four observations at the "center" of the experimental region (thereby providing both some coverage of the interior of the region, in addition to a view of some of its corners, and important replication as well).

TABLE 5.7 Data from an Initial Phase of a Process Improvement Study

x_1	x_2	x_3	x_4	Purity, y_1(%)	Yield, y_2(%)
1.00	1.0	1.00	2.0	62.1	35.1
2.50	1.0	1.00	5.0	92.2	45.9
1.00	1.8	1.00	5.0	7.0	4.0
2.50	1.8	1.00	2.0	84.0	46.0
1.00	1.0	2.50	5.0	61.1	41.4
2.50	1.0	2.50	2.0	91.6	51.2
1.00	1.8	2.50	2.0	9.0	10.0
2.50	1.8	2.50	5.0	83.7	52.8
1.75	1.4	1.75	3.5	87.7	54.7
1.75	1.4	1.75	3.5	89.8	52.8
1.75	1.4	1.75	3.5	86.5	53.3
1.75	1.4	1.75	3.5	87.3	52.0

The data in Table 5.7 are representative of what the group obtained.

The order in which the data are listed is simply a convenient systematic one, not to be confused with the order in which experimental runs were actually made. The table order is far too regular for it to constitute a wise choice itself. For example, the fact that all $x_3 = 1.0$ combinations precede the $x_3 = 2.5$ ones might have the unfortunate effect of allowing the impact of unnoticed environmental changes over the study period to end up being confused with the impact of x_3 changes. The order in which the 12 experimental runs were actually made was chosen in a "completely randomized" fashion. For a readable short discussion of the role of randomization in industrial experimentation, the reader is referred to Box.[20]

For purposes of this discussion, attention is focused on the yield response variable, y_2. Notice first that the four y_2 values from the center point of the experimental region have $\bar{y} = 53.2$ and $s = 1.13$ (which incidentally already appear to be an improvement over typical process behavior). As a partial indication of the logic that can be used to investigate whether the dependence of yield on the experimental factors is simple enough to be described adequately by the data of Table 5.7, one can compute some estimated "main effects" from the first eight data points. That is, considering first the impact of the variable x_1 (alone) on yield, the quantity

$$\bar{y}_{\text{high }x_1} - \bar{y}_{\text{low }x_1} = \frac{1}{4}(45.9+46.0+51.2+52.8)$$
$$- \frac{1}{4}(35.1+4.0+41.4+10.0)$$
$$= 26.35$$

is perhaps a sensible measure of how a change in x_1 from 1.00 to 2.50 is reflected in yield. Similar measures for the other variables turn out to be

$$\bar{y}_{\text{high }x_2} - \bar{y}_{\text{low }x_2} = -15.20$$

$$\bar{y}_{\text{high }x_3} - \bar{y}_{\text{low }x_3} = 6.10$$

and

$$\bar{y}_{\text{high }x_4} - \bar{y}_{\text{low }x_4} = 0.45$$

These measures provide some crude insight into the directions and magnitudes of influence of the experimental variables on y_2. (Clearly, by these measures $x_1 = 2.50$ seems preferable to $x_1 = 1.00$, and the run time variable x_4 seems to have little impact on yield.) But they also provide strong evidence that the nature of the dependence of yield on the experimental factors is too complicated to be described by the action of the factors individually. For example, if it *were* the case that the separate actions of the experimental factors were adequate to describe system behavior, then standard statistical theory and the data indicate that the mean response for the $x_1 = 1.00$, $x_2 = 1.0$, $x_3 = 1.00$, and $x_4 = 2.0$ set of conditions would be around

$$\hat{y} = \bar{y}_{\text{corners}} - \frac{1}{2}(-26.35) - \frac{1}{2}(-15.20)$$

$$- \frac{1}{2}(6.10) - \frac{1}{2}(0.45) = 27.45$$

(where $\bar{y}_{\text{corners}}$ is standing for the mean of the first eight yields in Table 5.7). But the observed yield of 35.1 is clearly incompatible with such a mean and the standard deviation value (of $s = 1.13$) derived from the repeated center point. Also, other simple evidence that (at least linear and) separate action of the four factors is not enough to describe yield adequately is given by the large difference between $\bar{y}_{\text{corners}} = 35.8$ and the observed mean from the center point $\bar{y} = 53.2$. (As it turns

TABLE 5.8 Data from a Second Phase of a Process Improvement Study

x_1	x_2	x_3	x_4	Purity, $y_1(\%)$	Yield, $y_2(\%)$
1.00	1.0	1.00	5.0	64.0	35.3
2.50	1.0	1.00	2.0	91.9	47.2
1.00	1.8	1.00	2.0	6.5	3.9
2.50	1.8	1.00	5.0	86.4	45.9
1.00	1.0	2.50	2.0	63.9	39.5
2.50	1.0	2.50	5.0	93.1	51.6
1.00	1.8	2.50	5.0	6.8	9.2
2.50	1.8	2.50	2.0	84.6	54.3

out, calculations that we will not show here indicate the *possibility* that individual action of the factors *plus joint action* of the Reactant A and Reactant B mole ratios is sufficient to describe yield. But in any case, the point is that the data of Table 5.7 provide evidence that the pattern of dependence of yield on the experimental variables is not simple, and thus that completion of the 2^4 factorial is in order.)

After a complete analysis of the first round of experimental data, the project team "ran the second half fraction" of the 2^4 factorial, and data similar to those in Table 5.8 were obtained. (Again, no significance should be attached to the order in which the observations in Table 5.8 are listed. It is not the order in which the experimental runs were made.)

The data from the second phase of experimentation served to complete the project team's 2^4 factorial picture of yield and confirm the tentative understanding drawn first from the initial half fraction. It is seen that the combinations listed in Table 5.8 are in the same order as the first eight in Table 5.7 as regards levels of experimental variables x_1, x_2, and x_3, and that the corresponding responses are very similar. (This, by the way, has the happy practical implication that run time seems to have little effect on final purity or yield, opening the possibility of reducing or at least not increasing the standard run time.) Thorough data analysis of a type not shown here left the project team with a clear (and quantified version of the) understanding that

Reactant A and B mole ratios have important individual and joint effects on the responses, and that, acting independently of the other two reactants, Reactant C also has an important effect on the responses. However, it did *not* yet provide a solution to the team's basic problem, which was to reach a 59 percent mean yield goal.

The data of Tables 5.7 and 5.8 do hold out hope that conditions producing the desired purity and yield can be found. That is, though none of the 16 corners of the experimental region nor the center point appeared to meet the team's yield goal, the data do show that there is substantial *curvature* in the yield response. (The joint effect of x_1 and x_2 amounts to a kind of curvature, and the non-linearity of response indicated by a large difference between $\bar{y}_{corners} \approx 35.8$ and $\bar{y} = 53.2$ at the center of the experimental region also is a kind of curvature.) If one could "map" the nature of the curvature, there is at least the possibility of finding favorable future operating conditions in the interior of the initial experimental region defined in (5-13).

It ought to be at least plausible to the reader that 2^4 factorial data (even supplemented with center points) are not really sufficient to interpolate the nature of a curved response over the experimental region. More data are needed, and a standard way of augmenting a 2^p design with center points to one sufficient to do the job is through the addition of so-called star points to produce a central composite design. Star points are points outside the original experimental region whose levels of all but one of the p experimental factors match those of the center point. Figure 5.17 shows graphical representations of central composite designs in $p = 2$ and $p = 3$ factors.

The project team conducted a third phase of experimentation by adding eight star points to their study and obtained data similar to those in Table 5.9.

The data in Tables 5.7–5.9 taken together turn out to provide enough information to enable one to rather thoroughly quantify the "curved" nature of the dependence of y_2 on x_1, x_2, x_3, and

x_4. A convenient and often successful method of accomplishing this quantification is through the least squares fitting of a general *quadratic response surface*. That is, central composite data are sufficient to allow one to fit an equation to a response that involves a constant term, linear terms in all the experimental variables, quadratic terms in all of the experimental variables, and cross-product terms in all pairs of the experimental variables. Appropriate use of a multiple regression program with the project

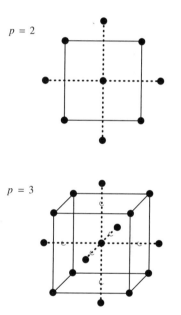

Fig. 5.17. $p = 2$ and $p = 3$ central composite designs.

TABLE 5.9   Data from a Third Phase of a Process Improvement Study

x_1	x_2	x_3	x_4	Purity, $y_1(\%)$	Yield, $y_2(\%)$
0.6895	1.4	1.75	3.5	20.8	13.0
2.8105	1.4	1.75	3.5	95.9	54.3
1.75	0.8344	1.75	3.5	99.9	62.4
1.75	1.9656	1.75	3.5	65.9	41.2
1.75	1.4	0.6895	3.5	64.4	32.7
1.75	1.4	2.8105	3.5	64.8	40.3
1.75	1.4	1.75	1.379	88.1	52.7
1.75	1.4	1.75	5.621	88.9	50.5

data represented here produces the fitted equation

$$y_2 \approx 15.4 + 37.9x_1 - 66.2x_2 + 48.8x_3$$
$$+ 0.97x_4 - 16.1x_1^2 - 0.03x_2^2$$
$$- 13.6x_3^2 - 0.046x_4^2 + 26.5x_1x_2$$
$$+ 0.344x_1x_3 - 0.217x_1x_4$$
$$+ 1.31x_2x_3 - 0.365x_2x_4 + 0.061x_3x_4$$

This may not seem to the reader to be a particularly helpful data summary, but standard multiple regression tools can be used to deduce that an essentially equivalent, far less cluttered, and more clearly interpretable representation of the relationship is:

$$y_2 \approx 13.8 + 37.8x_1 - 65.3x_2 + 51.6x_3$$
$$- 16.2x_1^2 - 13.6x_3^2 + 26.5x_1x_2 \quad (5\text{-}14)$$

Equation (5-14) provides an admirable fit to the data in Tables 5.7–5.9, is in perfect agreement with all that has been said thus far about the pattern of dependence of yield on the experimental factors, *and* allows one to do some intelligent interpolation in the initial experimental region. Use of an equation like (5-14) ultimately allowed the project team to determine that an increase of x_1 only would, with minimal change in the existing process, allow

them to meet their yield goal. (In fact, the single change in x_1 proved to be adequate to allow them to meet all of their yield *and* purity goals!)

Graphical representations similar to those in Figs 5.18 and 5.19 for (5-14) with $x_3 = 1.75$ (the standard operating value for x_3) were instrumental in helping the team understand the message carried by their data and how yield could be improved. Figure 5.18 is a so-called contour plot (essentially a topographic map) of the fitted equation, and Fig. 5.19 is a more three-dimensional-looking representation of the same surface. Both types of display are commonly used tools of modern statistical experiment design and analysis. The contour plot idea is particularly helpful where several responses are involved, and by overlaying several such plots one can simultaneously picture the various implications of a contemplated choice of process conditions.

SPECIAL STATISTICAL TOOLS FOR CHEMICAL APPLICATIONS

The statistical methods discussed thus far are of a quite general nature, routinely finding application beyond the bounds of the chemical industry. In this section, we will briefly highlight two statistical methodologies whose most important applications are to chemical

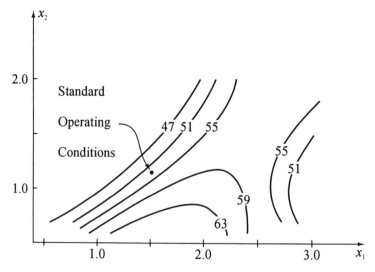

Fig. 5.18. A contour plot of fitted yield when $x_3 = 1.75$. (From *Statistics for Engineering Problem Solving (1st Ed.)* by S. B. Vardeman © 1994. Reprinted with permission of Brooks/Cole, a Division of Thomson Learning; www.thomsonlearning.com. FAX 800-730-2215.)

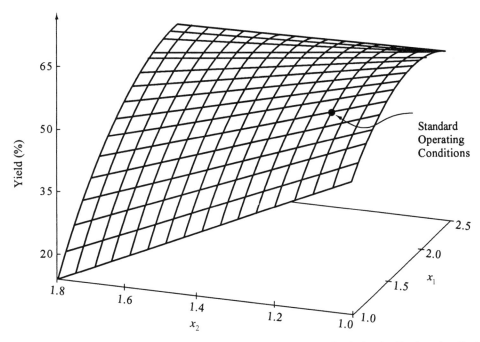

Fig. 5.19. A perspective graph of fitted yield when $x_3 = 1.75$. (From *Statistics for Engineering Problem Solving* by S. B. Vardeman © 1994. Reprinted with permission of Brooks/Cole, a Division of Thomson Learning; www.thomsonlearning.com. Fax 800-730-2215.)

problems. That is, we will touch on some of the ideas of mixture experiments and the role of statistics in mechanistic modeling.

Mixture Experiments

In many situations in industrial chemistry, some important measured property of a product is a function of the proportions in which a set of p ingredients or components are represented in a mixture leading to the product. For example, case studies in the literature have covered subjects ranging from octanes of gasoline blends, discussed by Snee;[21] to strengths of different formulations of ABS pipe compound, treated in Koons and Wilt;[22] to aftertaste intensities of different blends of artificial sweeteners used in an athletic sport drink, discussed by Cornell;[23] to moduli of elasticity of different rocket propellant formulations, considered by Kurotori.[24] For experimenting in such contexts, special statistical techniques are needed. These tools have been discussed at length by Cornell,[25,26] and our purpose here is not to attempt a complete exposition, but only to whet the reader's appetite for further reading in this area.

The goal of mixture experimentation is to quantify how proportions $x_1, x_2, x_3, \ldots, x_p$ of ingredients 1 through p affect a response y. Usually, the hope is to fit some kind of approximate equation involving some parameters b, say

$$y \approx f(x_1, x_2, \ldots, x_p | b)$$

to a set of n data points $(x_1, x_2, \ldots, x_p, y)$, for the purpose of using the fitted equation to guide optimization of y, that is, to find the "best" blend. The logic of data collection and equation fitting is complicated in the mixture scenario by the fact that

$$x_1 + x_2 + \cdots + x_p = 1 \qquad (5\text{-}15)$$

The linear constraint (5-15) means that (p way) factorial experimentation is impossible, and that special measures must be employed in order to use standard regression analysis software to do least squares equation fitting. We will briefly describe in turn some approaches to experimental design, equation fitting, and presentation of results for the

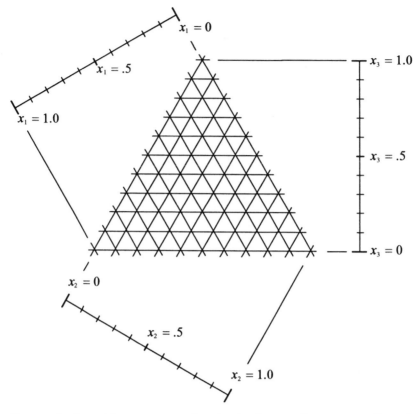

Fig. 5.20. The set of points with $x_1 + x_2 + x_3 = 1$ and a simplex coordinate system. (From *Statistics for Engineering Problem Solving* by S. B. Vardeman © 1994. Reprinted with permission of Brooks/Cole, a Division of Thomson Learning; www.thomsonlearning.com. Fax 800-730-2215.)

mixture problem under its fundamental constraint (5-15).

In the case of $p = 3$ (a three-component mixture problem), the set of all possible combinations of values for x_1, x_2, and x_3 satisfying (5-15) can be conveniently represented as an equilateral triangular region. Figure 5.20 shows such a region and the so-called simplex coordinate system on the region. The corners on the plot stand for cases where the "mixture" involved is actually a single pure component. Points on the line segments bounding the figure represent two-component mixtures, and interior points represent genuine three-component mixtures. For example, the center of the simplex corresponds to a set of conditions where each component makes up exactly one-third of the mixture.

One standard mixture (experimental) design strategy is to collect data at the extremes (corners) of the experimental region along with

TABLE 5.10 (x_1, x_2, x_3)
Points in a Particular $p = 3$
Simplex Lattice Design

x_1	x_2	x_3
1	0	0
0	1	0
0	0	1
$\frac{1}{3}$	$\frac{2}{3}$	0
$\frac{2}{3}$	$\frac{1}{3}$	0
$\frac{1}{3}$	0	$\frac{2}{3}$
$\frac{2}{3}$	0	$\frac{1}{3}$
0	$\frac{1}{3}$	$\frac{2}{3}$
0	$\frac{2}{3}$	$\frac{1}{3}$
$\frac{1}{3}$	$\frac{1}{3}$	$\frac{1}{3}$

collecting data on a regular grid in the experimental region. Figure 5.21 shows a $p = 3$ example of such a so-called simplex lattice design, and Table 5.10 lists the (x_1, x_2, x_3)

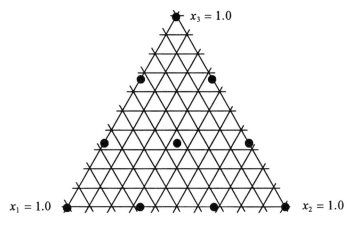

Fig. 5.21. A $p = 3$ simple lattice design.

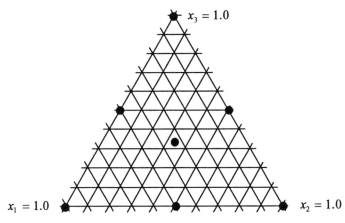

Fig. 5.22. A $p = 3$ simplex centroid design.

points involved. (As in the cases of the data in Tables 5.7–5.9, the order used in the listing in Table 5.10 is not one that would be used in sequencing data collection runs. Instead, a randomly chosen order often is employed.)

Another standard mixture experiment strategy is the so-called simplex centroid design, where data are collected at the extremes of the experimental region and for every equal-parts two-component mixture, every equal-parts three-component mixture, and so on. Figure 5.22 identifies the blends included in a $p = 3$ simplex centroid design.

Often, the space of practically feasible mixtures is smaller than the entire set of x_1, $x_2, \ldots, x_p$ satisfying (5-15). For example, in many contexts, "pure" mixtures do not produce viable product. Concrete made using only

water and no sand or cement obviously is a useless building product. One common type of constraint on the proportions $x_1, x_2, \ldots, x_p$ that produces quite simple experimental regions is that of lower bounds on one or more of the individual proportions. Cornell,[25] for example, discusses a situation where the effectiveness in grease stain removal of a $p = 3$ bleach mixture was studied. Past experience with the product indicated that the proportions by weight of bromine, x_1, of powder, x_2, and of HCl, x_3, needed to satisfy the constraints:

$$x_1 \geq 0.30, \ x_2 \geq 0.25, \ \text{and} \ x_3 \geq 0.02 \quad (5\text{-}16)$$

for effective action of the product (i.e., the mixture needed to be at least 30% bromine, at least 25% powder, and at least 2% HCl by weight.)

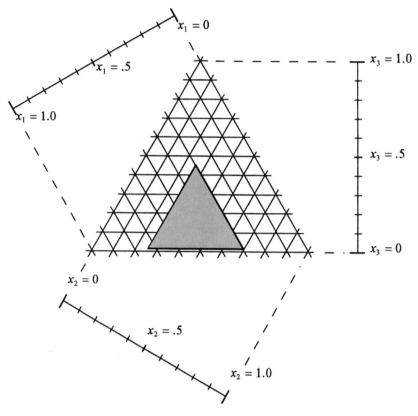

Fig. 5.23. The $p = 3$ simplex and a set of feasible bleach mixtures.

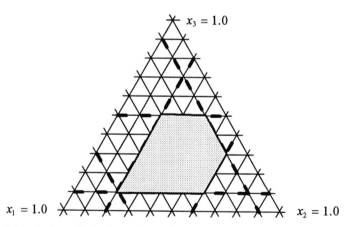

Fig. 5.24. An irregularly shaped experimental region in a $p = 3$ mixture study.

The effect of adding the lower bound constraints (5-16) to the basic mixture constraint (5-15) can be pictured as in Fig. 5.23. There, a triangular subregion of the basic $p = 3$ simplex depicts the feasible (x_1, x_2, x_3) points. The choice of experimental mixtures for such an experimental region can be made by direct analogy to or rescaling of designs such as the simplex lattice and simplex centroid designs illustrated above to cover the entire simplex. (It is common to refer to the rescaling process as the use of pseudo-components.)

Constraint systems more complicated than simple lower bounds produce irregularly shaped experimental regions and less obvious methods of choosing $(x_1, x_2, \ldots, x_p)$ points to cover the experimental region. When $p = 3$, it is possible to sketch the region of feasible points on a simplex plot and use it to help guide the choice of mixture experiment strategy. Figure 5.24 illustrates the kind of region that can arise with other than exclusively lower bound constraints.

When more than three components are involved in a mixture study, such plots are, of course, no longer possible, and other more analytic methods of identifying candidate experimental mixtures have been developed. For example, McLean and Anderson[27] presented an algorithm for locating the vertices of an experimental region defined by the basic constraint (5-15) and any combination of upper and or lower bound constraints

$$0 \le a_i \le x_i \le b_i \le 1$$

on the proportions x_i. Cornell[25,26] discusses a variety of algorithms for choosing good mixture experiment designs under constraints, and many of the existing algorithms for the problem have been implemented in the MIXSOFT software package developed by Piepel.[28]

Empirical polynomial descriptions of (approximately) how a response y depends upon proportions $x_1, x_2, \ldots, x_p$ are popular mixture analysis tools. The process of fitting polynomials to mixture experiment data in principle uses the same least squares notion illustrated in the fitting of a parabola to the data of Table 5.2. However, the mechanics of using standard multiple regression analysis software in the mixture context is complicated somewhat by the basic constraint (5-15). For example, in view of (5-15) the basic ($p + 1$ parameter) linear relationship

$$y \approx b_0 + b_1 x_1 + b_2 x_2 + \cdots + b_p x_p \quad (5\text{-}17)$$

is in some sense "overparameterized" in the mixture context, in that it is equivalent to the (p parameter) relationship

$$y \approx b_1 x_1 + b_2 x_2 + \cdots + b_p x_p \quad (5\text{-}18)$$

if one identifies the coefficients in (5-18) with the sums of the corresponding coefficients in

(5-17) and the coefficient b_0. As a result, it is the "no intercept" relationship (5-18) that is typically fit to mixture data when a linear relationship is used. In a similar way, when a second-order or (multivariable) quadratic relationship between the individual proportions and the response variable is used, it has no intercept term and no pure quadratic terms. For example, in the $p = 3$ component mixture case, the general quadratic relationship typically fit to mixture data is

$$y \approx b_1 x_1 + b_2 x_2 + b_3 x_3 + b_4 x_1 x_2$$
$$+ b_5 x_1 x_3 + b_6 x_2 x_3 \quad (5\text{-}19)$$

(Any apparently more general relationship involving an intercept term and pure quadratic terms can by use of (5-15) be shown to be equivalent to (5-19) in the mixture context.) Relationships of the type of (5-19) are often called Scheffé models, after the first author to treat them in the statistical literature. Other more complicated equation forms are also useful in some applications, but we will not present them in this chapter. The interested reader is again referred to Cornell[25,26] for more information on forms that have been found to be tractable and effective.

We should point out that the ability to fit equations of the form (5-18) or like (5-19), or of an even more complicated form, is predicated on having data from enough different mixtures to allow unambiguous identification of the parameters $\underline{b}$. This requires proper data collection strategy. Much of the existing statistical research on the topic of mixture experiment design has to do with the question of wise allocation of experimental resources under the assumption that a particular type of equation is to be fit.

One's understanding of fitted polynomial (and other) relationships often is enhanced through the use of contour plots made on coordinate systems such as that in Fig. 5.25. (This is even true for $p \ge 3$ component mixture scenarios, but the use of the idea is most transparent in the three-component case.) A plot like Fig. 5.25 can be a powerful tool to aid one in understanding the nature of a fitted equation

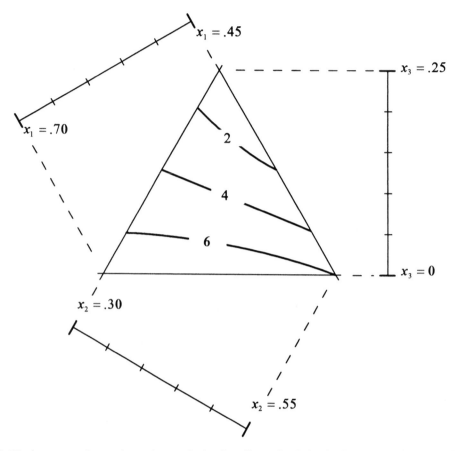

Fig. 5.25. A contour plot made on the $p = 3$ simplex. (From *Statistics for Engineering Problem Solving* by S. B. Vardeman © 1994. Reprinted with permission of Brooks/Cole, a Division of Thomson Learning: www.thomsonlearning.com. Fax 800-730-2215.)

and finding regions of optimum fitted response.

The mixture experiment counterpart to conventional screening/fractional factorial experimentation also is possible. So-called axial designs have been developed for the purpose of providing screening-type mixture data for use in rough evaluation of the relative effects of a large number of mixture components on a response variable. The same kind of sequential experimental strategy illustrated in the process improvement example is applicable in mixture contexts as well as contexts free of a constraint such as (5-15).

Mechanistic Model Building

The kinds of equations most easily fit to multi-factor data using standard (least squares) regression analysis techniques are polynomial equations such as (5-6), (5-14), (5-18), and (5-19). These are particularly convenient because they are linear in their parameters, b. But they are probably best thought of as empirical "mathematical French curve" descriptions of the relation of a response, y, to the explanatory variables, x. Polynomial equations function as extremely useful summaries of observed patterns in one's data, but they do not typically provide direct insight into chemical mechanisms that produce those patterns, and the fitted parameters, b, do not often have direct physical meanings. Their use is particularly appropriate where there is little a priori knowledge of mechanisms involved in a process that might aid in its description, and/or no such knowledge is really essential to achieving one's goals.

Sometimes, however, it is desirable (on the basis of possible reaction kinetics or for other reasons) to posit theoretical descriptions of process outputs in terms of explanatory variables. That is, physicochemical principles often lead (through differential or integral equation descriptions of a system) to equation forms for a response that, like

$$y = \frac{K_1 K_A K_B P_A P_B}{(1 + K_A P_A + K_B P_B)^2}$$

$$y = C_0 \exp(-Kt)$$

and

$$y = \frac{K_1 x}{1 + K_2 x}$$

are nonlinear in the parameters. Although such equations or models may be less tractable than empirical polynomial equations, the parameters involved more often than not *do have* direct physical interpretations. Further, when such a model can be verified as being an adequate description of a process (thereby confirming scientific understanding) and the parameters involved are estimated from process data, such mechanistic models can provide much safer extrapolations beyond an experimental region than the cruder empirical polynomial models.

The process of research in chemical systems is one of developing and testing different models for process behavior. Whether empirical or mechanistic models are involved, the discipline of statistics provides data-based tools for discrimination between competing possible models, parameter estimation, and model verification for use in this enterprise. In the case where empirical models are used, techniques associated with "linear" regression (linear least squares) are used, whereas in mechanistic modeling contexts "nonlinear" regression (nonlinear least squares) techniques most often are needed. In either case, the statistical tools are applied most fruitfully in iterative strategies.

Reilly and Blau[29] and Chapter 16 of Box et al.[19] provide introductions to the general philosophy of using statistical methods in mechanistic modeling contexts, as well as a number of useful references for further reading.

Fairly sophisticated and specialized statistical software is needed in the practical application of nonlinear regression methods to mechanistic modeling for industrial chemistry applications. The techniques implemented in such software are discussed in Seber and Wild,[32] Bates and Watts,[30] Bard,[31] and Riley and Blau.[29]

MODERN BUSINESS PROCESS IMPROVEMENT AND THE DISCIPLINE OF STATISTICS

The modern global business environment is fiercely competitive in all sectors, including the chemical sector. It is by now widely recognized that corporate survival in this environment depends upon constant improvement in *all* business processes, from billing to production. Companies have adopted a variety of programs and focuses aimed at facilitating that improvement. A decade ago, efforts organized around a *Total Quality Management* banner (with liberal references to emphases of consultants like W. E. Deming, J. M. Juran, and A. Feigenbaum) were popular. More recently, programs keyed to ISO 9000[33] certification criteria and Malcolm Baldridge Award[34] criteria have become prominent. And currently probably the most visible programs are the so-called *Six Sigma* programs.

In one sense there is nothing new under the sun, and all successful business process improvement programs (including those in the chemical sector) must in the end reduce to organized problem-solving disciplines. So it is not surprising that programs quite different in name are often very alike in fundamental content. And as they must necessarily make use of empirical information (data), they must have significant statistical components. To make this connection to statistics slightly more explicit, we proceed to provide a few additional details on the Six Sigma movement. (Further material on the subject is easy to find using an Internet search engine, as there are many consultants eager to sell their advice and Six Sigma training. The American

Society for Quality at www.asq.org offers many entries into the subject. And a search at amazon.com for "Six Sigma" books already produced 6666 hits in May 2004. Fashions change quickly enough that it seems pointless to provide more detailed recommendations for follow up on the subject.)

The phrase "Six Sigma" originated at Motorola in the late 1980s. Programs there and at General Electric in the mid-1990s are widely touted as important contributors to company profits and growth in stock values. The name is now commonly used in at least three different ways. "Six Sigma" refers to

- a goal for business process performance
- a discipline for improvement to achieve that performance
- a corporate program of organization, training, and recognition conceived to support the process improvement discipline

As a goal for business process improvement, "Six Sigma" is equivalent to "$C_{pk} = 2$." What is perhaps confusing to the uninitiated is that this goal has connections (through normal distribution tail area calculations) to small ("parts per million") fractions defective relative to two-sided specifications on y. Six Sigma proponents often move between the "small process variation" and "parts per million" understandings with little warning.

Six Sigma process improvement disciplines are typically organized around the acronym "**MAIC**." The first step in a MAIC cycle is a **M**easure step, wherein one finds appropriate process responses to observe, identifies and validates measurement systems and collects baseline process performance (process monitoring) data. The second step is an **A**nalyze step. This involves summarizing the initial process data and drawing appropriate inferences about current process performance. The third step in a MAIC cycle is an **I**mprove step, where process knowledge, experimentation, and more data analysis are employed to find a way to make things work better. Finally, the four-step cycle culminates in a **C**ontrol (process monitoring) effort. The object here is to see that the newly improved performance is maintained after a project team moves on to other problems.

Six Sigma corporate organization, training, and recognition programs borrow from the jargon and culture of the martial arts. People expert in the process improvement paradigm are designated "black belts," "master black belts," and so on. These people lead project teams and help train new initiates ("green belts") in the basics of the program and its tools (including statistical tools). The emphasis throughout is on completing projects with verifiable large dollar impact.

Having made the point that improvement in all business activities is of necessity data-driven, it is hopefully obvious that the emphases and methods of the subject of statistics are useful beyond the lab and even production. Of course, for broad implementation, it is the most elementary of statistical methods that are relevant.

CONCLUSION

We have tried in this chapter to give readers the flavor of modern applied statistical methods and to illustrate their usefulness in the chemical industry. Details of their implementation have of necessity been reserved for further more specialized reading, for which the interested reader is encouraged to consult the references given in this chapter.

REFERENCES

1. Albin, S., "The Lognormal Distribution for Modeling Quality Data When the Mean is Near Zero," *J. Qual. Technol.*, **22**, 105–110 (1990).
2. Chambers, J., Cleveland, W., Kleiner, B., and Tukey, P., *Graphical Methods of Data Analysis*, Duxbury, Boston, MA, 1983.
3. Odioso, R., Henke, A., Stauffer, H., and Frech, K., "Direct Hydration of Olefins," *Ind. Eng. Chem.*, **53**(3), 209–211 (1961).
4. Devore, J., *Probability and Statistics for Engineering and the Sciences* (3rd ed.), Brooks/Cole, Pacific Grove, CA, 1991.

5. Vardeman, S. B., and Jobe, J. M., *Basic Engineering Data Collection and Analysis*, Duxbury/Thomson Learning, Pacific Grove, CA, 2001.

6. Vardeman, S. B., *Statistics for Engineering Problem Solving*, PWS-Kent, Boston, MA, 1994.

7. Neter, J., Kutner, M., Nachsteim, C., and Wasserman, W., *Applied Linear Statistical Models* (4th ed.), McGraw-Hill, New York, 1996.

8. Hahn, G., and Meeker, W., *Statistical Intervals: A Guide for Practitioners*, Wiley, New York, 1991.

9. Wadsworth, H., Stephens, K., and Godfrey, B., *Modern Statistical Methods for Quality Control and Improvement*, Wiley, New York, 1986.

10. Burr, I., *Statistical Quality Control Methods*, Dekker, New York, 1976.

11. Duncan, A., *Quality Control and Industrial Statistics* (5th ed.), Irwin, Homewood, IL, 1986.

12. Grant, E., and Leavenworth, R., *Statistical Quality Control* (7th ed.), McGraw-Hill, New York, 1996.

13. Ott, E., and Schilling, E., *Process Quality Control*, McGraw-Hill, NY, 1990.

14. Vardeman, S., and Jobe, J. M., *Statistical Quality Assurance Methods for Engineers*, Wiley, New York, 1999.

15. Vander Wiel, S., Tucker, W., Faltin, F., and Doganaksoy, N., "Algorithmic Statistical Process Control: Concepts and an Application," *Technometrics*, **34**(3), 286–297 (1992).

16. Wernimont, G., "Statistical Quality Control in the Chemical Laboratory," *Qual. Eng.*, **2**, 59–72 (1989).

17. Hicks, C., and Turner, K., *Fundamental Concepts in the Design of Experiments* (5th ed.), Oxford University Press, New York, 1999.

18. Mason, R., Gunst, R., and Hess, J., *Statistical Design and Analysis of Experiments*, Wiley, New York, 1989.

19. Box, G., Hunter, W., and Hunter, J. S., *Statistics for Experimenters*, Wiley, New York, 1978.

20. Box, G., "George's Column," *Qual. Eng.*, **2**, 497–502 (1990).

21. Snee, R., "Developing Blending Models for Gasoline and Other Mixtures," *Technometrics*, **23**, 119–130 (1981).

22. Koons, G., and Wilt, M., "Design and Analysis of an ABS Pipe Compound Experiment," in *Experiments in Industry: Design, Analysis and Interpretation of Results*, R. Snee, L. Hare, and J. R. Trout (eds.), pp. 111–117, ASQC Quality Press, Milwaukee, WI, 1985.

23. Cornell, J., "A Comparison between Two Ten-Point Designs for Studying Three-Component Mixture Systems," *J. Qual. Technol.*, **18**, 1–15 (1986).

24. Kurotori, I., "Experiments with Mixtures of Components Having Lower Bounds," *Ind. Qual. Control*, **22**, 592–596 (1966).

25. Cornell, J., *How to Run Mixture Experiments for Product Quality*, Vol. 5 in the ASCQ "Basic References in Quality Control" series, American Society for Quality Control, Milwaukee, WI, 1983.

26. Cornell, J., *Experiments with Mixtures: Designs, Models, and the Analysis of Mixture Data* (2nd ed.), Wiley, New York, 1990.

27. MacLean, R., and Anderson, V., "Extreme Vertices Design of Mixture Experiments," *Technometrics*, **8**, 447–454 (1966).

28. Piepel, G., *Mixsoft and Misoft User's Guide, Version 1.0*, Mixsoft-Mixture Experiment Software, Richland, WA, 1989.

29. Reilly, P., and Blau, G., "The Use of Statistical Methods to Build Mathematical Models of Chemical Reacting Systems," *Can. J. Chem. Eng.*, **52**, 289–299 (1974).

30. Bates, D., and Watts, D., *Nonlinear Regression Analysis and Its Applications*, Wiley, New York, 1988.

31. Bard, Yonathan, *Nonlinear Parameter Estimation*, Academic Press, New York, 1974.

32. Seber, G., and Wild, C., *Nonlinear Regression,* Wiley, New York, 1989.

33. International Organization for Standardization (www.iso.ch).

34. National Institute of Standards and Technology (www.quality.nist.gov).

6

Green Engineering–Integration of Green Chemistry, Pollution Prevention, and Risk-Based Considerations

David Shonnard[1], Angela Lindner[2], Nhan Nguyen[3], Palghat A. Ramachandran[4], Daniel Fichana[5], Robert Hesketh[5], C. Stewart Slater[5], Richard Engler[6]

OVERVIEW

Literature sources on green chemistry and green engineering are numerous. The objective of this chapter is to familiarize readers with some of the green engineering and chemistry concepts, approaches and tools. In

[1]Department of Chemical Engineering, Michigan Technological University. Dr. Shonnard authors the section on Environmental Performance Assessment for Chemical Process Design
[2]Department of Environmental Engineering Sciences, University of Florida at Gainesville. Dr. Lindner authors the section on Understanding and Prediction of Environmental Fate of Chemicals.
[3]US EPA. The chapter does not represent the views of EPA or the U.S. Government. Nhan Nguyen authors the Introduction to Green Chemistry and Green Engineering section and coordinates development of the chapter.
[4]Department of Chemical Engineering, Washington University of St. Louis, MO. Dr. Ramachandran authors the section on P2 Heuristics.
[5]Department of Chemical Engineering, Rowan University. Mr. Fichana, Dr. Hesketh and Dr. Slater co-author the section on Life Cycle Assessment.
[6]US EPA. The chapter does not represent the views of EPA or the U.S. Government. Richard. Engler contributes to the Introduction to Green Chemistry and Green Engineering section.

order to do this, the chapter is organized into five sections as follows.

Section I provides an introduction to green chemistry and green engineering. Section II provides examples of pollution prevention heuristics for chemical processes. Heuristics of the two most important unit operations, reactors and separators, are covered. Section III introduces readers to the concept of environmental fate and transport and prediction of environmental fate properties. Understanding of environmental fate and transport is important for exposure assessment and also is essential for evaluating environmental performance of processes and products during process development and design.

Section IV covers the environmental performance assessment for chemical processes design and introduces a three-tier approach to green engineering design of processes incorporating green chemistry, pollution prevention, environmental fate and transport and life cycle approach. Finally, Section V provides more background and examples on life cycle assessment which is an essential principle of green engineering.

I. INTRODUCTION TO GREEN CHEMISTRY AND GREEN ENGINEERING

The Pollution Prevention Act of 1990 (42 U.S.C. 13101-13109) established a national policy to prevent or reduce pollution at its source whenever feasible. The Pollution Prevention Act also provided an opportunity to expand beyond traditional EPA programs and devise creative strategies to protect human health and the environment. The pollution prevention (P2) hierarchy established by this act is illustrated later in Table 6.1. Shortly after the passage of the Pollution Prevention Act of 1990, the EPA's Office of Pollution Prevention and Toxics (OPPT) launched a model research grants program called "Alternative Synthetic Pathways for Pollution Prevention." This program provided grants for research projects that include pollution prevention in the design and synthesis of chemicals. The grant program eventually resulted in the establishment of the Green Chemistry Program around 1991 and 1992. Over the years, this program has catalyzed the development of many green chemistry approaches and environmentally benign chemical syntheses.

Green chemistry is the use of chemistry to reduce pollution at the source. More specifically, green chemistry is the design of chemical products and processes that reduce or eliminate the use or generation of hazardous substances. It is an overarching philosophy of chemistry, rather than a discipline of chemistry such as organic or inorganic. In fact, green chemistry is usually multidisciplinary, drawing on a broad range of expertise. Green chemistry, also called *sustainable chemistry,* offers lower hazard alternatives to traditional technologies. Green chemistry and green engineering are partners in achieving sustainability.

Green engineering research was begun in the early 1990s to support environmental risk assessment of new chemicals. OPPT staff realized that the risk-based tools used to assess human health and environmental risk of new chemicals, when combined with traditional engineering design, could result in "greener" processes. This recognition eventually resulted in the establishment of the green engineering program which has catalyzed the green engineering movement, including many educational initiatives, research programs, and development of environmentally beneficial processes and products.

Environmental risk or environmental impact is an essential concept of green engineering (http://www.apa.goviogpugreenengmeaning).[1] There are numerous literature references on environmental risk assessment and some are included in the list of references in this chapter. Risk assessment considers the extent of harm a chemical and its uses pose to human health and the environment. Mathematically, it is expressed as a function of hazard and exposure:

$$\text{Risk} = f(\text{hazard and exposure})$$

A hazard is anything that will produce an adverse effect on human health and the environment. In environmental risk assessment, the hazard component generally refers to toxicity. Exposure is the quantitative or qualitative assessment of contact to the skin or orifices of the body by a chemical. Traditional pollution prevention techniques focus on reducing waste as much as possible; however, risk assessment methods used in pollution prevention can help quantify the degree of environmental impact for individual chemicals. This approach provides a powerful tool that enables engineers to better design processes and products by focusing on the most beneficial methods to minimize all aspects of risk.

By applying risk assessment concepts to processes and products, one can accomplish the following.

- Estimate the environmental impacts of specific chemicals on people and ecosystems.
- Prioritize chemicals that need to be minimized or eliminated.
- Optimize design to avoid or reduce environmental impacts.
- Assess feed and recycle streams based on risk and not volume within a chemical process.
- Design "greener" products and processes.

Historically, scientists, engineers and policy makers have focused their energy on minimizing exposure as the easiest way to minimize risk to humans or to the environment. Green chemistry focuses on the hazard or toxicity component of the equation instead of exposure. In reducing hazard, risk is reduced in a more reliable and frequently cost-effective manner. Exposure controls, such as personal protective equipment, thermal oxidizers, or treatment plants, are frequently expensive. Also, there is always a risk that an exposure control can fail. Such failure can be mitigated and minimized, but it cannot be eliminated. If instead, hazard is reduced, risk is minimized more reliably. To be sure, there are not yet technologies to eliminate the use or generation of all hazardous substances, so exposure controls are still needed, but to the extent that future technological developments can minimize hazard, the need for treatment or remediation will also be minimized.

Chemistry is an inherently creative discipline. Chemists routinely create new molecules and new methods to make molecules. Green chemistry taps this creativity. The 12 Principles of Green Chemistry,[2] originally published in *Green Chemistry: Theory and Practice*,[3] provide a roadmap for implementing green chemistry. Green chemistry techniques and principles are very powerful, especially for development of new chemicals and processes.

Green engineering (see Fig. 6.1) provides a system-based framework for evaluating and improving the environmental performance of chemical processes and products (both new and existing) by integrating consideration of health and environmental risk, green chemistry, and pollution prevention approaches into traditional engineering design.

As defined previously, risk is a function of hazard and exposure. In the environmental risk context, hazard is a function of toxicity which is affected by physical/chemical and environmental fate properties and hence chemistry. Green chemistry reduces the risk or environmental impact of processes or products by focusing on the hazard component.

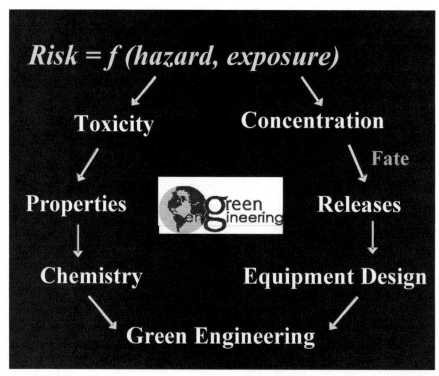

Fig. 6.1. Green engineering is a holistic approach to green process development and design.

Exposure is a function of concentration and also is affected by the environmental fate properties of a chemical. Concentration is a function of the environmental release that is affected by the equipment or unit operations.

Environmental fate is an important subject area of environmental engineering discipline, especially in the modeling and design of treatment systems. It is also an important component of green engineering. Environmental fate properties can be useful for designing greener chemicals and performing environmental performance assessment of processes.

Green engineering minimizes risk or environmental impact by addressing both the exposure component or unit operations and the hazard or chemistry of the process or product (Fig. 6.2).

Figure 6.3 provides an example for a green engineering process. The left-hand side is a conventional process for making methyl acetate which involves a reactor and a series of distillation steps. The right-hand side is a greener process which involves a combined unit operation, reactive distillation. The green process is superior to the conventional process in many aspects, including chemistry, exposure, release, energy, and economics.

There have been a number of green engineering concepts, approaches, and tools that have been developed since the introduction of green engineering in the mid-1990s. Many of the green engineering approaches, concepts, and tools have been compiled into a standard textbook, *Green Engineering: Environmentally Conscious Design of Chemical Processes* by Allen and Shonnard.[5] These concepts, approaches, and tools can also be found via accessing the EPA Green Engineering website at www.epa.gov/oppt/greenengineering. The Web site contains links to many computerized green design tools that can be downloaded free of charge.

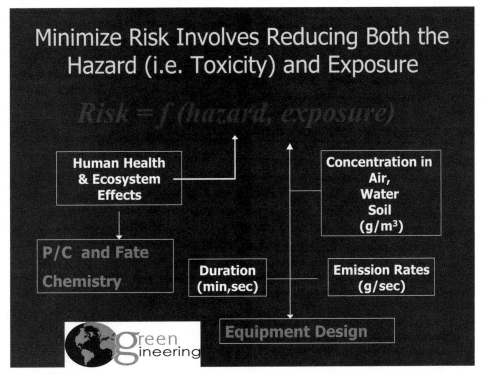

Fig. 6.2. Green engineering addresses both the hazard and exposure components of the risk equation. (P/C = physical chemical properties).

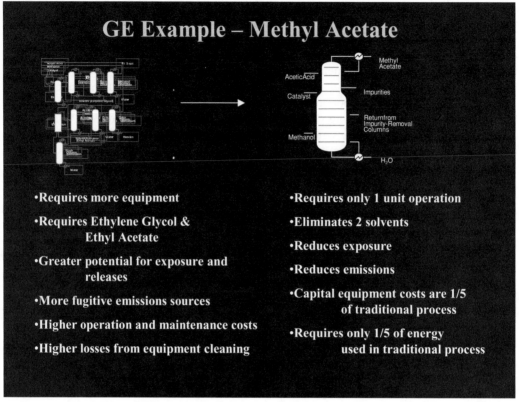

Fig. 6.3. Green engineering example: conventional versus greener process using reactive Distillation technologies for the production of high-purity metal acetate (Adapted from Malone and Russ[4] and modified.)

TWELVE PRINCIPLES OF GREEN CHEMISTRY*

1. *Prevent waste: Design chemical syntheses to prevent waste, leaving no waste to treat or clean up.* As commonsensical as this sounds, many chemists do not consider the waste generated by the syntheses they design. Coproducts (substances that are formed in stoichiometric amounts during a reaction) and byproducts (substances that are formed unintentionally, usually in side reactions) are an inconvenience with which others are expected to deal.

2. *Design safer chemicals and products: Design chemical products to be fully effective, yet have little or no toxicity.*

*Adapted from *Green Chemistry: Theory and Practice.*[3]

Although much is understood about the hazard of various substances, chemists frequently neglect toxicity as a design criterion when they are evaluating the performance of a chemical substance. Because hazardous materials are so routinely used in the laboratory, hazard becomes a trivial matter. Hazard should be explicitly considered and minimized during chemical design.

3. *Design less hazardous chemical syntheses: Design syntheses to use and generate substances with little or no toxicity to humans and the environment.* A holistic view of a synthetic pathway often allows a chemist to change factors and minimize hazard in a number of steps simultaneously. A safer synthesis may simply reduce the number of isolations and purifications,

or it may allow a cascade of changes where a new first step fosters changes down the line.

4. ***Use renewable feedstocks:*** *Use raw materials and feedstocks that are renewable rather than depleting.* Renewable feedstocks are often made from agricultural and forest products or are the wastes of other processes; depleting feedstocks are made from fossil fuels (petroleum, natural gas, or coal) or are mined. Another advantage of renewable feedstocks is that often the oxidation state of the feedstock is often close to that of the desired product. This simplifies and reduces the number and extent of chemical transformations necessary in a synthetic pathway.

5. ***Use catalysts, not stoichiometric reagents:*** *Minimize waste by using catalytic reactions.* Catalysts are used in small amounts and can carry out a single reaction many times. They are preferable to stoichiometric reagents, which are usually used in excess and work only once. Even oxidations and reductions, which require a change in oxidation state and therefore an ultimate electron sink or source can be improved by using a catalyst. If reduction using hydrogen gas or oxidation with oxygen can be selectively catalyzed, the hazard and amount of waste generated is minimized.

6. ***Avoid chemical derivatives:*** *Avoid using blocking or protecting groups or any temporary modifications if possible.* Derivatives use additional reagents and generate waste. By definition, any group that is added and removed is waste. Reuse of protecting groups is uneconomical, leading to substantial amounts of waste when they are used. Reactants or reaction conditions may be tailored to maximize selectivity for the desired moiety.

7. ***Maximize atom economy:*** *Design syntheses so that the final product contains the maximum proportion of the starting materials. There should be few, if any,*

wasted atoms. Reactions should also be designed minimize the *E*-factor (ratio of the mass of all reaction waste to the mass of the desired product). Being aware of all the reaction inputs and outputs helps the designer maximize the benefit of the new chemistry. Materials efficiency not only minimizes environmental impact, but maximizes cost efficiency.

8. ***Use safer solvents and reaction conditions:*** *Avoid using solvents, separation agents, or other auxiliary chemicals. If these chemicals are necessary, use innocuous chemicals.* More and more research demonstrates reactions may not require solvents to proceed in clean and quantitative yields. When solvents are required, water, CO_2, ethanol, or other low-toxicity alternatives are preferred to traditional organic solvents.

9. ***Increase energy efficiency:*** *Run chemical reactions at ambient temperature and pressure whenever possible.* Shorter reaction times also help minimize energy use. Designers should be cognizant of the conditions necessary to carry out their transformations.

10. ***Design chemicals and products to degrade after use:*** *Design chemical products to break down to innocuous substances after use so that they do not accumulate in the environment.* This is arguably one of the hardest principles to apply, despite the existence of sound experimental evidence of what groups degrade well in the environment. The trick, of course, is how the substance "knows" it is at the end of its useful life. Products must be stable long enough to be available before and during use. Degradation may be triggered by a change in the conditions: presence or absence of water, light, oxygen, microorganisms, or other environmental factors.

11. ***Analyze in real time to prevent pollution:*** *Include in-process real-time monitoring and control during syntheses to minimize or eliminate the formation of*

byproducts. It is possible to monitor reactions and to adjust the reaction parameters, such as feedstock ratios, temperature, pressure, etc., to maximize the selectivity for the desired product.

12. ***Minimize the potential for accidents:*** *Select chemicals and their physical forms to minimize the potential for chemical accidents, including explosions, fires, and releases to the environment.* Whenever possible, select lower energy forms of substances to minimize the energy needed as well as the energy content in case of accident. Avoid substances that are corrosive, highly reactive, or acutely toxic. Inherently safer chemistry reduces the risk from accidents as well as intentional harm.

PRINCIPLES OF GREEN ENGINEERING

Green engineering is the design, commercialization, and use of processes and products that are feasible and economical yet, at the same time, minimize generation of pollution at the source and risk to human health and the environment. Green engineering embraces the concept that decisions to protect human health and the environment can have the greatest impact and cost effectiveness when applied early to the design and development phase of a process or product.

There exist a number of green principles, including green engineering principles that have been developed. Most of these principles have overarching themes and objectives. For example, some were developed to help with design. The Green Engineering Principles[*] presented below were developed at the Green Engineering: Defining the Principles Conference held in Sandestin, Florida in May

---

[*]The preliminary principles forged at this multidisciplinary conference are intended for engineers to use as a guidance in the design or redesign of products and processes within the constraints dictated by business, government, and society such as cost, safety, performance, and environmental impact.

2003 and attended by more than 65 individuals, primarily engineers of various disciplines and scientists. The attendees used as a starting point various existing compilations of green and sustainability related principles in developing these nine Principles of Green Engineering. Examples of existing principles used in this exercise were the Hannover Principles, Twelve Principles of Chemistry, CERES Principles, and Twelve Principles of Green Engineering proposed (Anastas and Zimmerman, 2003). This list, along with the previously described Green Chemistry Principles, are intended to be used in tandem as guidance or rules for design and development of green processes and products. The Green Engineering Principles, described below, parallel many of the concepts and approaches covered in the textbook *Green Engineering: Environmentally Conscious Design of Chemical Processes.*[5]

1. ***Engineer processes and products holistically, use systems analysis, and integrate environmental impact assessment tools.*** These concepts resonated in a number of green and sustainable principles and are addressed at length in Allen and Shonnard.[5] Evaluate and reduce the environmental health and safety impacts of designs, products, technologies, processes, and systems on ecosystems, workers, and communities continually and "holistically." Avoid risk shifting (e.g., reducing releases to one environmental medium may increase risk to another medium and/or worker exposure and safety).

2. ***Conserve and improve natural ecosystems while protecting human health and well-being.*** [Reduce, reuse, and recycle the materials used in production and consumption systems, and ensure that residual waste can be assimilated by ecological systems. Rely on natural energy flows. Design processes and products to create cyclical material flows.]

3. ***Use life-cycle thinking in all engineering activities.*** [This is an important

concept of green engineering. Life-cycle approaches have been widely used by many companies to assess and improve the environmental performance of their products and processes. It is essential that one consider the environmental impacts throughout the product or process life cycle from extraction through manufacturing, use, and disposal.]

4. *Ensure that all material and energy inputs and outputs are as inherently safe and benign as possible.* [This principle addresses the hazard component of the risk equation. Risk will be minimized if material and energy sources are inherently benign and safe.]

5. *Minimize depletion of natural resources.* [This principle resonates in many existing lists of principles. It reminds the engineer of the need to reduce and recycle the materials used in production and consumption systems. Conserve energy and improve the energy of internal operations and of the goods and services. Make every effort to use environmentally safe and sustainable energy sources.]

6. *Strive to prevent waste.* [It is always better and more economical to prevent waste from occurring in the first place.]

7. *Develop and apply engineering solutions, while being cognizant of local geography, aspirations, and cultures.* [Successful implementation of green engineering solutions can be affected by such factors as availability of resources and geography. It is important that these factors are considered in selecting the green engineering solutions that are most effective for certain localities or regions.]

8. *Create engineering solutions beyond current or dominant technologies; improve, innovate, and invent (technologies) to achieve sustainability.* [This principle encourages engineers to be creative and innovative in design and development of green engineering solutions. Think "outside the box" for development of green and sustainable technologies.]

9. *Actively engage communities and stakeholders in development of engineering solutions.* [This is important to ensure that stakeholders are supportive of the engineering solutions.]

Engineers and chemists, as designers of products and processes, have a central role in designing chemical processes that have a minimal impact on the environment. Green chemistry and green engineering approaches and tools should be used to design new processes and modify existing processes. Green engineering broadens the scope of engineering design to encompass critical environmental issues and is an important framework for achieving goals of sustainable development.

II POLLUTION PREVENTION HEURISTICS FOR CHEMICAL PROCESSES

INTRODUCTION

This section discusses pollution prevention (P2) guidelines and heuristics in chemical process industries useful in process/product development and design stages. The heuristics can also be used for analysis of existing processes but it should be noted that the number of available options are somewhat limited at this stage because the design decisions are already locked in. Therefore, for existing processes, these rules provide some guidelines for maintenance procedures and for making retrofitting decisions.

The outline of this section is as follows: First is a listing and discussion of the general hierarchical rules to be evaluated during the process design stage. Then, because reactors and separators form the heart of a chemical process, specific aspects of P2 heuristics as applied to reactors and separators are discussed in somewhat more detail.

It is appropriate at this point to define what is meant by a heuristic and indicate the pattern of discussions in the other sections of this chapter. A heuristic is commonly defined as a general "rule of thumb," or procedure to arrive at a solution in the absence of a detailed analysis. More specifically, it is a set of rules, often based on common knowledge, used to guide thinking towards a final solution. In our case, the final goal is P2, and heuristics guide us towards achieving this goal.

In this section, the heuristics are often presented as questions rather than as affirmative rules. The rules are then often obvious answers to the questions, and, hence, often no explanation is necessary nor is any provided. However, if an elaboration or clarification is needed, this is provided immediately following the question. The question format of presenting the heuristics is very useful, for instance, in brainstorming sessions during process evaluation and design.

HIERARCHICAL RULES FOR WASTE MINIMIZATION

General rules for P2 follow the Pollution Prevention Act of 1990 (42 U.S.C. 13101-13109) and are applicable to any manufacturing activity. This act clearly identifies the waste management hierarchy, and we list them in Table 6.1 for completeness. The P2 Hierarchy specific to a chemical process is derived from this set of principles.

A simplified overall schematic of a chemical process is shown in Fig. 6.4 and forms a

Table 6.1   P2 Hierarchy According to Pollution Prevention Act of 1990

1. Source Reduction
2. In-process recycling
3. On-site recycling
4. Off-site recycling
5. Waste treatment to render the waste less hazardous
6. Secure disposal
7. Direct release to the environment

block diagram to analyze pollution prevention strategies. A flowsheet can be broken down into the following sections: (i) Pretreatment section, (ii) reactor section, (iii) separation section, and (iv) postprocessing section. The reactor and separator are usually coupled, and, hence, the P2 decisions are often to be made together for this section. Also it may be noted that the blocks cannot be treated in isolation and any design changes made in one section may affect the other. The blocks are, therefore, only convenient partitions for outlining a general strategy. Any decision made for one section has to be reexamined in the overall context, and reconsideration may often be needed. The rule "think outside the box" is very relevant here. With this background

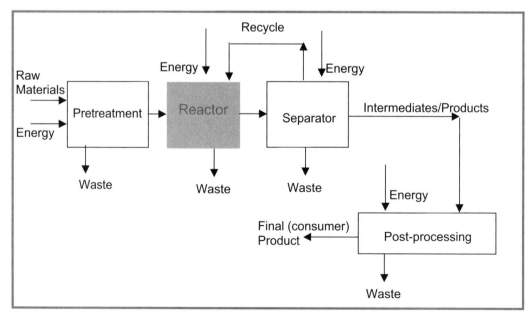

Fig. 6.4. Overall flowsheet showing the various sections for P-2 analysis.

TABLE 6.2   P2 Hierarchy for Flowsheet Analysis or Process Design

1. Batch or continuous mode
2. Input-output structure of the flowsheet
3. Recycle structure of the flowsheet
4. Reaction systems
5. Separation systems
6. Postprocessing and product sections (e.g., tableting, product drying, etc.)
7. Energy systems (boilers, cooling towers, heat exchangers, etc.)
8. Auxiliary equipment (piping, storage tanks, etc.)

setting, we can discuss the hierarchical approach to evaluate P2 opportunities.

The hierarchical rules for analyzing a flowsheet or designing a new process were first formulated by Douglas[6] and then modified by Rossiter and Klee,[7] and these form a starting point in this section. The steps are outlined in Table 6.2, which is an extended version of the original suggestions by Rossiter and Klee.[7] The key idea in forming such a hierarchy is that decisions made in earlier steps do not generally affect the design/changes to be made in the later steps and, hence, the rules can be followed on a "top to bottom" basis. This prunes the decision tree and makes the process synthesis task somewhat simpler. Each of the hierarchical steps is now discussed in detail together with the heuristics to be followed for each step.

Batch or Continuous?

The general guidelines are as follows.

1. *What is the annual rate of production?* For capacities of the order of 500 t/year or less, use batch. For capacities of more than the order of 5000 t/year, use a continuous process.
2. *Does the system involve RCRA hazardous chemicals?* If so, try to use continuous processes even if the scale is relatively small. Continuous processes generate waste at a constant rate. The composition of the waste is also constant with time and, hence, the treatment is easier. In contrast, batch processes generate wastes

intermittently with a large rate of generation for a short period (during the peak point in production). As a result, continuous processes are easier to monitor and control. Batch processes also generate additional waste due to the need for cleaning the equipment between two batches. Hence, the capacity alone may not be the deciding factor in choosing between batch and continuous operation.

It may be noted that raw material substitution is easier in batch processes. Therefore, if one anticipates a varying range of feed stocks then batch processes may have some merit. In a similar note, if the same equipment is used to make a wide range of products (e.g., dyestuffs) then a batch process may be more suitable.

Input-Output Structure

Input-output analysis focuses on the overall structure of the flowsheet, and the recycle streams are not considered here in as much as they do not appear in the overall flow streams. Hence, this structure is essentially an overall mass balance for the entire process. A simplified representation of the input-output structure is shown in Fig 6.5. The main purpose of this analysis is to identify the amount of raw materials used, useful products, and waste formed.

At this stage one can define a process efficiency parameter as:

Process efficiency = Mass of products formed/Mass of raw materials used which can be used as an indicator of the P2 success scorecard. The goal is to improve the process efficiency, and any changes can be benchmarked against the level of existing efficiency.

The heuristics to be considered at this stage are as follows:

Are any impurities in the feed tied to the waste streams? If so, try source reduction.
Is there a scope for better raw material selection? This would achieve a source reduction which is number 1 in the p2 list given in Table 6.1.

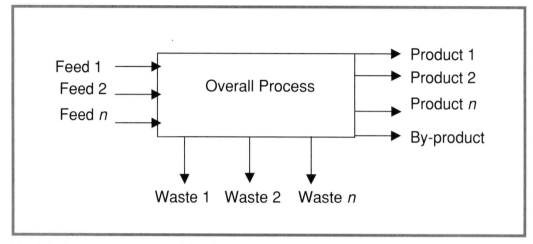

Fig. 6.5. Input-output structure of flowsheet.

Do the products have a toxic component?
Can this be minimized by proper raw
material selection?

*Can value be added to the waste? Are there
any raw materials that are suitable for in-
process recycling?* In-Process recycling
is covered under item 2 in the P2 list.

*Can wastes be used in the process recy-
cling? Are there waste agencies near by
to take these away?* This concern is
addressed in item 3 of the P2 hierarchy.

*Are wastes generated due to reactor ineffi-
ciency?* Some guidelines for improving
the reactor performance are discussed
below under "Heuristics for Green
Reactor Design."

*Are wastes generated as a result of separa-
tion inefficiency?* Some guidelines for
improving the separation processes are
discussed in a later section titled "The P2
Rules for Separations Devices."

Recycle Structure of the Flowsheet

At this level of analysis, the focus is on the
recycle streams. The pertinent questions to be
addressed at this stage are as follows.

*Is there a scope to redirect any output
waste streams back to the reactor feed
to reduce waste? Can the recycle to reactor be fed directly to the reactor or is a*
separate reactor needed for this purpose?
Often the streams leaving the reactor
have trace impurities (formed as byproduct in a side reaction) that can be harmful to the catalyst. Hence a separate
reactor may be more suitable for such
cases. In some cases, the streams can
be recycled back to the existing reactor
but only after a pretreatment step to
remove the trace contaminants and catalyst poisons.

Can any impurity be recycled to extinction?
This is possible if the impurities are
formed by a reversible reaction while the
main reaction is irreversible.

Is the recycle affecting the purge streams?
If so consider raw material pretreatment.
Nonreactive materials in the feed stream
are responsible for the purge and reducing their quantity would reduce the
purge. For example, if oxygen-enriched
air is used instead of air in a typical partial oxidation process, the quantity of the
purge and the associated waste can be
reduced.

*Can any waste from separation units be
reprocessed? Can any waste from separation be reused in any other part of the
process?*

*Has the scope for water recycle been exam-
ined completely?* Often the possibility of

reuse of stripped sour water, wastewater from utility blowdown, etc. exist, and these are often overlooked.

Reaction Systems

Having examined the recycle nature of the flowsheet and the relevant decisions made, the focus shifts to the reaction section. Some of the common issues related to reactor design to be addressed at this stage are as follows.

Are wastes formed from the side reactions? Often the side reactions can be minimized by simple changes in operational procedure. For example, temperature sensitivity of the side reactions can be examined and if the rate of the side reaction increases more with temperature than the main reaction, then lowering the reactor temperature will reduce the waste.

Are the side reactions leading to waste formed in a reversible step of the reaction? If so, there is scope to recycle wastes to extinction.

Is the catalyst used the best available or is there a scope for catalyst replacement?

Is there a scope for solvent substitution? If so, how would it affect the other steps? Additional information on solvent selection is discussed later.

Separation Systems

Having addressed issues related to the overall flowsheet and the reaction system, attention is now focused on the separation systems. Separation systems offer considerable scope for waste minimization. They can be generally classified into the following categories.

1. Gas–liquid separation
2. Gas–solid separation
3. Liquid–liquid separation
4. Liquid–solid separation

General heuristics related to separation systems are indicated here. Detailed heuristics with respect to each individual type of separation are addressed in "Separations Devices."

Some of the key questions to be addressed are the following.

1. *Are any waste streams produced as a result of poor separation?*

2. *Is the separation sequence optimum?* Separations that are easy are done first, followed by the less energy-intensive ones. The final separation step is usually assigned to the most difficult step because the overall quantity of stream to be treated would be minimum by then.

3. *Is the current separation the most suitable or should other methods of separation be considered?* For example, distillation is commonly the workhorse in the chemical industry and is often the immediate choice. Other separation methods, such as pervaporation, can be sometimes more effective especially if the volatility differences are small. Pervaporation is a membrane-based process with the difference that the permeate appears as a vapor, thus permitting solute recovery and recycle. For example, benzene can be recovered from hydrocarbon streams using this method in fairly high concentrations and in a usable form ready for recycle. Many alternative separation methods must be considered, and one should not simply bank on past experience or expertise.

4. *Is the current separation most efficiently designed or are there alternative designs that can reduce energy consumption and lead to less waste?* Examples may be in solvent extraction where a simple switch of solvent may lead to a nonhazardous waste. Also, conventional distillation is often used, but, currently, new design concepts, such as divided wall distillation,[8] may prove to be more beneficial for certain processes. Replacing a conventional packing by a more efficient, structured packing may improve the treatment of waste gases in absorption columns.

5. *Is the choice of the mass separation agent (MSA) both cost effective and*

environmentally benign? Can any improvements be possible in this choice? Caffeine extraction from coffee beans is an example. Chlorinated solvents were used as mass separating agents, which posed both a health risk to workers and global harm due to ozone depletion. Supercritical CO_2 and, more recently, supercritical water have replaced these as MSAs.

6. *Have mass integration possibilities been evaluated? Are mass exchange networks (MEN) in place in the flowsheet?* MEN synthesis has been addressed in the book by El-Halwagi,[9] and application of this method to water management has been addressed in a paper by Liu et al.[10] These aspects should be looked at carefully during the process synthesis stage, as they offer considerable scope for solvent or MSA reduction.

7. *Can the separation and reaction sections be suitably combined?* Combining reaction and separation in one unit can have some advantages. It often removes the equilibrium barrier, and, in some cases, complete recycle of unreacted feed is achieved in the same vessel. This question is more important in the design stage of new processes and is somewhat difficult to address in retrofitting existing processes. Membrane separations provide another area where reaction and separation can be combined.

8. *Can any hazardous waste be removed before discharging into a POWT (publically owned water treatment) facility?*

Postprocessing and Product Section

Postprocessing forms an additional important step in the manufacture of many consumer-oriented products, in contrast to commodity chemicals. Examples include tablets, pills, toothpaste, creams, and a wide variety of common products. The chemical industry has traditionally focused on commodity chemicals (process engineering), but, recently, the emphasis has shifted to product engineering. The postprocessing of

chemicals to make consumer-used products can often be a major source of pollutants. Hence, careful attention has to be paid to this part of the process as well. Also, this part of the overall plant may involve batch processing, whereas the rest of the process may be continuous processing. Some heuristics are listed below.

Are fines created as a result of product drying? If so, select a different type of dryer. High air velocities or thermal degradation of solids lead to fine formation in drying equipment. These can be avoided by choosing the proper type of dryer depending on the properties of the solid.

Is the blending operation optimized or is it creating some wastes? Can any waste from this operation be recycled?

Is the cleaning operation in between batches optimized? Is there a possibility of wash–solvent recycle here without affecting the overall product quality?

Can process equipments be coordinated to minimize vessel cleaning and to reduce the associated wastes?

How will the products be packaged? Will containers be available for recycle? Are any additional wastes generated as a result of the container reuse policy?

Energy Systems

Energy costs are a major part of the operating costs in chemical processes and, hence, any reduction in energy consumption leads to an increased profitability. Further increased energy use has a direct bearing on the greenhouse gas emissions and, thus, the optimization of the energy systems is also important from a P2 point of view. Some heuristics to be discussed at this stage of analysis are the following.

Is the process fully optimized for energy use?

Are the heat integration networks in place and is there further scope in reducing the energy costs?

Can cleaner fuels be used to reduce SO_2 emissions? If so, at what cost?

Do the steam systems operate at the needed pressure rather than at the available boiler pressure? Too large a pressure leads to a higher condensation temperature for steam. This causes an unnecessarily large temperature differential for heat transfer. Fouling and other problems arise as a consequence, reducing the energy efficiency. In other words, use utilities at the lowest practical temperature. A simple solution to reduce the steam temperature is the use of a thermocompressor. This device uses high-pressure steam to increase the pressure of low-pressure steam to form steam at a desired intermediate pressure.

Are the heat exchangers routinely maintained to reduce fouling? Are the state-of-the art cleaning methods used to reduce sludge formation?

Is the wastewater from high pressure cleaning of heat exchangers treated separately and not with all the other water streams? These often carry fine particles that provide a large surface area for oil and water to stabilize creating an oily sludge in the wastewater which is difficult to separate.

Can the cooling tower blowdown be reduced? Often pretreatment of fresh water to the cooling tower to reduce calcium salts can be helpful to reduce the scaling and thereby reduce the blowdown. Ion-exchange or even more expensive options, such as reverse osmosis, may prove to be beneficial in this regard depending on the quality (hard or soft) of the feed water.

Auxiliary Equipments

These include pumps, compressor, and storage tanks. The pumps, valves, flanges, etc., can often be a major source of pollution due to fugitive emissions (unintentional release of process fluids). Storage tanks also contribute to pollution by breathing and standing losses. Hence, sufficient attention should be paid to the design and maintenance of this equipment from a P2 standpoint, and this auxiliary equipment should not be taken for granted. The following guidelines are useful.

Can welded pipes be used instead of flanged pipes? This will reduce the fugitive emissions.

Can the total amount of equipment and number of connections be reduced?

Are tanks properly painted? A tank freshly painted will reduce the breathing losses by 50% as shown in an example by Allen and Rosselot.[11]

Has the EPA recommended estimates for fugitive emissions done on a periodic basis? Are any improvements noted over a period of time?

HEURISTICS FOR GREEN REACTOR DESIGN

Green design of a reactor can be approached in the following hierarchical manner with the top of the hierarchy being the least cost effective solution.

• Minor modifications in operating conditions and better "housekeeping" practices in existing processes.
• Additional "end-of-the-pipe" clean-up of wastes. (The goal here is to recycle and recover.)
• Major retrofitting of existing processes. Waste reduction and enhanced recycle. Goal is zero emissions and "total recycle."
• Development and installation of more efficient new process technologies that minimize waste and pollution.
• Process intensification concepts and new reactor design concepts.

The last two items in the above list involve considerable R&D and hence present a long-term strategy for waste minimization in reactors. These items should be looked into carefully at the early stage of process design. The first three can be attempted in existing processes.

The minor modifications can be addressed by looking at some of these issues.

A. Source reduction
    Use nonhazardous raw materials.
    Use renewable resources.
    Use benign solvents.
    Reduce use of solvents.

B. Routine maintenance
   Temperature control.
   Pressure control.
   Vent and relief system tuning.
   Routine calibration of instruments.
   Routine cleaning of steam jackets, coils, and other reactor cooling auxiliaries.
C. Operational changes
   Reduce byproducts and generate less waste.
   Produce products easy to separate.
   Use heat integration.
   Evaluate the effect of temperature on side-product formation rate. Can an increase in temperature be beneficial or should one decrease the temperature?
   Evaluate the role of mass transport for multiphase systems. In some cases an increased mass transfer may favor byproduct formation, and reducing the degree of agitation, for instance, may be helpful.
   Evaluate the role of mixing of reactants. Is a premixed feed better?
   Evaluate the role of catalyst. Is there a better or more stable catalyst?
   Is the temperature profile in batch or semi-batch the optimum? Can the temperature peak be minimized?
D. Design changes
   Recycle unreacted materials.
   Consider a separate reactor for recycle to minimize catalyst deactivation.
   Do not overdesign. Overdesign is not only capital consuming but also leads to increased energy costs that translate into greenhouse gases in the environmental context.

The above rules can be applied to an existing process. For new plants it may be necessary to revamp the entire process and use alternative benign production technologies. Some approaches to investigate here are as follows.

Improve atom efficiency.
Improve energy efficiency.
Novel chemistry, e.g., solid-catalyzed routes to replace liquid-phase routes or simple quantitative organic chemistry-based routes.
Novel solvents, e.g., CO_2 expanded systems, ionic liquids, etc.
Novel reactor concepts, e.g., periodic operation, membrane reactors, etc.
Reaction + Separation combination, e.g., catalytic distillation, extractive reactions.
Biphasic catalysis.

A more detailed description of the role of chemical reaction engineering is available in the paper by Tunca et al.,[12] along with some examples and recent trends. Another useful source is the EPA Green Chemistry Web site (www.epa.gov/oppt/greenchemistry).

THE P2 RULES FOR SEPARATIONS DEVICES

Some heuristics related to specific methods of separation are as follows.

Distillation Columns

Distillation columns are the workhorse of separations in the chemical industry, and these are very common and offer considerable scope for P2. Some important questions to be addressed in this context are listed below.

1. *Does the system form an azeotrope?* If so, is an entrainer used? Can one replace the entrainer by a more benign agent? Can the azeotrope be broken by other methods such as pervaporation (see paper by Wynn, 2001) or membrane separation and then continue with distillation to get the final purity?
2. *Is the reflux ratio in distillation column optimized?* Increase in reflux ratio increases the product purity. However, this causes a larger pressure drop in the column, and increases the reboiler temperature and the reboiler heat duty. Hence, there is a delicate balance among these factors, and often the reflux ratio has to be continuously adjusted to meet the change in feed composition and other day-to-day variations.

3. *Is the feed location optimum?* Simply relocating the feed may cause a more pure product and may be a simple strategy to reduce waste formation. It has both economic and environmental benefits.

4. *Is there a scope for combining reaction and distillation in one piece of equipment?*

Gas–Liquid Separation

These include absorption columns to remove a gaseous impurity as well as stripping columns to remove a VOC from a liquid (usually wastewater). Note that these involve a MSA (mass separation agent). For example, for stripping of VOC from wastewater, steam or air is used as the MSA. Hence, appropriate choice of MSA is an important consideration in the context of P2 for gas–liquid separations. Some important heuristics are as follows.

Are the off-gas specifications within the regulations? If not try to switch to more efficient structured packings.

Are the desorption systems at the optimum pressure?

Can the solvent losses be minimized? Should one use alternate solvents that have better heat stability over the repeated absorption–desorption cycle?

Gas–Solid Separations

Gas–solid systems are encountered in a number of processes, such as fluid bed dryers, fine capture from gaseous effluents to meet the PM_5 air quality criteria, etc. Efficient design of these systems will improve the air quality.

Another example of gas–solid separation is the adsorption process. The regeneration of adsorbent is often not complete due to pore diffusion limitations and other factors. Furthermore, the eventual replacement of spent adsorbent leads to solid wastes. Optimization of adsorbent pore structure is one option that can be examined here. The process of regeneration also leads to waste formation and needs to be set at optimum conditions.

Liquid–Liquid Separations

These systems are common in liquid extraction and also in a multiphase reactor with an organic and an aqueous phase. Common sources of pollution are incomplete separation and contamination due to trace organics in the aqueous phase. An example is in alkylation reactions (e.g., *n*-butane reaction with olefins to form isooctanes). Strong acids, such as sulfuric and hydrofluoric acids, are used as catalysts, and the recovery and the recycle of acid need to be optimized in order to reduce the waste generation.

Liquid–Solid Separations

These are encountered in separation such as filtration and involve flow through a membrane. The prevalence of this in the chemical industry has not, however, diminished the challenges in proper design of these systems. A wide range of membranes is available for separation of a solid from a liquid. Fouling is the major challenge in membrane processes and is the most common source of pollution. Membrane systems need periodic cleaning and are another source for pollution generation in terms of waste water. Some heuristics to consider at this stage of design are listed below.

Is the membrane hydrophobic or hydrophilic? Has the proper selection of membrane tied to the property of the slurry?

Can fouling be reduced by maintaining a higher shear rate at the interface?

In summary, the hierarchical process review and the various heuristics presented here provide a systematic method for identification of waste formation and the appropriate strategies to minimize waste in process industries. Maximum benefit of using this approach is realized at the process development and process synthesis stage followed by the process design stage. Some of the guidelines can also be used in the retrofitting stage for existing processes. Additional P2 methods, technologies and practices are found in the compilation by Mulholland and Dyer.[13]

ACRONYMS

MEN	Mass exchange network
MSA	Mass separation agent
P2	Pollution prevention
PM_5	Particulate matter <5 μm in diameter
POTW	Publicly owned (wastewater) treatment works
RCRA	Resource conservation and recovery act

ACKNOWLEDGMENT

P. A. Ramachandran would like to thank National Science Foundation for partial support of research in the general area of environmentally benign processing through the grant NSF-ERC center grant EEC-0310689.

III UNDERSTANDING AND PREDICTION OF THE ENVIRONMENTAL FATE OF CHEMICALS

INTRODUCTION

To understand the fate of a chemical once released into the environment requires knowledge of not only how the environment is modeled but also of the intimate connection among a chemical's structure, its physicochemical properties, and its behavior under a given set of conditions. The behavior of a chemical in the environment falls under one of two categories: *translocation*, resulting in the movement of the chemical either within a compartment or between compartments, and *transformation*, resulting in the alteration of the chemical's structure. In predicting the persistence of a chemical in the environment, three broad questions are asked:

1. Will it move, and, if so, where will it go?
2. Will its structure be altered?
3. How long will it persist?

The first question is answered by understanding the susceptibility of the chemical to translocation, whereas the second question is answered by understanding the chemical's susceptibility to transformation. Once these

two questions are answered, then the third question, requiring rates of translocation and/or transformation, can be answered, thus providing a foundation upon which subsequent assessments of risk to the environment and public health can be performed.

The goals of this section are to introduce methods of modeling chemical movement within and between environment compartments, to define specific translocation and transformation processes, to provide a basic understanding of the association among chemical structure, physicochemical properties, and susceptibility to specific translocation and transformation processes, and to provide methods of accessing and estimating physicochemical properties and environmental fate of chemicals.

TRANSLOCATION OF CHEMICALS IN THE ENVIRONMENT

Modeling the Environment

Before assessing how a chemical moves in the environment, the relevant media, or compartments, must be defined. The environment can be considered to be composed of four broad compartments—air, water, soil, and biota (including plants and animals)—as shown in Fig. 6.6. Various approaches to modeling the environment have been described.[14–16] The primary difference in these approaches is the level of spatial and component detail included in each of the compartments. For example, the most simplistic model considers air as a lumped compartment. A more advanced model considers air as composed of air and aerosols, composed of species such as sodium chloride, nitric and sulfuric acids, soil, and particles released anthropogenically.[17] A yet more complex model considers air as composed of air in stratified layers, with different temperatures and accessibility to the earth's surface, and aerosols segmented into different size classes.[16] As the model complexity increases, its resolution and the data demands also increase. Andren et al.[16] report that the simplest of models with lumped air, water, and soil compartments is suitable for

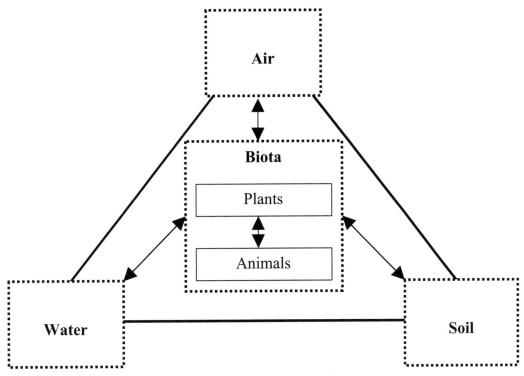

Fig. 6.6. Potential exchange routes of chemicals among the environmental compartments.

chemical fate evaluations at very large scales (global), whereas the more complex models, with greater levels of segmentation and spatial differences within each compartment, are best suited for problems involving regional to site-specific studies.

Regardless of the model complexity chosen, the transport and partitioning of a chemical between water and air, air and soil, and biota and all compartments must be assessed. Furthermore, each of these compartments may contain different air, liquid, and vapor phases, and chemical partitioning among these phases must be determined. The driving force of transport of chemicals within and between compartments is a difference between chemical potential, the tendency of a chemical to undergo physical or chemical change, in one region compared to the other. When thermal, mechanical, or material equilibrium has been upset in the compartment(s) in question, the chemical moves in response. Despite the awareness that nonequilibrium conditions

are at the heart of net chemical transport, modeling of partitioning of chemicals is commonly performed by assuming equilibrium conditions using estimates of liquid–vapor, solid–liquid, and solid–vapor partition coefficients.[18] These coefficients and other properties estimating environmental partitioning, along with properties enabling prediction of environmental transport of chemicals, are briefly discussed in the following sections. For greater detail on transport and fate of chemicals in the environment, the reader is encouraged to refer to numerous resources that are dedicated solely to this topic.[19–21]

Translocation Processes in Air

As characterized in Fig. 6.7, once a chemical is released into the atmosphere, it is rapidly transported by the average wind and subjected to dispersion, defined as spreading as a result of thermal or density gradients and/or turbulence, and advection, defined as movement as

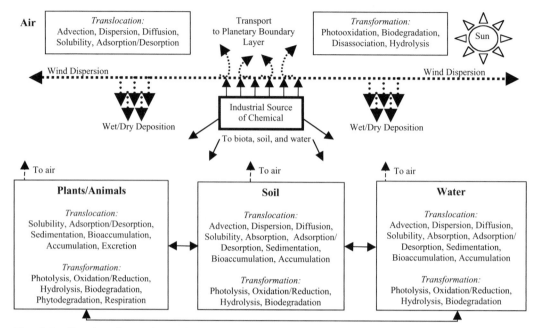

Fig. 6.7. Translocation and transformation processes possible in air, soil, water, and biota. (Adapted from Bishop.[21])

a result of mass flow in the wind. These processes occur within the planetary boundary layer up to approximately 1 km above ground level.[16] Individual contaminant molecules are also prone to move within air along high-to-low concentration gradients. This type of transport is known as diffusion. These three processes combined result in the same effects: movement of the contaminant away from the source by dilution of the contaminant concentration in the release plume and spreading of the contaminant over a larger area.

While in the air compartment, the contaminant "solubilizes" in the vapor–liquid phase or is associated with aerosol particles by adsorption. It is also prone to desorption from the aerosol particles into the vapor phase. Relevant properties of the air used to model transport of partitioning of a contaminant in the air compartment include temperature, turbulence, wind speed, size and composition of aerosol particles, etc.[16, 19] Relevant properties of the contaminant that measure its tendency to partition among the vapor, liquid, and solid phases in the air include its aqueous solubility (S_{aq}), vapor pressure (VP), Henry's constant

(K_H), and a variety of coefficients measuring sorption on solids. These properties will be discussed in more detail in subsequent sections.

The net dispersive, advective, and diffusive transport of contaminants in the air is greatly influenced by the degree of deposition to the soil, water, and biota. Deposition occurs in three steps: (1) turbulent diffusion through the surface layer of the atmosphere, (2) diffusion through a laminar sublayer just above the surface, and (3) the ultimate disposition of the chemical on the surface.[17] As discussed previously with the first step, the second step is also affected by properties of the atmosphere and terrain, including turbulence, wind speed, and temperature, along with the size and composition of the particles composing the aerosols.[16, 19] Dry deposition of the contaminant occurs in the absence of precipitation and involves contaminants with and without association with particles. Wet deposition results from condensation of aerosol particles or equilibrium partitioning of the "dissolved" contaminant from the air to the liquid phase. Relevant chemical properties that enable prediction of the tendency of the contaminant to undergo deposition include aqueous solubility (S_{aq}), vapor pressure (VP), Henry's constant (K_H), and various sorption coefficients on solids. The last step of deposition is the behavior of the chemical once it has reached the water, soil, or biota surfaces, where it may return to the atmosphere in its original form or in an altered structure.

Translocation Processes in Water

Like the transport of a chemical in air, movement of a chemical in water is governed by the flow characteristics of the water itself (advection) and by the degree of diffusion within the water body. Two different types of diffusion can exist in water bodies. Eddy diffusion (or eddy dispersion) results from the friction caused by the water flow over the sediment or soil bottom surfaces. The vertical and horizontal flow resulting from eddy diffusion is more random and temporal than advective flow and thus extends over a smaller region. Horizontal eddy diffusion tends to be greater than vertical flow, and the contaminant is therefore transported to a greater extent horizontally from a point of discharge.[22,23] A contaminant is also prone to transport by molecular diffusion, generated by concentration gradients. As a general rule, unless the water body is stagnant and uniform in temperature, molecular diffusion plays a minor role in transporting contaminants in comparison to eddy diffusion.[23,24] Temperature plays a significant role in determining the degree of eddy and molecular diffusion of a contaminant, particularly in water bodies with stratified layers, such as lakes. As described previously in air translocation processes, the net result of the advective and diffusive transport of a contaminant in water is dilution away from the point of discharge and spreading of the contaminant plume into regions of greater area.

A contaminant can partition from the aqueous phase to solid, air, and biota media, and the presence of each of these media can greatly influence the extent of transport of the contaminant. With a porous solid phase with which the contaminant does not interact, transport of the contaminant is governed by the same laws of mass transport, involving advection and diffusion, that apply in aqueous media free of solids. However, interaction of the contaminant with the solid medium greatly inhibits its movement in the aqueous phase. When contaminant mixtures are present, those with less interaction with the solid medium move along with the aqueous medium, whereas those with greater attraction to the solid medium are retained in proportion to their degree of interaction.[23]

As shown in Figure 6.7, partitioning between the aqueous and solid phases may result from absorption, adsorption/desorption, and sedimentation processes. The contaminant may be taken up into the interior of a solid by means of diffusion in a process known as absorption. As previously described in air–solid partitioning, the contaminant may also be taken up by the surface of the solid, known as adsorption, and its release from the surface of the solid is known as desorption. Adsorption of a chemical to soil or sediment particles may be a result of electrostatic or hydrophobic attraction between the contaminant and the solid surface.[23] Once associated with solid particles, the contaminant may also settle to the sediment surface. This process, known as sedimentation, typically occurs in water bodies with laminar flow, such as a wetland. The increase in contaminant concentration in the solid phases as a result of any of these partitioning processes is known as accumulation.[14]

The contaminant's aqueous solubility and density greatly influence its final disposition in water–solid systems. Dense nonaqueous phase liquids (DNAPLs) are chemicals with densities greater than water and typically low aqueous solubilities. DNAPLs naturally partition away from the aqueous phase and towards the solid phase, often pooling on top of an impermeable solid layer. On the other hand, light nonaqueous phase liquids (LNAPLs) possess densities less than water and have a tendency to pool on the water's surface where they may be prone to volatilization and photolysis reactions. Various liquid–solid partition coefficients have been developed to predict a contaminant's tendency to associate with the solid phase. These include the octanol–water partition coefficient (K_{ow}), the soil–water distribution ratio (K_d), the organic carbon partition coefficient (K_{OC}), the organic matter–water partition coefficient (K_{OM}). These will be discussed in detail in sections to follow.

Transport of a contaminant from water to air is influenced primarily by wind velocity.[16] The contaminant's density, vapor pressure, and aqueous solubility also factor into its tendency to be introduced into the air phase, and its Henry's constant (K_H) provides a good indication of this tendency. Biota have a strong attraction to hydrophobic contaminants, and, as a result, uptake of contaminants by partitioning into plants and animals, known as bioaccumulation, has been reported to be a dominant mechanism of removal.[16, 25] The tendency of a chemical to be taken up into biota is quantified by the bioconcentration factor (BCF), as measured by the ratio of its concentrations in biota and water.

Translocation Processes in Soil

Contaminants in the soil compartment are associated with the soil, water, air, and biota phases present. Transport of the contaminant, therefore, can occur within the water and air phases by advection, diffusion, or dispersion, as previously described. In addition to these processes, chemicals dissolved in soil water are transported by wicking and percolation in the unsaturated zone.[26] Chemicals can be transported in soil air by a process known as barometric pumping that is caused by sporadic changes in atmospheric pressure and soil–water displacement. Relevant physical properties of the soil matrix that are useful in modeling transport of a chemical include its hydraulic conductivity and tortuosity. The diffusivities of the chemicals in air and water are also used for this purpose.

As shown in Fig. 6.7, any of the translocation processes described previously in air and soil compartments may also occur in soil. Figure 6.8, adapted from Baum,[26] shows the potential intermedia exchanges that occur among the soil, air, water, and biota phases within the soil compartment. Exchange between air and water and air and solid phases may involve volatilization or deposition. Interchange between the water and solid phases may involve leaching (movement of the chemical in water through the soil column), absorption into the solid matrix, adsorption onto the solids, or desorption into the aqueous phase. Partitioning from any of the solid, water, or air phases to biota results in the bioaccumulation of the contaminant. Biota can use various means of elimination of the contaminant or its metabolites. Volatilization of the contaminant or its metabolites from plants to the air phase may occur, as well as excretion of the contaminant or its metabolites from roots or foliage to the surrounding soil and water phases in the root zone. The root zone may also serve to "stabilize" the contaminant in the soil, a process known as phytostabilization,[27] resulting in accumulation of the contaminant in the soil phase. Relevant physicochemical properties of the contaminant that provide an indication of the degree of these partitioning processes include aqueous solubility (S_{aq}), vapor pressure (VP), Henry's constant (K_H), soil partitioning coefficients, and the bioconcentration factor (BCF).

Translocation Processes Involving Biota

As mentioned previously and shown in Fig. 6.8, movement of a chemical in a system containing plants or animals may involve exchange with the air, soil, and water phases. Bioaccumulation results when the plant or animal uptakes the contaminant. In plants, the contaminant may be, in turn, released to the air by means of volatilization or to the soil with subsequent accumulation by phytostabilization, adsorption, and sedimentation or transport to the aqueous phase by advective, diffusive, or dispersive processes. Excretion of the contaminant or a metabolite from animals may also be received by any compartment.

Partitioning between a plant and air depends on the properties of the chemical,

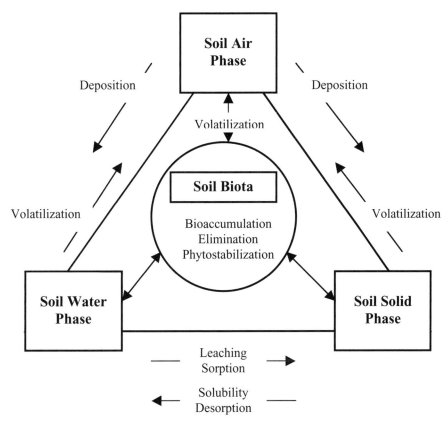

Fig. 6.8. Intermedia translocation processes involved in the soil compartment. (Adapted from Baum.[18])

such as vapor pressure (VP), properties of the plant, and the temperature. Andren et al.[16] report that partitioning from the air to plants occurs primarily through the foliage, and only compounds with mobility in the phloem can partition to the stem and trunk of the plant. As a general rule, hydrophilic compounds have higher phloem mobility, and, therefore, the octanol–water partition coefficient (K_{ow}) is a good indicator of the tendency of a chemical to be transported into a plant via its foliage. Uptake and release to/from the water and soil phases and the relevant parameters used to measure the tendency for these processes were discussed in the previous section. Of all the compartments of the environment, the least information is known concerning the translocation processes involving biota.[16] Also, the bioconcentration factors (BCFs) cannot be considered equilibrium constants

as the other partitioning coefficients are because biota are "reactive sorbents" where the contaminant may be transformed upon uptake. These and other transformation processes involved in each of the environmental compartments are discussed in the following sections.

TRANSFORMATION OF CHEMICALS IN THE ENVIRONMENT

The primary mechanisms of degradation of chemicals in soil, water, sediment, air, and biota environments are classified as biotic (biodegradation, phytodegradation, and respiration) or abiotic (hydrolysis, photolysis, and oxidation/reduction), as shown in Figure 6.7. Biodegradation, the transformation of chemicals by microorganisms, has potential to occur in any environmental compartment that contains moisture to support microbial processes; therefore, it can occur in all compartments, including in the root zone of plants. Phytodegradation, the transformation of chemicals by green plants, and respiration, the metabolism of chemicals by animals, occurs wherever these biota reside, and, as discussed previously, they interact with each of the other compartments of the environment. Hydrolysis, the reaction with water, can occur in any compartment containing water, whereas nonbiological oxidation/reduction reactions are most dominant in the air phase.[21] Photolysis, the reaction of a chemical with ultraviolet radiation from the sun, occurs on surfaces exposed to sunlight, including soil, water, air, and plants. Compounds not susceptible to any of these degradation processes will have a tendency to be persistent in the environment. These reactions and their kinetics are discussed in more detail below.

Biotic Transformation Processes

Biodegradation. Because of the abundance and species diversity of environmental microorganisms and their ability to adapt to many different conditions and to degrade a wide range of substrates, biodegradation plays a major role in the transformation of contaminants.[28] The rate of biodegradation is influenced by numerous factors, including availability of nutrients, pH, temperature, level of oxygen, moisture content, and chemical structure. The degradation of the contaminant is catalyzed by enzymes, and most biodegradation processes are composed of a series of steps mediated by different enzymes. If the chemical is completely degraded to carbon dioxide, methane, water, and other inorganic compounds, it is said to be mineralized. However, the contaminant is often converted to a chemical structure more complex than these mineralization products, and, in many cases, these incomplete degradation products, called daughter products or metabolites, are more toxic than the parent compound.

Some chemical compounds serve as a source of carbon and/or energy for the degrading microorganism, whereas in other cases they

are not but are, rather, "cometabolized." For example, methanotrophs, or methane-oxidizing bacteria, are known to oxidize aliphatic and aromatic compounds by cometabolism as long as methane is available to provide a source of carbon and energy. These bacteria are obligately aerobic because oxygen is necessary for their growth and metabolism; however, many other types of microorganisms cannot function in the presence of oxygen, and they are known as obligately anaerobic bacteria. Furthermore, bacterial populations exist that can function with or without oxygen, and they are known as facultative aerobes or facultative anaerobes.

Biodegradation reactions can be classified as aerobic or anaerobic, with oxidations and reductions dominating within these two classes, respectively. Oxidations occur in aerobic pockets in groundwater aquifers, near the soil surface, and aerobic zones of lakes and streams, for example. Dioxygenase and monooxygenase enzymes mediate oxidation reactions, yielding hydroxylated metabolites and possibly ring-cleavage products resulting from the activity of ring-cleaving dioxygenase enzymes. Figure 6.9 shows one of the known pathways of oxidation of benzene involving a consortium of microorganisms expressing diooxygenase enzymes that cleave the benzene ring between the two hydroxyl groups of the catechol, called *ortho*-cleavage. As implied by the TCA cycle endpoint in Fig. 6.9, the microorganisms involved in this pathway gain energy from the degradation of benzene. It is important to note that typical intermediates of biologically mediated oxidation reactions are more polar than the parent compound, thus with different environmental behavior.

More oxidized compounds, such as chlorinated benzenes, are susceptible to biologically mediated reduction in environments under anaerobic conditions, such as in lake and river sediments. It is known that highly polychlorinated biphenyl (PCB) congeners, for example, are susceptible to reductive dehalogenation, the result of the interaction of syntrophic microbial communities that are active under methanogenic and sulfate-reducing

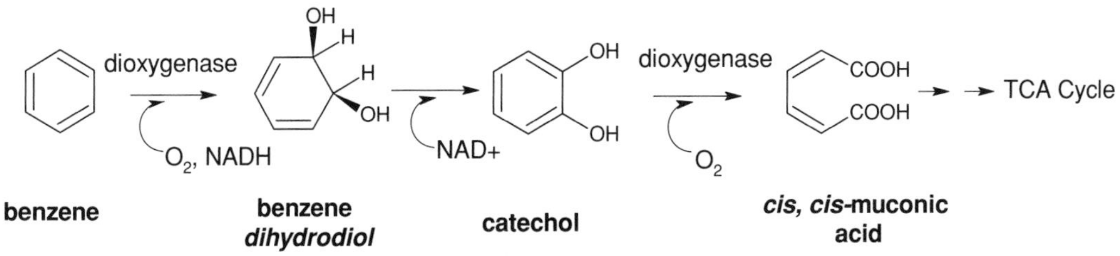

Fig. 6.9.   The *ortho*-cleavage pathway of benzene oxidation by dioxygenase enzymes.[29]

conditions.[30] Previous studies of Hudson River sediments have reported that the more chlorinated PCB congeners (with more than 3–4 chlorine atoms) are transformed to lesser chlorinated congeners by a series of dechlorination steps where the chlorine atoms are replaced by hydrogen atoms.[31] As shown in Fig. 6.10, preferential removal of the chlorine atoms in the *meta-* and *para-* positions on the ring occurs before the *ortho*-chlorines are removed, resulting in accumulation of the *ortho*-substituted congeners in the environment.[32]

As a result of a large body of studies, heuristics of biodegradation have been developed. Table 6.3 provides general guidelines for prediction of the tendency of a chemical to be biodegraded.[21] More detailed information about the specific biodegradation pathways and kinetics for individual groups of compounds and microorganisms can be found in numerous references.[33–35]

Phytodegradation. The transformation of contaminants by plants is believed to play a major role in contaminant removal from the biosphere, particularly considering that plants cover approximately 146,000,000 square kilometers of terrestrial surface.[28] Plants generally degrade chemicals by first uptaking them and subsequently transforming them to products that are conjugated or bound to the cell wall or stored in vacuoles.[36] Some plants use reductive dehalogenation and oxidation transformation pathways of degradation similar to those pathways followed by microorganisms in biodegradation. Compounds classes that have been reported to be degraded in aqueous systems by plants include halogenated hydrocarbons,[37,38] nitroaromatics,[39,40] organophosphate pesticides,[41] and polycyclic aromatic hydrocarbons.[42,43] At this time, there are few rules of thumb to enable prediction of phytodegradation of contaminants because the study of phytodegradation is relatively recent.

Abiotic Transformation Processes

Hydrolysis. Compounds that possess an electrophilic atom (electron-poor) have a tendency to undergo hydrolysis reactions with electron-rich water or hydroxide ion (OH^-). The reaction

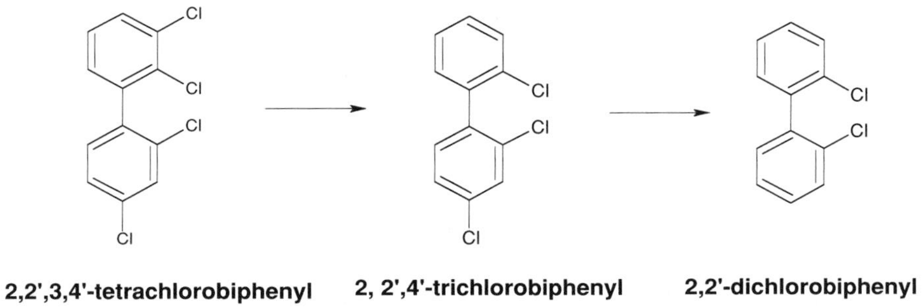

2,2',3,4'-tetrachlorobiphenyl   2, 2',4'-trichlorobiphenyl   2,2'-dichlorobiphenyl

Fig. 6.10.   Pathway of reductive dehalogenation of PCBs.[31]

TABLE 6.3   Chemical Characteristics Conferring Susceptibility to Biodegradation

Nonaromaticity
Unsaturated bonds
Straight chains of alkanes (greater than 9 carbons)
Soluble in water
Increased number of halogens (anaerobic conditions)
Decreased number of halogens (aerobic conditions)
Alcohol, aldehyde, and carboxylic acid functional
  groups
Ortho- or *para*-substitution patterns on a benzene ring

Adapted from Bishop.[21]

forms a product with a hydroxyl group replacing a group that leaves the parent compound. The products are more polar than the parent compound, thus indicating different environmental behavior that must be evaluated, as also reported for products of biological oxidation. Hydrolysis is observed anywhere water is present, thus indicating that it can occur in any environmental compartment, given the presence of compounds susceptible to this reaction. This is considered a dominant pathway of transformation of compounds with hydrolyzable groups in aquatic systems.[28] These reactions are extremely pH-sensitive, as they are often catalyzed by H^+ or OH^- ions. For example, Bishop[21] reports that the rate of hydrolysis of insecticide carbaryl increases logarithmically with pH, where the rate at pH 9 is 10 times greater than at neutral pH and 100 times greater than at pH 6.

Numerous functional groups have been reported to be hydrolyzable, and examples of compound classes that are susceptible to hydrolysis are provided in Table 6.4. The chemical structure of the compounds greatly affects the rates of hydrolysis as shown in Table 6.4, with half-lives measured at pH 7 and 25°C ranging from seconds to thousands of years. A detailed discussion of degradation of contaminants via hydrolysis can be found in Schwartzenbach et al.,[44] Larson and Weber,[23] and Wolfe and Jeffers.[45] Ney[14] has set a low range of hydrolysis half-lives falling below 30 days and a high range above 90 days. As shown in Table 6.5, compounds considered to be hydrolyzed rapidly are not prone to bioaccumulation, accumulation,

food-chain contamination, or adsorption, and are not generally considered persistent, whereas the opposite is true for compounds not hydrolyzed rapidly. Half-lives of various contaminants undergoing hydrolysis reaction have been reported to be significantly higher than those undergoing microbial transformation and phytotransformation, hydrolysis.[27] For example, tetrachloroethylene (PCE), a common solvent used in dry cleaning, possesses a half-life of over 10 years if subjected to hydrolytic conditions at neutral pH. In contrast, PCE's half-lives due to microbial transformation (using a density of 10^5 organisms per liter) and phytotransformation (using *Spyrogyra* spp. at a density of 200 grams wet weight per liter) have been reported to be 35 and 5 days, respectively.

Photolysis.   Photochemical transformations of chemicals, also known as photolysis, result from the uptake of light energy (quanta) by organic compounds. These reactions can occur in the gas phase (troposphere and stratosphere), the aqueous phase (atmospheric aerosols or droplets, surface waters, land–water interfaces), and the solid phase (plant tissue exteriors, soil and mineral surfaces).[23] Photolysis may be direct, where the structure of the chemical absorbing the light energy is "directly" transformed, or indirect, where photosensitizers, such as quinones and humic acids, absorb light energy and then transfer it to a contaminant whose structure is altered in the process.[46] Chemical reactions that affect structural change of photochemically excited contaminants include fragmentation, intramolecular rearrangement, isomerization, hydrogen atom abstraction, dimerization, and electron transfer from or to the chemical.[44]

Functional groups such as unsaturated carbon–carbon bonds and aromatic rings lend greater susceptibility to photolytic reactions, and compound classes such as nitrosamines, benzidines, and chlorinated organics are more readily photolyzed.[28,46] In order for photolysis to be considered to be a significant gas-phase destruction mechanism for a chemical, it must absorb light energy beyond a wavelength of 290 nm.[28] Specific compounds that

TABLE 6.4 Comparison of Hydrolysis Half-Lives of Representative Chemicals in Classes Susceptible to Hydrolysis at pH 7 and 25°C [14,23,44]

Hydrolyzable Compound Class	Representative Compound (s)	Half-Life, $t_{1/2}$
Monohalogenated hydrocarbons R—X	CH_3Cl	340 d
	$(CH_3)_2CHCl$	38 d
	$(CH_3)_3C—Cl$	23 s
Polyhalogenated hydrocarbons $R—X_n$	CH_2Cl_2	700 yr
	$CHCl_3$	3500 yr
	$CHBr_3$	700 yr
	$BrCH_2—CH_2Br$	4 yr
Carboxylic acid esters	$R_1 = CH_3–, R_2 = –CH_2CH_3$	2 yr
	$R_1 = CH_3–, R_2 = –C(CH_3)_3$	140 yr
	$R_1 = CH_3–, R_2 = C_6H_5–$	38 d
Dialkyl phthlates	$R = $ (2-ethylhexyl)	100 y (pH 8)
Amides	$R_1 = CH_3–, R_2 = –H, R_3 = –H$	4000 yr
	$R_1 = CH_3–, R_2 = CH_3–, R_3 = –H$	40,000 yr
	$R_1 = –CH_2Cl, R_2 = –H, R_3 = –H$	
Carbamates	$R_1 = CH_3–, R_2 = CH_3–, R_3 = –CH_2CH_3$	1.5 yr 50,000 yr
	$R_1 = –H, R_2 = C_6H_5–, R_3 = –C_6H_4NO_2$	25 s
Phosphoric Acid Triesters	$R_1 = CH_3–, R_2 = CH_3–$	1.2 yr
	$R_1 = C_6H_5–, R_2 = C_6H_5–$	320 d
Thiophosphoric Acid Triesters	$R_1 = CH_3CH_2–, R_2 = C_6H_5NO_2–$	89 d
	$R_1 = CH_3CH_2–, R_2 = $ (2-pyrimidinyl isopropyl)	23 d

Structures:

Carboxylic acid esters:
$$R_1—C(=O)—O—R_2$$

Dialkyl phthalates: ortho-benzene with two $C(=O)—O—R$ groups.

$$R = H_2C—CH(CH_2CH_3)... CH_3$$

Amides:
$$R_1—C(=O)—NH—R_2—R_3$$

Carbamates:
$$(R_1)(R_2)N—C(=O)—O—R_3$$

Phosphoric Acid Triesters:
$$(R_1O)_2—P(=O)—O—R_2$$

Thiophosphoric Acid Triesters:
$$(R_1O)_2—P(=S)—O—R_2$$

TABLE 6.5 Heuristics for hydrolysis half-lives and susceptibility of contaminants to translocation and transformation processes[14]

Process	Rapid Hydrolysis ($t_{1/2} < 30$ days)	Slow Hydrolysis ($t_{1/2} > 90$ days)
Accumulation	Not likely	Yes
Bioaccumulation	Not likely	Yes
Food-chain contamination	Not likely	Yes
Persistence	Negligible	Yes
Adsorption	Negligible	Maybe
Dissipation	Yes	Negligible to slowly

do absorb greater than 290 nm include 2,3,7,8-tetrachlorodibenzo-*p*-dioxin (2,3,7,8-TCDD), 1,2,3,7,8-pentachloro-*p*-dioxin, octachloro-*p*-dioxin, dibenzo-*p*-dioxin, hexachlorobenzene, and various polychlorinated dibenzofurans, and polychlorinated and polybrominated biphenyls.[47,48,49,40,51] The kinetics of photolysis of herbicides in natural waters and soil surfaces were compared by Konstantinou et al.,[52] who reported faster rates in soils than in lake, river, marine, and ground waters and significant enhancement of rates in the presence of increasing dissolved organic matter. Applying the same general rule as described for hydrolysis (Table 6.5), if the photolytic half-life of a contaminant is less than 30 days, it is not considered to be persistent; however, if the photolysis half-life is greater than 90 days, the contaminant poses risk to accumulation, bioaccumulation, and food-chain contamination.[14]

Oxidation/Reduction Reactions. Reactions of chemicals via abiotic oxidation or reduction involve a transfer of electrons and result in a change in oxidation of the state of the product compared to its parent compound. As a general rule, reduction reactions are prevalent in soil sediments, while oxidation reactions are more important in surface waters and in the atmosphere.[28]

The contaminant that is oxidized serves as the donor of electrons that are transferred to oxidizing agents including O_2, ferric (III) iron, and manganese (III/IV) in aquatic/soil systems and ozone (O_3), hydrogen peroxide (H_2O_2), and free radicals, such as the hydroxyl radical, in the atmosphere. Reduction reactions result in the transfer of electrons from an electron donor to the contaminant. Examples of electron donors present in the environment include pyrite (FeS), ferrous carbonates, sulfides, and natural organic matter. As described previously for biologically mediated reduction reactions, highly oxidized contaminants, such as PCE and highly chlorinated PCBs, can be reduced abiotically provided that they are in intimate contact with electron donors. Unfortunately, the kinetics of abiotic oxidation/reduction reactions have not been described for many contaminants.[28]

THE CONNECTION BETWEEN CHEMICAL PROPERTIES AND ENVIRONMENTAL FATE

As indicated previously, the final disposition of a chemical in the environment is dependent on the environmental conditions, characteristics of the media involved, and the various physicochemical properties of the contaminants. Table 6.6 provides a listing of properties describing the chemical, medium, and potential for translocation and transformation. This section focuses on chemical properties that are frequently used to assess the fate of a contaminant in air, water, soil, or biota. The goals of this section are to provide brief descriptions of relevant properties of a contaminant and to directly link specific ranges of these properties to predicted fate in the environment.

Traditional Chemical Properties

Traditional chemical properties including boiling point, T_b, melting point, T_m, density, ρ, surface tension, γ, vapor pressure, VP, and

TABLE 6.6 Useful Properties of the Chemical, the Medium, and Transport/Kinetics for Assessment of Environmental Fates of Contaminants[14,18,53]

Chemical	Medium	Transport and Kinetics
Traditional:	Temperature	Diffusivity in air and water
Boiling point, T_b	pH	
Melting point, T_m	Biota present	Phase transfer coefficients
Density, ρ	Light present	(air-water, air-soil)
Surface tension, ν		
Vapor pressure, VP	*Air:*	Half-lives of biotic and
Aqueous solubility, S_{aq}	Wind velocity, turbulence,	abiotic reactions
	stratification, composition, etc.	
Specialized:	*Water:*	
Henry's constant, K_H	Number of active cells, redox	
Octanol-water partition coefficient, K_{ow}	potential, activity, velocity, etc.	
Bioconcentration factor, BCF		
Soil-water distribution ratio, K_d	*Soil:*	
Organic matter-water partition coefficient, K_{om}	Hydraulic conductivity,	
Organic carbon-water partition coefficient, K_{oc}	tortuosity, composition, etc.	

aqueous solubility, S_{aq}, have use in many systems beyond environmental applications; however, they provide strong indication of partitioning among the air, water, soil, and biota compartments in the environment. When combined with more specialized physicochemical properties to be discussed below, these traditional chemical properties can provide a very powerful means of answering the basic questions of where will a chemical go, what reactions will it undergo, and how long will it persist in the environment.

A chemical's T_b, the temperature at which its vapor pressure equals the ambient pressure, and T_m, the temperature at which its solid and liquid forms are in equilibrium at ambient pressure, are easily located in references and databases. As a result, many of the correlations that have been constructed for property estimations use these parameters as independent variables. The T_b of a chemical can nonetheless provide an indication of the partitioning between gas and liquid phases,[53] with the higher values denoting a lower tendency to exist in the vapor phase. The surface tension, γ, of a chemical, the ratio of the work done to expand the surface divided by the increase in the surface area, is often used to estimate the VP of liquids in aerosols and in soil capillaries.[28] The VP of a chemical is the pressure of a pure chemical vapor that is in equilibrium with the pure liquid or solid, and

it provides an indication of the tendency of a chemical in its pure liquid or solid phase to volatilize. Chemicals with high vapor pressures will be likely to escape to air and thus exist in higher concentrations in the air phases of environmental compartments than those with low vapor pressures. Ney[14] suggests that the VP of a chemical is considered high if greater than 0.01 mm Hg, whereas low VP is considered to be below 10^{-6} mm Hg. The S_{aq} of a chemical is another property that can be directly used to assess translocation and transformation potential of a chemical. By definition, the S_{aq} of a chemical is its concentration in a saturated water solution. A chemical possessing high S_{aq}, defined by Ney[14] as greater than 1000 ppm, is more likely to be mobile in the aqueous environment than a chemical with a low S_{aq}, defined as less than 10 ppm.

Specialized Chemical Properties

Chemical properties that reflect the tendency of a chemical to partition between phases have been constructed specifically for environmental fate applications. These specialized properties are known as partition coefficients. The parameter that is a measure of a chemical's tendency to partition between water and air is known as the dimensionless Henry's constant (K_H), determined by the ratio of the equilibrium concentrations of the chemical in air and in water, respectively. The Henry's law constant (H), describing the ratio of the chemical's partial pressure in solution to its concentration in solution at equilibrium, is often used to describe chemical partitioning between air and water. K_H is simply the ratio of H to the product RT, where R is the gas constant and T is the temperature. Compounds with high H values (greater than 10^{-1} atm–m^3/mol) will prefer the air phase and tend to volatilize from the aqueous phase in water, soil, air, and biota compartments, whereas compounds with low H values (less than 10^{-7} atm–m^3/mol) prefer to escape the air phase into the aqueous phase.[53]

Partitioning of a chemical between the aqueous and nonaqueous phases is frequently measured by the octanol–water partition coefficient, K_{ow}. K_{ow} of a chemical is the ratio of its concentration in *n*-octanol, frequently used to represent a model phase of living and nonliving natural organic material, to its concentration in water. This parameter which ranges from 10 to 10^7 is a measure of the partitioning of a chemical between hydrophobic and hydrophilic phases in all environmental compartments. A high K_{ow} value (characterized as greater than 1000) indicates that a chemical has a tendency to be less mobile in the aqueous phase, more greatly adsorbed to solids, and bioaccumulate.[14,16,53] K_{ow} values are frequently used to determine another parameter, the bioconcentration factor (BCF), which is the ratio of a chemical's concentration in the tissue of a living organism to its concentration in water. A chemical with a high BCF (greater than 1000) will be more likely partitioned from water into the fatty tissues of fish and humans, for example, and thus is more likely to bioaccumulate and yield food-chain contamination effects. Chemicals with low potential for bioaccumulation typically have BCF values less than 250.[53]

Various coefficients are helpful in measuring the potential of a chemical to partition between the aqueous and solid phases. These parameters are valuable in predicting the potential of a chemical to adsorb to the solid phase. The soil–water distribution ratio, K_d, of a chemical is the ratio of its equilibrium concentration sorbed onto a solid phase to its equilibrium concentration in solution. Because most sorption of neutral, nonpolar chemicals occurs primarily on the organic matter in soil and sediments, the organic matter–water partition coefficient, K_{om}, is frequently used as a measure of a chemical's tendency to partition from water to solid phases. The useful parameter of K_{oc}, the organic carbon–water partition coefficient, has been introduced to describe the ratio of the equilibrium concentration of a chemical associated with the organic carbon content of soil to that associated with the water phase. K_{oc} is roughly 1.724 times greater than K_{om} because the organic-matter sediment is roughly 1.724 times larger than the organic carbon content.[28] With all of these partition coefficients, higher values indicate that a chemical has a greater tendency to leave the aqueous phase and sorb to soils and sediments. Log K_{oc} values of chemicals are considered high when greater than 4.5 and low when less than 1.5.[53]

Sources of Chemical Property and Fate Data

Handbooks. Tabulations of traditional properties of chemicals (T_b, T_m, ρ, S_{aq}) are often available in publications by fine-chemical manufacturers (e.g., Fisher Scientific, Web page: www.fishersci.com). Other collections of these parameters, surface tension, the specialized partition coefficient values, and kinetics of transformations include Winholz et al.,[54] Howard et al.,[55] Dean,[56] Lide,[57] Mackay et al.,[58] Howard and Meylan,[59] Tomlin,[60] Yaws,[61] and Verschueren.[62]

Software and Online Sources of Chemical Properties. Larson et al.[28] and Andren et al.[16] provide a thorough listing of electronic databases and online search engines for environmental fate properties. Very useful online databases for these parameters include Syracuse Research Corporation's (SRC's) (Syracuse, NY) Environmental Fate Database (EFDB) and the Hazardous Substances DataBank available at http://toxnet.nlm.nih.gov/. The SRC Web site also provides access to its DATALOG database that provides a literature search engine for numerous chemical properties. In addition, direct photolysis rates and half-lives of contaminants in the aquatic environment can be obtained from the U.S. EPA's GCSOLAR software, available at http://www.epa.gov/ceampubl/swater/gcsolar/. A very useful software for estimating physical/chemical and environmental fate properties is the EPI Suite software.[1]

Estimation Methods. Numerous references focus on the theory and application of estimating physicochemical properties of chemicals. Lyman et al.,[63] Neely and Blau,[64] Howard and Meylan,[59] Baum,[28] and Allen and Shonnard[53] provide thorough descriptions of the methods available for manually calculating chemical properties.

HEURISTICS FOR PREDICTING ENVIRONMENTAL FATE

Once the properties and fate data are obtained for a chemical, its general disposition in the environment can be estimated. Many of the properties described previously have been divided into ranges from low to high that enable estimation of tendencies of a chemical to undergo various translocation and transformation processes. Table 6.7 summarizes expected behavior of chemicals using values of VP, S_{aq}, K_{ow}, and K_{oc}. These findings are summarized below with general heuristics for predicting environmental fate of chemicals. These general trends in environmental fate can be combined with transformation kinetics to provide an assessment of exposure for any chemical.

IV ENVIRONMENTAL PERFORMANCE ASSESSMENT FOR CHEMICAL PROCESS DESIGN

INTRODUCTION

The chemical industry contributes significantly to economic development, yet faces many environmental and societal challenges that require a rethinking of traditional approaches in the commercialization of processes and products. The generation of toxic, hazardous, and global change byproducts of chemical processes are but a few examples of these challenges. A more comprehensive evaluation of economic, environmental, and societal consequences, at times spanning the entire product or process life cycle, is needed to achieve sustainable growth. Coincident with this will be an enlargement in the data and computational requirements for these assessments. Computer-aided analysis tools will therefore be needed to efficiently link process/product design with critical environmental and societal impacts in a larger systems analysis.

Fortunately, much progress has been made recently in developing environmental assessment methods and computer-aided tools to accomplish these goals. Table 6.8 shows a description of key factors for the environmentally conscious design of chemical processes for which computer-aided tools are useful. These tools fall into two categories: those providing information on environmental fate and impacts and those intended to improve process environmental performance. As will be shown, some tools provide information early in design whereas others are employed at later stages. These environmental factors and assessment tools should be incorporated into the design of chemical processes and products as illustrated in Fig. 6.11. The computer-aided tools in Fig. 6.11 include what has thus far been employed in traditional design: process simulation, design heuristics, and optimization. Added to these conventional design tools is a set of environmental evaluation methods and tools that inform the design activity on a range of potential impacts. The Green Engineering approach uses these environmental assessments in a hierarchical fashion during process and product design. A hierarchical approach for evaluating environmental performance during process design will be described in a later section.

OVERVIEW OF ENVIRONMENTAL ISSUES

Before beginning a series of case study evaluations, we will need to establish a set of

TABLE 6.7   Rules of Thumb of Chemical Property Ranges Guiding Prediction of Environmental Fates[14,53]

Property	Activity Predicted	Heuristic
VP (mm Hg) Low: $<10^{-6}$ Medium: 10^{-6}–0.01 High: 0.01	*Translocation:* Volatilization Accumulation Bioaccumulation Adsorption	1. Chemicals with high VP are more likely to volatilize into the air phase and less likely to accumulate, bioaccumulate, and adsorb to solids.
S_{aq} (ppm) Low: <0.1 Medium: 0.1–10,000 High: $>10,000$	*Translocation:* Mobility in water and soils Accumulation Bioaccumulation Volatilization Adsorption	2. High S_{aq} confers a greater tendency for a chemical to be mobile in the aqueous environment and is less likely to accumulate, bioaccumulate, volatilize, and be persistent. 3. Chemicals with high S_{aq} are prone to biodegradation and respiration processes.
	Transformation: Biodegradation Respiration	4. Chemicals with low S_{aq} have a greater tendency to be immobilized via adsorption and less likely to be leached in soil.
H (atm-m^3/mole) Low: $<10^{-7}$ Medium: 10^{-7}–10^{-1} High: $>10^{-1}$	*Translocation:* Mobility in air/water Volatilization	5. Chemicals with high H values have a greater tendency to escape the aqueous phase into the air phase, are less mobile in the aqueous environment and less biodegradable but more mobile in the air environment.
	Transformation: Biodegradation	6. Chemicals with higher molecular weight tend to experience a decrease in both VP and S_{aq}. Thus, VP is a better measure of tendency to volatilize for these chemicals.
K_{ow} Low: <3000 Medium: 3000–20,000 High: $>20,000$	*Translocation:* Accumulation Bioaccumulation Adsorption *Transformation:* Biodegradation Respiration	7. A chemical with a low K_{ow} indicates that it has high water solubility, aqueous mobility and is not susceptible to bioaccumulation, accumulation, or sorption to solids but is susceptible to biodegradation and metabolism by plants and animals.
K_{oc} Low: <30 Medium: 30–32,000 High: $>32,000$	*Translocation:* Accumulation Bioaccumulation Adsorption *Transformation:* Biodegradation Respiration	8. Chemicals with high K_{oc} will adsorb to organic carbon and is likely to bioaccumulate and accumulate and less likely to biodegrade or to be metabolized by plants and animals.
BCF Low: <250 Medium: 250–1000 High: >1000	*Translocation:* Accumulation Bioaccumulation Adsorption *Transformation:* Biodegradation Respiration	9. Chemicals with high BCF values will bioaccumulate and yield food-chain contamination. 10. A high BCF value implies a high K_{ow} value, and thus the same heuristics applying for K_{ow} apply for BCF.
Half-lives of hydrolysis (days) Low: <30 Medium: 30–90 High: >90	*Translocation:* Accumulation Bioaccumulation Adsorption	11. Compounds considered to be hydrolyzed rapidly are not prone to bioaccumulation, accumulation, food-chain contamination, or adsorption, and are not generally considered persistent, whereas the opposite is true for compounds not hydrolyzed rapidly.
	Transformation: Hydrolysis	12. In general, for all transformation reactions, if half-lives are high, the chemical is not considered to be susceptible to bioaccumulation, accumulation, food-chain contamination, or adsorption, and are not generally considered persistent.

TABLE 6.8   Description of Environmental Factors and Prediction/Analysis Methods

Environmental Factors	Description
Environmental properties of chemicals	Equilibrium distribution of chemicals among air, water, solid phases in the environment
	Degradation rates in air, water, and soil/ sediment
	Toxicological properties
	Structure–activity relationships based on chemical structure
	Online databases
Emission estimation from process units	Emission factors for major process units
	Emission correlations for fugitive sources, storage tanks, material transfer and handling
	Emissions for process heating and utilities
Environmental fate	Fate models for wastewater treatment
	Fate models in a multi-media environment
Environmental performance metrics	Models to characterize environmental impacts
Process intensification and integration	Integration of heat and power
	Mass integration to prevent waste
	Pinch analysis, source sink diagrams
Process optimization	Mixed integer nonlinear programming
	Multi-criteria optimization

environmental performance measures to use in design. These performance measures must reflect societal attitudes on the importance of several environmental impact categories. This is naturally a subjective judgment, but a consensus is emerging in the professional literature and in regulatory agencies on a set of environmental "midpoint" indicators for this purpose.[53, 65–69] Table 6.9 features several environmental impact categories, a description of the causes, and midpoint/endpoint effects.

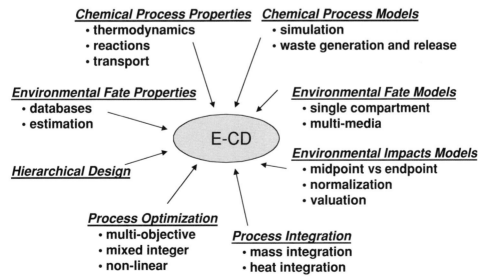

Fig. 6.11. The linking of environmental assessment methods and tools to the design of environmentally conscious chemical processes. E-CD = environmentally conscious design.

TABLE 6.9 Environmental Causes and Midpoint/Endpoint Effects

Impact Category	Initiating Event	Environmental Processes	Midpoint Effects	Endpoint Effects	Impact Indicator(s)
Global climate change	Emission of greenhouse gases (CO_2, CH_4, N_2O, CFCs)	Absorption of infrared radiation	Increase in atmospheric temperature, Sea level increase	Increases in human diseases and mortality Climate warming Ecological damage	Global Warming Potential (GWP)
Stratospheric ozone	Emission of ozone-depleting compounds (CFCs, Halons)	Chemical reaction release of •Cl and •Br in stratosphere	Catalytic destruction of ozone in stratosphere	Skin and crop damage, damage to materials	Ozone Depletion Potential (ODP)
Acidification	Release of oxides of sulfur and nitrogen (SOx, NOx)	Chemical oxidation and deposition of acid compounds	Acidic precipitation and lowering of pH in lakes and soils	Ecosystem changes and plant/animal death	Acidification Potential (AP)
Smog formation	Emission of smog precursors	Photochemical conversion to ozone and other species	Elevated ozone and aerosols in urban atmospheres	Reduced plant growth, human morbidity and mortality	Photochemical Ozone Creation Potential (POCP)
Human cancer and Noncancer effects	Release of toxic and carcinogenic compounds	Environmental fate and transport, exposure pathways	A dose to humans in excess of acceptable levels	Increases in human mortality and morbidity	Human Toxicity Potential (HTP)
Ecological toxicity	Release of toxic and carcinogenic compounds	Environmental fate and transport, exposure pathways	A dose to animals in excess of acceptable levels	Increases in animal mortality and morbidity	Ecological Toxicity Potential (ETP)
Resource consumption	Utilization of natural resources from the Earth	Extraction processes	Disruption of natural environments	Unsustainable depletion of resources	Fossil Energy, Land Use Area, Mineral Resource Use

ENVIRONMENTAL IMPACT ASSESSMENT: TIER I–TIER III

Figure 6.12 illustrates a design hierarchy in the synthesis and evaluation of chemical processes, integrating the design approach of Douglas[70] with environmental risk assessment methods.[53] Environmental evaluation progresses from simple ("Tier I") assessments early in design when process information is limited to detailed flowsheet evaluations later in the design process. Different assessment approaches are needed in early design as opposed to those applied later in flowsheet evaluation. In early design, a large number of design choices are considered, and a simple and efficient methodology is needed. Design choices in early design might include the selection of raw materials, solvents, reaction pathways, and heat or mass separating agents, and the number of choices may be very large. Later in design, we consider the environmental performance of a small number of process flowsheets. The purpose of this section is to present applications of this tiered approach in the comparison of alternative design choices, emphasizing the methodologies and computer-aided tools.

Early Process Design Evaluations: "Tier I" Assessment

Early design is arguably the most important stage in the hierarchy for conducting environmental assessment. Environmental burdens of chemical processes are largely "locked in" by choices made during early design and attempts to improve environmental performance after a chemical process is operational are expensive and disruptive. Given the importance of early design assessment and the need to provide information on a large number of choices, streamlined yet accurate assessment methods are a high priority.

Solvent Selection. Solvent selection is often conducted in early design of chemical processes. A method to match desirable solvent properties (solubility parameters, for example) while simultaneously avoiding undesirable environmental impacts (persistence, toxicity, volatility, etc.) would improve design performance. PARIS II is a program combining such solvent design characteristics. Solvent composition is manipulated by a search algorithm aided by a library of routines with the latest fluid property prediction techniques, and by another

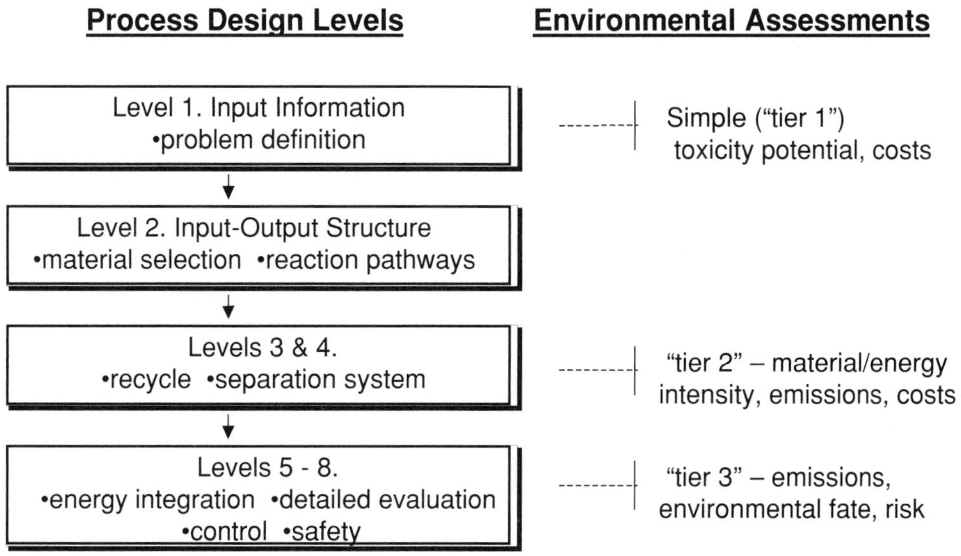

Process Design Levels   Environmental Assessments

Fig. 6.12. A hierarchical structure to environmental evaluation of chemical process design.

library of routines for calculating solvent environmental performance requirements.[71]

Reaction Pathway Selection. Reaction pathway selection is another very important early design activity. High conversion of reactants, selectivity to desired products, and avoiding byproduct reactions are among the 12 Green Chemistry principles from Anastas and Warner[3] described previously. Atom and mass efficiency are Green Chemistry performance measures that aid in early design assessment. Atom efficiency is the fraction of any element in the starting material that is incorporated in the product. The rationale is that reactions with high atom efficiencies will be inherently less wasteful. For example, in the reaction of phenol with ammonia to produce analine (C_6H_5-OH + NH_3 → C_6H_5-NH_2 + H_2O), atom efficiencies of C, H, O, and N are 100%, 77.8%, 0%, and 100%, respectively. Mass efficiency for this reaction is defined as the ratio of mass in the product with mass in the reactants. For this example mass efficiency is

Mass in Product = (6 C)(12) + (7 H) × (1) + (0 O) × 16) + (1 N) × (14) = 93 grams

Mass in Reactants = (6 C)(12) + (9 H) × (1) + (1 O) × 16) + (1 N) × (14) = 111 grams

Mass Efficiency = 93/111 × 100 = 83.8%

A higher atom and mass efficiency is desirable when comparing alternative reaction choices. A clear example of this is the production of maleic anhydride (MA) starting from either benzene or *n*-butane.[72] Benzene or *n*-butane is partially oxidized in the vapor phase in the presence of air and a solid catalyst at high temperature and pressure.

Benzene route: $2 C_6H_6 + 9 O_2 \rightarrow$ $2 C_4H_2O_3 + H_2O + 4 CO_2$

n-Benzene route: $C_4H_{10} + 7/2 O_2 \rightarrow$ $C_4H_2O_3 + 4 H_2O$

Mass efficiencies for these routes are 44.4% and 57.6% for the benzene and *n*-butane routes, respectively. This simplistic analysis

indicates that *n*-butane is the "greener" reaction route. But further analysis at this early design stage can shed more light on the differences in environmental performance. We will focus on raw material cost and CO_2 generation in this screening comparison because of concerns of economics and global climate change. It is necessary to bring in differences in conversion and selectivity at this stage. The benzene route has typical conversions of 95% and selectivity to MA of 70%, with approximately equimolar amounts of CO and CO_2 generated as byproducts. For the *n*-butane route these values are 85% and 60%, respectively. Assuming one mole of MA produced, 1/0.70 mole of benzene or 1/0.60 mole of *n*-butane is needed. The raw material costs are as follows:

Benzene: (1 mole/0.70 mole) × (78 g/mole) × (0.000280 $/g) = 0.0312 $/mole of MA

n-Butane: (1 mole/0.60 mole) × (58 g/mole) × (0.000214 $/g) = 0.0207 $/mole of MA

This result shows that the *n*-butane route costs less for the raw material than the benzene route due to the lower price of *n*-butane, even though the molar yield of MA from *n*-butane is less. This simplistic economic analysis is of relevance because raw material costs often dominate. The generation of CO_2 in the reactor is estimated as follows:

Benzene: [1 mole MA × (2 mole CO_2/mole MA)] + [(1 mole Benzene/0.70 mole MA) × (0.965 − 0.7) × 6/2] = 3.071 mole CO_2/mole of MA.

n-Butane: (1 mole *n*-Benzene/0.60 mole MA) × (0.85 − 0.6) × 4/2 = 0.833 mole CO_2/mole of MA.

The first [] term in the Benzene calculation accounts for the two carbons that are liberated when MA is formed, and the second [] term is for conversion to byproducts, with ½ going to CO_2 and the other to CO. MA from benzene generates almost four times as much CO_2 in the reactor than the *n*-butane route. Additional CO_2 is generated when unreacted benzene or *n*-butane and byproduct CO is incinerated in the pollution control equipment from the process.

The total CO_2 emission including pollution control is calculated as shown next.

Benzene: $3.071 + (1.071)(0.99) + (0.0714)(0.99)(6) = 4.595$ moles CO_2/mole of MA.

n-Butane: $0.833 + (0.833)(0.99) + (0.25)(0.99)(4) = 2.688$ mole CO_2/mole of MA.

The second term in each summation is CO converted to CO_2 in the pollution control device; the third is due to unconverted feedstock. These screening calculations verify that the n-butane route emits approximately ½ the CO_2 compared to the benzene pathway. Based on the economic and environmental screening, the benzene route would be excluded from further consideration. More detailed calculations based on optimized flowsheets confirm these screening calculations are accurate.[72]

Table 6.10 provides another example of the type of assessment that chemical engineers will need to perform with limited information.[73] A process engineer evaluating two alternative synthesis routes for the production of methyl methacrylate can use data on persistence, bioaccumulation, toxicity, and stoichiometry to quickly evaluate potential environmental concerns. These data can be estimated using group contribution methods when measured values are not available. The estimates of persistence, bioaccumulation, toxicity, and stoichiometry can then be combined to provide preliminary guidance. In this case concerns about the health and safety issues associated with sulfuric acid dominate, and the isobutylene route appears preferable because it requires less acid. Although more detailed data are available for these two processes, this level of data is typical of what might be available for new process chemistries.

Evaluations During Process Synthesis: "Tier II" Assessment

"Tier II" environmental assessment is employed for flowsheet synthesis on a smaller number of design alternatives. This provides an opportunity to evaluate the impacts of separation and other units in the process in addition to the reactor. Identification of emission sources and estimation of release rates are also part of this assessment. Inclusion of additional environmental and sustainability metrics (energy intensity [energy consumption/unit of product], water intensity, toxic release intensity, etc.) are hallmarks of "Tier II" assessment. More information on "Tier II" assessment is provided in the text by Allen and Shonnard.[53]

Detailed Evaluation of Process Flowsheets: "Tier III" Assessment

After a process flowsheet has been established, it is appropriate for a detailed environmental impact evaluation to be performed. The end result of the impact evaluation will be a set of environmental metrics (indexes) representing the major environmental impacts or risks of the entire process. A number of indexes are needed to account for potential damage to human health and to several important environmental compartments.

In quantitative risk assessment,[74] it is shown that impacts are a function of dose, dose is a function of concentration, and concentration is a function of emission rate. Therefore, emissions from a process flowsheet are a key piece of information required for impact assessment during process design. A number of computer-aided tools are available to generate pollutant emissions to air using process flowsheet information. Emission factors are used to calculate emission rates from major units (distillation columns, reactors, other columns, furnaces, boilers, etc.) based on process flows and utilities.[75] Fugitive sources (valves, pumps, flanges, fittings, sampling valves, etc.) number in the thousands for a typical chemical process and together contribute to facility air emissions. A compilation of emission estimation methods for chemical processes is found in Chapter 8 of Allen and Shonnard.[53]

As previously discussed, the concentrations in the relevant compartments of the environment (air, water, soil, biota) are dependent upon the emission rates and the chemical/physical properties of the pollutants. A fate

TABLE 6.10   Stoichiometric, Persistence, Toxicity and Bioaccumulation Data for Two Synthesis Routes for Methyl Methacrylate[53,73]

Compound	Pounds Produced or Pounds Required per Pound of Methyl Methacrylate[1]	Atmospheric Half-Life or Aquatic Half-Life[2]	1/TLV[4] (ppm)[1]	Bioconcentration Factor[5] (conc. in lipids/conc.in water)
Acetone-cyanohydrin route				
Acetone	−.68	52 days/weeks	1/750	3.2
Hydrogen cyanide	−.32	1 year/weeks	1/10	3.2
Methanol	−.37	17 days/days	1/200	3.2
Sulfuric acid[3]	−1.63		1/2(est.)	
Methyl methacrylate	1.00	7 hours/weeks	1/100	2.3
Isobutylene route				
Isobutylene	−1.12	2.5 hours/weeks	1/200 (est)	12.6
Methanol	−0.38	17 days/days	1/200	3.2
Pentane	−0.03	2.6 days/days	1/600	81
Sulfuric acid[3]	−0.01		1/2 (est)	
Methyl methacrylate	1.00	7 hours/weeks	1/100	2.3

Copyright American Institute of Chemical Engineers and reproduced by permission.

1. A negative stoichiometric index indicates that a material is consumed; a positive index indicates that it is produced in the reaction. A screening environmental index is used for comparison in "Tier I" assessment; Environmental index $= \Sigma \; |v_i| * (TLV_i)^{-1}$.
2. The atmospheric half-life is based on the reaction with the hydroxyl radical; aquatic half-life via biodegradation is based on expert estimates. From: EPISuite software (http://www.epa.gov/oppt/greenengineering/software.html) or ChemFate Database (http://www.syrres.com/eswc/chemfate.htm).
3. The lifetime of sulfuric acid in the atmosphere is short due to reactions with ammonia.
4. TLV is the threshold limit value, and the inverse is a measure of inhalation toxicity potential for a chemical. Values taken from NIOSH Pocket Guide to Chemical Hazards (http://www.cdc.gov/niosh/npg/npg.html), and the Specialized Information Service of the National Library of Medicine (NLM) (http://sis.nlm.nih.gov/).
5. Bioconcentration factor is an indicator of a chemical's potential to accumulate through the food chain. From the EPISuite software (http://www.epa.gov/oppt/greenengineering/software.html) or ChemFate Database (http://www.syrres.com/eswc/chemfate.htm).

and transport model transforms emissions into environmental concentrations. Although single compartment fate models are often used to predict concentrations downwind or downgradient from emissions sources—for example, an atmospheric dispersion model, a groundwater fate model, or river model—most applications of risk assessment to the design of chemical processes employ a multimedia compartment model (MCM). MCMs predict regional pollutant transport and fate, with a typical scale of 100 km × 100 km of the earth's surface. Mechanisms of diffusive and convective transport as well as degradation reactions are applied to separate well-mixed environmental compartments (air, water, soil, sediment). Steady-state concentrations in each compartment are expected to match actual environmental concentrations within

an order of magnitude. An illustrative example calculation using the "Level III" model of Mackay[76] is presented in Chapter 11 of Allen and Shonnard.[53]

Finally, information regarding toxicity or inherent impact is required to convert the concentration-dependent doses into probabilities of harm (risk). Based on this understanding of risk assessment, the steps for environmental impact assessment are grouped into three categories, (1) estimates of the rates of release for all chemicals in the process, (2) calculation of environmental fate and transport and environmental concentrations, and (3) the accounting for multiple measures of risk using toxicology and inherent environmental impact information. Computer-aided software packages integrating these three calculation steps are available and have been linked to commercial process simulation packages.[72,77,78] Figure 6.13 shows the information flows occurring in the software tool EFRAT (Environmental Fate and Risk Assessment Tool), which links with the simulation package HYSYS. Flowsheet stream and utility information from HYSYS is automatically transferred to EFRAT once the flowsheet is "synchronized" with EFRAT. Emission factors and correlations within EFRAT estimate release rates to the air, a MCM predicts environmental partitioning, and finally a relative risk assessment module

in EFRAT generates nine environmental risk indicators. A version of EFRAT is available free of charge for education purposes.[79]

Figure 6.14 provides an example of the type of decision that a process designer would face in flowsheet evaluation. In this simple example, absorption with a regenerable solvent is used to capture (and recycle or sell) toluene and ethyl acetate, which might otherwise be emitted into the atmosphere. To increase the fraction of the hydrocarbons that are absorbed, the circulation rate of the

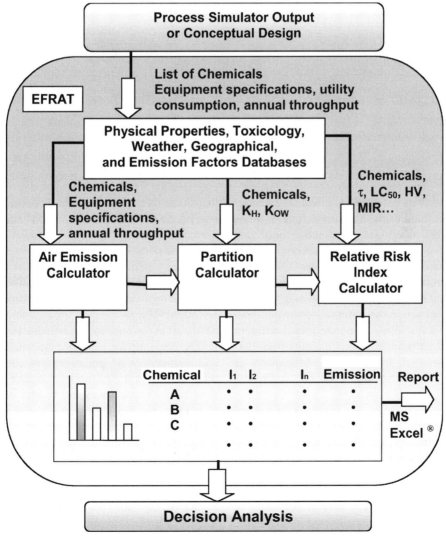

Fig. 6.13. Information flow diagram for EFRAT and relation to HYSYS[TM]. Shonnard and Hiew. *Environmental Science and Technology,* **34**(24), 5222–5228 (2000).[77] Copyright American Chemical Society and reproduced by permission.

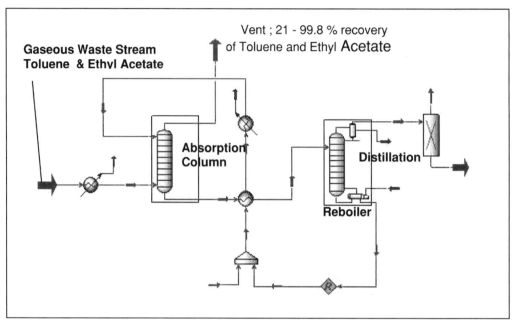

Fig. 6.14. Absorption with a regenerable solvent is used to capture toluene and ethyl acetate. Shonnard and Hiew, *Environmental Science and Technology,* 34(24), 5222–5228 (2000).[77] Copyright American Chemical Society and reproduced by permission.

solvent can be increased, but this will increase the duties of the reboiler, condenser, and pumps in the system, increasing energy use and increasing atmospheric emissions (primarily of criteria pollutants, sulfur and nitrogen oxides, particulates, carbon monoxide, and of the greenhouse gas CO_2). The process engineer will need emission estimation tools to evaluate such trade-offs, and will need to evaluate the potential environmental and economic costs associated with different types of emissions, in this example the relative costs of hydrocarbon emissions as opposed to emissions of criteria pollutants and CO_2.

Figure 6.15 shows the variation of several environmental indices as a function of absorption solvent flow rate. There is a sharp decrease in the global warming index (I_{GW}) with increasing solvent flow rate until about 50 kgmoles/hr, due mostly to toluene recovery. (Note that toluene and ethyl acetate have global warming impacts assuming that all emitted VOCs are oxidized to CO_2 in the environment and that the VOCs are of fossil origin.) Thereafter, increas-

ing utility-related emissions of greenhouse gases (primarily CO_2) drive this index up faster than its rate of decrease by further recovery of ethyl acetate. The smog formation index (I_{SF}) decreases sharply with absorber solvent flow rate in the range of 0–50 kgmoles/hr, again due to recovery of toluene. Afterwards, there is a slow decline in I_{SF} with increasing solvent flow rate above 50 kgmoles/hr as ethyl acetate is recovered. The acid rain index I_{AR} increases in nearly direct proportion to solvent flow rate. Utility consumption and its associated sulfur and nitrogen oxide emissions (precursors to acid rain) drive the I_{AR} up with higher solvent flow rates. These flowsheet results begin to reveal the complex trade-offs inherent in environmental assessment of chemical processes and products. Does the design engineer operate the process at 50 kgmoles/hr for the absorber solvent to minimize emission of CO_2 and global warming impacts or operate at higher values to reduce smog formation? These types of value judgments are commonly encountered in the decision-making process. Notwithstanding the complexity, environmental

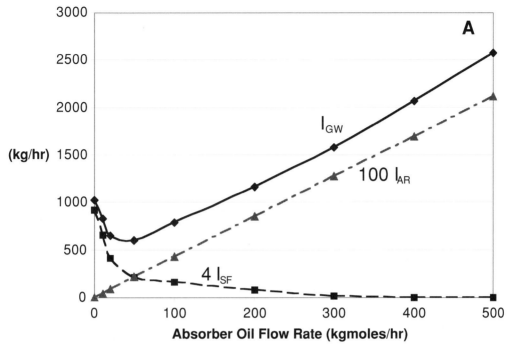

Fig. 6.15. Variation of environmental indexes with absorber oil flow rate for global warming (I_{GW}), smog formation (I_{SF}), and acid rain (I_{AR}). Shonnard and Hiew, *Environmental Science and Technology,* 34(24), 5222–5228.[77] Copyright American Chemical Society and reproduced by permission.

metrics inform the design activity on potential environmental impacts of the process.

Hybrid Screening Evaluations: Combining "Tier I"–"Tier III"-Life Cycle Assessment

Early design assessments similar to those presented in previous sections of this chapter have the following limitations: (1) they tend to focus on the reaction step and neglect the impacts of downstream units, (2) the assessment includes one or a small number of environmental indicators, and (3) the early assessment typically does not consider impacts beyond the process boundary, for example, the environmental burdens associated with the life cycle of materials used in the process. In this section we will explore some approaches to address these limitations.

Combining "Tier I" with Tier "III". In the next example, the maleic anhydride process

from an earlier section is reexamined by including the effects of units downstream from the reactor. Figure 6.16 shows the major emission sources from the process: the reactor, absorption unit, and distillation. Emission factors along with stream flow rates in the reactor are used to estimate releases of benzene, *n*-butane, CO, and MA. The emission from the reactor is estimated using an average emission factor (EF_{av}, 1.50 kg emitted/10^3 kg throughput; Allen and Shonnard 2002, chapter 8, eqn.8-4) using the equation

$$E = M_{voc}EF_{av}$$

In this equation, M_{voc} is the mass flow rate (kilograms per unit time) of the volatile organic compound in the reactor, and it is taken as the average mass flow rate through the reactor. The emissions from the absorber column originate from the offgas vent and this stream contains unreacted raw material, byproducts, and product. Raw material, especially benzene, and one of the byproducts,

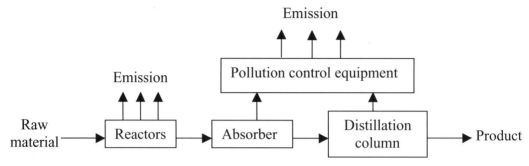

Fig. 6.16. Major emission sources from the maleic anhydride process.

CO, are toxic. To include the effects of pollution control at this early design stage, it is assumed that unreacted raw material, byproducts, and unrecovered product exiting the reactor are incinerated to CO_2 with 99% destruction efficiency with the remaining 1% released to the environment. The recovery of MA in the separation system is assumed to be 99%, with the remaining 1% going to pollution control.

Based on the process description above, Table 6.11 can be generated which contains the total emission of each pollutant. Emissions are converted to impact indicators using the environmental fate and impact assessment tool EFRAT.[79] A comparison of environmental indicators for MA production from either benzene or n-butane is shown in Table 6.12. All of the environmental indices in the n-butane process are less than or equal to those in the benzene process. This table reveals that the benzene route is estimated to have greater environmental impacts than the n-butane route. This early assessment is more rigorous than the approach in the previous section titled "Early Process Design Evaluations" because more indicators were used and the effects of units downstream of the reactor were included.

Including Life-Cycle Assessment at the Early Design Stage. In this example, a comparison is made between a new wood pulp bleaching process and a conventional process in order to uncover advantages and limitations of the "Green Chemistry" innovation.[80] In the production of most paper, wood is chemically treated with NaOH and Na_2S at high temperature ($170°C$[81]) to remove lignin and some hemicellulose, yielding pulp with a small residual of

TABLE 6.11 Stream Flow Rates and Emission Rates for Screening Evaluation of Environmental Impacts from Two Alternative Reaction Routes for Maleic Anhydride Production

	Benzene Process				
	(kg/mole of MA)				
	Benzene	*CO*	*CO_2*	*Maleic anhydride*	*Total*
Total process emission rate	1.4×10^{-4}	3.23×10^{-4}	2.03×10^{-1}	8.33×10^{-5}	2.04×10^{-1}
	n-Butane Process				
	(kg/mole of MA)				
	n-Butane	*CO*	*CO_2*	*Maleic anhydride*	*Total*
Separation unit (w/pollution control)	2.3×10^{-4}	2.51×10^{-4}	1.18×10^{-1}	8.33×10^{-5}	1.18×10^{-1}

TABLE 6.12   Environmental Indices of Both Routes (All the Values Are in Units of kg/mole of MA)

Chemical	I_{FT}	I_{ING}	I_{INH}	I_{CING}	I_{CINH}	I_{OD}	I_{GW}	I_{SF}	I_{AR}
Benzene	6.8×10^{-6}	3.3×10^{-3}	4.6×10^{-2}	1.4×10^{-4}	1.4×10^{-4}	0.0	2.0×10^{-1}	2.5×10^{-5}	0.0
n-Butane	3.0×10^{-6}	3.1×10^{-3}	3.8×10^{-2}	0.0	0.0	0.0	1.2×10^{-1}	4.4×10^{-6}	0.0

lignin that must be removed by bleaching to achieve a high brightness paper. Conventional bleaching utilizes ClO_2 and NaOH in sequential processes (Fig. 6.17). This process generates approximately 0.5 kg of chlorinated organics per ton of bleached pulp, and even though these water pollutants are less persistent, bioaccumulative, and toxic than those generated using Cl_2 (earlier bleaching agent), they are still of concern. Furthermore, the need to raise and lower the stream temperature consumes a large amount of energy and cooling water.

The 1999 Green Chemistry Challenge Award was given to Professor Terrence J. Collins of Carnegie Mellon University for the development of "Tetraamido-macrocyclic ligand" TAML™ activators.[82] These organic amine compounds contain a catalytic iron center that activates hydrogen peroxide at minute concentrations in the bleaching process (50 g/ton pulp). Although information is lacking about the toxicity of the TAML™ Oxidant Activators, it is known that the temperature of the TAML™ peroxide process is 50°C compared to 70°C for chlorine dioxide bleaching (Fig. 6.18). We will use these temperature differences as the basis of the green engineering analysis in comparing conventional ClO_2 bleaching with TAML™ process, with energy over the life cycle being the sustainability indicator.

Let us assume that the chlorine dioxide bleaching process has the sequence *DEDED*, (D = chlorination with ClO_2, E = extraction with NaOH) in which the pulp is cold-water drum washed between *DE* stages giving a drum exit temperature of 50°C. TAML™ peroxide beaching process is assumed to follow a sequence *PPPD* (P = perxide). Thus, two of the *DE* stages are replaced with two *PP* stages in which there exists a 20°C difference between the stream temperatures. Furthermore, assuming a 10% consistency for the pulp (9 tons water/ton pulp), we can estimate an energy savings of Energy Saved/ton pulp = 2 stages × (20 × 9/5)°F/stage × (2,200 Btu/ton water × 9 tons water/ton pulp) = 1.426×10^6 Btu/ton pulp. (Note, the contribution of the

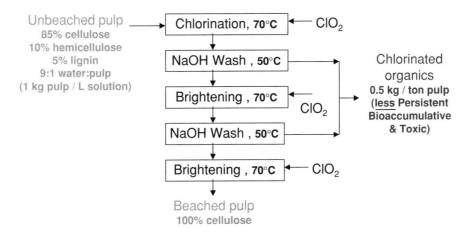

Fig. 6.17. Chlorine dioxide process for bleaching of unbleached wood pulp.

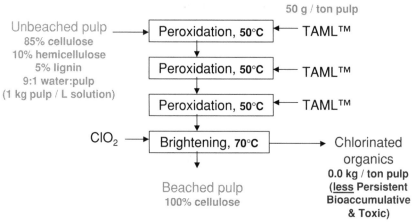

Fig. 6.18.   Wood pulp bleaching process using TAML™ activators and hydrogen peroxide.

pulp to the energy analysis is neglected, being a minor component.)

According to Department of Energy statistics the average energy consumed per ton of pulp is 39.4×10^6 Btu/tons pulp.[81] Given that roughly ½ of the pulp in the United States is bleached, this energy savings is about 2% of the annual energy consumed for production of pulp. This significant energy savings was cited in the application for the 1999 Green Chemistry Award.[82] This analysis included the energy differences between these bleaching processes in terms of the stream heating requirements. The next section will present a more complete energy efficiency analysis of the TAML™ peroxide beaching processes by including the effects of producing and delivering the bleaching agents to the process. Figure 6.19 shows the chain of material flows in the production of ClO_2, H_2O_2, NaOH, and TAML activator. Table 6.13 shows the primary energy intensity for each of the materials in the chain of materials for ClO_2 nd H_2O_2 bleaching of pulp.

The overall energy change for substituting H_2O_2 bleaching for ClO_2 bleaching is obtained by combining the net energy change from producing the chemicals and then subtracting the in-process energy savings. This yields a total decrease in energy consumption for the TAML™ H_2O_2 bleaching process of $(1.15 \times 10^7 - 1.87 \times 10^6$ Btu / ton pulp$)$ $- 1.426 \times 10^6 = -8.63 \times 10^6$ Btu / ton pulp.

The energy decrease as a percentage of the energy consumption rate of the pulp and paper industry is $0.5 \times 8.63/39.394 \times 100 = 11.0\%$

This Green Engineering screening level energy analysis of the Green Chemistry award for TCF (total chlorine free) bleaching of pulp points out a potential energy benefit by implementing the TAML™ process in the U.S. pulp and paper industry of approximately 11.0% over the "cradle to gate" life cycle for pulp. This conclusion is based upon a screening-level analysis that is uncertain, and more detailed analysis of the life-cycle energy burdens must be completed before the energy implications are fully understood. One step in the life cycle for ClO_2 production was not included for lack of data, the process energy for converting $NaClO_3$ to ClO_2. This screening assessment is expected to underestimate the energy savings for implementing this new technology. However, the analysis points to the need for including all relevant life-cycle stages in the evaluation, not only the in-process changes, as was done previously. The energy burdens of the chemicals in the supply chain for each bleaching alternative are more significant than the in-process energy improvements.

CONCLUSIONS

Environmental performance assessment is a powerful tool for moving the chemical industry

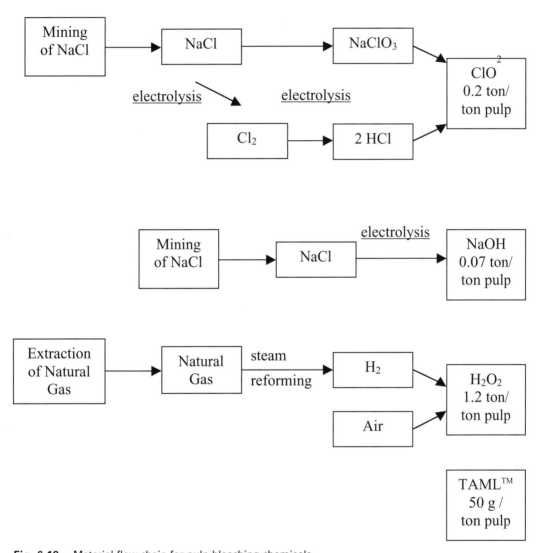

Fig. 6.19. Material flow chain for pulp bleaching chemicals.

toward sustainability. It provides important insights to the process designer on potential environmental impacts of alternative choices at several stages in the design process. Computer-aided tools further enhance the environmentally conscious engineering design activity. These tools for solvent selec-

tion, reaction pathway analysis, emission estimation, environmental property estimation, fate modeling, and life-cycle assessment will find increased use in the chemical industry. They will be particularly effective in decision making when coupled with economic and societal criteria. Application of these tools for

TABLE 6.13 The Subitems in Calculating the Energy Requirement in ClO_2 and H_2O_2 bleaching

	Energy in Original Units	Energy (Btu / ton ClO_2 or ton NaOH)
ClO_2 Bleaching		
Mining NaCl	1.3 MJ/kg NaCl	9.69×10^5 Btu/ton ClO_2
Electrolysis of NaCl to $NaClO_3$	17.1 kWh/kg $NaClO_3$	8.35×10^7 Btu / ton ClO_2
HCl production	2.5 MJ/kg HCl	2.32×10^6 Btu/ton ClO_2
HCl production electricity	0.043 kWh/kg HCl	1.44×10^5 Btu/ton ClO_2
Total from ClO_2		8.69×10^7 Btu/ton ClO_2
NaOH production	21.3 MJ/kg NaOH	1.83×10^7 Btu/ton NaOH
Total ClO_2 Bleaching		(1.87×10^7 Btu/ton pulp)
H_2O_2 / TAML Bleaching		
H_2O_2 Production		9.60×10^6 Btu/ton H_2O_2
TAML Production		1.0×10^8 Btu/ton TAML[TM]
Total H_2O_2 / TAML Bleaching		1.15×10^7 Btu/ton pulp

Energy intensities are taken from the SimaPro 6.0 software and other sources.[80]
1 Btu = .0010551 MJ
1 Btu = .0002931 kWh
1 kWh = 3.6 MJ

Example Calculation: The energy required to produce one ton of ClO_2 is therefore
(0.867 kg NaCl /kg ClO_2) × (1.3 MJ / kg NaCl) × (2,000/2.205 kg/ton) × (947.8 Btu/MJ) +
(24,478 kWh / ton of ClO_2) × (3,411.8 Btu/kWh) for electrolysis of NaCl solution +
(1.08 tons HCl/ton ClO_2) × (2.5 MJ/kg HCl) × (2,000/2.205 kg/ton) × (947.8 Btu/MJ) +
(1.08 tons HCl/ton ClO_2) × (0.043 kWh/kg HCl) × (2,000/2.205 kg/ton) × (3,411.8 Btu/kWh) =
$9.69 \times 10^5 + 8.35 \times 10^7 + 2.32 \times 10^6 + 1.44 \times 10^5 = 8.69 \times 10^7$ Btu/ton ClO_2.

comparison of traditional processes and products to innovative alternatives is already taking place in fields such as bio-based products, recycle of wastes, elimination of toxic compounds, green chemistry, benign solvents, and material flow analysis.

V LIFE-CYCLE ASSESSMENT

INTRODUCTION

Products and processes all have a natural life cycle. For example, the life cycle of a product starts from the extraction of raw materials for its production and ends when the product is finally disposed. In the production, use and disposal of this product, energy is consumed and wastes and emissions are generated. A life-cycle assessment is an analysis in which the use of energy and materials are quantified and the potential environmental and societal impacts are predicted. Life-cycle thinking is progressively being adopted by industry as an essential tool for analyzing processes and products. For example 3M has plans to perform life-cycle evaluations on all of their products.[83] BASF is using eco-efficiency analysis of a life-cycle assessment to compare products they produce and examine their economic and ecological characteristics.[84,85] This section presents a review of methods used for life-cycle assessment and then gives examples of assessments conducted on chemical products and processes.

Life-cycle assessment methodology has been used since the 1960s with early studies that focused solely on energy usage and solid waste issues. This focus continued in life-cycle assessments performed during the oil crisis in the 1970s.[86,87] The unique aspect of all of these initial studies was the early development and use of life-cycle data inventories with less emphasis on environmental risk impacts of the associated processes studied. A method published by the Royal Commission on Environmental Pollution in 1988, employing the "Best Practicable Environmental

Option," attempted to minimize the environmental burdens at the LCA practitioner's manufacturing plant, but this method did not analyze the entire lifecycle of the product or process in question.[88] However, even today, life-cycle assessments that focus primarily on mass and energy balances (e.g., Beaver et al.[89]) provide useful life-cycle information.

Life-cycle assessments capturing the full life of a process or product came into fruition in the late 1980s but were performed primarily by private companies, thus leaving into question the specific methodology used and the biased nature of the results. In 1993, the Society of Environmental Toxicology and Chemistry (SETAC) addressed the concern of no common framework for LCA methodology by introducing principles on how to conduct, review, use, and present the findings of a life-cycle assessment.[87] The International Organization of Standardization (ISO) introduced Standards 14040 and 14041 on life-cycle assessments in 1997 and 1998, respectively,[90,91] and Standards 14042 and 14043 in 1999 and 2000, respectively.[92,93] Figure 6.20 illustrates the Life-Cycle Principles and Framework according to ISO 14040-14043.[90–93]

Life-cycle impact assessments typically consist of four steps.[53] The first step includes the definitions of the system boundary, scope, and the functional unit. The second step consists of an inventory of the inputs and outputs of the system. Figure 16.21 depicts the various components of life-cycle inventory. As the earliest life-cycle assessments were previously described, many life-cycle assessments end at this stage, and conclusions are made on minimizing mass and energy usage. To increase the information gained by using life-cycle assessment, the relative hazards of the chemicals used in and emitted from the process must be included. The linking of the inputs and outputs determined in step 2 to their inherent risk to the environment and public health is the third step of LCA. The fourth and final step of LCA is interpreting the results and recommending process improvements.[94] SETAC has defined a different, but similar, fourth step which is termed "improvement assessment" which includes

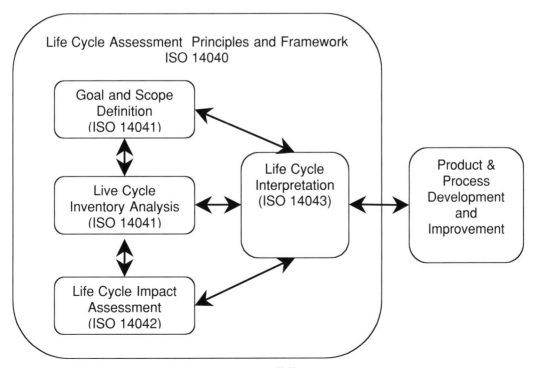

Fig. 6.20. ISO Standards for Life-Cycle Assessment.[90–93]

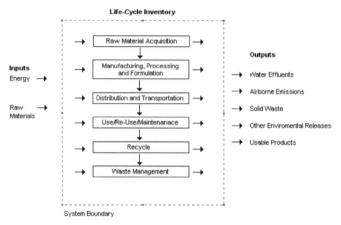

Fig. 6.21. Components of a Life-Cycle Inventory

ways to improve the impact assessment.[86] There are several good reference texts that give detailed procedures on life-cycle assessment.[95,96,97]

Life-cycle assessments are typically conducted on end products, but can also be used proactively for process selection, design, and optimization. Figure 6.22 illustrates the various steps involved in product and process life cycle. Some debate exists on the relative value of life-cycle assessments applied to products versus processes. Burgess and Brennan[87] state that a life-cycle assessment conducted on a product is also valid for the processing steps involved in the manufacture of the product and that the purpose of conducting an LCA on a process is different than on a product. On the contrary, Chevalier et al.[98] state that a life cycle conducted on a process is more thorough than a life cycle conducted on a product. A life-cycle assessment on a process is usually conducted in the research and development phase to determine if there are other options, such as replacing a solvent with a less environmentally harmful solvent, for comparison to processes used by competitors, to reduce liability, and for marketing and policy-making purposes.[87] Regardless of whether a process or product is evaluated, the basic steps of LCA remain the same. These steps are presented in more detail to follow.

GOAL AND SCOPE OF LCA

Defining a boundary for a life-cycle assessment varies depending on the methodology used. As previously described, ISO 14040 standardized life-cycle impact assessments in 1997[99,100] and established that a life-cycle assessment should be conducted in terms of elementary flows. This approach is typically understood as a cradle-to-grave approach, involving raw material extraction through the disposal of the product, including all of the relevant supply chains.[101] The standard was further refined by ISO 14041,[91] which established a "streamlining" approach to LCA that allowed omission of certain processes, inputs, and outputs if these processes are deemed insignificant. Burgess and Brennan[87] state that assumptions concerning the boundary conditions are necessary to maintain manageability and to ensure that the law of diminishing returns can be observed after three upstream processes. However, according to the ISO standards, these data must be collected before they can be disregarded.

Scoping and goal definition is often thought of as the most important process of an LCA, and care must be taken to establish appropriate boundaries that are consistent with the objectives of a study. Suh et al.[102] warn that including a scientific basis for excluding and including processes in system boundaries is essential and emphasizes the difficulty in

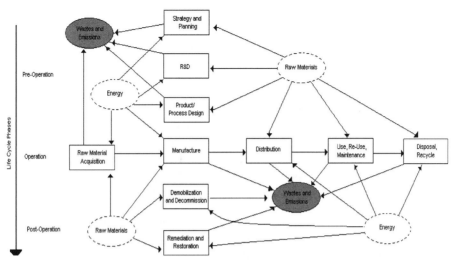

Fig. 6.22. Product and process life cycles.[53]

doing so, particularly if the practitioner is not familiar with the specific process under study.

The functional unit is defined as the basis upon which all data and impacts are presented and compared and typically relates to the "function" of the product studied. For example, the basis for comparing plastic and paper grocery bags can be the number of bags required to carry a certain amount and type of grocery items. Another example of a functional unit directly reflecting the product function is used when comparing cloth and disposable diapers, accounting for the number of diapers required for each diapering. A common functional unit used in LCAs involving product manufacture is "per mass of product"; however, there are other appropriate functional units that may be more relevant to the specific product effectiveness studied.

METHODS OF LIFE-CYCLE INVENTORY

The inventory stage involves the collection of all the data that will be used in the life-cycle analysis. The quality of the data is an important part of the life-cycle inventory (LCI) process, and, as with any model, the results of an LCA are only as good as the data inputs. There are two basic sources of data for an LCA, primary and secondary in nature. Primary data are derived directly from the

process in question. These are the most accurate data that can be applied to an LCA and, as a result, are the most desirable. However, in many cases, data are proprietary and are not readily available to the public,[87] thus necessitating the LCA practitioner to seek secondary sources of data, including databases, peer-reviewed literature, etc., that may not be as accurate and are not often accompanied with error estimates. Because of the frequent use of secondary data, debate within the LCA community has arisen on how to best capture and report error and subsequently interpret results of LCAs.

There are four main methods of conducting a life-cycle inventory, each differing in its degree of detail. These methods are economy scale, life-cycle scale, equipment scale, and a hybrid LCI,[101] listed from the most general to the most detailed in approach. The most general analysis, at the economy scale, contains all the information from cradle-to-grave of a product or process but is not specific for an individual process. On the other hand, the most detailed method, the hybrid LCI approach, is typically used for gate-to-gate (within the grounds of a manufacturing facility) analysis. A brief summary of each life-cycle inventory method, along with its advantages and disadvantages, is presented below.

The "economy scale" method of LCI typically uses national statistics about resource use and emissions from a specified sector and considers the entire economic system. With this method the boundary is defined from the cradle to the grave of the product. The most well-known method in this category is provided in the EIOLCA software developed at Carnegie Mellon University (2005). There are two disadvantages to this approach. First, an assumption must be made that there is a linear relationship between the dollars spent on the product in question and its environmental impact. In many cases, this is a good assumption, such as in processes involving obtaining a higher purity chemical. This assumption, however, is not valid for other industries with equivalent products having different perceived values, such as automobiles. There is also another drawback to using this type of method because the data typically used are a composite for the entire industry. This assumption may be valid assuming that the company is average in its performance; however, this assumption does not account for industries that are leaders or laggards in areas such as sustainability, operation efficiency, or whether they are using state-of-the-art processes.

The second LCI approach that can be used is at the life-cycle scale, involving focus on a specific sector. Industry uses this approach most frequently in developing LCIs. Although the data using the life-cycle scale approach are more detailed than the data offered in economic scale approach, they do not include specific information about individual processes, equipment, or other materials that could be considered negligible.

Another method for conducting an LCI is the equipment-scale approach, most frequently used in a "gate-to-gate" analysis of various processes. Full life-cycle assessments do not often include this LCI approach even though the most accurate data can be obtained from this method. The drawback of using this method is that it is very time and resource intensive. Another issue with this LCI approach is that a comparison with other products is relatively difficult to obtain in as

much as a gate-to-gate inventory does not account for how the given raw materials are being produced. This type of method is usually supplemented with data from the other LCI methods.

The hybrid LCI approach combines the features of the economic, life-cycle, and equipment-scale approaches in an attempt to overcome the shortcomings of the previous methods. This approach may involve combining the economy scale with the life-cycle scale to give a cradle-to-grave analysis with more detailed information on the specific industry. A hybrid LCI may also combine the life-cycle, scale LCI approach with the equipment-scale approach.[102] The advantage of this method is that it fills the gaps that are left from the life-cycle scale with data from the economic-scale, thus yielding a full "cradle to gate" life cycle. This type of life cycle also has the same limitations of the economy-scale approach, as described above.

There are several other methods that can be used for a life-cycle inventory which include a limited life-cycle inventory.[89] This method only considers metrics such as mass, water usage, energy, toxics emitted, and overall pollutants emitted. In-depth calculations are used, but all of the common metrics are not used as stated by Allen and Shonnard.[53] Another method for life-cycle assessment was proposed by Lei et al.,[103] coined by the authors as "the Most of the Most." This approach consists of finding the most significant impact factors and then selecting the most significant stages of the life cycle. Although this method was designed to consider the whole process life cycle, it severely limits the size of an LCA.

Life-cycle assessment can also be conducted using the concept of exergy, which is defined as the available energy for the specific process. This method reduces the amount of double counting involved in a process.[104] According to Cornelisen and Hirs,[105] an exergetic life-cycle assessment is a good tool to use in the area of depletion of natural resources, as life-cycle assessments are weak in evaluating this impact. An exergetic analysis compares the irreversibility of products, and the product with the lower irreversibility is more sustainable.[106] Cornelisen and Hirs[105] compared coal, green wood, and dried wood in an exergetic life-cycle assessment. According to this study, it appears as though using waste wood as chipboard rather than as an energy source results in less depletion. It was also found that using green wood for electricity production rather than for making chipboard gives less depletion of resources compared to using waste wood for electricity production. This difference is because the green wood needs to be dried prior to being used in chipboard production. Ukidiwe and Bakshi[101] stated that the exergetic life-cycle assessment does not account for various ecological resources and suggests that a thermodynamic life-cycle assessment can be used in place of an exergetic LCA.

If there is only limited information that can be obtained for a specific process, then this information must be estimated. Jimenez-Gonzalez et al.[107] have proposed several rules of thumb. This methodology includes determining the inlet temperature and pressure, reflux ratio, fugitive losses, and a way to account for any water in contact with other chemicals as contaminated. They mention that using this approach produces results within a 20% error.[108]

IMPACT ASSESSMENT AND ANALYSIS

There are three steps recommended in a complete impact analysis: classification, characterization, and valuation. Classification involves grouping of chemical emissions into impact categories. Typical impact categories include global warming, ozone depletion, smog formation, human carcinogenicity, atmospheric acidification, aquatic and terrestrial toxicity, habitat destruction, eutrophication, and depletion of nonrenewable resources. Xun and High[109] offer that these categories can be subdivided into four classes of metrics: generic for both chemical and site, chemical specific (not accounting for environmental conditions), chemical specific in a generic environment (such as a chemical's global effects), and

site and chemical specific (such as releases into a specific waterway).

The characterization step of impact assessment in LCA involves quantifying the specific impact category chosen. For example, the mass or dollar amount of material emitted is multiplied by the potential for the compound to cause the chosen impact. This provides a weighted factor for the specific impact category. When comparing two similar products and one product is higher for all impact categories, no further analysis is needed. This is rarely the case, however, thus necessitating valuation. The valuation step consists of determining which impact categories are the most significant. An example of a valuation step, the total amount of ocean and land that is needed to "buffer" the various environmental impacts may be calculated,[53] the so-called "critical volume" approach. It is also possible to conduct the valuation step in terms of a monetary value. Burgess and Brennan[87] state that developing a standard for assigning relative weights to the categories is difficult because there is no clear consensus on how to carry out the valuation process. One method to approach valuation uses weighting of impact categories based on their relevance to a specific country.[110,66] These authors also show how weighted factors can be used to determine which factor contributes most using societal factors. They use a process called Eco-Efficiency to determine which process causes the least environmental impact at the lowest normalized cost.

LCA IN PRACTICE

Typical life-cycle assessments are conducted during the product review stage of a process, after the plant, prototypes, and detailed designs of the product have been performed. However, Mueller et al.[111] state that life-cycle evaluation should be conducted starting at the planning stage of product development. They illustrate this using an example of multifunctional chip cards that are used in a wide variety of electronics. They determine the amount of material used for each board, how much is recycled, how much is incinerated, the toxic emissions, and the energy required to mine/produce the material. Their results show that the board is nearly 50% PVC, which accounts for two-thirds of the toxic emissions, but only 8.1% of the total energy required for its production. The most energy-intensive material in a multifunctional chip card is silver oxide, comprising only 2.5% of the product by weight, but requiring the most energy to produce. If LCA were applied at the product design stage of a chip card, alternatives to PVC and silver oxide could be explored and compared in terms of their contribution to emissions and energy requirements.

There have been a number of studies on fuels used for electricity generation. One of these studies focuses on the use of natural gas, heavy oil, or coal in cogeneration of electricity.[112] Using a numerical "eco-load total standardized" evaluation system, these authors found that coal had the lowest eco-load of all alternatives considered. In another LCA study, Goralczyk[113] compares hydroelectric, photovoltaic cells, wind turbines, oil, coal, and natural gas and quantifies that electricity from hydropower had the least environmental impact. Schleisner[114] focuses on wind farms in a life-cycle inventory study that focuses on the materials used to manufacture the windmills and reports that 2% of the electricity generated during the windmill's lifetime is used to manufacture the windmill components.

A life-cycle assessment for various forms of production of hydrogen has also been conducted. Koroneos et al.[115] examine six methods used to manufacture hydrogen: photovoltaic cells, solar thermal energy, wind power, hydroelectric power, biomass degradation, and natural gas steam reforming. These authors show that wind power yields the lowest environmental impacts for greenhouse gas formation, acidification, eutrophication, and smog formation, whereas photovoltaic cells result in the largest total environmental impact. As a result of this study, these authors recommend the use of wind power, hydropower, and solar thermal power to produce hydrogen because these are the "most environmentally friendly methods."

Life-cycle assessments have also been conducted on transportation fuels. Furuholt[116] has conducted a study comparing the production of diesel, gasoline, and gasoline with MTBE in Norway. The factors that were considered in this study were global warming, photo-oxidant formation, acidification, eutrophication, fossil energy, and solid waste. The impacts in the environmental categories listed above were conducted on the basis of 1 MJ of energy. In both analyses it was found that gasoline with MTBE contributes the most to the impact categories. It was also found that diesel fuel and gasoline have approximately the same scaled values for acidification, eutrophication, and solid waste, but gasoline without additives has 1.5 times the global warming, 2.6 times the photochemical oxidant formation, and uses 1.5 times the fossil energy as diesel.

Another study on automobile fuel options was conducted by MacLean and Lave.[117] This study focused on light-duty vehicles and the CO_2 equivalent gases released during manufacture, gasoline refining, operation, maintenance, and other services. The authors cite that 73% of the greenhouse equivalent gases are released during operation and propose viable alternatives to the use of gasoline in vehicles. They emphasize that, although battery-powered vehicles have zero emissions, there are other factors that give this alternative a negative environmental impact, such as the use of heavy metals. Hybrid vehicles are also discussed as an alternative, but they state that the higher sales price of the Toyota Prius is not justified by fuel savings, emissions reductions, or a combination of the two. Diesel fuel, another fuel alternative, has a well-to-tank efficiency of 24%, whereas gasoline only has a 20% efficiency. There are also many drawbacks to the use of diesel, including higher NO_x and particulate matter emissions, possible carcinogens. MacLean and Lave also cite that fuel cell vehicles are 20 years away from wide-scale use. Ethanol is a viable option as a fuel source, and there are two renewable processes that can be used to obtain ethanol. The first method is from plant cellulosic material. The well-to-tank efficiency for this material ranges from 80–95%, and the emissions are 15 g CO_2 equivalent gases/MJ. Ethanol can also be manufactured using corn. This process releases six times as much CO_2 equivalent gases/MJ as the previous process.

Other than fuels, life-cycle assessments have been performed on other sectors of the transportation industry. One study focused on the catalytic converters for passenger cars.[118] The goal of this life-cycle assessment was to compare the life-cycle impacts of a catalytic converter and the environmental benefits in terms of emission reductions through the exhaust pipe. The study on catalytic converters involved a cradle-to-grave approach, but excluded the mining and transportation of raw materials because no data were available. The criteria used for environmental loads were global warming potential, waste, eutrophication, acidification, resource use, and photochemical ozone creation potential. It was found from this study that waste and global warming are drastically increased, but acidification, eutrophication, and photochemical ozone creation potential are drastically decreased as a result of converter use. Auxiliary power units for diesel trucks were compared by Baratto and Diwekar.[119] The environmental criteria used were the same criteria as mentioned by Allen and Shonnard,[53] but also included the toxicity to humans, terrestrial species, and aquatic species. It was found the auxiliary power unit had the least impact for all the categories. An economic analysis was also conducted, and it was found that the payback period was a little over two years.

Eagen and Weinberg[120] conducted a life-cycle assessment on two different anodizing processes, differing in the mixture of boric and sulfuric acid or chromic acid used. Boric and sulfuric acid are shown to be a better choice than a mixture of boric and chromic acid. Tan et al.[121] have conducted a cradle-to-gate life-cycle assessment of an aluminum billet, which included the mining of bauxite, the processing of the alumina, and the final casting process for three plants located in Australia. Four different scenarios were analyzed—a base case, a reduction of scrap metal, a more sustainable practice for the smelter, and the latter with clean coal technology—and they conclude that implementation of the last case, clean coal technology, decreases all emissions considered and decreases the global warming potential by 21%.

A few life-cycle assessments have been conducted on processes involved in the pharmaceutical industry. A study conducted by Jodicke et al.[122] focuses on one processing step of an intermediate using either a metal catalyst or bio-catalyst. These authors show that the solvents used in the extraction of the product play a large role in the environmental impacts. Jimenez-Gonzalez et al.[123] conducted a cradle-to-gate life cycle for a pharmaceutical product of GlaxoSmithKline. The metrics used for this study were eutrophication, acidification, greenhouse gases, photochemical ozone creation, energy, and mass requirements. This study reports that eutrophication, ozone creation, total organic carbon, energy, and raw materials are most affected by this process, whereas greenhouse gas formation and acidification are most affected by energy use requirements. The manufacturing process is broken down further and shows the impact of solvents, chemicals, and internal drug manufacture on the environmental criteria listed above. Solvent selection also contributes significantly to the impact of the manufacture of a pharmaceutical product. Solvents contribute 75% to the energy use, 80% of the mass (excluding water), and 70% of the ozone depletion. Energy also contributes 70% to resource depletion and 90% to greenhouse gas emissions. Jimenez-Gonzalez and Overcash[124] compared two processes for making sertraline: the THF and TOL processes. These two processes were analyzed from the lab scale to the production scale. In comparing energy usage between the lab and production scales energy usage decreased by about 70%. It was also found that there is no significant energy difference during the final production stage of the product with regard to the two different processes. The results of an eco-efficiency analysis on Vitamin B2 production was reported by Wall-Markowski et al.[125] In this study they compared vitamin B2 produced from three fermentation processes and one chemical process. One of the three bioprocesses outperformed the others in both economic and ecological indicators. Based on the results of this study BASF recently started vitamin B2 production using a one-step fermentation from vegetable oil.

Life-cycle assessments have also been conducted on seawater desalination technologies. One study was conducted by Raluy et al.[126] used SimaPro 5.0 software[127] for the analysis portion. Three desalination technologies were compared: multi-effect distillation, multi-stage flash, and reverse osmosis. The study focused on the environmental criteria of CO_2, NO_x, nonmethane volatile organic compounds (NMVOC), and SO_x. The analysis focused on integrating the distillation process and flash process with a cogeneration plant and with reverse osmosis. This study also compared different regions with the primary difference related to the type of fuel used for electricity production. It was found that in models using the average of the European countries, as well as in models of Spain and Portugal alone, that the multi-stage flash had the least environmental burdens. However, the French and Norwegian models showed that reverse osmosis had the least environmental negative effects using Eco-indicator 99, Ecopoint 97, and CML 2 baseline impact-assessment methods that accompany the software. Raluy et al.[126] also stated that using a hybrid plant cuts down energy usage by 75%.

There have also been a number of LCA studies performed on industrial paint coatings. One study done by Shonnard et al.[110] compared five different coating processes for wooden doors. It appears from this analysis that the UV coating process has the least risk potential, raw material consumption, emissions, and energy consumption. Papasavva et al.[128] also conducted a life-cycle assessment on paints used in the automotive indus-

try. This study focused on three types of coating materials: primer, basecoat, and clear coat. Three primers were investigated, one solvent-based and two powders (acrylic and polyester). The two basecoats that were used were waterborne and white and pewter in color, respectively. The two clear coats were both acrylic, but one was a solvent-based and the other, powder-based. The white basecoat was chosen for the article because the energy required to produce either basecoat was approximately the same. The criteria used by Papasavva et al. were material requirements, energy consumption, atmospheric emissions, water emissions, solid waste emissions, particulate matter, SO_x, NO_x, CO, VOC, and CO_2. They show that there is a trade-off among environmental factors. These authors report that using the combination of powder primer, water basecoat, and powder clearcoat (PP2–WB1–PC2) results in the least energy requirement, water consumption, solid waste, and VOC emissions. However, this combination of surface coatings exceeded the other scenarios in PM, SO_x, and CO_2 equivalent air emissions. An interesting finding by Dobson[129] is that use of water-based and solvent-based paints yielded the same environmental impacts as incineration of the VOCs.

The pulp and paper industry is another industry where life-cycle assessment methodology has been applied. One example is a paper by Lopes et al.,[130] who compared the two major fuels used in the pulp and paper industry: fuel oil and natural gas. The environmental categories were the same categories listed by Allen and Shonnard.[53] The use of methane in place of fuel oil decreases all of the environmental parameters except photochemical ozone formation, which does not vary between fuel options.

Other life-cycle assessments have been conducted on postconsumer-use recycling. One of these studies by Rios et al.,[131] focused on the end-of-life recycling of plastics used in electronics. The focus of this paper was on the separation and sorting of various types of plastics for recycling. Another study[53] compared PET bottles to glass bottles with and without recycling. It was found that recycling the PET bottles has the least environmental impact and the same normalized cost as the glass bottle. Song and Hyun[132] focused on the recycling aspects of PET bottles for 11 different scenarios and showed that as the percent of bottles collected increases, the energy used for collection also increases. They cited that the recycle pathway which produces the least CO_2, SO_x, and NO_x was the closed loop and landfill pathway, and the pathway which produced the least solid waste was the pathway for pyrolysis and incineration. Ekvall[133] conducted a life-cycle analysis for recycling newsprint in Sweden. Shiojiri et al.[134] conducted a life-cycle assessment on sulfur hexafluoride. There were different ways to use and to recycle the sulfur hexafluoride. From the study conducted on sulfur hexafluoride "energy consumption as well as global warming risk can be reduced by using a mixture of SF_6 with nitrogen as an insulating gas compared to SF_6," but the other environmental impacts will increase due to transportation to the recycling plant.

There have been life-cycle assessments conducted on waste management. Jimenez-Gonzalez et al.[107] conducted a partial life-cycle inventory on three different waste treatments for pharmaceutical waste: wastewater treatment, incineration, and solvent recovery. Chevalier et al.[98] compared two flue gas treatment processes for waste incineration using life-cycle assessment. Other life-cycle assessments have been done on consumer products. One example is a life-cycle assessment on the production of clothing for the production of various indigo dyes for dying denim.[110] A 2-D plot was used with the axis being cost and normalized environmental impact. It was found that dying the denim electrochemically in a 40% vat solution has the least environmental impact and cost compared to dying the denim using indigo plants.

There have also been life-cycle assessments conducted on the food industry. In the study by Cederberg and Mattson[135] organic milk farms were compared to conventional milk farms. They show an increase in global warming in methane due to organic farming, but a decrease in other compounds such as carbon dioxide and N_2O. Most of the acidification potential for these farms was caused by the release of ammonia from manure. The eutrophication parameter was estimated for this study and was based on the manure application rate and a higher phosphorous surplus on conventional farms. They stated that organic farming reduces pesticide use, global warming, acidification, and eutrophication. Zabaniotou and Kassidi[136] conducted a study on two types of egg packaging material, recycled paper and polystyrene. The functional unit for this study was done on a packaging basis instead of a product mass basis. The environmental factors used in the study for egg containers were global warming, ozone depletion, acidification, eutrophication, particulate matter, heavy metals, carcinogenic substances, and photochemical ozone creation potential. This study concluded that polystyrene packages contribute more to acidification potential, winter and summer smog formation, whereas recycled paper egg packages contribute more to heavy metal and carcinogenetic substances impact. Another study on food packaging was conducted by Bohlmann,[137] which focused on a comparison of polypropylene and biodegradable packaging. The functional unit for this case study was the packaging required to fill 1000 kg of yogurt. The environmental criteria used for this comparison were energy and greenhouse gases. It appears as though the biodegradable package consumes less energy, but has slightly higher greenhouse gas emissions. Bohlmann cited that the greenhouse gas emissions are equivalent if the biodegradable packaging is fully decomposed in the landfill.

Anderson et al.[138] report a life-cycle assessment on tomato ketchup, in an attempt to identify "hot spots" in the life cycle of the product. In this study, it was found that the food-processing stage contributes the most to greenhouse gases, human toxicity, and acidification. They also suggest that the major contributor to eutrophication is the agriculture sector of the economy.

CONCLUSIONS

Life-cycle assessment is an important green engineering tool for analyzing processes and products. This tool is currently being used by industry to improve the environmental and economic performance of products and processes. Examples have been given in this chapter of life-cycle assessment methods as well as many studies done on chemical processes and products. Through the use of life-cycle assessment tools industry will move to sustainable production of current and new products.

REFERENCES

1. EPA Green Engineering Web site. www.epa.gov/oppt/greenengineering.
2. EPA Green Chemistry Web site, 12 Principles of Green Chemistry. www.epa.gov/oppt/greenchemistry/principles.html.
3. Anastas, P.T., and Warner, J.C., *Green Chemistry: Theory and Practice*, Oxford University Press, New York, 1998.
4. Malone, M. F. and Huss. R.S., "Green Chemical Engineering Aspects of Reactive Distillation," *Environ. Sci. Technol.* **37**, 23, 5325–5329, 2003.
5. Allen, D. T., and Shonnard, D.R., *Green Engineering: Environmentally Conscious Design of Chemical Processes* (3rd ed.), Prentice-Hall, NJ, 2004.
6. Douglas. J. M., *Conceptual Design of Chemical Processes,* McGraw-Hill, New York, 1988.
7. Rossiter, A. P. and Klee. H., *Hierarchical Review for Waste Minimization,* A. P. Rossetier ed., McGraw-Hill, New York, 1995.
8. Schultz, M. A., Stewart, D.G., Harris, J. M. Rosenblum, S. P., Shakur, M.S., and O'Brien, D. E., *Reduce costs with dividing-wall columns,* CEP, May pp 64–71, 2002.
9. El-Halwagi, M. M., *Pollution Prevention Through Process Integration*, Academic Press, San Diego, CA, 1997.
10. Liu, Y. A., Lucas, B., and Mann, J. M., "Up-to-date tools for water system optimization," *Chem. Eng*, **111** (1), 30–41, 2004.
11. Allen, D. T. and Rosselot, K. S., *Pollution Prevention for Chemical Processes,* John Wiley and Sons, New York, 1997.
12. Tunca, C., P. A. Ramachandran, and M. P Dudukovic, Role of chemical reaction engineering in sustainable development, Paper presented at AIChE session 7d, Austin, Texas, November 10, 2004.
13. Mulholland, K.L. and Dyer, J.A., *Pollution Prevention: Methodology, Technologies, and Practices,* American Institute of Chemical Engineers, New York, p. 214, 1999.
14. Ney, Ronald E. Jr., *Where Did That Chemical Go?* Van Nostrand Reinhold, New York, 1990.
15. Mackay, D., Patterson, S., diGuardo, A, and Cowan, CE. "Evaluating the environmental fate of a variety of types of chemicals using the EQC Model. *Environ. Toxicol. Chem.,* **15**, 1627–1637, 1996.
16. Andren, A.W., Mackay, D., Depinto, J. V., Fox, K., Thibodeaux, L.J., McLachlan, M., and Haderlein, S. Intermedia partitioning and transport. In: Klečka, G, Boethling, B, Franklin, J., Grady, L., Graham, D., Howard, P.H., Kannan, K., Larson, B., Mackay, D., Muir, D., and van de Meent, D., editors, *Evaluation of Persistence and Long-Range Transport of Organic Chemicals in the Environment.* SETAC Press, Pensacola, FL, pp. 131–168, 2000.
17. Franklin, J., Atkinson, R., Howard, P.H., Orlando, J. J., Seigneur, C., Wallington, T.J., and Zetzsch, C. Quantitative determination of persistence in air. In: Klečka G, Boethling, B., Franklin, J., Grady, L., Graham, D., Howard, P.H., Kannan, K., Larson, B., Mackay, D., Muir, D., and van de Meent, D., editors, *Evaluation of Persistence and Long-Range Transport of Organic Chemicals in the Environment.* SETAC Press, Pensacola, FL, pp. 7–62, 2000.
18. Thibodeaux, L.J. *Environmental Chemodynamics: Movement of Chemicals in Air, Water, and Soil*, 2nd edition. John Wiley & Sons, New York, 1996.
19. Clark, M.M. *Modeling for Environmental Engineers and Scientists*. John Wiley & Sons, New York, 1997.
20. Schnoor, J.L. *Environmental Modeling*. John Wiley & Sons, New York, 1997.
21. Bishop, P.L. *Pollution Prevention: Fundamentals and Practice*. Waveland Press, Long Grove, IL, 2004.
22. Thomas, R.G. Volatilization from water. In: Lyman, WJ, Reehl, WF, and Rosenblatt, D.H., editors. *Handbook of Chemical Property Estimation Methods.* American Chemical Society, Washington, DC, pp. 15-1-15-34, 1990.

23. Larson, R. A. and Weber, E. J. *Reaction Mechanisms in Environmental Organic Chemistry*. Lewis Publishers, Boca Raton, FL, 1994.

24. Lerman, A. Time to chemical steady states in lakes and oceans. In: Hem, J.D., editor. *Nonequilibrium Systems in Natural Water Chemistry*. Advan. Chem. Ser. #106, American Chemical Society, Washington, DC, pp. 30–76, 1971.

25. Thibodeaux, L.J, Valsaraj, K.T., and Reible, D.D., "Associations of polychlorinated biphenyls with particles in natural waters. "*Water Sci. Technol.* **28**(8):215–221, 1993.

26. Baum, E. J. *Chemical Property Estimation: Theory and Application*. Lewis Publishers, Boca Raton, FL, 1998.

27. McCutcheon, S.C., and Schnoor, J.L., Overview of phytotransformation and control of wastes. In: McCutcheon, S.C., and Schnoor, J.L., *Phytoremediation: Transformation and Control of Contaminants*. John Wiley & Sons, New York, pp. 3–58, 2003.

28. Larson, R., Forney, L., Grady Jr., L., Klečka, G. M., Masunanga, S., Peijnenburg, W., and Wolfe L. Quantification of Persistence in Soil, Water, and Sediments. In: Klečka, G, Boethling, B, Franklin, J, Grady, L., Graham, D, Howard, P.H., Kannan, K, Larson, B, Mackay, D, Muir, D, and van de Meent, D., editors, *Evaluation of Persistence and Long-Range Transport of Organic Chemicals in the Environment*. SETAC Press, Pensacola, FL, pp. 63–130, 2000.

29. Gibson D.T. and Subramanian, V., Microbial degradation of aromatic compounds. In: Gibson, D.T., editor *Microbial Degradation of Organic Compounds*. Marcel Dekker, New York, 1984.

30. Bedard D.L., and Quensen, J.F., III. Microbial reductive dechlorination of polychlorianted biphenyls. In: Young, L.Y., and Cerniglia, C.E., editors, *Microbial Transformation and Degradation of Toxic Organic Chemicals*, John Wiley & Sons, Inc., New York, 1995.

31. Fish K.M., and Principe, J.M., Biotransformations of Arochlor 1242 in Hudson River Test Tube Microcosms, *Appl. Environ. Microbiol.*, **60** (12), 4289–4296, 1994.

32. Ye, D., Quensen III, J.F., Tiedje, J.M, and Boyd, S.A., Evidence for para-dechlorination of polychlorobiphenyls by methanogenic bacteria. *Appl. Environ. Microbiol.* **61**:2166–2171, 1995.

33. Chakrabarty, A.M., *Biodegradation and Detoxification of Environmental Pollutants*. CRC Press, Boca Raton, FL, 1982.

34. Alexander, M., *Biodegradation and Bioremediation*. Academic Press, San Diego, 1994.

35. Young, L.Y., and Cerniglia, C.E., *Microbial Transformation and Degradation of Toxic Organic Chemicals*, Wiley-Liss, New York, 1995.

36. Burken, J.G., Uptake and metabolism of organic compounds: Green liver model. In: McCutcheon, S.C. and Schnoor, J.L., editors. *Phytoremediation: Transformation and Control of Contaminants*, John Wiley & Sons, New York, pp. 59–84, 2004.

37. Jeffers, P.M., and Wolfe, N.L., Degradation of methyl bromide by green plants. In: Seiber, J.N., editor. *Fumigants: Environmental Fate, Exposure and Analysis*. American Chemical Society, Washington, DC, 1997.

38. O'Neill, W, Nzengung, V., Noakes, J., Bender, J., and Phillips, P., Biodegradation of tetrachloroehtylene and trichloroethylene using mixed-species microbial mats. In: Wickramanayake, G.B., and Hinchee, R.E., editors. *Bioremediation and Phytoremediation*, Batelle, Columbus, WA, pp. 233–237, 1998.

39. Hughes, J.B., Shanks, J., Vanderford, M., Lauritzen, J., and Bhadra, R., "Transformation of TNT by aquatic plants and plant tissue cultures." *Environ. Sci. Technol.* **31**:266–271, 1997.

40. Vanderford, M., Shanks, J.V., Hughes, J.B., "Phytotransformation of trinitrotoluene (TNT) and distribution of metabolic products in myriphyllum aquaticum." *Biotechnol. Lett.* **199**:277–280, 1997.

41. Gao, J., Garrison, A.W., Hoehamer, C., Mazur, C., and Wolfe, N.L., Phytotransformations of organophosphate pesticides using axenic plant tissue cultures and tissue enzyme extract. In situ and on-site bioremediation. The Fifth International Symposium, San Diego, 19–22 April 1999.

42. Cunningham, S.D., and Berti, W.R., "The remediation of contaminated soils with green plants: An overview". *In vitro Cell Dev. Biol. Plant* **29**:207–212, 1993.

43. Banks, M.K., Schwab, A.P., Govindaraju, R.S., and Kulakow P. Phytoremediation of hydrocarbon contaminated soils. In: Fiorenza, S, Oubre, L.C., and Ward, C.H., editors. *Phytoremediation*, CRC Press, New York, 1999.

44. Schwartzenbach, R.P., Gschwend, P. M., and Imboden, D.M. *Environmental Organic Chemistry*. 1st edition. John Wiley & Sons, New York, 1993.

45. Wolfe, N.L., and Jeffers, P.M., Hydrolysis. In: Boethling. R.S., and Mackay, D., editors. *Handbooks of Property Estimation Methods for Chemicals: Environmental and Health Science*. CRC Press, Boca Raton, FL, pp. 311–334, 2000.

46. Zepp, R.G., Experimental approaches to environmental photochemistry. In: Hutzinger, O., editor. *The Handbook of Environmental Chemistry, Vol. 2, Part B*. Springer-Verlag, Berlin, Germany, pp. 19–41, 1982.

47. Alebić-Juretić, A., Güsten, H., and Zetzsch, C., "Absorption spectra of hexachlorobenzene adsorbed on SiO_2 powders." *Fresenius J. Anal. Chem.*, **340**:380–383, 1991.

48. Bermen, J.M., Graham, J.L., and Dellinger, B., "High temperature UV absorption characteristics of three environmentally sensitive compounds." *J. Photochem. Photobiol A: Chem*, **68**:353–362, 1992.

49. Tysklind, M., Lundgren. K., and Rappe, C., "Ultraviolet absorption characteristics of all tetra-to octachlorinated dibenzofurans," *Chemosphere*, **27**:535–546, 1993.

50. Kwok, E.S.C., Arey, J., and Atkinson, R., "Gas-phase atmospheric chemistry of dibenzo-p-dioxin and dibenzo-furan, *Environ. Sci. Technol.*, **28**:528–533, 1994.

51. Funk, D.J., Oldenborg, R.C., Dayton, D.P., Lacosse, J.P., Draves, J.A., and Logan, T.J., "Gas-phase absorption and later-induced fluorescence measurements of representative polychlorinated dibenzo-p-dioxins, polychlori-nated dibenzofurans, and a polycyclic aromatic hydrocbon," *Appl. Spectros.*, **49**:105–114, 1995.

52. Konstantinou, I.K., Zarkdis, A.K., and Albanis, T.A., "Photodegradation of selected herbicides in various natu-ral waters and soils under environmental conditions, *J. Environ. Qual.* **30**:121–130, 2001.

53. Allen, D.T., and Shonnard, D.R. *Green Engineering: Environmental Conscious Design of Chemical Processes*, Prentice Hall, Upper Saddle River, NJ, 2002.

54. Windholz, M., Budavara, S., Blumetti, R.F., and Otterbein, E.S., *The Merck Index: An Encyclopedia of Chemicals, Drugs, and Biologicals*. 6th Edition. Merck & Co., Rahway, NJ, 1983.

55. Howard, P.H., Boethling, R.S., Jarvis, W.F., Meylan, W.M., and Michalenko, E.M., *Handbooks of Environmental Degradation Rates*. Lewis Publishers, Chelsea, MI, 1991.

56. Dean, J.A., *Lange's Handbook of Chemistry*. 14th Edition. McGraw-Hill, New York, 1992.

57. Lide, D.R., *CRC Handbook of Chemistry and Physics*. 74th Edition. CRC Press, Boca Raton, FL, 1994.

58. Mackay, D., Shiu, W.Y., and Ma, K.C., *Illustrated Handbook of Physical Chemical Properties and Environmental Fate of Organic Chemicals, vols. 1–5*, Lewis Publishers, Boca Raton, FL, 1992–1997.

59. Howard, P.H., and Meylan, W.M., *Handbook of Physical Properties of Organic Chemicals,* CRC Press, Boca Raton, FL, 1997.

60. Tomlin, C.D.S., editor, *The Pesticide Manual*. 11th Edition, British Crop Protection Council, Farnham, Surrey, UK, 1997.

61. Yaws, C.L., *Chemical Properties Handbook*, McGraw-Hill, New York, 1999.

62. Verschueren, K., *Handbook of Environmental Data on Organic Chemicals*. 3rd and 4th editions. Van Nostrand-Reinhold, New York, 1996, 2001.

63. Lyman, W.J., Reehl, W.F., and Rosenblatt, D.H., *Handbook of Chemical Property Estimation Methods: Environmental Behavior of Organic Compounds*. McGraw-Hill, New York, 1982.

64. Neely, W.B., and Blau, G.E., *Environmental Exposure from Chemicals, Vols. I and II*. CRC Press, Boca Raton, FL, 1985.

65. Bare, J.C., Norris, G.A., Pennington, D.W., and McKone, T., "TRACI: The tool for the reduction and assess-ment of chemical and other impacts," *J. Indust. Ecol.*, **6**(3–4), 49–78, 2003.

66. Goedkoop, M., "The Eco-indicator 95, Final Report", Netherlands Agency for Energy and the Environment(NOVEM) and the National Institute of Public Health and Environmental Protection (RIVM), NOH report 9523, 1995.

67. Heijungs, R., Guinée, J.B., Huppes, G., Lankreijer, R.M., Udo de Haes, H.A., and Wegener Sleeswijk, "Environmental life cycle assessment of products. Guide and backgrounds", NOH Report Numbers 9266 and 9267, Netherlands Agency for Energy and the Environment (Nov.), 1992.

68. ISO 14040-14049, 1997–2002, *Environmental Management – Life Cycle Assessment*, International Organization for Standardization, Geneva, Switzerland.

69. SETAC, Society for Environmental Toxicology and Chemistry, "Guidelines for Life-Cycle Assessment: Code of Practice", Brussels, Belgium, 1993.

70. Douglas, J.M., 1992, *Ind. Eng. Chem. Res.*, **41**(25), 2522.

71. PARIS II, 2005, http://www.tds-tds.com/

72. Chen, H. and Shonnard, D.R. "A systematic framework for environmental-conscious chemical process design: Early and detailed design stages, *Indust. Eng. Chem. Res.*, **43**(2), 535–552, 2004.

73. Allen, D.T. and Shonnard, D.R., Green engineering: Environmentally conscious design of chemical processes and products, *AIChE J.*, **47**(9), 1906–1910, 2001.

74. NRC (National Research Council), 1983, *Risk Assessment in the Federal Government: Managing the Process*, Committee on Institutional Means for Assessment of Risks to Public Health, National Academy Press, Washington, DC.

75. Air CHIEF, accessed 2005, The **Air** **C**learing**H**ouse for **I**nventories and **E**mission **F**actors, CD-ROM, http://www.epa.gov/oppt/greenengineering/software.html.

76. Mackay, D., Shiu, W., and Ma, K., *Illustrated Hand book of Physical-Chemical Properties and Environmental Fate for Organic Chemicals*, 1ˢᵗ Edition, Vol. 1–4, Lewis ublishers, Chelsea, MI, 1992.

77. Shonnard, D.R. and Hiew, D.S., 2000, "Comparative environmental assessments of VOC recovery and recycle design alternatives for a gaseous waste stream," *Environ. Sci. Technol.*, **34**(24), 5222–5228.

78. WAR (**WA**ste **R**eduction Algorithm), http://www.epa.gov/oppt/greenengineering/software.html

79. SACHE, accessed 2005, Safety and Chemical Engineering Education, American Institute of Chemical Engineers, http://www.sache.org

80. Shonnard, D.R., Tools and Materials for Green Engineering and Green Chemistry Education, Green Chemistry and Engineering Education – a Workshop Organized by the Chemical Sciences Roundtable of the National Research Council, 7–8 November 2005.

81. Genco, J.M., Pulp. In *Kirk-Othmer Encyclopedia of Chemical Technology*, Vol. 20, J.I. Kroschwitz, Ed. John Wiley and Sons, New York, p. 493, 1991.

82. Collins, T.J., Horwitz, C., and Gordon-Wylie, S.W. Project Title: TAML™ Activators: General activation of hydrogen peroxide for green oxidation processes, provided by Mary Kirchhoff, Green Chemistry Institute, American Chemical Society, 1999.

83. Miller, K., Comments at the panel discussion in the session "Building the Business Case for Sustainability, AIChE Spring Meeting, Atlanta, 12 April, 2005.

84. Schrott, W. and Saling, P., "Eco-efficiency analysis—Testing products for their value to the customer. *Melliand Textilberichte* **81**(3),190, 192–194, 2000.

85. Landsiedel, R. and Saling, P. "Assessment of toxicological risks for life cycle assessment and eco-efficiency analysis." *Int. J. Life Cycle Assess.* **7**(5), 261–268, 2002.

86. Azapagic, A., "Life cycle assessment and its application to process selection, design, and optimization," *Chem. Eng. J.*, **73**, 1 (April), 1–21, 1999.

87. Burgess, A.A. and Brennan, D.J. "Application of life cycle assessment to chemical processes," *Chem. Eng. Sci.* **56**, 8 (April) 2589, 2609, 2001.

88. Royal Commission on Environmental Pollution, "Best practicable environmental option" Twelfth Report, Cm130, London, England, United Kingdom, 1988.

89. Beaver, E. R., Calculating metrics for acetic acid production, AIChE Sustainability Engineering Conference Proceedings, Austin, TX, November 2004, pp. 7–15.

90. International Organization of Standardization (ISO) 1997. *Environmental management – Life cycle assessment – Principles and framework*. International Organization of Standardization, Geneva, Switzerland (International Standard ISO14040:1997(E)).

91. International Organization of Standardization (ISO) 1998. *Environmental management – Life cycle assessment – Goal and scope definition and inventory analysis*. International Organization of Standardization, Geneva, Switzerland (International Standard ISO14041:1998(E)).

92. International Organization of Standardization (ISO). 2000. *Environmental management – Life cycle assessment – Life cycle impact assessment*. International Organization of Standardization, Geneva, Switzerland (International Standard ISO14042:2000(E)).

93. International Organization of Standardization (ISO). 2000. *Environmental management – Life cycle assessment – Life cycle interpretation*. International Organization of Standardization, Geneva, Switzerland (International Standard ISO14043:2000(E)).

94. Ukidwe, N. W. and Bakshi, B. R. Economic versus natural capital flows in industrial supply networks-Implications to sustainability, AIChE Sustainability Engineering Conference Proceedings, Austin, TX, November 2004, pp. 145–153.

95. Graedel, T.E., *Streamlined Life-Cycle Assessment,* Prentice Hall, Englewood Cliffs, NJ, 1998.

96. Curran, M., ed. *Environmental Life-Cycle Assessment*, McGraw Hill, New York, 1997.

97. Azapagic, A., Perdan, S., and Clift, R., *Sustainable Development in Practice: Case Studies for Engineers and Scientists*, John Wiley and Sons Ltd, New York, 2004.

98. Chevalier, J., Rousseaux, P., Benoit, V., and Benadda, B., "Environmental assessment of flue gas cleaning processes of municipal solid waste incinerators by means of the life cycle assessment approach," *Chem. Eng. Sci,* **58**, 10 (May) 2053–2064, 2003.

99. Ekvall, T. and Finnveden, G. "Allocation in ISO 14041 – A critical review," *J. Cleaner Prod.* **9**, 3 (June) 197–208, 2001.

100. Bakshi, B. R. and Hau, J.L., A multiscale and multiobjective approach for environmentally conscious process retrofitting, AIChE Sustainability Engineering Conference Proceedings, Austin, TX, November, pp. 229–235, 2004.

101. Ukidwe, N. W. and Bakshi, B.R., A multiscale Bayesian framework for designing efficient and sustainable industrial systems, AIChE Sustainability Engineering Conference Proceedings, Austin, TX, November, pp. 179–187, 2004.

102. Suh, S., Lenzen, M., Treloar, G. J., Hondom, H., Harvath, A., Huppes, G., Jolliet, O., Klann, U., Krewitt, W., Morguchi, Y., Munksgaard, J., and Norris, G., "System boundary selection in life-cycle inventories using hybrid approaches," *Environ. Sci. Technol.*, **38**, 3, 657–663, 2004.

103. Lei, L., Zhifeng, L., and Fung, R., "The Most of the Most"-Study of a New LCA Method, IEEE Proceedings, pp. 177–182, 2003.

104. Bakshi, B.R. and Hau, J. L., Using exergy analysis for improving life cycle inventory databases, AIChE Sustainability Engineering Conference Proceedings, Austin, TX, November pp. 131–134, 2004.

105. Cornelissen, R.L. and Hirs, G.G., "The value of the exergetic life cycle assessment besides the LCA," *Energy Conversion Manage.*, **43**, 1417–1424, 2002.

106. Becalli, G., Cellura, M., and Mistretta, M., "New exergy criterion in the "multi-criteria" context: a life cycle assessment of two plaster products," *Energy Conserv. Manage.* **44**, 2831–2838, 2003.
107. Jimenez-Gonzalez, C., Overcash, M.R., and Curzons, A., "Waste treatment modules- a partial life cycle inventory," *J. Chem. Technol. Biotechnol.* **76**, 707–716, 2001.
108. Jimenez-Gonzalez, C., Kim, S., and Overcash, M. R., "Methodology for developing gate-to-gate life cycle inventory information," *Int. J. Life Cycle Assess.*, **5**, 153–159, 2000.
109. Xun & High 2004.
110. Shonnard, D. R., Kichere, A., and Saling, P., "industrial applications using BASF eco-efficiency analysis: Perspectives on green engineering principles," *Environ. Sci. Technol.*, **37**, 5340–5348, 2003.
111. Mueller, J., Griese, H., Schischke, K., Stobbe, I., Norris, G.A., and Udo de Haes, H.A., Life cycle thinking for green electronics: Basics in ecodesign and the UNEP/SETAC life cycle initiative, International IEEE Conference on Asian Green Electronics, pp. 193–199, 2004.
112. Widiyanto, A., Kato, S., Maruyama, N., and Kojima, Y., "Environmental impact of fossil fuel fired co-generation plants using a numerically standardized LCA scheme," *J. Energy Resource Technol.*, **125**, 9–16, 2003.
113. Goralczyk, M., "Life-cycle assessment in the renewable energy sector," *Appl. Energy*, **75**, pages 205–211, 2003.
114. Schleisner, L., "Life cycle assessment of a wind farm and related externalities," *Renewable Energy*, **20**, 279–288, 2000.
115. Koroneos, C., Dompros, A., Roumbas, G., and Moussiopoulos, N., Life cycle assessment of hydrogen fuel production processes," *Int. J. Hydrogen Energy*, **29**, 1443–1450, 2004.
116. Furuholt, E., "Life cycle assessment of gasoline and diesel," *Resources, Conserv. Recycl.*, **14**, 251–263, 1995.
117. MacLean, H. L. and Lave, L. B., "Life cycle assessment of automobile/fuel options," *Environ. Sci. Technol.* **37**, 5445–5452, 2003.
118. Amatayakul, W. and Ramnas, O., "Life cycle assessment of a catalytic converter for passenger cars," *J. Cleaner Prod.*, **9**, 395–403, 2001.
119. Baratto, F. and Diwekar, U.M. "Life cycle of fuel cell-based APUs," *J. Power Sources*, **139**, pages 188–196, 2005.
120. Eagan, P. and Weinberg, L., "Application of analytic hierarchy process techniques to streamlined life-cycle analysis of two anodizing processes," *Environ. Sci. Technol.*, **33**, 1495–1500, 1999.
121. Tan, R., Khoo, B.H., and Hsien, H., "An LCA study of a primary aluminum supply chain," *J. Cleaner Prod.* **13**, 6 (May) 607–618, 2005.
122. Jodicke, G. Zenklusen, O., Weidenhaupt, A., and Hungerbuhler, K., "Developing environmentally sound processes in the chemical industry: A case study on pharmaceutical intermediates," *J. Cleaner Prod.* **7**, 2 (March) 159–166, 1999.
123. Jimenez-Gonzalez, C., Curzons, A.D., Constable, D.J.C., and Cunningham, V.L., "Cradle-to-gate life cycle inventory and assessment of pharmaceutical compounds," *Int Life Cycle Assess.*, **9**, 114–121, 2004.
124. Jimenez-Gonzalez, C. and Overcash, M. R., "Energy optimization during early drug development and the relationship with environmental burdens," *J. Chem. Technol. Biotechnol.* **75**, 983–990, 2000.
125. Wall-Markowski, C. A, Kicherer, A., and Saling, P., "Using eco-efficiency analysis to assess renewable-resource-based technologies," *Environ. Progress* **23**(4), 329–333, 2004.
126. Raluy, R.G., Serra, L., Uche, J., and Valero, A., "Life-cycle of desalination technologies integrated with energy production systems," *Desalination*, **167**, 445–458, 2004.
127. Pre Consultants, Amersfoot, The Netherlands.
128. Papasavva, Kia, S., Claya, J., and Gunther, R., "Characterization of automotive paints: An environmental impact analysis," *Progress Organic Coatings*, 43, 193–206, 2001.
129. Dobson, I. D., "Life cycle assessment for painting processes, putting the VOC issue in perspective," *Progress Organic Coatings*, **27**, pages 55–58, 2001.
130. Lopes, E., Dias, A., Arroja, L., Capela, I., and Pereira, F., "Application of life cycle assessment to the Portuguese pulp and paper industry," *J. Cleaner Prod.*, **11**, 51–59, 2003.
131. Rios, P., Stuart, J.A., and Grant, E., "Plastics disassembly versus bulk recycling: Engineering design for end-of-life electronics resource recovery," *Environ. Sci. Technol.* **37**, 5463–5470, 2003.
132. Song, H.-S. and Hyun, J.C., "A study on the comparison of the various waste management scenarios for Pet bottles using life-cycle assessment (LCA) methodology," *Resources, Conserv. Recycl.* **27**, 267–284, 1999.
133. Ekvall, T., "Key methodological issues for life cycle inventory analysis of paper recycling," *J. Cleaner Prod.*, **7**, 281–294, 1999.
134. Shiojiri, K., Yanagisawa, Y., Fujii, M., Kiyono, F., and Yamasaki, A., A life cycle impact assessment study on sulfur hexaflouride as a gas insulator, AIChE Sustainability Engineering Conference Proceedings, Austin, TX, November, 135–143, 2004.
135. Cederberg, C. and Mattson, B., "Life cycle assessment of milk production – A comparison of conventional and organic farming," *J. Cleaner Prod.*, **8**, 49–60, 2000.

136. Zabaniotou, A. and Kassidi, E., "Life cycle assessment applied to egg packaging made from polystyrene and recycled paper," *J. Cleaner Prod.* **11**, 549–559, 2003.
137. Bohlmann, G.M., "Biodegradable packaging life-cycle assessment," *Enviro Progress*, **23**, 4, 342–346.
138. Anderson, K, Ohlsson, T., and Olsson, P., "Screening life cycle assessment of tomato ketchup: A case study," *J. Cleaner Prod.* **6**, 277–288, 1998.

ADDITIONAL SUGGESTED READING

Introduction to Green Chemistry and Green Engineering

Allen, D. and Rosselot, K. *Pollution Prevention for Chemical Processes,* John Wiley and Sons, New York, 1997.

Allen, D.T and Shonnard, D.R. "Green Engineering: Environmentally Conscious Design of Chemical Processes and Products," *AIChE J.* **47**(9), 1906–1910, 2001.

Anastas, P.A. and Zimmerman, J.B. "Design through the twelve principles of green engineering," *Environ. Sci. and Technol.* 37(5) (March), 94A–101A, 2003.

Boethling, R. and Mackay, D. *Handbook of Property Estimation Methods for Chemicals.* Lewis Publishers, Roca Raton, FL, 2000.

Byrd, D. and Cothern, R. *Introduction to Risk Analysis.* Government Institutes, 2000.

Daugherty, J. *Assessment of Chemical Exposures.* Lewis Publishers, Roca Ratan, FL, 1998.

El-Halwagi, M. *Pollution Prevention through Process Integration.* Academic Press, San Aeogo, CA, 1997.

EPA Exposure Assessment Web site. www.epa.gov/oppt/exposure.

EPA Pollution Prevention Framework Web site. www.epa.gov/oppt/p2framework/.

Hesketh, R.P., Slater, C.S., Savelski, M.J., Hollar, K., and Farrell, S. "A program to help in designing courses to integrate green engineering subjects," *Intl. J. Eng. Educ.* **20**(1) 113–128, 2004.

Graedel, T.E. and Allenby, B. R. *Industrial Ecology*, Prentice Hall, Englewood Cliffs, NJ, 1995.

Martin, A. and Nguyen, N. "Green engineering: Definiting the principles – results from the Sandestin Conference." *Environ. Progress.* (December), 233–236, 2003.

Ritter, S. "A green agenda for engineering." *Chem. Eng. News*, **81**, 29 July 21, 30–32, 2003.

Shonnard, D.R., Allen, D.T., Nguyen, N., Austin, S.W., and Hesketh, R., "Green engineering education through a US EPA/academia collaboration," *Environ. Sci. and Technol.* 37(23) 5453–5462, 2003.

Slater, C. S. and R.P. Hesketh, "Incorporating green engineering into a material and energy balance course," *Chem. Eng. Educ.* **38**(1), 48–53, 2004.

Socolow, R., Andrews, F., Berkhout, F., and Thomas, V. *Industrial Ecology and Global Change.* Cambridge University Press, New York, 1994.

2.2 Pollution Prevention Heuristics for Chemical Processes

EPA Green Chemistry Web site, Green chemistry expert system: Analysis of existing processes, building new green processes, and design, (www.epa.gov/greenchemistry/tools.htm).

Freeman, H., ed., *Industrial Pollution Prevention Handbook*, McGraw Hill, New York, April 1994.

Allen, D. T. and Shonnard, D. R., *Green Engineering: Environmentally Conscious Design of Chemical Processes*, "Prentice-Hall, Upper Saddle River, NJ, 2002.

Dyer, J. A. and Mulholland, K. L., "Prevent pollution via better Reactor design and operation," *CEP* (Feb), 1998.

Wynn, C. "Pervaporation comes of age," *CEP* (October), 66–72, 2001.

2.3 Understanding and Prediction of the Environmental Fate of Chemicals

Crosby, D.G. and Wong, A.S., "Environmental degradation of 2,3,7,8-tetrachlorodibenzo-p-dioxin (TCDD)." *Science,* 195:1337–1338, 1977.

Hansch, C, and Leo, A. *Exploring QSAR: Fundamental and Applications in Chemistry and Biology*, American Chemical Society, Washington, DC, 1995.

Jeffers, P.M. and Wolfe, NL. "Green plants: A terrestrial sink for atmospheric methyl bromide." *Geophy. Res. Lett.* **25**:43–46, 1998.

Mackay, D., *Multimedia Environmental Models: The Fugacity Approach.* Lewis Publishers, Boca Raton, FL, 1991.

Meylen, W.M., and Howard, P.H., "Atom/fragment contribution method for estimating ocatnol-water partition coefficients." *J. Pharm. Sci.*, **84**:83–92, 1995.

Syracuse Research Corporation [SRC]. Syracuse, NY. Internet address: http://www.syrres.com, 2005.

2.4 Environmental Performance Assessment for Chemical Process Design

Cano-Ruiz, J.A. and McRae, G.J., "Environmentally conscious chemical process design." *Ann. Rev. Ener. Environ.*, **23**, 499, 1998.

2.4 Life-Cycle Assessment

Carnegie Mellon University, http://www.eiolca.net/index.html, developed by Green Design Initiative, Carnegie Mellon University, last logon March 21, 2005

Bauman, H. and A–M. Tillman, The Hitch Hiker's Guide to LCA: An Orientation in Life Cycle Assessment Methodology and Applications. Studentlitteratur AB, 2004.

Graedel, T. E., Streamlined Life-Cycle Assessment, Prentice Hall; 1998.

7

Industrial Catalysis: A Practical Guide

Robert J. Farrauto*

THE IMPORTANCE OF CATALYSIS

Every student of chemistry, material science, and chemical engineering should be schooled in catalysis and catalytic reactions. The reason is quite simple; most products produced in the chemical and petroleum industry utilize catalysts to enhance the rate of reaction and selectivity to desired products. Catalysts are also extensively used to minimize harmful byproduct pollutants in environmental applications. Enhanced reaction rates translate to higher production volumes at lower temperatures with smaller and less exotic materials of construction necessary. When a highly selective catalyst is used, large volumes of desired products are produced with virtually no undesirable byproducts. Gasoline, diesel, home heating oil, and aviation fuels owe their performance quality to catalytic processing used to upgrade crude oil.

Margarine, cakes, chocolate, salad oils, and other everyday edible products are produced from natural oils via catalytic hydrogenation. Polyethylene and polypropylene plastics, com-

monly used in packaging of foods, films, fibers, liquid containers, etc. require catalysts for cost-effective high volume production. Because of highly active and selective catalysts polyester fibers used in clothing can be produced at reasonable prices for the mass market. Catalysts enhance the production of ammonia-based fertilizers that enrich the earth's nutrient deficient soils for efficient agriculture. Catalytically produced ethylene oxide is a precursor to antifreeze. Formaldehyde is produced catalytically and used as a preservative and as a component in some polymer resins.

It is good to keep in mind the importance of catalysts in protecting the environment. They are frequently installed in the exhaust ducts from chemical operations to convert volatile organic compounds generated during manufacturing operations, into harmless products. Catalysts also provide the environmental benefit of clean air by abating pollutants generated from stationary and mobile combustion sources. In many locations around the industrialized world coal- and gas-fired power plants have special catalysts installed in the ducts to eliminate pollutants dangerous to our health.

*BASF Catalysts LLC and Columbia University

Many gas-fired compressors that pump natural gas through millions of miles of pipelines are also equipped with exhaust catalysts to clean emissions at moderate conditions. Even fast-food restaurants are being equipped with catalysts to eliminate odors from the cooking process. The most widely used treatment of exhaust pollutants is that of the catalytic converter present in the exhaust manifold that cleans emissions from the internal combustion engines of gasoline- and diesel-fueled automobiles and trucks. As modern commercial passenger jets fly above 30,000 feet there is a need to destroy the few ppm ozone that enters the airplane with make-up air to ensure passenger and crew comfort and safety. Radiators on select vehicles have a catalytic coating deposited on their surface that decomposes harmful ground-level ozone as the vehicle is driven.

All of this gives the consumer the benefits of readily available high quality products at reasonable prices. From food to clothing to medicines to clean energy, catalysts play a major role in products people use in everyday life.

The forthcoming description of catalysts and catalytic processes should only serve as a primer towards understanding the basic principles with some examples of applications in the field of petroleum processing, chemical production, and environmental air purification. Table 7.1 gives a list of some of the many commercial catalytic applications.

TABLE 7.1   Some Commercial Catalytic Reactions

Reaction Name	Example of Chemical Reaction	Major Catalyst Components	Commercial Applications
Petroleum Processing			
1. Cracking	$C_{16}H_{34} = C_8H_{18} + C_8H_{16}$	Faujasite Zeolite	Naphtha, heating oil
2. Hydrodesulfurization	$RS + H_2 = H_2S + RH$	Co, Mo/Al$_2$O$_3$ (S)	Sulfur free fuels
3. Naphtha reforming	$C_6H_{12} = C_6H_6 + 3H_2$	Pt,Re/Al$_2$O (Cl)	High octane gasoline
4. Alkylation	$C_3H_6 + C_4H_{10} = C_7H_{16}$	Liquid acids	Gasoline
Hydrogenation of Functional Groups			
1. Double bonds	$H_2R = R'-R''H_3 + H_2$ $= H_3R-R'HR''H_3$	Ni or Pd or Ru on carriers	Air stable compounds, edible/nonedible oils
2. Aldehydes	$RCHO + H_2 = RCH_2OH$	Pd or Ru or Ni on carriers	Alcohols
Selective Oxidations			
1. Nitrogen dioxide for nitric acid	$4NH_3 + 5O_2 = 4NO_2 + 6H_2O$	Pt, Rh gauze	Fertilizer, explosives
2. Sulfuric acid	$SO_2 + 1/2O_2 = SO_3$	V$_2$O$_5$ on TiO$_2$	Dissolving minerals
3. Ethylene oxide	$CH_2 = CH_2 + 1/2O_2 = C_2H_4O$	Ag on alpha Al$_2$O$_3$	Antifreeze, polyester, fibers, bottles
4. Formaldehyde	$CH_3OH + 1/2O_2 = CH_2O + H_2O$	Bi, Mo or Ag	Monomers, preservatives
Syntheisis Gas/H$_2$			
1. Syntheisis gas	$CH_4 + H_2O = 3H_2 + CO$	Ni/Al$_2$O$_3$	Production of chemicals
2. Water gas shift	$CO + H_2O = H_2 + CO_2$	Fe,Cr and Cu,Zn,Al	High purity H$_2$
Pollution Control			
1. Automotive (TWC)	$CO, HC + O_2 = CO_2 + H_2O$ $CO + NO = N_2 + CO_2$	Pt, Rh or Pd on Al$_2$O$_2$ on a Monolith	Clean emissions
2. Stationary Reduction of Nox	$2NO_2 + 4NH_3 + O_2 = 3N_2 + 6H_2O$	V$_2$O$_5$/TiO$_2$ on a monolith and metal exhanged zeolite	Reduction of NOx from power plants
Polymerization	$nCH_2 = CH_2 = (-CH_2-)_{n+1}$	Cr/SiO$_2$ or TiCl$_4$ + alkyl aluminum halide	Plastics

HOW DOES A CATALYST WORK?

A catalyst increases the reaction rate or activity relative to an uncatalyzed process by providing a less energetic pathway for conversion of reactants to products. In this regard the catalyst provides a chemical and energetic shortcut by lowering the energy barrier (i.e., activation energy) of reactants going to products. If no catalyst were present, higher temperature would be required to initiate the reaction. Higher temperatures often lead to undesirable byproducts and sometimes decomposition of one of the reactants. Therefore by initiating the reaction at a lower temperature the process is more controlled and the desired product can be produced. This is the most important advantage for catalytic processes that is exploited in many product applications.

The catalyst is not consumed in the process it accelerates but does undergo various chemical changes during the process by interacting with the reactants and products. Mechanistically some or all of the reactants adsorb onto active sites of the catalyst where bonds are rapidly made or broken. For a heterogeneous solid catalyst processing a liquid and/or gas, the adsorption of reactants is called chemisorption that has the kinetics and reaction energies equivalent to a chemical reaction. Frequently chemisorbed species decompose to an intermediate that is rapidly converted to other intermediates or the final product. After the reaction is complete the catalyst returns to its original state. In this regard there is no net change of the catalyst. Therefore a very small amount of catalyst can process many molecules.

WHAT ARE THE CATALYTIC METALS AND METAL OXIDES?

Most catalytic metals and metal oxides are derived from Group VIII of the periodic table. Of special importance are Fe, Co, Ni, Rh, Pd, and Pt but also of importance are Cu and Ag in Group 1b, V in Group Vb, and Cr and Mo in Group V1b. Three of the precious metals Rh, Pd, and Pt are extensively used in many industries due to their extremely high activity and selectivity. They are rare in nature and

very expensive, and thus spent catalysts are routinely recycled, purified, and reused. However, the so-called base-metal Fe, Co, Ni, Cu, V, Cr, and Mn but especially Ni and Cu are used for specialty chemical applications. Base-metal catalysts usually have modest activities but are much less expensive and in certain cases more selective than the precious metals. Therefore it is always desirable to search for less expensive base-metal catalysts whenever possible. This has been especially the case for replacing precious-metal-containing automotive emission control catalysts but because of lower activity and stability in the severe environment of an automobile exhaust they are only used as promoters.

More examples of the efficient use of catalytic metals and metal oxides will be given in the applications section of this brief review.

THE STRUCTURE OF HETEROGENEOUS CATALYSTS

The process of chemisorption of reactants requires adsorption on the surface of the catalyst. Therefore to maximize the rate the catalytic surface area should also be maximized. This is achieved by dispersing the catalytic species onto a high surface area inorganic carrier. An ideal dispersion of Ni on Al_2O_3 is shown in Figure 7.1.

Ideally every Ni atom should be accessible to the reactants for maximum efficiency in the conversion process. Although this is possible when the catalyst is first prepared, the dynamics of the catalytic reactions lead to some agglomeration. Catalyst scientists, however, have developed procedures and stabilizers to minimize the extent of agglomeration and therefore dispersed catalysts can be classified as nanomaterials with sizes only slightly greater than 1 nm or 10 Å.

The carrier can be thought of as a sponge with pores from 1 to 100 nanometers (10 to 1000 Å) in diameter. If one were to measure the internal surface area of just 20 grams with an internal surface area of 200–300 m^2/g it would be equivalent to about 1 football field. Carriers such as Al_2O_3, SiO_2, TiO_2, CeO_2, ZrO_2, C, and combinations of these materials

X = Ni ATOMS

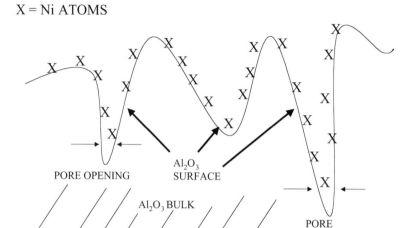

Fig. 7.1.   Ideal dispersion of an active catalyst species on a porous carrier. Example given is Ni on Al_2O_3.

are commonly used. All have different surface properties and are used in applications dependent on the requirement for acidity, inertness to solubility, interactions with reactants, affinity for catalytically active components and resistance to components in the gas phase. High surface area Al_2O_3 is not well suited for combustion reactions in which SO_2/SO_3 are present due to the formation of $Al_2(SO_4)_3$. In such cases high area TiO_2 and/or ZrO_2 are used because of their inertness. Carbons are mostly used for supporting precious metals in hydrogenation reactions. In addition to their chemical role precious metal recovery is achieved simply by burning the carbon.

The most common carrier is gamma alumina (γ-Al_2O_3). It has an internal area of >200–300 m^2/g. Its surface is highly hydroxylated (i.e., Al-$O^{-2}\,H^+$). The H^+ sites provide acidity required for many reactions and exchange sites for catalytic metal cations.

Zeolites are combinations of Al_2O_3 and SiO_2 that are crystalline structures with precisely defined pore structures in the molecular size range (0.4–1.5 nm or 4–15 Å). A related group of materials known as mesoporous silica–alumina has extended the range of pore sizes attainable in ordered SiO_2–Al_2O_3 supports to 4 nm (40 Å). They are commonly used in the chemical and petroleum industry due to their surface acidity and ability to exclude molecules

larger than the pore diameter. For this reason they are often referred to as molecular sieves. Their surfaces contain Al–OH groups with acidic and exchangeable H^+. In the application section some of these materials will be more thoroughly discussed.

Rate-Limiting Steps for a Supported Catalyst

Supporting a catalytic component introduces a physical size constraint dictated by the pore size of the carrier. Thus a key consideration is the accessibility of the reactants to the active catalytic sites within the high surface area carrier. Consider a hydrogenation reaction in which Ni is located in extremely small pores (i.e., 1 nm or 10 Å). The H_2 molecule has easy access but a large molecule, having a size comparable to the diameter of the pore, would experience great resistance moving towards the active sites. If large amounts of Ni are present in pores and are not accessible to the molecules to be hydrogenated the reaction rate will not be enhanced to its fullest potential. Thus the carrier with its geometric sizes and its pore size distribution must be carefully designed to permit the reagents and products to move in the pores with minimum resistance.

Following are the seven fundamental steps in converting a reagent molecule(s) to its product(s) using a supported heterogeneous catalyst.

1. Bulk diffusion of all reactants to the outer surface of the catalyzed carrier from the external reaction media
2. Diffusion through the porous network to the active sites
3. Chemisorption onto the catalytic sites (or adjacent sites) by one or more of the reactants
4. Conversion and formation of the chemisorbed product
5. Desorption of the product from the active site
6. Diffusion of the products through the porous network to the outer surface of the catalyzed carrier
7. Bulk diffusion of the products to the external fluid

Steps 1, 2, 6, and 7 depend on the physical properties of the catalyzed carrier and are not activated processes (no intermediate chemical complex is formed). For this reason we use the term apparent activation energy which is a term useful for comparing temperature dependence as will be described later. Steps 3 through 5 are chemically activated (with intermediate complexes formed during conversion to products) and depend on the chemical nature of the interaction of the reactants and products with the active sites.[1,2]

Step 1 is referred to as bulk mass transfer. It describes the transfer of reactants from the bulk fluid to the surface of the catalyzed carrier. When this step is rate limiting reactant molecules arriving at the external surface of the catalyst are converted instantaneously resulting in zero concentration of reactants at the surface. Thus the internal surface of the catalyst is not used. Such as mass transfer controlled process is nonactivated and we assign an apparent activation energy of less than 2 kcal/mol. Rates vary only slightly with temperature ($T^{3/2}$) which, as will be shown below, allows it to be distinguished from other rate-limiting steps. Step 7 is similar to Step 1 except that the products diffuse from the external surface of the catalyst particle into the bulk fluid. The temperature dependence of

this phenomena is relatively weak and has an apparent activation energy similar to that observed in Step 1 when it is rate limiting. When only the external surface of the catalyst particle is participating in the catalysis it is said to have a low effectiveness factor. The effectiveness factor is defined as the actual rate divided by the maximum rate achievable when all catalytically active sites participate in the reaction. In the case of bulk mass transfer the effectiveness factor approaches zero.

Steps 2 and 6 are both pore diffusion processes with apparent activation energies between 2 and 10 kcal/mol. This apparent activation energy is stated to be about 1/2 that of the chemical rate activation energy. The concentration of reactants decreases from the outer perimeter towards the center of the catalyst particle for Step 2. In this case some of the interior of the catalyst is being utilized but not fully. Therefore the effectiveness factor is greater than zero but considerably less than one. These reactions are moderately influenced by temperature but to a greater extent than bulk mass transfer.

Steps 3, chemisorption of the reactant(s), 4, chemical reaction forming the adsorbed product, and 5, desorption of the product(s) from the active site(s) are dependent on the chemical nature of the molecule(s) and the nature of their interaction with the active site(s). Activation energies are typically greater than 10 kcal/mol for kinetically or chemically controlled reactions. Chemical kinetic phenomena are controlling when all transport processes are fast relative to the reactions occurring at the surface of the active species so the effectiveness factor is one. All available sites are being utilized and the concentration of reactants and products is uniform throughout the particle. These reaction processes are affected by temperature more than either transport mechanisms. Fig. 7.2 shows the conversion versus temperature and concentration profiles of a reagent (R) for the three regimes of rate control.

Because of the significant differences in temperature dependence the kinetically limited reactions can be distinguished from pore diffusion that in turn can be differentiated from bulk mass transfer. This is shown in Fig. 7.2 in

CONVERSION

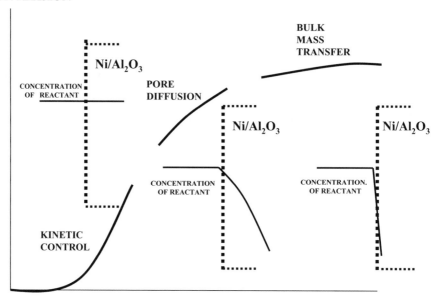

TEMPERATURE

Fig. 7.2. Conversion of a reactant vs. temperature. The concentration of reactants [*R*] within the porous catalyst structure. Concentration of *R* is (a) uniform for kinetic control, (b) decreasing within the catalyst for pore diffusion control, and (c) zero immediately at the surface of the catalyst for bulk mass transfer.

which conversion of reactants is measured against temperature. The first evidence of conversion is the sharply increasing slope that indicates kinetic control whereas pore diffusion shows a lower change in slope as the temperature increases. The bulk mass transfer process shows little change in conversion with increasing temperature. Thus at low temperature the reaction is controlled by chemical reactions (3, 4, or 5) and pore diffusion limited reactions exist when the supply of reactants to the active sites within the particle is limiting the rate (2 or 6). Finally, at the highest temperature, chemical reactions at the external surface are faster than bulk mass transfer (1 or 7) and the reaction is considered limited by bulk mass transfer.

The corresponding concentration of reactant R is also shown for each regime. The concentration of R is constant within the catalyst for kinetically limited reactions. The concentration of reactant gradually decreases within the catalyst particle for the pore diffusion limited case because the rate is limited by transport through the porous network. For bulk mass transfer limited cases the concentration of R is zero at the gas/solid interface.

Activation Energies and Rate-Limiting Steps. The heterogeneous catalyzed NH_3 synthesis from N_2 and H_2 will be used to illustrate the relationship between rate and activation energy. There are a series of steps in the Fe catalyzed process.

The process steps within the pore structure of the catalyst are

1. Diffusion of N_2 and H_2 to the active Fe site within the catalyst pore structure
2. Chemisorption of H_2 on the active Fe surface
3. Dissociation of chemisorbed H_2 to H atoms on the Fe site
4. Chemisorption of N_2 on the Fe site
5. Dissociation of N_2 to N atoms on the Fe surface
6. Surface reaction between adsorbed N and H atoms forming chemisorbed NH_3

7. Desorption of NH_3 from the surface
8. Diffusion of the NH_3 into the bulk gas

Dissociation of chemisorbed N_2 (Step 5) is the slowest and thus is rate limiting.

$$N_2\text{—Fe (active site)} \rightarrow 2N\text{—Fe}$$

The overall rate of reaction is determined by the slowest of these steps. In other words the overall reaction cannot be faster than the slowest step. The slow step and hence the overall reaction rate is characterized by the apparent activation energy.

An important detail is that an individual rate-limiting step may be endothermic whereas the overall reaction is exothermic as in this case. This is illustrated in Fig. 7.3. The chemisorption of N_2 is exothermic and its dissociation is endothermic (1A). However, the overall reaction of $N_2 + H_2$ to NH_3 is exothermic (1B). The overall activation energy and kinetics are dictated by the slow step. The reaction heat liberated ($\Delta H_{25}°C) = -11$ kcal/mole is the thermodynamic value associated with the overall reaction.

It is very important to understand that the catalyst only promotes the rate of a reaction and cannot change the equilibrium concentra-

tions of reactants and products. It cannot make a thermodynamically unfavorable reaction occur. It increases the rate at which equilibrium is achieved while always respecting the thermodynamics of the equilibrium constant and the enthalpy (ΔH) and free energy (ΔG) of the overall reaction. Process conditions (T&P) are changed to give more favorable thermodynamics.

Consider an everyday example of how we are all influenced by rate-limiting steps. If you are driving on a one-lane road behind a slow-moving truck your speed and those behind you is no greater than that of the truck although you certainly have the potential to increase your rate. Thus the time required to arrive at your destination is controlled by the speed of the truck. Taking an analogy, we can liken the truck to a slow chemical reaction step where the overall reaction rate, and the time required to achieve products, is no faster than the speed of the conversion at the surface of the catalyst. Returning to our highway story if you take a bypass road you can increase your rate of speed and decrease the time needed for you to arrive at your destination. The new road is analogous to a catalyst that provides a different pathway to the final product. It is likely, however, that

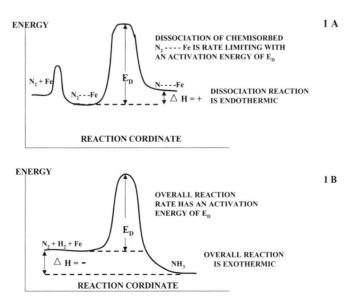

Fig. 7.3. Activation energy diagram for NH_3 synthesis: (1A) chemisorbed N_2 dissociation (rate-limiting step); (1B) overall reaction for NH_3 synthesis.

you will again be limited by another obstacle (narrowing of the new road due to construction) that will slow you and the others behind you, as you maneuver through it. This may be likened to pore diffusion where you are limited by the width of the passage. Mass transfer control can be thought of as reaching the maximum speed your vehicle can safely achieve within the local speed limits. The activation energy reflects the slow step and the kinetics of the overall reaction rate.

Selectivity

In many processes multiple reaction pathways are possible and it is the role of the catalyst and reaction conditions to maximize the rate along the desired path. Selectivity is defined as the amount of desired product divided by reactant converted. A practical example of the catalyst directing reactants to a selective product is shown by the oxidation of ammonia to nitric oxide, which is used in the production of fertilizers.

$$4NH_3 + 5O_2 \rightarrow 4NO + 6H_2O;$$
$$\Delta H_{25} = -54.2 \text{ kcal/mole } NH_3$$

The operating temperature of the process is 900°C and both the standard free energy of $\Delta G_{25} = -57.2$ kcal/mol of NH_3 and the equilibrium constant of $K_{NO} = 10^{10}$ are very favorable.

However, the decomposition pathway to N_2 is even more thermodynamically favorable with $\Delta G_{25} = -77.9$ kcal/mol of NH_3 and an equilibrium constant of $K_{N2} = 10^{15}$ at 900°C.

$$4NH_3 + 3O_2 \rightarrow 2N_2 + 6H_2O;$$
$$\Delta H_{25} = -75.5 \text{ kcal/mol of } NH_3$$

The presence of a PtRh gauze catalyst catalyzes the reactants along the NO pathway with a selectivity of 98%. Therefore although the free energy is more favorable and the equilibrium constant for the N_2 reaction is 10^5 times greater, the highly selective PtRh catalyst promotes the NH_3 oxidation reaction to NO. In contrast the presence of Pd favors the N_2 product. In each case the catalyst respects the equilibrium constant but directs the reactants to specific products.

A second reaction that is currently receiving a great amount of attention because of low-temperature fuel cells is the purification of traces of CO present in a H_2 stream. The fuel cell directly converts chemical energy (H_2 and O_2) to electricity bypassing the mechanical (pistons, turbines, etc.) and combustion steps associated with conventional power generation. The mechanical step limits efficiency and combustion generates pollutants (CO, HC, and NO_x). The heat and power generated from the fuel cell hold promise for powering vehicles and for providing heat and electricity to residential and commercial buildings with the only product being H_2O. H_2 and CO are produced by catalytic steam reforming of a hydrocarbon (e.g., natural gas). The subsequent water gas shift reaction generates more H_2 from the CO + H_2O reaction. Traces of CO exiting the shift reactor must be removed from the H_2 because it poisons the anode of the low-temperature fuel cell. The H_2 content of the gas is about 75%, and the CO is about 0.1% (balance is H_2O and CO_2). Although both standard state free energies are similar, a highly selective Pt containing catalyst promotes the oxidation of the CO with minimum oxidation of the H_2 purifying the latter for a low-temperature fuel cell.

$$CO + \tfrac{1}{2} O_2 \rightarrow CO_2 \quad \Delta G_{25} = -61 \text{ kcal/mole}$$
$$\Delta H_{25} = -68 \text{ kcal/mole}$$

$$H_2 + \tfrac{1}{2} O_2 \rightarrow H_2O \quad \Delta G_{25} = -57 \text{ kcal/mole}$$
$$\Delta H_{25} = -55 \text{ kcal/mole}$$

A small amount of air is injected into the reactor. The inlet H_2/CO ratio is 750 whereas the exit ratio must be 75,000. Thus the free energy for CO oxidation is becoming less favorable (more positive) as CO is reduced below 10 ppm. An effective catalyst currently in use commercially is Pt on Al_2O_3 promoted with a small amount of Fe. It operates at an inlet of 90°C and reduces the CO to less than 0.001% with a selectivity of well over 50% depending on management of the exothermic heat of reaction. This is quite remarkable given the increasingly large excess of H_2 as the reaction approaches completion. The same catalyst, but without the Fe, requires 170°C to achieve the same conversion of CO but with a selectivity less than 25%.

Catalyst Preparation

In the example given above a small amount of Fe is added to a Pt on Al_2O_3 catalyst. The catalyst is prepared by a very unique procedure that must be strictly adhered to in order to achieve the desired results. The Pt and Fe must be in such close proximity that the CO chemisorbs on the Pt and the O_2 on the Fe after which they react to form CO_2.[3] Simply reproducing the composition will not give acceptable performance. The specific details of catalyst preparation may be confidential and are most often covered by patents and trade secrets.

Some general guidelines for supported catalyst preparations are presented below, however, the reader should consult the many references and patents available on the subject.[4] Even by doing so the precise details used by industry to optimize the catalyst will often not be found.

Known amounts of salt(s) of catalytic metals are dissolved in aqueous solutions and impregnated into carrier materials. The wet mass is dried at 110°C and calcined in air at 300–500°C, releasing the decomposable salt components and depositing the metal oxide on the surface within the depths of the porous carrier. For many oxidation reactions the catalyst is now ready for use but for hydrogenation it is necessary to reduce the impregnated metal oxide or salt chemically. Usually this is accomplished by flowing H_2, under conditions consistent with the maximum temperature of use for the reaction of interest.

The carrier can be in the form of a powder used for slurry reactions or a particulate such as a sphere, cylinder, or tablet (typically a few mm in diameter) used in fixed bed reactors. The size and shape depend very much on what is anticipated to be the rate-limiting step. For example, for a reaction limited by pore diffusion it is customary to use a smaller particle in the shape of a star, trilobe, or wagon wheel to decrease the diffusion path while increasing the external geometric surface area. Mechanical strength and solubility

under reaction conditions must be considered in the selection. Although it is often stated that the carrier is inert there are many cases where this is not the case. Some carriers provide acid or basic sites that act as co-catalysts with the metal or metal oxides performing other functions. Petroleum reforming (discussed later) requires a hydrogenation function, provided by the metal, to dissociate H_2 and the carrier provides the acid site to enhance isomerization reactions.

Multi-channel ceramic monoliths (Fig. 7.4) are now the primary choice as support structures to carry the active catalytic species for cleaning emissions from various sources of pollution.[5] Figure 7.4 shows the shapes used for both automotive and stationary pollution abatement applications.

The largest application is the automotive catalytic converter that converts carbon monoxide (CO), hydrocarbons (HC), and nitric oxides (NO_x) to nontoxic compounds. The monolith structure offers high cell densities and thus geometric surface area upon which the catalyst is deposited permitting smaller reactor sizes, high mechanical strength, excellent thermal shock resistance, and low-pressure drop.[5] A powdered carrier, usually stabilized Al_2O_3, is impregnated with catalyst precursor salts. A slurry of the impregnated carrier is prepared and milled to some desirable particle size. The monolith is dipped into the slurry and "washcoated" onto the walls of all of the channel surfaces (see Fig. 7.4). Air is blown through the channels to remove excess slurry. It is then dried and calcined in air at about 500°C. The finished structure now contains the proper amount of catalyst uniformly deposited throughout the channel length. The washcoat thickness is greatest at the corners or fillets of the cell due to its sharp angle. The reactants flow through the channels and catalysis occurs on the washcoated walls. There are many other variations of preparing monolith catalysts with different carriers and compositions. There are monoliths made of metal, some of which have parallel channels and others with nonparallel channels designed for tortuous flow to enhance mass transfer.

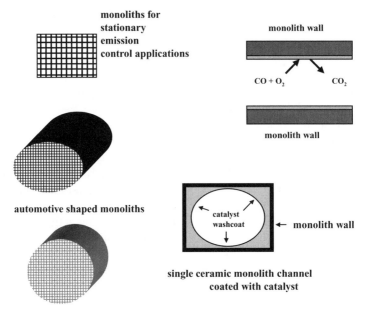

Fig. 7.4. Typical ceramic monolith geometries used for environmental emission control for vehicle and stationary applications.

A HETEROGENEOUS CATALYTIC REACTION: AN EXAMPLE

There is a great desire to use naturally occurring and renewable biomass for producing useful products. Furfural is extracted from cornhusks and contains an aldehyde functional group. If this group is selectively hydrogenated to the corresponding alcohol the product furfuryl alcohol can be used to make corrosion-resistant resins for preparing polymers to make molds for shaping products. This reaction is selectively carried out with a Cu, Cr_2O_3 catalyst (copper chromite powder) in a slurry phase stirred batch reactor (see "Reactor Types" below) at 3000 psig and 150°C.

$$C\text{-}C\text{-}C\text{-}C\text{-}CHO + H_2 \rightarrow C\text{-}C\text{-}C\text{-}C\text{-}CH_2OH$$
$$\backslash \, / \qquad\qquad\qquad \backslash \, /$$
$$O \qquad\qquad\qquad\qquad O$$

Hydrogen gas is dissociatively chemisorbed onto the surface of the Cu-containing catalyst producing highly active hydrogen atoms. The high pressure is needed to ensure adequate solubility of the H_2 in the furfural liquid. The aldehyde functional group forms a weak bond with these active adsorbed atoms and is hydrogenated to the finished product. In the absence of the catalyst the diatomic hydrogen molecule would have to be dissociated in the gas phase at a much higher temperature leading to the decomposition of the aldehyde group.

ACTIVE CATALYTIC SITES

Not all catalytic sites are equal. Ideally each catalytic site is an atom having equal activity. This is never the case for a supported heterogeneous catalyst. One of the great mysteries in catalysis is the exact nature of the active site. Some catalytic species may be so well dispersed that they have no defined structure or are amorphous (no long-range structural order) whereas others may be highly crystalline. Amorphous catalytic components have greater catalytic surface area because fewer atoms are buried within a large crystal. However, the nature of the carrier and the catalytic species and the method used to deposit it on the carrier gives rise to a very heterogeneous surface with different sites having different surface energies and different activities.

For example, it is believed that defects in the crystal structure produce highly energetic and active sites for catalytic reactions. This may be true but the more crystalline the catalytic site the lower is the number of surface atoms and the lower is its catalytic surface area. All this being said there are reactions that favor certain catalyst crystalline sizes and are said to be structure sensitive. The above discussion points to the mystery of catalysis. The goal of finding a universal model describing the nature of the active catalytic site still eludes us today and will undoubtedly be the subject of fundamental research for years to come.

Reactor Types

There are many different reactor designs but the two most commonly used are fixed bed and batch slurry phase. For a fixed bed reactor a given volume of solid particulate or monolith supported catalyst is fixed in a heated tube located within a furnace and liquid and/or gaseous reactants flow through the bed. This type of process is commonly used for large continuous-volume production where the reactor is dedicated to making only one product such as a bulk chemical or petroleum product.

Monolithic supports are commonly used for environmental applications and will be discussed in more detail later.[5] Batch reactors are used mostly for small-scale production such as the hydrogenation of intermediates in the production of medicines in the pharmaceutical industry. The catalyst powder is mixed in a precise amount of reactant in a pressurized-stirred autoclave. A gaseous reactant, usually H_2, is introduced at elevated pressures and the reaction proceeds with continuous monitoring of the H_2 consumed. The catalyst is separated from the product via filtration and is often used again depending on its retained activity and selectivity.

For the production of gasoline and other fuels by catalytic cracking of oils, a fluid bed reactor is used. This is a hybrid of a fixed bed and slurry phase reactor. The catalyst is fluidized as it interacts with the feed to be processed. This application is so important it

will be highlighted in the application section of this review.

Kinetics

The overall kinetics of a heterogeneous catalytic reaction can be controlled by any of the seven steps listed above.[6-9] We can distinguish which is rate controlling by determining the temperature dependence of the reaction. Once we know this we can design the catalyst to enhance the rate of the slowest step.

For example, bulk mass transfer (Steps 1 and 7) can be enhanced by increasing the turbulence of the reactants by increased stirring for a batch process or by increasing the linear velocity (see below) in the case of a fixed bed reactor. Increasing the geometric surface area of the catalyst also favors a reaction limited by bulk mass transfer. This is accomplished by decreasing the particle size of a particulate or by increasing the number of channels in a monolithic structure. Turbulence can be introduced in a monolith channel by modifying the surface from smooth to rough. Because kinetically controlled reactions have a stronger temperature dependence than transport controlled reactions they are affected the most by increasing temperature. Pore diffusion resistance is decreased by increasing the pore size of the carrier or by using a smaller diameter carrier. One may also deposit the active catalytic species nearer the surface of the carrier to decrease the diffusion path length. The rate of a reaction limited by pore diffusion is moderately enhanced with temperature.

For chemically controlled reactions one must modify the catalyst itself by increasing the number of active sites (increasing the catalytic surface area) or finding a crystal size that is more active for a given reaction. Often the activity is increased by the addition of promoters to the catalyst (i.e., Fe addition to Pt described under "Selectivity") that enhance the activity. Having the highest activation energy it is affected more than the transport regimes by increasing the temperature. Many examples of this will be given in the example section of this chapter.

General Kinetic Rate Equations.   The rate of a bi-molecular reaction is given by

$$\text{Rate} = dA/dt = -[k_F (A)^a (B)^b \\ - k_{REV} (D)^d (E)^e]$$

Rate is the disappearance of reactants with time expressed as the derivative $-d[A]/d(t)$. $[A]$ and $[B]$ are the concentrations of reactants and $[C]$ and $[D]$ are the concentrations of the products. The exponents a, b, c, and d are the reaction orders for each compound. The rate constants are k_F for the forward and k_{REV} for the reverse reaction. For those cases where the reaction is far away from equilibrium the reverse rate is negligible and this term is dropped from the rate expression.

To determine the rate constant and the reaction order at a specific temperature it is often convenient to increase the concentration of one reactant at least 20 times that of the other to maintain it relatively constant during the reaction. Thus with a high concentration of reactant B one my write $k_F [B]^b = k_F^*$

$$d[A]/dt = -k_F^* (A)^a$$

If the reaction order is to be determined one may take the natural log of the rate equation and obtain

$$-\ln(d[A]/dt) = \ln k_F^* + a \ln[A]$$

A plot of the $-\ln(d[A]/dt)$ versus $\ln[A]$ will produce a straight line with a slope equal to a and intercept $\ln k_F^*$

If one assumes $a = 1$ and integrates the rate expression

$$\int d([A]/[A]) = -k_F^* \int d(t)$$

Integration from the initial concentration A_o to A at anytime and from $t = 0$ to t

$$\ln([A_o]/[A]) = k_F^* t$$

$$[A] = [A_o - xA_o] \quad \text{where } x \text{ is the fraction converted}$$

Plotting $A_o \ln[1/(1 - x)]$ versus t will give a straight line with a slope equal to k_F^*

Kinetics for Fixed Bed Continuous Reactions. For continuous flow reactors we use the term space velocity (SV) defined as the volumetric

flow rate at *STP* divided by the volume of catalyst. That ratio yields the reciprocal of the residence or space-time (t)

$$SV(h^{-1}) = \frac{\text{flow rate cm}^3/\text{h}(STP)}{\text{catalyst volume (cm}^3)}$$

$$\frac{1}{SV} = \text{residence time or space-time}$$

Thus the rate equation for continuous-flow-reactions is

$$\ln([A_o]/[A]) = k_F^* \ t = k_F^*/SV$$

The linear velocity (LV) or superficial velocity is an important engineering term because it relates to pressure drop and turbulence. This parameter is often increased in fixed bed reactors to enhance bulk mass transfer and heat transfer.

$$LV \text{ (cm/h)} = \frac{\text{flow rate cm}^3/\text{h}(STP)}{\text{frontal area of catalyst (cm}^2)}$$

Kinetics of a Slurry Phase Reaction in a Batch Process.   This example is for the liquid phase hydrogenation of nitrobenzene to aniline with a powdered catalyst. These reactions typically are controlled by the supply of H_2 to the active sites.

$$3H_2 + C_6H_5NO_2 \rightarrow C_6H_5NH_2 + 2H_2O$$

H_2 must be

1. Transported from the bulk gas phase and dissolved in the liquid nitrobenzene.
2. Diffuse to the outside of the catalyst particle and into the pore structure.
3. H_2 and nitrobenzene react at the catalytic site.
4. Products diffuse through the pores and into the bulk liquid.

Steps 1, 2, and 4 are mass transfer phenomena while step 3 is kinetic.

At steady state the rate of mass transfer of reactants $(Rate)_M$ is equal to the kinetic rate $(Rate)_R$. This assumes Step 4 is fast and not rate limiting.

$$(Rate)_{net} = (Rate)_{MT} = (Rate)_R$$

$$(Rate)_{MT} = k_m (H_{2g} - H_{2s})/H_{2g} \\ = 1 - (H_{2s}/H_{2g})$$

where H_{2g} = H_2 concentration in the gas
H_{2s} = H_2 concentration at catalyst surface
k_m = Mass transfer rate constant

$$(Rate)_R = k_R (H_{2s}Q)/H_{2g}$$

where k_R = kinetic rate constant
Q = the amount of catalyst

Equating $(Rate)_{MT}$ and $(Rate)_R$ and rearranging one obtains

$$(Rate)_{net} = \frac{k_R k_m Q}{k_R Q + k_m}$$

Taking the inverse for the general rate equation and dividing both sides by $k_R k_m Q$ one obtains

$$\frac{1}{(Rate)_{net}} = \frac{1+1}{k_M k_R Q}$$

$$(Rate)_{net}^{-1} = k_M^{-1} + (k_R Q)^{-1}$$

A plot of inverse $(Rate)_{net}$ versus the inverse of Q yields a straight line with the slope equal to the inverse of k_R and the intercept the inverse of k_m. This is shown in Fig. 7.5.

When the amount of catalyst Q is large $k_R Q \gg k_m$

$$(Rate)_{net} = k_m$$

The rate is limited by mass transfer because the reactants are consumed immediately at the outer surface of the catalyst. For small amounts of catalysts $k_m \gg k_R Q$

$$(Rate)_{net} = k_R Q$$

The reaction is kinetically controlled limited by the amount of catalyst.

Arrhenius Equation. The general rate constant (k) is an exponential function of temperature as described by the Arrhenius equation

$$k = k_o \exp^{(-E/RT)}$$

where
E = Activation energy for chemical control ("apparent" for diffusion limited processes)
R = Universal gas constant
T = Absolute temperature
k_o = Absolute rate constant

Taking the natural log of the equation gives

$$\ln(k) = \ln(k_o) - E/RT$$

The plot of ln (k) versus T^{-1} gives a straight line with a slope equal to $-E/R$ and intercept the absolute rate constant k_o as shown in Fig. 7.6. The lowest slope represents reactions controlled by bulk mass transfer, and the largest is

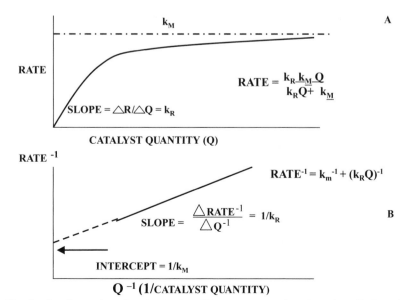

Fig. 7.5.   Kinetics for slurry phase/batch reaction: (A) rate vs. quantity of catalyst; (B) line plot of the rate expression.

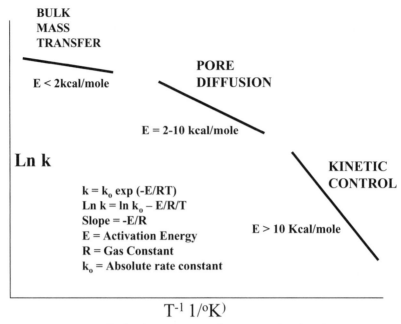

Fig. 7.6. Arrhenius profile for three regimes of rate control with activation energies (E).

for chemical or kinetic control. This method allows for the comparison of different rate-limiting steps but it must be clearly understood that diffusion processes are not activated and thus we use the term apparent activation energy for them only to allow comparisons to activated processes such as chemically controlled processes.

Rate Models

The Langmuir–Hinshelwood kinetic model describes a reaction in which the rate-limiting step is reaction between two adsorbed species such as chemisorbed CO and O reacting to form CO_2 over a Pt catalyst. The Mars–van Krevelen model describes a mechanism in which the catalytic metal oxide is reduced by one of the reactants and rapidly reoxidizd by another reactant. The dehydrogenation of ethyl benzene to styrene over Fe_2O_3 is another example of this model. Ethyl benzene reduces the Fe^{+3} to Fe^{+2} whereas the steam present reoxidizes it, completing the oxidation–reduction (redox) cycle. This mechanism is prevalent for many reducible base metal oxide catalysts. There are also mechanisms where the chemisorbed species reacts

with a gas phase molecule and the combination rapidly converts to the final product. There are many kinetic models that describe different mechanisms and the reader is directed to some outside references.[6–9]

Catalyst Deactivation

The first indication of catalyst deactivation is a significant change in the activity/selectivity of the process. Catalyst deactivation occurs in all processes but it often can be controlled if its causes are understood. This subject is very extensive and the reader is encouraged to seek additional information in references given here.[10,11] In the following we will present some of the most common deactivation modes especially for heterogeneous catalysts. These are pictorially shown in cartoon form in Fig. 7.7.

Sintering of the Active Components. Catalytic scientists go to great lengths to disperse the active catalytic species over the surface of a carrier to maximize the number of sites available to the reactants. Small particles or crystallites have a high surface-to-volume ratio that is a highly unstable thermodynamic

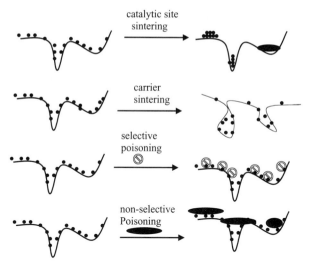

Fig. 7.7. Idealized poisoning mechanism: (a) sintering of the catalytic components, (b) sintering of the carrier, (c) selective poisoning, and (d) nonselective poisoning.

condition. The simple principle of Ostwald ripening indicates that small crystallites tend to grow to larger ones to bring the surface to volume condition to a favorable low free energy state. Thermal sintering occurs when small particles of active catalyst migrate over the surface of the carrier and agglomerate to form larger particles. There are other mechanisms of sintering but conceptually this is the easiest to understand. The net effect is the loss of catalytic surface area that leads to loss of activity. The most frequently encountered cause is high temperature. This condition is encountered in Pt-, Pd-, and Rh-containing catalytic converters present in automobile exhausts where temperatures close to 1000°C are commonly experienced. An oxidizing environment promotes the sintering of Pt by the formation of highly mobile or volatile Pt oxides. PdO on the other hand, tends to form a stronger bond with the Al_2O_3 surface and thus sintering is not significant at modest temperatures. In contrast it does sinter more readily in reducing environments. A catalytic species strongly bound to the surface is less likely to sinter. For this reason a carrier such as SiO_2, which contains few OH groups on the surface relative to Al_2O_3, leads to sintering of the supported metal or metal oxide more readily. Catalyst companies have incorporated "rare earth stabilizers" into

the formulations to minimize the rate of growth of the metal and metal oxide components. Stabilizers slow the rate of sintering but do not completely prevent it due to the thermodynamic nature of the phenomenon. The goal is to minimize the rate to acceptable levels to ensure acceptable life of the catalyst.

Carrier Sintering. The purpose of the carrier is to provide a high surface area upon which the catalytic components can be dispersed. The high surface area leads to sintering by collapse of the pore structure that subsequently blocks (or occludes) the active sites by preventing access of the reactant. For some carriers such as Al_2O_3 there are changes to the crystal structure that occur as the temperature is increased. The most common is the conversion of high surface area (gamma) γ-Al_2O_3 (200m²/g) to low area (alpha) α-Al_2O_3 (1–5 m²/g) at temperatures greater than about 800°C. This process occludes the catalytic components within the carrier and prevents the reactants from having access. The easiest analogy to understand is the truck that breaks down at the tunnel entrance; it prevents other vehicles from entering. High temperatures and steam are two of the most significant contributors to carrier sintering. Catalyst companies have incorporated metal oxides, such as Ba and La in precise percentages, into the carrier to minimize the sintering rate.

Poisoning. Specific components present in the reactant feed can adsorb selectively onto active catalytic sites rendering them inactive, in much the same way as CO can react with Fe-hemoglobin in the blood. For heterogeneous catalysts sulfur compounds are the most universal poisons for both base metal catalysts and to a lesser extent precious metals. Sulfur compounds present in petroleum, chemical, and environmental streams adsorb on the surface of Ni, Cu, Co, etc. forming metal sulfides that have little or no activity. In general poisoning by sulfur compounds is irreversible. For this reason upstream processes are used to reduce the sulfur to acceptable levels.

Sulfur oxides (SO_2 and SO_3) present in flue gases from upstream combustion operations adsorb onto the catalyst surface and in many cases form inactive metal sulfates. It is the presence of sulfur compounds in petroleum-based fuels that prevent the super-sensitive base metal catalysts (i.e., Cu, Ni, Co, etc.) from being used as the primary catalytic components for many environmental applications. Precious metals are inhibited by sulfur and lose some activity but usually reach a lower but steady state activity. Furthermore the precious metals are reversibly poisoned by sulfur compounds and can be regenerated simply by removing the poison from the gas stream. Heavy metals such as Pb, Hg, As, etc. alloy with precious metals and permanently deactivate them. Basic compounds such as NH_3 can deactivate an acidic catalyst such as a zeolite by adsorbing and neutralizing the acid sites.

Water is a reversible poison in that it will weakly adsorb (physically adsorb) on sites at low temperature but readily desorbs as the temperature is increased.

One interesting example of different selective poisoning mechanisms is that of SO_3 deactivation of Pt on Al_2O_3 used for abating emissions from combustion reactions. The Pt oxidizes the SO_2 to SO_3 and the latter adsorbs onto the Al_2O_3 forming a sulfate. Slowly the carrier surface becomes so sulfated that it occludes the Pt within the pores and the cata-

lyst slowly deactivates. By using a nonsulfating carrier such as TiO_2 or ZrO_2 deactivation can be prevented. In contrast SO_3 directly adsorbs on Pd sites and deactivation occurs rapidly.

Poisoning is not always bad. There are situations where a catalyst is intentionally poisoned to decrease activity towards an undesirable reaction. In the hydro-desulfurization and -demetallization of a petroleum feedstock the catalyst is presulfided prior to introducing the feed to decrease its activity and minimize cracking reactions that will produce unwanted gases. Another is the use of ammonia to slightly poison a Pt catalyst used in the hydrogenation of fats and oils to decrease undesirable oversaturation.

Nonselective poisoning or masking is caused by debris depositing on the surface of the catalyst physically blocking sites. Corrosion products from the reactor walls and contaminants such as dust, oil, etc. can be eliminated by careful filtration upstream, but this mechanism of deactivation is a constant problem in many applications. Regeneration is possible for precious metal oxidation catalysts designed to abate volatile organic compounds (VOC) from flue gases. The reactor is bypassed when the activity begins to decline to unacceptable levels. High-velocity air is passed through the catalyst bed and loosely held debris is dislodged. In some cases chelating solutions are used to solubilize the metal contaminants such as Fe without destroying the catalyst. Coking is a common phenomenon when petroleum and/or high molecular weight chemical compounds are processed. Hydrogen-deficient-hydrocarbons are formed from undesirable side reactions and block access to the catalytic sites deep within the pores of the catalyst. This deactivation mode has been positively integrated into the fluid catalytic cracking process for converting heavy oils to useful products. The coked catalyst is regenerated with air in a separate reactor and the heat liberated used to preheat the feed as it enters the cracker.

CATALYST CHARACTERIZATION

The goal of catalyst characterization is to relate the physical and/or chemical properties of the catalyst to performance. Some of the most important catalytic properties are physical surface area, pore size distribution, active catalytic surface area, the morphology or crystal structure of the carrier and active components, the location of the active components within the carrier, and the presence of surface contaminants or poisons on the surface. Fortunately there are many instrumental tools readily available in modern laboratories to measure these properties for fresh and spent catalysts. There are many reference books and monographs available that describe the strengths and limitations of the instrumental methods used in characterizing catalysts.[12,13]

The chemical composition can be measured by traditional wet and instrumental methods of analysis. Physical surface area is measured using the N_2 adsorption method at liquid nitrogen temperature (BET method). Pore size is measured by Hg porosimetry for pores with diameters larger than about 3.0 nm (30 Å) or for smaller pores by N_2 adsorption/desorption. Active catalytic surface area is measured by selective chemisorption techniques or by x-ray diffraction (XRD) line broadening. The morphology of the carrier is viewed by electron microscopy or its crystal structure by XRD. The active component can also be measured by XRD but there are certain limitations once its particle size is smaller than about 3.5 nm (35 Å). For small crystallites transmission electron microscopy (TEM) is most often used. The location of active components or poisons within the catalyst is determined by electron microprobe. Surface contamination is observed directly by x-ray photoelectron spectroscopy (XPS).

Making the characterization measurements is of critical importance in the diagnostics of the catalysts but interpreting those most responsible for changes in activity or selectivity requires experience and good comparative kinetics for fresh and aged materials. It should be standard practice to compare fresh and aged catalytic performance with the changes observed in your characterization diagnostics. Measuring rate-limiting steps and activation energies will provide invaluable insight into the major causes of deactivation.

HOMOGENEOUS CATALYTIC REACTIONS

In a homogenous catalytic reaction the reactants and catalysts are in the same phase. The catalyst is a metal (Rh, Co, Ni, etc.) chelated with organic ligands (often phosphine-containing) soluble in the reaction media and because no support is used, pore diffusion does not exist. However, bulk mass transfer is a concern especially when the reaction is a hydrogenation because H_2 must be dissolved in the liquid and make contact with the catalyst. This is accomplished by using high pressure and vigorous stirring. Homogeneous catalysis is most often used in the pharmaceutical industry where the desired selectivity can only be achieved with active complexes. A significant issue is separation of the catalyst from the final product to achieve the required purity. Furthermore, recovery of the catalyst is most often necessary especially for expensive precious metal containing complexes such as Rh. Distillation is sometimes used provided there is a significant difference in vapor pressure of the product from the catalyst. The catalyst is also recovered by ion exchange with a suitable sequestering agent such as an amine compound. The efficiency of the separation allows for catalyst reuse and is essential for an economic process.

An example will be given in "Commercial Applications."

Commercial Applications

There are literally hundreds of commercial catalytic processes carried out for high and low volume premium products. Only a few have been selected below as examples of everyday products essential for a high quality life. Table 1 also presents listings of some of the major catalytic processes but the reader is directed to references given in this review for a more complete listing.[9]

Petroleum Processing

Hydro-Demetallization (HDM) and -Desulfurization (HDS) of Heavy Oils. The hydrocarbon petroleum fractions contain varying amounts of inorganic impurities such as nickel, vanadium, and sulfur-containing compounds all of which must be removed to make high-quality products both functionally and environmentally.[14, 15] Metals, if present in gasoline or diesel, will create significant engine wear and the sulfur would produce sulfur oxides during combustion and ultimately sulfuric acid in the atmosphere. Furthermore they will deactivate the catalysts used in the petroleum upgrading processes and in their ultimate application as a fuel will damage the performance of the abatement catalyst.

Crude oil contains about 0.01% metals and up to 5% sulfur present in large aromatic structures. These levels are highly dependent on the origin of the crude. For example, California crude is relatively low in sulfur but higher in metals than crude from Kuwait. Any process to remove them must be economical with little destruction of the hydrocarbons and minimum consumption of H_2. The catalyst is Co, Mo/Al_2O_3 with particles a few mm in diameter. Although sulfur is usually a poison for catalytic reactions it is used here in a positive function to control selectivity. It is presulfided to decrease activity towards excessive consumption of H_2 that leads to unwanted saturation of aromatic molecules.

$$S\text{-Co, Mo}/Al_2O_3$$

$$R\text{-M} + H_2 \rightarrow M + R\text{-H}$$

$$R'S + H_2 \rightarrow R\text{-H} + H_2S$$

R and R′ = organic host of metals and sulfur
M = metal (Ni or V)

The hydrogenation process is carried out at 500°C and pressures in excess of 30 atm in fixed bed reactors containing catalysts with varying physical properties to accommodate the metal deposition that occurs during the reaction. In some cases moving bed reactors are used where spent catalyst is continuously removed and fresh catalyst added.

The first reactor contains the Co, Mo deposited on a low surface area Al_2O_3 with large pores to allow deep penetration of the metals into the particle. The second bed will treat a feed with less metal so its pore size is smaller and

surface area slightly larger. The metal penetration here is less deep than in the first bed and allows for some hydro-desulfurization. The final bed contains the highest surface area and smallest pores and is designed to perform most of the desulfurization.

The catalyst is regenerated frequently during its useful life but once spent it is leached and the metals recovered.

Catalytic Cracking for the Production of Useful Fuels. Gasoline and diesel fuel, home and commercial heating oil, kerosene, jet fuel, etc. are all produced by catalytically cracking fractions of distilled crude oil. Crude oil is distilled in large vertical towers where the various fractions present are separated according to their boiling ranges. The light gases (C_3 and C_4 propane and butane, respectively) are distilled first and the light/heavy naphtha fraction (C_5 to C_{10} pentanes to branched cyclopentanes), the precursors to gasoline, distill between roughly 70 and 200°C. Diesel fuel and heating oils (No. 1 and 2) are collected between 200 and 340°C. The remaining heavy hydrocarbons (called vacuum distillates) are used for lubricants and road construction.

The composition and molecular weight distribution of the crude oil depends on its origin but generally less than 50% is within the molecular and boiling range for transportation and heating fuels. Thus the role of the cracking process is to break or crack the higher molecular weight fractions into lower molecular weight compounds to be used for more useful products. Therefore the catalyst is at the heart of the refining industry.

Cracking Catalysts. The catalysts used for cracking are called zeolites.[16,17] They are SiO_2-Al_2O_3 materials in which Si, in its tetrahedral SiO_2 structure, is replaced with Al cations. They are produced by reacting sodium silicate with a water-soluble salt of Al followed by hydrothermal treatment in an autoclave. The zeolite is unique in that it has a well-defined crystal structure with a precise pore size (or aperture) ranging typically from 0.3–4 nm or 3–40 Å. This unique pore structure is responsible for separating molecules in accordance with their cross-sectional area. A molecule smaller than the aperture can enter the interior although a larger one cannot. Hence the term molecular sieve is often used to describe zeolites. The composition and pore size can be varied giving rise to a large number of different zeolites with different pore sizes and crystal structures. They are usually identified by the Si/Al ratio, the crystal structure, and the size and shape of the pore. The Si^{+4} is bonded to 4 O^{-2} and each is bonded to another Si^{+4} establishing charge neutrality. Substituting Al^{+3} for Si cation upsets charge neutrality and requires another positive charge to satisfy the oxygen ions.

$$\begin{matrix} | & | & | & | \\ O & O & O & O \\ \text{-O-Si-O-Si-O-Si-O-Al-O}^- \text{M}^+ \\ O & O & O & O \\ | & | & | & | \end{matrix}$$

Neutrality is satisfied by a cation (e.g., M^+) which is usually Na^+ derived from the salts used in the synthesis. When the cation is exchanged with a proton an acid site is created. This is the key active site for catalytic cracking reactions. The first exchange is with NH_4^+ which when heat-treated decomposes to NH_3 and the H^+ is retained on the zeolite. The acid zeolite is designated HZ

$$\begin{matrix} | & | & | & | \\ O & O & O & O \\ \text{-O-Si-O-Si-O-Si-O-Al-O}^- \text{H}^+ \\ O & O & O & O \\ | & | & | & | \end{matrix}$$

The active zeolite for cracking reactions is called Faujasite and is classified as an X zeolite (HX). It has a Si/Al ratio of 1.0–1.5 with a pore or aperture size of 0.74 nm or 7.4 Å forming an aperture composed of 12 oxygen anions as shown in Fig. 7.8. The midpoint of each line represents an O^{-2} bonded to either Si^{+4} or Al^{+3}. It is the AlO^- site that requires a metal cation for charge balance. For cracking catalysts these sites are H^+. The higher the Al content (lower Si/Al) the greater the number of acid sites, but the lower the thermal stability. The zeolite is embedded within an amorphous SiO_2-Al_2O_3 structure that initiates the cracking of the large molecules but also captures impurities

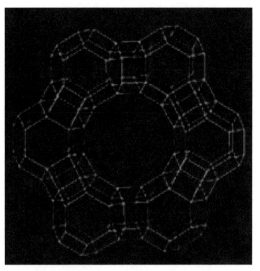

Fig. 7.8. Faujasite zeolite.

such as organic compounds containing Ni and V that will severely deactivate the zeolite. Having its own acid sites, it also functions to break large molecules into smaller sizes where the zeolite can polish them to desired products. Catalyst particle sizes vary between 50 and 100 microns depending on the fluidization dynamics of the process.

During the fluidized catalytic cracking (FCC) process a C–C bond is broken and a proton transferred from the catalyst to the molecule forming a positively charged carbo-cation. This ion can react with other hydrocarbons transferring its proton generating new carbo cations. Ultimately the large molecule is cracked to a smaller alkane and alkene with the regeneration of the protonated zeolite completing the catalytic cycle.

$$\text{Paraffin cracking: } C_{16}H_{34} \xrightarrow{\text{HX}} C_8H_{18} + C_8H_{16}$$
Catalyst HX

$$\text{Dealkylation: } C_6H_5\text{-}CH_2CH_3 \xrightarrow{\text{HX}} C_6H_5H + CH_2 = CH_2$$
Catalyst HX

Excessive extraction of H leads to the formation of hydrogen-deficient, high-boiling hydrocarbons called coke. Coking reactions are catalyzed by acid. The coke masks the surface

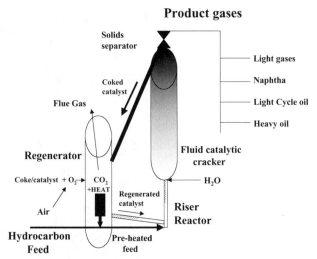

Fig. 7.9.   Schematic of Fluid Catalytic Cracking (FCC) reactor with catalyst regenerator.

and blocks the pore of the catalyst preventing access of the feed molecules leading to a loss in activity.

Cracking is carried out in a fluid bed process as shown in Fig. 7.9. Catalyst particles are mixed with feed and fluidized with steam upflow in a riser reactor where the reactions occur at around 500°C. The active life of the catalyst is only a few seconds because of deactivation caused by coke formation. The deactivated catalyst particles are separated from the product in a cyclone separator and injected into a separate reactor where they are regenerated with a limited amount of injected air. The regenerated catalyst is mixed with the incoming feed which is preheated by the heat of combustion of the coke.

Zeolites play a major role as catalysts and/or adsorbents in the petroleum, chemical, and lately in a growing number of environmental applications. The reader should consult references available.[17]

Naphtha Reforming for High-Octane Gasoline.   Gasoline is volatilized and injected into the cylinders of the internal combustion engine where it is ignited under compression by a spark plug in the power stroke. Maximum power occurs when the cylinder reaches top dead center (maximum compression) and the mixture ignited by the spark plug. A high-octane gasoline is formulated not to preignite

before reaching top dead center during compression to avoid the pinging or "knocking" sound that detracts from power. Before the mid-1970s tetraethyl lead was added to quench preignition reactions but because lead is no longer permitted the gasoline must be formulated to resist combustion until initiated by the spark. High-octane compounds such as aromatics and branched-paraffins are used in place of lead compounds. Today oxygenates are added to boost octane allowing decreases in carcinogenic aromatics.

Fuel-quality gasoline is made by a process called catalytic reforming[18,19] in which molecules in the gasoline boiling range (called naphtha) are isomerized, dehydrogenated, and aromatized to make high-octane products. The most widely used reforming catalyst is Pt, Re on chlorinated Al_2O_3 particles (3–5 mm diameter). The Pt is the active component primarily for dehydrogenation and aromatization reactions and the Cl adds to the acidity of the carrier and is the active site for isomerization. The Re is believed to minimize coke formation. Dehydroisomerization requires both metal and acid functions. Some reactions are endothermic (dehydrogenation and dehydroisomerization) and others are exothermic (isomerization and dehydroaromatization). One can see below that the reactions lead to an increase in octane number

The dehydrogenation of cyclohexane to benzene and H_2 increases the octane number from 75 to 106.

$$\text{Pt site}$$
$$C_6H_{12} + HEAT \rightarrow C_6H_6 + 3H_2$$

Isomerization of *n*-butane to *i*-butane increases octane from 94 to 101.

$$\text{acid site} \quad \overset{CH_3}{}$$
$$CH_3CH_2CH_2CH_3 \rightarrow CH_3CHCH_3 + HEAT$$

Heptane has a defined octane number of 0 and when dehydroaromatization occurs toluene is formed with an octane number of 116.

$$\text{Pt + acid site}$$
$$CH_3CH_2CH_2CH_2CH_2CH_2CH_3 \rightarrow C_6H_5CH_3$$
$$+ 4H_2 + HEAT$$

The formation of benzene by the dehydroaromatization coupled with isomerization of methyl cyclopentane also increases octane from 76 to 106.

$$\text{Pt + acid sites}$$
$$CH_3\text{-}C_5H_9 + HEAT \rightarrow C_6H_5H + 3H_2$$

The reforming process operates with three or four reactors in series. The feed is delivered to the first reactor at 500°C that is charged with the smallest amount of catalyst (5% of the total and the highest space velocity) to perform the easy but highly endothermic dehydrogenation reactions. To minimize coke formation a small amount of H_2 is recycled from the product. The products and unreacted feed are then reheated to 500°C and fed to a second bed containing about 15% of the total catalyst charge where isomerization reactions occur. The unreacted feed and product are then reheated to 500°C where the more difficult dehydroisomerization reactions take place with 20% of the total catalyst charge. The final reactor contains 60% of the total catalyst charge and performs dehydrocyclization. Swing reactors are in place to allow the process to continue as each bed is being regenerated by coke burn-off. After regeneration the catalyst must be rejuvenated by the addition of chloride. The final step is reduction of the metal to its active state.

CATALYSTS FOR CONTROLLING AUTOMOTIVE EMISSIONS

Oxidation Catalysts to Abate Unburned Hydrocarbon and CO Emissions

Catalytic converters were first installed in U.S. cars in 1976.[20, 21] They were passive devices in that they were simply placed in the exhaust with no communication with the engine or its control strategy. It catalyzed the oxidation of the unburned hydrocarbons (C_yH_n) and carbon monoxide (CO) emitted during the incomplete combustion of the fuel. In some vehicles excess air was pumped into the exhaust to ensure sufficient oxygen to complete the catalytic oxidation. This resulted in about a 90% reduction of these two pollutants relative to the uncontrolled uncatalyzed vehicle.

$$CO + 1/2O_2 \rightarrow CO_2 + HEAT$$

$$C_yH_n + (1 + n/4)O_2 \rightarrow yCO_2 + n/2H_2O$$
$$+ HEAT$$

The presence of the catalyst provides a lower-energy chemical path than that offered by a thermal reaction. A catalyst accelerates oxidation of hydrocarbon/carbon monoxide/air mixtures that lie outside the flammability range required for thermal reactions. In the exhaust of the automobile the composition of the pollutants is far below the flammability range yet the oxidation reactions occur by the catalyst providing a lower-energy chemical path to that offered by the thermal reaction. An excellent example is the oxidation of CO with and without a catalyst. Without a catalyst the rate-limiting step is O_2 dissociation at 700°C followed by reaction with gas phase CO. In the presence of the Pt catalyst O_2 dissociation is rapid and the rate-limiting step becomes the surface reaction between adsorbed O atoms and CO that occurs below 100°C.

Two approaches were used in the design of the converters both of which were positioned in the exhaust physically under the driver's seat. Both used precious metals (Pt and Pd) as the active catalytic components dispersed on Al_2O_3 (stabilized against carrier sintering with 1–2% CeO_2 and sometimes alkaline metal oxides). One major automobile company used catalyzed Al_2O_3 beads (4 mm in diameter) and a spring-loaded pancakelike vessel to decrease the linear velocity and thus pressure drop. This design decreases backpressure that detracts from the power by offering less resistance to flow. Another used a new ceramic monolithic structure with hundreds of parallel channels (see Fig. 7.4). Upon the walls was deposited a coating of stabilized Al_2O_3 containing the active precious metals.. The cordierite structure ($2MgO-5SiO_2-2Al_2O_3$) has a melting point over 1300°C sufficiently high to withstand the expected temperatures in the exhaust. It was extruded and had excellent resistance to breakage due to thermal shock experienced during the transient operation of the vehicle. The cellular structure had between 200 and 400 cells per square inch (cpsi) parallel to the flow. With channel diameters of 0.059 inches (200 cpsi) and 0.044 inches (400 cpsi) they had open frontal areas of about 70% offering little resistance to flow and thus low back-pressure. It was incorporated into the exhaust system with retainer rings and surrounded by layers of insulation to minimize breakage due to vibration and heat. The regulations required that the converter have a life of 50,000 miles. To ensure this life it was necessary to remove the tetraethyl lead, used to boost octane, in gasoline because the Pb poisoned the Pt and Pd by alloy formation.

Oxidation catalysts were used until 1979 in both the particulate (bead) form and monolith structure. Road experience demonstrated that the particulate beds were not mechanically stable and were breaking apart. In contrast the washcoated monoliths were found to be highly reliable so they became the structure of choice.

Three-Way Catalytic Conversion

In 1980 additional regulations imposed by the U.S. Environmental Protection Agency (EPA) required control of NO_x (NO, NO_2, N_2O) emissions. Its removal coupled with the continuing need to remove CO and C_yH_n proved to be quite challenging because the latter had to be oxidized and the former reduced. Thus it appeared two separate environments were needed. This problem was solved by the development of the three-way catalyst or TWC capable of catalyzing the conversion of all three pollutants simultaneously provided the exhaust environment could be held within a narrow air-to-fuel range. This is shown in Fig. 7.10.

This range was defined between the fuel-lean and fuel-rich sides of the stoichiometric point, where the amount of O_2 is precisely

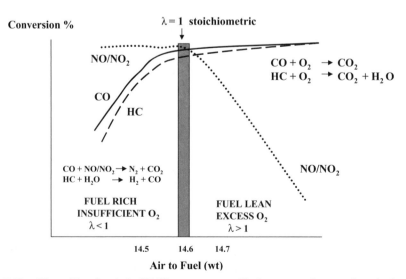

Fig. 7.10. Three-Way Catalytic (TWC) converter profile for conversion vs. air-to-fuel ratio.

sufficient for oxidizing both the CO and C_yH_n. This control required an O_2 sensor that is discussed below.

In the TWC the Pt functions primarily as the catalyst for the oxidation reactions and the Rh catalyzes the NO_x reduction.

$$CO + NO \text{ (or } NO_2) \to 1/2\ N_2 + CO_2$$

$$H_2 + NO \text{ (or } NO_2) \to 1/2\ N_2 + H_2O$$

$$C_yH_n + (2 + n/2)\ NO \text{ (or } NO_2) \to (1 + n/4)$$
$$N_2 + yCO_2 + n/2H_2O$$

The second reaction requires H_2 that is produced catalytically by the steam reforming reaction that occurs when excess C_yH_n is present.

$$C_yH_n + (n - 2)H_2O \to (n + 1)H_2 + yCO_2$$

O_2 Sensor. The control of the exhaust composition was essential to maintain the air-to-fuel ratio close to stoichiometric for simultaneous conversion of all three pollutants. This control came about with the invention of the O_2 sensor.[21,22] The sensor head of this device was installed in the exhaust immediately at the inlet to the catalyst and was able to measure the O_2 content instantly and precisely. It generates a voltage consistent with the Nernst equation in which the partial pressure of O_2 $(PO_2)_{exhaust}$ in the exhaust develops a voltage (E) relative to a reference. The exhaust electrode was Pt deposited on a solid oxygen ion conductor of yttrium-stabilized zirconia (ZrO_2). The reference electrode, also Pt, was deposited on the opposite side of the electrolyte but was physically mounted outside the exhaust and sensed the partial pressure $(PO_2)_{ref}$ in the atmosphere. E_o is the standard state or thermodynamic voltage. R is the universal gas constant, T the absolute temperature, n the number of electrons transferred in the process, and F the Faraday constant.

$$E = E_o + RT/nF\ [\ln(PO_2)_{ref}/(PO_2)_{exhaust}]$$

The CO and C_yH_n catalytically react with the O_2 at the surface of the Pt electrocatalyst. When the O_2 content is below stoichiometric the electrode surface is depleted of O_2 causing an increase in the $(PO_2)_{ref}/(PO_2)_{exhaust}$ generating

a large voltage. When the O_2 is higher than stoichiometric the voltage is decreased. Thus the electrodes must also function as catalysts. The voltage signal generated continuously fluctuates as the O_2 content is adjusted from sub to excess stoichiometric. Naturally the exhaust electrode had to be resistant to exhaust poisons and temperature variations so it was engineered with great care.

The voltage signal is fed back to the air/fuel intake system of the engine through an electronic control unit that controls the ratio necessary to maintain the proper window in the exhaust. Given the finite time necessary for the feedback system to function it creates a perturbation of the O_2 content in the exhaust. The TWC catalyst had to be engineered to respond to these changes. The catalyst was composed of Pt, Rh on stabilized Al_2O_3 on a ceramic monolith but an oxygen storage component (OSC) capable of storing and releasing O_2 was added. When the engine momentarily delivers less O_2 than the stoichiometric amount the hydrocarbons present reduce the oxygen storage component. During higher O_2 spikes the excess is stored on the OSC according to the fuel lean reaction below.

$$\text{Fuel rich:} \quad 2CeO_2 \to Ce_2O_3 + 1/2O_2$$

$$\text{Fuel lean} \quad Ce_2O_3 + 1/2O_2 \to 2CeO_2$$

The current OSC material is CeO_2–ZrO_2 (proprietary promoters are added to stabilize it against sintering) where the oxidation state of the cerium is sufficiently labile to respond to the requirements for the OSC. The ZrO_2 is added to enhance thermal stability of the OSC.

U.S. federal regulations require that the driver be alerted to a malfunctioning catalyst through a signal on the dashboard. Currently there is no instrumentation commercially available to sense the effectiveness of the catalyst to meet the onboard diagnostic requirement. An indirect solution is to place a second O_2 sensor at the exit of the catalytic converter. If the OSC in the catalyst is working properly its voltage signal would have virtually no fluctuations because the O_2 content would be always zero. If the OSC is not functioning properly O_2 will break through at the exit and

INTERNAL COMBUSTION ENGINE +
CLOSE COUPLED CATALYST + TWC
WITH FEEDBACK CONTROL

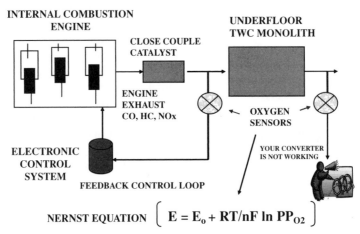

Fig. 7.11. Schematic of a exhaust system with a close couple and TWC catalyst. The system is equipped with dual O_2 sensors, one at the inlet and one at the outlet of the TWC converter.

the sensor would undergo similar fluctuations as the inlet sensor. Comparing these two signals generates the diagnostic that informs the operator of a malfunction.

A modern converter with dual O_2 sensors (one at the inlet and one at the outlet) is shown as Fig. 7.11.

Modern Catalytic Converter Systems

Modern TWC-equipped vehicles are required to meet minimum emission standards for 150,000 miles.[21,22] It should be understood that after this time period the catalyst is still extremely active but has lost sufficient activity that it no longer meets the stringent EPA standards. The source of deactivation is sintering of the catalytic metals, especially the OSC and the carrier due to the extremely high temperatures (900–1000°C) experienced in the exhaust. The steam produced from combustion enhances the degree of sintering. There are proprietary stabilizers added to the formulations that minimize the extent of sintering. Poisoning effects by sulfur and oil components (Zn, P, Ca, S, etc.), have been minimized by reductions in fuel sulfur and careful design of the washcoat to prevent

contact of the poisons with the catalytic components. These catalyst improvements, coupled with enhanced engine control, have resulted in lifetimes of at least 150,000 miles.

At start-up the catalyst is cold and there is a substantial emission of hydrocarbons. It is during the first one or two minutes of cold-start operation that the vehicle can fail the federal test procedure. Here the reaction is kinetically controlled. Once it gets sufficiently warm the reaction exotherm quickly raises the temperature and the reaction becomes limited by bulk mass transfer. The space velocity varies between 5000 (idle) and 75,000 h^{-1} at high speed. So manufacturers had to design the catalyst for kinetic control and bulk mass transfer conditions. The cold-start issue was addressed by positioning a small oxidation catalyst (close coupled) up against the exhaust ports of the engine to ensure rapid heat up and light-off. This is shown in Fig. 7.11. The newest cordierite monoliths have lower weights for faster light-off and high geometric areas (600–900 cpsi) to ensure adequate bulk mass transfer area and lower pressure drop to meet modern driving demands and ever-increasing regulations.

It is truly remarkable that catalysts can function so well in the exhaust of the modern high-speed vehicle. This fact has raised confidence in industry to use different monolithic (ceramic and metal) structures as supports for catalysts for other environmental applications such as diesel exhausts, power and chemical plants, restaurants, and even on widebody aircraft.

CATALYTIC HYDROGENATION OF VEGETABLE OILS FOR EDIBLE FOOD PRODUCTS

Triglycerides

Plant-derived oils such as soy, cottonseed, peanut, canola, corn, etc. are natural sources of edible products such as baking dough for cakes, cooking oils, salad dressing, chocolates, margarine, etc. Nonedible products such as lubricants, creams, lotions, etc. can also be produced depending on the processing of the oils. Natural oils are composed of long chains of fatty acid esters called triglycerides as shown in Fig. 7.12. The triglyceride chains are polyunsaturated, the degree of which influences their stability against oxidation in air.

Catalytic hydrogenation of the double bonds improves the stability against air and raises the melting point such that solids can be produced. Thus, the precursor to chocolate candy, margarine, or a cake mix is liquid oil that upon hydrogenation becomes an edible solid at room temperature. The more double bonds hydrogenated (the more saturated) the higher is the melting point, the lower is the reactivity towards air but often more injurious to our health by deposition of cholesterol in our blood vessels. The goal of a good catalyst coupled with the proper process conditions is to produce a reasonably healthy product with the desired melting point range with sufficient air stability to permit good shelf life.

Oils are classified by the length of the glyceride chain and degree of polyunsaturation. Typically nature produces oils with each chain length between 12 and 22 carbons with up to three unsaturated bonds usually all in the *cis*-form. Triglycerides with 18 carbons per length and three double bonds at positions 9,12, and 15 counting from the first carbon in the ester group are called linolenic and designated C18:3. This structure is shown in Fig. 7.12. The outermost double bond is so reactive towards air that oils with three double bonds in the alkyl chain are rare. Therefore the most prevalent in nature have double bonds at positions 9 and 12 and are referred to as linoleic (C18:2). Its reactivity is about half that of linolenic. The least reactive is oleic (1/20 that of linolenic) with only one double bond per length at position 9 (C18:1). Stearic is the term used for glycerides with all bonds saturated (C18:0). Not surprisingly this form has virtually no reactivity towards air, has a high melting point, and is unhealthy.

The source of the oils plays a major role in producing a product with the desired melting point, stability, and health consequences. Cotton, sunflower, corn, and soy bean oils are a mixture of the four basic triglycerides with 50–70% C18:2 being the most dominant

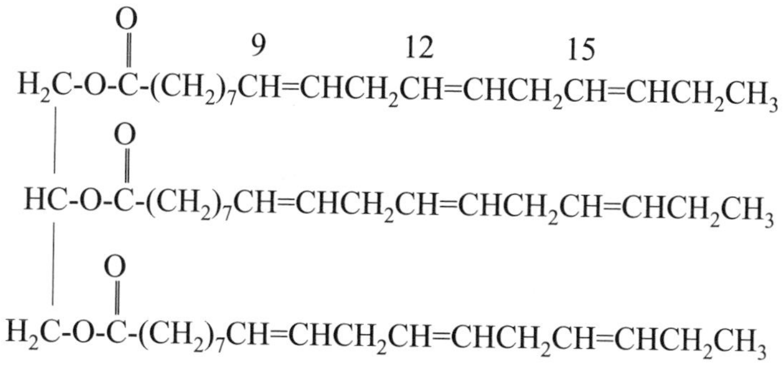

$$\begin{array}{c}
\overset{\displaystyle O}{\underset{\displaystyle |}{\|}} \qquad\;\; 9 \qquad\quad 12 \qquad\quad 15 \\
H_2C\text{-}O\text{-}C\text{-}(CH_2)_7CH\text{=}CHCH_2CH\text{=}CHCH_2CH\text{=}CHCH_2CH_3 \\
\overset{\displaystyle O}{\underset{\displaystyle |}{\|}} \\
HC\text{-}O\text{-}C\text{-}(CH_2)_7CH\text{=}CHCH_2CH\text{=}CHCH_2CH\text{=}CHCH_2CH_3 \\
\overset{\displaystyle O}{\underset{\displaystyle |}{\|}} \\
H_2C\text{-}O\text{-}C\text{-}(CH_2)_7CH\text{=}CHCH_2CH\text{=}CHCH_2CH\text{=}CHCH_2CH_3
\end{array}$$

Fig. 7.12.   Unsaturated fat molecule. Example shown is linolenic oil.

followed by 20–30% C18:1 with less than 1% C18:3. Less than 10% are other saturated oils such as C16:0. Palm kernel and coconut oils have almost 80% saturated triglycerides (C12:0, C14:0, and C16:0), have high melting points, and are stable against air but not healthy. They are used for protecting the skin against excessive sun exposure. Olive oil has up to 80% C18:1 and is therefore relatively healthy.

The most common oil hydrogenation catalysts are 20–25% Ni on Al_2O_3 and SiO_2. The nickel salts are either impregnated or co-gelled with a carrier precursor such as a soluble Al or Si salt. The catalytically active state of Ni is the reduced (metallic) form. The activation step is performed during manufacture at which time the catalyst is coated with a fatty gel to protect it from air oxidation during shipment.

Less than 1% by weight of catalyst is added to the batch reactor where the fatty protective gel slowly dissolves and the hydrogenation reaction commences. Temperatures of 100°C and H_2 pressures of 3–5 atmospheres are used to ensure adequate dissolution of the H_2 into the feed stream. Reactions are carried out between 100–160°C. Stirring is vigorous to maximize bulk gas diffusion to the catalyst surface. The catalyst particles are small to minimize pore diffusion resistance and increase liquid–solid mass transfer area.

Cu supported on Al_2O_3 is less active than Ni and this property is used to "brush" hydrogenate. Only minimum hydrogenation occurs maintaining the melting point but sufficient to improve stability against air. This is sometimes used for producing salad oils.

The reaction profile is generally sequential with hydrogenation first occurring on the most active double bonds followed by those less active.[22–24] Time distribution shows the linoleic form decreasing as the oleic form increases. After extended reaction time the stearic begins to form as the oleic is slowly hydrogenated. Thus one can design the process to control the product distribution in a predictable manner.

Hydrogenation of the linoleic form with a melting point of -13°C will produce a oleic product with a melting point of 5.5°C very suitable for consumption. During the partial hydrogenation process it is most desirable to minimize isomerization to the *trans*-isomer (the hydrocarbon groups are *trans* to each other across the remaining unsaturated bonds) because this structure raises the "bad" cholesterol or LDL (low density lipids). A partially hydrogenated *cis*-structure may have a melting point of 6°C whereas its *trans* isomer melts at 40°C. The *trans* isomer is more readily formed at high reaction temperatures, high Ni catalyst loadings, and at low hydrogen pressures (low concentration of H_2 at the catalyst surface). Pt containing oil hydrogenation catalysts produce considerably less *trans* than Ni, however, the high activity causes excessive hydrogenation of the double bonds. To minimize this effect NH_3 is intentionally added to the feed or catalyst to poison its activity towards hydrogenation. By so doing a low *trans* oil is produced without excessive saturation of the double bonds. In 2006 the U.S. Food and Drug Administration will require labels that report the amount of *trans* components present in edible products.

Catalyst deactivation is mainly caused by mechanical attrition due to the rigorous stirring. In most cases adsorption guard beds are used upstream to remove most of the impurities such as sulfur and phosphorous often found in the feed. Recognizing that some poisons may break through, the catalyst has an average pore size sufficiently large to admit the triglycerides but smaller than the average size of the organic compounds containing P and S. The spent catalyst is separated from the product by filtration. Given the increasing cost of Ni it is recovered, refined, and used to make fresh catalyst.

FERTILIZERS AND HYDROGEN GENERATION

General Reactions

Ammonium nitrate (NH_4NO_3) and urea ($CO(NH_2)_2$) are two major sources of the world's fertilizers. The nitrate is produced by reaction of ammonia and nitric acid.

$$NH_3 + HNO_3 \rightarrow NH_4NO_3$$

Urea is produced by the reaction of NH_3 with CO_2 and its subsequent decomposition.

$$2NH_3 + CO_2 \xrightarrow{\text{HEAT}} NH_2COONH_4 \rightarrow CO \\ (NH_2)_2 + H_2O$$

Ammonia is produced by the catalytic hydrogenation of N_2 with H_2

$$3H_2 + N_2 \rightarrow 2NH_3$$

Hydrogen is produced by a series of catalytic reactions the first of which is hydrocarbon reforming starting, and the second is water gas shift. Considering natural gas (CH_4) as the starting hydrocarbon

$$CH_4 + H_2O \rightarrow 3H_2 + CO$$

$$CO + H_2O \rightarrow H_2 + CO_2$$

Nitric acid is produced by the selective catalytic oxidation of NH_3 and its subsequent hydration

$$2NH_3 + 7/2O_2 \rightarrow 2NO_2 + 3H_2O$$

$$3/2NO_2 + 1/2H_2O \rightarrow 3/2HNO_3 + 1/2NO$$

Each catalytic step will be discussed in this section.

Hydrogen Generation for the Production of NH_3

Producing H_2 from hydrocarbons such as natural gas is currently practiced in the chemical industry [25-28] under steady-state conditions with carefully controlled catalytic unit operations. The overall process is as shown in Fig. 7.13.

Traces of organic sulfur compounds such as mercaptans, thiosulfides, and alkyl sulfides are added to natural gas to impart odor for safety detection of leaks. Because sulfur compounds are poisons to the downstream catalysts they must be removed. The

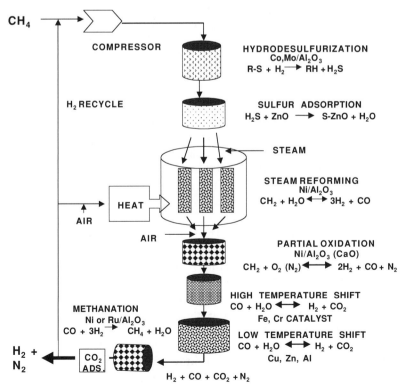

Fig. 7.13. The catalytic synthesis of hydrogen from natural gas for ammonia synthesis using hydrodesulfurization, steam reforming, partial oxidation, water gas shift, methanation, and CO_2 scrubbing.

technology of choice is hydrodesulfurization or HDS.

$$RH\text{-}S + H_2 \rightarrow H_2S + RH_2$$

The reaction is carried out at about 300–400 psig and 300–400°C but these conditions vary with the hydrocarbon feed. The catalyst is 3% Co, 15% Mo deposited on spheres (25–75 m^2/g) Al_2O_3 with diameters of 2–3 mm. The catalyst is presulfided to decrease its activity towards undesirable side reactions such as coking.

The H_2S produced is removed downstream from the HDS reactor by adsorption on ZnO particles at about 400–500°C.

$$H_2S + ZnO \rightarrow ZnS + H_2O$$

Primary reforming of the sulfur-free natural gas (i.e. CH_4) is the first step to produce a H_2-rich gas.

$$CH_4 + H_2O \rightarrow CO + 3H_2$$

The reaction is highly endothermic and thus is favored at high temperatures. The maximum temperature achievable is limited by metallurgy of the reactor. Given the increase in gas volume the reaction is favored by low pressures.

The catalyst is approximately 30% Ni with about 14% CaO on highly densified alpha alumina (α-Al_2O_3) with a surface area about 2–5 m^2/g. The CaO reacts with the Al_2O_3 forming $CaAl_2O_4$ for added mechanical strength under the severe operating conditions of 800°C and 300–400 psig and a steam environment up to 75%. The reaction rate is limited by a combination of heat transfer and pore diffusion, the latter due to the inadequate accessibility of the reactants to the catalyst interior. To counter pore diffusion limitations the catalyst is manufactured as a donut with two to three holes to increase the external contact area and decrease the diffusion path. The space velocity is between 1000 and 2000 h^{-1}. The active catalyst is Ni metal so it must be reduced carefully with H_2 prior to introducing the feed. The reaction is carried out in a series of tubular parallel reactors located in a large fired box furnace to provide the necessary heat.

Deactivation is due mostly to the slow accumulation of sulfur that breaks through the upstream HDS/ZnO guard beds. Sulfur irreversibly decreases the activity of the Ni that allows the methane decomposition rate to become significant leading to the accumulation of hydrogen-deficient carbon or "coke." This builds up within and between catalyst particles leading to its fracture and an increase in pressure drop. During process shut down the catalyst must be "passivated" to protect against air oxidation of the Ni and a subsequent fire due to its strong exotherm creating a safety hazard at the plant site. This is accomplished by periodically injecting small amounts of air and carefully oxidizing the surface of the Ni while monitoring the exotherm.

Equilibrium limits conversion of the CH_4 so partial oxidation or secondary reforming of the unreacted CH_4 is used to generate more heat and H_2 in a secondary reforming step. The addition of air also serves the purpose of providing the required N_2 for the subsequent ammonia synthesis reaction. Secondary steam reforming also uses a high-temperature resistant Ni containing catalyst that must retain its strength after prolonged exposure to close to 1200°C due to the oxidation in the front end of the bed. The catalyst used is α-Al_2O_3 impregnated with about 18–20% Ni and 15% CaO.

$$2CH_4 + N_2 + 2O_2 \rightarrow CO + 3H_2 + CO_2 + H_2O + N_2$$

The exit from the secondary reformer contains about 10–12% CO, is cooled to about 350°C and fed to a high-temperature water gas shift (HTS) reactor.

$$CO + H_2O \rightarrow H_2 + CO_2$$

The particulate catalyst is composed of 90% Fe and 10% Cr. The Cr minimizes sintering of the active Fe phase. The catalytic reaction is limited by pore diffusion so small particles are used. The exit process gas contains about 2% CO as governed by the thermodynamics and kinetics of the reaction. This reaction is slightly exothermic and thermodynamics favor low temperatures that decrease the reaction rate. It is therefore necessary to further cool the mix to about 200°C where it is fed to a low-temperature shift reactor (LTS) containing another particulate catalyst composed of 30–35% Cu, 45% ZnO, and 13–20% Al_2O_3. The catalyst is active in the reduced form so it must be carefully activated with H_2 avoiding excessive overheating which will cause sintering. The Zn and Al_2O_3 are added to stabilize the Cu because it is sensitive to sintering. The CO is decreased to its thermodynamic limit as imposed by the temperature and gas compositions. Typically the CO is reduced to less than about 0.5 %. The catalyst deactivates by traces of sulfur and sintering of the active Cu phase. Because of the necessity to operate at low temperatures the reaction rate is slow and large volumes and low space velocities (1500–2500 h^{-1}) are used.

The active Cu containing catalyst is also very air sensitive (like the Ni reforming catalyst) and will spontaneously oxidize generating uncontrolled reaction heats. Thus it must be passivated before discharged and exposed to air. A small amount of air is added to the reactor and the temperature monitored. This process is continued until the exotherm is small enough that the catalyst can be safely removed from the reactor.

The remaining CO, which poisons the downstream ammonia synthesis catalyst, is removed by methanation using either a Ni or Ru on Al_2O_3 catalyst at 300°C

$$CO + 3H_2 \rightarrow CH_4 + H_2O$$

The CO_2 is scrubbed in an amine solution.

Ammonia Synthesis

The Haber process for the synthesis of ammonia from H_2 and N_2 has been practiced since the beginning of the 20th century always with a massive Fe catalyst.[29]

$$N_2 + 3H_2 \rightarrow 2NH_3$$

It is mildly exothermic so the reaction is thermodynamically favored at lower temperatures but at high pressures due to the contraction of gaseous product volume. To obtain reasonable rates the process is operated at about 450°C and pressures approaching 5000 psig at a space velocity of 10,000–15,000 h^{-1}. The process is operated in a recycle mode so

ammonia is continuously removed aiding the equilibrium.

The active catalyst is 75–80% Fe metal, 10% Fe_2O_3, 4% Al_2O_3, less than 5% alkali and alkaline earth (Li, Ca, and Mg), with 1% SiO_2 added to minimize sintering of the Fe. The promoters are added to a melt of magnetite (Fe_2O_3). The solid mass is then ground to 1 mm particles, charged to the fixed bed reactor, and slowly reduced with H_2 at 500°C. The reduction generates active Fe metal with some porosity due to liberation of oxygen forming H_2O. The surface area is increased from about 1 m^2/g to about 20 m^2/g. The small particle size of the finished catalyst is necessary to minimize pore diffusion limitations. Special precaution is necessary during discharge from the reactor because air exposure will spontaneously oxidize the Fe surface generating large quantities of heat.

The catalyst is poisoned by CO, CO_2, and H_2O so they must be rigorously removed upstream in the hydrogen synthesis process. Oxygen molecules are permanent poisons. Other poisons such as sulfur, arsenic, halides, and phosphorous must be carefully removed upstream in as much as they too are permanent poisons.

Nitric Acid Synthesis

Nitric acid is produced by the selective oxidation of NH_3 over a gauze catalyst composed of 90%Pt, 10%Rh (some gauze is 90% Pt, 5% Rh and 5% Pd).[30] This reaction was used in the "Selectivity" section to demonstrate the high efficiency with which PtRh leads to NO production as opposed to more thermodynamically favored N_2.

$$\overset{PtRh}{4NH_3 + 5O_2 \rightarrow 4NO + 6H_2O}$$

$$4NH_3 + 3O_2 \rightarrow 2N_2 + 6H_2O$$

The low-pressure process (15–30 psig) produces NO with a selectivity of 98% and the high-pressure process (150 psig) has a selectivity of 94%. The high-pressure plant allows for a smaller reactor (and gauze) diameter (3 ft) compared to 12 ft for the low-pressure process.

The feed is composed of 10–12% NH_3 in air and is fed to the reactor at an inlet temperature of about 250°C and a space velocity approaching 50,000 h^{-1}. For a high-pressure plant the exotherm generates an outlet temperature of 900°C. The NO product is cooled and noncatalytically converted to NO_2, its thermodynamically favored state at low temperatures, which reacts with H_2O forming HNO_3.

The finished alloy catalyst is manufactured by a knitting process to form a wire gauze that looks like a door screen. The Rh is added to impart mechanical strength during the wire drawing operations. During reaction the catalyst undergoes an unusual morphology change. The Pt forms an oxy–nitrogen species and volatilizes from the gauze. The smooth wires become roughened and sprout resembling cauliflower. The surface area of the gauze increases by 20 times. The loss of Pt enriches the surface with Rh and the catalyst slowly loses activity. After approximately 90 days of operating a high-pressure plant, the Pt content of the gauze is reduced to 50%. The volatile Pt is captured

downstream on a "getter gauze" made of Pd. The Pd surface catalytically decomposes the gaseous oxy–nitro Pt species and a Pt–Pd alloy forms that allows for easy recovery of the precious metals. The spent catalyst is returned to the supplier where the precious metal is recovered for future use.

Another source of deactivation is Fe contamination originating from the corrosion of upstream equipment depositing on the gauze resulting in decomposition of the NH_3 to N_2. Another source is from the Fe-containing ammonia synthesis catalyst.

Although the largest use for nitric acid is NH_4NO_3 fertilizers it is also used for explosives and nylon polymers.

Pure Hydrogen Generation with Pressure Swing Adsorption (PSA) Purification

For applications in which N_2 is not needed, such as H_2 or alcohol production, pressure swing adsorption is used. The process flow diagram is shown in Fig. 7.14.

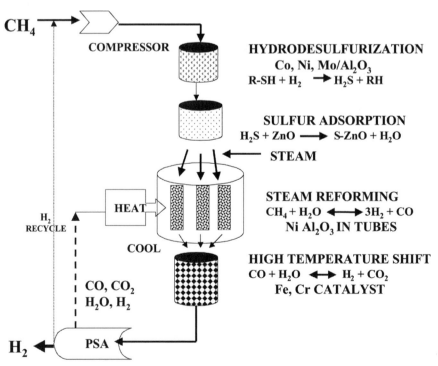

Fig. 7.14. The catalytic synthesis of hydrogen from natural gas using hydrodesulfurization, steam reforming, water gas shift, and pressure swing absorption (PSA).

There is a renewed interest in hydrogen generation for the developing hydrogen economy with the anticipated use of fuel cells as a power source for vehicles.[32] The fuel cell generates electricity by electrochemically oxidizing H_2 and reducing O_2. Because it directly converts chemical to electrical energy without using the traditional mechanical steps of piston-driven engines and turbines it promises to be more efficient, cleaner, decrease our dependence on oil, and generate less greenhouse gas. The small-scale generation of H_2 for cost-effective refueling stations is a major issue that is aggressively being studied. Ultimately it will be derived from water by electrolyzers using natural sources such as solar, wind, and geothermal energy. Until these technologies are available natural gas reforming is a likely source because infrastructures exist in many cities in the world. The fuel cell powered vehicle will require an infrastructure similar to gasoline and diesel service stations. High-pressure H_2 will have to be available to refuel the vehicles. Such demonstration stations are now being built in various parts of the world. Pressure swing adsorption (PSA) is used for final H_2 purification. The partial oxidation and the low-temperature shift are eliminated. There is therefore a loss of some hydrogen production but the final H_2 is not diluted with N_2. The PSA purification unit replaces the methanator and CO_2 scrubber and produces pure H_2. Some of the H_2 is recycled for HDS and some combusted to provide the heat for steam reforming.

Given the need for smaller size reformers to be operated in local communities safety will be an elevated concern. These reformers must contain nontoxic and air-insensitive catalysts thus eliminating Ni, Cu, and Cr from consideration. A major research effort is underway to redesign the entire H_2 generation process system using modern materials such as monoliths and precious metal catalysts.[32]

For fuel processors directly integrated to a residential fuel cell H_2 must be maximized and no pressure is available for PSA. In these cases a high- and low-temperature shift catalyst will be required. Hydrogen purification to reduce the CO to less than 10 ppm will be managed by preferential oxidation.[33]

PRODUCTION OF BUTYRALDEHYDE: A HOMOGENEOUS CATALYTIC REACTION

Butyraldehyde

The incorporation of an inner layer of poly(vinyl butyral) or (PVB) in the glass of an automobile windshield protects against serious head injuries when a passenger strikes it during an accident. The strongly adherent coating is optically transparent and maintains the glass intact (antishattering agent) when a foreign object hits the surface. Thus the glass does not shatter when a stone strikes its surface. PVB is produced by reaction of polyvinyl alcohol (PVA) with linear butylraldehyde ($CH_3CH_2CH_2CHO$).

$$-(-CH_2CHCH_2CH-)- + CH_3CH_2CH_2CHO$$
$$\quad\quad\ OH\quad\ OH$$

$$\downarrow$$

$$\begin{array}{c} CH_2 \\ / \quad \backslash \\ -(-CH_2CH \quad\quad CH-)- \\ | \quad\quad\quad\ / \\ O \quad\quad O \\ \backslash \quad / \\ CH \\ CH_2 \\ CH_2 \\ CH_3 \end{array}$$

Another important application of butyraldehyde is in the production of oxo-alcohols for use as plasticizers used to improve mixing of solid compounds that must be molded or extruded into specific shapes. The hydrogenation catalyst is Ni/Al_2O_3.

$$CH_3CH_2CH_2CHO + H_2 \xrightarrow{Ni/Al_2O_3} CH_3CH_2CH_2CH_2OH$$

Butylraldehyde is produced by a homogeneous catalytic process called hydroformulation in which CO and H_2 are added to liquid propylene using a soluble cobalt-containing complex catalyst $Co(CO)_6$. The reaction is carried out with a butanol solvent

$$CH_3CH = CH_2 + H_2 + CO \xrightarrow{Co(CO)_6} 80\%$$
$$CH_3CH_2CH_2CHO + 20\% \; CH_3CHCHO$$
$$+ \; CH_3CHCHOCH_3$$

The conditions needed to catalyze the reaction are very severe; pressure = 3000–4500 psig and 150°C. The high pressure maintains the propylene in solution, ensures sufficient solubility of the H_2 and CO and maintains the Co-carbonyl complex stable against decomposition. The product distribution is 4:1 linear to branch.

Much less severe conditions can be used with the Wilkinson homogeneous catalyst rhodium tricarbonyl triphenyl phosphate, $HRh(CO)_3(PC_6H_5)_3$. Pressures equal to 225 psig and temperatures of 100°C selectively produce the more useful linear form.[34] The milder conditions more than compensate for the more expensive Rh (1000 times that of Co). The aldehyde product is distilled leaving the catalyst in the solvent ready for reuse.

Homogeneous catalysts are structurally well-defined complexes and because they are soluble in the reaction mix are not subject to pore diffusion limitations as are heterogeneous catalytic materials. They are almost always highly selective towards desired products. The main consideration is that the complex be stable and reactor conditions chosen such that all the gaseous reactants are adequately dissolved and mixed in the liquid phase. Homogeneous catalysts are easily characterized by standard instrumental methods for compound identification such as XRD or spectroscopy. Deactivation is associated with attack by traces of carboxylic acidic byproducts and impurities in the feed such as O_2 and chlorides that attack the ligand groups.

POLYETHYLENE AND POLYPROPYLENE FOR THE PRODUCTION OF PLASTICS

Polyethylene

Specially prepared plastics are rapidly replacing traditional metal components because of their strength, transparency, resilence, lighter weight, and greater corrosion resistance. The largest volume products are polyethylene and polypropylene. Each has its own contributions to the marketplace where the former is primarily used for low-strength applications such as milk and food containers. Polypropylene is used when enhanced strength, higher melting temperatures, and greater resistance to chemicals such as chemical holding tanks and automobile bumpers are required.

There are two prevalent methods of producing polyethylene both of which involve heterogeneous catalysts. A slurry phase process utilizes chromium oxide deposited on SiO_2 dispersed in a solvent such as cyclohexane at 80–150°C and a pressure between 300 and 500 psig. The process operates in a recycle mode with a residence time of 2–3 hours. The product containing the solvent and polymer is flashed leaving the polymer. The catalyst is usually left in the polymer because its concentration is extremely low. The operating conditions are adjusted to produce both high- and low-density polyethylene. The active site is Cr^{+2} produced by the reduction of Cr^{+6} by ethylene. The reaction mechanism proposed is that the polymer coordinates with one of the Cr^{+2} sites and the incoming ethylene coordinates with another site. Insertion of the ethylene into the double bond of polymer propagates its growth.

A second method of production utilizes the Ziegler–Natta $TiCl_4$ catalyst with liquid co-catalysts such as an alkyl aluminum halide. This is a reactive catalyst that must be prepared at the exclusion of air and water. The alkyl group of the co-catalyst coordinates with the Ti^{+3} site. The polymer grows by insertion of the ethylene into the double bond of the adsorbed polymer on another site.

Polypropylene

The most modern production route for polypropylene (PP) is also the Ziegler–Natta catalyst.[35–37] The catalyst is $TiCl_4$ supported on $MgCl_2$ along with aluminum alkyl halide co-catalyst such as diethyl aluminum fluoride $(CH_3CH_2)_2AlF$. The $MgCl_2$ is milled to a very disordered but active structure and the $TiCl_4$ is added. Production of the PP is carried out in a fluidized gas phase reactor between 50 and 100°C and 100–600 psig. Ethyl benzoate is also used as part of the catalyst preparation and functions to reduce the $TiCl_4$ to active $TiCl_3$. The role of the alkyl component of the co-catalyst is to coordinate with the Ti^{+3} site where it inserts into the adsorbed polypropylene continuing the chain growth. The amount of catalyst used is so small it is retained in the final polymer product with no negative consequences. Unreacted gases are removed and recycled at the completion of the process. The most desirable product for the largest market is the isotactic form in which all CH_3 groups are on the same side of the polymer chain. Typically it has a density of $0.9g/cm^3$, a melting point of 170°C and an average molecular weight of 500,000. The polypropylene product is mixed in a separate reactor with ethylene to make a block polymer with enhanced mechanical properties.

Water, CO, and O_2 are the most significant poisons and are carefully removed upstream of the process.

The catalyst preparation and the process are far more complicated than presented here so the reader is encouraged to refer to more detailed references.[35–37]

CATALYST CHALLENGES

Catalysts will have additional challenges as we move forward in the 21st century. In this author's mind, one of the most critical is the need to balance our rapidly expanding energy needs with the environment. Catalysts are already playing a dominant role in pollution abatement and in the production of specialty petroleum and chemical products. The main challenge will be to use bio-renewable energy sources (Chapter 33) as well as solar, wind, geothermal, etc. with the specific goal of freeing us from the use of fossil fuels. The hydrogen economy coupled with the fuel cell holds great promise as one road to meet this challenge. Catalysts will play a key role in this pursuit although the road map is not yet complete. This will be an exhilarating ride as we find our way to clean energy.

Another area of great importance is the emerging application of bio-catalysis using nature's catalysts such as enzymes to produce a growing number of pharmaceutical and agricultural products. This subject is outside the scope of this review so other sections of this Handbook (see chapter 31) should be consulted.

REFERENCES

1. Smith, J., *Chemical Reaction Engineering*, McGraw-Hill, New York, 1981. Fogler, H., *Elements of Chemical Reaction Engineering*, 2nd edition, Prentice Hall, Englewood Cliffs, NJ, 1992.
2. Schweich, D., "Transport effects" in *Encyclopedia of Catalysis* I. Horvath, ed. Vol 6, 507, Wiley, New York, 2003.
3. Liu, X., Korotkikh, O., and Farrauto, R. "Selective Catalytic Oxidation of Co in H_2: Structural Study of Fe-oxide promoted Pt/alumina catalyst," in *Applied Catalysis A: General* **226**, 293, 2002.
4. Bartholomew, C. and Farrauto, R. *Fundamentals of Industrial Catalytic Processes*, 2nd edition, Chapter 2, Wiley, New York, 2006.
5. Heck, R. and Farrauto, R. with Gulati, S., *Catalytic Air Pollution Control: Commercial Technology,* Chapters 2, 4, 7, and 9, Wiley, New York, 2002.
6. Missen, R., Mims, C., and Saville, B., *Introduction to Chemical Reaction Engineering and Kinetics*, Chapter 4, Wiley, New York, 1999.
7. Broadbelt, L., "Kinetics of catalyzed reactions," in *Encyclopedia of Catalysis,* I. Horvath, Ed., Vol. 4, 472, Wiley, New York, 2003.
8. Chorkendorff, I. and Niemantsverdrietm, J., *Concepts of Modern Catalysis and Kinetics*, Chapters 2 and 9, Wiley-VCH, Weinheim, Germany, 2003.
9. Bartholomew, C. and Farrauto, R. *Fundamentals of Industrial Catalytic Processes*, 2nd edition, Chapter 1, Wiley, New York, 2006.
10. Bartholomew, C., "Catalyst deactivation/regeneration," in *Encyclopedia of Catalysis* I. Horvath, Ed. Vol. 2, 82 Wiley, New York, 2003.
11. Bartholomew, C. and Farrauto, R. *Fundamentals of Industrial Catalytic Processes*, 2nd edition, Chapter 5, Wiley, New York, 2006.

12. Knozinger, H. and Maximlians, L., Catalyst characterization in *Encyclopedia of Catalysis* I. Horvath, Ed., Vol. 2, Wiley, New York, 2003.
13. Bartholomew, C. and Farrauto, R. *Fundamentals of Industrial Catalytic Processes*, 2nd edition, Chapter 3, Wiley, New York, 2006.
14. Van Rysaselberghe, V. and Froment, G., "Hydrodesulfurization" *Encyclopedia of Catalysis* I. Horvath, Ed., Vol. 3, 667, Wiley, New York, 2003.
15. Brunet, S., Mey, D., Perot, G., Bauchy, C., and Diehl, F., *Applied Catalysis A: General* **278**, 2, 143, 2005.
16. Gates, B., "Catalysis by solid acid," in *Encyclopedia of Catalysis* I. Horvath, Ed., Vol. 2, 104, Wiley, New York, 2003.
17. Csicsery, S. and Kiricsi, I. "Shape selective catalysis," in *Encyclopedia of Catalysis* I. Horvath, Ed., Vol. 6, 307, Wiley, New York, 2003.
18. Moser, M., "Reforming," in *Encyclopedia of Catalysis* I. Horvath, Ed., Vol. 6, 1, Wiley, New York, 2003.
19. Antos, G. Aitani, A., and Parera, J. (Eds.), *Catalytic Naphtha Reforming, Science and Technology*, Marcel Dekker, New York, 1995.
20. Heck, R. and Farrauto, R., "Automotive catalysis," in *Encyclopedia of Catalysis,* I. Horvath, Ed., Vol. 1, 517, Wiley, New York, 2003.
21. Heck, R. and Farrauto, R. with Gulati, S., *Catalytic Air Pollution Control: Commercial Technology,* Chapter 6, Wiley, New York, 2002.
22. Albright, L., "Hydrogenation (partial) of triglycerides," in *Encyclopedia of Food Science* Peterson, S and Johnson, H., Eds., 398, Avi Publishing, Westport, CT, 1978.
23. Bartholomew, C. and Farrauto, R. *Fundamentals of Industrial Catalytic Processes*, 2nd edition, Chapter 7, Wiley, New York, 2006.
24. Albright, L. and Wisniak, J., *J. Amer. Oil Chem. Soc.* **39**, 14, 1962. "Selectivity and Isomerization during Partial hydrogenation of cottonseed oil and methyl oleate: Effect of operating variables."
25. Rostrup-Nielsen, J. and Nielsen, T. R., *CATTECH* **6.** 4, 150, 2002.
26. Osterkamp, P. "Synthesis gas" in *Encyclopedia of Catalysis* I. Horvath, Ed., Vol. 6, 456, Wiley, New York, 2003.
27. Nielsen, J. "Hydrogen generation by catalysis" in *Encyclopedia of Catalysis* I. Horvath, Ed., Vol. 4, 1, Wiley, New York, 2003.
28. Kondrantenko, E. and Baerns, M., "Synthesis gas generation," in *Encyclopedia of Catalysis* I. Horvath, Ed., Vol. 6, 424, Wiley, NY, 2003
29. Pesce, L and Jenks, W. "Synthetic nitrogen products," in *Riegels Handbook of Industrial Chemistry,* J. Kent, Ed., 11th Edition, Kluwer, New York, 2006.
30. Bartholomew, C. and Farrauto, R. *Fundamentals of Industrial Catalytic Processes*, 2nd edition, Chapter 8, Wiley, New York, 2006.
31. McNicol, B. and Williams, K. "Catalysts for fuel cells," in *Encyclopedia of Catalysis* I. Horvath, Ed., Vol. 2, 387 and 42, Wiley, New York, 2003.
32. Farrauto, R. J., Hwang, S., Shore, L., Ruettinger, W., Lampert, J., Giroux, T., Liu, Y., and Ilinich, O., *Ann. Rev. Mater. Res. Soc.* **33**, 1, 2003.
33. Shore, L. and Farrauto, "Preferential oxidation of CO in H2 streams," in Vielstich, W., Lamm, A., and Gasteiger, H., Eds., *Handbook of Fuel Cells*, **3**, Part 2, Wiley, West Sussex, England, 211, 2003.
34. Kohlpaintner, C., "Hydroformulation," in *Encyclopedia of Catalysis,* I. Horvath, Ed., Vol. 3, 787, Wiley, New York, 2003.
35. Vlad, G. "Polymerization," in *Encyclopedia of Catalysis,* I. Horvath, Ed., Vol. 5, 611, Wiley, New York, 2003.
36. Rodriguez, F., *Principles of Polymer Systems*, 4th Edition, Taylor & Francis, Washington, DC, 1996.
37. Knapczyk, J. and Simon, R., "Synthetic resins and plastics," in *Riegels Handbook of Industrial Chemistry*, J. Kent, Ed., 11th edition, Kluwer, New York. 2006.

8

Environmental Chemical Determinations

William L. Budde*

INTRODUCTION

Environmental chemical determinations are identifications and measurements of the concentrations of elements, compounds, or ions in environmental media. In a chemical determination equal importance is given to the correct identification of the substance, and to its accurate and precise measurement. There has been a tendency in some environmental work to place more emphasis on making accurate and precise measurements, and to give less attention to ascertaining the correctness of the identification of the substance being meaured.

Air, water, soil, and sediment are broad categories of environmental media and each of these can be divided into several subcategories, for example, ambient air, indoor air, industrial or workplace air, and vapor emissions from mobile or stationary combustion sources. Body fluids and tissue are also relevant environmental media because they are often analyzed to determine human, animal, and plant exposure to environmental chemicals.

SIGNIFICANCE OF ENVIRONMENTAL CHEMICAL DETERMINATIONS

Accurate and precise identifications and measurements of specific chemical substances are fundamental to environmental studies and protection programs. Determinations are required to understand natural background concentrations of chemicals in the environment, the nature and extent of environmental pollution by anthropogenic chemicals, trends in concentrations of these substances, the transport and fate of chemical substances, and the causes of variations of concentrations intime and space. Accurate and precise determinations are also required to assess human health and ecological risks caused by exposure to natural and anthropogenic substances, establish air and water quality standards, develop pollution control strategies, evaluate the effectiveness of pollution prevention and treatment technologies, and monitor compliance with and the effectiveness

*U.S. Environmental Protection Agency (retired), Cincinnati, Ohio. ohbudde@fuse.net.

of discharge and other government regulations. These data are also needed to develop, calibrate, and verify mathematical models used to predict the impact of changes in concentrations of specific substances in the environment. Environmental chemical deteminations are also required to set priorities and make cost estimates for the remediation of abandoned hazardous waste sites.

In addition to specific chemical substances, a variety of other determinations is very important and is often required for environmental studies and protection programs. These include identifications and measurements of bacteria, viruses, protozoa, and minerals such as asbestos fibers. Measurements of meteorological conditions, particulate matter in air by size, water turbidity, biological oxygen demand, chemical oxygen demand, and radioactivity are also very important. These and other similar determinations are beyond the scope of this chapter and the reader should consult other reference books for information about these topics.

CHEMICAL ANALYSIS STRATEGIES

A wide variety of chemical analytical strategies is used in environmental analysis programs. In order to develop an appropriate strategy for a specific program, the objectives of the program must be thoroughly defined and understood by all participants. An analytical strategy for the program can then be developed by selecting and combining the most appropriate analytical methods and other key elements that will provide the results needed to fulfill the objectives of the program. Some of the major issues and strategies that must be considered in developing a specific strategy for an analysis program are described in this section.

Samples and Sampling Strategies

The goals of the environmental program will usually define the specific types of samples that must be acquired and analyzed, for example, ambient air, drinking water, ground

water, soil, sediment, blood, sweat, urine, tissue, etc. The sampling strategy is a broad but detailed plan for the acquisition of the samples needed for the project. Sampling techniques and procedures are the actual physical processes used to acquire the samples. Sample preservation protects the integrity and validity of the samples before, during, and after the actual sampling process. Sampling strategies, techniques, and sample preservation are major factors contributing to the success or failure of environmental chemical determinations.

The sampling strategy defines in detail what, where, when, and how the environmental samples will be acquired. This plan should specify all details including the number of sampling stations for air, water, soil, sediment, or fish samples; the locations of the sampling stations; the time of the day of sampling; the frequency of sampling; the depth of sampling for water, soil, and sediment samples; the meteorological conditions and altitude for air sampling; and, the need for replicate samples taken at the same time and place. For body fluids and tissue samples myriad details must be considered including the number of persons contributing or the kinds of wildlife-related samples that must be collected. The plan should specify a series of discrete samples, a composite sample from contributions at various intervals, or a continuous flow of sample over a period of time. A sampling strategy that does not provide appropriate and representative samples seriously jeopardizes the value of the chemical determinations.

The sampling techniques and procedures define the type of sampling equipment, the containers used for the samples, the procedures used to clean the sample containers and sampling equipment, the calibration of sampling equipment, and sample compositing procedures if composite samples are required. Sample preservation must ensure that the chemical composition of the sample at the time of analysis is the same as it was at the time and place of sampling and is not the result of physical or chemical changes caused by the conditions of shipment and storage prior to chemical analysis. Significant research has been conducted to find sample containers that resist adsorption or degradation of sample components. Samples are often shipped and stored at low temperatures, and treated with various chemical reagents, to retard microbiological degradation of the analytes or to prevent chemical reactions that can change the composition of the sample. Appropriate sampling techniques and preservation procedures are required to provide valid samples and valid chemical determinations.

Determination of Total Elements or Total Related Substances.

A standard analysis strategy is the determination of the total amount of an element in a sample where the element is present in several or more elemental forms, compounds, ions, oxidation states, or physical phases. This strategy was developed before chromatographic and other techniques were available to separate the individual substances containing the element of interest. It is widely used in environmental studies and government regulatory programs. However, the meaning of *total* is variable because some analytical methods do not include all the physical phases or chemical forms in the determination. The sample preparation procedures in individual analytical methods define exactly which elemental forms, compounds, ions, oxidation states, and physical phases are included in a total measurement. A similar strategy is the determination of the total amount or total concentration of a group of closely related compounds or ions without specifying the relative or absolute concentrations of the individual substances. Closely related substances usually have some common physical or chemical properties, elemental composition, or structure.

The broadest definition of, for example, total mercury (Hg) in a sample is the sum of the elemental Hg^0, the Hg in all inorganic compounds of Hg^{+1}, the Hg in all inorganic compounds of Hg^{+2}, and the Hg in all organic compounds in which Hg is bonded to C, O, N, S, or some other element. All phases are included because some species may be present in the vapor phase (Hg^0), some soluble in water, and some insoluble in water or present in the particulate phase of an air sample. More limited definitions of total are often used in analytical methods that separate physical phases or chemical forms that contain the element of interest.

If a water sample contains both soluble and insoluble manganese (Mn) compounds and ions, and it is filtered to separate the dissolved and insoluble fractions, and the filtrate and insoluble residue are analyzed separately, the results can be expressed as total dissolved Mn and total suspended or insoluble Mn. Phosphorus (P) can be determined colorimetrically as the ortho-phosphate ion, PO_4^{-3}, in aqueous samples after a reaction that forms an intensely blue-colored derivative. However polyphosphate ions and other ions and compounds containing P do not form this derivative. Total P in a sample can be determined with the same colorimetric procedure after acid hydrolysis and oxidation of all ions and compounds containing P to PO_4^{-3}. In some elemental analyses the sample is treated with reagents designed to make available for measurement some fraction of an element or elements but not the total amount. For example, a soil sample may be treated with water at pH 3 to simulate the leaching process of acid rain. A total elemental analysis of the filtrate provides information about just those elements solubilized by the mild acid treatment. This can be called the determination of total mild acid leachable elements.

The measurements of total organic carbon (TOC) and total organic halogen (TOX) in a sample are used to assess many types of environmental samples. The analytical methods for TOC employ procedures to physically separate the inorganic carbon, that is, carbonate, bicarbonate, cyanide, and other inorganic substances containing C, from the organic compounds and ions in the sample. The TOC is then measured by oxidation of the organic compounds and ions to carbon dioxide which is determined by one of several techniques. The analytical methods for TOX use procedures to separate inorganic halogen-containing ions from halogen-containing organic compounds and then determine the total halogens in the organic compounds.

Some analytical methods have been developed for the determination of groups of closely related compounds or ions. The classic example is the measurement of combined phenolic materials colorimetrically after a reaction that forms a red derivative with many phenols. However, different phenols form derivatives with somewhat different visible absorption spectra and various phenols have different reactivities with the derivatizing reagent. Therefore this method is calibrated with pure phenol and the measurement gives just an estimate of the total phenols in the sample. For this reason this method is probably no longer widely used and phenols are usually determined as individual compounds with other analytical methods. Subgroups of complex mixtures of congeners are sometimes measured together for convenience of interpretation or for government regulatory programs. These determinations give concentrations of, for example, total tetrachlorobiphenyls, total pentachlorobiphenyls, etc. Similar determinations of chlorinated dibenzo-*p*- dioxins and chlorinated dibenzofurans at each level of chlorination are specified in some analytical methods.

Determination of Specific Substances

Before the development of efficient chromatographic separation techniques and selective and sensitive detectors, analytical methods for the determination of specific analytes in environmental samples were very limited. Those methods depended on highly selective chemical reactions that are relatively rare and difficult to discover, or on very selective physical measurements such as atomic absorption or emission techniques for elemental analytes. Therefore only a relatively few analytical methods for the most common and amenable organic and inorganic compounds or multi-element ions were developed. It would have been essentially impossible, and enormously costly, to develop, test, document, and implement a large number of nonchromatographic analytical methods for a wide range of often similar organic and inorganic analytes.

Vapor phase or gas chromatography (GC) and high-performance liquid chromatography (HPLC) provided capabilities for the separation of microgram (10^{-6}) and smaller quantities of often similar individual substances in complex mixtures. A variety of GC and HPLC detectors were developed and, depending on the detector, the separated substances could be determined with good to excellent sensitivity and with fair to excellent reliability. It was soon recognized that determinations of a broad range of specific chemical substances was not only practical but also essential to achieve the goals of many environmental studies and protection programs.

Gas chromatographic techniques were first applied during the 1960s to synthetic organic compounds and natural products which are often complex mixtures of organic compounds. The earliest applications of these techniques in environmental research were GC-based determinations of petroleum hydrocarbons in polluted air and chlorinated hydrocarbon pesticides in several types of samples. Analytical chemists and environmental scientists concerned about organic pesticides, industrial organic chemicals, and other organic compounds emphasized the development of analytical methods for the determination of specific organic compounds. In contrast, the emphasis in elemental and inorganic analyses during the 1960–1980s was, with a few exceptions, on determinations of the total amounts or concentrations of specific elements in a sample without regard to the specific compounds or ions containing those elements. With the recognition that toxicity and other environmentally significant properties varied widely with the specific compound or ion, inorganic analytical chemists later focused on chromatographic separations of individual species. The term *speciation* came into use, mostly by inorganic chemists, to distinguish this type of analysis from the conventional determinations of the total amounts or concentrations of the elements in a sample.

There are two general strategies for the determination of specific organic or inorganic compounds and ions in environmental samples.[1] The target analyte strategy dominated analytical chemistry before the development of chromatographic separation techniques and is by far the most commonly used with contemporary separation and detection techniques. The broad spectrum strategy became feasible with the development of high-resolution chromatographic separation techniques and spectroscopic detectors.

Target Analyte (TA) Strategy. Target analytes are known substances with known chemical, physical, and other properties. They are either known or thought to be in samples and they must be determined to meet the objectives of the environmental analysis program. The TA strategy is the analysis of the sample with an analytical method that is designed and optimized to determine the target analyte or a group of similar and separated target analytes. If the target analytes are a diverse group with sufficiently different chemical and physical properties, they are divided into subgroups according to their similar properties. Several optimized analytical methods and separate environmental samples are used to determine the analytes in the subgroups.

The TA strategy has many advantages that favor its widespread application. Sample preparation procedures can be designed to separate the target analytes from the sample matrix with maximum efficiency, and to concentrate them in a suitable solvent for further chromatographic separation. Some interferences can be separated from the analytes during sampling or sample processing by pH adjustments, chemical derivatization, evaporation of nontarget substances, or other techniques. Chromatographic separation and detection techniques can be selected to give the best practical resolution of analytes, selective detection, and the lowest detection limits. The detector can be calibrated for

quantitative analysis with standard solutions of the target analyte or analytes, and the analytical method tested, perfected, and validated using test sample matrices fortified with known concentrations of the target analyte or analytes.

The target analyte strategy is widely accepted, understood, and used in analytical chemistry, environmental research, environmental protection programs, and in many other fields of investigation. It is used in most environmental quality surveys and government regulatory compliance monitoring programs. Many analytical methods for a variety of target analytes have been developed, tested, documented, and implemented.[1-4] Cost estimates for analyses are readily made and analytical costs are not difficult to control. The target analyte strategy is used in the vast majority of chemical analyses reported in the scientific literature.

The Broad Spectrum (BS) Strategy. The objective of the BS strategy is to discover the substances present in the sample and to measure their concentrations without a predetermined list of target analytes.[1] As a starting point the BS strategy may follow the general scheme of a target analytical method, but with minimum sample processing to allow a broad variety of generally similar substances to reach the chromatographic separation and the detector. Procedures to remove target analyte interferences are minimized or not used to avoid discarding interesting and potentially important components. If the sample is very complex, it is usually divided into fractions that are likely to contain components with similar properties. Several or more different types of chromatographic separations may be required for the various fractions depending on the components of the sample and the breadth of information desired. The chromatographic detectors are generally spectroscopic detectors that can provide information about the composition and structure of the sample components. This information is used to identify known substances, unexpected substances, and even unknown substances. The most important and widely used detector for the broad spectrum strategy is the mass spectrometer[1], but infrared, nuclear magnetic resonance, and other spectroscopic techniques are sometimes used.

The identification of all or most substances in one or more chromatograms, even with extensive spectroscopic data, can be a challenging and difficult process. The available data may not be sufficient to even tentatively identify all the components, especially if pure authentic samples of suspected substances are not available in the laboratory. Calibration of a broad spectrum method for quantitative analysis is delayed until the desired components are identified. For these reasons, and the general preoccupation with target analytes, the BS strategy is much less common than the target analyte strategy.

The BS strategy is obviously important for the discovery of unknown naturally occurring substances and anthropogenic chemicals in the environment. However, BS strategies are more difficult than TA strategies to develop, document, and implement. Cost estimates for analyses are difficult to make because the number of substances found, identified, and measured is not known until after the samples are analyzed. Therefore analytical costs are difficult to control. Because of the potential costs, the instrumentation requirements, the technical skills needed, and time required, the BS strategy is not often employed especially by programs with limited objectives and budgets. A strategy sometimes used in environmental studies is to develop a target analyte method that can meet the objectives of the study and give some attention to other chromatographic peaks to identify potentially new or unexpected substances.

Single-Analyte and Multi-Analyte Methods

Traditional analytical methods, which were generally developed prior to the widespread application of GC and HPLC techniques, were nearly always designed for a single target analyte, for example, the colorimetric determination of PO_4^{-3}. With the development of GC and HPLC separation techniques, the

determination of several or more similar compounds or ions in a sample was feasible and multi-analyte analytical methods were developed and documented. However, analytical chemists did not generally attempt to include more than about 10–20 target analytes in a method. This strategy was necessary because early GC and HPLC columns were not very efficient and most GC and HPLC detectors were either not selective or had limited selectivity. Sample preparation procedures designed to reduce or eliminate interferences continued to be very important. Analytical methods for chlorinated hydrocarbon pesticides with a GC separation and an electron capture detector require appropriate sample preparation. These multi-analyte pesticide methods are sometimes called multi-residue methods because the pesticides are residues in crops and other samples.

As more efficient high-resolution chromatographic separation techniques were developed, and spectroscopic detectors came into widespread use, the number of target analytes in multi-analyte methods was increased. The separation of most or all target analytes is often feasible and spectroscopic detectors usually provide sufficient information to make correct identifications of target analytes even when some are not fully separated. Well-tested and documented analytical methods for 80–100 or more analytes in some types of samples are presently available.[1–2] Multi-analyte methods significantly reduce the cost of an analysis on a per analyte basis and add support to the already strong justification for high resolution chromatographic separations and spectroscopic detectors.

Remote Laboratory Analyses and Field Analyses

Chemical analyses can be conducted in a laboratory remote from the locations where the samples are taken or in the field near the sampling sites (on-site). On-site analyses can be conducted in a field laboratory which may be a temporary building or a truck trailer, van, or recreational vehicle equipped with utilities services and analytical equipment. Another type of field laboratory is within a materials or fluids processing facility. Alternatively, field analyses can be conducted with mobile or portable instrumentation carried in a small van, sport utility vehicle, moved with a hand cart, or carried by a person. Each of these strategies has some advantages and some disadvantages.

The remote laboratory has the advantages of providing carefully controlled temperature, humidity, ventilation, and background conditions with adequate space and utilities to support a large array of major analytical instrumentation and a staff of skilled analytical chemists and technicians working in a convenient, comfortable, and safe environment. The major disadvantage is that environmental samples must be carefully preserved, shipped, and stored prior to analysis. Furthermore, the analytical results may not be available for several days or weeks because of the time required to transport the samples to the laboratory, incorporate the analyses into work schedules, and service the multiple clients of a remote laboratory.

Field determinations have the major advantage of greatly reducing the time between acquisition of the samples and the availability of the analytical results. This may allow utilization of the results quickly which can provide significant cost savings in the field operations that utilize the analytical information. For example, the rapid availability of results from field determinations can be used to fine-tune a sampling strategy to obtain the most significant samples for detailed remote laboratory analysis. Similarly, the results from field determinations can be used to direct the work of construction crews or well drillers using heavy and costly equipment. Mobile or portable instrumentation can provide rapid and low-cost results from a large number of samples taken over a broad geographic area in a short period of time.

The analytical methods that are feasible in the field may be significantly limited compared to what is feasible in a remote permanent laboratory. If a field laboratory is located in a temporary building or a large truck trailer, and adequate utilities and personnel are available, many of the kinds of analytical methods that are routinely implemented in a remote laboratory may be feasible in the field. However, because of space and power limitations, a broad variety of instrumentation is usually not available and the number of different analytical methods that can be implemented is smaller than in a remote permanent laboratory. Mobile or portable instrumentation is usually more limited and generally sample analyses are less complete and detailed than in a field or remote laboratory.

Discrete Samples and Continuous Monitoring

Field analyses can be conducted with discrete or composite samples similar to the samples used in a remote laboratory, or by continuous monitoring of substances in a flowing stream of gases or liquids. Continuous monitoring is required when the results are needed within a period of time that is shorter than the time required to acquire and to analyze conventional discrete samples. Continuous monitoring is often required to determine substances in a processing facility and provide rapid feedback of results that are used in process control strategies. If continuous determinations are made with sufficient speed to permit changes in sampling or other strategies while the determinations are in progress, these are called real-time analyses.

Continuous environmental monitoring in the field is needed when the substances present and their concentrations are changing rapidly. Discrete samples taken at inappropriate times will give results that do not correctly assess the variable conditions in the atmosphere or a flowing stream. Sample compositing may provide a better assessment, but continuous monitoring and integration of the determinations over time gives the best assessment. The determination of the sources of fugitive emissions that are rapidly dispersed in the atmosphere or in a flowing water stream requires continuous and sometimes mobile continuous monitoring. The concentrations of air pollutants such as CO, NO_x, SO_2, and O_3, that vary widely with sunlight, automobile traffic, wind speed, wind direction, and other meteorological conditions are monitored continuously.

Sampling and analytical equipment used in processing facilities and in the field have some of the same requirements. However, equipment used in a facility may have fewer constraints in regard to size, weight, and power requirements than mobile and portable field equipment. Standard chromatographic separation techniques are generally too slow for continuous measurements although GC separations, especially fast GC, are used for process control.

Analytical Quality Assurance and Control

Analytical quality assurance (QA) is a broad program of actions designed to ensure that the chemical determinations are of known and acceptable quality. The QA program encompasses all aspects of the chemical analysis from the design of the sampling strategy to the documentation of the results. Analytical quality control (QC) is the implementation of specific actions designed to control the quality of the determinations at some defined level of acceptance. All physical measurements have some degree of variability and uncertainty and the QA/QC program should define the limits of these. The quality of analytical determinations is judged by the attributes, or figures of merit, of an analytical method, which include selectivity, sensitivity, detection limits, signal/noise, recovery, accuracy, bias, precision, and validation. These attributes are determined using a variety of special measurements and fortified samples (QA/QC samples) which are described in the analytical method. The time relationship between the analyses of the environmental samples and the determination of the quality attributes is critical in estimating the quality of the environmental determinations.

Published environmental chemical determinations are often used in ways that are not anticipated by the original investigators. The determinations may be used by other investigators to estimate human or ecological exposure, assess the level of industrial discharges, or to develop environmental quality standards or discharge limitations. Therefore the results of the QA/QC measurements, and their time relationships to the analysis of environmental samples, should be permanently attached to the environmental sample results. These QA/QC data should demonstrate that the environmental data were obtained under controlled conditions that provide credibility to the results. Other users of the data, sometimes years in the future, should be able to assess the quality and applicability of the information and take into account the variability and uncertainty in the determinations.

DEVELOPMENT AND DOCUMENTATION OF ANALYTICAL METHODS

The complete development and documentation of an analytical method usually occurs over a period of several years or more. The germ of most new methods is often a research project in which a determination is needed. Organizations that specialize in developing and manufacturing analytical instrumentation frequently play a major role in supporting the development of new analytical methods. But instrumentation alone does not constitute an analytical method. The general acceptance and widespread use of methods depends on many factors including the needs for research or environmental monitoring of various substances, the cost and complexity of instrumentation, the required laboratory or field skills, and especially government regulations. There is a high degree of variability in the completeness, documentation, testing, and validation of published analytical methods.

Research Methods

The basic concepts and fundamentals of most analytical methods are usually first published in scientific journals, for example, The

American Chemical Society publications *Analytical Chemistry, Environmental Science and Technology,* and the *Journal of the American Chemical Society.* These descriptions are usually brief and often just summarize the techniques and procedures of the method. Research reported in scientific journals is often focused on a detailed investigation of a narrow subject area, and new analytical techniques and procedures developed for the research are rarely tested in a broader context. Analytical quality assurance and control are often minimal or not described. Techniques and procedures described in scientific research journals are the beginnings of analytical methods, however, they usually require considerable development, modifications, and testing before they become widely accepted analytical methods.

Methods in Development

Academic research groups, industrial laboratories, and government agencies interested in conducting environmental surveys or monitoring programs often adapt research techniques and procedures to the broader needs of the survey or program. During this stage considerable experimentation is underway to evaluate equipment, instrumentation, chemical reagents, and other materials used in the method. The incipient method may be tested with many potential analytes and sample matrices. Modifications are implemented to minimize or eliminate problems discovered during the development process. Techniques and procedures are developed for identifications of analytes, calibration of instruments for quantitative analysis, and analytical quality assurance and control. Methods in development may be described in a series of draft versions that are not usually formally published but may appear in bound or unbound technical reports. These methods are described in more detail than in scientific journal articles, and descriptions often contain detailed information about required equipment, supplies, reagents, instrumentation, and personnel skills. The technical reports and draft method descriptions are often distributed

informally through personal contacts, scientific conferences, or the Internet.

Methods Published by Standard-Setting Organizations

Standard-setting organizations that publish analytical methods for environmental analyses include The American Society for Testing and Materials (ASTM),[4] The American Public Health Association (APHA),[3] The American Water Works Association (AWWA), The Water Environment Federation (WEF), The International Standards Organization (ISO), and the Association of Official Analytical Chemists (AOAC). These organizations assemble working committees of experts that consider developed, widely used, and generally accepted analytical methods for publication as standardized analytical methods. Requirements of individual organizations vary and these may include a description of the method in a specific editorial format and the availability of multi-laboratory validation data for the proposed analytical method. Some of these organizations also sponsor or participate in multi-laboratory validation studies of proposed analytical methods. When an analytical method is published by one of these organizations it usually has been thoroughly tested and used by many laboratories and analysts, and is widely known and accepted by specialists in that type of determination. Published methods are reviewed periodically, updated as needed, and may eventually be replaced by entirely new methods.

Methods Published or Referenced in Government Agency Regulations

The U. S. Environmental Protection Agency (USEPA) and other federal, state, and local government agencies in the United States may require chemical analyses to determine compliance with air quality, water quality, liquid waste discharge, solid waste disposal, and other environmental regulations. The analytical methods specified for these purposes may be methods in development, methods published by standard-setting organizations, or new methods documented in the regulatory proposals.[2] The USEPA has promulgated two types of analytical methods. Some regulations, for example, some drinking water and waste water regulations, require either analytical methods designated in the regulations or approved alternative test methods for compliance monitoring. Other regulations, for example, some USEPA solid waste regulations, include or reference analytical methods that are suggested or optional but allow any other appropriate analytical methods.

CHARACTERISTICS OF ANALYTES, SAMPLES, AND SAMPLING TECHNIQUES

The physical and chemical properties of analytes and the nature of the sample have a major impact on, and often limit, the sampling and other procedures and techniques that can be employed in an analytical method. Major issues that must be considered when developing an analytical method are the volatilities, thermal stabilities, photochemical stabilities, polarities, water solubilities, and chemical reactivities of the sample components or target analytes; the physical state of the sample; and the nature of the sample matrix. Analytes, whether organic or inorganic, can be broadly divided into three categories based partly on vapor pressure, or volatility, at ambient temperature and on some other physical and chemical properties. There are major differences in the procedures and techniques used to acquire and process condensed-phase and vapor-phase samples.

Sampling ground water requires expert selection of procedures and techniques to avoid significant analytical errors.

Volatile Analytes

Volatile analytes are usually defined as those having vapor pressures (VP) greater than about 0.1 Torr at 25° C and an external pressure of 760 Torr (1 Torr = 1 mm of Hg or 133 pascals (Pa)). Figure 8.1 shows the structures, molecular weights (MW), boiling points (BP), and VPs of three representative volatile compounds of environmental interest. Most volatile analytes have MWs below 200 but a low MW does not guarantee that an analyte will be volatile. Many substances have MWs below 200 but they are not volatile because they are ionic or have polar groups of atoms or engage in hydrogen bonding with other molecules. Some compounds have high VPs, for example, MTBE in Fig. 8.1, but reduced volatility in water because of their high water solubilities. Boiling points of volatile substances range from below 0° C to above 200° C. Some compounds with BPs at the upper end of this range are surprisingly volatile. For example, nitrobenzene in Fig. 8.1 has a BP of 211° C but still has a VP of 0.245 Torr at 25 C. At normal room temperatures the distinctive odor of nitrobenzene vapor is readily detected by most people.

Semivolatile Analytes

Semivolatile analytes are usually defined as those having VPs in the range of 0.1 Torr to

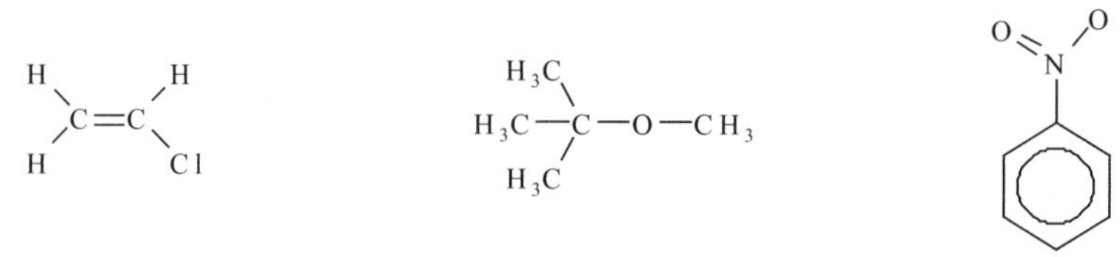

Vinyl chloride
MW 62.5
BP -13 °C
VP 2980 Torr @ 25 °C

Methyl *tert.*-butyl Ether (MTBE)
MW 88
BP 55 °C
VP 249 Torr @ 25 °C

Nitrobenzene
MW 123
BP 211 °C
VP 0.245 Torr @ 25 °C

Fig. 8.1.   The structures, molecular weights (MW), boiling points (BP), and vapor pressures (VP) of three representative volatile compounds of environmental interest.

about 10^{-9} Torr, but this range is approximate and some substances with VPs in this range are considered volatiles or even nonvolatiles (next section). For example, Hg^0 has a VP of 2×10^{-3} Torr at 25° C, but is often considered a volatile analyte. Most semivolatile analytes have molecular weights in the range of 100–500, but a MW in this range does not guarantee that a compound is a semivolatile. However, because semivolatiles nearly always have higher MWs and contain more atoms than the volatiles, they have a significantly larger number of isomers, congeners, and chiral forms. Semivolatiles are usually devoid of structural groups that are susceptible to thermal decomposition below about 300 C, or cause high polarity, or impart high water solubility, or are very chemically reactive. Figure 8.2 shows the structures, MWs, some melting points (MPs), and VPs of four representative semivolatile compounds of environmental interest. Both benzo[a]pyrene

and 2,3,7,8-tetrachlorodibenzo-*p*-dioxin have VPs at the low end of the semivolatile range, but they have little or no polarity and tend to behave as do other semivolatiles with higher VPs. The VPs of semivolatile compounds are generally insufficient to give vapor concentrations at ambient temperatures that can be detected by a distinctive odor.

The large number of potential congeners and isomers of some semivolatile compounds is illustrated in Table 8.1 which shows the numbers of possible chlorinated biphenyl, chlorinated dibenzo-*p*-dioxin, and chlorinated dibenzofuran congeners and isomers. The total number of chlorinated congeners of each parent compound is the sum of the possible isomers at each level of chlorination. Dibenzofuran has more possible chlorinated congeners and isomers than chlorinated dibenzo-*p*-dioxin (Fig. 8.2) because it has just one ring oxygen and a less symmetrical structure. If F, Br, I, CH_3, or any other uniform

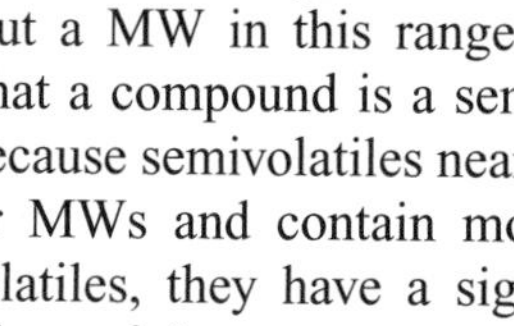

Benzo[α]pyrene
MW 252
MP 179 °C
VP 5.49 x 10^{-9} Torr @ 25 °C

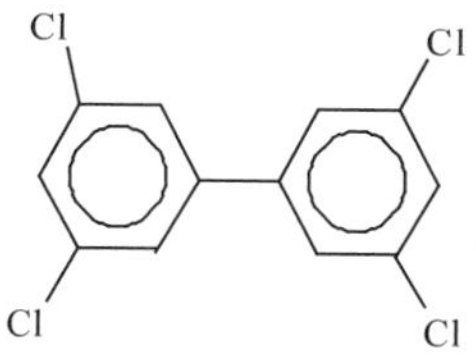

3,3',5,5'-Tetrachlorobiphenyl
MW 292
VP 8.45 x 10^{-6} Torr @ 25 °C

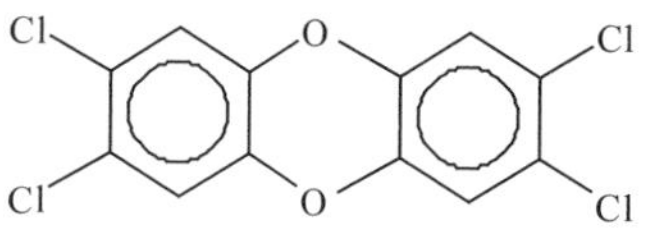

2,3,7,8-Tetrachlorodibenzo-*p*-dioxin
MW 322
MP 305 °C
VP 1.5 x 10^{-9} Torr @ 25 °C

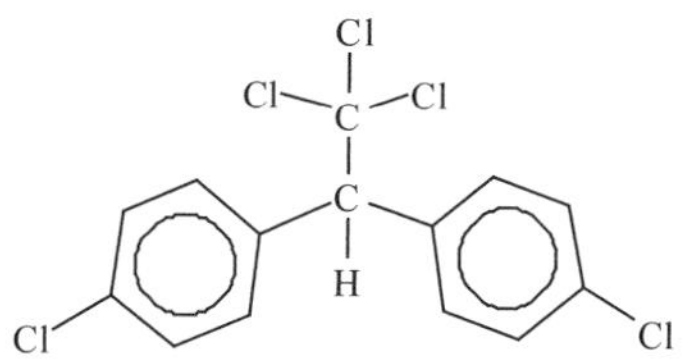

1,1-bis(4-Chlorophenyl)-2,2,2-
trichloroethane (4,4'-DDT)
MW 354.5
MP 109 °C
VP 1.60 x 10^{-7} Torr @ 20 °C

Fig. 8.2. The structures, MWs, some melting points (MP), and VPs of four representative semivolatile compounds of environmental interest.

TABLE 8.1.   The Number of Possible Chlorinated Biphenyl, Chlorinated Dibenzo-*p*-dioxin, and Chlorinated Dibenzofuran Congenors and Isomers

Parent Compound	Total Congeners	Cl_1	Cl_2	Cl_3	Cl_4	Cl_5	Cl_6	Cl_7	Cl_8	Cl_9	Cl_{10}
Biphenyl	209	3	12	24	42	46	42	24	12	3	1
Dibenzo-*p*-dioxin	75	2	10	14	22	14	10	2	1		
Dibenzofuran	135	4	16	28	38	28	16	4	1		

substituent replaces Cl, the same numbers of possible congeners and isomers would exist. If different atoms or groups of atoms were mixed as substituents on these parent compounds, a significantly larger number of different substances, congeners, and isomers are possible.

Nonvolatile Analytes

All other compounds and essentially all ions are classified as nonvolatile. Substances in this category have VPs lower than about 10^{-9} Torr or have structural groups that are susceptible to thermal decomposition below about 300 C, or that cause high polarity, or that impart high water solubility, or that are very chemically reactive. Figure 8.3 shows the structures, MWs, some MPs, and VPs of six representative compounds of environmental interest from this group. All these examples except the herbicide glyphosate have reported VPs in the semivolatile range, but they do not behave as typical semivolatile compounds. Nitroglycerine and N-nitrosodiphenylamine undergo thermal decomposition at temperatures below well below 300° C. The pesticide carbofuran and the herbicide diuron are more thermally stable, but they also tend to decompose below 300° C and on hot surfaces. The polarities of the functional groups of this class of compounds cause adsorption on polar surfaces and dipole–dipole interactions with other molecules. Nearly all carboxylic acids, for example, the herbicide 2,4-D in Fig. 8.3, are susceptible to decarboxylation at elevated temperatures and they interact with basic substances or basic surfaces which reduces their volatility. Similarly, basic substances such as amines and some other nitrogen

compounds interact with acids and acidic surfaces which reduces their volatility.

Condensed-Phase Samples

If only semivolatile and nonvolatile components are of interest, or the sample only contains these categories of analytes, sampling procedures for most condensed-phase samples are not difficult and require only a few precautions. These include the materials used for sample containers, the cleaning of sample containers, and the preservation of the sample. Water samples for elemental analyses are treated with acid at the time of sampling, or well before analysis, to reduce the pH to $< \sim 2$. This ensures the solubility of metal-containing ions which can precipitate or adsorb on container walls at a pH $> \sim 2$. Glass containers are not used for elemental analysis samples because aqueous acid solutions can leach trace elements from glass. Samples are taken in plastic containers which provide the added benefits of reduced weight compared to glass, little or no breakage during handling and shipping, and single-use containers that can be disposed of at a plastic recycling facility. However, samples for the determination of organic analytes are taken in glass containers to prevent background contamination of the samples by organic compounds that can leach from plastic materials. These samples are often preserved by addition of acid to make the sample pH $< \sim 2$ to retard microbiological degradation of some analytes. Acid leaching of glass containers generally has no effect on the concentrations of organic analytes.

Condensed-phase samples containing volatile analytes require special techniques because the volatile components are elusive

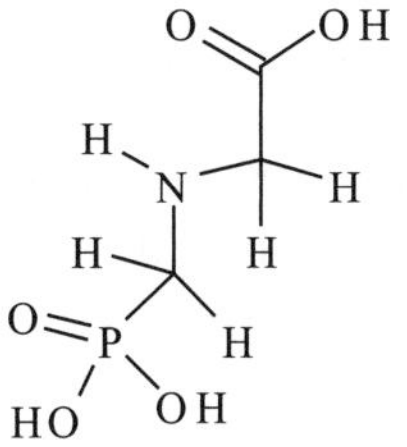

Glyphosate
MW 169
MP 200 °C
VP 2.89 x 10^{-10} Torr @ 25 °C

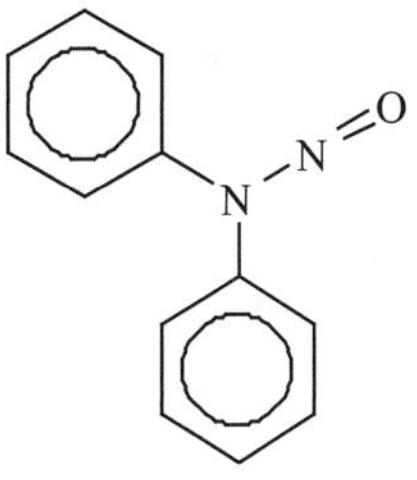

N-nitrosodiphenylamine
MW 198
VP 1.98 x 10^{-3} Torr @ 25 °C

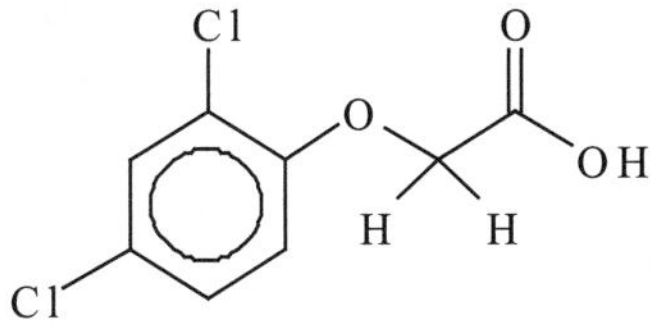

2,4-Dichlorophenoxyacetic
acid (2,4-D)
MW 221
MP 141 °C
VP 6 x 10^{-7} Torr @ 25 °C

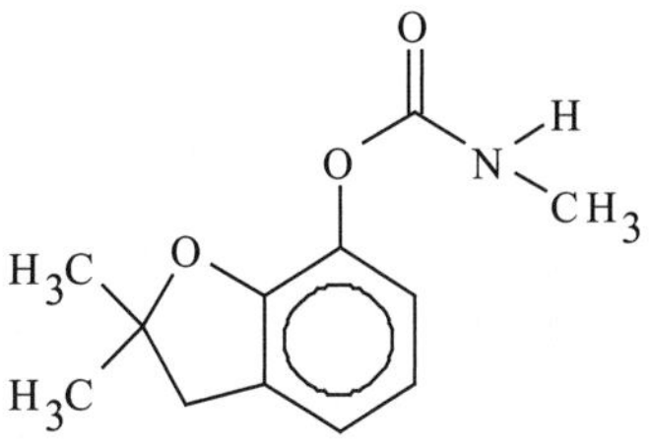

Carbofuran
MW 221
MP 151 °C
VP 4.85 x 10^{-6} @ 19 °C

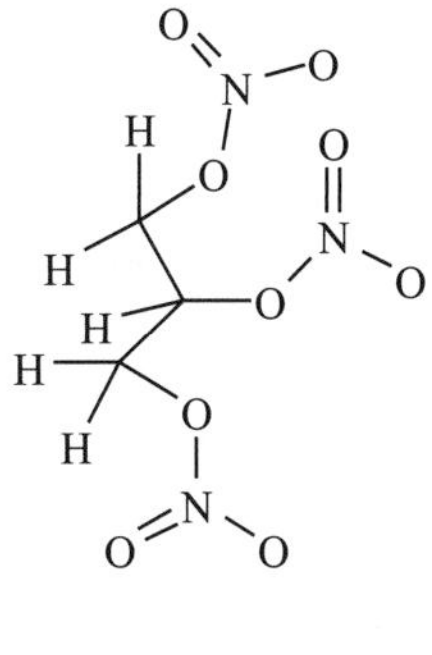

Nitroglycerine
MW 227
Decomposes 50-60 °C
VP 2.6 x 10^{-4} Torr @ 20 °C

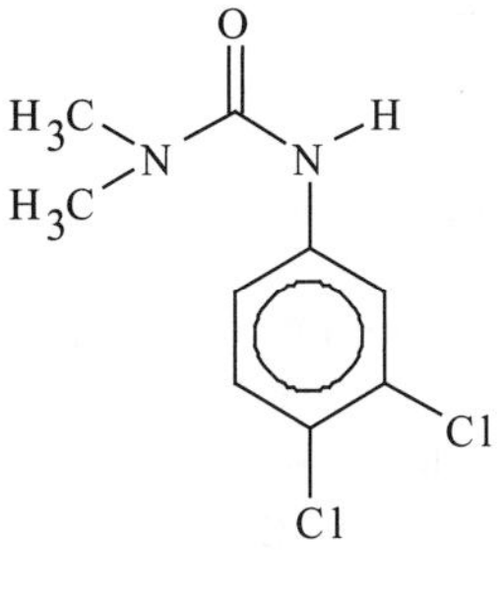

Diuron
MW 233
MP 158 °C
VP 2.7 x 10^{-6} Torr @ 30 °C

Fig. 8.3. The structures, MWs, some MPs, and VPs of six representative nonvolatile compounds of environmental interest.

and can be readily lost during sampling and sample processing. Water samples must be taken with little or no agitation of the sample and poured into the container so there is no air space, often called a headspace, between the top of the sample and the air-tight seal of the container. This prevents vaporization of volatile components into the headspace during shipment and storage which reduces the concentration in the aqueous phase. Aliquots of water samples are removed from sample containers by inserting a syringe needle through the inert air-tight septum seal and drawing water into a gas-tight syringe. Solid samples, for example, soils and sediments, may also contain volatile components trapped in pores or dissolved in associated water and these components are easily lost during sample handling. Solid and semisolid samples are carefully and quickly placed in wide-mouth sample containers that are sealed in the field and not opened during any subsequent processing of the sample. Volatile components are usually partitioned into water, organic solvents, or the vapor phase and the extracts are analyzed using techniques developed for liquid- or vapor-phase samples. Water or an organic solvent are added to the sample containers through inert and air-tight septum seals with a syringe.

Ground water sampling is a significant challenge because water must be lifted to the surface through a bore hole from depths ranging from a few meters to 75 meters or more. The lifting process can disturb the equilibrium between dissolved analytes and analytes associated with particulate matter, and significantly change the temperature and pressure of the water sample. These factors can cause changes in concentrations of analytes and raise questions about the representativeness of the sample. Volatile components of ground water samples are particularly susceptible to losses caused by temperature and pressure changes and degassing of the sample. Several types of down-hole water samplers are available and several types of pumps are used to purge a well and lift a stream of ground water to the surface. The materials of construction of the down-hole

samplers, pumps, tubing used with pumps, and well casings can also have an impact on the concentrations of analytes in the sample. Considerable research has been conducted to evaluate the materials and techniques used to sample ground water, however, uncertainties remain because there is no satisfactory ground water standard of reference for comparison of various techniques.

Semivolatile analytes in vapor-phase samples are often associated with particulate matter or aerosols that are collected on glass or quartz fiber filters in a flowing air stream.[1] However some semivolatile analytes that have higher VPs, for example, 3,3',5,5'-tetrachlorobiphenyl in Fig. 8.2, can slowly vaporize from a particle trapped on a filter in the flowing air stream, and will be lost unless captured by an in-stream sampling device. Small glass or metal tubes containing polyurethane foam are often used to capture vaporized semivolatile analytes. Other-solid phase adsorbents, which are described in the next section, are also used to trap semivolatile analytes vaporized from particulate filters.

Vapor-Phase Samples

There are two general approaches to sampling air, or vaporous emissions from stationary (stack) and mobile (automobile, truck, etc.) sources, for the laboratory determination of volatile analytes.[1] Bulk vapor-phase samples can be taken in the field in various containers and transported to a remote or field laboratory for analysis. Containers used for bulk vapor-phase samples include flexible polyvinyl fluoride (Tedlar™) bags, evacuated glass or metal reservoirs, and thermally insulated cryogenic collection vessels. Alternatively, the volatile analytes can be separated from the main components of air in the field and just the analytes and their collection devices transported to the laboratory. The principal techniques used to separate volatile analytes from air in the field are cryogenic traps, impingers, and solid-phase adsorbents.

Bulk Vapor-Phase Samples. Flexible plastic sample bags are generally limited to vapor-phase samples that can be analyzed within a short time after sample collection. This limitation is due to potential losses of analytes by surface adsorption and surface chemical reactions. Transportation of inflated bags over long distances to a remote laboratory is cumbersome and can result in total losses of samples due to punctures and other accidents. Plastic bags are used in some laboratory operations, for example, to collect automobile exhaust and vaporized fuel, and for samples that can be conveniently transported to a laboratory.

Evacuated stainless steel canisters are widely used collection devices for ambient air samples.[6] Sample canisters have smooth and inert internal surfaces and few or no active sites that adsorb volatile analytes or catalyze chemical reactions. The 1–6 L canisters are easily transported to a remote or a field laboratory. Canisters are leak tested and cleaned in the laboratory prior to use, evacuated to about 5×10^{-2} Torr or less, and transported to the sampling site where samples are taken by opening the sampling valve. Composite samples can be taken over time and/or space and an in-line pump can be used to pressurize the container with either additional sample air or pure air if sample dilution is required. Pressurized samples are useful when longer-term composite samples are taken or when larger samples are needed to lower detection limits.

Condensation of an entire air sample with liquid nitrogen or liquid helium has been used for many studies. However, this technique is expensive to implement and requires specialized portable equipment for handling cryogenic fluids in the field.

Separated Analytes. Cold trapping is used to separate volatile analytes from the main components of air in the field. Air is drawn by a pump through an inert, often nickel, metal tube immersed in a fluid at a very low temperature, for example, $-150°$ C. The tube may be packed with some inert material such as Pyrex™ glass beads and the temperature is sufficient to condense most analytes but insufficient to condense oxygen or nitrogen.

Plugging of the condensation tube with ice or other solids can be a major problem when sampling large volumes of humid air. This problem is addressed by the employment of air dryers that trap moisture but allow the nonpolar volatile analytes to pass into the cold stage of the trapping system. However, the more polar and water-soluble analytes are also removed by efficient air drying systems. Another potential problem is that some analytes may react during subsequent processing with trapped ozone, nitrogen oxides, or other substances present in the air. A practical limitation of the cold trapping technique is the requirement for liquid nitrogen or liquid argon in the field during extended sampling periods.

Impingers are used to extract various substances from vapor-phase samples.[1] An impinger is a closed glass or metal vessel with an inlet tube that extends to near the bottom of a liquid and an outlet tube well above the surface of the liquid. The impinger may contain various aqueous or nonaqueous liquids including solutions of derivatizing agents. Vapor is drawn by a pump into the inlet tube and bubbled through the liquid which dissolves soluble analytes or the analytes react with reagents in solution to form soluble compounds. The liquid may also condense various substances, including water vapor, and collect fine particulate material that passes through a coarse filter or another separation device. A important advantage of an impinger compared to some bulk air sampling techniques is that hundreds of liters of air can be drawn through the device over a period of several hours. Sampling trains with multiple collection devices in series are used to collect different fractions of a vapor-phase sample. Components of sampling trains can include particulate filters of several types and sizes, particle-size separation devices, multiple impingers, and solid-phase adsorbents.

Solid-phase adsorbents are simple and inexpensive devices used to separate volatile analytes from the principal components of air in the field.[1] A porus solid-phase adsorbent is placed in a glass or metal tube which is taken to the field where air is drawn through the adsorbent to trap the analytes. The adsorbent tube is then sealed and returned to the laboratory for analysis. Many types of solid-phase adsorbents are used including alumina, activated carbons of various types, charcoal, graphitized carbon black (Carbopack™ B and C), carbon molecular sieves, GC packing materials such as Chromosorb™ 101 and 102, ethylvinylbenzene-divinylbenzene copolymer (Porapak Q™), styrene-ethylvinylbenzene-divinylbenzene terpolymer (Porapak P), silica gel, 2,6-diphenyl-*p*-phenyleneoxide (Tenax-GC™), styrene-divinylbenzene copolymers (XAD-1™, XAD-2™, and XAD-4™), and acrylic ester polymers (XAD-7™ and XAD-8™). Adsorbent trapping is applicable to a wide variety of nonpolar and some polar volatile analytes.

Some adsorbents do not bind analytes strongly, which is an advantage for subsequent processing, but may result in the gradual release of some volatile analytes during long vapor sampling periods. Long sampling periods are often required, for example, with ambient air, because of generally lower analyte concentrations compared to vapor-phase samples from stationary or mobile sources of air pollutants. Breakthrough of the more volatile polar and nonpolar analytes from the adsorbent is controlled by placing additional adsorbents in the tube or additional sampling tubes in a sampling train. Analytes adsorbed on solid-phase materials may also react with substances in the air, for example, nitrogen oxides and ozone, to produce products not in the original sample. Long air sampling times also risk oxidation of some adsorbents and the production of background substances and other artifacts. These potential problems are usually evaluated during analytical method development and controlled by using appropriate quality control procedures

PROCESSING OF SAMPLES BEFORE DETERMINATION OF THE ANALYTES

Some processing of samples before the determination of the analytes is often required to achieve optimum analytical method performance. Sample processing also defines exactly what elemental forms, compounds, ions, oxidation states, and physical phases are included in the determination of specific substances, or in the total amount of an element or a group of related substances in a sample. The degree of sample processing that is either appropriate or needed depends on a variety of considerations including:

- The chemical analysis strategy, that is, a total, target analyte, or broad spectrum determination
- The physical phases of the sample that are included in the determination
- Whether the samples will be analyzed in a remote or a field laboratory or with laboratory, mobile, or portable instrumentation
- The complexity of the sample matrix
- The number of target analytes
- The diversity of properties of the target analytes
- The nature and concentrations of analyte interferences
- The detection limits required
- The efficiency of the chromatographic separation used in the analytical method
- The selectivity of the chromatographic detector or the nonchromatographic measurement technique
- The willingness to risk severe contamination of a GC or HPLC column, or some other critical instrument component, by a concentrated or high background environmental sample

Some sample matrices, for example, ambient air or drinking water, are analyzed without sample processing with some analytical methods. Volatile compounds in ambient air collected in canisters or volatile compounds trapped from ambient air on solid-phase adsorbents are determined directly. Similarly, the total concentrations of some elements in low turbidity drinking or surface water samples preserved at a pH <2 can be determined without further sample processing. Common inorganic anions, for example, Cl^-, Br^-, NO_3^-, etc., are determined in surface, ground, drinking, and some other water samples without sample processing. However, many sample matrices and analytical methods require at least some, and sometimes considerable, sample processing prior to the determination of the analytes.

Semivolatile organic analytes trapped on air filters or on solid-phase adsorbents or in water, soil, and other solid samples are extracted with an organic solvent, or a solvent mixture, and the extracts are concentrated by evaporation of the solvent before the determination of the analytes. Solid samples are often extracted with the classical Soxhlet apparatus or with a variety of other techniques including several that use organic solvents at elevated temperatures and pressures. If a sample is highly concentrated, for example, an industrial wastewater or a soil from a hazardous waste land fill, fractionation of the solvent extract before determination of the analytes is usually required. Fractionation is often accomplished with open column liquid chromatography or preparatory scale high-performance liquid chromatography. The determination of total elements in particulate matter collected on air filters, in most water samples, or in solids requires an acid digestion of the sample to ensure that the elements are in water-soluble chemical forms for the measurement. The vigor of the digestion also determines the recovery of the elements from sediment and other insoluble material.

Invariably the analytical method developer is required to make compromises between the amount and complexity of the sample processing and the separating power, selectivity, and other attributes of the chromatographic or nonchromatographic determination. These compromises are often strongly influenced by the projected cost and time required for various method options and by the desired quality, detail, and reliability of the results. Major issues usually are the availability of laboratory or field equipment and instrumentation, the experience and skill of the staff in using the equipment, and other laboratory or field infrastructure required to complete the analyses of the samples. Most research and standard analytical methods contain many compromises that may not be clearly defined in the method description, but should be understood by the user.

CHROMATOGRAPHIC ANALYTICAL METHODS

Nearly all contemporary analytical methods designed for specific compounds or ions in environmental samples employ some type of chromatographic separation. Gas chromatography is the dominant technique for the separation of volatile and semivolatile analytes (Figs. 8.1 and 8.2 and Table 8.1). Reverse-phase high-performance liquid chromatography is used for the separation of nonionic analytes that are thermally unstable, nonvolatile, or reactive and not amenable to GC (Fig. 8.3). High-performance ion exchange chromatography and capillary electrophoresis are used for the separation of ionic substances. Analytes, either fully or partially separated, are sensed by in-line chromatographic detectors that produce electronic signals that are usually converted into digital form and stored in computer data systems. The computer systems, which are used with nearly all contemporary chemical analytical instrumentation, typically have software to control the operation of the instrument, acquire and store raw data, and reduce the data to more usable analytical information.

Chromatographic Separation Techniques

Gas Chromatography. The basic components of a gas chromatograph are a carrier gas system, a column, a column oven, a sample injector, and a detector. Very pure helium is the near-universal carrier gas for environmental and many other analyses. Open tubular GC columns are constructed of fused silica with low-bleed stationary phases of varying polarity chemically bonded to the silica surface. Columns are typically 30–75 m in length and have inside diameters (ID) in the range of about 0.25–0.75 mm. The column oven is capable of precise temperature control and temperature programming at variable rates for variable times.

Analytes are introduced into GC columns with several techniques. An aliquot of a relatively concentrated vapor or air sample, for example, from a plastic bag or a canister, can be introduced into a short section of tubing of known volume, called a sample loop, and subsequently purged with carrier gas into the GC column. Volatile analytes in ambient air samples in a canister or trapped on a solid-phase adsorbent are usually concentrated and focused in a cryogenic trap or a secondary adsorbent trap, then thermally vaporized into the GC carrier gas stream. However, in some analytical methods, volatiles trapped on an absorbent are thermally desorbed directly into the GC column. Aliquots of organic solvent extracts from various aqueous and solid samples are usually injected with a syringe into the carrier gas stream in a heated injection port. Both manual and automated syringe injection systems (autoinjectors) are used and the latter are generally very reliable, precise, and have the capacity to process many samples unattended.

Mixtures of analytes are separated by repeated equilibrations between the vapor state, where the analytes are entrained in the flowing carrier gas toward the detector, and the absorbed or dissolved state, where they are attracted to the stationary phase on the wall of the column by generally weak molecular forces. Analytes that tend to favor the absorbed state move more slowly through the column to the detector than do analytes that favor the vapor state. Complete separations of a few nanograms (10^{-9}) or less of each of 40 to 50 or more analytes in 30 min or less is not uncommon.

High-Performance liquid Chromatography. The basic components of a high-performance liquid chromatograph are a high-pressure mobile-phase delivery system, a metal column packed with fine particles containing the stationary phase, a sample injector, and a detector. A high-pressure pump is used to force the mobile-phase solvent or solvent mixture through the packed column. The term *high performance* is often used to distinguish this technique from open-column liquid chromatography conducted at atmospheric pressure with gravity flow of the mobile phase. Columns vary in diameter from <1–5 mm ID, or larger, and from a few cm in length to 30 cm or more. Column packings consist mainly of silica particles, usually 3–10 μm in diameter, coated with low-bleed stationary phases of varying polarity chemically bonded to the silica surface. Organic polymers are also used as stationary phases in analytical separations. Automated or manual syringe injections of aliquots of liquid samples and known-volume sample loops are used for sample injection.

The dominant HPLC technique is the reverse-phase configuration in which a nonpolar or slightly polar stationary phase is used with a more polar mobile phase that is often water, methanol, acetonitrile, or mixtures of these solvents. The normal-phase configuration, which was developed before reverse phase, uses a more polar stationary phase and a nonpolar mobile phase, but it is not often used in contemporary HPLC. A vast array of organic compounds containing a variety of functional groups are retained on nonpolar or slightly polar stationary phases, and are sufficiently soluble in more polar mobile phases to give excellent separations. Analytes are retained on the column by an equilibrium process in which the dissolved molecule is alternately associated with the stationary phase, through weak noncovalent bonding interactions, and the mobile phase where it is transported toward the detector.

Isocratic elution is the use of a mobile phase that has a constant composition throughout the elution of analytes from the column. Gradient elution is the gradual changing of the mobile phase composition as the analytes elute from the column. Gradient elution is frequently used to enhance analyte resolution and shorten the time required for a separation. The pH of typical mobile phases can be adjusted over a reasonable range to improve resolution and selectivity. Various buffer substances, salts, and ion pair reagents can be used to control pH, ionic strength, and to facilitate the separation of some analytes. Reverse-phase HPLC is more complex than GC because of the number of operational parameters including column dimensions, column packings, mobile phases, gradient elution, and various mobile phase additives.

High-Performance Ion Exchange Chromatography. Ion exchange is another form of HPLC that uses a stationary phase consisting of a cross-linked synthetic organic polymer, often called a resin, with $-SO_3H$ or $-NH_2$ groups attached to phenyl or other aromatic rings on the polymer backbone. Mobile phases are usually water or water and a miscible organic solvent. In basic solution the SO_3H groups of a cation exchange resin are ionized and consist of $-SO_3^-$ groups and associated counter ions, for example, Na^+. Cationic analytes are retained on the cation exchange column by displacing the resin counter ions in an equilibrium process. The cationic analytes are subsequently eluted with a mobile phase containing a high concentration of counter ions or counter ions of a higher charge, for example, Ca^{++}. Isocratic and gradient elutions are used with gradients in both solvent composition and counter ion concentration. Mobile-phase pH is a very important operational parameter that can have a significant impact on the retention or elution of various ionizable analytes from a resin.

In an acid solution the amino groups of an anion exchange resin are protonated and exist as $-NH_3^+$ groups and associated counter ions, for example, Cl^-. Analyte anions are retained on the column by displacing the resin counter ions in an equilibrium process. The anion analytes are subsequently eluted with a higher concentration of counter ions or some type of gradient elution. Anion exchange chromatography is widely used for the separation of inorganic anions and chelated metal anions in aqueous samples. With some natural or industrial water samples that contain high concentrations of ions such as Na^+, Ca^{++}, Mg^{++}, Cl^-, and SO_4^{--}, cation and anion analytes may not be retained on the column and this can result in poor analyte recoveries.

Other Chromatographic and Related Techniques. Supercritical fluid chromatography (SFC), capillary electrophoresis (CE), and several related separation techniques are occasionally used in environmental chemical determinations. The CE technique is very important in biochemistry and molecular biology because of the very high resolving power that can be achieved, the high speed of separations compared to HPLC, and its ability to separate charged species such as proteins, peptides, and deoxyribonucleic acid fragments. These techniques have considerable potential for environmental analyses and may emerge as very important techniques in the future.

Chromatography Detectors

Online detectors for GC, HPLC, and other chromatographic separation techniques are conveniently divided into three general classes:

- Nonselective detectors
- Selective detectors
- Spectroscopic detectors

Nonselective detectors respond to most or all changes in the composition of the carrier gas or mobile phase and are capable of detecting nearly all entrained or dissolved analytes. The GC flame ionization and thermal conductivity detectors and the HPLC refractive index detector are examples of nonselective detectors. Nonselective detectors provide no direct information about the identity of the analytes except the time of arrival, the time of maximum concentration, and the time of return to pure carrier gas or mobile phase. Thus the peak shape of the analyte is usually well defined, the measurement of peak area or height can be precise, but the information needed to identify the analyte that caused the change is weak. Identifications are based on comparisons of measured retention times of separated analytes with retention times of expected analytes measured under the same chromatographic conditions. Although measured retention times can be very precise, especially with fused silica capillary GC columns and multiple internal standards, there is a high probability of coelution of two or more analytes in most environmental samples. This is because the peak capacity of a chromatogram, which is the number of analytes that can be fully separated, is limited. For most environmental samples the peak capacity is much smaller than the number of possible analytes that respond to a nonselective detector. Therefore, there is a significant risk of misidentifications or false positives.

Selective detectors respond to only certain classes of analytes and they are often used in environmental chemical determinations. The electron capture, photoionization, electrolytic conductivity, and flame photometric GC detectors are selective for limited groups of analytes. Selective HPLC detectors include electrochemical detectors and the single wavelength ultraviolet-visible (UV-VIS) absorption and fluorescence detectors. Preinjection or postcolumn online chemical reactions are used to convert nonresponsive analytes into derivatives that respond to a selective detector. The information produced by a selective detector about the nature of the analyte is improved, compared to the nonselective detector, but still is not strong because the most significant piece of information obtained is the retention time. Although the range of potential analytes that responds to a selective detector is much narrower than the range that responds to a nonselective detector, multiple potential analytes still can have the same retention time. The nature of the sample is a factor in considering whether multiple analytes could be present that may have the same retention time.

Spectroscopic detectors measure partial or complete energy absorption, energy emission, or mass spectra in real-time as analytes are separated on a chromatography column. Spectroscopic data provide the strongest evidence to support the identifications of analytes. However, depending on the spectroscopic technique, other method attributes such as sensitivity and peak area measurement accuracy may be reduced compared to some nonselective and selective detectors. The mass spectrometer and Fourier transform infrared spectrometer are examples of spectroscopic detectors used online with GC and HPLC. The diode array detector, which can measure the UV-VIS spectra of eluting analytes is a

selective spectroscopic detector because only some analytes absorb in the UV-VIS region of the spectrum. The mass spectrometer, which can use a variety of ionization techniques, is probably the most widely used GC and HPLC detector for environmental chemical determinations.[1] All detectors respond, to some extent, to natural background substances in environmental samples and sample extracts. Background chemical noise is highest in the most contaminated environmental samples and at the highest instrument sensitivities, but spectroscopic detectors are best equipped to distinguish chemical noise from environmental analytes.

NONCHROMATOGRAPHIC ANALYTICAL METHODS

Analytical methods that do not employ a chromatographic separation of analytes are widely used for some environmental determinations. These methods generally depend on highly selective physical measurements or selective chemical reactions. Atomic absorption spectrometry (AAS), atomic emission spectrometry (AES), and mass spectrometry (MS) are most often used for determinations of the elements, but electrochemical and colorimetric techniques are sometimes used. Derivatives of analytes, which are sometimes called complexes, that strongly absorb in the UV-VIS spectrum are employed in colorimetric methods for the elements and some compounds and multi-element ions. Ozone, nitrogen oxides, and some other substances in air are determined by selective chemical reactions that produce measurable light emission (chemiluminescence). Fluorescence spectrometry is used in some methods, especially with fiber optics technology, for remote monitoring. Immunoassays have been developed for a few analytes and they are especially useful in field analyses.

Elemental Analysis

Atomic emission and atomic absorption techniques are highly selective because the number of elemental analytes is small and each has a unique atomic spectrum. Frequencies are selected for measurements to maximize selectivity and sensitivity, minimize interferences, and correct for interferences when necessary. Mass spectrometry is the separation and measurement of ions in the gas phase by their mass-to-charge ratios. Elemental analyses by mass spectrometry provides high selectivity, sensitivity, and the ability to correct for interferences when necessary.[1] Colorimetric and electrochemical techniques were used more frequently for elemental determinations before the development of the three major spectroscopic techniques. Their use has decreased significantly because of inherent limitations in selectivity, sensitivity, susceptibility to interferences, and the need for chemical processing to prepare energy-absorbing derivatives.

Atomic Absorption. During the 1960s flame AAS became the dominant technique for determinations of the elements in low concentrations in water samples and aqueous extracts of other samples. Aqueous sample aerosols are injected directly into the flame and precise determinations of many elements can be made. However, instrument detection limits for some important elements, for example, As, Cr, Pb, Sb, Se, and Tl, in water and other matrices are >50 μg/L, and often much greater, and AAS is limited to measuring one element at a time. The high-temperature graphite furnace sample introduction system provided instrument detection limits of <5 μg/L for most elements, and allowed automation of sample processing, but was still limited to measuring one element at a time. Other AAS sample introduction techniques, especially chemical reduction and elemental Hg vaporization (cold-vapor) and conversion of As and Se to volatile hydrides, provided similarly low detection limits. The graphite furnace and Hg cold-vapor techniques are often referred to in the scientific literature as flameless AAS methods.

Atomic Emission. Elemental analysis with AES had been practiced since the mid-1930s using flames and arc or spark discharges to vaporize and atomize samples and excite the atoms for optical emission. The AES technique provided rapid simultaneous or sequential multi-element determinations, but the flames and arc or spark discharges had significant limitations. Interferences from electrode and other sample components were not uncommon, elemental measurements were often imprecise, detection limits were not sufficiently low, and liquid samples were difficult to analyze. The increasing demand for rapid, selective, and sensitive multi-element determinations led to the development of the inductively coupled plasma (ICP) sample vaporization, atomization, and excitation source which met the analytical requirements and gave new life to AES. The ICP allowed the direct injection of aqueous aerosols into a $5500°K–8000°K$ argon ion plasma and elemental measurements with minimal or no interferences from background components or other analytes. Analytical methods utilizing ICP/AES are widely used for the determination of multiple elements in environmental samples.

Mass Spectrometry. The ICP technique also produces gas-phase elemental ions and the ICP was quickly adapted as an ion source for mass spectrometry. This combination became one of the most useful and important techniques for rapid multi-element analyses of gases, liquids, and solids.[1] The argon ICP is a very efficient ion source that produces mainly singly charged ions. It is estimated that 54 elements, all metals, are ionized with 90% or greater efficiency. Only C, H, N, a few electronegative elements, and the noble gases are ionized with efficiencies less than 10%. A small number of elements, for example, As, B, Be, Hg, I, P, S, Se, and Te have estimated ionization efficiencies in the 10–90% range. Although Ar has a low ionization efficiency, it is present in the ion source in great quantity and gives a few significant ions at m/z 40 ($Ar^{+\cdot}$) and m/z 80 ($Ar_2^{+\cdot}$). Other background ions, for example, m/z 41 (ArH^+) and m/z 56 ($ArO^{+\cdot}$), are formed from sample components, usually water, and mineral acids that are used to ensure dissolution of some analytes.

These ions obscure ions from some elements, but techniques are available to circumvent most of these interferences.

Organic and Inorganic Compounds and Ions

Colorimetric determinations are selective for specific substances, for example, the PO_4^{-3} ion, because of selective color-forming chemical reactions between reagents and target analytes. Several of these techniques continue to be important for the determination of a few inorganic anions. However, ion-exchange chromatographic techniques are increasingly used for determinations of multiple ions in environmental samples. Some electrochemical techniques, for example, ion-selective electrodes, are selective because the materials of construction, for example, ion-selective membranes, and operating parameters are carefully chosen so the devices respond only to specific analytes. Ion-selective electrodes are widely used for the determination of pH and a few other inorganic ions, for example, F^-. Immunoassay methods, which are very widely used and enormously important in clinical analyses, have been developed for a small number of organic compounds, for example, several triazine pesticides. However, many chemicals of environmental interest have MWs and shapes that are too small and similar for effective application of immuno-chemical techniques.

The further development of nonchromatographic analytical methods based on these and other techniques has been impeded by fundamental limitations in selectivity, and sometimes sensitivity, for a wide variety of similar substances, for example, those in Table 8.1 and Figs. 8.1–8.3. The cost of developing, testing, and documenting a large number of specialized methods, often for just one or a few analytes, is generally prohibitive compared to the cost of chromatographic methods that allow the determination of multiple generally similar analytes in an environmental sample.

GLOSSARY

Accuracy: the degree of agreement between the measured concentration of a substance in a sample and the true value of the concentration in the sample.

Analysis: the process of investigation of a sample of the physical world to learn about its chemical components, composition, structure, or other physical or chemical characteristics. Generally only samples are analyzed and individual elements, compounds, and ions are separated from one another, identified, measured, or determined. A pure compound or multi-element ion is analyzed only when it is investigated to determine its components, composition, structure, or other physical or chemical characteristics.

Analyte: a general term for any element, compound, or ion that is present in a sample or is targeted for determination in a sample.

Analytical method: the complete process used to determine an analyte or analytes in a sample. The analytical method documents all the individual steps in the process from sampling to reporting the results.

Analytical method attributes: measures of the quality, reliability, and uncertainty of the determinations obtained with an analytical method. Typical analytical method attributes are selectivity, sensitivity, detection limits, signal/noise, recovery, accuracy, bias, precision, and validation. *Analytical method attributes* are sometimes called *figures of merit.*

Bias: the systematic error in a measurement of the amount or concentration of an analyte in a sample.

Congeners: compounds or ions that are members of a series of related substances that differ only by the number of hydrogens that have been substituted by the same atom.

Derivatives: compounds or ions that are produced by chemical reactions of analytes. An analytically useful derivative has physical or chemical properties that are not possessed by the analyte but that can be employed to determine the analyte.

Detection limit: the minimum quantity or concentration of an analyte that can be detected with an analytical method or technique. There are no generally accepted standard criteria for detection and detection limits often depend on the sample matrix. Therefore detection limits must include the criteria for detection and the nature of the sample matrix. A technique with a higher analyte sensitivity does not always provide a lower detection limit because interfering sample matrix components may also be observed with higher sensitivities.

Determination: the identification and the measurement of the concentration of an analyte in a sample.

Isomers: two or more compounds or multi-element ions that have the same elemental composition but different structures.

Precision: the degree of random variation in repetitive measurements of the concentration of an analyte in a sample. Precision is usually measured by the standard deviation or the relative standard deviation of the measurements.

Procedure: a specific part of an analytical method that is concerned with one aspect of the method, for example, the liquid–liquid extraction of groups of similar analytes from a water sample.

Qualitative analysis: the process of only identifying the analytes in a sample.

Quantitative analysis: the process of both identifying and measuring the concentrations of the analytes in a sample.

Recovery: the amount of analyte measured in a sample matrix as a fraction of the amount of the same analyte that was added to the sample. If the analyte is present in the sample before the addition, the native amount is subtracted from the measured quantity before calculating the recovery.

Sample matrix: the general nature of the sample and its components that can have a significant impact on the performance of an analytical method. For example, sea water and fresh water sample matrices are significantly different and this difference can affect the performance of an analytical method.

Selectivity: is a qualitative estimate of how well the analyte identification procedure is able to distinguish an analyte in a sample from one or many similar analytes with similar, or even some of the same, physical or chemical properties.

Sensitivity: is the electronic or other measurable signal produced by the analytical method or measurement technique per unit amount of analyte.

Signal/Noise: is the ratio of analyte electronic or other measurable signal to the mean background matrix signal.

Technique: is a specific way of manipulating a sample or substance or measuring a substance. One or more techniques may be used within each procedure and several procedures may be used within an analytical method.

Validation: is the determination of the attributes, or figures of merit, of an analytical method for one or more analytes in one or more sample matrices by one or more analysts in one or more analytical laboratories and the acceptance of the attributes as reasonable and useful by the users of the data. There are many levels of analytical method validation ranging from the validation of a method for a single analyte in a single matrix by a single analyst in a single laboratory to a multi-analyte, multi-matrix, multi-analyst, and multi-laboratory validation.

REFERENCES

1. Budde, W. L., *Analytical Mass Spectrometry: Strategies for Environmental and Related Applications*. Oxford University Press, New York, 2001.
2. http://www.epa.gov/nerlcwww/methmans.html
3. Clesceri, L. S., Greenberg, A. E., and Eaton, A. D. (Eds.), *Standard Methods for the Examination of Water and Wastewater*, (20th ed.), American Public Health Association, Washington, DC, 1998 and later editions.
4. *Annual Book of ASTM Standards, Section 11, Water and Environmental Technology*. American Society for Testing and Materials, West Conshohocken, PA.
5. Parker, L. V. "The Effects of Ground Water Sampling Devices on Water Quality: A Literature Review." *Ground Water Monitoring Remediation* **14**, 130–141, 1994.
6. *Standard D5466-95, Standard Test Method for Determination of Volatile Organic Chemicals in Atmospheres (Canister Sampling Methodology), Annual Book of ASTM Standards*, Vol. 11.03, American Society for Testing and Materials, W. Conshohocken, PA, **1998** and later editions.

SUGGESTED ADDITIONAL READING

Bloemen, H. J. Th. and Burn, J. (Eds.), *Chemistry and Analysis of Volatile Organic Compounds in the Environment*. Blackie Academic & Professional, London, 1993.

Keith, L.H. (Ed.), *Principles of Environmental Sampling*, 2nd ed., American Chemical Society, Washington, DC, 1996.

Schwarzenbach, R. P., Gschwend, P. M., and Imboden, D. M.. *Environmental Organic Chemistry*, Wiley, New York, 1993.

Taylor, J. K., *Quality Assurance of Chemical Measurements,* Lewis, Chelsea, MI, 1987.

9

Nanotechnology: Fundamental Principles and Applications

Koodali T. Ranjit and Kenneth J. Klabunde*

INTRODUCTION

Nanotechnology research is based primarily on molecular manufacturing. Although several definitions have been widely used in the past to describe the field of nanotechnology, it is worthwhile to point out that the National Nanotechnology Initiative (NNI), a federal research and development scheme approved by the congress in 2001 defines nanotechnology only if the following three aspects are involved: (1) research and technology development at the atomic, molecular, or macromolecular levels, in the length scale of approximately 1–100 nanometer range, (2) creating and using structures, devices, and systems that have novel properties and functions because of their small and/or intermediate size, and (3) ability to control or manipulate on the atomic scale. Nanotechnology in essence is the technology based on the manipulation of individual atoms and molecules to build complex structures that have atomic specifications. In the year 2004 alone, 22 U.S. government agencies are expected to spend a total of $1 billion in research devoted to nanoscience and nanotechnology.

Nanotechnology is a future manufacturing technology that is expected to make most products lighter, stronger, less expensive, and more precise. As stated earlier, at the core of nanotechnology is what is called "molecular technology" or "molecular manufacturing." Molecular technology refers to manufacturing processes using molecular machinery, i.e., giving molecule-by-molecule control of products and byproducts via positional chemical synthesis. The ancient style of technology dating back from chipping stones to the modern technology of making Si chips for microprocessors contains trillions of atoms and molecules and hence the methods of making them belong to a class called bulk technology. In contrast, nanotechnology will handle individual atoms or molecules with control and precision and hence it is widely believed that nanotechnology will change our world more than can be imagined.

---

* Ranjit T. Koodali, Department of Chemistry, University of South Dakota, Vermillion, SD 57069 Kenneth J. Klabunde, Department of Chemistry, Kansas State University, Manhattan, KS 66506.

The dawn of nanotechnology was envisioned by the late Nobel laureate Richard Feynman as far back as 1959.[1] In his seminal talk at the annual meeting of the American Physical Society titled, *"There's Plenty of Room at the Bottom—An Invitation to Enter a New Field of Physics,"* Feynman discussed the advantages that could be obtained through precise control of atoms. Feynman, the visionary, had stated in his talk, "But I am not afraid to consider the final question as to whether, ultimately—in the great future—we can arrange the atoms the way we want; the very *atoms*, all the way down!" There seems to be no doubt that Feynman originated the idea of nanotechnology.

The revolutionary idea proposed by Feynman generated a substantial volume of technical literature in the emerging field of nanotechnology. Notable among them was a book written by Eric Drexler in 1986 titled, *Engines of Creation—The Coming Era of Nanotechnology.*[2] In his book, Drexler had stated that "Our ability to arrange atoms lies at the foundation of technology and for better or for worse; the greatest technological breakthrough in history is still to come." Throughout the late eighties and early nineties, research in the area of nanoscience and nanotechnology blossomed and scientists were able to manipulate individual atoms and molecules. For example, in 1989, a team at IBM showed that they could use a Scanning Tunneling Microscope (STM) tip to move atoms, and they spelled out the letters "I B M" with 35 individual xenon atoms on a nickel surface.[3]

In the United States in 1998, several federal agencies formed a group called Inter-agency Working Group on Nanotechnology (IWGN) to discuss future plans in the area of nanoscience and nanotechnology. The IWGN was the forerunner to the current NNI initiative to foresee research and development in nanotechnology and to forecast future directions. The federal government initiative includes activities ranging from basic and fundamental nanoscience to development of specific nanotechnology devices and applications. They include the design and manufacture of nanostructured materials that are correct and precise at the atomic and single-molecule level. These advances are aimed at applications such as cost-effective manufacture of nanoscale microelectronics, more efficient and cost-effective energy conservation and storage devices, and biological sensors with applications to both health care and chemical and biological threat detection. Indeed "nanomania" is sweeping through all fields of science and technology and even day-to-day products. For example, Dockers, manufacturers of the popular Khakis brand, recently came out with an advertisement for their new brand of pants, called the "Go Khakis", which promises to keep ones' legs stain-free using revolutionary nanotechnology! Aside from stain-free pants and nanotech tennis and golf balls (that are claimed to correct their own flight!), there are few commercial products based on nanotechnology. However, that is all about to change in the coming decade.

A NEW REALM OF MATTER

Chemistry is the study of atoms and molecules whose dimensions are generally less than one nanometer, whereas the majority of physics, particularly the area of condensed matter physics, deals with solids of essentially an infinite array of bound atoms or molecules greater than 100 nm. A very big gap exists between these two extreme regimes. The regime called the nanoscale regime lies intermediate between the realms of quantum chemistry and solid state or condensed matter physics and deals with particle sizes of 1 to 100 nm. In this regime, neither classical laws of physics nor quantum chemistry rules are applicable. Hence this regime represents a new realm of matter. This new field of nanoscale materials is a multidisciplinary area that touches the fields of chemistry, physics, metallurgy, biology, electronics, computers, medicine, and mathematics. Therefore, interdisciplinary research is required for substantial progress to be made in the field of nanotechnology.

The field of nanomaterials is not unique to manmade substances; nature has been utilizing nanomaterials for millions of years. Biological systems contain several nanoscale materials. For example, bones, teeth, and shells are molecular composites of proteins and biominerals that have superior strength and toughness. Human bones contain mineral materials with particle sizes in the nanoscale regime. Another example is found in certain aquatic bacteria that are able to orient themselves using the Earth's magnetic field. They are able to do this because they contain chains of nanosized, single-domain magnetite (Fe_3O_4) particles. A further example is the teeth of herbivorous mollusks. Their teeth have a complex structure that contains nanocrystalline needles of goethite (FeO(OH)). It is quite a marvel that nature produces such tough materials out of protein constituents as soft as human skin and mineral constituents as brittle as a classroom chalk. In fact nature can be considered as providing model systems for developing technologically useful nanomaterials. According to Dickson, "Life itself could be considered as a nanophase system!"[4]

The three most important aspects concerning nanoscale materials are synthesis, physical, and chemical properties. However, the most important of these is synthesis. Several synthetic methods developed during the last decade have allowed scientists to produce large quantities of nanomaterials. The nanoparticles prepared are usually reactive with oxygen and water, and it is quite difficult to prepare in monodisperse form (one size). Generally, it is best to produce nanomaterials with a narrow size distribution. It is quite challenging to prepare kilogram quantities of pure nanomaterials that are monodisperse. Thus, innovative and creative synthesis methods have to be developed so that the useful properties of the nanomaterials can be used for the benefit of mankind. In this chapter, we shall discuss two methods that have been developed in the authors' laboratory for the production of nanoparticles in large quantities. The first method, called the Solvated Metal Atom Dispersion (SMAD) method, allows the preparation of gram scale and higher quantities of metal nanoparticles such as Au, Ag, Cu, Pd, etc., metal-supported catalysts, bimetallic alloy materials such as Pt-Sn, Ag-Au, Mn-Co, etc., and also semiconductors such as CdS, CdSe, PbS, etc. that have particle sizes in nanoscale dimensions. A further important aspect of the SMAD method is that it leads to the production of high-purity materials (with no formation of byproducts) and most important, the materials are highly monodisperse.

SOLVATED METAL ATOM DISPERSION (SMAD) METHOD FOR THE PREPARATION OF NANOPARTICLES

Metal-supported catalysts are among the important classes of synthetic materials developed in the past 50 years. The importance of these materials can be gauged from the fact that about 20% of the U.S. national gross product is dependent on such catalysts. Thus, the preparation of zero valent metal particles supported catalysts is very vital to the economy of any nation. Before discussing the advantages of the SMAD method, it is worthwhile to briefly review the traditional methods of preparation of metal-supported catalysts.

Metal ion reduction has been the principal procedure for preparing metal-supported catalysts. In this method, metal nitrates or metal halides are dissolved in water and the solid support such as alumina or silica is slurried in the aqueous metal salt solution. Then, the water is evaporated and the dispersed metal salt solution is converted into the oxide form by heating the dry slurry at high temperature in air or oxygen. Finally the metal oxide is reduced to the metallic state by passing a stream of hydrogen gas to give the final metal/support catalyst. During the reduction step, metal atom clustering or agglomeration (sintering) occurs and thus the size of the metal particles obtained varies considerably. The agglomeration of the metal particles is quite difficult to control because of the high temperatures involved in the reduction process. The metal particles obtained are usually crystalline and spherical in shape although other shapes have also been reported. The advantages of this method of preparation are the ease of preparation, possibility of scale-up for industrial applications, and the relative low cost involved. However, there are several disadvantages and they are: (1) incomplete reduction of the metal ions to the zero valent state, (2) sintering of the metal atoms during the high-temperature treatment and (3) difficulty in reducing two metal particles simultaneously to make bi-metallic particles supported oxide catalysts.[5,6]

Another method of preparation of metal-supported catalyst is decomposition of organometallic reagents on catalyst supports. In this procedure, stable organometallic compounds are adsorbed from either solution or gas phase onto a catalyst support. Controlled thermolysis of the organometallic compounds leads to loss of ligand molecules and the metal particles continue to grow on the support surface. Theoretically, by using an organometallic cluster compound, it is possible to form a well-defined cluster of metal atoms on the support surface; however, in practice, the removal of the ligands in an organized manner has not been possible. Cluster aggregation and decomposition are frequently encountered and they constitute the major disadvantages of this approach. A related method that has been employed has been the use of plasma discharge to decompose the organometallic compound. However, this method suffers from the disadvantage that only a limited number of organometallic compounds can be utilized for the preparation of metal supported catalysts.

A novel approach developed in 1976 makes use of weakly stabilized metal atoms or "solvated metal atoms."[7] The nature of the solvent and the metal determine the stability of the metal solvate. Such metal solvates are normally thermally unstable and hence can be used as precursors to metal particles. On warming the metal solvates, metal cluster nucleation takes place and in the presence of a support, both nucleation and particle growth occur that leads to the formation of a highly dispersed metal-supported heterogeneous catalyst. This has been termed by the authors as the "Solvated Metal Atom Dispersed" or SMAD procedure. The numerous advantages of this method are: (1) unusual metal morphologies that are highly reactive are often encountered because the metal particle deposition occurs at low temperature, (2) very high dispersion of the zero valent metal is achieved, (3) no reduction step is necessary and sintering or agglomeration of metal particles is avoided, (4) many metal–solvent combinations can be used, (5) bi-metallic metal

or alloy particles can be prepared, (6) metal particles are securely anchored to catalyst support by the reaction of solvated atoms with surface hydroxyl groups, and (7) handling of toxic organometallic compounds can be avoided. However, skillful operation of the experimental setup is required, and although a number of metal–solvent combinations can be used, only some of the combinations have been found to be of synthetic utility.

The SMAD catalysts have been generally prepared in the following manner. An aluminum oxide coated tungsten crucible is first degassed by placing it into a reaction vessel (Fig. 9.1) and heating to $\sim 1400°C$ in vacuum overnight. Then the reactor is filled with air and a gram scale quantity of metal is placed in the crucible. About 100 mL of organic solvent (toluene or pentane) is degassed and placed in Schlenk tubes and attached to the SMAD reactor. This whole system is sealed to vacuum and the reactor flask is surrounded by liquid nitrogen, and evacuated for a few hours (~ 3 h) typically reaching pressures of 1×10^{-3} Torr. After approximately 30 min, the solvent is deposited on the cold walls of the reactor flask forming a layer of frozen solvent. At the same time, the crucible is slowly heated to the evaporation temperature of the metal. The metal vapor and solvent are co-deposited for a period of 3 h. After the deposition of the solvent layer and the evaporation of the metal are completed, the frozen matrix is warmed from $-196°C$ until the solvent melts to form a solvent-solvated metal solution. The solution is stirred and warmed further in the presence of a support (silica or alumina, ~ 20 g preheated at 500°C for 3 h in dry air and cooled and transferred to the reaction chamber and placed in vacuum prior to the metal vapor reaction). After reaching room temperature, the mixture is vacuum siphoned and placed under an atmosphere of nitrogen in a glassware. The solvent is then slowly removed under vacuum and the resulting powder is then outgassed at room temperature to 1×10^{-5} Torr and used without any further treatment as catalyst. The metal concentration, temperature, and time of impregnation have an effect

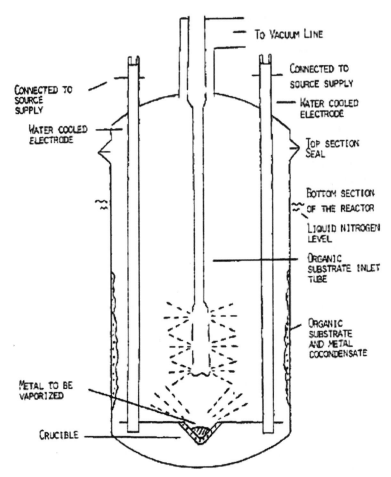

Fig. 9.1. Apparatus used for the preparation of nanoparticles by the Solvated Metal Atom Dispersion Method (SMAD). (Reprinted with permission from Klabunde, K. J., *Acc. Chem. Res.*, 1975, Fig.1, copyright © (1975) American Chemical Society.)

on the size and morphology of the metal particle size. Support effects are not pronounced, but presence of acid sites on the support and a higher specific surface area of the support tend to allow better dispersion of the metal particles. The following section describes some recent results obtained for the preparation of Au, Ag, and semiconductor nanoparticles (CdS, PbS, CdSe, etc.) using the SMAD method.

Gold Nanoparticles

Gold has received the attention of mankind since ancient times. During the past several years there has been a dramatic increase in research pertaining to the synthesis and appli-cation of Au nanoparticles in the emerging fields of nanoscience and nanotechnology.[8] Au nanoparticles exhibit many fascinating properties such as electronic, magnetic, and optical properties (quantum size effect) that are strongly dependent on the size and shape of the nanoparticle. Nanoparticles of gold and other noble metals have attracted the attention of not only chemists and physicists, but also biologists, computer scientists, electronic engineers and metallurgists. These nanoparticles have potential applications in the development of optical and electronic devices, magnetic recording media, micro-electronics, sensors, and catalysis.[9] Another feature of the nanoparticles that is important for future technological applications is the arrangement of nanoparticles into regular and periodic two- or three-dimensional structures called nanocrystalline superlattices (NCSs). The formation of such two- dimensional (2D) or three-dimensional (3D) nanocrystalline superlattices leads to the formation of materials characterized by very different properties compared to those of the discrete species. The collective interactions of the individual nanoparticles in the NCS and the individual particles themselves lead to the manifestation of such unique properties compared to the bulk particles.[10]

Several methods have been reported for the synthesis of 2D and 3D NCSs.[11] The most common method is to reduce the metal salt in the presence of stabilizing agents.[12] However, they invariably end up in the formation of byproducts and an additional step is required for the isolation of pure metal nanoparticles. Thus, the traditional methods of preparation are quite tedious and time consuming and the possibility of scale-up is fraught with more difficulties and challenges. An important advantage of the SMAD method is the possibility for scaling up the process. Industrial applications of nanoparticles require their preparation in large quantities and reproducible quality. Thus, the SMAD method seems to be a very facile method for the preparation of large-scale quantities of metal nanoparticles.

Colloidal solutions of gold in acetone have been well documented in the literature. A combination of two solvents such as acetone and toluene were used in the preparation of Au nanoparticles by the SMAD method.[13] Acetone, being a polar solvent, solvates the Au metal atoms and clusters during the warming stage of the SMAD method and acts as a stabilizing agent. After the removal of acetone from the system, the Au nanoparticles are stabilized by dodecanethiol molecules, which enable their dispersion in nonpolar solvents such as toluene. A second step called the digestive ripening is carried out in the presence of toluene as the solvent. The digestive ripening step is the key step for the formation of almost monodisperse Au nanoparticles from the polydisperse Au–toluene–thiol colloid. In this step, the Au–toluene–thiol colloid is heated to reflux under Ar atmosphere for a couple of hours. This leads to a narrowing of the particle size distribution and monodisperse colloids are obtained. This particle narrowing is caused by the transfer of gold atoms/clusters from particle to particle until a thermodynamic equilibrium is achieved between the formation of more gold–sulfur bonds and the size of the particles as shown by the following equations.

$$\text{Au vapor + acetone} \xrightarrow{77K} \text{(Au)(acetone) solvate} \quad (9\text{-}1)$$

$$\text{Au (acetone)} + C_{12}H_{25}SH \longrightarrow (Au)_x(HSC_{12}H_{25})_y \quad (9\text{-}2)$$

$$(Au)_x(HSC_{12}H_{25})_y \xrightarrow[\text{reflux}]{\text{toluene}} (Au)_{5000}(HSC_{12}H_{25})_{300} \quad (9\text{-}3)$$

Although the exact mechanism of the digestive ripening is unknown at this moment, the process probably involves dissolution of surface atoms or clusters of atoms by the ligand (dodecanethiol) molecules. So a dissolution and reprecipitation process is believed to happen in which the reactive sites such as corners and edges are probably the first atoms that are prone to this dissolution–reprecipitation process. The dry product is then obtained by cooling the colloidal solution to room temperature and precipitating with absolute ethanol. The final dry product is shiny dark crystals consisting of Au nanoparticles stabilized by dodecanethiol molecules that are readily soluble in nonpolar solvents such as toluene or hexane. Treatment with a long chained thiol in refluxing toluene causes a remarkable particle size narrowing so that each gold nanoparticle is 4.6 nm $\pm$ 0.1 nm (about 5000 atoms/particle and 300 $C_{12}H_{25}SH$ ligands/particle). A remarkable property of these colloidal solutions is their tendency to organize into 2D and 3D NCSs.[13] The TEM figures of the Au colloids obtained after digestive ripening and cooling to room temperature show remarkable features. After just 15 minutes, these Au nanoparticles organize and from predominantly 3D NCSs and some 2D NCSs when deposited on a support. Interestingly, these NCSs are stable and even after two months, distinct 2D and 3D structures can be seen.

High-resolution TEM show that the Au nanoparticles prepared by the SMAD procedure lack faceting, have defective core structures, and behave as do hard spheres. They self-assemble and organize into hcp-type superlattices with long-range translational order. Figure 9.2 shows hcp nanocrystal superlattice of Au nanoparticle oriented along the [0001] direction. The Au nanoparticles have predominantly spherical shapes and the TEM of the Au nanoparticles showed reproducibly hcp structures indicating the versatility of the SMAD method.[14]

Silver Nanoparticles

Silver nanoparticles are of great importance due to their ability to efficiently interact with light because of plasmon resonances.[15] These are collective oscillations of the conducting electrons in the metal. Indeed, Ag nanoparticles are envisaged to be vital components of optical and photonic devices in the future. Over

the years, several methods have been reported for the synthesis of Ag nanoparticles. These include the Creighton method[16] in which sodium borohydride is used for the reduction of silver nitrate, or the Lee–Meisel method[17] which is considered to be reduction of silver nitrate by sodium citrate. The most common source of silver ions has been silver nitrate; however, other sources such as silver perchlorate, silver sulfate, and silver 2-ethylhexonate have also been used.[17] The high reduction potential of Ag ions (0.799 V) allows it to be reduced by several inorganic and organic compounds. Several reducing agents have been employed for the reduction of silver ions, $NaBH_4$, sodium citrate, ascorbic acid, ethanol, pyridine, N,N-dimethylformamide, hydrazine hydrochloride, poly(ethylene glycol) have all been used as reducing agents.

For the long-term stability of the Ag nanoparticles, surfactants have been used as protecting agents.[18] Stabilizers such as sodium 2-diethylhexyl sulfosuccinate (AOT), cetyltrimethylammonium bromide (CTAB), and poly(vinyl pyrrolidone) (PVP) have been effectively employed for stabilization and also to direct the aspect ratio of the Ag nanoparticles producing different shapes such as plates, rods, discs, etc.[19] All the above methods in general produce nanoparticles of Ag, but very

Fig. 9.2. High Resolution Transmission Electron Microscopic (HRTEM) image of Au nanoparticles stabilized by dodecanethiol ligand molecules after SMAD and digestive ripening procedure. (Reprinted from Stoeva, S. et al *J. Phys. Chem. B,* 2003, 107, 7441–7448, Fig. 11(c), by permission of the American Chemical Society, copyright © 2002, American Chemical Society.)

careful control of conditions is necessary in order to obtain Ag nanoparticles with narrow size distribution. Scalability and the purity of the Ag nanoparticles is often a problem. The preparation of zero valent silver is a challenge because of its relative ease of oxidation compared to Au nanoparticles. More often, after several days of standing under ambient conditions, the surface of the Ag nanoparticles becomes oxidized and hence a thin oxide layer coating is formed.

A key requirement for applications is a reproducible and reliable method that can produce large quantities of Ag nanoparticles with a narrow size distribution. The SMAD method has the potential for preparation of large-scale quantities; a 5000fold scale-up is possible by the SMAD method compared to the inverse micelle method of production of metal nanoparticles. Gram-scale quantities of monodisperse spherical Ag nanoparticles have been successfully prepared using the SMAD method followed by digestive ripening.[20] Two different capping agents or stabilizing agents were evaluated and it was found that the use of dodecanethiol led to Ag nanoparticles with a mean diameter of 6.6 +/− 1 nm, and trioctyl phosphine capped particles were 6.0 +/− 2 nm.

As with the Au nanoparticles, these Ag nanoparticles too were found to organize into two- and three-dimensional superlattices with a well-defined geometry through self-assembly in solution. 2D arrays were found when either dodecanethiol or trioctyl phosphine were used as ligands; interestingly, only circular 3D superlattices were observed when trioctyl phosphine was used as the protecting ligand and triangular 3D superlattices were observed when dodecanethiol was used as the protecting ligands. The synthetic procedure for the preparation of Ag nanoparticles was similar to that used for Au, however, some changes were needed to prevent the oxidation of Ag. Ag nanoparticles are normally resistant to the formation of 3D structures and structures reported earlier in the literature are usually amorphous. Thus, the observation of crystalline 3D Ag superlattice structures formed via the SMAD method and subsequent digestive ripening is clearly remarkable. Efforts are underway to extend this synthetic procedure for the preparation of other elements of the periodic table to produce new monodisperse metal nanoparticles such as Pb and Pd that can self-assemble and form hitherto unknown superlattice structures.

Semiconductor Nanoparticles

The size-dependent properties of semiconductor nanoparticles were first reported by several groups in the early 1980s. In their pioneering works, Brus, Henglein, and Grätzel, independently developed liquid-phase synthesis of stable semiconductor nanoparticles suspended in solvents such as acetonitrile and water.[21–23] Several nanocrystalline semiconductors such as II–VI materials (CdS, HgS, and CdSe), I–VII materials (AgBr, AgCl), and III–V materials (GaAs) that are narrowly dispersed have been prepared. The strategy adopted by Brus was to separate the temporal growth and nucleation sites so that a highly crystalline and monodisperse semiconductor could be obtained. Rapid injection of a room-temperature solution containing the inorganic precursors into a preheated solvent (350°C) results in rapid nucleation of the semiconductor nanoparticle. This is followed by lowering the temperature to around 300°C and this allows for a slow growth of the nanocrystals until all the reactants are consumed. The size and shape of the semiconductor nanocrystals can be controlled by adjusting the ratio of the concentration of the inorganic precursors and the concentration of the stabilizing or capping agents (surfactants, dendrimers, and polymers). This type of synthesis is called the bottom-up method. In contrast, in the top-down method, the macroscopic or bulk material is "machined" down to nanometer-length scale by laser ablation–condensation or lithographic techniques.

In bulk semiconductors, the conduction band and the valence band are separated by a band gap. On excitation, electrons are promoted to the conduction band and holes are produced in the valence band. The charge carriers (electrons and holes) are separated by distances that encompass a number of ions or molecules that constitute the semiconductor material. This separating distance is called the Bohr radius and normally has dimensions on the nanometer scale. When the size of the semiconductor material itself becomes similar to the Bohr radius or smaller, it leads to a situation in which the charge carrier excitons have a restricted space or volume to move and thus their motion is confined. Like the motion of an electron in a box, the kinetic energy and the excitation energy of the electron increase as the size of the box decreases. Similarly, the band gap of the semiconductor increase as the size of the particles becomes smaller than their Bohr radius. Along with the band gap, the color and emission of the nanoparticle changes with change in the size of the nanoparticle. The physical, chemical, and optical properties of semiconductor nanoparticles thus are sensitive to both the size and shape of the nanoparticles. The surface-to-volume ratio increases as the semiconductor nanoparticle decreases in size and because the properties of the semiconductor depend on the size and shape of the nanoparticle, semiconductor nanoparticles exhibit very different properties compared to bulk semiconductors.[24] Several applications are envisaged for the use of these semiconductor particles; these include photovoltaics, nanoelectronics, and optoelectronics.[25] Other applications where these semiconductor nanoparticles find use are in the fields of photocatalysis, photodegradation, and detoxification of pollutants.[26]

The SMAD method has been successfully developed to synthesize gram scale quantities of high-purity metal sulfide semiconductor nanoparticles such as ZnS, PbS, CdS, and SnS.[27] The advantages of the SMAD method compared to other methods reported in the literature are: (1) gram scale quantities of stable nanocrystalline (1–10 nm) particle sizes can be prepared without any stabilizers with relative ease, (2) the synthesis is very reproducible and yields are typically >90%, (3) the nanoparticles have fairly uniform particle size

distribution and hence are highly monodisperse, (4) no purification step is necessary and hence easy work-up affords high purity metal sulfide nanoparticles, and (5) the surfaces of the metal sulfide nanoparticles are "clean"; that is, no oxide coating is found on the surface of the nanoparticles. In addition, the SMAD method for the synthesis of metal sulfide nanoparticles allows one to tune the particle size and hence its band gap by appropriate choice of solvents and warm-up time of the matrix consisting of frozen solvent and metal sulfide vapors. The textural properties such as specific surface area and pore volume of the nanoparticles are much higher compared to the commercial samples. For example, ZnS prepared by the SMAD method using pentane as the solvent leads to the production of ZnS nanoparticles with surface area as high as 237 m^2/g whereas a commercial ZnS (Fluka) has a surface area of only about 10 m^2/g. Similarly PbS and CdS nanoparticles prepared by the SMAD method have surface areas ~120 m^2/g. Thus, the SMAD method of preparation of metal sulfide nanoparticles leads to the production of high surface area metal sulfides and hence these metal sulfide nanoparticles are excellent candidates as photocatalysts.

Dielectrics (Insulator Nanoparticles)

Metal oxides are ubiquitous in catalysis and are key components in several catalytic reactions. They function directly as catalytic reactive centers or serve as high surface area supports to disperse active metal centers or as promoters to enhance the rate of catalytic reactions. Many commercial catalysts consist of zero valent metal atoms dispersed finely on a high surface area metal oxide support such as silica or alumina.

Surface chemistry is of importance in numerous processes such as catalysis, corrosion and adsorption. When the particles sizes are in the 1–10 nm range, a whole new field of surface chemistry is realized. In recent years it has been possible to successfully produce nanocrystalline metal oxides of MgO, CaO, ZnO, TiO_2, CuO, CeO_2 and other binary metal oxides by a specially designed sol-gel or aerogel process.[28] Based on the results obtained so far, one can conclude that: (1) intrinsic reactivities are higher per unit surface area for nanocrystals compared to microcrystals, and (2) consolidation of the nanoparticles into pellets does not significantly lower the surface reactivity and the surface area when moderate pressures are used.[29] The high surface area of the nanomaterial ensures that a high percentage of the atoms are on the surface. Thus, reactant–surface interactions can reach stoichiometric ratios. For spherical iron atoms it has been calculated that for a 3 nm particle, 50% of the atoms lie on the surface whereas a 20 nm particle has less than 10% of the atoms on the surface. Another important feature that has larger practical implications is that the nanoparticles exhibit higher intrinsic chemical reactivities as the particle size becomes smaller. The reason for the dramatic increase in the activities is most probably due to changes in crystal shape.

Until a few years back, the crystal shape of solid materials was of academic curiosity only and shape was not considered to have an effect on the chemical properties and reactivities of a material. However, recent studies clearly indicate that the shape of nanocrystals does indeed affect the chemistry. For example, it has been shown that 4 nm nanocrystalline MgO particles adsorb six molecules of SO_2 per nm^2 at room temperature and 20 Torr pressure.[30] However microcrystalline MgO adsorbs only 2 molecules of SO_2 per nm^2 under similar conditions. Similarly, the nanocrystalline aerogel prepared, AP–MgO material adsorbs four times as much CO_2 as the microcrystals. There are not only differences in the amounts of gaseous molecules adsorbed on these surfaces, but also the mode of surface binding can also be different. SO_2 binds more predominantly as a monodentate species on the AP–MgO crystal but favors a bidentate geometry on conventionally prepared, CP–MgO microcrystals. Clearly, these results indicate that the shape and size of the crystals affect the adsorptive properties of the MgO surfaces. The high reactivities of the polyhedral-shaped MgO and CaO are attributed to the higher percentage (~20%) of defect sites such as corner and edge sites on the surfaces of AP–MgO and AP–CaO. In contrast, a conventionally prepared CP–MgO possesses less than 0.5% of defect sites whereas a commercial sample of MgO and CaO essentially contains 0% of defect sites. Thus, the aerogel procedure leads to the production of a new family of inorganic porous metal oxides that exhibit unique properties, and the modified aerogel procedure is discussed in the next section. Figure 9.3 shows the various shapes that MgO crystals adopt when the particle size is changed.

MODIFIED AEROGEL PROCEDURE (MAP)

Aerogels are mesoporous materials that have nanoscale dimensions and have low density and high surface area.[31] They are widely used

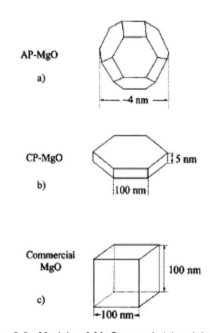

Fig. 9.3. Models of MgO crystal: (a) polyhedral-shaped, aerogel prepared AP–MgO nan crystals; (b) hexagonal-shaped, conventionally prepared CP–MgO microcrystals; and (c) cube-shaped commercial polycrystalline MgO. (Reprinted with permission from Klabunde, K.J. et al., *Chem. Eur. J.,* 2001, 7, 2505–2510. Fig. 2, by permission of Wiley-VCH, Publishers.)

as insulators and increasingly as heterogeneous catalysts. An aerogel procedure involving supercritical drying was first reported by Kistler in 1932 for the preparation of high surface area silica. However, the procedure developed by Kistler was time consuming. Teichner and his co-workers used organic solvents that dramatically decreased the processing time from weeks to hours.[32] A further modification of the aerogel method involved the addition of large amounts of aromatic hydrocarbons such as toluene to the alcohol–alkoxide mixture.[33] The addition of toluene dramatically increased the surface area of the resulting metal oxide and although the exact reason is not clear, it is believed that incorporation of a less polar solvent such as toluene could reduce the surface tension at the pore walls.

The modified aerogel procedure for the preparation of inorganic metal oxides has four steps. Let us consider the preparation of MgO by MAP. The first step is the preparation of Mg alkoxide by the reaction of Mg metal with alcohols such as methanol or ethanol. The next step is the hydrolysis of the Mg alkoxide in the presence of toluene. The third step is the supercritical or hypercritical drying procedure. In this step the gel obtained by the hydrolysis is placed in an autoclave and the wet gel is heated so that the pressure and the temperature exceed the critical temperature (T_c) and the critical pressure (P_c) of the liquid entrapped inside the pores of the gel. The supercritical conditions vary depending on the nature of the solvent employed in the preparation of the gel. The autoclave is slowly heated from room temperature to the critical temperature and after about 15 minutes of standing at T_c, the solvent is vented off to give a dry solid. This drying procedure preserves the texture of the wet gel by avoiding the collapse of the pores and hence a low-density material having very high surface area and large pore volume is obtained. In the final and fourth step the metal hydroxide, $M(OH)_2$ obtained is heated under vacuum (typically at 550°C) to convert it into MO and also to get rid of surface carbonates and other species such as methoxy groups adsorbed on the surface. The following section describes some of the properties of the metal oxide nanoparticles prepared by the modified aerogel procedure.

Metal Oxide Nanoparticles

The crystallite sizes of the metal oxide nanoparticles prepared by MAP are remarkable. For example MgO have crystallite sizes ~4 nm, AP–CaO ~7 nm, AP–TiO$_2$ ~10 nm, AP–Al$_2$O$_3$ ~2 nm and AP–ZrO$_2$ ~8 nm.[34–38] The surface area of AP–MgO is ~500 m^2/g, whereas a MgO sample conventionally prepared (referred to as CP henceforth) has surface area of ~200 m^2/g, and the surface area of a commercial (referred as CM) MgO sample is only around 30 m^2/g. The surface areas of AP–CaO, CP–CaO, and CM–CaO are ~150 m^2/g, ~100 m^2/g and ~1 m^2/g, respectively. Similarly, AP–Al$_2$O$_3$ possesses surface areas as high as 810 m^2/g, whereas a commercial Al$_2$O$_3$ prepared by the high-temperature method has a surface area of only about 100 m^2/g. These examples clearly indicate that the modified aerogel procedure results in the formation of ultrafine particles that have very high surface areas compared to commercial samples. Also, as the particle size decreases, the surface area increases and the reactivity is considerably enhanced. The increased reactivity is not simply due to the increased surface areas alone. These nanoparticles contain numerous defect sites such as crystal corners, edges, kinks, and ion vacancies. In addition, surface hydroxyl groups which can be either isolated or lattice bound add to the rich surface chemistry exhibited by these metal oxide nanoparticles.

The MAP process leads to the formation of free-flowing ultrafine powders rather than monoliths. The low-magnification TEM picture (not shown) of AP–MgO shows porous weblike aggregates in the range of about 1400 nm. These are formed by the interaction of the 4 nm (average) polyhedral crystallites and their overall size distribution is narrow. A high-resolution TEM image of AP–MgO is shown in Fig. 9.4. The TEM picture clearly shows that the cubelike crystallites aggregate

Fig. 9.4. High-resolution transmission electron microscope image of aerogel prepared AP–MgO. (Reprinted with permission from Richards, R. et al., *J. Am. Chem. Soc.* 2000, 122, 4921–4925, Fig. 2, copyright © (2000) American Chemical Society.)

into polyhedral structures that have numerous edge and corner sites. Also interesting is the formation of pores between the crystalline structures that are visible at this magnification. Essentially, the TEM picture is an atomic resolution image of MgO nanocrystallites with lattice planes of Mg and O ions clearly revealed.

More interesting behavior is exhibited by the metal oxides when the loose powders are consolidated into pellets. Compression is carried out by application of pressure using a hydraulic press. For example, AP–MgO powder has a low density of 0.30 cm^3/g. Upon compression at low pressure (3000 psi) this increases to 0.58 cm^3/g, and at 20,000 psi to 1.0 cm^3/g. The surface area of AP–MgO powder is 364 m^2/g. Upon compaction at a low pressure of 1000 psi, the surface area was found to reproducibly increase to ~370 m^2/g.[29] On compaction at higher pressures of 5000 psi and 20,000 psi, the surface area was found to be 366 m^2/g and 342 m^2/g, respec-

tively. Another interesting aspect is that these powders exhibit adsorption/desorption isotherms typical of bottleneck pores. However, after compression at 1000 psi, the samples of Mg(OH)$_2$ and MgO exhibit bottleneck and cylindrical pores, open at both ends. For comparison purposes the conventionally prepared (CP) samples too were subjected to compaction and their textural properties measured. The CP samples were more susceptible to collapse of pore volume, especially the calcium-containing samples. However, the AP–calcium samples exhibited more resistance. The AP–oxides, due to their polyhedral nanocrystal shapes and their tendencies toward forming porous weblike aggregates, are the most resistant to collapse under pressure. The CP–oxides, due to their more ordered hexagonal platelet shapes, more easily compress into denser structures.

The aerogel-prepared metal oxide nanoparticles constitute a new class of porous inorganic materials because of their unique morphological features such as crystal shape, pore structure, high pore volume, and surface areas. Also, it is possible to load catalytic metals such as Fe or Cu at very high dispersions on these oxide supports and hence the nanocrystalline oxide materials can also function as unusual catalyst supports. Furthermore, these oxides can be tailored for desired Lewis base/Lewis acid strengths by incorporation of thin layers of other oxide materials or by preparation of mixed metal oxides.

MIXED METAL OXIDE NANOPARTICLES

Many methods have been reported for the synthesis of mixed metal oxides.[39] Generally they have been prepared by co-precipitation of metal hydroxides followed by high-temperature treatment: solid-state physical mixing of hydroxides, oxides, or nitrates followed by high-temperature treatment or by hydrolysis of bimetallic bridged alkoxides. However, all these methods of preparation lead to partial segregation of the individual oxides as final products. This is because the starting precursors tend to hydrolyze at different rates leading to phase segregation.

Furthermore, the very high-temperature treatment leads to sintering of the particles resulting often in low surface area oxide materials. The MAP method avoids the problems inherent in the synthetic methods and provides an ideal way to prepare intermingled mixed metal oxide nanoparticles that are essentially mixed at the molecular level. In the MAP method, a large portion of a spectator solvent (toluene) is added, which greatly enhances gelation rates, thereby increasing the chances that the two hydrolyzing metal alkoxides gel together. This results in the formation of intermingled mixtures or gels that are essentially molecular in nature. This technique was used to prepare a series of intimately intermingled mixed metal oxide nanoparticles.[40,41] These mixed oxides were composed of alkaline earth oxide and alumina. It was found that only aerogel-prepared, AP–MgAl$_2$O$_4$ and AP–BaAl$_2$O$_4$, showed any peaks in the XRD. In the XRDs for AP–MgAl$_2$O$_4$ and AP–BaAl$_2$O$_4$, it was possible to identify peaks corresponding to MgO and BaO, respectively. This clearly indicates that the samples are intimately mixed. The intimately intermingled mixed metal oxide samples retain the high surface area and relatively large pore volumes when compared to those of the individual metal oxides. HRTEM studies clearly indicate the extent of the intermingling of the two oxides as is illustrated in Fig. 9.5 which shows the HRTEM image of AP–MgAl$_2$O$_4$. Individual and aggregated Al$_2$O$_3$ Boehmite planes mixed with AP–MgO nanocrystals can clearly be observed. This gives a clear indication that the MgO and Al$_2$O$_3$ are intimately mixed throughout the entire material resulting in some unique structures. The presence of MgO "guest" planes in the "host" AlO(OH) planes has led to an increase in the distance between the planes. Typically in AP–Al$_2$O$_3$ the spacing between planes is 6 Å. MgO is now sandwiched between the boehmite planes and the spacing has increased to 15 Å. The ability to disperse the basic nature of MgO throughout the high surface area framework of Al$_2$O$_3$ is clearly shown in these images and is an advantageous feature when dealing with surface adsorption behavior.

Thus, an interesting facet of the MAP procedure is that one can engineer and tune acid/base sites in them. The intimately intermingled MgO–Al$_2$O$_3$ nanostructures discussed earlier exhibit enhanced capacity/activity over the pure forms of either AP–Al$_2$O$_3$ or AP–MgO. The enhanced reactivity is attributed to the Lewis base nature of the very small and very well dispersed MgO crystallites that are "housed" within the large pores of Al$_2$O$_3$. In addition, such intimately intermingled mixed oxides are highly thermally stable with minimum sintering after heating up to 700°C.

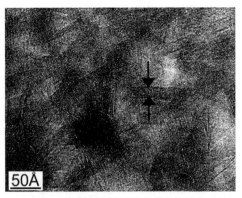

Fig. 9.5. High-resolution transmission electron microscopic image of AP–MgAl$_2$O$_4$. The arrows indicate the MgO sandwiched between Boehmite planes. (Medine, G.M. et al., *J. Mater. Chem.*, 2004, 14, 757–763, Fig. 2. Reproduced by permission of the Royal Society of Chemistry.)

APPLICATIONS

As stated earlier, catalytic processes contribute at least 20% of the GDP of the United States. Thus, one of the most important applications of nanostructured materials in chemistry lies in the field of heterogeneous catalysis. It is beyond the scope of this chapter to discuss the numerous applications of nanoparticles as catalysts. However, we confine our discussion to a select few examples that demonstrate the potential application of nanoparticles in the field of catalysis.

Catalysis—Dechlorination and Dehydrochlorination

The disposal and destruction of chlorinated compounds is a subject of great importance. The cost of complete elimination of chlorinated compounds is quite staggering and it is estimated to be over $100 billion/year. In 1993 some environmental agencies had proposed the need for a "chlorine–free economy". The most widespread group of chlorocarbons includes polychlorinated biphenyls (PCBs) and cleaning solvents such as CCl_4 and C_2Cl_4. The most common method used to destroy chlorocarbons is high-temperature thermal oxidation (incineration). The toxic chlorinated compounds seem to be completely destroyed at high temperatures; however, there is great concern about the formation of toxic byproducts such as dioxins that are more toxic than the chlorinated compounds themselves. The desired reaction is the complete oxidation of chlorocarbons to water, CO_2 and HCl without the formation of any toxic byproducts.

Chlorocarbons are destroyed during reactions with nanocrystalline metal oxides by becoming mineralized as environmentally benign metal chlorides and CO_2 gas.[36,42,43] The process can be written as

$$2MO + CCl_4 \rightarrow CO_2 + 2MCl_2 \quad (4)$$

Considering thermodynamics alone, the reaction of metal oxides with chlorocarbons is energetically favorable for both MgO and CaO,

$$2MgO_{(s)} + CCl_4 \rightarrow 2MgCl_{2\,(s)} + CO_{2\,(g)}$$
$$\Delta H° = -334 \text{ kJ/mol} \quad (5)$$

$$2CaO_{(s)} + CCl_4 \rightarrow 2CaCl_{2\,(s)} + CO_{2\,(g)}$$
$$\Delta H° = -573 \text{ kJ/mol} \quad (6)$$

CaO and MgO destroy chlorocarbons such as CCl_4, $CHCl_3$, and C_2Cl_4 at temperatures around 400–500°C in the absence of an oxidant, yielding mainly CO_2 and the corresponding metal chlorides. Also, it has been demonstrated that if MgO (CaO) particles were coated with transition metal oxides (e.g., Mn_2O_3, Fe_2O_3, CoO, or NiO), the reactivity could be enhanced substantially and a kinetic advantage could be obtained.[44]

The dehydrochlorination can be expressed as shown by Equation (7) below:

$$R\text{-}CH_2\text{-}CH_2\text{-}Cl \rightarrow R\text{-}CH=CH_2 + HCl \quad (7)$$

Nanoscale AP–MgO and AP–CaO were found to be superior to conventionally prepared CP–CaO, CP–MgO, and commercial CaO/MgO catalysts for the dehydrochlorination of several toxic chlorinated substances. The interaction of 1-chlorobutane with nanocrystalline MgO at 200–350°C results in both stoichiometric and catalytic dehydrochlorination of 1-chlorobutane to isomers of butene and simultaneous topochemical conversion of MgO to $MgCl_2$.[45] During the course of the study it was found that the surface of MgO was converted to $MgCl_2$, and a steady performance of the catalyst was observed only after the formation of the surface $MgCl_2$ phase. Also, a decrease in the surface area during the transformation of MgO to $MgCl_2$ was found due to aggregation of the nanoparticles. The study demonstrates that multi-component systems comprising nanoparticles of $MgO/MgCl_2$ have unusual catalytic properties in acid-base reactions.

Destructive Adsorption of Chemical Warfare (CW) Agents

Chemical warfare (CW) agents constitute one of the biggest threats to mankind along with biological warfare agents such as bacteria, fungi, and viruses. The huge stockpiles of CW nerve agents in the United States, Russia, and other nations such as North Korea are a matter of grave concern to the public. The CW agents are known to irreversibly react with the enzyme acetylcholinesterase (AChE) which prevents the enzyme from controlling the central nervous system and hence cause irreversible damage to humans. Hence, many methods have been attempted to decontaminate or neutralize these CW agents.[46] However, only a few reactions are feasible for practical neutralization. Examples of CW agents include VX [O-ethyl S-(2-diisopropylamino) ethylmethyl-phosphonothioate], GD (pinacolyl methylphosphono-fluoridate), and HD [bis(2-chloroethyl)sulfide].

The traditional method decontaminating CW agents is to neutralize them in basic media in solution phase. An approach in this direction has been to use oxidizing reagents such as aqueous bleach ($NaOCl$ or $Ca(OCl)_2$), or aqueous hypochlorite (ClO^-), or hypochlorous acid ($HClO$). However, these reactions require careful control of pH and some times solubility problems arise; for example, in basic bleach, VX has solubility problems. Also, large quantities of liquids have to be used for complete neutralization and hence aqueous phase methods are cumbersome. Dry powders have numerous advantages: (1) they are nontoxic, (2) easy to handle and store, (3) waterless and hence have low logistical burden, and finally (4) there is no liquid waste stream. Nanosize MgO, CaO, and Al_2O_3 can be used in room temperature reactions with chemical warfare agents such as VX, GD, and HD.[47–49] Reaction with nerve agents VX and GD and metal oxides results in the nerve agents being hydrolyzed. There are two important differences between their solution behavior and their destructive adsorption on nanoscale metal oxides. The first difference is that the nontoxic phosphonate products reside as surface bound complexes. The second important difference is that toxic EA-2192, which is known to form under basic hydrolysis in aqueous phase, is not observed on the surface of either MgO or CaO.

MISCELLANEOUS APPLICATIONS

The way to describe how nanotechnology is going to revolutionize our lives in the coming decade is to list several other applications. Some of the applications of nanostructured materials are already found in commercial products but many other applications are still in their developmental stages.

Environmental Chemistry

Nanoparticles of reactive metal fine powders such as Fe and Zn show high reactivity for chlorocarbons. Metal powder-sand membranes have been demonstrated to be effective for cleaning up groundwater contamination.[50]

Photocatalysis

Thin-film nanostructured titania films have been successfully developed that have self-cleaning and anti-fog properties. Pilkington and PPG glasses contain about 15 nm thick coatings of transparent and crack-free titania film integrated into glass that absorbs UV light. The UV light causes a photocatalytic reaction that leads to the production of highly reactive hydroxyl radicals which can break down dirt on glass. The coating of TiO_2 is also hydrophilic and so water forms a continuous sheet rather than drops and the sheeting action of water rinses dirt off the window.

Plastics

Incorporation of nanopowders into polymer matrices imparts several unique properties to the polymers. One can envisage several applications such as wear-resistant tires, replacement of metal body parts for vehicles, flame-retardant plastics, tougher coatings, and so on. Stronger and lighter polymers will allow further replacement of metals. Nanostructured polymer films of organic light-emitting diodes or OLEDs are already used in cell phones, laptop computers, digital cameras, TV displays, and computer monitors.

Medicine

The genetic material DNA is in the 2.5 nanometer range. Nanoparticle assay of DNA has been possible by coating gold nanoparticles with DNA strands. Medical researchers are working to develop new drug delivery methods. In nanoparticle form the drugs can be solubilized into the bloodstream and hence can effectively be used to treat several diseases. Quantum dots of semiconductor nanocrystals can be used to identify and locate cells and record biological activities. These nanocrystals offer optical detection about 100 times brighter than conventional dyes and MRI scans. Implantable devices that can automatically deliver drugs, and monitor and sense vital body functions such as heartbeat and glucose level in the blood, are being developed.

Electrochemistry

Nanostructured Li and Ni containing nickel-metal hydride batteries are widely used in cell phones, video camcorders, quartz watches, and pacemakers to name a few uses. Electrically conducting nanostructured mesoporous materials are envisaged as new materials for fuel cell applications, batteries, and ultracapacitors.

CONCLUSIONS

It is quite clear from the various applications listed above that nanotechnology has the potential to produce innumerable benefits for mankind. Also, it is expected that productivity can be increased through efficient molecular manufacturing. From the several applications listed, it is clear that nanotechnology and nanoscience will touch almost all aspects related to daily life. Hence, research and development in nanotechnology has also broad societal implications. Scientists and researchers thus have an important role in educating the public about the larger implications of nanotechnology and effectively communicating the goals and potential risks of this new emerging technology. It is worth concluding this chapter by quoting the Nobel laureate Richard Smalley,[51] "Just wait—the next century is going to be incredible. We are about to be able to build things that work on the smallest possible length scales, atom by atom. These little nanothings will revolutionize our industries and lives."

REFERENCES

1. Feynman, R.P., *Annual Meeting of the American Physical Society,* California Institute of Technology, Pasadena, CA, Dec 29, 1959.
2. Drexler, K.E., *Engines of Creation,* Anchor, Garden City, NY, 1986.
3. Eigler, D.M., and Schweizer, E.K., *Nature,* **344**, 524–526 (1990).
4. Dickson, D.P.E., in *Nanomaterials: Synthesis, Properties and Applications,* Edelstein, A.S., and Cammarata, R.C., (Eds.), pp. 459–476, Institute of Physics Publishing, Bristol, UK, 1996.
5. Klabunde, K.J., Li, Y.X., and Tan, B.J., *Chem. Mater.,* **3**, 30–39, (1991).
6. Li, Y.X., Klabunde, K.J., *J. Catal.,* **126**, 173–186, (1990).
7. Klabunde, K.J., Efner, H.F., Murdock, T.O., and Ropple, R., *J. Am. Chem. Soc.,* **98**, 1021–1023 (1976).
8. Daniel, M.-C., and Astruc, D., *Chem. Rev.,* **104**, 293–346, (2004).
9. Andres, R.P., Bielefeld, J.D., Henderson, J.I., Janes, D.B., Kolagunta, V.R., Kubiak, C.P., Mahoney, W.J., Osifchin, R., *Science,* **273**, 1690–1693 (1996).
10. Schmid, G., Chi, L., *Adv. Mater.,* **10**, 515–525, (1998).
11. Bethell, D., and Schiffrin, D.J., *Nature,* **382**, 581 (1996).
12. Green, M., and O'Brien, P., *J. Chem. Soc., Chem. Commun.,* 2235–2241, (1999).
13. Stoeva, S., Klabunde, K.J., Sorensen, C.M., and Dragieva, I. *J. Am. Chem. Soc.,* **124**, 2305–2311 (2002).
14. Stoeva, S.I., Prasad, B.L.V., Uma, S., Stoimenov, P.K., Zaikovski, V., Sorensen, C.M., and Klabunde, K.J., *J. Phys. Chem. B, 107,* 7441–7448 (2003).
15. Malynych, S., and Chumanov, G., *J. Am. Chem. Soc., 125,* 2896–2898 (2003).
16. Creighton, J.A., Blatchford, C.G., and Albrecht, M.G., *J. Chem. Soc., Faraday Trans. 2,* **75**, 790–798 (1979).
17. Lee, P.C., and Meisel, D., *J. Phys. Chem.,* 86, 3391–3395 (1982).
18. Leopold, N., and Lendl, B., *J. Phys. Chem. B,* 107, 5723–5727 (2003).
19. Caswell, K.K., Bender, C.M., and Murphy, C. J. *Nano Lett.,* **3**, 667–669 (2003).
20. *J. Colloid Interface Sci., 284,* 521–526, (2005).
21. Rossetti, R., Ellison, J.L., Gibson, J.M., and Brus, L.E., *J. Chem. Phys.,* 80, 4464–4469 (1983).
22. Fojtik, A., Weller, H., Koch, U., and Henglein, A. *Ber. Bunsen-Ges. Phys. Chem.,* **88**, 969–977 (1984).
23. Ramsden, J.J., and Grätzel, M., *Faraday Trans. 1,* **80**, 919–933 (1984).

24. El-Sayed, M.E., *Acc. Chem. Res.*, **37**, 326–333 (2004).
25. Davies, J.H., and Long, A.R., (Eds.), *Physics of Nanostructures*, Institute of Physics, Philadelphia, 1992.
26. Schiavello, M., *Photocatalysis and Environment: Trends and Applications*, Kluwer Academic, Boston, 1987.
27. Heroux, D.S., and Klabunde, K.J., Abstracts of the 38th Midwest Regional Meeting of the American Chemical Society, Columbia, MO, 339 (2003).
28. Lucas, E., Decker, S., Khaleel, A., Seitz, A., Fultz, S., Ponce, A., Li, W., Carnes, C., and Klabunde, K.J. *Chem. Eur. J.* **7**, 2505–2510 (2001).
29. Richards, R., Li, W., Decker, S., Davidson, C., Koper, O., Zaikovski, V, Volodin, A., Rieker, T., and Klabunde, K.J., *J. Am. Chem. Soc.,* **122**, 4921–4925 (2000).
30. Stark, J.V., Park, D.G., Lagadic, I., and Klabunde, K.J., *Chem. Mater.* **8**, 1904–1912 (1996).
31. Kistler, S.S., *J. Phys. Chem.* **36**, 52–64 (1932).
32. Teichner, S.J., in *Aerogel*, Fricke, J., (Ed.) Proceedings of the First International Symposium, Wurzburg, Springer-Verlag, Berlin, 1986, p 22.
33. Utamapanya, S., Klabunde, K.J., and Schlup, J.R., *Chem. Mater.*, **3**, 175–181 (1991).
34. Koper, O.B., Lagadic, I., Volodin, A., and Klabunde, K.J. *Chem. Mater.*, **9**, 2468–2480 (1997).
35. Klabunde, K.J., Stark, J.V., Koper, O., Mohs, C., Park, D.G., Decker, S., Jiang, Y., Lagadic, I., and Zhang, D., *J. Phys. Chem.,* **100**, 12142–12153 (1996).
36. Koper, O., and Klabunde, K.J., *Chem. Mater.*, **5**, 500–505 (1993).
37. Carnes, C.L., Kapoor, P.N., and Klabunde, K.J., *Chem. Mater.,* **14**, 2922–2929 (2002).
38. Bedilo, A., and Klabunde, K.J. *Nanostructured Mater.*, **8**, 119–135 (1997).
39. Kapoor, P.N., Heroux, D., Mulukutla, R.S., Zaikovskii, V., and Klabunde, K.J., *J. Mater. Chem.,* **13**, 410–414 (2003).
40. Carnes, C.L., Kapoor, P.N., Klabunde, K.J., and Bonevich, J., *Chem. Mater.,* **14**, 2922–2929 (2002).
41. Medine, G.M., Zaikovskii, V., and Klabunde, K.J. *J. Mater. Chem.* **14**,757–763 (2004).
42. Koper, O., Lagadic, I., and Klabunde, K.J., *Chem. Mater.*, **9**, 838–848 (1997).
43. Koper, O., and Klabunde, K.J., *Chem. Mater.*, **9**, 2481–2485 (1997).
44. Decker, S., and Klabunde, K.J., *J. Am. Chem. Soc.*, **118**, 12465–12466 (1996).
45. Fenelonov, V.B., Mel'gunov, M.S., Mishakov, I.V., Richards, R.M., Chesnokov, V.V., Volodin, A.M., and Klabunde, K.J., *J. Phys. Chem. B.*, **105**, 3937–3941 (2001).
46. Yang, Y.C., *Acc. Chem. Res.*, **32**, 109–115 (1999).
47. Wagner, G.W., Bartram, P.W., Koper, O., and Klabunde, K.J., *J. Phys. Chem. B.,* **103**, 3225–3228 (1999).
48. Wagner, G.W., Koper, O.B., Lucas, E., Decker, S., and Klabunde, K.J., *J. Phys. Chem. B.*, **104**, 5118–5123 (2000).
49. Wagner, G.W., Procell, L.R., O'Connor, R.J., Munavalli, S., Carnes, C.L., Kapoor, P.N., and Klabunde, K.J., *J. Am. Chem. Soc.*, **123**, 1636–1644 (2001).
50. Kastanek, F.; Kastanek, P., *J. Hazard. Mater.* **117**, 185–205 (2005).
51. Smalley, R., Nanotechnology, Congressional Hearings—*Emerging Technologies in the New Millenium*, The U.S. Senate Committee on Commerce, Science and Transportation, May 12, 1999.

10

Synthetic Organic Chemicals

Guo-Shuh J. Lee,* James H. McCain** and Madan M. Bhasin**

Synthetic organic chemicals can be defined as products derived from naturally occurring materials (petroleum, natural gas, and coal), which have undergone at least one chemical reaction, such as oxidation, hydrogenation, or sulfonation.

The volume of synthetic organic chemicals produced in the United States increased from about 42 billion lb in 1958 to more than 270 billion lb in 1990 and continues to increase steadily. The growth in production for the past 35 years is shown in Fig. 10.1. The effect of the economic slowdowns in 1974/75 and in the early 1980s on chemical output is reflected very clearly. Although the total volume increased from 1989 to 1990, the sales value actually decreased, presaging the slowdown of the early 1990s. Up until the mid-1960s, most of the phenomenal growth reflected the replacement of "natural" organic chemicals. Since that time, growth for synthetic materials has been dictated by the expansion of present markets and the development of new organic chemical end uses. It is not certain that this rate of growth can be

maintained as the uses of many products become strongly affected by environmental concerns. In fact, the future may see more emphasis on chemicals that can be produced from renewable raw material sources.

Those synthetic organic chemicals (excluding polymers) having production volumes of greater than four billion pounds in 2000, according to the Stanford Research Institute (SRI) Chemical Economics Handbook, are listed below. Note that the production of 5 chemicals alone totaled more than 100 billion lb, and 15 totaled 190 billion lb/year.

Ethylene dichloride	30.0 billion lb/year
MTBE	19.0
Methanol	18.7
Vinyl chloride	18.1
Ethanol	16.9
Styrene	12.6
p-Xylene	12.3
Terephthalic acid/DMT	10.1
Formaldehyde (37%)	10.0
Ethylene oxide	8.8
Cumene	8.6
Ethylene glycol	7.5
Oxo chemicals	7.1
Linear alfa olefins	5.2
Propylene oxide	5.2
Total	190.1

*The Dow Chemical Co., Midland, MI.
**The Dow Chemical Co., South Charleston, WV.

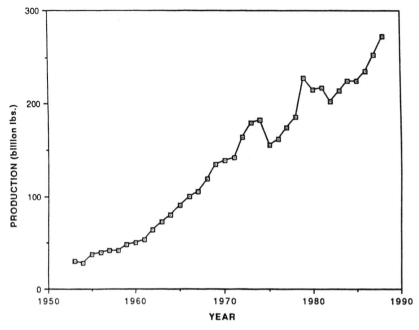

Fig. 10.1. Total production of synthetic organic chemicals, 1953–1990. (*U.S. International Trade Commission, December 1991.*)

Almost three thousand organic chemical products are currently derived from petrochemical sources. The commercial syntheses for all these products, however, can be traced back to one of six logical starting points. Consequently, this chapter has been subdivided according to the six major raw material sources: chemicals derived from methane, those from ethylene, propylene, "C_4s," higher aliphatics, and the aromatics.

A number of general references at the end of the chapter may be consulted for an overview of the subject.

CHEMICALS DERIVED FROM METHANE

Methane is readily available as the major component of natural gas, and its primary use is as fuel. It is also found in coal mines and as a product of anaerobic biological decay of organic materials in swamps and landfills. It is possible to make many chemicals from methane in a laboratory. However, methane is relatively inert chemically and is truly useful as a raw material for only a few commercial chemicals. These conversions generally require high temperatures and pressures or very aggressive chemicals such as chlorine, and usually are operated on a very large scale. Here, only those materials that are currently made from methane in commercial quantities are considered. The most important of these are shown in Fig. 10.2.

Synthesis Gas

The most important route for the conversion of methane to petrochemicals is via either hydrogen or a mixture of hydrogen and carbon monoxide. The latter material is known as synthesis gas. The manufacture of carbon monoxide–hydrogen mixtures from coal was first established industrially by the well-known water–gas reaction:

$$C + H_2O \rightarrow CO + H_2$$

Two important methods presently are used to produce the gas mixture from methane. The first is the methane–steam reaction, where methane and steam at about 900°C are passed through a tubular reactor packed with a promoted iron oxide catalyst. Two reactions are

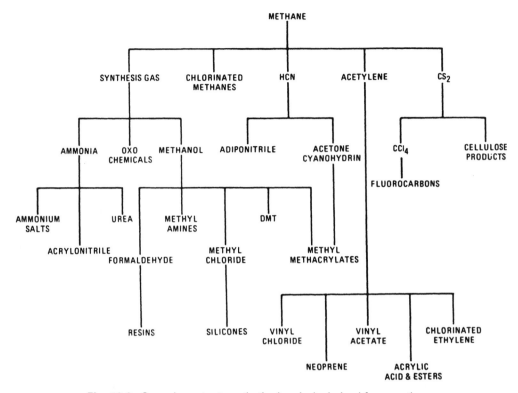

Fig. 10.2. Some important synthetic chemicals derived from methane.

possible, depending on the conditions:

$$CH_4 + H_2O \rightarrow CO + 3H_2$$

$$CO + H_2O \rightarrow CO_2 + H_2$$

The second commercial method involves the partial combustion of methane to provide the heat and steam needed for the conversion. Thus the reaction can be considered to take place in at least two steps, the combustion step:

$$CH_4 + 2O_2 \rightarrow CO_2 + 2H_2O$$

followed by the reaction steps:

$$CH_4 + CO_2 \rightarrow 2CO + 2H_2$$

$$CH_4 + H_2O \rightarrow CO + 3H_2$$

The process usually is run with nickel catalysts in the temperature range of 800–1000°C. Steam reforming usually is used on the lighter feedstocks, and partial oxidation is used for the heavier fraction.

Synthesis gas is the starting material for the manufacture of ammonia and its derivatives and also for methanol, as well as for other oxo-synthesis processes. It also is a source of carbon monoxide in the manufacture of such chemicals as acetic acid. And it is also a source of hydrogen for petroleum refining processes. However, shortages and corresponding high prices of natural gas and naphtha have generated interest in other synthesis gas feedstocks such as coal and residual oil.

Ammonia. Although ammonia is not an organic chemical, it is one of the largest-volume synthetic petrochemicals. From it many reactive organic chemicals are derived such as urea, acrylonitrile, caprolactam, amines, and isocyanates. Almost all of the 40.6 or more billion lb of ammonia produced in 1999 in the United States was based on hydrogen from petroleum and natural gas. Detailed descriptions of ammonia processes are found in Chapter 29.

Methanol. Methanol was once known as wood alcohol because it was a product of the destructive distillation of wood. All American methanol was produced in that way before 1926. That year, however, marked the first appearance in the world of German synthetic methanol. Today, almost all of the approximately 10 billion lb/year of methanol made in the United States comes from large-scale, integrated plants for conversion of natural gas to synthesis gas to methanol (Fig. 10.3). World consumption has reached 57 billion lb in 2000.

In principle, methanol, because it is derived from synthesis gas, can be made not only from convenient natural gas but also from any source of reduced organic carbon such as coal, wood, or cellulosic agricultural waste. It then can be used as a readily stored fuel or shipped for use as fuel or raw material elsewhere. As petroleum and natural gas become more difficult to recover, alternative carbon sources such as these will be used more. An example involving methanol is a coal-based acetic anhydride facility started in the 1980s in the United States. Another example is the use of some of the natural gas formerly wasted during recovery of Middle Eastern oil to make methanol and other chemicals. Direct use of methanol as a motor vehicle fuel is being studied, but it is not known when or even if such use will be significant in terms of methanol usage or decreased petroleum usage.

Methanol synthesis resembles that of ammonia in that high temperatures and pressures are used to obtain high conversions and rates. Improvements in catalysts allow operation at temperatures and pressures much lower than those of the initial commercial processes. Today, "low-pressure" Cu–Zn–Alminium oxide catalysts are operated at about 1500 psi and 250°C. These catalysts must be protected from trace impurities that the older "high-pressure" (5000 psi and 350°C) and "medium-pressure" (3000 psi and 250°C) catalysts tolerate better. Synthesis gas production technology has also evolved so that it is possible to maintain the required low levels of these trace impurities.

Methanol is used as a solvent, an antifreeze, a refrigerant, and a chemical intermediate. The greatest chemical uses for methanol as of 1998 were formaldehyde, 33 percent; MTBE, 27 percent; acetic acid, 7 percent; and chloromethane, 5 percent. Other chemicals derived from methanol include methyl methacrylate, methylamines, and dimethyl terephthalate.

Formaldehyde. Formaldehyde may be made from methanol either by catalytic vapor-phase oxidation:

$$CH_3OH + \tfrac{1}{2}O_2 \rightarrow CH_2O + H_2O$$

or by a combination oxidation–dehydrogenation process:

$$CH_3OH \rightarrow CH_2O + H_2$$

It also can be produced directly from natural gas, methane, and other aliphatic hydrocarbons, but this process yields mixtures of various oxygenated materials. Because both gaseous and liquid formaldehyde readily polymerize at room temperature, formaldehyde is not available in pure form. It is sold instead as a 37 percent solution in water, or in the polymeric form as paraformaldehyde $[HO(CH_2O)_nH]$, where n is between 8 and 50, or as trioxane $(CH_2O)_3$. The greatest end use for formaldehyde is in the field of synthetic resins, either as a homopolymer or as a copolymer with phenol, urea, or melamine. It also is reacted with acetaldehyde to produce pentaerythritol $[C(CH_2OH)_4]$, which finds use in polyester resins. Two smaller-volume uses are in urea–formaldehyde fertilizers and in hexamethylenetetramine, the latter being formed by condensation with ammonia.

U.S. production of formaldehyde in 2000 was approximately 10 billion lb of 37 percent formaldehyde, amounting to about 85 percent of capacity. Usage is expected to grow at about 3 percent/year through 2005.

Methyl Methyacrylate. Methyl methacrylate is formed in a three-step process from methanol, acetone, and HCN:

$$CH_3\overset{\overset{O}{\|}}{C}CH_3 \xrightarrow{\ HCN\ } CH_3\underset{\underset{CH_3}{|}}{\overset{\overset{OH}{|}}{C}}CN \xrightarrow{\ H_2SO_4\ }$$

$$CH_2{=}\underset{\underset{CH_3}{|}}{C}CONH_2 \ H_2SO_4 \xrightarrow{\ CH_3OH\ }$$

$$CH_2{=}\underset{\underset{CH_3}{|}}{C}COOCH_3 \ + \ NH_4HSO_4$$

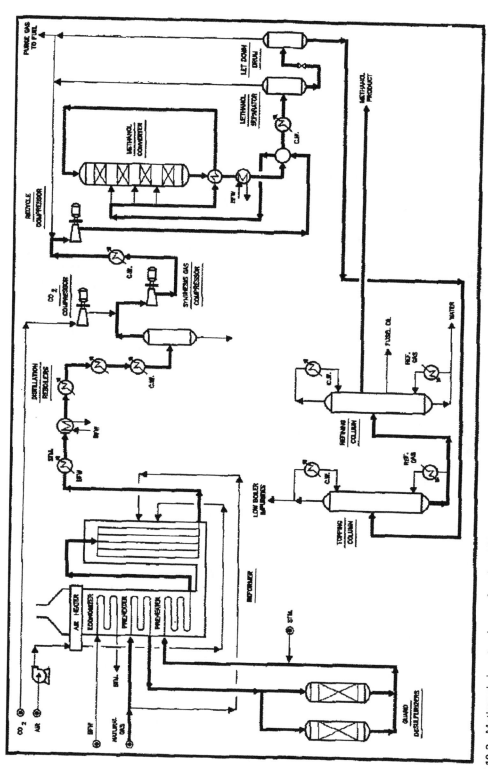

Fig. 10.3. Methanol via natural gas reforming. (*Chem Systems Report No. 93-1. Copyright Nexant Chem Systems, Inc. and used by permission of the copyright owner.*)

Although this is the major process in operation, there also is at least one commercially operated process for converting isobutylene and/or tert-butanol to methacrylic acid followed by esterification with methanol. A new process based on acetone cyanohydrin also has been reported. It avoids corrosive sulfuric acid and byproduct salts while coproducing formamide, which can be converted to HCN for recycle to the process.

U.S. production of methyl methacrylate in 1999 totaled 1.4 billion lb, which is about 82 percent of world capacity. Its uses are almost exclusively based on polymerization to poly(methyl methacrylate), which, because of its physical strength, weathering resistance, optical clarity, and high refractive index, has major uses in cast and extruded sheet (33%), molding powders and resins (16%), and surface coatings (22%).

Acetic Acid. The worldwide production of acetic acid was reported to be 15.7 billion lb in 1998. Acetic acid is a global product with about one third of production capacity now outside the United States, Western Europe, and Japan. The majority buildup is in Asia. In the future, the capacity in Asia will continue to increase substantially.

Acetic acid is produced by methanol carbonylation (the dominant process) as well as by acetaldehyde oxidation, ethanol oxidation, and light hydrocarbon oxidation. When methanol carbonylation was first practiced in the United States and West Germany, a cobalt iodide catalyst was used, and the process required up to 10,000 psia pressure. The technological breakthrough that allowed methanol carbonylation to become the leading acetic acid process was the discovery of rhodium–iodine catalysts, which can be operated at moderate pressure (500 psia) and at a methanol selectivity of 99 percent to acetic acid. Figure 10.4 is a schematic of the process. The recent advancement is the implementation of the low water technology. It significantly reduces the production cost by increasing productivity and lowering utility and capital cost.

In the United States, 5.3 billion lb of acetic acid was consumed in 1998. The applications, in decreasing order were: vinyl acetate (which alone accounts for more than 40 percent of U.S. acetic acid consumption), dimethyl terephthalate/terephthalic acid, acetate esters, cellulose acetate, other acetic anhydride uses, textiles, monochloroacetic acid, and several smaller uses. Growth projections are close to 3 percent from 1999 to 2003. The growth is tied largely to vinyl acetate monomer manufacture and to a lesser extent terephthalic acid manufacture.

Methyl t-Butyl Ether (MTBE). In 1980, MTBE was the fastest-growing derivative of methanol. This is a result of its only significant use, which is as an antiknock agent replacing lead in gasoline. In 1990, it was the fastest-growing chemical in the world. The world production has reached 48 billion in 1999. The U.S. production was about 30 billion lb, which was 63 percent of world capacity.

Because of its high miscibility in water and its increased use over the last several years, MTBE is now being found in many areas of the United States in groundwater reservoirs. This problem received national attention in June 1996, after MTBE was discovered in the drinking water supply of the city of Santa Monica, California. This incident led to many legislative initiatives in California and culminated in an Executive Order issued by California Governor Davis on March 25, 1999 to remove MTBE from all gasoline sold in California at the earliest possible date, but not later than December 31, 2002. Large declines in MTBE production and use are expected in the next few years.

MTBE is made by reacting methanol with the isobutylene contained in mixed-C_4 refinery streams. This is possible because butanes, the other butanes, and butadiene are inert under the mild conditions used. The process is catalyzed by acidic ion exchange resins.

Oxo Chemicals. The so-called oxo process combines carbon monoxide and hydrogen with olefins to make saturated aldehydes having one more carbon atom than the olefins have. The earliest such reaction studied used ethylene to produce both an aldehyde and a ketone.

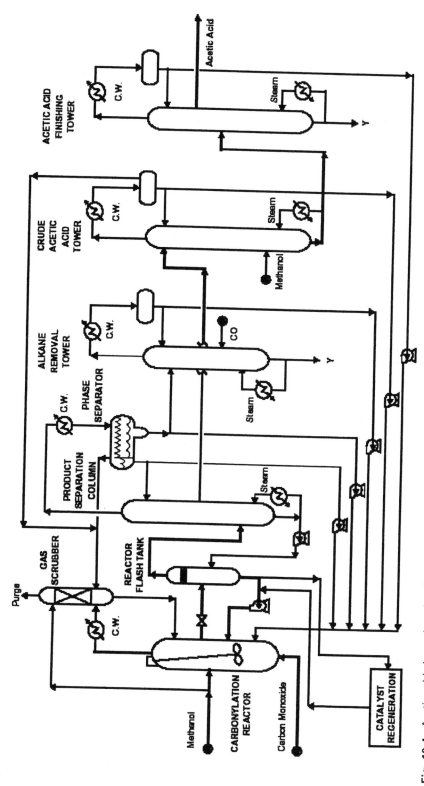

Fig. 10.4. Acetic acid via methanol carbonylation. (*Chem Systems Report No. 99/00 S5. Copyright Nexant Chem Systems, Inc. and used by permission of the copyright owner.*)

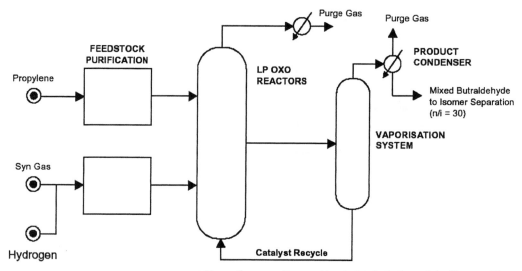

Fig. 10.5. Low-pressure oxo process. (*Chem Systems Report No. 98/99 S13. Copyright Nexant Chem Systems, Inc. and used by permission of the copyright owner.*)

Thus, the reaction was named "oxo" after the German *oxierung*, meaning "ketonization."

The low-pressure oxo process based on rhodium complex catalysts has largely replaced the older, high-pressure process, which used cobalt carbonyls as catalyst. The low-pressure process is operated at about 100°C and 200 psig. A new generation oxo process with bisphosphite modified rhodium catalyst is shown schematically in Fig. 10.5.

Oxo chemicals include butyraldehyde (*normal-* and *iso-*) and the corresponding alcohols, 2-ethylhexanol (from n-butyraldehyde), propionaldehyde, and n-propyl alcohol, and lesser amounts of higher aldehydes and alcohols derived from C_5 through C_{17} olefins. The total volume of products derived from oxo chemistry exceeds a billion pounds a year. Volumes and applications are given later in this chapter for the most important products.

Chloromethanes

The four chlorinated methanes are methyl chloride (CH_3Cl), methylene dichloride (CH_2Cl_2), chloroform ($CHCl_3$), and carbon tetrachloride (CCl_4). The U.S. production levels were about 520, 500, 450, and 750 million lb respectively, in 1990. Due to environmental regulation, there were some major changes.

The 2000 levels were 1267, 290, 792, and 40 million lb, respectively.

Methyl chloride is the only chlorinated methane with good growth. The principal use for methyl chloride is in the manufacture of chlorosilanes (89%) for the silicone industry. Other smaller uses are for methyl cellulose ether, quaternary ammonium compounds, herbicides, and butyl rubber.

Methyl chloride is produced by two methods: by the reaction of hydrogen chloride and methanol and by the chlorination of methane. Due to increasing demand for methyl chloride, the more selective methanol hydrochlorination has become increasingly important, whereas the nonselective methane chlorination route has declined. The hydrochlorination process also has the advantage that it utilizes, instead of generating, hydrogen chloride, a product whose disposal has become increasingly difficult. The methanol hydrochlorination process can be carried out in either liquid or gas phase. The gaseous phase reaction is carried out at 250–280°C. It uses a smaller reactor but requires extra energy to vaporize aqueous HCl.

$$CH_3OH + HCl \rightarrow CH_3Cl + H_2O$$

Methylene chloride and chloroform can be made, along with the other products, by the direct chlorination of methane. It is much more common, however, to produce them by the chlorination of methyl chloride. This can be done either thermally (350–450°C) or photochemically. The HCl byproduct can be recycled back into a hydrochlorination process for production of the methylene chloride starting material.

Methylene chloride is used primarily as a solvent for degreasing and paint removal, and it is also used in aerosols and foam-blowing agents. Since 1985, new environmental regulation has had a major impact on this chemical. The consumption of methylene chloride has been reduced by 60 percent due to recycling and product substitutions.

Chloroform is used to produce chlorodifluoromethane (HCFC-22), which is used as a refrigerant (70%) and to synthesize the monomer tetrafluoroethylene (30%). Fluoropolymers that use HCFC-22 as a feedstock are strong. But the uses for refrigerant will be phased out beginning 2010. The major uses for carbon tetrachloride were to make aerosol propellants such as dichlorodifluoromethane (CFC-12) and trichlorofluoromethane (CFC-11). The volume of carbon tetrachloride decreases to almost zero as CFC-11 and 12 will be phased out.

Acetylene

In the early days of the chemical industry, acetylene was a key starting material for many important products. Initially it was obtained for chemical purposes by reaction of calcium carbide with water; but that practice has given way to acetylene recovery from hydrocarbon cracking, so that now 86 percent of acetylene used in chemical manufacturing is made in this way. Owing to difficulty in its safe collection and transport, it is almost always used where it is prepared.

Acetylene still is a preferred raw material for some products, but it has been largely replaced by ethylene for many others. Chemicals once produced from acetylene by processes now considered outdated include: vinyl chloride, vinyl acetate, acetaldehyde, acrylonitrile, neoprene, and chlorinated solvents.

1,4-Butanediol. In 2000, 164 million lb of 1,4-butanediol were made from acetylene in the United States. It is the largest consumer of acetylene.

The Reppe process is used to make 1,4-1 butanediol from acetylene. In this process, acetylene and formaldehyde are reacted in the presence of a copper–bismuth catalyst. The resulting intermediate, 2-butyne-1,4-diol is hydrogenated over a Raney nickel catalyst:

$$HC{\equiv}CH + 2HCHO \rightarrow HOCH_2C{\equiv}CCH_2OH$$
$$\text{2-butyne-1,4-diol}$$

$$HOCH_2C{\equiv}CCH_2OH + H_2 \rightarrow HO(CH_2)_4OH$$
$$\text{2-butyne-1,4-diol} \qquad \text{1,4-butanediol}$$

The use of acetylene to make vinyl chloride (VCM) is now considered outdated and it is a minor process compared to the production of VCM from ethylene. Only 120 million lb of 16 billion lb VCM was made from acetylene.

Applications of 1,4-butanediol include the manufacture of THF (tetrahydrofuran), used as a solvent, and of poly(butylene terephthalate), used in thermoplastic resins.

Hydrogen Cyanide

Most of the hydrogen cyanide used today is prepared, as illustrated in Fig. 10.6, by ammoxidation of methane over a platinum catalyst. Absorption in water and distillation give pure hydrogen cyanide. Although all new hydrogen cyanide capacity is based on this technology, it also can be made from coke-oven gas, from acidification of inorganic cyanides, and by dehydration of formamide. A significant amount of material is obtained as a byproduct of ammoxidation of propylene in acrylonitrile manufacture. This amounts to about 25 percent of the demand. Because hydrogen cyanide is very toxic, the producers use much of it on-site to minimize the potential for human exposure during shipping.

The U.S. demand in 2000 was about 1.6 billion lb. Uses for hydrogen cyanide include: adiponitrile (for nylon 6/6), 47 percent; methyl methacrylate, 27 percent; sodium cyanide, 8 percent; methionine, 6 percent; and chelating agents, 2 percent.

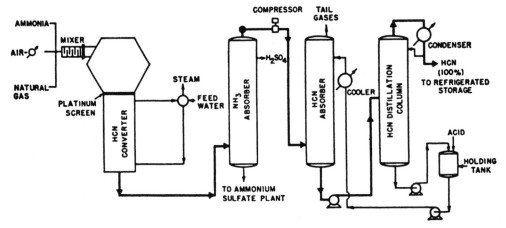

Fig. 10.6. The hydrogen cyanide process. (*Reprinted from Ind. Eng. Chem., 51, no. 10, 1235, 1959; Copyright 1959 by the American Chemical Society and reprinted by permission of the copyright owner.*)

Carbon Disulfide

Carbon disulfide is made by the catalytic reaction of methane with sulfur vapor. This can be thought of as analogous to oxidation of methane with oxygen, giving carbon dioxide. Carbon disulfide is used as raw material for making rayon (43%), agricultural chemicals (36%) and cellophane (3%). The total U.S. consumption in 2000 was about 160 million lb.

CHEMICALS DERIVED FROM ETHYLENE

Ethylene surpasses all other organic petrochemicals in production and in the amount sold. It is used as raw material for a greater number of commercial synthetic organic chemical products than is any other single chemical. Figure 10.7 shows the more important derivatives of ethylene. Ethylene consumption has grown explosively since 1940 when 300 million lb were used, mostly for making ethanol and ethylene oxide. During World War II, styrene use grew markedly, and polyethylene was developed as insulation for the then-new radar electronics. These materials later found a multitude of applications, which were responsible in large part for ethylene consumption reaching nearly 5 billion lb in 1960. Strong growth in ethylene dichloride and ethylene oxide contributed to over 18 billion lb of ethylene consumption in 1970. Continued growth raised this figure to 27 billion lb in

1978 and over 58 billion lb in 2000, which is almost two hundred times the 1940 volume. The overall growth for 48 years averaged a remarkable 10 percent/year.

Ethylene is manufactured by cracking hydrocarbons. A discussion of hydrocarbon cracking as a route to ethylene is found in Chapter 18.

Polyethylene

The largest consumers of ethylene are the various types of polyethylene: Low Density Polyethylene (LDPE), High Density Polyethylene (HDPE), and Linear Low Density Polyethylene (LLDPE). Chapter 15 gives detailed discussions of preparation of the various types of polyethylene.

LDPE is produced by high-pressure, high-temperature radical polymerization of pure ethylene. When improved properties are required, copolymers with one or more other vinyl monomers such as ethyl acrylate, vinyl acetate, or acrylic acid are used. LDPE has a relatively branched molecular structure, and the branches are relatively long. It is used for a multitude of purposes because of its properties and economics. Some important uses include: films for packaging of food and other merchandise; shipping trays and pallets; lightweight, flexible water- and chemical-resistant containers or barriers; and temporary coverings as in construction and agriculture.

ETHYLENE

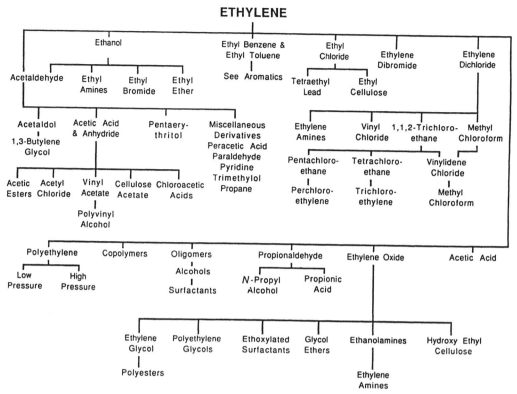

Fig. 10.7. Important derivatives of ethylene.

HDPE is produced by a low-pressure polymerization process in which highly pure gaseous ethylene is converted by proprietary catalysts to solid polyethylene particles. It has a very linear molecular structure. HDPE is stronger, tougher, and more rigid than LDPE, so it is used where such properties are advantageous. The major uses for HDPE are in blow-molded bottles, cans, and tanks for products such as milk, bleach, detergent, and fuel and in grocery sacks and other paper-replacement markets.

LLDPE is made by a catalytic process very similar to that for HDPE, but it is a softer polyethylene than HDPE with properties similar to those of LDPE. Its properties are achieved by inclusion of comonomers such as butene or hexene. A relatively disordered crystalline state is obtained by introducing many short branches into an otherwise highly linear molecule. Thus, the less expensive equipment of the HDPE process can be used to make a product having the greater flexibility and impact strength characteristic of LDPE.

In 2000, world capacity for polyethylene was nearly 112 billion lb.

Ethylene Oxide

Ethylene oxide was discovered in 1859 by Wurtz. He stated that ethylene oxide could not be made by direct oxidation of ethylene, and it was nearly 80 years before this was disproved. Wurtz made ethylene oxide by the method known today as the chlorohydrin process, in which ethylene is reacted in turn with hypochlorous acid and base. This process was commercialized during World War I in Germany, and until 1985 was still used commercially in the United States.

$$CH_2=CH_2 + HOCl \longrightarrow CH_2ClCH_2OH$$

$$2CH_2ClCH_2OH + Ca(OH)_2 \longrightarrow$$

$$CaCl_2 + 2H_2O + 2\overset{O}{\overset{/\backslash}{CH_2CH_2}}$$

Since 1985, processes for the direct oxidation of ethylene using either air or oxygen and

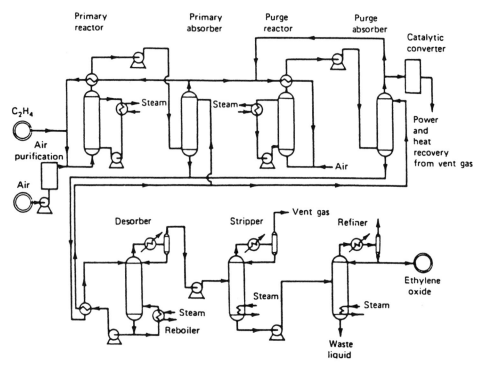

Fig. 10.8. Air-based direct oxidation process for ethylene oxide. (*Encyclopedia of Chemical Technology, Kirk and Othmer, Web site ed., ethylene oxide, manufacture, 2002. Copyright by John Wiley & Sons, Inc. and reproduced by permission of the copyright owner.*)

a silver catalyst have been the only remaining commercial processes for ethylene oxide production in the United States. Figure 10.8 illustrates an air-based process, and Fig. 10.9 an oxygen-based process. In this very exothermic conversion, oxygen and ethylene combine on a silver catalytic surface to make ethylene oxide. Oxygen and ethylene concentrations are controlled at low levels to avoid creating explosive mixtures.

The competing reactions of total combustion to carbon dioxide and isomerization must be avoided. Ethylene oxide plants in which air is used as source of oxygen require additional investment for purge reactors and associated absorbers. This investment is offset by the need, in the oxygen-based process, for an oxygen production plant and a carbon dioxide removal system. In general, the oxygen-based process is thought to be more economical, and all the plants built since the mid-1970s have been oxygen-based.

In 1999, U.S. ethylene oxide capacity was 9.1 billion lb with production of 8.2 billion lb.

Major uses in that year were: ethylene glycol, 57 percent; nonionic surfactants, 11 percent; ethanolamines, 11 percent; glycol ethers, 7 percent; diethylene glycol, 5 percent; and triethylene glycol, 2 percent. The remaining 7 percent of ethylene oxide consumption included PEGs (poly(ethylene glycol)), urethane polyols, and exports. In the following sections, several of these derivatives are discussed in more detail.

Ethylene Glycols. Monoethylene glycol or ethylene glycol is the major derivative of ethylene oxide. Ethylene glycol was initially made commercially by hydrolysis of ethylene chlorohydrin. Today, hydrolysis of ethylene oxide is the preferred route.

$$\underset{\displaystyle CH_2CH_2}{\overset{\displaystyle O}{\diagup\!\diagdown}} + H_2O \longrightarrow HOCH_2CH_2OH$$

Reaction of ethylene oxide with water is accomplished using a large molar excess of water to favor ethylene glycol formation over

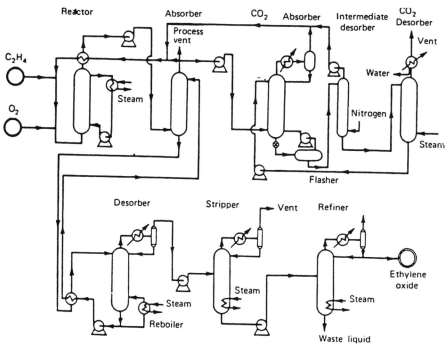

Fig. 10.9. Oxygen-based direct oxidation process for ethylene oxide. (*Encyclopedia of Chemical Technology, Kirk and Othmer, Web site ed., ethylene oxide, manufacture, 2002. Copyright by John Wiley & Sons, Inc. and reproduced by permission of the copyright owner.*)

that of diethylene glycol and triethylene glycol, as the glycols are more reactive with ethylene oxide on a molar basis than is water. A multiple-effect distillation scheme is used to recycle the unreacted excess water. The major product, ethylene glycol, is recovered between light impurities and the heavier glycol byproducts.

In 1999, U.S. consumption of ethylene glycol totaled 5.5 billion lb. Of that, 1.5 billion lb (28%) was used in the production of polyester bottles, primarily for soft drinks. Polyester fiber applications accounted for 1.4 billion lb (26%), primarily for the textile industry. Polyester film and miscellaneous applications consumed another 0.4 billion lb (7%). Antifreeze applications have held steady at approximately 1.6 billion lb over the last 20 years, and have become relatively less important with time than the polyester applications. This trend is expected to hold in the future. Increased demand for polyester bottles is expected to fuel growth in the United States, and bottle and textile applications are expected to fuel growth in other areas of the world.

Diethylene glycol usage is about 800 million lb/year in the United States. Major uses are unsaturated polyester resins (21%), polyurethane resins (21%), and antifreeze blending (10%). Other applications include use as raw materials for triethylene glycol (7%) and for morpholine (7%). Diethylene glycol is also used for dehydration of natural gas and in textile conditioning.

Triethylene glycol consumption is approximately 115 million lb/year in the United States. The major use, natural gas drying, depends on the low volatility and strong affinity of triethylene glycol for water. Lesser amounts are used as intermediate for vinyl plasticizers, polyester resins, and polyols. Additional direct applications include solvent and humectant uses.

Polyethylene glycols are produced by base-catalyzed addition of ethylene oxide to a low molecular weight glycol such as diethylene glycol. These glycols are higher-molecular-weight analogs of mono-, di-, and triethylene glycol but differ from the latter compounds in that they are not pure substances but rather consist of distributions of low molecular weight polymers. With average molecular weights beginning at about 200 (that of tetraethylene glycol) and going up to about 1000, these materials are liquids at ambient temperature. They are used as plasticizer intermediates, dispersant media, lubricants, and humectants. Above an average molecular weight of 1000, the polyglycols become waxy solids and find use in ointments, cosmetics, and lubricants taking advantage of their oil and water compatibility and low toxicity. At very high molecular weights, homopolymers of ethylene oxide are used for thickening, for water-soluble films, and for reducing friction in, for example, water delivery in fire hoses. Their value in this last application is that a given size hose can be made to deliver a greater flow of water.

Surfactants. Ethylene oxide-containing surfactants are generally of nonionic or anionic classes. The nonionic materials are made by base-catalyzed addition of ethylene oxide to either fatty alcohols or alkylphenols. Sulfation can be used to convert these compounds to the sulfated anionic surfactants. The products contain from a few to many ethylene oxide molecules per alcohol. The chain of poly(ethylene oxide) in a nonionic product acts as the hydrophile, and the alkyl or alkaryl residue is the hydrophobe. A sulfate salt group adds to the hydrophilicity of an anionic surfactant.

Surfactants based on aliphatic alcohols are used as cleaners in both domestic and industrial applications. They provide excellent properties such as wetting, dispersion, and emulsification. The ethoxylates derived from alkylphenols are chemically stable and highly versatile, finding more use in industrial practice than in domestic applications. They are used both as processing aids and as components in various products. Their applications include metal cleaning, hospital cleaners and disinfectants, agricultural chemical formulation surfactants, insecticides and herbicides, oil-well drilling fluids, and many others.

In the United States in 1999 for surfactant applications, 340 million lb of ethylene oxide were consumed in the production of alkylphenol ethoxylates, and 600 million lb were consumed

in the production of ethoxylates of aliphatic alcohols.

Ethanolamines. Ethanolamines are manufactured by reacting ethylene oxide and ammonia. The

$$NH_3 \xrightarrow{C_2H_4O} HOC_2H_4NH_2$$
$$\xrightarrow{C_2H_4O} (HOC_2H_4)_2NH$$
$$\xrightarrow{C_2H_4O} (HOC_2H_4)_3N$$

relative amounts of the three amines will depend primarily on the ammonia-to-oxide feed ratio. The three products are separated by distillation. Over the years, the relative demand for the three products has varied greatly. Thus, operational flexibility must be maintained.

The ethanolamines are water-miscible bases from whose properties stem their major uses as neutralizers in aqueous formulations such as metalworking fluids. Monoethanolamine is used in detergents, in "sweetening" (removing carbon dioxide and hydrogen sulfide from) natural gas, for removing carbon dioxide from ammonia during its manufacture, and as a raw material for producing ethyleneamines by reductive amination. Diethanolamine finds use in detergents and as an absorbent for acidic components of gases, as well as its major use as a raw material for surfactant diethanolamides of fatty acids. Triethanolamine's main end uses are in cosmetics and textile processing.

In the United States in 1999, ethylene oxide consumed in the production of ethanolamines totaled approximately 900 million lb. Approximately 300 million lb went into each of monoethanolamine, diethanolamine, and triethanolamine.

Glycol Ethers. In the same way that water reacts with one or more molecules of ethylene oxide, alcohols react to give monoethers of ethylene glycol, producing monoethers of diethylene glycol, triethylene glycol, and so on, as by-products.

$$ROH \xrightarrow{C_2H_4O} ROC_2H_4OH$$
$$\xrightarrow{C_2H_4O} RO(C_2H_4O)_2H$$
$$\xrightarrow{C_2H_4O} RO(C_2H_4O)_3H$$

Since their commercial introduction in 1926, glycol ethers have become valuable as industrial solvents and chemical intermediates. Because glycol monoethers contain a $-OCH_2CH_2OH$ group, they resemble a combination of ether and ethyl alcohol in solvent properties. The most common

$$2ROC_2H_4OH$$
$$\xrightarrow{H_2SO_4} ROC_2H_4OC_2H_4OR + H_2O$$

alcohols used are methanol, ethanol, and butanol. Principal uses for the glycol ethers are as solvents for paints and lacquers, as intermediates in the production of plasticizers, and as ingredients in brake fluid formulations. Condensation of the monoethers produces glycol diethers, which are also useful as solvents.

Solvent characteristics of glycol ethers are enhanced by esterifying with acetic acid. The resulting acetate esters are used extensively in coating formulations, especially those formulations in which their high solvent power allows a decreased total solvent usage in compliance with volatile organic compound (VOC) emission standards.

In the United States in 1999, ethylene oxide consumption for production of glycol ethers was approximately 560 million lb.

Other Uses of Ethylene Oxide. About 2 percent of ethylene oxide is consumed in miscellaneous applications, such as its use as a raw material in manufacture of choline, ethylene chlorohydrin, hydroxyethyl starch, and hydroxyethyl cellulose and its direct use as a fumigant/sterilant. Production of 1,3-propanediol via hydroformylation of ethylene oxide was begun on a commercial scale in 1999. 1,3-Propanediol is a raw material for polytrimethylene terephthalate, which finds uses in fibers, injection molding, and in film. Use of ethylene oxide in making 1,3-propanediol is expected to be as much as 185 million lb by 2004, up from 12 million lb in 1999.

Chlorinated Ethanes and Ethylenes

A number of important large-volume petrochemicals are obtained through the chlorination of ethane and ethylene. The largest-volume chlorinated derivative is 1,2-dichloroethane (18 billion lb/year); most of it is used to make vinyl chloride. It has about 4 percent growth rate in the past decade. Because of their unique solvent and chemical intermediate properties, the market for chlorinated ethanes and ethylenes (exclude 1,2-dichloroethane) grew steadily until it reached a peak in 1980. Owing to environmental problems, particularly in the solvent area, the demand for some of the end uses has been declining steadily since then. The 1999 annual production rates for tetrachloroethylene, trichloroethylene, and 1,1,1-trichloroethane are at about 300, 210, and 250 million lb. Figure 10.10 shows the possible production routes to the major chlorinated derivatives.

Chlorinated Ethanes. Of the nine possible chlorinated derivatives of ethane, only three are of commercial importance: ethyl chloride, 1,2-dichloroethane (ethylene dichloride), and 1,1,1-trichloroethane (methyl chloroform). The other compounds have no important end uses and are produced either as intermediates or as unwanted by-products. They normally are converted to useful materials by a cracking process (for trichloroethylene) or by perchlorination (for carbon tetrachloride and tetrachloroethylene).

Ethyl Chloride. Most of the ethyl chloride is made by the exothermic hydrochlorination of ethylene, in either the liquid or the vapor phase:

$$CH_2=CH_2 + HCl \xrightarrow{AlCl_3} CH_3CH_2Cl$$

A much smaller amount is produced by the thermal chlorination of ethane. This direct chlorination may be run in conjunction with another process, such as oxychlorination, which can use the byproduct HCl as feed.

Ethyl chloride rose to commercial importance because of the automotive industry. It was the starting material for tetraethyllead, at one time the most commonly used octane booster. Demand has been cut drastically because of the conversion from leaded to unleaded gasoline for environmental reasons. Other uses for ethyl chloride are in the production of ethyl cellulose, as an ethylating agent, as a blowing agent, and in solvent extraction.

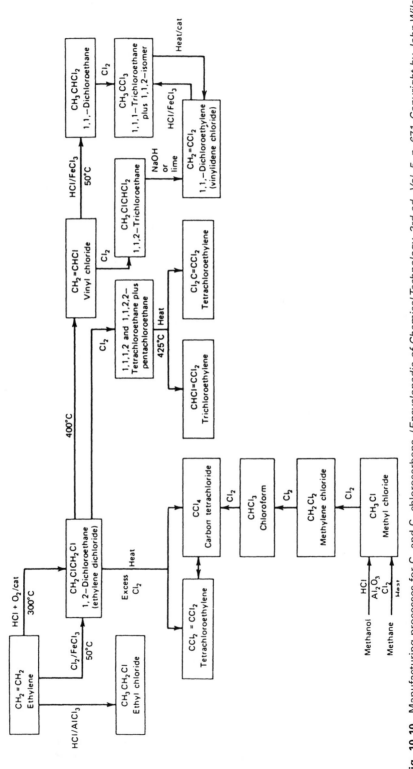

Fig. 10.10. Manufacturing processes for C_1 and C_2 chlorocarbons. (*Encyclopedia of Chemical Technology, 3rd ed., Vol. 5, p. 671. Copyright by John Wiley & Sons, Inc. and reproduced by permission of the copyright owner.*)

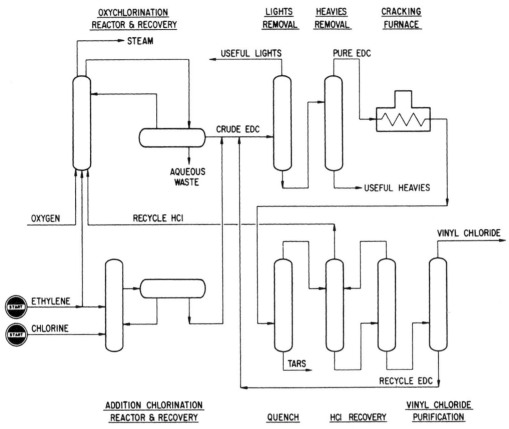

Fig. 10.11. Integrated EDC/vinyl chloride process. (*Hydrocarbons Processing, p. 174, 1985 November. Copyright Gulf Publishing Company and used by permission of the copyright owner.*)

1,2-Dichloroethene (Ethylene Dichloride). Ethylene dichloride (EDC) is one of the truly major petrochemicals. The U.S. production exceeded 18 billion lb in 1999. Almost 95 percent of this went into the manufacture of vinyl chloride monomer.

Figure 10.11 shows an integrated plant for producing EDC and vinyl chloride from ethylene, chlorine, and air. In this process, vinyl chloride (VCM) is produced by the thermal cracking of EDC. The feed EDC may be supplied from two sources. In the first source, ethylene and chlorine are reacted in essentially stoichiometric proportions to produce EDC by direct addition. In the second source, ethylene is reacted with air and HCl by the oxychlorination process. Ideally, both processes are carried out in balance, and the oxychlorination process is used to consume the HCl produced in the cracking and direct chlorination steps. The chemical reactions are

as follows.

$$CH_2{=}CH_2 + 2HCl + 1/2O_2 \rightarrow ClCH_2{-}CH_2Cl$$

$$CH_2{=}CH_2 + Cl_2 \rightarrow ClCH_2{-}CH_2Cl$$

$$ClCH_2{-}CH_2Cl \rightarrow CH_2{=}CHCl + HCl$$

Thus the overall reaction for the integrated plant is:

$$4CH_2{=}CH_2 + 2Cl_2 + O_2$$
$$\rightarrow 4CH_2{=}CHCl + H_2O$$

The direct chlorination of ethylene usually is run in the liquid phase and is catalyzed with ferric chloride. High-purity ethylene normally is used to avoid product purification problems. The cracking (pyrolysis) of EDC to VCM typically is carried out at temperatures of 430–530°C without a catalyst. The hot gases are quenched and distilled to remove HCl and then VCM. The unconverted EDC is returned to the EDC purification train. The oxychlorination step is the heart of the process and has two major variables, the type of reactor and the oxidant. The reactor may be either a fixed bed or a fluidized bed, and the oxidant is either air or oxygen. The temperature is in the range of 225–275°C with a copper chloride-impregnated catalyst.

Profitable disposal of the byproduct HCl once was the major restriction to the growth of EDC. The advances in the oxychlorination, which uses the HCl and air to produce ethylene dichloride, opened the door for the rapid replacement of the acetylene-based routes.

Almost 95 percent of all EDC goes to make VCM. Of that, less than 20 percent actually is isolated as EDC. Smaller uses are as a solvent and as a raw material for other chlorinated hydrocarbons such as trichloroethylene and perchloroethylene. Also a small amount is used to produce ethylene diamines.

Vinyl Chloride. Approximately 16.5 billion lb of VCM were produced in the United States in 1999, making it one of the largest-volume petrochemicals. It has been reported that more than 35 percent of the global production of chlorine goes to the manufacture of VCM. Although most of the VCM comes from EDC by the route described previously, it can be obtained from other sources, including its production in the catalytic hydrochlorination of acetylene and as a byproduct in the synthesis of other chlorinated hydrocarbons.

More than 95 percent of all VCM is used to produce polyvinyl chloride (PVC), an important polymer for the housing and automotive industries. (A detailed description of PVC is included in Chapter 15.) The rest of the VCM goes into the production of chlorinated solvents.

1,1,1-Trichloroethane (Methyl Chloroform). 1,1,1-Trichloroethane was a major solvent, particularly for cold and vapor degreasing. It was phased out for emissive uses in the United States in 1996 because of its ozone depletion potential. The only application left is as chemical precursor for HCFC-141b and HCFC-142b. However, both are subject to phaseout schedule of the Montreal Protocol,

and their production has been frozen at the 1996 level. The U.S. consumption has fallen from 700 million lb in 1988 to about 200 million lb in 1999.

1,1,1-Trichloroethane can be produced by three methods: by chlorination of 1,1-dichloroethane, from 1,1,2-trichloroethane via 1,1-dichloroethylene, and by direct chlorination of ethane. In the United States the first route produces about 70 percent. In this process the EDC feedstock is rearranged to 1,1-dichloroethane via cracking to VCM, followed by addition of HCl in the presence of a catalyst. For the final step, the dichloroethane is thermally or photochemically chlorinated. The reactions are as follows:

$$ClCH_2\text{–}CH_2Cl \rightarrow CH_2\text{=}CHCl + HCl$$

$$CH_2\text{=}CHCl + HCl \rightarrow CH_3\text{–}CHCl_2$$

$$CH_3\text{–}CHCl_2 + Cl_2 \rightarrow CH_3\text{–}CCl_3 + HCl$$

Chlorinated Ethylenes. VCM is by far the largest-volume chlorinated ethylene derivative. The others of commercial interest are tetrachloroethylene (perchloroethylene), trichloroethylene, and 1,1-dichloroethylene (vinylidene chloride).

Tetrachloroethylene (Perchloroethylene). Perchloroethylene historically has been the dominant solvent in the dry-cleaning industry because of its good stability and low flammability. Environmental concerns reduced its usage in dry-cleaning from 500 million lb in 1988 to less than 100 million lb in 1999. However, increasing quantities of perchloroethylene are being used to make alternative chlorofluorohydrocarbons, such as HCFC-123 and HCFC-134a. The total production volume in 1999 was 318 million lb.

Most perchloroethylene has been coproduced with carbon tetrachloride by the chlorination of propylene and/or chloropropanes. After the phaseout of CFC-11 and -12, the market for carbon tetrachloride disappeared. Producers have modified their units to shift the production to perchloroethylene.

An oxychlorination/oxyhydrochlorination process for the production of perchloroethylene and trichloroethylene is shown in Fig. 10.12.

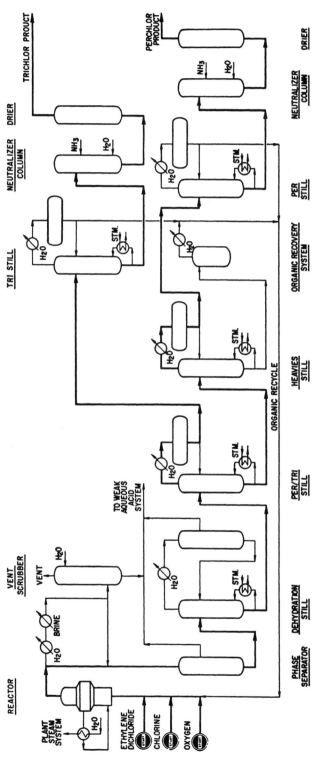

Fig. 10.12. Perchloroethylene/trichloroethylene plant. (Reproduced from Hydrocarbon Processing, p. 154, 1985 November. Copyright by Gulf Publishing Co. and used by permission of the copyright owner.)

The process can accept a wide range of low-cost feedstocks, such as ethylene, chlorinated C_2 hydrocarbons, and by-product streams from VCM chloromethanes, methyl chloroform, and EDC plants. The product ratio of trichloroethylene to perchloroethylene can be adjusted over a wide range.

Other Chlorinated Ethylenes. Trichloroethylene was a major solvent for degreasing in the late 1960s and early 1970s. Since that time, its production has decreased from 500 million lb to 100 million lb in 1993 because of environmental pressures on the solvent users and replacement by 1,1,1-trichloroethane. Recently, trichloroethylene has recovered market share in metal cleaning due to the phasing out of 1,1,1-trichloroethane in 1996. Also, the use as precursor for HFC-134a synthesis continues to increase. The production volume in 1998 was 245 million lb.

Although 1,1-dichloroethylene (vinylidene chloride) is a relatively small-volume product, it provides a way of upgrading the unwanted 1,1,2-trichloroethane by-product from the manufacture of EDC and 1,1,1-trichloroethane. Its major use is as an intermediate for polyvinylidene chloride and its copolymers, which are important barrier materials for food packaging.

Ethanol

Ethanol is made by both ethylene hydration and fermentation of starches and sugars. In this section the synthetic route will be discussed. The fermentation route is covered in Chapters 32 and 33.

In the World War II era, 72 percent of U.S. ethanol was derived from molasses fermentation. By 1978 the balance was 90 percent from direct catalytic hydration and the rest from fermentation. In 1998 the balance had returned to the dominance of fermentation, with 83 percent of the 10 billion lb of U.S. ethanol made in this way. The recent swing toward fermentation is due to the use of 90 percent of the fermentation ethanol as motor fuel, as a result of post-oil-embargo U.S. government policy.

Direct hydration of ethylene is by far the major route to synthetic ethanol. It is accomplished under pressure at 250–300°C over an acidic catalyst. Ethylene and high temperature steam are mixed and passed over an acidic catalyst, usually phosphoric acid on a support. A modest conversion is achieved even with the severe conditions. Cooling of the exit stream and passage through a separations system give ethylene and water for recycle. Ethanol is made either as a 95 percent azeotrope with water or as an anhydrous material from a drying system.

Synthetic ethanol has the following uses: as a chemical intermediate (for ethyl acetate, ethyl acrylate, glycol ethers, ethylamines, etc.), 30 percent; in toiletries and cosmetics, 20 percent; as a coatings solvent, 15 percent; as a raw material for vinegar, 10 percent; in household cleaners, 7 percent; in detergents, 5 percent; in pharmaceuticals, 5 percent; in printing inks, 3 percent; and in miscellaneous uses, 5 percent.

Ethylbenzene

Ethylbenzene is used almost exclusively (99%) as a raw material for producing styrene. The remainder is used as solvent and in the manufacture of diethylbenzene. The world and U.S. demand were 44.7 and 12.6 billion lb, respectively. A growth of about 3 percent/year was expected for the next few years.

Over 90 percent of all ethylbenzene is produced by alkylation of benzene with ethylene in the presence of an acidic catalyst such as aluminum chloride or an acidic zeolite. Figure 10.13 shows a liquid phase alkylation process with zeolite catalyst.

$$C_6H_6 + CH_2{=}CH_2 \rightarrow C_6H_5CH_2CH_3$$

Conversion to styrene is accomplished either by dehydrogenation:

$$C_6H_5CH_2CH_3 \rightarrow C_6H_5CH{=}CH_2 + H_2$$

or by a sequence of oxidation to ethylbenzene hydroperoxide, reduction to methyl phenyl carbinol (by a process that also oxidizes propylene to propylene oxide), and dehydration of the methyl phenyl carbinol to styrene.

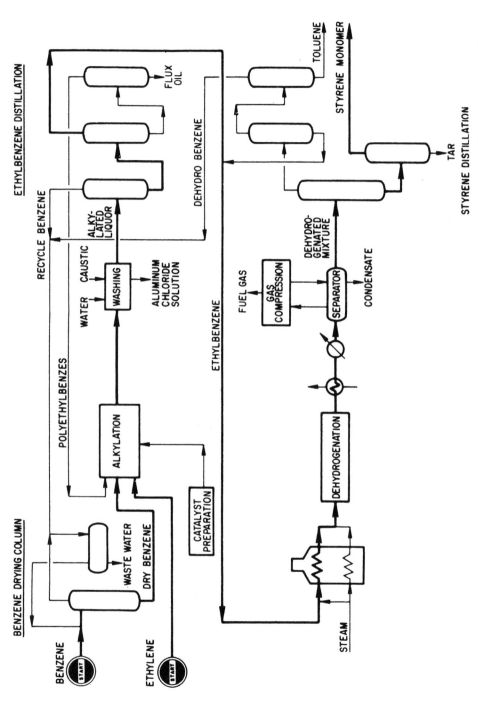

Fig. 10.13. Integrated plant for manufacture of ethylbenzene and styrene. (*Reproduced from Hydrocarbons Processing, Petrochemical Handbook, p. 169, 1985 November. Copyright Gulf Publishing Co. and used by permission of the copyright owner.*)

Acetaldehyde, Acetic Acid, Acetic Anhydride, Vinyl Acetate

Acetaldehyde. Acetaldehyde has been made from ethanol by dehydrogenation and by catalytic hydration of acetylene. Today direct oxidation of ethylene in the liquid phase catalyzed by palladium and copper has replaced these earlier methods. Figure 10.14 shows an ethylene-to-acetaldehyde unit based on this last route.

Acetaldehyde once was widely used as raw material for a variety of large-volume chemical products such as acetic acid and butanol. U.S. usage peaked in 1969 at 1.65 billion lb. Today, most of the former uses have been superseded by routes based on C_1 or other chemistry such as methanol carbonylation to acetic acid and butanol from propylene by oxo chemistry. Of the remaining uses, which totaled about 400 million lb in the United States in 2000, pyridine and substituted pyridines are the major consumers at 40 percent. It is also used as a raw material for peracetic acid, pentaerythritol, and 1,3-butylene glycol.

Acetic Acid. Acetic acid used to be derived from ethylene with acetaldehyde as an intermediate. The relatively high price of acetaldehyde compared to methanol and carbon monoxide, however, caused a shift away from this route. Although most acetic acid is currently produced by methanol carbonylation, as discussed earlier, a new route directly from ethylene was commercialized in 1997. This route employs a palladium-containing catalyst and combines ethylene directly with oxygen to produce acetic acid with approximately 86 percent selectivity. Figure 10.15 is a schematic diagram of the process.

Acetic Anhydride. A total of 1.9 billion lb of acetic anhydride was produced in the United States in 1999. Commercial production of acetic anhydride is currently accomplished through two routes, one involving ketene and the other methyl acetate carbonylation. A former route based on liquid phase oxidation of acetaldehyde is now obsolete.

In the ketene process, acetic acid is thermally dehydrated at 750°C to ketene. The ketene is separated from byproduct water and reacted with another mole of acetic acid to produce acetic anhydride. Figure 10.16 is a schematic diagram of this process.

The methyl acetate carbonylation process was successfully started and operated in the early 1980s. In this process, methyl acetate, itself the product of a one-step esterification

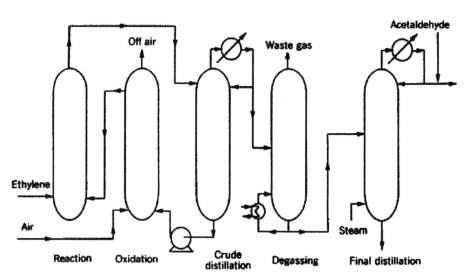

Fig. 10.14. Two-stage acetaldehyde process. (*Encyclopedia of Chemical Technology, Kirk and Othmer, Web site ed., acetaldehyde, manufacture, 2002. Copyright by John Wiley & Sons, Inc. and reproduced by permission of the copyright owner*)

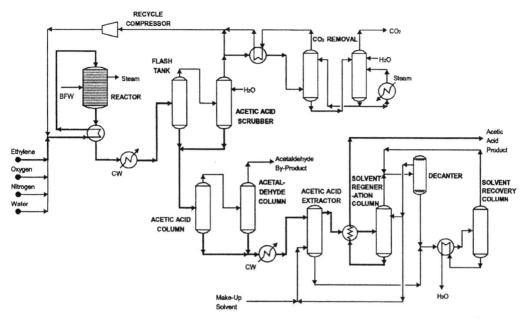

Fig. 10.15. Acetic acid via direct ethylene oxidation. (*Chem Systems Report No. 99/0055. Copyright Nexant Chem Systems, Inc. and used by permission of the copyright owner.*)

of acetic acid and methanol, is reacted with carbon monoxide in the presence of a promoted rhodium-iodide catalyst. Figure 10.17 illustrates this process exclusive of esterification to make methyl acetate.

The greatest use of acetic anhydride is in esterifying cellulose to cellulose acetate for application as cigarette filter tow and in textiles. In the United States, acetic anhydride is manufactured by cellulose acetate manufacturers and largely used internally. Other products using acetic anhydride as a raw material are mostly mature with low growth rates; they include triacetin, plastic modifiers and intermediates for pharmaceuticals, herbicides, pesticides, and dyes for polyolefins. Growth in production was projected to be 0.8 percent/year in the United States through 2003.

Vinyl Acetate. Vinyl acetate (VAM, for vinyl acetate monomer) production is the largest consumer of acetic acid worldwide. In North America, vinyl acetate production in 2000 was 1.7 billion lb. Growth in North America in the period 2000–2005 is expected to be 1.0 percent/ year and for the world, 2.4 percent/year.

Production of vinyl acetate is based primarily on vapor phase oxidative addition of acetic acid to ethylene. Figure 10.18 illustrates the process.

VAM finds exclusive use as a monomer or raw material for polymers and copolymers; and latex paints are the largest use for poly(vinyl acetate) (PVA) emulsions. Because latex paints cure without appreciable solvent emissions, regulatory pressures against such emissions favor the use of latex paints over solvent-based coatings. Adhesives are the second largest consumers of PVA emulsions, with a range of applications from packaging and wallboard to consumer "white" glue.

In the second major use of VAM, PVA is converted to poly(vinyl alcohol) (PVOH) by a transesterification reaction with methanol, giving methyl acetate as coproduct. PVOH finds its major end use in textile sizing and adhesives. Further reaction of PVOH with butyraldehyde or formaldehyde gives polyvinyl butyral (PVB) or polyvinyl formal, which together constitute the third largest consumption of VAM. PVB is used almost exclusively in the adhesive laminating inner layer in safety glass.

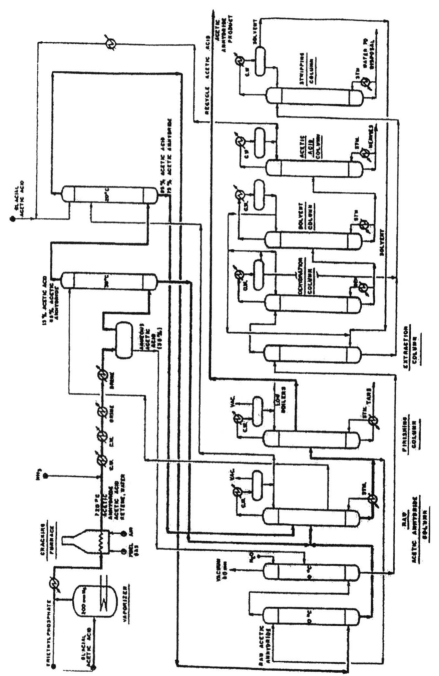

Fig. 10.16. Acetic anhydride from acetic acid ketene process. (*Chem Systems Report No. 97/98-1. Copyright Nexant Chem Systems, Inc. and used by permission of the copyright owner.*)

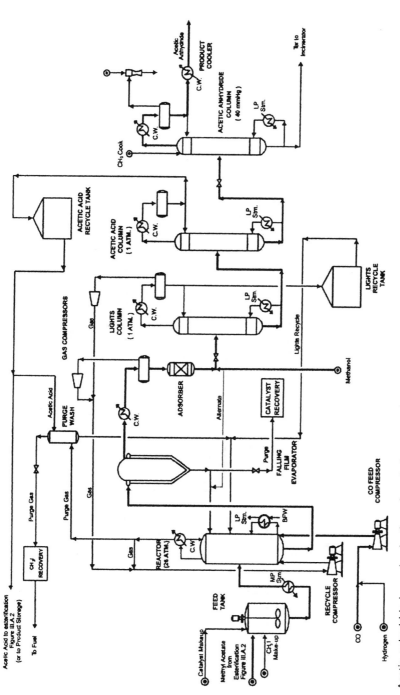

Fig. 10.17. Acetic anhydride by carbonylation of methyl acetate. (*Chem Systems Report No. 97/98-1. Copyright Nexant Chem Systems, Inc. and used by permission of the copyright owner.*)

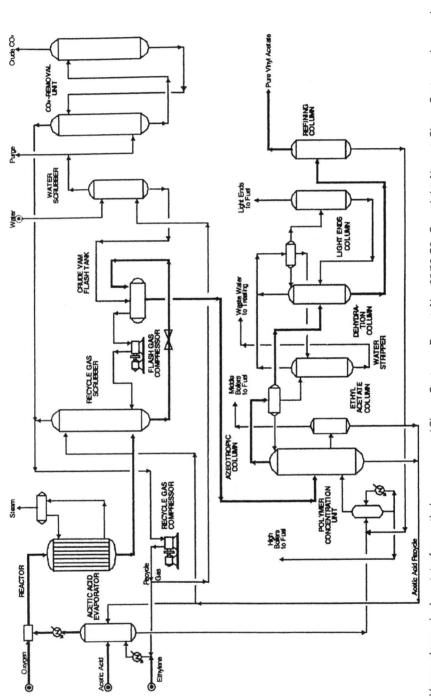

Fig. 10.18. Vapor phase vinyl acetate from ethylene process. (*Chem Systems Report No. 98/99-S3. Copyright Nexant Chem Systems, Inc. and used by permission of the copyright owner.*)

Ethylene Oligomers (Alpha Olefins) and Linear Primary Alcohols

Linear primary alcohols and alpha olefins in the C_6–C_{18} range have enjoyed remarkable growth in the last three decades. As esters, the C_6–C_{10} alcohols are used for plasticizing PVC. In the C_{12}–C_{18} range, the alcohols are used to make readily biodegradable surfactants of various types such as ethoxylates (nonionic), alcohol sulfates, and sulfates of ethoxylates (anionic). Alpha olefins are used as polyethylene comonomer (33%) and as raw materials for detergent alcohols (22%), oxo alcohols (10%), and lubricants and lube oil additives (18%).

Production of linear primary alcohols and production of alpha olefins are accomplished by similar reactions in which ethylene is oligomerized by organometallic catalysts based on aluminum alkyls such as triethylaluminum. The two processes are distinguished by the way in which the growing hydrocarbon chain is removed from the catalyst center. In the case of the alpha olefin products, ethylene growth to a hydrocarbon chain of a few or many two-carbon units is interrupted when the hydrocarbon on aluminum is displaced as an alpha olefin by exchange with fresh ethylene. Thus the aluminum alkyl is regenerated, and chain growth starts again. In production of alpha alcohols, the hydrocarbon group on the aluminum catalyst is cleaved by oxygen at the sensitive carbon–aluminum bond to give aluminum oxide and an alcohol. Figures 10.19 and 10.20 illustrate production schemes for alpha alcohols and alpha olefins, respectively.

Demand for alpha olefins in North America in 1999 totaled nearly 2.7 billion. The rate of growth through 2004 is expected to be 5.7 percent/year.

Ethylene–Propylene Elastomers

Ethylene propylene copolymer and terpolymer rubbers (EPRs) are produced at the rate in excess of two billion pounds per year worldwide. Of this, 41 percent is in North America, 27 percent in Western Europe, and 23 percent in Japan.

EPR is produced by polymerization of a mixture of ethylene and propylene and optionally a small amount of a nonconjugated diene such as ethylidene norbornene, norbornene, 1,4-hexadiene, or dicyclopentadiene. Two processes, one a solution and the other a suspension process, are employed. They use organometallic catalysts, the most common being products of combining (organo)vanadium halides with alkyl aluminum or alkyl aluminum halides. The resulting catalysts are deactivated by water and alcohols.

The comonomer diene confers sulfur vulcanizability on the elastomer. Otherwise, a peroxide cure is required for cross-linking. The polymers are readily oil-extended with 20–50 percent oil for many applications. Some uses result from the ability of these products to resist oxidation by ozone.

Applications of ethylene–propylene copolymers and terpolymers include: automotive (the major use area), thermoplastic olefin elastomers, single-ply roofing, viscosity index improvers for lube oils, wire and cable insulation, hose, appliance parts, and polymer modification.

Propionaldehyde

Propionaldehyde is produced by the oxo reaction of ethylene with carbon monoxide and hydrogen. n-Propyl alcohol is produced by hydrogenation of propionaldehyde, and propionic acid is made by oxidation of propionaldehyde.

n-Propyl alcohol is used as solvent in printing inks and as an intermediate in the preparation of agricultural chemicals. Propionic acid is used as a grain preservative as, for example, in preventing spoilage of wet corn used as animal feed. The use of propionic acid as a grain preservative is an alternative to drying by heating, which consumes fuel, and is considered mostly when fuel is expensive.

Other Ethylene Uses

Some lesser-volume ethylene uses are in:

- *Agriculture*, as a ripening agent for fruits and vegetables
- *Vinyl toluene* for use in unsaturated polyester resins

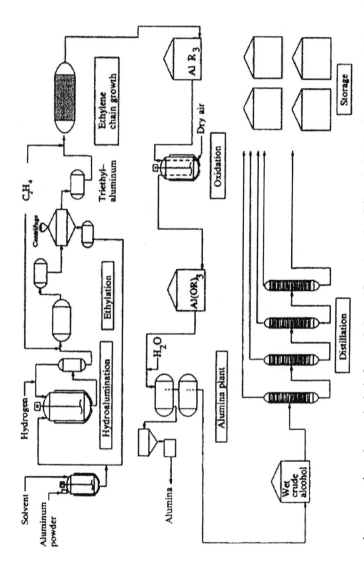

Fig. 10.19. Flow diagram of a process for primary alcohols from ethylene. (*Encyclopedia of Chemical Technology, Kirk and Othmer, Web site ed., alcohols, higher aliphatic, synthetic processes, manufacture, 2002. Copyright by John Wiley & Sons, Inc. and reproduced by permission of the copyright owner.*)

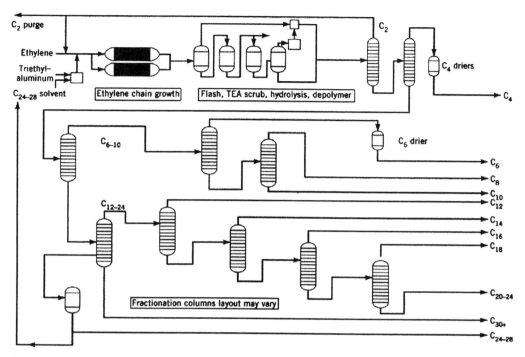

Fig. 10.20. Flow scheme of a process for alpha-olefins from ethylene. (*Encyclopedia of Chemical Technology, Kirk and Othmer, Web site ed., olefins, higher, manufacture, 2002. Copyright by John Wiley & Sons, Inc. and reproduced by permission of the copyright owner.*)

- *Aluminum alkyls* used in making organometallic catalysts and as initiators for processes such as ethylene–propylene rubber, polybutadiene, low-pressure polyethylene, and ethylene oligomerization to make alpha-olefins and C_6–C_{18} alcohols
- *Diethyl sulfate* made from sulfuric acid and ethylene and used as an alkylating agent in many applications
- *Alkylation of anilines* for chemical intermediates used in pesticides, pharmaceuticals, dyes, and urethane comonomers

CHEMICALS DERIVED FROM PROPYLENE

Propylene consumption for chemical synthesis in 1998 in the United States was 30 billion lb. This demand was exceeded by that of only one other synthetic organic chemical, ethylene. The demand was projected to grow at about 4.7 percent/year through 2003. Major uses of propylene are in polypropylene, acrylonitrile, propylene oxide, and cumene. A breakdown of propylene consumption by product is:

Polypropylene	45%
Acrylonitrile	12
Propylene oxide	11
Cumene	8
Oxo alcohols	8
Isopropyl alcohol	4
Oligomers	4
Acrylic acid	5
Export, other	3

Propylene is produced as a coproduct of ethylene cracking and is a product of petroleum refinery operations.

Polypropylene

More than 40 years after its introduction, polypropylene is the largest chemical consumer of propylene in the United States. It is produced primarily by a bulk or gas phase process, with the older slurry process still used by some. Because of its greater stiffness relative to polyethylene, polypropylene is used for more demanding applications. This stiffness is not without a drawback in the form of increased brittleness, which can be moderated through incorporation of ethylene by copolymerization. Its major uses are in fibers and injection molding. Chapter 15 reviews polypropylene in some detail.

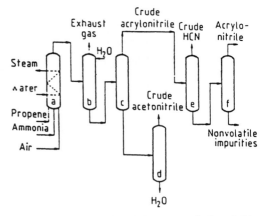

Acrylonitrile

In 1960, almost all of the 260 million lb annual production of acrylonitrile was based on acetylene. Ten years later, the volume had risen to 1.1 billion lb, which was based almost entirely on an ammoxidation process with ammonia, propylene, and air as feeds. However, in the latter 1980s the growth rate had slowed considerably.

Fig. 10.21. Simplified diagram of the Sohio acrylonitrile process: (a) fluidized-bed reactor; (b) absorber column; (c) extractive distillation column; (d) acetonitrile stripping column; (e) lights fractionation column; (f) product column. (*Patrick W. Langvardt, Ullmann's Encyclopedia of Industrial Chemistry, W. Gerhartz (Ed.), 5th ed. Vol. A1, p.179, 1985. Copyright Wiley-VCH Verlag GmbH & Co KG. Used with permission of the copyright owner and the author.*)

$$H_2C=CHCH_3 + HOCl \longrightarrow ClH_2CCHOHCH_3$$

$$ClH_2CCHOHCH_3 + NaOH \longrightarrow H_2C\overset{O}{\overset{/\backslash}{-}}CHCH_3 + NaCl$$

The air, ammonia, propylene process for acrylonitrile is shown in Fig. 10.21. The main reaction is given below.

$$2H_2C=CHCH_3 + 2NH_3 + 3O_2$$
$$\rightarrow 2H_2C=CHCN + 6H_2O$$

In this process the highly exothermic oxidation is performed in a fluidized bed to facilitate heat removal. Note also that acetonitrile and hydrogen cyanide are byproducts. In the case of hydrogen cyanide, this source is of major commercial importance.

Acrylic fibers are by far the major end use for acrylonitrile. They find use primarily in fabrics for clothing, furniture, draperies, and carpets. The second largest consumer of acrylonitrile is acrylonitrile–butadiene–styrene (ABS) and styrene acrylonitrile (SAN) resins. ABS is useful in industrial and construction applications, and the superior clarity of SAN makes it useful in plastic lenses, windows, and transparent household items.

Propylene Oxide

Propylene oxide (PO) is one of the most important organic chemicals from the propylene family. The global capacity was almost 11 billion lb in 1998. PO production in the United States reached 4.2 billion lb in 1998. Polyurethane polyether polyols are the largest usage, which contributed to about 60 percent of domestic PO consumption. It is also used to make propylene glycol, glycol ethers, polyglycols, glycerine, surfactants, and amino propanols.

PO was manufactured by the chlorohydrin route first during World War I in Germany by BASF and others. This route (below) involves reaction of propylene with hypochlorous acid followed by treatment of the resulting propylene chlorohydrin with a base such as caustic or lime. The products of the second reaction are PO and sodium or calcium chloride (Fig. 10.22).

Until 1969, the chlorohydrin process was the only PO process, and The Dow Chemical

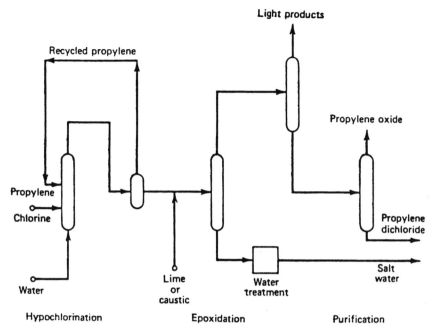

Fig. 10.22. Chlorohydrin process. (*Encyclopedia of Chemical Technology, Kirk and Othmer. 3rd ed., Vol. 19, p. 255, 1980. Copyright by John Wiley & Sons, Inc. and used by permission of the copyright owner.*)

Company was the largest producer. In that year, Oxirane brought on stream the first peroxidation process involving catalyzed epoxidation of propylene with tert-butyl hydroperoxide. In 1977, Oxirane (later Arco Chemical) commercialized a process which employed ethyl-benzene hydroperoxide as the epoxidizing agent and produced PO and styrene (Fig. 10.23).

conversion of ethyl benzene to styrene than is direct dehydrogenation.

Propylene Glycols. PO is converted to mono-, di-, and tri-glycols by a hydrolysis. It is similar to the hydrolysis of ethylene oxide to mono- and di-ethylene glycol. The propylene glycols are used for many of the same applications as the corresponding products derived from ethylene

$$RH + O_2 \longrightarrow ROOH$$

$$ROOH + H_2C{=}CHCH_3 \longrightarrow ROH + H_2C\overset{O}{\overset{\triangle}{-}}CHCH_3$$

The peroxide processes convert propylene to its epoxide while reducing the hydroperoxide to the corresponding alcohol (e.g., tert-butyl alcohol or phenyl methyl carbinol). Because the processes produce the alcohols in larger amounts than PO, their success depends upon finding uses for the alcohols. tert-Butyl alcohol can be dehydrated to isobutylene and hydrogenated to isobutane for recycle to the PO process. It can also be converted to MTBE. Phenyl methyl carbinol can be dehydrated to styrene, making this process a more involved

oxide. Because of their very low toxicity, they also can be used for pharmaceutical, cosmetic, food applications, liquid detergent, tobacco humectant, deicing fluid, and antifreezes. In 1999, the U.S. consumption for mono-, di-, and tri-propylene glycols were 1.08 billion lb, 125, and 16 million lb, respectively.

Isopropyl Alcohol

Isopropyl alcohol (IPA) has been called the first petrochemical. Both historically and today, it is

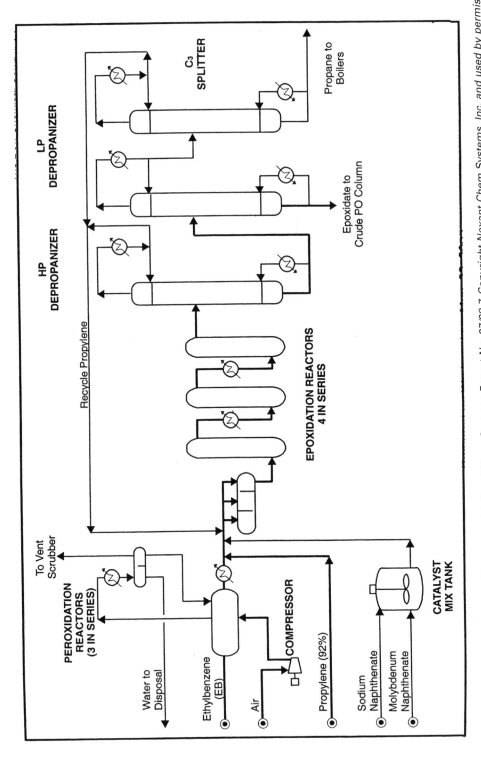

Fig. 10.23. Styrene/PO via peroxidation and epoxidation. (*Chem Systems Report No. 97/98-7. Copyright Nexant Chem Systems, Inc. and used by permission of the copyright owner.*)

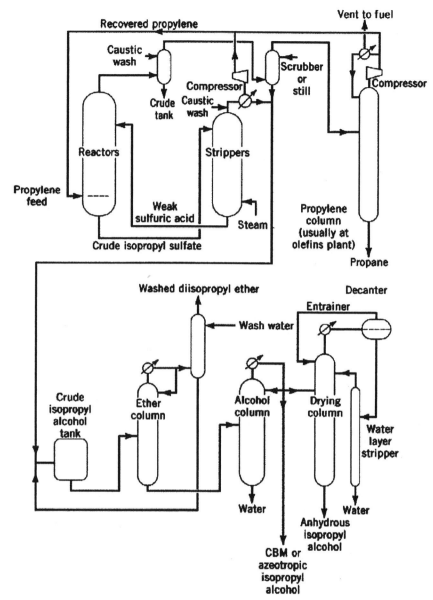

Fig. 10.24. Flow scheme of a process for isopropanol. (*Encyclopedia of Chemical Technology, Kirk and Othmer, Web site ed., isopropyl alcohol, manufacture, 2002. Copyright by John Wiley & Sons, Inc. and reproduced by permission of the copyright owner.*)

prepared by sulfuric acid-mediated indirect hydration of propylene (see Fig. 10.24). Originally it was the source of most of the acetone used in the world. Now, this route must compete with acetone derived from the cumene oxidation process, in which cumene is converted to equimolar amounts of phenol and acetone. The amount of IPA used for producing acetone declined from 47 percent in 1978 to 7 percent in 2001. IPA can also be made by hydrogenation of acetone, but the large capacity for IPA by indirect hydration is a disincentive for this application. Direct hydration of propylene in a vapor-phase, catalytic process is also commercially practiced. This is similar to hydration of ethylene to make ethanol. Relative to the sulfuric acid-mediated process, it offers the advantage of decreased corrosion. However, it suffers from a requirement for a pure propylene feed, whereas the former process can be used with a dilute, refinery stream.

Isopropyl alcohol is an excellent solvent with a blend of polar, nonpolar, and hydrogen-bonding character that makes it useful in a broad spectrum of applications. Its moderate volatility makes it convenient for uses involving evaporation or recovery by distillation. Thus, it is no surprise that much of IPA's consumption is for solvent uses. In North America in 1999 about 1.2 billion lb of IPA were consumed. Major uses were: solvent applications, 47 percent; isopropylamines, 15 percent, esters and ketones, 20 percent; and others, including pharmaceuticals, 18 percent. The total demand in 1999 was significantly less than the 1.9 billion lb recorded for 1978. This downward trend is not unique to IPA and is primarily a result of regulatory pressure in the United States to decrease emissions of VOCs in coating and other applications.

Cumene

Cumene manufacture consumed about 10 percent (2.2 billion lb) of the propylene used for chemicals in the United States in 1998. It is prepared in near stoichiometric yield from propylene and benzene with acidic catalysts (scheme below). Many catalysts have been used commercially, but most cumene is made using a "solid phosphoric acid" catalyst. Recently, there has been a major industry shift to zeolite-based catalyst. The new process has better catalyst productivity and also eliminates the environmental waste from spent phosphoric acid catalyst. It significantly improves the product yield and lowers the production cost. Cumene is used almost exclusively as feed to the cumene oxidation process, which has phenol and acetone as its coproducts.

$$C_6H_6 + H_2C{=}CHCH_3 \longrightarrow C_6H_5CH{\Large\langle}{\scriptstyle CH_3 \atop CH_3}$$

Acetone. Acetone in commerce is derived mostly from cumene oxidation. This is a two-step process involving oxidation of cumene to

the hydroperoxide followed by acid catalyzed decomposition to acetone and phenol:

$$C_6H_5CH{\Large\langle}{\scriptstyle CH_3 \atop CH_3} + O_2 \longrightarrow C_6H_5\overset{\displaystyle CH_3}{\underset{\displaystyle CH_3}{C}}{-}OOH$$

$$C_6H_5\overset{\displaystyle CH_3}{\underset{\displaystyle CH_3}{C}}{-}OOH \xrightarrow{\;H^+\;} C_6H_5OH + H_3C\overset{\displaystyle O}{\overset{\|}{C}}CH_3$$

The 1998 U.S. use of acetone was about 2.6 billion lb. Its major uses are methacrylic acid and esters (44%); solvent (17%); bisphenol-A (20%); and aldol chemicals (such as methyl isobutyl ketone) (13%).

Oxo Chemicals

About 8 percent of the propylene converted into chemicals is used to make oxo alcohols such as 1-butanol and 2-ethylhexanol, which are called oxo alcohols because they are derived from olefins by the oxo process, which converts them to aldehydes. (The oxo process was described earlier in this chapter.)

Butyl Alcohols and Aldehydes. Hydroformylation of propylene gives a mixture of n-butyraldehyde and isobutyraldehyde. This mixture is formed approximately in the ratio of 2 : 1 from the high-pressure, cobalt-catalyzed oxo process. There has always been a much greater demand for the linear n-butyraldehyde than the iso product, so it has been necessary to find uses for the latter.

The low-pressure, rhodium-catalyzed oxo process has made the product mix conform to the relative demand for the two aldehydes. This process gives a 10 : 1 ratio of n-butyraldehyde to isobutyraldehyde.

Each aldehyde can be hydrogenated to the corresponding alcohol for use as a solvent or an intermediate for plasticizers and resins. n-Butyraldehyde is also converted to 2-ethylhexanol by sequential condensation and hydrogenation. 2-Ethylhexanol is used to make the phthalate ester, which finds wide use as a plasticizer of PVC.

Propylene Oligomers: Dodecene and Nonene

The manufacturing processes for these materials are very similar to the one for cumene. When nonene is the desired product, additional fractionation is required, the extent of which is determined by product specifications.

In the reactor portion of this process, the olefin stock is mixed with benzene (for cumene) or recycle lights (for tetramer). The resulting charge is pumped to the reaction chamber. The catalyst, solid phosphoric acid, is maintained in separate beds in the reactor. Suitable propane quench is provided between beds for temperature control purposes because the reaction is exothermic.

Dodecene is an intermediate for surfactants, mainly through two routes. One, the larger user, produces dodecylbenzene sulfonate for anionic detergents. The other goes through the oxo process to tridecyl alcohol, which then is converted into a nonionic detergent by the addition of alkylene oxides.

Nonene has two major outlets, the larger being the oxo production of decyl alcohol which is used in the manufacture of esters, and so forth, for plasticizers. The other significant use for nonene is in the manufacture of nonylphenol, an intermediate for the important series of ethoxylated nonylphenol nonionic surfactants.

Acrylic Acid and Esters

Acetylene once was the raw material for commercial production of acrylic acid and esters, but in 1970 production of acrylic acid by oxidation of propylene was first practiced commercially. In a few years, the new process had essentially replaced the old. In 2000, acrylic acid production in the U.S. was of the order of 2.0 billion lb, and that of acrylate esters was of the order of 1.8 billion lb.

The oxidation of propylene is carried out in two stages (Fig. 10.25). Acrolein exiting the first-stage converter can be isolated, or it can be further oxidized to acrylic acid in the second converter. The process is operated with two reaction stages to allow optimum catalyst and process conditions for each step.

$$H_2C=CHCH_3 + O_2 \rightarrow H_2C=CHCHO + H_2O$$

$$H_2C=CHCHO + \tfrac{1}{2}O_2 \rightarrow H_2C=CHCO_2H$$

$$H_2C=CHCO_2H + ROH$$
$$\rightarrow H_2C=CHCO_2R + H_2O$$

Acrolein is very reactive and has some use as a chemical intermediate, as well as direct use

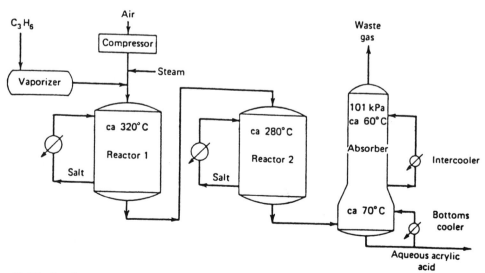

Fig. 10.25. Acrylic acid from oxidation of propylene. (*Encyclopedia of Chemical Technology*, Kirk and Othmer, 3rd ed., Vol. 1, p. 339, 1980. Copyright by John Wiley & Sons, Inc. and reproduced by permission of the copyright owner.)

as an aquatic pest control agent. Most acrolein is converted without isolation to acrylic acid.

Acrylic acid is almost exclusively used directly, or after conversion to an ester, as a monomer. Acrylate esters are produced by normal esterification processes. However, in dealing with acrylic acid, acrolein, or acrylates, unusual care must be taken to minimize losses due to polymerization and other side reactions such as additions of water, acids, or alcohols across the reactive double bond. Polyacrylic acids find use in superabsorbers, dispersants, and water treatment. The polyesters are used in surface coatings, textile fibers, adhesives, and various other applications.

Epichlorohydrin

Epichlorohydrin (ECH) is made from propylene, the majority is via allyl chloride intermediate. Total consumption in 1999 was about 600 million lb. Uses for ECH include epoxy resins (65%), synthetic glycerin (22%), and others (paper treatment, specialty ionic exchange resin, glycerol, and glycidol derivatives).

The key reaction in this manufacturing process is the hot chlorination of propylene, which fairly selectively gives substitution to methyl group rather than the addition to the double bond. In this chlorination step, fresh propylene is first mixed with recycle propylene. This mixture is dried over a desiccant, heated to 650–700°F, and then mixed with chlorine (C_3H_6 to Cl_2 ratio is 4 : 1) and fed to a simple steel tube adiabatic reactor. The effluent gases (950°F) are cooled quickly to 120°F and fractionated. The yield of allyl chloride is 80–85 percent.

Hypochlorous acid is then reacted with the allyl chloride at 85–100°F to form a mixture of dichlorohydrins. The reactor effluent is separated, the aqueous phase is returned to make up the hypochlorous acid, and the non-aqueous phase containing the dichlorohydrins is reacted with caustic or a lime slurry to form ECH which is steam-distilled out and given a finishing distillation.

ECH is used to manufacture epoxy resins for surface coating, castings, and laminates. It is hydrolyzed in 10 percent caustic to make synthetic glycerin (see "Glycerin"). ECH is also employed as a raw material for the manufacture and glycidol derivatives used as plasticizers, stabilizers, surface active agents, and intermediates for further synthesis. The polyamide/ECH resin (which is used in the paper industry to improve the wet strength) has had very good growth in the past few years. The average growth rate for ECH for the period 1990–1999 is about 3 percent and the trend is expected to continue.

Glycerin

Glycerin can be prepared from propylene (via ECH) or as a byproduct from fat and oil hydrolysis of the soap industry. Before 1949 all glycerin was obtained from hydrolysis of fatty triglycerides. In the past 50 years, the synthetic glycerin is to serve the portion of demand not satisfied by natural glycerin. In 1998, the U.S. production for natural versus synthetic is about 2.4 : 1. The diagram of different routes for the manufacture of glycerin is in Fig. 10.26.

In 1998, glycerin consumption in the United States was about 380 million lb. The average annual growth for 1990–1998 is about 2.3 percent/year. About 80 percent of U.S. usage of glycerin is in foods, pharmaceuticals, personal care, cosmetics, tobacco, and similar applications. This reflects its extremely low toxicity, sweet taste, and moisturizing and lubricating properties. Chemical uses for glycerin include use as a "starter" alcohol for polyols made by alkoxylation with propylene oxide and ethylene oxide, and as raw material for alkyd polymers, plasticizers, and explosives.

Glycerin by the Epichlorohydrin Process. In the ECH process, synthetic glycerin is produced in three successive operations, the end products of which are allyl chloride, ECH, and finished glycerin, respectively. Glycerin is formed by the hydrolysis of ECH with 10 percent caustic. Crude glycerin is separated from this reaction mass by multiple-effect evaporation to remove salt and most of the water. A final vacuum distillation yields a 99+ percent product.

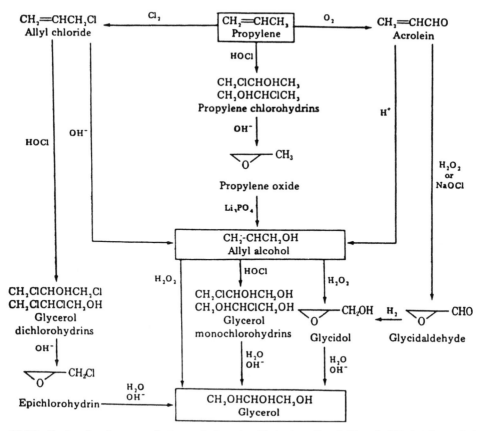

Fig. 10.26. Routes for the manufacture of glycerin. (*Encyclopedia of Chemical Technology, 3rd ed., Vol. 11, p. 923, 1980. Copyright by John Wiley & Sons, Inc. and reproduced by permission of the copyright owner.*)

CHEMICALS DERIVED FROM BUTANES AND BUTYLENES

Saturated four-carbon hydrocarbons (butanes) occur in natural petroleum products such as crude oil and the heavy vapors in wet natural gas. The saturated C_4s are also produced from other hydrocarbons during the various petroleum refining processes. The butylenes—unsaturated C_4s—do not occur in nature, but are derived from butanes or other hydrocarbons either deliberately or as by-products. The complex interrelationships of C_4 hydrocarbons, including their production and use, are described in Fig. 10.27.

The chemical uses of the C_4 hydrocarbons still account for only a small fraction of the available material. To put the volume of C_4s used in chemical manufacture in perspective with the amount used for fuel, one finds that

approximately 12 percent of the butanes and about 30 percent of the butylenes were used as chemical raw materials. The trends that affect availability of C_4 hydrocarbons for chemical and energy end uses are determined by the natural gas processors, petroleum refiners, and, to a growing extent, ethylene manufacture.

Changes in technology and in the availability of optimum feedstocks have far-reaching effects on the entire product mix. For example, when the availability of LPG and ethane for ethylene manufacture has decreased, n-butane and the higher crude cuts have been used, and the proportion of by-product butadiene has increased.

The spectrum of products which can be derived from the four-carbon hydrocarbons is shown in Fig. 10.28. Several of these can also

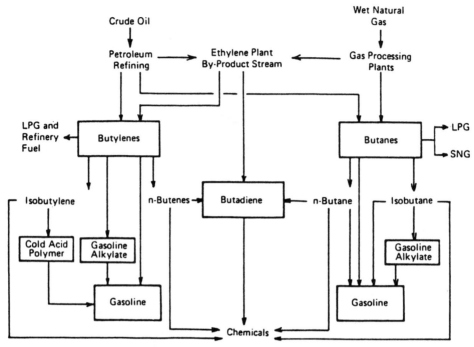

Fig. 10.27. Origins, interrelationships, and end uses of C$_4$ hydrocarbons. (*Reproduced from Chemical Economics Handbook, p. 300200A*, Stanford Research Institute, Menlo Park, CA, 1980, March.)

be produced from other raw materials and are described elsewhere in this chapter.

n-Butane Derivatives

n-Butane can be obtained from natural gas and from refinery hydrocracker streams. Most of the n-butane goes into fuel additive uses. The major chemical use is as a feedstock for ethylene production by cracking. The other important chemical uses for butane are in oxidation to acetic acid and in the production of maleic anhydride. In the past, butane also was the main feedstock for the production of butadiene by dehydrogenation, but it has been replaced by coproduct butadiene obtained from ethylene production.

Ethylene. The largest potential chemical market for n-butane is in steam cracking to ethylene and coproducts. n-Butane is a supplemental feedstock for olefin plants and has accounted for 1–4 percent of total ethylene production for most years since 1970. It can be used at up to 10–15 percent of the total feed in ethane/propane crackers with no major modifications. n-Butane can also be used as a supplemental feed at as high as 20–30 percent in heavy naphtha crackers. The consumption of C$_4$s has fluctuated considerably from year to year since 1970, depending on the relative price of butane and other feedstocks. The yield of ethylene is only 36–40 percent, with the other products including methane, propylene, ethane, and butadiene, acetylene, and butylenes. About 2–3 billion lb of butane are consumed annually to produce ethylene.

Acetic Acid. Acetic acid is the most important carboxylic acid produced industrially. The annual production in the United States in 1999 was almost 15.7 billion lb. As with many compounds produced on a large scale, acetic acid has several different commercial processes. The carbonylation of methanol is now the dominant route. (This process was described earlier in this chapter in the section "Methanol".) The oxidation of acetaldehyde, ethanol, and butane are also important. The percent world capacity for virgin acetic acid

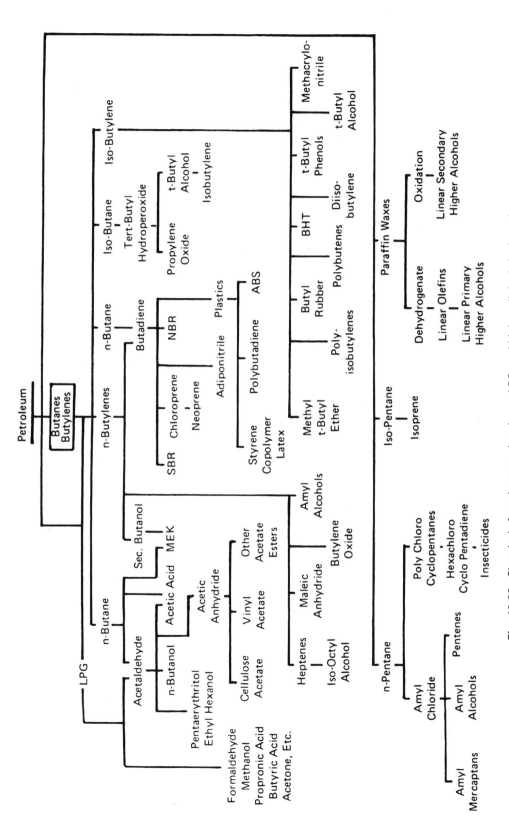

Fig. 10.28. Chemicals from butanes, butylenes, LPG, and higher aliphatic hydrocarbons.

by different starting materials are: methanol (60%), acetaldehyde (18%), ethanol (10%), and butane (8%).

The liquid-phase oxidation (LPO) of light saturated hydrocarbons yields acetic acid and a spectrum of coproduct acids, ketones, and esters. Although propane and pentanes have been used, n-butane is the most common feedstock because it can ideally yield two moles of acetic acid. The catalytic LPO process consumes more than 500 million lb of n-butane to produce about 500 million lb of acetic acid, 70 million lb of methyl ethyl ketone, and smaller amounts of vinyl acetate and formic acid. The process employs a liquid-phase, high-pressure (850 psi), 160–180°C oxidation, using acetic acid as a diluent and a cobalt or manganese acetate catalyst.

Figure 10.29 shows the flowsheet for this process. From the reactor, the product mixture is passed through coolers, and then through separators where the dissolved gases are released. The major components of the oxidized crude are acetic acid, methyl ethyl ketone, and various alcohols and acids. The initial reaction involves formation of butane-2 hydroperoxide, which is not isolated. Further oxidation and decomposition of the resultant radicals produce acetic acid.

Concurrently, the hydroperoxide may be converted to methyl ethyl ketone (MEK). If the initial radical attack is at the primary rather than the secondary carbon, the process makes propionic and formic acids. Reaction conditions can be changed to produce more MEK at the expense of some acetic acid. The maximum acetic acid/MEK ratio is 6.5–7 on a weight basis. If ethyl acetate is also formed, the ratio can go down to acetic acid/(ethyl acetate + MEK) of 3.6–4, with MEK being about 55 percent of the byproduct.

A portion of the acetic acid, which is the major product, can be converted in a separate unit to acetic anhydride. Acetic anhydride may be produced from acetic acid, acetone, or acetaldehyde. With both acetic acid and acetone the initial product is ketene. The ketene is highly reactive and reacts readily

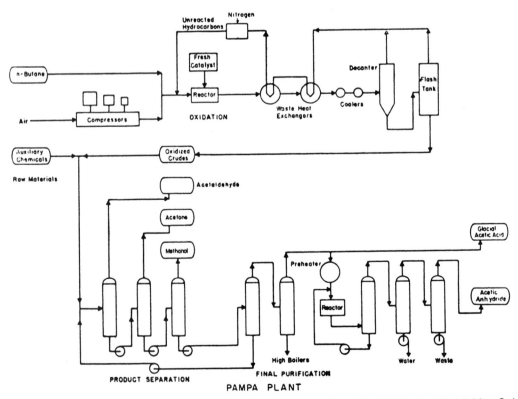

Fig. 10.29. Oxidation of butane. (*Pet. Ref. **38**, no. 11, 234, 1959. Copyright 1959 by Gulf Publishing Co.*)

with acetic acid to form acetic anhydride. All this takes place at 700–800°C in the presence of a triethyl phosphate catalyst. With acetic acid, the reactions are:

$$CH_3COOH \rightarrow CH_2=C=O + H_2O$$

$$CH_3COOH + CH_2=C=O \rightarrow (CH_3CO)_2O$$

Acetic anhydride is used to make acetic acid esters. It is especially effective in difficult acetylations, such as in the manufacture of aspirin and cellulose acetate.

Maleic Anhydride. Maleic anhydride is one of the fastest-growing chemical end-uses for butane. The demand in the United States was about 500 million lb in 2000. About 60 percent of the maleic anhydride produced goes into the manufacture of unsaturated polyester resins, used primarily in fiber-reinforced plastics for construction, marine, and transportation industries. It is also used to make lube oil additives, alkyd resins, fumaric and malic acids, copolymers, and agricultural chemicals.

Essentially all maleic anhydride is manufactured by the catalytic vapor-phase oxidation of hydrocarbons. Prior to 1975, benzene was the feedstock of choice. By the early 1980s, however, many producers had

Until the late 1990s nearly all butane oxidation to maleic anhydride was conducted in a fluid bed or in a fixed bed multitubular, tubeshell heat exchanger type of reactors. After over a decade of intensive R&D efforts, DuPont was successful in commercializing a Circulating Fluid Bed Reactor (CFBR) catalyst system. CFBR has the advantage of providing 10–15 percent higher selectivity even at higher conversions, thereby significantly reducing raw material costs. A diagram of the fluid bed process in presented in Fig. 10.30.

Isobutanes

t-Butyl Alcohol/Propylene Oxide. An important use for isobutane is in the peroxidation of propylene with t-butyl hydroperoxide. The feedstocks are propylene and isobutane, and the process is similar to the PO/styrene plant. (See the section "Chemicals from Benzene" below.) In the two-stage conversion route, oxidation of isobutane with air yields a mixture of t-butyl hydroperoxide and t-butanol in a liquid-phase reaction at 135–144°C. After separation of products, a molybdenum-catalyzed reaction of the hydroperoxide with propylene at 110°C yields PO and t-butyl alcohol (TBA).

$$2(CH_3)_3CH + 3/2O_2 \longrightarrow (CH_3)_3-C-O-OH + (CH_3)_3COH$$

isobutane   t-butyl hydroperoxide   TBA

$$CH_2=CHCH_3 + (CH_3)_3C-O-OH \longrightarrow H_2C\overset{O}{-}CHCH_3 + (CH_3)_3COH$$

propylene   t-butyl hydroperoxide   propylene oxide   TBA

switched to n-butane for economic and environmental reasons. Although benzene as a feedstock for maleic anhydride is no longer used in the United States, it is still used in older plants in Latin America, Europe, and East Asia. The oxidation reaction to produce maleic acid from n-butane is as follows:

$$2C_4H_{10} + 7O_2 \longrightarrow 2 \underset{O}{\overset{O}{\|}} + 8H_2O$$

The ratio of TBA to PO, and thus the isobutane requirement, can be adjusted from approximately 2:1 to 3:1. The U.S. capacity of PO in 1999 was about 1715 million lb. (The capacity for PO/styrene and chlorohydrin routes was 1120 and 2150 million lb, respectively). At this time, only one domestic PO producer is not using this process.

When this process was first introduced, TBA had a low value, so the TBA/PO ratio was kept to a minimum. Since the use for MTBE as oxygen enhancer in gasoline has became important, a higher TBA/PO value is used. Note, however, that changes in the

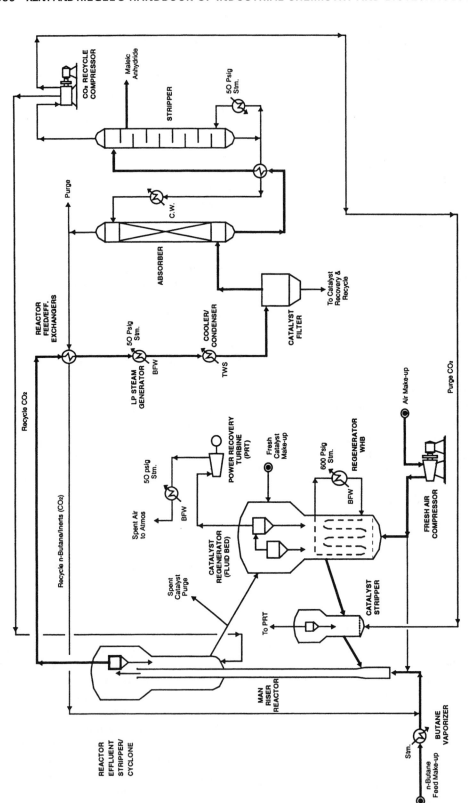

Fig. 10.30. Maleic anhydride DuPont process. (*Chem Systems Report No. 99/00-5. Copyright Nexant Chem Systems, Inc. and used by permission of the copyright owner.*)

MTBE picture also apply here—as discussed earlier. Dehydration of TBA yields high-purity isobutylene, which can be converted into MTBE with methanol. Today, almost 900 million lb of isobutylene is produced from TBA.

Other Isobutane-Based Chemicals. Isobutane can be directly dehydrogenated to isobutylene by a modification of the Houdry process. This can then be converted to MTBE. The estimated use is over 1 billion lb of isobutane. Because of their inertness and higher vapor pressures, high-purity propane and butanes have become the important substitutes for fluorocarbons as aerosol propellants. Isobutane can also be used as a solvent in polymer processing, and as a blowing agent for foamed polystyrene.

Butylenes

Butylenes are four-carbon monoolefins that are produced by various hydrocarbon processes, principally catalytic cracking at refineries and steam cracking at olefins plants. These processes yield isomeric mixtures of 1-butene, *cis-* and *trans-*butene-2, and isobutylene. Derivatives of butylenes range from polygas chemicals and methyl t-butyl ether, where crude butylenes streams may be used, to polybutene-1 and LLDPE, which require high-purity 1-butene. In 1997, the estimated consumption of butylenes (in billions of pounds) was: alkylation, 32.0; MTBE, 12.0; other, including polygas and fuel uses, 0.5.

The major chemical uses for n-butylenes are sec-butyl alcohol (and MEK), butadiene, butene-1, heptenes, and octenes. In 1978, butadiene accounted for almost 70 percent of the demand. At the present time sec-butanol and 1-butene are the largest chemical enduses of butylene, consuming about four fifths of the total.

1-Butene. The largest chemical use of n-butenes is 1-butene used in production of LLDPE, which requires alpha-olefin comonomers. Various processes for the production of LLDPE and HDPE incorporate 1-butene as a comonomer. This accounts for about 70 percent of 1-butene use. The alpha-olefin comonomers control the density and physical properties of the polymer. About 20 percent of the 12,500 million lb of HDPE production in the United States in 1997 utilized 1-butene as comonomer, as did 45 percent of the 6900 million lb of LLDPE. Most of the remainder was split between 1-hexene and 1-octene.

The remaining 30 percent of 1-butene is divided among several uses. About 10–15 percent of the 1-butene is polymerized in the presence of a Ziegler-type catalyst to produce polybutene-1 resin. The markets for this resin are pipe, specialty films, and polymer alloys. Approximately the same volume of 1-butene is reacted with synthesis gas in an oxo reaction to produce valeraldehydes. These C_5 aldehydes are then hydrogenated to amyl alcohols or oxidized to valeric acid. Amyl alcohols are consumed in the production of lube oil additives and amyl acetate and in solvent uses. Valeric acid goes into lubricant base stocks and specialty chemicals.

Smaller uses of 1-butene are in 1,2-butylene oxide, butyl mercaptan, and butyl phenols. Butylene oxide, produced by the chlorohydrin process, is used as a corrosion inhibitor in chlorinated solvents. Butyl mercaptan is a precursor for organophosphate herbicides, pharmaceutical intermediates, and is used as a gas odorant.

sec-Butanol and Methyl Ethyl Ketone. The next-largest use for n-butenes is in the manufacture of sec-butanol. A refinery butanes–butylenes stream, usually rich in butene-2, is contacted with 80 percent sulfuric acid to produce the sec-butyl hydrogen sulfate. Dilution with water and steam stripping produce the alcohol. MEK is obtained in high yields at catalytic dehydration of the alcohol at 400–500°C.

Solvent applications account for almost 95 percent of all MEK consumption; the rest goes to chemical uses such as MEK peroxide and methyl ethyl ketoxime. The solvent applications include surface coatings, adhesives, lube oil dewaxing, magnetic tape manufacture, and printing inks. Production of MEK in 1999 totaled almost 690 million lb.

Heptenes and Octenes. Heptene and octene are oligomers produced by the polymerization of refinery streams containing C_3 and C_4 hydrocarbons. Originally these polygas units were developed to provide a source of high-octane blending components from refinery gases, but many have since been adapted to produce heptenes and octenes for chemical uses. Heptenes are used primarily to produce isooctyl alcohol which is a precursor for lube oil additives, diisooctyl phthalate (DIOP), other plasticizers, and herbicide esters. Isooctyl alcohol was once the predominant plasticizer alcohol, but development of plasticizers from competing alcohols such as 2-ethylhexanol and linear alcohols has eroded its market. Octenes are precursors of isononyl alcohol, which in turn is the raw material in the manufacture of diisononyl phthalate (DINP) plasticizers. These plasticizers compete directly with dioctyl phthalate in many applications. The total U.S. consumption of butylenes for production of heptenes and octenes was approximately 240 million lb in 1997.

Butene-2. Most butene-2 in the United States goes into production of gasoline alkylate. Some butene-2 is used in solvent applications, and it is also the intermediate in the disproportionation process for producing propylene from ethylene.

Isobutylene

Methyl t-Butyl Ether. By far the largest use of isobutylene is in the manufacture of methyl t-butyl ether (MTBE). Since its introduction in 1979, the demand for MTBE as an octane improver in gasoline has grown phenomenally. By 2000, production had reached 46 billion lb/year. MTBE use is exclusively as an octane booster/combustion promoter in gasoline. Use of MTBE in gasoline, at least in the United States, is expected to decline in the coming years for the reasons discussed earlier in this chapter. The decline of MTBE production is expected to have a significant impact on C_4 uses in the future.

MTBE is produced by reacting methanol and isobutylene under mild conditions in the presence of an acid catalyst. The isobutylene feed is either mixed butylenes, a butylenes stream from catalytic cracking, or a butylenes coproduct from ethylene production. The reaction conditions are mild enough to permit the n-butenes to pass through without ether formation. Figure 10.31 shows a typical process for making MTBE.

Another approximately 1.5 billion lb of isobutylene goes into other chemical uses. These applications include polybutenes and derivatives of high-purity isobutylene such as butyl rubber, polyisobutylenes, and substituted phenols. Isobutylene is more reactive than the n-butenes, but many of its reactions are readily reversible under relatively mild conditions.

Polybutenes. More than 900 million lb of butylenes are consumed in the production of polybutenes. The process involves the

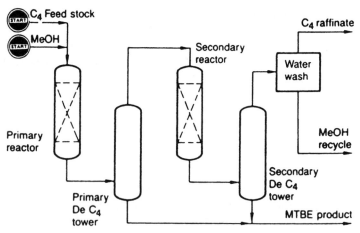

Fig. 10.31. Two-stage MTBE process. (*Bitar, L. S., Hazbun, E. A. and Piel, W. J., Hydrocarbons Processing, 63, no. 10, 54. 1984, October. Copyright Gulf Publishing Company and reproduced by permission of the copyright owner.*)

Friedel–Crafts polymerization of desulfurized C_4 refinery streams. Although the feed is a mixed butylenes stream, the polybutene product is essentially a polymer of isobutylene, with a few n-butene units occurring in the polymer chain. The optimum isobutylene concentration in the feed stream is reported to be in the 20–25 percent range. The polymerization reaction yields products with a molecular weight range of 300–3000.

The lowest-molecular-weight polymers (mol. wt. 300–350) are used for dielectric fluids and specialty lubricants. They comprise a very small part of the market. Polymers in the 700–750 range are precursors for caulks, sealants, and gasoline additives. Polymers with molecular weights of 900–1500 constitute the segment with by far the largest volume. They are used in the manufacture of lube oil additives via modification with phosphorus, amine, or succinimide groups. The resulting dispersants reduce oil consumption, restore compression, and improve oil viscosity. They are also used as gasoline additives and specialized plasticizers. The higher polymer fraction (mol. wt. 1500–3000) mainly is formulated into adhesives, caulks, and sealants. The use of polybutenes in gasoline detergents is expected to show continued growth.

Butyl Rubber. Almost two-thirds of the demand for high-purity isobutylene is for production of butyl rubber, which is produced by the cationic polymerization of high-purity isobutylene with isoprene (2–3% by weight) at low temperatures in the presence of a Friedel–Crafts catalyst. The isobutylene must be pure in order to obtain a high molecular weight product. The elastomer's outstanding property of impermeability to air and gas makes it particularly suitable for tire liners and tire bladders and valves. Other important butyl elastomers are the halogenated products, chlorobutyl and bromobutyl. Halogenated butyl rubbers, which are more compatible with other tire elastomers and can be cured faster than butyl rubber, have grown to be of more importance in recent years.

Polyisobutylenes. Polyisobutylenes are produced by the low temperature polymerization of high-purity isobutylene. The main commercial products have two molecular weight ranges: a 40,000–50,000 molecular weight polymer, and a polymer with molecular weight two to four times that of butyl rubber. The lower-weight polymer is used in binder systems, and in plasticizers and tackifiers for adhesives and sealants for electrical applications. The largest application for the higher-weight polymer is in the production of lube oil viscosity improvers. In 1997, about 50 million lb of high-purity isobutylene went to make polyisobutylenes.

Other Derivatives of Isobutylene. The production of many smaller-volume chemicals is based on high-purity isobutylene. The major chemicals are para-t-butylphenol, di-t-butyl-p-cresol (butylated hydroxytoluene, BHT), 2,6-di-t-butylphenol, t-butylamine, t-butyl mercaptan, and isobutyl aluminum compounds. Isobutylene usage by each of these six ranged from 8 to 18 million lb. The total volume of isobutylene that went for small-volume chemical production was about 140 million lb in 1997.

The substituted phenols and cresols constitute about half the total volume of this group. Para-t-butylphenol is produced by the alkylation of phenol with isobutylene. The principal applications for this derivative are in the manufacture of modified phenolic resins for the rubber industry and in surface coatings. BHT is obtained from isobutylene and p-cresol. Technical-grade BHT is an antioxidant for plastics and elastomers, and is a gum inhibitor in gasoline. Food-grade BHT is an antioxidant in edible oils, preserves, and many other foods. 2,6-Di-t-butylphenol is used to produce a wide range of plastics additives, antioxidants, and gasoline additives.

t-Butylamine is formed by the reaction of isobutylene with HCN in the presence of strong sulfuric acid. The intermediate t-butyl formamide is then hydrolyzed to form the amine and formic acid. This amine is used mainly to synthesize sulfonamide rubber accelerator compounds. The major use of n-butyl mercaptan is in odorant formulations for natural gas. The distributors of natural gas inject about one pound of odorant per million cubic feet. Isobutyl aluminum compounds are produced from high-purity isobutylene, hydrogen, and aluminum. Five distinctly different isobutyl aluminum compounds are produced in the United States. The principal end use for these compounds is as polymerization cocatalysts in the manufacture of polybutadiene, polyisoprene, and polypropylene. Other minor uses for high-purity isobutylene are in the manufacture of neopentanoic acid, methallyl chloride, and miscellaneous butylated phenols and cresols.

Butadiene

In the mid-1970s there were several major processes for making butadiene in the United States: steam cracking of naphtha, catalytic dehydrogenation of n-butene, dehydrogenation of n-butane, and oxidative dehydrogenation of n-butene. By 2000, more than 90 percent of all the butadiene was made as a coproduct with ethylene from steam cracking, and the only "on-purpose" production came from the dehydrogenation of butylene.

Thermal cracking of hydrocarbon feedstocks in the presence of steam at 700–900°C produces ethylene and several coproducts, including butylenes and butadiene. The yield of butylenes varies widely (to as high as 30%), depending on the feedstock and the severity of the cracking. The yield of butadiene is particularly high with naphthas and heavier feedstocks. Most ethylene producers recover a raw C_4 stream that contains butanes, butylenes, and butadiene. The butadiene is recovered by extraction, and the raffinate (containing butanes and butylenes) is used for gasoline blending or the production of chemicals.

The dehydrogenation process feed can be refinery streams from the catalytic cracking processes. This mixed C_4 stream typically contains less than 20 percent n-butenes. For use in dehydrogenation, however, it should be concentrated to 80–95 percent. The isobutylene generally is removed first by a selective extraction-hydration process. The n-butenes in the raffinate are then separated from the butanes by an extractive distillation. The catalytic dehydrogenation of n-butenes to 1,3-butadiene is carried out in the presence of steam at high temperature (>600°C) and

reduced pressure. A typical catalyst could be a chromium-promoted calcium nickel phosphate. The oxidative dehydrogenation process for butadiene reacts a mixture of n-butenes, compressed air, and steam over a fixed catalyst bed of tin, bismuth, and boron.

Of the 17 billion lb of butadiene consumed in 1999, almost two thirds went into the production of elastomers (styrene–butadiene latex rubber (SBR), polybutadiene, nitrile, and polychloroprene). Adiponitrile, ABS resins, styrene–butadiene latex, styrene block copolymers, and other smaller polymer uses accounted for the remainder. The largest single use was for styrene–butadiene copolymers (SBR and latex). Most of it was made by an emulsion process using a free-radical initiator and a styrene–butadiene ratio of about 1 : 3. More detailed description of the rubber and polymer used can be found in Chapters 16 and 15.

HIGHER ALIPHATIC HYDROCARBONS

Cyclopentadiene

Cyclopentadiene is a product of petroleum cracking. It dimerizes exothermically in a Diels–Alder reaction to dicyclopentadiene, which is a convenient form for storage and transport. Dicyclopentadiene plus cyclopentadiene demand in the United States amounted to 270 million lb in 1998.

Dicyclopentadiene can be converted back to cyclopentadiene by thermally reversing the Diels–Alder reaction. Cyclopentadiene also undergoes the Diels–Alder reaction with other olefins, and this chemistry has been used to make highly chlorinated, polycyclic hydrocarbon pesticides. These pesticides are so resistant to degradation in the biosphere, however, that they are now largely banned from use. It is also used as a monomer and a chemical intermediate.

Isoprene

Isoprene is the basic repeating unit in natural rubber and in the naturally occurring materials known as terpenoids. It is a diene like butadiene and is useful as a building block for synthetic polymers. The most frequently used synthetic procedure for making isoprene is acid-catalyzed reaction of formaldehyde with isobutylene, giving a dioxolane intermediate that is thermally cracked to isoprene. Isoprene also can be recovered from petroleum refinery streams. A total of 360 million lb of isoprene was used in the United States in 2000.

Isoprene is converted to elastomers such as poly(*cis*-1,4-isoprene), which is tough, elastic, and resistant to weathering and is used mainly for vehicle tires. Recently, block copolymers of isoprene with styrene have been finding use as thermoplastic elastomers and pressure-sensitive adhesives.

n-Paraffins and Olefins

n-Paraffins are the unbranched fraction of hydrocarbons found in petroleum. They can be separated from the branched and aromatic hydrocarbons by a process using a shape-selective, controlled-pore-size adsorbent. In this process, the small pores exclude branched or aromatic materials and allow them to flow through a column more readily than the linear hydrocarbons, which become adsorbed into the narrow pore structure of the adsorbent. Linear, internal monoolefins are produced by dehydrogenation of n-paraffins.

Primary and Secondary Higher Alcohols

Linear internal monoolefins can be oxidized to linear secondary alcohols. The alpha (terminal) olefins from ethylene oligomerization, described earlier in this chapter, can be converted by oxo chemistry to alcohols having one more carbon atom. The higher alcohols from each of these sources are used for preparation of biodegradable, synthetic detergents. The alcohols provide the hydrophobic hydrocarbon group and are linked to a polar, hydrophilic group by ethoxylation, sulfation, phosphorylation, and so forth.

CHEMICALS DERIVED FROM BENZENE, TOLUENE, AND XYLENE

Until World War II, most of the commercial aromatic chemicals in the United States and other countries were derived from the coal industry. The high-temperature carbonization of coal to produce coke for the steel industry also generated a liquid stream that was rich in aromatics. Benzene, toluene, and xylene (BTX) and other aromatics could be recovered from the coal tar by extraction and distillation. Recently, the importance of this source has greatly diminished; now almost all of the BTX in the United States is based on petroleum (17 billion lb of benzene, 12.6 billion lb of toluene, and 12.3 billion lb of xylene).

There are two major sources for petroleum-based aromatics, catalytic reformate and pyrolysis gasoline. The catalytic reformate is a refinery product that occurs in the catalytic reforming of naphthenes and paraffins in low-octane naphtha to produce a high-octane product. The pyrolysis gasoline stream that results from the steam cracking of hydrocarbons to produce ethylene and propylene is a very large (and growing) source of benzene and other aromatics. Because the demand for toluene is considerably less than that for benzene, some toluene is converted back to benzene by high-temperature hydrodealkylation (HDA) or by catalytic toluene disproportionation (TDP). The amount of benzene produced by HDA at any given time depends on the relative economics of benzene and toluene. Recently, the ratio of benzene produced from HDA versus TDP process is about 1 : 3. The specific process details of these production routes can be found in Chapter 18.

Compared to the amount of BTX that goes into fuel, the volume of aromatic chemicals used as chemical building blocks is relatively small. About half of the benzene and more than 90 percent of the toluene and xylenes end up in the gasoline pool.

Chemicals from Benzene

Benzene is by far the most important aromatic petrochemical raw material. During 1999, some 2.8 billion gal were consumed in the United States. This ranks it close to propylene as a chemical building block. Benzene has a broad end-use pattern. Its most important uses are for: ethylbenzene (styrene), 55.6 percent; cumene (phenol), 22.4 percent; cyclohexane

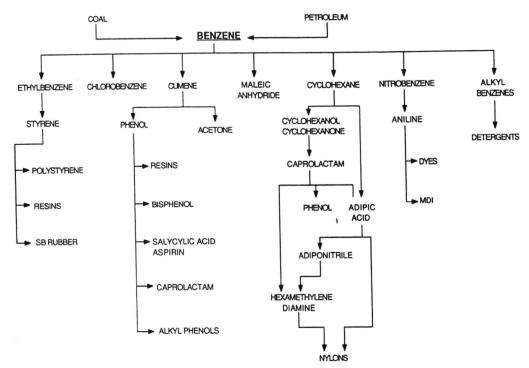

Fig. 10.32. Chemicals derived from benzene.

(nylon), 13.5 percent; nitrobenzene (aniline), 5 percent; and detergent alkylate, 3 percent. Other significant uses are for chlorobenzenes and maleic anhydride. Benzene is an excellent solvent, but it has been almost entirely replaced by less toxic materials. Some of the major end products are shown in Fig. 10.32.

Styrene. Styrene is the largest benzene derivative with annual consumption about 11.5 billion lb in the United States. It is produced mainly by catalytic dehydrogenation of high-purity ethylbenzene (EB) in the vapor phase. The manufacture process for EB is based on ethylene alkylation with excess benzene. This can be done in a homogeneous system with aluminum chloride catalyst or a heterogeneous solid acid catalyst in either gas or liquid-phase reaction. In the past decade, the liquid-phase alkylation with zeolite catalyst has won acceptance. Those processes have advantages of easier product separation, reducing waste stream, and less corrosion. In addition, it produces less xylene due to lower

process temperature. Since certain xylene isomers have very similar b.p. as styrene, it is difficult to be separated.

The EB dehydrogenation can be done with various commercially available styrene catalysts. The fractionation train separates high-purity styrene, unconverted EB, and minor reaction byproducts such as toluene.

Styrene is also produced as coproduct from a PO process. In this route, EB is oxidized to its hydroperoxide and reacted with propylene to yield propylene oxide. The coproduct methyl phenyl carbinol is then dehydrated to styrene. For every pound of PO produced, up to 2.5 lb of styrene can be produced. In 2000, about 25 percent of styrene was produced by this process in the United States.

The largest use for styrene (over 70%) is to make homopolymer polystyrene. The U.S. production volume reached 6.3 billion lb in 1998. Other major uses are in plastics, latex, paints, and coatings, synthetic rubbers, polyesters, and styrene-alkyd coatings. In these applications styrene is used in copolymers such as ABS (8%), styrene–butadiene latex (8%) rubber (SBR) (4%), unsaturated polyester resins (6%), and other polymer applications.

Cumene (Phenol). Cumene has become the second largest chemical use for benzene. It is produced by alkylating benzene with propylene at elevated temperature and pressure in the presence of a solid acid catalyst. The U.S. production was more than 6.9 billion lb in 1999. Of this, about 96 percent then was converted to phenol.

Before 1970, there were five different processes used to make phenol in the United States: the sulfonation route, chlorobenzene hydrolysis, the Raschig process, cumene oxidation, and the benzoic acid route. By 1978, the first three processes had essentially disappeared, and 98 percent of the remaining plant capacity was based on cumene oxidation. The oxidation process is shown in Fig. 10.33.

In this process, cumene is oxidized to cumene hydroperoxide by air at about 100°C in an alkaline environment. The oxidation products are separated, and the bottoms are mixed with a small amount of acetone and sulfuric acid and held at 70–80°C while the hydroperoxide splits into phenol and acetone. Total domestic phenol capacity with this process is about 4.8 billion lb/year. In the much smaller-volume benzoic acid process, toluene is air-oxidized to benzoic acid with a cobalt catalyst. The benzoic acid then is converted to phenol by an oxidative decarboxylation reaction with air at about 240°C.

The three major uses for phenol are in the manufacture of phenolic resins, bisphenol A, and caprolactam.

Phenolic resins. Resins such as those made from phenol and formaldehyde now account for about one third of the phenol consumed in the United States. They are widely used in construction related use such as plywood adhesives, foundry resins, thermoformed plastics, and surface coatings.

Bisphenol. Bisphenol-A (4,4'-isopropylidene-diphenol) accounts for 35 percent of phenol consumption and is used mainly in the production of polycarbonates (55%) and epoxy resins (25%), two of the fastest-growing families of plastics. Other uses are in the manufacture of flame retardant such as tetrabromobisphenol-A, polysulfone resins, and polyacrylate resins. The consumption of bisphenol in the United States in 1999 topped 2.1 billion lb.

Bisphenol is obtained by the reaction of phenol and acetone with HCl or acid resin as catalyst. In the HCl catalyzed process (Fig. 10.34), phenol and acetone in a molar ratio of about 3:1 are charged to an acid-resistant stirred reactor. A sulfur-containing catalyst is added, and then dry HCl gas is bubbled into the reaction mass. The temperature is maintained at 30–40°C for 8–12 hr. At the end of the reaction, the mixture is washed with water and treated first with enough lime to neutralize the free acid. Vacuum and heat

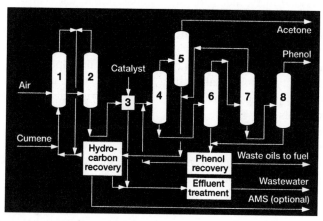

Fig. 10.33. Manufacture of phenol and acetone by oxidation of cumene. (*Hydrocarbon Processing.* p. 117, 2001, March. Copyright 2001 by Gulf Publishing Co.)

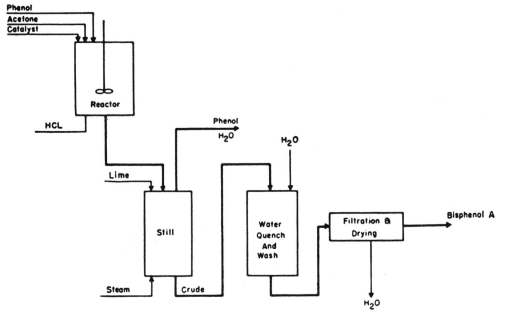

Fig. 10.34. Manufacture of bisphenol A. (*Pet. Ref.* **38**, *no. 11, 225, 1959. Copyright 1959 by Gulf Publishing Co.*)

are applied, and water and phenol are distilled separately from the mixture. The batch is finished by blowing the molten product with steam under vacuum at 150°C to remove the odor of the sulfur catalyst. The molten bisphenol is quenched in a large volume of water, filtered, and dried. All new bisphenol plants are based on the acid resin catalyst technology. The total capital investment for a resin plant is about half of a plant using HCl catalyst. The higher investment for the HCl-catalyzed process is due to the need for the corrosion resistant materials in the process equipment handling HCl and removing it after the reaction.

Other phenolics. Other major derivatives of phenol include caprolactam (14%), salicyclic acid (aspirin), alkylated phenols (3.6%), aniline (3.5%), and xylenols (3.5%).

The uses of caprolactam are described in the section "Cyclohexane." Almost 60 percent of the 27 million lb of salicyclic acid went to aspirin, with remainder mainly going to salicylate esters and phenolic resins. Aspirin (acetyl salicylic acid) has long been recognized for its analgesic and antipyretic properties. It must now share this market with the competing pain relievers such as acetaminophen and ibuprofen.

Cyclohexane. Cyclohexane is the basic starting material for nylon fibers and resins via the intermediates adipic acid, caprolactam, and hexamethylenediamine. The world consumption was about 10 billion lb (with 3.5 billion lb in the United States) in 2000. Of these three derivatives, adipic acid and caprolactam account for over 90 percent of cyclohexane consumption. Cyclohexane is also used as a solvent and as a starting material for cyclohexanol and cyclohexanone. Although cyclohexane can be recovered from natural gasoline, most is made by liquid or vapor-phase hydrogenation of benzene. A nickel or platinum catalyst is generally used at elevated temperature and pressure.

Adipic acid. The world consumption for adipic acid was 4.6 billion lb in 1999. Nylon 66, produced from adipic acid and hexamethylenediamine (HMDA), is currently the largest-volume domestic nylon. About 86 percent of all adipic acid goes to make nylon 66 fibers and resins. Although HMDA can be made from adipic acid, a major source is from adiponitrile. The commercial synthesis of adipic acid is a two-step reaction starting with either cyclohexane or phenol. In both cases, a cyclohexanone/cyclohexanol mixture is formed as an intermediate. This mixture is then catalytically oxidized to the adipic acid product with nitric acid. It can also be manufactured as a byproduct of the caprolactam process.

various other routes. For example, adipic acid can be made by butylene oxidation as well as by cyclohexane oxidation; HMDA can be made starting with butadiene, adipic acid, or acrylonitrile; and caprolactam is also produced from phenol. See Chapter 12 for a more detailed discussion of nylon fibers and resins.

Caprolactam. Essentially all caprolactam is used in the manufacture of nylon 6 fibers. In 1998, global demand reached nearly 7.3 billion lb with 1.7 billion lb used in North America. This is a fast-growing nylon with applications in carpets, textiles, and tires. Caprolactam can be produced from cyclohexane, phenol, and toluene via cyclohexanone. It is then reacted with hydroxylamine to give an oxime. The oxime undergoes an acid-catalyzed rearrangement to give caprolactam.

Although these nylon intermediates are derived mainly from cyclohexane, there are

Maleic Anhydride. Prior to 1975, benzene was the feedstock of choice for maleic anhydride manufacture. By the early 1980s, for economic reasons, many producers had switched to the n-butane process described in the section "n-Butane Derivatives". By 1988, all of the maleic anhydride produced in the United States came from that process. However, about half of the maleic anhydride produced abroad still comes from benzene oxidation, with a small amount being recovered as a coproduct in phthalic anhydride manufacture.

Detergent Alkylate. Alkylbenzenes are major intermediates in the manufacture of synthetic detergents. If a straight-chain alkyl group is used, the resulting product is linear alkylbenzene (LAB), a "soft" degradable alkylate. Using a branched olefin (from propylene tetramer) gives branched alkylbenzene (BAB), a "hard" nondegradable alkylate. Approximately 960 million lb (mainly dodecyl- and tridecyl-benzene) of LAB were produced in 1998. The production of BAB was about 188 million lb in 1985; it declined to zero in 1998.

The production of LAB involves the liquid-phase alkylation of benzene with linear monoolefins or alkyl chlorides. Liquid HF is used as catalyst for linear monoolefins. And the $AlCl_3$ is used as the catalyst for alkyl chlorides. Nowadays, acidic zeolite catalyst is used for olefin alkylation which generates less waste and reduces manufacture cost. The alkylate is then sulfonated to produce linear alkylbenzene sulfonate for biodegradable detergents. The manufacture of detergents is described in detail in Chapter 27.

The majority of LAB has a chain length of 10–13. This method is also used to make longer-chain (C_{20}–C_{22}) alkyl derivatives of benzene. Those alkylates are used as lubricants.

Nitrobenzene (Aniline). The U.S. nitrobenzene production was about 2 billion lb in 1999. Two types of manufacturing processes were used: the direct nitration and the adiabatic nitration process. In the direct nitration system, benzene is mixed with a mixture of nitric/sulfuric acid. The reaction can be carried out in either a batch or a continuous system. Those reactors require a cooling system to keep it at constant temperature. It also requires a separate system for sulfuric acid reconcentration. In the adiabatic process, water is flashed off under vacuum before the sulfuric acid/nitrobenzene separation. The advantage of the adiabatic process is to eliminate a separated sulfuric acid reconcentration unit. This also will provide a better heat integration. Recently, the disposal of nitrophenols has become a major issue for aniline manufacture. Small amounts of nitrophenols are always made during the benzene nitration. It is more of a problem for the adiabatic process due to higher process temperature. Now an improved adiabatic process has been developed. By using an enhanced mixing and a higher benzene/acid ratio it produces less nitrophenols and it still keeps the advantages of the adiabatic process.

In 2000, the aniline production in the United States was about 1.9 billion lb. Almost 98 percent of nitrobenzene is used for the production of aniline. Consequently many nitrobenzene plants are integrated with facilities for aniline production. The hydrogenation of nitrobenzene can be done in either the vapor over a copper–silica catalyst or in liquid phase over platinum–palladium catalyst. One of the smaller uses for nitrobenzene is the production of the pain reliever, acetaminophen.

Aniline can also be made by two other methods. In the first, nitrobenzene is reduced by reaction with scrap iron in the presence of a hydrochloric acid catalyst. The iron is oxidized to the ferrous state, and the coproduct aniline is separated. This route accounts for less than 5 percent of the current aniline production. The other process avoids nitrobenzene entirely and involves the vapor-phase ammonolysis of phenol, using an alumina catalyst. Aniline is formed with diphenylamine as a by-product. About 20 percent of the aniline is produced by this route.

Aniline is consumed as a raw material in the manufacture of a number of chemicals: *p,p*-methylene diphenyl diisocyanate (MDI), 65 percent; rubber-processing chemicals, 15 percent; herbicides, 5 percent; dyes and pigments, 4 percent; specialty fibers, 2 percent. Other uses are in pharmaceuticals and photo chemicals. Principal growth is occurring from demand for MDI and the small, rapidly growing specialty fibers.

Diisocyanates (MDI). The first step in the production of MDI (and polymeric "PMDI") is the condensation of aniline and formaldehyde to form diphenylmethylenediamine. The reaction conditions can be varied to change the isomer distribution of the product. This is followed by phosgenation to give an aromatic isocyanate product mix that corresponds to the starting polyaromatic amine.

$$2 \bigcirc{-}NH_2 + HCHO \xrightarrow{HCl} H_2N{-}\bigcirc{-}CH_2{-}\bigcirc{-}NH_2 + H_2O$$

$$H_2N{-}\bigcirc{-}CH_2{-}\bigcirc{-}NH_2 \xrightarrow{2 \, COCl_2} OCN{-}\bigcirc{-}CH_2{-}\bigcirc{-}NCO$$

Typically a mixture containing MDI and its dimer, trimer, and some tetramer is produced. Pure MDI can be separated by distillation. MDI is supplied in several grades, depending on the number of reactive units (–NCO groups) per molecule. The most common grade is polymeric MDI with a functionality of 2.3–3.0. The grades used in rigid foam production typically contain 40–60 percent pure MDI, with the balance being dimer and other isomers. Pure MDI is used mainly for RIM (reaction injection molding) systems.

The U.S. production of MDI was about 1.34 billion lb in 1998. Rigid polyurethane foams constitute the largest single use for MDI and its polymers, with total consumption of more than 850 million lb. Typical laminate and board foams contain more than 60 percent MDI, whereas pour systems and spray systems contain somewhat less. The main applications of rigid foam are in construction and in the manufacture of refrigerators and water heaters. Smaller uses are packaging, tank and pipe insulation, and transportation. A minor amount of polymeric MDI is used to make foundry sand binders. Pure MDI finds use in RIM systems, specialty coatings, thermoplastic resins, high-performance casting elastomers, and spandex fibers. A more detailed description of polyurethane polymers can be found in Chapter 15.

Other Uses for Aniline. Aniline currently is used as a raw material in most of the major groups of rubber-processing chemicals: accelerators, antioxidants and stabilizers, and anti-ozonants. The most important of these are the thiazole derivatives and substituted *para*-phenylenediamines. The demand for aniline in these rubber-processing uses is expected to grow at less than 1 percent/year. In agricultural chemicals, the major use for aniline is as a raw material in the manufacture of amide herbicides for controlling annual grasses and broadleaf weeds in various crops. More than 175 commercial dyes can be made from aniline, and many others are produced from aniline derivatives. This market, however, is not expected to show further growth. Since their introduction in 1979, the use of aniline in polyaramid specialty fibers has shown rapid growth. This trend is expected to continue in the future. Among the important pharmaceutical derivatives of aniline are the sulfonamides, a group of compounds used to combat infections.

Chlorobenzenes. Of the 12 different chlorobenzenes that can result from the chlorination of benzene, three are of most commercial importance: monochlorobenzene (MCB), *o*-dichlorobenzene (ODCB), and *p*-dichlorobenzene (PDCB). Chlorination of benzene can be done either batchwise or continuously in the presence of a catalyst such as ferric chloride, aluminum chloride, or stannic chloride. It is usually run as a three-product process; the current product distribution is about 52 percent to MCB, 17 percent ODCB and 31 percent PDCB. The pure compounds are separated from the crude by distillation and crystallization.

Production of monochlorobenzenes peaked in the 1960s with production volume at about 600 million lb. It was down to 152 million lb in 1998. The most significant cause for the decline is the replacement of monochlorobenzene by cumene as the preferred raw material for phenol manufacture. Other reasons include the elimination of the herbicide DDT, the change of diphenyl oxide process from chlorobenzene to phenol and a significant drop in solvent use. The production volume for ODCB and PDCB were 50 and 91 million lb, respectively, in 1998.

Monochlorobenzene. The largest use for monochlorobenzene, accounting for about 59 percent of the consumption, is in the production of chloronitrobenzenes. *p*-Nitrochlorobenzene (NCB) is converted into *p*-phenylenediamine for use as antioxidants in rubber processing. A smaller use for NCB is in the synthesis of the pain reliever, acetaminophen. *Ortho*-nitrochlorobenzene is a raw material for producing insecticides and several azo pigments. A large number of dyes also can be derived from either chlorobenzene or nitrochlorobenzene. About 13 percent of the monochlorobenzene is used as a solvent for pesticide formulation and in MDI processing. About 18 percent is used to make dichlorodiphenylsulfone, an intermediate in the manufacture of sulfone polymers.

Dichlorobenzenes. In 1998, approximately 16 million lb of the *o*-dichlorobenzene was converted into 3,4-dichloroaniline, the raw material for several major herbicides. Also a small amount (3 million lb) goes to various solvent applications. The major demands for *p*-dichlorobenzene come from uses in polyphenylene sulfide resins (50 million lb), room deodorants (16 million lb), and moth-control agents (11 million lb). Any future growth will have to come from the phenylene sulfide resins.

Trichlorobenzenes. A mixture containing trichlorobenzene is always obtained when chlorinating benzene. It can also be made through further chlorination of dichlorobenzene. Most trichlorobenzenes are produced as a mixture of 1,2,3-/1,2,4-trichlorobenzene. The 1,2,3-trichlorobenzene is sold as a raw material for pesticides. The 1,2,4-trichlorobenzene is used for the manufacture of the herbicide Banvel (dicamba). The estimated U.S. consumption was about 15 million lb in 1997.

Derivatives of Toluene

Although the bulk of the toluene is never isolated from the gasoline pool, approximately 12.6 billion lb of toluene was produced for nonfuel consumption in the United States in 2000. Approximately 80 percent of this toluene is used as feed stock for benzene and xylene. The second largest end-use of toluene is as a solvent for coatings, paints, and lacquers. Also

of importance is the use of toluene as an intermediate in the manufacture of other chemicals, mainly toluene diisocyanate, and also benzoic acid and benzyl chloride.

Toluene is converted into benzene by a catalytic hydrodealkylation (HDA) process at elevated temperature and pressure. The importance of this process is influenced by the relative value and demand for benzene, as benzene from this source is normally more costly than that isolated directly from refinery reformate streams. Benzene (along with xylenes) can also be obtained by the catalytic TDP. It has became favorable in recent years. Toluene consumption for toluene disproportionation versus HDA has changed from about 1/5 in 1990 to 2/1 in 2000. The volume of toluene that finds use as a solvent is expected to show a continued decline because of regulations controlling the emission of VOCs.

Toluene Diisocyanate (TDI). TDI is manufactured from toluene by the route indicated in the following equations:

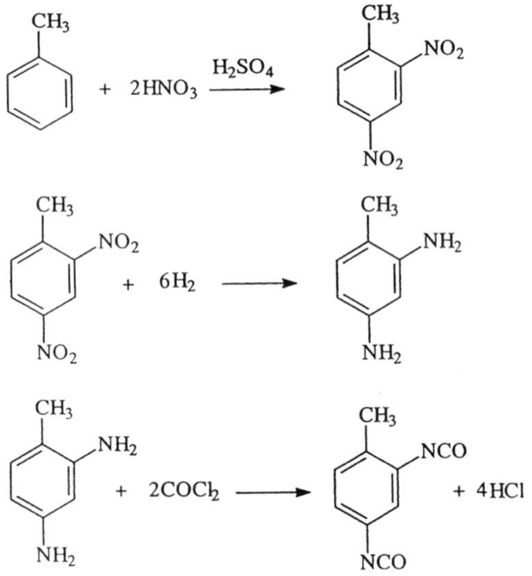

The synthesis of TDI begins with the nitration of toluene, using a nitric acid–sulfuric acid mixture. The nitration product typically contains at least 75 percent 2,4-dinitrotoluene with the balance mostly 2,6-dinitrotoluene, which is catalytically reduced to toluene diamine. Lastly, the diamine mixture is dissolved in chlorobenzenes and reacted with phosgene to produce the TDI. After phosgenation, the mixture is stripped of the solvent and separated by distillation. The final product is an 80 : 20 isomer mixture.

The annual U.S. production of TDI was 960 million lb in 1999. Most of the TDI is reacted with polyols to produce flexible polyurethane foams. These foams are widely used as cushioning materials in furniture, automobiles, carpets, and bedding. A small amount of TDI is used to make polyurethane coatings. Polyurethanes are discussed in detail in Chapter 15.

Benzoic Acid. Benzoic acid can be produced by the LPO of toluene using a catalyst such as cobalt or manganese. Domestic production of benzoic acid was about 130 million lb in 2000. Of this amount, about one half went to make phenol or phenolic derivatives. Other uses are in the synthesis of caprolactam and terephthalic acid, and as food additive, and as a plasticizer and resin intermediate.

Benzyl Chloride. The principal method for producing benzyl chloride involves the photochlorination of toluene, followed by neutralization and distillation. In 1999, 75 million lb of benzyl chloride were produced in the United States. About two-thirds was used to manufacture benzyl phthalates (mainly butyl benzyl phthalate), which are widely used as plasticizers. The other use was to make benzyl quarts. Benzyl chloride can also be used as raw material in the manufacture of benzyl alcohol, for use in photography, perfumes, and cosmetics. The production has increased considerably in Western Europe because of the greater use in solvents such as benzyl esters. But the U.S. production was stopped in 1999.

Chemicals from Xylene

Xylenes are obtained mainly (80%) from petroleum reformate streams in the form of "mixed xylenes." A typical composition of this stream is about 18 percent *p*-xylene, 40 percent *m*-xylene, 22 percent *o*-xylene, and 20 percent ethylbenzene. The major chemical uses of xylene, however, require the pure isomers. The purification process involves a number of steps. First the *o*-xylene is separated from the other aromatics by distillation, with the *meta* and *para* isomers going overhead along with the ethylbenzene. *p*-Xylene can be recovered by either adsorption or crystallization processes. The flow diagram in Fig. 10.35 depicts a two-stage crystallization process for recovery of high-purity *p*-xylene from mixed xylenes. In the adsorption process, the stream is charged to a fixed bed of molecular sieves, and the selectively adsorbed *p*-xylene is recovered by washing the bed with solvent. Because the demand for *p*-xylene is far greater than that for m-xylene, the raffinate usually is isomerized to form more of the *para* isomer. Toluene may also be disproportionated to form equivalent amounts of benzene and xylenes without any ethylbenzene. Recently, this TDP process became more favorable. Today, it contributes to 18.5 percent of xylene production.

In 1999, the total demand for the xylenes (12.3 billion lb) was roughly comparable to that for toluene. The volume of *o*-, *m*- and *p*-xylene were approximately 1.1, 0.27, and 9.9 billion lb, respectively. The principal uses of the three xylene isomers are the production of terephthalic acid (or di-methyl terephthalate), phthalic anhydride, and isophthalic acid, respectively.

Terephthalic Acid (Dimethyl Terephthalate). Terephthalic acid (TPA) and dimethyl terephthalate (DMT) are precursors for polyethylene terephthalate (PET), which in turn is used in the production of polyester fibers and film polyester thermoplastic PET bottles, and other resins. In 1999 the total U.S. production was more than 9 billion lb. In the past, the relative ease of producing high-quality DMT gave it the largest share of the terephthalate market. The trend is now toward TPA, as the result of technological advances that permit better purification of TPA and the use of the acid directly in polymer formation. The capacity is about 3 to 1 split in favor of TPA process.

One process for making TPA involves the air oxidation of a solution of *p*-xylene in acetic acid in the presence of a catalyst containing

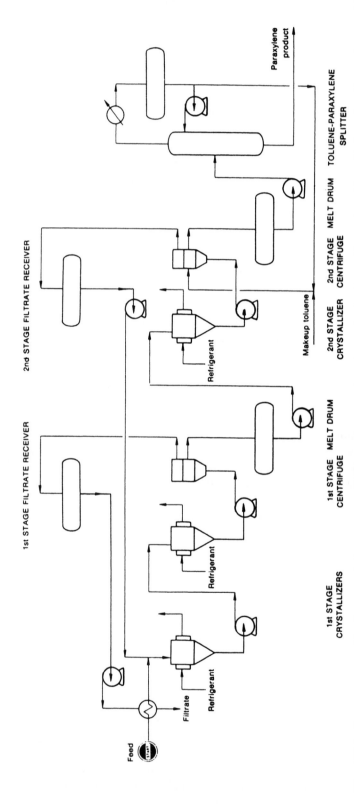

Fig. 10.35. Two-stage crystallization process for recovery of high-purity *para*-xylene from mixed xylenes. (*Reproduced from Hydrocarbon Processing, p. 175, Nov. 1985. Copyright 1985 by Gulf Publishing Co.*)

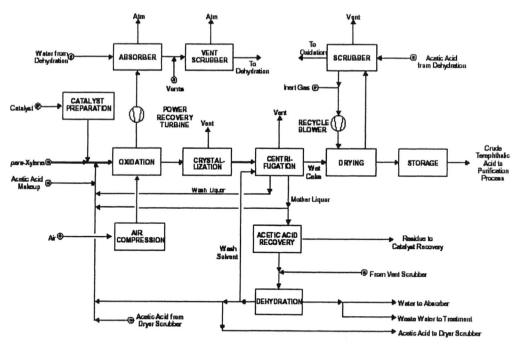

Fig. 10.36. Block flow diagram of generic crude TA process. (*Chem Systems Report No. 97/98-5. Copyright Nexant Chem Systems, Inc. and used by permission of the copyright owner.*)

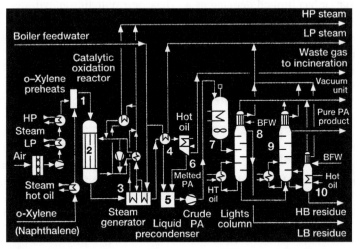

Fig. 10.37. Production of phthalic anhydride. (*Hydrocarbons Processing. p. 118, March, 2001.* Copyright 2001 by Gulf Publishing Co.)

cobalt, manganese, and bromide (Fig. 10.36). The liquid-phase reaction is conducted at about 200°C and 20 atm. pressure. The mixture is cooled to recover TPA by crystallization, dissolved in hot water, and hydrogenated to remove aldehydic by-products. Fiber-grade TPA is recovered by recrystallization.

Phthalic Anhydride. Although phthalic anhydride first was made commercially from the oxidation of naphthalene, by 1999, nearly 90 percent of the production had been converted to *o*-xylene as the feedstock. This is now essentially the only major use for the *o*-xylene. The flowsheet in Fig. 10.37 shows a typical process for making phthalic anhydride. The o-xylene is vaporized by injection into the hot gas stream and then passes through a catalyst-filled multitube reactor. The crude phthalic anhydride is desublimated, and any acid present is dehydrated in the predecomposer vessel. The crude is finally purified in two distillations. Although the fixed-bed process currently is important, there are a number of plants in which a fluidized-bed reactor is used.

About 1.1 billion lb of phthalic anhydride are produced annually in the United States. The major uses are in plasticizers, alkyd resins, and unsaturated polyester resins. The plasticizers are esters made by reacting two moles of an alcohol, such as 2-ethylhexanol, with one mole of phthalic anhydride. These plasticizers find major use in vinyl chloride polymers and copolymers. Alkyd resins are a type of polyester resin used in surface coatings. The most rapidly growing end use is in unsaturated polyester resins for reinforced plastics.

Isophthalic Acid. Although m-xylene is an abundant material, it has limited demand as a chemical raw material. The only major outlet is in the manufacture of isophthalic acid. More than 220 million pounds were produced in the United States in 1992, primarily for use in preparing alkyd resins and unsaturated polyester resins. Small amounts also are used in PET bottle applications.

Naphthalene Derivatives

The use of naphthalene for production of chemicals has been declining steadily, as it is being replaced by other petroleum-based materials. Current domestic consumption is only about 235 million lb. There are two commercial processes for producing naphthalene. One is the recovery of naphthalene from coal tar, and the other involves its recovery from certain aromatized petroleum fractions. Until the end of the 1950s, the only commercial source of naphthalene in the United States was coal tar. At that time petroleum-derived naphthalene became a commercial product and was quickly established as a desirable source of raw material for phthalic anhydride because of its quality, low sulfur content, and stable supply.

The oxidation of naphthalene to phthalic anhydride has long been its principal end use. In the last 30 years, however, o-xylene has supplanted naphthalene as the preferred raw material for phthalic anhydride, and it now accounts for about 90 percent of the PA production. The manufacture of phthalic anhydride is described above in the section "Chemicals from Xylene."

Presently the major end uses for naphthalene are phthalic anhydride (62%), carbaryl insecticide (15%), surfactants and dispersants (20%), and synthetic tanning agents (3%). Carbaryl (1-naphthyl, n-methylcarbamate) is a broad-spectrum insecticide. The major surfactants and dispersants are derivatives of naphthalene sulfonates; their principal applications are as commercial wetting agents, concrete additives, and rubber dispersants, and in agricultural formulations. The synthetic tanning agent are derivatives of naphthalene sulfonic acid and formaldehyde. Other smaller uses of naphthalene are in beta-naphthol and as a moth-repelling agent.

REFERENCES

1. **Chem System Inc, Process Evaluation/Research Planning:** Extending the Methane Value Chain 99-S9 (2000, October), Acetic Acid/Acetic Anhydride 97/98-1 (1999, January), Acetic Acid from Ethane Oxidation 99-S5 (2001, January), Methanol 98/99-4 (2000, May), Developments in Methanol Production Technology 96/97-S14 (1998, August), Methyl Methacrylate 99-2 (2001, September), Formaldehyde 00-8 (2001, April), Ammonia 97/98-6 (l998, August), Impact of MTBE Phaseout on Chemical Market 00/01 S2 (2001, June), Ethanol 99/00-8 (2001, August), Ethylbenzene/Styrene 99/00-6 (2000, August), Maleic Anhydride 99/00-5 (2000, May), Caprolactam 99/00-4 (2001, March), Adipic Acid 98/99-3 (1999, March), Nylon 6/Nylon 6,6 99-S1 (2000, March), Acrylic Acid/Acrylates 00/01-7 (2001, May), Development in Propylene Oxide Technology 00-S12 (2001, November), Glycerin 00-S4 (2001, November), Ethylene/Propylene 00-4 (2001, January), Acrylic Acid/Acrylates (2001, May), Epichlorohydrin 99-S11 (2000, August), Vinyl Chloride/Ethylene Dichloride

99-3 (2000, May), TDI/MDI 98/99-S8 (1999, October), Benzene/Toluene 98/99-6 (1999, May), Terephthalic Acid 97/98-5 (1999, February), Bisphenol A 97/98-4 (1998, May), Naphthalene and Derivatives 96/97-S9 (1998, March), Phenol/Acetone/Cumene 96/97-2 (1997, November), Development in p-Xylene Technology 96/97-S7 (1997, July), meta-Xylene/Isophthalic Acid 94/95 S-14 (1997, February).

2. **Chemical Economic Handbook: SRI International.** Menlo Park, CA, Ethylene Oxide (2000, August), Acetaldehyde (2001, September), Linear Alpha-Olefins (2000, September), Ethylene-Propylene Elastomers (2000, February), Ethylene (1999, April), Isopropyl Alcohol (IPA) (2000, May), Propylene (2000, March), Detergent Alcohols (2000, May), Plasticizer Alcohols (C_4-C_{13}) (1998, June), Propylene (2000, March), Butylenes (1998, August), 1,4-Butanediol (2001, March), Nonene (Propylene Trimer) and Tetramer (2001, August), Butadiene (2000, November), Methyl Ethyl Ketone (2000, April), Butanes (1999, July), Isoprene (2001, October), Cyclopentadiene/Dicyclopentadiene (1999, August), Benzene (2000, October), Toluene (2000, December), Xylene (2001, March), Bisphenol A (2001, February), Benzoic Acid (2001, September), Benzyl Chloride (2001, July), Aniline (2001, May), Chlorinated Methanes (2001, December), C2 Chlorinated Solvents (1999, January), Chlorobenzene (1999, December), Cyclohexane (2000, April), Cyclohexanol and Cyclohexanone—United States (1998, May), Dimethyl Terephthalate (DMT) and Terephthalic Acid (TPA) (2001, February), Epichlorohydrin (2000, December), Epoxy Resins (2001, January), Ethylbenzene (1999, September), Styrene (1999, August), Glycerin (1999, August), Isophthalic Acid (1999, March), Naphthalene (2000, December), Petrochemical Industry Overview (2001, April), Phenol (1999, March), Phthalic Anhydride (2001, June), Propylene Oxide (1999, December), Propylene Glycols (2000, December), Ethylene Dichloride (2001, January), Vinyl Chloride Monomer (2001, December), Natural Gas (1999, January), Ammonia (2001, March), Acetylene (2001, August), Acetic Acid (2000, February), Adipic Acid (2000, April), Caprolactam (2000, April), Carbon Disulfide (2001, December), Ethanol (1999, April), Formaldehyde (2000, July), Hydrogen Cyanide (2000, June), Ketene (1999, January), Methanol (1999, August), Methyl Methacrylate (2000, March), Maleic Anhydride (1999, August), Urea (2001, January), Vinyl Acetate (2001, March).

3. **Process Economics Program Report: SRI International.** Menlo Park, CA, Isocyanates IE, Propylene Oxide 2E, Vinyl Chloride 5D, Terephthalic Acid and Dimethyl Terephthalate 9E, Phenol 22C, Xylene Separation 25C, BTX, Aromatics 30A, o-Xylene 34 A, m-Xylene 25 A, p-Xylene 93-3-4, Ethylbenzene/Styrene 33C, Phthalic Anhydride 34B, Glycerine and Intermediates 58, Aniline and Derivatives 76C, Bisphenol A and Phosgene 81, C1 Chlorinated Hydrocarbons 126, Chlorinated Solvent 48, Chlorofluorocarbon Alternatives 201, Reforming for BTX 129, Aromatics Processes 182 A, Propylene Oxide Derivatives 198, Acetaldehyde 24 A2, 91-1-3, Acetic Acid 37 B, Acetylene 16A, Adipic Acid 3 B, Ammonia 44 A, Caprolactam 7 C, Carbon Disulfide 171 A, Cumene 92-3-4, 22 B, 219, MDA 1 D, Ethanol 53 A, 85-2-4, Ethylene Dichloride/Vinyl Chloride 5 C, Formaldehyde 23 A, Hexamethylenediamine (HMDA) 31 B, Hydrogen Cyanide 76-3-4, Maleic Anhydride 46 C, Methane (Natural Gas) 191, Synthesis Gas 146, 148, 191 A, Methanol 148, 43 B, 93-2-2, Methyl Methacrylate 11 D, Nylon 6-41 B, Nylon 6,6-54 B, Ethylene/Propylene 29 A, Urea 56 A, Vinyl Acetate 15 A.

4. **Chemical Profile of the Chemical Marketing Reporter, Schnell Publishing. (http://www.chemexpo.com)**, Nonylphenol (2001, July 9), Ethanolamines (1998, May 4), Ethylene Oxide (2001, August 27), Vinyl Acetate (2000, August 21), Propylene Glycol (1998, July 20), Isopropanol (1998, December 28), Cumene (1999, March 22), Acetone (1999, April 5), Acrylic Acid (1999, May 24), Butene-1 (1999, August 16), Butadiene (2000, March 6), Dicyclopentadiene (2001, July 23), Aniline (1999, February 8), Benzene (1999, December 6), Benzoic Acid (1996, November 4), Bisphenol A (2001, November 5), Cyclohexane (2001, May 28), Dipropylene Glycol (1998, July 27), Ethylbenzene (2001, April 30), Ethylene Dichloride (2001, March 19), Glycerine (2001, December 16), Linear Alkyl Benzene (2001, July 30), Methyl Chloride (2000, October 2), Methylene Chloride (2000, October 9), Chloroform (2000, October 16), Monochlorobenzene (1999, May 17), Naphthalene (1999, May 31), Nitrobenzene (1999, February 15), o-Xylene (1998, May 18), p-Xylene (1998, May 11), o-Dichlorobenzene (1996, September 9), p-Dichlorobenzene (1999, June 7), Perchloroethylene (2000, October 30), Phenol (1999, March 29), Phthalic Anhydride (2001, February 12), Propylene Oxide (1998, July 13), Propylene Glycol (1998, July 20), PTA/DTM (1998, July 6), Styrene (2001, May 14), Toluene (2000, October 23), Toluene Diisocyanate (TDI) (1999, April 12), Methyl Diphenyl Diisocyanate (MDI) (2000, January 3), Trichloroethylene (2000, September 25), Vinyl Chloride (1998, February 9), Formaldehyde (2001, May 21), Hydrogen Cyanide (2001, November 26), Methanol (2000, July 31), Methyl Methacrylate (1999, June 28), Maleic Anhydride (2001, February 5), Ammonia (1999, November 29), Acetylene (2001, December), Acetic Acid (2001, February 26), Adipic Acid (1998, June 15), Caprolactam (2001, April 16), Carbon Disulfide (2000, November 20), Cyclohexane (2001, May 28), Ethanol (2000, March 13).

11

Chemistry in the Pharmaceutical Industry

Graham S. Poindexter,* Yadagiri Pendri,**
Lawrence B. Snyder,* Joseph P. Yevich* and
Milind Deshpande***

INTRODUCTION

This chapter will discuss the role of chemistry within the pharmaceutical industry. Although the focus will be upon the industry within the United States, much of the discussion is equally relevant to pharmaceutical companies based in other first world nations such as Japan and those in Europe. The major objective of the pharmaceutical industry is the discovery, development, and marketing of efficacious and safe drugs for the treatment of human disease. Of course drug companies do not exist as altruistic, charitable organizations but like other share-holder owned corporations within our capitalistic society must achieve profits in order to remain viable and competitive. Thus, there exists a conundrum between the dual goals of enhancing the quality and duration of human life and that of increasing stock-holder equity. Much has

been written and spoken in the lay media about the high prices of prescription drugs and the hardships this places upon the elderly and others of limited income. Consequently, some consumer advocate groups support governmental imposition of price controls, such as those that exist in a number of other countries, on ethical pharmaceuticals in the United States.

However the out-of-pocket dollars spent by patients on prescription drugs must be weighed against the more costly and unpleasant alternatives of surgery and hospitalization, which are often obviated by drug therapy. Consideration must also be given to the enormous expense associated with the development of new drugs. It can take 10 years or more from the laboratory inception of a drug to its registrational approval and marketing at an overall cost which is now $600–800 million dollars and increasing. Only 1 out of 10 to 20,000 compounds prepared as drug candidates ever reach clinical testing in man and the attrition rate of those that do is >80 percent. The expense of

*Bristol-Myers Squibb Company, Wallingford, CT.
**Expicor, Inc., Hauppauge, NY
***Achillion pharmaceuticals, New Haven, CT.

developing a promising drug grows steadily the further through the pipeline it progresses; clinical trials can be several orders of magnitude more costly than the preclinical evaluation of a compound. While the sales of successful drugs that run the gauntlet and reach the shelves of pharmacies can eventually recoup their developmental expenses many times over, the cost of the drugs that fail is never recovered.

To a large extent, the difficulties associated with bringing a drug to market have arisen from the increasingly stringent but appropriate criteria that have been imposed by the Food and Drug Administration (FDA) in the United States and analogous regulatory agencies in other countries. It is unlikely that an occurrence like that of the thalidomide disaster, which resulted in horrible birth defects several decades ago, would happen again today. Furthermore the era of easy approval of "me-too" drugs is long past. During this era, which prevailed until the final two decades of the past century, it was possible to gain approval for drugs which, although they fell outside the scope of the patents covering a particular marketed drug, offered little advantage over the marketed agent. It is now necessary for a company to demonstrate that a drug, for which a New Drug Application (NDA) is submitted to the FDA, affords significant benefits in terms of efficacy and/or safety relative to the existing drug therapy. The approvability bar may be lowered for agents aimed at the treatment of life-threatening maladies such as cancer and AIDS or for those such as Alzheimer's disease where no effective therapy currently exists; but even in these cases it is incumbent upon the sponsoring company to provide compelling empirical evidence that their drug is safe and effective. The restrictions imposed by Health Maintenance Organizations (HMOs) can also have significant impact on the sales of any given drug. Most HMOs list only a select few drugs, for which they will cover costs, within any given category, such as antidepressants, antihypertensives, or cholesterol-lowering agents.

A major consequence of the financial and logistical impediments to the successful introduction of new drugs has been the high incidence of mergers and acquisitions among U.S.-based pharmaceutical companies in the recent past. These events have not occurred because bigger is necessarily better but because the critical mass of internal resources required to bring a drug from the test tube to the pharmacy continues to grow. In contrast to this trend among the major drug companies (often dubbed "big pharma") there has been a proliferation of start-up companies often founded by entrepreneurial scientists with "big pharma" or academic experience and financed by venture capital investment. While many such start ups are strictly bio-techs, others function as mini drug companies and are staffed by both chemists and biologists. Unlike their much larger brethren, the small companies cannot attempt to cover the breath of drug research but instead focus upon a particular therapeutic area and perhaps even a particular disease. Their mission is to discover drug candidates, which a large company may be interested in licensing and developing. The "big pharma" companies do not rely exclusively upon filling their developmental pipelines with drug candidates that have been discovered in-house but often enter into collaborations and licensing agreements to acquire the rights to promising agents from the labs of smaller companies or academic researchers.

MEDICINAL CHEMISTRY

Chemistry has long been an integral part of the pharmaceutical industry and its importance should not diminish. Many currently marketed drugs such as the antineoplastic agent, paclitaxel, and the antibiotic, vancomycin, are natural products. The extracts of plants and marine organisms and the products of soil bacteria fermentation will continue to be investigated as potential sources of powerful new drug substances. Chemists are certainly involved in this arena of drug discovery as they conduct the painstaking isolation, purification, and structural characterization of pharmacologically active components which most often are present in minute amounts in the natural source and which have extremely complex chemical structures. The enormous advances in molecular biology have resulted in the successful development of bio-engineered therapeutic agents, for example, human insulin, Herceptin (Genentech drug for breast cancer), and Enbrel (Immunex drug for rheumatoid arthritis). It is anticipated that many other biomolecules may be forthcoming for the treatment of human disease.

However the great majority of existing drugs are small organic molecules (MW ~200–600) that have been synthesized by medicinal chemists. There is no reason to doubt that most drugs of the future will also fall in this category. It is thus important to define what is meant by "medicinal chemist" and what role is played by the practitioners of this sub-discipline in the pharmaceutical industry. A traditional and perhaps somewhat narrow definition of medicinal chemist is that of a researcher engaged in the design and synthesis of bioactive molecules. As part of their academic training many medicinal chemists carried out doctoral and postdoctoral work that involved the total synthesis of natural products and/or the development of synthetic methodology. They are hired by pharmaceutical companies because of the skills they have gained in planning and conducting the synthesis of organic compounds. While such skills can remain important throughout chemists' careers, they alone are insufficient for the challenging task of drug discovery in which, unlike the academic environment, synthetic chemistry is just a means to an end rather than an end in itself. Thus, the enterprising young chemical researcher who enters the industry must be able and willing to undergo an evolution from that of pure synthetic chemist who knows how to make compounds to that of medicinal chemist who also has an insight into what to make and why.

Such insight is gained by acquiring an expanded knowledge base. It is important for the medicinal chemist to know what structural components act as pharmacophores in existing drugs. Pharmacophores, which can be of varying complexity, comprise the essential structural elements of a drug molecule that enable it to interact on the molecular level with a biological macromolecule such as a receptor or enzyme and thus impart a pharmacological effect. The medicinal chemist must become skilled at analyzing the structure activity relationships (SAR) that pertain to the series of compounds on which he/she is working. That is, how does the activity in a biological test of analogs within the series change depending on the introduction of substituents of various size, polarity, and lipophilicity at various domains of the parent drug molecule? Elucidation of the SAR within a series of active compounds is the key to optimizing the potency and other desirable biological properties in order to identify a new chemical entity (NCE) as a bona fide drug candidate. Quantitative structure activity relationships (QSAR) are often employed in this effort; analyses employing linear free energy relationships, linear regression, and other techniques can be utilized to correlate biological activity with the electronic, steric, polarizability, and other physical/chemical parameters of the substituent groups on members of a series of structurally related compounds.

The synthesis and isolation of pure enantiomers has become increasingly important. In the past chiral drugs were most often marketed as racemic mixtures since it was not deemed cost-effective to provide them in enantiomercially pure form. However, in many cases one or the other enantiomers of an optically active drug may have a significantly greater level of the desired biological activity and/or less side effect liability than its antipode. Regulatory agencies such as the FDA now routinely require that each enantiomer of a chiral drug be isolated and evaluated in tests of efficacy, side effects, and toxicity. If one of the enantiomers is shown to be clearly superior then it is likely that it is the form that will be developed as the drug candidate. Thus enantioselective chemical reactions which can afford a high enantiomeric excess(ee) of one or the other of a pair of enantiomers are valuable components of the medicinal chemist's synthetic tools. Enzyme chemistry plays a prominent role in drug R&D since isolated enzymes or microorganisms can often achieve an enantiospecific chemical transformation much more efficiently and economically than conventional synthetic methods. Many "big pharma" companies now have dedicated groups that exclusively study enzymatic reactions.

Research Strategies

The discovery of new drugs may occur by luck or serendipity or as the result of some brilliant insight. However pharmaceutical companies cannot depend on chance occurrences as a research strategy. The aforementioned "me-too" approach has hardly been abandoned and it is likely that the marketing of a novel drug will soon be followed by a number of competitors' agents but with the caveat that the latter offer some therapeutic advantage over the prototype.

The most scientifically sound approach is that of rational drug design, which is based on an understanding of the biochemical mechanisms underlying a particular disease. If, for example, overactivity or underactivity of a certain neurotransmitter system is believed to be responsible for a central nervous system (CNS) disorder such as depression, then medicinal chemists can endeavor to design agents capable of normalizing neurotransmission by their action upon the receptor proteins through which interneuronal communication is mediated. Cloning and expression of human genes to afford functional receptors and enzymes that can be studied in cell culture has been a tremendous advance in the ability to evaluate drug action at the molecular level. Likewise, molecular biology has afforded macromolecules that are essential to the life cycle of pathogens such as bacteria and viruses, thus enabling novel mechanistic strategies for the treatment of infectious disease. In many cases, X-ray crystallography has provided a detailed three-dimensional structure of a macromolecule such as an enzyme with and/or without a bound substrate. Researchers having expertise in computer assisted drug design (CADD) can depict the determined structure on silicon graphics terminals and in collaboration with medicinal chemists can propose drug molecules to fit the active site. Such detailed analysis of protein structure was instrumental in the design of a number of drugs that inhibit HIV protease, an enzyme essential to the integrity of the AIDS virus.

Up until now there have been approximately 1000 human proteins identified as potential targets for drug intervention in various diseases. It is estimated that the determination of the human genome will increase this number by at least tenfold. Therefore, it seems safe to predict that the rational approach to drug discovery will grow accordingly and with it the role of synthetic/medicinal chemistry. There will be intense competition within the pharmaceutical industry to determine the functional relevance of this multitude of new targets in the absence and presence of disease and a close nexus to this quest will be the search for compounds that can impart selective pharmacological effects upon the target proteins. But it is not likely that these goals can be met by employing only the classical iterative approach which entails one-compound-at-a-time synthesis and low volume testing. Instead the challenges of this exciting new era of research must be met by methodologies that can synthesize and test large numbers of compounds in a short period of time—that is, combinatorial chemistry and high-throughput screening (HTS). In the context of its application within pharmaceutical research, combinatorial chemistry should not be regarded as a separate discipline but instead as a technologically specialized part of medicinal chemistry. This topic will be discussed in detail in a later section of the chapter.

Another important interface occurs with chemists in process research and development.

In most cases medicinal chemists are not overly concerned with the cost, toxicity, or environmental impact of the starting materials, reagents and solvents they employ to synthesize target compounds since they are dealing with relatively small quantities of materials. Neither are reaction conditions employing very low or elevated temperatures and pressures problematic on the discovery scale. However these and other pragmatic considerations must be taken into account for the bulk scale preparation of experimental drugs. Process chemists must very often modify the synthetic procedures of their medicinal chemistry colleagues and in many cases devise an entirely new synthetic pathway. Process chemistry will also be discussed in an ensuing section.

Pharmacodynamics

Medicinal chemists must be generally knowledgeable about pharmacodynamics, that is, the effect of drugs upon biological systems. In addition to being aware of the state-of-the-art understanding of the biological mechanisms that underlie the particular diseases for which they are endeavoring to discover drug therapy they should know the basis of the various in vitro and in vivo tests that the biologists employ to evaluate both the potential efficacy and side-effect liability of the synthesized compounds. Because drug research covers a plethora of human diseases, each with its own unique combination of etiology and biochemical mechanisms, the number and diversity of biological tests are far too great to discuss in this chapter. Suffice it to say that in a general sense the primary and often even the secondary biological tests of drugs for a particular disease target are in vitro tests that can be run rather quickly, inexpensively, and on small amounts of compound. For example, these can be receptor binding assays for CNS drugs, enzyme assays for antihypertensive agents, inhibition of bacterial colony growth by antibiotics, and the killing of cultured cancer cells by oncolytic drugs. Encouraging in vitro results lead to

in vivo testing in some appropriate animal model. In vivo tests are more laborious and costly but are necessary to establish that a drug is effective in an intact living organism; they can range from complex behavioral paradigms for CNS drugs to enhancement of survival time of tumor-implanted mice by experimental cancer drugs. Evaluation of a NCE's propensity to cause side effects is as important as efficacy testing. Even if a compound shows an encouraging level of the desired activity, a lack of selectivity can cause it to induce a number of undesirable pharmacological effects thus precluding its further development. The medicinal chemist must be able to interpret the results of the tests run on his/her compounds and use this information as a guide to further synthetic work.

Pharmacokinetics and Toxicity

It is also necessary that chemists are attuned to various aspects of pharmacokinetics (PK), that is, the effects of biological systems upon drugs. These aspects—absorption, distribution, metabolism and excretion (ADME)—are as critical as biological activity in determining whether a NCE is a viable drug candidate. A compound may exhibit high affinity for a biological receptor or potent inhibition of an enzyme in an in vitro assay but if it is poorly absorbed or rapidly metabolized to inactive species then it will be ineffective as a drug. For example, the empirically based Lipinsky's rules of five (Table 11.1) define the limits of such physical/chemical parameters as molecular weight, lipophilicity,

TABLE 11.1   Lipinski's Rules for Drug Absorption

Absorption of a drug following oral administration is favored by:
  Molecular weight is <500
  The drug molecule has <5 hydrogen bond donors
  The drug molecule has <10 hydrogen bond acceptors
  The distribution coefficient, log P, is <5

Source: Lipinski, C. A., Lombardo, F., Dominy, D. W., and Feeny, P. J., *Adv. Drug Delivery Rev.*, **23**, 3–25 (1997).

and hydrogen bond forming moieties that must be considered for the absorption of orally administered drugs. A compound with potent intrinsic activity can be rendered ineffective in vivo by its rapid conversion to inactive metabolites. The susceptibility of compounds to metabolic conversion can be assessed by incubating them with liver homogenates from various species including rodent, dog, monkey, and man or with cloned, expressed human hepatic enzymes. Analysis of the incubates by liquid chromatography/ mass spectometry (LC/MS) can quantify the extent of metabolism and even identify some specific metabolites. In vivo adminstration of a NCE to one or several animal species is required to determine its oral bioavailability, half-life, and other PK properties such as distribution and elimination. If an unsatisfactory PK profile threatens to be the demise of an otherwise promising drug candidate, it falls upon the medicinal chemist to make structural permutations aimed at correcting the problem. If poor absorption is the problem this may entail modifying the lipophilicity of the drug molecule to render it more membrane permeable. A metabolic liability might be rectified by blocking the site of biotransformation with a metabolically inert atom or group.

Toxic effects upon blood or organs or the potential to cause gene aberrations will red flag a compound regardless of its having both excellent biological activity and PK properties. Promising lead compounds are screened in in vitro tests in bacteria and mammalian cells to determine whether they cause gene mutations and DNA damage. If they pass this hurdle the compounds are dosed on a daily basis for several weeks to several months in both a rodent and nonrodent (usually dog or monkey) species and the animals are observed for any adverse effects; the test animals are necropsied following conclusion of the study to ascertain whether any organ or tissue damage occurred. Unacceptable toxicological findings will invariably kill a drug candidate and again it is the medicinal chemist who will be called upon to save the day by devising and implementing structural modifications to eliminate the toxicity. This may be a more daunting task than overcoming a side effect or metabolic issue, especially if the toxicity is mechanism-based.

Drug Delivery

Drugs can be administered to patients in many ways. The most common and preferred route is oral administration and oral drugs are generally formulated as tablets or capsules in which a specific dose of the drug substance is homogeneously mixed with some inert filler or excipient. Some oral medications, such as pediatric formulations of antibiotics, are in solution form, as are injectable drugs. Obviously this requires satisfactory solubilization of the drug, preferably in aqueous medium. Compounds bearing some ionizable group such as a basic amine or an acidic function can usually be converted into water-soluble salts but neutral molecules present greater difficulties. In some cases the results of clinical trials will indicate that an experimental injectable drug shows promise of efficacy but does not elicit a robust response because its poor solubility limits the amount that can be administered and thus does not allow adequate plasma levels to be attained. Inadequate membrane permeability can restrict the absorption and bioavailability of an orally administered drug.

Medicinal chemists can respond to such findings by investigating the feasibility of preparing a suitable prodrug. A prodrug is a derivative in which a cleavable solubilizing group is covalently appended to the parent drug molecule, most often via a hetero atom such as oxygen or nitrogen. An effective prodrug is one which has much higher solubility than the parent drug and which following its administration is rapidly cleaved in vivo to achieve a therapeutically beneficial plasma concentration of the parent drug.

Patents

Patent protection on both its approved and experimental drugs is of critical importance to a pharmaceutical company. Issued patents provide the company with exclusivity for the manufacture, use, and sale of its drug products and it is highly unlikely that a company would undertake the risks and costs of developing an agent for which it had no patent protection. There are several types of patents of which the "composition of matter" (COM) or "product" patent may be deemed to have the greatest value. An approved COM patent covers specifically claimed compounds of a certain structural chemotype and provides empirical evidence that the claimed compounds have been prepared, characterized, and found to have some utility. In order to be patentable the compounds must have structural novelty and cannot have been publicly disclosed either in the scientific or patent literature or by a presentation. But structural novelty alone is not sufficient grounds for a patent; it must be demonstrated that the compounds are useful and in the context of a drug patent the proposed utility is for the treatment of some disease. The basis of such utility is activity in appropriate and relevant biological tests. Clinical data may also be used in support of a patent application although in the great majority of cases the applications are filed well before any compound within the application reaches clinical trials.

Medicinal chemists are closely affiliated with the patent process and are most commonly the inventors listed on COM patents covering drug substances. The chemists and other researchers with whom they collaborate must provide the chemical and biological data for the patent and the chemists will also provide input as to the scope and claims of the patent. Since patents are legal documents that provide the assignee exclusive proprietary rights to the covered subject matter for 20 years from the date of the patent's issue, it is essential that all supportive data be accurate and instructive. If a patent is ever challenged by another party and is found to contain erroneous information then it could be invalidated. Moreover, in the United States, patents are granted on a "first-to-invent" basis. Thus if two or more parties submit applications on identical subject matter to the U.S. Patent Office then the

patent will be awarded to the party that can prove that it had the earliest conception and reduction to practice of the subject matter. Therefore it is imperative that chemists maintain accurate records of all experimental work in a bound notebook and that such records are dated, signed, and witnessed.

Other types of drug-related patents include process, use, and formulation patents. Chemists are responsible for process patents, which describe an improved method of preparation of some drug substance but are minimally involved with the others. Use patents are based on the discovery of some unobvious utility of a compound that is either part of the public domain or covered by an existing patent; such discoveries are most likely to be made by biologists. Formulation patents disclose a preferred means of drug delivery of a known drug substance.

Clinical Trials

Even though there is no involvement of chemistry in the clinical evaluation of drugs, any discussion of the pharmaceutical industry must include clinical trials for the results of such trials determine whether or not an experimental drug has the combination of efficacy, safety, and tolerability which will allow it to achieve registrational approval and reach the market. If a drug candidate survives the hurdles of pharmacological, pharmacokinetic, and toxicological testing, the next customary step in the United States is the sponsoring firm's filing of an Investigational New Drug (IND) application with the FDA. This is a formal request to initiate clinical investigation in man and is accompanied by a detailed description of the planned studies and clinical protocols. Upon approval of the IND, Phase I clinical studies are initiated.

Phase I studies are conducted in healthy volunteers in order to establish the drug's safety and to determine appropriate dosage levels. If the drug is found to have an acceptable human pharmacokinetic profile and to be free of untoward side effect liabilities, it is advanced into Phase II trials, which are typically carried out in several hundred patients and may last from six months to two years. Phase II trials are designed to ascertain the appropriate dosing regimen for the drug and whether it is effective in treating the target disease. Only about one third of drugs pass Phase II trials, most failing because of the lack of efficacy. Those that pass are advanced into Phase III trials which may involve from several hundred to several thousand patients and which can last from one to three years or even longer depending on the type of drug under study and the complexities of the study design. Phase III trials provide the ultimate test of an experimental drug since they are designed to verify the drug's effectiveness against the target disease as well as its safety. For agents that are intended for chronic use, studies also monitor adverse reactions that may develop only after long-term use and the development of tolerance. Clinical studies of many drug classes will commonly employ several patient groups of approximately equal size with one group receiving the experimental drug, another placebo (nondrug), and another a positive control, that is, a marketed drug used to treat the same disease for which the experimental agent is being evaluated. In order to minimize the possibility of bias in favor of the test drug, such studies are most often run in a double-blinded manner with neither patients nor clinical investigators knowing which group is receiving which treatment until the conclusion of the trial.

 If a drug candidate is among the one in four to five that gets through Phases I–III and if statistical analysis of the clinical data supports its efficacy then the sponsoring firm will assemble the voluminous data into the NDA which is submitted to the FDA. Review of the NDA can take one to two years and often the FDA may request that additional information be provided or even that some additional studies be done. When approval is granted the company is then free to market the drug.

 The results of clinical evaluation of an experimental drug can feed back into medicinal chemistry. For example, if a drug is found to fail because of poor bioavailability in humans then medicinal chemists will endeavor to design and prepare an analog with improved pharmacokinetic properties.

Summary

The preceding sections present what is an admittedly superficial overview of the very extensive and complex topic of medicinal chemistry, its role in the pharmaceutical industry, and its interface with other disciplines. An acquired understanding of relevant biology, pharmacology, toxicity, and so on is not just of heuristic value but is necessary for the chemist to engage in meaningful dialogue with their colleagues who work in these specialties. Successful drug discovery and development cannot be done by individuals working in isolation but requires the interactive collaboration of many researchers representing a multiplicity of scientific disciplines as depicted in Fig. 11.1. It may be argued that medicinal chemists are the most versatile generalists among these researchers in that they must have primary expertise in chemistry along with extensive knowledge of numerous other areas.

 The following section presents examples of marketed drugs in a number of different therapeutic categories.

CARDIOVASCULAR AGENTS

Hypertension

A variety of agents of several mechanistic types are currently available for the treatment of hypertension (elevated blood pressure). The dihydropyridine derivative amlodipine (Norvasc®/Pfizer) is a receptor-operated, calcium entry blocker that prevents Ca^{++} entry into vascular smooth muscle cells. Amlodipine is also useful for the treatment of angina. Losartan (Cozaar®/Merck) and irbesartan (Avapro®/Bristol-Myers Squibb) are angiotensin receptor antagonists that inhibit the action of angiotensin II on the AT_1 receptor. Metoprolol (Toprol®/AstraZeneca) is a cardioselective, β_1-adrenergic receptor blocking agent and is also useful in the treatment of angina.

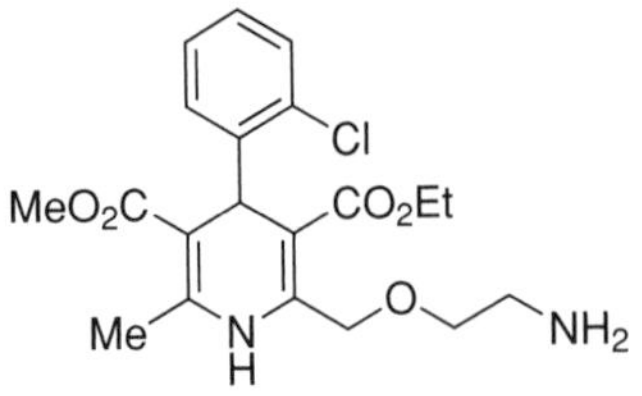

amlodipine

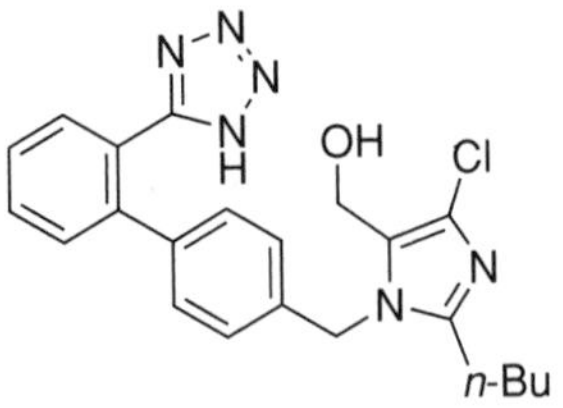

losartan

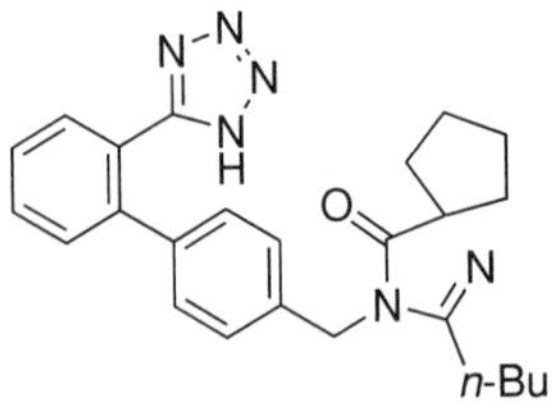

irbesartan

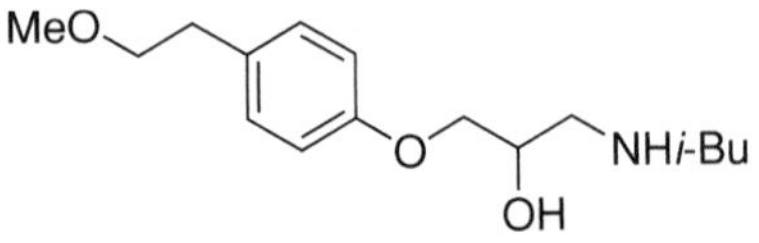

metoprolol

Congestive Heart Failure, Migraine, and Thrombolytic Agents

Enalapril (Vasotek®/Merck) and lisinopril (Zestril®/AstraZeneca and Prinvil®/Merck) are angiotensin-converting enzyme (ACE) inhibitors, useful in the treatment of congestive heart failure and hypertension by suppression of the renin–angiotensin–aldosterone system. Enalapril is an ethyl ester prodrug that is hydrolyzed in the liver to the active carboxylic acid enalaprilat. Sumatriptan is a selective agonist of serotonin (5-hydroxytryptamine) type-1 receptors (most likely the 5-HT$_{1B}$ and 5-HT$_{1D}$ subtypes) in the vasculature. It is thought to exert its beneficial effects on migraine headaches by selective constriction of certain large cranial blood vessels and/or possibly through suppression of neurogenic inflammatory processes in the central nervous system. Clopidogrel (Plavix®/Bristol-Myers Squibb, Sanofi-Synthelabo) is an inhibitor of ADP-induced platelet aggregation and is useful in the treatment of various thrombolytic events such as stroke and myocardial infarction.

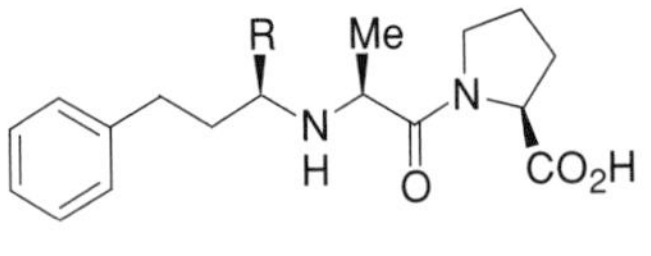

enalapril, R = CO$_2$Et
enalaprilat, R = CO$_2$H

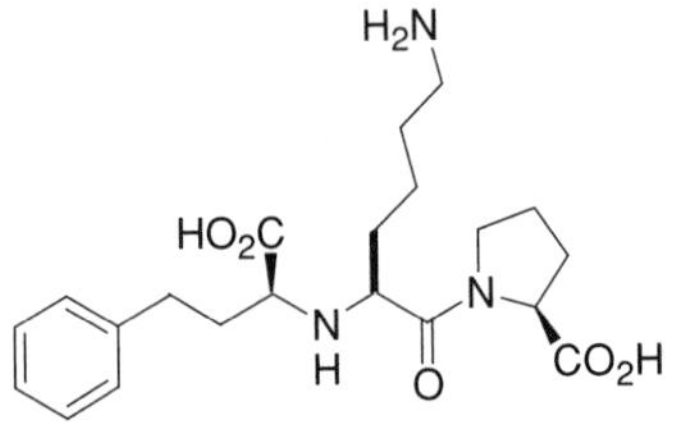

lisinopril

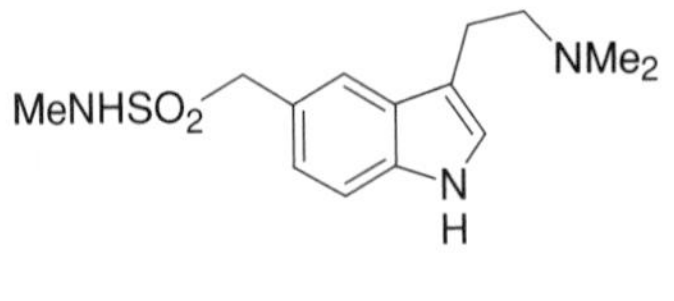

sumatriptan

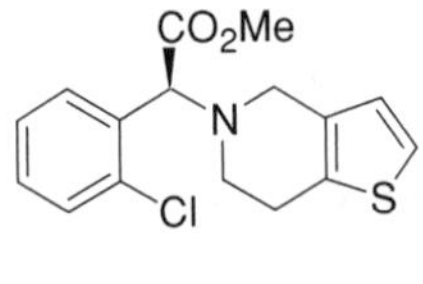

clopidogrel

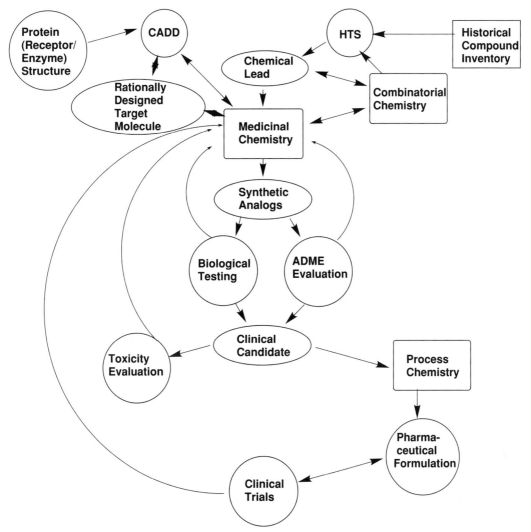

Fig. 11.1. Drug discovery and development is a complicated process that involves the interaction of researchers in various disciplines. Medicinal chemists may synthesize analogues based on chemical leads arising from the high throughput screening of combinatorial libraries or historical compound inventories. Alternatively, analogue synthesis can be based on the collaboration between medicinal chemistry and computer-assisted drug design to rationally design small molecules capable of interacting with a macromolecular biological target (receptor or enzyme). Subsequent biological, pharmacokinetic, and toxicological avaluations lead to identification of a drug candidate that, following development of a suitable bulk scale synthesis by process chemistry and pharmaceutical formulation, is advanced into clinical trials. Feedback to medicinal chemistry from any of these developmental steps can give rise to further synthetic modifications and refinements.

METABOLIC AGENTS

Hyperlipidemia

Simvastatin (Zocor®/Merck), pravastatin (Pravachol®/Bristol-Myers Squibb), atorvastatin (Lipitor®/Pfizer), and rosuvastatin (Crestor®/ AstraZeneca) are hydroxymethyl-glutaryl-CoA (HMG-CoA) reductase inhibitors (statins) that lower serum lipid levels by inhibiting cholesterol biosynthesis. Simvastatin and pravastatin are semi-synthetic, mevinic acid-derived antilipidemic agents whereas atorvastatin is a wholly synthetic, pentasubstituted pyrrolo heptanoic acid. Unlike pravastatin, atorvastatin, and atorvastatin, simvastatin is a lactone prodrug which must be converted to the corresponding, ring-opened δ-hydroxy acid in vivo. A newer agent with a novel mechanism of action is ezetimibe (Zetia®/Merck, Schering Plough). Ezetimibe does not inhibit cholesterol biosynthesis in liver as do the statins but rather inhibits cholesterol absorption in the intestine. This novel action is complementary to the HMG-CoA reductase mechanism displayed by the statins. An innovative new product for the treatment of hyperlipidemia is Vytorin®. It was developed by Merck and Schering Plough and consists of a mixture of simvastatin and ezetimibe in one pill.

Diabetes

A variety of mechanistic agents are currently available for the treatment of type 2 (noninsulin-dependent) diabetes mellitus (NIDDM)]. Rosiglitazone (Avandia®/GlaxoSmithKline) is a thiazolidinedione (glitazone) antidiabetic agent and an agonist at the peroxisome proliferator-activated receptor$_{gamma}$ (PPAR$_{gamma}$). Activation of this receptor enhances insulin sensitivity in target tissues by increasing

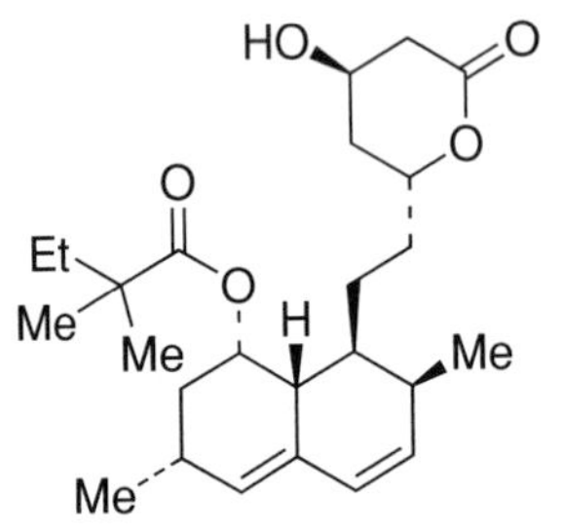

simvastatin

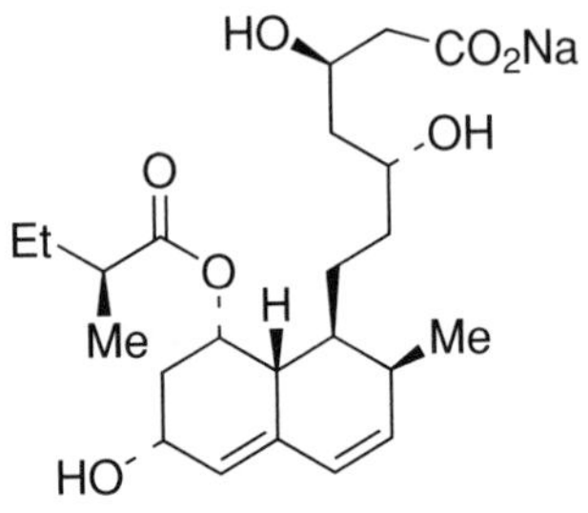

pravastatin

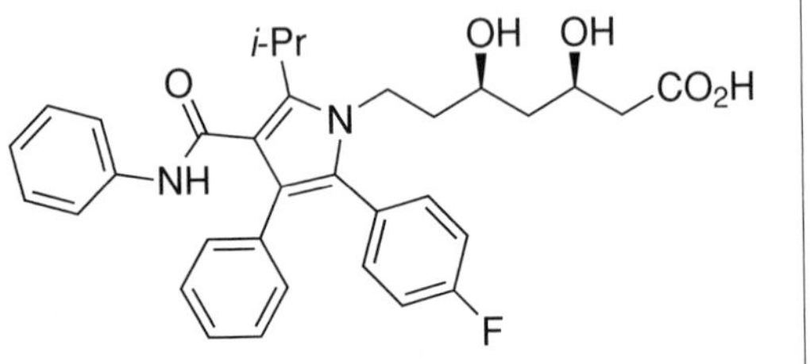

atorvastatin

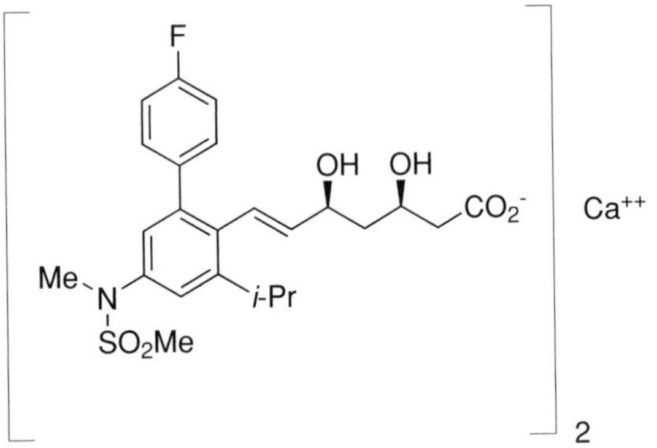

rosuvastatin

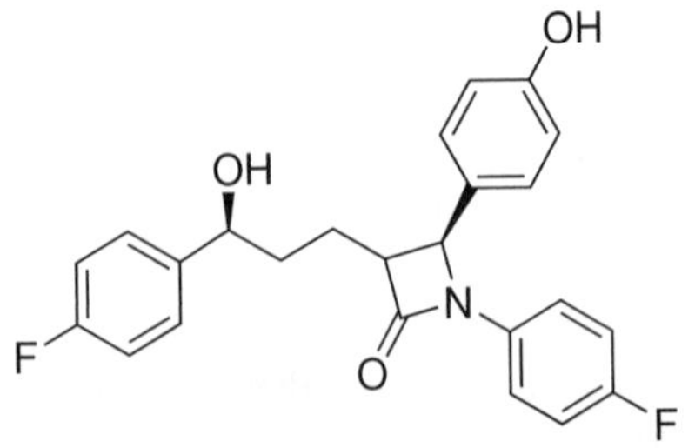

ezetimibe

insulin-responsive gene transcription. Metformin (Glucophage®/Bristol-Myers Squibb) is an antihyperglycemic agent that improves glucose tolerance in patients with type 2 diabetes. The compound acts by decreasing both hepatic glucose production and intestinal absorption of glucose, and improves insulin sensitivity by increasing peripheral glucose uptake and utilization. Glimepiride (Amaryl®/Aventis) is in the sulfonylurea class of antidiabetic agents. Glimepiride is thought to lower blood glucose concentration by stimulating insulin secretion in pancreatic beta cells.

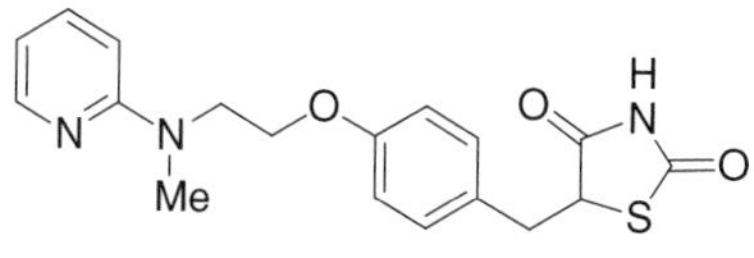

GASTROINTESTINAL AND GENITOURINARY AGENTS

Antisecretory

Ranitidine (Zantac®/GlaxoSmithKline) is a histamine H_2-receptor antagonist that inhibits the release of gastric acid and is useful in the treatment of a variety of hypersecretory conditions [dyspepsia, heartburn, duodenal and gastric ulcers, and gastroesophageal reflux (GERD)]. Lansoprazole (Prevacid®/TAP), omeprazole (Prilosec®/AstraZeneca), and esomeprazole (Nexium®/AstraZeneca) are benzimidazole

rosiglitazone

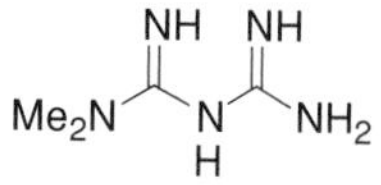

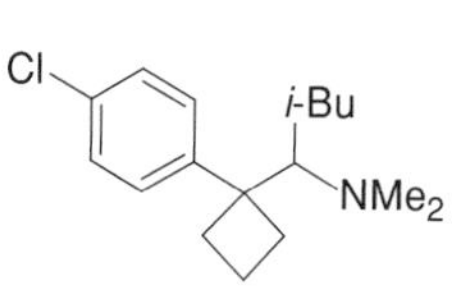

metformin

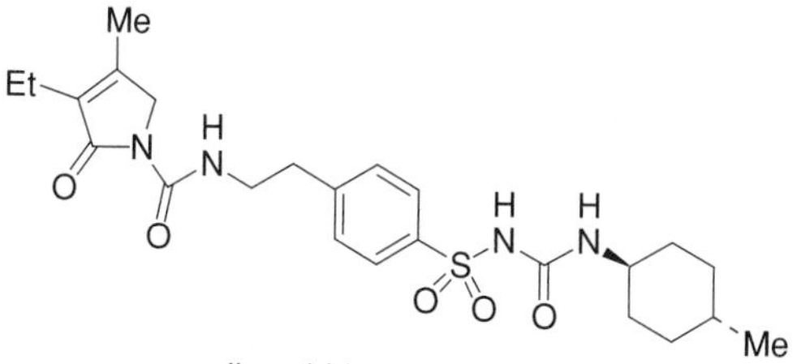

glimepiride

Obesity

Orlistat (Xenical®/Roche) is a reversible gastric and pancreatic lipase inhibitor. The compound has no effect on appetite suppression but rather acts by inhibiting dietary fat absorption from the GI tract. Sibutramine (Meridia®/Abbott) and its major active metabolites are re-uptake

gastric antisecretory agents and unrelated both chemically and pharmacologically to the H_2-receptor antagonists. These agents are known as proton pump inhibitors due to their ability to inhibit the H^+K^+-ATPase (the proton pump) in gastric parietal cells thereby blocking the secretion of hydrochloric acid. Esomeprazole is the S-enantiomer of omeprazole which is racemic and thus a mixture of both its R-and

orlistat

sibutramine

inhibitors of norepinephrine, serotonin, and dopamine and exert their beneficial effect through appetite suppression.

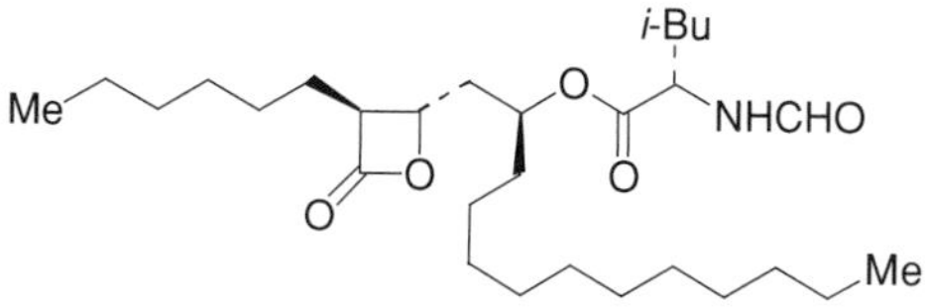

S-enantiomers. Lansoprazole and omeprazole are also useful in the management of duodenal and gastric ulcers, and GERD.

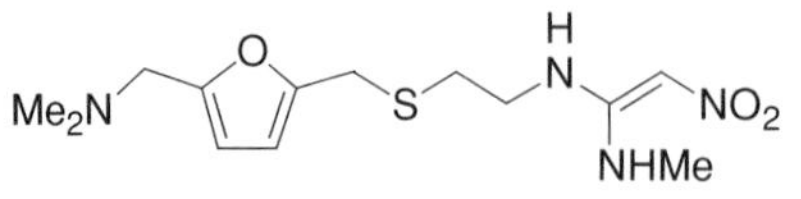

ranitidine

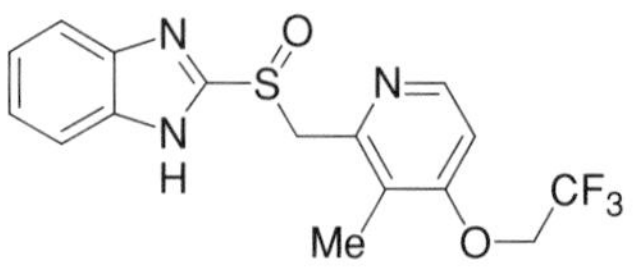

lansoprazole

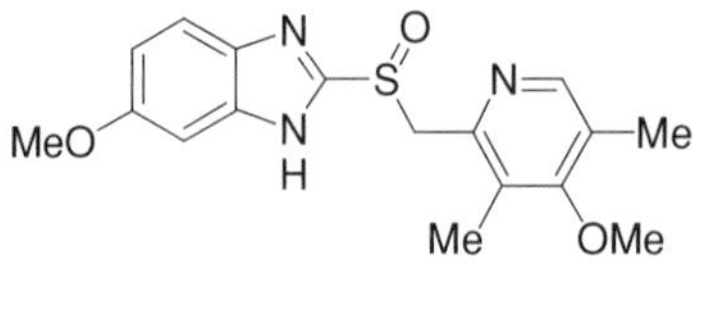

omeprazole

Benign Prostatic Hyperplasia and Urinary Urge Incontinence

Doxazosin (Cardura®/Pfizer), tamsulosin (Flomax®/Boehringer Ingelheim), and alfuzosin (Uroxatral®/Sanofi-Synthelabo) are used in the management of benign prostatic hyperplasia (BPH). The compounds are postsynaptic, α_1-adrenergic blocking agents that relax prostatic tissue and increase urinary outflow in men. Because tamsulosin demonstrates selectivity for the α_{1A}-adrenergic receptor subtype located in prostate over that of α_{1B}-subtype located in vascular tissue, there is a reduced incidence of cardiovascular side effects (hypotension, dizziness, and syncope). Because doxazosin is not selective for the α_{1A}-subtype, it is also useful in the treatment of hypertension. Finasteride (Proscar®/ Merck) and dutasteride (Avodart®/ GlaxoSmithKline) are 5α-reductase inhibitors that block the conversion of testosterone to 5α-dihydrotestosterone (DHT). Because DHT is an androgen responsible for prostatic growth, inhibition of the 5α-reductase enzyme is beneficial in reducing prostatic enlargement.

Erectile Dysfunction

The pyrazolopyrimidinone derivative sildenafil (Viagra® /Pfizer), the indolopyrazinone

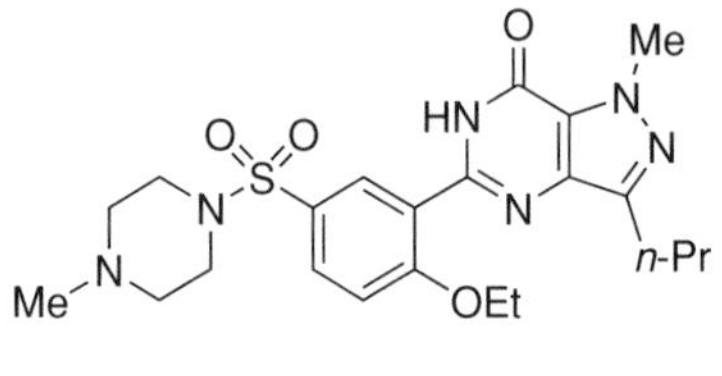

sildenafil

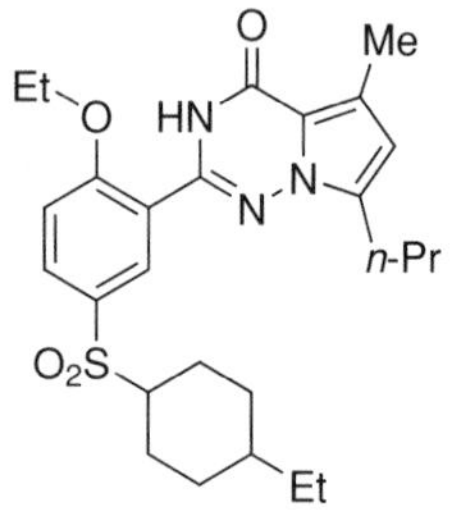

vardenafil

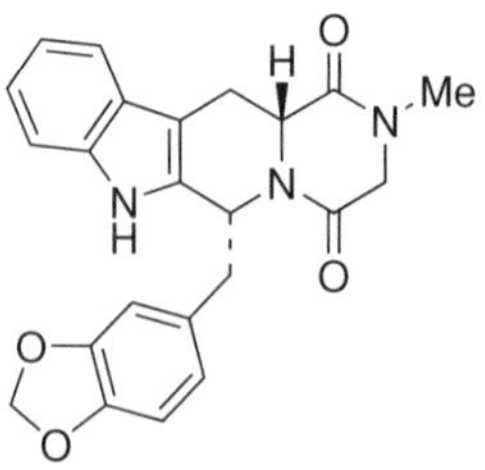

tadalafil

derivative tadalafil (Cialis®/Lilly ICOS), and the imidazotrizinone derivative vardenafil (Levitra®/Bayer) are selective inhibitors of the phosphodiesterase (PDE) type 5 enzyme. They act by selectively blocking the PDE type 5 isoenzyme ultimately causing vascular vasodilation in corpus cavernosal tissue which, in turn, leads to penile tumescence and rigidity.

PULMONARY AGENTS

Asthma and Allergic Rhinitis

Fluticasone (Flovent®/GlaxoSmithKline) is a synthetic corticosteroid derivative that is a selective agonist at the human glucocorticoid

blood–brain barrier, it is considered a "nonse-dating" antihistamine. More recently, the quaternary ammonium tricyle tiotropium bromide (Spiriva®/Boehringer-Ingelheim/Pfizer) has been introduced. It is a long-acting bronchodilator useful in the treatment of asthma and exerts its pharmacological effect through inhibition of the muscarinic M_3 receptor.

INFLAMMATION AND OSTEOPOROSIS

Arthritis

The diaryl pyrazole derivative celecoxib (Celebrex®/Pharmacia, Pfizer), the furanone derivative rofecoxib (Vioxx®/Merck), and the isoxazole derivative valdecoxib (Bextra®/

fluticasone   fexofenadine   tiotropium bromide

receptor and useful in the treatment of asthma. Although the precise mechanism of fluticasone in asthma is unknown, it is believed it's anti-inflammatory property contributes to its beneficial effect. The butyrophenone derivative fexofenadine (Allegra®/Aventis) is an antihistamine and used in the treatment of seasonal allergic rhinitis. Because fexofenadine does not readily cross the

Pfizer) are selective cyclooxygenase type 2 (COX-2) inhibitors and are useful in the treatment of arthritis. The compounds exert their pharmacological effect by selectively blocking the COX-2 enzyme to produce an anti-inflammatory effect without adverse gastrointestinal side effects. In addition, they also display analgesic and antipyretic activities in animal models.

celecoxib   rofecoxib   valdecoxib

Osteoporosis

The benzothiophene derivative raloxifene (Evista®/Lilly) is a selective estrogen receptor modulator (SERM). Raloxifene produces its biological actions via modulation (both activation and blockade) of estrogen receptors that ultimately results in decreased resorption of bone. The bisphosphonate derivative alendronate (Fosamax®/Merck), an inhibitor of osteoclast-mediated bone resorption, is also useful in the treatment of osteoporosis. Both raloxifene and alendronate are useful in the treatment of osteoporosis in postmenopausal women.

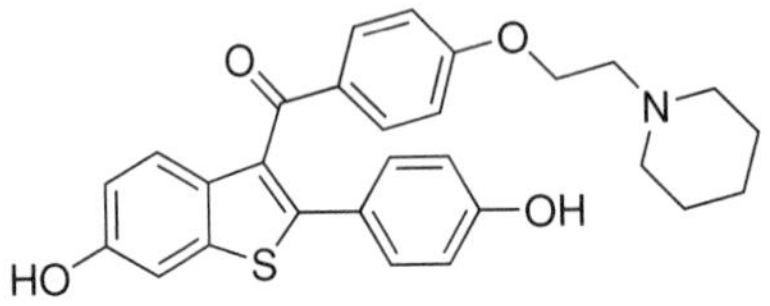

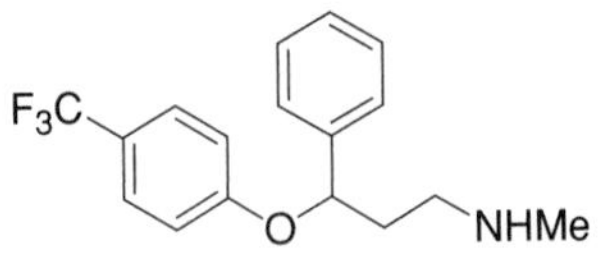

fluoxetine

sertraline

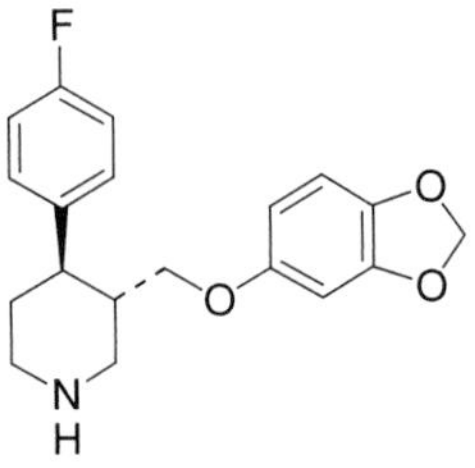

raloxifene

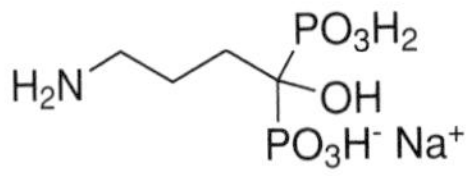

alendronate

paroxetine

CENTRAL NERVOUS SYSTEM AGENTS

Antidepressants

Fluoxetine (Prozac®/Lilly), paroxetine (Paxil®/GlaxoSmithKilne), and sertraline (Zoloft®/Pfizer) are selective serotonin reuptake inhibitors (SSRIs) and are useful in the treatment of depression. These agents potentiate the pharmacological actions of the neurotransmitter serotonin by preventing its reuptake at presynaptic neuronal membranes. In addition to its SSRI properties, venlafaxine (Effexor®/Wyeth-Ayerst) also appears to be a potent inhibitor of neuronal norepinephrine reuptake and a weak inhibitor of dopamine reuptake thereby enhancing the actions of these neurotransmitters as well. Venlafaxine is indicated for use in anxiety and depression.

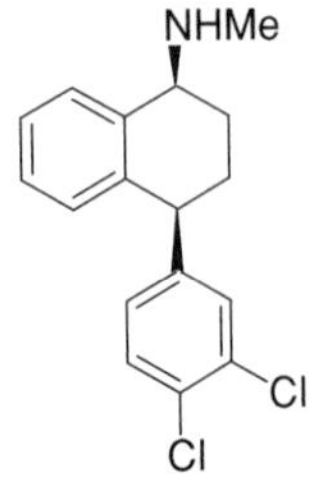

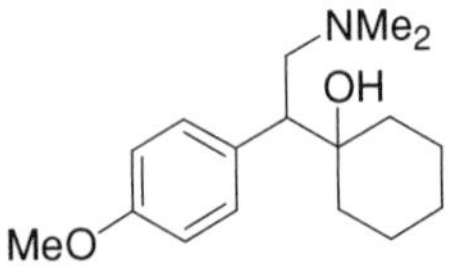

venlafaxine

Anxiolytics

Alprazolam (Xanax®/Pharmacia), a benzodiazepine derivative is used for the treatment of both anxiety and panic disorder and buspirone (Buspar®/Bristol-Myers Squibb) is indicated for the treatment of anxiety disorders. The mechanism of action of buspirone is distinct from that of the benzodiazepines and is believed to be mediated mainly through modulation of serotonergic neurotransmission via its interaction with the 5-HT$_{1A}$ serotonin receptor subtype.

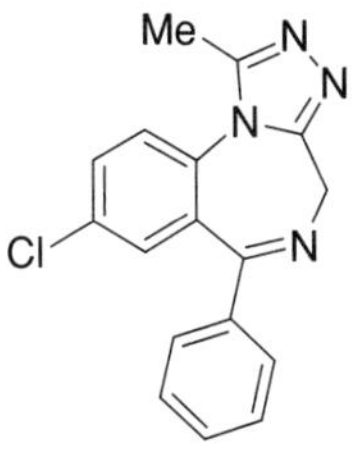

alprazolam

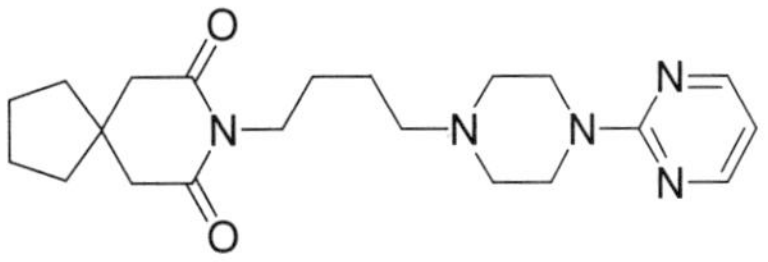

buspirone

Bipolar Disorders, Schizophrenia, and Epilepsy

The thienobenzodiazepine derivative olanzapine (Zyprexa®/Lilly), and benzisoxazole risperidone (Risperidal®/Janssen) are atypical antipsychotic agents. Olanzapine is used in the treatment of bipolar disorder and risperidone is useful in the management of schizophrenia. It is believed that both compounds exert their beneficial effects through antagonism of serotonergic and dopaminergic receptors. A newer agent for the treatment of schizophrenia is aripiprazole (Abilify®/ Bristol-Myers Squibb, Otsuka). It is believed the pharmacological effects are mediated through a combination of partial agonist activity at the dopamine D_2 and serotonin 5-HT_{1a} receptors and antagonism at the serotinergic 5-HT_2 receptor. The γ-aminobutyric acid derivative (GABA) gabapentin (Neurontin®/ Pfizer) is useful in the treatment

of epilepsy. Although structurally related to GABA, it has no GABA-ergic activity. The mechanism for its anticonvulsive actions is currently unknown.

Alzheimer's Disease

The indanone derivative donepezil (Aricept®/ Pfizer, Eisai) is an acetycholinesterase inhibitor and is structurally unrelated to other cholinesterase inhibitors. Because it increases the concentration of the neurotransmitter acetycholine at cholinergic sites, it is useful in the treatment of Alzheimer's disease (dementia). Another agent useful in the treatment of Alzheimer's disease is the adamantlyl amine derivative memantine (Namenda®/ Forest). Memantine is a N-methyl-D-aspartate (NMDA) receptor antagonist and is thought to exert its pharmacological effect by blocking the

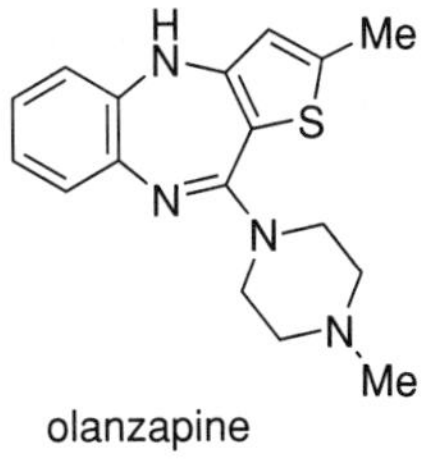

olanzapine

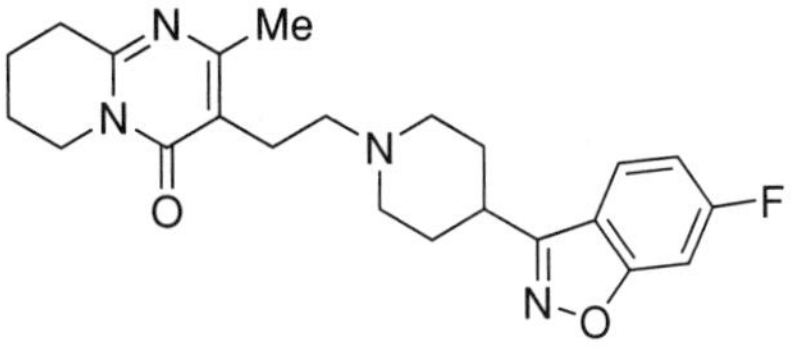

risperidone

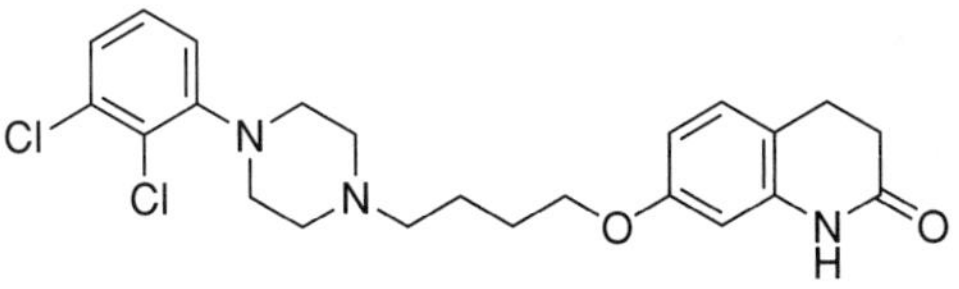

aripiprazole

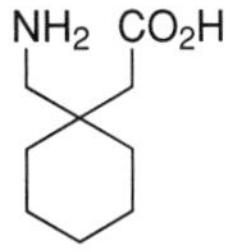

gabapentin

excitatory action of the amino acid gluta-mate on the receptor. Memantine has shown no evidence of preventing or slowing neu-rodegenaration in Alzheimer's patients.

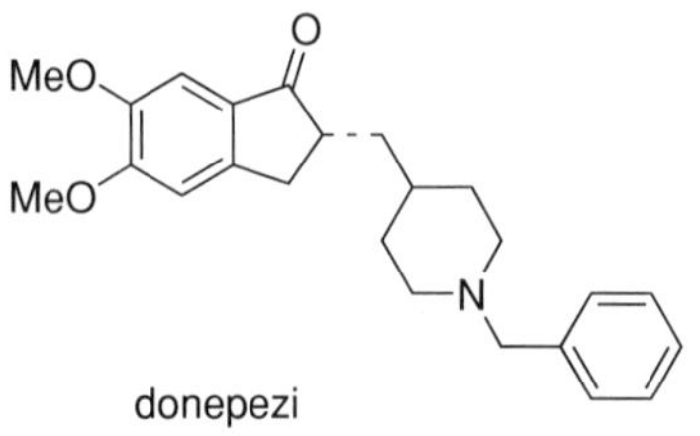

many bacterial infections. Amoxicillin/Clavulanate is one of the few approved drug mixtures and is a drug of choice for the treat-ment of otitis media. It is also an alternative

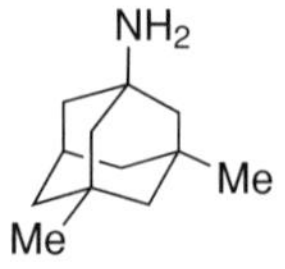

donepezi

memantine

INFECTIOUS DISEASES

Antibacterials

The primary driver for research in the anti-bacterial area over the past decade has been the emergence of resistant organisms. Important members of the ever-growing armamentarium of antibacterials include azithromycin (Zithromax®/Pfizer), linezolid (Zyvox®/Pharmacia), amoxicillin / clavulanate potassium (Augmentin®/GlaxoSmithKline), ciprofloxacin (Cipro®/Bayer) and daptomycin (Cubicin/Cubist). Azithromycin is a semisyn-thetic 9α-azalide analog of erythromycin pos-sessing improved resistance to acid-mediated degradation, increased activity against gram-negative organisms, and improved pharmaco-kinetics. It's indications include the treatment of mild to moderate upper and lower respira-tory tract infections and otitis media in pedi-atrics. Interestingly, azithromycin tends to concentrate in lung tissue which is the site of

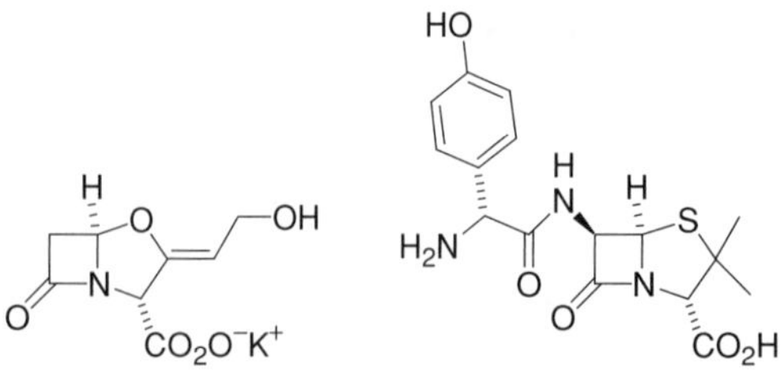

treatment for anthrax exposure in pediatrics. Ciprofloxacin is a totally synthetic antibacter-ial that acts as a DNA gyrase inhibitor. It is active against a broad range of pathogens including both gram-positive and gram-negative aerobic bacteria and is effective against urinary tract and lower respiratory tract infections. Linezolid is a totally syn-thetic oxazolidinone derivative which has a unique mechanism of action resulting in a low potential for cross resistance to other antibac-terials. Linezolid is indicated for the treat-ment of community acquired pneumonia, MRSA, and VRE infections and has the dis-tinctive characteristic of being 100% orally bioavailable. Daptomycin (Cubicin®/Cubist), a cyclic lipopeptide of molecular formula $C_{72}H_{101}N_{17}O_{26}$, is a bactericidal antibacterial agent used for the treatment of infections caused by gram-positive bacteria including those that are resistant to standard antibacter-ial regimens.

amoxicillin / clavulanic Acid

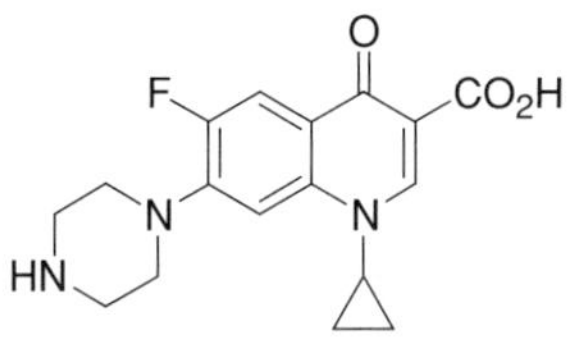

ciprofloxacin

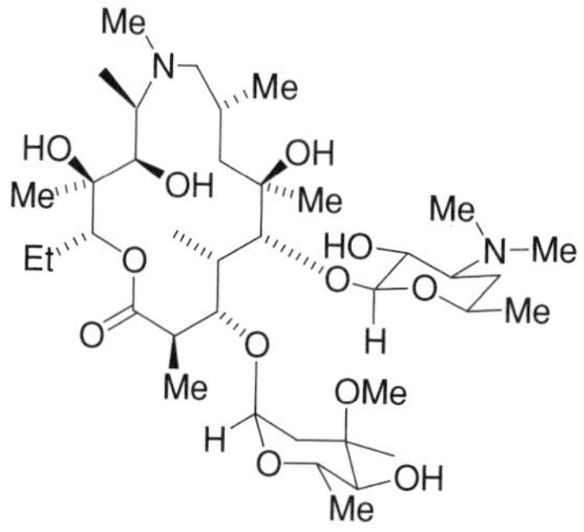

azithromycin

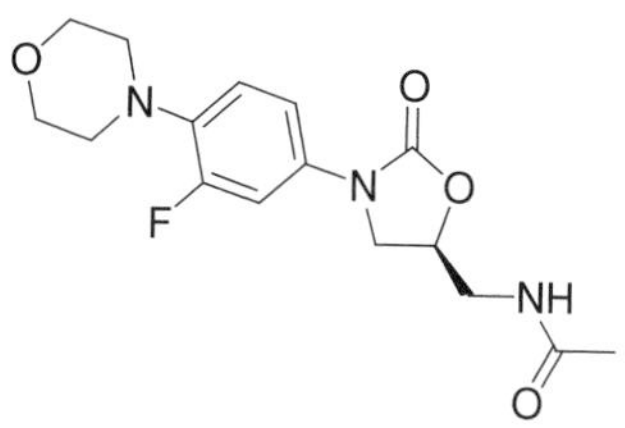

linezolid

Antifungals

The increasing immunocompromised patient population has exacerbated the need for effective antifungal agents to combat opportunistic fungal infections that arise in these patients. Fluconazole, an achiral triazole derivative, is indicated for the treatment of systemic candidiasis as well as meningitis caused by *Cryptococcus neoformans*. Itraconazole (Sporanox®/Janssen, Ortho Biotech), a mixture of four diastereomers, is used to treat aspergillosis, oral candidiasis, and histoplasmosis. These agents are structurally related to other imidazole-based antifungals such as ketoconazole and miconazole but have better antifungal activity and broader coverage.

HIV/AIDS, Hepatitis B and C, and RSV. Indinavir (Crixivan®/Merck) is one of a group of HIV protease inhibitors and is used in conjunction with other antiretroviral chemotherapeutic agents for the treatment of AIDS in adults and adolescents. It is a Phe-Pro scissile bond peptidomimetic with a hydroxyindane moiety that was optimized for selectivity and potency. More recently, atazanavir sulfate (Reyataz®/Bristol-Myers Squibb) was introduced as the latest protease inhibitor. Clinical data suggests that atazanavir may have a more favorable hypertriglyceridemia profile as compared to other protease inhibitors. Ribavirin (Rebetron®/Schering Plough and Virazole®/ICN) is a synthetic nucleoside

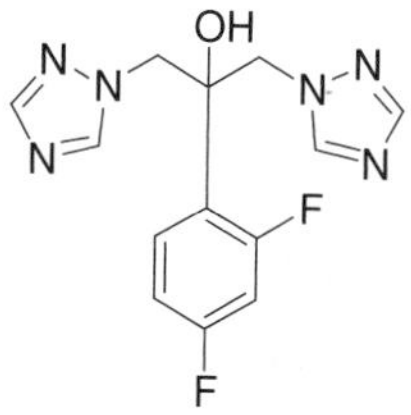

fluconazole

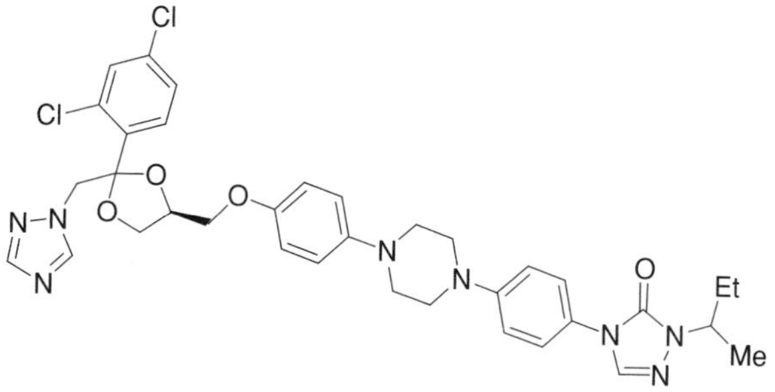

itraconazole

Antivirals

Antiviral research has become a major focus in the pharmaceutical industry over the past decade as evidenced by the marketing of a plethora of antiviral agents active against

used to treat respiratory syncytial virus (RSV) in hospitalized infants and is also used in combination therapy with interferon for the treatment of chronic hepatitis C. Efavirenz (Sustiva®/Bristol-Myers Squibb) is a synthetic nonnucleoside reverse transcriptase inhibitor (NNRTI) used in conjunction with other antiretroviral agents for the treatment of HIV.

Tamoxifen (Nolvadex®/AstraZeneca), a nonsteroidal antiestrogen chemotherapeutic possessing both agonistic and antagonistic properties, is used for the treatment and preven-

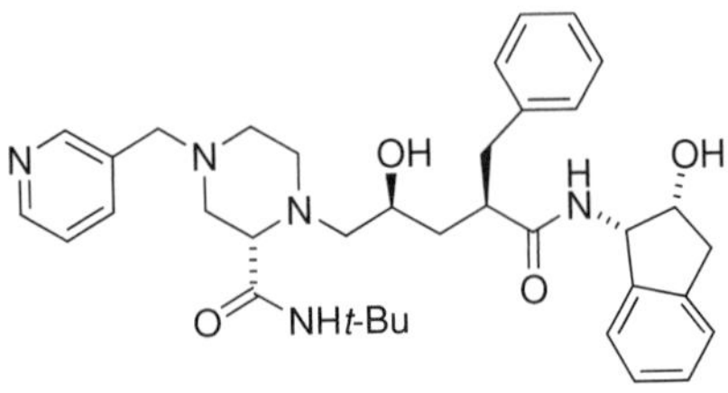

indinavir

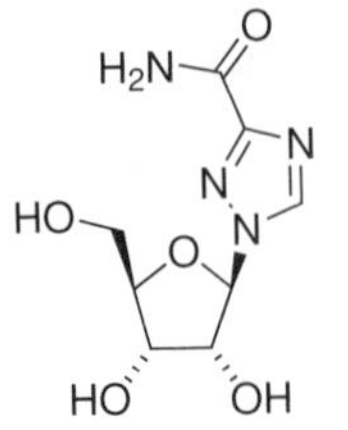

ribavirin

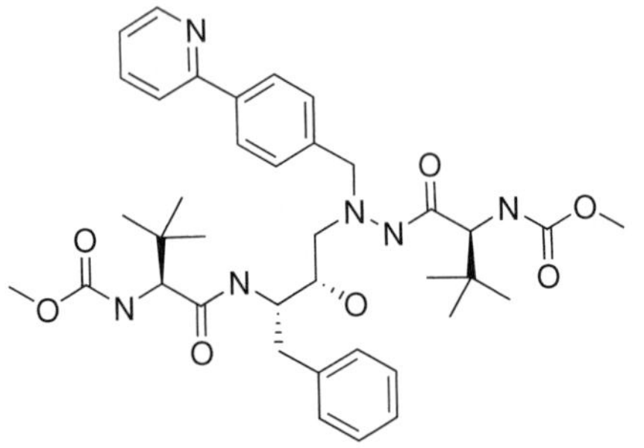

atazanavir

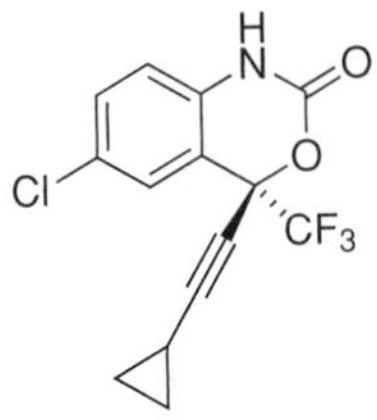

efavirenz

ANTINEOPLASTICS

Paclitaxel (Taxol®/Bristol-Myers Squibb) and irinotecan (Camptosar®/Pharmacia) were discovered as a result of natural product extract screening done at the NIH in the late 1960s by Monroe Wall and Mankush Wani. Paclitaxel is a naturally occurring diterpene that exerts its antineoplastic effect via stabilization of the mitotic spindle during cell replication. It is used for the treatment of nonsmall cell lung, breast, ovarian, and esophageal carcinomas as well as Kaposi's sarcoma. Irinotecan is a prodrug that upon release of the piperidinylpiperidine carbamate moiety reveals the pharmacologically active parent SN-38 which is itself a derivative of the naturally occurring camptothecin. Irinotecan exerts its antineoplastic activity via the inhibition of Type I DNA topoisomerase and stabilization of the transiently formed Topoisomerase I/DNA cleavable complex.

tion of breast cancer. Imitanib (Gleevac®/Novartis), an inhibitor of Bcr-Abl tyrosine kinase recently received FDA approval for the treatment of chronic myelogenous leukemia. Bortezomib (Velcade®/Millenium and Ortho Biotech), an iv ubiquitin proteosome inhibitor, is used for the treatment of multiple myeloma in pateints who have been refractory to other chemotherapeutic regimens. Cetuximab (Erbitux®/ImClone, Merck KGaA, and Bristol-Myers Squibb) a human–murine chimeric monoclonal antibody that blocks the epidermal growth factor receptor (EGFR), was developed for the treatment of irinotecan-refractory colorectal cancer. This agent is also used in patients who are irtolerant of irinotecan-based therapy. The small molecule EGFR tyrosine kinase inhibitor gefitinib (Iressa®/AstraZeneca) is used to treat nonsmall cell lung cancer.

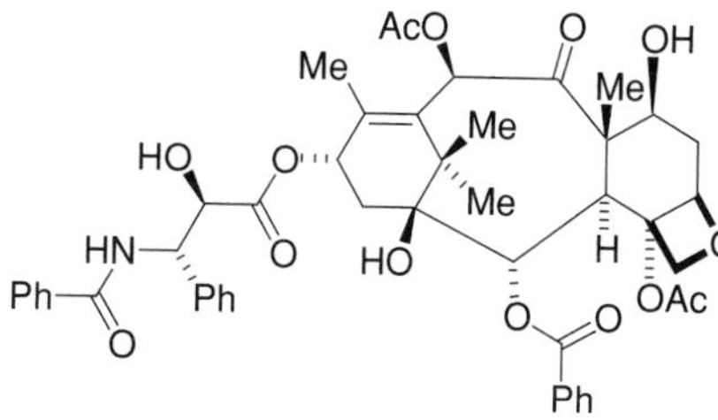

paclitaxel

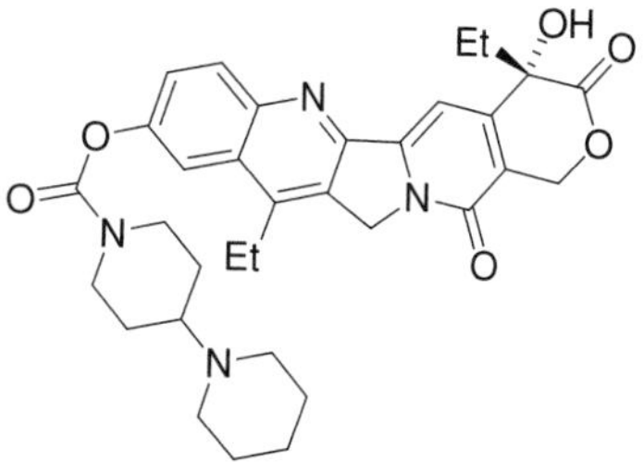

irinotecan

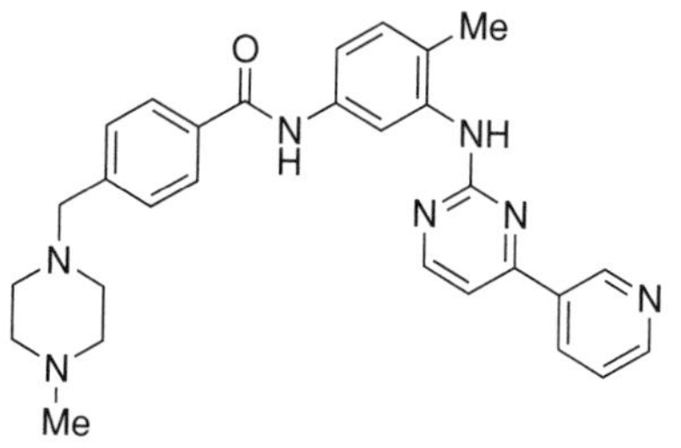

imatinib

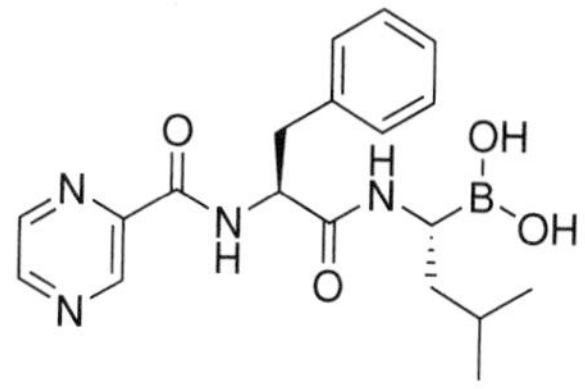

bortezomib

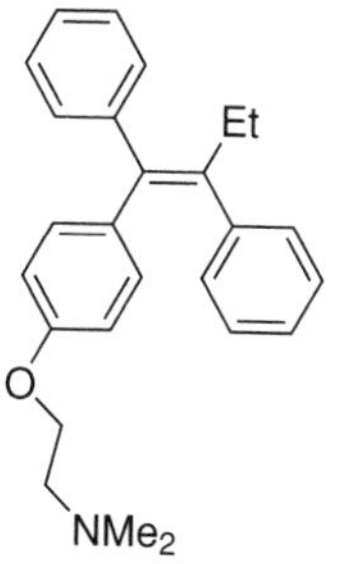

tamoxifen

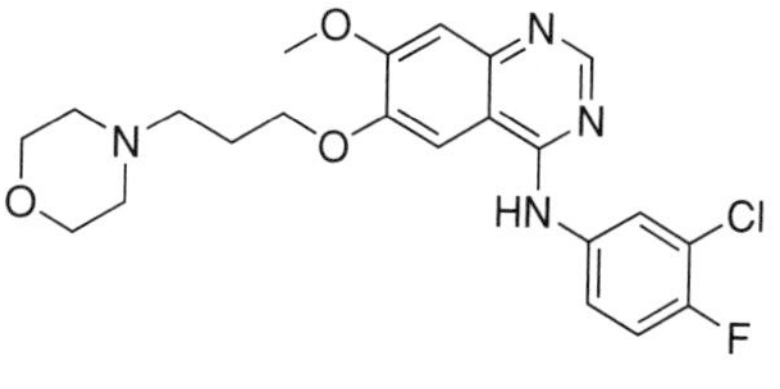

gefitinib

MISCELLANEOUS AGENTS

Glaucoma and Nausea

Latanoprost (Xalatan$/Pharmacia & Upjohn) is a topical, ocular hypotensive agent used to treat glaucoma. The compound is a synthetic analogue of the naturally occurring prostaglandin PGF2α and is thought to reduce intraocular pressure by increased outflow of the aqueous humor. Odansetron (Zofran®/GlaxoSmithKline) is a selective, serotonergic, 5-HT$_3$ receptor antagonist and is used to ameliorate nausea and vomiting associated with chemotherapy-induced emesis.

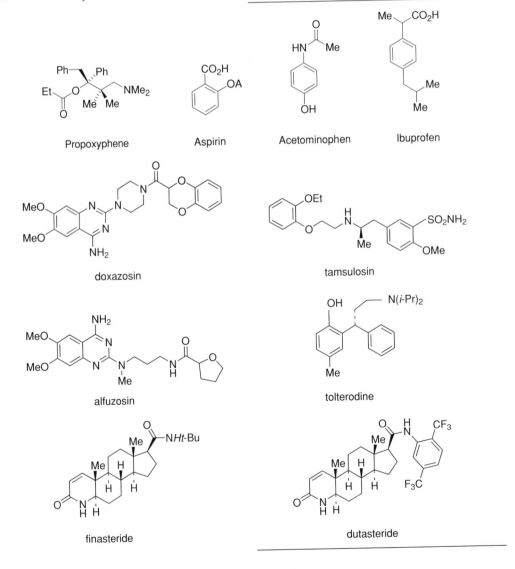

latanoprost

odansetron

Analgesics

Propoxyphene, along with aspirin, acetominophen, and ibuprofen are among the most widely used agents for the treatment of mild to moderate pain.

SMALL MOLECULE HIGH THROUGHPUT SYNTHESIS

The field of nonoligomeric, small molecule high throughput synthesis came into existence

Propoxyphene

Aspirin

Acetominophen

Ibuprofen

doxazosin

tamsulosin

alfuzosin

tolterodine

finasteride

dutasteride

in 1992. Since then, small molecule library synthesis has affected drug discovery efforts in lead identification, as well as lead optimization. In a recent review, R. Dolle has categorized synthetic libraries as follows: (1) discovery libraries: libraries synthesized with no preconceived notion about which molecular target it may be active against. These libraries tend to be large in size, typically >5000 compounds.(2) targeted libraries: these libraries are biased in their design and contain a pharmacophore known to interact with a specific target, or a family of targets.(3) optimization libraries: libraries are constructed around an existing lead with the intent to improve potency, selectivity, pharmacokinetic profile, etc. These libraries tend to be smaller in size, usually ranging from tens up to a few hundred compounds.

Discovery Libraries

Researchers have employed several different strategies to create populations of molecules that are used for broad-based screening. One strategy is to synthesize libraries of "privileged pharmacophores" such as benzodiazepines (1), triazines (2), and so on. A second strategy is to design scaffolds or templates for library synthesis that are based on important molecular recognition. Libraries of β-turn mimetics (3) synthesized by Ellman et al. are examples of templates for molecular recognition.

gies utilize resin-based split-pool synthesis to prepare large arrays of compounds. Libraries of >50K members were prepared by using chemically encoded beads. Chemical encryption, in the form of unique chemical markers (tags), is associated with synthetic identity of the library member tethered to the resin bead. The technology for chemical encoding was pioneered by W. Clark Still and subsequently commercialized by Pharmacopeia, Inc. Restricted amount (200–300 μg), lack of analytical characterization of library members, and the requirement of a specialized screening format for chemically encoded libraries have limited the utility of this technology. Radio-frequency encoded synthesis, developed and commercialized by IRORI, Inc. overcomes the afore mentioned limitations while retaining the efficiency of split-pool synthesis. Libraries of 10–15K members can be prepared, with individual members quantitated and characterized by LC/MS. Most pharmaceutical companies have utilized Rf-encoded synthesis in their lead identification efforts.

Targeted Libraries

Libraries targeted towards proteolytic enzymes, nonproteolytic enzymes, G-protein coupled receptors (GPCRs) and ion-channels have been very successful in lead identification. Libraries of hydroxamates (4), hydroxy ethylenes (5), boronic acids (6) and αketo sulfon-

1

2

3

The discovery of chemical encoding technologies and radio-frequency (Rf) encoded synthesis have had a major impact on synthesis of lead discovery libraries. Both technolo-

amides (7) have been prepared as inhibitors of metallo-, aspartyl, serine and cysteine proteases respectively, using either solid phase or solution phase synthesis.

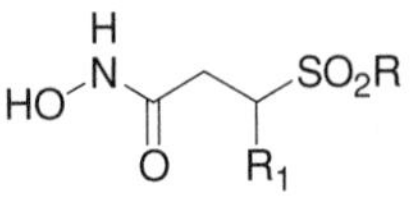

4

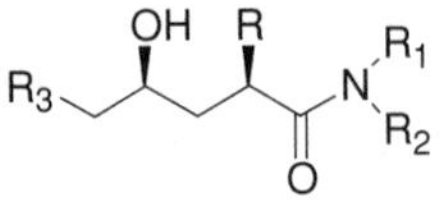

5

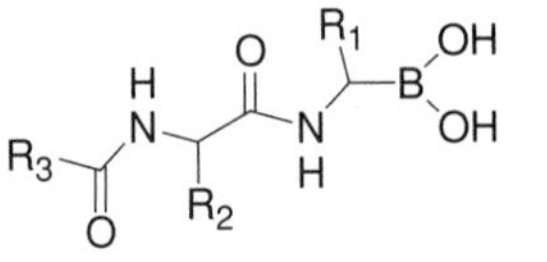

6

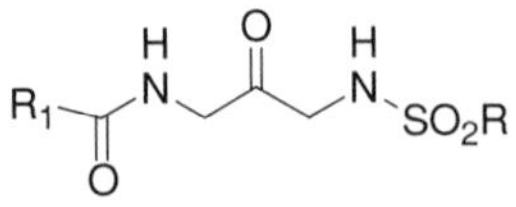

7

Structure-based design has been effectively utilized in synthesis of inhibitors of non-proteolytic enzymes. Inhibitors of MurB, an essential bacterial enzyme required for biosynthesis of peptidoglycan, were identified using the X-ray structure of the enzyme for library design. Thiazolidinone inhibitors (8) thus identified are the first examples of small molecule inhibitors of MurB.

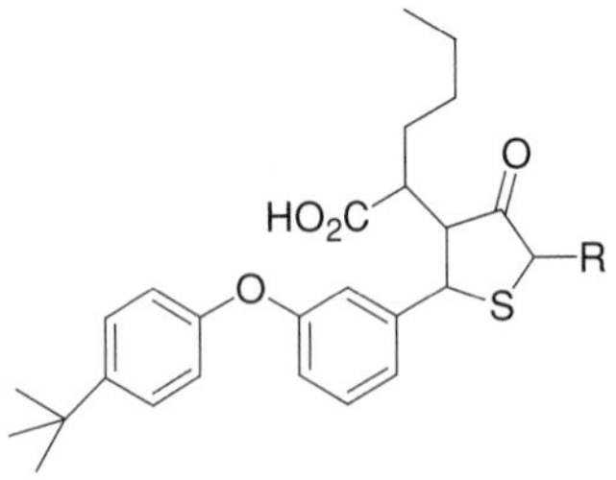

8

Substituted indoles ($5HT_{2a}$; D4 and α_{2a} receptor antagonists) and piperazines (δ opiod antagonists) are representative chemotypes targeted towards GPCRs.

Advances made in solid phase extraction (SPE) and in development of resin-based scavengers have increased the versatility of chemistries implemented for synthesis of

targeted libraries. A combination of cation exchange (SCX) and anion exchange (SAX) resin was effectively utilized to prepare libraries of highly substituted amides.

Scavenger resins and polymer-bound reagents are routinely used to prepare medium-sized (500–1000 member) libraries. Polymer-bound isocyanates (9) and aldehydes (10) are used to remove amines from reaction mixtures, while polymer-bound thiols (11) are used to scavenge halides.

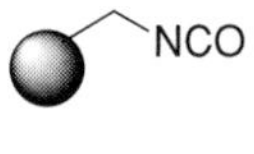

9

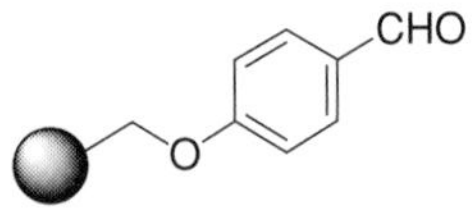

10

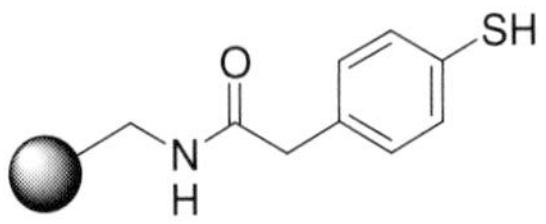

11

Optimization Libraries

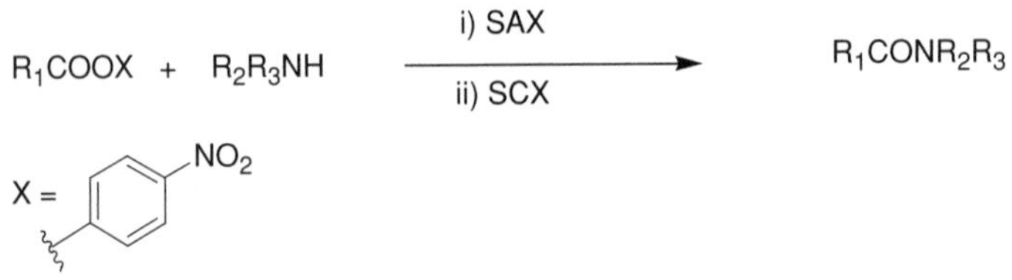

Starting with a lead structure, researchers have demonstrated that parallel synthesis can be effectively utilized to optimize activity, as well as reduce timelines for optimization. Parallel synthesis strategy was implemented to identify more potent analogs of influenza hemagglutinin inhibitor (12) (IC$_{50}$ = 4 μg/ml). Solid phase extraction was used to automate preparation of >400 analogs resulting in identification of compounds (13) (IC$_{50}$ = 20 ng/ml) and (14) (IC$_{50}$ = 20 ng/ml).

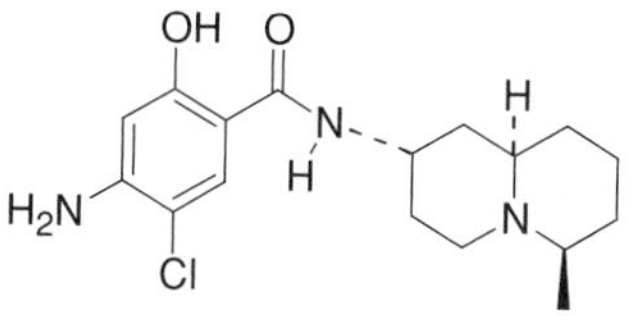

CHEMICAL PROCESS R&D IN THE PHARMACEUTICAL INDUSTRY

Most of the active pharmaceutical ingredients (APIs) of commercially available pharmaceuticals are manufactured either by chemical syntheses or microbial fermentations. However, some of the active ingredients are directly obtained from natural sources. This section addresses the development and manufacture

12

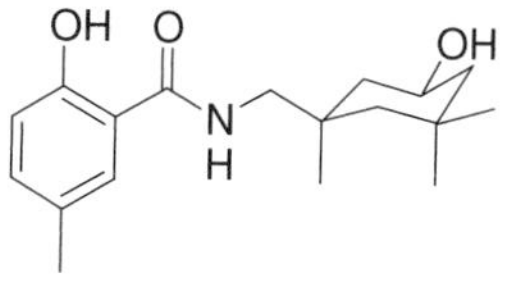

13

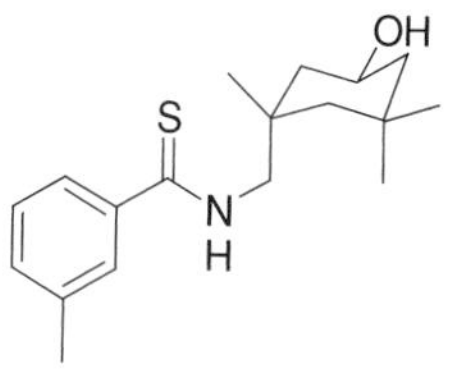

14

During the past 10 years, the pharmaceutical industry has expended significant resource in developing and assimilating technologies to increase synthesis throughput and decrease preclinical time lines. There are numerous examples in the literature demonstrating effective use of high throughput synthesis for lead discovery and optimization. There are two publicly known examples of clinical candidates that have emerged directly from optimization libraries. Ontogen Corporation identified OC144-093, (15) (IC$_{50}$ = 50 nM) as a P-glycoprotein modulator and Agouron Pharmaceuticals reported identification of AG-7088, (16) (k$_{obs}$/I = 1,470,000 M^{-1} S^{-1}), a clinical candidate for treatment of rhinovirus infection.

of APIs. Recent trend shows that >75 percent of the drug candidates in development are chiral and of complex structure. Incessant demand to shorten the timelines for the discovery, development and launch of NCEs coupled with environmental concerns has necessitated the development of higher yielding, more robust and environmentally friendly processes in shorter times. The success of a pharmaceutical company greatly depends not only on discovering blockbuster NCEs but also on its ability to design, optimize and scaleup a chemical process to commercial manufacturing with increasing rapidity. The chemical manufacturing process must be a robust procedure capable of operating routinely in a manufacturing environment.

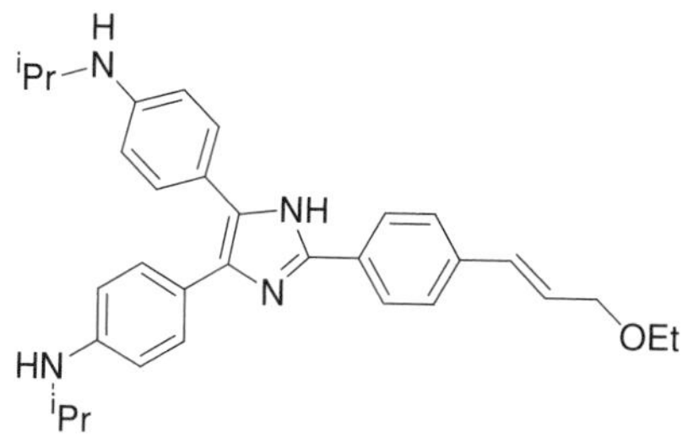

15

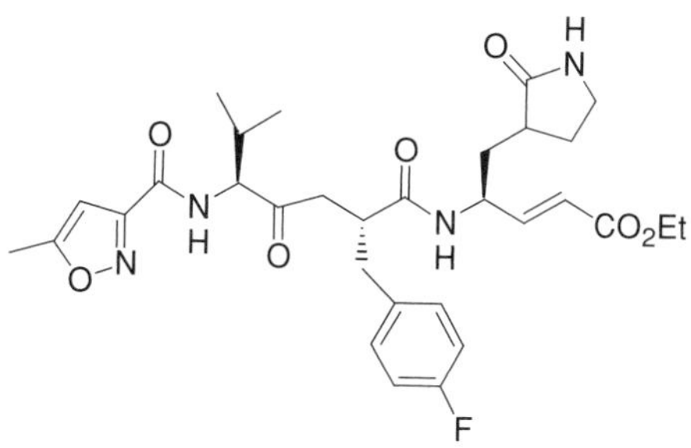

16

Considerable attention has to be given to various parameters in developing a manufacturing process for an API, including for example: efficiency of the synthesis, availability and cost of starting materials, toxicity of the reagents, stability and toxicity profiles of intermediates, formation of byproducts, and safe disposal of waste materials. Data from various aspects of chemical process development, including process structure and flowsheet, operational guidelines, optimization, process management, process control, fault diagnostics and equipment management need to be in place in order to support a smooth transition from laboratory to manufacturing plant. Safety is another critical factor requiring consideration for large-scale manufacture. All reactions should undergo a process hazard analysis for incident-free and successful plant implementation before scale-up. The use of automation in accelerating the design of cost-effective and well-understood synthetic processes has been demonstrated over the past few years by pharmaceutical companies and a few research groups in academia and is now beginning to grow very rapidly. Automation concepts and tools such as statistical design of experiments and parallel experimentation using in-house built reactor blocks or commercially available systems such as Zymark robots, ReactArray, Bohdan, Argonaut's Surveyor, or Mettler Toledo MultiMax will play a major role in increasing the productivity of process R&D with respect to speed and economics, as well as obtaining process knowledge. The application of microreaction technology (micropiloting) is another area that is growing rapidly to understand the chemical engineering aspects of process development. Some beneficial features of microreaction technology include mixing efficiency, enhanced heat transfer, and more uniform residence time distributions.

Production and logistical processes are becoming more complex due to an increasing number of products and smaller batch sizes. To manage this, supply chain optimization and production planning activities need to be addressed. Production simulation can be used for performance measurements and capacity assessments of manufacturing as well as material and information flow processes. Some applications of production simulation include bottleneck analysis, examination of process alternatives, assessment of investment decisions and solution of sequencing problems. Batch process development is a fairly complex series of engineering tasks. In the pharmaceutical industry, the production of a majority of APIs is based on a batch concept. This concept offers many advantages with respect to quality assurance as an individual batch can be accepted or rejected. However, the scale-up of the batch size without proper controls may lead to problems. The variety of the equipment involved often does not facilitate the scale-up process. In order to avoid scale-up problems, continuous or semi-continuous processes need to be adopted as alternatives to a batch production.

Crystallization, filtration, drying and milling (if required) are other important factors that need to be defined well before a process to manufacture solid APIs is finalized. Physicochemical properties of APIs play a vital role in providing the pharmaceutical drug products with desired bioavailability, manufacturing properties, and good final product quality. Particle size, density, flowability, polymorphism, hygroscopicity, and stability are critical properties for solid APIs in the formulation development. Polymorphism is very important in determining the physical properties of various crystal forms of a drug for optimal chemical and formulation processing, as well as for satisfying regulatory and patent issues for producing consistent solid forms of a drug.

The following flow diagrams show the preparation of APIs of some widely used pharmaceutical drugs in today's market. Scheme-1 shows the preparation of sildenafil. This route has a greater synthetic convergency than other published routes.

Scheme-1

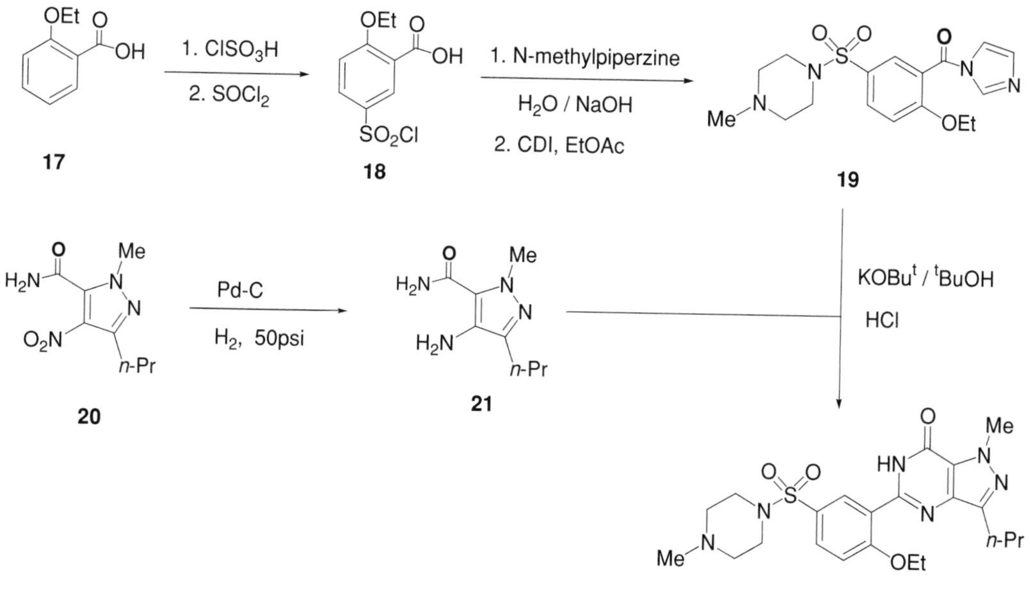

sildenafil

Synthesis of fluoxetine as a racemic mixture is shown in scheme-2. Recently several patents and publications have appeared in the literature describing the synthesis of (S)- and (R) enantiomers.

The single enantiomer of indinavir has five stereogenic centers, four of which are derived either directly or indirectly from epoxide (27). Synthesis of indinavir sulfate developed by Merck is shown in Scheme-3.

Scheme-2

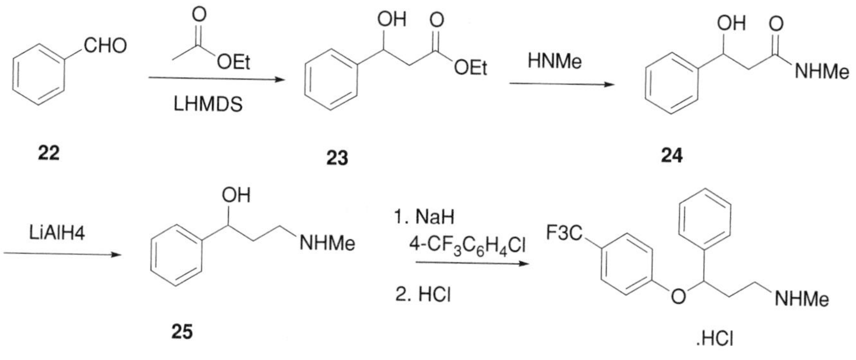

fluoxetine

Scheme-3

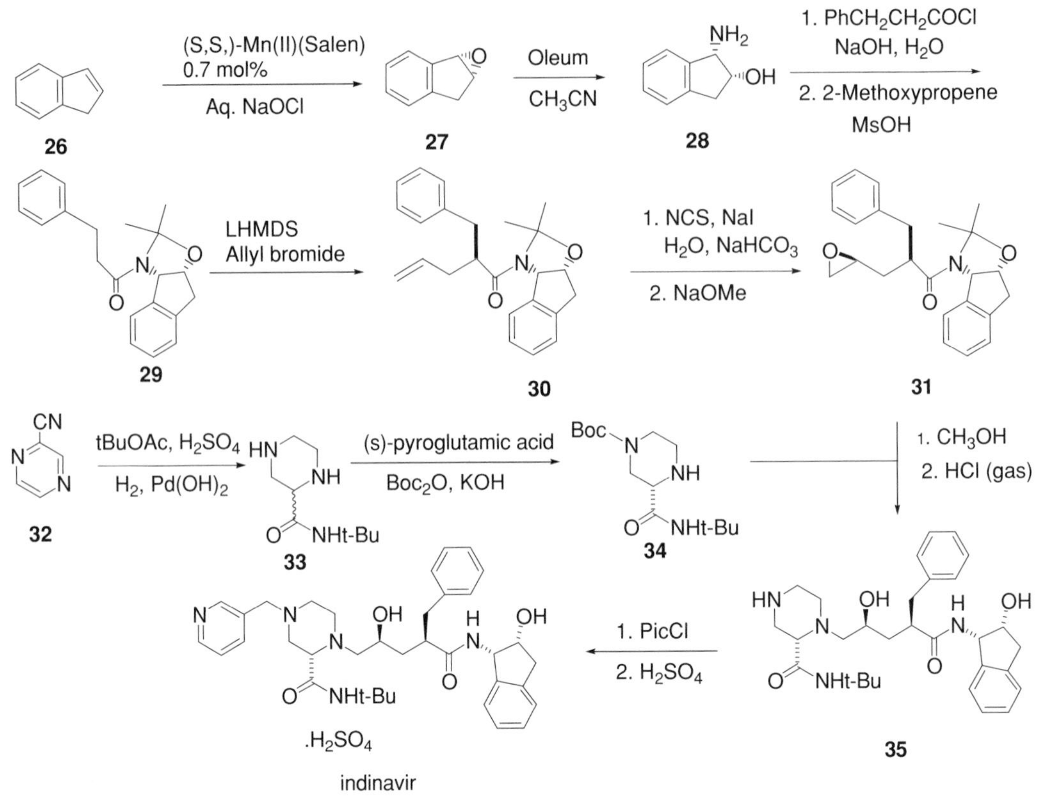

indinavir

CONCLUSION

The discovery and development of novel therapeutic agents by the pharmaceutical industry has afforded physicians an extensive armamentarium to fight a wide range of human disease. Of course there remains the opportunity for even more effective drugs with greater benefit-to-risk ratios than those currently available. The elucidation of the human genome will eventually lead to the identification of many new macromolecular targets for drug intervention. Chemistry has been and will likely continue to remain at the forefront of pharmaceutical research which will afford the drugs of the future.

REFERENCES

1. Krogsgaard-Larsen, P., Liljefors, T., and Madsen, U. (Eds.), *A Textbook of Drug Design and Development,* 2nd. ed., Harwood Academic Publishers, Amsterdam, (1996).
2. Spilker, B., *Multinational Drug Companies; Issues in Drug Discovery and Development,* Raven Press, New York, (1989).
3. Wermuth, C.G. (Ed.), *The Practice of Medicinal Chemistry,* Academic Press, San Diego, (1996).
4. Lipinski, C.A., Lombardo, F., Dominy, D.W., and Feeny, P. J., *Adv. Drug Delivery Rev.,* **23**, 3–25 (1997).
5. Testa, B., and Mayer, J. M., *Drug Metab Rev.,* **30**, 787–807 (1998).
6. Wess, G., Urmann, M., and Sickenberger, B., *Angew. Chem. Int. Ed.,* **40**, 3341–3350 (2001).
7. Miertus, S., and Fassina, G. (Eds.), *Combinatorial Chemistry and Technology; Principles, Methods and Applications,* Marcel Dekker, Inc., New York, (1999).
8. Dolle, R., *J. Combinatorial Chem.,* **3**, 477–518 (2001).
9. Anderson, N.G., *Practical Process Research and Development,* Academic Press, New York, (2000).

12

Manufactured Textile Fibers

Bhupender S. Gupta*†

TEXTILE BACKGROUND

The first conversion of naturally occurring fibers into threads strong enough to be looped into snares, knit to form nets, or woven into fabrics is lost in prehistory. Unlike stone weapons, such threads, cords, and fabrics—being organic in nature—have in most part disappeared, although in some dry caves traces remain. There is ample evidence to indicate that spindles used to assist in the twisting of fibers together had been developed long before the dawn of recorded history. In that spinning process, fibers such as wool were drawn out of a loose mass, perhaps held in a distaff, and made parallel by human fingers. (A maidservant so spins in Giotto's *The Annunciation to Anne*, ca. A.D. 1306, Arena Chapel, Padua, Italy.[1]) A rod (spindle), hooked to the lengthening thread, was rotated

so that the fibers while so held were twisted together to form additional thread. The finished length then was wound by hand around the spindle, which, in becoming the core on which the finished product was accumulated, served the dual role of twisting and storing, and, in so doing, established a principle still in use today. (Even now, a "spindle" is 14,400 yards of coarse linen thread.) Thus, the formation of any threadlike structure became known as spinning, and it followed that a spider spins a web, a silkworm spins a cocoon, and manufactured fibers are spun by extrusion, although no rotation is involved.

It is not surprising that words from this ancient craft still carry specialized meanings within the textile industry and have entered everyday parlance, quite often with very different meanings. Explanations are in order for some of the words used in the following pages. For example, as already indicated, "spinning" describes either the twisting of a bundle of essentially parallel short pieces of wool, cotton, or precut manufactured fibers into thread or the extrusion of continuous long lengths of manufactured fibers. In the former case, the short lengths are known as "staple" fibers, and

*College of Textiles, North Carolina State University.
†The author dedicates this chapter to the memory of late Dr. Robert W. Work, Professor Emeritus, a longtime friend and mentor.

The author gratefully acknowledges the assistance he has received from associates both from within the College of Textiles and from outside, including several fiber producing companies, in preparing this chapter.

Riegel's Handbook of Industrial Chemistry, 10th Edition
Edited by Kent. Kluwer Academic/Plenum Publishers. New York 2003

TABLE 12.1 Typical (Average) Values of Tensile and Physical Properties of Some Textile Fibers

Fiber	Breaking Stress (cN/tex)	Strain to Fail (%)	Moisture Regain (%)	Density (g/cc)
Natural				
Cotton	40	7	8	1.52
Flax	54	3	12	1.52
Silk	38	23	11	1.33
Wool	14	40	14	1.30
Regenerated				
Acetate	13	25	6	1.32
Rayon	25	20	14	1.51
Synthetic				
Acrylic	26	25	1.5	1.18
Modacrylic	28	32	1.5	1.32
Nylon	50	25	4	1.14
Polyester	50	15	0.4	1.38
Polypropylene	53	17	0	0.92

the resulting product is a "spun yarn," whereas the long lengths are called "continuous filament yarn," or merely "filament yarn." Neither is called a "thread," for in the textile industry that term is reserved for sewing thread and rubber or metallic threads. Although to the layperson "yarn" connotes a material used in hand knitting, the term will be used in the textile sense hereinafter.

Before manufactured fibers are discussed, it is necessary to define some terms.* The "denier" of a fiber or a yarn defines its linear density, that is, the mass in grams of a 9000 m length of the material at standard conditions of 70°F and 65 percent relative humidity. Although denier is actually a measure of linear density, in the textile industry, the word connotes the size of the filament or yarn. Fibers usually range from 1 to 15 denier, yarns from 15 to 1650. Single fibers, usually 15 denier or larger, used singly, are termed "monofils." The cross-sectional area of fibers of identical deniers will be inversely related to their densities, which range from 0.92 g/cc for polypropylene to 2.54 g/cc for glass. The approximate densities of some of the other commonly used

fibers are given in Table 12.1. Because by definition denier is measured at standard conditions, it describes the amount of "bone-dry" material plus the moisture regain, which ranges from zero for glass and polypropylene to 14 percent for rayon. It should be mentioned that some years ago scientific organizations throughout the world accepted the word "tex," this being the mass (g) of 1 km of the material, as a more useful term than denier. "Tex" is an accepted adjunct to the SI, or International System of Units, but it has received only limited acceptance in commerce, whereas the SI units are being employed increasingly in scientific organizations. Furthermore, the sizes of cotton, wool, and worsted yarns, and yarns containing manufactured fibers but produced by the traditional cotton, wool, or worsted systems, still are expressed in the inverse-count system that has been used for centuries.

The "breaking tenacity" or more commonly, "tenacity," is the breaking strength of a fiber or a yarn expressed in force per unit denier, that is, in grams per denier, calculated from the denier of the original unstretched specimen. "Breaking length" expresses the theoretical length of yarn that would break under its own weight, and is used mostly in Europe. "Elongation" means "breaking elongation" and is expressed in units of increase in length

*Each year the ASTM publishes in its *Book of Standards*, the most recent and accepted definition and test methods used in the textile and fiber industries.

to break calculated as a percentage of the original specimen length.

Typical force–elongation curves of some manufactured and natural staple fibers and textile-type manufactured filaments are shown in Figs 12.1 and 12.2. Table 12.1 gives the values of some of the physical and tensile properties of textile fibers.

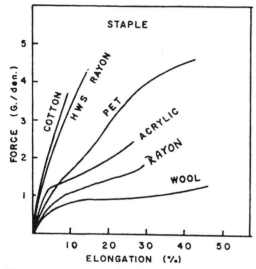

Fig. 12.1. Force–elongation curves of natural and manufactured staple fibers at standard conditions of 70°F and 65 percent humidity.

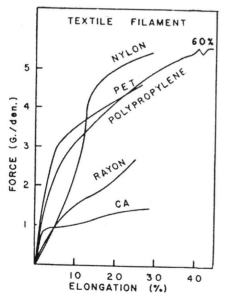

Fig. 12.2. Force–elongation curves of manufactured textile continuous-filament yarns at standard conditions of 70°F and 65 percent relative humidity.

HISTORY

Early humans, over time, became aware of the presence and usefulness of fibrous materials available from such sources as seeds, leaves, stems, animal coverings, and cocoons. They learned to spin, weave, knit, felt, or braid these fibers to protect their bodies and improve their lifestyle. A few hundred years ago, it has been suggested, someone took a clue from a busy worm and thought that it should be possible for humans to make a silk-like fiber that would be of commercial value. Curiosity combined with simple experiments strengthened that premise and much later led to the development of viscose rayon, only partially a manufactured fiber, in the 19th century. Further curiosity, war-time need, and superior commitment by modern researchers led to the synthesis of a totally synthetic fiber in the 1930s. The knowledge gained and the spark thus ignited resulted in the development of other fibers to such an extent that manufactured fibers now dominate the market in the industrialized nations in all major categories of apparel, home furnishing, and industrial end uses.

The story of the development of manufactured fibers is of great historical interest, beginning in 1664 when Robert Hooke, an Englishman, suggested that it should be possible to make a fiber much like silk that could be of value in the market place. Andemars, a Swiss chemist, received the first patent for making silklike fiber in 1855. He drew fibers by dipping a needle and pulling it out from a solution of cellulose nitrate containing some rubber.

The credit for using a spinnerette and forcing a solution through it for producing a fiber, however, goes to the English scientist Sir Joseph W. Swan, in the early 1880s. The first person to put the idea into commercial practice was the French chemist Count Hilaire de Chardonnet, who built the first plant to commercially produce a fiber based on regenerated cellulose, called "artificial silk," at Besançon, France, in 1891. During the last years of the 19th century and the beginning of the 20th century, progress was so rapid that the production of this fiber increased from several thousand pounds in 1891 to over two million in 1910. Commercial production of the fiber in the United States began in 1910 with the opening of the first plant. In 1924, the industry gave the "artificial silk" fiber a new name, rayon.

By the year, 1910, the brothers Camille and Henry Dreyfus had discovered a practical method for producing cellulose acetate polymer and were making plastic film and toilet articles in Basel, Switzerland. During World War I, they built a plant in England to produce acetate dope for painting airplane wings to render them air-impervious. The success of the product led the U.S. government to invite the Dreyfus brothers to build a plant in the United States, which started commercial production in 1924.

The successful manufacture of these two fibers, although based on fibrous materials available in nature, marked the beginning of the development of manufactured fibers in the 1930s. This effort, initiated by a technological breakthrough, was marked by the work of W. H. Carothers, aimed at learning how and why certain molecules joined to form large molecules, or polymers.[2] Fibers were described as being composed of high molecular weight linear polymers; and the first one to be manufactured, nylon 66, was synthesized and produced on a commercial scale in 1939. It was quickly followed by nylon 6, the second most widely used nylon, and modacrylic (1949), olefin (1949), acrylic (1950), polyester (1953), and triacetate (1954). Glass had joined this group of large-production items earlier in 1936.

Several other fibrous materials have been produced, but they are regarded as fibers with special performance characteristics, used either in limited textile or in specialized industrial applications. Some of these worth noting are Spandex (1959), Aramid (1961), polybenzimidazole (PBI) (1983), and Sulfar (1983).

Thus, the period from the 1930s to the 1960s can be considered as a time of discoveries and innovations for manufactured fibers, when the majority of the basic fibers were developed. The years since then may be thought of as a period of modification of

performance characteristics. The basic generic materials have been manipulated both chemically and physically to produce a wide variety of different fibers, tailored to secure the desired characteristics for specific end-use products. Thus, fibers can be extruded in different shapes and sizes for special purposes. They can be modified to offer greater comfort, flame resistance, or static-free behavior in apparel; they can offer soil-release and other desirable characteristics for carpets; they can be developed with unique surface characteristics, easier dyeability, or better blending qualities. The industry has begun to discover many possibilities for modifying the behavior of a given fibrous material. It has learned how to produce new fibers with greater strength, greater thermal resistance, or other special qualities.

Essentially, then, no new, large-volume, highly profitable fibers have been developed since the mid-1950s. Instead, the existing ones have become commodities with all the economic impact thereby implied. No major chemical engineering processes have been added, although the previously described ones have been modified to allow for spinning of liquid crystalline polymers or the formation of gel spun fibers. Research activity has been reduced and centered essentially on modifications of fiber size, shape, and properties, and many variants now are successfully marketed. Production volumes have increased enormously for nylon, polyester, and polyolefin.

FIBER CONSUMPTION

Figure 12.3 compares population growth with the production of manufactured fibers and the mill consumption of natural fibers in the United States. Per capita consumption of all fibers, starting at a level in the 1920s of about 30 lb, rose to approximately 40 lb following World War II and reached a level at or about 50 lb in the 1970s. It topped 60 lb in 1973, dropped below 45 lb in 1982, and in 1991 was at 55 lb. But clearly overshadowing the increases resulting from population growth and a higher standard of living are the volumes produced of, first, the cellulosic manufactured

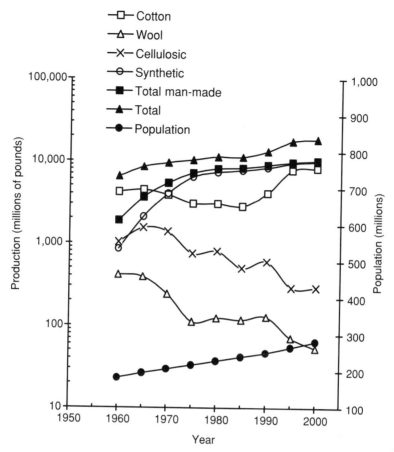

Fig. 12.3. Comparative growth of population and fiber consumption in the United States, 1960–2000.

fibers and, second, the noncellulosic or completely synthetic fibers. The consumption of manufactured fibers increased from about 2 billion lb in 1960 to nearly 9 billion lb in 1990. During the 1980s, the increase was about 11 percent, made up of a decrease in cellulosics of about 200 million lb (25%) and an increase in synthetics of about 1.2 billion lb (17%). Of the 13.2 billion lb of fibers used in the United States in 1990, 4.6 percent was the manufactured cellulosics, 63.3 percent was the synthetics (67.9% manufactured), 31.1 percent was cotton, and less than 1 percent was wool. A significant increase was noted in the use of cotton during the 1980s, from 3.0 billion lb in 1980 to 4.1 billion lb in 1990, believed to be in large measure due to advances in chemical finishes that made caring for the natural fibers easier than it was with the older technology.

To some extent, this increase also may have occurred because people were willing to accept a more wrinkled look in order to gain the comfort of hydrophilic fibers. Still, the general figures represent quite a reversal in the consumption of fiber types seen three or more decades ago. For example, in 1960, of the 6.5 billion lb of fibers used, manufactured fibers accounted for only 29 percent of the total, with cotton 65 percent and wool 6 percent.

Major applications of fibers lie in apparel, home furnishing, and industrial products. In each of these, manufactured fibers have made large inroads, and currently their usage dominates. As an illustration, consider the changes that have taken place in the use of the materials required in the manufacture of tire cords. Originally made from cotton, rayon took a commanding position during World War II. But as late as 1951, cotton comprised about 40 percent of the total output of tire cords of approximately half a billion lb, and nylon was at a negligible level of 4 million lb. By 1960, however, cotton had all but disappeared; nylon represented about 37 percent of the total (on a weight basis), even though only about 0.8 lb of nylon is needed to replace 1.0 lb of rayon. Whereas rayon for several years had dominated the so-called original-equipment tire market and nylon had held a corresponding position for replacement tires, more recently, glass and polyester have made heavy inroads into both—especially in belted constructions. The situation continued to change in favor of noncellulosic manufactured fiber usage in tires, so that by 1972, rayon was down to 14 percent, nylon up to 42 percent, polyester up to 32 percent, glass up to 7 percent, and steel at 5 percent, all on a weight basis. By the late 1970s, tire markets were dominated wholly by manufactured fibers with polyester holding over 90 percent of the passenger car original-equipment market and nylon commanding over 90 percent of the truck original-equipment market. This division of markets is a direct result of the performance characteristics of the two fibers. Polyester-containing tires are free of "flat spotting" or cold-morning thump, and so are preferred in passenger cars for their smooth ride. On the other hand, nylon-containing tires are tougher and more durable, and so are the choice for trucks and off-road vehicles.

The production of manufactured fibers throughout the world has developed in a manner that rather parallels the situation in the United States, as may be seen in Fig. 12.4. There are some expected differences, and obviously the data for world usage are strongly influenced by the large components attributable to the United States, which currently accounts for about 23 percent of the manufactured fiber and about 15 percent of the total fiber consumption. The output of the world cellulosics has leveled off, but expansion of the noncellulosics has continued unabated. The use, or at least the recorded use, of the natural fibers, cotton and wool,

rose rapidly in the 1950s, as the world economy rebounded at the conclusion of World War II. Since then (1960–1970), a modest increase has continued, essentially parallel to the growth of world population. But, in comparison with population trends, it appears that the great demand has been for manufactured textile fibers. Much of this increase has resulted from an improved standard of living and the absence of major wars.

A detailed economic examination of the processing of fibers and the changes that have taken place during the last half century would show two rather vivid occurrences. The first of these is a rapid decrease in the prices of the newer fibers as they became established, followed by a leveling out and stabilization. The second is the relative stability of prices of the manufactured fibers on short-term and even long-term bases, as compared with fluctuations in the prices for the natural fibers where governmentally imposed stability has not been in effect. Data are not presented about it in this text, but in the first half of the 20th century there was a saying in the textile industry that the person who made or lost money for the company was the one who was responsible for buying cotton and wool "futures."

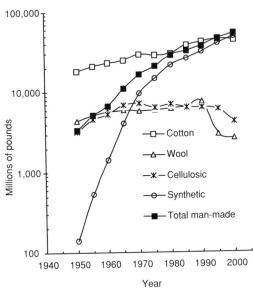

Fig. 12.4. Production of fibers in the world, 1950–2000.

However, it should also be emphasized that list prices of manufactured fibers are ceiling prices and do not reflect the short-term discounts, allowances, and special arrangements that are given in a free marketplace when the demand for any manufactured fiber softens.

A presentation of complete information about the consumption of raw materials, chemical reactions, reagents and catalysts used, and efficiencies of operation in the production of manufactured fibers undoubtedly would contribute to a better understanding of the industrial chemistry involved. Several factors have prevented this, however. In the first half of the 20th century, a historical belief in the efficacy of trade secrets still permeated the chemical industry. Even with the increased mobility of technical and scientific personnel during and following World War II, the idea still prevailed that if nothing other than patents was allowed to become public knowledge, so much the better. The situation has changed considerably since about 1960, as can be noted from the availability of information contained in the list of suggested readings that follows this chapter; yet, secrecy tends to be maintained despite the fact that key employees move from company to company, and the chemical engineering knowledge available in chemical companies that produce large volumes of fibers permits an almost complete appraisal of a competitor's activities.

In general, in the early period of production of a fiber, the cost of the original raw material may have had very little bearing on the selling price of the final fiber. A most important factor is the action of the producer's competitors and the conditions of the market and the demand that can be developed. But the complexity of the processes involved in conversion determines the base cost of the fiber at the point of manufacture. As the process becomes older, research reduces this complexity; with simplification, there may be rapid drops in plant cost. If demand remains high, such reductions will not be expected to be reflected in selling prices; rather, profits are high. As more producers enter the field in order to share in those profits, output capacity surpasses demand, and in accordance with classical economic theory, major selling price reductions result. This was happening, in general, in the 1960s and for cellulose-based manufactured fibers in the 1970s. But beginning in 1973, the cost of petroleum-based products started to rise steeply and erratically. This rise was based not on economic considerations alone but, on political considerations among the oil producing and exporting countries (OPEC) as well. Further upward pressure on manufactured fiber prices has resulted from governmental limitations placed on chemical usage and exposure and on amounts of chemicals that can be discharged into the air and water. To meet these limitations, the manufactured fiber industry has had to supply large infusion of capital. In some instances, such expenditures could not be justified, and plant capacity was shut down permanently. This was particularly true in the case of filament rayon. In recent years, the factors of rapidly rising raw material/energy prices and the costs of meeting environmental regulations have not allowed the prices of manufactured fibers to fall as production experience has been gained and technological advances have been introduced. Instead, selling prices have been continually adjusted upward in an effort to pass along unavoidable cost increases so as to maintain profitability. In areas of application in which a manufactured fiber replaced a natural one because of lower prices and stable availability, swings in fashion and increases in imports sometimes have caused a reduction in fiber utilization. This scenario, combined with environmental concerns, is believed to have particularly applied to acrylic fiber, whose production has decreased recently in both Europe and the United States.

The great importance of manufactured fibers in the chemical industry and in the overall economy of the United States (and, in general, the developed countries) becomes apparent when the volume of production of these materials is considered and compared with the market value of even the least expensive of the raw materials used by them. The amounts of oil and natural gas consumed by the manufactured fiber industry represent around 1 percent of national annual usage.

Of this amount, about one half is used to produce raw materials from petrochemicals, with the other half used for energy to convert trees to wood pulp for cellulose-based fibers and to convert the wood pulp and petrochemical-derived raw materials to fibers.

RAYON

Chemical Manufacture

Rayon, the first of the manufactured fibers produced in large volume, is based on the natural polymer cellulose, a repeat unit of which is shown below:

Two anhydroglucose units

Although in the early days the main source of this raw material was cotton linters, a combination of improved technologies for obtaining alpha cellulose from wood and the shortage of cotton linters used for the manufacture of cellulose nitrate during World War II resulted in a shift in raw material to wood pulp. Only certain trees constitute the most economical supply of dissolving pulp, as the final product is called; the process economics depend upon the cost of logs delivered at the pulping mill and the relative yield of alpha cellulose after the unusable lignin and the other components of the wood are discarded.

A general flow diagram for the manufacture of rayon is given in Fig. 12.5. The dissolving pulp is received by the rayon manufacturer in sheet or roll form. In the manufacturing process, impurities are removed, with special attention being given to removal of traces of such metallic elements as manganese and iron, the former having an effect on the manufacturing process (as will be noted later) and the latter an effect on the color of the final product. The production of dissolving pulp involves drastic chemical action at elevated temperatures, which substantially reduces the originally very high molecular weight of the cellulose. The portion not soluble in 17–18 percent aqueous caustic, known as alpha cellulose, remains, and the lower-molecular-weight beta and gamma fractions are largely soluble and lost. The composition of the pulp is aimed at high alpha content. A typical economic trade-off is involved. The pulp producers can secure an alpha content of up to 98 percent by means of a cold caustic extraction, or, on the other hand, the rayon manufacturer can use a less expensive, lower alpha content pulp (90–96%) and expect to secure a lower yield. The sellers have numerous grades available to meet the specific process needs and end-product requirements of each of the buyers.

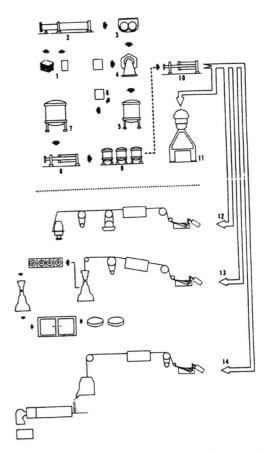

Fig. 12.5. Flow diagram for manufacture of viscose yarn: (1) cellulose sheets and caustic soda; (2) steeping press; (3) shredder; (4) xanthating churn; (5) dissolver; (6) caustic supply; (7) ripener; (8) filtration; (9) deaeration; (10) filtration; (11) continuous process; (12) tire cord; (13) pot spinning; (14) staple spinning.

In the manufacture of rayon, it is the usual practice to begin "blending" at the first step, which involves steeping the pulp. Further blending proceeds throughout successive steps. The warehouse supply of pulp consists

loses (also called hemicelluloses). The exact chemical composition of the soda cellulose is not known, but there is evidence that one molecule of NaOH is associated with two anhydroglucose units in the polymer chain.

$$(C_6H_{10}O_5)_n \text{ (cellulose)} + 18\% \text{ aqueous NaOH} \rightarrow [(C_6H_{10}O_5)_2 \cdot \text{NaOH}]_n \text{ (swollen, insoluble,}$$
$$\text{soda cellulose I)} + \text{soluble soda cellulose from } \beta \text{ and } \gamma \text{ celluloses}$$

of numerous shipments, and in making up the batches for the conventional process, a few sheets are taken from each of several shipments. This serves two purposes. It prevents a slight variation in a single pulp lot from unduly affecting any given volume of production, and it provides a moving average so that changes with time are reduced to a minimum.

The cellulose sheets are loaded vertically, but loosely, into a combination steeping bath and press (Fig. 12.6), which is slowly filled with a solution of 17–19 percent caustic, where they remain for about 1 hr. In the steeping, the alpha cellulose is converted into alkali or "soda" cellulose; at the same time, as already mentioned, the caustic solution removes most of the beta and gamma cellu-

The excess caustic solution is drained off for reuse. Additional amounts are removed by forcing the sheets through a press. The sheets are still in a swollen state and retain from 2.7 to 3.0 parts of the alkali solution. The spent steeping solution squeezed out of the pulp is processed for recovering the caustic from the organic materials.

The sheets of soda cellulose are discharged into a shredder. If blending is desired, the charges from two or more steeping presses are mixed in a single shredder, where the already soft sheets are torn into crumbs; cooling is provided to prevent thermal degradation. Shredding is controlled to produce crumbs that are open and fluffy, and that will allow air to penetrate the mass readily; this is essential in aging.

Fig. 12.6. Steeping of cellulose in the manufacture of viscose rayon. (*Courtesy Avtex Fibers, Inc.*)

Soda cellulose is aged by holding it at a constant temperature in perforated containers. The oxygen in the air produces uniform aging accompanied by a reduction in molecular weight and an increase in the number of carboxyl groups present. The target of aging is an average molecular weight high enough to produce satisfactory strength in the final fiber but low enough so that the viscosity of the solution will not be excessively high at the desired concentration for spinning. Each of the various rayon end products has its optimum degree of polymerization or chain length, ranging from about one fourth the original length for regular rayon to one half for certain high-performance fibers. As noted earlier, this optimum size is generally established by effecting a compromise between process economics and desired end-product properties. The aging proceeds for periods of up to two or three days, although the tendency is to speed up the operation by using higher temperatures and traces of metal ions, such as manganese or cobalt, to catalyze the reaction. A combination of experience and constant quality-control testing guarantees that the material will reach the correct point for conversion to cellulose xanthate.

Cellulose xanthate, or more exactly, sodium cellulose xanthate or sodium cellulose dithiocarbonate, is obtained by mixing the aged soda cellulose with carbon disulfide in a vapor-tight xanthating churn. Based upon weight of cellulose, the amount of carbon disulfide used will be in the range of 30 percent for regular rayon to 50–60 percent for modified varieties.

viscous, honey-colored liquid—hence the word "viscose." At this stage, the viscous solution may contain about 7.25 percent cellulose as xanthate in about a 6.5 percent solution of sodium hydroxide, although concentrations of both vary, depending on what end products are desired. The solution is ready for mixing with other batches to promote uniformity, to be followed by filtration, ripening, deaeration, and spinning. The filtration process usually involves several stages so that filters of decreasing pore size may be used to secure a balance of throughput and stepwise particle and gel removal.

Such an operation is a straightforward one for "bright" rayon, but only in the days of "artificial silk" did the shiny fiber alone satisfy the market. After a few years, a dull-appearing fiber also was demanded. At first, fine droplets of oil in the filaments were used to produce dullness until it was discovered that titanium dioxide pigment having a particle size smaller than 1 μm in diameter was even more satisfactory. The latter has since become the universal delustrant for all manufactured fibers. With the use of pigments of any type, problems of dispersion and agglomerate formation must be faced. The usual practice has been to add this pigment when mixing the cellulose xanthate into the dilute solution of caustic.

However, there are many other chemicals and additives that a producer may be required to add to the solution, including: (1) a few parts per million of a tracer element for later identification of the product; (2) coloring pigments for "dope dyed" rayon ("dope dyeing"

$$[(C_6H_{10}O_5)_2 \cdot NaOH]_n + CS_2 \text{ (30–60\% based on weight of cellulose in soda cellulose)} \rightarrow (C_6H_{10}O_5)_n[C_6H_7O_2(OH)_x(O\!-\!\underset{\underset{S}{\|}}{C}\!-\!S \cdot Na^+)_{3-x}]_m$$

or, for simplicity,

$$(\text{cellulose}\!-\!O\!-\!\underset{\underset{S}{\|}}{C}\!-\!SNa)$$

The xanthate is soluble in a dilute solution of sodium hydroxide—a characteristic discovered by Cross and Bevan in 1892—and this property makes the spinning of rayon possible. It is a yellow solid; when dissolved in a dilute solution of alkali, it becomes a

will be discussed in greater detail under another heading); (3) chemicals for controlling the rate of precipitation and regeneration for obtaining rayon with so-called high performance; and (4) polymers and chemicals to impart specific properties to the fiber. From the standpoint of chemical processing, it is obvious that these additives may also be added when the sodium cellulose xanthate is dissolved in dilute caustic solution, or may be injected into the solution before it enters the spinnerette prior to being extruded. To keep the operations as flexible as possible, the additives should be injected at the last possible moment so that when a changeover is desired, there will be a minimum amount of equipment to be cleaned. On the other hand, the farther along in the operation that additives are placed into the stream, the greater the problem of obtaining uniformity in an extremely viscous medium, and the greater the difficulty in maintaining exact control of proportions before the viscous solution is passed forward and spun. Furthermore, all insoluble additives must be of extremely small particle size, and all injected slurries must be freed of agglomerates by prefiltration; if not, the viscous solution containing the additives must be filtered. Each manufacturer of viscose rayon develops the particular conditions for making additions, depending on a multitude of factors, not the least of which is the existing investment in equipment. All manufacturers must face the universal necessity of filtering the solution with or without pigments or other additives, so that all impurities and agglomerates that might block the tiny holes in the spinnerette are removed.

Although it was known in the years following the discovery by Cross and Bevan that a viscose type of solution could be used in the preparation of regenerated cellulose, the conversion of this solution into useful fibers was not possible until the discovery that the solution required aging until "ripe." Ripening is the first part of the actual chemical decomposition of cellulose xanthate, which, if allowed to proceed unhampered, would result in gelation of the viscose solution.

Experience has taught the manufacturer the correct time and conditions for the aging operation, but the requirement of aging itself demands that the entire process be so planned that the viscose solution will arrive at the spinnerette possessing, as nearly as possible, the optimum degree of ripeness, to produce fibers having the desired characteristics. This degree of ripeness is determined by an empirical test made periodically, which is a measurement of the resistance of the solution to precipitation of the soda cellulose when a salt solution is titrated into it. Thus, it is known as the "salt index" or "Hottenroth number" after its originator. An additional step in the overall ripening operation involves the removal of dissolved and mechanically held air by the use of a vacuum on a moving thin film of the viscose solution.

It should be mentioned that so inevitable is decomposition of cellulose xanthate and consequent gelation of the contents of pipes and tanks that all viscose rayon plants must be prepared to pump in-process viscose solutions to other spinning machines or to a waste receiver, purging the entire system with dilute caustic solution, in the event of a long delay in spinning.

Wet Spinning

Spinning a viscose solution into rayon fibers (wet spinning) is the oldest of the three common ways of making manufactured fibers. In this method, the polymer is dissolved in an appropriate solvent, and this solution is forced through fine holes in the face of the spinnerette, which is submerged in a bath of such composition that the polymer precipitates. The pressure necessary for this extrusion is supplied by a gear pump, which also acts as a metering device; the solution is moved through a final or "candle" filter before it emerges

$$\text{Cellulose} - \text{O} - \underset{\underset{\text{S}}{\|}}{\text{C}} - \text{SNa} \underset{}{\overset{\text{H}_2\text{O}}{\rightleftharpoons}} \text{Cellulose} - \underset{\underset{\text{S}}{\|}}{\text{OC}} - \text{SH} + \text{NaOH} \xrightarrow{\text{H}_2\text{O}}$$

$$\text{Cellulose} + \underset{\underset{\text{S}}{\|}}{\text{HOC}} - \text{SH} \longrightarrow \text{Cellulose} + \text{CS}_2 + \text{H}_2\text{O}$$

from the holes of the spinnerette. There is immediate contact between these tiny streams and the liquid or "wet" bath. As the bath solution makes contact with the material extruded from the holes, chemical or physical changes take place. These changes, whether of lesser or greater complexity, convert the solution of high molecular weight linear polymer first to a gel structure and then to a fiber. As will be observed in what follows, it is an interesting fact that the spinning of viscose rayon, with all of the ramifications made possible by variations in the composition of the solution and the precipitating bath, as well as in the operating conditions, presents the chemist and the chemical engineer with both the oldest and the most complex wet-spinning process.

The formation of rayon fibers from viscose solution is far from being simple, from either a physical or a chemical standpoint. The spinning bath usually contains 1–5 percent zinc sulfate and 7–10 percent sulfuric acid, as well as a surface-active agent, without which minute deposits will form around the holes in the spinnerette. Sodium sulfate (15–22 percent) is present, formed by the reactions, and as sulfuric acid is depleted and sodium sulfate concentration builds up, an appropriate replenishment of the acid is required. There is a coagulation of the organic material as the sulfuric acid in the spinning bath neutralizes the sodium hydroxide in the viscose solution; at the same time, chemical decomposition of the sodium cellulose xanthate takes place to regenerate the cellulose. If zinc ions are present, which is the usual situation in the production of the improved types of rayon, an interchange takes place so that the zinc cellulose xanthate becomes an intermediate. It reacts at a slower rate, causing slower decomposition to cellulose. This provides conditions for more effective stretching or drawing of the fiber. Chemical additives usually are present to repress hydrogen ion action. The gel-like structure, the first state through which the material passes, is not capable of supporting itself outside the spinning bath. As it travels through the bath, however, it quickly becomes transformed into a fiber that can be drawn from the spinning bath and that can support itself in subsequent operation (Fig. 12.7). The

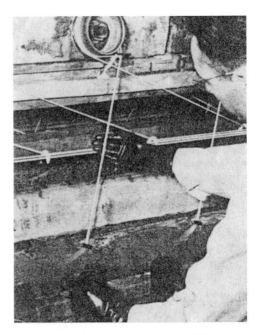

Fig. 12.7. Spinning of viscose rayon. (*Courtesy of Avtex Fibers, Inc.*)

reaction between the bath and the fiber that is forming are paramount in determining the characteristics of the final product; it is for this reason that additives (previously mentioned), as well as zinc ions, may be used to control both the rate of coagulation and regeneration. In this manner, the arrangement of the cellulose molecules may be controlled to produce the conformational structure desired. A practical application of this will be discussed later.

(Rapid reaction)

$$2\,\text{Cellulose} -\!\!\text{O}-\!\!\underset{\underset{\text{S}}{\|}}{\text{C}}-\!\!\text{SNa} + \text{H}_2\text{SO}_4$$

$$\longrightarrow \text{Cellulose} + \text{Na}_2\text{SO}_4 + \text{CS}_2$$

(Slow reaction)

$$2\,\text{Cellulose} -\!\!\text{O}-\!\!\underset{\underset{\text{S}}{\|}}{\text{C}}-\!\!\text{SNa} + \text{ZnSO}_4$$

$$\longrightarrow (\text{Cellulose O}-\!\!\underset{\underset{\text{S}}{\|}}{\text{C}}-\!\!\text{S})_2\,\text{Zn} + \text{Na}_2\text{SO}_4$$

$$(\text{Cellulose O}-\!\!\underset{\underset{\text{S}}{\|}}{\text{C}}-\!\!\text{S})_2\text{Zn} + \text{H}_2\text{SO}_4$$

$$\longrightarrow \text{Cellulose} + \text{ZnSO}_4 + \text{CS}_2$$

Because of hydraulic drag, stretching occurs in the bath and also in a separate step after the yarn leaves the bath. In both cases, the linear molecules of cellulose are oriented from random positions to positions more parallel to the fiber axis. If a rayon tire cord is to be the final product, the fibers must be severely stretched to produce a very high orientation of the molecule. This is the basis of the tire cord's high strength and ability to resist stretching, without which growth of the tire body would occur. For regular textile and nonwoven uses, such high strengths are not desired, and the spinning and stretching conditions are controlled to produce rayon of lower strength and greater stretchability under stress.

In order to stretch the yarn uniformly during the manufacturing process, two sets of paired rollers or "godets" are employed, each of the two sets operating at different rotational speeds. The yarn is passed around the first set of godets several times to prevent slippage and is supplied to the stretching area at a constant speed. A second set of godets moves it forward at a more rapid rate, also without slippage. Stretching may range from a few to 100 or more percent. Spinning speeds are of the order of 100 m/min, but may vary with both the size of the yarn and the process used.

Spinning conditions, composition of the spinning bath, and additives to the viscose solution determine the physical characteristics of the rayon—its breaking strength and elongation, modulus, ability to resist swelling in water, and characteristics in the wet state as compared with those of the dry material. Not only must the chemical composition of the spinning bath be carefully controlled, but the temperature must be regulated at a selected point, somewhere in the range of 35–60°C, to ensure those precipitation and regeneration conditions that are essential to the manufacture of any particular viscose rayon having the properties needed for a selected end use.

After precipitation and regeneration of cellulose have been completed and raw rayon fiber has been formed, the subsequent steps must be controlled so that differences in treatment are minimized; otherwise such sensitive properties

as "dye acceptance" will be affected, and the appearance of the final product will vary.

Minute traces of suspended sulfur resulting from the chemical decomposition of cellulose xanthate must be removed by washing with a solution of sodium sulfide. It is expedient to bleach the newly formed fibers with hypochlorite to improve their whiteness; an "antichlor" follows. The chemicals originally present and those used to purify the fibers must be removed by washing. As a final step, a small amount of lubricant is placed on the filaments to reduce friction and improve processability in subsequent operations.

Several different processes are used for the steps involved in spinning and purifying continuous filament rayon. One of the most common involves the formation of packages of yarn, each weighing several pounds, for separate treatment. After it has been passed upward out of the spinning bath and stretched to the desired degree, the yarn is fed downward vertically into a rapidly rotating canlike container called a spinning pot or "Topham" box (after the man who invented it in 1900). It is thrown outward to the wall of the pot by centrifugal force and gradually builds up like a cake, with excess water being removed by the same centrifugal force. This cake is firm, although it must be handled with care, and is sufficiently permeable to aqueous solution to permit purification.

In another method of package-spinning, the yarn is wound onto a mandrel from the side at a uniform peripheral speed. With this process, the yarn may be purified and dried in the package thus formed. In any of these systems, the spinning and stretching, as well as subsequent steps, may involve separate baths.

The continuous process for spinning and purifying textile-grade rayon yarn merits particular mention from the standpoint of industrial chemistry, as it is rather an axiom that a continuous process is to be preferred over a batch or discontinuous operation. This method employs "advancing rolls" or godets that make it possible for the yarn to dwell for a sufficient length of time on each pair, thus allowing the several chemical operations to take place in a relatively small area. Their operation depends on the geometry existing when the shafts of a pair of adjacent cylindrical rolls are oriented slightly askew. Yarn led onto the end of one of these and then around the pair will progress toward the other end of the set with every pass, the rate of traversing, and therefore the number of wraps, being determined by the degree of skewness.

The production of rayon to be converted to staple fiber also is amenable to line operation. Here, the spinnerette has many thousands of holes, and a correspondingly large number of filaments are formed in the precipitation bath. The resulting tow then is stretched to the desired degree and immediately cut in the wet and unpurified condition. The mass of short lengths can be conveyed through the usual chemical treatments, after which it is washed and dried. It is fluffed to prevent matting and is packaged for shipment in large cases.

Cuprammonium, Nitrocellulose, and Cellulose Acetate Processes for Rayon

Cuprammonium Cellulose. Cellulose forms a soluble complex with copper salts and ammonia. Thus, when cellulose is added to an ammoniacal solution of copper sulfate that also contains sodium hydroxide, it dissolves to form a viscous blue solution, and in this form it is known as cuprammonium cellulose. The principles on which the chemical and spinning steps of this process are based are the same as those for the viscose process. Cellulose is dissolved, in this case in a solution containing ammonia, copper sulfate, and sodium hydroxide. Unlike the viscose solution, the cuprammonium solution need not be aged and will not precipitate spontaneously on standing except after long periods. It is, however, sensitive to light and oxygen. It is spun into water and given an acid wash to remove the last traces of ammonia and copper ions. Although this rayon was never manufactured in a volume even approaching that achieved by the viscose process, the smaller individual filaments inherent in it made it useful in certain specialty markets. It no longer is manufactured in the United States but continues to be made abroad.

Nitrocellulose and Cellulose Acetate. Although nitrocellulose and cellulose acetate intermediates have been made and regenerated to form cellulose fibers, neither of these historical processes is still in operation.

Textile Operations

After the filament rayon fiber has been spun and chemically purified, much of it passes through what are known as "textile operations" before it is ready to be knitted or woven. Because these steps of twisting and packaging or beaming are common to the manufacture of all manufactured fibers, it is advisable to review briefly the background and the processes.

Rayon, the first manufactured fiber, not only had to compete in an established field, but also had to break into a conservative industry. Silk was the only continuous filament yarn, and products made from it were expensive and possessed of high prestige, so that they offered a tempting market for rayon. Thus, the new product entered as a competitor to silk and, as already noted, became known as "artificial silk." Under the circumstances, it was necessary for rayon to adapt itself to the then existing silk processing operations and technologies. It was customary to twist several silk filaments together to secure a yarn of the desired size, strength, and abrasion resistance. Because rayon was weaker than silk and its individual filaments were smaller, it required as much twisting as silk or even more.

This twisting could have been carried out in the same plant where the yarn was spun, but the existence of silk "throwsters" (from the Anglo-Saxon *thrawan*, to twist or revolve) made that unnecessary. However, as the rayon industry developed, the amount of yarn twisted in the producing plant or sent forward to throwsters decreased. Over the years, the trend has been to use less twist and to place, instead, several thousand parallel ends directly on a "beam," to form packages weighing as much as 300–400 lb, which are shipped directly to a weaving or knitting mill. The advent of stronger rayons, as well as other strong fibers, and a diminishing market for crepe fabrics which required highly twisted yarns, accelerated the trend away from twisting.

In all twisting and packaging operations, the yarn makes contact with guide surfaces and tensioning devices, often at very high speeds. To reduce friction, it is necessary to add a lubricant as a protective coating for the filaments. This is generally true of all manufactured fibers, and it is customary to apply the lubricant or "spinning finish" or "spinning lubricant" as early in the manufacturing process as possible. For those materials that develop static charges in passing over surfaces, this lubricant also must provide antistatic characteristics.

It is difficult to overstate the importance of fiber lubricants to the successful utilization of manufactured fibers. Few problems can be more damaging to a fiber-handling operation than a lubricant upset. A separate chapter could be written on lubricant usage, but some of its more important aspects will be mentioned here. Obviously, lubricants must reduce friction between the fiber and various surfaces to allow movement without excessive damage to the fiber or the surface contacted, the latter being any one of a variety of metals or ceramics. Similarly, a fiber comes in contact with the surface of other fibers in staple fiber processing and in packaging. The lubricant composition must be stable under a variety of storage conditions, without decomposing or migrating within the package or being lost from the fiber surface by adsorption into the fiber, and must be nontoxic and nondermatitic. It also must be compatible with other materials added during textile processes, such as the protective size coat applied to warp yarns before weaving or the wax coat often applied to yarns before circular knitting. Possible metal corrosion must be evaluated for each lubricant composition. Finally, after having performed its function, the fiber finish or lubricant must be completely scoured from the fabric to permit uniform adsorption of dyes and fabric finishes. The application of sewing lubricants to fabric to be cut and sewn is yet another area requiring attention.

Spun Yarn. After rayon became established in the textile industry, where it could be used as a silklike fiber, and its selling price was greatly

reduced, other markets for it were developed. The cotton, wool, worsted, and linen systems of converting short discontinuous fibers to yarns were well established, and their products were universally accepted. Here again it was necessary to make rayon fit the requirements of existing equipment and historically acceptable operations. The first of these was that it be cut into the same lengths as those found in cotton and wool. Fortunately, the viscose rayon process was and is eminently suited to the production of tows containing thousands of filaments. The pressure required to force the solution through the holes is so low that neither thick metal sections nor reinforcement of the surface is necessary to prevent bulging, and large spinnerettes containing several thousand holes can be used. Furthermore, the spinning bath succeeds in making contact with all the filaments uniformly. As a result, the spinning of viscose rayon tow is very similar in principle to the production of the smaller continuous filament yarns.

Because both cotton and wool possess distortions from a straight rodlike structure, machinery for their processing was designed to operate best with such crimped fibers. Thus, it was necessary for rayon staple to possess similar lengths and crimpiness in order to be adapted to existing equipment. The crimp, that is, several distortions from a straight path per inch, is produced in rayon "chemically" by modification of the structure. The precipitation-stretching step in spinning is carried out so that the skin and the core of the individual filaments are radially nonuniform and constantly changing over very short lengths along the filaments. Because the skin and the core differ in sensitivity to moisture, the two components shrink differentially, leading to the development of permanent distortions of the filaments. The latent chemical crimp may be enhanced by a thermomechanical step. In this, the tow is fed between two wheels, which in turn force it into a chamber, called a "stuffer box," heated with steam, in which it is forced against the compacted material ahead of it, causing the straight filaments to collapse immediately. As the mass of material is pressed forward, it becomes tightly compacted, and it tends to remain axially distorted after it escapes through a pressure-loaded door at the opposite end of the chamber.

Modified Viscose Rayon Fibers

The variations of chemical ingredients, their concentrations, and the temperature of the spinning bath determine the rates of coagulation and regeneration, and thus the relative amounts of "skin" and "core" in the cross-section of the fiber. The skin is known to possess a higher degree of order and better mechanical properties than the core, so an increase in its proportion is desired in higher performance fibers. The degree of orientation is determined by the stretch imparted. Inherent in process variations such as these is what may be called the "art" of viscose rayon manufacturing, whereby great diversity in rayon properties can be obtained. The cross-sectional morphologies of some rayons are illustrated in Fig. 12.8.[3,4]

High-Wet-Modulus Rayon. One of the important innovations in the rayon industry has been the development of High-Wet-Modulus (HWM) rayon. For its manufacture, the cellulose molecules require a higher degree of polymerization (DP) than regular rayon; so aging and ripening times are decreased, and the processing temperature is

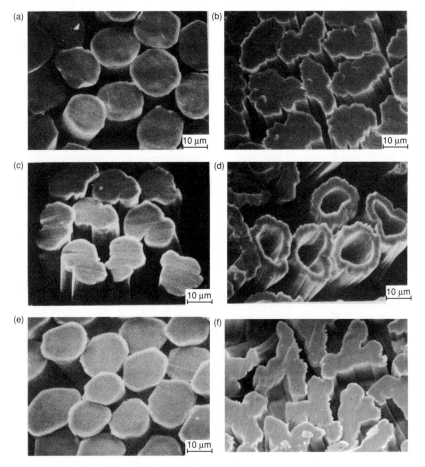

Fig. 12.8. Cross-sectional morphologies of some of the rayon fibers. (a) High wet modulus; (b) regular rayon; (c) crimped HWM; (d) hollow; (e) cuprammonium; (f) trilobal. (Sources: All except *trilobal*: Turbak, A., "Rayon," in *Encyclopedia of Polymer Science and Engineering*, 2nd ed., Vol. 14, p. 55, copyright John Wiley & Sons, Inc., New York, 1985 and used with permission of the copyright owner; trilobal photo: Gupta, B. S. and Hong, C. T., *INJ*, **7**(1), 38 (1995).)

lowered. As the viscose solution flows through the spinnerette into the bath, coagulation takes place to form the needed skin. However, in order for an increase in tenacity to occur, regeneration should proceed slowly; this is aided by using a lower concentration of acid in the spin bath. Also, zinc is added because zinc ions in the bath slow down regeneration by forming zinc cellulose xathate, which is more resistant to acid decomposition than is sodium xanthate. Because of the slow generation process, what is actually formed is nearly an "all-skin" rayon with a round cross-section. Because the structure also is stretched before it crystallizes, a higher stretch is possible; this gives a higher orientation. The fiber thus has significantly higher strength than regular rayon. This higher tenacity exists not only when the material is dry, but also when it is wet; hence the name high-wet-modulus rayon. Because of its high structural orientation and greater order, the fibers have fewer physically accessible sites for water molecules; thus it is less susceptible to swelling and to the adverse effects of basic cleaning solutions, so that the fiber's launderability is improved. The HWM rayons are used extensively for blending with cotton, wool, silk, and all other manufactured fibers.

High Absorbency Rayons. Over the past years, disposable products have become commonplace, especially in the United States and Europe. Cellulosic fibers, particularly rayons, have served the needs of the disposables industry because of their absorbent qualities. The most useful fibers for disposable/absorbent applications are the rayons with crenulations, crimp, and hollow regions, all of which add to the absorbency of the fiber. These characteristics are achieved in varying degrees by physical and chemical alterations in the spinning process. Crenulations, or random irregularities in the shape of the cross-section, typical for most rayon fibers, are caused by the rapid formation of skin before the dehydration is complete. As the fiber interior loses solvent, it collapses in certain areas and produces the crenulated shape. Furthermore, fabricators have learned how to control the cross-sectional shape of filaments by using spinnerettes containing other than round holes. One example is fibers having Y or trilobal-shaped cross-sections, which have been found to be capable of picking up more water and at a faster rate than possible with fibers of round cross-sections.[4]

Hollow viscose fibers contain gas pockets produced by adding "blowing" agents, such as sodium carbonate, to the viscose. When carbon dioxide is released during regeneration, the fibers inflate, leading to the formation of hollow filaments. The added free volume and decrease in molecular order, increase the ability of the fibers to pick up water.

Other New Developments

A number of other developments are taking place in the rayon industry, the target of one being the manufacture of lint-free rayon for use in products such as circuit boards. In another, graphite particles are blended in with the viscose to reduce static buildup. Production of flame-retardant rayon has received increased attention, in one case being achieved by the addition of phosphorus compounds to the spinning dope. The advanced technology now can produce flame-resistant fibers that, when exposed to high temperatures, will not shrink or emit toxic gases. Other developments are in the area of finding better and environmentally safer solvents for the cellulosic raw material. Searches are under way for solvents that may lead to lyotropic liquid crystalline polymer solutions from which ultrahigh-strength and high-modulus fibers can be spun.

Environmentally Friendly High Wet Strength Rayon—Lyocell

Low wet strength of rayon in general has restricted the application of the fiber to disposable and semi-durable materials. Additionally, the environmental concerns associated with its manufacture have resulted in significant curtailment in production of the fiber during the recent past. Search for ways and means to produce a high wet strength fiber using an environmentally acceptable process have occupied much of the research effort during the past two decades. One of the latest additions to the family of rayon fibers is Lyocell. The fiber has wet strength comparable to that of the natural cellulosic fibers and is manufactured with a solvent that is essentially totally recovered and recycled.[5] The solvent used is N-methyl morphine oxide, $O(C_4H_8)NOCH_3$, popularly known as amine oxide. The manufacturing process involves the dissolving of pulp in hot amine oxide, filtration of the solution, and then spinning into a bath containing a dilute solution of the solvent. The bath removes the amine oxide from the fibers, which are washed and dried, and the removed solvent is almost totally reclaimed for further use. The final fiber is said to have a different molecular structure from that of normal rayon, and a smooth surface and a round cross-section. The fiber is noted to be stronger than cotton and normal rayon in both the dry and the wet states.[6]

CELLULOSE ACETATE

Historical

Cellulose acetate was known as a chemical compound long before its potential use as a plastic or fiber-forming material was recognized. The presence of hydroxyl groups had made it possible to prepare cellulose esters from various organic acids, as cellulose consists of a long molecular chain of beta-anhydroglucose units, each of which carries three hydroxyl groups—one primary, the other two secondary. The formula for cellulose (already noted) is $[C_6H_7O_2(OH)_3]_n$; when this is fully esterified, a triester results. It was learned quite early that although cellulose triacetate is soluble only in chlorinated solvents, a product obtained by partial hydrolysis of the triester to a "secondary" ester (having about 2.35–2.40 acetyl groups per anhydroglucose unit) was easily soluble in acetone obtaining a small amount of water. Many other cellulose esters have been prepared, but only the acetate has been commercialized successfully as a manufactured fiber. Propionates and butyrates, and mixed esters of one or both the

acetate, have applications as plastics. The first acetate fibers were produced in 1921 in Europe and in 1924 in the United States.

Manufacture of Cellulose Secondary Acetate

Cellulose acetate originally was made from purified cotton linters, but this raw material has been entirely replaced by wood pulp. The other raw materials used are acetic acid and acetic anhydride.

Cellulose acetate is manufactured by a batch process (see Fig. 12.9). There has been mention in the patent literature of a continuous system, but its utilization as a production process has not been announced. The "charge" of cellulose, purified, bleached, and shredded, is of the order of 800–1500 lb. It is pretreated with about one third its weight of acetic acid and a very necessary amount of water, about 6 percent of its weight. If it is too dry at the time of use, more H_2O must be added to the acetic acid. A small amount of sulfuric acid may be used to assist in swelling the cellulose and to make it "accessible" to the esterifying mixture.

Although there has been much discussion of the chemistry of cellulose acetylation, it is now generally agreed that the sulfuric acid is not a "catalyst" in the normal sense of the word, but rather that it reacts with the cellulose to form a sulfo ester. The acetic anhydride is the reactant that provides the acetate groups for esterification. The acetylation mixture consists of the output from the acetic anhydride recovery unit, being about 60 percent acetic acid and 40 percent acetic anhydride, in an amount 5–10 percent above the stoichiometric requirement, to which has been added 10–14 percent sulfuric acid based on the weight of cellulose used. The reaction is exothermic and requires that the heat be dissipated.

In preparing for acetylation, the liquid reactants are cooled to a point (0°C) where the acetic acid crystallizes, the heat of crystallization being removed by an appropriate cooling system. The slush of acetic acid crystals in the acetic anhydride–sulfuric acid mixture is pumped to the acetylizer, a brine-cooled mixer

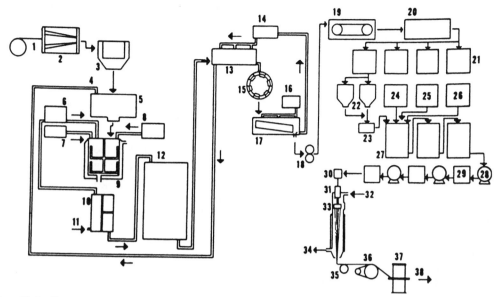

Fig. 12.9. Flow diagram for manufacture of cellulose acetate yarn: (1) wood pulp; (2) attrition mill; (3) cyclone; (4) 35% acetic acid; (5) pretreater; (6) magnesium acetate solution; (7) precooled acetylation mix; (8) sulfuric acid; (9) acetylator; (10) ripener; (11) steam; (12) blender; (13) precipitator; (14) dilute acetic acid; (15) hammer mill; (16) water; (17) rotary screen washer; (18) squeeze rolls; (19) drying oven; (20) blender; (21) storage bins; (22) silos; (23) weight bins; (24) acetone; (25) wood pulp; (26) pigment; (27) mixers; (28) hold tanks; (29) filter press; (30) pump; (31) filter; (32) air; (33) jet; (34) acetone recovery; (35) oiling wheel; (36) feed roll; (37) bobbin; (38) inspection.

Fig. 12.10. Process vessel for acetylation of cellulose. (*Courtesy Celanese Fibers Co.*)

of heavy construction (see Fig. 12.10). The pretreated cellulose is dropped in from the pretreating unit located above. The reaction is highly exothermic, and at the start large amounts of heat are produced. As the temperature of the reaction mixture rises to the melting point of the acetic acid (16.6°C), its large heat of fusion (45.91 cal/g) prevents a dangerous rise in temperature that would degrade the molecular weight of the cellulose chain. As the reaction proceeds, brine in the jacket of the acetylizer provides additional cooling.

$$\text{Cellulose} + (CH_3CO)_2O + H_2SO_4 \ (10\text{--}15\%$$

$$\text{based on weight of cellulose}) \ \xrightarrow[\text{Anhydrous}]{CH_3COOH}$$

$$[C_6H_7O_2(OSO_3H)_{0.2}(CH_3COO)_{2.8}]_n$$

The reaction product is soluble in the acetylation mixture; as it is formed and dissolved, new surfaces of the cellulose are presented to the reagents. One variation of this procedure uses methylene chloride, rather than an excess of acetic acid in the reaction mixture. This chemical is used both to dissipate the heat by refluxing (boiling point, 41.2°C) and to dissolve the cellulose ester as it is formed. As the reaction proceeds, the temperature is allowed to rise. Because cellulose is a natural product obtained from many sources, it varies slightly in composition, and at the end of the reaction cannot be predicted exactly; the disappearance of fibers as determined by microscopic examination thus is the usual means of following its progress.

During the acetylation operation, a certain amount of chain fission is allowed to take place in the cellulose molecule. This is to ensure that the viscosity of the cellulose acetate spinning solution will be low enough for ease of handling but high enough to produce fibers with the required strength. The temperature of the reaction controls the rates of both acetylation and degradation of molecular weight.

The next step in the manufacture is "ripening," whose object is to convert the triester, the "primary" cellulose acetate, to a "secondary" acetate having an average of about 2.35–2.40 acetyl and no sulfo groups (if any sulfuric acid is used in pretreatment) per anhydroglucose unit. While the cellulose sulfo-acetate is still in the acetylizer, sufficient water is added to react with the excess anhydride and start the hydrolysis of the ester. Usually the water is used as a solution of sodium or magnesium acetate, which increases the pH and promotes hydrolysis. The temperature is raised to about 70–80°C, by direct injection of steam to speed up the reaction. Hydrolysis is continued until the desired acetyl content is obtained. When this value is reached, an aqueous solution of magnesium or sodium acetate is added to cool the batch and stop the hydrolysis. It is then ready for precipitation. For example,

$$[C_6H_7O_2(OSO_3H)_{0.2}(CH_3COO)_{2.8}]_n$$
$$+ (CH_3COO)_2Mg \xrightarrow{\text{Aqueous conc. } CH_3COOH}$$
$$[C_6H_7O_2(OH)_{0.65}(CH_3COO)_{2.35}]_n + MgSO_4$$

The solution is carried to the verge of precipitation by adding dilute acetic acid. Then it is flooded with more dilute acetic acid and mixed vigorously, so that the cellulose acetate comes out as a "flake" rather than a gelatinous mass or fine powder. The flake then is

washed by standard countercurrent methods to remove the last traces of acid, and is dried in a suitable dryer.

Manufacture of Cellulose Triacetate

To obtain completely acetylated cellulose, the reaction requires the use of perchloric acid rather than sulfuric acid as the catalyst. In the presence of 1 percent perchloric acid, a mixture of acetic acid and acetic anhydride converts a previously "pretreated" cellulose to triacetate without changing the morphology of the fibers. If methylene chloride rather than an excess of acetic acid is present in the acetylation mixture, a solution is obtained. However, usually a degree of substitution between 92 and 100 percent is acceptable. For obtaining such a triester, it is possible to use about 1 percent sulfuric acid instead of perchloric acid. When the sulfoacetate obtained from such a reaction is hydrolyzed with the objective of removing only the sulfo-ester groups, the resulting product has about 2.94 acetyl groups per anhydroglucose unit. The preparation, hydrolysis, precipitation, and washing of "triacetate" are in all other respects similar to the corresponding steps in the manufacture of the more common secondary acetate. Cellulose triacetate, formerly produced under the trade name Arnel® by Celanese Corporation, is no longer in production in the United States.

Acid Recovery. In the manufacture of every pound of cellulose acetate, about 4 lb of acetic acid is produced in 30–35 percent aqueous solution. The accumulated acid contains a small amount of suspended fines and some dissolved cellulose esters. To remove the suspended material, the acid is passed slowly through settling tanks. Then it is mixed with organic solvents, so that the acid becomes concentrated in an organic layer, which is decanted. Distillation separates the acid from the organic solvent.

To produce the acetic anhydride, the acid is dehydrated to ketene and reacted with acetic acid using a phosphate catalyst at 500°C or higher in a tubular furnace.

results of the analyses determine how much further blending is necessary. After blending and mixing of portions of selected batches, the lot is air-conveyed to large storage bins or "silos," which are filled from the center of the top and emptied from the center of the bottom, thus bringing about further mixing.

Spinning Cellulose Acetate

Acetone is metered into a vertical tank equipped with a stirrer, and the cellulose acetate flake and filter aid are weighed in an automatic hopper; all operations are controlled by proportioning methods common to the chemical industry. The ratio of materials is about 25 percent cellulose acetate, 4 percent water, less than 1 percent ground wood pulp as a filter aid, and the remainder acetone. The mixture moves forward through two or three stages at the rate at which it is used, the hold time being determined by experience. After dissolution is completed, filtration is carried out in batteries of plate and frame filter presses in three or even four stages, the passage of the "dope" being through presses of decreasing porosity.

Much of the cellulose acetate is delustered by the addition of titanium dioxide pigment, as with viscose rayon. Between filtrations (and after the last filtration), the dope goes to storage tanks that serve to remove bubbles; in this case, a vacuum is not necessary. From the final storage tank, it is pumped into a header located at the top of each spinning machine; then it is directed to a series of metering gear pumps, one for each spinnerette. Because the holes in the cellulose acetate spinning spinnerette are smaller (0.03–0.05 mm) than those in the corresponding viscose devices, great care must be taken with the final filtration. An additional filter for the removal of any small particles that may have passed through the large filters is placed in the fixture, sometimes called the "candle," to which the spinnerette assembly is fastened. A final filter is placed in the spinnerette-assembly unit over the top of the spinnerette itself.

The method used for spinning cellulose acetate is "dry" spinning. The dope is heated

Fig. 12.11. Recovery of acetic acid. (*Courtesy Celanese Fibers Co.*)

$$CH_3COOH \xrightarrow[\text{Catalyst}]{\text{Heat}} H_2O + CH_2{=}C{=}O$$

$$CH_2{=}C{=}O + CH_3COOH \rightarrow (CH_3CO)_2O$$

The mixture of unreacted acid, water, and anhydride is fed to a still, which yields dilute acetic acid overhead and an anhydride–acetic acid mixture at the bottom (see Fig. 12.11). Conditions are controlled in such a way that the raffinate is about 40 percent anhydride and 60 percent acetic acid. As already mentioned, this is the desired ratio for the reaction mixture used for acetylation of cellulose.

Blending of Flake. As in the manufacture of viscose, the products of batch operations are blended to promote uniformity in the manufacture of cellulose acetate. Although a blend of different celluloses is selected in the beginning, the pretreatment, acetylation, and ripening are batch operations with little or no mixing. Before precipitation, a holding tank provides an opportunity for mixing; then precipitation, washing, and drying—all continuous—promote uniformity. The dried cellulose acetate flake moves to holding bins for analysis—the moisture content, acetyl value, and viscosity being especially important. The

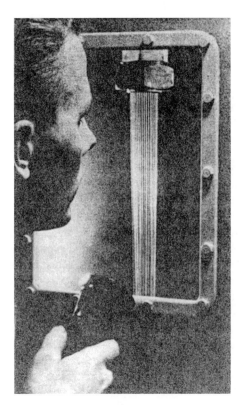

Fig. 12.12. Dry spinning of cellulose acetate. (*Courtesy Tennessee Eastman Co.*)

(in some cases above the boiling point of acetone, 56.5°C) to lower its viscosity and thus reduce the pressure required to extrude it, and to supply some of the heat needed for evaporating the acetone solvent (see Fig. 12.12).

The spinnerette is stainless steel, and because the filaments must be heated and prevented from sticking together, and because space must be allowed for the escape of acetone vapor, the holes must be kept farther apart than those of the spinnerettes used for wet spinning. As the hot solution of cellulose acetate in acetone emerges downward into the spinning cabinet, an instantaneous loss of acetone takes place from the surface of the filaments, which tend to form a solid skin over the still liquid or plastic interior. A current of air, either in the direction the filaments are moving or countercurrent, heats the filaments, and as the acetone is diffused from the center through the more solid skin, each filament collapses to form the indented cross-sectional shape typical of cellulose

acetate. The heated air removes the vaporized acetone. Each manufacturer uses a preferred updraft, downdraft, or mixed-draft operation, as needs dictate.

The cabinet through which the yarn passes vertically downward must be long enough to allow sufficient acetone to diffuse outward and evaporate from the surfaces of the filaments so that the latter will not stick to the first surface contacted or fuse to each other. The temperature of the air in the cabinet, the rate of flow, the length of the cabinet, the size and number of filaments, and the rate of travel are all interrelated in the spinning process. Because it is desirable to increase spinning speeds to the limit of the equipment, the tendency has been to construct longer spinning cabinets as each new plant is built. Present spinning speeds are of the order of 600 or more meters per minute, measured as the yarn emerges from the cabinet.

Other dry-spinning operations have followed essentially the same pattern. For example, the dry spinning of cellulose triacetate was identical to that for secondary acetate except that the acetone solvent had been replaced by a chlorinated hydrocarbon such as methylene chloride, the solubility of which was improved by the addition of a small amount of methanol (5–15%).

The acetate yarn emerging from the cabinet makes contact with an applicator that provides the lubricant required to reduce both friction and static formation in subsequent operations. With its surface lubricated, the yarn passes around a "feed" roll that determines the rate of withdrawal from the spinning cabinet, and then to any of several desired packaging devices.

Unlike the packaging of rayon yarn, cellulose acetates are either "ring" spun or wound into a package called a "disc," "zero twist," or "cam wound." In the ring-spun package, the yarn carries a slight twist of less than one turn per inch, but it requires a relatively expensive bobbin. Since the trend is toward less twisting, such acetate yarn is "beamed" in the producer's plant after little or no twisting, the heavy beams being shipped directly to knitters or weavers (see Fig. 12.13).

Fig. 12.13. Beaming cellulose acetate yarn from a reel holding about 800 packages of yarn. (*Courtesy Tennessee Eastman Co.*)

Filament yarns are twisted for two reasons. One is to supply certain esthetic characteristics such as touch, drapability, and elasticity. The other more fundamental reason is to provide physical integrity to the filament bundle so that it can be warped, woven, and knitted without excessive breakage or fraying of individual filaments.

The yarn just mentioned as having no twist imparted before beaming may have been subjected to intermingling just prior to windup after extrusion. In the intermingling process, yarn with no twist, and usually under low tension, is passed through a zone where it is impinged upon by a jet stream of compressed air. This causes the filaments to interlace or intermingle with each other, and they can become metastable in this configuration when tension is reapplied. In this condition, the yarn has the integrity of twisted yarn and will

pass through several textile processing steps without difficulty; but with each handling, some of the intermingling is worked out.

Solvent Recovery. The air containing the acetone vapor is drawn out of the spinning cabinet and passed through beds of activated carbon that sorb the organic solvent. The acetone is recovered by steaming and then by separating it from the water by distillation. The efficiency of recovery is about 95 percent.

Dope-Dying. As with viscose rayon, colored pigments or dyestuffs may be added to the spinning solution so that the yarn will be colored as it is produced, thus eliminating the need for dyeing the final fabric. As mentioned earlier, even in using titanium dioxide, a compromise must be made on the basis of two competing needs. Complete mixing, uniformity, and filtration require that the addition be made early in the operation; minimal cleaning problems during changeovers require just the opposite. There exist two solutions to the problem. If a manufacturer must produce a multitude of colors in relatively small amounts, it is desirable to premix individual batches of spinning dope. Each batch should be pretested on a small scale to ensure that the desired color will be acceptable when it is produced. Facilities must be provided to allow each batch of colored dope to be cut into the system very close to the spinning operation in order to minimize pipe cleaning. Permanent piping must be flushed with solvent or the new batch of colored dope; some of the equipment may be disassembled for mechanical cleaning after each change of color.

Another method of producing spun-dyed yarn involves using a group of "master" dopes of such color versatility that when they are injected by appropriate proportioning pumps into a mixer located near the spinning operation, they will produce the final desired color. The advantages of such an operation are obvious; the disadvantage lies in the public demand for an infinite number of colors. No small group of known pigments will produce final colors of every desired shade.

PROTEIN FIBERS

As previously mentioned, the use of naturally existing polymers to produce fibers has had a long history. In the case of cellulose the results were fabulous. An initial investment of $930,000 produced net profits of $354,000,000 in 24 years for one rayon company.[7] On the other hand, efforts to use another family of natural polymers—proteins—have thus far resulted in failure or at best very limited production.

These regenerated proteins are obtained from milk (casein), soya beans, corn, and peanuts. More or less complex chemical separation and purification processes are required to isolate them from the parent materials. They may be dissolved in aqueous solutions of caustic, and wet-spun to form fibers, which usually require further chemical

treatment as, for example, with formaldehyde. This reduces the tendency to swell or dissolve in subsequent wet-processing operations or final end uses. These fibers are characterized by a wool-like feel, low strength, and ease of dyeing. Nevertheless, for economic and other reasons they have not been able to compete successfully with either wool (after which they were modeled) or with other manufactured fibers.

NYLON

Historical

Nylon was the first direct product of the technological breakthrough achieved by W. H. Carothers of E. I. duPont de Nemours & Co. Until he began his classic research on high polymers, the production of manufactured fibers was based almost completely on natural linear polymers. Such materials included rayon, cellulose acetate, and the proteins. His research showed that chemicals of low molecular weight could be reacted to form polymers of high molecular weight. By selecting reactants that produce linear molecules having great length in comparison with their cross-section, fiber-forming polymers are obtained. With this discovery, the manufactured fiber industry entered a new and dramatic era.

Manufacture

Nylon 66. The word "nylon" was established as a generic name for polyamides, one class of the new high molecular weight linear polymers. The first of these, and the one still produced in the largest volume, was nylon 66 or polyhexamethylene adipamide. Numbers are used following the word "nylon" to indicate the number of carbon atoms contributed by the diamine and dicarboxylic acid constituents, in this case hexamethylenediamine and adipic acid, respectively.

To emphasize the fact that it does not depend on a naturally occurring polymer as a source of raw material, nylon often has been called a "truly synthetic fiber." To start the synthesis, benzene may be hydrogenated to cyclohexane:

$$C_6H_6 + 3H_2 \xrightarrow{\text{Catalyst}} C_6H_{12}$$

or the cyclohexane may be obtained by fractionation of petroleum. The next step is oxidation to a cyclohexanol–cyclohexanone mixture by means of air:

$$xC_6H_{12} + O_2(\text{air})$$
$$\xrightarrow{\text{Catalyst}} yC_6H_{11}OH + zC_6H_{10}O$$

In turn, this mixture is oxidized by nitric acid to adipic acid:

$$C_6H_{11}OH + C_6H_{10}O + HNO_3$$
$$\xrightarrow{\text{Catalyst}} (CH_2)_4(COOH)_2$$

Adipic acid so obtained is both a reactant for the production of nylon and the raw material source for hexamethylenediamine, the other reactant. The adipic acid first is converted to adiponitrile by ammonolysis and then to hexamethylenediamine by hydrogenation:

$$(CH_2)_4(COOH)_2 + 2NH_3$$
$$\xrightarrow{\text{Catalyst}} (CH_2)_4(CN)_2 + 4H_2O$$
$$(CH_2)_4(CN)_2 + 4H_2 \xrightarrow{\text{Catalyst}} (CH_2)_6(NH_2)_2$$

Another approach is through the series of compounds furfural, furane, cyclotetramethylene oxide, 1,4-dichlorobutane, and adiponitrile, as illustrated below. The furfural is obtained from oat hulls and corn cobs.

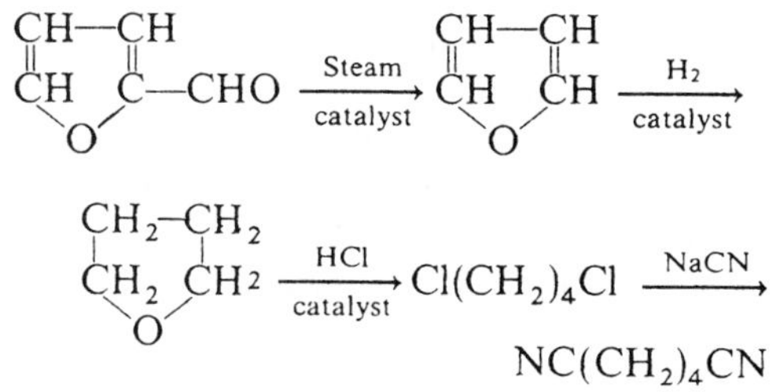

Or, 1,4-butadiene obtained from petroleum, may be used as a starting raw material to make

the adiponitrile via 1,4-dichloro-2-butene and 1,4-dicyano-2-butene:

$$CH_2{=}CHCH{=}CH_2 \rightarrow ClCH_2CH{=}CHCH_2Cl$$
$$\xrightarrow[\text{Catalyst}]{\text{HCN}} NCCH_2CH{=}CHCH_2CN$$
$$\xrightarrow[\text{Catalyst}]{H_2} NC(CH_2)_4CN$$

When hexamethylenediamine and adipic acid are mixed in solution in a one-to-one molar ratio, the "nylon salt" hexamethylenediammoniumadipate, the direct progenitor of the polymer, is precipitated. After purification, this nylon salt is polymerized to obtain a material of the desired molecular weight. It is heated to about 280°C under vacuum while being stirred in an autoclave for 2–3 h; a shorter holding period follows; and the process is finished off at 300°C. The molecular weight must be raised to a level high enough to provide a fiber-forming material, yet no higher. If it is too high, the corresponding viscosity in the subsequent spinning operation will require extremely high temperatures and pressure to make it flow. Accordingly, a small amount of acetic acid is added to terminate the growth of the long-chain molecules by reaction with the end amino groups.

The polymerized product is an extremely insoluble material and must be melt-spun, as discussed later. Therefore, should a delustered or precolored fiber be desired, it is necessary to add the titanium dioxide or colored pigment to the polymerization batch prior to solidification. For ease of handling, the batch of nylon polymer may be extruded from the autoclave to form a thin ribbon, which is easily broken down into chips after rapid cooling. But, whenever possible, the liquid polymer is pumped directly to the fiber melt spinning operation (see Fig. 12.14).

Nylon 6. Nylon 6 is made from caprolactam and is known as Perlon® in Germany, where it was originally developed by Dr. Paul Schlack.[8] Its production has reached a very large volume in the United States in recent years.

Like nylon 66, nylon 6 uses benzene as raw material, which is converted through previously mentioned steps to cyclohexanone.

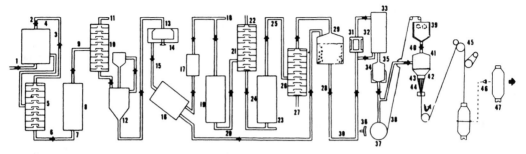

Fig. 12.14. Flow diagram for the manufacture of nylon 66 yarn: (1) air; (2) cyclohexane from petroleum; (3) reactor; (4) recycle cyclohexane; (5) still; (6) cyclohexanol–cyclohexanone; (7) nitric acid; (8) converter; (9) adipic acid solution; (10) still; (11) impurities; (12) crystallizer; (13) centrifuge; (14) impurities; (15) adipic acid crystals; (16) dryer; (17) vaporizer; (18) ammonia; (19) converter; (20) crude adiponitrile; (21) still; (22) impurities; (23) hydrogen; (24) converter; (25) crude diamine; (26) still; (27) impurities; (28) nylon salt solution; (29); reactor; (30) stabilizer; (31) calandria; (32) evaporator; (33) excess water; (34) autoclave; (35) delustrant; (36) water sprays; (37) casting wheel; (38) polymer ribbon; (39) grinder; (40) polymer flake; (41) spinning machine; (42) heating cells; (43) spinnerette; (44) air; (45) draw twisting; (46) inspection; (47) nylon bobbin. (Note: Whenever the demand for liquid polymer at a spinnerette is large, as, for example, in the spinning of tire yarn, it is pumped directly from the autoclave.)

This compound is in turn converted to the corresponding oxime by reaction with hydroxylamine, and cyclohexanone oxime is made into caprolactam by the Beckmann rearrangement.

$$\underset{\substack{H_2 \; H_2 \\ H_2 \; H_2}}{H_2 \langle \bigcirc \rangle C{=}O} + H_2NOH \longrightarrow$$

$$\underset{\substack{H_2 \; H_2 \\ H_2 \; H_2}}{H_2 \langle \bigcirc \rangle C{=}NOH} + H_2O$$

$$\underset{\substack{H_2 \; H_2 \\ H_2 \; H_2}}{H_2 \langle \bigcirc \rangle C{=}NOH} \xrightarrow{H_2SO_4} \underset{CH_2(CH_2)_4\overset{\displaystyle|}{C}{=}O}{\overset{\displaystyle \overline{\qquad}NH\overline{\qquad}}{}}$$

After purification, the lactam is polymerized by heating at elevated temperatures in an inert atmosphere. During self-condensation, the ring structure of the lactam is opened so that the monomer acts as an epsilon-aminocaproic acid radical. Unlike that of nylon 66, the polymerization of caprolactam is reversible; the polymer remains in equilibrium with a small amount of monomer. As with nylon 66, nylon 6 is extruded in thin strands, quenched, and cut into chips for subsequent spinning, or the molten polymer is pumped directly to the spinning equipment.

Melt spinning

Because of its extremely low solubility in low-boiling and inexpensive organic solvents, nylon 66 required a new technique for converting the solid polymer into fibers; hence the development of "melt" spinning, the third basic method for producing manufactured fibers. The following description refers essentially to nylon 66 because it was the first to use the method, but the process applies, in general, to all melt-spun manufactured fibers.

In the original production of nylon fiber by melt spinning, the chips of predried polymer were fed from a chamber onto a melting grid whose holes were so small that only passage of molten polymer was possible. Both solid and liquid were prevented from contacting oxygen by maintaining an inert nitrogen atmosphere over the polymer supply. The polymer melted in contact with the hot grid and dripped into a pool where it became the supply for the spinning itself. This melting operation has been entirely replaced by delivery of the molten polymer pumped directly from the polymerization stage or by "screw" melting. In the latter process, the solid polymer in chip form is fed into an extrusion-type screw contained in a heated tube. The depth and the helix angle of the grooves are engineered in such a way that melting takes place in the rear section, and the molten polymer is moved forward under increasing pressure to a uniformly heated chamber preceding the metering pump.

Whatever means is used to secure the molten polymer, it is moved forward to a gear-type pump that provides both high pressure and a constant rate of flow to the final filter and spinnerette. The filter consists of either sintered metal candle filters, several metal screens of increasing fineness, or graded sand arranged in such a way that the finest sand is at the bottom. After being filtered, the molten polymer at a pressure of several thousand pounds per square inch is extruded through the small capillaries in the heavily constructed spinnerette. It is necessary to maintain the temperature of the pool, pump, filter, spinnerette assembly, and spinnerette at about 20–30°C above the melting point of the nylon, which is about 264°C for nylon 66 and 220°C for nylon 6. Fibers having desired cross-sectional shapes can be produced by selecting spinnerettes containing holes of appropriate configuration. An example of a trilobal spinnerette capillary and the shape of the resulting trilobal fiber is given in Fig. 12.15.

The nylon production process requires that the extruded fibers emerge from the spinnerette face into a quench chamber where a cross current of relatively cool quench air is provided to promote rapid solidification. The solid filaments then travel down a chimney to cool further, and a lubricant is applied before they make contact with the windup rolls in order to prevent static formation and to reduce friction in subsequent textile operations. The freshly spun yarn from the spinning chamber is taken up by a traversing winder onto a yarn package and "drawn" in a separate operation. In modern high-speed processes, drawing still is required, but in many cases this is combined with spinning in a single operation, as will be described in what follows.

Drawing

It was learned early that the "as-spun" fibers made from nylon 66 could be extended to about four times their original length with very little effort, but that thereafter a marked resistance to extension took place. It was discovered that during this high extension, the entire length of fiber under stress did not

(a)

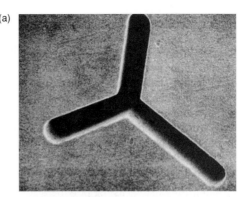

(b)

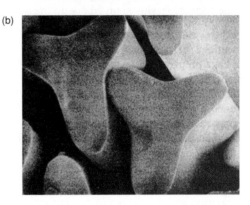

Fig. 12.15. Scanning electron micrographs of: (a) trilobal-shaped spinerette, and (b) resulting cross-sections of nylon fibers. The pictures were taken at different magnifications. (*Courtesy BASF Corporation*)

extend uniformly. Rather, a "necking down" occurred at one or more points, and when the entire length under tension had passed through this phenomenon, a high-strength fiber was obtained. It also was found that when more than one necking down was allowed to take place in a given length of fiber, a discontinuity occurred at the point where the two came together. Accordingly, the drawing operation was aimed at forcing the drawing to occur at a single point as the yarn advanced from the supply to the takeup package.

Where still used, cold drawing consists essentially of removing the yarn from the package prepared in the melt spinning operation and feeding it forward at a uniformly controlled rate under low tension. It is passed around a godet or roller that determines the supply rate and prevents slippage; for nylon 66, it then is wrapped several times around a stationary snubbing pin. From there it goes to a second roller that rotates faster than the supply roller to produce the desired amount of stretch, usually about 400 percent. The necking down occurs at the pin. In the case of nylon 6, drawing may be effected satisfactorily without passing the yarn around such a snubbing pin.

The long molecules of the nylon 66 or 6 polymer, which are randomly positioned in the molten polymer, when extruded from the spinnerette tend to form "crystalline" areas of molecular dimensions as the polymers solidify in the form of freshly spun fibers. In the drawing operation, both these more ordered portions as well as the amorphous areas tend to become oriented so that the lengthwise dimensions of the molecules become parallel to the long axis of the fiber, and additional intermolecular hydrogen bonding is facilitated. It is this orientation that converts the fiber having low resistance to stress into one of high strength.

By controlling the amount of drawing as well as the conditions under which this operation takes place, it is possible to vary the amount of orientation and the degree of crystallization. A minimal amount is preferable in the manufacture of yarns intended for textile applications wherein elongation of considerable magnitude and low modulus or stiffness is required rather than high strength. On the other hand, strength and high modulus are at a premium when fibers are to be used in tire cords and other industrial applications. High resistance to elongation is imperative if the tire is not to grow under conditions of use. In this connection, it should be noted that nylon tire cord that has been produced by twisting the original tire yarn

and plying the ends of these twisted yarns together is hot-stretched just before use at the tire plant to increase strength and reduce even further the tendency to elongate under tension.

The separate operations of spinning and drawing nylon presented a challenge whose object was combination of the two operations into a single continuous step. But the problem was obvious, for the operating speeds of the two separate steps already had been pushed as high as was thought to be possible. How then would it be possible to combine them into a continuous spin-draw, wherein a stretching of about 400 percent could take place? The answer lay in the manner in which the cooling air was used and in the development of improved high-speed winding devices. By first cooling the emerging fibers by a concurrent flow of air and then cooling them further by a countercurrent flow, the vertical length of the cooling columns can be kept within reason. In-line drawing may occur in one or two stages, and relaxation may be induced if needed. The final yarn is said to be packaged at speeds of 6000 m/min.

Other Nylons, Modifications, and New Developments

Although nylon 66 and 6 account for most of the polyamide fibers produced, a great many others have been experimentally synthesized and have been developed and manufactured in commercial amounts. Of these, some have been made into fibers, some with limited economic success. These nylons are identified by either the same numbering system used for nylon 66 or 6 or by a combination of numbers and letters, as follows:

Nylon 3	$-\!\!\left[NH-(CH_2)_2CO\right]_n\!\!-$
Dimethyl nylon 3	$-\!\!\left[NH-C(CH_3)_2-CH_2-CO\right]_n\!\!-$
Nylon 4	$-\!\!\left[NH-(CH_2)_3-CO\right]_n\!\!-$
Nylon 6T	$-\!\!\left[NH-(CH_2)_6-NH-CO-\bigcirc-CO\right]_n\!\!-$
Nylon 7	$-\!\!\left[NH-(CH_2)_6-CO\right]_n\!\!-$
Nylon 12	$-\!\!\left[NH-(CH_2)_{11}-CO\right]_n\!\!-$
Nylon PACM-12	$-\!\!\left[NH-\bigcirc-CH_2-\bigcirc-NH-CO-(CH_2)_{10}-CO\right]_n\!\!-$
Nylon 46	$-\!\!\left[NH-(CH_2)_4-NH-CO-(CH_2)_4-CO\right]_n\!\!-$
Nylon 610	$-\!\!\left[NH-(CH_2)_6-NH-CO-(CH_2)_8-CO\right]_n\!\!-$

Dimethyl nylon 3 is solution-spun because it tends to decompose during melt spinning. Nylon 4 has a moisture regain (mass water per unit mass of dry fiber, under standard atmospheric conditions of 20°C and 65% RH) of 6–9 percent[9] and therefore is superior to other nylons for textile usages, being comparable to cotton. Nylon 11 was developed in France and has been trademarked as Rilsan®. It has a moisture regain of 1.8 percent and density of 1.04 g/cc as compared with 4 percent and 1.14 g/cc, respectively, for nylon 66. Nylon 7 is made in the former Soviet Union and marketed under the name Enant. The fiber has better stability to heat and ultraviolet light than nylon 66 and 6. Nylon 6T, an aromatic polymer, has a much higher melting point (370°C), a higher density (1.21 g/cc), and slightly higher moisture regain (4.5%) than nylon 66. It also has superior resistance to nylon 66 against heat. This fiber has served as a precursor to the development of aramid fibers. Nylon PACM-12, formerly produced under the trade name Qiana® in the United States, is no longer in production.

A *Chemical and Engineering News* report[10] suggests that the most serious competition to nylon 66 and 6 will be provided by a new, still experimental fiber, nylon 46, being developed by DSM in the Netherlands, Trade-named Stanyl®, this fiber results from the interaction of 1,4-diaminobutane and adipic acid. Better order in the structure in the fiber leads to greater crystallinity and, thus, to greater density (1.18 g/cc). The fiber has a melting point of about 300°C, and a breaking stress or tenacity of 9.5 gram force/denier (~1 GPa), modulus at 120°C of 20 gram force/denier (2.1 GPa), and shrinkage in gas at 160°C of 3 percent.

Some of the outstanding characteristics of nylon that are responsible for its many uses in apparel, home furnishing, and industrial products are its high strength and toughness, elastic recovery, resilience, abrasion resistance, and low density. Among many applications of the fiber are such products as intimate apparel and foundation garments, sportswear, carpets, parachutes, tents, sleeping bags, and tire cords.

The world production of nylon has doubled in recent years, increasing from 3.8 billion lb in 1970 to about 7.5 billion lb in 1990. The fiber accounts for about 24 percent of the synthetic fibers produced worldwide. In the United States, the production of the fiber also doubled in two decades, increasing from 1.2 billion lb in 1970 to about 2.4 billion lb in 1990.

The newest activity connected with nylon is the effort at developing a more highly oriented stronger nylon than possible by the current technology. Because of the formation of hydrogen bonds between the chains, the normal polymer is restricted in terms of the maximum draw ratio by which it can be oriented. In the new technique,[11] nylon 66 is dissolved in an agent such as gallium trichloride, which effectively breaks the hydrogen bonds. The solution is spun by the dry-jet wet spinning method. The GaCl$_3$/Nylon 66 complex so obtained can now be stretched to very high draw ratios, levels as high as 40X have been possible. Once drawn, the structure is soaked in water to remove gallium trichloride, which allows the hydrogen bonds to reestablish and link the chains. In preliminary work done thus far, the strength and the modulus obatined exceed the values usually found in nylon 66.

POLYESTERS

Historical

The stimulus for the development of polyester, as for nylon, was provided by the fundamental work of Carothers. Although his team's initial work was directed toward this material, because of greater promise shown by polyamides at the time, the developmental work on polyesters was temporarily set aside. The polymer, however, attracted interest in Great Britain, where J. T. Dickson and J. R. Whinfield experimented with it and developed a successful polyester fiber.[7] They found that a synthetic linear polymer could be produced by condensing ethylene glycol with terephthalic acid or by an ester-exchange between the glycol and pure dimethyl terephthalate. The polymer thus obtained could be converted to fibers having valuable properties, including the absence of color. Like nylon, this material has been popularized under its generic name, polyester or just "poly." Those persons working with it commonly refer to it as PET. It first appeared under the trade name Terylene® (Imperial Chemical Industries, Ltd.,) in England, and was first commercialized in the

United States in 1953 as Dacron® (E. I. duPont de Nemours & Co).

Manufacture

When the development of polyethylene terephthalate (PET) occurred, ethylene glycol already was being produced in large amounts from ethylene, a by-product of petroleum cracking, by the oxidation of ethylene to ethylene oxide and subsequent hydration to ethylene glycol, which, in a noncatalytic process, uses high pressure and temperature in the presence of excess water.

$$CH_2{=}CH_2 + O_2 \longrightarrow \overset{\displaystyle O}{\overset{\diagup\diagdown}{CH_2{-}CH_2}}$$
$$\xrightarrow{H_2O} HOCH_2CH_2OH$$

methanol. This process then is repeated with the second methyl group to secure the dimethyl ester of terephthalic acid.

Either phthalic anhydride or toluene, both in ample supply as raw materials, can be used in the Henkel processes. Use of phthalic anhydride depends only upon dry isomerization of the potassium salt of the *ortho* derivative to the *para* form at about 430°C and 20 atmospheric pressure; or toluene is oxidized to benzoic acid, whose potassium salt can be converted to benzene and the potassium salt of terephthalic acid by disporportionation.

The first step in the reaction of dimethyl-terephthalate and ethylene glycol is trans-esterification to form bis(*p*-hydroxyethyl) terephthalate (bis-HET) and eliminate methanol.

$$OOC\langle\bigcirc\rangle COOCH_3 + 2HOCH_2CH_2OH$$

$$\xrightarrow{\sim 200°C} HOCH_2CH_2OOC\langle\bigcirc\rangle COOCH_2CH_2OH + 2CH_3OH\uparrow$$

On the other hand, although *o*-phthalic acid, or rather its anhydride, had long been produced in enormous amounts for use in the manufacture of alkyd resins, the *para* derivative was less well known and not available on a large scale. The synthesis is a straightforward one, however, from *p*-xylene, which is oxidized to terephthalic acid, either by means of nitric acid in the older process or by air (catalyzed) in the newer one. In the early years this compound then was converted to the easily purified dimethyl ester in order to obtain a colorless polymer adequate for the manufacture of commercially acceptable fibers.

Several other methods were developed for producing the desired dimethyl terephthalate. The Witten (Hercules) process goes from *p*-xylene to toluic acid by oxidation of one of the methyl groups on the ring, following which the carboxyl group is esterified with

This product then is polymerized in the presence of a catalyst to a low molecular weight compound and the by-product glycol is eliminated. In a second stage, at a temperature of about 275°C and under a high vacuum, the molecular weight is raised to secure the melt viscosity desired for the particular material involved. Like nylon, this final material may be extruded, cooled, and cut into chips for storage and remelting, or it may be pumped directly to the spinning machines.

From the beginning, it was obvious that there would be considerable progress in industrial chemistry, to say nothing of cost reduction, if the process could be simplified by making it unnecessary to go through the dimethyl derivative to secure a product of adequate purity. This was accomplished in the early 1960s when methods of purifying

$$CH_3\langle\bigcirc\rangle CH_3 \xrightarrow[190°C]{O_2} HOOC\langle\bigcirc\rangle CH_3 \xrightarrow[150°C]{CH_3OH} CO_3OOC\langle\bigcirc\rangle CH_3$$

$$\xrightarrow[\text{steps}]{\text{Same two}} CH_3OOC\langle\bigcirc\rangle COOCH_3$$

the crude terephthalic acid were developed, and conditions and catalysts were found that made possible the continuous production of a color-free polymer. It is said that the selection of the catalyst is especially aimed at the prevention of ether linkages in the polymer chain due to intracondensation of the glycol end groups.

Two additional rather similar routes are known. Both depend upon the reaction between ethylene oxide, rather than ethylene glycol, and terephthalic acid to form the bis-HET monomer already mentioned. The difference between the two methods lies in the point where purification is done: in one case, it is the crude terephthalic acid; in the other, it is the bis-HET monomer. In both cases this monomer is polymerized by known procedures to form a fiber-grade polyester. The titanium dioxide delustrant is added, as might be expected, early in the polymerizing process.

Another polyester that has reached long-term commercialization is now produced in limited volume as Kodel 200® by Tennessee Eastman Co. and is considered to be 1–4 cyclohexylene dimethylene terephthalate. The glycol that is used instead of ethylene glycol in this process exists in two isomeric forms, one melting at 43°C and the other at 67°C. This makes possible their separation by crystallization, to secure the desired ratio of the two forms for conversion to the polymer. This ratio determines the melting point of the polymer, a most important property for a material that is to be melt-spun. The polymer from the 100 percent *cis* form melts at 275°C, and that from the 100 percent *trans* form at 318°C. Indications are that the commercial product is about 30/70 *cis-trans*.

In 1973, the Federal Trade Commission modified the generic definition of polyester to include in the polyester category materials that previously were polyester ethers or benzoate polyesters. As a result, the fiber known as poly (ethylene oxybenzoate) or PEB and manufactured under the trade name A-Tell in Japan came to be known as polyester. This material is made by reacting parahydroxybenzoic acid and ethylene oxide to give

paraoxyethylenebenzoic acid, which is then polymerized to obtain PEB:

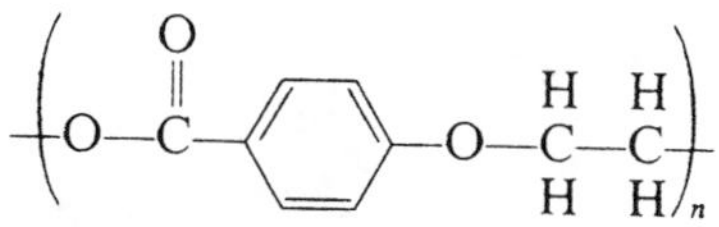

The fiber softens at about 200°C and melts at 225°C. It is said to have a silk-like hand and appearance and other properties comparable to those of other polyesters.

Polyesters are melt-spun in equipment essentially the same as that used for nylon, already described. Wherever the volume is large and the stability of demand is adequate, the molten polymer is pumped directly from the final polymerization stage to the melt-spinning machine. The molten polymer is both metered and moved forward at high pressure by use of an extruder coupled with a gear-type pump, through filters to the spinnerette, which contains capillaries of about 9 mils (230 μm) diameter. Great care is taken to eliminate moisture and oxygen from the chips, if they are used, and from the spinning chamber. When the polyester fibers are destined to become staple, the emerging filaments from a number of spinnerettes are combined to form a tow, which can be further processed as a unit. Continuous filament yarn is packaged for further processing such as drawing or texturing. Spin-drawing, described later, has become commonplace today and represents major cost savings to the fiber manufacturer.

Drawing

Unlike nylon, which in the as-spun state contains a high amount of crystalline component, PET fibers are essentially amorphous as spun. In order to secure a usable textile yarn or staple fiber, this product must be drawn under conditions that will result in an increase in both molecular orientation and crystallinity. This is done by drawing at a temperature well above the glass transition point, T_g, which is about 80°C. Conditions of rate and temperature must be selected so that the amorphous areas are oriented, and crystallization will take place as the temperature of the drawn

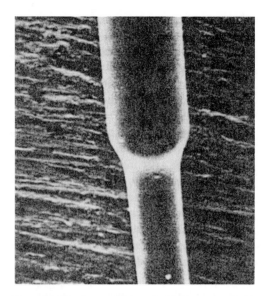

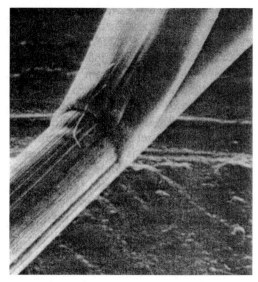

Fig. 12.16. Drawnecking in polyester single filament. (*Courtesy E. I. DuPont de Nemours & Co.*)

Fig. 12.17. Skin-peeling in polyester showing fibrillar structure. (*Courtesy E. I. DuPont de Nemours & Co.*)

fibers drops to room temperature. An appropriate contact-type hot plate or other device is used, and about 300–400 percent extension is effected.[10] Figure 21.16 shows a filament with drawn and undrawn segments. As with nylon, the conditions of draw, especially the amount, determine the force—elongation properties of the product. Industrial-type yarns, such as those intended to be used as tire cord, are more highly drawn than other yarns and have higher strength with less elongation. The fibers develop the much desired fibrillar morphology for such applications (Figure 12.17).

Heat Setting

The ability of textile fibers to be "set" is not characteristic of manufactured fibers alone. Aided in many cases by the presence of starch, cotton fabrics can be ironed to a smooth and wrinkle-free condition; also, a sharp crease in wool trousers has been commonplace for generations. In other words, these fabrics were exposed to moisture at elevated temperatures while being held or pressed into desired geometrical configurations and then allowed to cool before being released from constraint. Such fabrics tend to

remain unchanged while cool and dry, even though the fibers from which they are formed carry internal stresses; but reversion takes place upon washing or exposure to high relative humidity.

With the development of nylon, and especially polyesters, a durable kind of setting has become possible. When fabrics made from these fibers are shaped and then exposed to elevated temperatures either in the dry condition or, in the case of nylon particularly, in the presence of water vapor, thermoplastic relaxation of induced stresses in the fiber takes place and configurations at the molecular level adjust to a new and lower energy level. This depends on not only the temperature used but also the duration of the exposure. Thus a few seconds at 230°C will produce the same results as exposure for a considerably longer period at a temperature 50–75° lower. The permanency of the setting, that is, the ability of a fabric or garment to return to its original configuration after temporary distortion even while exposed to moisture and raised temperatures, is a function of the severity of the heat setting. To impart true permanence, it is essential that the internal crystalline structure be annealed.

It is this property of polyamides and polyesters that has been the main factor contributing to "ease of care" and the "wash and wear" characteristics of garments made from these polymers. In turn, these garments have revolutionized both the textile and the apparel industries.

Textured Yarns

Fundamentally, the manufacture of "textured" yarns is closely related to the heat setting of fabrics, which must be composed of thermoplastic fibers such as nylon or polyester, the difference being that the individual filaments or bundle of filaments in textured yarns are distorted from an essentially straight rodlike form and then heat-set. In some instances, the fibers are distorted in a more or less random way; at other times, a regular pattern is introduced.

The first commercially successful textured yarn was produced by highly twisting nylon 66, heat-setting it as a full package of yarn, and then untwisting it through zero and a small amount of twist in the opposite direction. This process changed yarn from a close-packed structure to one that was voluminous because of mutual interference of distorted filaments. The technique of heat-setting the twisted yarn as a batch-unit operation now has been replaced by a continuous operation, using what is known as a "false twisting" process. This is based upon the principle that if a length of yarn is prevented from rotating at both ends but is rotated on its axis at its center point, the resulting two sections will contain both "Z" and "S" twists in equal amounts. When this occurs with a moving yarn, any element in it will first receive a twist in one direction, but after passing the false twisting point must revert to zero twist. If it is then made to pass over a hot plate while in the twisted state and is heat-set in that configuration, even after returning to the untwisted condition, the individual filaments will tend to remain distorted when lengthwise stress is released. Because of the low mass and diameters of textile yarns or monofilaments, it is possible to false-twist them at extremely high rotational speeds. Yarn forward speeds of about 1000 m/min are currently obtainable by passing the yarn between, and in contact with, high-speed-friction twisting discs. (When attempts are made to secure higher rates, problems of twist control develop.) The same technique is now more commonly applied to unoriented (undrawn) or partially oriented yarn (POY) at the draw-texturing machine. The resulting yarn may be heat-set as part of the same continuous operation by passing it through a second heater under conditions of overfeed or little or no tension in order to secure both thermally stable geometric configurations in the individual distorted filaments that comprise the yarn, and the degree of "stretchiness" and bulk desired in the final product.

Because these yarns are being made in one less step and also within the plants spinning the parent product, this latest development may be said to constitute another advance in the industrial chemical technology of manufactured-textile products. This draw-texturing appears to be especially applicable to polyester yarns intended for fabrics known as "double knits" and "textured wovens."

Yarn can be forced forward by means of "nip" rolls, although this may seem to be quite contrary to the old adage that one cannot push on an end of string. When this is done so that the yarn is jammed into a receiver (stuffer-box) already full of the preceding materials, it collapses with sharp bends between very short lengths of straight sections. In this condition heat is applied, usually in the form of superheated steam, to set it. In practice, the mass of such yarn is pushed through a heated tube until it escapes at the exit past a spring-loaded gate. During this passage it is heat-set in a highly crimped configuration; then it is cooled before being straightened and wound onto a package. In another continuous process, the yarn or monofilament is pulled under tension over a hot sharp edge so that it is bent beyond its elastic limit and is heat-set in that condition. The process is known as "edge crimping," and the result is not unlike that produced by drawing a human hair over the thumbnail. The process is not used much today, but a yarn with similar crimp is produced by bicomponent spinning.

When such yarns are knitted or woven into fabric, the filaments tend to return to the configurations in which they were originally

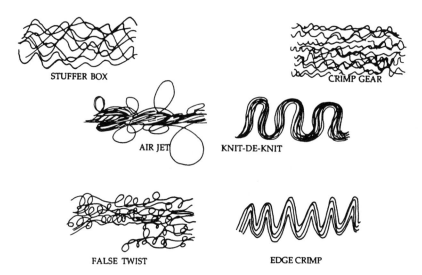

Fig. 12.18. Schematic comparison of the shapes of some textured yarns.

heat-set. Contraction takes place in the direction of the yarn axis, and this in turn converts the smooth flat fabric into a "stretch" fabric and gives the surface a textured appearance. These fabrics or the garments made from them, whatever the process used to produce the yarns, may be given additional heat treatment to secure stability in a desired geometric configuration. A degree of stretch may be retained, or a flat and stable textured surface may be produced. There are a number of variations of the texturing process, which, combined with the many possibilities of heat setting, impart considerable versatility to the final product (see Fig. 12.18). The growth in the use of these products in the 1960s is well known. Carpeting also provides a significant market for them, as texture is one of the most important characteristics of soft floor coverings. Such products have been important to the successful use and expanded development of nylon and polyester yarns.

Staple Process

Unlike nylon, which is used mostly in the form of continuous filament yarn, polyester is utilized both in staple and in continuous filament form in large volumes. For producing staple fiber, the spinning machines consist of a series of packs, 10–30, each with 1000–3000 holes.

The extrudates from different packs after solidification and application of finish are combined to form a subtow and collected in a can. Then several such cans are brought together in a creel area, and subtows from these cans are combined to feed a staple drawline. The latter may involve one or more stages of drawing and relaxation, one or more stuffer-box type crimpers, a drying unit, and a cutter. A baling unit following the cutter collects and bales the cut fiber.

Continuous Filament Yarn Process Variants

For producing continuous filament yarn, several process routes are available,[12] each of which involves the basic processes of polymer synthesis, extrusion, quenching, stretching, and winding. In one process, specially meant for textile-type uses, spinnerettes with 20–100 holes typically are used. After solidification and finish application, the filaments may be split into smaller bundles, depending upon end-use applications, for downstream processing. The drawing can also be done in a single integrated process (spin-draw) immediately after finish application to yield a fully oriented yarn (FOY), which having no bulk or texture, is referred to as a flat yarn. Spinning speeds of the order of 1000 m/min or more and winding speeds of about 4000 m/min are used. Similar yarn may be produced on a separate drawing unit, but this process—called draw-twisting because a certain amount of twist is also inserted prior to winding—is less commonly used today. It is more usual to see the separate drawing process done in conjunction with a texturing process—most commonly false twisting, and called draw-texturing—to give a textured yarn. The process leads to orientation and crystallization of the filament structure and bulking of the yarn. The feed yarn for such processes usually is produced at 3000–4000 m/min spinning speed as POY. The latter is designed to be run on draw-texturing machines at speeds ranging from 800 to 1200 m/min.

For industrial applications, polyester filaments having high strength low shrinkage (HSLS), low creep properties, and high glass transition temperatures are targeted. To produce such a filament, more severe processing conditions and higher molecular weight polymer are generally used. The filament is spun at low speeds (500–1000 m/min), sometimes with retarded quench, to obtain minimum orientation. The drawing can be achieved in the more common integrated, or spin-draw, process or in a separate draw-twist operation. High strength is achieved by drawing the filament to several times its length over very high-temperature rolls and then heat-setting and relaxing the structure prior to winding. Low shrinkage properties are obtained with a relaxing step at high temperature.

Modifications and New Developments

As was the case with the nylon fibers, the potential polyester fibers offered in apparel, home furnishing, and industrial applications was judged to be enormous. For this potential to be realized in practice, however, some characteristics had to be improved, and others had to be engineered for specific end uses. Thus, fibers of different cross-sectional shapes were developed in order to impart anti-soiling, reflective, and resilient characteristics for rug and carpet applications. A difficulty associated with the early polyester fiber that restricted its applications was its lack of ability to take on dyes through one of many methods available for dyeing. This problem was overcome by introducing chemicals that added sulfonate groups to the molecule and by substituting in some cases isophthalic acid for a small portion of terephthalic acid. These changes allowed fibers to be dyed by cationic and disperse dyes, the dyes most frequently used for polyester. Another area of modification has been the development of inherently flame-resistant fiber. One process involves copolymerizing a derivative of phosphoric acid with PET. An exciting new development in polyester filament yarn for apparel uses is the production of microdenier fiber (denier per filament less than one), discussed later in a separate section. The introduction of finer-denier yarns opened up a whole new field for developing fabrics with special esthetic and performance characteristics that were not possible earlier.

With the use of fibers in conveyor belts, tires, and composites, fibers of greater strength and modulus and lower extensibility have been needed. Much effort was directed in the 1970s, and later, to developing such fibers from polyester. The composition and the properties of wholly aromatic polyamides or aramids are discussed in a later section. When both the diacid and diamine components are *para*-substituted aromatic compounds, the resulting polymer is capable of forming lyotropic liquid–crystalline solutions. These solutions can be dry- or wet-spun into fibers with unusually high tensile strength and tensile modulus. When a similar strategy is tried to make polyester fiber from a homopolymer of a *para*-substituted aromatic diacid and a *para*-substituted aromatic diol, only infusible and intractable materials are obtained. A solution to this problem has been found in the development of polyester copolymers that give thermotropic liquid–crystalline melts over a useful temperature range and have viscosities suitable for melt extrusion into fibers or films having high levels of orientation. Spin-line stretch factors of the order of several hundred percent are used to achieve orientation, and physical properties are developed further by heat treatment at temperatures approaching melting conditions.

The first fibers from a thermotropic liquid crystalline melt whose properties were reported were spun from a copolyester of *para*-hydroxybenzoic acid (PHB) and PET by workers at Tennessee Eastman Co. The preparation of the copolymer proceeds in two stages. First, *para*-acetoxybenzoic acid is reacted with PET in an acidolysis step to give a copolyester prepolymer, which in the second step is condensed further to a higher degree of polymerization suitable for fiber formation.

ponents. Some frequently used constituents, in addition to those mentioned above, are 2,6-naphthalene-dicarboxylic acid, hydroquinone, 4,4'-biphenol, isophthalic acid, and 4,4'-dihydroxy-diphenyl ether.

ACRYLICS

Polymer Manufacture

Acrylic fibers are spun from polymers that are made from monomers containing a minimum of 85 percent acrylonitrile. This compound may be made from hydrogen cyanide

When the mol. percent of PHB in the copolymer exceeds about 30–40 percent, a liquid–crystalline melt is obtained. Up to about 60 mol. percent, order in the melt increases and melt viscosity decreases. Compositions containing about 60 mol. percent PHB can be melt-spun into fibers using standard extrusion techniques. It is the unusual combination of properties that makes this class of materials valuable for the formation of high-strength fibers and plastics.

Among melt-spun fibers, those based on thermotropic liquid–crystalline melts have the highest strength and rigidity reported to date, and appear comparable to polyamides spun from lyotropic liquids–crystalline solutions. This was a very active field of research in the 1970s and later, and many comonomers have been reported. Obviously, these compositions must contain three components at a minimum, but many have four or five com-

and ethylene oxide through the intermediate ethylene cyanohydrin:

$$CH_2CH_2 + HCN \longrightarrow HOCH_2CH_2CN \xrightarrow[-H_2O]{catalyst}$$

$$CH_2{=}CHCN$$

It also may be made directly from acetylene and hydrogen cyanide:

$$CH{\equiv}CH + HCN \rightarrow CH_2{=}CHCN$$

But the reaction that currently is preferred uses propylene, ammonia, and air:

$$3CH_2{=}CHCH_3 + 3NH_3 + 7O_2\,(air)$$

$$\xrightarrow[<500°C]{Catalyst} CH_2{=}CHCN + 2CO + CO_2$$

$$+ CH_3CN + HCN + 10H_2O$$

Pure acrylonitrile may polymerize at room temperature to polyacrylonitrile (PAN), a compound that, unlike polyamides and polyesters, does not melt at elevated temperatures but only softens and finally discolors and decomposes. Nor is it soluble in inexpensive low-boiling organic solvents. Because fibers made from it resist the dyeing operations commonly used in the textile industry, the usual practice is to modify it by copolymerization with other monomers, for example, vinyl acetate, styrene, acrylic esters, acrylamide, or vinyl pyridine in amounts up to 15 percent of the total weight (beyond which the final product may not be termed an acrylic fiber). The choice of modifier depends on the characteristics that a given manufacturer considers important in a fiber, the availability and cost of the raw materials in the manufacturer's particular area of production, and the patent situation.

In copolymerizing acrylonitrile with another monomer, conditions must be controlled in such a way that the reaction produces a polymer having the desired chain structure and length. The reaction takes place in the presence of substances capable of producing free radicals. In addition, certain trace metals that have been found to increase reaction rates offer a means of controlling chain length. When polymerization is carried out in solution, after an induction period, the reaction is rapid and liberates a considerable amount of heat. Furthermore, because the polymer is not soluble in the monomer, a thick paste is formed. These facts limit the usefulness of such a process. Carrying out the polymerization in the presence of a large amount of water (water/monomer of 2/1 to 3/1) is a convenient method and the one most generally used. In this case the polymer forms a slurry, and the water provides a means for removing the heat from the site of the reaction. Moreover, most of the common redox-catalyst systems are water-soluble. Polymerization may be carried out batchwise or by a continuous process.

In the batch method, the monomers and catalyst solutions are fed slowly into an agitated vessel containing a quantity of water. The heat of reaction is removed either by circulating cold water through the jacket surrounding the vessel or by operating the reaction mixture at reflux temperature and eliminating the heat through the condenser water. The monomer and catalyst feeds are stopped when the desired amounts have been added, and polymerization is allowed to continue until there is only a small amount of monomer remaining in the reaction mixture. Then the slurry is dumped from the reaction vessel, filtered, washed, and dried (see Fig. 12.19).

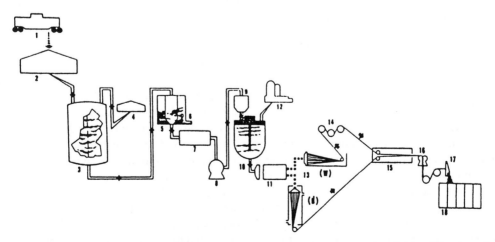

Fig. 12.19. Flow diagram for the manufacture of acrylic fiber: (1) acrylonitrile; (2) tank farm; (3) polymerizer; (4) comonomer and catalyst; (5) centrifuge; (6) waste liquid; (7) dried polymer; (8) grinding; (9) polymer storage; (10) dissolver; (11) filter; (12) solvent plant; (13) spinnerette; (13w) wet spinning; (13d) dry spinning; (14) roller dryer; (15) additional treatment; (16) crimper; (17) cutter; (18) acrylic fiber bale.

In the continuous-overflow method, rather than stopping the monomer and catalyst feed when the reaction vessel is full, the slurry is simply allowed to overflow; the solids are removed by filtration, washed, and dried. The filtrate contains a certain amount of unreacted monomer, which is recovered by steam distillation after the trace metal present has been chelated to stop the redox reaction and thus further polymerization. The dried polymer is the raw material from which fibers are spun.

As mentioned, PAN polymer has two major weaknesses: (1) extremely poor dyeing characteristics due to highly ordered structure, and (2) very low solubility in most of the solvents. To overcome these problems, comonomers are incorporated into the chains. Comonomers have been used to improve hydrophilicity, dyeability, flame retardancy, heat resistance, and so on. For obtaining hydrophilic acrylic fiber, comonomers containing hydrophilic functional groups such as hydroxyl, ester, carboxyl, amide, and substituted amide are used. To make the fiber base-dyeable, comonomer containing carboxylic and sulfonic acid groups is introduced; and to obtain acid-dyeable fiber, comonomers containing pyridine, aliphatic amine, or quaternary ammonium salt are used. Antistatic acrylic fibers can be made by incorporating in the chain polyethylene oxide, polyalkylene derivatives, polyethylene glycol, acrylates, or methacrylates as the comonomers. For improved flame retardancy, halogen-containing vinyl comonomers are used.

Spinning

As already indicated, pure PAN softens at elevated temperatures, and thermal decomposition starts before the molten state is reached. The same is true of the copolymers commonly used to produce fibers. Accordingly, melt spinning is impossible; spinning must be done from a solution of the polymer. Both dry and wet spinning are carried out in current commercial operations.

The operations used to either wet- or dry-spin acrylics are essentially the same as those already described for rayon and acetate, respectively. The polymer must be completely dissolved in solvent and the solution filtered to remove any impurities that would cause spinnerette blockage. Because acrylic polymers are not soluble in common nonpolar solvents, polar substances such as dimethylformamide, dimethylacetamide, or aqueous solutions of inorganic salts such as zinc chloride or sodium thiocyanate are required. Only wet spinning is possible with the latter. Dimethyl formamide boils at $152.8°C$ and exerts a vapor pressure of 3.7 mm of Hg at $25°C$ compared with acetone (used in dry spinning of cellulose acetate), which has a vapor pressure of 228.2 mm of Hg at $25°C$. It follows that, unlike acetone which requires an activated-carbon system for recovery, dimethylformamide may be condensed directly from the gas stream used to evaporate the solvent from the forming fiber.

In order to obtain the desired characteristics of modulus, rupture tenacity, and rupture elongation, acrylics, like rayon, require stretching which is usually carried out after the fiber has been spun, either as part of the fiber washing operation or after the fiber has been dried. These same properties are influenced by spinning speeds, and the temperature of the drying air, if they are dry-spun, or the temperature and the composition of the bath, if wet-spun. The multitude of combinations made possible by the use of various comonomers and the flexibility of the fiber-forming operations furnish the different manufacturers with versatility and the users with a variety of acrylic fibers. Figure 12.20 shows a wet-spinning operation for acrylic tow.

Acrylic fibers possess a property that made it possible for them in the late 1950s and early 1960s to find immediate, even spectacular, acceptance in the knitted sweater field, until then dominated by wool. When acrylic fibers, normally in the form of a heavy tow, are hot-stretched (e.g., by being drawn over a hot plate and then cooled under tension), they are converted to a labile state. Upon immersion in hot water, such fibers will contract considerably, but not to their prior unstretched length. In practice, this characteristic is used to

Fig. 12.20. Wet spinning of acrylic tow. (*Courtesy Monsanto Co.*)

produce a bulky yarn resembling the woolen yarns long accepted for use in sweaters. The process is described briefly below.

Using "stretch-break" equipment, the stretched labile fibers are further cold-stretched to the breaking point so that the fiber breaks at different points leading to a distribution of fiber lengths, similar to the lengths found in wool. These are crimped and then mixed with thermally stable acrylic fibers that have been stretched and relaxed and have about the same length and degree of crimp. The blend is converted to a spun yarn by the same process used in making woolen yarns, and in turn this yarn is knitted into sweaters and other similar products. When such garments are dyed in hot water, the labile fibers, intimately blended with stable ones, contract lengthwise individually. In the process, segments of the stable units tend to be carried along physically by entrapment and friction; but because such fibers do not change their overall length, the yarn as a whole decreases in length. Lateral displacement of the large volume of stable fibers results in the formation of a more voluminous structure known as "hi-bulk" yarn.

Bicomponent or Conjugate Spun Fibers

As will be shown, it should be theoretically possible to make any of the common manufactured fibers in bicomponent forms. However, acrylics have received the most attention for quite good reasons. Their general characteristics have tended to make them competitive with wool. This means that they should be processible on machinery developed for handling wool, as well as capable of being accepted into markets previously dominated by an animal hair fiber. It follows that because the natural fiber possesses crimp which produces the cohesion that determines its behavior in processing and in part its appearance and "hand" in usage, a similar crimp was desired for acrylics.

The principle that is the basis for bicomponent fibers usually is likened to that which underlies the bicomponent metal strips often used in temperature controllers. With the latter, differential-thermal expansion of the two joined components results in a bending of the thermal element. With fibers, moisture usually is the agent that acts upon the two side-by-side portions. Differential swelling or shrinkage causes the fiber to be brought into a crimped, or preferably, a spirally distorted condition. As such, the side-by-side structure exists naturally in wool.

The combination of small size and large number of holes in a spinnerette might lead one to conclude that it would be almost impossible to design a spinnerette assembly that could bring two streams of polymer or polymer solutions together at each such hole and extrude them side-by-side to form a single filament. Such designs have, in fact, been made; but solutions of fiber-forming polymers fortunately possess properties that encourage laminar flow and thus make other approaches possible. This phenomenon was remarked upon earlier in connection with dope dyeing; when a suspension of a colored pigment is injected into a dope stream, a considerable problem must be overcome to achieve adequate mixing so as to secure "dope dyed" fibers of a uniform color. Thus, it was known that when two streams of essentially the same solution of a fiber-forming polymer are brought together, side-by-side, and moved forward down a pipe or channel by the same amount of pressure behind each, virtually no mixing takes place. By bringing these streams to each spinnerette hole in such an individual side-by-side arrangement and using appropriate mechanical separators, the extruded filament from each hole will have a bicomponent structure. In addition to producing fibers in which the two components form a bilateral symmetrical structure, an ingenious arrangement of predividers of the two streams can produce from the full complement of holes in a single spinnerette a selected group of fibers wherein the amount and the position of each of the two components are randomly distributed throughout their cross-sections. It follows that curls of uniform or random geometry may be produced to meet the required needs.

The worldwide production of acrylic fiber has declined significantly over the years because of the environmental concerns associated with the solution-spun process. In view of this, scientists have sought over the years a method that could render the high acrylics melt-spinnable. Such a method would not only be economical and environmentally friendly but also allow for engineering the fiber with a wider range of morphologies and properties. In 1997, British Petroleum patented a polymerization process in which the two components usually used in developing spinnable acrylic copolymer were redistributed to allow the resulting material to be melt processable.[13,14] Preliminary findings show that the polymer can be melt-spun into reasonable fine denier fibers with mechanical properties expected of the usual solution-spun material.[15]

VINYL AND MODACRYLIC FIBERS

Vinyls

When nylon 66 was developed, it was described as being "synthetic" or "fully synthetic" in order to differentiate it from rayon and acetate. This was no small act of courage, as the word "synthetic," in that period just following the repeal of Prohibition in the United States, was often associated in the public mind with the least palatable kind of alcoholic beverages. In due time, what is known in the advertising business as "puffing" led it to be known as the "first fully synthetic fiber," which was an anachronism. It so happens that fibers based upon polyvinyl chloride (PVC) predated nylon by several years.

About 1931, the production of fibers from PVC was accomplished by dry spinning from a solution in cyclohexanone. But by chlorinating the polymer, it was possible to secure solubility in acetone, which has the advantage of possessing a boiling point about 100°C lower than that of cyclohexanone. Several million pounds per year of this fiber were produced in Germany during World War II to relieve the shortages of other materials. Unfortunately, PVC begins to soften at about 65°C, and in the fibrous state, it shrinks disastrously upon heating. Because of its low softening point, it cannot be dyed at the temperatures commonly used for this purpose, and, furthermore, it resists dyeing.

Modifications of PVC have been produced by copolymerization with other monomers.

The first successful one consisted of 90 percent vinyl chloride copolymerized with 10 percent vinyl acetate. It was dry-spun from acetone and given the trade name Vinyon by its producer, Union Carbide Corporation. (In 1960, vinyon was accepted as a generic name for fibers containing not less than 85 percent vinyl chloride.) It has never been produced in large volume; it is used for heat-sealable compositions.

A copolymer of vinyl chloride with vinylidene chloride was used for a number of years to produce melt-spun, heavy monofilaments, which found use in heavy fabrics, where the chemical inertness of the polymer was needed, in outdoor furniture, and in upholstery for seats in public-transportation vehicles.

Another vinyl-based fiber, polyvinyl alcohol, or vinal, was developed in Japan but has not been produced or used in the United States. As such, it illustrates the importance of both relative availability of raw materials and differences in markets, in the success of a chemical product. Acetylene made from calcium carbide is converted to vinyl acetate, which, following polymerization, is saponified to polyvinyl alcohol.

$$CH\equiv CH + CH_3\overset{\overset{\displaystyle O}{\|}}{C}\!-\!OH \longrightarrow$$

$$CH_2\!=\!CH\overset{\overset{\displaystyle O}{\|}}{O}CCH_3 \xrightarrow[\text{catalyst}]{\text{heat}}$$

$$\underset{\underset{\underset{\displaystyle O}{\overset{\displaystyle |}{}}}{}}{+CH_2CH+}_{n} \xrightarrow{H_2O} +CH_2\underset{\underset{\displaystyle OH}{|}}{CH}+_{n} + CH_3COOH$$

$$CH_3\overset{\overset{\displaystyle }{C}}{\underset{\displaystyle O}{}}$$

The polyvinyl alcohol is soluble in hot water, and the solution is wet-spun into a coagulating bath consisting of a concentrated solution of sodium sulfate. The fibers are heat-treated to provide temporary stability so that they may be converted to the formal derivative by treatment with an aqueous solution of formaldehyde and sulfuric acid. This final product resists hydrolysis up to the boiling point of water. It seems reasonable to assume that it contains hemiacetal groups and some unreacted hydroxyls on the polymer chain as

well as cross-linking acetyl groups between the adjacent molecules.

Under the trade name Kuralon® (Kuraray Co., Ltd), it achieved a production level of about 180 million lb in 1970, but production dropped to 16 million lb of continuous filament and 87 million lb of staple in 1980. The former has been mainly used in industrial rubber products, and the latter has been used mostly for uniforms, nonwoven and coated fabrics, and filters.

Modacrylics

In the United States, the modification of PCV has moved in the direction of copolymerizing vinyl chloride with acrylonitrile, or perhaps it should be said that PAN has been modified by copolymerizing the acrylonitrile with chlorine-containing vinyl compounds. In any case, one modacrylic fiber is currently produced in the United States, a modacrylic being defined as containing at least 35 percent but not over 85 percent acrylonitrile.

The first two modacrylic fibers ever introduced in the United States were Dynel® (by Union Carbide) in 1949 and Verel® (by Tennessee Eastman) in 1956. The former was a copolymer of 60 percent vinyl chloride and 40 percent acrylonitrile, and the latter was said to be a 50–50 copolymer of vinylidene chloride and acrylonitrile with perhaps a third component graft-copolymerized onto the primary material to secure dyeability. SEF® and its version for wigs, Elura®, were introduced by Monsanto Fibers in 1972. A few foreign manufacturers are making modacrylic fibers, but the only modacrylic fiber currently in production in the United States is SEF®.

Modacrylic fibers, like acrylic, require after-stretching and heat stabilization in order to develop the necessary properties. It is thought that the stretching is of the order of 900–1300 percent, and that, in a separate operation, shrinkage of about 15–25 percent is allowed during the time that the fibers are heat stabilized.

The modacrylic fibers, like vinyon and unlike the acrylic fibers, have not become general purpose fibers. They can be dyed satisfactorily and thus are acceptable in many normal textile products; but their non-flammability tends to place them in uses where that property is important, even vital. Blended with other fibers, they are used in carpets; but their largest market is in deep-pile products, such as "fake furs," or in doll hair, where a fire hazard cannot be tolerated.

ELASTOMERIC FIBERS

The well-known elastic properties of natural rubber early led to processes for preparing it in forms that could be incorporated into fabrics for garments. One such process uses standard rubber technology. A raw rubber of high quality is compounded with sulfur and other necessary chemicals, calendered as a uniform thin sheet onto a large metal drum, and vulcanized under water. The resulting skin is spirally cut into strips that may be as narrow as they are thick, for example, 0.010 in. by 0.010 in.^{2} in cross-section. These strips are desulfurized, washed, dried, and packaged. Larger cross-sections are easier to make. This product, coming out of the rubber rather than the textile industry, is known as a thread.

Another method produces a monofilament known as a latex thread. As the name would indicate, rubber latex is the raw material, and because extrusion through small holes is required, the purity of the material must be of a high order. With proper stabilization, the latex solution may be shipped from the rubber plantation to the plant, where it is compounded with sulfur and other chemicals needed for curing, as well as with pigments, antioxidants, and similar additives. This is followed by "precuring" to convert the latex to a form that will coagulate upon extrusion into a precipitating bath of dilute acetic acid and will form a filament having sufficient strength for subsequent operations. It passes out of the bath and is washed, dried, vulcanized in one or two stages, and packaged.

The rubber threads manufactured by either process can be used as such in combination with normal nonelastomeric yarns in fabrics made by weaving or knitting; but most of them, especially those made by a latex process, first are covered by a spiral winding of natural or manufactured yarns. Often two layers are applied in opposite directions to minimize the effects of torque. Such coverings have two purposes. The first is to replace the less desirable "feel" of rubber on human skin by that of the more acceptable "hard" fiber. The second concerns the engineering of desired properties into the product to be woven or knitted into fabric. As an elastomeric material begins to recover from a state of high elongation, it supplies a high stress; but as it approaches its original unstretched condition, the stress drops to a very low order. When wound in an elongated state with a yarn having high initial modulus and strength, the elastomeric component cannot retract completely because its lateral expansion is limited, and jamming of the winding yarn occurs. Thus, the combination of such materials can be made to provide stretch and recovery characteristics needed for a broad spectrum of applications.

The traditional elastomeric threads have been subject to certain inherent limitations, however. The presence of unreacted double bonds makes them sensitive to oxidation, especially with exposure to the ultraviolet radiation of direct sunlight. They also have low resistance to laundry and household bleaches and dry-cleaning fluids.

During recent years, elastomeric yarns or threads have been used to impart comfort, fit, and shape retention to a variety of garments such as women's hosiery and swimwear. Such garments must be thin and highly effective per unit of weight. The materials of which they are composed must be compatible with these requirements. Thus, it was not unexpected that the producers of manufactured fibers, already eminently successful in meeting the needs of the marketplace, should look to the field of elastomeric fibers for new possibilities. Given the limitations of rubbers, both natural and synthetic, as well as the relationships between molecular structure and behavior of fiber-forming linear polymers, the scientists faced new challenges.

As an oversimplification, it can be said that within limits a rubberlike material can be stretched relatively easily but reaches a state where crystallization tends to occur. The structure produced in this manner resists further extension, and the modulus rises sharply. In contrast to the conditions that occur when the manufactured fibers discussed earlier such as nylon or polyester are drawn to form fibers of stable geometry in the crystalline and oriented states, the crystalline state of the elastomeric fibers is labile unless the temperature is lowered materially. Thus, to improve on the chemical sensitivity of rubber, new approaches were necessary. The solution was found in developing linear block copolymers containing "soft" liquidlike sections that impart elasticity, connected with "hard" components that act as tie points to hold the structure together.

The soft, flexible, and low-melting part is commonly an aliphatic polyether or a polyester with hydroxyl end groups and molecular weight in the range of 500–4000. The hard portion is derived from an aromatic diisocyanate supplied in an amount that will react with both end groups of the polyether or polyester to form urethane groups. The product, an intermediate known as a pre-polymer, is a thick liquid composed essentially of molecules carrying active isocyanate groups at each end. For example:

$$HO(RO)_n + m\ OCNR'NCO \rightarrow$$

$$OCNR'N\overset{HO}{\underset{|}{C}}O(RO)_n\overset{OH}{\underset{|}{C}}NR'NCO$$

where—(RO)—is an aliphatic polyether chain, R' is one of several commonly available ring structures, $n \sim 10$–30, and $m \sim 1.5$–2.

The elastomeric polymer is obtained by "extending" the prepolymer through its reaction with short-chain diols such as butanediol or diamines such as ethylene diamine, thus completing the formation of hard groups between soft, flexible chains. When amines are used, the final step is typically done in a polar solvent such as dimethyl acetamide. The conversion of these polymers into usable fibers may be accomplished by wet-, dry-, or melt-spinning operations, depending on the polymer. Additives to impart whiteness or improve resistance to ultraviolet radiation and oxidation may be incorporated in the spinning solutions or in the melts.

The development of elastomeric fibers has resulted in a variant of wet spinning called "reaction" or "chemical" spinning. In point of fact, rayon, the first wet-spun material, might properly be said to be produced by "reaction wet spinning" or "chemical wet spinning" because complex chemical reactions always have been involved in that operation. In any case, it has been found that the prepolymer of an elastomeric fiber may be extruded into a bath containing a highly reactive diamine so that the chemical conversion from liquid to solid occurs there.

The elastomeric fibers produced in this fashion are based upon segmented polyurethanes and by definition are known generically as spandex yarns. Each manufacturer uses a trade name, for the usual commercial reasons. Perhaps the most noteworthy aspect from the standpoint of industrial chemistry is the multitude of options available to the manufacturer through the ingenious use of various chemicals for soft segments, hard units, chain extenders, and conditions of chemical reaction, followed by numerous possibilities for extrusion and after-treatments. In the United States, there are two main producers of spandex fibers: DuPont (Lycra®) and Globe Rubber Co. (Cleerspan®, Glospan®). There are numerous worldwide producers, including: Bayer, Germany (Dorlastan®); Asahi, Japan (Roica®); Nisshinbo, Japan (Mobilon®); and Tae Kwang, Korea (Acelan®).

POLYOLEFIN FIBERS

Polypropylene

Although polyethylene was considered a source of useful fibers at an early date, its low melting point (110–120°C) as well as other limitations precluded active development during the period when production of other fibers based upon the petrochemical industry expanded enormously. The higher melting point of high-density polyethylene gave some promise, but it was overshadowed by the introduction of polypropylene (PP) around

1958–1959. Great expectations were held for the latter as a quick competitor with the polyamides and the polyesters, already successful, as well as the acrylics, which then were entering the fiber field in volume. PP was thought to have several advantages. The raw material costs were low, only a few cents a pound; also there was a high level of sophistication in the spinning and processing of fibers, and a presumption that this would readily lead to the development of means for converting the polymer to fibers; and, finally, there was the belief that the American consumer would be ready to accept, and perhaps even demand, something new and different, which this polymer offered. However, the limitations of PP fibers, such as lack of dyeability, low melting temperature, low heat stability, and poor light stability, combined with the lower prices and the greater versatility of the already established fibers, dashed the hopes for quick success. However, all of these deficiencies except the low melting temperature and lack of dyeability now have been overcome. The fiber has found an increasingly important place, and its properties have led to new techniques of manufacture and specialized uses.

The structural formula of PP is as given below, where $100,000 < n < 600,000$ for chips or granules, and $50,000 < n < 250,000$ for fibers:

$$\left[\begin{array}{c} -CH_2-CH- \\ | \\ CH_3 \end{array}\right]_n$$

The steric configuration is extremely important in the polymer. Only isotactic polypropylene (iPP) has the properties necessary for forming fibers. The molecules are crosslinked only by Van der Waals forces, so it is important that they pack as closely as possible. The isotactic molecules form a 3_1 helix, as shown in Fig. 12.21,[16] and exhibit a high crystallization rate. The atactic molecules, shown in the figure, do not pack well, and although the syndiotactic molecules can pack better and crystallize, this configuration is not a normal product of commonly used catalyst systems.

Some properties of isotactic, syndiotactic, and atactic PP are listed in Table 12.2.[17] The insolubility of iPP in hydrocarbon solvents at room temperature often is used to separate iPP from atactic polypropylene (aPP).

Early in the manufacture of PP, a concept was developed for dry spinning directly from the solution obtained in the polymerization operation. Had it been feasible, it would have been the realization of a chemical engineer's dream: the gaseous olefin fed into one end of the equipment, and the packaged fiber, ready for shipment to a textile mill, coming out the other end. But it did not turn out that way, and today melt spinning is the accepted technique for the production of staple fibers, monofilament, and multifilament yarns. To this usual method have been added the fibrillation and the "slit film" procedures for producing yarns.

The PP materials are completely resistant to bacterial attack, are chemically inert, and are unaffected by water. Monofilaments can be produced that possess high strength, low elongation under stress, and dimensional stability at normal atmospheric temperatures.

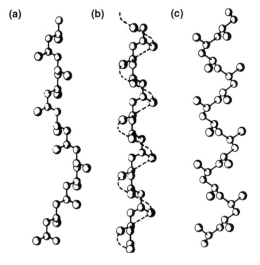

(a)   (b)   (c)

Fig. 12.21. Diagrams showing (a) irregular atactic, (b) stereoregular isotactic, and (c) stereoregular syndiotactic configurations in polypropylene polymer. (Source: Ahmed, M., "Polypropylene Fibers–Science and Technology," *Textile Science and Technology*, **5**, 16, Elsevier Science Publishers SV, Academic Publishing Div., New York, Amsterdam (1982).) Copyright © M. Ahmed. By permission.

TABLE 12.2 Properties of Isotactic, Syndiotactic, and Atactic Polypropylene

Property	Isotactic	Syndiotactic	Atactic
Density, g/cm³	0.92–0.94	0.89–0.91	0.85–0.90
Melting Point, °C	165	135	
Solubility in hydrocarbons at 20°C	None	Medium	High
Yield strength	High	Medium	Very low

Source: Lieberman, R. B., and Barbe, P. C., "Propylene Polymers," in *Concise Encyclopedia of Polymer Science and Engineering*, J. I. Kroschwitz (Ed.), p. 916, 1990. Copyright © John Wiley and Sons and reproduced by permission of the copyright owner.

PP monofilaments have found broad application in cordage and fishing nets (which float), and if highly stabilized they are woven into fabrics used for outdoor furniture, tarpaulins, and similar applications. Large filament denier staple is used widely in "indoor–outdoor" carpets. Also, staple fibers have found major applications in tufted indoor carpets and nonwovens used for diaper, filtration, and civil engineering fabrics.

Synthesis. The early PP plants used a slurry process adopted from polyethylene technology. An inert liquid hydrocarbon diluent, such as hexane, was stirred in an autoclave at temperatures and pressures sufficient to keep 10–20 percent of the propylene monomer concentrated in the liquid phase. The traditional catalyst system was the crystalline, violet form of $TiCl_3$ and $AlCl(C_2H_5)_2$. Isotactic polymer particles that were formed remained in suspension and were removed as a 20–40 percent solid slurry while the atactic portion remained as a solution in the liquid hydrocarbon. The catalyst was deactivated and solubilized by adding HCl and alcohol. The iPP was removed by centrifuging, filtration, or aqueous extraction, and the atactic portion was recovered by evaporation of the solvent. The first plants were inefficient because of low catalyst productivity and low crystalline yields. With some modifications to the catalyst system, basically the same process is in use today.

In 1963, liquid polymerization was introduced in which liquid propylene, catalysts, and hydrogen were pumped continuously into the reactor while polypropylene slurry was transferred to a cyclone separator. The unconverted monomer gas was removed, compressed, condensed, and recycled, and the polymer was treated to reduce the catalyst residue. This system also suffered from a poor catalyst yield, and the polymer produced lacked the required stereospecificity, so that it was necessary to remove the atactic portion of the polymer.

In the mid-1960s, a gas phase process was introduced for production of the polymer. The monomer was pumped over adsorbing beds and entered the reactor with the catalyst system. These feed streams of monomer and catalyst, together with a mechanical stirrer, created a turbulent bed of powdered polymer. Periodically the polymer powder was vented off in a carrier gas to extrusion storage hoppers. Meanwhile, the heat of polymerization was removed by condensing the unreacted monomer in a cooling loop and returning it to the reactor, where it immediately vaporized. This process eventually led to the production of highly crystalline products and was adopted by several companies in the United States.

Most processes in use today rely on a combination of these technologies. Montedison's introduction in 1975 of third-generation catalysts gave high yields and allowed polymerization to take place at 60–80°C and 2.5–3.5 MPa (362–507 psi). This was welcome news during the energy crisis, but the resulting polymer was not stereospecific enough to eliminate the need for removal of aPP. Real progress came with the discovery of superactive third-generation catalysts, which gave both the optimal yield and stereospecificity.[18]

Production. Classical melt spinning, which was developed for the production of nylon filaments, is widely used to produce PP fibers today. It involves a high-speed process (2000–3000 m/min) that is particularly suitable for long production runs. The average molecular weight of polypropylene polymer, like that of other addition (olefin) polymers, is relatively high compared to that of other polymers. This results in a high melt viscosity; so, unlike the case of other polymers, its extrusion temperatures are 70–100°C above its melting point. Single-screw extruders are used for melting and homogenizing the polymer. The screw diameters are from 45 to 200 mm and screw lengths are 24, 30, or 36 times the diameter. The polymer granules are fed into the extruder hopper, where they are melted and homogenized. Chips carrying pigments can be fed into and blended with the main charge of the extruder if colored fibers are desired. The molten polymer is forced through the spinnerette via a screen pack to eliminate any contaminant particles. The spinnerette hole determines the shape of the filaments, and the flow rate and the takeup speed determine the size. The polymer has high specific heat and low thermal conductivity, so the cooling zones must be longer than those for polyester and nylon. For filament yarn production, the filaments are drawn at high speed and wound on packages. For staple fiber production, the filaments are collected in the form of tow and then are drawn, crimped, and cut. The multifilament yarns are often textured to improve bulk and appearance. The false twist method is generally used to texture finer yarns, with the stuffer box used for coarser yarns.

The short spinning method used to produce staple fiber is considerably slower than high-speed spinning. The lower spinning speeds (30–150 m/min) would have a negative effect on productivity, but this is counteracted in industry by the use of spinnerettes with a large number of holes (up to 55,000). The required cooling zones are much shorter because of the lower speeds and the use of higher volumes of quench air, which gives this method its name. Because the drawing units can match these low speeds, the two can be fed directly and continuously from the spinning machine to the draw frame to the

texturing chamber or the crimper, to produce bulked continuous filament (BCF) yarn or staple fiber, respectively. The short spinning method is used to produce high-tenacity fibers. Many PP yarns are produced using the slit film method. The film extruder is almost identical to a filament extruder. In it, the molten polymer is forced through a film die that converts the melt into film, where the thickness of the film can be controlled by adjusting one of the die lips. The takeoff unit is either a chill roll that removes the film uniformly and cools it below T_m or a water bath followed by nip rolls. The cooled film is slit into separate tapes using a slitter bar that contains a large number of special knives separated by spacers. The film tapes are heated and drawn to their final length.

Several methods are used to produce fibrillated film. They are produced in much the same way as slit film, but these techniques take advantage of the tendency of PP to fibrillate. In one method, a profiled tape is extruded and hot-drawn. It is drawn again to achieve a 10:1 ratio, and the film splits into separate filaments. In another method, called roll embossing, the film is hot-drawn, and then embossed using profiled rollers. The profiled film is drawn again, and fibrillation results. In the pin-roller technique, the film is drawn and cut by knives or pins on a rotating cylinder. This method can produce individual fibers or a controlled web network.

PP nonwovens are created by forming a staple fiberweb and then consolidating it into a fabric. The fibers can be entangled by a needling machine, an air jet, or a water jet. Another method uses a single-stage process in which melt-spun fibers are drawn through an air aspirator jet and deposited randomly on a conveyor. The fibers then are bonded by fusion under heat and pressure to give a "spunbonded" nonwoven fabric (see Fig. 12.22). Yet another method, known as melt blowing, can be used in producing a nonwoven web having special characteristics. The melt-blown (MB) fibers are characterized as ultrafine fibers because of their size relative to other fibers. In the process, molten polymer is forced through a melt blowing die and die tip orifice that are

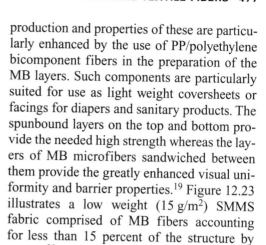

production and properties of these are particularly enhanced by the use of PP/polyethylene bicomponent fibers in the preparation of the MB layers. Such components are particularly suited for use as light weight coversheets or facings for diapers and sanitary products. The spunbound layers on the top and bottom provide the needed high strength whereas the layers of MB microfibers sandwiched between them provide the greatly enhanced visual uniformity and barrier properties.[19] Figure 12.23 illustrates a low weight (15 g/m^2) SMMS fabric comprised of MB fibers accounting for less than 15 percent of the structure by weight.[20]

The production figures of PP show impressive growth. The volume, which was less than 10 million lb in 1965, reached over 1.5 billion lb in 1990. This included approximately 17 percent filament yarn, 23 percent staple fiber, 18 percent spunbonded fabrics, and 38 percent split film products.

It is recognized that despite its many advantages compared with other synthetic fibers, the use of the fiber is restricted by the fact that most dyestuffs lack affinity for the material. This makes the fiber unsuitable for conventional exhaustion dyeing techniques using disperse or acid dyes. A number of attempts have been made in recent years to improve dyeability of the material.[21] In one of the most recent studies an alloy fiber of polypropylene (PP) and nylon 6 (N6) has been produced by melt spinning. The fiber contained PP grafted with maleic anhydride (MAH) as the third component. The MAH units reacted with the amine end groups of N6 to form block or graft copolymers that acted as a compatibilizer for the blend. Addition of 10% N6 produced a fiber that could be dyed conventionally using disperse dyes.[22]

High Molecular Weight

Polyethylene is probably the material one encounters most abundantly in daily life, such as in the form of grocery bags, shampoo bottles and toys, but now one also finds it as the material in such super high performance products as bullet-proof vests. For such a

Fig. 12.22. Spunbonded polypropylene showing interfiber bonding, which binds the structure. (*Courtesy E. I. DuPont de Nemours & Co.*)

finer than most orifices used to make manufactured fibers. As the polymer emerges, it is attenuated by a jet of high-velocity hot air. This allows the polymer to stay in a molten state but attenuate until broken. The fibers then come in contact with cool quench air, which causes the former to solidify. The fibers deposit on a collector screen and form a MB web. These webs, composed of very fine size fibers, can be engineered for applications in areas where barrier (such as against bacteria), insulative, and absorptive (such as for oil) characteristics are important.

One of the most recent developments in the use of polyolefin fibers is in composites of spunbond (SB) and melt-blown systems. Some examples of structures being made are SB/MB, known as SM, SB/MB/SB, or SMS, and other combinations of SB and MB. The

Fig. 12.23. Spunbond/melt-blown composite, SMMS, containing two layers of melt-blown microfibers sandwiched between two layers of spunbonded material and bonded by hot calendaring. (*Source:* Madsen, J. B., *Nonwovens World*, p. 69, 2001, August–September.)

versatile fibrous material, it has the simplest structure of all known polymers:

$$\left(\begin{array}{cc} \overset{\text{H}}{\underset{\text{H}}{\text{C}}} & \overset{\text{H}}{\underset{\text{H}}{\text{C}}} \end{array}\right)_n$$

However, during the general addition polymerization process, some of the carbon atoms, instead of having hydrogen attached to them, have segments of polyethylene chains grown on them, leading to a branched or low-density polyethylene (LDPE). If branching is eliminated or greatly minimized, such as found with the use of the Ziegler–Natta polymerization process, utilizing special catalysts, one can get linear chains capable of greater packing. This provides a high-density polyethylene (HDPE) material suitable for fiber use. Molecular weights of the order of one million have been achieved resulting in ultrahigh molecular weight polymer (UHMWPE). Accordingly, a recent addition to the group of high performance fibers is the ultrahigh molecular weight, extended-chain, linear polyethylene. Although fundamental work in the area of developing fully oriented and crys-talline structures in polyethylene polymer had been going on since the mid-1960s, it was not until the late 1970s that the possibility of producing such materials on a commercial scale became evident. Presently, three companies worldwide are manufacturing extended-chain polyethylene fibers, which have very high strength and high modulus characteristics.

Routes to High Performance. The achievement of ultimate strength and high modulus in fibers has been a subject of great interest to material scientists and fiber producers. In an attempt to identify the preferred structures for high performance, many theoretical analyses have been conducted to calculate the limiting values. The basis of such calculations is the assumption that in a fiber in which all polymer chains of infinitely long dimensions are extended and oriented parallel to the axis, rupture will occur only when the stress exerted exceeds the intramolecular bond strength. Such calculations show that the limiting values of conventional fibers usually are several times higher than those obtained in actual practice. The reasons for this difference lie in one or more of the following explanations: (1) the molecular weights are not high enough; (2) the chains are not fully extended; (3) the chains are not fully oriented. For a given polymer system and molecular weight, tensile strength and modulus can be enhanced by extending and orienting the chains. Practically, many modern techniques, including zone stretching, multiple-step stretching, and state-of-the-art high-speed extrusion methods, have been adopted to achieve such results.

Still, the results of studies on conventional fibers show that modifications in physical processing alone cannot lead to values that even approach the theoretical maximum within an order of magnitude. Flexible chains, which characterize the bulk of the commercial polymers, tend to conform to a random coil or folded chain structure in an as-spun material, and are very difficult, to reorganize into an extended-chain structure by known methods. Accordingly, in order to achieve ultimate properties, either novel spinning methods are needed, which allow flexible chains to be fully extended, or the chains used must be so configured that they have high intrinsic stiffness and would remain extended in solution or a melt. Both of these goals have been met and are exemplified in the production of (1) high-strength polyethylene fibers, discussed in this section, and (2) aramid fibers, discussed in the next section.

Extended-chain polyethylene fiber became available commercially in 1984 when DSM, a Dutch firm in the Netherlands, introduced Dyneema®, and Mitsui Petrochemicals in Japan announced Tekmilon®. Allied Signal of the United States entered the field in 1985 when it introduced Spectra® fibers. These materials are characterized by very high strength and modulus, which are achieved by the use of ultrahigh molecular weight polyethylene spun by the gel spinning method into fibers having extended-chain structures and near perfect orientation.

Gel Spinning. In general, the purpose for which the gel spinning method is used is to produce an as-spun fiber that contains a loose network of chains with few entanglements, which then can be drawn out to ultrahigh levels to yield a highly oriented structure. The surface-growth method of Pennings,[23] which uses a Couette type apparatus, and the gel spinning method of Smith–Lemstra,[24] which uses a more conventional spinning apparatus, led to the achievement of this goal. In the former method, a polymer solution is stirred between two counterrotating cylinders that provides the elongational flow necessary for initial chain alignment. Essentially the same result is achieved by passing the polymer solution through a constriction prior to spinning in the latter method.

In the surface-growth technique, polymer solution between the inner rotating cylinder and the outer stationary cylinder is maintained within a certain temperature range above the polymer crystallization temperature. A fiber seed is immersed through an opening in the outer cylinder, and its tip is made to attach to the polymer layers absorbed on the inner rotating cylinder. By pulling on the other end of the fiber and winding it onto a bobbin, while at the same time replenishing the solution in the gap between the cylinders, a fiber can be produced continuously. This process results in a fibrous precipitate with a "shish-kebab" morphology (Fig. 12.24). One can conclude, then, that the flow field that is developed extends the chains, which then crystallize in a fibrillar form. Chains that are left unextended, those that are below a certain length for the speed, use the preformed clusters (shishes) as nuclei and crystallize as overgrowths in the form of chain-folded lamellae (kebabs). Fibers formed in this way showed remarkable mechanical properties with modulus in excess of 1200 gram force/denier (102 GPa) and tensile strength in the neighborhood of 30 gram force/denier (2.6 GPa).

Production. Recognition that the shish-kebab fibers produced by the surface-growth procedure result from the deformation of a gel-like entangled network layer at the rotor surface led to the development of gel-spun polyethylene fibers. The fiber is made by the solution spinning method. The polymer is

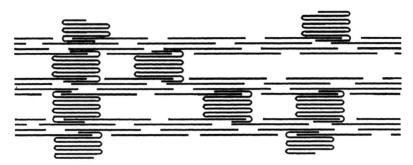

Fig. 12.24. "Shish-kebab" structure showing extended-chain crystals with lamellar overgrowths.

dissolved in a high-temperature solvent, such as decalin, at a low concentration, 10 percent or less, and extruded into cold water. Initially the fiber is formed as a gel that still contains the spinning solvent. It can be oriented by drawing at remarkably high draw ratios of 30 or greater, either before or after solvent removal. A method of spinning polyethylene fiber is described in the patent by Kavesh and Prevorsek.[25] The factors that govern the properties of fibers produced by this method are the polymer molecular weight, the concentration of polymer in the solvent, the type of solvent, the solution temperature, and the stage and the extent of drawing.

The high-performance behavior in these fibers is obtained from their having a very high molecular weight (1–5 million) combined with a very high degree of extended-chain crystal continuity. The fiber can aproach a 100 percent crystalline structure with a theoretical maximum density ($\sim$0.97 g/cc). Their use of course, is, limited by the melting point of the polymer, which even in the extended form is only about 150°C. Although higher than that of ordinary polyethylene, it is still much lower (80°C or more) than those of commercial textile and other higher performance industrial fibers. There also has been concern about the creep that occurs in these fibers, although significant improvement has been made in this regard since the introduction of the first fiber. However, these fibers have such a unique combination of strength and lightness that they have proved highly successful in a number of applications, such

as sailcloth, body armor, medical implants, fishing net, and sports equipment.

ARAMIDS

Introduction

As pointed out in the preceding section, a second route for developing fibers having properties approaching the ultimate is the use of polymer chains that have high intrinsic stiffness and will remain extended in solution or melt. The development of aramid organic fibers based on aromatic polyamides met these requirements and added another chapter to the history of the development of synthetic fibers. Nomex® aramid, a thermally resistant fiber based on a *meta*-oriented structure, was commercialized by the DuPont company in 1962.

Following the technological breakthroughs which led to the discovery of (1) the liquid crystalline behavior of *para*-oriented aramids[26] and (2) a novel method for spinning anisotropic liquid crystalline polymer solutions,[27] Kevlar® aramid fiber was produced and commercialized by the DuPont company in 1972. Other fibers based on aromatic polyamide compositions, which were produced and commercialized by other companies, were Technora® (Teijin, Japan), Teijinconex® (Teijin, Japan), and Twaron® (Akzo, The Netherlands). Additionally, SVM is a fiber produced in the Former Soviet Union and it was announced in 1990 that a new aramid fiber had been introduced by Hoechst, in Germany.

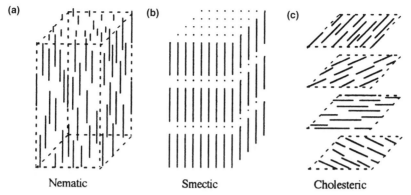

Fig. 12.25. Structure of liquid crystalline polymers showing (a) nematic, (b) smectic, and (c) cholesteric orders.

The rigid rod chains in *para*-aramids tend to form so-called liquid crystals when they are dissolved in polar solvents or heated to certain temperatures. The polymers showing liquid crystalline behavior in melts are called the thermotropic liquid crystalline polymers, and those showing similar properties in solution are called the lyotropic liquid crystalline polymers. These liquid crystals exist in three distinct phases according to their specific structures (Fig. 12.25).

A one-dimensional arrangement of rigid rods is called a nematic structure; a two-dimensional ordered arrangement represents a smectic structure; and the cholesteric structure is formed in terms of rotating oriented sheets so that rigid rods align parallel to each other in every layer, but the directional vectors in each layer are different. The preferential phase considered for fiber spinning is nematic. Nematic solutions or melts are easy to develop into oriented structures through shearing and elongational flow during extrusion because of the rigidity of the polymer chains. Thus high orientation can be obtained in the as-spun fibers without much post-treatment.

Manufacture

In 1973, the Federal Trade Commission recognized aramid as a distinctly different generic material and defined it as "a long-chain synthetic polyamide in which at least 85 percent of the amide linkages are attached directly to two aromatic rings." This distinguishes aramids from nylon, which was redefined as a polyamide with less than 85 percent of the amide linkages attached to two aromatic rings. The first aramid fiber produced in the United States was Nomex®, the reactants being *m*-phenylene diamine and isophthaloyl chloride to give poly(*m*-phenylene-isophthalamide) (MPD-I).

MPD-1

This polymer could not be melted without decomposition, so the preferred fiber formation route was solution spinning. Patent literature suggests that the fiber is spun from a solvent system composed of dimethylformamide and lithium chloride. The final properties are achieved by stretching in steam after washing to remove residual solvent.

The physical and chemical properties of this fiber are not remarkably different from those of other strong polyamides, but it does have excellent heat and flame resistance that makes it particularly suited for use in protective clothing and in specified industrial end uses. Military flight suits, fire-fighter uniforms, and hot gas filtration are a few of its many possible applications.

The other important fiber in this category, which also was first produced commercially in the United States, is Kevlar®, introduced in

1971 as fiber B and later coded as Kevlar® 29. It was produced from poly(p-phenylene terephthalamide).

poly(p-phenylene terephthalamide)

Later, a higher modulus version, Kevlar® 49, believed to be made by heat anealing of Kevlar® 29, was introduced.

Poly(p-phenylene terephthalamide) PPD-T, also called para-aramid, can be polymerized to a fiber-forming molecular weight by poly-condensation of terephthaloyl chloride and 1,4-phenylene diamine.

higher molecular weight is reached. Large-scale manufacture of a polymer requires continuous polymerization to minimize cost. In the case of PPD-T, special problems that had to be accommodated included the rapid gelation of the reaction solution with increasing molecular weight, the need to control the temperature of a vigorous exothermic reaction, and the handling of solvent HMPA, which is suspected to be a carcinogen.

Although melt-spinning would be preferred from the standpoint of process simplicity and conversion cost, aramids must be spun from solutions, by wet, dry, or dry jet–wet methods because they decompose before or during melting, ruling out melt spinning. Dry-spinning

p-phenylenediamine
(PPD)

terephthaloyl chloride
(TCL)

poly(p-phenylene terephthalamide)
(PPD-T)

Poly(1,4-phenylene terephthalamide) (PPD-T) of high molecular weight can be prepared by low-temperature solution polymerization techniques. This polymer is less soluble in amide-type solvents than is poly(p-benzamide). The most successful conditions required hexamethylene phosphoramide (HMPA), the original solvent, alone or mixed with N-methyl-2-pyrrolidone (NMP), although other mixtures such as NMP containing CaCl$_2$ also could produce a fiber-forming polymer. During polymerization, the molecular weight increases rapidly within the first few seconds of the reaction. The critical molecular weight or viscosity is exceeded, and the stir opalescence typical of lyotropic solutions is observed. Although gelation of the reaction mixture occurs quickly, polymerization continues, but at a greatly reduced rate. With the choice of a suitable solvent system, gelation can be delayed until the desired

is used to produce Nomex® fiber, where a dope (20%) of the polymer in solvent is converted to yarn. In wet spinning, the polymer dope is extruded into a nonsolvent where the fiber coagulates. The coagulated fiber then is washed and often drawn to develop desired fiber properties. In wet spinning, the spinnerette is in the coagulation bath, hence the dope temperature and the coagulation temperature are the same. In dry jet–wet spinning (Fig. 12.26), as used for Kevlar®, on the other hand, the spinnerette is separated from the coagulation bath, allowing independent control of the dope extrusion and coagulation temperatures. The extrusion jet is placed a small distance above the coagulation bath, and the nascent fibers descend into the liquid, pass under a guide, and proceed in the bath while undergoing stretch; then they are withdrawn from the bath and wound up. A subsequent washing step may be required to

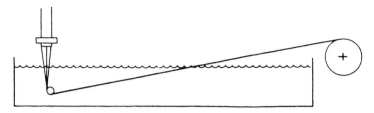

Fig. 12.26. Schematic of dry jet–wet spinning method.

remove residual acid solvent. In systems where the dope must be hot to lower viscosities to extrudable levels, and the bath cold for developing specific structures during coagulation, as is preferred with Kevlar®, dry jet–wet spinning is the option. For production of Kevlar® fibers, a PPD-T/concentrated sulfuric acid solution, containing approximately 20 percent polymer, is extruded at 90°C into a cold water bath (~1°C). Following extrusion, the fiber is washed, dried, and given post-treatment, depending upon the properties desired.

Kevlar® is reported to have about twice the breaking strength (23–27 gram force/denier or 2.9–3.4 GPa) of high-tenacity nylon and polyester, but its most outstanding physical property is its high stiffness (550–1000 gram force/denier or 70–127 GPa), which can range up to an order of magnitude greater than that of standard polyester. This property has led to high-volume usage of the fiber as reinforcement in composite materials such as belts in radial tires and aerospace structures. The world production figure (1999) of this fiber was estimated to be over 60 million lb.

HIGH-TEMPERATURE-RESISTANT FIBERS

The need for high-temperature-resistance fibers has arisen from demands of a number of industrial applications, as well as applications in aerospace programs. In many of these applications, the usual characteristics of organic-based fibers are desired, but the high temperature resistance of inorganic fibers is required. Thus, the fibers are expected to retain their structural integrity at temperatures of 300°C and above for considerable periods of time, but otherwise their properties should resemble those of the more common manufactured textile fibers. The two leading groups of fibers in this area are the *meta*-aramids and PBI.

Meta-Aramid

The major fibers in this group, based on sales volume, are the products made from poly(*m*-phenylene-isophthalamide), which were introduced by DuPont in 1962 as HT-1 nylon (later known as Nomex®) and by Teijin in 1972 as Conex®. General manufacture of the fiber was described earlier under the heading "Aramids." Other manufacturers now are entering the field with products of similar chemical structure. Although these products are made and spun by different processes, their chemical and physical properties are similar. The fiber is usually utilized in the form of cut staple, which are amenable to conversion on traditional spinning and weaving machinery. Its mechanical properties (tenacity 4–5 gram force/denier, ultimate strain 25–30 percent and initial modulus 90–100 gram force/denier) compare favorably with those of other textile fibers. The main utility of the fiber lies in the resistance it offers to combustion: it has a limiting oxygen index (LOI) of about 0.29, a melting point above decomposition temperature, an ignition point above 600°C and a flash point about 800°C.[28] The LOI gives a relative measure of flame resistance; the higher the number, the lower the flammability. The fiber Kermel® from Rhone Poulenc also is classified as an aramid. It is, chemically, a polyamide-imide fiber and has an LOI of about 0.31. A major application of these fibers is in protective clothing. In order to reduce their cost, they often are blended with other, less expensive, flame-retardant fibers such as those based on cotton, rayon, and wool.

PBI

Poly-2-2′-(*m*-phenylene)-5,5′-bibenzimidazole, commonly called polybenzimidazole (PBI), was developed under the aegis of the U.S. Air Force Materials Laboratory in cooperation with the then-existing Celanese Corporation. The fiber went into commercial production in the United States in 1983. It is a condensation polymer obtained from the reaction of tetra-aminobiphenyl and diphenylisophthalate in a nitrogen atmosphere at temperatures that may reach 400°C in the final stages.[29] The structure of a repeating unit is shown below.

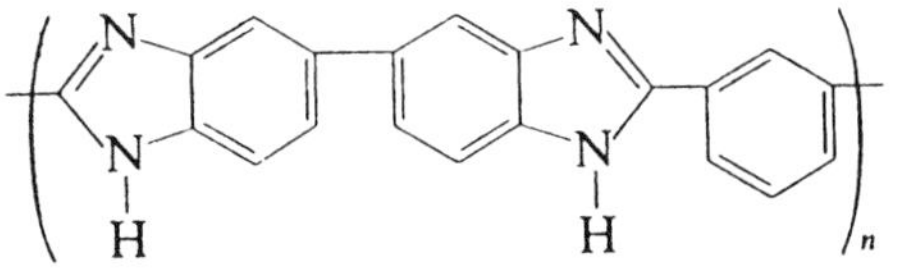

The polymer is dissolved at a high temperature under nitrogen pressure in dimethylacetamide, to which a small amount of lithium chloride may be added to increase the stability of the solution. Then, it is dry-spun in an atmosphere of heated nitrogen (about 200°C), from which the solvent is recovered; next it is stretched slightly in steam and washed. Drawing and relaxing are done in an inert atmosphere, as might be expected, because temperatures up to 250°C or higher are used. The fiber then is given a stabilization treatment in a sequence of steps and made into staple fibers using conventional crimping and cutting techniques. The stabilization treatment involves reaction with sulfuric acid and heating at high temperatures (~475°C) for short periods of time. The process, known as sulfonation, yields a product that has a significantly lower shrinkage than the unstabilized material.

The final yarn is golden yellow, and because this color appears to be an intrinsic property of the polymer, it may have some limitation as far as the civilian market is concerned. This material originally suffered from high shrinkage on exposure to flame; however, further developments including the sulfonation treatment have reduced the shrinkage to only about 5–10 percent at 600°C.[30] The fiber is capable of retaining about one half of its original strength (~3 gram force/denier) upon exposure to air for 18 hr at 350°C or 1 hr at 425°C, it has an LOI of 0.41, which is well above that of the aramids, and it has high resistance to inorganic acids and bases and organic chemicals. Further, with tensile and moisture regain properties comparable to those of many textile fibers, the PBI fiber is well suited for blending with other fibers and conversion into final products using conventional spinning and weaving or non woven equipment. Applications include high-performance protective apparel, flight suits, and aircraft furnishings.

POLYTETRAFLUOROETHYLENE

Historical

Polytetrafluoroethylene (PTFE) was discovered in 1938 when Dr. Roy J. Plunkett and his assistants working on new nontoxic, nonflammable refrigerants at E. I. duPont de Nemours and Co. found that one cylinder, which was supposed to contain tetrafluoroethylene (TFE), ceased to release the gaseous material. Upon opening the cylinder, they discovered that the inside was covered with a white powder. Polymerization of TFE had taken place. The result was PTFE or Teflon®, which is the trade name applied to that polymer by duPont. The company tested the polymer and found that it was virtually inert to all known solvents, acids, and bases, a characteristic that was unique at that time. It also found that this material was resistant to high temperature and had the lowest coefficient of friction of any known solid. The research on PTFE was intensified during World War II because of military demands for improved materials for products such as gaskets, packing, and linings for containers for handling corrosive materials. When the war was over, the release of classified information caused a booming interest in the polymer. Since that time, PTFE has found itself in many different applications.

Manufacture

In one process, the manufacture of the monomer, TFE, involves the following reactions: Hydrogen fluoride is made by reacting calcium fluoride with sulfuric acid:

$$CaF_2 + H_2SO_4 \rightarrow CaSO_4 + 2HF$$

Chloroform is reacted with hydrogen fluoride in the presence of antimony trifluoride as the catalyst:

$$CHCl_3 + 2HF \rightarrow CHClF_2 + 2HCl$$

TFE is obtained by the thermal decomposition of this monochlorodifluoromethane (known as Freon) in a continuous noncatalytic gas-phase reaction, carried out at or below atmospheric pressure at temperatures from 600 to 900°C:

$$2CHClF_2 \rightarrow C_2F_4 + 2HCl$$

Numerous side-products are generated in this process. Many of them are present in trace amounts, but the highly toxic perfluoroisobutylene, $CF_2 = C(CF_3)_2$, requires special precautions.

TFE also may be manufactured by the reaction of zinc and tetrafluorodichloroethane:

$$ClF_2C\text{–}CF_2Cl + Zn \rightarrow C_2F_4 + ZnCl_2$$

or by the reaction of tetrafluoromethane molecules in an electric arc furnace:

$$2CF_4 \rightarrow C_2F_4 + 2F_2$$

TFE is a colorless, tasteless, odorless, and nontoxic gas. To avoid any undesired reactions during storage, inhibitors must be added. The polymerization is carried out by an addition-type reaction in an aqueous emulsion medium and in the presence of initiators such as benzoyl peroxide, hydrogen peroxide, and persulfates. The monomer is fed into a cooled emulsion medium and then heated to a temperature of 70–80°C, at which the polymerization takes place.

$$CF_2 = CF_2 \rightarrow (-CF_2\text{–}CF_2-)$$

The pressure may range from 40 to 100 atmospheres. After removal of the unreacted material, the polymer is washed, pressed, and dried. The degree of polymerization can be quite high, of the order of 50,000. PTFE is manufactured in four different forms: granular, fine powder, aqueous dispersion, and micro powder, and in a variety of grades, each differing in properties. Fillers such as glass fibers, asbestos, graphite, or powdered metals may be added to the granules in order to modify properties.

Fiber Manufacture

Because Teflon® is not soluble, it cannot be wet- or dry-spun, and because it is thermally unstable at its melting point of about 400°C, this combination would seem to pose an impossible problem for the production of fibers. Research into the fundamental characteristics of the polymer, however, revealed that the submicroscopic particles precipitated from the polymerization reaction were about 100 times as long as they were thick.

In one manufacturing process, an aqueous dispersion of PTFE is mixed with a solution of ripened cellulose xanthate, from which a fiber is obtained by a wet-spinning process, after which the cellulose is completely decomposed by heating. The remaining PTFE is sintered into continuous fibers by transporting them over heated metal rolls, followed by stretching to achieve the desired diameters and physical structure. In a process known as "paste extrusion," the powder is mixed with an organic plasticizer and compressed at 300–500 psi to make a preform or billet. The latter is extruded into filaments which are then dried to evaporate the lubricant. In another process, films are produced that are slit into strips of very small widths, which then are stretched and sintered. In yet another, tapes are extruded, which are converted to fibers by stretching while being twisted to a very small cross-sectional area.

Properties

The unique combination of properties, including chemical resistance, thermal oxidation resistance, high lubricity, electrical and thermal insulation, low flammability and excellent weatherability, are derived primarily from two factors, namely the molecular structure and the molecular weight of PTFE. The structure consists of a core chain of carbon atoms with a fluorine sheath, which essentially completely shields the core. The fluorine atoms are so tightly packed that steric interactions cause a slight rotation of the carbon chain from the normal planar zig-zag to a helical conformation. This dense shield has a low surface energy and a very smooth surface with no side chains or imperfections. As a result, the interchain forces are low and individual molecules are able to slide past each other with relative ease. Compared with other polymers, therefore, creep tends to be high.

The polymer chains pack themselves very closely and regularly to give cylindrical packing which consequently leads to very high crystallinity (~90%). The material has a high melting point which is about 330°C. The smooth surface when combined with low surface energy makes the surface so neutral that it resists sticking to any material. Accordingly, PTFE has the unique property of the lowest coefficient of friction (~0.007), with essentially no stick-slip character. The material has no affinity for water and, therefore, is totally hydrophobic.

The chemical inertness and the thermal stability of this polymer are so great that in spite of its high price ($4–10/lb, depending on the resin type) it is used in chemical operations where drastic conditions exist and no other organic material is suitable. Its low friction allows it to be used as non-stick coating for metals, work surfaces, and cooking utensils. The polymer is regarded as biocompatible and tends to be accepted by the body. This has allowed PTFE to be explored as a material for surgical implants.

Expanded PTFE (ePTFE)

In early 1970s, Dr. Robert W. Gore invented a process by which PTFE could be expanded and gave the trade name of Gore-Tex® to the product obtained. In this process, the specific gravity is reduced by the introduction of micropores but much of the original properties of the polymer are retained. The new material is a hydrophobic but porous membrane of PTFE that is used as a protective layer in a number of applications.

In the process used, in general, a paste is formed of the PTFE polymer and a plasticizer and shaped into an article. It is then expanded by stretching in one or more directions, and while it is held in the stretched form, it is stabilized by heating to high temperature (~327°C) and cooling. The porosity that is produced by the process is retained in the final product.[31] The structure formed by the process consists of "nodes" and "fibrils." The nodes, that vary in size from 5 to 500 μm and are always found perpendicular to the direction of expansion, are interconnected by fibrils. This is seen in Fig. 12.27, which shows a micrograph of Mikrotex®, an ePTFE product.[32] An ePTFE may have as many as a billion or more randomly spaced pores per square centimeter. These pores are unique in size: they are three or more orders of magnitude smaller than the size of a water droplet but two or more orders of magnitude larger than a water vapor molecule. Thus, when used as a rain wear, the product allows the perspiration vapors to escape but blocks out the liquid water from penetrating. The product can likewise serve as a barrier against chemicals and microbes.

GLASS AND CARBON FIBERS

Glass

Among the manufactured inorganic fibers, glass is produced in by far the largest volume. There has been a rapid increase in the use of textile grades of these fibers, and outside the textile field enormous quantities of glass fibers are used in air filters, in thermal insulation (glass wool), and for the reinforcement of plastics.

Glass possesses obvious and well-known characteristics which have largely determined the methods used to form it into large objects. It flows readily when molten and can be

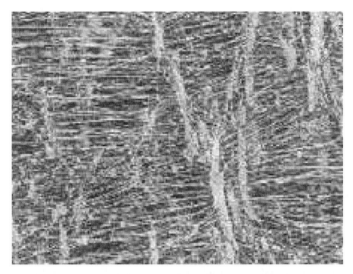

Fig. 12.27. Expanded PTFE micrograph (Microtex® Membrane at 500×) showing a porous structure containing nodes interconnected by fibrils. (*Courtesy Menardi-MikroPul, LLC.*)

drawn into filaments, whose extreme fineness appears to be limited only by the drawing speed. The method used in producing textile-grade glass fibers follows this principle (see Fig. 12.28.)

In the commercial operation, the molten glass, produced either directly from raw materials or by remelting of marbles, is held at a uniform temperature in a vessel, whose bottom carries a bushing containing small uniform holes. The molten glass flows through these holes as tiny streams that are attenuated into filaments at speeds on the order of 3000 m/min; these flaments are coated with a lubricant, gathered into groups to form yarns, and wound up. For a particular glass viscosity,

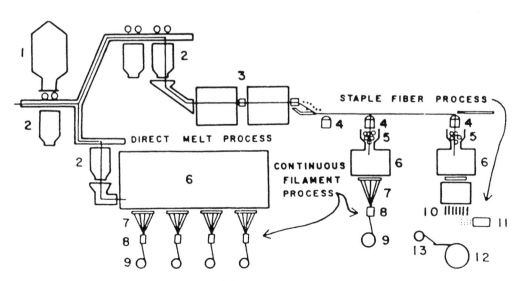

Fig. 12.28. Flow diagram for manufacture of textile glass fiber: (1) glass batch; (2) batch cans; (3) marble forming; (4) cullet cans; (5) marbles; (6) melting furnaces; (7) filament yarn formation; (8) gathering and sizing; (9) yarn packaging; (10) air jets; (11) lubricant spray; (12) collection for staple fibers; (13) staple fiber packaging. (*Courtesy Owens-Corning Fiberglass Corp.*)

the size of the individual filaments is determined by the combination of the hole size and the speed of attenuation.

Because of the inherently high modulus of glass, very fine filaments are required in order to approach the required properties of textile materials. Thus, the diameter of glass filaments falls in the range of 3.8–7.6 μm whereas the average diameter of the finest organic fibers is about twice as great. The fiber and yarn numbering system is based on nomenclature used in the glass industry and differs from the traditional systems accepted in the textile and organic fiber industries.

The method of manufacture of glass staple fibers differs from those used to produce the corresponding organic materials, all of which are based upon cutting the continuous filament product. Air jets, directed in the same line of flow as the emerging streams of glass, attenuate the streams, and break the solid glass into the lengths desired for further processing, which are gathered on an appropriate vacuum drum and delivered as slivers or a matte. To produce fibers that may be coarser and considerably less uniform in length, to be used for the production of filters, paper, or thermal insulation, large streams of molten glass are cross-blown by blasts of hot air, steam, or burning gas.

As might be expected from the nature of glass, the conversion of glass fibers into final products has required the development of new lubricants, finishes, and processing techniques. For example, because glass fabrics cannot be dyed directly or printed with the colors demanded for their acceptance as draperies, the colorant must be applied to a resin coating. But before applying the coating it is necessary to remove the lubricant that was placed on the fibers to permit their conversion into a fabric, which is done by burning. The elevated temperature resulting from this operation also relaxes the internal strains developed in the glass fibers during the steps of the textile operations and sets the yarns in the required geometry. The fabric then is resin-treated, cured, and dyed or printed.

Another inherent property of glass is the tendency of unprotected fiber surfaces to abrade each other to destruction under the action of very little mechanical working. When it was first considered for rubber-reinforcing purposes, its poor adhesion to rubber and the inadequacy of the then-existing bonding agents frustrated attempts by manufacturers to take advantage of the very high tensile strength, completely elastic behavior, high modulus, and lack of moisture sensitivity of glass fibers. However, it has been possible to modify the fiber surfaces so that satisfactory adhesion is achievable, and the impregnant can be applied in such a way that fiber-to-fiber contact is prevented. With the adhesion problems solved, glass in cord form could effectively enter markets in belt-type tire construction and in all kinds of power-transmitting rubber belts. It is estimated that the total glass fiber production in the United States in 2000 was approximately 2.5 billion lb.

Carbon and Graphite Fibers

Following World War II, the development of jet aircraft and rockets created demands for fibers having thermal resistance, strength, and modulus far beyond what could be obtained in existing organic fibers. Much of this need was for reinforcing materials that could be embedded in matrices of one type or another. As a result, techniques have been developed for preparing fibers from a good many metals and refractory inorganic compounds. Although these materials are essential for certain uses, the volume of production still is low, and the prices are correspondigly high (as much as $1000/lb).

Carbon and graphite fibers are made from rayon and acrylic precursors by driving off virtually all of the hydrogen and oxygen contained in them. The principle is essentially the same as that which brought about the formation of coal, or, citing a more recently discovered and dramatic example, the conversion of the original wooden beams of buildings in Herculaneum, buried by a flow of mud from Vesuvius in A.D. 67, to what appears to be charcoal. In the present commercial process, the starting material is selected so as to produce a final product of the desired size and properties. Cotton, bamboo, and other natural fibers were the earliest materials used as precursors. In the

1950s, rayon was used for this purpose, but the first attempts led to fibers with poor strength. Union Carbide, in the mid-1960s, made the first strong carbon fiber from rayon and extended its work to using polyacrylonitrile as the pecursor. The use of mesophase pitch as the starting material began in the 1970s. Carbon fibers may be divided into three types: (1) highly graphitized with a high modulus, (2) heat-treated at a lower temperature to produce fibers with lower modulus but high strength, and (3) randomly oriented crystallites with relatively low modulus, and low strength, but most important, low cost.

The preparation of fibers generally consists of heat-treating the precursor at a low temperature (200–350°C), usually in air, which gives a stable fiber for higher temperature processing. During this step, extensive decomposition occurs, and a percentage of the initial weight is lost, which is related to the cleavage of the C–C and C=O bonds and expulsion of H_2O, CO, and CO_2 (additionally HCN in the case of PAN precursor) as gases. This step is followed by carbonization in an inert atmosphere at 1000–2000°C, which is said to collapse the cyclized structure into a stacked ring carbon fiber structure. Almost all noncarbon elements are evolved as volatiles. This is followed by graphitization, which usually is carried out at temperatures above 2500°C for short periods in argon or nitrogen. The process increases the purity, removes the defects, and further improves the order in the structure. Thus, the difference between the so-called carbon and graphite fibers lies in differences in the ranges of temperatures at which the last step is carried out, the degree of carbon content (97% for carbon and 99.6% for graphite), and the mechanical properties, which are superior in the graphite fibers. In usual discussions, the term "carbon fibers" covers both materials. The majority of carbon fibers produced today are made from a PAN precursor. PAN fibers are fine, and have a higher degree of molecular order and a higher decomposition point than those from rayon. This precursor also leads to a greater carbon yield (45%, as opposed to 24% from rayon), but the fibers are more expensive to produce than those based on rayon. Recent

commercial developments have allowed the production of carbon fibers from low-cost petroleum or coal-tar pitch, instead of synthetic fibers. These precursors lead to a higher yield (90%) of carbon, improved lubricity of fiber products, and higher production rates. However, the pitch-based fibers may be more brittle and harder to handle, have a higher specific gravity (2.0, as opposed to 1.8 for those from PAN and 1.66 from rayon) and lower compatibility with some matrix materials.

The fibers have a diversity of applications. One major application is in composites, where they are used for reinforcing resins and metals to provide structural materials with high strength, high modulus, and light weight. The resulting composites are used in the aircraft and aerospace (the largest users), automotive, and sports industries. The fibers also find uses in protective garments, electrical devices, insulation, and filtration. The use of the fibers in the world was about 25 million lb in 2000. The prices dropped significantly during the past decades, but in 2000 the majority remained in the $15–70/lb range. The price of the ultrahigh-modulus carbon fibers, however, can be as high as $1500/lb, or greater.

SULFAR

Historical

Sulfar fibers are extruded from poly(phenylene sulfide) or PPS by the melt-spinning process. The first PPS polymer was made in 1897 by the Friedel–Crafts reaction of sulfur and benzene. Researchers at Dow Chemical, in the early 1950s, succeeded in producing high-molecular weight linear PPS by means of the Ullmann condensation of alkali metal salts of *p*-bromothiophenol.

In 1973, Phillips Petroleum Company introduced linear and branched products under the trade name Ryton® by reacting 1,4-dichlorobezene with sodium sulfide in a dipolar aprotic solvent. In 1983, the same company succeeded in stable melt-spinning of PPS. In 1986, the Federal Trade Commission gave the fiber the generic name Sulfar, defined as "a manufactured fiber in which the fiber forming substance is a long chain synthetic polysulfide in which at least 85 percent of the sulfide (–S–) linkages are attached directly to two aromatic rings."

Manufacture

In one process, synthesis was carried out by self-condensation of a metal salt of a *p*-halothiophenol:

$$n\,X\!-\!\!\langle\bigcirc\rangle\!-\!SM$$

$$\xrightarrow{200°C}\left(\!\langle\bigcirc\rangle\!-\!S\right)_{\!n} + n\,MX$$

where X is a halogen, preferably Br, and M is a metal (Na, Li, K, or Cu, preferably Cu). This reaction was carried out under nitrogen in the solid state or in the presence of materials such as pyridine as reaction media. Considerable difficulty was encountered in removing the by-product, copper bromide, from polymers made by this process. The current commercial synthesis of the polymer is carried out by reaction between *p*-dichlorobenzene and sodium sulfide in a polar solvent.

The process discovered by workers in the laboratories of Phillips Petroleum Co. marked a significant departure from prior processes, and made it possible to prepare a variety of arylene sulfide polymers from the readily available starting materials.[33]

$$n\,Cl\!-\!\!\langle\bigcirc\rangle\!-\!Cl + 2n\,Na_2S \xrightarrow[solvent]{heat}$$

$$\left(\!\langle\bigcirc\rangle\!-\!S\right)_{\!n} + n\,NaCl$$

Melt spinning of PPS involves problems such as plugging of the filter and the spinnerette. In order to prevent cross-linking and gel formation during the process of spinning, the company introduced cure retarders comprising Group IIA or Group IIB metal salts of fatty acids, which improved the heat stability of PPS:

$$(CH_3(CH_2)_nCOO)_{\overline{2}}M$$

where M is a Group IIA or IIB metal, and n is an integer from 8 to 18. Representative compounds of the type described above include calcium stearate, calcium laurate, calcium caparate, and calcium palmitate. In general, the cure-retardant additives are employed in an amount within the range, of about 0.1–5, preferably about 0.5–2, weight percent based on the weight of the PPS.

Sulfar fibers, sold under the trade name Ryton® in the United States, are characterized by high heat resistance, inherent flame retardancy, excellent chemical resistance, low friction coefficient, good abrasion resistance, and good electrical properties.[34] Physical characteristics include medium tenacity (3.5 gram force/denier or 423 MPa) and elongation (25–40%) and low shrinkage (<5% at 100°C). The fibers find application in a number of industrial products, including filter fabric for coal-fired boiler bag houses, paper maker's felts, materials for electrical insulation, high-performance composites, gaskets, and packings.

Several Japanese and European companies have begun the production of PPS, some with a U.S. partner. The new decade should see the introduction of a number of new fibers and fiber products based on the PPS polymer.

MICRODENIER FIBERS

One of the more important developments in the field of fiber technology in recent years has been the production of fine denier fibers, with worldwide activity in developing new products and outlets for them. Improvements in the quality of polymers, coupled with new technology for extrusion of fibers, have led to the production of fibers with sizes ranging from 1 to 0.1 denier or even lower. Fabrics produced from such fibers have novel and unique properties, and they are finding applications in a wide variety of apparel, and industrial products, including high-fashion fabrics with silk-like texture, synthetic suede, breathable porous but rainproof or bacteria-proof fabrics, wipes for oil and other spills, clean room materials, cloths for cleaning camera, microscope, and spectacle lenses, and for compact discs.

Unfortunately, there is no universally accepted definition for microdenier, and companies have been free to use terms they choose. In general, there seems to be a consensus that the term "fine" may be used for a denier of 1 or less and "micro" for a denier less than 0.5. The production of such fibers is difficult and expensive because the throughput rate must be reduced in order to obtain fine denier filaments, and there is a limit to how fine in denier a regular process could be made to go. Generally, the technology of production of fine and microdenier filaments can be divided into five categories. First, conventional spinning technology can be used, by using fine-size dies and adjusting the throughput rate and quenching and drawing parameters to obtain fine denier polyester filaments. This method has been used by most fiber manufacturers to produce such fibers in limited quantities. In another method, alkali reduction or surface etching is used to dissolve the surface layers, in an effort to reduce the cross-sectional size of filaments in a polyester fabric. A weight loss of as much as 25 percent has been achieved by this process. In the MB process, molten polymer is forced through a melt blowing die and die tip orifices, and the emerging stream is attenuated by a jet of high-velocity hot air until broken. Then the broken fiber is forced into a stream of cold air, where the fiber solidifies. The fiber is collected on a wire screen or apron with other fibers, and a homogeneous MB web is produced. This method is used extensively with PP materials, but other polymers such as nylon, polyester, and polyethylene also have been used. Fibers as fine as 0.5 μm and finer have been produced.

In another approach, filaments containing two polymers that do not adhere to each other are spun and then split. One may, for example, spin a bicomponent fiber of nylon that has several filaments of polyester embedded. After a fabric containing bicomponent filaments has been woven, it is treated to split the components, thus converting the original filament to several smaller filaments. Deniers of the order of 0.1 can be achieved by this ingenious method. In the last method, instead of splitting the two components as in the previous example, one component is dissolved away chemically, leaving bundles of very fine fibers in the fabric.

The majority of the technology for producing fine and microdenier fibers is new and thus expensive; so efforts in the future can be expected to be directed toward optimizing process parameters in existing methods and discovering faster and cheaper methods for manufacturing these esthetically very pleasing and functionally very promising fibers.

NANOFIBERS

An extreme example of the microdenier-size fibers discussed above is the nanofibers that are one or more orders of magnitude smaller in diameter. The fiber with diameter at submicron or nanometer level is spun by a process known as electrospinning, in which the fiber is spun in a field involving high electrostatic forces.[35]

The electrospinning set-up essentially consists of a capillary tube or a needle attached to a syringe filled with a polymer solution or melt, a grounded collector screen, and a high-voltage power supply (Figure 12.29). The collector typically is a metal plate, an aluminum foil or a metal grid. Other forms of the collector, such as a rotating drum, have been used. When the potential (1–30 kV), is applied, the pendant droplet at the end of the needle becomes charged and two opposite forces, namely the surface tension of the droplet itself and the electrostatic force due to applied field, act on it. As the potential is increased, at a certain point, electrostatic force overcomes the surface tension of the droplet and a conelike structure, commonly known as a "Taylor cone," is formed at the tip of the droplet and a jet of polymer solution or melt emanates from it which is accelerated towards the collector.[36] In this process, the polymer jet gets drawn to submicron level while the solvent evaporates or the melt cools down to form the fibers that deposit on the collector and form a nonwoven web.

The electrospun nanofiber webs, with very high surface area to volume ratio and high

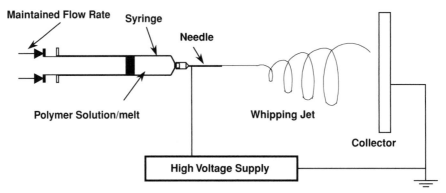

Fig. 12.29. Electrospinning set-up in horizontal configuration.

porosity with interconnected pore networks, have been explored for applications in a number of highly specialized areas such as the scaffolding for tissue engineering, wound dressing, drug delivery, nanosensors, and nanoelectronics. More than 50 different polymers have been electrospun successfully in a solution or melt form that include synthetic nondegradable (polyesters, polyamides, polyurethanes, polycarbonates, etc.), natural (collagen, elastin, chitosan, silk, etc.), and synthetic biodegradable (polyglycolic acid, polycaprolactone, polylactic acid, polylactide-co-glycolide etc.) polymers.[37]

FIBER VARIANTS

Introduction

In a previous section, data and plots were given showing the rapid rise in consumption and production of manufactured fibers at the expense of natural fibers. The principal reason for this has been the wide range of manufactured fiber variants that can be produced from a single fiber-forming polymer. The wide range of polymers available, each with its particular properties, adds yet another dimension. This is not to say that there is only one type of cotton, wool, silk, or asbestos fiber; there are many varieties of natural fibers, but their supply is limited by natural factors such as climate and genetics. The relative availabilities of manufactured fiber types can be altered by controlled chemical-process changes, whereas the amount and the quality of a desired cotton type that can be grown is determined to a great extent by climatic conditions, which humans have not yet learned to control. Another factor that has aided the growth of manufactured fibers is their consistent quality and properties. Again, the grade and the quality of natural fibers are subject to the vagaries of nature.

For the purposes of this discussion, fiber variants will be divided into two types: chemical and physical. Chemical variants will be those involving a small but significant change in composition, whereas physical changes will be those involving a change in either the dimensions of the fiber or its stress/strain or stability features. The definitions of the two variants also could be based on modification of either esthetics or functionality.

Physical Variants

Most manufactured fibers are available as staple, tow, and filament. Natural fibers are available only in the characteristic forms in which they occur, with filament silk and cotton staple as examples. All manufactured fibers are formed initially as filament yarns. The German adjective *endlos* (literally translated "endless") is very descriptive, as filament yarns are continuous strands consisting of one or more members that for most practical purposes are infinite in length. Fine filament yarns (40–100 denier) are used in producing lightweight apparel fabrics, whereas coarse filament yarns (800–1200 denier) are found as reinforcement in tires or conveyor belts. These examples are chosen as extremes to show the range or applicability of manufactured fibers, and represent only a small fraction of the actual range of end uses.

When many filament yarns are collected into a bundle immediately after formation, the resulting structure is called a tow. Tows may range from 10,000 to over a million in total denier. In a next step, the tow may be crimped by the process previously described, which imparts what is usually a sawtooth appearance along the length of all the filaments. For some end uses, the crimped tow itself is provided by the fiber maker, as are, for example, the previously discussed acrylic tows, which are converted to staple as part of the spinning of yarns with a woollike character. Another example is the cellulose acetate tows used to form cigarette filters or the ink reservoirs for marker pens. In the latter case, the compact tow bundle first is treated to separate the individual filaments, giving a voluminous structure, which then is gathered into a continuous rod, wrapped in paper, and cut into appropriate lengths.

In the fiber-making plant, tow may be cut into short lengths of staple, ranging in length from $\frac{1}{16}$ to 6 in., depending upon the end use. For instance, the very short staples are used either in making flocked structures or in the production of papers containing blends of natural cellulose and manufactured fiber. The longest lengths are used in spinning heavy yarns for carpets or cordage. But most staple is $1\frac{1}{2}$–3 in. length and is used to form blends with cotton, rayon, or wool in the yarns employed in standard apparel fabrics. The staple length of the manufactured fiber is chosen to match that of the other blend component; otherwise, uneven yarn of poor quality results.

The size of a manufactured fiber can be altered by changing the size of the hole through which it is extruded and maintaining a constant takeup speed. Commercial fibers range from about 1.25 to 25 denier, corresponding to average diameters of about 5–50 μm. The very large ones would be used to make doll's hair or wigs, but the majority of manufactured fiber staples are made in the 1.5–6.0 denier range, corresponding to cotton blending fiber at the low denier end and coarse wool blending fiber at the high end. Staple and filament yarns used in carpets are in the 12–16 denier range, whereas industrial filament yarns such as tire cord are about 6 denier per filament. The size of a fiber is a determinant of its stiffness, which in turn influences the draping, quality, and the surface feel (often called hand in the textile industry) of a fabric made from it.

Although fibers normally are extruded through circular jet holes, the use of noncircular holes in the jets has led to the availability of a wide variety of cross-sectional shapes. In the case of fibers dry- or wet-spun from solutions, most of the mass exiting from the jet hole is not the polymer. For example, cellulose acetate fiber is made from a 25 percent solution of cellulose acetate in acetone containing a small amount of water. After the fiber leaves the jet face, solvent begins to evaporate, and as a result the area of the fiber cross-section decreases. The final result is a fiber of roughly circular cross-section, but with a serrated edge, and much smaller in area than the parent jet hole. When cellulose acetate is extruded through a triangular jet hole, the end result is a fiber of "Y" cross-section, due to shrinkage from the original triangular shape.

In the melt-spinning process, there is no solvent loss to influence final cross-sectional shape. In the case of a triangular jet hole in melt spinning, the molten fiber leaves the jet face with a triangular cross-section but, being molten, immediately tends to return to a circular cross-section due to surface tension forces. It is necessary, therefore, to quench or cool the fiber as soon as possible in order to maintain the desired cross-sectional definition. Some loss of this definition is unavoidable during the drawing step that normally follows extrusion in a melt-spinning process.

Fibers of noncircular cross-section can modify and change both functional and esthetic properties in textile structures. The triangular cross-section is typical in those respects; its shape leads to a stiffer fiber than circular fiber of the same cross-sectional area, and in a fabric this results in less drapability and a crisper surface feel. Also, the flat surfaces reflect light in a different way than do curved surfaces and can create desirable lustrous effects. These optical effects are subject to many subtle influences having to do with the size of the reflecting surface and the amount of internal reflection that takes place. Triangular or "Y" cross-sectional fibers have greater specific surface areas per unit weight than their circular counterparts, which accounts for their use in aerosol filtration, where surface area is a major factor in efficiency. Yarns made with triangular cross-section fibers are more voluminous than those from round cross-section fibers. Thus when the two types are made into fabrics of equal weight, the variant cross-sectional fabric will transmit less light and be less permeable to air than that made with round cross-section fibers. In the spinning of blended staple yarns, maximum strength is obtained if the blend components have similar load/elongation characteristics. The stiffness of polyester staple fibers can be varied mainly by changing the draw ratio used to orient the fiber after extrusion and thus producing fibers suitable for blending with cotton, rayon, acrylic, or wool fibers, which possess markedly different properties. In many industrial applications, fiber and yarns are used under conditions where they bear a load while hot, as for example in tires and power transmission belts, where it is important that they do not grow or stretch significantly under these conditions. Accordingly, industrial yarns are drawn to a greater extent than other yarns to reduce as much as possible the stretch of the final products. Relatively, these yarns have high strengths, high stiffnesses, and low elongations at rupture.

Although in most instances it is desirable to have fibers that are dimensionally stable, in some structures it is advantageous to use mixtures of fibers that are stable with fibers that do shrink upon exposure to heat or steam. High shrinkage potential usually is built into a fiber by stretching it and not giving it a stabilization treatment. The yarn bulking that occurs with blend yarns of high- and low-shrinkage acrylic staples has already been described, in the section on that fiber. Using the same principle, feltlike structures can be made by heat treatment of nonwoven battings containing low and high-shrink staple.

Chemical Variants

The use of titanium dioxide as an additive in the delustering or dulling of manufactured fibers was discussed in the section on viscose rayon. The addition of pigment particles influences the processing and performance of the fiber, along with changing its appearance. Because of the whiteness of the delustered fiber, it requires more dyestuff to reach a given shade than that needed in the case of a bright fiber. The sliding friction of a delustered fiber is lower because the pigment particles protruding from the fiber surface reduce the contact area between a fiber and the surface it slides against—a guide, for example. By the same token, some pigments can accelerate the wear of contacted surfaces. It has been suggested that the drawing and consequent orientation of delustered fibers proceeds more smoothly because the pigment particles act as nucleation sites where molecular motion is initiated. Unless specially treated, the surface of anatase—one of the crystalline forms of titanium dioxide—can accelerate the ultraviolet light degradation of acetate or nylon fibers. It is postulated that the crystal surface catalyzes the formation of peroxides from the water and oxygen under the influence of ultraviolet light, and that peroxides are the active species in the resulting polymer degradation. For this reason, rutile—the other common crystalline form of titanium dioxide—is used to deluster fibers when improved sunlight resistance is needed.

The degree of polymerization or the molecular weight of the polymers on which manufactured fibers are based can be controlled as part of one of the early steps in the process. The polymer molecular weight chosen for a fiber has a strong influence on process economics, ease of conversion to fiber, and end-use performance. Commercially acceptable fibers are based on the best balance of these factors. However, in fabrics that are open in texture and made from standard polyester staple spun yarns of low twist, a condition known as "pilling" will develop as a result of wearing. The pills are made up of fiber ends that have worked loose from the yarn bundles as a result of surface rubbing and have wrapped around themselves. In the case of fabrics from natural fibers, which are generally less wear-resistant than manufactured fibers, these pills or fiber bundles will be lost by attrition with continuing wear. Because the wear resistance of a manufactured fiber is related to its molecular weight in a general way, the pilling tendency of a polyester staple can be reduced by lowering its degree of polymerization. This compromises the tensile characteristics of the fiber only to a small extent, and all the other desirable properties such as minimum care characteristics are essentially unaffected. When the molecular weight of a polyester or a nylon polymer is increased above the standard, the resultant fibers will have increased tensile properties and fatigue resistance. The filament polyester and the nylon yarns used in end products such as ties and conveyor belts are based on such polymers.

The technology for "dope dyeing" or mass coloration of fiber as part of the fiber manufacturing operation was described in the sections on viscose rayon and cellulose acetate. The use of this technology has decreased, largely because of problems of profitably managing the required inventory of colors in rapidly changing fashion markets, and for this reason it is used extensively only if the fiber cannot be dyed by any other means. PP fiber for use in outdoor carpets is a good example of such a situation. However, if one regards "white" as a color, one finds that a substantial portion of the polyester staple fiber produced for blending with cotton or rayon contains an optical brightener or a fluorescing agent. This is needed to overcome the yellowing tendency of polyester following the absorption of hydrophobic soils. The cotton or rayon fibers

in a blend are continually rewhitened by the fluorescing agents added to laundry detergents for that purpose, but these agents are without effect on the polyester blend component. Because the polyester component usually is at least half of the blend, the spun-in optical brightener it contains is vital for the maintenance of overall whiteness.

In the cases of polyester, nylon, and acrylic fibers, their manufacturers have developed fiber variants with a wide range of dyeing behavior, referred to as dye variant fibers. Polyester fibers usually are dyed with what are described as disperse dyes. These dyes are only slightly soluble in boiling water and are used in the form of dispersions. The dye in aqueous solution is assumed to be in monomolecular form, and is absorbed from the dyebath into the polyester fiber by a process often called solid solutioning. As this occurs, more solid dye is dissolved to replenish that entering the fiber. The overall rate of dyeing is a very complex phenomenon but in part is determined by the molecular structure of the fiber through which the dye must diffuse. Generally, the more that the polymer molecules have been organized by drawing and annealing into more geometrically perfect domains or crystallites, the slower is the dyeing rate. The ability of the molecules to be thus organized into compact ordered structures can be reduced by polymerizing a small amount of a foreign dibasic acid or glycol into the polymer. Usually, 5–10 mol. percent is sufficient to prevent this regularity. Adipic acid, isophthalic acid, and polyethylene glycols are used to produce some of the comonomers for fast- or deep-dyeing polyesters. In this way, these fibers are made more economical to dye or print because special dyebath additives, high dyeing temperatures, and high-pressure steam-print fixation usually are not required.

Polyester fibers can be given an additional mechanism for dyeing if an ionic comonomer is added during polymerization. A common additive is an alkali metal salt of dimethyl-5-sulfo-isophthalate, which gives sulfonic (anionic) groups as part of the polymer structure. These groups allow the fiber to absorb basic (cationic) dyes by a specific ionic mechanism. The amount of cationic dye that can be absorbed by the fiber is stoichiometrically related to the number of anionic sites present in the fiber; this is quite distinct from the general solid solutioning that takes place with disperse dyes and polyester. A cationic dyeable polyester is useful for two main reasons in fabric coloration. First, cationic dyes give brighter, clearer shades than disperse dyes, and this can be important for both solid-dyed shades and prints during ever changing fashion cycles. Second, fabrics containing arrangements of unmodified and cationic dyeable polyesters can be dyed in the piece to a variety of color/white combinations by selection of dyestuffs, as day-to-day changes in demand may require. This is more economical than dyeing yarns to different colors, holding them in inventory, and then weaving them into fabrics, if and when they are required.

Nylon 66 can be dyed either with disperse dyes or with acid (anionic) dyes, the former carried out by essentially the same steps as described above for polyester, except that nylon absorbs these dyes much more readily. The ability of nylon to absorb acid (anionic) dyes is the result of a significant number of accessible free amine (cationic) end groups being present in the polymer. The dyeing of nylon with acid dyes is analogous to the dyeing of wool or to the dyeing of modified polyester with basic dyes, except that the polarities of the interacting groups are reversed in the latter case.

For best fastness to light and washing, nylon 66 is dyed with acid dyes, and nylon dye variants thus are based on manipulation of the level of acid dye uptake. By adding a monobasic acid such as acetic acid to the reaction mix near the end of the polymerization process, the amino end groups are converted to amide groups which have no affinity for acid dyes under normal dyeing conditions.

This technique creates light acid dyeing or acid reserve dye variants that can have some capacity to absorb basic dyes at the carboxyl end groups. Nylon dye variants with increased acid dye uptake can be made by using a slight excess of diamine in the polymerization. In this way, there are no free carboxylic acid end groups. Nylon dye variants have found the greatest acceptance in floor coverings where attractive patterns can be piece-dyed using controlled dyebath conditions and selected acid dyestuffs.

Acrylic fibers are dyed most frequently with basic dyes. This is made possible by copolymerizing acrylonitrile with an acidic monomer such as styrene-*para*-sulfonic acid. Acrylic fiber suitable for acid dyeing can be made by using a basic comonomer such as a vinyl pyridine or a vinyl pyrrolidone.

Fibers not having inherent flame resistance often can be given this property by incorporation of a suitable additive. This may be done by copolymerization of the additive into the polymer reaction of the additive with the polymer after polymerization, or by applying a polymeric or monomeric noninflammable finish to the surface as a coating. These additives usually contain bromine, nitrogen, or phosphorus, or a combination of these elements. Great care must be taken in choosing the additive and its level of addition in order to prevent loss of other desirable fiber properties and to avoid any harmful effects to processors or ultimate consumers.

A wide variety of special durable surface treatments have been used on manufactured fibers. These include treatments for imparting such characteristics as soil resistance, antistatic behavior, and wearer comfort through moisture wicking and transport. Fiber finishes also have been used successfully in promoting adhesion between two materials, as, for example, between polyester tire cord and rubber, and between glass fiber and polyester resin.

REFERENCES

1. Time-Life Books, *Seven Centuries of Art*, New York, 1970.
2. Mark, H., and Whitby, G. S., *Collected Papers of W H. Carothers*, John Wiley & Sons, New York, 1940.
3. Turbak, A., "Rayon," in *Encyclopedia of Polymer Science and Engineering*, 2nd ed., Vol. 14, A. Klingsberg and T. Baldwin (Eds.), p. 55, John Wiley & Sons, New York, 1985.

4. Gupta, B. S., and Hong, C. J., *INJ*, **7**(1), 38 (1995).
5. Davis, S., *Textile Horizons*, **9**(2), 62 (1989).
6. Albrecht, W., Reintjes, M., and Wulfhorst, B., *Chem. Fibers Int.*, **47**, 298 (1997).
7. Markham, J. W., *Competition in the Rayon Industry*, p. 16, Harvard University Press, Cambridge, MA, 1952.
8. Robinson, J. S., *Fiber-Forming Polymers: Recent Advances*, Noyes Data Corp., Park Ridge, NJ, 1980.
9. Barnes, C. E., "Nylon 4-Development and Commercialization," *Lenzinger Berichte*, **62**, 62–66, March 1987.
10. O'Sullivan, D., *Chemical and Engineering News*, **62**(21), 33 (1984).
11. Jung, Dong-Wook, Kotek, R., Vasanthan, N., and Tonelli, A. E., "High modulus Nylon 66 fibers through Lewis acid-base complexation to control hydrogen bonding and enhance drawing behavior," *Am Chem. Soc., Polymeric Materials: Science and Engineering Division Preprints* (2004), 91, 354–355.
12. Davis, G. W., Everage, A. E., and Talbot, J. R., *Fiber Producer*, **12**(6), 45 (1984).
13. Smierciak, R. C., Wardlow, E., and Lawrence, B. U.S. Patent 5,602,222, 1997.
14. Smierciak, R. C., Wardlow, E., and Lawrence, B. U.S. Patent 5,618,901, 1997.
15. Hutchinson, S. R., "Thermoplastic Polyacrylonitrile," North Carolina State University, M.S. Thesis, Raleigh, 2005.
16. Ahmed, M., *Polypropylene Fibers-Science and Technology, Textile Science and Technology,* Vol. 5, p. 16, Elsevier, New York, 1982.
17. Lieberman, R. B., and Barbe, P. C., "Propylene Polymers," in *Concise Encyclopedia of Polymer Science and Engineering*, J. I. Kroschwitz (Ed.), p. 916, John Wiley & Sons, New York, 1990.
18. Hogan, J. P., and Banks, R. L., "History of Crystalline Polypropylene," in *History of Polyolefins*, R. B. Seymour and T. Cheng (Eds.), p. 103, D. Reidel, Boston, 1986.
19. Gupta, B. S., and Smith, D. K., "Nonwovens in Absorbent Materials," in *Absorbent Technology*, P. K. Chatterjee and B. S. Gupta (Eds.), p. 378, Elsevier, Amsterdam, 2002.
20. Madsen, J. B., *Nonwovens World*, 69 (2001, August–September).
21. N. Sekar, *Colourage*, 47 (2), 33, 2000.
22. R. Kotek, Afshari, M., Gupta, B., Kish, M. H., and Jung, D., *Color. Technol*, 120, 26, 2004.
23. Zwijnenburg, A., and Pennings, A. J., *Colloid Polym. Sci.*, 259, 868 (1978).
24. Smith, P., and Lemstra, P. J., *J. Mater. Sci.*, 15, 505 (1980).
25. Kavesh, S., and Prevorsek, D., U.S. Patent 4,413,110, to Allied Chemical, 1983.
26. Kwolek, D. L., U.S. Patent 3,600,350, to E. I. duPont de Nemours and Co., 1971.
27. Blades, H., U.S. Patent 3,767,756, to E. I. duPont de Nemours and Co., Inc., 1973.
28. McIntyre, E., *Textile Horizons*, **8**(10), 43 (1988).
29. Chenevey, E. C., and Conciatori, A. B., U.S. Patent 3,549,603, to Celanese Corp., 1970.
30. Coffin, D. R., Serad, G. A, Hicks, H. L., and Montgomery, R. T., *Textile Res. J.*, **52**, 466 (1982).
31. Gore, R. W., U.S. Patent 3,953,566, to W. L. Gore & Associates, Inc., 1973, April 27.
32. Menardi-MikroPul L. L. C., www.mikropul.com/products/media/mikrotex.html
33. Edmonds, J. T., Jr., and Hill, H. W., Jr., U.S. Patent 3,354,129, to Phillips Petroleum Co., 1967.
34. Scruggs, J. G., and Reed, J. O., "Polyphenylene Sulfide Fibers," in *High Technology Fibers, Part A*, M. Lewin and J. Preston (Eds.), Marcel Dekker, Inc., New York, 1985.
35. Formhals, A., *Process and Apparatus for Preparing Artifical Threads*, U.S. Patent 1,975,504 (1934).
36. Reneker, D.H., and Chun, I., "Nanometre Diameter Fibres of Polymer, Produced by Electrospinning," *Nanotechnology,* 7: 216–223 (1996).
37. Huang, Z.M., Zhang, Y.Z., Kotaki, M., and Ramakrishna S., "A Review on Polymer Nanofibers by Electrospinning and Their Applications in Nanocomposites," *Compo. Sci. and Tech.,* 63: 2223–2253 (2003).

SUGGESTED READING

The reader is referred to the four encyclopedias listed below for additional information. They contain enormous quantities of information on manufactured fibers as well as comprehensive bibliographies.

Concise Encyclopedia of Polymer Science and Engineering, John Wiley & Sons, New York, 1990.
Encyclopedia of Polymer Science and Engineering, 2nd ed., John Wiley & Sons, New York, 1985. (17 volumes, index volume, and supplement volume.)
Encyclopedia of Polymer Science and Technology, Interscience Publishers, New York. (16 volumes.)
Kirk-Othmer Encyclopedia of Chemical Technology, 3rd ed., Interscience Publishers, New York, (21 volumes and a supplement, 3rd ed.; to date, 16 volumes.)

The following books contain broad discussions of manufactured textile fibers.

Baer, E. and Moet, A. (Eds.), *High Performance Polymers*, Hanser, New York, 1991.
Billmeyer, F. W., *Textbook of Polymer Science*, John Wiley & Sons, New York, 1984.

Ciferri, A., and Ward, I. M. (Eds.), *Ultra-High Modulus Polymers*, Applied Science, London, 1979.

Datye, K. V., *Chemical Processing of Synthetic Fibers and Blends*, John Wiley & Sons, New York, 1984.

Hearle, J. W. S., and Peters, R. H. (Eds.), *Fibre Structure*, The Textile Institute, Manchester, Butterworths, London, 1963.

Lewin, M., and Preston, J. (Eds.), *Handbook of Fiber Science and Technology: High Technology Fibers*, Vol. III, Marcel Dekker, New York, 1983.

Mark, H. F., Atlas, S. M., and Cernia, E. (Eds.), *Man-Made Fibers; Science and Technology*, Vols. I, II, and III, John Wiley & Sons, New York, 1967, 1968, and 1968.

Moncrieff, R. W., *Man-Made Fibres*, 6th ed., John Wiley & Sons, New York, 1975.

Morton, W. E., and Hearle, J. W. S., *Physical Properties of Textile Fibres*, The Textile Institute, Manchester, Butterworths, London, 1993.

Peters, R. H., *Textile Chemistry; The Chemistry of Fibers,* Vol. I and *Impurities in Fibers; Purification of Fibers*, Vol. II, Elsevier, New York, 1963 and 1967.

13

Dye Application, Manufacture of Dye Intermediates and Dyes

H. S. Freeman* and G. N. Mock**

INTRODUCTION

Dyeing

It is difficult if not impossible to determine when mankind first systematically applied color to a textile substrate. The first colored fabrics were probably nonwoven felts painted in imitation of animal skins. The first dyeings were probably actually little more than stains from the juice of berries. Ancient Greek writers described painted fabrics worn by the tribes of Asia Minor. But just where did the ancient craft have its origins? Was there one original birthplace or were there a number of simultaneous beginnings around the world?

Although it is difficult to determine just when each respective civilization began to use dyes, it is possible to date textile fragments and temple paintings, which have survived the ensuing centuries. The ancient Egyptians wove linen as early as 5000 BC, and paintings on tomb walls infer that colored wall hangings were in use by 3000 BC. By 2500 BC, dyer's thistle and safflower were used to produce red and yellow shades. Egyptian dyers developed a full range of colors by 1450 BC.

Another cradle of civilization was the Indian subcontinent where religious and social records dating to 2500 BC refer to dyed silk and woven brocades of dyed yarn. Cotton, first cultivated in the Indus valley of Pakistan was woven as early as 2000 BC. A book written around 300 BC included a chapter on dyes. It is believed that systematic dyeing occurred in China as early as 3000 BC near the city of Xian in the Hoang (Yellow) River Valley, although there is no conclusive proof. Empress Si-Ling-Chi is credited with the discovery of silk about 2640 BC. Kermes and indigo were used as dyes as early as 2000 BC. Fragments of silk have been found in the corrosive patina of bronze swords of the Shang dynasty (1523–1027 BC), but most assuredly these samples are not the oldest.[1]

The New World was similarly active in developing the textile art. With help from the desert climate in the high Andes of Peru, dyed samples of wool have been preserved and

*Ciba Professor of Dyestuff Chemistry, College of Textiles, North Carolina State University.
**Professor Emeritus of Textile Engineering, College of Textiles, North Carolina State University.

recovered from burial sites. These fragments have been dated to the millennium before the Christian era. The western and southwestern regions of the United States provided homes for the Anasazi, or ancient ones, who dwelt in the region of Mesa Verde National Park in southwestern Colorado, northern Arizona and New Mexico, and eastern Utah. Again the dry climate has helped to preserve samples from these early civilizations.

Very little in the nature of large, intact textile samples has survived in Europe. Remains of a large woolen robe, the Thorsberg Robe, found in northern Germany and dated prior to 750 BC, indicate a highly developed dyeing and weaving technology.[2] Indirect evidence is more plentiful: for example, a tombstone of a purpurarius, a Roman purple dyer, was found near Parma, northern Italy, and a dyer's workshop excavated in Pompeii. This great center of the Roman Empire was destroyed by the eruption of Mount Vesuvius in 79 AD. Similar stone vats for dyeing have been excavated in the tells of Israel and in present-day Turkey.

The Dark Ages following the fall of the Roman Empire were dark indeed, with little development of the dyer's art. The robes of a number of the monastic orders were brown and black, surely a dark age. By the end of the 1300s, however, civilization began making the swift and certain strides that have led to our present level of development. In 1371, the dyers of Florence, a city famous for its Renaissance art, formed a guild, or association of like merchants and craftsmen, which lasted for eleven years. Other guilds were being formed in other centers across Europe. Some of these guilds exist to this day. The Worshipful Company of Dyers was formed in 1471 in London. One of the legacies of this guild is a Publications Trust, which has underwritten the publication of a number of books on dyeing in cooperation with The Society of Dyers and Colourists.

The art and craft of dyeing was largely passed down from father to son or from craftsman to apprentice by word of mouth and example until the early 1500s. The *Plictho* of Gioanventura Rosetti, a Venetian armory superintendent, is believed to be the first published book on dyeing. It certainly is the oldest surviving European text to have come down to us in the twenty-first century. Five known Italian editions were published between 1548 and 1672. A French edition appeared in 1716. It is interesting that no known English translation was made until 1968[3] when Sidney Edelstein of the Dexter Chemical Company and Hector Borghetty collaborated to reproduce a facsimile of the original 1548 edition along with a complete translation into English. During his extensive travels, Rosetti collected dyeing recipes and processes used in the flourishing city states of Venice, Genoa, and Florence. He published: *Plictho de L'arte de Tentori che insegna tenger pani telle banbasi et sede si per larthe magiore come per la comune* or *Instructions in the Art of the Dyers, Which Teaches the Dyeing of Woolen Cloths, Linens, Cottons, and Silk by the Great Art as Well as by the Common* or simply *Instruction in the Art of Dyeing*. The book was divided into four sections: the first and second sections were devoted to the dyeing of wool, cotton, and linen; the third to the dyeing of silk and the use of fugitive colors; and the fourth to the dyeing of leather and skins. Approximately 160 complete recipes were preserved in the first three sections. Edelstein and Borghetty labored diligently in determining the meanings of terms in recipes written in the dialect of 16th Century Italy.

Synthetic Dyes

The father of modern synthetic dyes was William Henry Perkin (1838–1907), who synthesized mauve, or aniline purple, in 1856. The story behind this great story bears telling. William's father was a builder who wanted him to become an architect, but like many others, Perkin did not follow his father's chosen profession. Perkin studied at the City of London School where he became interested in chemistry at the age of 12. A teacher, Mr. Hall, gave him work in the laboratory, which in turn, inspired Perkin to follow his natural curiosity. At age 15, Perkin entered the Royal College of Science and listened to the lectures of the great German chemist, August Wilhelm von Hofmann (1818–1892). He was granted an assistantship under von Hofmann at age 17. Because his work did not allow time for his own research, he set up a separate laboratory at home and it was there that he discovered aniline purple, the first dyestuff to be commercially produced. Another dye, based on naphthalene, and prepared in collaboration with Arthur H. Church, actually preceded aniline purple, but was not commercially produced before aniline purple. Aniline purple was discovered at this home during Easter vacation while looking for quinine, an antimalarial drug. After oxidizing aniline with potassium dichromate and getting a black precipitate, extraction with ethanol gave a brilliant purple solution. Almost immediately, he sent a sample of this dye to a dyer in Perth with a request to dye silk fabric. The dyer's report read: "If your discovery does not make the goods too expensive, it is decidedly one of the most valuable that has come out for a long time." Trials on cotton were not as successful because the need for a mordant was not realized. Perkin later reported, "The value of mauve was first realized in France, in 1859. English and Scotch calico printers did not show any interest in it until it appeared on French patterns, although some of them had printed cloth for me with that colour."[4]

Since that beginning, thousands of dyes have been synthesized; some 1500 to 2000 are commercially successful today. Until 1884, however, all synthetic dyes required a mordant to give acceptable wash fastness on the textile substrate. In 1884, Böttiger produced Congo Red, which could dye cotton directly without a mordant. These dyes were commonly called direct dyes. In order to improve washfastness, the path taken in some synthetic dye chemistry was to build the dye from two or more components, directly in the fiber, or in situ. By building a large molecule without solubilizing groups within the fiber, washfastness was markedly improved. The first practical development along these lines was by A. G. Green who synthesized primuline, a dye that because of poor light fastness was not commercially important but later led

TABLE 13.1 Dyes in Order of Discovery

Basic: Mauve or Aniline Purple, Perkin, 1856; Fuchsin, Verguin, 1859
Acid: Alkali Blue, Nicholson, 1862
Vat: Alizarin, Gräbe & Liebermann, 1868
   Indigo, von Bayer, 1880, discovered structure and synthesized indigo
   Indanthrene, Böhn, 1901
Direct: Congo Red, Böttiger, 1884
Direct Developed: Primuline coupled with beta-naphthol, Green, 1887
Sulfur: Vidal, 1893
Azoics: Zitscher & Laske, 1911
Disperse: A simple azo dye, 1920s
Phthalocyanine: Linstead & Diesbach, 1928–1929
Reactive: commercialized, I.C.I., Rattee and Stevens, 1956

the way to many important commercial dyes. Table 13.1 lists a number of classes of dyes along with the date of discovery. Worldwide, 80% of all dyes go into textiles and 20% into paper, leather, food, and the like.

The Development of the U.S. Dyestuff Industry

The natural dyes industry was more than just a cottage industry in Colonial America. Indigo was a very important cash crop in South Carolina among the coastal islands and for some distance inland. Plantations existed well into the early 1900s despite the growth of the synthetic dyestuff industry.

The modern synthetic dye industry in the United States dates from World War I. However, in 1864, Thomas Holliday of Great Britain, and in 1868, the Albany (NY) Aniline Company with participation of Bayer of Germany began coal-tar dye manufacture. In the early 1900s, most synthetic dyes used in the United States were imported from Germany and Switzerland. With the outbreak of World War I, the British naval blockade of Germany prevented export of dyes from Europe. In spite of the blockade, the German submarine, *Deutschland*, ran the British blockade and sailed into American ports twice with dyestuffs and drugs. The Germans needed critical war material and export moneys; the United States and others needed dyes. Ironically, in 1914, German dyes were used by French dyers to dye the official French Army uniforms. The outbreak of war and ensuing blockade showed the United States how important dyes were to the American economy. Several companies began investigative work, that would lead to dye synthesis; they found that dyestuffs were very difficult to make; the chemistry was much more complex than imagined. A real boost to the U.S. industry came after World War I, when the German patents were given over to the Allies via the Alien Property Custodian. According to Lehner, DuPont reportedly spend $43 million, a tremendous sum of money in the early twentieth century, before ever showing a profit.[5] Obviously, only financially strong companies could afford to enter the business. The early pioneers included Allied (formed by merging five companies), American Cyanamid, and DuPont, to name only a few who survived to become major factors later in the twentieth century. In 1938, others included Dow, German-owned General Aniline and Film (GAF), and Swiss-owned Cincinnati Chemical Company (Ciba, Geigy, and Sandoz). In the 1960s, 50 to 60% of all U.S. manufacturing was in the hands of four principal U.S. companies:

- Allied Chemical (later sold to Bayer of Germany and to independent investors as Buffalo Color, 1977)
- American Cyanamid
- GAF, the result of the break-up of the German cartel, I. G. Farben, which was nationalized during World War II and sold to BASF in 1978
- DuPont (sold in 1980–1981 with various lines going to Crompton & Knowles, Ciba, and Blackman-Uhler)

Today, there are at least 42 dyestuff manufacturers, distributors, and repackaging agents in the United States.[6] Of the major companies, Swiss and German-based companies tend to dominate the U.S. market. Those companies include:

- Ciba, formerly Ciba-Geigy, Switzerland
- Clariant (split off from Sandoz in 1995), Switzerland

- DyStar, formed by the merger of Hoechst and Bayer (1995), acquisition of BASF textile colors which include the former Zeneca, and Mitsubishi of Japan, Germany
- Crompton & Knowles sold to Yorkshire Group Plc and is now Yorkshire Americas

There is no major surviving U.S.-based company. Raghavan[7] and Mock[8] give an interesting description of these mergers and the reasons behind them.

Most of the international companies have limited manufacturing facilities in the United States and major facilities in other countries where environmental laws are not as stringent or where the parent companies have a modern integrated low-pollution facility. These facilities in the U.S. minimize the tariffs paid and also allow quicker response to the marketplace. Ciba has a manufacturing facility in St. Gabriel, LA; Clariant in Martin, SC; DyStar near Charleston, SC; and Yorkshire Americas at Lowell, NC.

Today well over 1500 dyes are produced in commercial quantities, although only a select handful in each class are the true "workhorse" colors found in virtually every dyehouse dyeing a particular substrate for a particular enduse. Approximately two-thirds of the dyes and pigments consumed in the United States are used by the textile industry. One-sixth of the dyes and pigments are used for coloring paper, and the rest are used chiefly in the production of organic pigments and in the dyeing of leather and plastics.

Dyes are catalogued and grouped under a set of rules established by the Colour Index committee, consisting of representatives from the Society of Dyers and Colourists (SDC), Bradford, England, and the American Association of Textile Chemists and Colorists (AATCC), Research Triangle Park, NC. Table 13.2 shows how over 9000 dyes are enumerated in the current Index.[9] The Colour Index, now in its fourth edition, is updated periodically with newly released information and is available in book form and on CD-ROM. Volumes 1 to 3, published in 1971, contain the C. I. name and number, chemical class, fastness properties, hue indication, application, and usage. Volume 4, also published in 1971,

TABLE 13.2 Dyes Listed in the Colour Index

Shade	Yellow	Orange	Red	Violet	Blue	Green	Brown	Black	Sum
Acid	255	178	439	131	356	121	453	236	2169
Direct	171	121	263	104	303	100	242	189	1493
Disperse	243	155	371	102	371	9	27	31	1309
Reactive	188	123	252	44	250	27	48	46	978
Pigment	194	69	262	50	74	52	42	32	775
Basic	108	68	115	50	164	6	23	11	545
Vat	48	29	61	17	74	44	84	65	405
Sulfur	23	5	14	—	20	37	96	18	213
Mordant	65	47	95	60	82	36	92	96	573
Natural	26	4	34	—	—	5	13	—	82
Solvent	174	105	232	49	134	33	58	52	837
Food	15	8	17	—	5	4	3	3	55
Total									9475

contains the structures of all disclosed structures. Volume 5, last published as part of Volume 9 in 1993, contains the commercial names of all known dyes and pigments. Volumes 6 to 8 are supplements with updates to information in Volumes 1 to 4 up to 1976, 1981, and 1987, respectively. An online version of the Colour Index containing a significant number of new entries was made available in 2002.

In addition to the Colour Index, AATCC publishes a *Buyer's Guide* annually in July.[6] Part A lists dyes, pigments, and resin-bonded pigment colors available from companies who choose to list this information.

The textile industry uses a large number of dyestuffs from each of the dye categories, the choice depending on the shade, fiber, and dyeing process, end-use of the textile product, requirements for fastness, and economic considerations. To provide an understanding of the interrelationships that exist among the various dye classes and fiber types, a brief survey of the major fibers follows.

TEXTILE FIBERS

In this survey, commercially important textile fibers are grouped by their origin. First there are the natural fibers from plant sources, cotton and flax, and those from animal sources, wool and silk. A second group consists of those fibers that are regenerated or chemically modified natural materials, the rayon and acetate fibers. The final group consists of synthetic fibers, which include polyester, nylon, acrylics, polyolefins, and elastane.

Natural Fibers

Cotton. Cotton fibers are comprised mainly of cellulose, a long-chain polymer of anhydroglucose units connected by ether linkages. The polymer has primary and secondary alcohol groups uniformly distributed throughout the length of the polymer chain. These hydroxyl groups impart high water absorption characteristics to the fiber and can act as reactive sites. The morphology of the cotton fiber is a complex series of reversing spiral fibrils. The fiber in total is a convoluted collapsed tube with a high degree of twist occurring along the length of the fiber. This staple fiber occurs in nature in lengths of ½ to 2 inches, depending on the variety and growing conditions. The diameter ranges from 16 to 21 microns (one micron is 1×10^{-6} meter).

Flax. Flax is also a cellulosic fiber but has a greater degree of crystallinity than cotton. The morphology of flax is quite different from that of cotton. Flax fibers have a long cylindrical shape with a hollow core. The fibers range in length from ½ to 2½ inches, with a diameter of 12 to 16 microns. Flax staple is comprised of bundles of individual fibers. Historians believe that flax was among the first fibers to be used as a textile fiber. In recent years, its commercial importance as a textile fiber has decreased significantly.

Wool. Wool fibers are comprised mainly of proteins: the polypeptide polymers in wool are produced from some 20 alpha–amino acids. The major chemical features of the polypeptide polymer are the amide links, which occur between the amino acids along the polymer chain, and the cystine (sulfur to sulfur) crosslinks, which occur in a random spacing between the polymer chains. The polymer contains many amine, carboxylic acid, and amide groups, which contribute in part to the water-absorbent nature of the fiber.

The morphology of wool is complex. There is an outer covering over the fiber, the cortical. There are also overlapping scales having a ratchet configuration that causes shrinkage and felting. The coefficient of friction in wool fibers is vastly different between the tip and the root, depending on which way the scales point. Wool can be made washable by chemically abrading the scales or coating the fibers with another polymer.

Wool fibers are not round but are oval in cross-section. The cortical cells constitute the major component of the fiber, and are aligned along the axis of the fiber. There is a medulla section at the center region of the fiber. Each fiber has a bicomponent longitudinal crystalline arrangement. One side of the fiber contains alpha–keratin crystalline regions, and the other contains beta–keratin crystalline regions. Alpha–keratin and beta–keratin have different moisture absorption characteristics, and this difference is what gives wool fibers crimp and springiness. It is also the reason why wool fibers kink in conditions of high humidity.

Wool fibers are sheared from about 30 major sheep breeds. The length of the wool fibers varies from 1 to 14 inches and depends on the breed, the climate, and the location on the sheep's anatomy. The fibers can be very fine to very coarse, ranging from 10 to 50 microns in diameter. The longer, coarser fibers normally are used for woolen fabrics, whereas the shorter finer ones are used for worsted fabrics.

Silk. Silk, like wool, is a protein fiber, but of much simpler chemical and morphological makeup. It is comprised of six alpha–amino acids, and is the only continuous-filament natural fiber. Historians claim that silk was discovered in China in 2640 BC. Silk fiber is spun by the silkworm as a smooth double-strand, each part having a trilobal cross-section. This configuration helps give silk its lustrous appearance. The fiber is unwound from the cocoon the silkworm spins as it prepares its chrysalis. The filaments are smooth and have no twists in their length, which can vary from 300 to 1800 yards. The diameter of silk is very fine, ranging from 2 to 5 microns. Because of the labor-intensiveness of sericulture and subsequent preparation of the fiber, silk remains a luxury fiber.

Regenerated Fibers

Rayon. Viscose rayon, like cotton, is comprised of cellulose. In the manufacturing process, wood pulp is treated with alkali and carbon disulfide to form cellulose xanthate. Subsequently, the reaction mass is forced through a spinneret and precipitated in an acid coagulation bath as it is formed into a continuous filament. The fiber has a round striated cross-section. Rayon staple is made by "breaking" the continuous strands into staple-length fibers. Viscose rayon is conventionally produced in diameters varying from 9 to 43 microns.

Acetate. Triacetate and diacetate fibers are manufactured by the chemical treatment of cellulose obtained from refined wood pulp or purified cotton lint. Most of the hydroxyl groups are acetylated (esterified) by treating the cellulose with acetic acid. This determines the chemical configuration of triacetate. Acetate or diacetate is made by the saponification of one of the acetylated groups, thus restoring a hydroxyl to each cellulosic monomer unit. Theoretically, then, diacetate has two acetylated groups in each glycoside unit. The conversion of the hydroxyl groups causes these fibers to be hydrophobic and changes the dyeing characteristics drastically from those of the normal cellulosic fibers. Triacetate fibers are spun by mixing the isolated reaction product (flake) with methylene chloride and alcohol. The spinning solution (dope) is forced through a spinneret and dry-spun into continuous filaments.

An alternate way of wet spinning is also possible. Acetate fibers are spun by mixing the isolated reaction product with acetone and water. The spinning solution is formed into filaments by evaporating the solvent and coagulating the acetate in a manner similar to that for triacetate (i.e., by the dry-spinning method).

Synthetic Fibers

Nylon. In 1939 the DuPont Company introduced the first truly synthetic textile fiber. Dr. Wallace Carothers invented nylon as a result of his basic research into polymer science. Chemically, nylon is a polyamide fiber. The two major types of nylon polymer are used in textiles: type 6,6 which is made by using hexamethylene glycol and adipic acid, and type 6, which is made by polymerizing ε-caprolactam. Nylon fibers are made by melt-spinning the molten polymer. The result is a continuous filament fiber of indeterminate length. It is spun in many deniers, with its diameter varying from 10 to 50 microns. The cross-section usually is round, trilobal, or square with hollow channels when used as carpet fiber.

Polyester. Polyester is made by the polymerization reaction of a diol and a diester. The main commercial polymer is formed by a condensation reaction using ethylene glycol and terephthalic acid. Fibers are formed by melt-spinning. Commercially introduced in 1953 by the DuPont Company as Dacron, polyester fibers have high strength, and very low moisture absorbance. The fiber is usually spun with a round cross section. Polyester is the most-used synthetic fiber around the world.

Acrylics. The DuPont Company introduced the first commercial acrylic fiber, Orlon, in 1950. Acrylics are made from the polymerization of acrylonitrile and other co-monomers to allow dyeability and to open the internal structure. The fibers are produced by either solvent-spinning (Orlon), or wet-spinning (Acrilan). In the solvent-spinning process, the polymer is dissolved in a low-boiling liquid solvent such as dimethyl formamide and extruded in a warm air chamber. In wet-spinning, the polymer is dissolved in a suitable solvent, extruded into a coagulation bath, dried, crimped, and collected. Although the acrylic fibers are extruded as continuous filaments, they subsequently are cut into staple-length fibers. Acrylics have found a niche market as a substitute for wool or in wool blends (blankets, sweaters, etc.) and in awnings and boat covers. The cross-section of the filament varies among manufacturers, Orlon having a dog-bone configuration and Acrilan having a lima-bean shape. Acrylic fibers are quick drying and wrinkle resistant.

Polyolefins. Polyolefin fibers are produced from the polymerization of ethylene or propylene gas. The catalysis research of Ziegler and Natta led to the development of these polymers to form crystalline polymers of high molecular weight. Hercules Inc. produced the first commercial fibers in 1961. The fibers made from these polymers are melt-spun. The cross-sections are round, and the fibers are smooth. They have extremely low dye affinity and moisture absorbance. Colored fiber is normally produced by mixing pigments in the melt polymer prior to extrusion.

Elastane. The DuPont Company commercialized the first manufactured elastic fiber, Lycra, in 1958. Originally categorized as a spandex fiber, the name "elastane" has become more common around the world. This specialty fiber is described as a segmented polyurethane that contains "hard" and "soft" segments; their ratio determines the amount of stretch built into the fiber. Elastane fibers are formed by dry-spinning or solvent-spinning. The continuous filaments can be coalesced multifilaments or monofilaments, depending on the manufacturer.

Because most dyeings are applied from water solutions or dispersions, the effect of water absorption by the fiber is an important criterion. Table 13.3 shows the hydrophobic/hydrophilic characteristics of the important fibers. The cellulosic and natural fibers are the most hydrophilic, and polyolefin is the most hydrophobic.

TABLE 13.3 Hydrophobic / Hydrophilic Characteristics of Various Fibers

Fiber	Moisture Content, % (at 65 % R.H.–70°F	Water Retention, % (Weight Change in Water)	Swelling, % (Volume Change in Water)
Acetate	6.5	20–25	NA
Acrylic	1.0–1.5	4.5–6	2
Cotton	7	45	45
Elastane	0.3–1.5	low	low
Nylon	3.0–5.0	9–12	13
Polyester	0.3–0.5	3–5	0.5
Olefin	0.01–0.1	Very low	Very low
Triacetate	3.2	12–18	NA
Viscose	13	90–100	95
Wool	13–15	42	42

Microdenier Fibers. The first commercial production of microfiber in the United States was in 1989 by the DuPont Company. Today microfibers are produced in a variety of synthetic fibers (i.e., polyester, nylon, acrylic, etc.) A microfiber is a fiber that is less than one denier per filament. Yarns made from microdenier filaments are able to give silklike hand to fabrics.

DYE CLASSIFICATION

This section covers structural features that govern the classification and application of various dye classes. In this regard, the chemistry of acid, azoic, basic, direct, disperse, reactive, sulfur, and vat dyes is presented. With regard to the application of synthetic dyes to textiles, it is well known that dyeing of textile fibers from an aqueous dyebath involves four steps: exhaustion, diffusion, migration, and fixation. In step 1, individual dye molecules move from the dyebath to the fiber surface and in step 2, dye molecules move from the fiber surface into the amorphous regions of the fiber. In step 3, dye molecules move from regions of high concentration to regions of low concentration

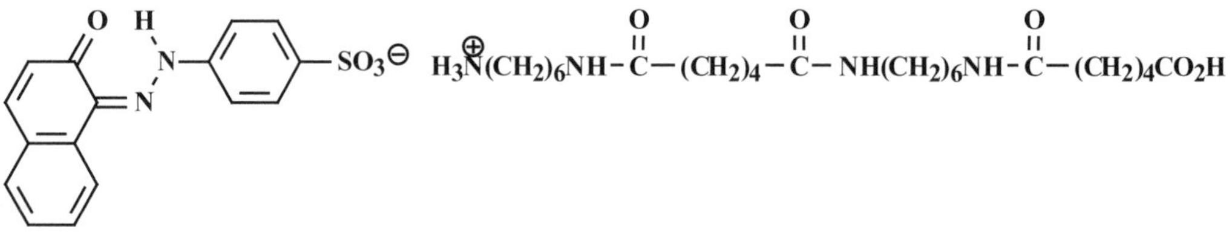

Fig. 13.1. Ionic bond formation between nylon 66 and C.I. Acid Orange 7.

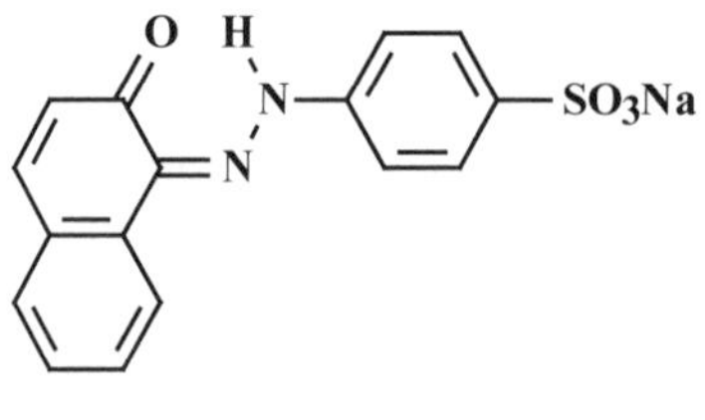

Acid Orange 7 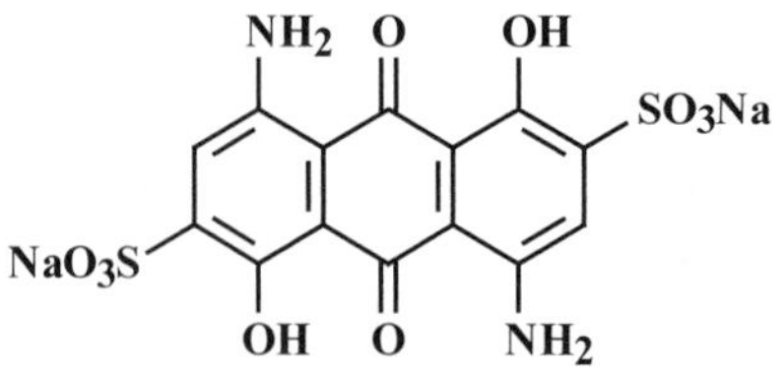   Acid Blue 45

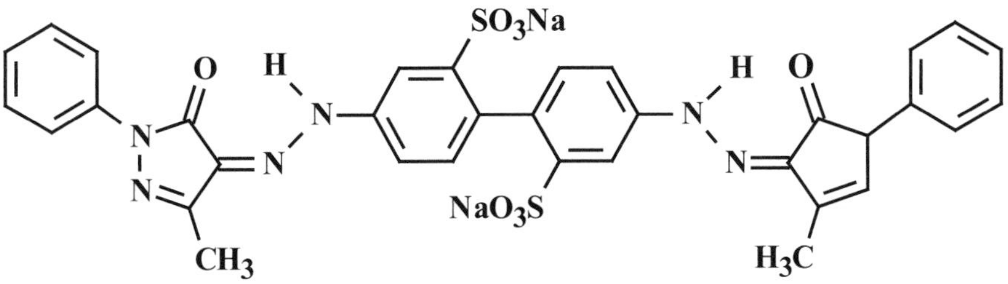

Acid Yellow 42

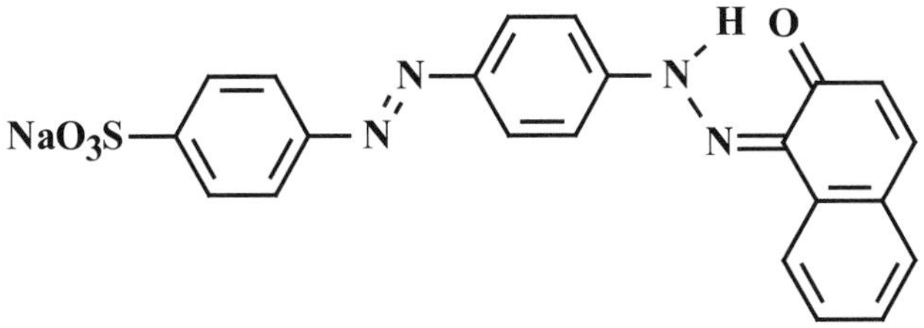

Acid Red 151

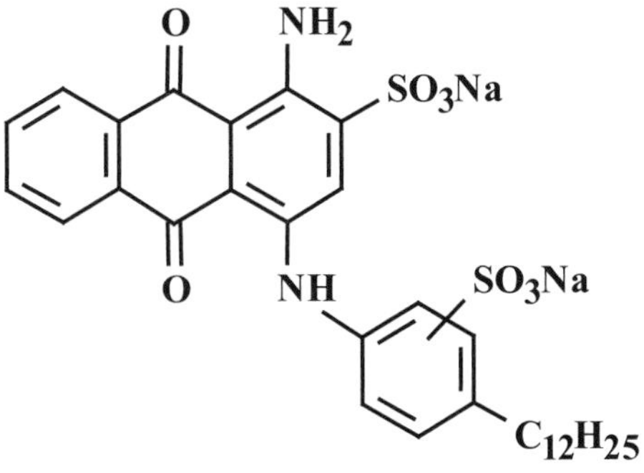

Acid Blue 138

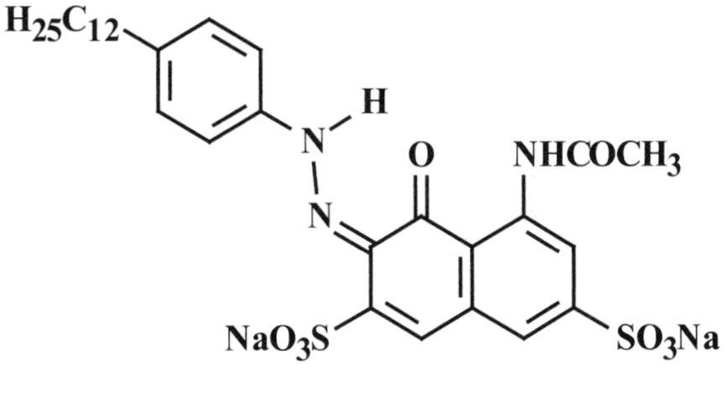

Acid Red 138

(i.e., migrate) to become uniformly distributed within the polymer matrix. In step 4, dye molecules interact with groups along the polymer chain via primary or secondary valency forces. Dye-polymer interactions can involve ionic bonding (e.g., acid dyes on nylon or wool), covalent bonding (e.g., reactive dyes on cotton), mechanical entrapment (e.g., vat dyes and sulfur dyes on cotton), secondary valency forces (direct dyes on cotton), or solid–solid solution (e.g., disperse dyes on polyester).

Acid Dyes

Acid dyes derive their name from the conditions associated with their application, in that they are typically applied to textile fibers from dyebaths containing acid.[9] Most acid dyes have one or two sodium sulfonate ($-SO_3Na$) groups and, therefore, are water soluble and capable of bonding with fibers having cationic sites (cf. Fig. 13.1). They give a wide range of bright colors on textiles, especially when monoazo and anthraquinone structures are used.

Acid dyes vary widely in molecular structure and in the level of acid required for dye application. They include relatively low molecular weight dyes such as C.I. Acid Orange 7 and C.I. Acid Blue 45, both of which are readily applied to polyamide and protein fibers and are known as *level dyeing acid dyes*. As the name suggests, these dyes are characterized by good migration and, therefore, readily produce level dyeings with time. In addition they give reasonably good lightfastness and barré coverage. The application of level dyeing acid dyes to nylon and wool utilizes weak acid and strong acid, respectively. For applications requiring good washfastness, *milling acid dyes* or *supermilling acid dyes* are employed. Both of these dye types afford relatively poor barré coverage, however. The former type dyes are applied from weakly acidic dyebaths whereas the latter are generally applied at neutral pH, with molecular size increasing as acid strength decreases. Examples of milling acid dyes are C.I. Acid Yellow 42 and C.I. Acid Red 151, and supermilling acid dyes include C.I. Acid Blue 138 and C.I. Acid Red 138. Because it is well known that azo dyes derived from naphthol and pyrazolone intermediates exist predominantly in the hydrazone form, this tautomeric form is given for Acid Yellow 42, Acid Red 151, Acid Red 138, and for the appropriate dyes that follow in this chapter.

Acid dyes include metal-complexed azo structures, where the metals used are cobalt, chromium, and iron.[10] Examples are 1:1 and 2:3 chromium complexes and 1:2 cobalt complexes, where the numbers employed represent the ratio of metal atoms to dye molecules. Metal-complexed dyes can be formed inside textile fibers by treating suitably dyed fibers with a solution containing metal ions.[11] In this case, the metal-free forms of these azo dyes are known as *mordant dyes* and contain mainly *ortho, ortho'*-bis-hydroxy or *ortho*-carboxy, *ortho'*-hydroxy groups (e.g., C.I. Mordant Black 11, Mordant Yellow 8, and Mordant Orange 6). When the metal complexes are formed prior to the dye application process, the resultant dyes are known as

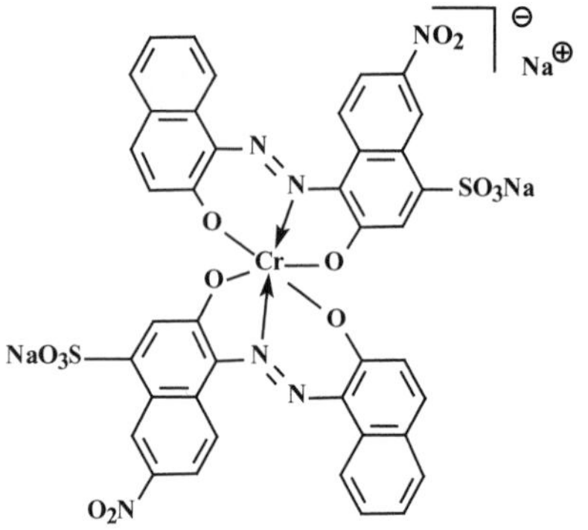

Acid Black 172

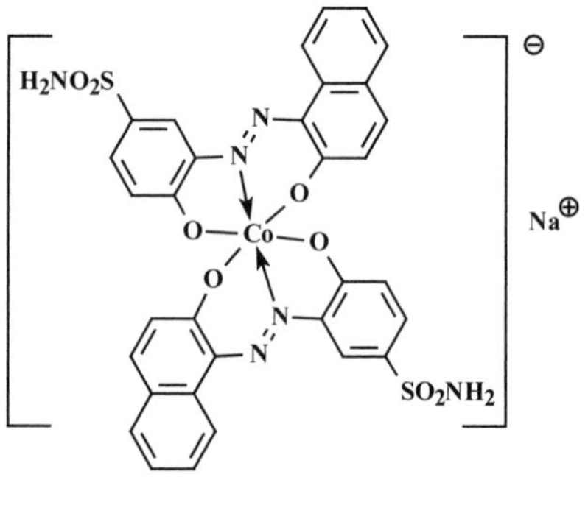

Acid Red 182

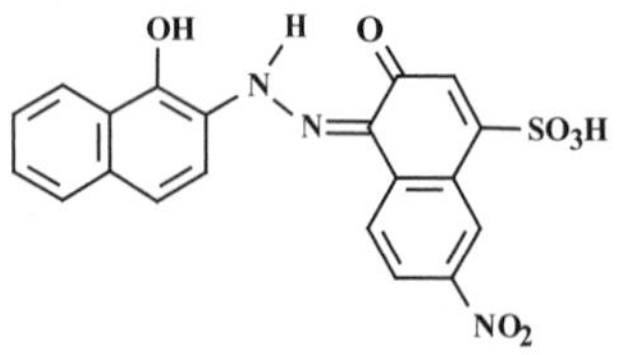

Mordant Black 11

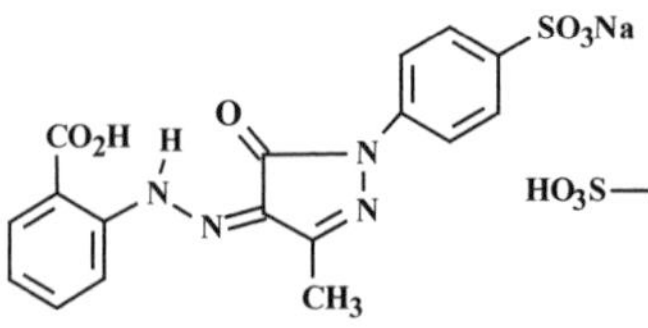

Mordant Yellow 8

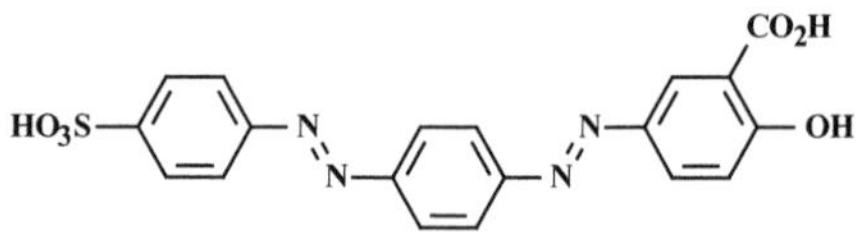

Mordant Orange 6

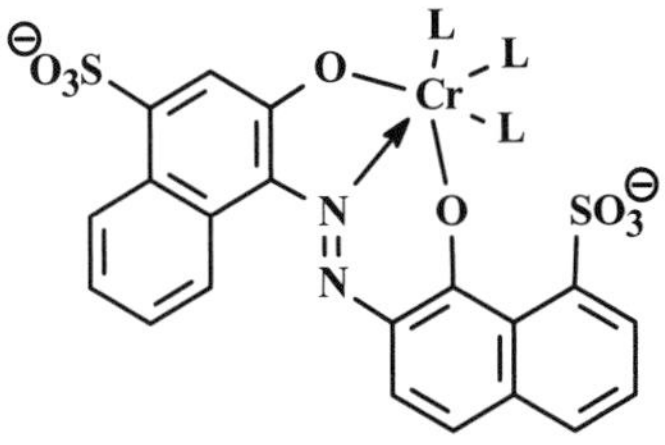

Acid Blue 158

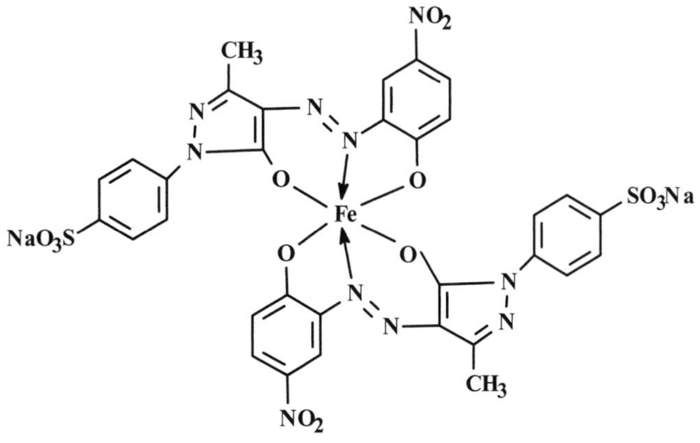

Acid Brown 98

premetallized acid dyes and vary in the acid strength required in the application step.[12] The 1:1 chromium complexes (e.g., C.I. Acid Blue 158) are stable only in very strong acid, making them suitable for wool but not nylon. Neutral dyeing premetallized acid dyes contain $-SO_2NH_2$ or $-SO_2CH_3$ groups in lieu of $-SO_3Na$ groups (see Acid Black 172 vs. Acid Red 182). In this case, dye–fiber fixation occurs because the combination of trivalent metal ion (e.g., Cr^{3+}) and four attached negatively charged ligands gives the complex a net negative charge.

Metallization of azo dyes enhances lightfastness, reduces water solubility, causes a bathochromic shift in color, and dulls the shade. Iron complexes generally give brown shades (e.g., C.I. Acid Brown 98) and are most often used to dye leather.

Azoic Dyes

Azoic dyes are mainly bright orange and red monoazo dyes for cotton, with dull violet and blue colors also possible.[13] They are water insoluble and consequently give high washfastness. They are also referred to as azoic "combinations" rather than "dyes" because they do not exist as colorants until they are formed inside the pores of cotton fibers.[14] They are quite important for printing on cotton and often give good lightfastness in heavy depths. Their bleachfastness is better

than direct and sulfur dyes and good crockfastness requires efficient soaping after the application step. The formation of these dyes requires two constituents: an azoic coupling component and an azoic diazo component, examples of which are shown in Figs. 13.2 and 13.3. The azoic coupling components are beta-naphthol and BON acid derivatives and the azoic diazo components are substituted anilines.

Azoic dyes are also known as naphthol dyes, because all employ a naphthol component in their formation, and they can be produced in batch or continuous processes. Because they have a limited shade range, they are best known for their ability to provide economical wetfast orange and red shades on cotton. A generic azoic dye structure is shown in Fig. 13.4.

Basic or Cationic Dyes

Basic dyes were developed to dye negatively charged acrylic fibers, forming ionic bonds in the fixation step (Fig. 13.5).[15] They owe their name to the presence of aromatic amino (basic) groups, and in this case a cationic amino group is present. Generally, they have excellent brightness and color strength, especially among the triarylmethane types. However, their lightfastness is often low, when they are applied to fibers other than acrylics. Basic dyes include those containing a fixed cation, examples of which are C.I.

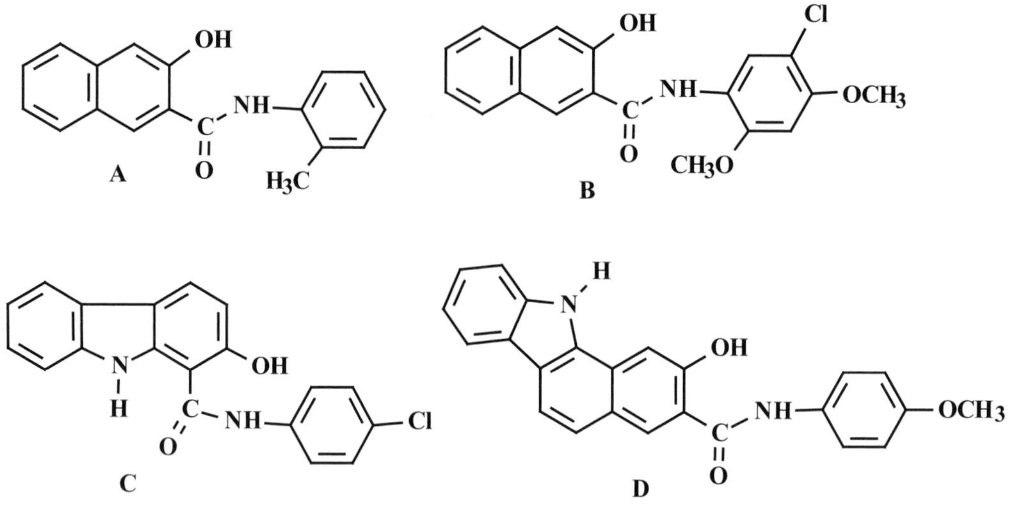

Fig. 13.2. Structures of C.I. Azoic Coupling Components 18 (A), 12 (B), 15 (C), and 13 (D).

Basic Blue 22, and C.I. Basic Red 18. The tri-arylmethane dye C.I. Basic Violet 3 has a mobile cation that produces resonance structures of comparable energy.

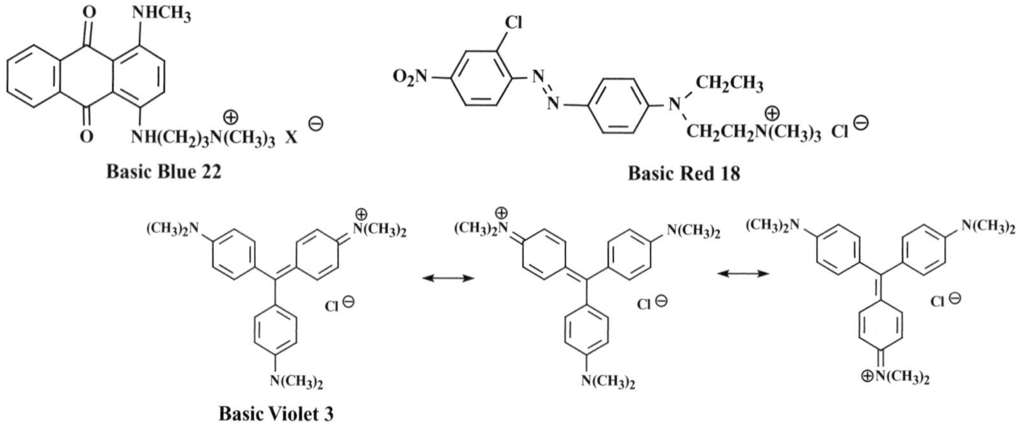

retarding agents are used, the dye–retarder bond is broken by increasing the dyebath temperature, giving controlled release of dye molecules to facilitate levelling.

Basic Blue 22

Basic Red 18

Basic Violet 3

Basic dyes are applied from weakly acidic dyebaths (pH = 4.5–5.5) and often require the use of anionic or cationic retarding agents to control the rate of dye strike and give level dyeings. Suitable retarding agents either form a weak bond with dyesites along the polymer chain or interact with the dye in the dyebath. In the former case, a significant fraction of the cationic retarder employed is displaced by the dye as dyeing progresses because the dye has higher affinity for the fiber. When anionic

To help determine which basic dyes can be combined for shade matching, key dyebath parameters have been developed.[16] The first parameter pertains to the dyes themselves and is known as the *combinability constant* (k). This value provides a measure of how fast a basic dye will dye the fiber, and the dyes are rated on a scale of 1 (fast) to 5 (slow). The second parameter pertains to the fiber type involved and is known as the *fiber saturation value* (S_F). This value provides an indication

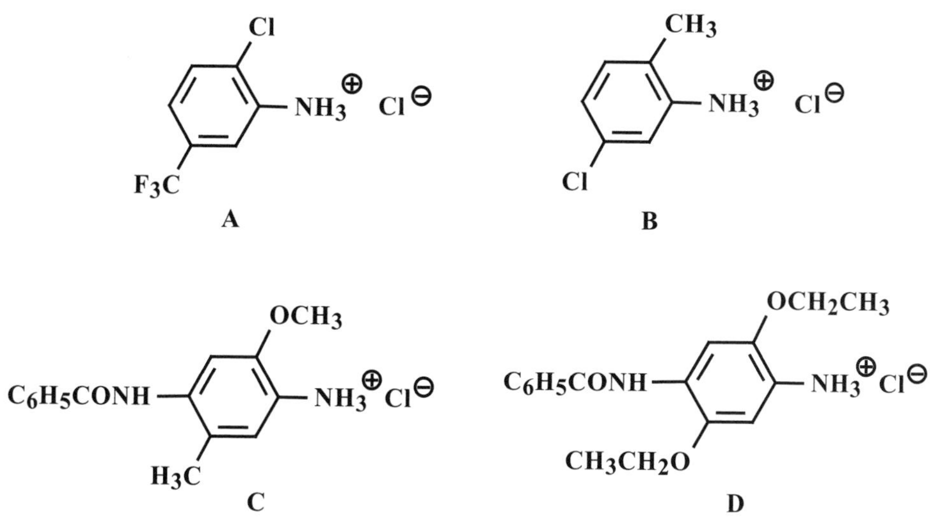

Fig. 13.3. Structures of C.I. Azoic Diazo Components 49 (A), 32 (B), 41 (C), and 20 (D).

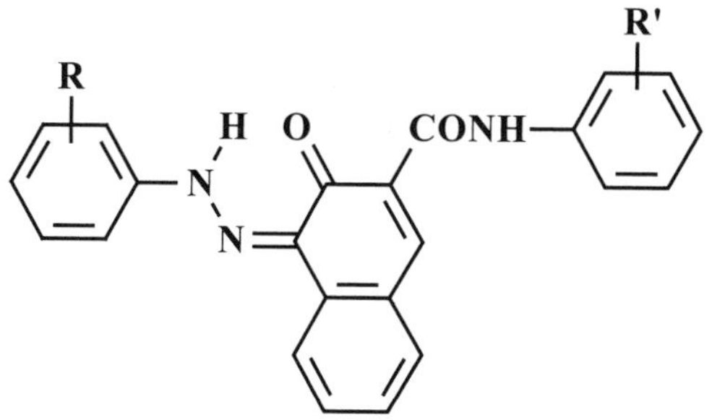

Fig. 13.4. Generic structure for azoic dyes, where R and R′ = alkyl, alkoxy, halo, and nitro groups.

of how much dye the fiber will hold at the saturation point. In this regard, the dye used is C.I. Basic Green 4 and typical saturation levels are 1.0 to 4.0% based on the weight of the fibers (*owf*) for light to deep dyeings. The third parameter is the *dye saturation factor* (f), which is a measure of the capacity of a basic dye for saturating a fiber. This factor is influenced by the molecular size and purity of the dye. In this case, the goal is to avoid placing more dye on the fiber than the number of dye sites, and the standard is C.I. Basic Green 4 ($f = 1\%$).

Direct Dyes

Direct dyes are anionic colorants that have affinity for cellulosic fibers.[17] They were the first dyes that could be used to dye cotton in the absence of a mordanting agent, giving rise to the term *direct-cotton dyes*. Like acid dyes, direct dyes contain one or more $-SO_3Na$ groups, making them water-soluble. Unlike acid dyes, they interact with cellulose (Cell–OH) chains via secondary valency forces (e.g., H-bonding and dipole–dipole interactions), as illustrated in

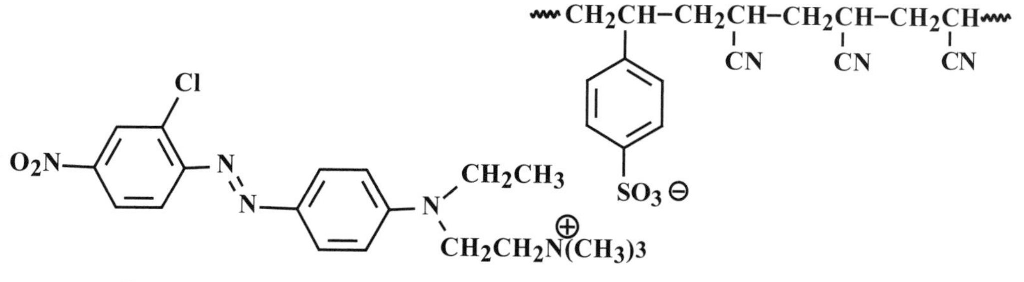

Fig. 13.5. Ionic bond formation between polyacrylonitrile and C.I. Basic Red 18.

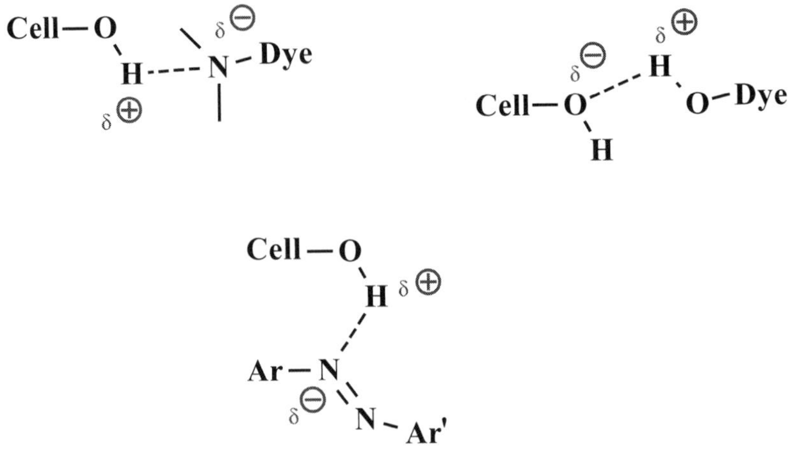

Fig. 13.6. H-bonding interactions between cellulose and polar groups in direct dye molecules.

Fig. 13.6. The combined effects of these rather weak forces and sulfonated structures cause direct dyes to have low intrinsic washfastness.

Direct dye structures are based on four main chromophores: azo (e.g., C.I. Direct Red 81, C.I. Direct Yellow 28, and C.I. Direct Black 22), stilbene (e.g., C.I. Direct Yellow 12 and C.I. Direct Yellow 11), oxazine (e.g., C.I. Direct Blue 106 and C.I. Direct Blue 108), and phthalocyanine (e.g., C.I. Direct Blue 86 and C.I. Direct Blue 199). About 82% of all direct dyes have disazo or polyazo structures, with stilbene and monoazo structures occupying about 5% each and thiazole, phthalocyanine, and dioxazine structures covering the remaining few percent.[18]

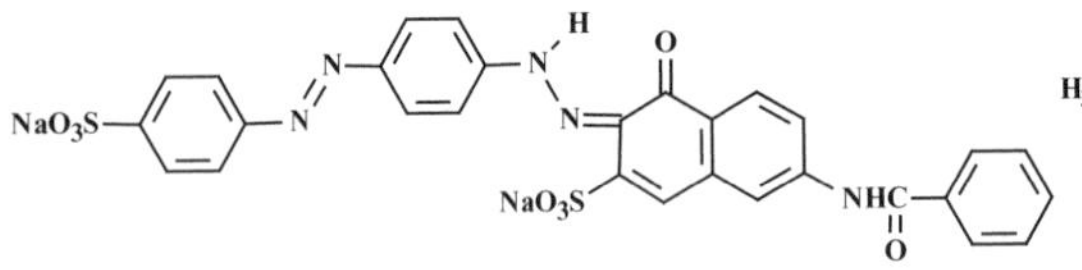

Direct Red 81

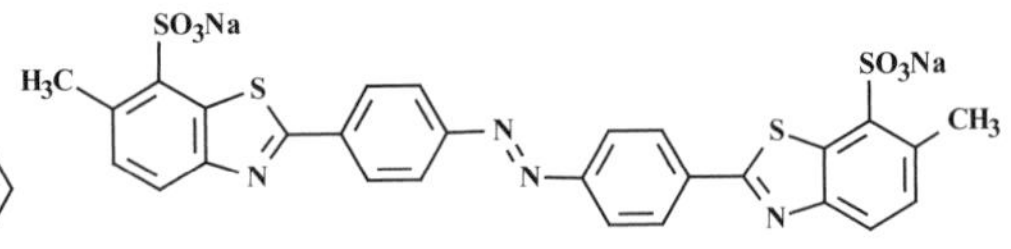

Direct Yellow 28

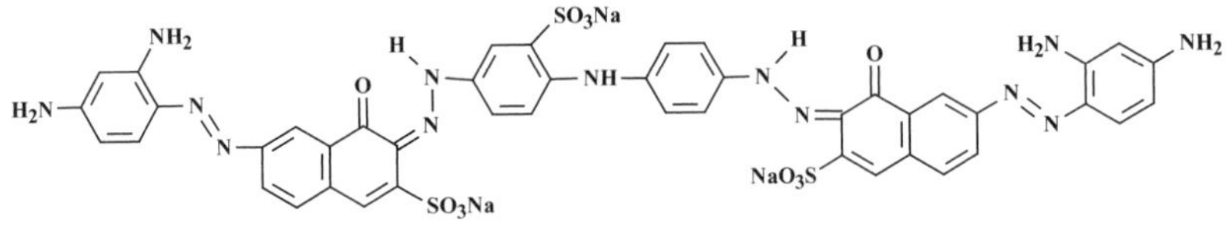

Direct Black 22

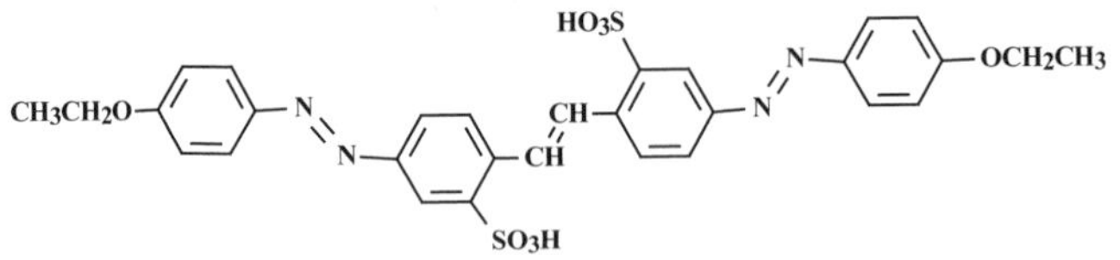

Direct Yellow 12

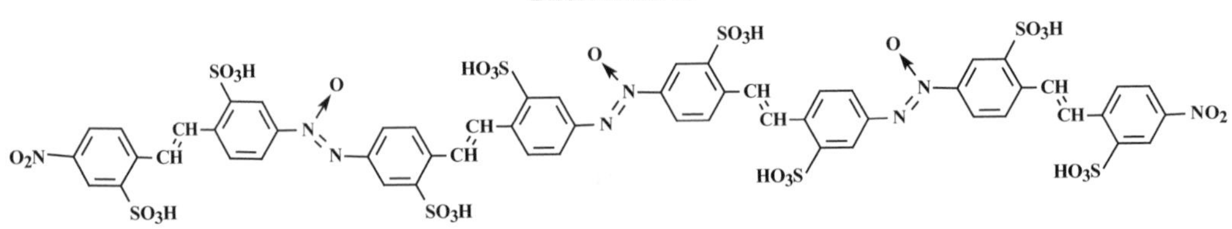

Direct Yellow 11

Suitably substituted direct dyes can be converted to metal complexes. In this regard, Cu is the metal of choice and examples are C.I. Direct Blue 218, C.I. Direct Red 83, and C.I. Direct Brown 95. About 5% of all azo direct dyes are metal complexes and unlike most direct dyes, these dyes have good lightfastness, as would be anticipated.

Direct dyes are subdivided into three classes (A, B, and C), to assist the dyer in selecting appropriate combinations for color matching.[19] Class A direct dyes give good migration and leveling with time. The dyer employs 5–20% salt for their application and in this case all of the salt may be added at the beginning of the dyeing cycle. An example of

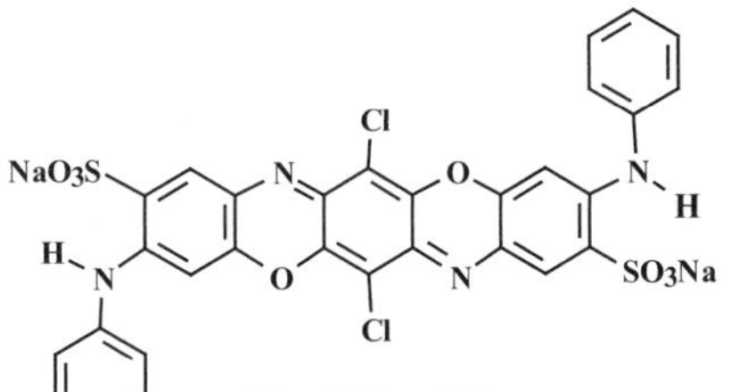

Direct Blue 106

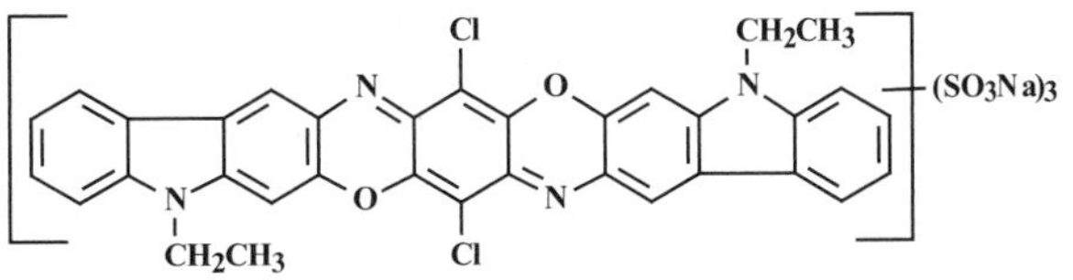

Direct Blue 108

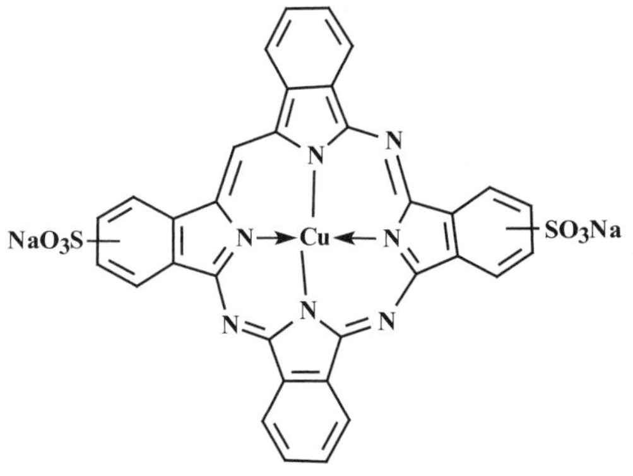

Direct Blue 86

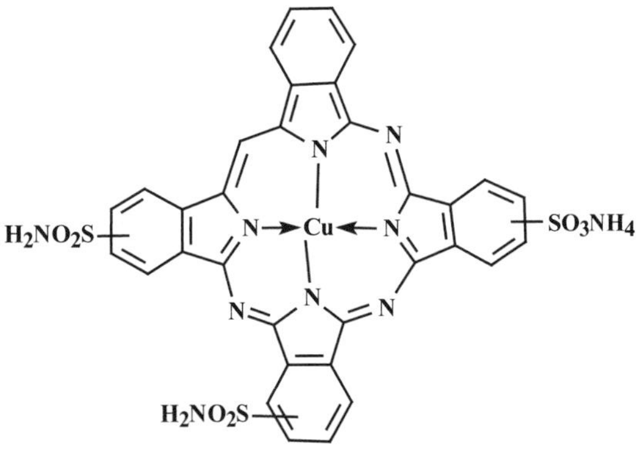

Direct Blue 199

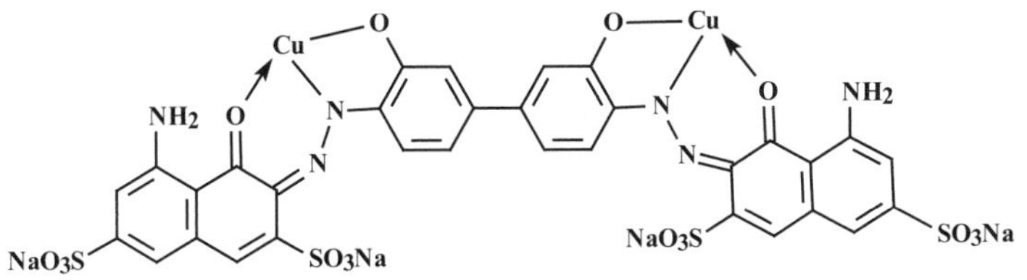

Direct Blue 215

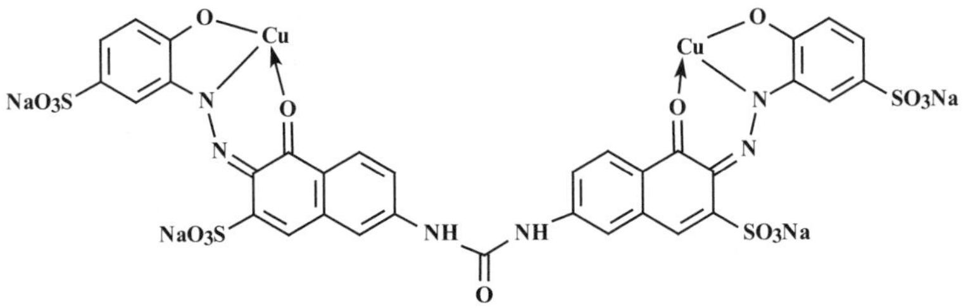

Direct Red 83

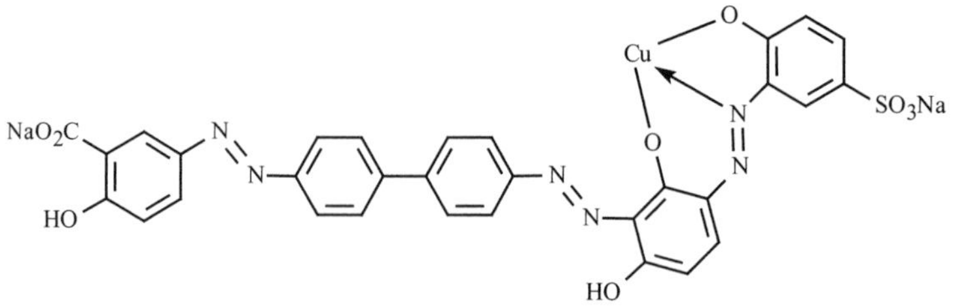

Direct Brown 95

this direct dye class is C.I. Direct Yellow 12. Class B direct dyes have poor migration and leveling properties and require the controlled addition of salt to afford level dyeings. They are larger than the former types and have better washfastness. An example is C.I. Direct Blue 1. Class C dyes are the largest of the direct dyes and, consequently, have the best washfastness but poorest leveling properties. Leveling requires careful control of the rate of temperature rise during the dyeing process. Some salt may be added but less than the amount used with classes A and B. An example of this dye class is C.I. Direct Black 22.

of the unsuitable dyes reveals that they have groups which are subject to hydrolysis. In the case of Direct Red 83, hydrolysis essentially cuts the molecule in half, eliminating fiber affinity (Fig. 13.7).

Because many direct dyes do not have good washfastness and lightfastness, their dyeings on cotton are often treated with a chemical agent, in what is commonly known as an aftertreatment process. The most widely used aftertreatment methods involve (1) cationic fixatives, (2) copper sulfate, or (3) diazotization and coupling reactions. The first and third methods are designed to enhance wash-

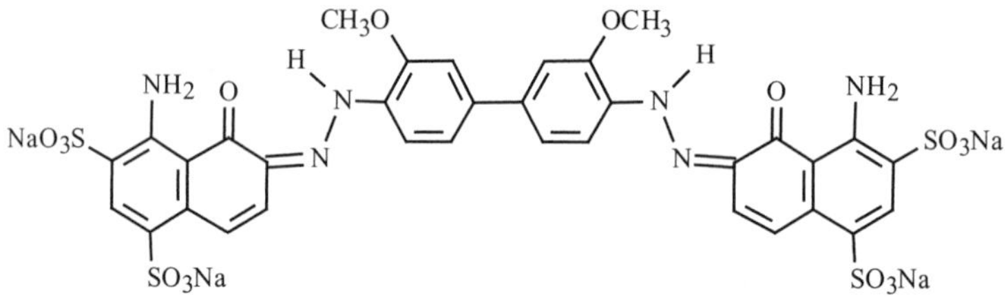

Direct Blue 1

The high temperature stability of direct dyes is an important consideration if one wishes to use these dyes as the colorant for cotton when dyeing a polyester/cotton blend at 130°C.[20] The key to success is to choose dyes that are resistant to hydrolysis. Suitable dyes include C.I. Direct Yellow 105, C.I. Direct Orange 39, and C.I. Direct Blue 80, whereas unsuitable dyes include C.I. Direct Yellow 44, C.I. Direct Red 80, and C.I. Direct Red 83. A quick examination of the structures

fastness and are illustrated in Figs. 13.8 and 13.9. The use of cationic fixatives ties up sodium sulfonate groups, reducing the water solubility of the treated dye. Diazotization and coupling enlarges the size of the dye, making desorption more difficult, and simultaneously makes the dye less hydrophilic. This process requires the presence of at least one diazotizable primary arylamino (Ar–NH$_2$) group in the dye structure. In this two-step process, the amino group is diazotized by

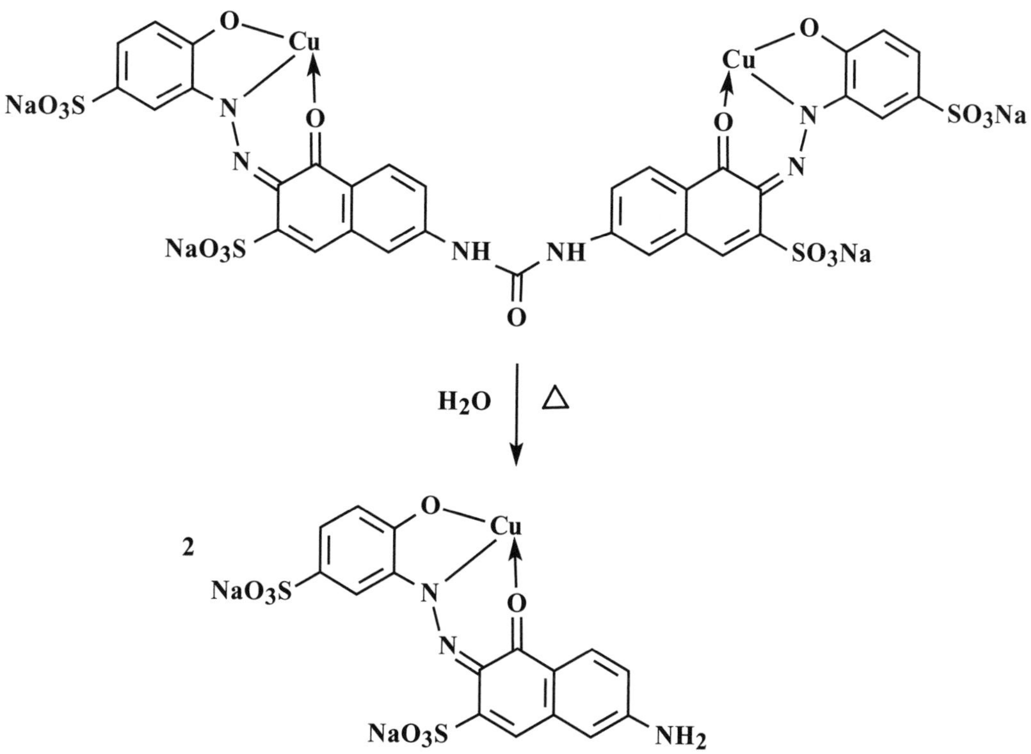

Fig. 13.7. High temperature hydrolysis of Direct Red 83.

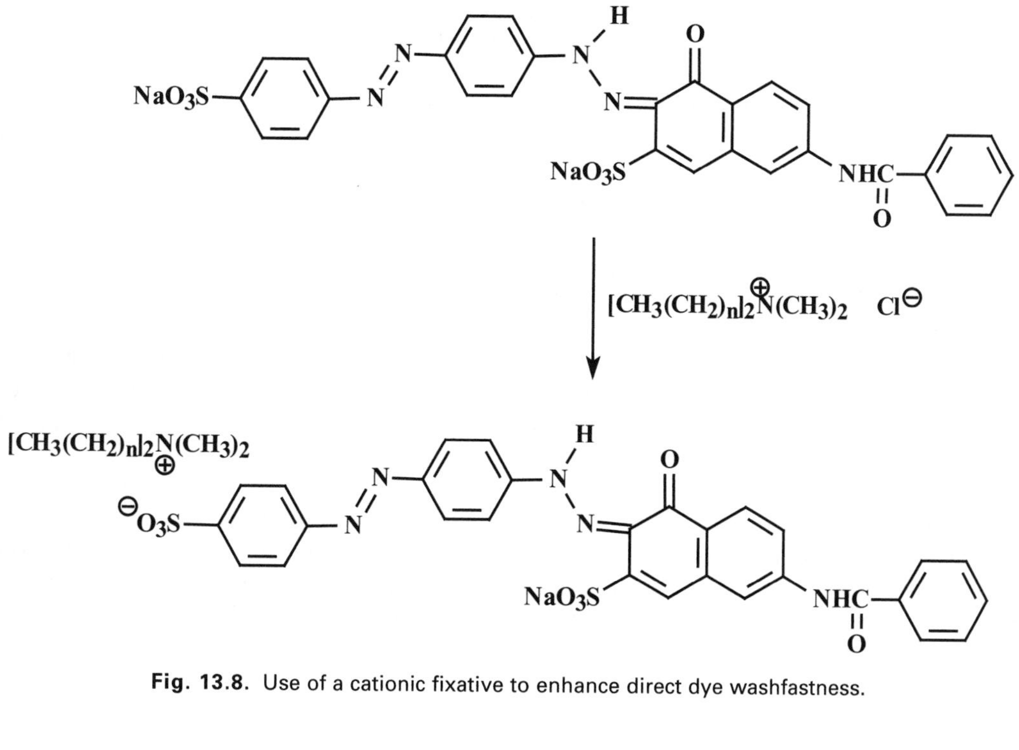

Fig. 13.8. Use of a cationic fixative to enhance direct dye washfastness.

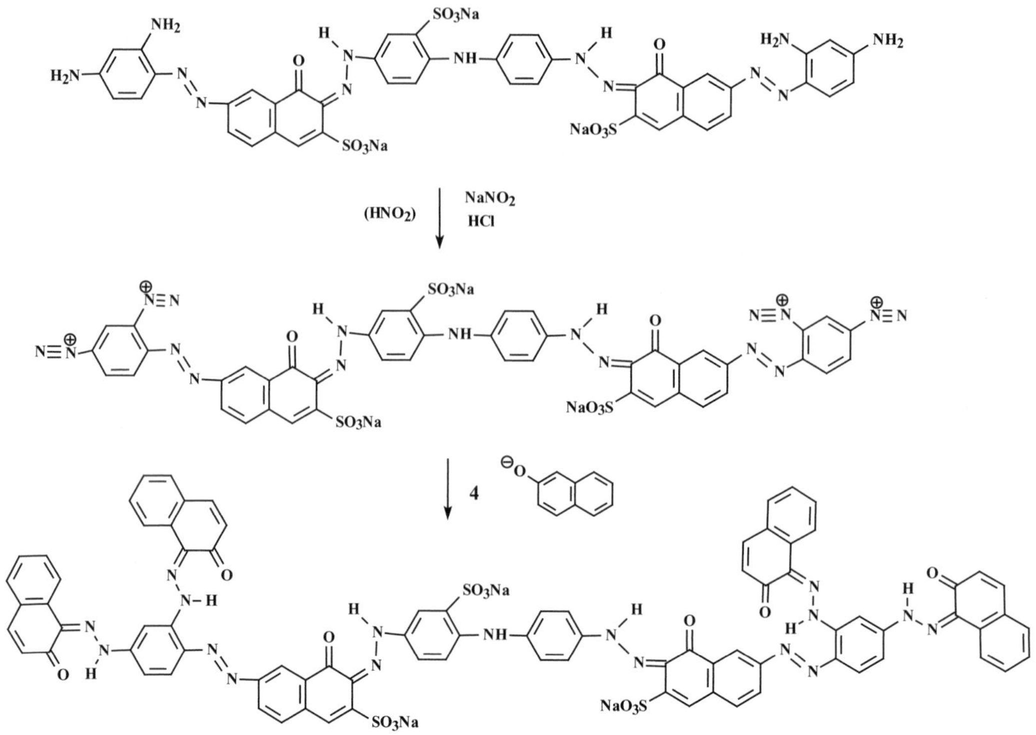

Fig. 13.9. Use of the two-step development process to enhance direct dye washfastness.

treatment with nitrous acid (HNO_2) and the resultant diazonium groups are coupled with a naphthoxide to give new azo groups. It should be pointed out that the addition of new azo groups can also affect dye color. Therefore, this process is most often used for navy and black shades, where the differences in shade variations from batch to batch are less objectionable.

Although copper sulfate aftertreatments are designed mainly to enhance lightfastness, the reduction in water solubility that accompanies Cu-complex formation can have a beneficial effect on washfastness. This treatment also dulls the fabric shade and causes a shift in dye color, so that the resultant color must be the one the dyer is seeking.

The dye used as an example in Fig. 13.10 is C.I. Direct Black 38. It is worthwhile to note that this dye is one of many that were synthesized from benzidine, an established human carcinogen.[21] Nowadays, such dyes are regarded as cancer-suspect agents because of their potential to generate free benzidine upon metabolic breakdown.[22] With this point in mind, regulations preventing the use of azo dyes derived from benzidine and 20 other aromatic amines in textiles have appeared.[23] This requires dye chemists to consider the genotoxicity of potential metabolites in the design of new azo dyes.[24]

Disperse Dyes

Disperse dyes were invented to dye the first hydrophobic fiber developed, namely cellulose acetate, and were initially called acetate dyes.[25] The term disperse dyes is more appropriate, because these dyes are suitable for a variety of hydrophobic fibers and it is descriptive of their physical state in the dyebath. Disperse dyes have extremely low water solubility and to be applied from this medium they must be (1) dispersed in water using a

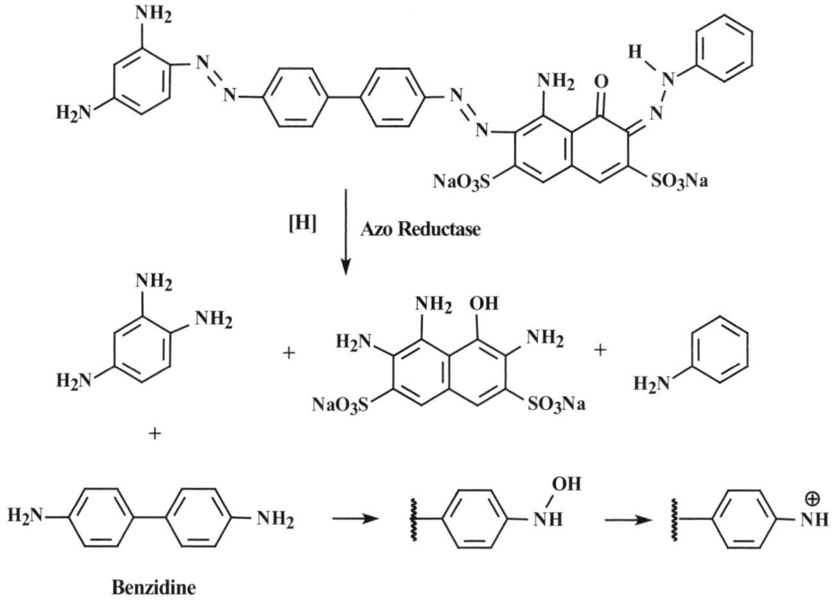

Fig. 13.10. Metabolic breakdown of a direct dye by azo reductase enzymes.

surfactant (dispersing agent) and (2) milled to a very low particle size (1–3 microns). These nonionic hydrophobic dyes can be used on acetate, triacetate, polyester, nylon, acrylic, and polyolefin fibers, and their mechanism of fixation involves solid–solid solution formation.

Disperse dyes provide a wide range of bright colors on textiles and many have excellent build-up and barré coverage properties. In addition, they have good washfastness properties but their lightfastness varies with structure. They are suitable for continuous dyeing, a process that takes advantage of their sublimation properties. Disperse dye end-use applications are often based on their classification. The classification system employed is shown in Table 13.4. Low-energy disperse dyes are the easiest to exhaust under atmospheric dyeing conditions but have the lowest thermostability, with the latter property making them unsuitable for automotive applications. They are used to dye acetate, triacetate, and nylon fibers. On the other hand, the high energy dyes are best applied under pressure

($T = 130°C$) and are most appropriate for polyester body cloth for automobile interiors. Medium-energy dyes are also used to dye polyester and can be applied at atmospheric pressure using a carrier.

Disperse dyes vary in the type of chromophore present and include azo, anthraquinone, nitro, methine, benzodifuranone, and quinoline based structures. Examples of the first three types are given in Table 13.4, and representative of the latter three types are C.I. Disperse Blue 354, C.I. Disperse Yellow 64, and C.I. Disperse Red 356. Most disperse dyes have azo (~59%) or anthraquinone (~32%) structures. Azo disperse dyes cover the entire color spectrum, whereas the important anthraquinone disperse dyes are mainly red, violet, and blue. The azo types offer the advantages of higher extinction coefficients (ϵ_{max} = 30,000–60,000) and ease of synthesis, and the anthraquinones are generally brighter and have better photostability (lightfastness). The key weaknesses associated with the anthraquinone dyes are their low extinction

TABLE 13.4   Disperse Dye Classification and Examples

Class	Molecular Mass	Sublimation Temperature
Low Energy	<300	<150°C

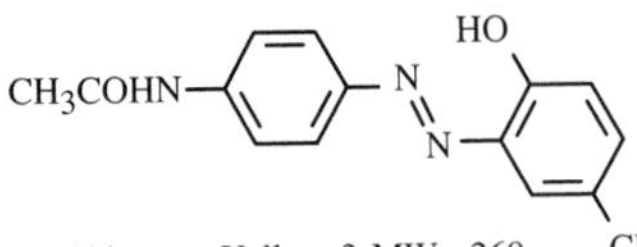

Disperse Yellow 3 MW = 269

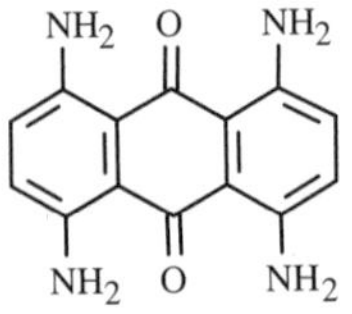

Disperse Blue 1 MW = 269

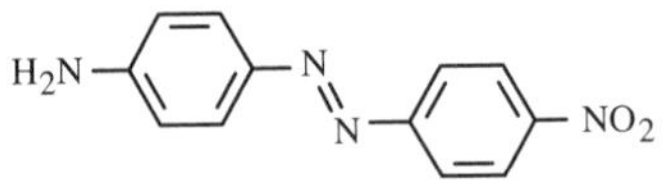

Disperse Orange 3 MW = 242

Medium Energy	300–400	150–210°C

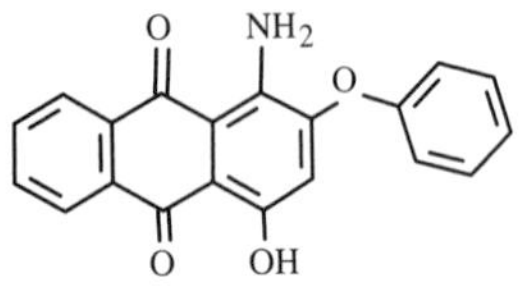

Disperse Red 60 MW = 331

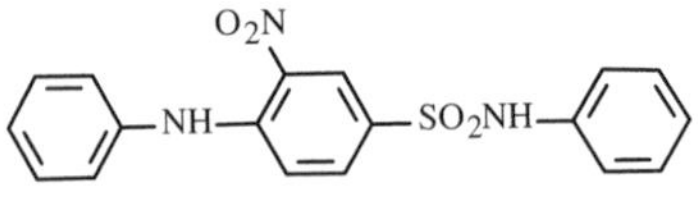

Disperse Yellow 42 MW = 369

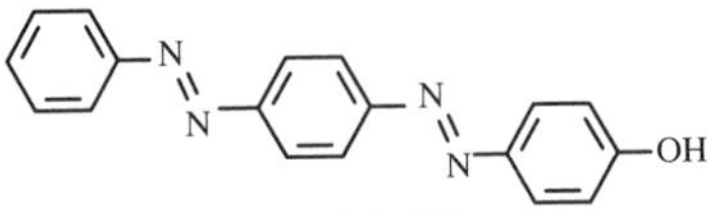

Disperse Yellow 23 MW = 302

High Energy	>400	>210°C

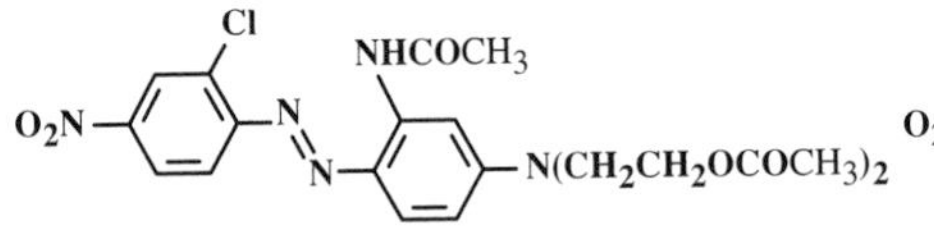

Disperse Red 167:1 MW 507

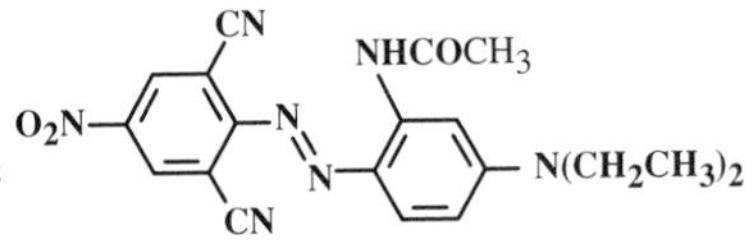

Disperse Blue 165 MW 405

coefficients (ϵ_{max} = 10,000–15,000) and less environmentally friendly synthesis.

synthesis. They have ϵ_{max} = 50,000–80,000, good brightness, and good washfastness.

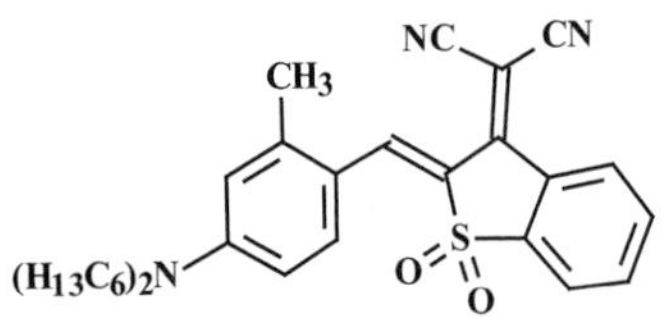

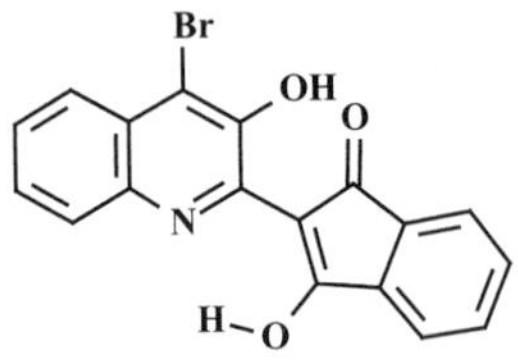

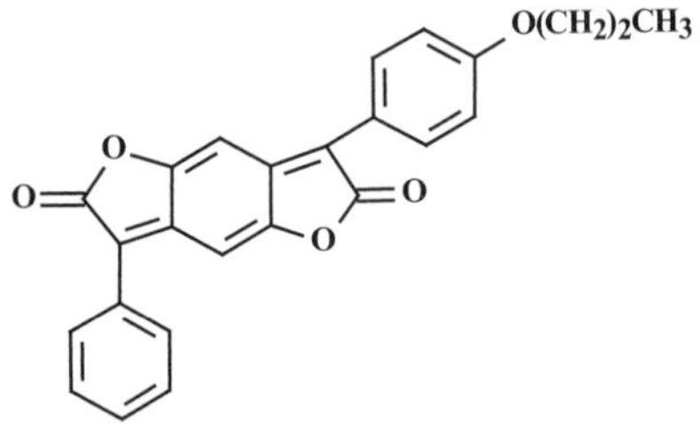

To produce disperse dyes having the brightness of the anthraquinone system and the color strength of the azo system, azo dyes based on heteroaromatic amines were developed.[26–28] Examples are C.I. Disperse Red 145, Disperse Blue 148, Disperse Red 156, and C.I. Disperse Blue 339. These dyes employ aminated thiazoles, benzothiazoles, benzisothiazoles, and thiadiazoles in their

Another key feature of disperse dyes with heteroaromatic systems is their less complex structures. Compare, for example, the fewer number of substituent groups in the diazo component (left side of the azo bond) of Disperse Red 167 versus Disperse Red 156 and Disperse Blue 165 versus Disperse Blue 102. However, these dyes are more expensive than disperse dyes derived from benzeneamines, owing to their low reaction yields, and have lower lightfastness than the anthraquinone dyes.

The use of disperse dyes in applications requiring high lightfastness involves the coapplication of photostabilizers. These agents

dyes have very high washfastness and are used for leisure wear and other applications requiring stability to repeating laundering. Each dye is composed of five basic parts:

$$SG–C–B–RG–LG$$

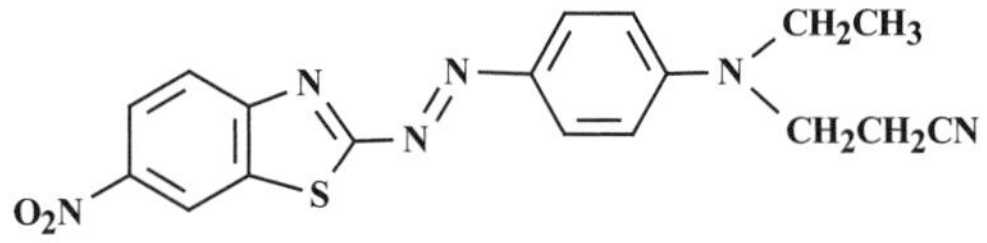

Disperse Red 145

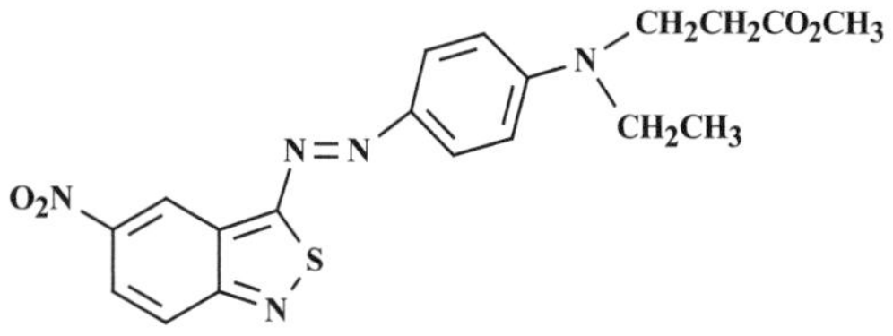

Disperse Blue 148

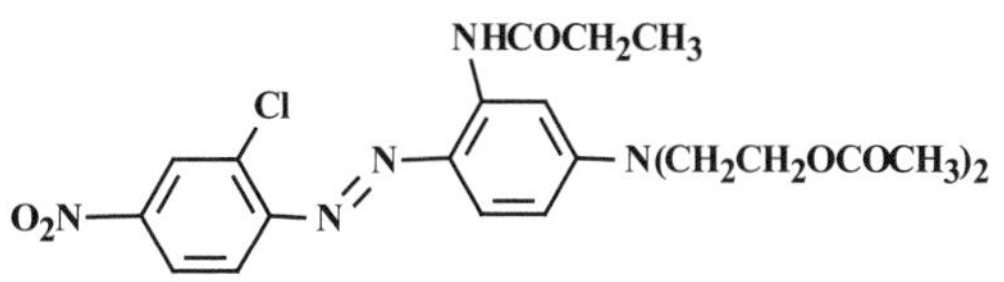

Disperse Red 167

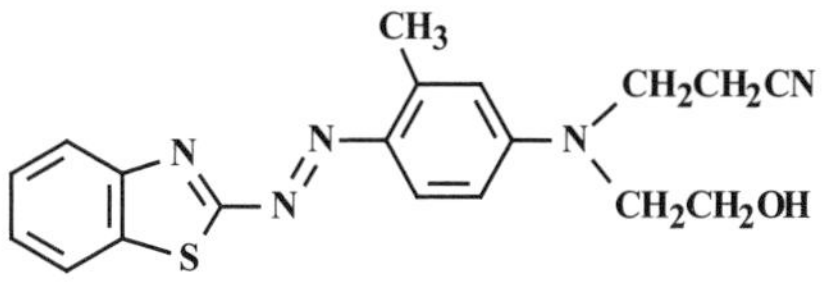

Disperse Red 156

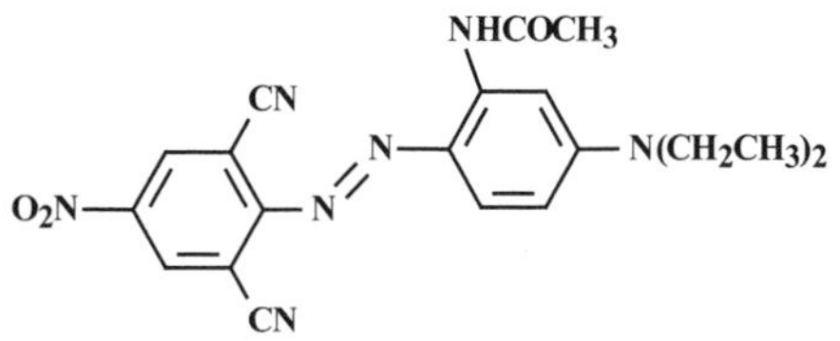

Disperse Blue 165

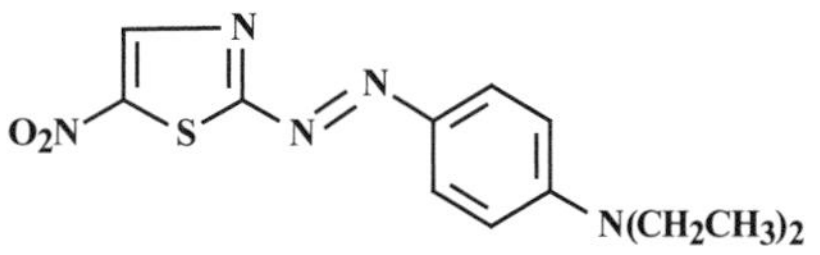

Disperse Blue 102

enhance dye stability by quenching the excited states of disperse dyes, probably via energy transfer, or by preferential absorption (screening) of UV radiation. They are also known as UV absorbers and exhaust from the dyebath like disperse dyes. They encompass benzophenone, benzotriazole, oxalanilide, and hindered amine/phenol structures (see Fig. 13.11).[29]

Reactive Dyes

Reactive dyes are used mainly as colorants for cotton, although they are also suitable for nylon and wool.[30] They are water soluble, due to the presence of one or more $-SO_3Na$ groups, and undergo fixation to polymer chains via covalent bond formation. Reactive

In this regard, SG = water solubilizing group ($-SO_3Na$), C = chromogen (e.g., azo, anthraquinone), B = bridging or linking group (e.g., $-NH-$), RG = reactive group (e.g., chlorotriazine, vinylsulfone), and LG = leaving group (e.g., $-Cl$, $-F$, $-SO_4H$). These parts are illustrated for the structure in Fig. 13.12. This structure also shows that reactive dye structures can be quite small, much smaller in fact than those characterizing direct dyes. As a consequence, reactive dyes have significantly lower inherent affinity for cotton and can require high levels of salt (200–300 g/L) in their dyebaths to promote exhaustion.[31]

In addition to giving high washfastness on cotton, reactive dyes usually give bright shades. The latter property arises from the fact that reactive dyes are often acid dye structures

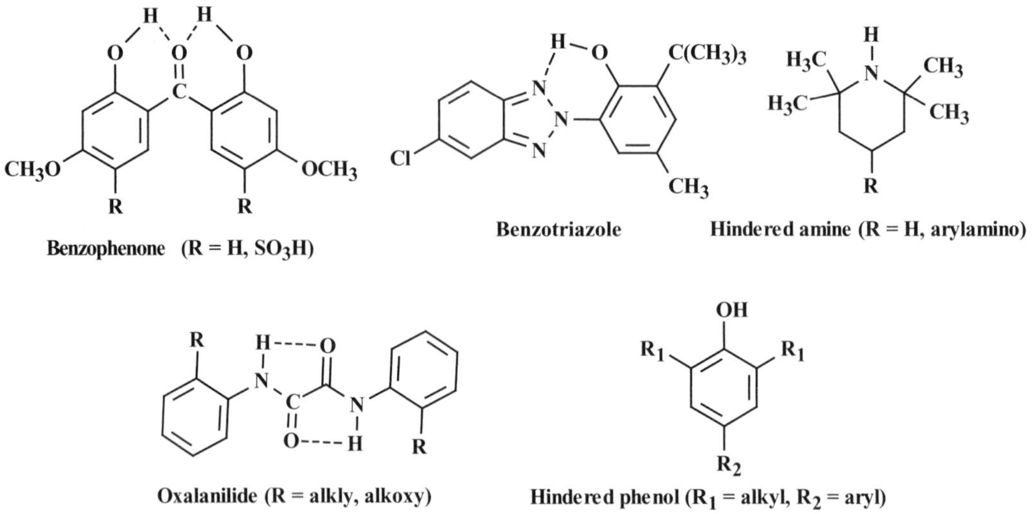

Fig. 13.11. Structures of disperse dye photostabilizers.

linked to reactive groups, as shown in Fig. 13.13. Reactive dyes have moderate-to-good lightfastness and fair-to-poor chlorine fastness.

Although the most commonly used reactive systems involve the halotriazine and sulfatoethyl sulfone (vinyl sulfone) groups, halogenated pyrimidines, phthalazines, and quinoxalines are also available (Fig. 13.14). For all of these systems, alkali is used to facilitate dye–fiber fixation, and fixation occurs either by nucleophilic substitution or addition (Figs. 13.15–13.16).

The requirement for alkali in the application of reactive dyes to cotton leads to an undesirable side reaction, namely hydrolysis of the reactive groups before dye–fiber fixa-

tion can occur (Fig. 13.17). Because the hydrolyzed dye cannot react with the fiber, this leads to wasted dye and the need to treat the residual color in the wastewater prior to dyehouse discharges. To improve percentage fixation, dyes with two or more reactive groups were developed (Fig. 13.18). This makes it possible for dye–fiber fixation to occur even when one reactive group undergoes hydrolysis.[32]

Sulfur Dyes

Sulfur dyes are water-insoluble dyes that are applied to cotton.[33] They are used primarily for their economy and high washfastness, are

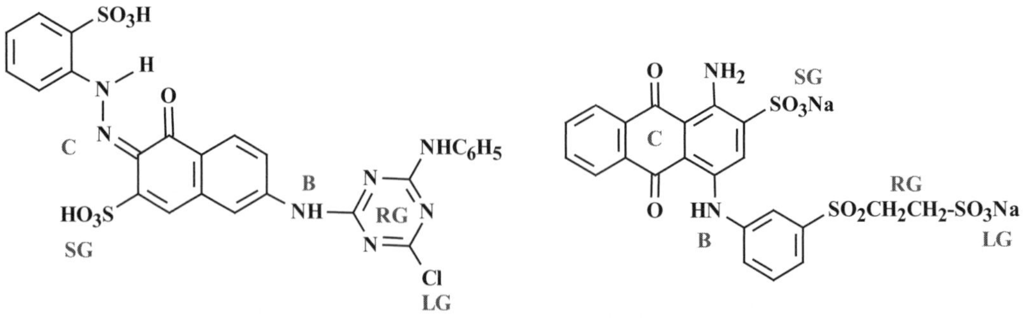

Fig. 13.12. Structures showing the basic parts of two fiber-reactive dyes.

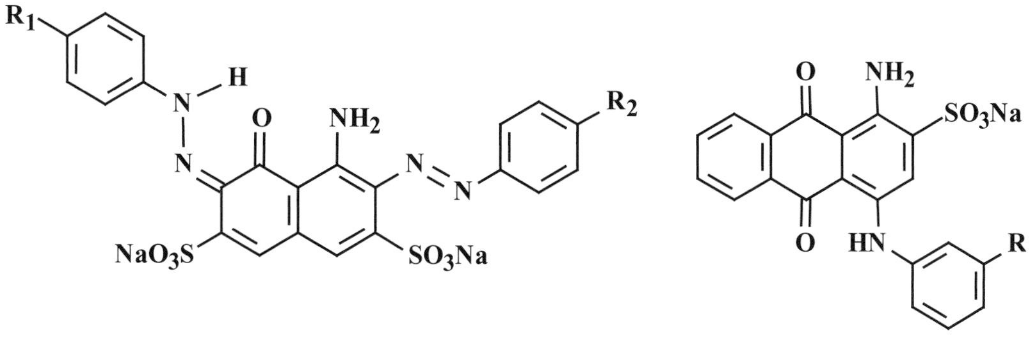

Fig. 13.13. Comparison of some acid dye and fiber-reactive dye structures. Acid Black 1: R_1 = H; R_2 = NO_2; Acid Blue 25: R = H; Reactive Black 5: R_1 = R_2 = $SO_2(CH_2)_2OSO_3Na$; Reactive Blue 19: R_1 = $SO_2(CH_2)_2$ OSO_3Na.

easy to apply, and give mainly dull shades. Yellow, red, brown, olive, and blue colors can be produced, however, sulfur dyes are most important for their ability to delivery washfast black shades on cotton. In this regard, C.I. Sulfur Black 1 is the main dye used commer-cially. Sulfur dyes have acceptable lightfast-ness but poor bleachfastness.

Due to extremely low solubility, the precise structures of most sulfur dyes remain unknown. Much of what we know about sulfur dye structures arises from the characterization of

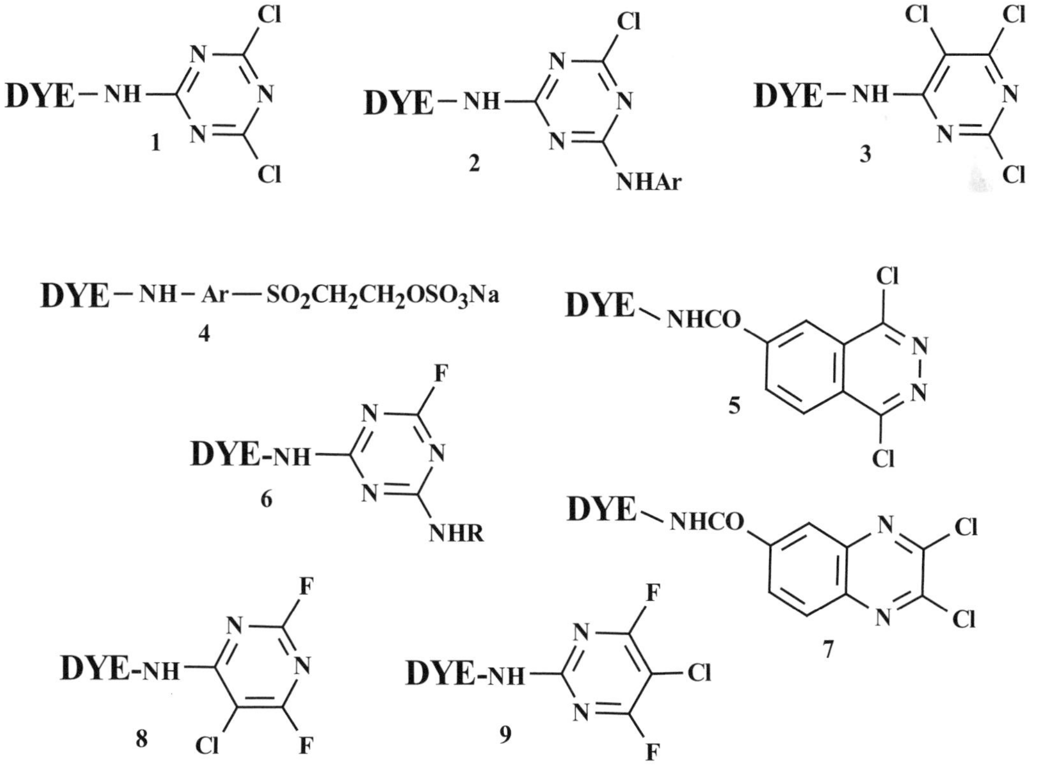

Fig. 13.14. Examples of reactive groups found in fiber-reactive dye chemistry, where 1 = dichlorotri-azine, 2 = monochlorotriazine, 3 = trichloropyrimidine, 4 = sulfatoethyl sulfone, 5 = dichlorophthalazine, 6 = monofluorotriazine, 7 = dichloroquinoxaline, 8,9 = difluorochloropyrimidine.

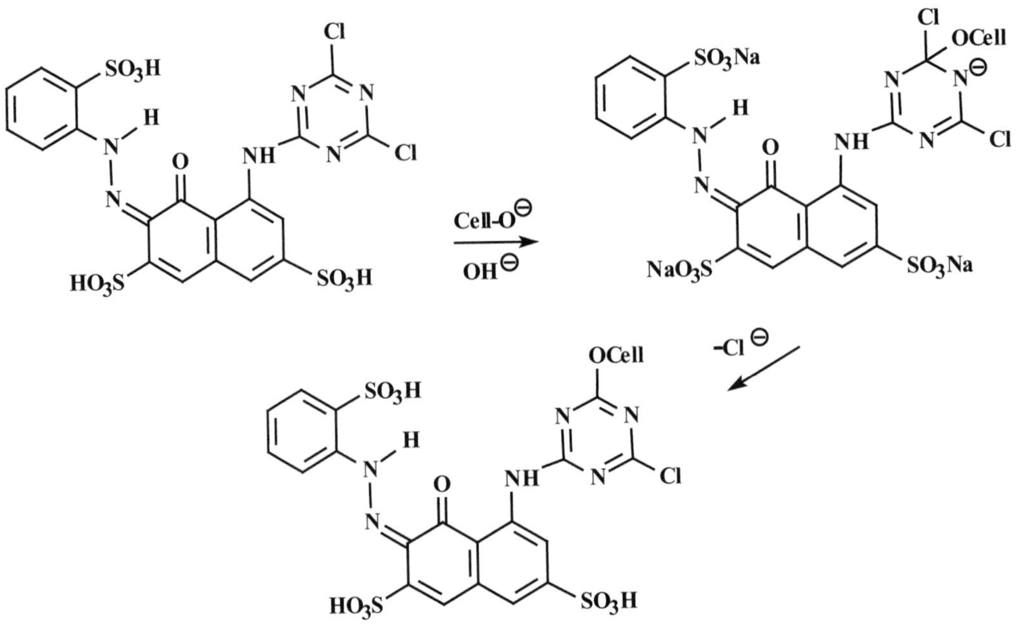

Fig. 13.15. Reactive dye fixation to cellulose via nucleophilic substitution.

certain degradation products or reaction precursors.[34] Based on such work, it has been possible to determine that structures of the type shown in Fig. 13.19 are covered in this dye class. A key common feature of sulfur dyes is the presence of sulfide $(-S_n-)$ bonds,

and it is this feature that makes dye application from an aqueous medium possible.

The reaction of sulfur dyes with sodium sulfide (Na_2S) at pH >10 effects the reduction of the sulfide bonds, giving their water-soluble (leuco) forms. The reduced forms

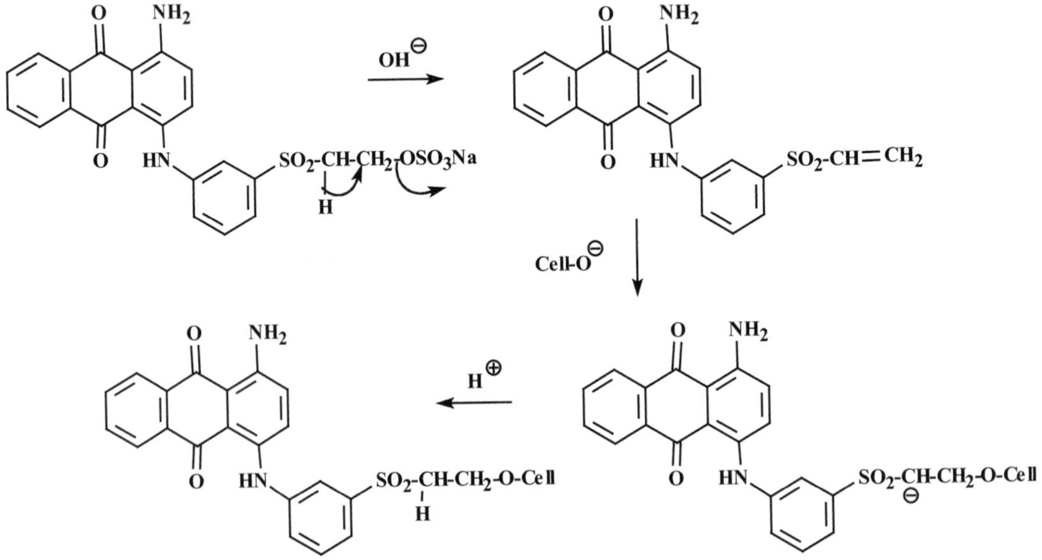

Fig. 13.16. Reactive dye fixation to cellulose via nucleophilic addition.

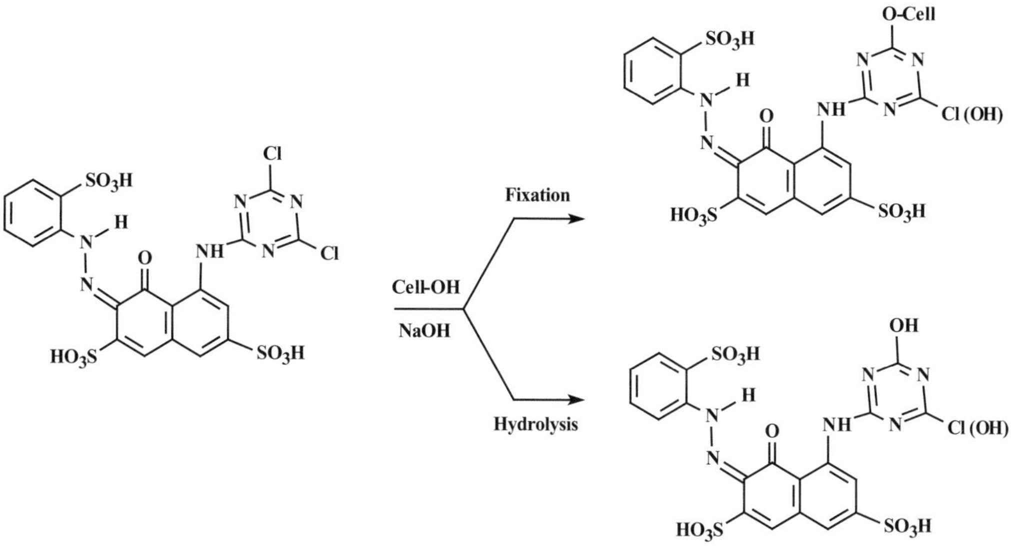

Fig. 13.17. Competing reactions when reactive dyes are applied to cotton.

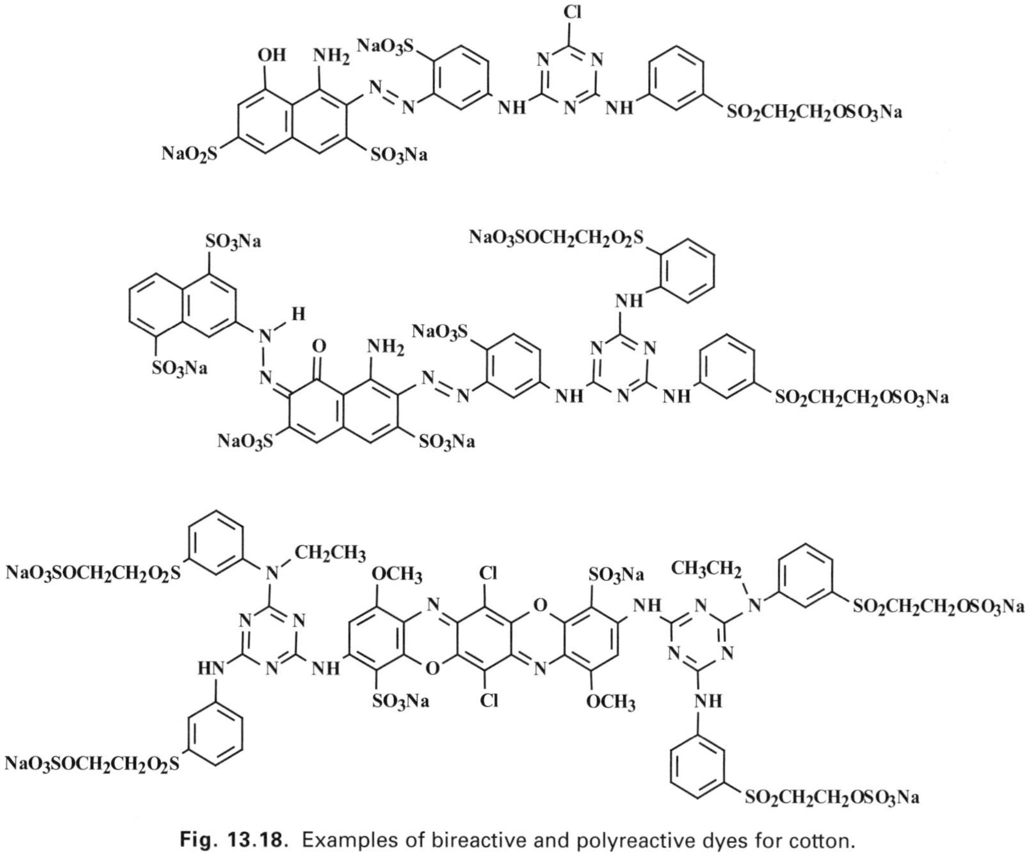

Fig. 13.18. Examples of bireactive and polyreactive dyes for cotton.

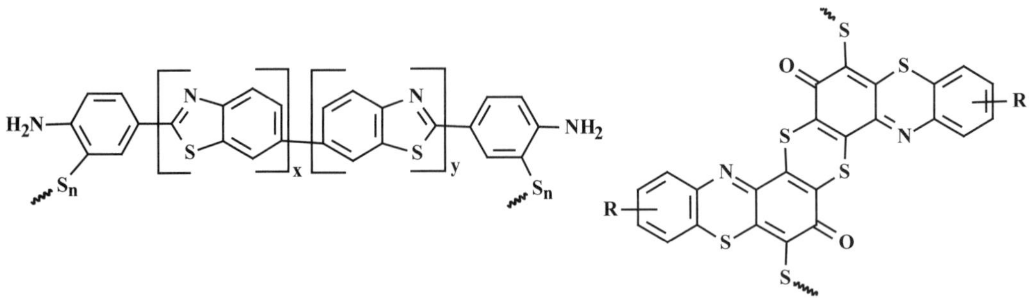

Fig. 13.19. Examples of sulfur yellow (left) and red (right) dye structures.

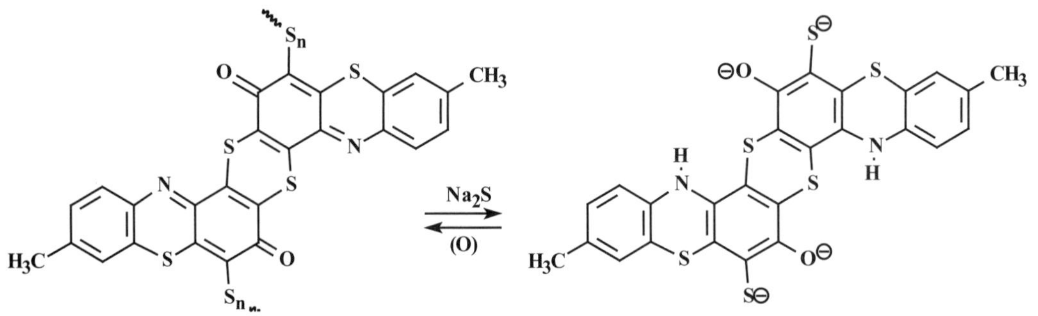

Fig. 13.20. Steps involved in the application of sulfur dyes to cotton.

behave like direct dyes, in that they exhaust onto cotton in the presence of salt. Once applied, the reduced dyes are reoxidized to their water-insoluble forms, giving dyeings with good washfastness. This chemistry is illustrated in Fig. 13.20. Although the oxygen in air can be used for the oxidation step, an agent such as hydrogen peroxide is used because it works faster. Sulfur dyes have also been marketed in their prereduced form (Dye-S⁻Na⁺), as ready-to-use C.I. Leuco Sulfur dyes. Dye exhaustion in the present of salt is followed by oxidation. Similarly, water-soluble sulfur dyes containing thiosulfate groups are sold as C.I. Solubilized Sulfur dyes. They are known as "Bunte salts,"[35] have better leveling properties than the C.I. Sulfur dyes, and are attractive for package dyeing. The C.I. Solubilized sulfur dyes are applied with Na_2S and the chemistry associated with their two-step application is summarized in Fig. 13.21, along with a representative dye structure.

Vat Dyes

Like sulfur dyes, vat dyes are water-insoluble colorants for cotton that must be reduced to their soluble "leuco" forms to be applied from an aqueous dyebath.[36] Their name originates from their early application from wooden vessels known as vats. The term "vatting" is used to refer to the application of these dyes via chemical reduction followed by oxidation. Vat dyes are easier to reoxidize than sulfur dyes and the oxygen in air is often the agent used. As would be anticipated, most vat dyes display high washfastness. As a class, they have the best lightfastness and bleach fastness among the dyes families suitable for cotton. Some cause catalytic fading or phototendering on cotton.[37]

Vat dyes have mainly anthraquinone (82%) or indigoid/thioindigoid (9%) structures, with the former having much better fastness properties. The anthraquinone vat dyes exhibit a bathochromic color shift (λ_{max} of higher wavelength) upon reduction to their leuco forms, whereas the indigoids exhibit a hypsochromic shift. Examples of the two structural types are shown in Figs. 13.22 and 13.13. Anthraquinone vat dyes having a single anthraquinone unit exist; however, those with the best fastness properties seem to have the equivalent of two anthraquinone units.

Step 1: $\text{Dye-S-SO}_3^{\ominus}$ + $\text{S}_2^{\ominus}$ → $\text{Dye-S}^{\ominus}$

Step 2: $\text{Dye-S}^{\ominus}$ + $\text{Dye-S-SO}_3^{\ominus}$ → Dye-S-S-Dye

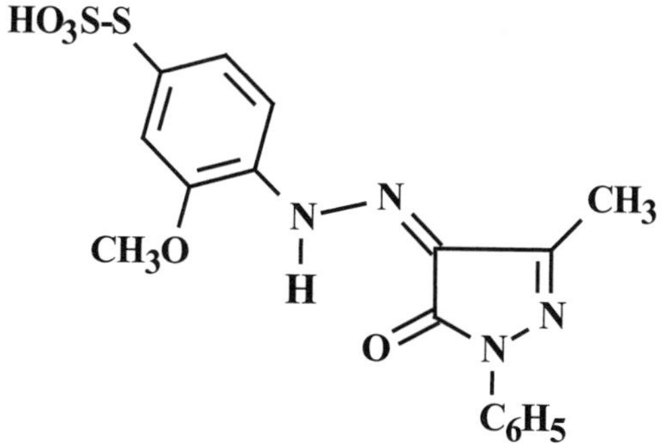

Fig. 13.21. Two-step chemistry employed in the application of C.I. Solubilized Sulfur dyes (top) and a representative dye structure (bottom).

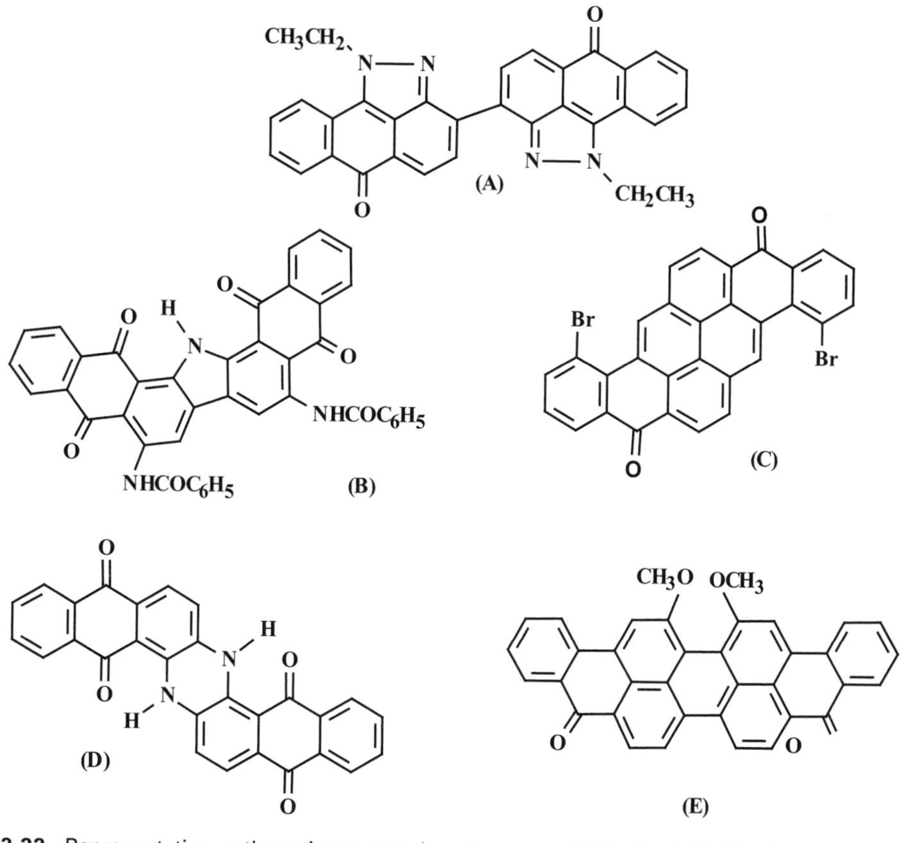

Fig. 13.22. Representative anthraquinone vat dye structures: CI Vat Red 13 (A), Vat Black 27 (B), Vat Orange 2 (C), Vat Blue 4 (D), and Vat Green 1 (E).

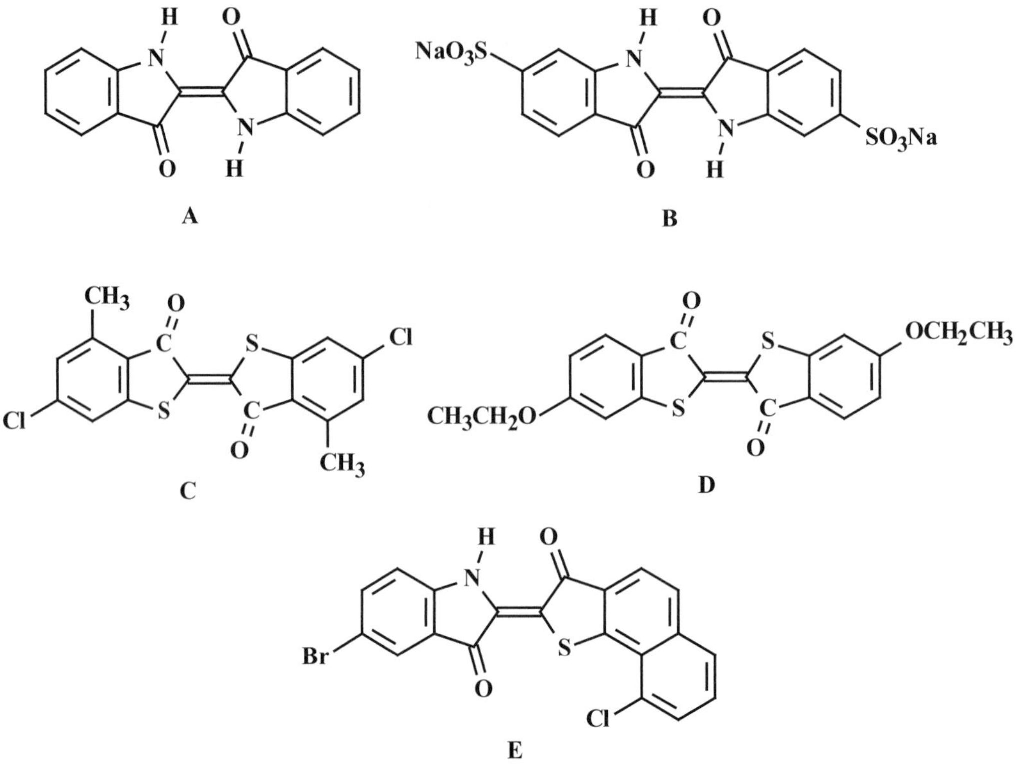

Fig. 13.23. Representative indigoid and thioindigoid vat dye structures: CI Vat Blue 1 (A), Food Blue 1 (B), Vat Red 1 (C), Vat Orange 5 (D), and Vat Black 1 (E).

No doubt the best-known and biggest volume vat dye is C.I. Vat Blue 1, indigo, the denim blue dye. Closely related structures are the thioindigoids (4%), which have a sulfur atom in lieu of the –NH– group (Fig. 13.23). The thioindigoids are used mainly as colorants for printing and give orange and red hues. A few dyes having the features of both indigoid types are also known (e.g., Ciba Violet A).

The chemistry associated with the vatting process is illustrated in Fig. 13.24. For the reduction step, a mixture of sodium hydroxide (caustic) and sodium hydrosulfite (hydro, $Na_2S_2O_4$) is used. Depending upon the amount of caustic and hydro employed one or both of the anthraquinone rings may undergo reduction.

Vat dyes are also available in prereduced forms (3%), an example of which is the leuco sulfuric acid ester C.I. Solubilized Vat Blue 4.

These water-soluble forms have affinity for cellulose and exhaust like direct dyes. They are oxidized to the insoluble form using hydrogen peroxide.

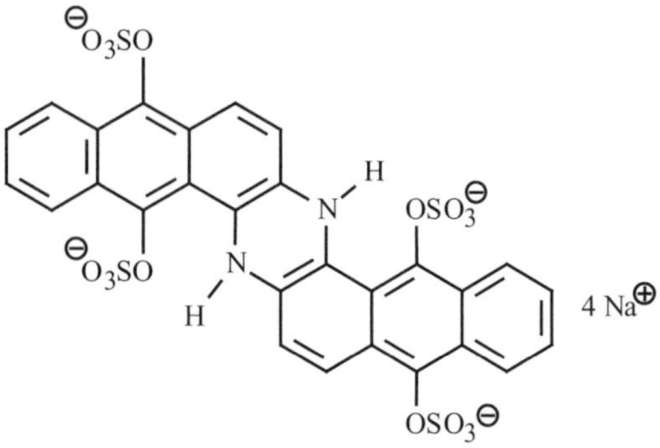

C.I. Solubilized Vat Blue 4

Vat dyes are brighter than direct and sulfur dyes but less so than reactive dyes. They are the colorants of choice when dye bleachfastness on cotton is important. They span the

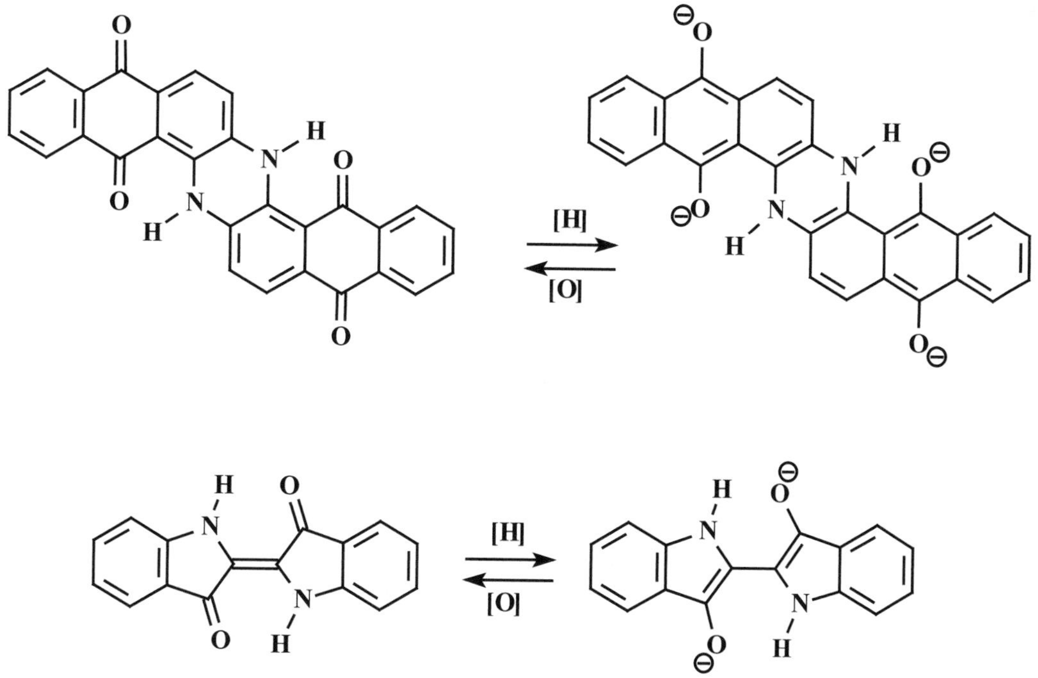

Fig. 13.24. Chemistry involved in the vatting of the CI Vat dyes Blue 4 (top) and Blue 1 (bottom).

entire color spectrum and can be applied to cotton using a variety of methods. With regard to the latter point, they can be further classified based on the temperatures involved in their application. Accordingly, there are hot (50–60°C), warm (40–50°C), and cold (25–30°C) dyeing vat dyes. The hot dyeing types are large planar leuco forms having high affinity and no salt is required for their application. The cold types are small molecules with low affinity and require repeated application to get good build-up. Indigo falls into the cold dyeing category.

THE APPLICATION OF DYES

The process of dyeing may be carried out in batches or on a continuous basis. The fiber may be dyed as stock, yarn, or fabric. However, no matter how the dyeing is done, the process is always fundamentally the same: dye must be transferred from a bath—usually aqueous—to the fiber itself. The basic operations of dyeing include: (1) preparation of the fiber, (2) preparation of the dye bath, (3) application of the

dye, and (4) finishing. There are many variations of these operations, depending on the kind of dye. The dyeing process is complicated by the fact that single dyes seldom are used. The matching of a specified shade may require from two to a dozen dyes.

Fiber Preparation

Fiber preparation ordinarily involves scouring to remove foreign materials and ensure even access to dye liquor. Some natural fibers are contaminated with fatty materials and dirt, and synthetic fibers may have been treated with spinning lubricants or sizing that must be removed. Some fibers also may require bleaching before they are ready for use.

Dye-Bath Preparation

Preparation of the dye bath may involve simply dissolving the dye in water, or it may be necessary to carry out more involved operations such as reducing the vat dyes. Wetting agents, salts, "carriers," retarders, and other dyeing assistants also may be added. Carriers are swelling agents that improve the dyeing rate of very hydrophobic fibers such as the polyesters. Examples are *o*-phenylphenol and biphenyl. Retarders are colorless substances that compete with dyes for dye sites or form a complex with the dye in the bath and act to slow the dyeing rate. Their use is necessary when too-rapid dyeing tends to cause unevenness in the dyeings.

Finishing

The finishing steps for many dyes, such as the direct dyes, are very simple: the dyed material merely is rinsed and dried. Vat-dyed materials, on the other hand, must be rinsed to remove the reducing agent, oxidized, rinsed again, and soaped before the final rinsing and drying steps are carried out. Generally, the finishing steps must fix the color (if fixation has not occurred during application) and remove any loose dye from the surface of the colored substrate. Residual dyeing assistants such as carriers also must be removed.

The types of textile structures that lend themselves to continuous dyeing methods are woven and tufted carpets. Continuous dyeing is designed for long runs of similar product; it is a high-output method of dye application.

The first volume-yardage continuous process was the continuous pad-steam process for vat dyes on cotton. The vat dye dispersion was padded onto the cloth and dried; this was followed by passage through a reducing bath, steaming for 30 seconds, passage through an oxidizing bath and, finally, washing. When it was discovered that disperse dyes could be thermosoled into polyesters by treatment with dry heat for 60 seconds and 400°F, this procedure was readily adapted to continuous processing. The advent of large volumes of dyed polyester-cotton-blend fabrics in the late 1960s made it possible to combine these two processes into one thermosol pad-steam system.

Tufted nylon carpet grew to be the number-one floor covering in the United States in recent decades. Continuous open-width ranges were developed but not without a great deal of ingenuity to deliver the precise loading of liquid to the tufted surface. This was accomplished by a dye applicator that flooded the dye solution onto the carpet surface. The advancing technology in continuous, metered dyeing systems has created a need for dyes in liquid form, both dispersions and solution. The dyes used in carpet dyeing, for the most part, are supplied by the dye manufacturers as liquids. See Fig. 13.25.

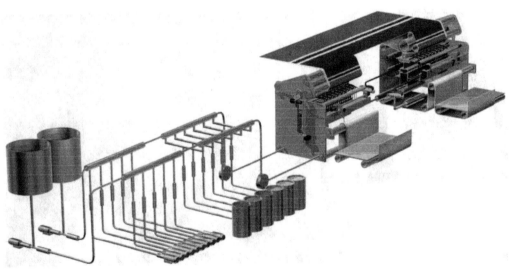

Fig. 13.25. Fluidyer Carpet Range Dye Applicator. (Courtesy of Zima Corporation.)

Dyeing Methods/Batch

Virtually all types of textile structures are dyed by batch (or exhaust) methods of dyeing, such as stock, yarn, circular knits, warp knits, woven fabrics, garments, carpets, and so on. Batch methods include beck dyeing, jig dyeing, pad-batch, beam dyeing, and others. These methods are dictated primarily by the physical structure of the textile product and the type of fiber(s) it contains. Each of these batch methods employs a different type of machine. As an example, a circular knit fabric comprised of cotton could be dyed in a beck, whereas the same structure comprised of polyester most likely would be dyed in a high-pressure jet machine, and a garment constructed from the circular knit cotton likely would be dyed in a garment machine.

Stock dyeing often is carried out in large heated kettles made of stainless steel or other corrosion-resistant metal. These kettles can be sealed and used for dyeing at temperatures somewhat above the boiling point of water at atmospheric pressure.

Yarns are dyed in package machines. In this arrangement the yarn is wound onto perforated dye tubes and placed on spindles that are fit into a closed kettle. The dye solution is heated and pumped through the spindle and yarn package. A cycle of inside-outside flow usually is used to provide level dyeing by equal exposure of the dye to yarns. Although the basis of package dyeing has not changed, a number of refinements have been introduced in recent years. Precision winding of the yarn has improved quality by giving a more uniform package density. Horizontal machines and valving between chambers to allow reconfiguration of the dye machine to control the size of the dyeing have changed the way package dye houses are built. Robotization has been widely utilized to load and unload machines. Also lower-ratio dyebaths with higher flow rates have improved the energy efficiency of the newer machines.

Fabrics are dyed in machines that move them through the dye liquor either under tension (jig) or relaxed (beck). Fabrics also can be dyed in full width by winding them on a perforated beam through which hot dye liquor is pumped. This is the principle of the beam dyeing machine.

The pressure-jet dyeing machine is unique in that it has no moving parts. The cloth, in rope form, is introduced into a unidirectional liquid stream enclosed in a pipe. Liquor is pumped through a specially designed venturi jet imparting a driving force that moves the fabric. The two fabric ends are sewn together to form a continuous loop.

The first jet machine was introduced in 1965. There are two major types of jet dyeing machines: the vertical kier and the elongated horizontal kier (see Fig. 13.26). In general, the kier uses small water volumes, whereas the elongated types use larger-volume ratios in dyeing. The kier types normally are used for more substantial fabrics, and the elongated types are suited for fine or delicate fabric styles. Important features in today's machines are improved corrosion-resistant alloys and the ability to operate at higher efficiencies with minimum energy consumption. The control systems have been refined; there is simultaneous loading and unloading. Larger-capacity machines also are being built; a jet dye machine has been developed for carpet dyeing.

PRINTING

Printing is a special kind of localized dyeing that produces patterns. Four kinds of printing have long been recognized: (1) direct, (2) dyed, (3) discharge, and (4) resist. In direct printing, a thickened paste of the dye is printed on the fabric to produce a pattern. The fabric then is steamed to fix the dye and is finished by washing and drying. Dyed printing requires that the pattern be printed on the fabric with a mordant. The entire piece then is placed in a dye bath containing a mordant dye, but only the mordanted areas are dyeable. Washing then clears the dye from the unmordanted areas, leaving the pattern in color.

In discharge printing, the cloth is dyed all over and then printed with a substance that

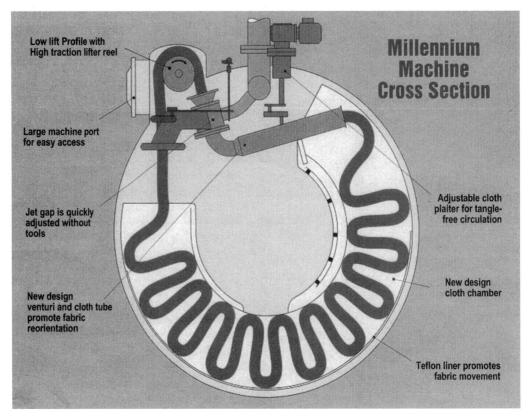

Fig. 13.26. Millennium Jet Dyeing Machine. (Courtesy of Gaston County Dyeing Machine Co.)

can destroy the dye by oxidation or reduction, leaving the pattern in white. When a reducing agent such as sodium hydrosulfite is used to destroy the dye, the paste may contain a reduced vat dye. Finishing the goods by oxidation and soaping then produces the pattern in color. In resist printing, certain colorless substances are printed on the fabric. The whole piece then is dyed, but the dye is repelled from the printed areas, thus producing a colored ground with the pattern in white.

Printing is most often done with rotary screens etched in the design to be printed. Printing paste is fed constantly to the center of the rotating screen from a nearby supply, and a squeegee pushes the colored paste through the holes in the screen, leaving the dye paste only in the intended areas, a separate screen is required for each color in the pattern. See Fig. 13.27.

An important recent advance in the pattern-coloring of textiles is ink-jet or digital printing. Milliken's Millitron and Zimmer's ChromoJet

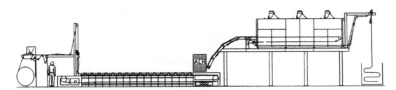

Fig. 13.27. Rotascreen V Rotary Screen Print Machine. (Courtesy of Zimmer Machinery Co.)

have been successfully used for carpet and upholstery markets for over two decades. Finer-resolution machines began to emerge in the late 1980s when Stork introduced a prototype machine. During the past five years a number of manufacturers have introduced digital ink-jet printers that use either CYMK (Cyan, Yellow, Magenta, and Black) to make a composite color or true-color machines that use mixed pigment systems. The current machines are very successful at furnishing one of a kind and for use in rapid prototyping.

PIGMENT DYEING AND PRINTING

Pigment dyeing and printing are processes that compete with the more conventional means of dyeing and printing described above. These processes use water-insoluble dyes or pigments that are bound to the surfaces of fabrics with resins. A paste or an emulsion, containing pigment and resin or a resin-former, is applied to the fabric. The goods then are dried and cured by heat to produce the finished dyeing or print. During the heating or curing, fabric, resin, and pigment become firmly bonded together. This method of color application is economical and produces good results. It should be noted that the pigment is confined to the surface of the fabric and can be selected without regard for fiber affinity.

NONTEXTILE USES OF DYES

Colorants for nontextile use have been developed mainly for use in hair dyeing, photography, biomedical application, and electronics and reprographics.[33–39] Dye application areas involving the latter areas include ink-jet printing, thermal or pressure dye transfer, laser printing, liquid crystal displays, optical data storage, and nonlinear optics. In several nontextile applications, dyes are not used for their ability to deliver color. Instead, they are used because of their potential electrical properties, such as photoconduction and electrostatic charging of toners, and in some cases they are used because they absorb IR

radiation, which induces heating effects. The latter property is important in optical data recording.

Liquid Crystal Dyes

Dyes for liquid crystalline media typically have (1) nonionic structures, (2) high purity, (3) solubility and compatibility with the medium, (4) a transition dipole that is parallel with the alignment axis of the molecular structure, and (5) good alignment with the liquid crystal molecule.[40] Examples include the disazo and anthraquinone dyes in Fig. 13.28.

Ink-jet Dyes

Dyes used in this area must have the following properties: (1) very good water solubility; (2) low toxicity; (3) good stability to UV light, heat, and moisture; (4) quick fixation to paper following application (deposition); and (5) good color strength.[41] To achieve high resistance to removal by water (wetfastness), ink-jet dyes often contain fewer sulfonate groups and one or more carboxylate groups. This change in structural features allows the dyes to have good solubility in alkaline ink formulations but high wetfastness following deposition. This change in solubility behavior is known as *differential solubility*.[42] Structures in Fig. 13.29 illustrate the type changes made to the early ink-jet dye C.I. Food Black 2 to enhance wetfastness.

New water-soluble yellow dyes for ink-jet printing are similar to the initially used dye C.I. Direct Yellow 86, except that they are smaller (Fig. 13.30). The size change is designed to provide the solubility needed for high throughput ink cartridge systems, without clogging the ink-jet nozzles.

Most of the new water-soluble magenta dyes are based on H-acid. Examples include dyes that contain a fluorocyanophenyl group (Fig. 13.31). To improve the lightfastness of magenta dyes, gamma acid can be used as the coupling component. For very bright magenta prints, dyes based on xanthene structures can be used. Examples include C.I. Acid Red 52,

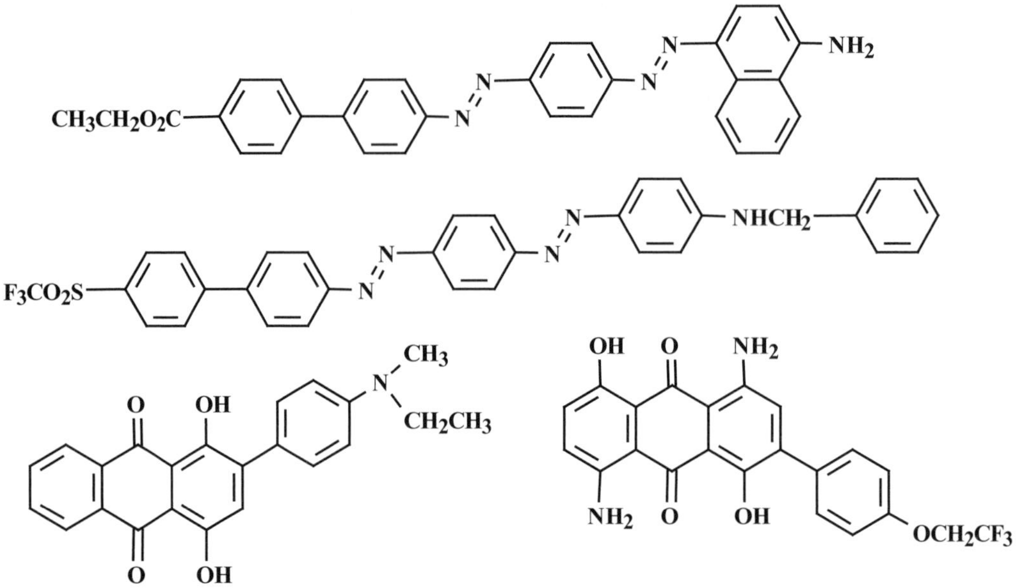

Fig. 13.28. Azo and anthraquinone dyes suitable for liquid crystalline media.

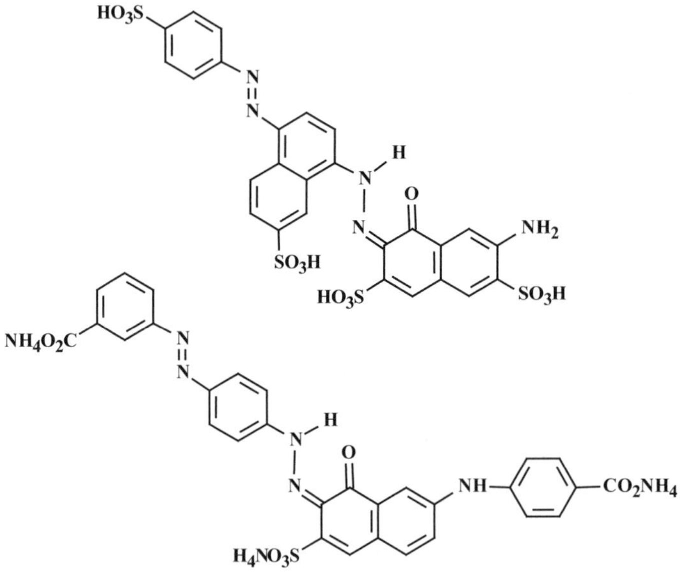

Fig. 13.29. Food Black 2 (top) and its wetfast analog (bottom).

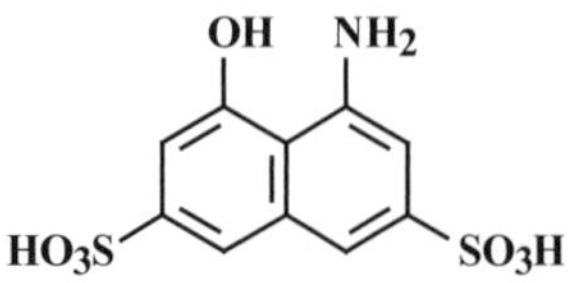

H-acid

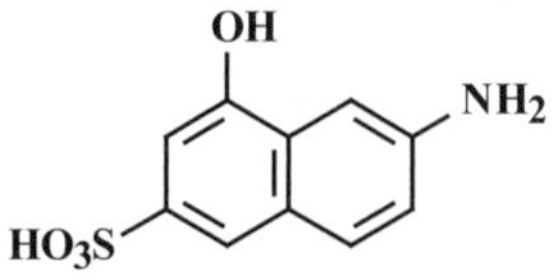

Gamma acid

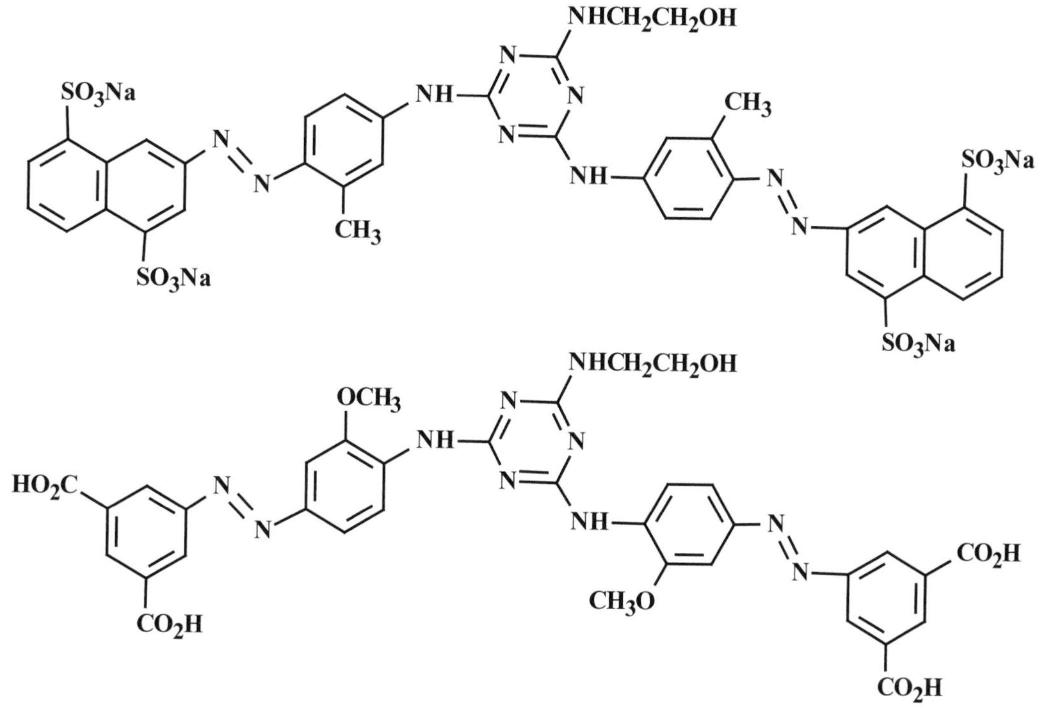

Fig. 13.30. Examples of yellow ink-jet dyes (upper dye is Direct Yellow 86).

which has low photostability, and a carboxy-lated analog, which has better photostability and wetfastness (Fig. 13.32).

The water-soluble cyan dyes continue to be based on the copper phthalocyanine system. In this regard, C.I. Direct Blue 199 has proved effective, due to its good color strength and photostability. Carboxylated analogs of this type of dye have been developed to enhance wetfastness on paper (Fig. 13.33).

Thermal and Pressure-Sensitive Printing

In direct thermal printing, a color former (colorless) and a developer (acidic) are brought into contact in the presence of heat, to

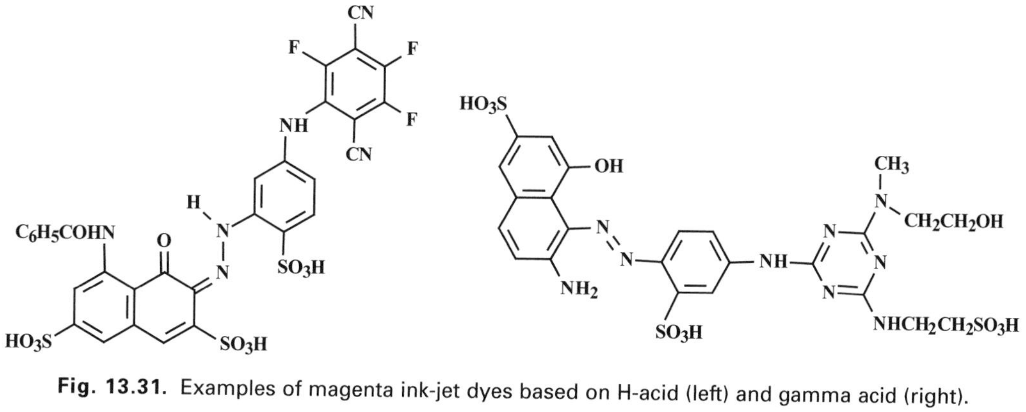

Fig. 13.31. Examples of magenta ink-jet dyes based on H-acid (left) and gamma acid (right).

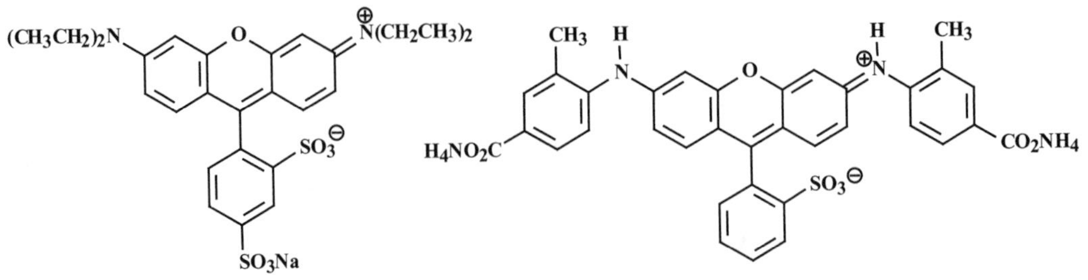

Fig. 13.32. Acid Red 52 (left) and a more lightfast analog (right).

produce color on paper.[43] The most important color in thermal printing is black and the majority of the color formers are fluorans (Fig. 13.34). The most important application of direct thermal printing is in facsimile machines.

In pressure-sensitive printing technology the color former is dissolved in a solvent and encapsulated.[44] The use of pressure (pen, typewriter key) ruptures microcapsules containing the color former, which generates color upon contacting a developer. Black prints are usually obtained either from fluorans or from color former mixtures. Compounds of the type shown in Fig. 13.35 can be used in two- and three-component mixtures.

Organic Photoconductors and Toners

Photoconductors and toners are used in photocopiers and laser printers to produce images.[45] Organic photoconductors consist

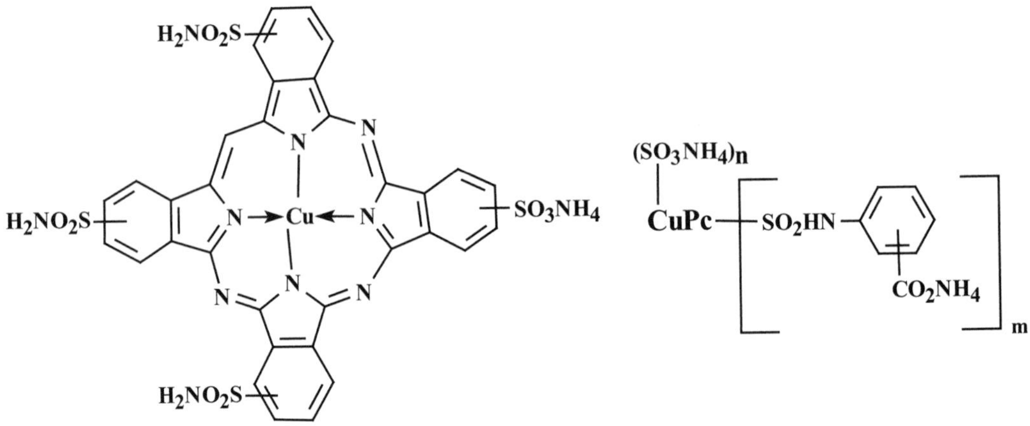

Fig. 13.33. Direct Blue 199 (left) and a wetfast analog (right), where Pc = phthalocyanine.

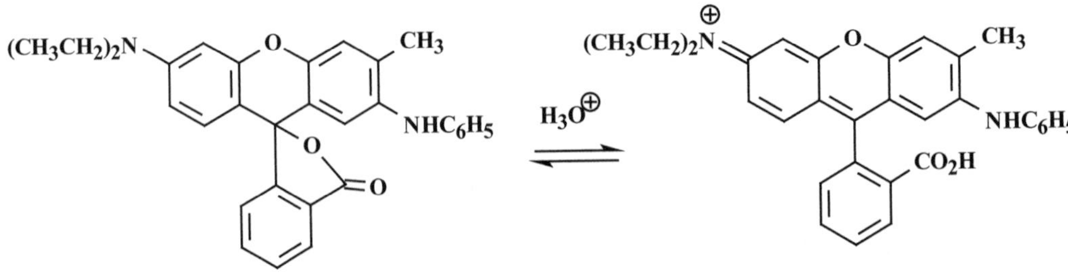

Fig. 13.34. Acid-induced conversion of a color former to its colored form.

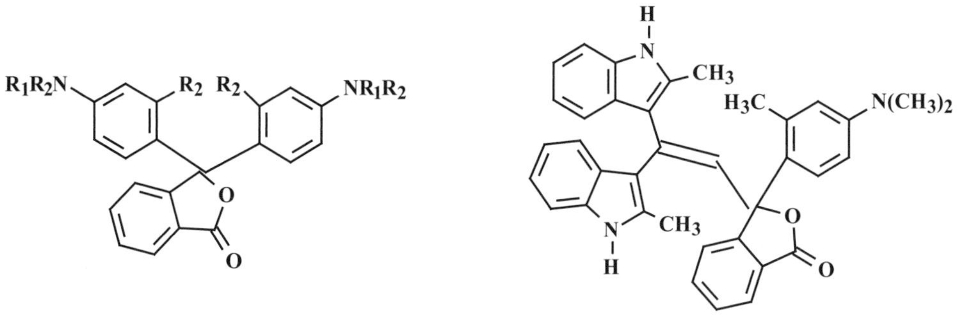

R$_1$: C$_{1-6}$ alkyl, R$_2$: C$_{1-6}$ alkyl, alkoxyalkyl

Fig. 13.35. Color former combinations suitable for pressure-sensitive printing.

of a charge-generating layer and a charge-transporting layer. The former is comprised of pigments and the latter is comprised of electron-rich organic compounds that are usually colorless. Suitable organic pigments for charge generation include azo pigments, tetracarboxydiimides, polycyclic quinones, phthalocyanines, perylenes, and squarylium compounds (e.g., Fig. 13.36).

Colorants are used in toners to provide color and control the electrostatic charge on toner particles. Diarylides and monoarylides have been used as the yellow pigments in colored toners. The magenta pigments are often quinacridones and the cyan pigments are copper phthalocyanines.

Infrared Absorbing Dyes

Infrared dyes include indoleninecyanines and azulenium compounds, both of which are used in optical recording materials.[46] Other examples are metal (Mn, Fe, Co, Cd, Al, Cu, Pd)-complexed phthalocyanines, quinones, quinonoids, and imminium and diiminium compounds (Fig. 13.37).

Laser Dyes

Lasers in which dyes comprise the active medium have become one of the most widely used types.[47] The key virtue of these systems is their ability to cover virtually the entire fluorescence spectral region. Accordingly, the most

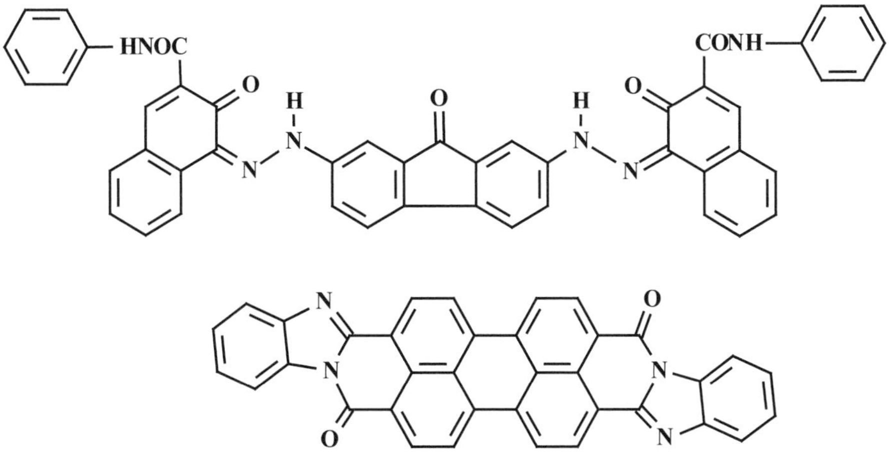

Fig. 13.36. Azo (top) and perylene (bottom) pigments for charge generation.

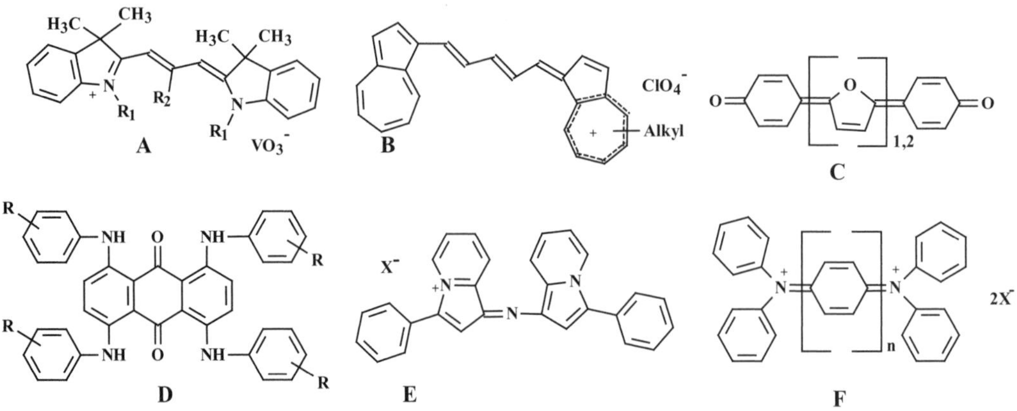

Fig. 13.37 Structures of indoleninecyanine (A), azulenium (B), quinonoid (C), anthraquinone (D), and imminium (E) and diiminium (F) IR-absorbing dyes, where R = alkyl or alkoxy.

commonly used dyes are highly fluorescent and include coumarin, rhodamine, oxazine, and *syn*-bimane structures (Fig. 13.38). Dye lasers are employed in liquid form, which allows them to dispel excessive heat by recirculating the dye solution. Good photostability and efficient laser action under flashlamp excitation are important properties.

near infrared (NIR) and IR-absorbing dyes can be used in automated DNA sequencing (dye A), fluorescent dyes can be used in cancer detection (dye B), and certain azo and heterocyclic dyes can be used in virus (dye C), cell (dye D), and bacteria (dye E) detection. An in-depth summary of dyes in this area has been published recently.[48]

Biomedical Dyes

Dyes can be used clinically in bioanalysis and medical diagnostics and in the treatment of certain diseases (cf. Fig. 13.39). For instance,

Hair Dyes

About 80% of the dyes used in hair coloring are known as oxidation hair dyes.[49–50] The remaining 20% of the available hair dyes

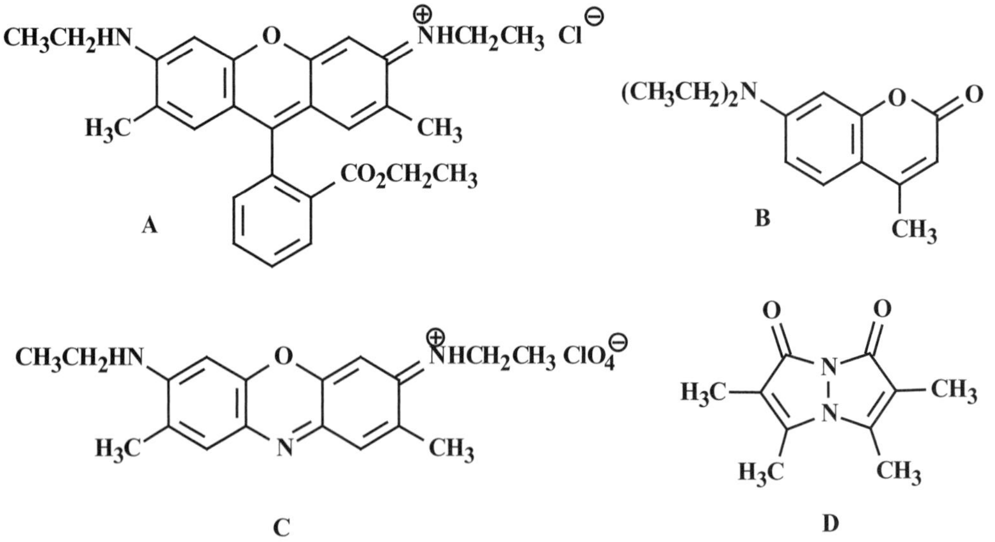

Fig. 13.38. Rhodamine (A), coumarin (B), oxazine (C), and *syn*-bimane (D) laser dye structures.

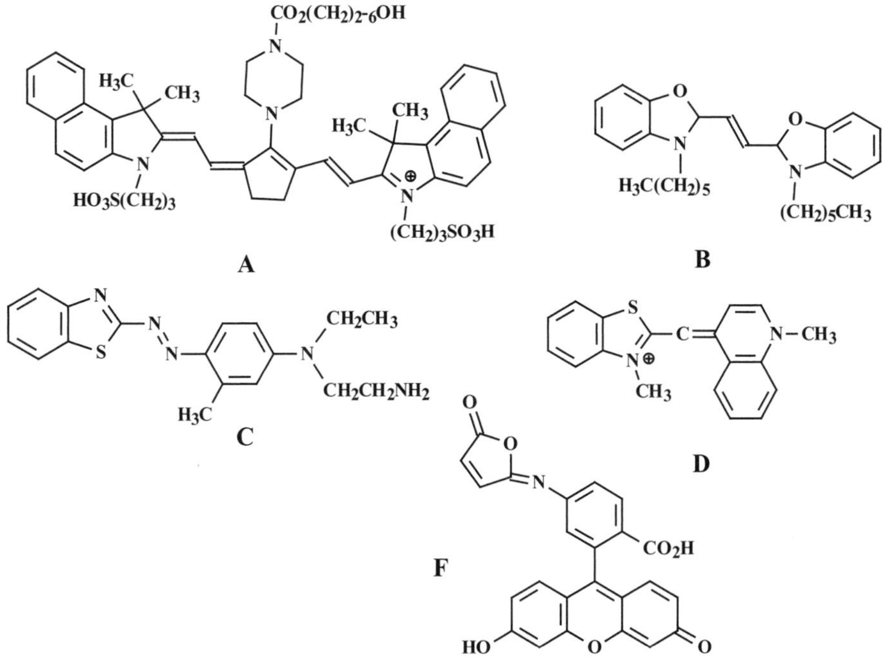

Fig. 13.39. Examples of experimental dyes used in biomedical applications.

are mainly synthetic dyes that have affinity for protein substrates. Oxidation dyes are produced directly on hair by oxidizing aromatic diamines (e.g., *para*-phenylenediamine or 2,5-diaminotoluene) with a suitable oxidizing agent. In this regard, the diamines have been referred to as "primary intermediates" and the oxidizing agents (e.g., hydrogen peroxide) as "developers". Other suitable primary intermediates are aminodiphenylamines, aminomethylphenols, and *para*-aminophenol.

When used alone, the primary intermediates give a quite limited shade range following oxidation on hair. To enhance the range of available hair colors, the primary intermediates are oxidized in the presence of suitable "couplers.") Whereas most couplers do not produce colors when exposed to developers alone, they give a wide array of hair shades in combination with primary intermediates. Suitable couplers include 3-aminophenol, resorcinol, and α-naphthol.

The chemistry associated with the oxidation of primary intermediates is now reasonably well known. For *para*-phenylenediamine and *para*-aminophenol, this involves the process outlined in Fig. 13.40. It can be seen that dye formation is a two-step process involving oxidation and self-coupling.

C.I. Basic dyes such as Yellow 57, Red 76, Blue 99, Brown 16, and Brown 17 have been used in color refreshener shampoos and conditioners. Similarly, C.I. Acid dyes such as Yellow 3, Orange 7, Red 33, Violet 43, and

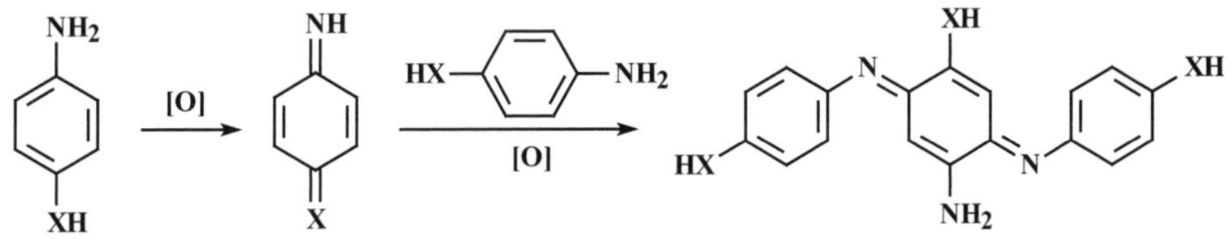

Fig. 13.40. Oxidation of primary intermediates, where X = O, NH.

Blue 9 have been used in shampoos, in this case to deliver highlighting effects.[51]

Photographic Dyes

Color photography is still one of the most important and interesting nontextile uses for synthetic dyes. The chemistry employed is comparable to that described above for oxidation hair dyes, in that an oxidizable substrate (e.g., phenylenediamine) is combined with a coupler to produce the target colorant. In this case the diamine is referred to as the "developer," and it is oxidized by silver halide in the photographic film. The oxidized developer then reacts with the coupler to form the dye. This process produces a negative dye image consisting of yellow, magenta, and cyan dyes in proportion to the amount of red, blue, and green light absorbed by the film.[52]

Some widely used developers are shown in Fig. 13.41. They can be used to produce the yellow, magenta, and cyan dyes shown in Fig. 13.42. These dye structures demonstrate that acetoacetanilide, pyrazolone, and indoaniline intermediates are useful for producing yellow, magenta, and cyan colors, respectively.

DYE INTERMEDIATES

The raw materials used to synthesize organic dyes are commonly referred to as dye intermediates. Largely, they are derivatives of aromatic compounds obtained from coal tar mixtures. The majority of these derivatives are benzene, naphthalene, and anthracene based compounds. This section provides an overview of the chemical reactions used to prepare the key intermediates employed in dye synthesis. In this regard, emphasis is placed on halogenated, aminated, hydroxylated, sulfonated, and alkylated derivatives of benzene, naphthalene, and anthraquinone.

Most dye intermediates are prepared by reactions involving electrophilic or nucleophilic substitution processes. The electrophilic processes include nitration, sulfonation, and halogenation reactions, and the nucleophilic processes include hydroxylation and amination reactions. Electrophilic substitution reactions are of the form shown in Fig. 13.43. In this regard, the incoming electrophile (electron-seeking species) reacts with the more electron-rich positions. When the aromatic ring contains ring-activating groups (e.g., hydroxy, alkoxy, amino, alkyl), the incoming

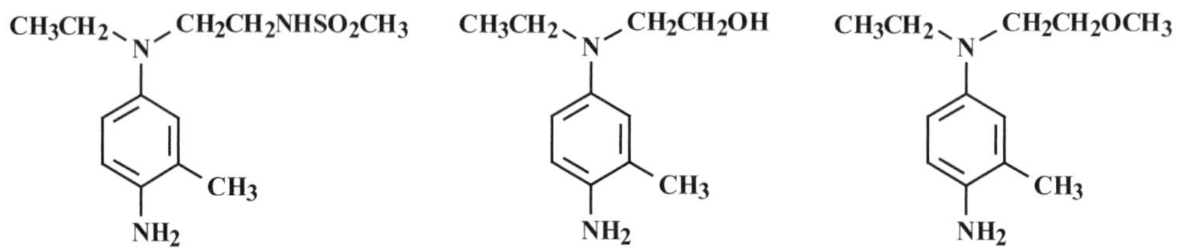

Fig. 13.41. Structures of some developers used in color photography.

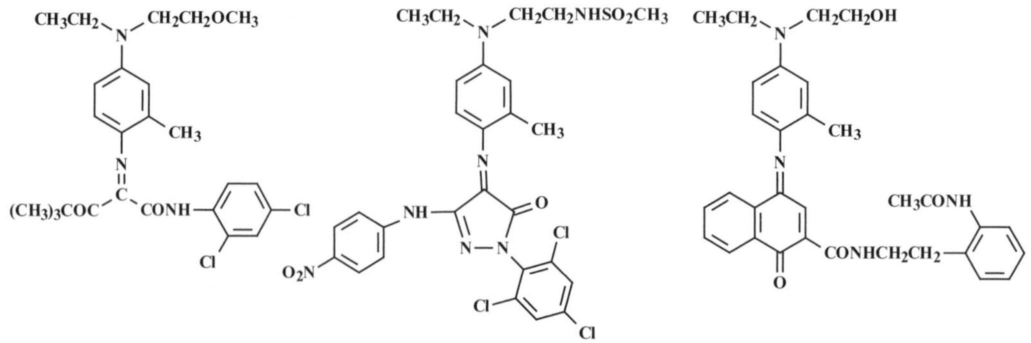

Fig. 13.42. Structures of yellow (left), magenta (center), and cyan (right) photographic dyes.

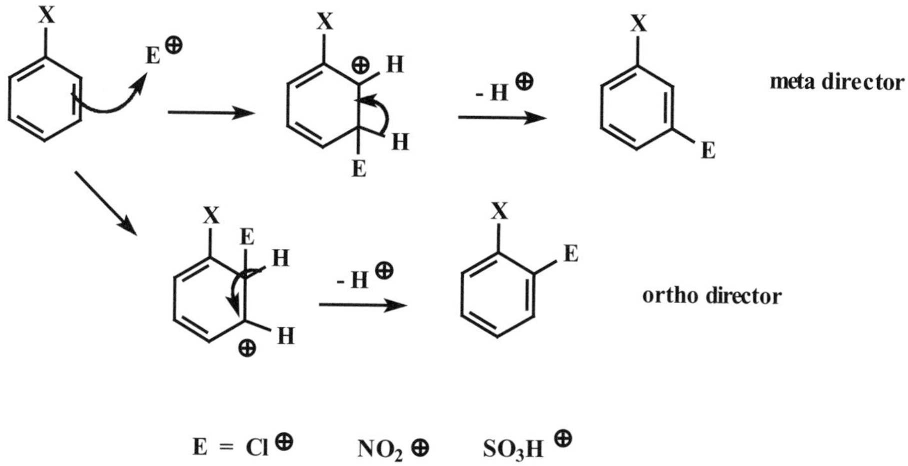

Fig. 13.43. Electrophilic attack of an aromatic ring containing deactivating and activating groups.

group will attack *ortho/para* positions. If ring-deactivating groups (e.g., nitro, sulfonic acid, carboxylic) are present, the positions *meta* to the deactivating groups will be attacked.

Other key dye intermediates are prepared by oxidation and reduction processes. Examples of each of these processes are covered in the sections that follow.

Nitration

For dye intermediates, this process involves the introduction of one or more nitro (NO_2) groups into aromatic ring systems. Nitro groups serve as chromophores (color bearers, precursors for amino groups, and as auxochromes (color aiders). Because they are *meta*-directing groups they are also useful in the strategic placement of another incoming group.

Nitric acid (HNO_3) is the chemical agent commonly used in nitration reactions. Depending upon the degree of ring activation, HNO_3 may be used in combination with other acids. In fact, nitrations are often conducted by using a mixture of HNO_3 and sulfuric acid (H_2SO_4). This combination is known as "nitrating mixture" or "mixed acid," and it is especially effective when deactivated ring systems are to be nitrated. Dilute HNO_3 or a HNO_3/acetic acid (CH_3CO_2H) mixture can be used for nitrating very reactive ring systems. When the former is used there is also the potential for ring oxidation to occur rather than the desired nitration, depending upon the actual compound undergoing nitration. Examples of nitration reactions are shown in Figs. 13.44–13.46. The nitration of toluene (Fig. 13.44) is selected because it illustrates what can happen when monosubstituted benzenes having a ring-activating group are

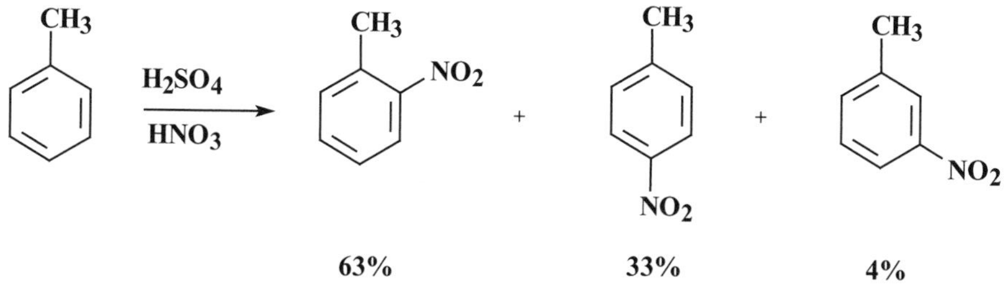

Fig. 13.44. Nitration of toluene using mixed acid.

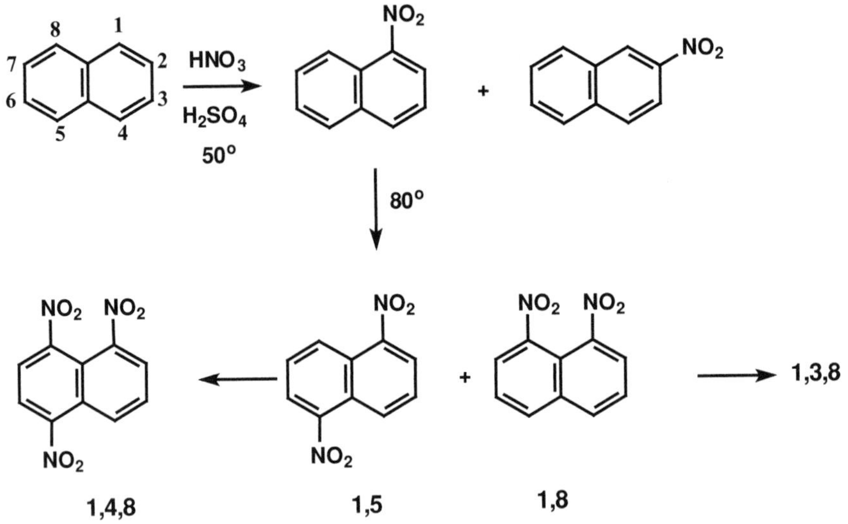

Fig. 13.45. Nitration of naphthalene using mixed acid.

used. In this case, the principal products reflect a statistical mixture of *ortho* and *para* isomers, with only a small amount of the *meta* isomer obtained. Nitration is conducted near 20°C and the products are separated by distillation.

Nitration of naphthalene gives mostly the 1-nitro isomer (~90%), initially. Introduction of a second nitro group takes place in the opposite ring because the existing nitro group reduces the reactivity of the ring to which it is attached. Although, nitro groups are *meta* directors, in this case they can also direct the incoming second (or third) nitro group to a *peri* position. In the naphthalene ring system, the *peri* positions are those that are 1,8 and 4,5 to each other (Fig. 13.45).

The nitration of anthraquinone at 50°C gives, initially, the 1-nitro isomer, and if nitration continues at 80–90°, the 1,5 and 1,8 isomers

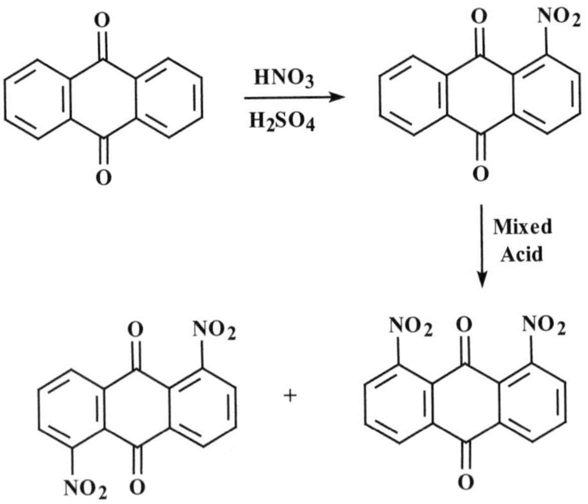

Fig. 13.46. Nitration of anthraquinone using mixed acid.

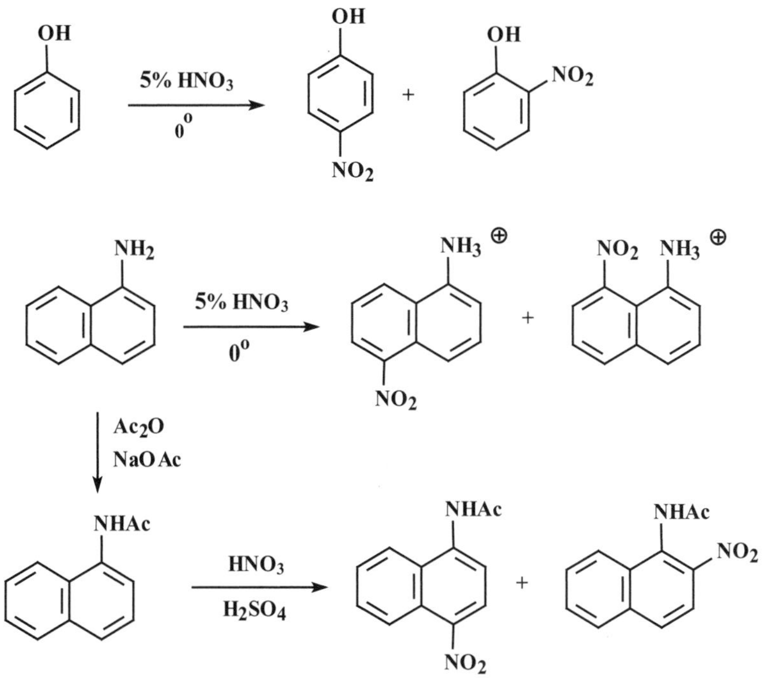

Fig. 13.47. Nitration of phenol and 1-naphthylamine.

are obtained. Further nitration is impractical and serves to point out that the anthraquinone ring is appreciably less reactive than the naphthalene system. This will be more evident as the chemistry reported in this section continues to unfold.

The nitration of phenols and amines must be conducted with care, as these systems are subject to ring oxidation if the temperature gets too high. For instance, the nitration of phenol itself is conducted near 0°C using 5% HNO_3. This gives a mixture of *ortho* and *para* isomers that can be separated by steam distillation (Fig. 13.47). Aromatic amines are often protected by *N*-acetylation prior to nitration. This reduces both the potential for ring oxidation and the amount of *meta* isomer that forms when the amino group undergoes protonation. The protonated amino group ($-NH_3^+$) is a *meta* director, unlike the free

amino (NH_2) and the acetylated amino (NHAc) groups. This chemistry is illustrated in Fig. 13.47 for 1-naphthylamine. Following nitration, the acetyl group can be removed by hydrolysis.

Reduction

The most important reduction reactions are those leading to aromatic amines that are suitable for azo dye formation. Although this usually involves the reduction of a nitro group to an amino (NH_2) group, the reduction of azo groups to amino groups is also an important process. Agents that are commonly used to effect chemical reductions include: Fe + HCl or H_2SO_4; Na_2S; NaSH; Zn + NaOH; H_2 + transition metal catalysts; and $Na_2S_2O_4$. Examples of these reductions are given in Figs. 13.48 to 13.51. While the reduction of

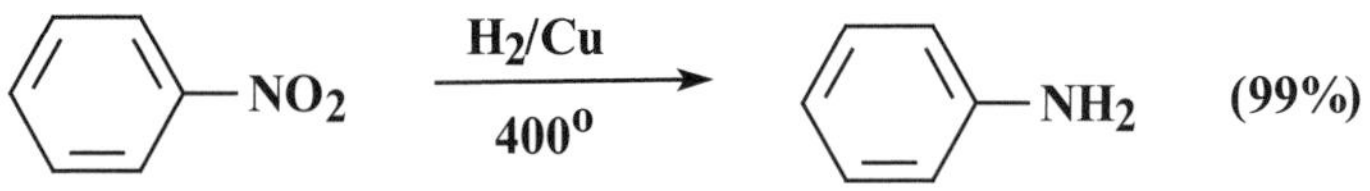

Fig. 13.48. Commercial process for the reduction of nitrobenzene.

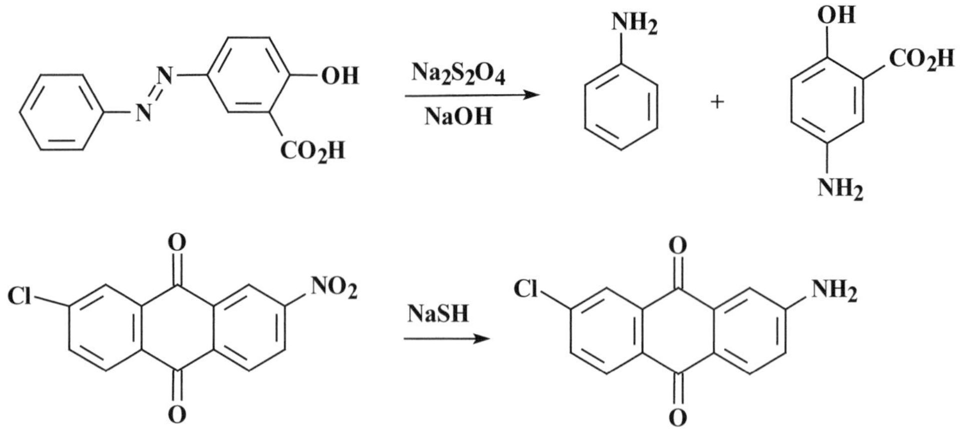

Fig. 13.49. Reduction of azo (upper) and nitroanthraquinone (lower) compounds.

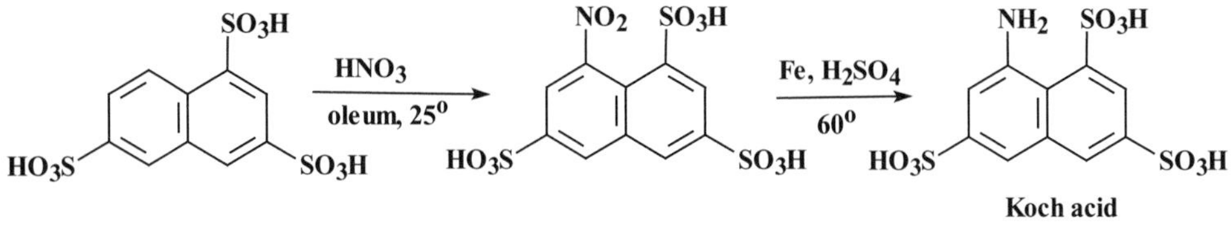

Fig. 13.50. Formation of Koch acid via nitration and reduction steps.

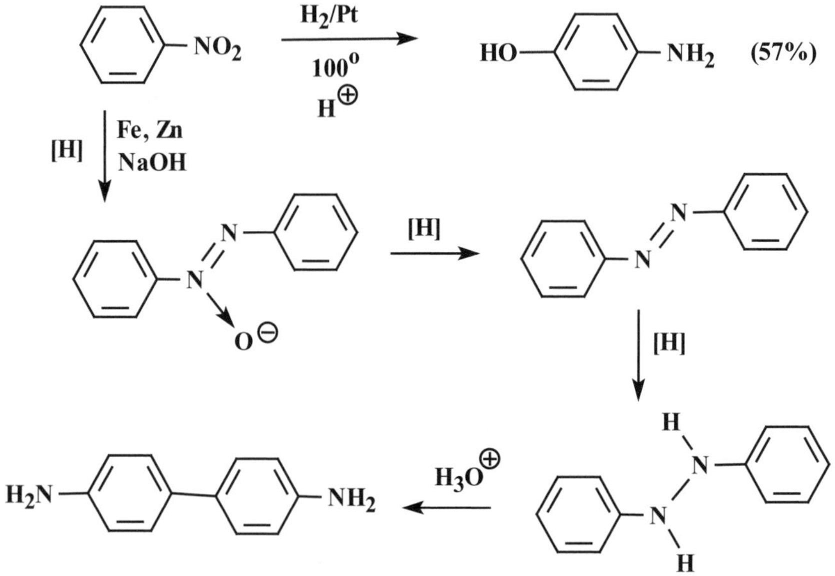

Fig. 13.51. Reduction of nitrobenzene in acidic and alkaline media.

nitrobenzene can be conducted in a number of ways, a key commercial process involves the method in Fig. 13.48, where high-temperature hydrogenation is used.

The reduction of azo compounds using sodium hydrosulfite ($Na_2S_2O_4$) and NaOH is an important reaction, as it provides an indirect method for the amination of phenols and naphthols (Fig. 13.49). The reduction of nitro groups in anthraquinone compounds works best when a mild reducing agent (e.g., sodium hydrosulfide, NaSH) is used. In this way one avoids reducing the quinoid system.

An example of an important reduction reaction involving Fe + H^+ is shown in Fig. 13.50. In this case the sequential use of nitration and reduction is illustrated.

It must also be pointed out that the medium employed in the reduction process can play a major role in the outcome of the reaction. A good example is the reduction of nitrobenzene in the presence of acid or alkali. One should expect the reduction to follow the course shown in Fig. 13.48 under normal conditions, however, in acidic media the product obtained is mainly *para*-aminophenol. In fact, this has long been the key step in the commercial route to acetaminophen,[52] which is obtained by *N*-acetylation of the reduction product. When the reduction is conducted in the presence of alkali and Zn, the nitro compound is converted to a hydrazo compound via azoxy and azo intermediates. The hydrazo compound is important because it can be treated with acid to form diaminobiphenyls known as benzidines. These reactions are shown in Fig. 13.51. Because benzidine (4,4'-diaminobiphenyl) itself is known to be a human carcinogen, its use as a dye intermediate is substantially curtailed in the western world.

Amination

In as much as the previous section covers the reduction of nitro and azo compounds as a method for introducing amino groups, the focus of this section is direct aminations involving replacement reactions and examples of indirect amination. In the former case, amination via the replacement of activated halogens using an alkyl or arylamine is widely used. The examples given in Fig. 13.52 show that halogens positioned *ortho* to a nitro group or in an α-position on the anthraquinone ring can replaced by amino groups. The former reaction also works well when the groups are *para*. However, the reaction is difficult and usually impractical when electron-donating rather than electron-attracting groups are situated *ortho* and/or *para* to the halogen. In the case of the anthraquinone system, α-sulfonic acid and α-nitro groups can also be replaced.

An important amination reaction involves hydroxy-substituted naphthalenes (Fig. 13.53). In a process known as the Bucherer reaction, naphthols are heated under pressure with a mixture of ammonia and sodium bisulfite. As

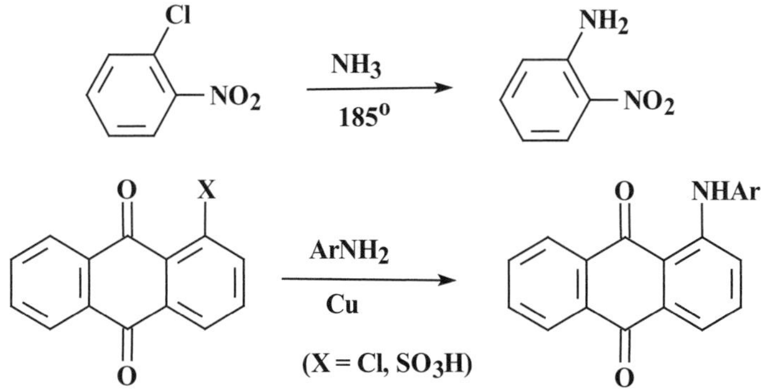

Fig. **13.52.** Amination reactions involving benzene and anthraquinone compounds.

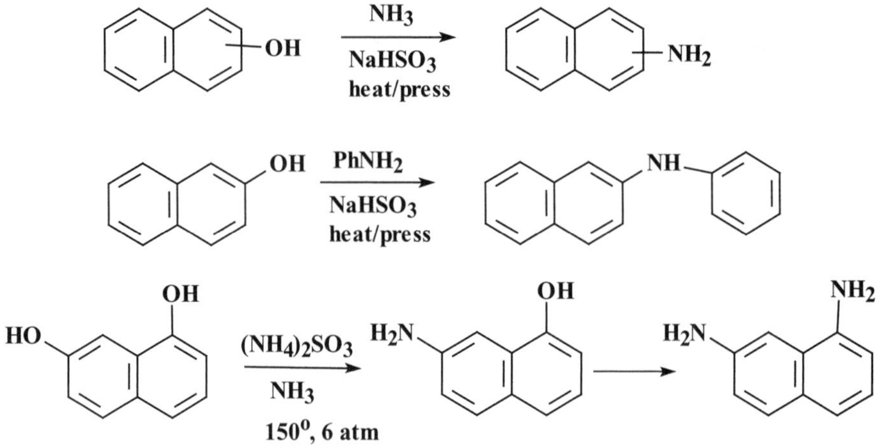

Fig. 13.53. Amination of naphthalene compounds via the Bucherer reaction.

the second and third examples indicate, the reaction works with aromatic amines and is selective. Note that the β-hydroxy group reacts preferentially when an a-hydroxy group is also present, and that two hydroxy groups in the same compound can be replaced.

An alternative route to the synthesis of aminoanthraquinones is the two-step sequence shown in Fig. 13.54. In this case, amination occurs via the condensation of *para*-toluenesulfonamide with chloroanthraquinone followed by hydrolysis of the sulfonamide bond. This method provides a way to introduce an –NH₂

group without the use of ammonia gas and the associated high temperatures and pressures.

Another interesting reaction is shown in Fig. 13.55. In this example, amination and sulfonation occur when a-nitronaphthalene is reduced by heating it under pressure with NaHSO₃.

Sulfonation

The introduction of one or more sulfonic acid groups (sulfonation) into dye intermediates is often conducted to confer water solubility, to provide fiber affinity, and to direct other

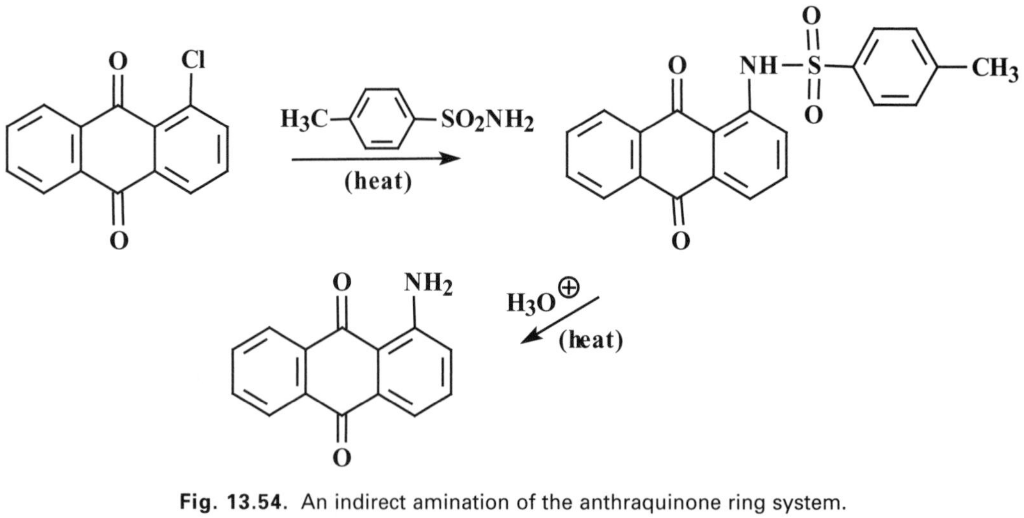

Fig. 13.54. An indirect amination of the anthraquinone ring system.

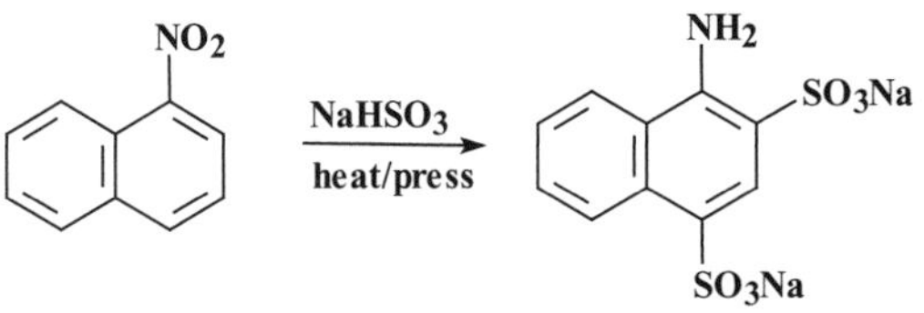

Fig. 13.55. A one-step amination and sulfonation of naphthalene.

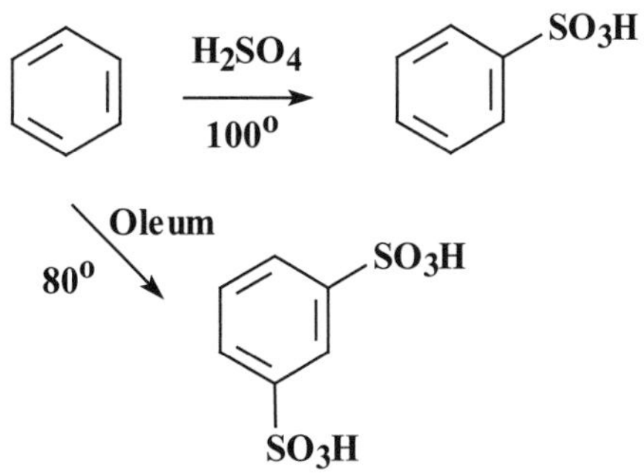

Fig. 13.56. The mono and disulfonation of benzene.

incoming groups in the steps that follow sulfonation. In most cases this process employs sulfuric acid but in difficult cases, for example, deactivated ring systems, oleum (an SO_3/H_2SO_4 mixture) is used. This chemistry is illustrated in Fig. 13.56 for benzene. Here we see that benzene can be sulfonated using sulfuric acid and that the introduction of a second sulfonic acid group requires oleum. When a more reactive system is sulfonated, less stringent conditions are required. For example, naphthalene (Fig. 13.57) is readily sulfonated up to four times without using

oleum. It is important to note that it is not possible to have sulfonic acid groups that are *ortho*, *para*, or *peri* to each other in the naphthalene system

By contrast, the sulfonation of anthraquinone requires oleum and no more than two sulfonic acid groups can be introduced. In this system, sulfonation in the α-position requires the use of HgO as a catalyst. Examples of the

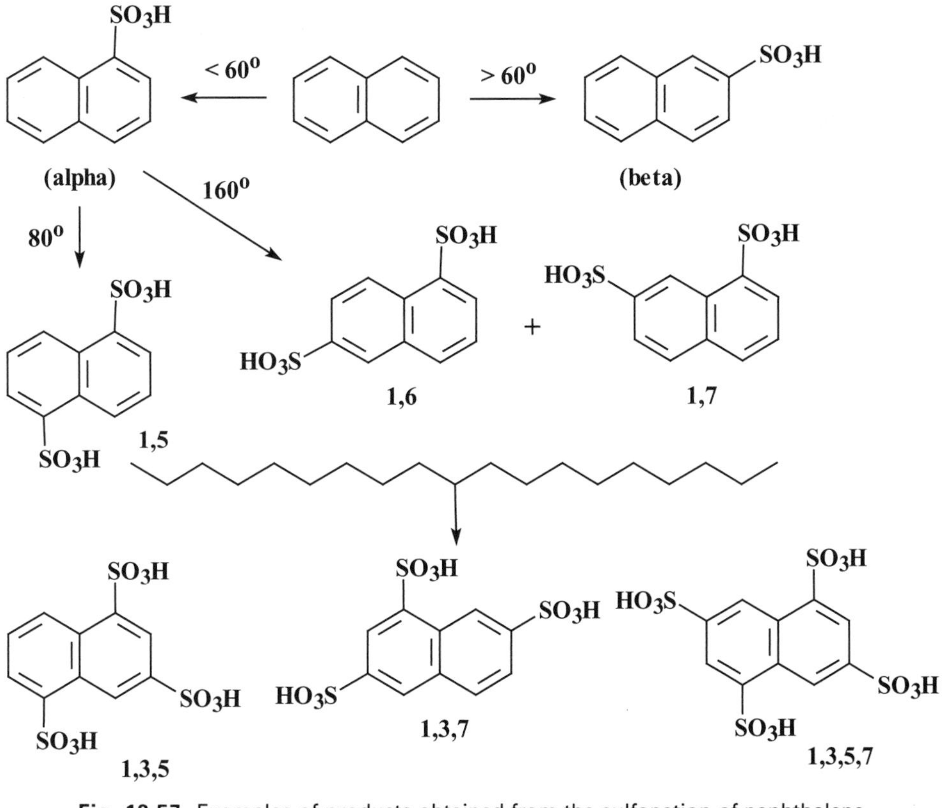

Fig. 13.57. Examples of products obtained from the sulfonation of naphthalene.

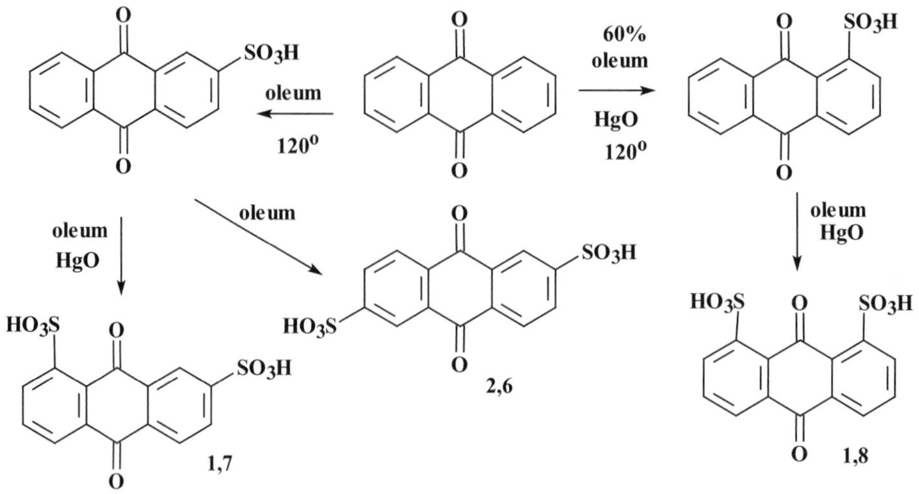

Fig. 13.58. Typical products produced from the sulfonation of anthraquinone.

possible products are shown in the scheme in Fig. 13.58.

The sulfonation of β-naphthol produces several important dye intermediates, the nature of which depends upon the conditions employed (Fig. 13.59). At low temperatures, sulfonation occurs in the α-position to give oxy-Tobias acid. Under ambient conditions Crocein acid is produced and at elevated temperatures three other products are obtained, including two that are disulfonated.

The sulfonation of aromatic amines such as aniline can give a mixture of products that must be separated prior to dye synthesis.

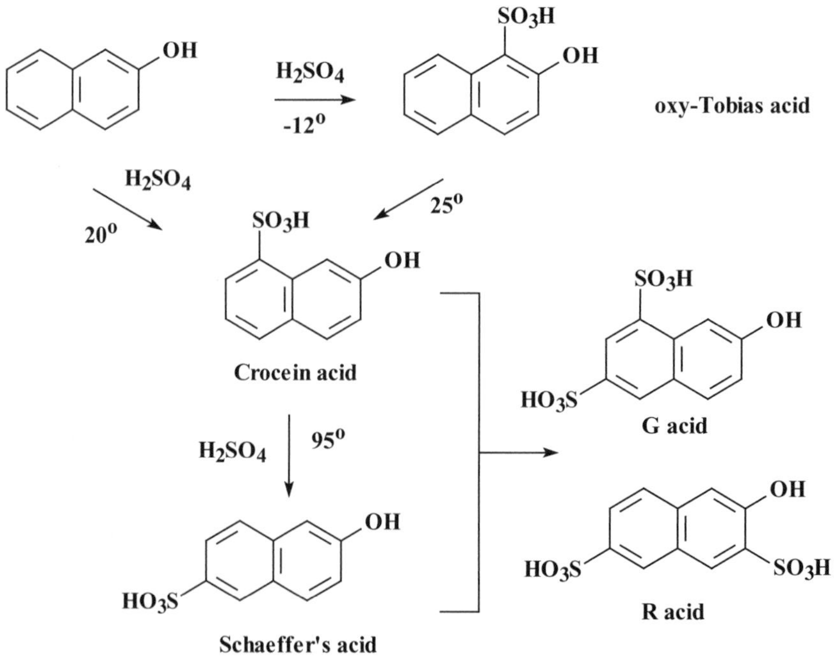

Fig. 13.59. Dye intermediates prepared from the sulfonation of β-naphthol.

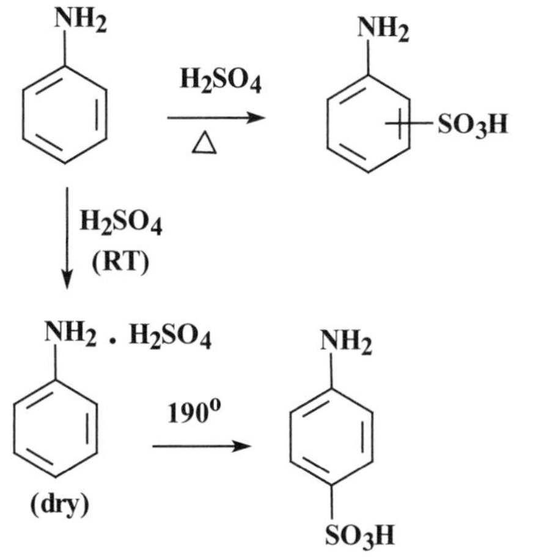

Fig. 13.60. Direct sulfonation of aniline (upper) versus the baking reaction (lower).

When a single product is sought, the "baking" reaction is often employed (Fig. 13.60). In this process, the sulfate salt of aniline is prepared, dried, and then "baked" in an oven under vacuum. The product in this case is the important dye intermediate, sulfanilic acid. Similarly, naphthylamine sulfonic acid can be produced, and if the *para*-position is occupied, sulfonation of an *ortho*-position occurs (Fig. 13.61).

Halogenation

For dye intermediates, halogenation most often involves the incorporation of chloro groups.

As pointed out earlier, halogens are important as leaving groups in the amination process, but they can also be used to enhance brightness and influence color. Later, we show that halogens are important as leaving groups in reactive dye chemistry, and in this regard chloro and fluoro groups are used.

Figure. 13.62 to 13.65 provide examples of chlorination reactions. In the first example, the commonly used agent $FeCl_3/Cl_2$ is employed for the chlorination of benzene and naphthalene rings. This method is not practical for the chlorination of anthraquinone. In this case the most important reaction is the tetrachlorination process shown in Fig. 13.63.

Because the chlorination of phenols and aromatic amines can be difficult to control, chlorination of these systems usually employs agents that will give a single chloro group when this is the desired outcome. In this regard, $NaOCl$ and SO_2Cl_2 are quite useful chlorinating agents (see Fig. 13.64). In cases involving amines, often such compounds are protected by acetylation prior to chlorination. If the reactivity of the ring has been reduced by the presence of a deactivating group (e.g., $-NO_2$), acetylation may not be needed.

Halogens are also introduced via indirect methods, three examples of which are shown in Fig. 13.65. In the first case (sequence "A"), aniline is diazotized and the resultant diazonium compound is heated with cuprous

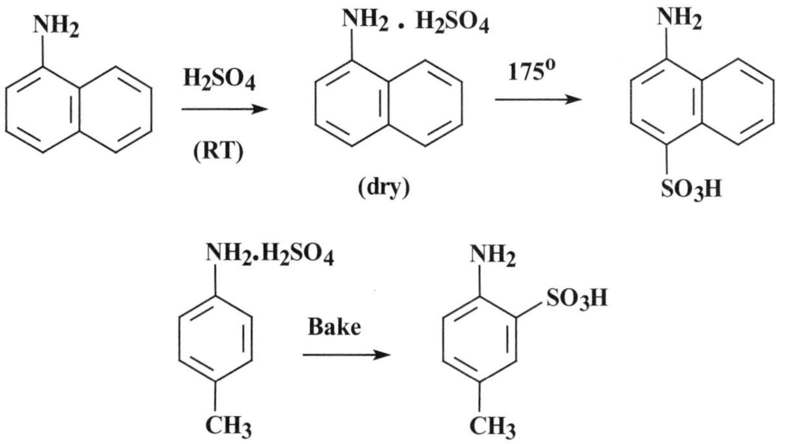

Fig. 13.61. Additional examples of the baking reaction.

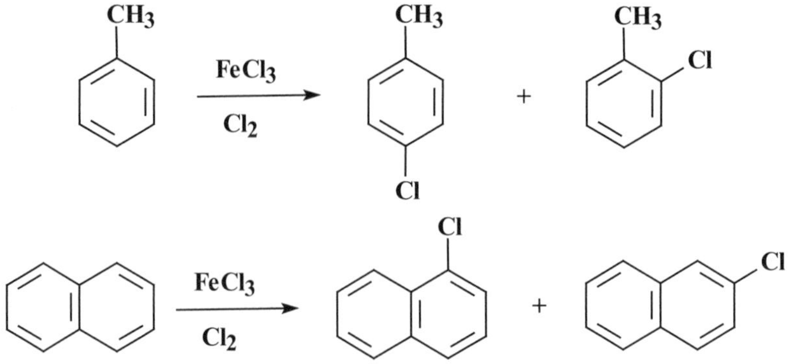

Fig. 13.62. Halogenation of benzene and naphthalene rings.

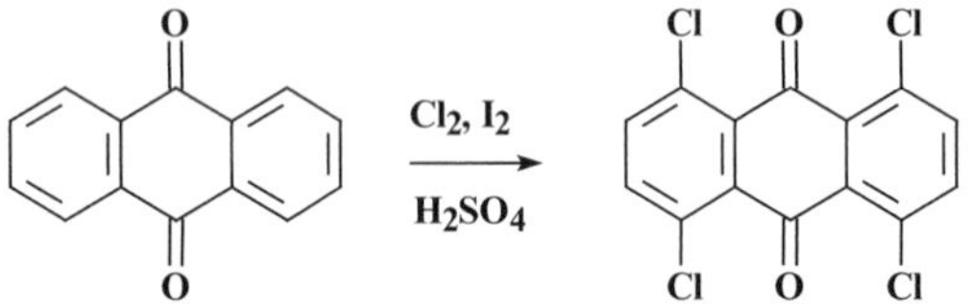

Fig. 13.63. Synthesis of the dye intermediate 1,4,5,8-tetrachloroanthraquinone.

chloride to give chlorobenzene, in a process known as the Sandmeyer reaction.[53] Alternatively, the diazonium compound can be converted to the tetrafluoroborate salt, which in turn is heated to give fluorobenzene. In sequence "B", anthraquinone-2-sulfonic acid is converted to the corresponding chloro compound by treatment with $NaClO_3/HCl$. All three reactions can be used to prepare a wide array of halogenated aromatics.

Chlorination is also an important step in the synthesis of oxygenated aromatic compounds. In this case, chlorination takes place at alkyl groups attached to the rings and is conducted in the absence of iron. The use of UV light

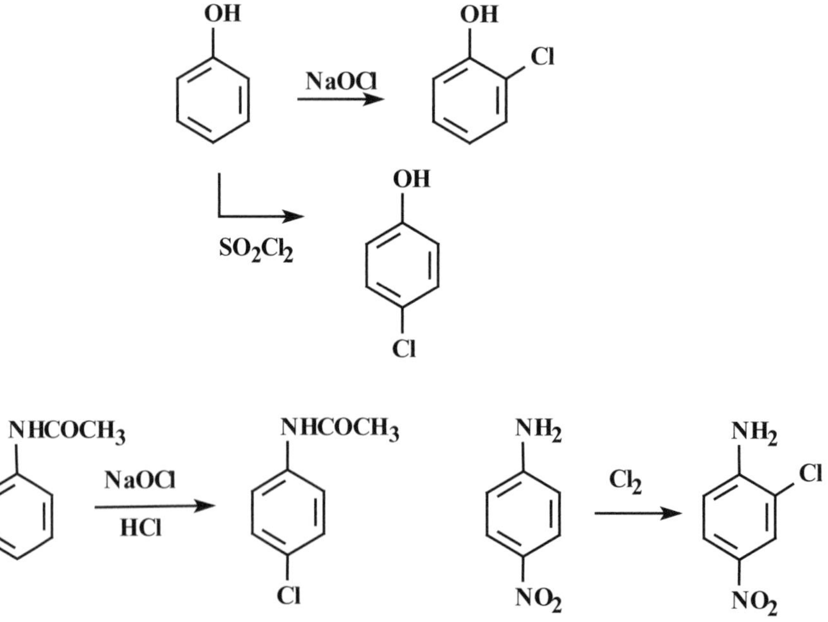

Fig. 13.64. Chlorination of phenol and anilines.

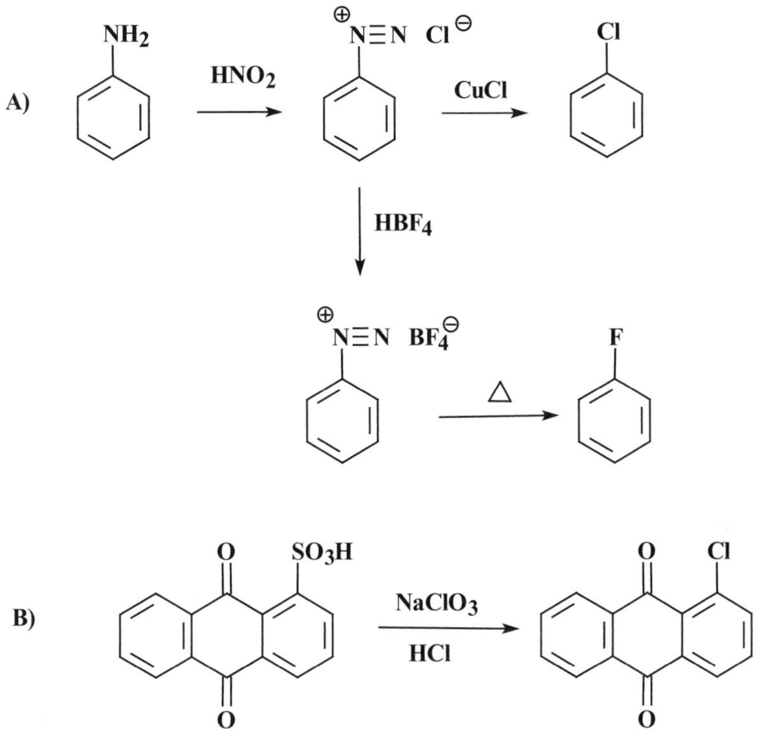

Fig. 13.65. Three important indirect halogenation reactions.

speeds up this reaction, which is illustrated for toluene in Fig. 13.66. This free radical chlorination of toluene gives a mixture of benzyl chloride, benzal chloride, and benzotrichloride, which in turn can be hydrolyzed to benzyl alcohol, benzaldehyde, and benzoic acid.

Hydroxylation

The introduction of hydroxy groups is important in dye chemistry because it opens the door to azo dye formation, using phenols and naphthols, and provides an important auxochrome. Hydroxylation methods include alkali fusion, replacement of labile groups,

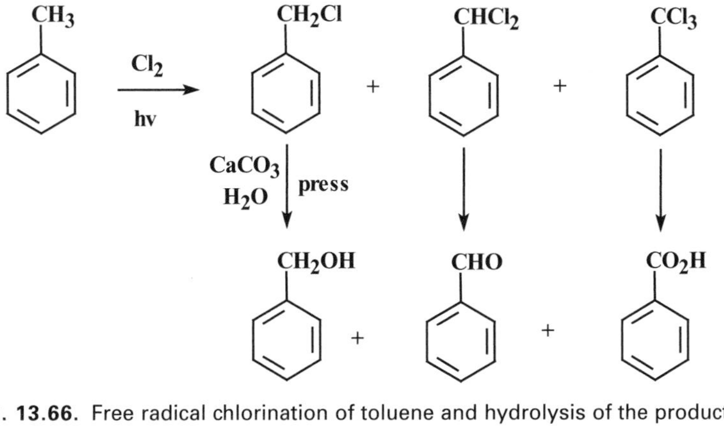

Fig. 13.66. Free radical chlorination of toluene and hydrolysis of the products.

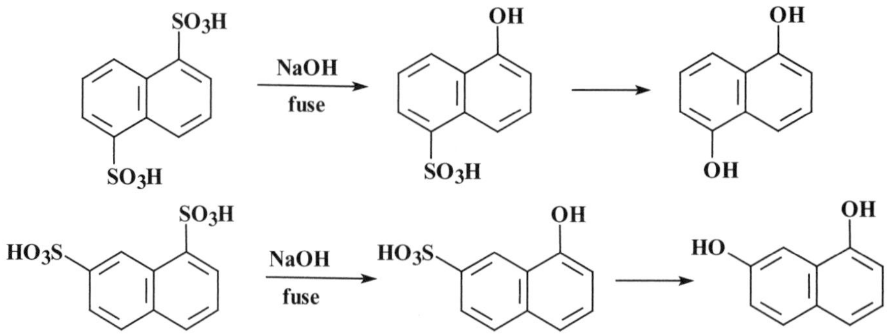

Fig. 13.67. Sodium hydroxide fusion as a hydroxylation process.

and the reverse Bucherer reaction. In the alkali fusion reaction, naphthalene sulfonic acids are reacted with molten NaOH, KOH, or combinations of the two, as illustrated in Fig. 13.67. When disulfonated naphthalenes are used, the reaction can be stopped at the mono-hydroxylation stage if this is the desired outcome. The second example shows that the α-sulfonic acid group reacts faster.

When sulfonated anthraquinones are used, hydroxylation is conducted with $Ca(OH)_2$ to avoid over oxidation that occurs when hot NaOH is used. Example reactions are shown in Fig. 13.68.

The short sequence in Fig. 13.69 shows that aqueous alkali can also be used in hydroxylation reactions. In both cases, however, elevated temperatures are required.

Other important hydroxylation reactions are shown in Fig. 13.70. Here it can be seen that the Bucherer reaction is reversible, that the fusion reaction works for sulfonated benzene compounds, and that diazonium compounds undergo hydrolysis to produce phenols/naphthols.

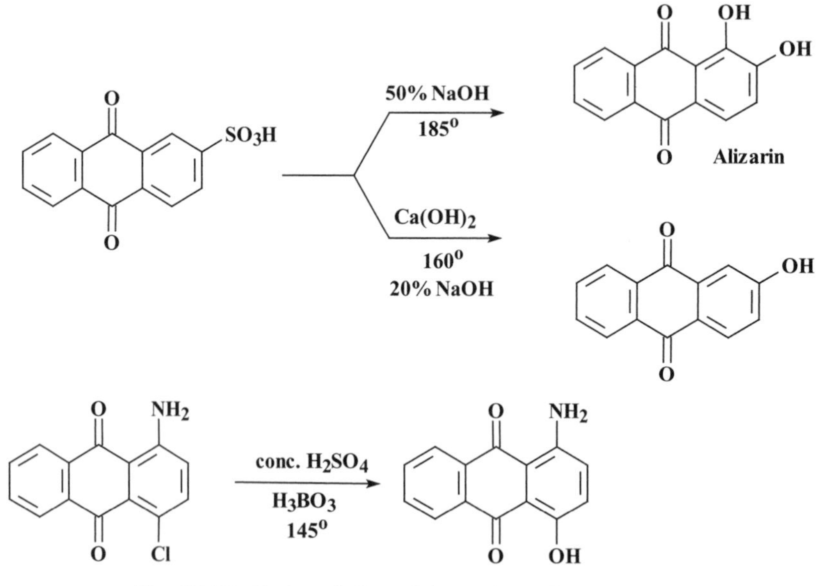

Fig. 13.68. Hydroxylation of the anthraquinone system.

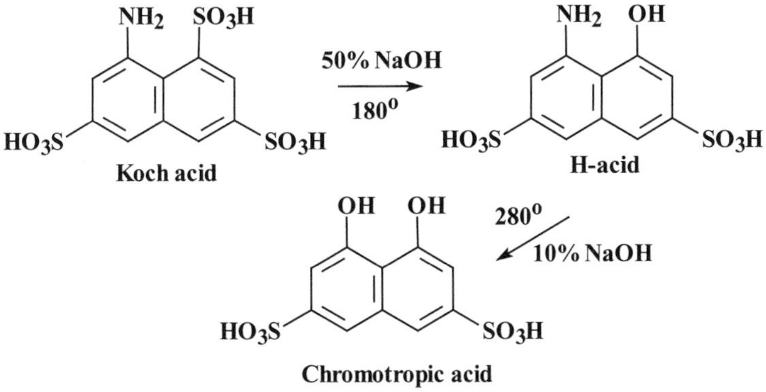

Fig. 13.69. Synthesis of two key dye intermediates using hydroxylation steps.

Oxidation

Although the oxidation of aromatic methyl groups can be conducted via the two-step sequence shown in Fig. 13.66, a convenient alternative process involves potassium dichromate. In this case, the ring system involved must be stable to the conditions of the reactions. Another important oxidation reaction involves the conversion of naphthalene to phthalic anhydride, which can be accomplished using hot $KMnO_4$ or V_2O_5. These two reactions are illustrated in Fig. 13.71. Later we show that the oxygen in air can be used as the oxidant for certain organic dyes.

Other Important Reactions

Diazotization. The conversion of a primary aromatic amine to a diazonium compound is known as diazotization. Although this process is covered in more detail in our discussion of azo dye synthesis, it is worthwhile to point out that the diazonium group ($-N_2^+$) is used to produce a wide range of intermediates. As indicated in Fig. 13.72 diazotization is often achieved through the action of nitrous acid (HNO_2) and the resultant diazonium group can be replaced by various groups or reduced to give arylhydrazines.

Carboxylation. The introduction of carboxyl groups into the structures of phenols and naphthols produces some important dye intermediates, including salicylic acid and β-oxynaphthoic acid (BON acid). This process is conducted under pressure at elevated temperatures using the sodium salts of phenols/naphthols and in the case of β-naphthol, the carboxyl group enters

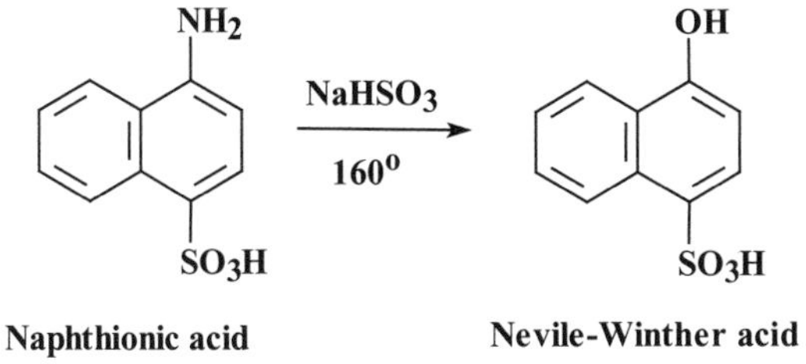

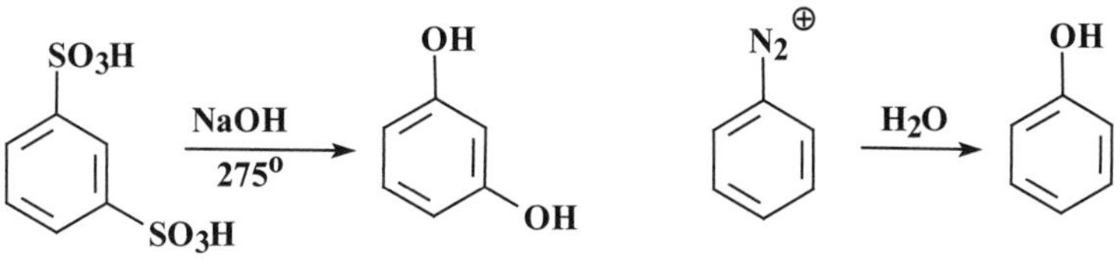

Fig. 13.70. Other examples of hydroxylation reactions.

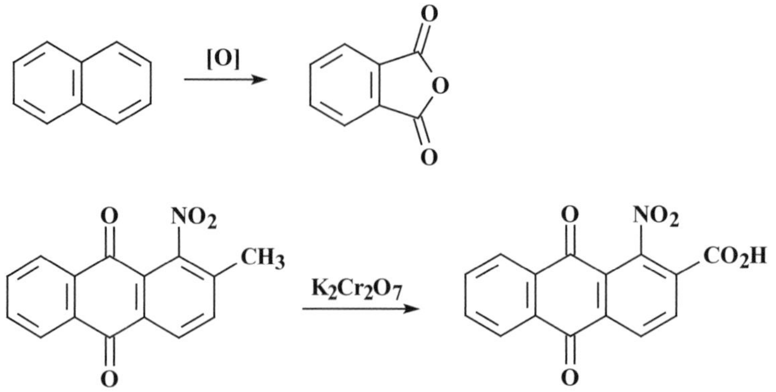

Fig. 13.71. Oxidation of naphthalene (top) and a methylated anthraquinone (bottom).

the 3-position (cf. Fig. 13.73). The free acid ($-CO_2H$) group is produced by acid treatment in the final step.

DYE MANUFACTURE

In this section, we summarize the principal methods of synthesis for different dye classes.

Emphasis is placed on dyes presently in commerce and the industrial methods suitable for making them. Before doing so, we review the important principles that set dyes apart from other classes of organic compounds.

Unlike other organic compounds dyes possess color because they (1) absorb light in the visible spectrum (400–700nm), (2) have at

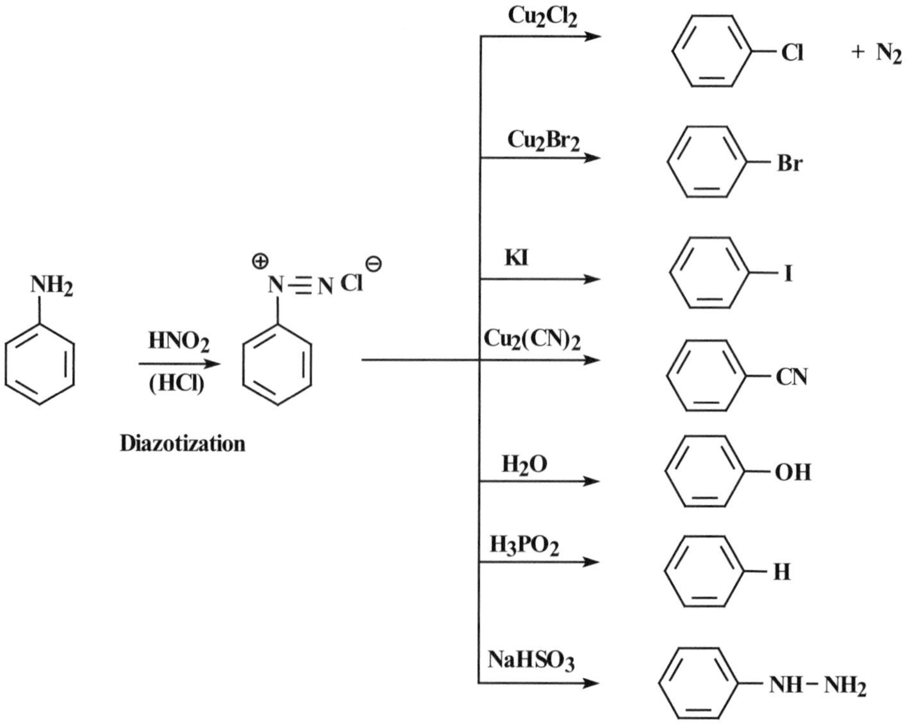

Fig. 13.72. Diazotization of aniline and its conversion to other types of intermediates.

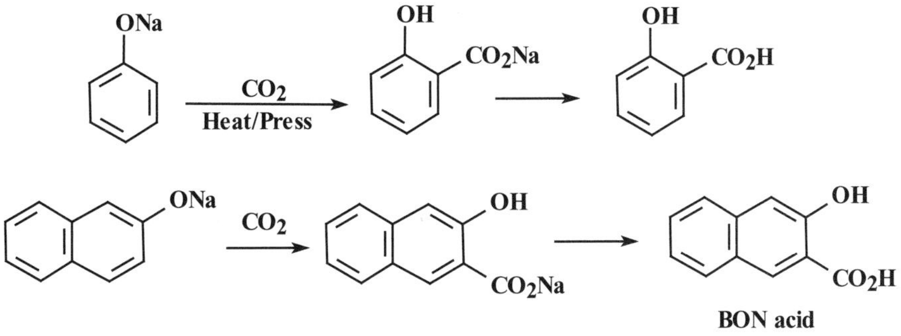

Fig. 13.73. Carboxylation of phenol and β-naphthol.

least one chromophore (color bearing group), (3) have a conjugated system (system of alternating double and single bonds), and (4) exhibit resonance (a stabilizing force in organic compounds). Table 13.5 shows the relationships between wavelength of visible light and color absorbed/seen and the other three factors are illustrated in Fig. 13.74 to 13.76.

Concerning the various factors responsible for color in organic compounds, it is worthwhile to point out that the chromophore must be part of a conjugated system. This is illustrated through the examples in Fig. 13.77. When the azo group is connected to methyl groups the resultant compound is colorless. When it is attached to aromatic rings, the compound

TABLE 13.5. Wavelength of Light Versus Color

Wavelength Absorbed	Color Absorbed	Color Seen
400–435	Violet	Yellow-Green
435–480	Blue	Yellow
480–490	Green-Blue	Orange
490–500	Blue-Green	Red
500–560	Green	Purple
560–580	Yellow-Green	Violet
580–595	Yellow	Blue
595–605	Orange	Green-Blue
605–700	Red	Blue-Green

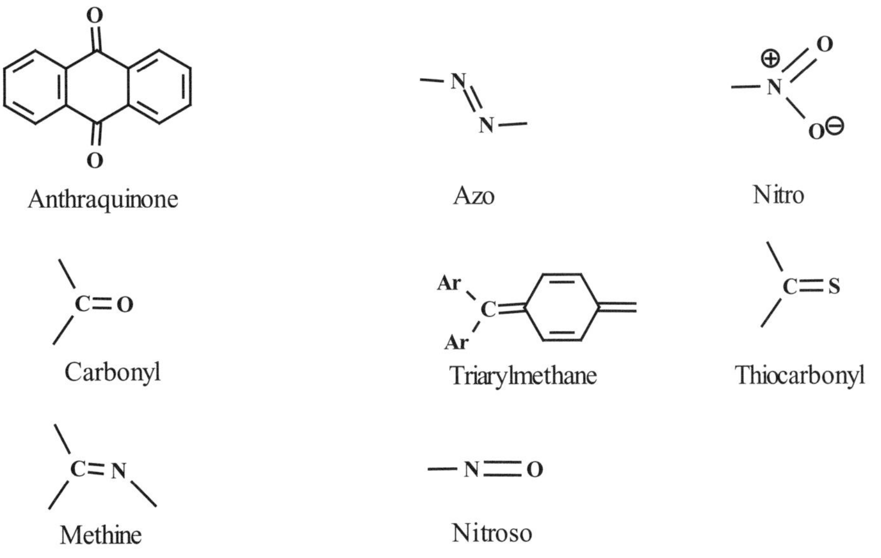

possesses color. Similarly, the structures in Fig. 13.75 illustrate the importance of having an extended conjugated system. In this case,

Fig. 13.74. Chromophores commonly found in organic dyes.

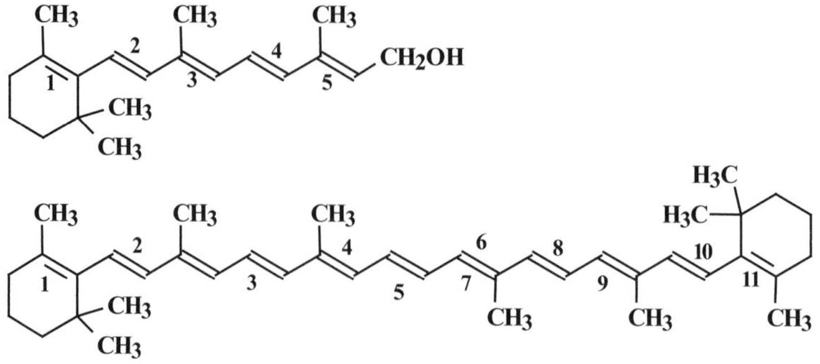

Fig. 13.75. Comparison of the conjugated systems in Vitamin A (top) and β-carotene (bottom).

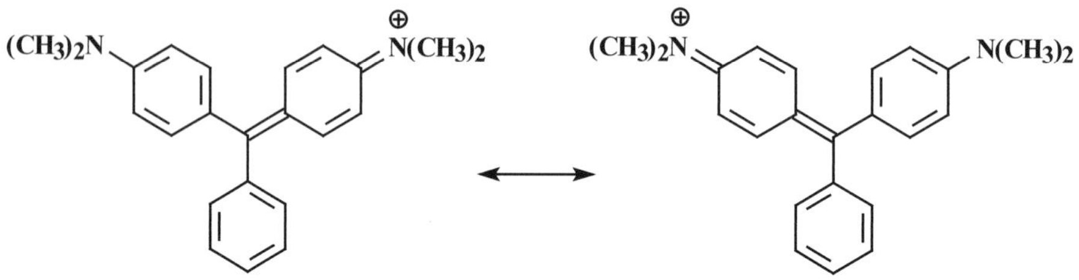

Fig. 13.76. Resonance structures for Malachite Green (C.I. Basic Green 4).

doubling the length of the conjugated system for Vitamin A to give β-carotene causes the λ_{max} value to shift from 325nm to 466 and 497nm.

Nitro Dyes

As the name suggests, this very small class of organic dyes has at least one nitro group as the chromophore. Nitro dyes invariably are yellow or orange and are important for their economical cost and good lightfastness. Examples include the dyes shown in Fig. 13.78–C.I. Acid Orange 3 (A), C.I. Disperse Yellow 42 (B), C.I. Acid Yellow 1 (C), and

C.I. Disperse Yellow 70 (D). A key disadvantage of nitro dyes is their low color strength (ϵ_{max} = 5000–7000). Improvements in color strength have been achieved by incorporating an azo group, as illustrated in dye D.

Representative syntheses are shown in Figs. 13.79 and 13.80. In the first example, C.I. Disperse Yellow 42 is prepared by condensing two molecules of aniline with one molecule of 4-chloro-3-nitrobenzenesulfonyl chloride, using ethanol as the solvent. In the second example, C.I. Acid Orange is prepared in a 3-step synthesis, starting from 2-chloro-5-nitrobenzenesulfonic acid.

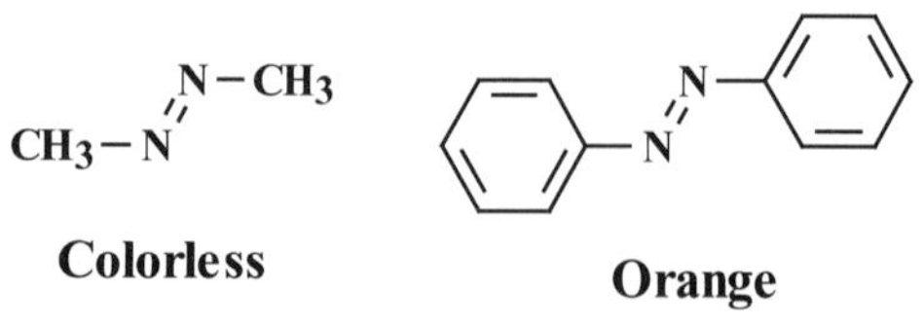

AZO DYES

Azo dyes are by far the largest family of organic dyes. They play a prominent role in acid, direct, reactive, azoic, and disperse dye structures, as shown previously, and include structures that cover the full color spectrum. Generally, the synthesis of azo dyes involves a process known as diazo coupling. In this

Colorless

Orange

Fig. 13.77. Impact of having a chromophore apart from (left) or part of (right) a conjugated system.

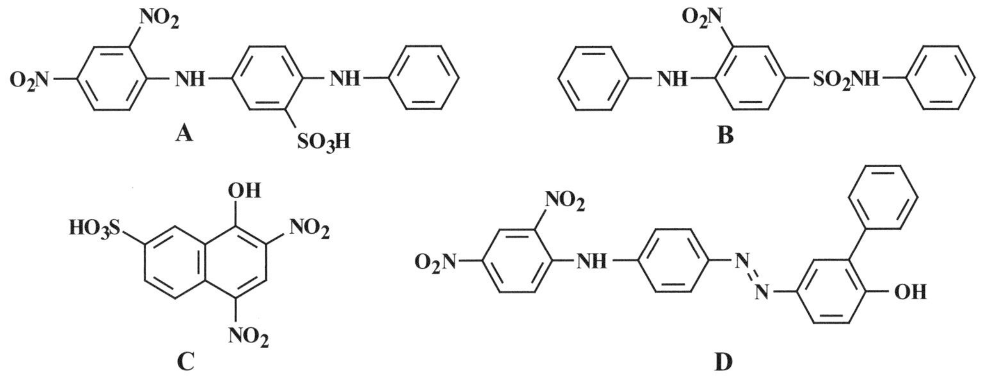

Fig. 13.78. Examples of nitro dye structures.

process, a diazotized aromatic amine is coupled to a phenol, naphthol, aromatic amine, or a compound that has an active methylene group, as illustrated in the two-step synthesis in Fig. 13.81. Step 1 is the conversion of aniline to benzenediazonium chloride, a process known as diazotization, and step 2 is the reaction of the diazo compound with phenol to produce the corresponding azo dye, a process known as diazo coupling.

Diazotizations are normally conducted in an aqueous medium containing nitrous acid, generated *in situ* from HCl + NaNO$_2$, and a primary aromatic amine. When weakly basic or heteroaromatic amines are used in azo dye synthesis, H$_2$SO$_4$ is often used as the reaction medium, forming H(NO)SO$_4$ (nitrosylsulfuric acid) as the diazotizing agent.[54] The stoichiometry associated with this reaction is given in Fig. 13.82, and although only 2 moles of acid per mole amine are required, in practice 2.2 to 2.5 moles are used. Diazotizations are most often conducted at 0–10°C because the resultant diazo compounds are usually unstable at higher temperatures.

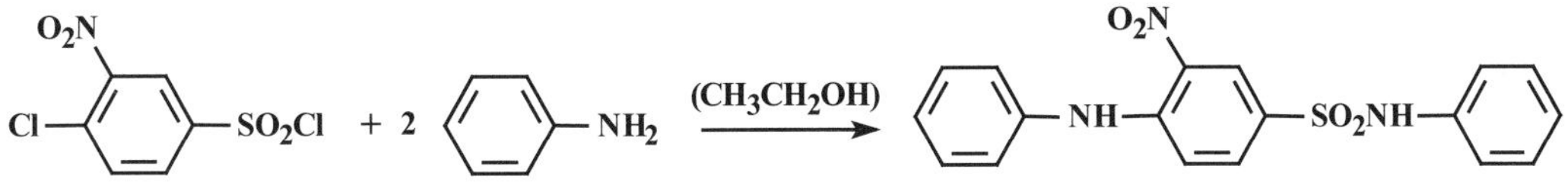

Fig. 13.79. Synthesis of Disperse Yellow 42.

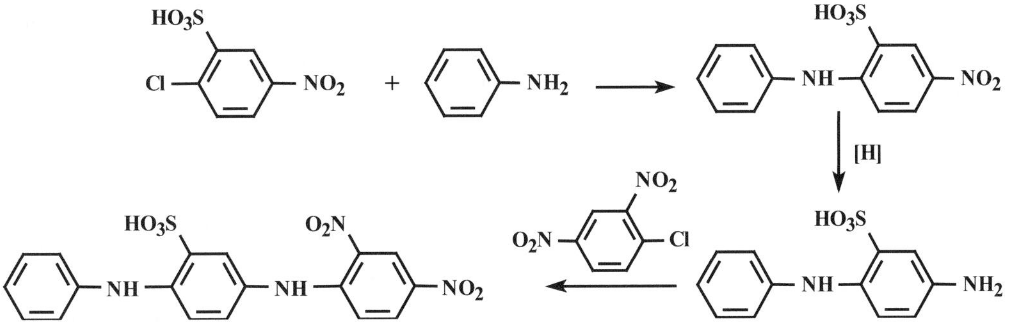

Fig. 13.80. Synthesis of Acid Orange 3.

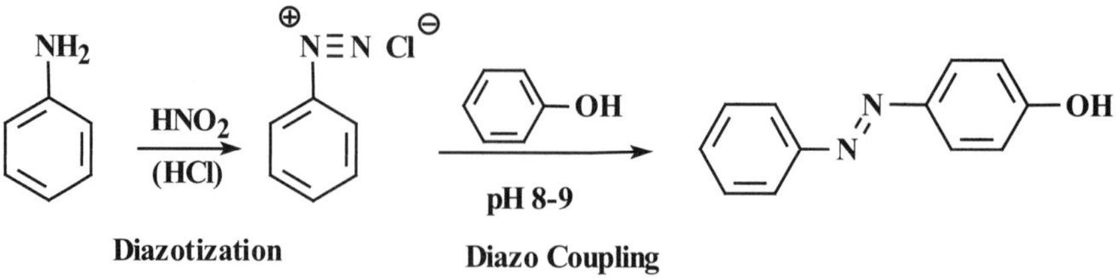

Fig. 13.81. Two-step synthesis of an azo dye from aniline and phenol.

Fig. 13.82. Summary of the diazotization process.

Examples of aromatic amines that can be diazotized are shown in Fig. 13.83. This extremely abbreviated list is designed to show that a wide variety of amines can be used, including hydrophobic, weakly basic, hydrophilic, and heterocyclic compounds. *ortho*-Diamines are not typically used because of their propensity to undergo triazole formation (Fig. 13.84).

Examples of compounds that can be used as coupling components in azo dye synthesis are shown in Figs. 13.85 to 13.87. The first group is comprised of phenols and naphthols, the second group is comprised of amines that couple, and the third contains couplers that have an active methylene group (see Fig. 13.87). Compounds in the first and third groups require ionization using alkali, to give sufficient reactivity for diazo coupling, and

the pH employed is usually 8–9. Because aromatic amines are appreciably more reactive, they couple at pH 5–6. Arrows have been used to indicate the coupling positions for the various couplers. Compounds such as 1-naphthol or 1-naphthylamine give a mixture of monoazo dyes by coupling in the 2-position or the 4-position. When couplers containing –OH and –NH$_2$ groups are employed (see Fig. 13.85), coupling may occur twice, giving disazo dyes. In such cases, coupling is first conducted in acid, *ortho* to the –NH$_2$ group, and then in alkali. This is important because the introduction of the first azo group decreases the reactivity of the coupler. The ability to ionize the –OH group provides sufficient ring activation for the second coupling. In the case of gamma acid, one has the lone option of coupling under acidic or alkaline conditions.

When primary amines are used as couplers, coupling can occur on the ring or at the amino group itself unless the amino group is blocked. One good way to block this group is by converting it to the *N*-sulfomethyl group, as illustrated in Fig. 13.86. The products formed are also known as omega salts.[55] The

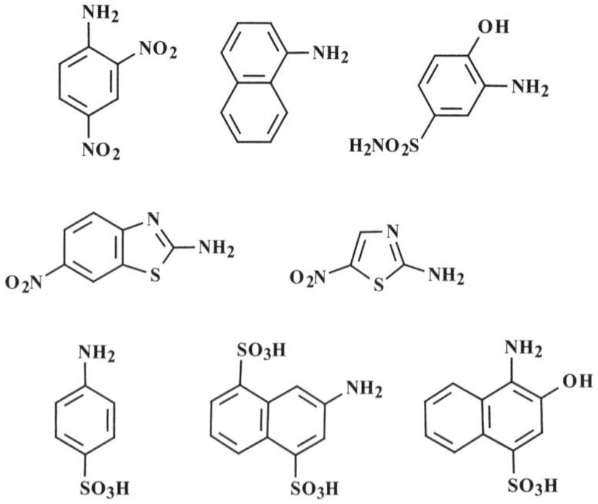

Fig. 13.83. Representative aromatic amines used in azo dye synthesis.

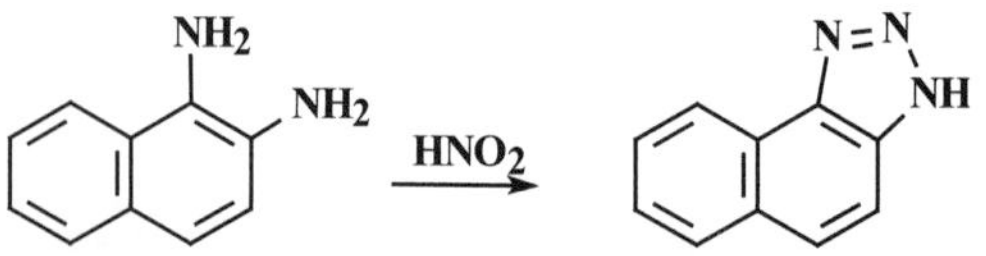

Fig. 13.84. Triazole formation from the diazotization of an *ortho*-diamine.

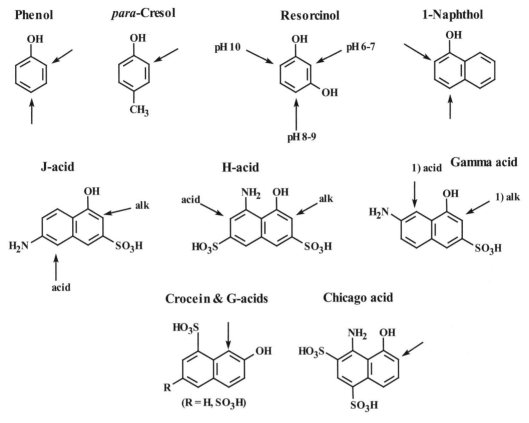

Fig. 13.85. Examples of phenol and naphthol-based couplers.

blocking group can be removed following the coupling step, by treating the resultant azo dye with an alkaline solution.

There are also important examples of phenolic compounds that do not couple (see Fig. 13.88). In these examples, the required coupling positions are blocked, the ring is too deactivated, or the compounds undergo oxidation in the presence of the diazo compound.

The synthesis of azo dyes can be illustrated using the following letter designations:

- **A** = Diazotizable amine
- **D** = Tetrazotizable diamine
- **E** = Coupler that couples once
- **M** = 1° Amine that couples once and is diazotized and coupled again
- **Z** = Coupler that couples twice
- **Z · X · Z** = Binuclear coupler that couples twice

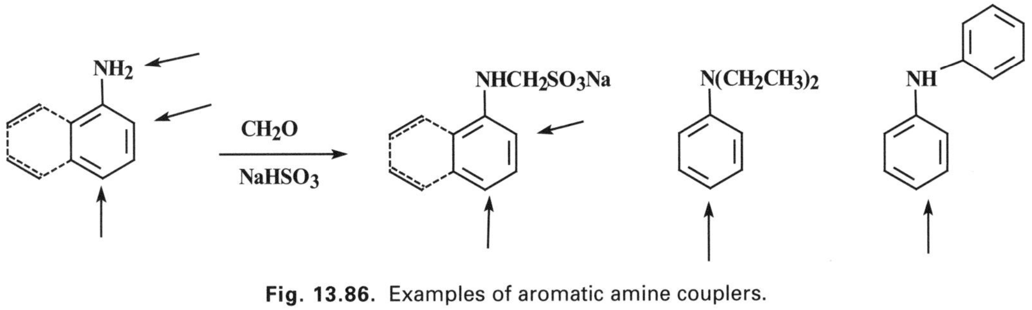

Fig. 13.86. Examples of aromatic amine couplers.

Acetoacetanilide

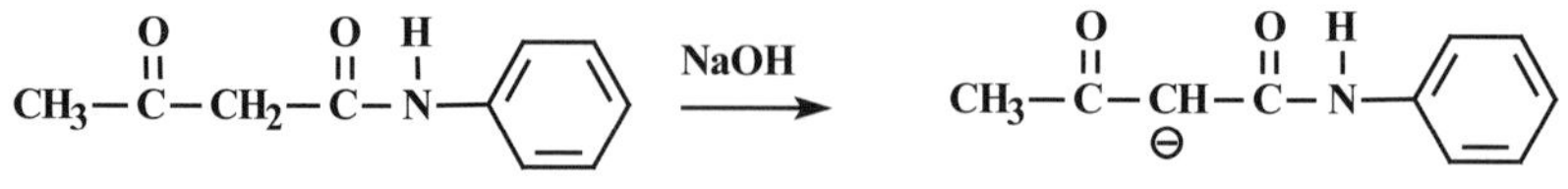

An *N*-Methylphenylpyrazalone 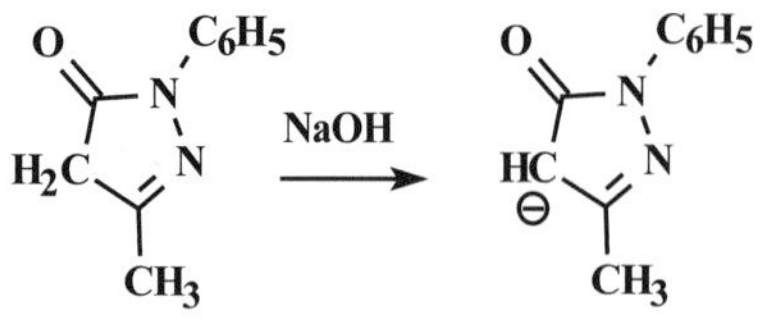   **A Pyridone Compound**

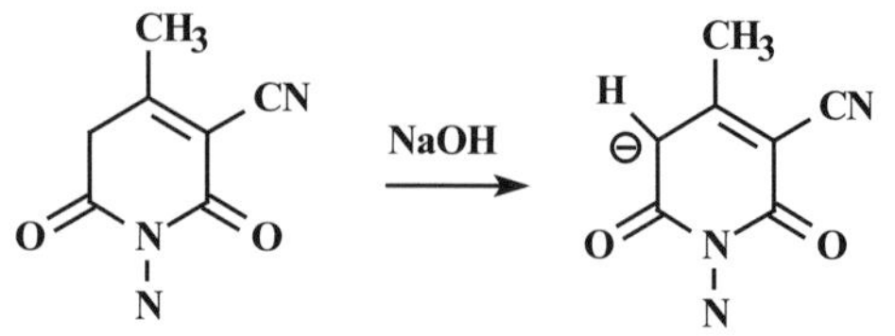

Fig. 13.87. Examples of couplers having an active methylene (–CH$_2$–) group.

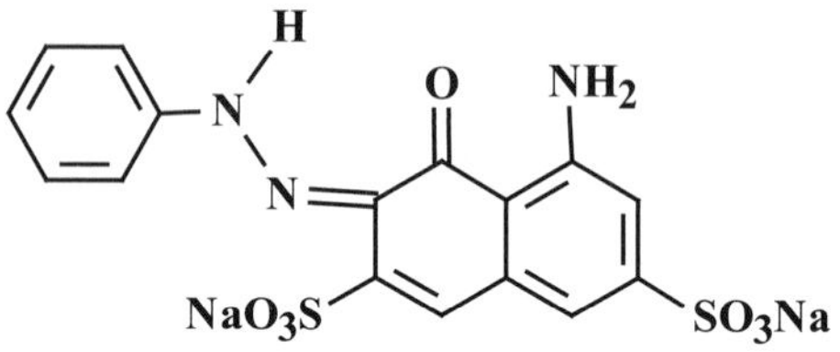

Acid Red 33

These designations are used to provide an indication of how a given dye has been assembled, and will be used in describing the azo dye syntheses covered in the subsections that follow.

Monoazo Dyes

Azo dyes of this type are manufactured predominantly by the reaction between a diazotized amine ("A") and a type "E" or "Z" coupler. The synthesis can be as simple as coupling diazotized aniline to H-acid, in an A → E process, to produce C.I. Acid Red 33. An example of a reactive dye that is manufactured via an A → E process is C.I. Reactive Red 1. In this case, the target dye is manufactured as shown in Fig. 13.89, which shows that the reactive group can be introduced prior to (sequence 1) or after (sequence 2) the

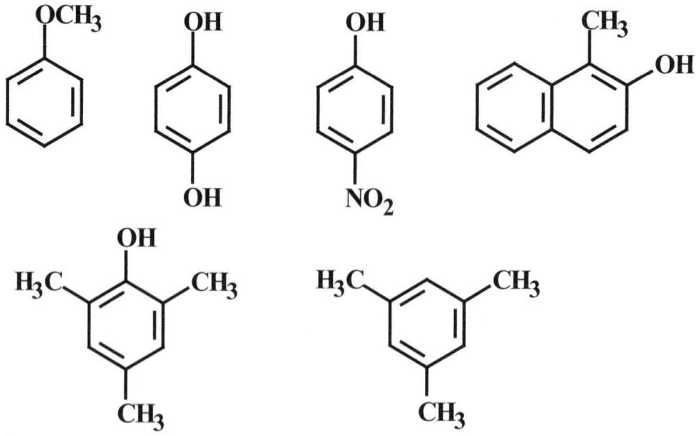

Fig. 13.88. Examples of compounds that do not couple.

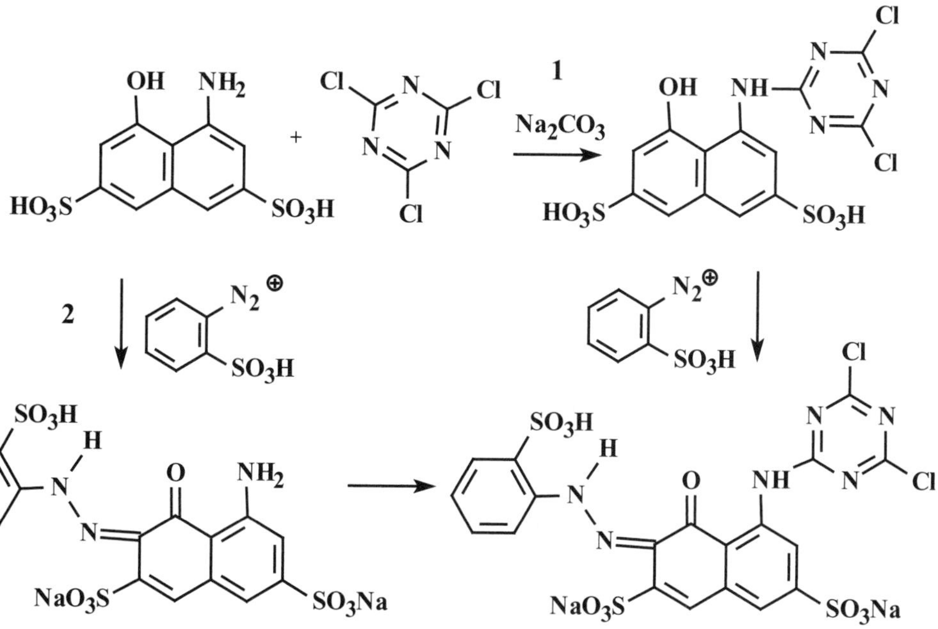

Fig. 13.89. Two approaches to the synthesis of Reactive Red 1.

coupling step. Similarly, monoazo bireactive dyes are made by this process (Fig. 13.90). This illustrates that a quite complex arylamine can be used as the diazo compound.

Other examples of monoazo dyes that are synthesized via an A → E process are shown in Fig. 13.91, further illustrating the wide range of structural types that can be manufactured this way.

Disazo Dyes

There are four often-used methods for synthesizing dyes containing two azo linkages,

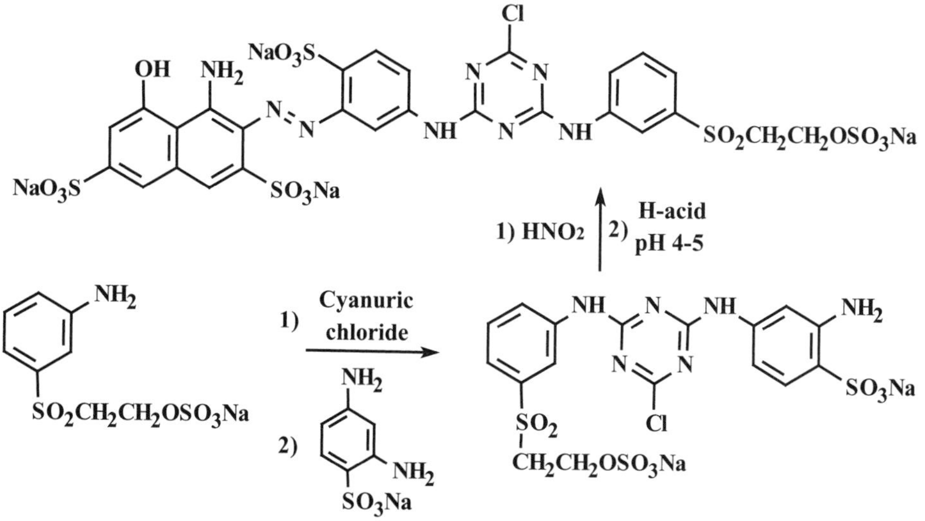

Fig. 13.90. Synthesis of a monoazo dye containing two different reactive groups.

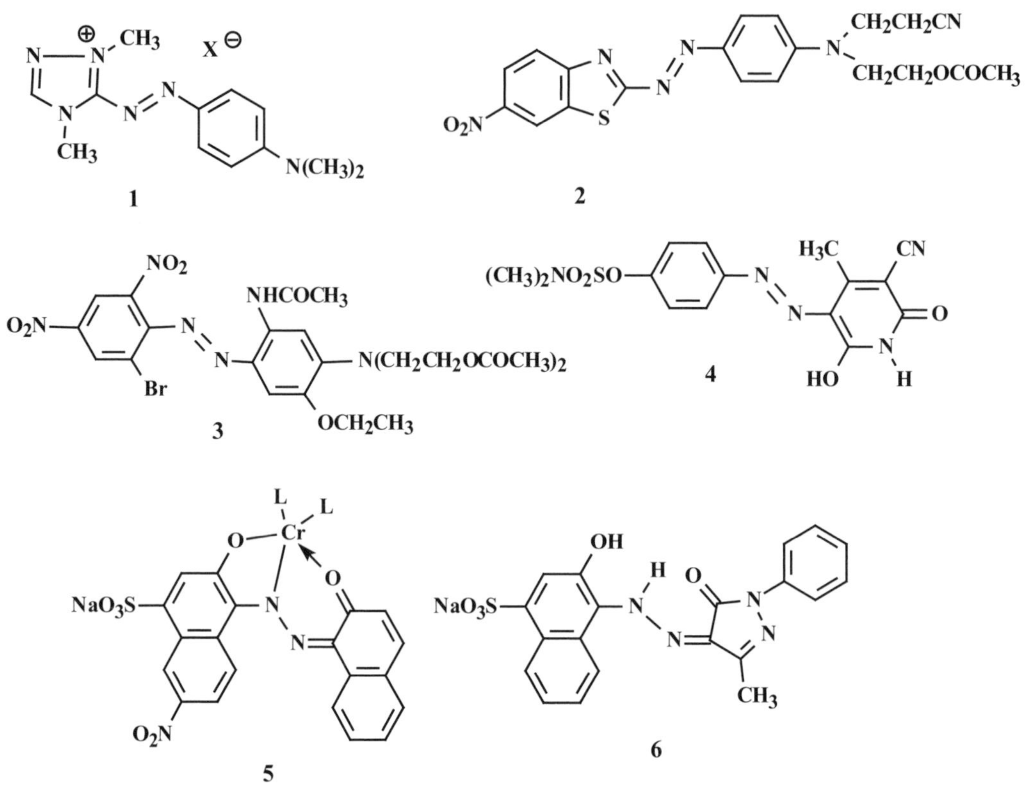

Fig. 13.91. Monoazo cationic (**1**), disperse (**2-4**), 1:1 chromium complexed (**5**), and mordant (**6**) dyes prepared via an A → E process.

each of which requires two diazo coupling reactions. A nontraditional "disazo" dye involves 1:2 metal complex formation.

Type $A^1 \rightarrow Z \rightarrow A^2$ Synthesis. Dyes of this type include those shown in Fig. 13.92 (C.I. Acid Black 1 (**7**), C.I. Mordant Brown 1 (**8**), and C.I. Acid Black 17 (**9**), C.I. Direct Orange 18 (**10**)), in which couplers such as H-acid, resorcinol, and *meta*-phenylenediamine are coupled twice. Although A^1 and A^2 are different in the present examples, they need not be different. As pointed out above, coupling *ortho* to the amino group of H-acid is usually conducted first, under weakly acidic conditions, followed by coupling with diazotized aniline under alkaline conditions. This is also true for the structurally similar dye **9**, which is prepared from S-acid. In the case of dye **10**, however, coupling with aniline under slightly acidic conditions is the second step. For dye **8**, both couplings are

conducted under acidic conditions, with 2-amino-4-nitrophenol introduced first.

Type $E^1 \rightarrow D \rightarrow E^1$ Synthesis. Dyes of this type require the conversion of an aryldiamine to a tetrazonium compound (one that has two diazonium groups), in a process know as tetrazotization. See Fig. 13.93, which involves environmentally friendly alternatives to benzidine. Following tetrazotization, one tetrazonium molecule reacts with two coupler molecules to produce the target dye, examples of which are provided in Fig. 13.94 (**11**: C.I. Direct Red 28 (Congo Red), **12**: C.I. Direct Yellow 12, and **13**: C.I. Acid Yellow 42). Disazo dyes prepared this way include dye **14** (C.I. Direct Blue 15), which is converted to the important bis-copper complex, C.I. Direct Blue 218 (see Fig. 13.95).

Type A → M → E Synthesis. This is one of the largest groups of disazo dyes, as they

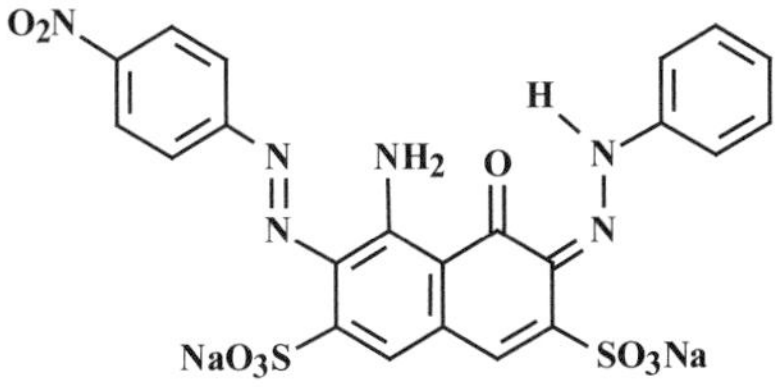

Fig. 13.92. Disazo dyes prepared by coupling twice to H-acid (**7**), *meta*-phenylenediamine (**8**), S-acid (**9**), and resorcinol (**10**).

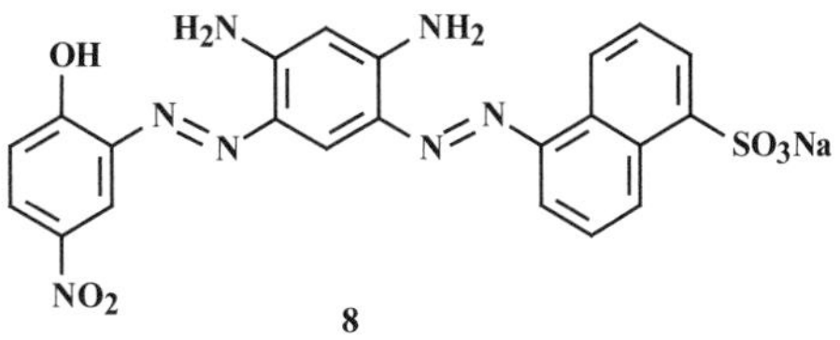

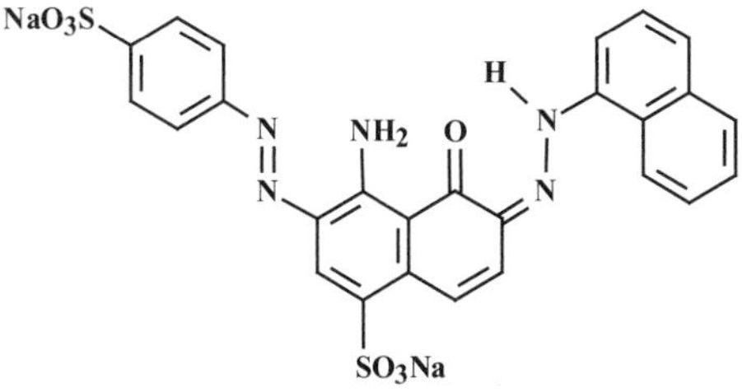

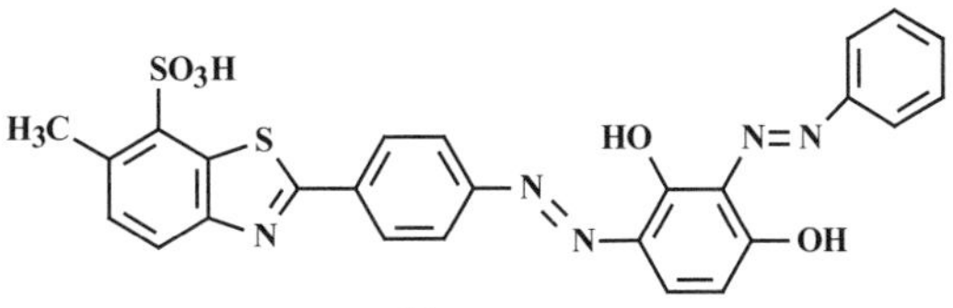

include acid, disperse, direct, and reactive dye structures. A representative synthesis is shown in Fig. 13.96. The second diazotization and coupling steps can be conducted inside certain textile fibers. For instance, disperse black dyes are produced in the presence of cellulose acetate by conducting the chemistry shown in Fig. 13.97 after dyeing cellulose acetate with the monoazo dye.

Examples of dyes made via an A → M → E synthesis are shown in Fig. 13.98. Although most azo disperse dyes are based on monoazo structures, disazo structures such as **15** (C.I.

Disperse Orange 13) and **16** (C.I. Disperse Orange 29) are manufactured. An important direct dye of this type is **17** (C.I. Direct Red 81), a reactive dye is **18** (C.I. Reactive Blue 40), and acid dyes include **19** (C.I. Acid Red 151) and **20** (C.I. Acid Blue 116).

Type $A^1 \rightarrow Z \cdot X \cdot Z \leftarrow A^1$ Synthesis. Disazo dyes of this type are produced from coupling twice to dye intermediates such as those shown in Fig. 13.99, and are largely direct dyes for cotton. A representative synthesis is shown in

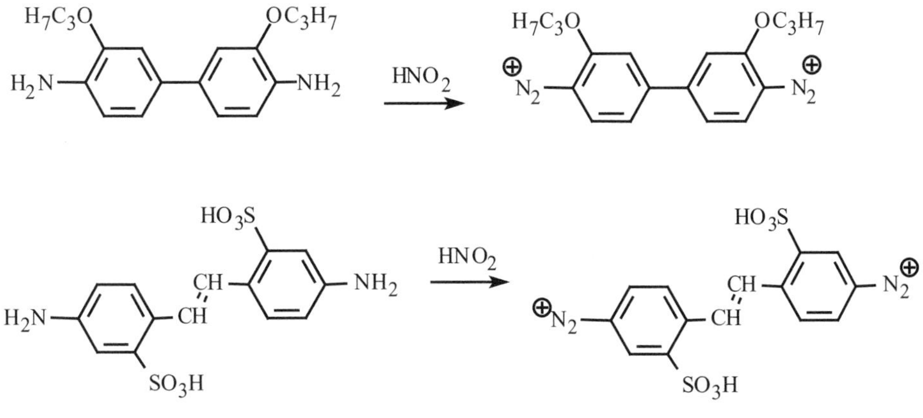

Fig. 13.93. Tetrazotization of a di-n-propoxybenzidine (top) and a diaminostilbene disulfonic acid (bottom).

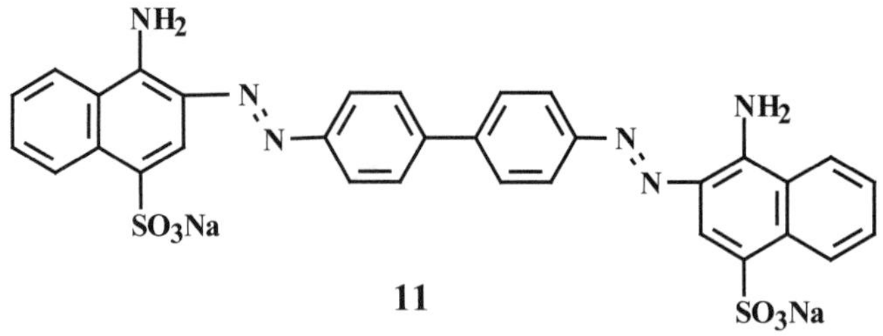

Fig. 13.94. Examples of disazo dyes (**11–13**) prepared via a type $E^1 \leftarrow D \rightarrow E^1$ synthesis.

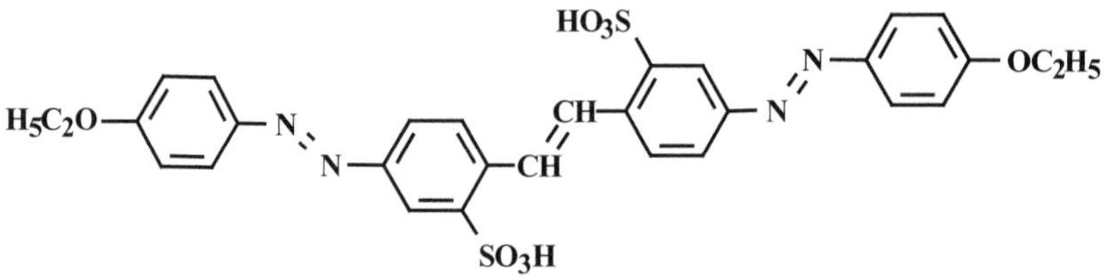

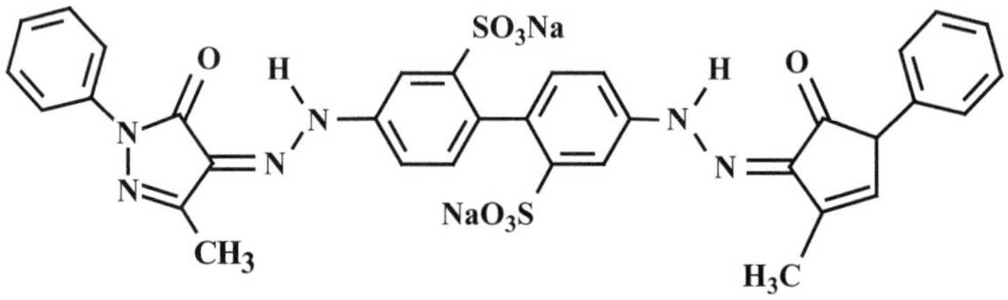

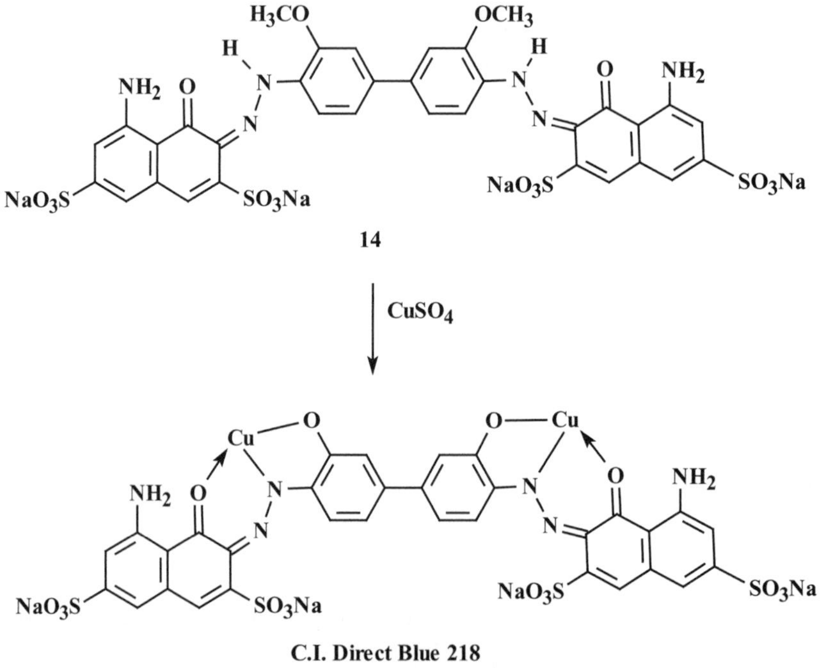

C.I. Direct Blue 218

Fig. 13.95. Formation of Direct Blue 218.

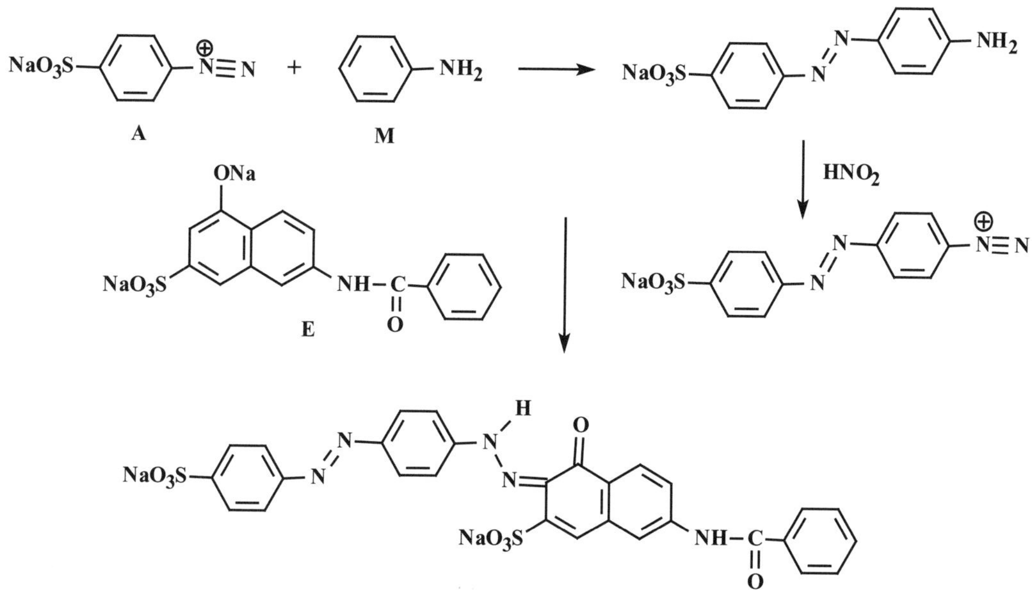

Fig. 13.96. Disazo dye synthesis via an A → M → E process.

Fig. 13.100), for C.I. Direct Red 83. In this case the target dye is prepared by metallization after the coupling step.

Disazo dyes such as C.I. Direct Yellow 44 are prepared according to the sequence shown in Fig. 13.101. In this example, a pair of monoazo dyes is reacted with phosgene.

1:2 Metal Complexes. Although somewhat different from the previous examples and methods, dyes containing two azo groups can also be synthesized by forming 1:2 metal complexes of suitably substituted monoazo dyes. The resultant

dyes are mostly acid dyes for protein and polyamide substrates and the metals employed are Cr, Co, and Fe. Examples shown in Fig. 13.102 are for C.I. Acid Black 172 (**21**) and C.I. Acid Yellow 151 (**22**). In these examples, the corresponding monoazo dye is treated with one-half the molar amount of $Cr_2(SO_4)_3$ or $CoCl_3$, respectively.

Polyazo Dyes

In this section, we cover the synthesis of dyes containing three or more azo linkages. In

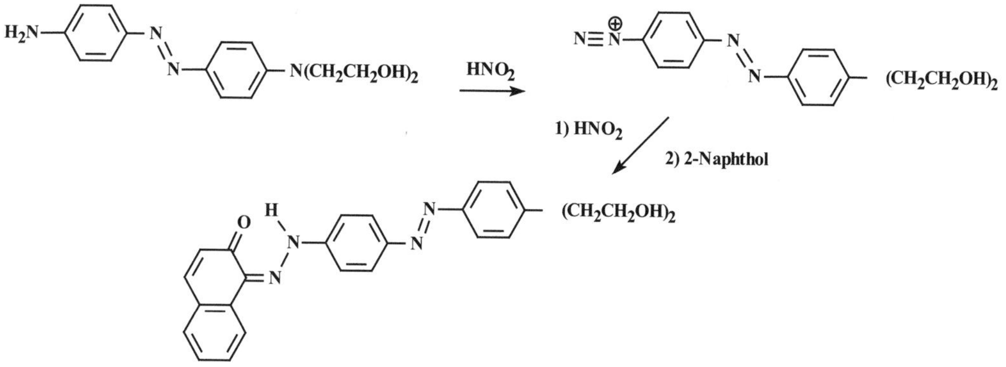

Fig. 13.97. Disazo disperse black dye synthesis conducted inside cellulose acetate fibers.

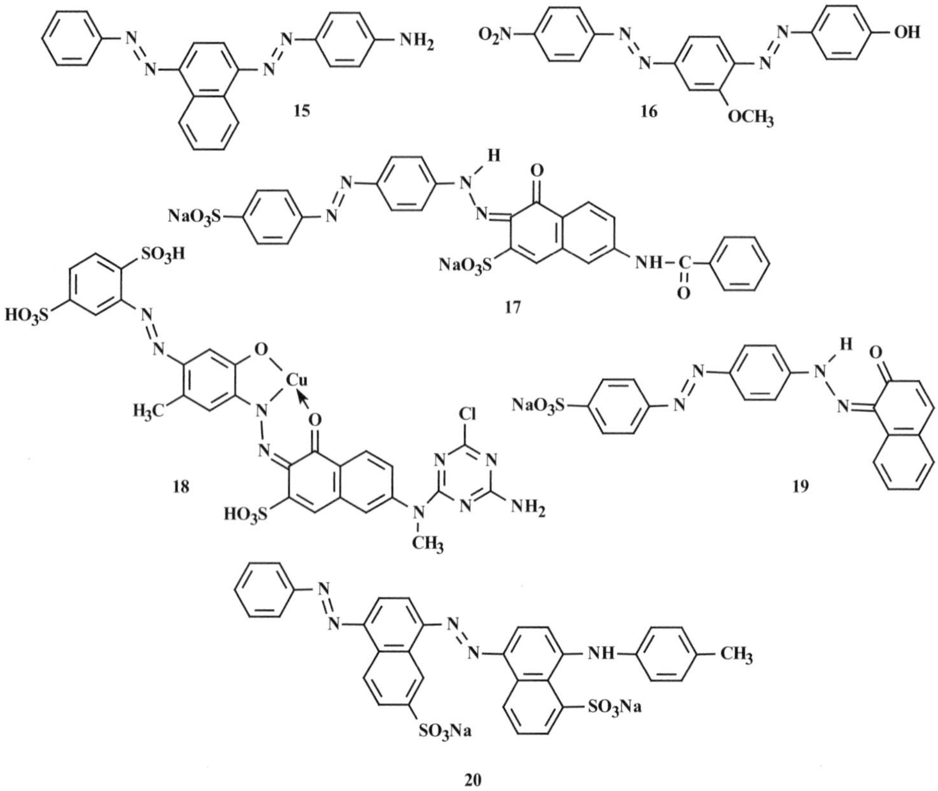

Fig. 13.98. Examples of disazo dyes prepared by a type A → M → E synthesis.

this regard, methods for producing trisazo dyes (those having three azo linkages) include E ← D → Z ← A and A → M^1 → M^2 → E syntheses. Examples are shown in Fig. 13.103 for C.I. Acid Black 234 (**23**) and C.I. Direct Blue 71 (**24**). In the synthesis of dye **23**, an unsymmetrical dye can be made from

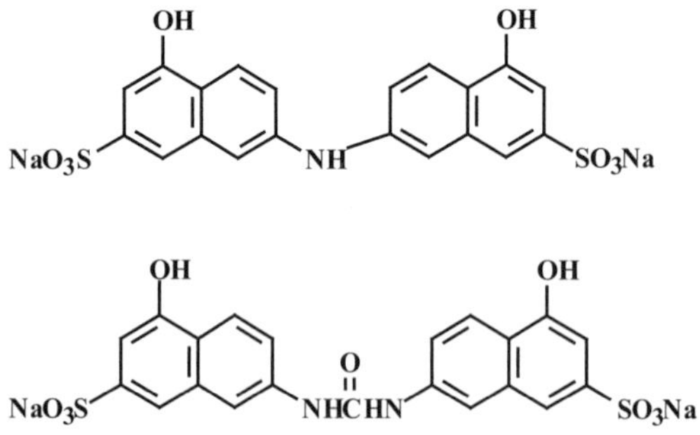

Fig. 13.99. Structures of J-acid imide (top) and J-acid urea (bottom).

diamine **25** because the end of the tetrazonium compound (cf. **26**) that is *para* to the –SO_2 moiety is more reactive than the one that is *para* to the –NH moiety (Fig. 13.104).

Dyes containing four azo linkages are direct dyes for cotton and can be prepared in several ways, including via A → M → Z ← D → E, A^1 → Z^1 ← D → Z^2 ← A^2, E^1 ← M^1 ← D → M^2 → E^2, E^1 ← D^1 → Z ← D^2 → E^2, and E^1 ← D → M^1 → M^2 → E^2 sequences. Examples of the second and third methods are shown in Fig. 13.105. Note that both are symmetrical molecules, the first of which (C.I. Direct Brown 44) employs *meta*-phenylenediamine as a type "Z" coupler and a type "D" diazo component. In the second example (C.I. Direct Black 22), gamma acid is twice used as the "M" moiety, and the dye is synthesized by (1) coupling tetrazotized benzidine disulfonic acid to two molecules of gamma acid, (2) diazotizing the amino groups

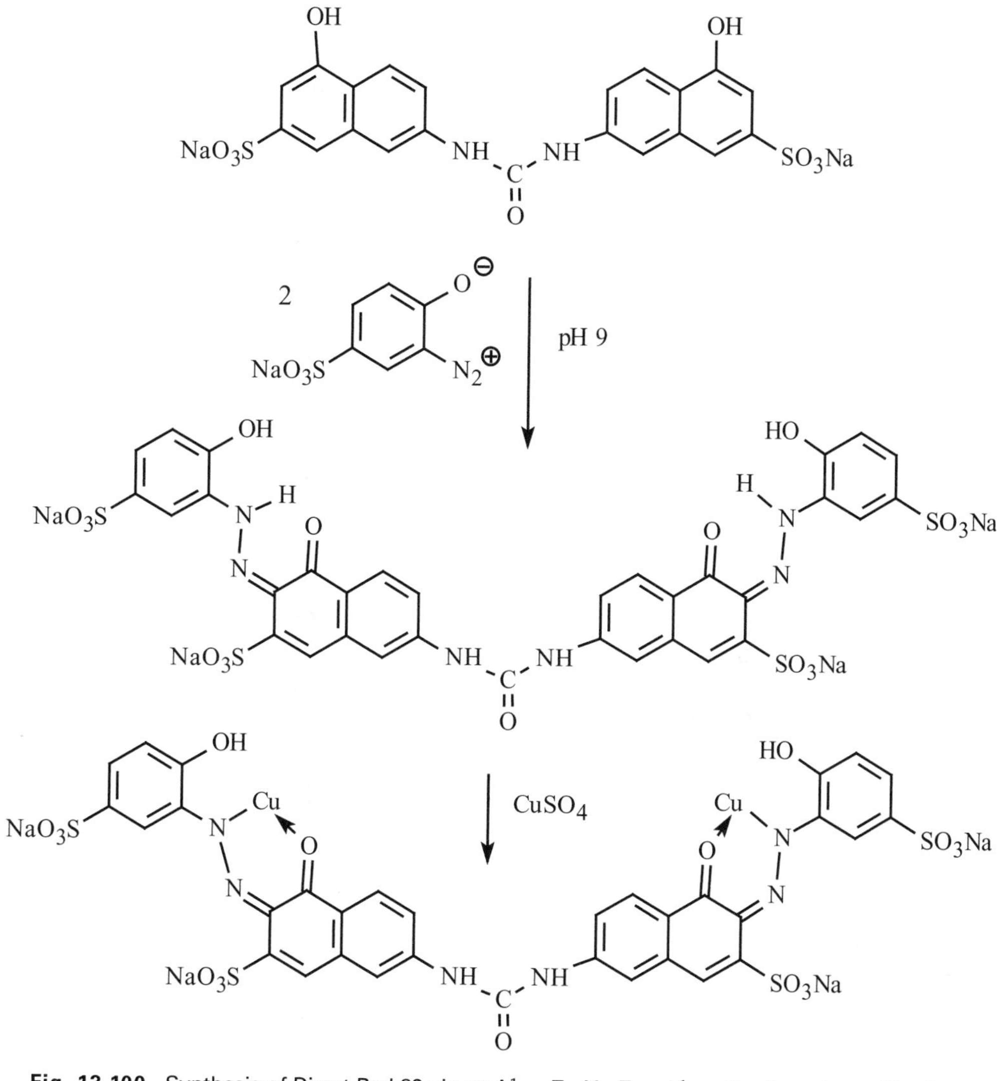

Fig. 13.100. Synthesis of Direct Red 83 via an $A^1 \rightarrow Z \cdot X \cdot Z \leftarrow A^1$ synthesis and metallization.

on the gamma acid moieties, and 3) coupling to two molecules of *meta*-phenylenediamine.

TRIPHENYLMETHANE DYES

Triphenylmethane dyes are usually prepared in two steps: 1) condensation of an *N,N*-dialkylaniline with a benzaldehyde compound and 2) oxidation of the resultant leuco base (**27**). The synthesis of C.I. Basic Green 4 (Malachite Green) is given as an example in Fig. 13.106. Alternatively, C.I. Acid Green 50 is prepared in three steps: 1) condensation of

N,N-dimethylaniline and *para*-(*N,N*-dimethylamino)benzaldehyde to produce Michler's hydrol (**28**), condensation with R-acid to give an intermediate leuco base (**29**), and (3) oxidation to give the target dye. Historically, PbO$_2$ has been used as the oxidizing agent. However, concerns about its toxicity have led to the use of a more environmentally friendly agent such as tetrachloro-*para*-benzoquinone (chloranil).

In another synthetic variation, C.I. Acid Violet 17 is prepared in the four steps shown in Fig. 13.107. The different steps in this

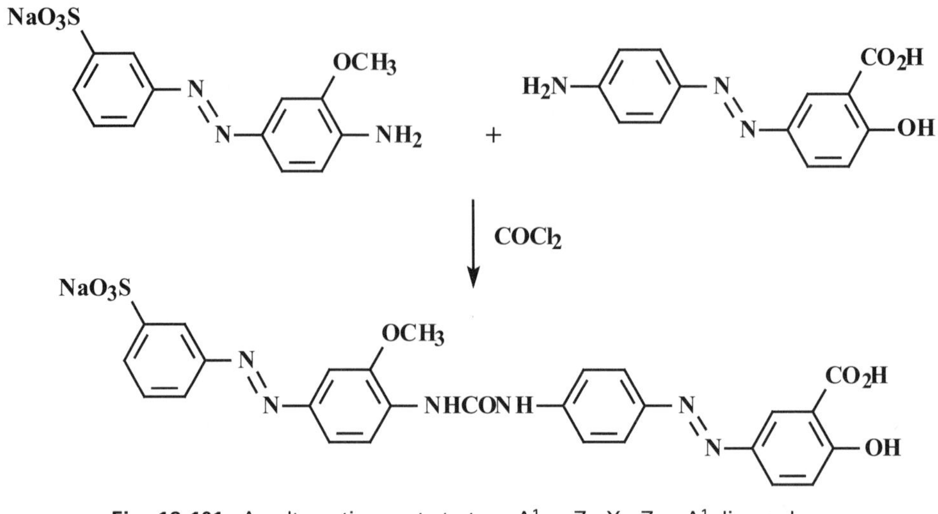

Fig. 13.101. An alternative route to type $A^1 \rightarrow Z \cdot X \cdot Z \leftarrow A^1$ disazo dyes.

process are the synthesis of the N-arylmethyl intermediate **30** and the diphenylmethane intermediate **31**. Oxidation to the intermediate hydrol and condensation with N,N-dimethylaniline produce the target dye.

Structurally related dyes are synthesized by condensing phenols with phthalic anhydride to give a colorless intermediate lactone (**32**) that reacts with alkali to give the colored form. An example of this dye type is phenolphthalein, the synthesis of which is shown in Fig. 13.108.

XANTHENE DYES

Like phenolphthalein, xanthene dyes are prepared in a condensation reaction involving phthalic anhydride. However, resorcinol is employed instead of phenol. The simplest representative of this family is C.I. Acid Yellow 73 (fluorescein), which is made via the sequence of steps shown in Fig. 13.109. Similarly, C.I. Acid Red 92 is made by the condensation of tetrachlorophthalic anhydride and resorcinol followed by bromination.

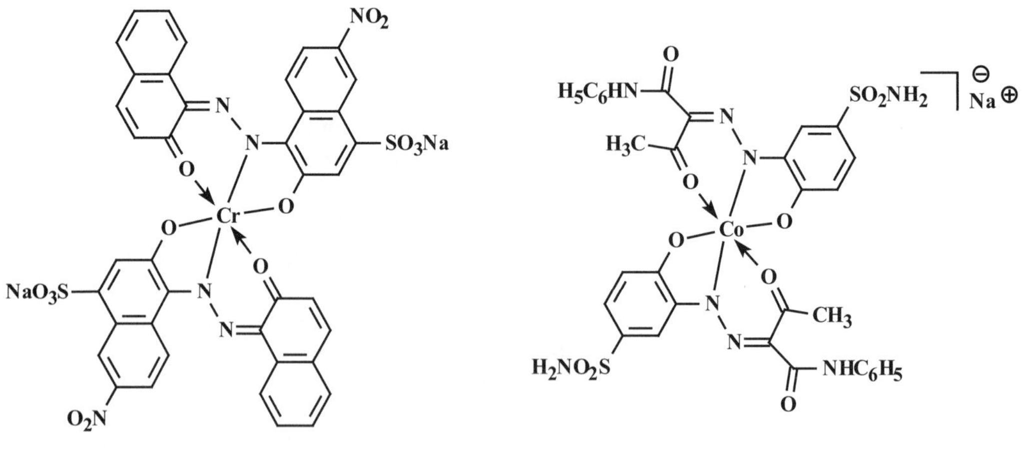

21   22

Fig. 13.102. Representative 1:2 bisazo metal complexed dyes.

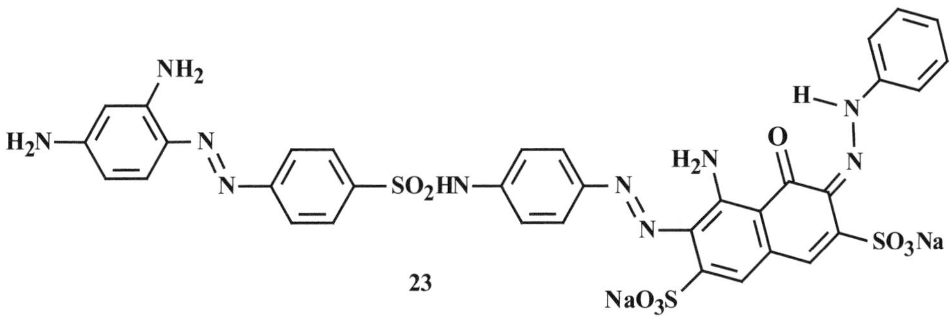

23

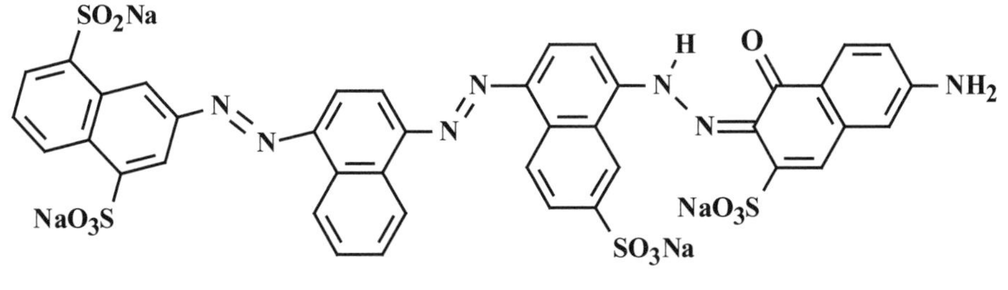

24

Fig. 13.103. Examples of trisazo dyes.

ANTHRAQUINONE AND RELATED DYES

The commercial preparation of anthraquinone dyes begins with the synthesis of anthraquinone itself. In this regard, the three-step synthesis involves: (1) the oxidation of naphthalene to phthalic anhydride, (2) Friedel–Crafts acylation of benzene to give a keto acid, and (3) cyclodehydration using H_2SO_4. See Fig. 13.110. The preparation of 1,4-disubstituted anthraquinones utilizes the intermediates

25

HNO$_2$

26

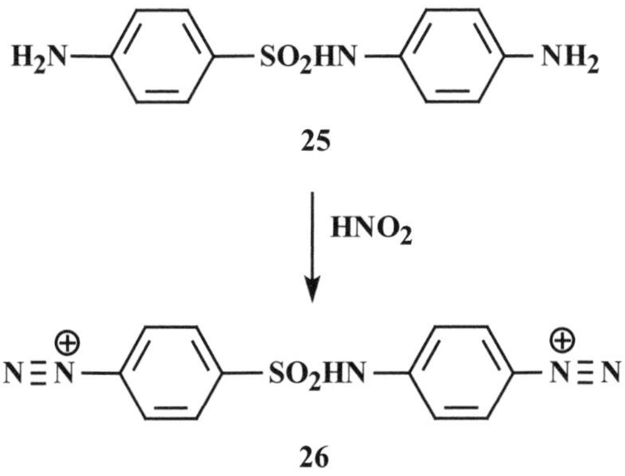

Fig 13.104. Structures of compounds **25** and **26**.

prepared in Fig. 13.111, where R = OH corresponds to quinizarin.

The reduction of quinizarin using sodium hydrosulfite produces leuco quinizarin, which, in turn, undergoes condensation with alkyl- or arylamines and reoxidation to produce blue and green disperse and solvent dyes. Although chemical oxidation can be used, air oxidation is normally sufficient. See steps "A" and "B" in Fig. 13.112 for the general reaction scheme. The use of boric acid in the reduction step follows the course outlined in Fig. 13.113, where the synthesis of C.I. Solvent Green 3 is given as an example.[56]

Anthraquinone Disperse Dyes

Examples of dyes prepared using the above methods are shown in Fig. 13.114. The C.I. disperse dyes Red 15, Violet 1, Blue 3, Violet 27, Blue 19, and Blue 23, are prepared from leucoquinizarin. When unsymmetrical dyes such as Disperse Blue 3 are made, the use of a mixture of two amines in the condensation step gives the corresponding symmetrical dyes as by-products. In this

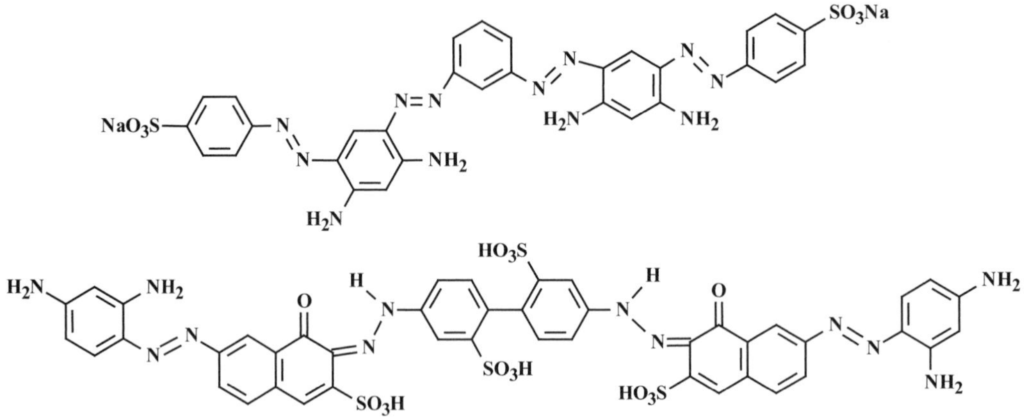

Fig. 13.105. Representative polyazo dyes prepared via the $A^1 \rightarrow Z^1 \leftarrow D \rightarrow Z^2 \leftarrow A^2$ (top) and $E^1 \leftarrow M^1 \leftarrow D \rightarrow M^2 \rightarrow E^2$ (bottom) methods.

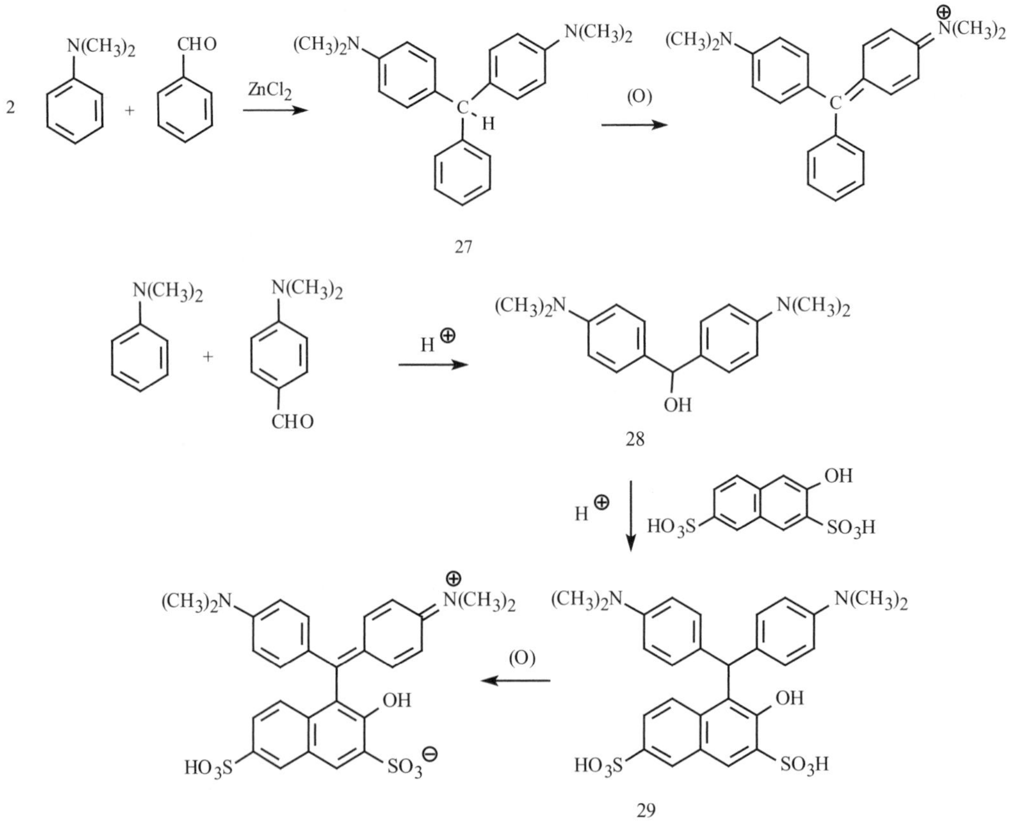

Fig. 13.106. Two-step synthesis of Malachite Green (Basic Green 4).
Three-step synthesis of Acid Green 50.

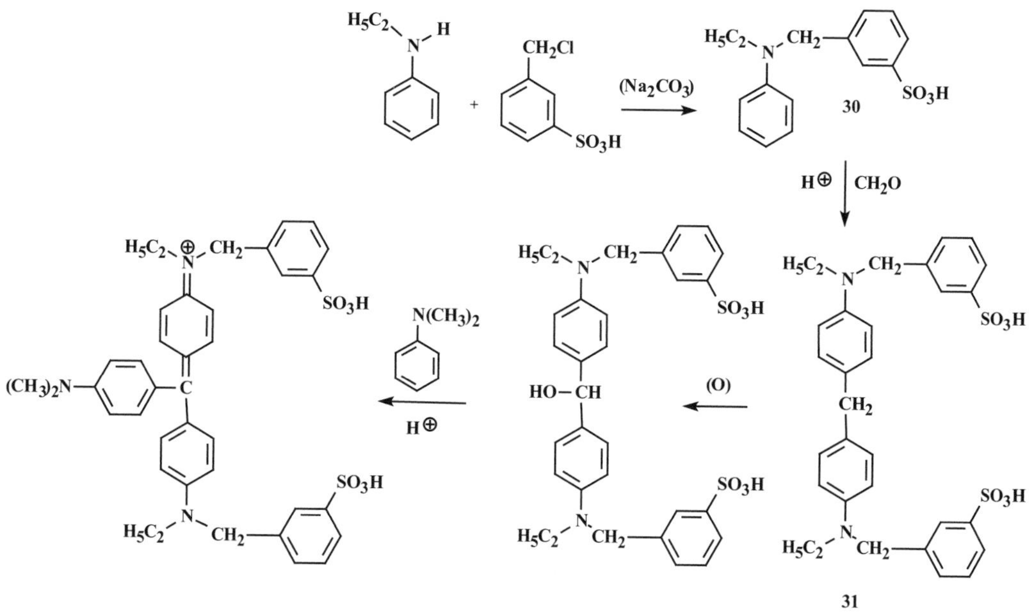

Fig. 13.107. Synthesis of Acid Violet 17 via diarylmethane intermediate **31**.

case, Disperse Blue 23 would be one of the byproducts.

The synthesis of Disperse Red 4 employs the dibromoanthraquinone intermediate **33**, which is hydrolyzed to compound **34** and converted to the target dye upon alcoholysis. See Fig. 13.115. The synthesis of Disperse Violet 26 is conducted in two steps: (1) chlorination of Disperse Violet 1 in the 2, 3-positions using SO_2Cl_2 and (2) condensation with phenol.

Disperse dyes containing substituents in both anthraquinone rings are often prepared from dinitroanthrarufin (DNA) and dinitrochrysazin (DNC), the structures of which are shown in Fig. 13.116. Examples of these dyes are C.I. Disperse Blue 56 and Blue 77. The former dye is made by reduction of

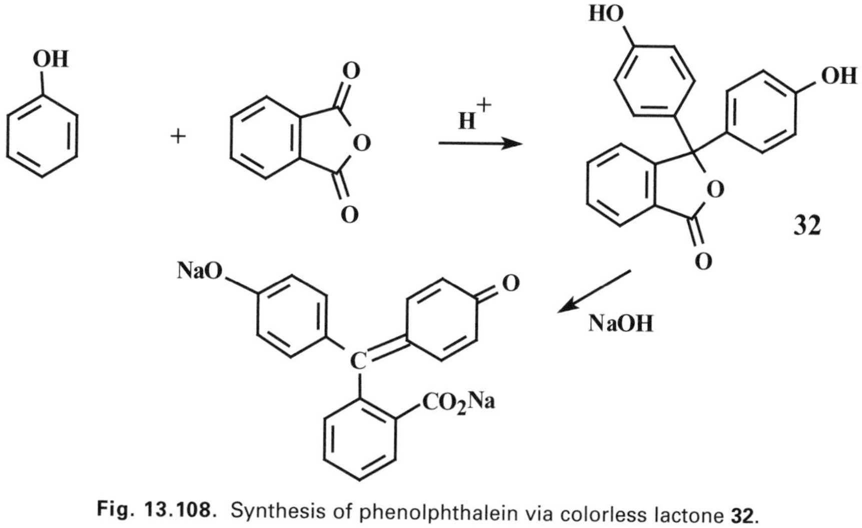

Fig. 13.108. Synthesis of phenolphthalein via colorless lactone **32**.

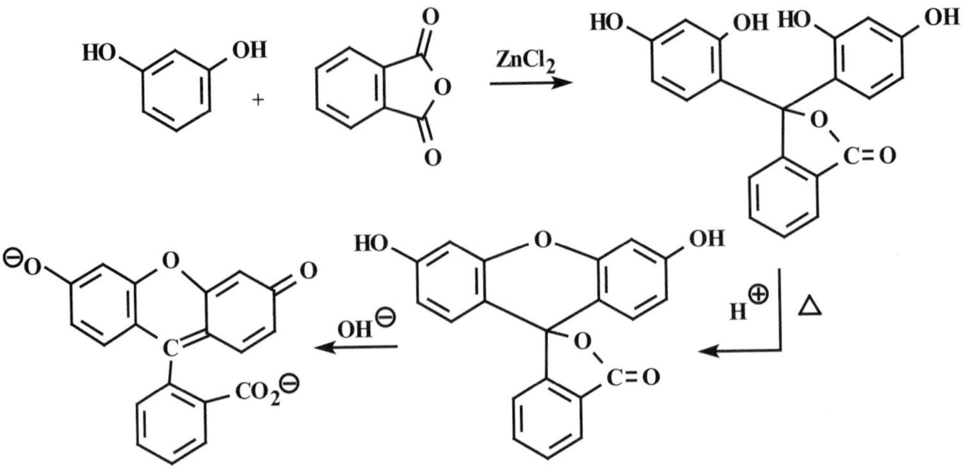

Fig. 13.109. Synthesis of Acid Yellow 73, a xanthene dye.

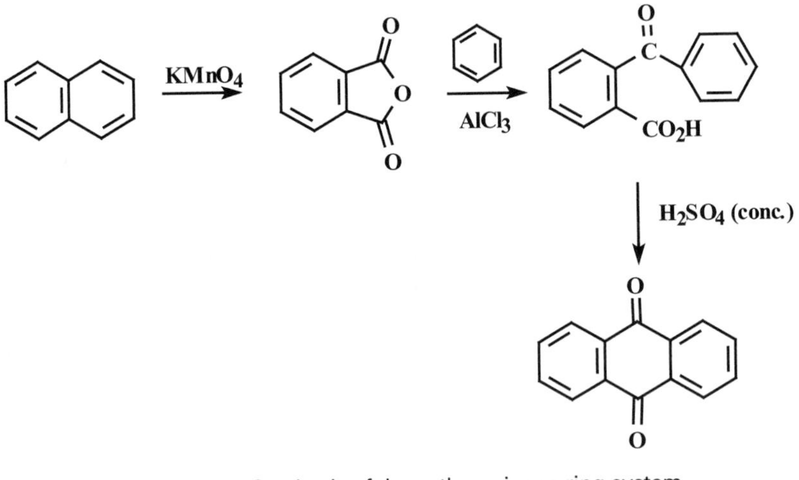

Fig. 13.110. Synthesis of the anthraquinone ring system.

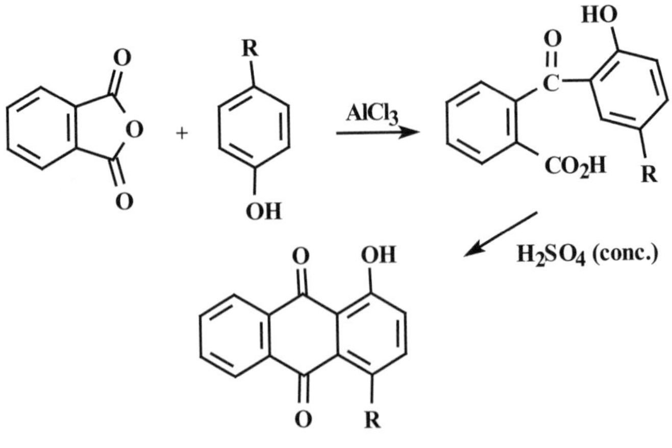

Fig. 13.111. Synthesis of key substituted anthraquinone intermediates, where R = Cl, OH.

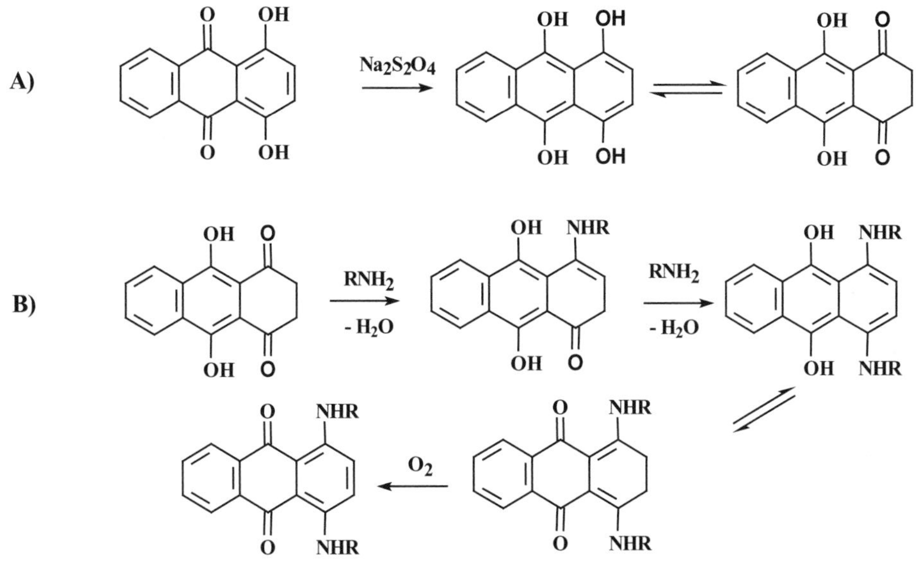

Fig. 13.112. General reaction scheme for the synthesis of 1,4-diaminoanthraquinone dyes.

dinitroanthrarufin followed by bromination, and the latter is made by condensing aniline with DNC. The DNC condensation shows that nitro groups in the α-position can be displaced like a halogen.

The dichlororinated precursor for Disperse Violet 26 can be used to make turquoise blue dyes such as C.I. Disperse Blue 60, as shown in Fig. 13.117. In this sequence, the chloro groups are replaced by cyano groups, using NaCN, and the resultant intermediate (35) is hydrolyzed to give the corresponding imide (36), which in turn is alkylated to give the target dye.

Anthraquinone Acid Dyes

A key intermediate in the synthesis of anthraquinone acid dyes is bromamine acid.

This compound is made via the sequence shown in Fig. 13.118. Acid dyes made from this intermediate include C.I. Acid Blue 25 and C.I. Acid Blue 40, and C.I. Acid Blue 127.

The synthesis of C.I. Acid Blue 127 takes place according to the route shown in Fig. 13.119. A key step in the synthesis is the formation of diamine 37, which is produced in two steps from N-sulfomethylaniline: (1) condensation with acetone and (2) hydrolysis to remove the protecting group. At this point, one molecule of diamine 37 is condensed with two molecules of bromamine acid to form the dye.

Another important dye is C.I. Acid Green 25. This dye is made by the sulfonation of C.I. Solvent Green 3 (Fig. 13.113). Because the

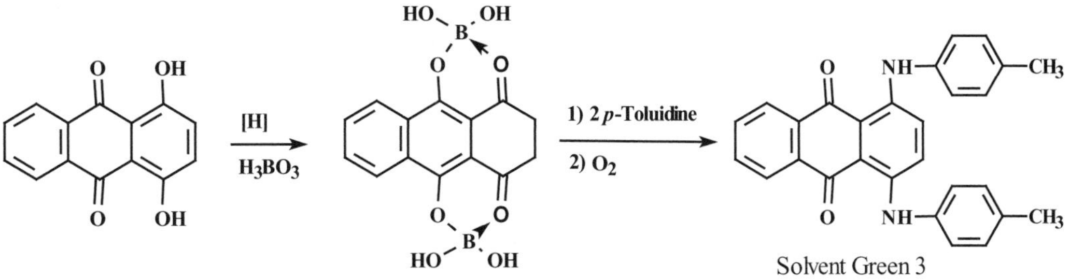

Fig. 13.113. Boric acid catalyzed synthesis of Solvent Green 3.

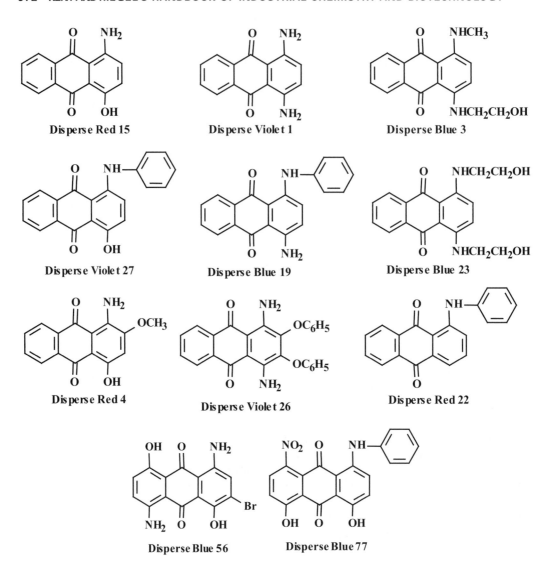

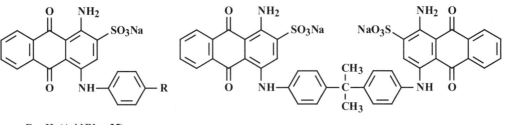

Fig. 13.114. Representative 1,4-disubstituted anthraquinone dyes.

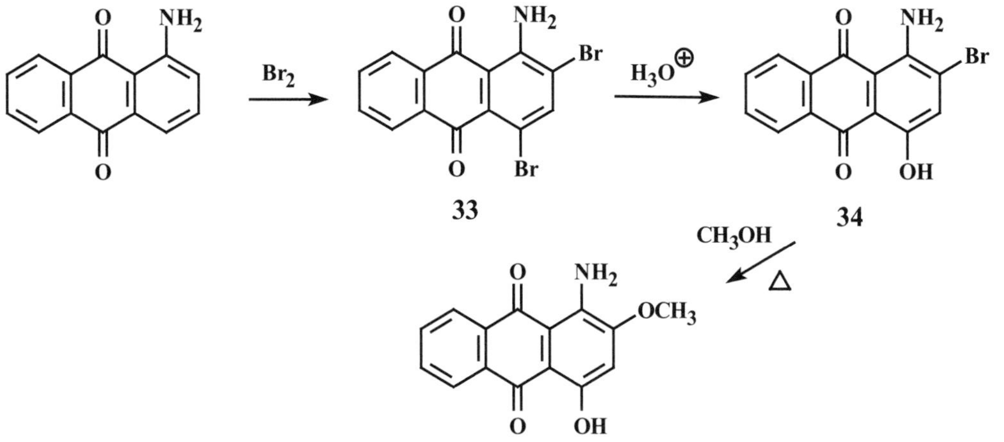

Fig. 13.115. Synthesis of Disperse Red 4.

benzene rings are more reactive than the anthraquinone system, sulfonation occurs there preferentially.

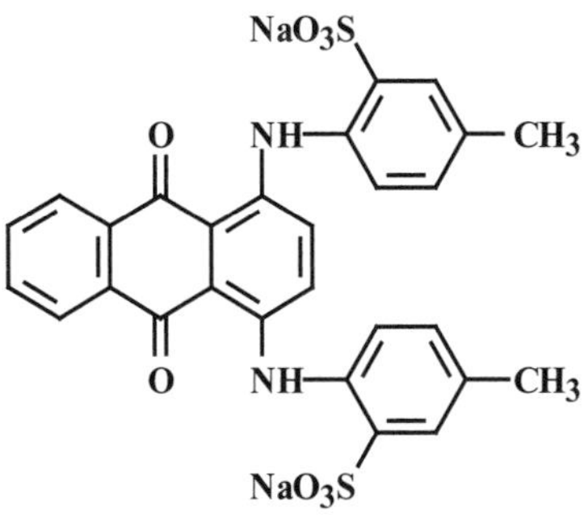

Acid Green 25

Anthraquinone Basic Dyes

Dyes of this type include C.I. Basic Blue 22 and Basic Blue 47. The synthesis of Basic Blue 22 is shown in Fig. 13.120, as an example of the type of chemistry required. The sequence begins with the preparation of N,N-dimethylpropylenediamine, which in turn is combined with methylamine and condensed with leucoquinizarin. Oxidation gives the key intermediate **38**, which is alkylated using methyl chloride to produce the dye.

Anthraquinone Reactive Dyes

Three examples of dyes of this type are C.I. Reactive Blue 19 (**39**), Reactive Blue 2 (**40**),

and Reactive Blue 4 (**41**). All three dyes can be synthesized by condensing the appropriate arylamine with bromamine acid. In the case of the high-volume dye Reactive Blue 19, arylamine **44** is the key intermediate, and its synthesis is shown in Fig. 13.121. Chlorosulfonation and then reduction of the intermediate sulfonyl chloride produce the sulfinic acid **42** Alternatively, the reduction step can be conducted with $Na_2S_2O_4$. Alkylation of the sulfinic acid with 2-chloroethanol or ethylene oxide (a more

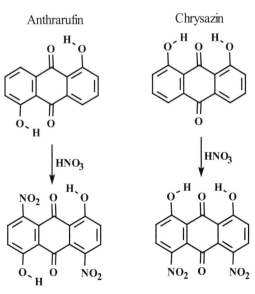

Fig. 13.116. Formation of DNA and DNC

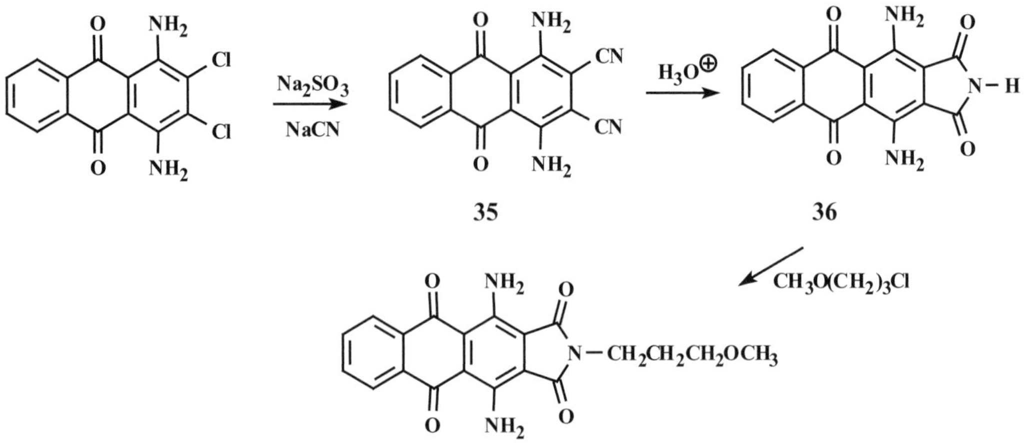

Fig. 13.117. A three-step synthesis of Disperse Blue 60.

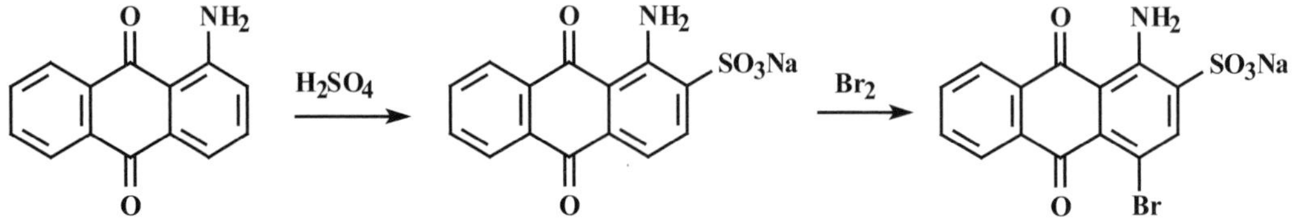

Fig. 13.118. A two-step synthesis of bromamine acid.

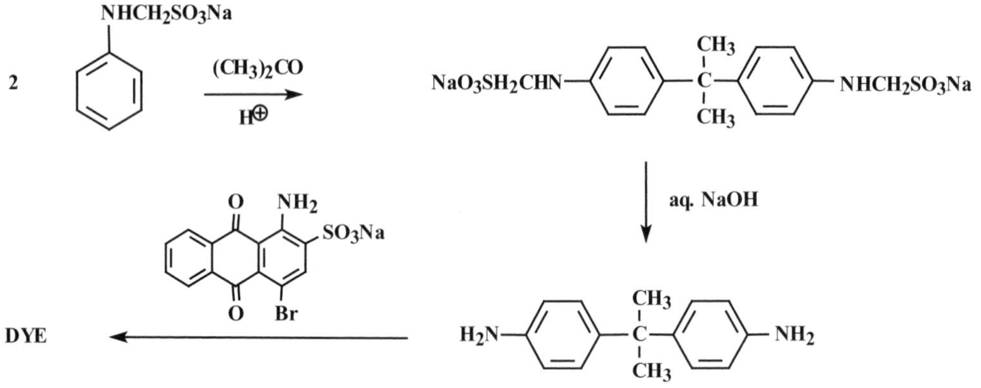

Fig. 13.119. Synthesis of Acid Blue 127.

toxic agent) produces compound **43**. Treatment of this compound with hot H_2SO_4 gives simultaneous hydrolysis of the acetamido (–NHAc) group and sulfonation of the hydroxyethyl (–CH$_2$CH$_2$OH) group to give key intermediate **44**.

Similarly, dyes **40** and **41** are prepared by condensing 2,5-diaminobenzenesulfonic acid with bromamine acid, which reacts first at the less hindered amino group, followed by a reaction with cyanuric chloride to introduce the reactive group. These steps produce dye **41** and

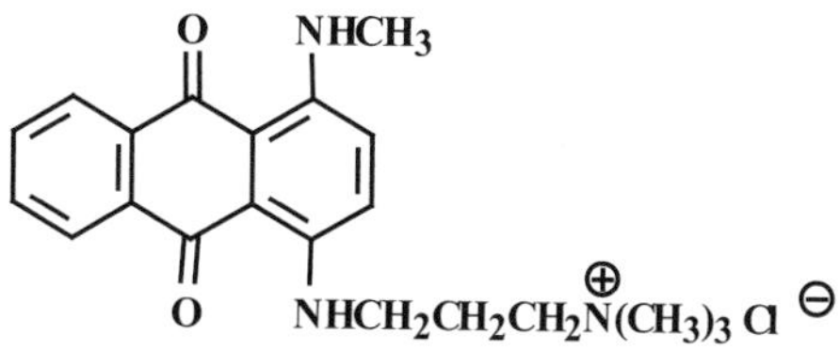

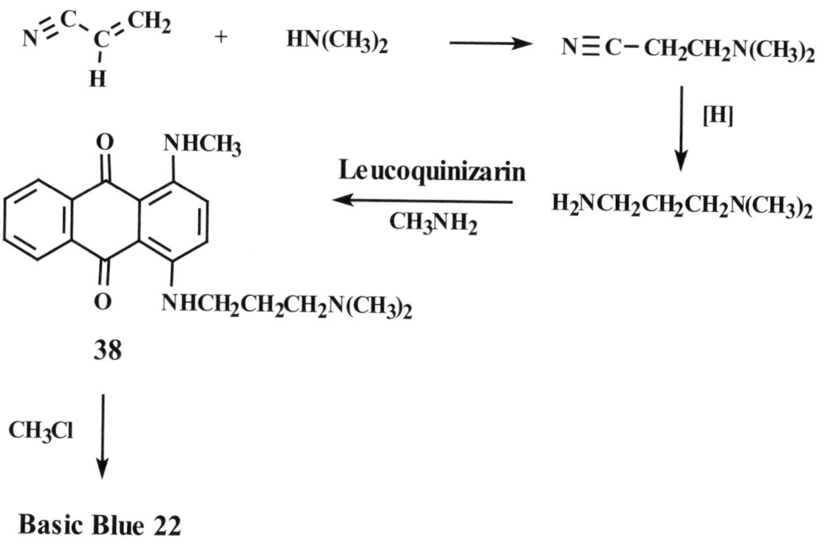

Fig. 13.120. Synthesis of Basic Blue 22.

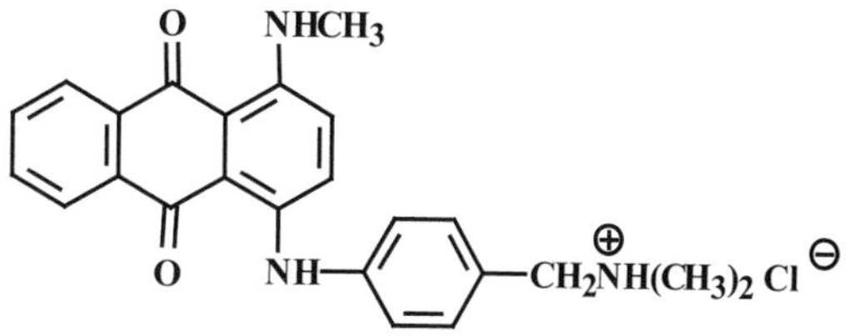

dye **40** is formed by reacting **41** with a mixture of sulfonated anilines. See Fig. 13.122.

VAT DYES

The synthesis of vat dyes covers the full gamut of simple to complex chemistry. We have chosen examples to illustrate the broad spectrum of possible structures and synthetic methods. Emphasis is placed on anthraquinone vat dyes, because they dominate the number of commercial dyes.

Anthraquinone

The simplest anthraquinone vat dyes are benzoylated amines such as C.I. Vat Yellow 3 (**45**) and Vat Yellow 33 (**46**). The syntheses are shown in Figs. 13.123 and 13.124.

Anthraquinone vat dyes containing a thiazole ring include C.I. Vat Yellow 2, the synthesis of which is shown in Fig. 13.125. In this case, at least two approaches are possible. In the first, 2,6-diaminoanthraquinone is condensed with benzotrichloride in the presence of sulfur and the initial product is oxidized without isolation to give the target dye. Alternatively, the starting diamine can be chlorinated and converted to the corresponding dithiol (**47**). At this point condensation with benzaldehyde followed by oxidation (e.g. air or dichromate) gives the dye.

Important vat dyes containing a carbazole moiety include C.I. Vat Brown 3 and Vat Black 27. These dyes are made according to the method shown in Fig. 13.126 for Vat Brown 3. The synthesis employs an Ullmann-type condensation reaction between compounds

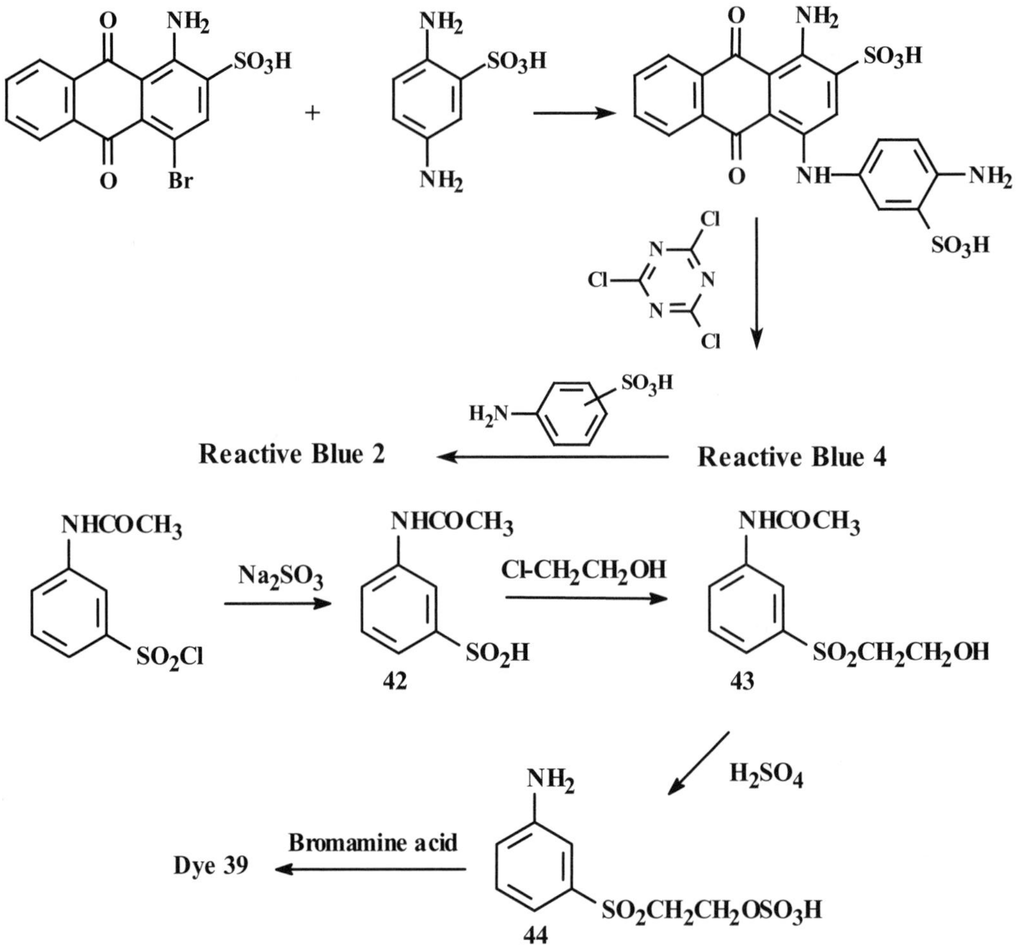

Fig. 13.121. Synthesis of three reactive blue dyes from bromamine acid.

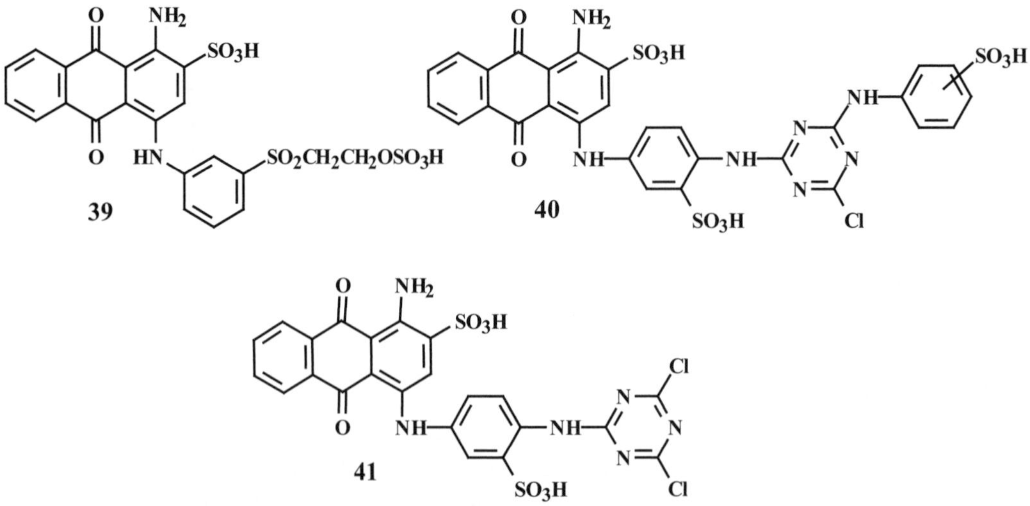

Fig. 13.122. Structures of Reactive Blue 19 (**39**). Reactive Blue 2 (**40**), and Reactive Blue 4 (**41**).

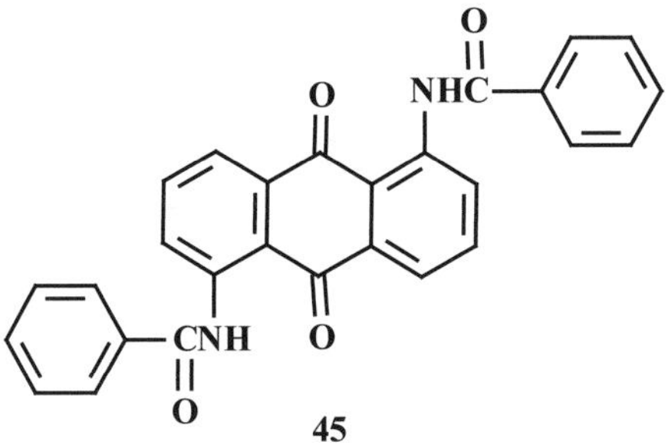

45

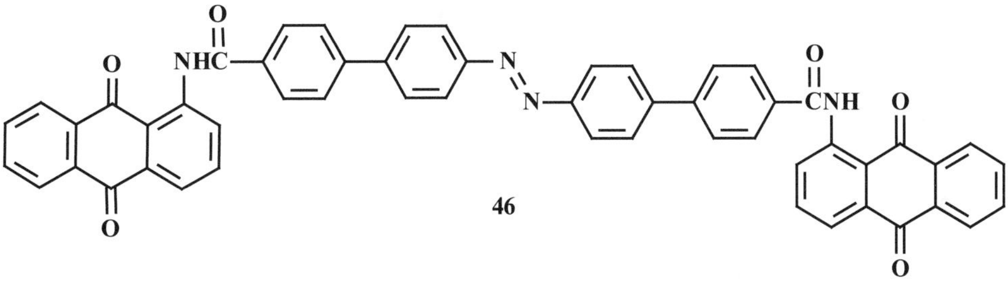

46

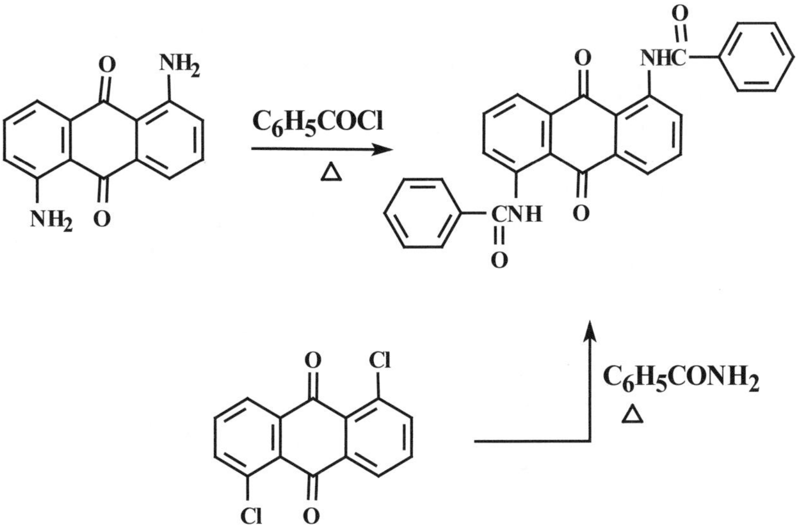

Fig. 13.123. Two approaches to the synthesis of Vat Yellow 3.

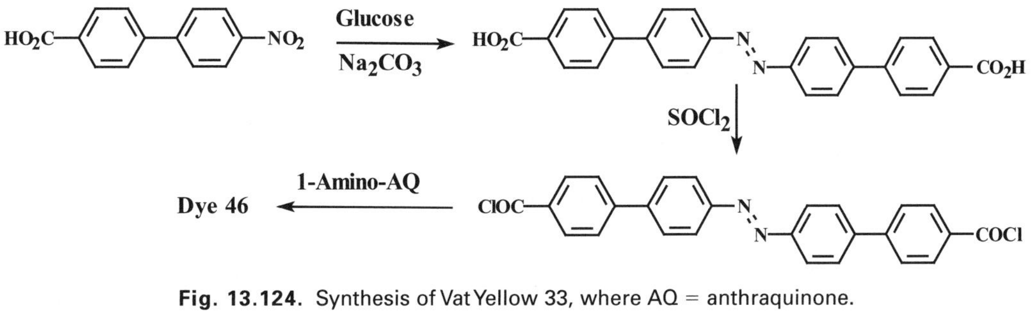

Fig. 13.124. Synthesis of Vat Yellow 33, where AQ = anthraquinone.

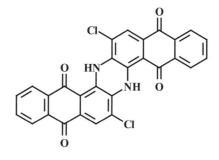

Fig. 13.125. Alternatives to the synthesis of Vat Yellow 2.

48 and **49** followed by acid-induced cyclization using H_2SO_4.

Vat dyes that do not contain all of the elements of the anthraquinone moiety include benzanthrone-based vat dyes such as C.I. Vat Orange 1 and Vat Green 1, which are made according to the routes shown in Figs. 13.127 and 13.128. The first synthesis is a three-step process: (1) dibenzoylation of naphthalene, (2) Lewis acid-induced cyclization to the benzanthrone system, and (3) dibromination. The

Other important anthraquinone vat dyes belong to the family known as indanthrones. Important examples of this structural type are C.I. Vat Blue 4 and Vat Blue 6. Vat Blue 4 is made by heating 1-amino or 2-aminoanthraquinone at 220–230°C in a KOH/H_2O mixture. The Vat Blue 6 synthesis is a much longer process that requires the synthesis of 2-chloro-3-aminoanthraquinone.[57] The resultant amine is brominated and converted to the target dye via an Ullmann reaction.

Vat Blue 4

Vat Blue 6

second synthesis is a four-step process: (1) oxidative-coupling of benzanthrone in the presence of alkali to give compound **50**, (2) H_2SO_4-induced ring closure to give compound **51**, (3) reduction to compound **52**, and (4) methylation to give the target dye.

INDIGOID AND THIOINDIGOID

By far the most important member of these vat dye families is C.I. Vat Blue 1 (indigo). Its synthesis can be achieved via the four-step method shown in Fig. 13.129. The method shown is known as the Heumann–Pfleger synthesis,[58] where the key intermediate, *N*-carboxymethylaniline, is fused with NaNH$_2$. The cyclic product of the fusion step undergoes air oxidation to give indigo.

Thioindigoids are similarly prepared, in that the synthesis of carboxymethyl intermediates is conducted. The resultant cyclic ketones are much less air sensitive, making oxidation with a chemical agent important. However, this also means that unsymmetrical thioindigoid systems can be synthesized (see Fig 13.130). Although many have been made, few are in commerce today. Examples are C.I. Vat Red 1 and Vat Red 41.

SULFUR DYES

Earlier we mentioned that sulfur dye chemistry, although quite old, is still much less well defined than for the other classes of dyes. It is clear, however, that many sulfur dyes are produced by the sulfur bake process and that compounds containing the benzothiazole group (e.g. **53**) are formed in route to the final dyes. For instance, the synthesis of C.I. Sulfur Yellow 4 follows a course of the type outlined in Fig. 13.131. In this regard, heating a mixture of *para*-toluidine and sulfur produces a 2-(*para*-aminophenyl) benzothiazole. The

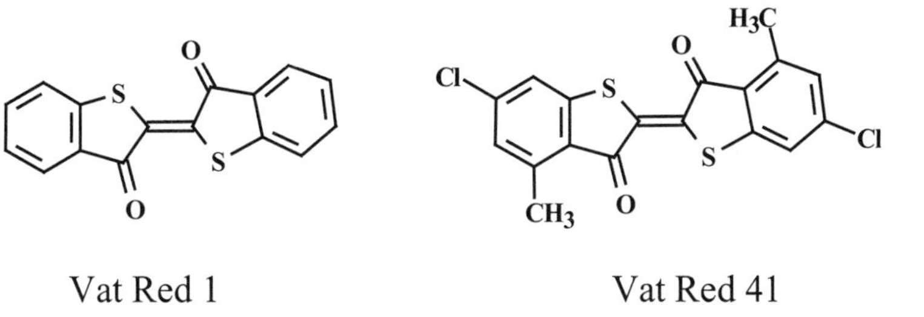

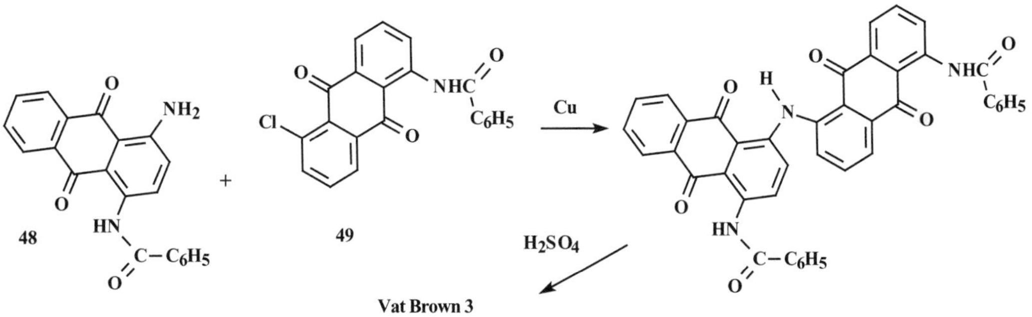

Vat Red 1

Vat Red 41

Vat Brown 3

Fig. 13.126. Synthesis of carbazole-based anthraquinone vat dyes.

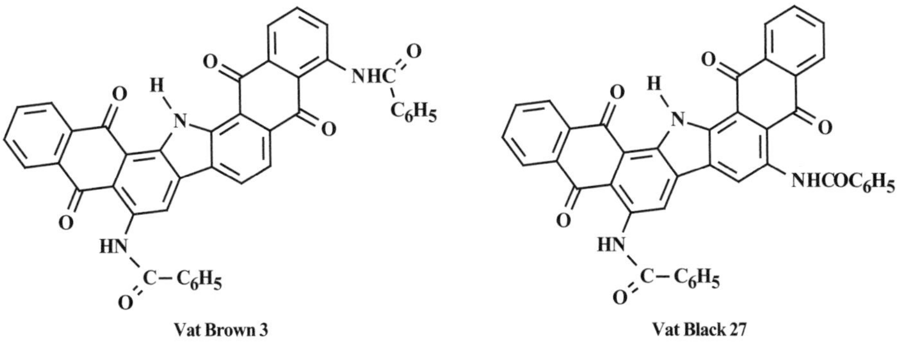

Vat Brown 3

Vat Black 27

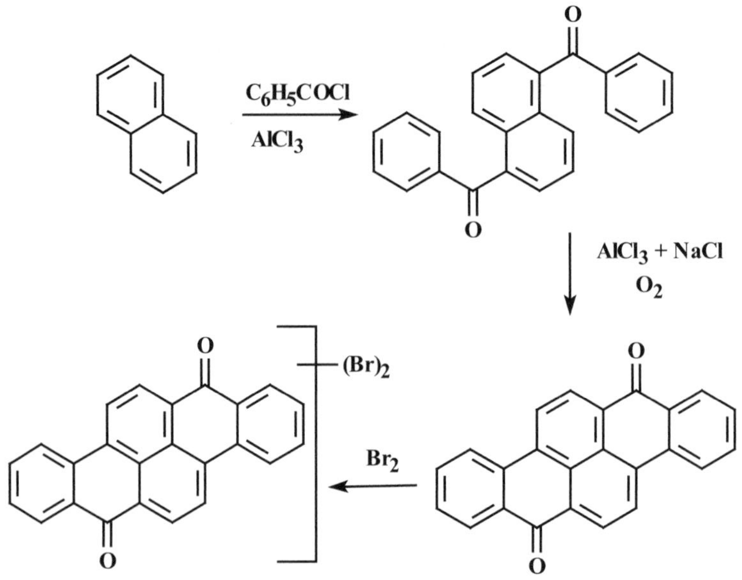

Fig. 13.127. Three-step synthesis of Vat Orange 1.

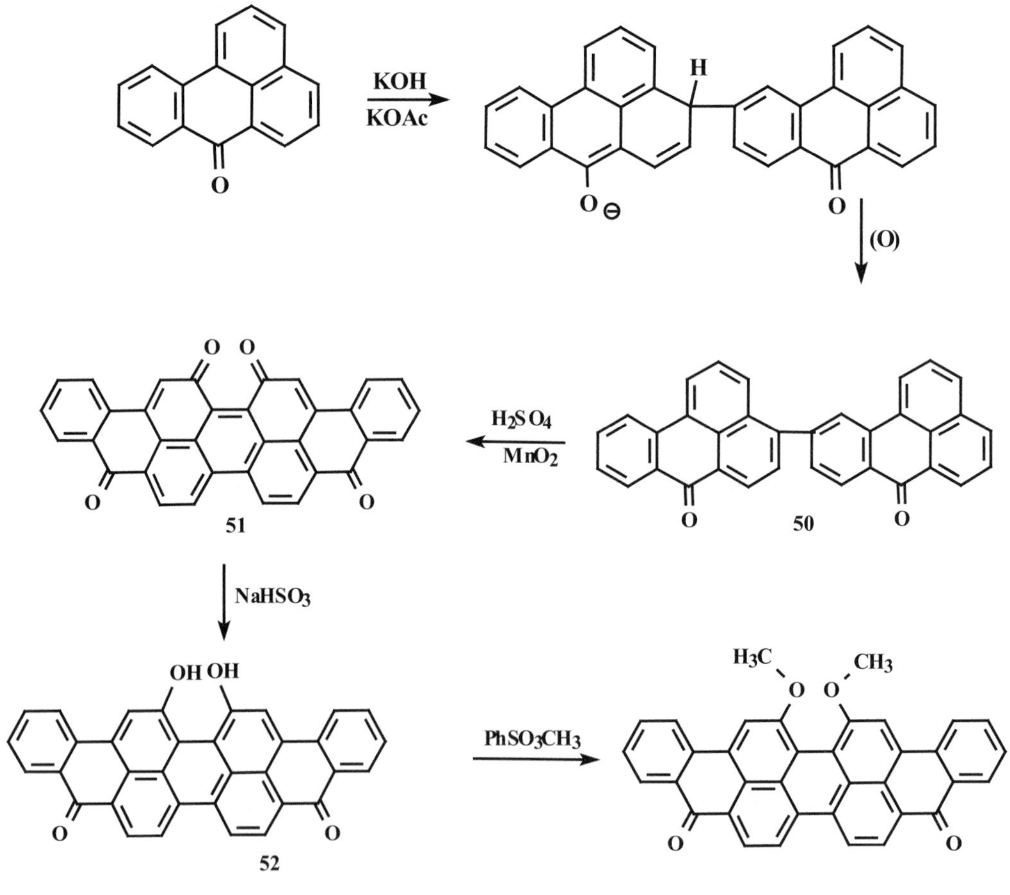

Fig. 13.128. Synthesis of Vat Green 1 from benzanthrone.

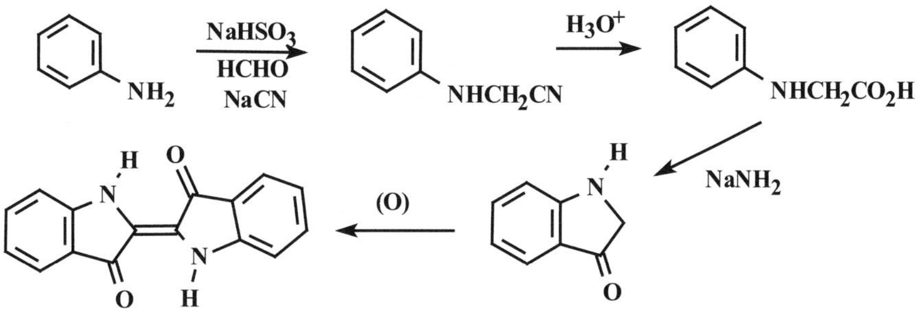

Fig. 13.129. Commercial preparation of synthetic indigo.

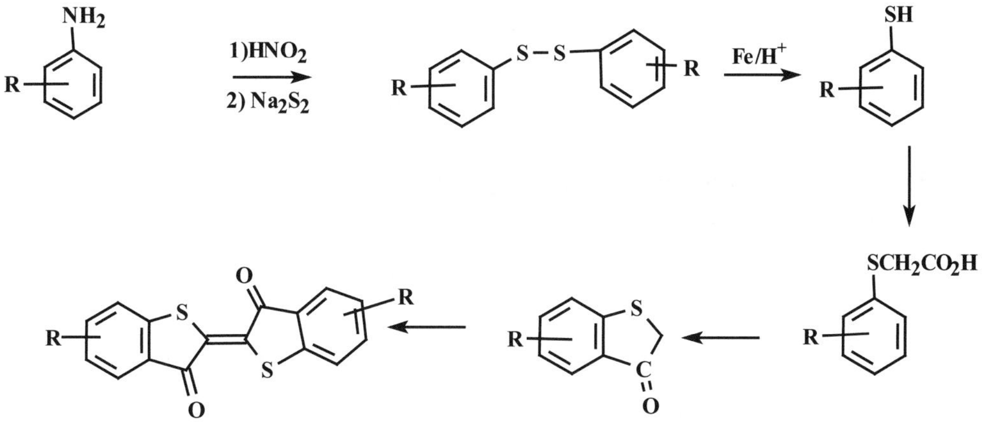

Fig. 13.130. Synthesis of thioindigoid dyes.

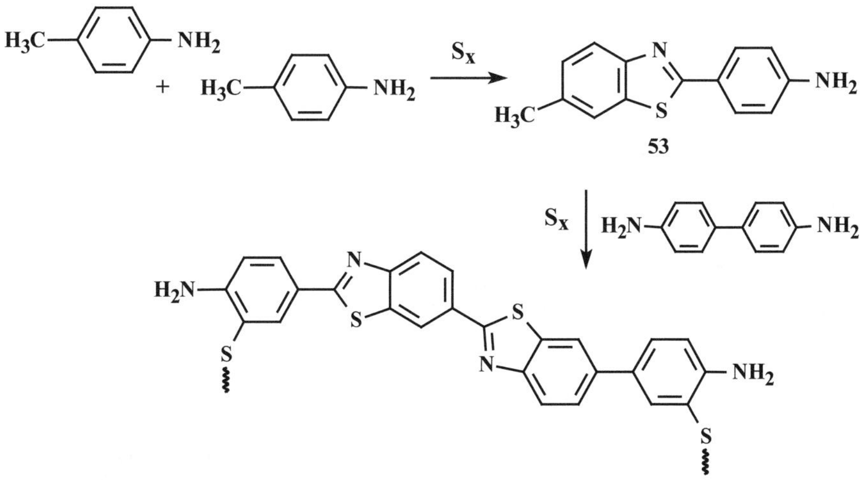

Fig. 13.131. Synthesis of Sulfur Yellow 4 by the sulfur bake process.

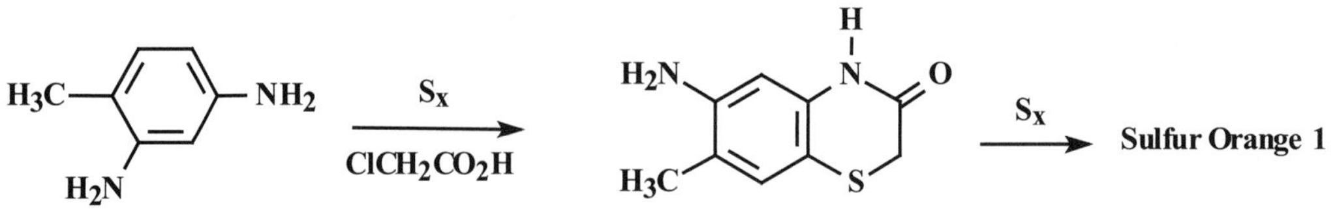

Fig. 13.132. Synthesis of Sulfur Orange 1 by the sulfur bake process.

sulfur bake process has also been used to make C.I. Sulfur Orange 1, where benzothiazone intermediate **54** is produced along the way.[59] See Fig. 13.132.

Sulfur blue dyes are often made using an organic solvent such as n-butanol, in what is known as the solvent reflux process. Examples are C.I. Sulfur Blue 9 and Sulfur Blue 13. In this case, intermediate structures are indophenols (e.g., **55**). See Fig. 13.133. Similarly, sulfur dyes containing benzothiazine groups can be made from

tetrahalogenated benzophenones. See Fig. 13.134.

Sulfur black dyes are synthesized according to the methods shown in Fig. 13.135. In these examples sodium polysulfide is the sulfurizing agent employed.

PHTHALOCYANINE DYES

The synthesis of the copper phthalocyanine (CuPc) system is achieved as shown in Fig. 13.136. Here it can be seen that any of

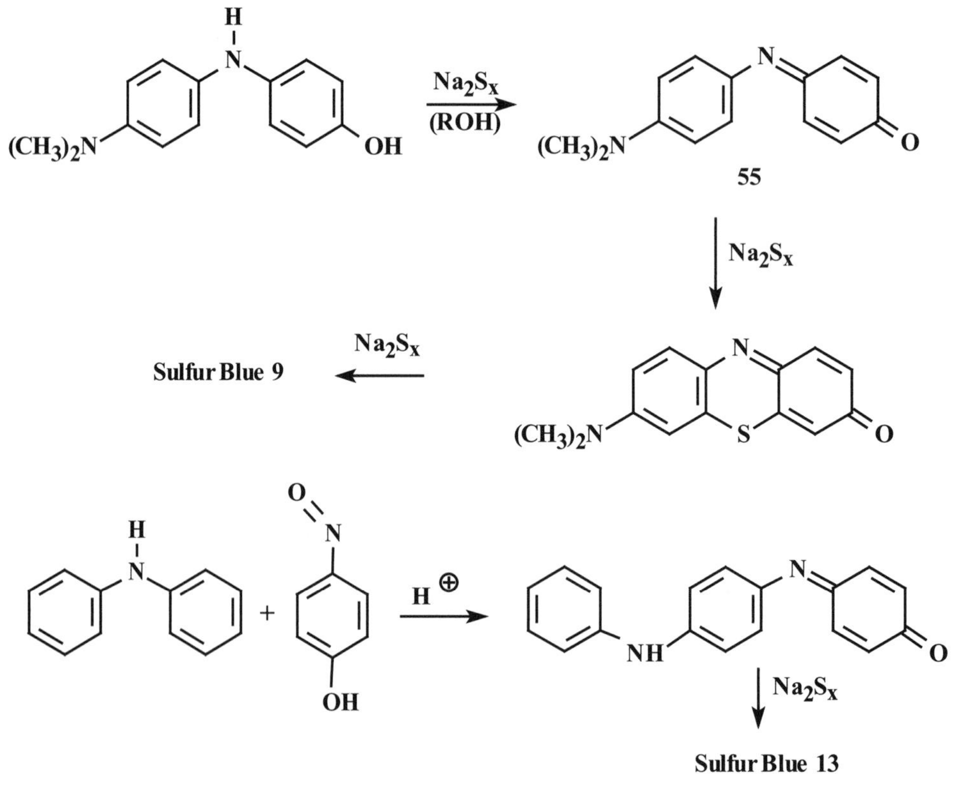

Fig. 13.133. Synthesis of blue dyes by the solvent reflux process.

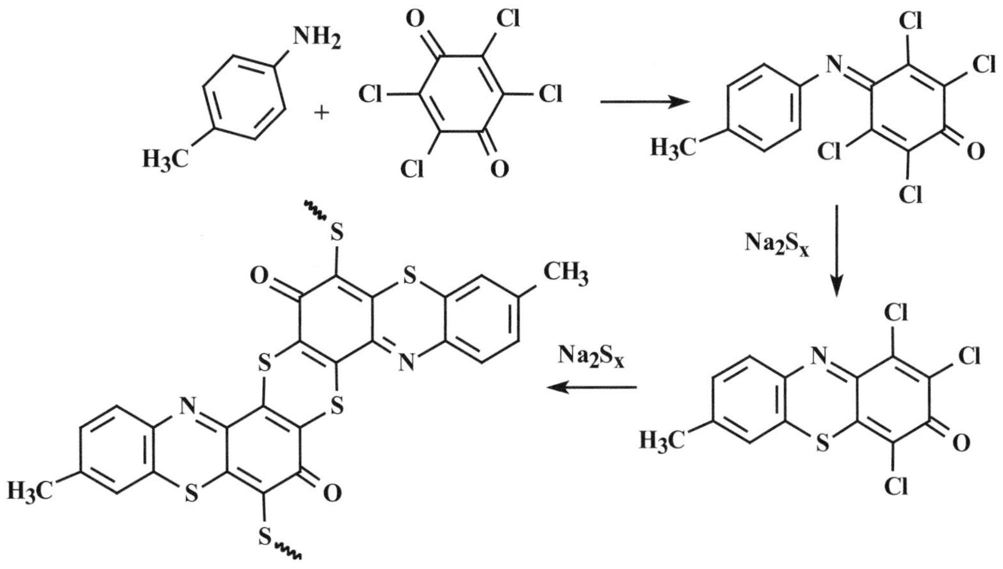

Fig. 13.134. Sulfur dye synthesis from tetrachlorobenzoquinone.

four precursors can be used. Disulfonation gives C.I. Direct Blue 86 and tetrasulfonation gives C.I. Acid Blue 249.

The chlorosulfonation of the CuPc system opens the door to the synthesis of reactive dyes, as shown in Fig. 13.137. In this case, aminochlorotriazine **56** reacts with a CuPc–SO$_2$Cl intermediate to give a monochlorotriazine reactive dye (**57**), which in turn can be used to make the cationic reactive dye **58**.

FLUORESCENT BRIGHTENERS (COLORLESS "DYES")

Many fluorescent brighteners are derivatives of 4,4'diamino-stilbene-2,2'-disulfonic acid (**59**), an example of which is C.I. Fluorescent Brightener 32 (Fig. 13.138). In this case, successive reactions involving diamine **59** with two molecules of cyanuric chloride and two molecules of aniline followed by hydrolysis of the final chloro groups give the target compound.

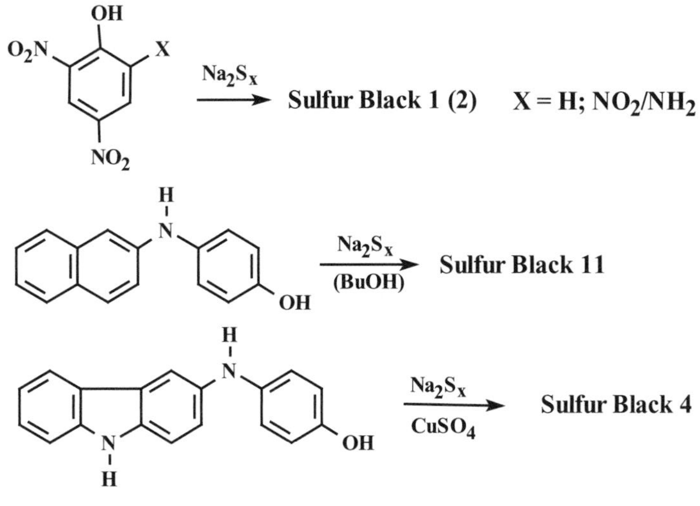

Fig. 13.135. Synthesis of sulfur black dyes.

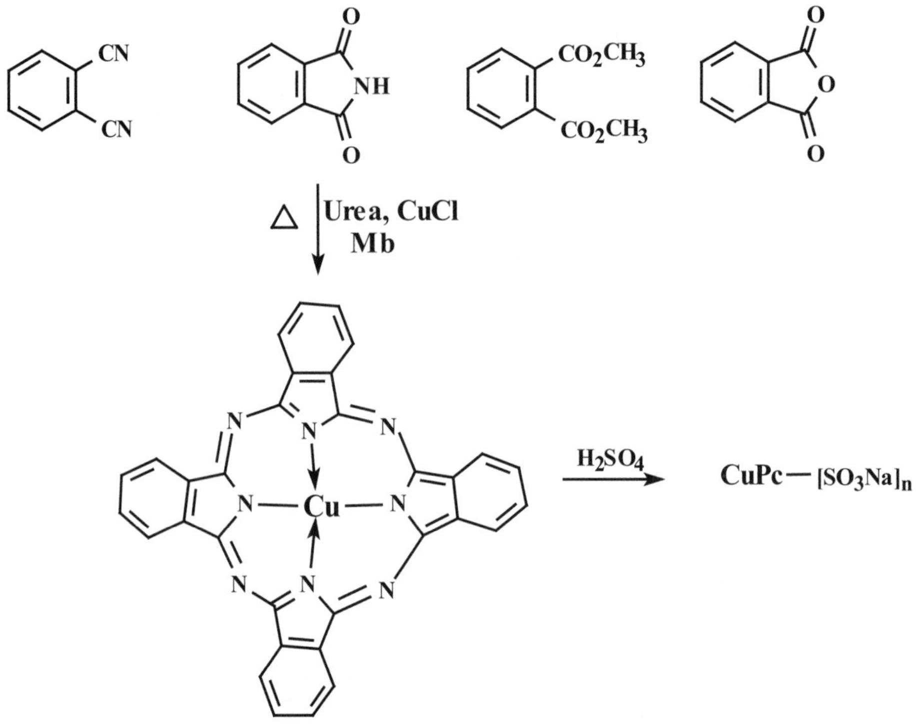

Fig. 13.136. Synthesis of phthalocyanine and sulfonated derivatives.

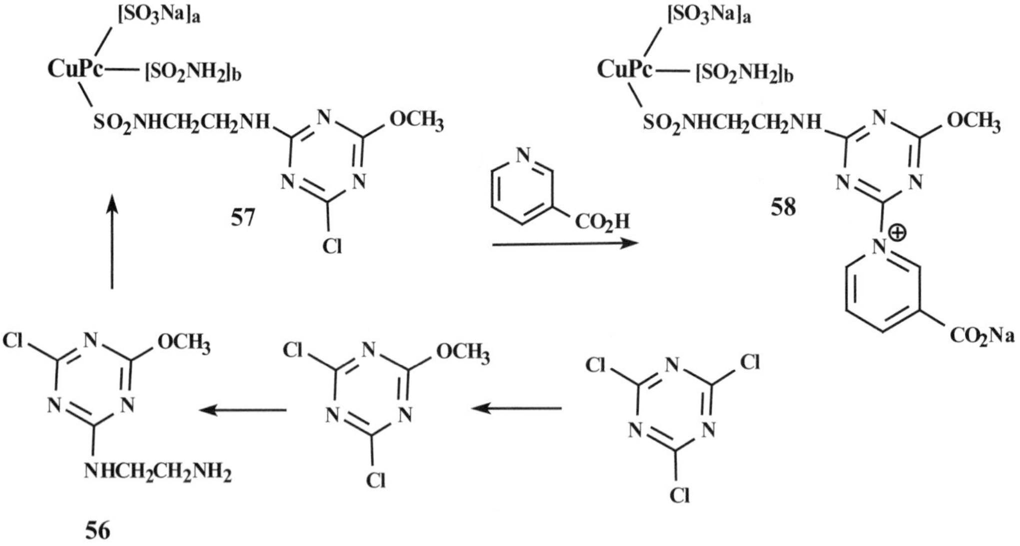

Fig. 13.137. Synthesis of CuPc-based reactive dyes.

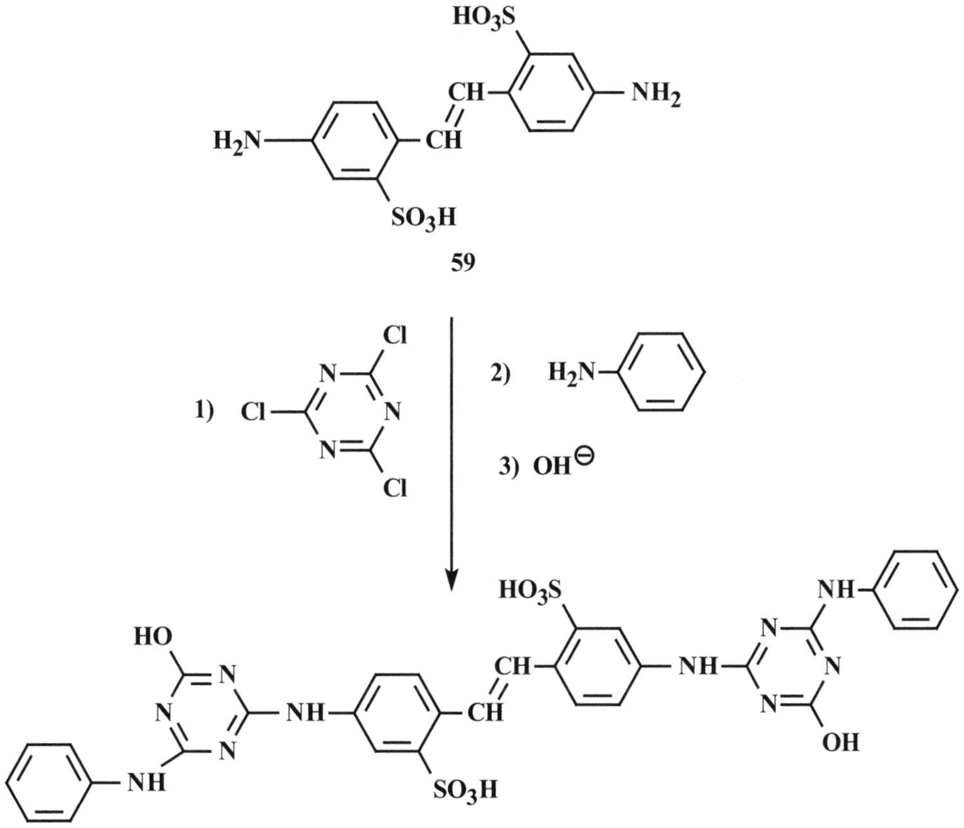

Fig. 13.138. Synthesis of Fluorescent Brightener 32.

Structurally related fluorescent brighteners containing a benzotriazole moiety are made according to the route shown in Fig. 13.139. In this case, diamine **59** is tetrazotized, coupled to 2 molecules of 1,6-Cleve's acid, and the intermediate disazo stilbene structure (**60**) is oxidized to C.I. Fluorescent Brightener 40. Nowadays, monosulfonated benzotriazole brighteners are more important.[60] The synthesis of one example is shown in Fig. 13.140 for C.I. Fluorescent Brightener 46.

Examples of hydrophobic fluorescent brighteners include C.I. Fluorescent Brighteners 199, 130, 236, and 162. The synthesis of these compounds is shown in Figs. 13.141 to 13.144. In the first of these examples, a *bis*-stilbene structure is made in two steps from *bis*-chloromethyl-xylene, using the traditional reaction of a phosphorus ylide with an aldehyde as the key step in the sequence.

In the second example, the synthesis of a coumarin-type fluorescent brightener is illustrated. Here, *meta*-hydroxy-*N*,*N*-diethylaniline is condensed with ethyl acetoacetate followed by cyclization of the intermediate keto ester **61**. The latter compound undergoes acid-catalyzed cyclization and dehydration to give C.I. Fluorescent Brightener 130. See Fig. 13.142.

A fluorescent brightener containing coumarin and triazole groups is made according to the method shown in Fig. 13.143. The synthesis begins with the preparation of amino-coumarin **62**, which in turn is coupled to Tobias acid with concomitant loss of the SO₃H group and then oxidized to give C.I. Fluorescent Brightener 236.

The final example is for a naphthalimide structure that is made from acenaphthene (**63**) in the four-step sequence shown in

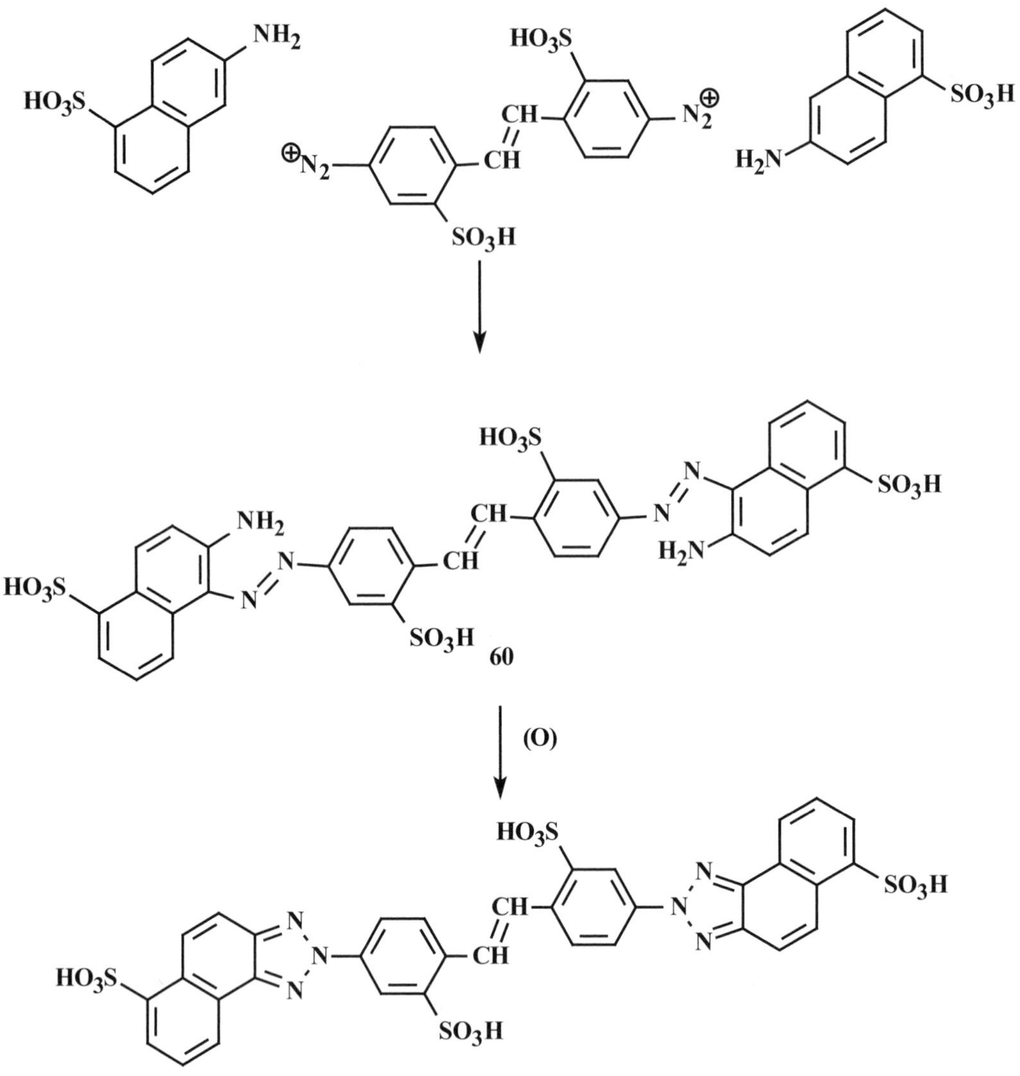

Fig. 13.139. Synthesis of Fluorescent Brightener 40.

Fig. 13.144: (1) sulfonation, (2) chromate oxidation to give the naphthalic anhydride (**64**), (3) condensation with *N*-methylamine, and (4) replacement of the sulfonic acid group in a reaction with methoxide. This process gives C.I. Fluorescent Brightener 162.

PRODUCTION AND SALES

During the 1990s, the large international companies began to form alliances with producers around the world. Hoechst AG, which had done little research on disperse dyes since the 1970s, signed an agreement in 1990 with Mitsubishi of Japan and gained access to a strong line of disperse dyes. BASF AG and Mitsui signed agreements for vat dyes. ATIC resulted from a joint venture between ICI and Atul of India. Finally, a major break came in January 1995, when Bayer AG and Hoechst AG, the parent companies in Germany, announced the formation of DyStar, a worldwide consolidation of their textile dye businesses, which included the US Hoechst Celanese, and Bayer. Within a short time, BASF acquired the textile dyes business of ICI/ Zeneca. Swiss companies Ciba and Clariant

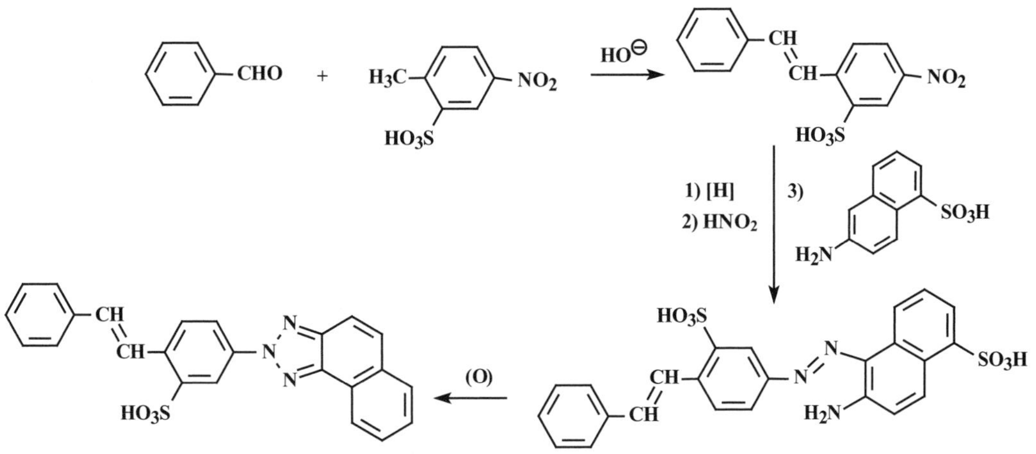

Fig. 13.140. Synthesis of Fluorescent Brightener 46.

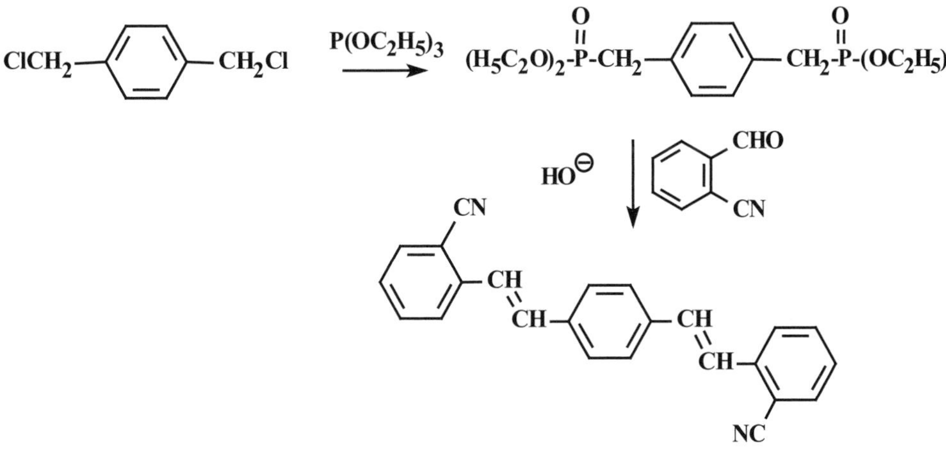

Fig. 13.141. Synthesis of Fluorescent Brightener 199.

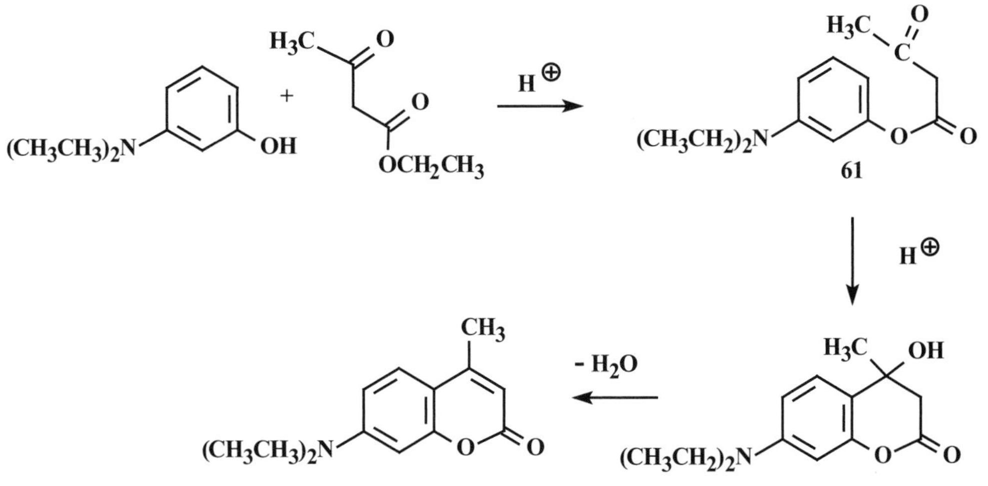

Fig. 13.142. Synthesis of Fluorescent Brightener 130.

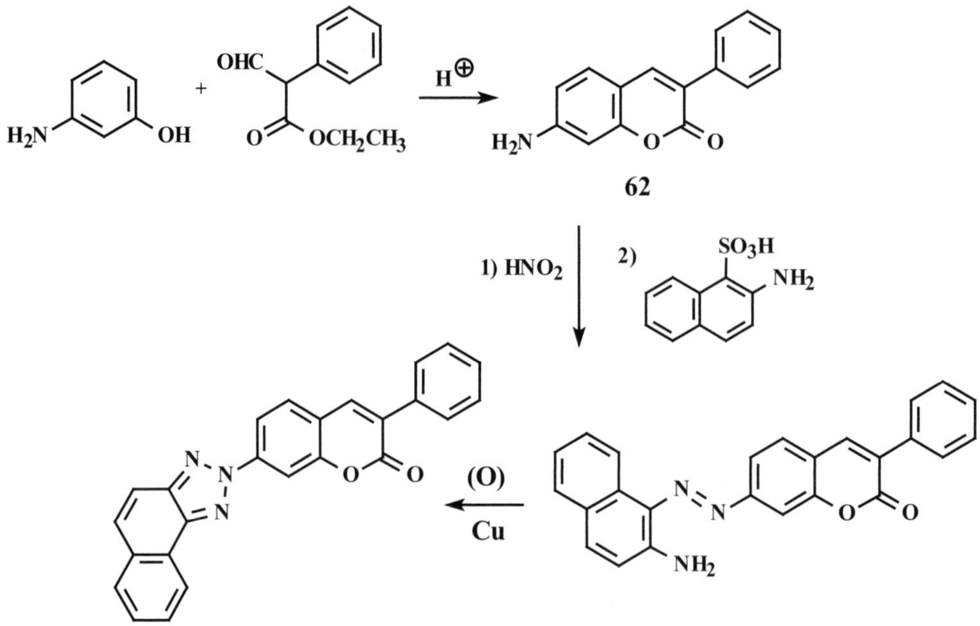

Fig. 13.143. Synthesis of Fluorescent Brightener 236.

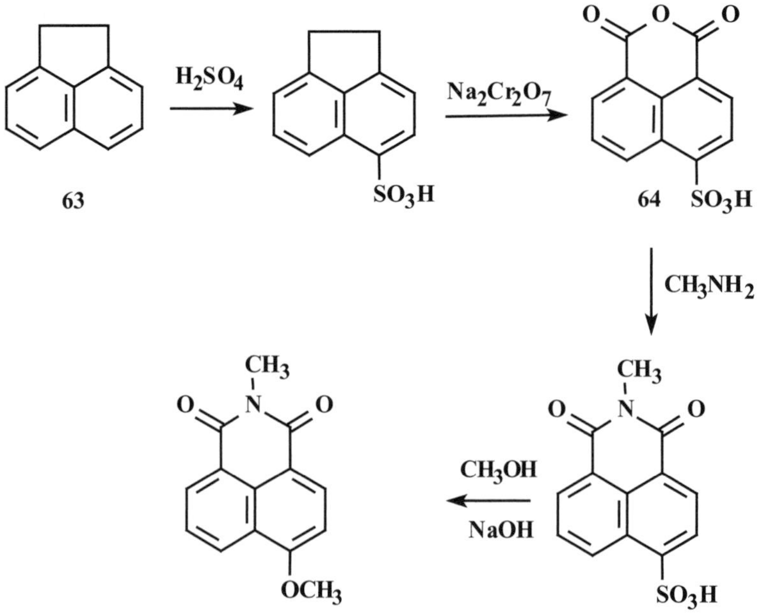

Fig. 13.144. Synthesis of Fluorescent Brightener 162.

(derived by consolidating Sandoz and portions of Hoechst in 1995) announced a merger of the textile dyes business but cancelled the venture in 1998. Crompton & Knowles (C&K) emerged as the sole U.S.-based major company, but the company struggled during the late 1990s and was sold to Yorkshire Group PLC of the United

Kingdom. Yorkshire Pat-Chem and C&K became Yorkshire Americas.

Globalization and establishment of NAFTA meant fewer textile dyes were needed and manufactured in the United States during the late 1990s. The market shrank from 232 million pounds ($955 million) in 1994 to 214 million pounds ($689 million) in 1998 with further cuts expected. Imported dyes expanded but prices fell. Some 1.1 million pounds of disperse dyes were brought in with a value of $5 million in 1992. In 1999, 5.7 million pounds with a value of $10 million were imported. For each class of dyes, you can find expansion of imports for fewer and fewer dollars. The latest year when consumption was publicly revealed is given in Table 13.6.

Table 13.6. World market Textile Dyes 2002 Volume*

	Volume tons	%
Reactive	179,381	28
Direct	15,986	3
Vats	18,663	3
Indigo	35,159	6
Sulfur	92,873	15
Disperse	175,845	28
Acid	23,257	4
Metal complex	17,202	3
Chrome	5,519	1
Cationic	18,568	3
Naphtol	8,942	1
Phthalogen	591	0
Pigment prep.	46,885	7
TOTAL	**638,871**	**100**

*Market Survey DyStar 2002.

REFERENCES

1. Edelstein, S. M., *Historical Notes on the Wet-Processing Industry*, Dexter Chemical Co., Bronx, NY, no date.
2. Zahn, J., *Bayer Farben Revue*, **12**, 32; **11**, 75 (1967).
3. Edelstein, S. M., and H. C. Borghetty, *The Plictho of Gioanventura Rosetti*, M. I. T. Press, Cambridge, MA, 1969.
4. Robinson, R., "The Life and Work of Sir William Henry Perkin," *Proceedings of the Perkin Centennial 1856–1956*, AATCC, 41 (1957).
5. Lehner, S., "America's Debt to Perkin," ibid., 269.
6. "Buyer's Guide," *AATCC Review*, **1**, (7), 17 (2001).
7. Raghavan, K. S. S., *Textile Dyer and Printer*, **XV**, (20), 21, (1982).
8. Mock, G. N., *AATCC Review*, **1** (12), 18 (2001).
9. Compiled from: *The Colour Index*, **5**, 4th edition (1992).
10. (a) Crossley, M.L., *American Dyestuff Reporter*, **27**(3), 124 (1938); (b) Crossley, M.L., *American Dyestuff Reporter*, **28**(3), 487 (1939).
11. Welham, A.C., *Journal of the Society of Dyes & Colourists*, **102**, 126 (1986).
12. Beffa, F. and Back, G., *Review of Progress in Coloration*, **14**, 33 (1984).
13. Allen, R.L.M., *Colour Chemistry*, Chapter 7, Thomas Nelson & Sons, London, 1971.
14. Hueckel, M., *Textile Chemist & Colorist*, **1**(11), 510 (1969).
15. Raue, R., *Review of Progress in Coloration*, **14**, 187 (1984).
16. Aspland, J.R.., *Textile Chemist & Colorist*, **25**(6), 21 (1993).
17. Allen, R.L.M., *Colour Chemistry*, Chapter 5, Thomas Nelson & Sons, London, 1971.
18. Shore, J., *Colorants and Auxiliaries–Organic Chemistry, Properties and Applications*, Vol. 1 (Colorants), Society of Dyers and Colourists, Manchester, England, p. 20, 1990.
19. Aspland, J.R.., *Textile Chemist & Colorist*, **21**(11), 21 (1991).
20. Shore, J., *Review of Progress in Coloration*, **21**, 23 (1991).
21. Haley, T.J., *Clinical Toxicology*, **8**, 13 (1975).
22. Prival, M.J., Bell, S.J., Mitchell, V.D., Peirl, M.D., and Vaughan, V.L., *Mutation Research*, **136**, 33 (1984).
23. Ecological and Toxicological Association of Dyes and Organic Pigments Manufacturers, ETAD Information Notice No. 6, *Textile Chemist & Colorist*, **28**(6), 11 (1996).
24. Freeman, H.S., Esancy, M.K., Esancy, J.F., and Claxton, L.D., *CHEMTECH*, **21**(7), 438 (1991).
25. Dawson, J.F., *Review of Progress in Coloration*, **14**, 90 (1984).
26. Annen, O., Egli, R., Hasler, R., Henzi, B., Jakob, H., and Matzinger, P., *Review of Progress in Coloration*, **17**, 72 (1987).
27. Weaver, M.A. and Shuttleworth, L., *Dyes & Pigments*, **3**, 81 (1982).
28. Egli, R., "The Chemistry of Disperse Blue Dyes, Past and Present," Chapter 1, in *Colour Chemistry*, A.T. Peters and H.S. Freeman, Editors, Elsevier Applied Science, London, 1991.
29. Reinert, G., *Review of Progress in Coloration*, **27**, 32 (1997).

30. Renfrow, A.H.M. and Taylor, J.A., *Review of Progress in Coloration*, **20**, 1 (1990).
31. Aspland, J.R.., *Textile Chemist & Colorist*, **24**(5), 31 (1992).
32. Carr, K., "Reactive Dyes, Especially Bireactive Molecules," in *Modern Colorants: Syntheses and Structures*, Volume 3, Chapter 4, A.T. Peters and H.S. Freeman, Editors, Blackie Academic and Professional, London, 1995.
33. Wood, W.E., *Review of Progress in Coloration*, **7**, 80 (1976).
34. Guest, R.A. and Wood, W.E., *Review of Progress in Coloration*, **19**, 63 (1989).
35. Aspland, J.R.., *Textile Chemist & Colorist*, **24**(4), 27 (1992).
36. a. Baumgarte, U., *Review of Progress in Coloration*, **17**, 29 (1987). b. Jones, F., *Review of Progress in Coloration*, **19**, 20 (1989).
37. Egerton, G.S., *Journal of the Society of Dyes & Colourists*, **65**, 764 (1949).
38. Freeman, H.S. and Peter, A.T., Editors, *Colorants for Non-textile Applications*, Elsevier Science, Amsterdam, 2000.
39. Gregory, P., "Colorants for High Technology," in *Colour Chemistry*, A.T. Peters and H.S. Freeman, Editors, Elsevier Applied Science, London, 1991.
40. Freeman, H.S. and Sokolowska, J., *Review of Progress in Coloration*, **29**, 8 (1999).
41. Bauer, W. and Ritter, J., "Tailoring Dyes for Ink-Jet Applications", in Z. Yoshida and Y. Shirota, Editors, *Chemistry of Functional Dyes*, Chapter 8.1, p. 649, Mita, Tokyo, 1993.
42. Carr, K, "Dyes for Ink Jet Printing", Chapter 1, in Freeman, H.S. and Peter, A.T., Editors, *Colorants for Non-textile Applications*, Elsevier Science, Amsterdam, 2000.
43. a. Gregory, P. *Dyes & Pigments*, **13**, 251 (1990). b. Bradbury, R, "Thermal Transfer Printing," Chapter 2, in Freeman, H.S. and Peter, A.T., Editors, *Colorants for Non-textile Applications*, Elsevier Science, Amsterdam, 2000.
44. Bamfield, P. *Chromic Phenomena: Technological Applications of Colour Chemistry*, Chapter 1, p. 50, The Royal Society of Chemistry, Cambridge, UK, 2001.
45. Inokuchi, H., "Organic Semiconductors: A Still Fashionable Subject," in Z. Yoshida and Y. Shirota, Editors, *Chemistry of Functional Dyes*, Chapter 7.1, p. 521, Mita, Tokyo, 1993.
46. Bamfield, P. *Chromic Phenomena: Technological Applications of Colour Chemistry*, Chapter 4, pp. 245–256, The Royal Society of Chemistry, Cambridge, UK, 2001.
47. Pavlopoulos, T., "Laser Dyes," Chapter 7, in Freeman, H.S. and Peter, A.T., Editors, *Colorants for Non-textile Applications*, Elsevier Science, Amsterdam, 2000.
48. Moura, J.C.V.P., "Biomedical Applications of Dyes," Chapter 5, in Freeman, H.S. and Peter, A.T., Editors, *Colorants for Non-textile Applications*, Elsevier Science, Amsterdam, 2000.
49. Corbett, J.G., *Review of Progress in Coloration*, **15**, 52 (1985).
50. Corbett, J.G., *Journal of the Society of Dyes & Colourists*, **83**, 273 (1967).
51. Corbett, J., "Hair Dyes," Chapter 10, in Freeman, H.S. and Peter, A.T., Editors, *Colorants for Non-textile Applications*, Elsevier Science, Amsterdam, 2000.
52. Waller, D., Zbigniew, H.J., and Filosa, M., "Dyes used in Photography," Chapter 3, in Freeman, H.S. and Peter, A.T., Editors, *Colorants for Non-textile Applications*, Elsevier Science, Amsterdam, 2000.
53. Moury, D.T., *Chemical Reviews*, **42**, 213 (1948).
54. Zollinger, H. *Color Chemistry*, 2nd edition, Chapter 7, VCH, Weinheim, 1991.
55. Fierz-David, H.E. and Blangey, L., *Fundamental Processes in Dye Chemistry*, 5th edition, Interscience, NY, p.250, 1979.
56. Gordon, P.F. and Gregory, P., *Organic Chemistry in Colour*, Chapter 2, Springer-Verlag, Berlin, 1983.
57. Zollinger, H. *Color Chemistry*, 2nd edition, Chapter 8, VCH, Weinheim, 1991.
58. Püntener, A. and Schlesinger, U., "Natural Dyes," Chapter 9, in Freeman, H.S. and Peter, A.T., Editors, *Colorants for Non-textile Applications*, Elsevier Science, Amsterdam, 2000.
59. Hallas, G., "Chemistry of Anthraquinoid, Polycyclic, and Miscellaneous Colorants, Chapter 6, in Shore, J., *Colorants and Auxiliaries–Organic Chemistry, Properties and Applications*, Vol. 1 (Colorants), Society of Dyers and Colourists, Manchester, England, p. 20, 1990.
60. Siegrist, A.E., Heffi, H., Meyer, H.R., and Schmidt, E., *Review of Progress in Coloration*, **17**, 39 (1987).

14

The Chemistry of Structural Adhesives: Epoxy, Urethane, and Acrylic Adhesives

Denis J. Zalucha, Ph.D. and Kirk J. Abbey, Ph.D.

INTRODUCTION

Adhesives have been used successfully in a variety of applications for centuries. Today, adhesives are more important than ever in our daily lives, and their usefulness is increasing rapidly. In the past few decades there have been significant advances in materials and in bonding technology. People now routinely trust their fortunes and their lives to adhesively bonded structures and rarely think about it.

At the same time, the subject of adhesives and adhesion continues to receive much attention by both industrial and academic researchers as evidenced by many measures including the continued growth of membership and attendance at the annual meeting of The Adhesion Society.[1] The scientific literature continues to grow at a rapid pace. New books devoted to general and specialized aspects of adhesion and adhesives continue to appear.[2–23] New patents on adhesive compositions and processes are granted almost daily. Figure 14.1 shows U.S. Patent activity for the past ten years. The overwhelming majority of these refers to uses of adhesive compositions to produce commercially useful structures

and products. Many refer to novel application methods. A significant number do refer to chemical innovations in adhesive compositions to provide improved or specialized products. Figure 14.2 shows a sample of the most active world patenting organizations irrespective of adhesive chemistry. There are scientific journals devoted to the science and technology of adhesives and their use (Table 14.1). Specialized Internet sites have also appeared in recent years (Table 14.2). A significant portion of the publications dealing with adhesives is concerned with epoxy, urethane, and acrylic structural adhesives as they are used in a wide variety of commercially important applications. This chapter reviews some of the chemistry of these adhesive types as it applies to structural applications, that is, to those applications where the adhesive bond must carry a load while being resistant to dimensional changes, also known as creep.

Adhesion

Materials are generally defined as adhesives by what they do. Almost any organic polymer and even many inorganic materials can function as

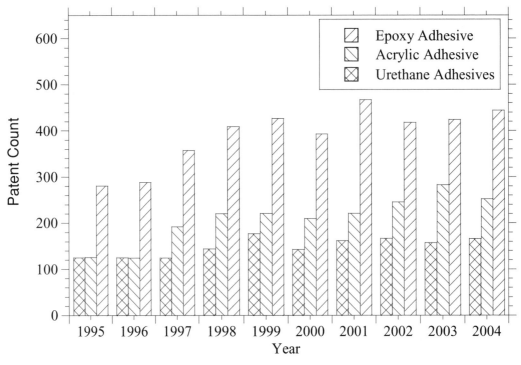

Fig. 14.1. U. S. patents issued from 1995 through 2004. Search terms used were "epoxy adhesive," "urethane adhesive" summed with "polyurethane adhesive," and "acrylic adhesive."

adhesives in some situations. However, whether they are organic polymers or inorganic, or perhaps even solders or brazing alloys, all adhesives share common traits in performing their functions.

1. An adhesive, by surface attachment only, transfers and distributes mechanical loads among the components of an assembly.
2. At some time in the course of the bond formation, the adhesive must be liquid or behave as a liquid in order to wet the adherends.
3. An adhesive carries some continuous, and often variable, load throughout its life.
4. An adhesive must work with the other components of the assembly to provide a durable product that is resistant to degradation by elements of the environment in which it will be used.

The expectations of the user are extremely important in determining whether an adhesive is "good" or "bad." Adhesives are judged on the ability of the whole assembly to meet the user's expectations, which will, in turn, depend on the way the assembly is loaded and tested and on what and where the weakest points of the assembly are located.[13] Adhesion is not an intrinsic property of any polymer but is rather a property of the whole assembly. Structural adhesives are distinguished from nonstructural adhesives by the magnitude of the load that they carry.

Curing

The chemistry of a structural adhesive is designed to do at least two important things. First, the adhesive must at some time pass through a fluid state in order to wet the adherends. Second, the adhesive in its final state in the bond line must be a solid, high-molecular-weight polymer that is able to carry and transfer mechanical forces. In almost all cases, the polymer matrix of a structural adhesive will be crosslinked. The chemistry must

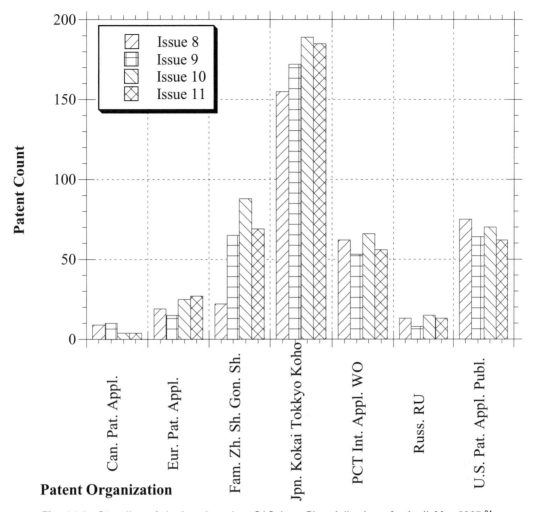

Fig. 14.2. Sampling of citations based on CASelects Plus: Adhesives, for April–May 2005.[24]

make possible some manufacturing process for the assembly that allows for the liquid state, and there must be some mechanism for passing from the liquid to the solid, load-bearing state. The process of going from the liquid to the solid state is usually termed "curing."

There are three general ways in which adhesives are cured. In the first "hot-melt" method, the adhesive can be applied in a molten state and allowed to cool and solidify in the bond line. In the second method, the adhesive can be applied as a solution or dispersion and the

TABLE 14.1.   Journals Devoted to Adhesion and Adhesives

Title	Publisher
The Journal of Adhesion	Taylor & Francis, Inc.
Journal of Adhesion Science and Technology	Brill Academic Publishers
International Journal of Adhesion and Adhesives	Elsevier
CA Selects Plus: Adhesives	American Chemical Society
Adhesives & Sealants Industry	Business News Publishing Co.

TABLE 14.2.   Internet Web Sites Focused on Adhesives

Web Address	Publisher
www.adhesivesandsealants.com	VertMarkets, Inc.
www.adhesiveshq.com	Verdex Group, Thomas Publishing
www.specialchem4adhesives.com	SpecialChem, S. A.
www.adhesionsociety.org	The Adhesion Society

carrier liquid allowed to evaporate, leaving behind the high-molecular-weight polymer. In the third method the adhesive consists of a low viscosity fluid containing reactive groups that undergo polymerization in the bond line to build the molecular weight sufficiently to carry a load, entailing in most cases forming a crosslinked network. This method is the one most commonly used with epoxy, urethane, and acrylic structural adhesives. The chemistry, methods, and mechanisms for accomplishing this polymerization will be covered in more detail in the following sections.

Adhesion Mechanisms

Focusing on the third cure method, once the liquid adhesive has been applied to the adherends and intimate contact and wetting have been established, the liquid mass is cured by polymerizing it to a solid, high-molecular-weight, load-bearing state. It now transfers load among the components of the assembly. The mechanisms of adhesion can be grouped into three or four categories. Kinloch identifies four categories and devotes an entire chapter to elaborate these in detail.[12] In this review, three mechanisms will be invoked to explain the adhesion of one material to another: (1) mechanical interlocking, (2) electrostatic attraction, and (3) the formation of chemical bonds across the interface. All three mechanisms may play some role in any given bonding situation although often one contributes much more than the others.

Mechanical interlocking is usually invoked when describing the adhesion of ice to glass. Silver amalgam dental fillings are held in place largely by mechanical interlocking. Although there probably are some exceptions, mechanical interlocking usually is not a major factor in bond formation with structural adhesives because, unlike water freezing, most materials contract on cooling or curing. Interdiffusion of polymer chains (i.e., entanglements) may also be considered to be mechanical interlocking at the atomic scale, but it requires mutually compatible and essentially uncrosslinked compositions to occur. This is likely to be most important in the welding of plastics.

The other extreme in bonding is the formation of direct covalent chemical links across the interface. These bonds would be expected to be quite strong and durable, but they require special attention not only to the chemistry of the adhesive, but also to that of the substrate. It is necessary that there be mutually reactive chemical groups tightly bound on the adherend surface and in the adhesive, and there is evidence that such bonds can be formed under controlled conditions. Silane coupling agents are one example of using specific reactive groups to promote the formation of direct chemical bonds.

X = -OH, -NH$_2$, -SH, etc.

By far the dominant adhesion mechanism, particularly in the absence of covalent linkages, is the electrostatic attraction of the polar groups of the adhesive to polar groups of the adherends. These are mainly forces arising from the interaction of permanent dipoles, including the special cases of hydrogen bonding (10–25 kJ/mol) and Lewis acid-base interactions (<80 kJ/mole).[25,26] These forces provide much of the attraction between the adhesive and the adherend and also provide a significant portion of the cohesive strength of the adhesive polymer. These interactions are generally classified as attractive long-range forces that drop off inversely to the sixth power of the distance, $1/r^6$, when r is very large compared to the dipole charge separation distance; that is, $r > 1.5$ nm. However, at shorter distances attractive interactions rise more rapidly especially when dipoles are aligned.

In two articles on the cohesive and adhesive strengths of polymers, Mark[27,28] derives some estimates of what adhesive bond strengths might be achievable with covalent bonds or polar forces across an interface. He concludes that the bonds actually achieved in real life are only a small fraction of what he estimates for the situation in which covalent chemical bonds are the main contributors to adhesion. He further proposes that even if there are a significant number of covalent chemical bonds across the interface, the failing strength of the bond still will depend on the strength of the polar bonds. The polar bonds will fail at a lower strength than the covalent bonds, and the applied load then will be concentrated on the covalent bonds. The measurable mechanical strength of a partially covalent adhesive bond still will be dominated by the polar forces. The implication is that although increasing the proportion of covalent bonds across the interface can enhance durability, the ultimate load-carrying capacity probably will not be significantly affected.

Surfaces

Adhesives must function solely through surface attachment. Therefore, the nature of the condition of the adherend surface is crucial to the formation of strong and durable bonds. By "surface" we usually mean that region of a material which interacts with its surroundings. There is some region of a bonded assembly where the adhesive and the adherend interact, but only rarely is this a sharp boundary. Usually it is a very diffuse, somewhat ill-defined region of interaction that has become

known as an interphase rather than an interface. For example, the interaction with a freshly cleaved single crystal of zinc might occur over only a few atomic layers or a few nanometers. Rough or porous surfaces present more surface area than smooth ones of the same dimensions, and the adhesive might reach a depth of several hundred nanometers on a porous adherend such as wood or paper.

One very important aspect of surfaces is that they rarely have the same chemical composition as the bulk material and often seem to be entirely unrelated to the bulk. The surface usually consists of several regions having no clear boundaries. A metal alloy might have a well-defined bulk composition, but at the surface there probably will be a region that is still metallic but is of different chemical composition because of alloying elements or impurities that have segregated at the surface. On top of this region there probably will be a layer of oxides and hydroxides formed by reaction with the atmosphere. There also will be many other contaminants such as nitrogen, sulfur, and halogen compounds formed by interaction with the pollutants in the atmosphere. Finally there will be several layers of adsorbed water. The surface of a metal also might be contaminated with rolling oils, cutting lubricants, drawing compounds, or corrosion inhibitors. Mechanical working of the metal might even mix these contaminants with the other surface materials to create something like an inhomogeneous "frosting" on the surface.

Engineering plastics display some of the same surface phenomena as metals, in that the surface is very different from the bulk. The manufacturing process often introduces anisotropy so that the mechanical properties of the material are different in different directions. In addition, it is common to find that components of the plastic have accumulated at the surface. Low molecular weight polymers or oligomers, plasticizers, pigments, mold release agents, shrink control agents, and other processing aids as well as adsorbed contaminants often are present.

The nature of the surface of an engineering plastic can change rapidly in response to its surroundings. The bulk of the material might be in the glassy state, but because of the concentration of low-molecular-weight material and contaminants, the surface region can be quite mobile. Exposure of the surface to a polar environment, such as by wiping with a polar solvent, can cause polar groups in the plastic surface to preferentially orient themselves outward. Exposure to a nonpolar medium can bring out the nonpolar nature of the surface.

The cured adhesive itself can be expected to be inhomogeneous particularly when arising from a formulated composition containing fillers, rubber-toughening agents, and other additives. Low-molecular-weight materials can be drawn into the surface of a porous adherend, leaving higher-weight polymer and fillers behind. The polar or nonpolar nature of the adherend influences the orientation and morphology of adhesive polymers in the interphase. Compounds at the adherend surface can catalyze or inhibit polymerization. Solvents in the adhesive can swell the adherend or dissolve portions of the adherend surface.

The interphase region is complex, and its composition is usually unknown. Primers or surface treatments often are used to improve control of the interphase and provide increased adhesion, durability, and resistance to aggressive environments. The chemistry of primers and surface treatments is as varied as the chemistry of the adhesives, but they will not be considered further in this chapter.

Any bonded construction consists of at least two adherends and one adhesive and contains at least two interphase regions. It is important to remember that the performance of the construction, its durability, its mechanical properties, and its response to tests and challenges, are all properties of the entire assembly. The successful use of adhesives depends on taking account of all parts of the construction and the process. Whereas the adhesive is just one part of the assembly, its chemistry plays an important role in the bonding process.

The following sections discuss the chemistry of some major classes of thermosetting, structural adhesives.

EPOXY STRUCTURAL ADHESIVES

Introduction

Epoxy adhesives command a large portion of the structural adhesives market. Many people are probably familiar with epoxy structural adhesives, as these are the typical two-part adhesives found in hardware stores and supermarkets. Epoxy adhesives owe their popularity with both the general public and industry to their ease of use, their relative safety, and their compatibility with many adherends (Fig. 14.3). The various chemical reactions involving the epoxy ring provide a fertile field for the development of a wide range of properties.[29] Even a half century after the first epoxy patents were issued,[30] new patents on epoxy adhesive technology continue to appear every month (Fig. 14.1.)

Epoxy adhesives get their name from the portion of the adhesive containing 1,2-epoxy, epoxide, or oxirane ring. This three-member ring consists of two carbon atoms joined to an oxygen atom. The highly strained geometry of this moiety with a strain energy of 114 kJ/mole accounts for its reactivity with many nucleophilic or electrophilic compounds. Typical

Epoxy ring

epoxy resins used to formulate epoxy adhesives have at least two epoxy rings, usually at the ends of a relatively short-chain prepolymer. The epoxy groups then are reacted with other epoxy groups in a chain-growth polymerization or with another curative in a step-growth polymerization to produce a polymer network, which can be either thermoplastic or thermoset. The polymer linkages created by reaction of the epoxy ring are polar and provide adhesion to a variety of polar surfaces. With the proper backbone polymers and curing agents, cured epoxy adhesives can be very tough and resistant to chemical degradation. Bonded assemblies can have a high degree of durability and environmental resistance. Epoxy adhesives are widely used in the transportation industry where they can greatly reduce the amount of welding required.

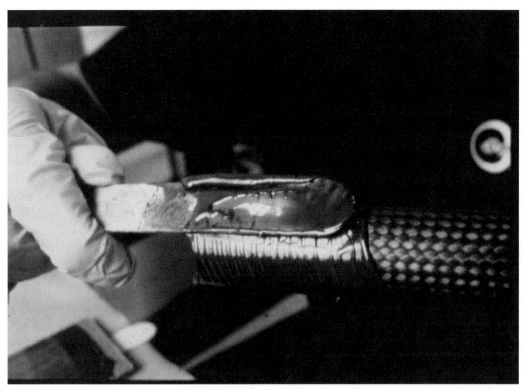

Fig. 14.3. Construction of a bicycle frame from carbon fiber composite tubing using an Araldite™ epoxy adhesive. (Courtesy of Huntsman Advanced Materials.)

Commercial Epoxy Resins

Epichlorohydrin, or 1-chloro-2,3-epoxypropane, is the key raw material in the manufacture of the most widely used epoxy resins known as glycidyl ethers, amines, and esters. The most commercially important epoxy resins for structural application, for coatings, or for adhesives are the diglycidyl ether of bisphenol A (DGEBA, also referred to as bisphenol A diglycidyl ether, BADGE) and its higher

Epichlorohydrin derived epoxy resins

homologues. Pure DGEBA crystallizes after a short time in storage, melting point ~42°C. Because of this, most liquid grades used in epoxy adhesives contain 10–15% higher oligomers, average $n \approx 0.15$–0.20 as well as some side products. Typical commercial resins have molecular weights of about 370 and epoxide equivalent weights in the range of about 180–190, and will have some small percentage of material that is less than difunctional.[31,32] Various grades of liquid diglycidyl ether of bisphenol A resins are manufactured by several companies including Dow Chemical (D.E.R.™),[31] Resolution Performance Products (formerly Shell Chemical, EPON™),[32] Reichhold (Epotuf™),[33] Cognis(Chem Res™),[34] and Huntsman (Araldite™).[35]

The higher-molecular-weight epoxy resins can be produced commercially either through the direct reaction of bisphenol A, NaOH, and epichlorohydrin or by chain extension of the DGEBA with bisphenol A. For resins with $n > 2$, the number of secondary hydroxyls exceeds the number of oxirane groups. They are often used as the reactive site for coatings.

are polymers made by condensing phenol or cresol with formaldehyde in the presence of an acid catalyst. The result is a chain of phenol or cresol groups held together by methylene bridges in a mixture of *ortho* and *para* isomers. The hydroxyl groups then can be reacted with epichlorohydrin and dehydrohalogenated with base to produce an epoxy functional novolac. The simplest, containing only two rings, is the diglycidyl ether of bisphenol F. It is a liquid resin having an epoxy equivalent weight of about 165.

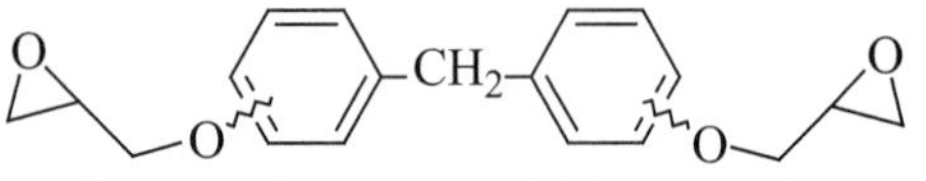

Diglycidyl ether of bisphenol F

Unlike that of the higher molecular weight bisphenol A resins, the epoxy equivalent weight of the epoxy novolac resins remains relatively constant with increasing molecular weight because all of the aromatic rings contain hydroxyl groups that can be epoxidized. Higher-molecular-weight epoxy novolac resins can produce adhesives with higher crosslink densities. Value of n for epoxy novolac resins

Epoxy phenol novolac

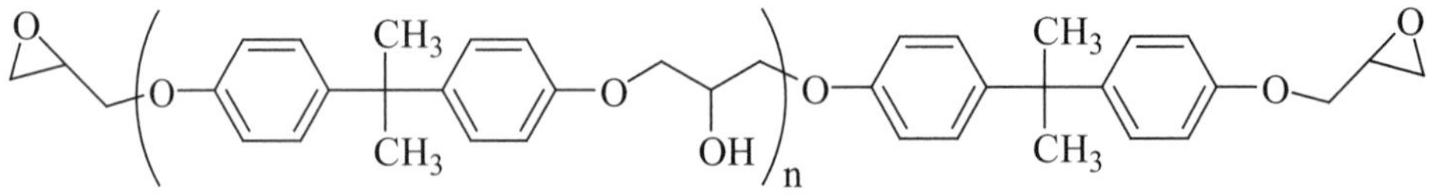

The value of n can be as high as 90.[36] The higher-weight polymers give greater toughness but with a greater tendency for swelling by solvents. Their high T_gs and higher melt viscosity make them less suitable for most adhesive applications.

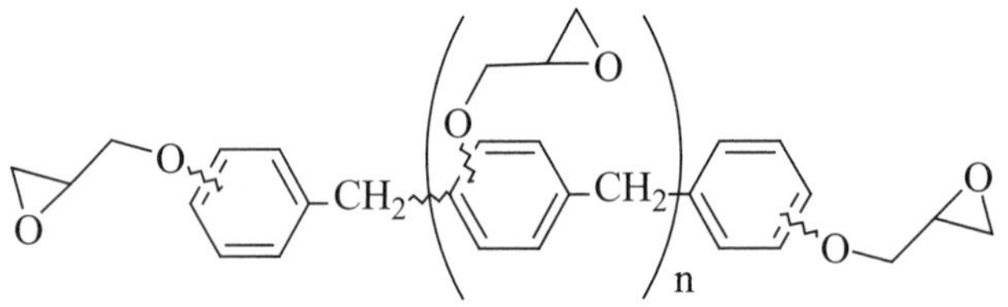

used in adhesives are usually in the range of about 0.2–3.5.

Various aliphatic and aromatic monoglycidyl ethers are available that can be used as reactive diluents of many formulated epoxy structural adhesives. They are

Bisphenol A resin

Another important class of aromatic epoxy resins is the epoxy novolac resins. Novolacs

used to lower the viscosity of the composition without introducing low molecular

TABLE 14.3.   Select Examples of Monofunctional Epoxy Diluents

Composition	Nominal Structures	Tradename
n-Butyl glycidyl ether		Heloxy Modifier 61[a] 6741[b]
2-Ethylhexyl glycidyl ether		Heloxy Modifier 116[a] 6746[b]
o-Cresyl glycidyl ether		Heloxy Modifier 62[a] 6742[b] Araldite® DY-K[c]
Pheny glycidyl ether		Heloxy Modifier 63[a] 6740[b]
Glycidyl ester of neodecanoic acid		Cardura™ Glycidyl Ester[a] 6770[b] Glydexx™ N 10[d]

[a] Resolution Performance Products.
[b] Pacific Epoxy Polymers, Inc.
[c] Huntsman.
[d] ExxonMobil Chemical.

weight unreactive species that might migrate out of the adhesive after it is cured. These compounds are made in much the same way that the polyfunctional epoxy resins are made. Some of the common reactive diluents are given in Table 14.3.

Butyl glycidyl ether has the lowest viscosity and gives the greatest viscosity reduction for the same weight concentration. However, it has a higher vapor pressure than either cresyl glycidyl ether or the higher alkyl glycidyl ethers, and is more likely to cause problems with skin sensitivity and toxicity. The cresyl glycidyl ether is particularly effective at reducing the tendency of liquid epoxy resins to crystallize on storage. For safety reasons the higher-molecular-weight, lower-vapor-pressure materials find more use even though they might be less efficient than those with a higher vapor pressure.

Aliphatic and other specialty aromatic poly-funtional glycidyl ethers are also widely used either for viscosity reduction or improving one or more particular properties. Table 14.4 lists examples of some of these.

The diglycidyl ether of 1,4-butanediol is a low-viscosity difunctional epoxy that is a somewhat less efficient diluent than the monoepoxides but does offer the advantage of having two reactive sites. Neopentylglycol diglycidyl ether and cyclohexane dimethanol diglycidyl ether impart greater rigidity than the 1,4-butanediol diglycidyl ether. Resorcinol diglycidyl ether is a very reactive material useful for reducing viscosity and gives higher crosslink densities than bisphenol A resins themselves.

Epoxidized oils such as soybean oil or lin-seed oil, which are made by oxidizing the internal unsaturation in the fatty acid chain, find much use as plasticizers, particularly in vinyl resins. An internal epoxy ring is much less reactive than a terminal epoxy ring and usually does not take part in typical epoxy curing reactions under mild conditions. These materials can be used as plasticizers in epoxy adhesives but cannot really be considered reactive diluents.

The cycloaliphatic compounds such as 3',4'-epoxycyclohexylmethyl 3,4-epoxycy-clohexane carboxylate (Cryacure UVR-6110 from Dow Chemical or Uvacure 1500 from Cytec) are made by direct epoxidation of the corresponding cyclohexene with peracetic

TABLE 14.4.   Select Polyfunctional Diluent Resins

Composition	Nominal Structures	Trade Name
Trimethylolpropane triglycidyl ether		6752[a] Heloxy™ Modifier 48[b] Araldite DY-T[c]
Neopentylglycol diglycidyl ether		6749[a] Heloxy™ Modifier 68[b]
1,4-Butanediol diglycidyl ether		6749[a] Heloxy™ Modifier 67[b] Araldite® RD-2[c]
Cyclohexane dimethanol diglycidyl ether		6757[a] Heloxy™ Modifier 107[b]
Resorcinol diglycidyl ether		6769[a]

[a] Pacific Epoxy Polymers, Inc.
[b] Resolution Performance Products.
[c] Huntsman.

acid. They produce high T_g polymers too brittle for use as the sole epoxy compound in adhesives. They are marketed towards the ultraviolet cationic cure coatings applications.

3',4'-Epoxycyclohexylmethyl
3,4-epoxycyclohexane carboxylate

Tougher polymers of somewhat lower T_g are obtained from glycidyl esters such as the diglycidyl ester of hexahydrophthalic acid (Araldite CY-184 from Huntsman). The glycidyl esters are prepared by reacting the corresponding cycloaliphatic carboxylic acid with epichlorohydrin and dehydrohalogenating with NaOH.

Diglycidyl ester of hexahydrophthalic acid

There is also a variety of other specialty epoxy resins used in limited quantity for applications demanding some specific performance property. Tri- and tetra-functional materials such as triglycidyl-p-amino phenol, triglycidyl isocyanurate, or N,N,N',N' tetraglycidyl-4,4'-diaminophenyl methane (tetraglycidyl methanedianiline) have been used in adhesives requiring high heat resistance and good chemical resistance.

Tetrglycidyl methanedianiline

A typical sales specification for an epoxy resin used in adhesives will include the epoxy equivalent weight (EEW), also sometimes called the weight per epoxy (WPE), viscosity, and density at some specified temperature and the average functionality or number of epoxy groups per molecule. Sometimes a specification on total chlorine is included, which gives some indication of bound chlorine not removed by the NaOH in the dehydrohalogenation process. This bound chlorine is the result of epichlorohydrin side reactions during the epoxy synthesis. A high chlorine content indicates that there will be a large number of molecules of low functionality because each chlorine remaining represents one epoxy ring not formed in the dehydrohalogenation step.

Epoxy Cure Chemistry

Epoxy structural adhesives rely on the chemical reactions of the epoxy group with other reactants to pass from a liquid, wetting state to a solid, load-bearing state. There are a number of ways in which this is done, but all fall under one, or some combination, of three general schemes: step-growth polymerizations through reaction with curing agent, chain-growth polymerizations initiated by Lewis acids, or chain-growth polymerizations initiated by Lewis bases.[§] Often the cure times of the slower step-growth curing adhesives are shortened by including Lewis acid or Lewis base catalysts.

It would be tempting to consider the step-growth and chain-growth polymerization reactions as if they were independent and one could have the choice of either in any particular situation. The truth is that there are aspects of both types of polymerization in the cure of almost every epoxy structural adhesive. Such multiple-cure reactions often make it difficult to calculate the stoichiometry of an epoxy adhesive formulation. One type might predominate, depending on the formulation and cure conditions, but the effects of the other could not be completely discounted. The significance of this statement can be seen by looking at the two generalized reactions.

A typical step-growth reaction is one in which a single epoxy ring reacts with the active hydrogen of the curing agent. The general reaction is:

$$R'XH + \underset{\text{H}_2\text{C}-\text{CHR}}{\overset{\text{O}}{\triangle}} \longrightarrow \underset{\text{R'XH}_2\text{C}-\text{CHR}}{\overset{\text{OH}}{|}}$$

Model epoxy reaction

In order to form a high molecular weight polymer, both the epoxy-containing material and the curative must be at least difunctional. If both behave as difunctional materials, the resulting polymer is linear, and then it is necessary to drive the reaction nearly to completion to obtain a high-enough molecular weight to be useful. Neither of the reactants can contain a significant amount of monofunctional material. Monofunctional impurities or reactive diluents will act as chain terminators and limit the ultimate molecular weight obtainable.

Chain-growth polymerizations occur through the reaction of epoxy rings with the active site on a growing chain and not with each other or with a second curing agent. After being initiated by a Lewis acid or a Lewis base, the growing chain will continue to consume epoxy groups and can reach a high molecular weight very rapidly. Because the epoxy groups are reacting with the growing chain, even a monofunctional epoxy compound can be polymerized. If A* is the initiator and M the monomer unit, the general reaction is:

$$A^* + n\text{M} \rightarrow \text{A} - \text{M}_{n-1}\text{M}^*$$

Chain-growth reaction

where the asterisk indicates the active site of the growing chain.

Lewis bases initiate anionic chain growth polymerizations, the generalized reaction being that of a propagating alkoxide anion.

$$A^- + n\left(\underset{\text{H}_2\text{C}-\text{CHR}}{\overset{\text{O}}{\triangle}}\right) \longrightarrow A\left(\underset{\text{R}}{\overset{\text{R}}{\text{O}}}\right)_n\text{O}^-$$

Anionic chain growth

Lewis acids initiate cationic chain-growth polymerizations. There are several possible chain propagation reactions, and the mechanism of cationic chain growth is still open to

Lewis Acid

[§]Lewis acids and Lewis bases are discussed below in the section on "Chain-Growth Polymerizations."

some debate. However, the propagating species is likely to be an alkylated epoxy cation, or oxonium ion.

In chain-growth polymerizations, epoxy reactants containing more than one epoxy ring per molecule can form tightly crosslinked, three-dimensional networks, as each epoxy group acts as a difunctional reactant.

Step-Growth Polymerization. Only a relative few of the dozens of active hydrogen compounds that undergo reactions with the epoxy ring find widespread use in epoxy structural adhesives. The most common are amines, acid anhydrides, phenols, thiols, and carboxylic acids.

Primary and secondary amines react with epoxy groups to form secondary or tertiary amine linkages. The uncatalyzed reactions proceed at room temperature with the glycidyl ethers and glycidyl esters over many hours with handling strength of >50 psi being achieved in about 2–3 hours. The resulting polymer depends on the structure of the reactants and the degree of cure, and thus a variety of final adhesive properties can be had. The reaction is susceptible to general acid catalysis, that is, by hydrogen bonding. Indeed, the reaction is autocatalytic as secondary alcohols are generated during the ring opening reaction. Better hydrogen-bond donors, such as phenols, are commonly used.[37–39] The reaction of the secondary amine with the epoxy group produces a tertiary amine, which can in turn be the Lewis base that initiates an anionic chain growth polymerization of the remaining epoxy groups, depending on cure conditions.

Epoxy amine reaction

The simple linear aliphatic diamines, $H_2N(CH_2)_n NH_2$, can be used as curatives in adhesives. For small values of n, the short distance between the amine groups can hinder the reaction of the second amine and slow the

cure process. Also, the resulting products tend to be brittle for values of n less than about 6.

More flexible, tougher products can be obtained by using liquid diamines or polyamines having more flexible backbones. For example, Jeffamine T.403 (Huntsman) is a low-viscosity liquid (70 mPa/sec) having a molecular weight of about 440 and an amine hydrogen equivalent weight of about 81. It is a poly(propylene oxide) triamine made from the polyether initiated by trimethylol propane, and is promoted as a flexible crosslinker for epoxy systems. Their lower reactivity relative to other aliphatic amine may be a consideration in their usage.

Various polyamines also are useful as curatives for epoxy adhesives. An example of a simple polyamine is diethylenetriamine (DETA).

$$H_2NC_2H_4NHC_2H_4NH_2$$
Diethylenetriamine

Its higher-weight homologues also are quite useful and are made by adding (CH_2CH_2NH) groups, leading to a homologous series of the form.

$$H_2N-(\ C_2H_4NH\ )_n-C_2H_4NH_2$$

where n is two, the material is triethylenetetraamine (TETA), and for $n = 3$ it is tetraethylenepentamine (TEPA), and so on. These three are the most important members of this series for adhesive applications, and are used more often than the simple aliphatic diamines. They often are used in combination with some other curative. Calculating stoichiometry can be difficult with these polyamines. All the active hydrogens might not be available because of steric factors introduced by the first reactions, or because once the first one or two active hydrogens are reacted with epoxy rings, the molecule is anchored into the chain and cannot readily diffuse to other epoxy groups. Therefore, a modest excess of polyfunctional amine is often added to the adhesive formulation. If the amine is used in too much excess, as with any other imbalanced stoichiometry, the final composition may have too many dangling chain ends, that is, be soft and cheesy, or even have unreacted amine that can leach from the product.

Other series of polyamines can be made too. For example, propyl groups can replace some or all of the ethyl groups, or the compound might be modified by reaction with an excess of a monoepoxide to give a hydroxy functional amine. Such modifications are made to improve adhesive properties and sometimes to lower toxicity of the curing agent or make it easier to handle.

Among the most important amine functional epoxy curing agents are the polyamidoamine resins. These are made from dimerized unsaturated fatty acids by reaction of the dimer acid with a polyamine such as diethylenetriamine.

The polyamidoamines are very high-viscosity liquids, some having viscosities over 50,000 mPa/sec. Typical amine equivalent weights are 100–150. The polyamidoamines react with bisphenol A epoxy resins at room temperature although the adhesives usually require several hours to reach sufficient molecular weight to carry a load. Cure times can be shortened to a few minutes at about 150°C.

Epoxy adhesives cured with polyamidoamines are flexible, tough, durable adhesives useful on a wide variety of adherends. They probably have contributed heavily to making the words "epoxy" and "adhesive" equivalent for many people.

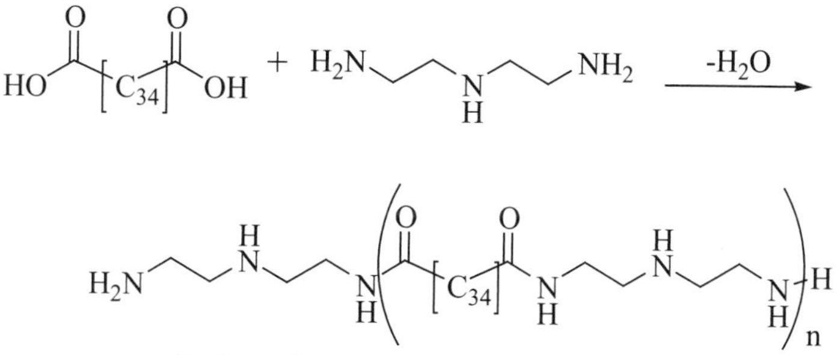

Polyamide Polyamine synthesis

These polyamidoamines are available from several suppliers worldwide (Cognis (Veramide®), Arizona Chemical (Uni-Rez®), Air Products (Ancamide™), and others) and are among the most common curatives in the general-purpose, "do-it-yourself " two-package epoxy adhesives. They have a distinctive odor somewhat like popcorn and are easily recognized in adhesive formulations. The polyamide backbone does contribute to the overall good mechanical properties of the polyamide amine cured adhesives.

A variety of aromatic polyfunctional amines is also used in curing epoxy adhesives. They generally are slower to react than the aliphatic amines and require a heat cure to be practical. They do provide generally better high-temperature properties than the aliphatic amines. The most commonly used aromatic amine curatives are m-phenylenediamine (MPDA, DuPont), methylenedianiline (MDA, Bayer), and diaminodiphenyl sulfone (DDS, Aceto). Albemarle Corporation's Ethacure® 100 is claimed to be a more user-friendly

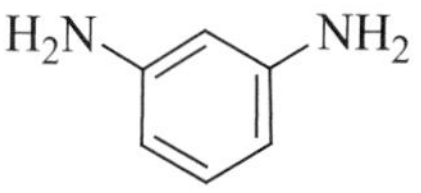

m-Phenylenediamine

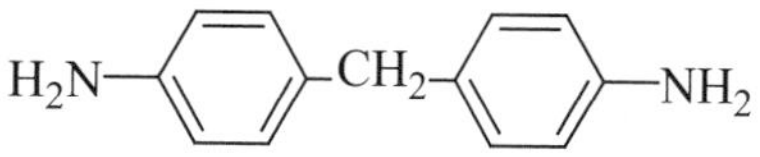

Methylenedianiline

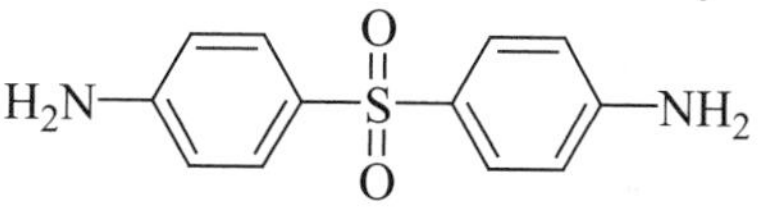

Diaminodiphenyl sulfone

curative that can be used in place of MPDA or MDA.

All of the curatives described so far are used to make two-package adhesives in which the curing agent is packaged separately from the epoxy resin. Once they are mixed, they have a limited pot life, usually less than a few hours.

It is possible to make one-package epoxy adhesives that can have very long shelf lives at room temperature but cure rapidly when heated. One amine curative widely used to make single-package heat-cured epoxy adhesives is dicyandiamide (cyanoguanadine), commonly known as dicy.

Dicyandiamide

It is made by dimerizing cyanamide in basic aqueous solution, and is a colorless solid melting at 208°C. Dicyandiamide is soluble in polar solvents, but at room temperature is insoluble in bisphenol A epoxy resins. It can be made into a very fine powder and milled into epoxy resins to form stable dispersions. Because the dicy is insoluble in the epoxy, the only possible reaction sites are at the particle surfaces. Although some reaction certainly occurs over a short time, the adhesives easily can have a useful shelf life of six months. On heating to about 150°C, the dicyandiamide becomes soluble in the epoxy resin, and the adhesive polymerizes rapidly. Cure can be accelerated by incorporation of tertiary aromatic amines or substituted ureas.

Carboxylic acids can be used to cure epoxy adhesives or otherwise modify epoxy adhesives. The reactions can be complex. If no hydroxyl groups are present initially, the first reaction will be that of the active hydrogen with the epoxy ring to form an ester. This will produce an hydroxyl group on the backbone and allows for competing reactions. The organic acid can catalyze the etherification reaction with the hydroxyl group or undergo a condensation esterification reaction directly with the hydroxyl group.

Epoxy acid reaction

Etherification reaction

Condensation reaction

These reactions usually are slow at room temperature, and the adhesives must be cured with heat. Tertiary amines or amidines can be used to catalyze the reaction whereby the carboxylate anion is the nucleophile and the ammonium or amidinium ion act as hydrogen-bond donors. Dusek et al. have shown that transesterification only occurs after essentially complete consumption of the carboxylic acid in compositions initially having equivalent amounts of acid and epoxy groups.[40,41] The transesterification leads to an equilibrium sol–gel composition.

Acid anhydrides also can be used to cure epoxy adhesives although they usually are used only where good service at high temperatures is required. Most of the anhydride-cured epoxy adhesives are cured at high temperature. Because most of the anhydrides are relatively small molecules, the products tend to be tightly crosslinked and can be somewhat brittle.

The first step in the anhydride cure in the absence of a tertiary amine catalyst is ring opening of the anhydride by an active hydrogen, perhaps from water or hydroxyl groups already present on the epoxy resin.

The resulting acid then reacts as a typical organic acid. In the presence of a tertiary amine, the initial formation of a zwitterionic acylammonium carboxylate salt is possible. The carboxylate anion then reacts with the epoxide as a nucleophile.

$$R_3N + H_2C\!-\!CHR' \longrightarrow R_3\overset{\oplus}{N}\diagdown\diagup\!\!\overset{R'}{\underset{O}{\diagdown}}{}^{\ominus}$$

Formation of anionic chain initiator

This anion can continue to react with epoxy rings, adding them to the chain until the anion is destroyed in some side reaction.

Anhydride ring opening by hydroxyl

Anionic chain propagation

An interesting class of curatives, but with more limited usage, is that of thiols (also known as mercaptans). Thiols react with terminal epoxide groups quite rapidly when a tertiary amine catalyst is present even at temperatures below 0°C. The reaction is similar to that found with hydroxyl groups and produces a polythioether product. These are the familiar "five-minute" epoxy adhesives and have the characteristic odor of thiols. The rapid cure can be controlled so as to give very good open, or handling, time by the incorporation of very minor amounts of weakly acidic components such as chlorophenol, paraben esters, or carboxylic acids.[42] One limitation on wider application is the commercial availability of suitable polythiols. CapCure™ trifunctional thiols (Cognis) are widely used, but give cured products with low T_gs unsuitable for many structural applications.

Chain-Growth Polymerizations. Chain-growth polymerizations are very important to many commercially successful epoxy structural adhesives. They can be extremely rapid and contribute to the fast cure times needed for high productivity in many manufacturing operations.

A Lewis base is a compound that contains an unshared pair of electrons capable of undergoing chemical reactions. Tertiary amines are examples of Lewis bases, and often are used in epoxy curing agents. In an anionic epoxy polymerization the propagating species is the alkoxide anion generated by the reaction of the Lewis base with an epoxy ring.

The product is a polyether, which can be tightly crosslinked when polyfunctional epoxides are used, as each epoxy ring can become part of a different chain.

Lewis acids, compounds with empty orbitals capable of accepting electron pairs, initiate cationic polymerization of epoxy resins. In this case the propagating species is a positive ion. The most commonly used Lewis acids are the boron trihalides, particularly BF_3 and BCl_3. They usually are used in the form of complexes because both are gases at room temperature and are so reactive with epoxy resins that they can be difficult to control. Lewis base complexes with the boron trihalides have much lower reactivity at room temperature but can react quickly on relatively mild heating, depending on the particular complex. Boron trifluoride readily forms complexes with ethers, alcohols, and amines, and several of these complexes are commercially available. A boron trifluoride ethanolamine complex can be included in the curative portion of a two-package epoxy adhesive. When the two packages are mixed at the time of use, the mixture can have a pot life of hours at room temperature but polymerize in minutes at temperatures of 100–150°C.

One very interesting new application of Lewis acids in curing epoxy adhesives has appeared within the last 25 years. The Lewis acid initiator for the cationic polymerization is formed by the heat or ultraviolet light-induced decomposition of Lewis acid:Lewis base salts. Several patents by Crivello and coworkers[43–47] describe compounds containing aromatic onium salts such as iodonium or sulfonium, in which the cation is stabilized by the aromatic rings. The field is covered in more detail in a recent review.[48]

Diphenyl iodonium salt

The counterion generally is a large stable anion such a $[SbF6]^-$, $[AsF6]^-$, $[PF6]^-$, or, $[BF4]^-$. The more stable anions are less likely to terminate the growing cationic chains than are typical anions. The salts can be dissociated with heat to release the cation, which appears capable of initiating cationic polymerization of many materials in addition to epoxy rings. If the proper dye sensitizers are added, the cation can be liberated by ultraviolet light to initiate the polymerization. The cations persist for quite some time after the light source is removed. Acids of the form H^+SbF5X^-, where X is a halogen, also have been used to catalyze epoxy reactions.[49]

Evolution

There are many examples of the evolution of epoxy structural adhesives. The development of heat activated epoxy adhesives for use with induction heating apparatus[50] can lead to increased assembly line productivity. Epoxy adhesives suitable for use on oily metal[51] can reduce the number of manufacturing operations. Adhesion promoters such as dithiooxamides can be included in epoxy formulations to improve adhesion and durability.[52] Epoxy resins have been modified with phosphorous[53] to introduce flame retardance. Siloxanes have been used to modify both epoxy adhesives and adherends in order to improve adhesion and durability.[54] Epoxy adhesives are being packaged in novel ways such as forming the two parts of the adhesive into sheets and interleaving them to produce a room-temperature stable, heat-curable construction.[55]

Summary

Epoxy structural adhesives have proven to be versatile and reliable compositions. Their widespread use and acceptance is in part due to the varied chemistry of the epoxy ring and the skill of scientists and adhesive formulators in developing high-quality compositions that can produce reliable, reproducible structural joints even when applied by relatively unskilled users. New compositions and chemical reactions continue to be disclosed, and it is certain that the knowledge of epoxy chemistry will continue to grow. Many of the new discoveries will find their way into new epoxy structural adhesives.

URETHANE STRUCTURAL ADHESIVES

Introduction

The term "urethane adhesive" as it is generally used encompasses a lot of chemistry that is not necessarily urethane chemistry. "Urethane" is the common name for the compound ethyl carbamate. In common usage, "urethane adhesive" generally means an adhesive polymer derived from isocyanate chemistry and reactions of isocyanates with active hydrogen compounds. However, isocyanate reactions do not always lead to urethane linkages, and there are ways of arriving at urethane linkages without involving isocyanates. In this section the common approach is taken; that is, an adhesive that uses reactions of the isocyanate group to bring about polymerization in the bond line is considered a urethane adhesive.

The study of industrial applications of isocyanate chemistry and polymers derived from isocyanates received much attention in Europe, particularly Germany, in the 1930s and during World War II. Patents on aspects of urethane chemistry appeared as early as 1937.[55] The effort going into understanding isocyanate chemistry and commercializing urethane products continues.[56–60]

The isocyanate group consists of a linear arrangement of nitrogen, carbon, and oxygen atoms.

$$-N=C=O$$

Several possible electronic configurations can be drawn, most of which involve a positive charge on the central carbon atom. This partial positive charge on the carbon atom accounts for much of the reactivity of the isocyanate group with nucleophilic groups. Not only does the isocyanate group react with a variety of potential curing agents, but also it is very reactive with many of the adherend surfaces on which urethane adhesives are used.

The same high reactivity that makes the isocyanate a desirable reactant for structural adhesives also renders the unreacted isocyanate more acutely toxic than, for instance, epoxy adhesives. Only a few isocyanates are safe enough and easy enough to handle that they find widespread use in urethane structural adhesives.

Isocyanate Preparation

Organic isocyanates are the major building blocks of urethane structural adhesives. They can be synthesized by a variety of routes, but most of the commercially available isocyanate compounds used in adhesives are made by the reaction of a precursor primary amine or amine salt with phosgene, followed by dehydrohalogenation. The reaction with phosgene usually is carried out at a relatively low temperature, less than 60°C, and then the temperature is raised to 100–200°C to remove the HCl.

$$RNH_2 + COCl_2 \longrightarrow$$
$$[RNHCOCl + HCl] \xrightarrow{\text{heat}} RNCO + 2HCl$$

intermediate   isocyanate

Isocyanate synthesis by phosgenation

There are many possible side reactions, and the yield depends on the reaction conditions, which usually are specific to the starting materials and desired product. Much work has gone into determining the reaction conditions for manufacturing as clean a product as possible.

The amount of isocyanate in a commercial isocyanate or isocyanate-containing formulation usually will be specified as weight percent isocyanate (as NCO) or as an amine equivalent weight, which is the weight of material containing sufficient isocyanate to react with one mole of amine hydrogen.

Isocyanate Reactions

Organic isocyanates can undergo a large number of reactions with active hydrogen compounds. One test commonly used to determine the presence of active hydrogen atoms and the number of active hydrogens per molecule is the Tschugaeff–Zerewitinoff analysis or, more commonly, the Zerewitinoff test. An excess of a Grignard reagent, methyl magnesium iodide, is added to the sample to be tested, and the amount of methane evolved in measured.

$$CH_3MgI + RH \rightarrow CH_4 + RMgI$$

Zerewitinoff reaction

Urethane adhesives take their name from the product of the most common step-growth polymerization reaction used to generate the adhesive polymers. Isocyanates react with hydroxyl groups to create urethane (or carbamate) linkages:

$$R-NCO + R'OH \longrightarrow R-\overset{H}{N}-\overset{O}{\overset{\|}{C}}-OR'$$

Urethane formation

If both the isocyanate and the hydroxyl-containing material are difunctional, if the mixture is made up to have one isocyanate per hydroxyl, if there are no side reactions, and if the reaction can be driven to completion, a single linear thermoplastic polymer should result. If one or more of the reactants is more than difunctional, it is possible to create an infinite three-dimensional network.

In general, primary hydroxyl groups are faster to react than secondary hydroxyls, which are in turn faster than tertiary hydroxyls, absent catalysts. When reacted with hydroxyl-containing compounds, aliphatic isocyanates tend to be more sluggish than their aromatic counterparts. Urethane linkages made with tertiary hydroxyls tend to be less stable and at high temperature can dissociate into an olefin and an amine with loss of carbon dioxide. Metal compounds, particularly tin compounds such as dibutyl tin dilaurate, and various amines catalyze the isocyanate–hydroxyl reaction.

Isocyanates will react with amines to produce substituted ureas, primary amines being faster than secondary amines.

$$R{-}NCO \quad + R'NH_2 \longrightarrow RHN{-}\overset{\displaystyle O}{\overset{\|}{C}}{-}NHR'$$

Urea formation from 1° amine

$$R{-}NCO \quad + R'R''NH \longrightarrow RHN{-}\overset{\displaystyle O}{\overset{\|}{C}}{-}NR'R''$$

Urea formation from 2°C amine

The ureas are more rigid linkages than the urethane structure, but they also are generally more resistant to heat and chemical degradation. The reaction of isocyanates with amines is generally so rapid that it is nearly impossible to control the reaction well enough to make it useful in formulating urethane adhesives. What usually happens is that the reaction takes place faster than it is possible to mix the adhesive. If it is necessary to include urea linkages in the final production, that can be done by making a prepolymer that has the urea linkages in it already or by taking advantage of the slower reaction of isocyanates with water.

Small (much less than stoichiometric) amounts of diamines sometimes are added to the hydroxyl portion of the adhesive to provide a rapid but limited molecular weight increase as soon as the components are mixed.[61,62] In this way a two-package adhesive can be made that will flow easily before mixing but will not readily flow after mixing. Such adhesives can be applied to vertical surfaces or overhead, and will remain in place until the bonds are closed and cured.

The reactions of amines and isocyanates are important in adhesives because of the possible reaction of isocyanates with water. Because isocyanates react readily with water, raw materials used in formulating adhesives must be dry, and the compositions must be protected from moisture, including atmospheric humidity, during storage. The first reaction with water is the formation of a carbamic acid, which rapidly loses carbon dioxide to form a primary amine.

$$RNCO + H_2O \longrightarrow$$

$$\left[R{-}\overset{\displaystyle H}{\underset{}{N}}{-}\overset{\displaystyle O}{\overset{\|}{C}}{-}OH \right] \longrightarrow$$

$$RNH_2 + CO_2$$

Water NCO reaction

The primary amine then can react with another isocyanate to produce a urea. Thus one mole of water consumes at least two moles of isocyanate, builds molecular weight, and liberates carbon dioxide in the process. If this happens in a closed container such as a drum of adhesive, the result can be explosive, particularly as the reaction mass rises in temperature because of the exothermic reactions. On the other hand, these reactions can be useful in a bond line because under the proper conditions desirable urea linkages can be introduced into the curing adhesive through the reactions of small amounts of water normally present on the adherend surfaces.

The hydrogen atom attached to the nitrogen atom of the urethane group is active enough that it can react with another isocyanate group to produce an allophanate. This is an additional crosslinking mechanism for urethane polymers and can disturb the stoichiometry of the system by consuming an additional isocyanate group for each allophanate formed. Elevated temperatures usually are needed to produce allophanates in uncatalyzed systems; allophanation reactions can be catalyzed by tertiary amines.

$$R''NCO + R{-}\overset{\displaystyle H}{\underset{}{N}}{-}\overset{\displaystyle O}{\overset{\|}{C}}{-}OR' \longrightarrow {}''R{-}\overset{\displaystyle H}{\underset{}{N}}{-}\overset{\displaystyle O}{\overset{\|}{C}}{-}\underset{\displaystyle R}{\overset{}{N}}{-}\overset{\displaystyle O}{\overset{\|}{C}}{-}OR'$$

Allophanate formation

In a similar fashion, a urea hydrogen atom can react with an additional isocyanate group to produce a biuret.

$$R''NCO + R{-}\overset{\displaystyle H}{\underset{}{N}}{-}\overset{\displaystyle O}{\overset{\|}{C}}{-}\overset{\displaystyle H}{\underset{}{N}}{-}R' \longrightarrow R''{-}\overset{\displaystyle H}{\underset{}{N}}{-}\overset{\displaystyle O}{\overset{\|}{C}}{-}\underset{\displaystyle R}{\overset{}{N}}{-}\overset{\displaystyle O}{\overset{\|}{C}}{-}\overset{\displaystyle H}{\underset{}{N}}{-}R'$$

Biuret formation

Isocyanate groups also will react with themselves to form a variety of compounds. Two isocyanate groups can react to form a dimer or uretidinedione.

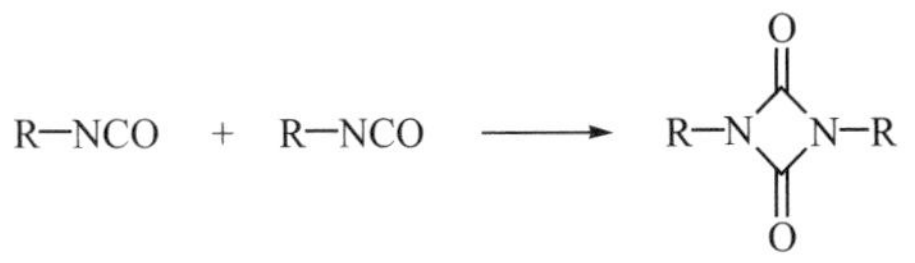

R—NCO + R—NCO $\longrightarrow$ R—N$\overset{\overset{O}{\parallel}}{\underset{\underset{O}{\parallel}}{\diamondsuit}}$N—R

Isocyanate dimerization

These dimers can be dissociated to regenerate the original isocyanates with heat, and some of the dimers, such as the dimer of toluene diisocyanate, have become commercially important.

Three isocyanate groups can react to form a trimer or substituted isocyanurate ring. Phosphines or bases such as sodium acetate or sodium formate can catalyze this reaction. The isocyanurate ring is thermally stable, has good chemical resistance, and can enhance the resistance of a urethane adhesive to aggressive environments.

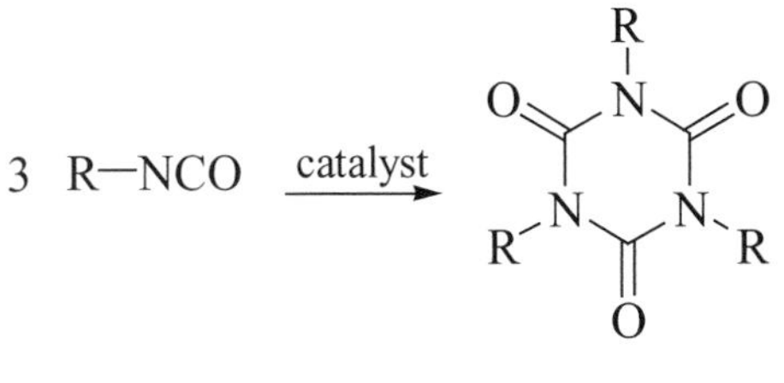

3 R—NCO $\xrightarrow{\text{catalyst}}$

R
|
N
O⟋ ⟍O
| |
R—N   N—R
   ⟍N⟋
    |
    O

Isocyanates also can react with each other to produce carbodiimides with the loss of carbon dioxide. This reaction requires high temperatures unless catalyzed by specific phosphorus compounds. Formation of carbodiimides normally is not an important cross-linking mechanism in polyurethane adhesives. However, carbodiimides are sold by Dow Chemicals (Ucarlnk™), Nisshinbo Industries (Carbodilite™), and Stahl USA (XR-2569). They have been recommended as water scavengers, crosslinkers, and stabilizers for carboxyl functional polyurethanes. The carbodiimide can react with water to give a urea, which still can react with additional isocyanate to produce a biuret.

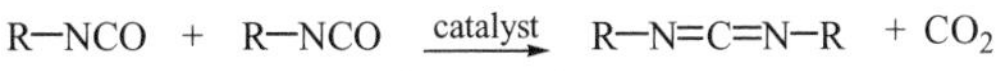

R—NCO + R—NCO $\xrightarrow{\text{catalyst}}$ R—N=C=N—R + CO$_2$

Carbodiimide formation

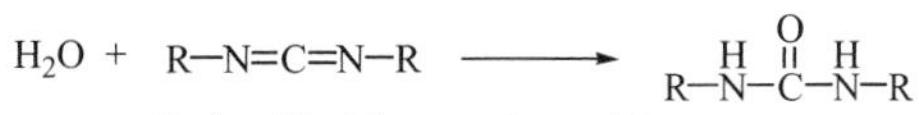

H$_2$O + R—N=C=N—R $\longrightarrow$ R—N$\overset{H}{}$—C$\overset{\overset{O}{\parallel}}{}$—N$\overset{H}{}$—R

Carbodiimide reaction with water

Carbodiimides can react with additional isocyanate groups to form uretone imines, which sometimes are used to modify polyisocyanates used in urethane structural adhesives.

R—N=C=N—R + R'—NCO $\longrightarrow$ R'—N$\diamondsuit$N—R

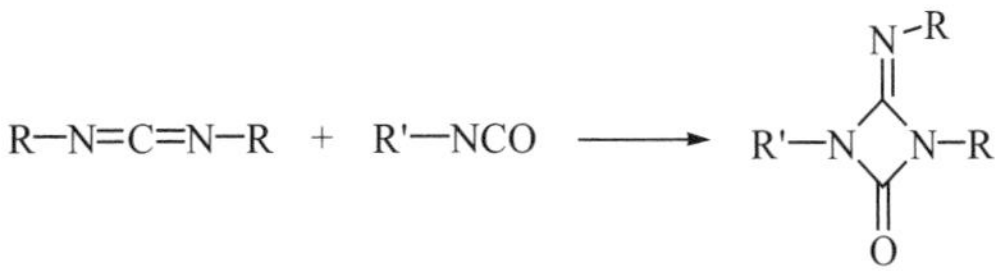

Uretone imine formation

Important Isocyanates

Dozens of isocyanate functional compounds have been synthesized, but only a few find much use in urethane structural adhesives. The choices are largely dictated by a combination of performance, price, and safety considerations. Most of the materials used in adhesives are derived from the aromatic isocyanates, toluene diisocyanate (TDI) and 4,4'-diphenyl-methane diisocyanate (MDI).

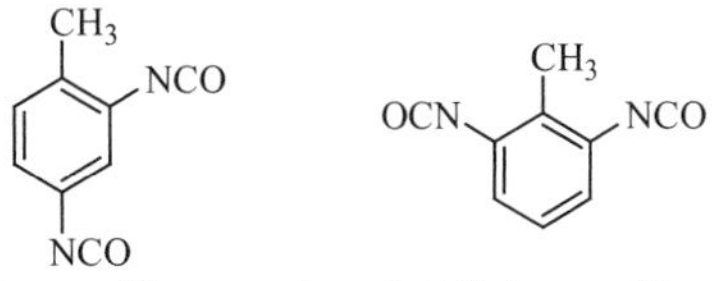

2,4-Toluene diisocyanate   2,6-Toluene diisocyanate
(TDI)

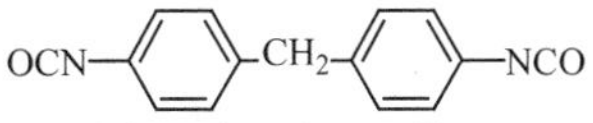

4,4'-Diphenylmethane diisocyanate
4,4-Methylene bis(phenyl isocyanate) (MDI)

Where color and light stability of the adhesive are important, and cure speed or cost is less important, aliphatic isocyanates are frequently used. Adhesives derived from isophorone diisocyanate (IPDI), hexamethylene diisocyanate (HDI), or 4,4'-dicyclohexylmethane diisocyanate (H$_{12}$MDI) are available.

Isophorone diisocyanate (IPDI)

Hexamethylene diisocyanate (HDI)

4,4'-Dicyclohexylmethane diisocyanate (H₁₂MDI)

The common commercial TDI is an 80/20 mixture of the 2,4- and 2,6-isomers. The pure 2,6-isomer is available also and is sometimes called TDS. To make toluene diisocyanate, toluene first is nitrated to produce a mixture of the 2,4- and 2,6-dinitro isomers. The dinitrate is reduced to the diamine and reacted with phosgene, which is followed by dehydrohalogenation to give the diisocyanate. Because of its relatively high vapor pressure and toxicity, adhesives rarely contain toluene diisocyanate monomer. Typically, excess TDI is reacted with another material such as trimethylolpropane or a polyester diol or polyether diol to produce a higher-molecular-weight isocyanate functional compound, which is safer and easier to handle than the free TDI. Still, there will be some free TDI present in the adduct, and much work has gone into finding synthesis schemes to minimize the free TDI in prepolymers and adducts.[63]

MDI can be considered the first member of a series of polyisocyanates of the general form:

Aromatic polyisocyanates

MDI is the most important member of the series although materials with n of 1 or more also are available. The precurser amine is made by condensation of aniline hydrochloride with formaldehyde, followed by reaction with phosgene and dehydrohalogenation. When the aniline is present in excess, the diamine and consequently the diisocyanate are produced in greatest yield, with nearly all of this being the 4,4'-isomer.

MDI is a solid at room temperature with a melting point of about 38°C, which usually is stored and shipped in the molten state for convenience. However, on standing in the liquid state, the MDI slowly dimerizes, and the liquid MDI becomes saturated with dimer at about 1 percent dimer by weight. The dimer then begins to precipitate. The isocyanate content of the remaining liquid will then remain constant.

Hexamethylene diisocyanate (HDI) can be used as an aliphatic crosslinker in urethane adhesives but presents significant health risks because of its toxicity and high vapor pressure. HDI is more commonly used in the form of its biuret, which is much safer and easier to handle than HDI.

Biuret of hexamethylene diisocyante

Isophorone diisocyanate (IPDI) is another low-viscosity aliphatic diisocyanate that is useful in formulating light-stable polyurethane adhesives, and is somewhat lower in cost than hexamethylene diisocyanate. IPDI has a low vapor pressure at room temperature but, like most low-weight isocyanates, still can present a health hazard. Higher-molecular-weight adducts of IPDI, such as its isocyanurate trimer, are available. The trimer is a high-melting solid (100–115°C), and one loses the advantage of the low viscosity liquid in using it but gains in safety and easy handling.

In IPDI, the two isocyanate groups are not equivalent. One is attached directly to the aliphatic ring, and a methylene group separates the other from the ring. Because their environments are different, the reactivity of the two groups are different; and the reactivity can depend on the choice of catalyst. For instance, Ono and coworkers[64] have shown that when IPDI is reacted with a primary hydroxyl group, the primary NCO is most reactive when the catalyst is 1,4-diazabicyclo-[2.2.2]octane. The secondary NCO attached to the ring is the most reactive when dibutyl tin dilaurate is the catalyst.

The saturated analogue of MDI, 4,4'-dicyclohexyl methane diisocyanate, has found limited use as an aiphatic isocyanate in adhesives. This material is known by a variety of names including Desmodur W™ (Bayer), hydrogenated MDI (or HMDI or H_{12}MDI), reduced MDI (RMDI), and saturated MDI (SMDI). It is a low-viscosity liquid with a fairly high vapor pressure, so it too must be handled with care. In adhesive compositions, the diisocyanate usually is used to make an isocyanate functional prepolymer by reacting excess diisocyanate with a hydroxyl or amine functional polymer such as a polyester diol.

Because it contains two saturated six-member rings, 4,4'-dicyclohexyl methane diisocyanate can exist in three isomeric forms, in which the orientations of the NCO groups with respect to the rings and each other are different. The *trans–trans* isomer is a solid at room temperature so that over time this isomer can precipitate from the remaining liquid isomers. Samples of 4,4'-dicyclohexyl methane diisocyanate that have been stored at cool temperatures for a while often have a solid layer of *trans–trans* precipitate in the bottom.

An offering by Cytec Specialty Chemicals, the meta isomer of tetramethyl xylene diisocyanate (TMXDI) is interesting because it contains an aromatic ring, but the NCO groups themselves are aliphatic isocyanates and have reaction characteristics typical of aliphatic diisocyanates. It reacts even more sluggishly than the more standard aliphatic isocyanates because of steric interactions, making the reactions easier to control. Compounds such as dimethyl tin dilaurate, lead octoate, or tetrabutyl diacetyl distannoxane have been shown to be effective catalysts for the isocyanate–hydroxyl reaction. The manufacturer claims that it is less toxic than many other isocyanates.

m-TMXDI

Blocked Isocyanates

Blocked isocyanates are compounds formed by the reaction of an organic isocyanate with an active hydrogen compound where the reaction is reversible with moderate heat. The blocked isocyanate can be used in formulating adhesives or other reactive compositions even in the presence of materials that normally would react rapidly with the isocyanate. Phenol is one example of a blocking agent used with isocyanates. There are many other blocking agents in use, including lactams, oximes, and malonates. Even isocyanate dimers such as TDI dimer could be considered blocked isocyanates because they will dissociate with heat to regenerate the isocyanates.

Phenol blocking reaction

There are two major limiting factors to the use of blocked isocyanates in urethane structural adhesives. First, the adherends must be sufficiently heat resistant to withstand the temperatures needed to cause rapid dissociation of the blocked isocyanate. Second, the blocking agent is present to continue competition with the intended curing agent for reaction with the isocyanate. After the adhesive is cured, the blocking agent remains trapped in the bond line and could contribute to poor mechanical properties in the bond, poor resistance to harsh environments, or exude or extract into the environment. Applications on porous substrate, such as wood, or where de-blocking occurs before closure of the bond, such as laminating, are compatible with some of these limitations.

Evolution

Urethane adhesives have also received much attention. New processes are being developed for the production of isocyanate containing

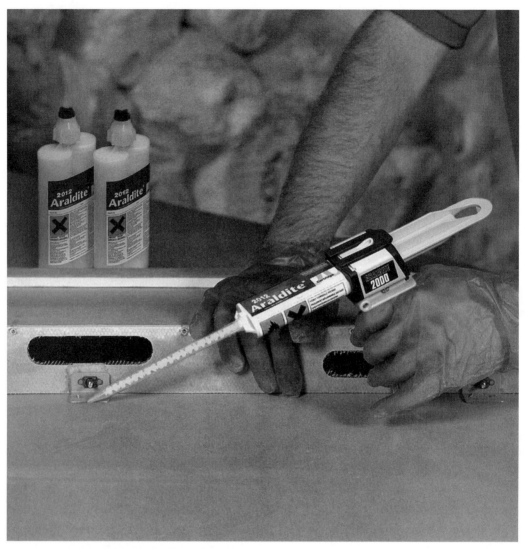

Fig. 14.4.   Assembly of an aircraft headliner using an Araldite™ two-part urethane adhesive. (Courtesy of Huntsman Advanced Materials.)

raw materials[65] and methods of stabilizing them.[66] Formulators continue to produce new mixtures and intermediates from standard materials (Fig. 14.4).[67,68] New blocking agents for isocyanates are being developed to facilitate the formulation of stable urethane adhesives compositions rapidly curable when heated.[69,70]

Summary

Urethane structural adhesives have proved quite successful in bonding applications that

take advantage of their toughness, particularly when the adhesive must function at low temperatures. The reactivity of the isocyanate group may require some extra care in handling and storing the adhesives but the same reactivity provides many opportunities for the polymer chemist and adhesive formulator. The principal reaction used in curing urethane structural adhesives is the reaction of the isocyanate group with primary and secondary hydroxyl groups. Both the isocyanate groups and the hydroxyl groups can be carried on a wide variety of low-molecular-weight oligomers such as polyesters, polyethers, polybutadienes, polyurethanes, and polymer blends so that many of the desired properties of the final adhesive can be built into the polymer before the adhesive is cured.

ACRYLIC STRUCTURAL ADHESIVES

Introduction

Acrylic structural adhesives are more recently developed products compared to epoxy or urethane adhesives. They first appeared in Europe in the mid-1960s and were commercialized in the United States a few years later.[71] They share many similarities to anaerobic adhesives that preceded them by a few years.[†] The first offerings were rather brittle products of use in limited applications where toughness and flexibility were not critical. However, because these new acrylic structural adhesives polymerized in the bond line through a free radical chain growth polymerization mechanism, they did have the big advantage of curing rapidly at room temperature. This polymerization mechanism offered a variety of potential advantages in manufacturing. Bonded structures could be assembled relatively rapidly without the need of curing ovens. Assemblies did not need to be stored for long periods while room-temperature cures were completed. The adhesives could be made at very low viscosities for easy handling and dispensing. In the past 35 years there has been much creative work done with the chemistry of acrylic adhesives so that now they are available in many forms with a wide range of properties.

Acrylic Monomers

All acrylic structural adhesives consist basically of a solution or a mixture of polymers and unsaturated, low-molecular-weight, free-radical-polymerizable monomers with other materials added as needed for the particular intended use. In addition, some precursor of the polymerization initiator will be present.

The most common monomers used are methacrylic acid (2-methyl propenoic acid) and its esters or, less commonly, acrylic acid (propenoic acid) and its esters.

Methacrylic acid   Acrylic acid

Several processes for making acrylic acid have been developed.[72] Since the 1980s manufacturers have almost exclusively used a two-step, vapor-phase oxidation of propylene that proceeds through acrolein. The acrylic acid then can be esterified with the appropriate alcohols. If current shifts in petroleum prices continue, one of the alternative routes to acrylic acid may become economically preferable.

Unlike acrylic acid, methacrylic acid continues to be produced by a variety of processes including oxidation of ethylene, propylene, or isobutylene.[73] The older commercial process for making methacrylic acid based on propylene proceeded through acetone cyanohydrin. This process is steadily declining in usage because of high waste costs. Rapid development of specialty catalysts in the 1980s by several Japanese companies has moved isobutylene-based production to the forefront. Also, for companies with internal synthesis gas (syngas) production, routes based on ethylene are economically viable.

Direct esterification of methacrylic acid with alcohols using sulfuric acid or other catalysts can be used to prepare methyl methacrylate (MMA) and other esters. Commercial routes for the direct preparation of MMA and some lower alkyl esters also exist. In the 1990s, researchers at Shell developed a direct route to MMA from propyne (methylacetylene), carbon monoxide, and methanol using a Pd(II) catalyst. The limited availability of propyne may slow the expansion of this highly efficient route to high purity MMA. Transesterification of MMA is often the preferred route for the preparation of other esters.

[†]See the discussion in [11, pp. 217].

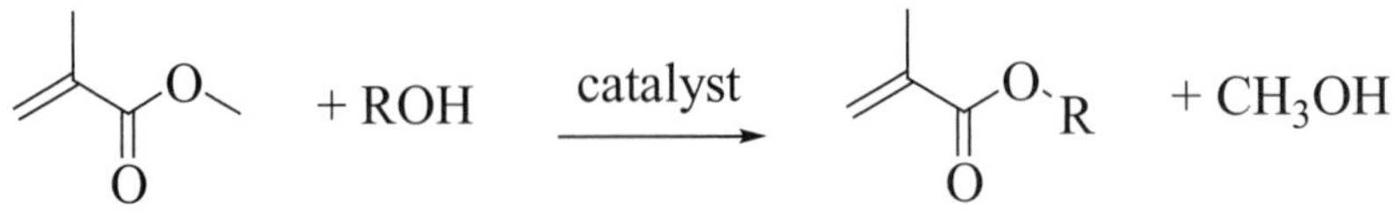

Transesterification of MMA to other methacrylate esters

The choice of monomers that are useful in acrylic structural adhesives is rather limited. Cost always is an important factor, and because acrylic structural adhesives consume only a very small portion of the world's output of acrylic monomer, the formulator usually must rely on acrylic monomers that are made in large quantity for other uses. In addition, the monomers must polymerize readily at room temperature. If a mixture of monomers is to be used, the monomers must copolymerize easily. Finally, the monomers must be good solvents or dispersants for the polymers used in formulating the adhesive. The acrylic monomers finding most use in acrylic structural adhesives are methyl methacrylate and tetrahydrofurfuryl methacrylate. The later, albeit more expensive, has a much higher flashpoint and a generally perceived low odor, yet maintains good solvency.

Curing

The curing reaction of an acrylic structural adhesive is the chain-growth polymerization of the acrylic monomer. The monomer units are not reactive with each other but react only with a growing chain having an active site on one end. In order to begin a chain, one must generate an initiator in the monomer solution. In the case of acrylic structural adhesives, this initiator nearly always is a free radical, a species having an unpaired, reactive electron.

After chains have been initiated, there are three general types of reaction that can occur: propagation, chain transfer, and chain termination. If "*" represents the active site, "A*" represents the initiator, and M is a monomer unit, then the four reaction types are:

Initiation: $A^* + M \rightarrow AM^*$
Propagation: $AM^* + nM \rightarrow AM_nM^*$
Chain transfer: $AM_nM^* + M \rightarrow AM_nM + M^*$

Termination: $RM^* + R'M^* \rightarrow RMMR'$
                (combination)
$RM^* + R'M^* \rightarrow RMC{=}CH_2 + RMCHCH_3$
(disproportionation)

In addition, a growing chain might undergo chain transfer or be terminated by reaction with a variety of unknown impurities invariably present in any mixture.

The initiating radical is usually created in a redox reaction. Common reactions involve the reduction of an organic peroxide by some reducing agent such as an amine or an ion capable of undergoing a one-electron transfer reaction. Anaerobic adhesives generally rely on metal ions derived from the surface to be bonded as part of the redox system. Acrylic adhesives generally embody the reducing agent in the monomer mixture and place the oxidizer in a monomer free package.

One example of an efficient free radical initiator generating reaction is the reduction of

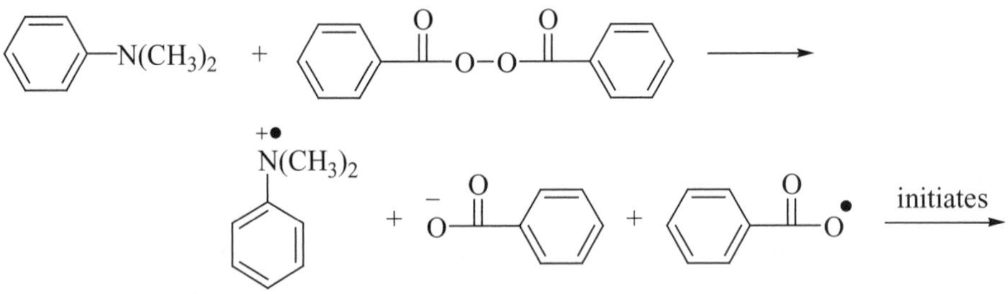

Benzoyl peroxide reaction with N,N-dimethyl aniline

diacyl peroxides such as benzoyl peroxide (BPO) by tertiary aromatic amines such as N,N-dimethyl aniline (DMA).

The condensation products of amines and aldehydes have often been used as the reductant. The most common commercial example is that arising from aniline and butyraldehyde for which the active ingredient has been identified as 3,5-diethyl-1,2-dihydro-1-phenyl-2-propyl-pyridine (DHP). A high purity grade, >85% active, has become available in recent years (Reilly Industries, PDHP™ Adhesive Accelerator; Vanderbilt, Vanax™ 808 HP).

DHP

The free-radical polymerization of methacrylate adhesives may show four stages, Figure 14.5: inhibition, solution polymerization, "gel" polymerization, and glassy polymerization. All commercial monomers have inhibitors added to help prevent premature polymerization during storage. During the inhibition stage, the redox system generates radicals and some minor addition to monomer may occur, but the inhibitors and any dissolved oxygen, a very potent inhibitor of free-radical polymerization, will prevent the formation of polymer and any significant consumption of monomer. Once the oxygen and inhibitors are consumed, very-high-molecular-weight polymer will form, $0.1-1 \times 10^6$, if no efficient chain transfer agents are present. As the reaction progresses, the polymer–monomer solution becomes increasingly viscous and the termination process is retarded. At about 25–30% monomer consumption, the polymer entanglements (gel) become so profound that a

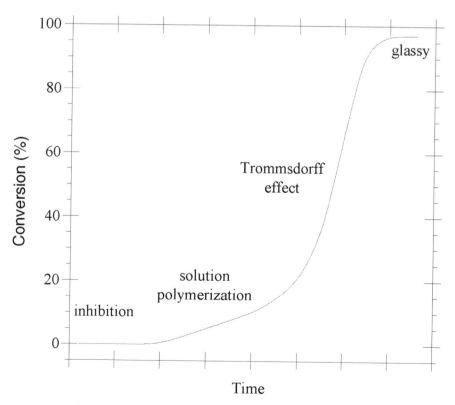

Fig. 14.5. Four stages of free-radical polymerization in bulk.

rapid rise in observed rate of polymerization is noted as the rate of termination reactions plummets. This rapid acceleration is referred to as the Trommsdorff effect. Because the diffusion of monomer molecules is not appreciably retarded by the "gel," there is no significant reduction in the propagation rate. Eventually, the polymer–monomer solution will approach the glassy state and even small molecules such as monomer and the redox components cannot move easily. For all practical purposes, the reaction comes to a halt soon thereafter as the smaller, more rapidly diffusing oxygen molecules reenter the bondline and quench the remaining polymer radicals.

Another consequence of oxygen inhibition of free-radical reactions is that the adhesive surface exposed to air will contain a higher concentration of lower-molecular-weight species than the bulk of the adhesive. This can result in the surface being soft, tacky, or even remaining fluid.

Formulation

Although the free radical chain growth polymerization brings processing advantages, it also brings limitations. The product of the polymerization of the monofunctional acrylic monomers is a linear thermoplastic polymer. In the case of a copolymer that is largely methyl methacrylate, the glass transition temperature of the polymer will be well above room temperature if the polymer has been driven to a high-enough molecular weight. The resulting adhesive is brittle, and although its shear strength might be quite high, its ability to withstand peeling forces or impacts will be low. One could use monomer mixtures yielding copolymers having lower glass transition temperatures and expect that the adhesive might be somewhat tougher. However, this parameter is difficult to control because when curing the acrylic adhesive in the bond line, one really is doing a bulk polymerization under uncontrolled conditions and trying to produce a specific polymer with a specific set of material and performance properties. The approach taken by most adhesive manufacturers or formulators is to use "prepolymers,"

high-molecular-weight oligomers that can be made under controlled conditions and then dissolved or dispersed in the acrylic monomer. By forming key elements of the polymer chains under controlled conditions, many of the desired properties of the final product can be built into the adhesive when it is formulated and before the material is cured. The prepolymer can cause the solution polymerization stage of polymerization to be skipped, thereby decreasing the cure time substantially.

The simplest approach to improving toughness is to dissolve or disperse a nonreactive rubbery polymer in the monomer mixture that is to be polymerized. Examples of such materials are nitrile rubbers, polyethers, and acrylic rubbers. The choice will depend on cost, desired properties, and the solubility of the polymer in the monomer. As the monomer polymerizes, it will lose its solvating ability for the dissolved polymer. Eventually, when enough of the monomer has been consumed, it can no longer hold the polymer in solution, and the polymer precipitates. The polymer then can segregate as a totally separate phase or as distinct domains heterogeneously dispersed throughout the acrylic polymer matrix, or it can be homogeneously trapped in the acrylic matrix, depending on the rates of reaction, the compatibility with the acrylic polymer, and the relative mobility of the dissolved polymer.

One way of influencing the way in which the added oligomer is distributed in the final cured produce is to provide reactive sites on the oligomer so that it can be incorporated into the acrylic matrix as it forms. For instance, if the added oligomer contains terminal, active, acrylic, or methacrylic unsaturation, it can be easily incorporated into the growing acrylic polymer chains as the adhesive cures. Then it is unable to precipitate as a separate phase and must remain more or less uniformly distributed throughout the matrix. On the other hand, incompatible polymers can be added to deliberately encourage the formation of reinforcing domains. If the oligomer has two or more functional groups capable of participating in the polymerization, there is at least the possibility of forming crosslinks in the polyacrylate matrix. Particular advantage has been claimed for combinations of reactive and nonreactive rubbers with attention to their molecular weights.[74]

A technology developed at Du Pont[75] combines the use of reactive sites on the oligomers with the initiation reaction. The resulting family of acrylic structural adhesives has become popularly known as "second generation acrylics." They consist essentially of solutions of chlorosulfonated polyethylene (Du Pont Hypalon™) in acrylic or methacrylic monomers. The chlorosulfonyl groups present on the polymer will react with amine aldehyde condensation products, DHP above, to generate free radical initiators, at least some of which are claimed to be on the oligomer backbone. The speculation is that the adhesive polymer chains then grow by graft polymerization from various sites on the rubber oligomer. Peroxides and metal ions can be added to accelerate the initiation and polymerization. One of the significant advantages of this type of acrylic structural adhesive is the ability to bond oily or dirty metals with only minimal surface preparation, although some surface preparation usually is necessary to improve bond durability (Fig. 14.6).

Fig. 14.6. Applying Fusor™ adhesive for repair of a steel automobile hood. (Courtesy of Lord Corporation.)

Other acrylic structural adhesives can be used on poorly prepared metal surfaces. Inclusion of reactive monomers having phosphate groups can lead to significant improvements in primary bonds to many unprepared metals.[76] Kowa American supplies one such monomer, 2-hydroxyethylmethacrylate acid phosphate.

Many other modifications to acrylic adhesives are useful. Improved heat resistance of acrylic adhesives can be achieved by incorporating small amounts of epoxy resin into the formulations,[77] or by including cyanate esters and organometallic compounds to provide an additional crosslinking mechanism.[78] Lower sensitivity of oxygen can be achieved by

polyolefin bonding.[85] A commercial product appeared in the mid- 1990s that was later twice improved, Scotch-Weld™ DP-8010.[86] By 2005, more than 30 patents had since appeared covering particular blocking groups, deblocking agents, and additives to improve adhesive performance. Later Dow Chemical made a push into this field with more than 10 patents on adhesive composition as well as others in specific automotive applications.[87] It has been commercialized under the LESA trademark.[88] The amine-blocked organoborane adhesives generally require refrigeration for good shelf live. A family of more robust, internally coordinated organoboranes has been disclosed.[89,90]

Organoborane–amine complex and activation

incorporation of a wax that migrates to the monomer–air interface providing a diffusion barrier or by using more efficient initiators.[76,79] Innovations in acrylic adhesive have led to new initiator reactions that provide rapid initiation and cure while at the same time allowing for improved shelf life.[80] Acrylic adhesives cured by ultraviolet light have been made from mixtures of acrylic functional urethane prepolymers mixed with free radically polymerizable monomers and photo initiators.[81,82]

Organoborane initiators for acrylic adhesives have received much attention arising from the observation that they give exceptional adhesion to low energy surfaces, such as polyethylene and polypropylene without surface preparation.[83] Earlier use of tributylborane–amine complexes in methyl methacrylate and activated with isocyanates, acid chlorides, or sulfonyl chlorides for a dental resin was reported in 1969.[84] In the mid-1990s, 3M became active in the area of amine-blocked organoboranes with a focus on

Summary

Acrylic structural adhesives have the advantage of easy handling and processing and rapid cure at room temperature. The ability of some compositions to adhere to unprepared metal surfaces or low surface energy substrates can allow substantial cost savings in manufacturing processes. The brittleness of the polymers made from the monomers often can be overcome by formulating to include tough reactive oligomers in the liquid adhesive. The variety of possible initiators and mechanisms for generating them has brought many innovative compositions to market.

HYBRID ADHESIVES

Adhesives usually are classified as epoxy, urethane, or acrylic, based upon the chemistry used to bring about the polymerization of the liquid adhesive. However, the distinction among the various types of adhesives is not always clear, and many hybrid adhesives have been developed and marketed. The objective usually is to take advantage of the desirable mechanical properties or chemical resistance of one polymer while retaining the processing attributes of a different cure system.[91]

It is common in acrylic structural adhesives to use oligomers that have a desirable backbone and are terminated with free-radical-polymerizable bonds. A variety of isocyanate-terminated polyurethanes can be adapted to use in acrylic structural adhesives by reacting the terminal isocyanates with a hydroxy functional acrylic monomer such as 2-hydroxyethyl methacrylate.[76,92,93]

Sometimes the system is formulated so that multiple cure mechanisms are possible and can occur sequentially or simultaneously. Compositions that rely on both epoxy and urethane chemistry are examples.[62,94,95] These are compositions containing the diglycidyl ether of bisphenol A, an isocyanate or isocyanate-terminated prepolymer, amines or other reactants for either epoxy or isocyanates, and catalysts.

Some of the more interesting and innovative work has occurred in areas combining aspects of more than one chemistry type. For instance, moisture-curable thermoplastic adhesives have received much attention. Hot melt adhesives have been developed that contain active, moisture-curable isocyanate groups. The compositions provide rapid processing on assembly lines because a reasonable bond is formed as soon as the thermoplastic adhesive cools from the melt. However, bond strength and performance improve with time as the composition is slowly crosslinked to a thermoset by reaction of the isocyanates with atmospheric moisture.[96,97]

EVOLUTION

The fundamental chemistry of the structural adhesives described here can change very little. Vinyl and acrylic monomers polymerize by chain growth polymerization initiated by free radicals or ions. Isocyanate and epoxy compounds react with compounds containing active hydrogen in step growth polymerization. Epoxy-containing compounds undergo chain growth polymerization initiated by certain Lewis acids and Lewis bases. These reactions will remain the most important ones for the polymerization of these raw materials.

Most of the raw materials used in large quantities in structural adhesives are used because they are widely available, relatively safe, and inexpensive. Quite often they are made in large quantities for uses other than adhesives and the adhesive manufacturers have taken advantage of supply and price. Few new basic raw materials are being developed specifically for the adhesive industry although the traditional raw materials are being combined in new ways to enhance desired adhesive properties. New initiators, adhesion promoters, primers, and specialty chemicals are being developed for use in small quantities to provide wider application latitude and improved performance.

The evolution of structural adhesives will certainly continue. Each increment in strength, durability, processing speed and ease, safety, reliability and reproducibility opens new commercial markets, not only to displace older joining methods but also to allow for the manufacture of new structures not possible without adhesives.

CONCLUSION

Epoxy, urethane, and acrylic structural adhesives have been commercially successful because each can be used, under the proper conditions, to make reliable, durable, and useful adhesively joined assemblies. The adhesives are classified according to the polymerization reactions used to bring the liquid adhesive to a high-molecular-weight load-bearing state. The reactive sites on the uncured adhesive and the overall polarity of the cured adhesive are important in the adhesion of the polymer to any specific material. The polymerization reactions determine the processing requirements and are important to the mechanical properties and environmental resistance of the cured adhesive. Through the use of specific reactive oligomers it is possible to build certain molecular structures into the composition that will remain in the cured adhesive to provide desired mechanical or chemical properties.

The successful use of adhesive joining in producing any assembly depends on viewing the assembly as a whole from the very beginning of its design. The key is remembering that the adhesive is only one component of the assembly. Adhesion is a property of the whole assembly.

REFERENCES

1. The Adhesion Society, *www.adhesionsociety.org*, 2 Davidson Hall – 0201, Blacksburg, VA 24061, phone: 540-231-7257, fax: 540-231-3971, e-mail: adhesoc@vt.edu
2. Patrick, R. L. (Ed.), *Treatise on Adhesion and Adhesives, Vol. 1, Theory*, Marcel Dekker, New York, 1967.
3. Patrick, R. L. (Ed.), *Treatise on Adhesion and Adhesives, Vol. 2, Materials*, Marcel Dekker, New York, 1969.
4. Kaelble, D. H., *Physical Chemistry of Adhesion*, Wiley-Interscience, New York, 1971.
5. Patrick, R. L. (Ed.), *Treatise on Adhesion and Adhesives, Vol. 3*, Marcel Dekker, New York, 1973.
6. Skeist, L. (Ed.), *Handbook of Adhesives*, 2nd ed., Van Nostrand Reinhold, New York, 1977.
7. Wake, W. C., *Adhesion and the Formulation of Adhesives*, 2nd ed., Applied Science, New York, 1982.
8. Lee, L.-H., *Adhesive Chemistry, Developments and Trends*, Plenum, New York, 1984.
9. Panek, J. R. and Cook, J. P., *Construction Sealants and Adhesives*, 2nd ed., Wiley-Interscience, New York, 1984.
10. Adams, R. D. and Wake, W. C., *Structural Adhesive Joints in Engineering*, Elsevier, New York, 1984.
11. Hartshorn, S. R. (Ed.), *Structural Adhesives Chemistry and Technology*, Plenum Press, New York, 1986.
12. Kinloch, A. J., *Adhesion and Adhesives Science and Technology*, Chapman and Hall, London, 1987.
13. Tong, L. and Steven, G. P., *Analysis and Design of Structural Bonded Joints*, Kluwer Academic, Boston, 1999.
14. Ward, N. and Young, T., *The Complete Guide to Glues and Adhesives*, Krause, Iola, WI, 2001.
15. Gierenz, G. (Ed.), *Adhesives and Adhesive Tapes*, John Wiley & Sons, New York, NY, 2001.
16. Moore, D. R. (Ed.), *Fracture Mechanics Testing Methods for Polymers, Adhesives and Composites*, Elsevier, Amsterdam, 2001.
17. Veselovskii, R. A., Kestelman, V. N., and Veselovsky, R. A., *Adhesion of Polymers*, McGraw-Hill, New York, 2001.
18. Pocius, A. V., *Adhesion and Adhesives Technology*, Hanser Gardner, 2002.
19. Pocius, A. V, and Dillard, D., and Chaudhury, M., *Surfaces, Chemistry and Applications: Adhesion Science and Engineering*, Elsevier Science & Technology, Amsterdam, 2002.
20. Pizzi, A. (Ed.), *Handbook of Adhesive Technology*, Marcel Dekker, New York, 2003.
21. Swanson, D. W., *Adhesive Materials for Electronic Applications: Polymers, Bonding, and Reliability*, William Andrew, Norwich, NY, 2005.
22. Fisher, L. W., *Selection of Engineering Materials and Adhesives*, CRC, Boca Raton, FL, 2005.
23. Packham, D. E., *Handbook of Adhesion*, John Wiley & Sons, Hoboken, NJ, 2005.
24. Chemical Abstracts Service, American Chemical Society, 2005.
25. Fowkes, F. M., and Mostafa, M. A., *Ind. Eng. Chem. Prod. Res. Dev.* 37, 605 (1978).
26. Fowkes, F. M., *Physicohemical Aspects of Polymer Surfaces*, Vol. 2, Mittal, K. L. (Ed.), Plenum, New York, p. 583.
27. Mark, H. F., "Future Improvements in the Cohesive and Adhesive Strength of Polymers-Part I," *Adhes. Age*, 22(7), 35–40 (1979).
28. Mark, H. F., "Future Improvements in the Cohesive and Adhesive Strength of Polymers-Part II," *Adhes. Age*, 22(9), 45–50 (1979).
29. May, C.A. (Ed), *Epoxy Resins: Chemistry And Technology*, 2nd ed., Marcel Dekker, New York, 1988.
30. Castan, P., *Process of Preparing Synthetic Resins*, U.S. Patent No. 2,324,483, 1943.
31. Dow Chemical, Dow Liquid Epoxy Resins, Form No. 296-00224-0199 WC+M, 1999, (http://epoxy.dow.com/index.htm).
32. Resolution Performance Products, EPON Resins and Modifiers SC:3059-01, 2002, (http://www.resins.com/resins/am/products/Epon.html).
33. Reichhold, Inc, (http://www.reichhold.com/coatings/products/brand.cfm?ID=24).
34. Cognis Corp., (http://www.cognis.com/framescout.html?/ProductCatalog/FindYourProduct.html).
35. Huntsman Corp., (http://www.huntsman.com/structural-composites/).
36. Resolution Performance Products, Eponol™ Resin 53-BH-35.
37. Partansky, A. M., *Amer. Chem. Soc., Div. Org. Coatings Plast. Chem.* 28(1), 366 (1968).
38. Partansky, A. M., *Advan. Chem. Ser., Epoxy Resins*, 92, 29 (1970).
39. Hine, J., Linden, S.-M., and Kanagasabapathy, V. M., *J. Org. Chem.* 50(25), 5096 (1985).
40. Matejka, L., Pororny, S., Dusek, K., *Polym. Bull.* (Berlin) 7(2–3), 123–8 (1982).

41. Dusek, K. and Matejka, L., "Transesterification and Gelation of Polyhydroxy Esters Formed from Diepoxides and Dicarboxylic Acids," in *Rubber Modified Thermoset Resins*, Riew, C. K. and Gillham, J. K. (Eds), Am. Chem. Soc., Adv. Chem. Series 208 (1984), pp. 15–26.

42. Abbey, K. J., Pressley, M. W., and Durso, S. R., "Controlled Cure of Thiol-Epoxy Systems," in *Proceedings of the 22nd Annual Meeting of The Adhesion Society*, Panama City Beach, FL, February 21–14, 1999, David R. Speth (Ed.).

43. Crivello, J. V., *Heat Curable Compositions*, U.S. Patent No. 4,173,551, 1979.

44. Crivello, J. V., *Heat Curable Cationically Polymerizable Compositions and Method of Curing Same with Onium Salts and Reducing Agents*, U.S. Patent No. 4,216,288, 1980.

45. Crivello, J. V., *UV Curable Compositions and Substrates Treated Therewith*, U.S. Patent No. 4,319,974, 1982.

46. Crivello, J. V. and Ashby, B. A., *Methods of Adhesive Bonding Using Visible Light Cured Epoxies*, U.S. Patent No. 4,356,050, 1982.

47. Crivello, J. V. and Lee, J. L., *Photocurable Compositions*, U.S. Patent No. 4,442,197, 1984.

48. Yagci, Y.; Reetz, I., "Externally stimulated initiator systems for cationic polymerization," *Progr. Polym. Sci.* 23(8), 1485–1538 (1998).

49. Tarbutton, K. S. and Robins, J., *Acid Catalyzed, Toughened Epoxy Adhesives*, U.S. Patent No. 4,846,905, 1989.

50. Jorissen, S. A., Ferguson, G. A., and Imirowicz, K., *Epoxy Compound Blend with Di(aminoalkyl) Ether of Diethylene Glycol*. U.S. Patent No. 5,548,026, 1996.

51. Baldwin, J. M. and Robins, J., *Epoxy Adhesive Composition Comprising a Calcium Salt and Mannich Base*, U.S. Patent No. 5,629,380, 1997.

52. Markevka, V. C., Griggs, A. L., and Tarbutton, K. S., *Epoxy Adhesives with Dithiooxamide Adhesion Promoters*, U.S. Patent 5,712,039, 1998.

53. Harold, S. and Schmitz, H.-P., *Phosphorus-modified Epoxy Resins Comprising Epoxy Resins and Phosphorus-containing Compounds*, U.S. Patent 5,830,973, 1998.

54. Mowrere, N. R., Kane, J. F., and Hull, C. G., *Siloxane-modified Adhesive/Adherend Systems*, U.S. Patent 5,942,073, 1999.

55. Rijsdijk, H., Overbergh, N., DeBlick, G., Miles, G., and Kennan, A., *Curable Adhesive System*, U.S. Patent No. 5,952,071, 1999.

55. Bayer, O., German Patent No. 728.981, 1937.

56. Saunders, J. H. and Frisch, K. C., *Polyurethanes, Chemistry and Technology*, Vols. 1 and 2, Interscience, New York, 1962, 1964.

57. Oertel, G. (Ed.), *Polyurethane Handbook*, 2nd E., Hanser, Munich, 1994.

58. Randall, D., Lee, S. (Eds.), *The Polyurethane Book*, John Wiley & Sons, New York, 2003.

59. Dunn, D. J., *Engineering and Structural Adhesives*, Rapra Review Reports, 15(1), Report 169, Rapra Technology Ltd., Shropshire, UK, 2004.

60. *Polyurethane Adhesives*, Rapra Published Search Number 114, Rapra Technology Ltd., Shropshire, UK.

61. Fabris, H. J., Maxey, E. M., and Uelzmann, H., *Urethane Adhesive Having Improved Sag Resistance*, U.S. Patent No. 3,714,127, 1973.

62. Goel, A. B., *Sag Resistant Urethane Adhesives with Improved Antifoaming Property*, U.S. Patent No. 4,728,710, 1988.

63. Baueriedel, H., *Adhesives Based on Polyurethane Prepolymers Having a Low Residual Monomer Content*, U.S. Patent No. 4,623,709, 1986.

64. Ono, H.-K., Jones, F. N., and Pappas, S. P., "Relative Reactivity of Isocyanate Groups of Isophorone Diisocyanate. Unexpected High Reactivity of the Secondary Isocyanate Group," *J. Polym. Sci., Polym. Lett. Ed.*, 23, 509–515 (1985).

65. Okawa, T., *Process for Producing Isocyanate Compound*, U.S. Patent No. 5,166,414, 1992.

66. Nagata, T., Yamashita, H., Kusumoto, M., and Okazaki, K., *Stabilizing Method of Isocyanate Compounds and Isocyanate Compositions Stabilized Thereby*, U.S. Patent No. 5,302,749, 1994.

67. Ohashi, Y., Matsuda, H., Nishi, E., and Nishida, T., *Moisture Curing Urethane Adhesive Composition*, U.S. Patent No. 5,698,656, 1997.

68. Fukatsu, S. and Hattori, Y., *Moisture Curable Polymer Composition and Process for Production Thereof*, U.S. Patent No. 5,767,197, 1998.

69. Schoener, T. E. and Housenick, J. B., *Polyurethane Reaction System Having a Blocked Catalyst Combination*, U.S. Patent No. 6,348,121, 2002.

70. Ambrose, R., Retsch, W. R., Jr., and Chasser, A., *Blocked Isocyanate-based Compounds and Compositions Containing the Same*, U.S. Patent No. 6,288,199, 2001.

71. Bader, E., U.S. Patent 33,333,025 (1967).

72. Bauer, W. Jr., "Acrylic Acid and Derivatives," in *Kirk-Othmer Encyclopedia of Chemical Technology*, John Wiley & Sons, New York, 2003.

73. Wilczynski, R. and Juliette, J. J., "Methacrylic Acid and Derivatives," in *Kirk-Othmer Encyclopedia of Chemical Technology*, John Wiley & Sons, New York, 2003.

74. Huang, J.-P., Righettini, R. F, and Dennis, F. G., *Adhesive Formulations*, U.S. Patent 6,225,408 (2001).

75. Briggs, P. C. and Muschiatti, L. C., U.S. Patent 3,890,407 (1975).

76. Zalucha, D. J., Sexsmith, F. H., Hornaman, E. C., and Dawdy, T. H., *Structural Adhesive Formulations*, U.S. Patent No. 4,223,115, (1981).

77. Dawdy, T. H., *Epoxy Modified Structural Adhesives Having Improved Heat Resistance*, U.S. Patent 4,467,071 (1984).

78. McCormick, F. B., Drath, D. J., Gorodisher, I., Kropp, M. A., Palazzotto, M. C., and Sahyun, M. R. V., *Energy-curable Cyanate/Ethylenically Unsaturated Compositions*, U.S. Patent No. 6,029,219 (2000).

79. Righettini, R. R. and Dawdy, T. H., *Free Radical Polymerizable Compositions Including Para-Halogenated Aniline Derivatives,"* U.S. Patent 5,932,638 (1999).

80. Edelman, R. and Catena, W., *Rapid Curing Structural Acrylic Adhesive*, U.S. Patent No. 5,865,936 (1999).

81. Usifer, D. A. and Broderick, I. C., *Urethane Adhesive Compositions*, U.S. Patent No. 5,426,166 (1995).

82. Usifer, D. A. and Broderick, I. C., *Urethane Adhesive Compositions*, U.S. Patent No. 5,484,864 (1996).

83. Imai, Y., Fujisawa, S., Matsui, H., Yamazaki, H., Masuhuara, E. Japanese Kokai 69-100477, 1973.

84. Fujisawa, S.; Imai, Y.; Masuhara, E. Iyo Kizai Hokoku, Tokyo Ika Shika Daigaku 1969, 3, 64–71.

85. Zharov, J. V.; Krasnov, J. N., *Polymerizable Compositions Made with Polymerization Initiator Systems Based on Organoborane Amine Complexes*, U.S. Patent 5,539,070, 1996.

86. 3M Technical Literature for DP-8010; January 2002.

87. Sonnenschein, M. F., Webb, S. P., and Rondan, N. G., *Amine Organoborane Complex Polymerization Initiators and Polymerizable Compositions*, U.S. Patent 6,706,831, 2004.

88. Leaversuch, R., "Long-Glass PP Makes InroadsIn Automotive Front Ends," *Plastics Technol.*, online article, http://www.plasticstechnology.com/articles/200207cu1.html, Gardner.

89. Kendall, J. L. and Abbey, K. J., *Internally Coordinated Organoboranes*, U.S. Patent Application 20040242817.

90. Abbey, K. J. and Kendall, J. L., "Internally coordinated organoboranes: Stability and activation in polyolefin adhesives," *Polym. Mater. Sci. Eng.* 2004(2).

91. Pohl, E. and Osterholz, F. D., *Novel Vulcanizable Silane-Terminated Polyurethane Polymers*, U.S. Patent 4,645,816, 1987.

92. Brownstein, A. M., *Anaerobic Adhesive*, U.S. Patent 3,428,614, 1969.

93. Su, W.-F., A., *UV Curable High Tensile Strength Resin Composition*, U.S. Patent 4,618,632, 1986.

94. Hawkins, J. M., *Epoxy Resin Adhesive Compositions Containing an Isocyanate Terminated Polyurethane Prepolymer and a Chain Extender*, U.S. Patent No. 3,636,133, 1972.

95. Trieves, R. and Pratley, K. G. M., U.S. Patent No. 4,623,702, 1986, November.

96. Anderson, G. J. and Zimmel, J. M., *Thermally Stable Hot Melt Moisture-cure Polyurethane Adhesive Composition*, U.S. Patent No. 5,939,499, 1999.

97. McInnis, E. L., Santosusso, T. M., and Quay, J. R., *Hot Melt Adhesives Comprising Low Free Monomer, Low Oligomer Isocyanate Prepolymers*, U.S. Patent No. 6,280,561, 2001.

15

Synthetic Resins and Plastics

Rudolph D. Deanin* and Joey L. Mead**

INTRODUCTION

Definition

Plastic (*adj.*) is defined by Webster as "capable of being molded or modeled (e.g., clay)...capable of being deformed continuously and permanently in any direction without rupture." Plastic (*n.*) in modern industry covers high-molecular-weight organic compounds that can be formed into any desired shape and then solidified into a useful product that can withstand the mechanical stresses normally applied to it.

History

Commercial plastics began in 1868 when John and Isaiah Hyatt plasticized cellulose

*Professor, Plastics Engineering Department, University of Massachusetts at Lowell. (Sections on Commercial Plastic Materials; Plastics Processing.)
**Associate Professor, Plastics Engineering Department, University of Massachusetts at Lowell. (Section on Polymer Chemistry.)

nitrate and molded it into billiard balls, and later into combs, brushes, and other useful articles. Commercial *synthetic* plastics began in 1908 when Leo Baekeland reacted phenol, formaldehyde, and wood flour and molded them into electrical insulators and a growing variety of other product specialties. During the next 30 years, a variety of plastics appeared as interesting specialties. During World War II the U.S. and German governments met the severe demands of advanced military technology by supporting vast research, development, plant construction, and manufacturing of a number of major polymers for plastics and rubber applications. With the end of the war in 1945, all of this technology and production capacity converted to civilian products, and commodity plastics began their tremendous growth, first polyvinyl chloride and polystyrene, then polyethylenes, polypropylene, polyesters, and polyurethanes, along with a constantly expanding range of more specialized polymers for more demanding applications.

Advantages of Plastics over Conventional Materials

We have had conventional structural materials—metals, ceramics, glass, wood, leather, textiles, paper—for thousands of years, during which time we have been able to explore and exploit them thoroughly to full maturity. By comparison, plastics are so new that we are still learning to develop and use them. Their exponential growth is due to the fact that they offer many advantages over conventional materials.

Processability and Product Design. It is much easier to convert plastic materials into an almost unlimited range of products.

Modulus. Plastics cover an extremely broad range from extremely rigid to stiffly flexible to extremely soft and rubbery, sometimes even within a single chemical family.

Elasticity. Their recovery from deformation is superior to almost all conventional materials.

Impact Strength. Plastics are much less brittle than ceramics, glass, and paper, and some families can be made almost unbreakable.

Lubricity and Abrasion Resistance. Specific plastic materials offer outstanding self-lubricating performance and abrasion resistance.

Thermal Insulation. Plastics offer very good insulation against heat and cold, and can be foamed to increase their insulating qualities much further.

Flame-Retardance. Many plastics are less flammable than wood and paper, and most can be formulated to make them much more resistant to burning.

Electrical Insulation. Plastics are excellent electrical insulators. They can also be formulated to provide semi-conductivity, and high or low dielectric constant and loss.

Transparency/Opacity. Some plastics can approach or equal the transparency of glass. All can be formulated to a wide range of translucency to opacity.

Color and Appearance. Most conventional materials have a very limited color range, and they simply look like what they are. Plastics can be produced in an almost infinite range of colors, transparent or opaque, and surfaces to simulate all of the conventional materials or create totally new ones.

Chemical Resistance. Plastics are generally far superior to metals in corrosion resistance. Among the 100 families of commercial plastics, they offer a wide variety of chemical resistance, or solubility/reactivity, as desired for different types of products.

Water Resistance. It is far superior to wood and paper.

Permeability. Different families of plastics offer a wide range from highly impermeable barrier materials to membranes of different separation abilities.

Weathering. Some plastics are very resistant to weather. Others have moderate resistance. Still others can be designed for self-destruction to alleviate collection of solid waste.

Cost. Contrary to some popular belief, plastics are not "cheap." But their superior processability often makes the finished product less expensive.

This economic advantage, plus all their other advantages listed above, accounts for their success in replacing conventional materials, and in leading to new products which were not even possible before.

Markets for Plastics

The U.S. plastics market has grown past 100 billion pounds/year. The largest share (24%) goes to packaging, both rigid packages and film. Close behind is building and construction (20%), mainly piping, plywood, siding, insulation, and flooring. Smaller, more specialized amounts go into agriculture, aircraft, appliances, autos and trucks, electrical and electronics, furniture, glazing, housewares, luggage, marine, medical, office equipment, optical, tools, toys, and miscellaneous industrial and consumer products. Thus, taken as a whole, plastics is one of the largest and fastest-growing industries in the United States.

Major Classes of Plastic Materials

The primary binary classification is the distinction between thermoplastics and thermosets. Thermoplastics are stable large molecules, typically molecular weights 10^4–10^6, which soften on heating to permit melt processing, and solidify on cooling to give solid finished poducts; the process is reversible, so they are essentially recyclable within the limits of their thermal stability. Thermosetting plastics are reactive low-molecular-weight polymers, which may be melted or even poured, shaped into final products, and then reacted further into cross-linked molecules of essentially infinite molecular weight; the process is essentially irreversible, so they are difficult or impossible to recycle. Thermoplastic processing is simpler and more economical, so it accounts for about 85 percent of the plastics market. On the other hand, thermosetting plastics permit many special processes and offer outstanding final properties, which accounts for the health of their share of their more specialized markets.

Another important distinction is based on: (1) filled and reinforced plastics, and (2) foams. (1) When any family of polymers is combined with particulate inorganic fillers, this produces major increase in density, modulus, dimensional stability, heat transfer, dielectric constant, and opacity, and frequently a decrease in cost. When the fillers are reinforcing fibers, they can further produce a great increase in strength, impact resistance, and dimensional stability. Thus, these properties may depend more upon the use of fillers and fibers, than upon the choice of the particular polymer in which they are used.

(2) When a polymer is liquified, foamed, and solidified to trap the air spaces within it, air contributes so much to the final properties

that it may be more important than the particular polymer in which it is dispersed. The most outstanding effects are flotation by closed-cell foams, softness in open-cell foams, impact cushioning, thermal and electrical insulation, and permeability in open-cell foams.

Following these introductory remarks, we turn to a study of the fundamental aspects of Polymer Chemistry (Part I), followed with a discussion of Commercial Plastic Materials (Part II), and conclude with Plastic Processing (Part III).

PART I. POLYMER CHEMISTRY

Materials are often classified as metals, ceramics, or polymers. Polymers differ from metals and ceramics, by their lower densities, thermal conductivities, and moduli. A vast array of products utilize plastic materials. For example, in applications requiring lighter weight, plastics offer an advantage over other choices as a result of their lower density. Polymeric materials are used in automotive, packaging, and consumer goods, just to name a few. The requirements for these diverse applications vary greatly, but through proper control, plastic materials can be synthesized to meet these varied service conditions.

MOLECULAR WEIGHT

A polymer is prepared by linking a low molecular weight species, called a monomer (such as ethylene), into an extremely long chain, called a polymer (such as polyethylene), much as one would string together a series of beads to make a necklace (see Fig. 15.1). As molecular weight increases, the properties of the material change. Looking at the alkane hydrocarbon series with the general structure $H–(CH_2)_n–H$, we can see the

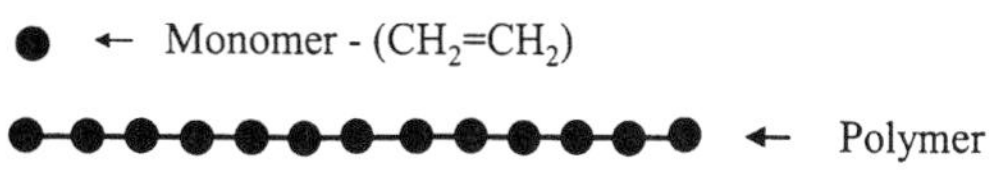

● ← Monomer - $(CH_2=CH_2)$

●●●●●●●●●●●● ← Polymer

Fig. 15.1. Polymerization.

material change from a gas for values of $n = 1$–4, a liquid for $n = 5$–11, a high viscosity liquid from $n = 16$–25, a crystalline solid for $n = 25$–50, to a tough plastic solid for $n = 1000$–5000.[1] The molecular weight affects both the mechanical and processing behavior of the polymer. In general, higher molecular weights result in improved mechanical properties, but face more difficulty in processing. Unlike low-molecular-weight species, polymeric materials do not possess one unique molecular weight, but rather a distribution of molecular weights as depicted in Fig. 15.2. Molecular weights for polymers are usually described by two different average molecular weights, the number average molecular weight, $\overline{M}_n$, and the weight average molecular weight, $\overline{M}_w$. These averages are calculated using the equations below:

$$\overline{M}_n = \sum_{i=1}^{\infty} \frac{n_i M_i}{n_i}$$

$$\overline{M}_w = \sum_{i=1}^{\infty} \frac{n_i M_i^2}{n_i M_i}$$

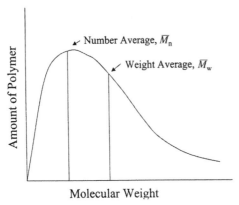

Fig. 15.2. Molecular weight distribution.

where n_i is the number of moles of species i and M_i is the molecular weight of species i.

CHAIN STRUCTURE

Polymerization can produce linear chains, but other structures can exist as well. As shown in Fig. 15.3 branched and crosslinked structures can be formed. Linear and branched structures can be shaped and reshaped simply by heating and are called *thermoplastics*. In the case of a crosslinked structure a three-dimensional network is formed that cannot be reshaped by heating. This type of structure is called a *thermoset*.

Macromolecular conformations describe the positions of the atoms that occur due to rotation about the single bonds in the main chain.[2] Polymer chains in solution, melt, or amorphous state exist in what is termed a random coil. The chains may take up a number of different conformations, varying with time. Figure 15.4 shows one possible conformation for a single polymer chain. In order to describe the chain, polymer scientists utilize the root mean square end-to-end distance ($\langle r^2 \rangle^{1/2}$), which is the average over many conformations. This end-to-end distance is a function of the bond lengths, the number of bonds, and a characteristic ratio, C, for the specific polymer.

CHEMICAL STRUCTURE

The chemical characteristics of the starting low-molecular-weight species will help determine the properties of the final polymer. Along the chain axis primary bonds hold the atoms together and determine molecular

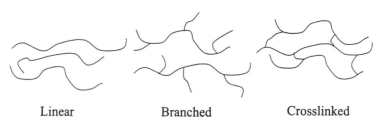

Linear   Branched   Crosslinked

Fig. 15.3. Polymer structures.

properties such as flexibility and glass transition temperatures. Flexibility of the chain is governed by the ease of rotation about main chain bonds. The presence of methylene units or carbon–oxygen single bonds act to increase the flexibility of the chain. Groups or interchain interactions (described below) that tend to restrict rotation will decrease the flexibility of the molecule. An extreme example of this principle is the rigid rod polymers. Figure 15.5 shows several examples of these types of structures. The aromatic groups along the backbone act to stiffen the polymer chain and restrict rotation, causing the polymer to remain straight, much like a log.[3] Materials such as these may exhibit liquid crystalline behavior.

The forces holding the many individual chains together (interchain forces) are determined by secondary bonds, except in the case of thermosets where primary bonds hold the chains together. The type and strength of the secondary bonds (often termed van der Waals forces)[4] will depend on the structure of the

polymer. In the case of hydrocarbon polymers, such as polyethylene, the secondary bonds are dispersion forces. For polymers containing carbon and oxygen groups, such as the polyesters, the presence of the $-C=O$ bond results in a dipole due to the different electronegativities of the carbon and oxygen atoms. The presence of polarity in the polymer will act to increase the intermolecular forces. In the case of polyamides, hydrogen bonding between the polymer chains leads to high intermolecular forces. The strength of the intermolecular forces will affect the properties of the polymer such as viscosity, solubility, miscibility, surface tension, and melting point. As discussed above, interchain forces will also affect the flexibility of the chain if they restrict free rotation.

MORPHOLOGY

In its solid form a polymer can exhibit different morphologies depending on processing conditions and the structure of the polymer chain. Amorphous polymers show no order to the arrangement of the chains. The chains are entangled with each other, much like the strands of spaghetti on one's plate. An example of an amorphous polymer is polystyrene. If the polymer backbone has a regular, ordered microstructure, then the polymer can pack tightly into an ordered crystalline structure, although the material will generally be only semicrystalline. Examples of semicrystalline polymers are polyethylene and polypropylene. The amorphous and

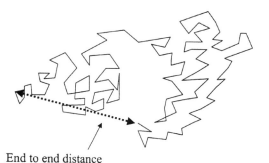

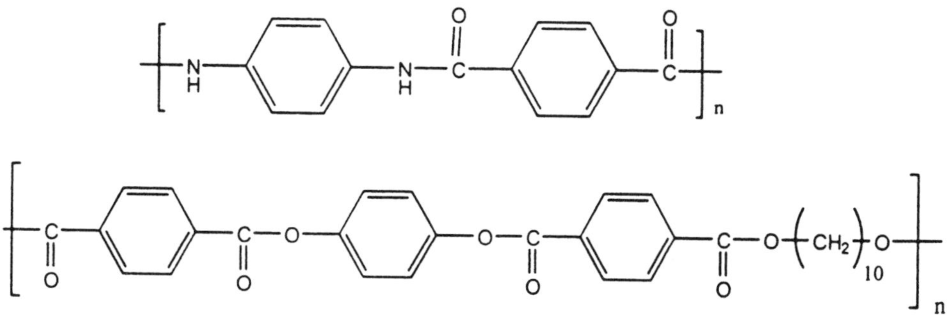

End to end distance

Fig. 15.4. Random coil chain.

Fig. 15.5. Rigid rod-type polymers.

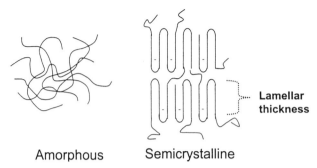

Amorphous   Semicrystalline

Fig. 15.6. Amorphous and semicrystalline structures.

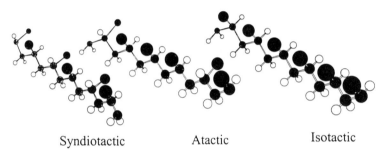

Syndiotactic   Atactic   Isotactic

Fig. 15.7. Syndiotactic, atactic, and isotactic structures.

crystalline chain structures are depicted in Fig. 15.6.

The exact make-up and architecture of the polymer backbone will determine the ability of the polymer to crystallize. Figure 15.7 shows the different types of microstructure that can be obtained for a vinyl polymer. Isotactic and syndiotactic structures are considered stereospecific polymers, and their highly regular backbone structure allows them to crystallize. The atactic form is irregular and would produce an amorphous material. This nature of the polymer microstructure can be controlled by using different synthetic methods. As will be discussed below, the Ziegler–Natta catalysts are capable of controlling the microstructure to produce different types of stereospecific polymers.

TRANSITION TEMPERATURES

Glass Transition Temperature (T_g)

Chain flexibility is governed by molecular structure, but is also affected by temperature.

As the temperature is reduced, amorphous polymers reach a temperature where large-scale (20–50 chain atoms) segmental motion ceases.[5] This temperature is called the glass transition temperature or T_g. Volume–temperature plots can be used to indicate the T_g. At the T_g transition temperature a several decade change in the modulus occurs and the material changes from a rigid solid to a rubbery material. At still higher temperatures, provided the material is a thermoplastic, it becomes a liquid, which can flow and be processed. The behavior in the glass transition region is depicted in Fig. 15.8.

The location of the glass transition temperature will depend on the nature of the polymer. Generally, a plastic differs from a rubbery material due to the location of its glass transition temperature. A plastic has a T_g above room temperature, while a rubber has a T_g below room temperature. As previously mentioned the flexibility of the chain will affect the value of T_g. Flexible groups will tend to lower the T_g, while stiffening groups will act to increase it. Side groups can also affect the value. The

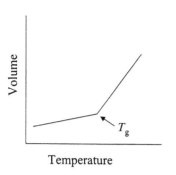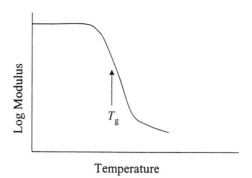

Fig. 15.8. Behavior at the glass transition temperature.

TABLE 15.1   Effect of Chain Structure on T_g [6]

	T_g (°C)
Side groups	
Polypropylene	−10
Polystyrene	100
Polarity	
Polypropylene	−10
PVC	85
Polyacrylonitrile	101
Symmetry	
Polypropylene	−10
Polyisobutylene	−70
PVC	87
Polyvinylidene chloride	−19

effect of various factors on the value of T_g is shown in Table 15.1. Aliphatic side groups will tend to have decreasing T_gs as the length of the side chain increases; however, rigid side groups will tend to increase the T_g. Increased polarity of the polymer will also increase the glass transition temperature as shown by the polymers polypropylene, polyvinyl chloride, and polyacrylonitrile. Symmetry about the backbone can act to decrease the T_g as illustrated by the pairs, polypropylene and polyisobutylene, and polyvinyl chloride (PVC), and polyvinylidene chloride.

Increases in number average molecular weight, $\overline{M}_n$, and cross-link density will both act to increase the T_g. The addition of plasticizers, such as in the case of PVC, will decrease the value of the glass transition temperature. The glass transition temperature of the plasticized material can be estimated if the glass transition temperature of the two components (A and B) and their weight fractions are known.[7]

$$\frac{1}{T_g} = \frac{W_A}{T_{gA}} + \frac{W_B}{T_{gB}}$$

where T_{gA} and T_{gB} are the glass transition temperatures of components A and B, respectively, and W_A and W_B are the weight fractions.

One difficulty in obtaining values for the glass transition temperature is its dependence on measurement rate. When experiments are conducted at slow rates, the measured values will be lower than those measured at more rapid rates. Other difficulties include experimental problems and many definitions and interpretations on how to measure the values. Measurement of the temperature at which a step change in the volumetric thermal expansion coefficient occurs, when heating and cooling rates are 1°C/min, is perhaps one of the less ambiguous methods.[8] The measured value of the glass transition temperature will increase approximately 3°C (volumetric measurements) to 7°C (maximum in tan δ from dynamic mechanical analysis) for a decade change in rate.

Crystallization and Melting Points (T_M)

A number of polymers may exhibit the ability to crystallize. As previously mentioned, such polymers are semicrystalline, meaning they will have regions of amorphous and crystalline material dispersed throughout the part. Both the morphology and degree of crystallinity can be affected by the processing conditions used to manufacture the part. These changes can greatly affect the mechanical behavior of the material. The degree of crystallinity and the melting point are often measured using differential scanning calorimetry.

The general structure of the crystalline regions is rather complex. Polymer chains appear to fold back and forth into a lamellar structure as shown in Fig. 15.6. These lamellae form layers of ribbonlike structures, tied together with amorphous regions.

Semicrystalline polymers do not exhibit a single sharp melting point, but rather melt over a range of temperatures. In addition, the melting point of the material will be higher the greater the lamellar thickness, which can be controlled by the crystallization temperature.[9] Higher crystallization temperatures generally lead to greater lamellar thickness.

Polymer structure will affect the melting point of the polymer in a number of ways. Polymers containing polar groups will have higher melting points. Polymers with hydrogen bonding would be expected to have even higher melting points. For example, polyethylene has a melting point of 135°C, while nylon 6 has a melting point of 265°C. The presence of chain stiffening groups in the backbone will act to raise the melting point of the polymer; contrast the melting point of polyethylene (135°C) with poly p-xylene (400°C). In general, polymers with high interchain forces and rigid chains will tend to have the highest melting points.

In addition to the melting point, the degree of crystallinity will influence the behavior of the polymer. Differential scanning calorimetry (DSC) can be used to determine both the melting point and the percent crystallinity. Figure 15.9 shows a representative DSC curve for a melting point. The area under the melting peak is related to the percent crystallinity of the polymer sample. The percent crystallinity may have a dramatic effect on the mechanical behavior of the material. Above the T_g of the material, the presence of crystalline regions will act to increase the rigidity of the polymer, resulting in an increasing modulus with increasing percent crystallinity.[10]

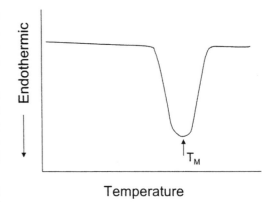

Fig. 15.9. Differential scanning calorimetry—melting point.

The kinetics of crystallization are a function of the crystallization temperature.[11] As the crystallization temperature is decreased the rate of crystallization increases, up to a maximum. When the crystallization temperature is lowered beyond this point the rate begins to slow down as the molecular motion of the chains is reduced. At temperatures below the T_g crystallization is suppressed.

POLYMERIZATION

There are two primary polymerization approaches: step-reaction polymerization and chain-reaction polymerization.[12]

Step-Reaction Polymerization

In step-reaction polymerization (also referred to as condensation polymerization) any two polyfunctional (typically bifunctional) monomers of the correct chemical species can react, often liberating a small molecule such as water. As the reaction continues, higher molecular weight species are produced as longer and longer polyfunctional groups react together. For example, two monomers can react to form a dimer, then react with another monomer to form a trimer. The reaction can be described as n-mer + m-mer → $(n + m)$mer, where n and m refer to the number of monomer units for each reactant. Monomer functionalities of two will produce linear polymers. With functionalities greater than two, branching or crosslinking can occur. Molecular weight of the polymer builds up gradually with time and in order to produce high molecular weight polymers high conversions are usually required. Polymers synthesized by this method typically have atoms other than carbon in their backbone. Examples of polymers produced by step-growth polymerization include polyesters and polyamides. A typical step growth reaction is shown in Fig. 15.10.

Molecular weight control can be accomplished by stopping the reaction; however, further heating may result in changes in the molecular weight. A stoichiometric imbalance (excess of one of the reactants over another) can also be used to control the molecular weight. Finally, a monofunctional reagent can be used to control the molecular weight.[13]

Chain-Reaction Polymerization

In contrast to the slow step-reaction polymerizations, chain-reaction polymerizations are fairly rapid.[14] Chain-reaction polymerizations (often referred to as addition polymerizations) require the presence of an initiator for polymerization to occur. Initiation can occur by a free radical, an anionic, or a cationic species, which open the double bond of a vinyl monomer and the reaction proceeds as shown in Fig. 15.11 where * may be a radical,

cationic, or anionic species. Chain-reaction polymers typically contain only carbon in their backbone and include such polymers as polystyrene and polyvinyl chloride. Once initiated, the chain will continue to add monomer until a termination reaction stops the growth. Termination generally occurs in radical type reactions, but in anionic and some cationic polymerizations termination reactions may not be present and the polymerization is termed a "living polymerization." This has important commercial implications in that a second monomer may be added to the living end to form block copolymers.

Radical Polymerization. Free radical polymerization consists of three steps, initiation, propagation, and termination or chain transfer. Initiation consists of two steps, decomposition of the initiator to form a radical species, followed by addition of the initiator to the first monomer unit.

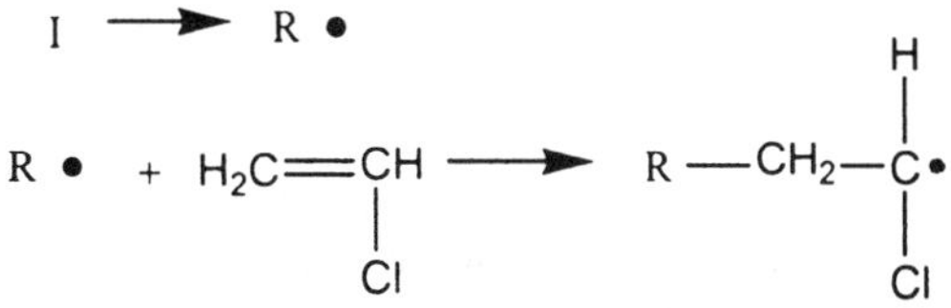

$$I \longrightarrow R\bullet$$

$$R\bullet + H_2C{=}CH \longrightarrow R{-}CH_2{-}\overset{H}{\underset{Cl}{C}}\bullet$$

Typical initiators include benzoyl peroxide or azobisisobutyronitrile, but other radical generating methods, such as high-energy

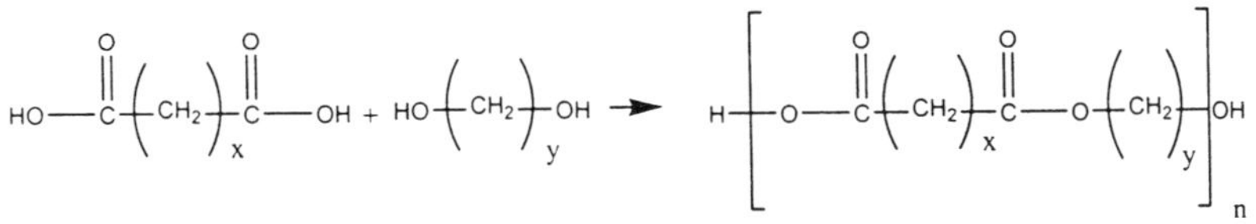

Fig. 15.10. Step-growth polymerization.

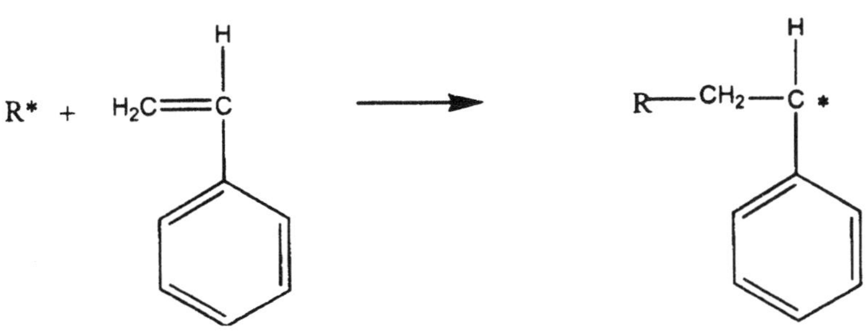

Fig. 15.11. Addition polymerization.

radiation or photolytic decomposition of compounds, may also be used.[12]

The next step in the reaction is propagation.

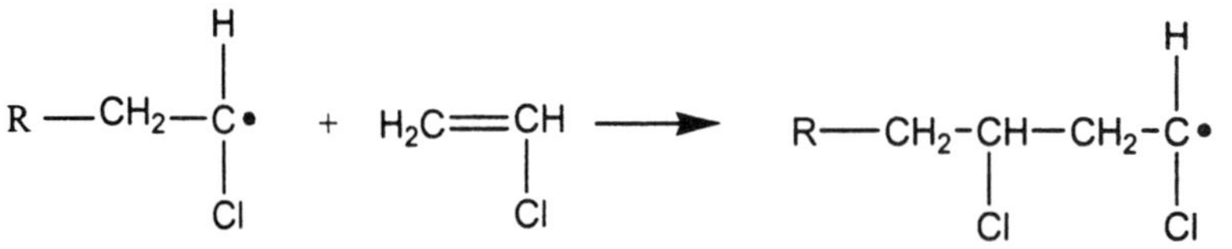

Propagation continues until the radical is terminated. Termination occurs when two radical species meet and react either by coupling or by disproportionation as shown in Fig. 15.12.

In radical polymerizations other reactions occur to prematurely stop chain growth. These reactions are termed chain transfer and when they occur they reduce the molecular weight of the chain. Chain transfer acts to transfer the radical from the growing end of the chain to another species. This may include solvent, initiator, polymer, or a deliberately added chain transfer agent. Propagation now begins from the new radical generated, while the original chain is terminated. When chain transfer to polymer occurs this produces branching along the polymer backbone. Low-density polyethylene is produced via free radical processes, with considerable transfer to polymer. Branching along the polyethylene backbone suppresses the degree of crystallinity, resulting in a lower density material.

Cationic Polymerization. This follows similar steps as with free radical polymerization. Initiator types are typically acids, such as H_2SO_4 and H_3PO_4, and Lewis acids, such as $AlCl_3$, BF_3, $TiCl_4$, and $SnCl_4$. Lewis acids generally require the presence of a proton source, such as trace amounts of water.[15] Monomers exhibit considerable selectivity in their ability to be polymerized via ionic mechanisms. The nature of the substituent (Y) will influence the electron density of the double bond and thus its ability to polymerize.[16] Substituents that are electron donating, such as alkoxy or alkyl, will increase the electron

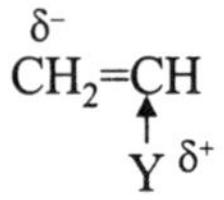

$$CH_2\!=\!CH \atop \underset{Y \; \delta^+}{\uparrow}$$

density of the double bond and allow for polymerization by cationic methods.

Typical monomers that may be polymerized by cationic methods include styrene, isobutylene, and vinyl ethers. Unlike radical polymerizations, solvent polarity can influence the rate of polymerization. This is due to the presence of the counterion (see Fig. 15.13). For example, more polar solvents can increase the degree of separation between the growing end and the counterion during the propagation step, increasing the rate of propagation.[17]

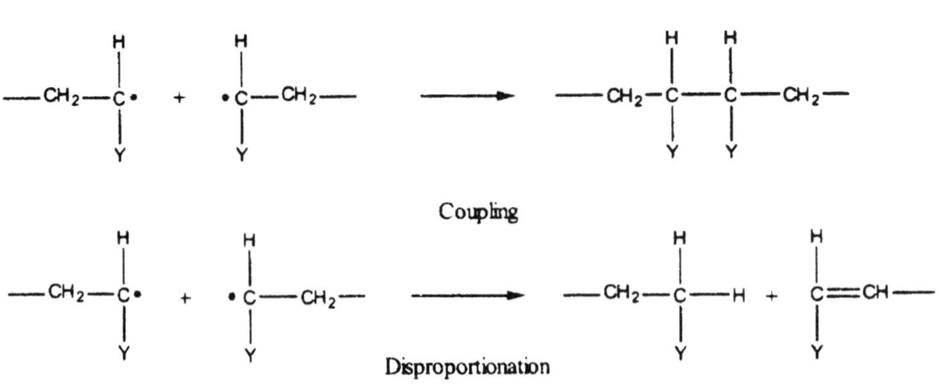

Fig. 15.12. Termination mechanisms.

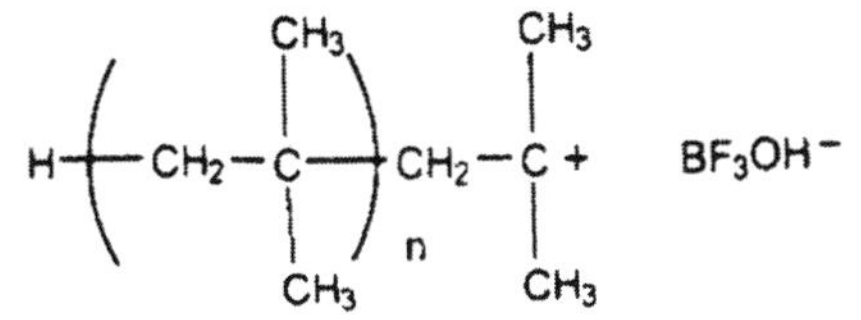

Fig. 15.13. Active cationic propagating species.

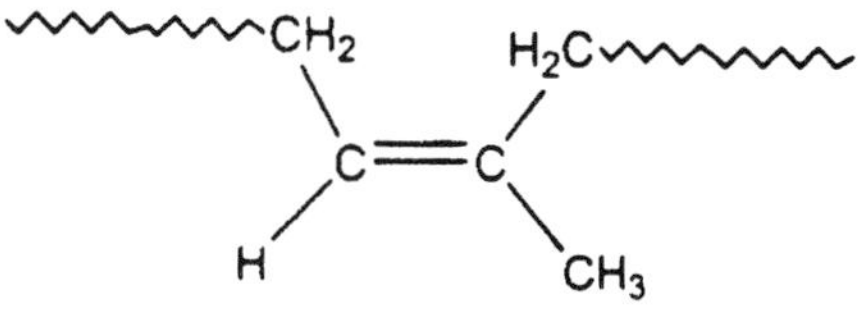

Fig. 15.14. *Cis* 1,4 polyisoprene.

Chain transfer is the most common chain terminating reaction in cationic polymerization and can include transfer to monomer, solvent, and polymer. Termination by combination with the counterion can also occur in some systems. In some cases, cationic polymerization may be used to prepare stereoregular polymers. Although the exact mechanism is unclear, it is known that stereoregularity varies with initiator and solvent.[18] Lower temperatures also tend to favor more stereoregular polymers.

Anionic Polymerization. This is similar to cationic polymerizations, except that the propagating species is anionic. Initiator types are typically alkali metals or their compounds. In the case of anionic polymerizations, electron withdrawing substituents, such as cyano, nitro, carboxyl, and vinyl, facilitate polymerization by anionic means. Termination in anionic polymerization is generally by chain transfer. If the system is purified so that chain transfer is suppressed, the propagating species may remain active resulting in what is termed "living polymerization".[19] This allows for the preparation of block copolymers, where one monomer is polymerized, followed by addition of a second monomer to the living end.

As with cationic polymerization the propagation rate is highly dependent on the solvent. Solvents that reduce the association between the growing chain end and the counterion result in faster rates of propagation.[20]

In anionic polymerization of vinyl monomers (nondiene), low temperatures and polar solvents favor the preparation of syndiotactic polymers.[21] Nonpolar solvents tend to favor isotactic polymerization. In the case of diene monomers such as butadiene and isoprene, the use of lithium based initiators in nonpolar solvents favors preparation of the *cis* 1,4 polymer (see Fig. 15.14).

Coordination Polymerization

In the 1950s, Karl Ziegler discovered a way to polymerize ethylene in a linear structure to produce high-density polyethylene (HDPE) using transition metal compounds and organometallic compounds. Using similar catalysts, Giulio Natta polymerized alpha olefins, for example propene, with controlled stereoregularity. These catalyst systems are called Ziegler–Natta catalysts, and are widely used for the synthesis of a number of commodity plastics, such as high-density polyethylene and polypropylene.[22] Ziegler–Natta catalysts may be either insoluble (heterogeneous) or soluble (homogeneous) systems.

Ziegler–Natta Catalysts (Heterogeneous). These systems consist of a combination of a transition metal compound from groups IV to VIII and an organometallic compound of a group I–III metal.[23] The transition metal compound is called the catalyst and the organometallic compound the cocatalyst. Typically the catalyst is a halide or oxyhalide of titanium, chromium, vanadium, zirconium, or molybdenum. The cocatalyst is often an alkyl, aryl, or halide of aluminum, lithium, zinc, tin, cadmium, magnesium, or beryllium.[24] One of the most important catalyst systems is the titanium trihalides or tetrahalides combined with a trialkylaluminum compound.

The catalyst system is prepared by mixing the two compounds in the solvent, usually at low temperatures. Polymerization occurs at specific sites on the catalyst surface. There are several proposed mechanisms for polymerization, but the important aspect of both is that the polymerization occurs in coordination

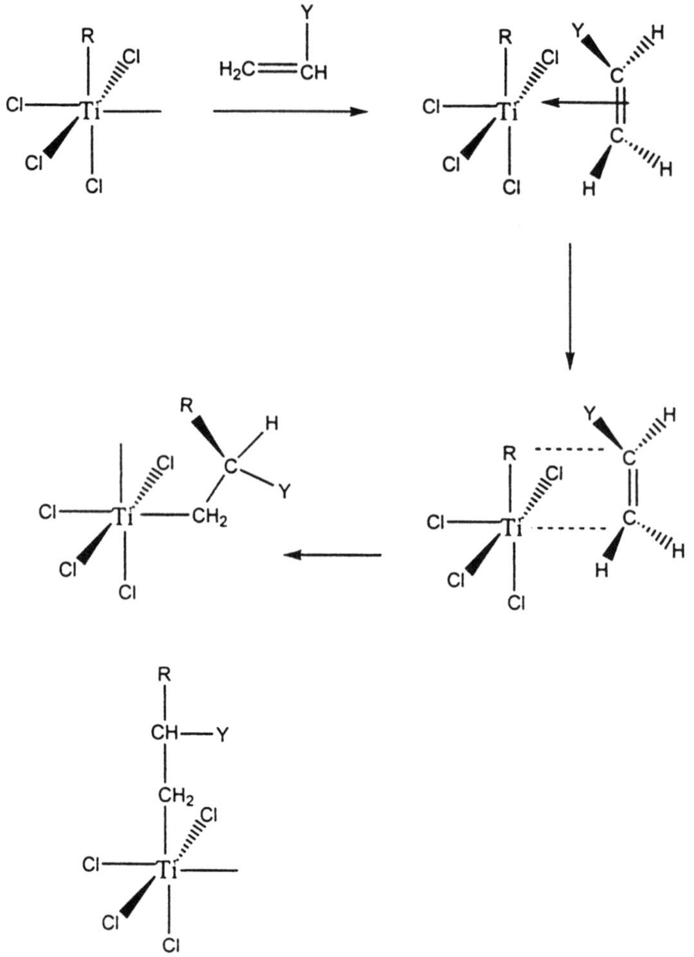

Fig. 15.15. Ziegler–Natta polymerization (nonmetallic mechanism).

with the catalyst. The pi bond of the monomer complexes with the transition metal and is then inserted in between the transition metal and the carbon of the coordinated polymer chain. One of the proposed mechanisms is shown in Fig. 15.15. Isotactic polymers are generally formed with the insoluble catalysts. Syndiotactic polypropylene has been formed with both heterogeneous and homogeneous catalysts.

Metallocene Catalysts (Homogeneous Ziegler–Natta). Solid Ziegler–Natta catalysts suffered from several problems, including the presence of multiple polymerization sites on the catalyst surface and catalyst residue in the final polymer, requiring a secondary purifica-

tion step.[25] The use of soluble catalysts offered an answer to some of these problems. These catalysts are composed of a metal atom (the active site), a cocatalyst, and a ligand system. Zirconium is the most commonly used metal although other metals such as Ti, Hf, Sc, and Th have been used. The most commonly used ligand is cyclopentadienyl. Methylalumoxane is typically the counterion. Figure 15.16 shows a proposed structure for methylalumoxane and a generalized metallocene structure.

The soluble catalysts can prepare polymers with very good stereospecificity and narrow molecular weight distributions, as a result of the uniformity of the active sites. In fact, these catalysts are often referred to as "single site catalysts".[26] The polydispersity ($\overline{M}_w/\overline{M}_n$),

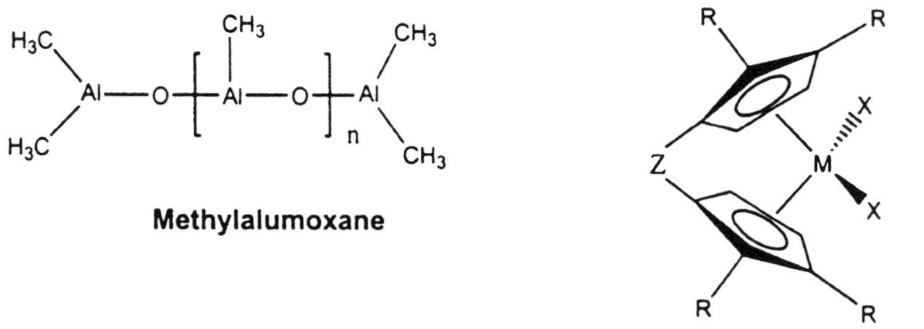

Methylalumoxane

Metallocene

M = metal (Zr, Ti, or Hf)   X = Cl or alkyl

Z = optional bridging group   R = H or alkyl

Fig. 15.16. Metallocene catalysts.

which is a measure of the molecular weight distribution, is about 2–2.5 for the soluble catalysts and 5–6 for the heterogeneous Ziegler–Natta systems. By proper selection of the catalyst, syndiotactic, atactic and isotactic polypropylene and higher alpha-olefins can be synthesized. In addition, it has been possible to prepare polypropylene with alternating blocks of isotactic and atactic chains. The resulting material exhibits elastomeric properties.

POLYMERIZATION METHODS

While polymerization by ionic methods is usually performed in solution, free radical polymerizations can be performed in solution, bulk, suspension, or emulsion.[27] Each of these methods is described below.

Bulk Polymerization

This is one of the simplest methods of polymerization. It is often used in the polymerization of step-growth polymers.[28] In these types of systems the viscosity remains low for a large portion of the reaction and heat transfer is easily controlled. Chain-growth polymers are more difficult to polymerize by this method due to the rapid and highly exothermic reactions. As the viscosity increases, thermal control becomes more difficult and may result in thermal runaway or localized hot spots. Commercial use of bulk polymerization for vinyl polymers is rather limited for

those reasons. Bulk polymerization may be either homogeneous or heterogeneous, as in the case where the polymer is insoluble in the monomer.[29]

Solution Polymerization

Solution polymerization offers improved heat transfer over bulk polymerizations. Proper selection of the solvent is critical to avoid chain transfer reactions. Coupled with environmental concerns over organic solvents, the complete removal of solvents from the polymer also poses a potential problem. Recent work has been performed on the use of supercritical carbon dioxide as a solvent, which is easy to remove and poses less environmental concerns.[30]

Suspension Polymerization

From an environmental standpoint, the use of water as a solvent is desirable. Unfortunately, many of the monomers of interest are insoluble in water, but suspension polymerization offers a way to utilize water. Suspension polymerization is performed by mechanically dispersing a monomer in an incompatible solvent, most often water. The system is heterogeneous and when polymerization is complete the polymer is collected as granular beads. This method is not suitable for tacky materials, such as elastomers, as the beads will tend to clump together.[31]

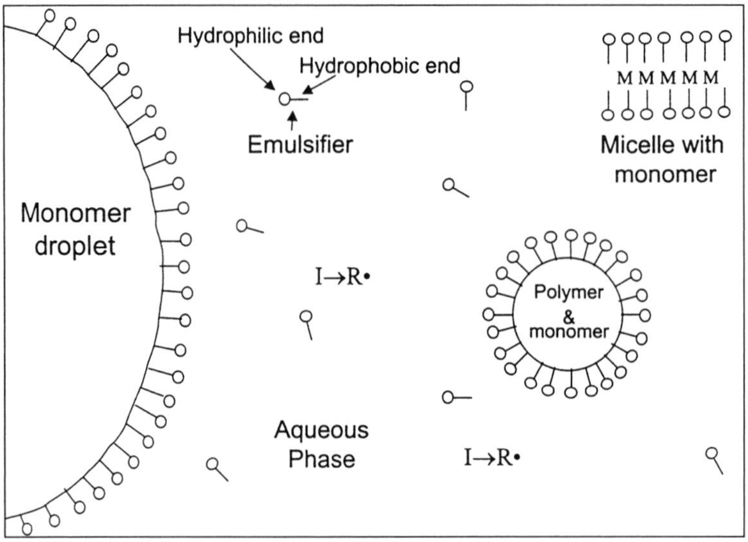

Fig. 15.17. Emulsion polymerization.

Monomer droplets are suspended in the water through the use of agitation and stabilizers, such as methyl cellulose, gelatin, polyvinyl alcohol, and sodium polyacrylate.[32] Typical droplet sizes are 0.01–0.5 cm. A monomer soluble initiator is added to begin the polymerization. The kinetics of suspension polymerization are the same as for bulk polymerization, but suspension polymerization offers the advantage of good heat transfer. Polymers such as polystyrene, PVC, and polymethyl methacrylate are prepared by suspension polymerization.

Emulsion Polymerization

Superficially, emulsion polymerization resembles suspension polymerization, but there are a number of important differences. Water is used as the continuous phase and heat transfer is very good for both suspension and emulsion polymerization. In contrast to suspension polymerization, the polymer particles produced in emulsion polymerization are on the order of 0.1 μm in diameter.[33] Another important difference is the presence of an emulsifying agent or soap. At the beginning of polymerization the soap molecules aggregate together in a group of about 50–100 molecules to form what is called a micelle. Some of the

monomer enters the micelles, but most of it is contained in monomer droplets. A water-soluble initiator is added, which migrates to the micelles as a result of their large surface to volume ratio and initiates polymerization primarily in the micelles. As polymerization continues, the micelles grow by addition of monomer from the water. The monomer droplets provide additional monomer to the aqueous phase as polymerization continues. Polymerization continues in the micelles until a second radical enters to terminate the reaction. As a result, very high molecular weight polymers may be synthesized.[34] Figure 15.17 shows an overall view of the emulsion polymerization process. Emulsion polymerization is a widely used technique, especially useful for making synthetic rubber, latex paints, and adhesives.

COPOLYMERIZATION

In many cases the properties of a single type of polymer cannot meet the demands of a particular application. One approach to solving this problem is to combine two monomers into a single polymer through copolymerization. The properties of the resulting copolymer will then depend on the chemical nature of the monomers used and the microstructure

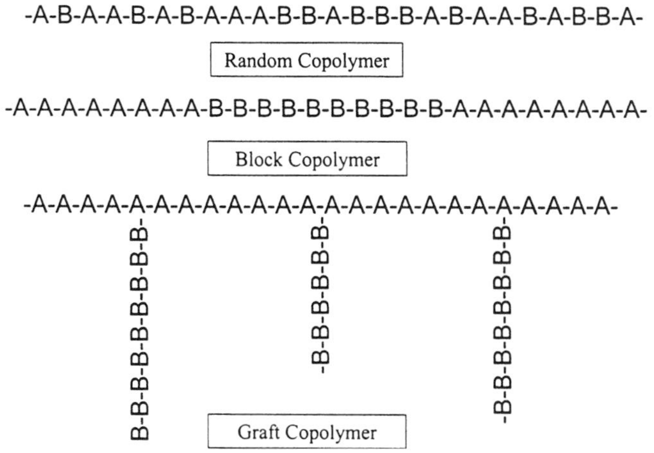

-A-B-A-A-B-A-B-A-A-A-B-B-A-B-B-B-A-B-A-A-B-A-B-B-A-

> Random Copolymer

-A-A-A-A-A-A-A-A-A-B-B-B-B-B-B-B-B-B-B-A-A-A-A-A-A-A-A-

> Block Copolymer

-A-A-A-A-A-A-A-A-A-A-A-A-A-A-A-A-A-A-A-A-A-A-A-A-A-A-A-A-A-A-

> Graft Copolymer

Fig. 15.18. Copolymer structures.

of the chain. Monomers may be placed in the chain in a variety of ways, including random, block, and graft copolymers as depicted in Fig. 15.18.

Random Copolymerization

Copolymerization can occur through any of the chain reaction polymerization mechanisms described above; however, the reactivity of a given monomer toward the second monomer can vary. Thus, not all combinations of monomers may be copolymerized. Each active end will exhibit different reactivity toward each monomer, which can be expressed as reactivity ratios, r_1 and r_2.[35] These reactivity ratios (r_1 in this example) show the tendency of a given active end, for example M_1^*, to add its own monomer (M_1) over the other monomer (M_2). The copolymer composition at any instant can be determined by the composition of the feedstock and the reactivity ratios by

$$F_1 = \frac{r_1 f_1^2 + f_1 f_2}{r_1 f_1^2 + 2f_1 f_2 + r_2 f_2^2}$$

where F_1 is the mole fraction of monomer 1 in the copolymer, f_1 and f_2 are the mole fractions of monomer 1 and 2, respectively, in the feedstock, and r_1 and r_2 are the corresponding reactivity ratios. In general, the copolymer composition will not be the same as the feed composition.

The properties of random copolymers are often a weighted average of the two polymers. For example, the T_g of a single-phase copolymer typically falls somewhere in between the T_gs of the two homopolymers. This can be estimated using[36]

$$a_1 c_1 (T_g - T_{g1}) + a_2 c_2 (T_g - T_{g2}) = 0$$

where T_{g1} and T_{g2} are the glass transition temperatures for the pure homopolymers, a_1 and a_2 depend on the monomer type, and c_1 and c_2 are the weight fractions of monomers 1 and 2, respectively. For crystalline polymers, the degree of crystallinity and melting point decrease as the second monomer is added.[37]

Block and Graft Copolymers

As mentioned above, the ability to have "living polymerizations" offered the potential to make block copolymers. In the preparation of a block copolymer the sequence of addition can be important to ensure that the second monomer is capable of adding to the living end. An example is the formation of a polystyrene—polymethyl methacrylate block copolymer.[38] In this case polystyrene is polymerized first, followed by addition of the methyl methacrylate. The block copolymer could not be formed if methyl methacrylate were polymerized first, as styrene will not add to the methyl methacrylate living end. An important example of block copolymerization is the synthesis of ABA triblock copolymers from styrene and diene monomers to form thermoplastic elastomers. They may be prepared by sequential polymerization from one end or from the middle using a difunctional initiator. Block copolymers may also be prepared with polymers containing functionalized end groups that are linked together.

There are three basic methods to produce graft copolymers.[39] In the first method a monomer is polymerized in the presence of a polymer and chain transfer provides the branching. Initiator, monomer and polymer are combined to prepare the copolymer. The initiator may function by polymerizing the monomer, which then reacts with the polymer to form the graft, or by forming a reactive site on the polymer, which then polymerizes the monomer onto the polymer.

The second method of forming a graft copolymer is to polymerize a monomer with a polymer containing a reactive functional group (or sites that may be activated). Irradiation is commonly used to form active sites for graft copolymerization, but other methods may also be used. When irradiation is used the grafting mechanism is free radical in nature. In cases where the monomer is present when the polymer is irradiated, homopolymerization may also occur. If homopolymerization is a concern, then the polymer may be irradiated in the presence of air (or oxygen) to form hydroperoxide groups. The pretreated polymer is then mixed with monomer and heated to initiate the grafting reaction by decomposition of the peroxide groups.

The third method of preparing a graft copolymer is to combine two polymers with functional groups that can react together. An example is the reaction of oxazoline-substituted polystyrene with polymers containing functional groups such as alcohols, amines, and carboxylic acids.

Unlike the random copolymers, block and graft copolymers separate into two phases, with each phase exhibiting its own T_g (or T_M).[40] The modulus-temperature behavior of a series of

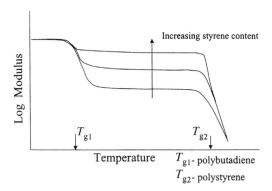

Fig. 15.19. Modulus-temperature behavior of a triblock copolymer.

triblock copolymers prepared from styrene and butadiene (SBS) is shown in Fig. 15.19. With increasing styrene content the modulus in the plateau region (between the two glass transition temperatures) increases. Mixing of the two phases may shift the values of the two transition temperatures.

MECHANICAL PROPERTIES

Many factors influence the mechanical behavior of polymers including polymer type, molecular weight, and test procedure. Modulus values may be obtained from a standard tensile test with a given rate of crosshead separation. In the initial linear region, the slope of a stress–strain curve will give the elastic or Young's modulus, E. Unlike many other materials, polymeric material behavior may be affected by factors such as test temperature and rates. This can be especially important to the designer when the product is used or tested at temperatures near the glass transition temperature, where dramatic changes in properties occur as depicted in Fig. 15.8. The time-dependent behavior of these materials is discussed below.

Viscoelasticity

Polymer properties may exhibit considerable time dependent behavior, depending on the polymer type and test conditions. Increases in testing rate or decreases in temperature cause the material to appear more rigid, while an increase in temperature or decrease in rate

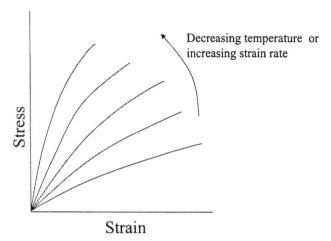

Stress

Decreasing temperature or
increasing strain rate

Strain

Fig. 15.20. Viscoelastic behavior of polymers.

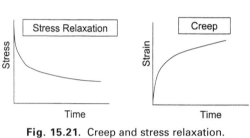

Fig. 15.21. Creep and stress relaxation.

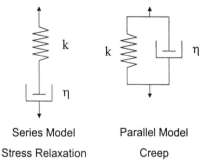

Series Model

Parallel Model

Stress Relaxation

Creep

Fig. 15.22. Spring and dashpot models for stress relaxation and creep.

will cause the material to appear softer. Viscoelastic response of a polymer to changes in testing rate or temperature is shown in Fig. 15.20. This time-dependent behavior can also result in long-term effects such as stress-relaxation or creep.[41] Creep and stress relaxation are shown in Fig. 15.21. A specimen held under a fixed load will continue to elongate with time, a process called creep. A spring and dashpot in parallel is often used as a simple model to predict the creep behavior (see Fig. 15.22). This model predicts the time-dependent strain to be

$$\varepsilon(t) = \varepsilon_o e^{-t/\tau}$$

where τ is the characteristic relaxation time (η/k).

Stress relaxation is the decrease in load of a material held at a fixed displacement. Figure 15.22 shows the spring and dashpot in series that can be used to model the stress relaxation behavior. Using this model one obtains the following equation for the stress behavior.

$$\sigma(t) = \sigma_o e^{-t/\tau}$$

These models are useful for understanding the general concept of viscoelasticity, but are typically unable to accurately model the time dependent behavior. More accurate prediction can be obtained by using models with more elements.

If stress relaxation curves are obtained at a number of different temperatures, it is found that these curves can be superimposed by horizontal shifts to produce what is called a "master curve".[42] This concept of time–temperature equivalence is very important to understanding and predicting polymer behavior. As an example, a polymer at very low temperatures will behave as if it were tested at higher temperatures at much higher testing rates. This principle can be applied to predict material behavior under testing rates or times that are not experimentally accessible, through the use of shift factors (a_T) and the equation below

$$\ln a_T = \ln\left(\frac{t}{t_o}\right) = -\frac{17.44\,(T - T_g)}{51.6 + T - T_g}$$

where T_g is the glass transition temperature of the polymer. This has practical applications where one may be interested in the mechanical properties of a material at low temperatures under high rate (perhaps impact rate) conditions. It is also important to recognize that if material properties are measured at room temperature, they may not adequately reflect the material behavior at much lower temperatures or higher rates.

Failure Behavior

Part design requires the avoidance of failure without overdesign of the part, which leads to increased part weight and cost. Failure behavior depends on material type, service temperature, and rates. A tensile stress–strain test can be used to gather some information on material strength and behavior, for example, stress (or strain) at break, σ_B. Figure 15.23 shows some different types of failure behavior. Materials failing at rather low elongations (1% strain or less) are considered to have undergone brittle failure.[43] General purpose polystyrene and acrylics are polymers that show this type of failure. Failure usually starts at a defect where stresses are concentrated. Once a crack is formed it will grow as a result of stress concentrations at the crack tip. Many amorphous polymers will also exhibit what are called "crazes." Crazing is a form of yielding and can enhance the toughness of a material. Although crazes appear to look like cracks, they are load bearing, with fibrils of material bridging the two surfaces as shown in Fig. 15.24.

Polymers also exhibit what is termed ductile failure by yielding of the polymer or slip of the molecular chains past one another. This is most often indicated by a maximum in the tensile stress–strain test or the yield point, σ_Y (see Fig. 15.23). Above this point the material may exhibit lateral contraction upon further extension, called necking.[44] In the necked regions, molecules are oriented in the direction of deformation, resulting in increased stiffness in the necked region. As a result of this localized stiffness increase, material adjacent to the neck is deformed and the neck region continues to grow. This process is known as cold-drawing (see Fig. 15.25), which may result in elongations of several hundred percent.

As might be expected, temperature will influence the behavior of the material. The effect of temperature on modulus has been discussed above. Very different behavior may be seen in a single polymer simply by changing

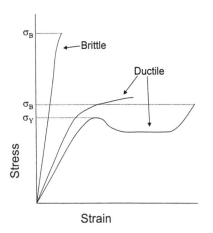

Fig. 15.23. Types of stress–strain behavior.

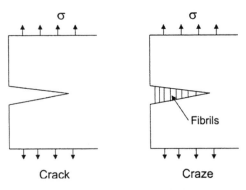

Fig. 15.24. Crazes and cracks.

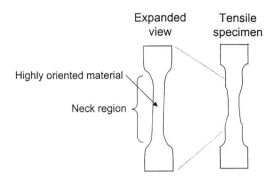

Fig. 15.25. Cold-drawing.

the temperature.[45] Beginning at temperatures well below the T_g, the stress–strain behavior will show very low elongations, with no yield point. As the temperature is raised, the material will exhibit a yield point and elongation increases. The yieldpoint is typically seen at temperatures near the glass transition temperature of the polymer. The higher the testing rate, the higher the temperature must be for yielding to occur.

Under repeated cyclic loading, a material may fail at stresses well below the single-cycle failure stress found in a typical tensile test, a process called fatigue.[46] Fatigue behavior is usually described by plotting the maximum stress versus the number of cycles to failure. The exact conditions of the fatigue test can be varied and are specified by the service requirements. Type of loading (compression, tension, shear), load or displacement control, frequency, waveform (typically sinusoidal), and ratio of maximum to minimum stress may all be varied. Thermal effects and the presence or absence of cracks are other variables that may be included when evaluating the fatigue life of a material.

PART II. COMMERCIAL PLASTIC MATERIALS

CLASSES OF FAMILIES OF COMMERCIAL PLASTICS

Commercial plastic materials may conveniently be divided into five major classes. Commodity thermoplastics are families which are produced in volumes of a billion pounds or more per year. Engineering thermoplastics

are generally designed to withstand use at higher temperatures. Specialty thermoplastics are designed for a greater variety of performance and applications. Thermoplastic elastomers can be processed like thermoplastics, and give many of the properties of conventional thermoset vulcanized rubbers. And thermoset plastics are permanently crosslinked during processing, generally to give higher performance than would be possible with linear thermoplastics.

Commodity Thermoplastics

Seven families of thermoplastics exceed a billion pounds per year in the United States: polyethylenes, polypropylene, PVC, polystyrene and its copolymers, polyethylene terephthalate, acrylonitrile-butadiene-styrene (ABS) and nylon (Table 15.2).

Polyethylenes. The major members of the polyethylene family are distinguished primarily by the amount of side-branching, which decreases regularity, crystallinity, density, melting point, modulus (rigidity), and strength (Table 15.3).

TABLE 15.2 Commodity Thermoplastics in the United States and Canada

	Million Metric Tons[a]
Polyethylenes	15.7
Polypropylene	8.0
PVC	6.7
Polyethylene terephthalate	3.4
Polystyrene and its copolymers	2.9
ABS	0.6
Nylon	0.6

[a]To convert MMT to pounds, multiply by 2200.

TABLE 15.3 Major Members of the Polyethylene Family: Typical Properties

Polyethylene	*HDPE*	*LLDPE*	*LDPE*
Density	0.96	0.93	0.92
Melting point, °C	130	124	108
Modulus, MPa[a]	1400	450	180
Tensile strength, MPa[a]	26	14	12

[a]To convert MPa to psi, multiply by 145.

High-Density Polyethylene (HDPE). Polymerization of ethylene to polyethylenes is most often carried out at low temperature and pressure, using either the Ziegler aluminum triethyl plus titanium tetrachloride catalyst system, the Phillips chromic oxide plus silica plus alumina system, or more recently the newer metallocene single-site catalyst systems.

$$nCH_2{=}CH_2 \rightarrow -(CH_2{-}CH_2)_n-$$

$$CH_2 = CH_2 \longrightarrow -(CH_2 - CH_2) - (CH_2 - \underset{|}{CH}) - (CH_2 - \underset{|}{CH}) - (CH_2 - \underset{|}{CH}) -$$

with pendant CH_2/CH_3, $(CH_2)_3/CH_3$, $(CH_2)_n/CH_3$

This produces a regular, linear polymer which is 50–90 percent crystalline, making it rigid and strong. Its largest use is blow-molded bottles for milk, water, juice, noncarbonated drinks, laundry detergent, antifreeze, and motor oil. Other large uses are large molded containers, pipe and conduit, and shopping bags. For detergent bottles, environmental stress-crack resistance is increased by increasing molecular weight, adding ethylene/propylene rubber, or copolymerization with a small amount (1–2%) of l-butene to reduce crystallinity. For auto gasoline tanks,

Low-Density Polyethylene (LDPE). Polyethylene was originally made by polymerization at high temperature and pressure; during the polymerization, side-reactions produced several percent of short-chain branching (ethyl and butyl groups) which reduced regularity/crystallinity (e.g., to 30–40%) and gave a stiffly flexible plastic material; and chain-transfer reactions also

produced some long-chain branching which gave pseudoplastic melt processability, important in the production of extruded blown film. Largest use is in packaging film; second largest is for coating and laminating on cardboard, paper, and foil.

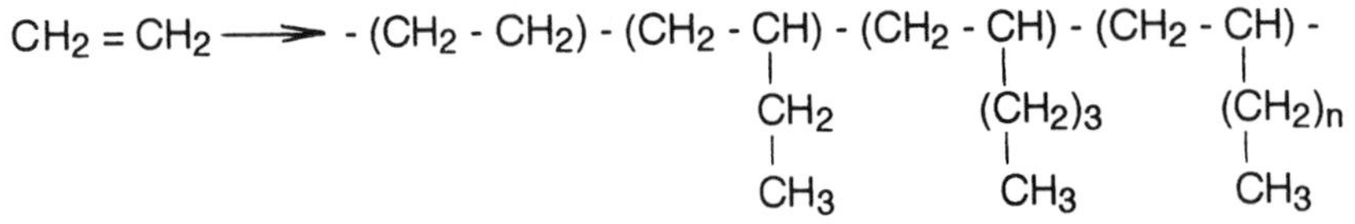

Linear Low-Density Polyethylene (LLDPE). Ziegler and metallocene polyethylene can be modified by copolymerization with several percent (e.g., 10%) of butene, hexene, or octene, to reduce regularity/crystallinity and modulus, and thus produce

$$CH_2 = CH_2 + CH_2 = \underset{|}{CH} \longrightarrow -(CH_2 - CH_2) - (CH_2 - \underset{|}{CH}) -$$

with C_2H_5 substituent

Ethylene   Butene

$$CH_2 = CH_2 + CH_2 = \underset{|}{CH} \longrightarrow -(CH_2 - CH_2) - (CH_2 - \underset{|}{CH}) -$$

with C_4H_9 substituent

Ethylene   Hexene

$$CH_2 = CH_2 + CH_2 = \underset{|}{CH} \longrightarrow -(CH_2 - CH_2) - (CH_2 - \underset{|}{CH}) -$$

with C_6H_{13} substituent

Ethylene   Octene

impermeability may be increased by dispersing nylon in the polyethylene. For extreme abrasion resistance in materials-handling machinery, and in hip and knee bone replacement, Ultra-High Molecular Weight (UHMWPE: MW $3–6 \times 10^6$) is outstanding.

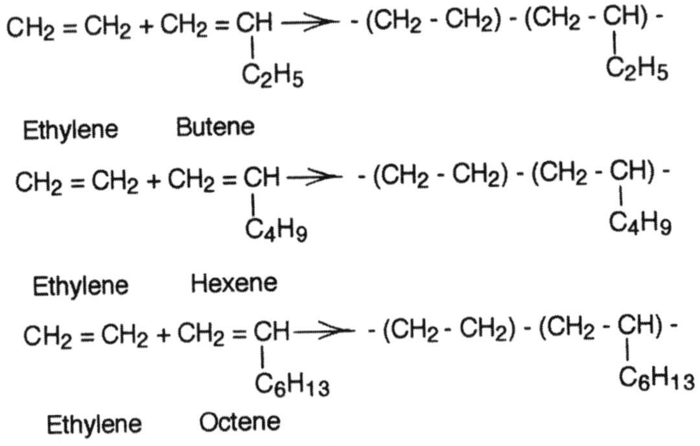

low-density polyethylene at low temperature and pressure. The resulting LLDPE is stronger than the original LDPE, and therefore preferred for packaging film and trashcan liners. Lack of long-chain branching makes it difficult for extrusion-blown film, so processors add enough LDPE to improve processability.

Very Low Density (VLDPE) and Ultra Low Density (ULDPE) Polyethylenes. These are made by copolymerization with increasing amounts of comonomers, especially 1-octene, reducing regularity/crystallinity (density 0.91– 0.86) down toward ethylene/propylene rubber. These are soft and flexible enough to compete with plasticized polyvinyl chloride and thermoplastic elastomers in some applications.

Ethylene/vinyl Acetate Copolymers (EVA). LDPE is easily internally plasticized by copolymerization with increasing amounts (2–20%) of vinyl acetate comonomer.

Polypropylene. Polymerization of propylene to polypropylene (Table 15.4)

$$CH_2 = CH \longrightarrow -(CH_2 - CH) - $$
$$\qquad\quad | \qquad\qquad\qquad | $$
$$\qquad\quad CH_3 \qquad\qquad\quad CH_3$$

produces an asymmetric carbon atom, which can be either right-handed or left-handed. Peroxide initiation of free-radical polymerization produces a random ("atactic") mixture of right- and left-handed carbon atoms, which is too irregular to crystallize, so the resulting polymer is a soft gummy rubber. Ziegler triethyl aluminum plus titanium chloride catalysis of polymerization pre-orients each monomer unit before inserting it into the polymer chain, and produces all-right-handed

$$CH_2 = CH_2 + CH_2 = CH \longrightarrow -(CH_2 - CH_2) - (CH_2 - CH) -$$
$$\qquad\qquad\qquad\qquad | \qquad\qquad\qquad\qquad\qquad\qquad | $$
$$\qquad\qquad\qquad\qquad O \qquad\qquad\qquad\qquad\qquad\qquad O$$
$$\qquad\qquad\qquad\qquad | \qquad\qquad\qquad\qquad\qquad\qquad | $$
$$\qquad\qquad\qquad\qquad C = O \qquad\qquad\qquad\qquad\qquad C = O$$
$$\qquad\qquad\qquad\qquad | \qquad\qquad\qquad\qquad\qquad\qquad | $$
$$\qquad\qquad\qquad\qquad CH_3 \qquad\qquad\qquad\qquad\qquad CH_3$$

This reduces regularity/crystallinity (melting points 108–70°C), increasing softness/flexibility (modulus 138–19 MPa), transparency, and polar adhesion, again competing with plasticized PVC and thermoplastic elastomers, particularly in packaging film, coatings, and adhesives.

Ionomers are generally ethylene copolymers with 5–10 percent of methacrylic acid, half-neutralized by sodium or zinc.

or all-left-handed ("isotactic") polymer which is very regular and therefore able to crystallize (e.g. 60%). This is a high-melting (165°C), rigid, strong, low-cost commodity plastic, very competitive with more costly polystyrene and ABS. Another unique quality is its ability to form an "integral hinge," so that rigid packaging container plus lid can be molded as a single part.

$$\qquad\qquad\qquad CH_3 \qquad\quad CH_3$$
$$\qquad\qquad\qquad | \qquad\qquad\quad | $$
$$-(CH_2 - CH_2) - (CH_2 - C) - (CH_2 - C) -$$
$$\qquad\qquad\qquad\qquad\qquad | \qquad\qquad\quad | $$
$$\qquad\qquad\qquad\qquad\quad CO_2H \qquad CO_2^- \, Na^+ \text{ or } Zn^{++}$$

This reduces regularity/crystallinity, increasing softness/flexibility (modulus 250 MPa) and transparency. The ionic groups increase strength (20 MPa) and polar adhesion, again competing with plasticized PVC and thermoplastic elastomers, particularly in packaging films and sporting goods.

It has two major weaknesses: (1) Methyl branching leaves tertiary hydrogen atoms,

$$\qquad\qquad\qquad H^*$$
$$\qquad\qquad\qquad | $$
$$\qquad\quad -CH_2 - C -$$
$$\qquad\qquad\qquad | $$
$$\qquad\qquad\qquad CH_3$$

TABLE 15.4 Typical Properties of Polypropylenes

Type	General Purpose	High-Impact	40% Talc	40% Glass Fiber
Modulus, MPa	1450	1100	3100	7000
Tensile strength, MPa	36		32	83
Ultimate elongation, %	500		10	2
Notched Izod impact strength, J/cm[a]	1.1	5.6	1.3	3.9
Heat deflection temperature, °C at 1.82 MPa	57	53	79	149

[a]To change J/cm to FPI, multiply by 1.87.

TABLE 15.5 Typical Properties of Rigid PVC

Type	General Purpose	High-Impact	30% Glass Fiber	Chlorinated PVC
Modulus, MPa	3,300	2,900	8,300	2,800
Tensile strength, MPa	48	43	90	56
Ultimate elongation, %	145	130	2	35
Notched Izod impact strength, J/cm	0.32	8.56	0.70	0.91
Heat deflection temperature, °C at 1.82 MPa	67	66	76	103

which are sensitive to thermal oxidative aging. (2) Bulky methyl groups cause steric hindrance and stiffen the polymer molecule; while this increases modulus, it also causes embrittlement below room temperature. The polypropylene industry has successfully retarded aging by adding phenolic and aliphatic sulfide antioxidants; and it has reduced low-temperature embrittlement by adding or grafting ethylene–propylene–diene (EPDM) rubber. A third common improvement is the addition of fillers to increase rigidity, making it even more competitive with polystyrene.

With these problems solved, polypropylene use has been growing rapidly. Extrusion and stretch orientation produces excellent synthetic fiber for carpeting and synthetic turf, and excellent film for packaging. Other large uses are rigid packaging (replacing polystyrene), automotive parts (replacing ABS), and a great variety of injection-molded consumer products.
Polyvinyl Chloride. (Table 15.5) this is the most versatile of the commercial thermoplastic polymers. It is used mainly for rigid and flexible plastics, for rubberlike products, for coatings on steel, cloth, and paper, and in smaller amounts for specialty fibers. It is processed mainly by extrusion and calendering, and in smaller amounts by injection, compression, and

blow molding, thermoforming, rotomolding, plastisol/organosol technology, casting, pouring, and foaming. This extreme versatility comes primarily from the use of plasticizers, aided by a host of other types of additives.

Vinyl chloride is polymerized primarily by peroxide-initiated polymerization in aqueous suspension,

$$CH_2 = CH \longrightarrow -(CH_2 - CH) - $$
$$\qquad\quad | \qquad\qquad\qquad | $$
$$\qquad\quad Cl \qquad\qquad\qquad Cl $$

Producing a fine porous powder which is easy to blend with compounding ingredients. A smaller amount is polymerized in emulsion and spray-dried for plastisols and organosols. And an even smaller amount is copolymerized with vinyl acetate in organic solution, to produce a uniform copolymer which precipitates at a constant composition and molecular weight.

Rigid PVC is polymerized at 55–70°C and then compounded with organotin stabilizer, acrylic processing aid, lubricants, and optionally with rubbery impact modifier.

It has high modulus, strength, and resistance to creep, weathering, and chemicals; and compounding with rubbery impact modifier produces high resistance to brittle failure. It is mainly extruded into pipe, house-siding, and

TABLE 15.6 Typical Properties of Flexible PVC

Plasticizer	Dioctyl Phthalate			Dioctyl Adipate		
Parts/hundred of resin	30	50	75	30	50	70
Shore a hardness	95	79	62	90	74	60
Tensile strength, MPa	31	21	14	27	20	14
Ultimate elongation, %	220	300	400	270	395	410
Flex temperature, °C	+20	0	−26	−15	−48	−75

TABLE 15.7 Typical Properties of Polystyrenes

Polymer	Polystyrene	Impact Styrene		Styrene/Acrylonitrile
		Moderate	Very High	
Modulus, MPa	2900	1700	1100	3500
Tensile Strength, MPa	46	21	13	73
Notched Izod impact strength, J/cm	0.17	1.28	3.70	0.26
Heat deflection temperature, °C at 1.82 MPa	83	81	81	100

window and door frames, while smaller amounts are injection molded into pipe fittings and blow-molded into bottles for detergents and other chemicals.

Flexible PVC (Table 15.6) is made by polymerizing at 40–55°C and then compounding with 20–80 PHR (parts per hundred of resin) of dioctyl phthalate and/or other monomeric liquid plasticizers (e.g., dioctyl adipate for low-temperature flexibility, oligomeric polyesters for permanence, organic phosphates for flame-retardance), plus a synergistic stabilizer system usually composed of barium or calcium soap, zinc soap, epoxidized fatty ester, and organic phosphite.

This flexible-rubbery material is most often calendered into film and sheet for clothing, luggage, raincoats, upholstery, and flooring, or extruded into garden hose, wire and cable insulation, and medical tubing.

For plastisol/organosol technology, high-molecular-weight PVC powder is slurried in liquid plasticizer, and poured, dip- or spread-coated onto metal, fabric, or paper to produce dishwasher racks, upholstery, and wallpaper, respectively.

Polystyrene and its Copolymers. Polystyrene (Table 15.7) is made by continuous bulk polymerization, initiated by peroxides and

$$CH_2 = CH \longrightarrow -(CH_2 - CH)-$$

heat, with temperature rising continuously to keep the system molten, and optionally adding solvent ("solution polymerization") to facilitate the process. It was the first commodity thermoplastic, combining easy melt processability, rigidity, and glass-like transparency for major applications in packaging, toys, and housewares. When swollen with 10 percent pentane and heated, it expands to rigid closed-cell foams with densities as low as 0.01 (expanded polystyrene or EPS), which are popular for packaging hot and cold foods, delicate instruments and appliances.

The major weaknesses of polystyrene are brittleness, and softening in hot water. Brittleness is remedied by dissolving 2–10 percent of rubber in styrene monomer before polymerization, producing "high-impact styrene" (HIPS), in which 10-μm rubber particles improve impact strength by an order of magnitude, with some sacrifice of other mechanical properties and transparency; this accounts for more than half of the total "polystyrene" market.

Heat resistance is improved by copolymerization with 15–30 percent of acrylonitrile, producing styrene/acrylonitrile (SAN).

TABLE 15.8 Typical Properties of ABS

Type	Moderate Impact	Very High Impact
Modulus, MPa	2,500	2,000
Tensile strength, MPa	46	38
Notched Izod impact strength, J/cm	3.5	9.0
Heat deflection temperature, °C at 1.82 MPa	94	88

Polarity and hydrogen-bonding stiffen the polymer molecule, improving mechanical properties and especially resistance to hot water. This improvement is useful in household products, autos, and appliances.

Acrylonitrile/Butadiene/Styrene (ABS). The benefits of impact styrene and SAN are

equipment cabinets, and drain-waste-vent (DWV) pipe.

Polyethylene Terephthalate. PET (Table 15.9) is produced by continuous melt condensation polymerization of ethylene glycol plus terephthalic acid,

combined in ABS (Table 15.8). Typically, 15–30 percent acrylonitrile and 45–75 percent styrene are copolymerized in the presence of 5–30 percent of polybutadiene rubber, producing some graft terpolymer, and dispersing 1-μm rubber domains in an SAN matrix.

This balance of properties has found particular usefulness in appliances, autos, electronic

followed by solid-state finish to reach high molecular weight and high purity. It was originally developed for synthetic fibers, replacing cotton as the leading textile fiber worldwide. Later it became popular in packaging film (e.g., boil-in-bag) and magnetic tape because of its high strength, cling, melting point, impermeability, and clarity. Finally, it became

TABLE 15.9 Typical Properties of PET

| | | | | Reinforced Molding | |
				15% Glass	55% Glass
Form	Fiber	Film	Bottles		
Modulus, MPa		3600	3200	5900	18000
Tensile strength, MPa	550	200	120	110	200
Ultimate elongation, %	29	125		2	1
Notched Izod impact strength, J/cm				0.64	1.07
Heat deflection temperature, °C at 1.82 MPa				210	229

the basic material for blow-molding bottles for carbonated beverages, after which it proved completely recyclable into bulk fiber for bedding, furniture, and clothing. More specialized engineering grades for solid molding applications are generally reinforced by short glass fiber.

In summary, commodity thermoplastics are manufactured readily at low cost, and offer a combination of processability, mechanical, thermal, optical, and chemical properties that are useful in a wide range of mass markets and products.

Engineering and Specialty Thermoplastics

Many present and future applications of thermoplastics make greater demands for higher properties, and especially combinations of properties, than are available from the commodity materials. To satisfy these demands, organic polymer chemists and chemical engineers have developed and commercialized over four dozen major types of polymers, offering many improved properties to meet these demands. They may be listed as follows, and then compared in their abilities to satisfy these requirements.

Acrylonitrile/Styrene/Acrylic Rubber (ASA)

Acrylonitrile/Styrene/Olefin Rubber (ASO)

Poly(1-Butene)

Poly(4-Methyl Pentene-1) (TPX)

Poly-p-Xylylene (Parylene) $- CH_2 -$ ⬡ $- CH_2 -$

Poly(Vinylidene Chloride) (PVDC)
$$- CH_2 - \underset{\underset{Cl}{|}}{\overset{\overset{Cl}{|}}{C}} -$$

Poly(Tetrafluoroethylene) (PTFE)
$$- \underset{\underset{F}{|}}{\overset{\overset{F}{|}}{C}} - \underset{\underset{F}{|}}{\overset{\overset{F}{|}}{C}} -$$

Fluorinated Ethylene/Propylene (FEP)
$$- (CF_2 - CF_2) - (CF_2 - \underset{\underset{CF_3}{|}}{CF}) -$$

Perfluoroalkoxy Ethylene (PFA)
$$- (CF_2 - CF_2) - (CF_2 - \underset{\underset{OC_3F_7}{|}}{CF}) -$$

Ethylene/Tetrafluoroethylene (ETFE) $- (CH_2 - CH_2) - (CF_2 - CF_2) -$

Chlorotrifluoroethylene (CTFE)
$$- (\overset{\overset{Cl}{|}}{CF} - CF_2) -$$

Ethylene/Chlorotrifluoroethylene (ECTFE)
$$- (CH_2 - CH_2) - (\overset{\overset{Cl}{|}}{CF} - CF_2) -$$

Poly(Vinylidene Fluoride) (PVDF) $- (CH_2 - CF_2) -$

Poly(Vinyl Fluoride) (PVF)
$$- (CH_2 - \underset{\underset{F}{|}}{CH}) -$$

Ethylene/Vinyl Alcohol (EVAL or EVOH)
$$- (CH_2 - CH_2) - (CH_2 - \underset{\underset{OH}{|}}{CH}) -$$

Poly(Vinyl Formal)
$$- (CH_2 - \underset{\underset{O}{|}}{CH} - CH_2 - \underset{\underset{O}{|}}{CH}) -$$
$$\underset{CH_2}{\diagdown \diagup}$$

Poly(Vinyl Butyral)
$$- (CH_2 - \underset{\underset{O}{|}}{CH} - CH_2 - \underset{\underset{O}{|}}{CH}) -$$
$$\underset{\underset{C_3H_7}{|}}{\underset{CH}{\diagdown \diagup}}$$

Poly(Methyl Methacrylate) (PMMA)
$$- (CH_2 - \underset{\underset{CO_2CH_3}{|}}{\overset{\overset{CH_3}{|}}{C}}) -$$

Poly(Acrylonitrile) (PAN)
$$- (CH_2 - \underset{\underset{CN}{|}}{CH}) -$$

Phenoxy Resin

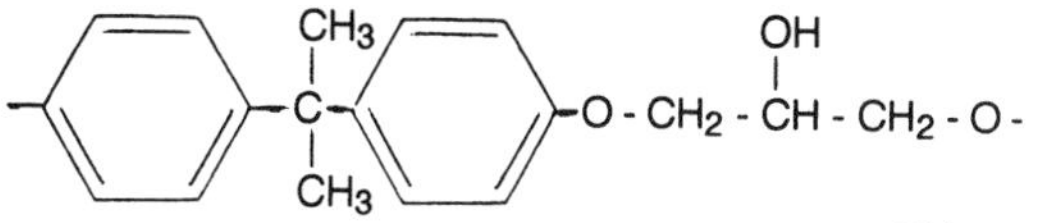

$$CH_3$$
$$OH$$

Phenoxy Resin — [benzene] — C — [benzene] — O - CH₂ - CH - CH₂ - O -

Poly(Phenylene Ether) (PPE or PPO)

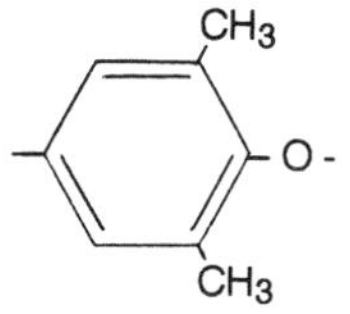

Poly(Phenylene Sulfide) (PPS)

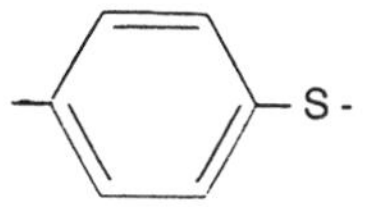

Poly(Ether Ether Ketone) (PEEK)

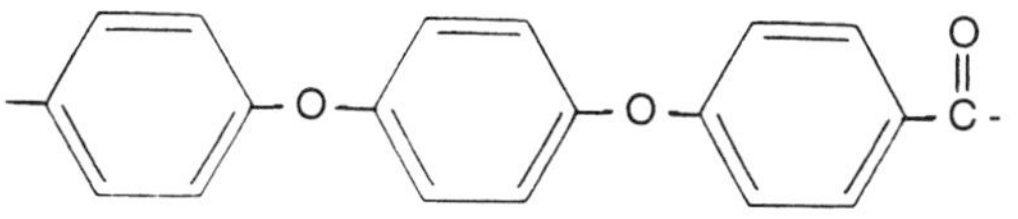

Polysulfone (Udel)

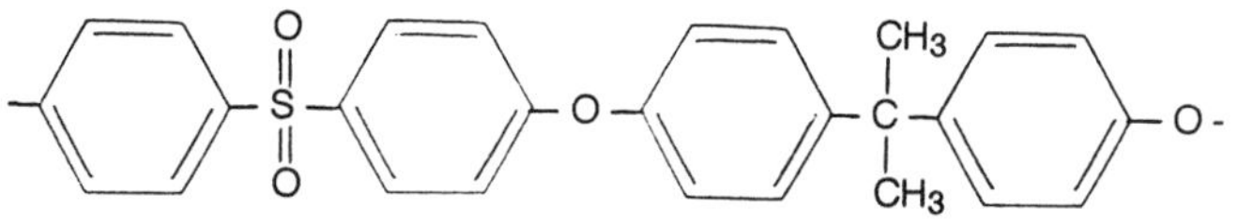

Polyaryl Sulfone

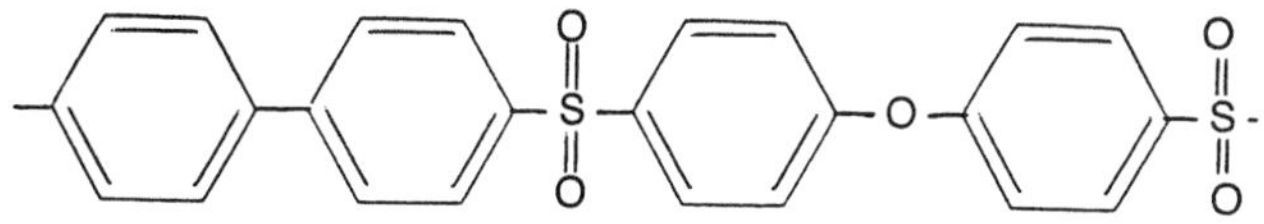

Polyether Sulfone

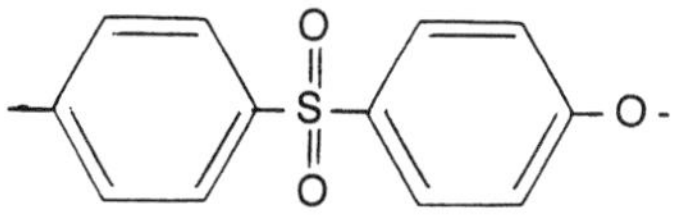

Polyphenyl Sulfone (Radel)

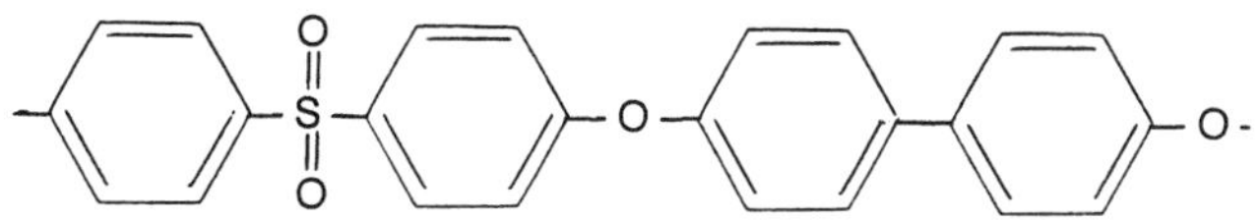

Poly(Oxymethylene) (Polyacetal)   - (CH₂ - O) -

Cellulose Acetate (CA)

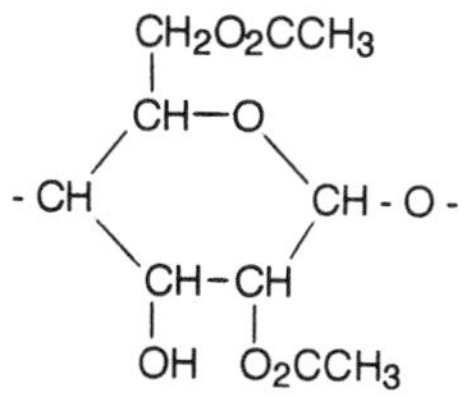

$$CH_2O_2CCH_3$$
$$CH-O$$
- CH   CH - O -
$$CH-CH$$
$$OH \quad O_2CCH_3$$

Cellulose Acetate/Propionate (CAP)

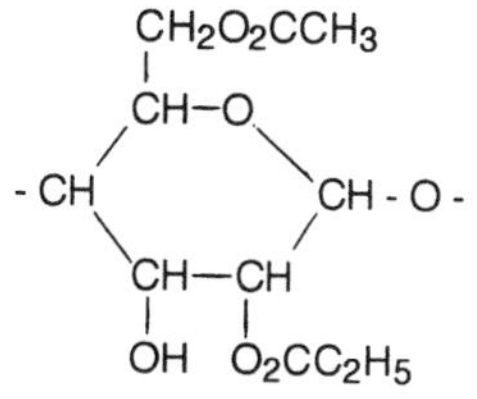

$$CH_2O_2CCH_3$$
$$CH-O$$
- CH   CH - O -
$$CH-CH$$
$$OH \quad O_2CC_2H_5$$

Cellulose Acetate/Butyrate (CAB)

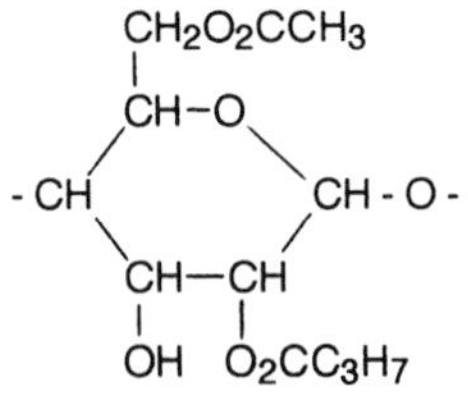

$$-CH \underset{\underset{\displaystyle OH \quad O_2CC_3H_7}{|}}{\overset{\overset{\displaystyle CH_2O_2CCH_3}{|}}{\overset{\displaystyle CH-O}{}}}\!\!\!\!\!\!\!\!\!\!\!\! CH-O-$$

Cellulose Nitrate

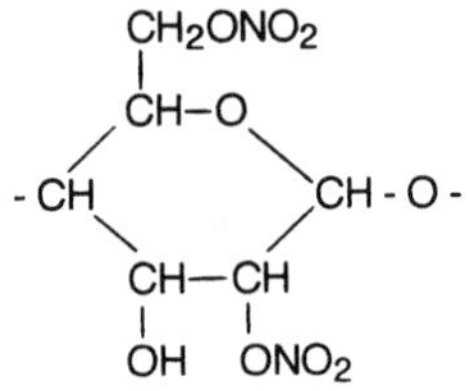

$$-CH \underset{\underset{\displaystyle OH \quad ONO_2}{|}}{\overset{\overset{\displaystyle CH_2ONO_2}{|}}{\overset{\displaystyle CH-O}{}}}\!\!\!\!\!\!\!\!\!\!\!\! CH-O-$$

Ethyl Cellulose (EC)

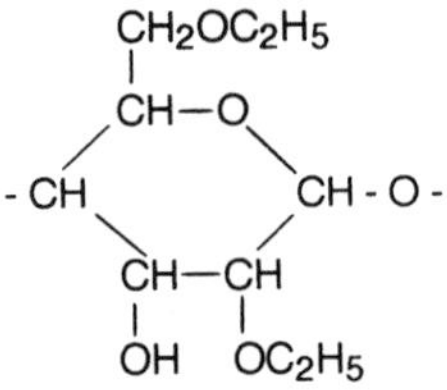

$$-CH \underset{\underset{\displaystyle OH \quad OC_2H_5}{|}}{\overset{\overset{\displaystyle CH_2OC_2H_5}{|}}{\overset{\displaystyle CH-O}{}}}\!\!\!\!\!\!\!\!\!\!\!\! CH-O-$$

Polycarbonate (PC)

$$-\!\!\!\!\bigcirc\!\!\!\!-\overset{\overset{\displaystyle CH_3}{|}}{\underset{\underset{\displaystyle CH_3}{|}}{C}}-\!\!\!\!\bigcirc\!\!\!\!-O-\overset{\overset{\displaystyle O}{\|}}{C}-O-$$

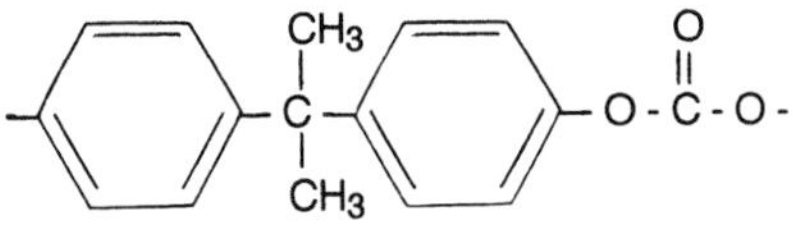

Poly(Butylene Terephthalate) (PBT)

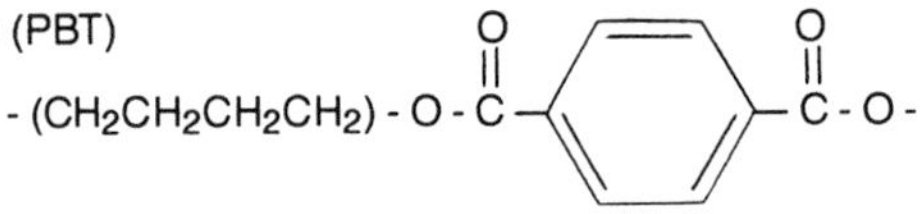

$$-(CH_2CH_2CH_2CH_2)-O-\overset{\overset{\displaystyle O}{\|}}{C}-\!\!\!\!\bigcirc\!\!\!\!-\overset{\overset{\displaystyle O}{\|}}{C}-O-$$

Polyarylate

$$-\!\!\!\!\bigcirc\!\!\!\!-\overset{\overset{\displaystyle CH_3}{|}}{\underset{\underset{\displaystyle CH_3}{|}}{C}}-\!\!\!\!\bigcirc\!\!\!\!-O-\overset{\overset{\displaystyle O}{\|}}{C}-\!\!\!\!\bigcirc\!\!\!\!-\overset{\overset{\displaystyle O}{\|}}{C}-O-$$

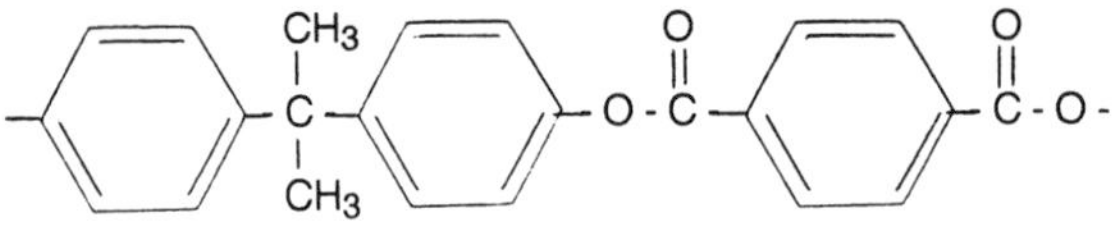

Liquid Crystal Polyesters (LCP)

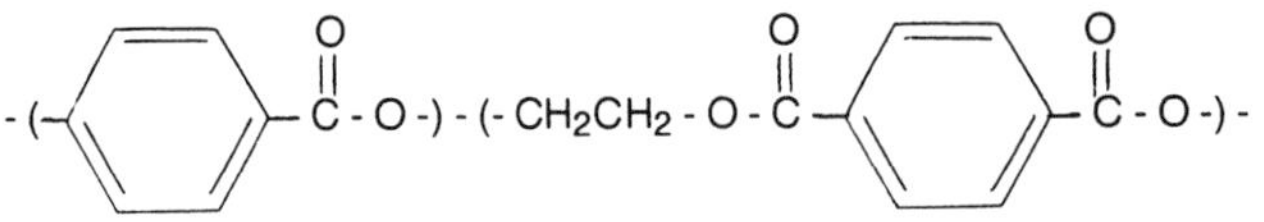

$$-(-\!\!\!\!\bigcirc\!\!\!\!-\overset{\overset{\displaystyle O}{\|}}{C}-O-)-(-CH_2CH_2-O-\overset{\overset{\displaystyle O}{\|}}{C}-\!\!\!\!\bigcirc\!\!\!\!-\overset{\overset{\displaystyle O}{\|}}{C}-O-)-$$

Nylon 6 (PA6)

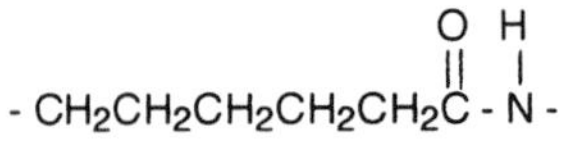

$$-CH_2CH_2CH_2CH_2CH_2\overset{\overset{\displaystyle O}{\|}}{C}-\overset{\overset{\displaystyle H}{|}}{N}-$$

Nylon 66 (PA66)

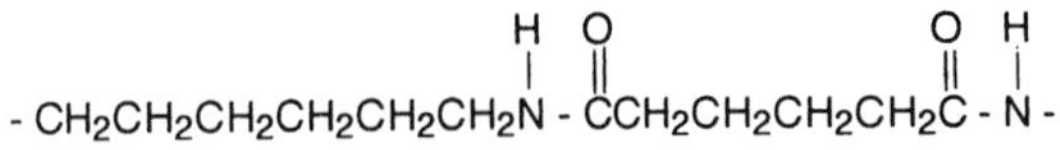

$$-CH_2CH_2CH_2CH_2CH_2CH_2\overset{\overset{\displaystyle H}{|}}{N}-\overset{\overset{\displaystyle O}{\|}}{C}CH_2CH_2CH_2CH_2\overset{\overset{\displaystyle O}{\|}}{C}-\overset{\overset{\displaystyle H}{|}}{N}-$$

Nylon 69 (PA69)

$$-(CH_2)_6\overset{\overset{\displaystyle H}{|}}{N}-\overset{\overset{\displaystyle O}{\|}}{C}(CH_2)_7\overset{\overset{\displaystyle O}{\|}}{C}-\overset{\overset{\displaystyle H}{|}}{N}-$$

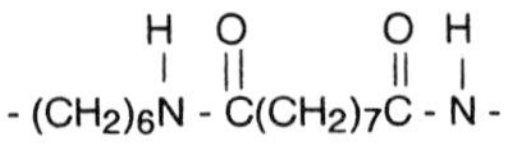

Nylon 610 (PA610)

$$-(CH_2)_6\overset{\overset{\displaystyle H}{|}}{N}-\overset{\overset{\displaystyle O}{\|}}{C}(CH_2)_8\overset{\overset{\displaystyle O}{\|}}{C}-\overset{\overset{\displaystyle H}{|}}{N}-$$

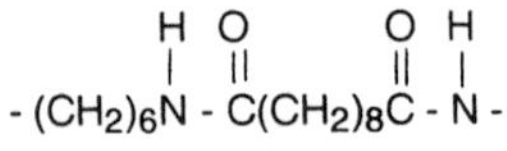

Nylon 612 (PA612)

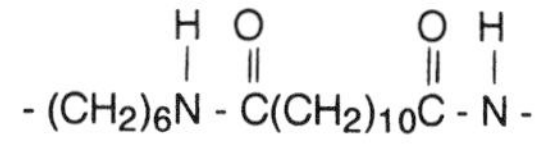

$$- (CH_2)_6 N - \overset{\overset{O}{\|}}{C}(CH_2)_{10}\overset{\overset{O}{\|}}{C} - \overset{\overset{H}{|}}{N} -$$

Nylon 11 (PA11)

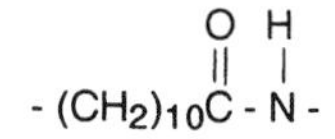

$$- (CH_2)_{10}\overset{\overset{O}{\|}}{C} - \overset{\overset{H}{|}}{N} -$$

Nylon 12 (PA12)

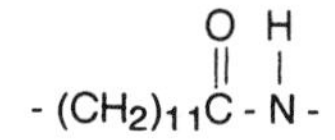

$$- (CH_2)_{11}\overset{\overset{O}{\|}}{C} - \overset{\overset{H}{|}}{N} -$$

Poly(Phthalamide) (PPA)

$$- R - \overset{\overset{H}{|}}{N} - \overset{\overset{O}{\|}}{C} - \boxed{} - \overset{\overset{O}{\|}}{C} - \overset{\overset{H}{|}}{N} -$$

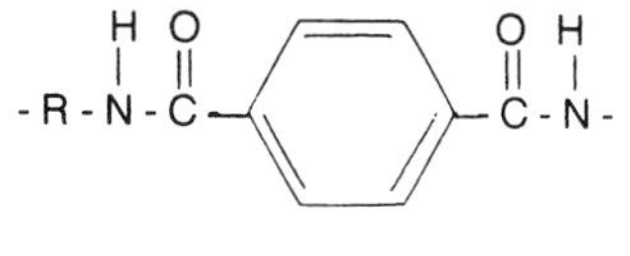

Poly(Amide Imide) (PAI)

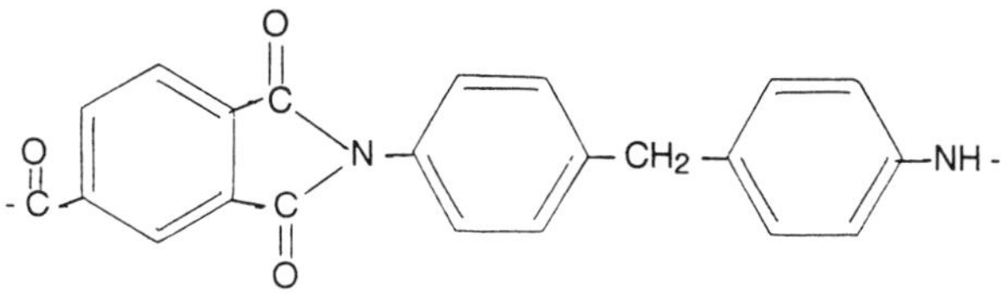

Poly(Ether Imide) (PEI)

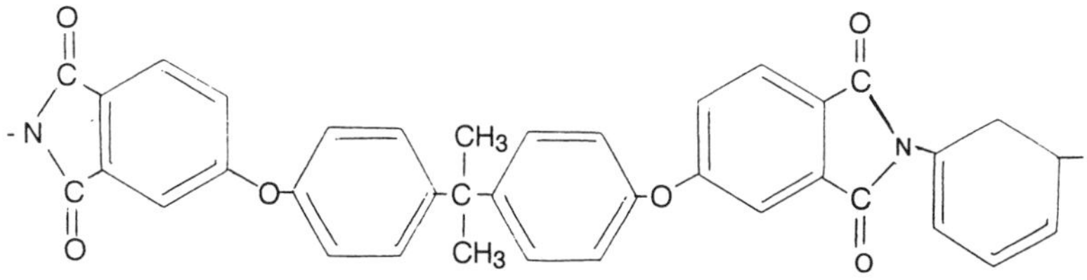

The leading materials, in terms of market volume, are: nylon, polycarbonate, polybutylene terephthalate, polyphenylene ether, polyoxymethylene, and polyethylene terephthalate.

Perhaps even more important than their structures, most of these polymers are frequently reinforced by glass fibers or even carbon fibers, which contribute tremendously to their properties, and must be considered in any comparison of their practical performance. Reinforcing fibers generally raise modulus 2–4 fold and usually increase breaking strength somewhat. In crystalline plastics, they often raise maximum use temperatures dramatically.

Maximum Use Temperature. The most frequent requirement for higher engineering performance is retention of properties at higher temperatures. Whereas most commodity thermoplastics soften and distort in boiling water, engineering thermoplastics are most often characterized by their ability to stand much higher temperatures. This is measured most often by the short-term Heat Deflection Temperature (HDT) under a load of 1.82 MPa, less often but perhaps more practically by an estimated Continuous Service Temperature (CST) in long-term use (Table 15.10).

Major applications are primarily in electrical and electronic products, auto parts, industrial products, and appliances.

Impact Strength. When plastics are compared with metals and wood, they often fail under high-speed impact. Flexible molecules such as nylons have some inherent impact strength. Of the rigid molecules, only polycarbonate and polyphenyl sulfone combine inherent rigidity and high impact strength, and we still do not understand the secret of their success. Most plastics can be reinforced with long fibers to increase their impact strength, but processing becomes much more difficult. Some plastics have been toughened by

TABLE 15.10   Typical Temperature Resistance of Engineering Thermoplastics

Polymer	Reinforcing Fiber	HDT (°C)	CST (°C)
Liquid crystal polyester		347	355
Polyetheretherketone	30% Glass	316	250
Polyamideimide	Filled	280	80
Polyphthalamide	15% Glass	277	
Polyphenylene sulfide	30% Glass	265	220
Perfluoroalkoxy ethylene	20% Carbon	260	260
Nylon 66	33% Glass	250	130
Nylon 610	30% Glass	215	110
Polyaryl sulfone	30% Glass	213	
Nylon 612	33% Glass	210	110
Nylon 6	33% Glass	210	121
Polyetherimide		210	170
Polyethylene terephthalate	30% Glass	210	140
Ethylene/tetrafluoroethylene	25% Glass	210	177
Polyether sulfone		204	179
Polybutylene terephthalate	33% Glass	204	
Polyphenyl sulfone		190	
Polysulfone	10% Glass	183	160
Nylon 11	23% Glass	176	90
Fluorinated ethylene/propylene	20% Glass	176	204
Polyvinylidene fluoride	20% Carbon Fiber	173	121
Polyarylate		172	
Polyoxymethylene	30% Glass	160	105
Nylon 12	23% Glass	160	100
Polyphenylene ether		146	105
Ethylene/Chlorotrifluoroethylene	20% Glass	136	150
Polycarbonate		130	

dispersing tiny rubber domains in them, and this technique is currently expanding with the development of compatibilization technology. One unique material is poly(1-butene), whose high creep- and puncture-resistance make it particularly desirable in pipe and tubing.

Transparency. Some applications of plastics require transparency. Amorphous plastics should be able to transmit light. Some factors which prevent transparency include unsaturation/light absorption, crystallinity, fillers and reinforcing fibers, and use of rubber particles to increase impact strength. The plastics most often used for their transparency are poly(4-methylpentene-1) (TPX), poly(methyl methacrylate) (almost equal to glass), cellulose acetate, propionate, and butyrate, polycarbonate, and polysulfones (slightly yellow). As a research challenge, it is quite possible that fillers and rubber particles could

be modified to match the refractive index of the matrix polymers, and thus retain transparency while offering their reinforcing effects on mechanical properties.

Processability. Engineering performance generally requires rigid molecules to give maximum rigidity, strength, and high-temperature performance. Melt processability, on the other hand, generally requires flexible molecules to give a fluid melt. Several polymers which manage to combine easy melt processability, with high rigidity/strength/heat resistance, include nylons, polyoxymethylene (polyacetal), and liquid crystal polymers. The first two combine flexible molecules in the melt plus high crystallinity in the solid form. The molecular rigidity of liquid crystal polymers explains their high modulus and strength, but their easy melt processability remains something of a mystery.

Lubricity. Plastic gears and bearings are less polar than metals, and therefore are relatively self-lubricating, without the need for lubricating oil. Ultrahigh molecular weight polyethylene, fluoropolymers, polyoxymethylene (polyacetal), and nylons are the best. To improve other plastics even further, adding powdered polytetrafluoroethylene to them can produce an even greater increase in lubricity, and simultaneously also an increase in abrasion resistance.

Barrier Plastics. When plastics replace metals and glass in packaging, their permeability is often a limiting property. Barrier performance generally increases with density and crystallinity. The most promising barrier plastics include ethylene/vinyl alcohol, polyvinylidene chloride, polyacrylonitrile, and polyethylene naphthoate. These are used most efficiently by laminating them to commodity plastics such as polyethylene and polyethylene terephthalate.

Weather-Resistance. For long-lived outdoor products, most plastics can be stabilized somewhat by opaque UV reflectors or at least dissolved UV stabilizers. For inherent resistance to sunlight, rain, and other components of weather, some preferred plastics include acrylonitrile/styrene/acrylic rubber, acrylonitrile/styrene/ethylene-propylene rubber, polyvinyl chloride, fluoropolymers, and polymethyl methacrylate.

Chemical Resistance. Plastics are generally superior to metals in resistance to aqueous inorganic environments. Their resistance to organic solvents depends mainly on crystallinity and difference in polarity: nonpolar polymers are more resistant to polar organics, while polar polymers are more resistant to nonpolar organics.

Individual Specialties. There are a number of special plastics which are used for their individual special properties and applications. Poly(4-methylpentene-1) combines rigidity, impact resistance, heat resistance, transparency, and chemical resistance, making it a unique replacement for glass in chemical equipment. Poly-*p*-xylylene is an extreme specialty coating, applied by vapor deposition to produce thin uniform films for electrical insulation. Polyvinyl formal is used in specialty wire-coating. Polyvinyl butyral offers the high toughness, adhesion, and clarity which make it the critical component in safety glass. Phenoxy resin was one of the first engineering plastics, offering a combination of processability, rigidity, strength, toughness, transparency, adhesion, and chemical resistance; its present use is primarily in coatings and adhesives. Cellulose nitrate was the first commercial plastic; its present uses are primarily as high-quality coatings on wood furniture and leather goods. Ethyl cellulose is a very tough transparent adhesive material, mainly used in coating bowling pins and specialty papers. Many other commercial polymers are more important in nonplastic applications such as rubber, textiles, paper, coatings, and adhesives (see Table 15.11).

TABLE 15.11 Major Markets for Thermoplastics

Market	Million Metric Tons
Packaging	12.5
Building & Construction	6.6
Consumer & Institutional	6.5
Transportation	2.2
Furniture and Furnishings	1.5
Electrical & Electronic	1.3
Adhesives, Inks, & Coatings	0.5
Industrial & Machinery	0.4
All other	5.0
Total	36.5

Thermoplastic Elastomers

Soft flexible rubbery behavior depends on long flexible polymer molecules in the form of random coils. Strength, heat and chemical resistance depend on attachment between the coils. Conventional rubber chemistry uses vulcanization, permanent thermoset primary covalent cross-links, usually by sulfur plus metal oxides, to hold the coils together; but this makes processing more difficult, and recycling very difficult. In the past 40 years, this technology has been supplemented by the

TABLE 15.12   Typical Property Ranges of Thermoplastic Elastomers

Polymer System	Shore Hardness	Tensile Strength (MPa)	Ultimate Elongation (%)	Low Temperature Limit (°C)	High Temperature Limit (°C)
Styrene-Diene	A23-D65	2–34	200–1750	−75	100
TPOlefin	A67-D72	3–36	20–900	−60	120
TPVulcanizate	A40-D60	3–19	300–650	−63	135
Polyurethane	A60-D78	14–47	250–966	−50	120
Polyetherester	D35-D82	9–47	170–600	−50	137
Polyetheramide	D25-D69	6–62	250–760	−40	150

development of thermoplastic elastomers. These are based mainly on block copolymers, in which long flexible blocks form the continuous rubbery matrix, and short glassy or crystalline blocks form the thermoplastic "cross-links," secondary attractions that give strength, heat, and chemical resistance. These are much more attractive to the plastics processing industry, offering easy thermoplastic processing and good recyclability. They now account for about 10 percent of the total rubber market. There are six families of commercial thermoplastic elastomers (Table 15.12).

The primary variable in each of these families is the ratio of rubbery soft block to glassy or crystalline hard block, thus offering a wide range in balance of soft flexible properties vs. strength, heat, and chemical resistance.

Styrene-Diene.   These ("styrenic") thermoplastic elastomers are block copolymers of styrene with butadiene (SBS) or isoprene (SIS) in about 30/70 monomer ratio.

SBS $(CH_2 - CH) - (CH_2 - CH = CH - CH_2) - (CH_2 - CH)$

SIS $(CH_2 - CH) - (CH_2 - C = CH - CH_2) - (CH_2 - CH)$, with CH_3

They have all the rubberiness of the butadiene rubber matrix, and the glassy polystyrene domains hold them together up to the softening point of polystyrene. Since

their unsaturation is sensitive to oxygen and ozone aging, SBS is often saturated by hydrogenation (SEBS) to improve age-resistance.

$- (CH_2 - CH) - [(CH_2 - CH_2) - (CH_2 - CH)] - (CH_2 - CH) -$, with C_2H_5

They are the leading class of thermoplastic elastomers, 45 percent of the total market, used mainly in adhesives, shoe soles, wire and cable insulation, kitchen utensils, medical products, and auto parts.

Thermoplastic Olefin.   These thermoplastic elastomers are primarily blends, or block or graft copolymers, of ethylene/propylene rubber with polypropylene.

$- (CH_2 - CH_2) - (CH_2 - CH) - + - (CH_2 - CH) -$, with CH_3 and CH_3

They have all the rubberiness of the ethylene/propylene (EPR) rubber matrix, and the crystalline polypropylene (PP) domains hold them together. As saturated elastomers, they have natural resistance to oxygen and ozone aging. They are the second largest class of thermoplastic elastomers, 25 percent of the total market, used mainly in mechanical rubber parts.

Thermoplastic Vulcanizates.   These are a surprising improvement over conventional thermoplastic olefins. Vulcanized ethylene/

propylene/diene (EPDM) rubber particles are dispersed in a thermoplastic polypropylene (PP) matrix. The vulcanized EPDM has higher heat and chemical resistance than ordinary EPR, and the PP matrix provides thermoplastic processability. In more specialized grades, the vulcanized rubber may be nitrile rubber for greater oil resistance, or butyl rubber for impermeability. They are used where greater heat and/or chemical resistance are required, for example, in oil wells, mechanical goods, and building and construction.

be offered by various manufacturers. For the most part they are balanced towards less rubbery block and more crystalline block, combining moderate flexibility with greater strength, heat, and chemical resistance. They are used mainly in automotive and other mechanical parts requiring this combination of properties.

Polyetheramide. These thermoplastic elastomers are typically block copolymers of polyether rubber with nylon crystalline domains.

$$- (CH_2CH_2CH_2CH_2O)_n - (\overset{\overset{\displaystyle O}{\|}}{C} - CH_2CH_2CH_2CH_2CH_2 - \overset{\overset{\displaystyle H}{|}}{N})_m -$$

Polyurethane. This rubber is mainly thermoset, but thermoplastic processability can be achieved by block copolymers of amorphous polyurethane rubber with strongly hydrogen-bonded crystalline polyurethane blocks.

A number of such combinations are mentioned in the literature, and may be offered by various manufacturers. They combine the soft flexible rubberiness of polyether or polyester elastomers with the high strength, heat-, oil-,

They combine the high strength, oil- and gas-resistance of polyurethanes with the advantage of thermoplastic processability. They are about 15 percent of the total thermoplastic elastomer market, used in auto parts, wire and cable, medical products, and fuel hose.

Polyetherester. These thermoplastic elastomers are typically block copolymers of polyoxybutylene rubber with polybutylene terephthalate crystalline domains.

and gas-resistance of nylons, and thus find use in auto parts, wire and cable, and sporting goods.

Thermoset Plastics

Whereas difunctional monomers produce linear thermoplastic polymers, monomers with higher functionality can react further during processing, cross-linking up to infinite molecular weight. Such thermosetting processing may be more difficult, but infinite

A number of other rubber and crystalline blocks are mentioned in the literature, and may

cross-linking produces extreme increases of rigidity, creep-resistance, dimensional stability, heat-resistance, and chemical resistance, which are valuable in many demanding engineering applications. Thus thermoset plastics account for about 15 percent of the total plastics market. The major thermoset plastics families may be ranked in order of market size as shown in Table 15.13.

Polyurethanes. Most polyurethane chemistry may be simplified down to three basic reactions:

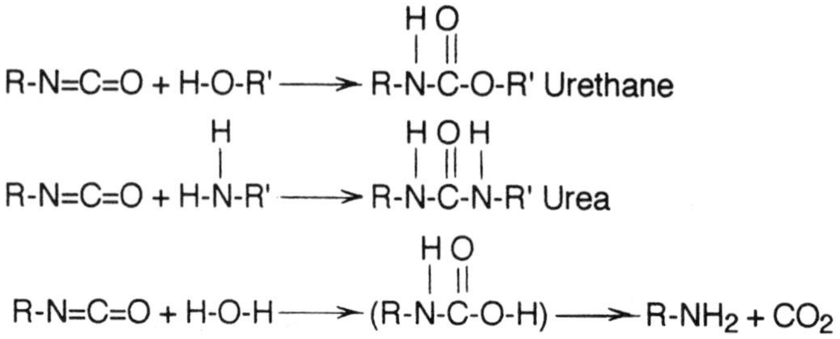

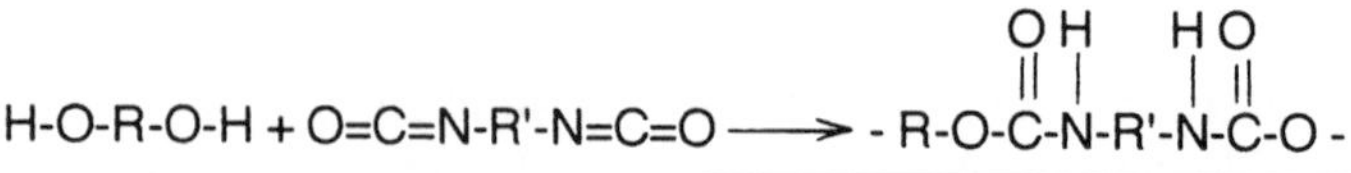

When the polyols are trifunctional or higher, they form thermoset polyurethanes.

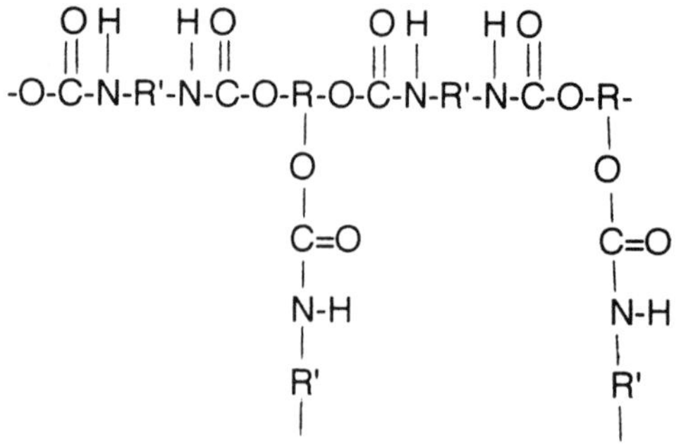

$$-O-\overset{\overset{\displaystyle O}{\|}}{C}-\overset{\overset{\displaystyle H}{|}}{N}-R'-\overset{\overset{\displaystyle H}{|}}{N}-\overset{\overset{\displaystyle O}{\|}}{C}-O-R-O-\overset{\overset{\displaystyle O}{\|}}{C}-\overset{\overset{\displaystyle H}{|}}{N}-R'-\overset{\overset{\displaystyle H}{|}}{N}-\overset{\overset{\displaystyle O}{\|}}{C}-O-R-$$

with pendant chains:

$$\begin{array}{c} O \\ | \\ C=O \\ | \\ N-H \\ | \\ R' \\ | \end{array}$$

$$R-N=C=O + H-O-R' \longrightarrow R-\overset{\overset{\displaystyle H}{|}}{N}-\overset{\overset{\displaystyle O}{\|}}{C}-O-R' \quad \text{Urethane}$$

$$R-N=C=O + H-\overset{\overset{\displaystyle H}{|}}{N}-R' \longrightarrow R-\overset{\overset{\displaystyle H}{|}}{N}-\overset{\overset{\displaystyle O}{\|}}{C}-\overset{\overset{\displaystyle H}{|}}{N}-R' \quad \text{Urea}$$

$$R-N=C=O + H-O-H \longrightarrow (R-\overset{\overset{\displaystyle H}{|}}{N}-\overset{\overset{\displaystyle O}{\|}}{C}-O-H) \longrightarrow R-NH_2 + CO_2$$

Polyols are usually aliphatic polyethers or polyesters.

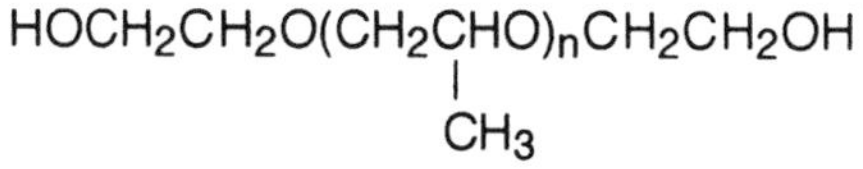

$$\underset{\underset{CH_3}{|}}{HOCH_2CH_2O(CH_2CHO)_nCH_2CH_2OH} \quad \text{Typical Polyether}$$

$$HOCH_2CH_2O(\overset{\overset{\displaystyle O}{\|}}{C}CH_2CH_2CH_2CH_2\overset{\overset{\displaystyle O}{\|}}{C}OCH_2CH_2O)_nH \quad \text{Typical Polyester}$$

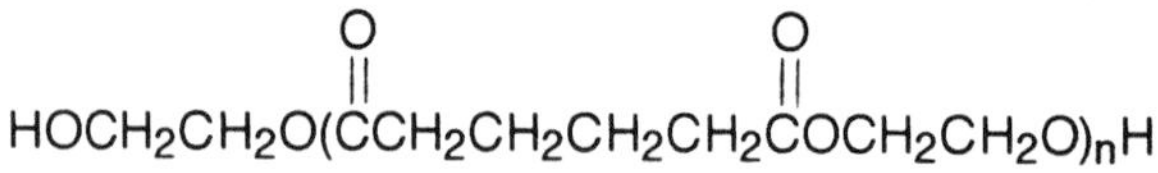

Polyisocyanates are usually toluene diisocyanate or diphenylmethane diisocyanate.

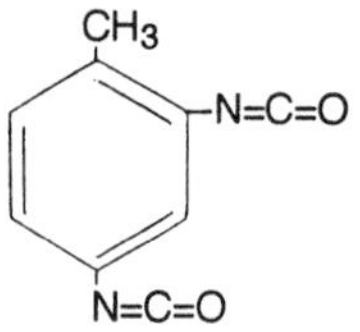

Toluene Diisocyanate (TDI)

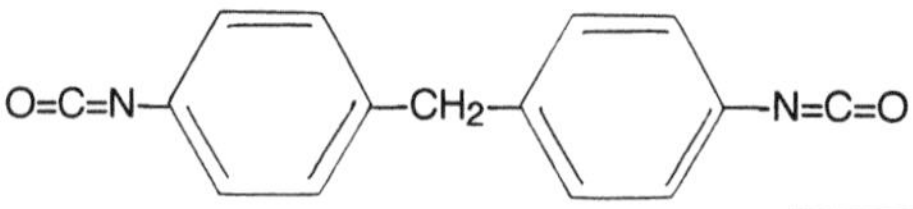

Diphenylmethane Diisocyanate

When the polyols are linear (difunctional) they form thermoplastic polyurethanes.

$$H-O-R-O-H + O=C=N-R'-N=C=O \longrightarrow -R-O-\overset{\overset{\displaystyle O}{\|}}{C}-\overset{\overset{\displaystyle H}{|}}{N}-R'-\overset{\overset{\displaystyle H}{|}}{N}-\overset{\overset{\displaystyle O}{\|}}{C}-O-$$

TABLE 15.13 Thermoset Plastics Market

Family	Percent of Thermoset Market
Polyurethane	36
Phenol–formaldehyde (Phenolic)	29
Urea–formaldehyde	19
Polyester	10
Epoxy	4
Melamine–formaldehyde	2

TABLE 15.14 Polyurethane Markets

Form	Percent of Polyurethane Market
Flexible foam	48
Rigid foam	28
Reaction injection molding (RIM)	6
Rubber, spandex, sealants, adhesives, coatings	17

TABLE 15.15 Phenolic Plastics Markets

Market	Percent of Market
Adhesive for plywood	51
Binder for Fibrous & Granulated Wood	17
Binder for Glass Wool Insulation	15
Molding Powders	8
Paper Laminate Board	4
Foundry Resins	3
Binder for Abrasive Products	2

In most processes, the reactive liquids are mixed and poured, and polymerized and cured rapidly to the final products (Table 15.14).

Most polyurethane is foamed during the polymerization/cure reaction.

Flexible foam. This is made by mixing long trifunctional polyol with isocyanate to form the polyurethane, and adding a little excess isocyanate and water to the reaction to produce carbon dioxide which produces the foam. The largest use is in furniture, with smaller amounts in auto seating, mattresses, rug underlay, textiles, and packaging.

Rigid Foam. This is made by mixing short polyfunctional polyol with di- or higher polyisocyanate, and foaming either with volatile liquid or with isocyanate and water. The largest use is in building insulation, with smaller amounts in refrigeration, industrial insulation, packaging, autos, and marine flotation.

Reaction Injection Molding. RIM mixes polyol, polyamine, polyisocyanate, and strong catalyst, and injects the mixture rapidly into a mold, where it cures rapidly to form large parts very economically. It is used primarily for producing front ends and other large parts of autos.

Phenol–Formaldehyde. Phenolic plastics were the first commercial synthetic plastics in 1908, and were the leading commodity plastic for 40 years, until the growth of vinyl and styrenic thermoplastics (Table 15.15). Now quite mature, they remain the second largest family of thermoset plastics.

While their largest use is as adhesives for outdoor plywood and glass wool insulation, they provide a group of compression-molding fiber-reinforced plastics which meet high engineering performance requirements (Table 15.16).

For performance under severe conditions, they compare very favorably with more expensive engineering thermoplastics as shown in Table 15.17.

The chemistry of phenolic molding powders begins with the reaction of phenol with formaldehyde.

Phenol Formaldehyde Methylolphenol Dimethylolphenol

TABLE 15.16   Typical Properties of Molded Phenolic Plastics

Grade	General-Purpose	Engineering
Fibrous filler	Wood flour	Glass
Modulus, MPa	9,000	18,000
Flexural strength, MPa	76	275
Notched Izod impact strength, J/cm	0.20	5.3
Heat deflection temperature, °C at 1.82 MPa	168	240
Maximum continuous use temperature, °C	149	194

TABLE 15.17   Phenolics as Engineering Plastics

Property	Phenolic	Engineering Thermoplastics
Price, in cents/cubic inch	2.7–8.0	9.5–18.4
Creep modulus, 100 hr/14 MPa, in MPa	28	5.5–10
Compressive creep, 14 MPa/50°C, in %	0.02	0.1–1.4
Heat resistance, in °C	315	121–260

The high reactivity of the methylol groups makes it easy to polymerize and cure phenolic polymers,

$$P-CH_2OH + HOCH_2-P \rightarrow P-CH_2OCH_2-P$$
$$\rightarrow P-CH_2-P$$

$$P-CH_2OH + H-P \rightarrow P-CH_2-P$$

and also to copolymerize them with melamine and furan plastics,

$$P-CH_2OH + HOCH_2-M \rightarrow P-CH_2-M$$

and with cellulosic reinforcements such as wood, cotton, and paper.

$$P-CH_2OH + HO-Cellulose$$
$$\rightarrow P-CH_2-O-Cellulose$$

It is generally a three-stage process: (1) initial reaction to A-stage resin produces low-molecular-weight oligomers, which are still soluble, fusible, and reactive. (2) Melt compounding advances them to B-stage resins which are fairly fusible to doughy melts, and still reactive. (3) Molding them into finished products advances them to C-stage resins which are fully cross-linked to stable thermoset plastics of three-dimensional infinite molecular weight, high modulus and strength, and very resistant to heat and chemical environments.

Urea–Formaldehyde. These plastics became commercial about 1929. Urea and formaldehyde react very readily to form methylol compounds, mainly dimethylol urea.

The methylol groups are very reactive, condensing with each other, with the N–H groups in urea, and with the –OH groups in cellulose.

Thermosetting cure produces highly cross-linked three-dimensional molecules of infinite size.

The largest use is for binding fibrous and granulated wood into indoor composition board. Smaller uses are for wet-strength paper and permanent-crease textiles. About 4 percent of urea–formaldehyde resin is combined with alpha-cellulose to make molding powders (Table 15.18). These find use mainly in electrical parts such as switches, wall plates and receptacles, circuit breakers, electric blankets, handles and knobs.

Unsaturated Polyesters. The chemistry of unsaturated polyesters was developed in the 1930s, and manufacture of glass–fiber-reinforced polyesters began in the early

1940s. Their outstanding performance was recognized early in the typical consumer compliment: "That's not cheap plastic, that's high-performance fiberglass." They matured early, and form about 10 percent of the present thermosetting plastics market (Table 15.19).

Their chemistry is fairly complex. The most common material is made from propylene glycol plus maleic anhydride plus phthalic anhydride.

When these are cooked together, maleic anhydride isomerizes to fumaric acid, and they condense to form low-molecular-weight propylene fumarate phthalate copolyester oligomers. These are mixed with styrene monomer, reinforced by glass fibers, usually extended with low-cost fillers, and cured by peroxide to form rigid strong products which are very resistant to impact and heat (Table 15.20).

TABLE 15.18   Typical Properties of Urea–Formaldehyde-Alpha-Cellulose Moldings

Flexural modulus, MPa	10,000
Flexural strength, MPa	100
Notched Izod impact strength, J/cm	0.16
Heat deflection temperature, °C at 1.82 MPa	132

TABLE 15.19 Thermoset Polyester Markets

Market	Percent of Market
Building & Construction	57
Industrial	13
Tanks & Containers	9
Marine	5
Auto	4
Other	12

The mechanics of processing are carried out by a number of techniques. *Hand layup or sprayup* produces large shapes such as boat hulls, recreational vehicles, mobile homes, truck cabs, and tub-shower units. *Continuous panel processing* produces room dividers and skylights. *Compression molding* of sheet molding compound (SMC) and bulk molding compound (BMC) produces autobody parts, bathtubs, septic tanks, trays, tote boxes, and equipment housings. *Pultrusion* produces flagpoles, archery bows, and park benches. And *filament winding* is highly engineered to produce maximum strength in pipes and storage tanks.

Epoxy Resins. These were developed in the 1940s and offered a unique combination of engineering performance which made them a popular family of thermoset plastics: fast low-temperature cure with low pressure and low shrinkage; high adhesion to polar surfaces; hardness, heat- and chemical-resistance (Table 15.21). Their largest use is in coatings for corrosion protection and electronic equipment. The second largest is in printed circuit boards. And three other important uses are adhesives, flooring, and high-performance fiber-reinforced plastics. See Table 15.22.

Epoxy chemistry is complex. Most epoxy resins are made by reaction of epichlorohydrin with bisphenol A.

TABLE 15.20 Typical Properties of Glass–Fiber-Reinforced Thermoset Polyesters

Process	BMC	SMC	Woven Cloth	Filament Wound
Modulus, MPa	14,000	11,000	14,000	6000–24,000
Strength, MPa	121	159	414	283–586
Notched Izod impact strength, J/cm	4.0	7.8	9.4	—
Heat deflection temperature, °C at 1.82 MPa	182	225	205+	—

TABLE 15.21   Property Ranges for Cured Epoxy Resins

Flexural modulus, MPa	14,000–34,000
Flexural strength, MPa	55–655
Notched Izod impact strength, J/cm	0.16–21.0
Heat deflection temperature, °C at 1.82 MPa	93–288

TABLE 15.23   Typical Properties of Cellulose-Filled Melamine–Formaldehyde

Flexural modulus, MPa	9000–11,000
Flexural strength, MPa	70–124
Notched Izod impact strength, J/cm	0.13–0.21
Heat deflection temperature, °C at 1.82 MPa	127–143

While they are called "resins," they are really monomers to low-molecular-weight oligomers, liquids to soluble fusible solids, with high reactivity in the epoxy rings, and fair reactivity in the internal hydroxyl groups. They are cured most often by room-temperature reaction with polyamines or polyamide amines.

For higher-temperature cure and heat-resistance, they are cured most often by acid anhydrides.

TABLE 15.22   Epoxy Resin Markets

Market	Percent of Market
Coatings	53
Printed Circuit Boards	13
Adhesives	9
Flooring & Paving	8
Reinforced Plastics	7
Tooling & Molding	3
Other	7

There are also many more types of epoxy resins and curing agents for more specialized applications.

Melamine–Formaldehyde.   These resins became commercial in the 1930s. Their combination of high thermosetting reactivity, cured hardness, and resistance to heat, weather, and chemical environments made them particularly valuable for their good appearance and durability (Table 15.23).

Their largest use is in coatings, where they are used to cure acrylic automotive coatings and polyester appliance coatings. Their second largest use is in countertops, where they protect the decorative surfaces against abrasion, heat, and chemical attack. Their third use is in dinnerware, where their light weight, impact resistance, and attractive appearance are very competitive with china.

Melamine chemistry begins with the addition of 2–3 mols of formaldehyde to form methylol melamines.

These are very reactive with each other, with the remaining N–H bonds on melamine, with the hydroxyl groups in acrylic and polyester coatings, and with the hydroxyl groups in paper for countertops and in alpha-cellulose for molded dinnerware.

$$RNHCH_2OH + HOCH_2NHR' \rightarrow$$
$$RNHCH_2OCH_2NHR' \rightarrow RNHCH_2NHR'$$

$$RNHCH_2OH + R'NH_2 \rightarrow RNHCH_2NHR'$$

$$RNHCH_2OH + HOR' \rightarrow RNHCH_2OR'$$

The combination of resonance stabilization in the melamine heterocycle, and the high cross-linking between methylol melamines and with the other polymers, all produce the outstanding properties which make it a valuable specialty member of the thermoset plastics spectrum.

GENERAL CONSIDERATIONS

Structure–Property Relationships

When plastics engineers want to improve properties in an existing product, or when they want to select the optimum material for a new product, a routine search of existing tables of properties may sometimes be sufficient. For more professional judgment and problem-solving, however, and for planning development of new materials, they need to understand the basic relationships between polymer structure and practical properties. Here are some starting guidelines.

Molecular Weight. Low molecular weight gives lower melt viscosity for injection molding, and easier solution processing in general. High molecular weight is preferred for extrusion, and particularly for blow molding, thermoforming, stretch orientation, and thermoplastic foaming. In finished products, high molecular weight generally gives higher mechanical strength and chemical resistance.

Molecular Flexibility/Rigidity. Flexible molecules generally give lower melt viscosity for easier processing; softer, more flexible, more extensible products; and higher impact strength, friction, and acoustic absorption. Rigid molecules generally give higher rigidity, strength, creep resistance, heat deflection temperature, and impermeability; and lower coefficient of thermal expansion, and dielectric constant and loss.

Crystallinity. When polymers crystallize, their melting points are much higher and sharper than the softening points of amorphous plastics. This requires higher processing temperatures, but gives higher maximum use temperatures in the final products. Increasing crystallinity generally increases rigidity, strength, creep resistance, dimensional stability, impermeability, and chemical resistance; but decreases impact strength and transparency.

Orientation. Stretch orientation of extruded fibers and films greatly increases modulus, strength, transparency, and impermeability. It is also useful for producing shrink-packaging; conversely, a disadvantage is thermal dimensional instability. When orientation occurs accidentally in injection molding, calendering, thermoforming, and other processes, it generally produces undesirable anisotropy of final structure and properties.

Polarity and Hydrogen-Bonding. Whereas hydrocarbon polymers are nonpolar and have weak intermolecular attraction, introduction of negative atoms into the polymer molecule—oxygen, nitrogen, chlorine—produces permanent polarity, giving strong intermolecular attractions. Increasing polarity generally requires higher processing temperatures, and gives higher modulus, strength, creep resistance, heat deflection temperature, crystalline melting point, dielectric constant and loss, and gasoline and oil resistance.

When the polymer contains oxygen and especially nitrogen, polarity also produces hydrogen-bonding, which is an even stronger intermolecular attraction, and produces all the same effects to an even greater extent. Another effect of hydrogen-bonding is water absorption. In fabrics this produces greater comfort; but in plastics it decreases modulus, strength, and dimensional and chemical stability. These effects are most noticeable in nylons and cellulosics.

Cross-Linking. Thermoplastics are stable linear molecules which are softened by heat and soluble in solvents of similar polarity; this makes for easy processability. Primary covalent cross-linking in thermosets converts them into three-dimensional molecules of infinite size, with tremendous changes in properties: insolubility and infusibility; higher modulus, creep resistance, maximum use temperature, and chemical resistance; and lower extensibility, impact strength, thermal expansion, dielectric constant and loss, solvent swelling, and permeability. Cross-linking produces shrinkage strains which embrittle the polymer, so most thermoset plastics must be reinforced with fibers; the result is synergistic improvement of modulus,strength, impact resistance, and dimensional stability, producing enhanced engineering performance.

Additives

Polymers are rarely used in pure form. They are almost always improved by use of additives to enhance specific properties. The major classes of additives may be briefly summarized as follows.

Stabilizers. Organic polymers are not perfectly stable. Specific polymers and specific products require additives to improve their stability during processing and/or long-term use of finished products. *Antioxidants* are added to polyolefins and rubber-modified impact plastics to protect against atmospheric oxygen; these are primarily hindered phenols and polyphenols, sometimes synergized by aliphatic sulfides or organic phosphites, used in fractions of a percent up to several percent. *Thermal stabilizers* must be added to PVC to prevent dehydrochlorination, discoloration, and cross-linking during melt processing: organotin esters are strongest, used as a fraction of a percent for processing rigid PVC; barium/zinc soaps plus epoxidized fatty esters plus organic phosphites, total concentration several percent, form a synergistic stabilizer system for plasticized flexible PVC; and basic lead oxide compounds, several percent, are best for wire and cable insulation. *Ultraviolet light stabilizers* are needed in products for outdoor applications: cyclic hindered amines and *o*-hydroxy benzophenones and benzotriazoles are used at a fraction of a percent in clear products, while carbon black and especially aluminum flake are extremely effective in opaque products. *Biostabilizers* are used to protect natural polymers and monomeric additives against attack by microorganisms; these are chemicals which require a delicate balance between toxicity to microorganisms vs. safety for macroorganisms like ourselves.

Fillers are inorganic powders added in large amounts to increase modulus, dimensional stability, and opacity, and often to reduce cost. *Reinforcing fibers* are mostly glass, occasionally carbon or organic fibers, typically added in concentrations of 10–40 percent, to increase modulus, strength, impact strength, creep resistance, and dimensional stability; long and continuous fibers give the greatest improvement in properties, while short chopped fibers $(\frac{1}{16}-2$ in.$)$ permit fairly conventional melt processing.

Coupling agents are chemical surface treatments applied to fibers, and sometimes to fillers, to strengthen the interface between inorganic fillers and fibers and organic polymer matrixes, to improve dispersion and stress transfer across the interface. Most common are organosilicon compounds of the type $(RO)_{2-3}SiX_{2-1}$, where RO is typically methoxy or ethoxy to react with the silanol surface of glass fibers, and X is an organic group designed to react with a thermosetting polymer matrix, or at least to be attracted toward a thermoplastic polymer matrix.

Plasticizers are typically organic liquids of very low volatility, which are miscible with a polymer, and are added to it to improve processability and, in larger amounts (20–80 parts per hundred of resin), to make it soft and flexible, or even rubbery and/or adhesive. The major portion of the plasticizer market (80%) is aliphatic and aromatic esters, which go to convert rigid PVC into flexible PVC. The remainder goes to improve the processability of cellulosics, and for a variety of specialized uses in other polymers.

Lubricants are a variety of proprietary additives, which are used either to improve melt flow, release from steel process equipment, or self-lubricity in final products such as gears and bearings.

Flame-retardants may be built into the polymer during polymerization or cure, or they may be physical additives to the finished composition. Organic phosphorus compounds are the most effective, typically requiring about 2 percent of phosphorus to prevent burning. Organobromine (10–20%) and chlorine (10–40%) are effective when used in larger amounts. Antimony oxide, and some other metal oxides, can synergize the action of bromine and chlorine, reducing total flame-retardant concentrations to 5–10 percent. And alumina trihydrate and magnesium hydroxide, which release water when heated, are becoming increasing popular when used in large enough amounts to be effective (40–60%).

Colorants. One third of plastic materials are used in their natural color. The other two thirds are colored for esthetic and/or functional reasons, typically using about 1 percent of colorant. Inorganic minerals and synthetic colors give greater stability and opacity; while organic colors are available in greater variety, miscibility, and efficiency, and less likely to raise questions of toxicity. Thus use of inorganics is decreasing, while use of organics is increasing.

Organic peroxides are used to initiate free-radical polymerization of ethylene, butadiene, styrene, vinyl chloride, vinyl acetate, and methyl methacrylate. They are also used to cure unsaturated polyesters, occasionally to cross-link thermoplastics such as polyethylene and polyacrylates, and increasingly for grafting and compatibilization of polymer blends. A variety of organic peroxides offer useful reactivity over a temperature range from 0 to 130°C or more, for different polymers and different processes.

Polymer Blends. Blending of polymers with each other accounts for approximately 40 percent of the present plastics market, and the practice is growing continually, because it permits the development of improved properties without the cost of inventing new polymers. When polymers are fairly miscible, as in the polyethylenes, and in polyphenylene ether plus polystyrene, blending can be used to produce intermediate properties and balance of properties. Most polymer blends are not miscible, and separate into microphases; if these can be strongly bonded at the interface, it is often possible to produce synergistic improvement of properties, particularly the balance between rigidity and heat deflection temperature on the one hand, plus ductility and impact strength on the other; or between soft, flexible, rubbery properties on the one hand, plus the strength of "thermoplastic cross-linking" on the other. Some of the newer blends are used to produce barrier properties and other valuable improvements.

Critical Properties: Challenges to the Plastics Industry

It is easy to be positive and proud of the accomplishments of the plastics industry. On the other hand, there are a number of areas in which it is obvious that plastics are not yet perfect, areas where major breakthroughs could open major new markets and uses for plastics in the future. It may be stimulating to explore some of these here.

One-step conversion of monomer to finished product could reduce processing steps and costs dramatically. Epoxy cure and polyurethane RIM are examples of very fast reactions producing finished products. Monomer casting of acrylics and nylon 6 are commercial one-step processes. Polymer chemists have many more polymerization reactions which can rapidly convert monomers into high-molecular-weight or even thermoset polymers. What is needed are strong cooperative programs between polymerization chemists and plastics process engineers to develop these possibilities into commercial realities.

Continuous-fiber reinforcement gives plastic products which are not simply quantitatively, but often qualitatively, superior to most present commercial practice. Most plastic processing is limited to conventional melt flow of short-fiber reinforcements, which sacrifices much of the potential benefits of reinforcement. There are a few processes for incorporating continuous fiber reinforcement—filament-winding, pultrusion, swirl conformation of fibers in polymer sheets, and mixed fabrics of reinforcing fibers and plastic fibers. More vigorous development of such techniques could rapidly produce plastics products with far superior properties.

Modulus and creep-resistance of plastics are still inferior to metals, ceramics, and glass, which means there are areas where they cannot compete. We know that molecular rigidity, crystallinity, polarity, and reinforcing fibers can all go a long way toward closing the gap.

Abrasion-resistance of transparent plastics still cannot equal that of glass. We have ways of improving or coating the polymer, but customers keep telling the industry that there is still a long way to go.

Coefficient of thermal expansion of organic polymers is 1–2 orders of magnitude higher than metals, ceramics, and glass, which gives serious difficulties in product design, and especially in mating plastics parts with inorganic parts in an assembled product. Molecular rigidity, crystallinity, fillers, and especially fibrous reinforcement can go a long way toward bridging the gap, but there still is a gap to be overcome.

Thermal conductivity of plastics is very low, which makes them excellent insulators against heat and cold. On the other hand, there are times when high thermal conductivity is preferred, for example, in processing, cooking, and heating equipment. Inorganic fillers can help, in proportion to their volume concentration. Perhaps this approach can be carried further; or perhaps there are totally different mechanisms waiting to be discovered.

Heat resistance of organic polymers is far lower than that of metals, ceramics, and glass. There have been major improvements, based on aromatic and heterocyclic resonance, ladder structures, and other mechanisms, and we may see further improvement in the future. Perhaps more serious limitations are the high cost of synthesis and the difficulty of processing these polymers into the desired final products. This is an area where the polymer chemist could use more help from the plastics engineer.

Electrical conductivity of polymers is very low, making them very useful as insulation.

On the other hand, there are products in which conductivity would be very desirable. Semiconductivity is fairly easily achieved by adding semi-compatible hydrophilic organic compounds, and fairly high conductivity can be achieved by metallic fillers, especially fibers. Research is developing polymers which are inherently conductive due to conjugated unsaturation plus doping with inorganic electron donors. This is an area where research support and activity are making good progress at present.

Dielectric breakdown occurs when high voltage drop across an insulator causes some current to leak through, turning to heat, and ultimately decomposing the polymer and burning a conductive carbon track right through the insulator. Practically, some polymers are more resistant than others; but more theoretical understanding is needed in order to design polymer structures which will offer superior resistance to dielectric breakdown.

Outdoor weathering of plastic products has been the subject of both theoretical and practical study. Some plastics can last for many years, others for more limited times, and a few can actually be designed to self-destruct rapidly. Mechanisms involve ultraviolet light, atmospheric oxygen, water, transition metals, acid rain, wind-blown dust, and microbiological action. Polymer structure and additives respond to these mechanisms in various ways. Whereas there presently exist a fair theoretical understanding and practical control measures, there remains much to learn in order to achieve the ultimate goals of long-term weather stability and efficient control of solid waste.

Solvent-resistance of organic polymers varies with polarity, crystallinity, and cross-linking, so it is usually possible to solve solvent-attack problems by proper choice of polymer; but it is important to remember that they do not easily compete with metals, ceramics, and glass.

Barrier properties of organic polymers cannot equal metals and glass. Plastics offer so many other advantages that we often try to compromise or laminate to optimize overall balance of properties. It is known that molecular rigidity and crystallinity improve barrier performance, and from practical experience we can identify several very high-barrier polymers; but the details that make one much better than another are not understood so discovery of superior barriers is slow and uncertain.

Cost of engineering performance is a major factor retarding growth. Organic polymer chemists can easily design and synthesize polymer structures with higher and higher properties, but the cost of synthesis and difficulty of processing often inhibit their use for many years. Closer cooperation between chemists and engineers should be the optimum route to more efficient development.

Fire Performance

Wood burns. Most of the fires throughout history, causing death and destruction, have been caused by wood. It is rare that anyone will say "Ban wood." Plastics are also carbon compounds, and they also burn. When plastics are involved in a fire, even to the slightest degree, there is often an outcry, "Ban plastics." Thus the use of plastics in building and transportation is seriously restricted by this prejudice.

Plastics are not all equally flammable. Some burn as readily as fuels. Some do not burn spontaneously; but when exposed to a severe fire, they can be burned. And some require enriched or pure oxygen environment before they will burn. Most plastics can be made more resistant to burning, by incorporating flame-retardant elements—phosphorus, bromine, chlorine, antimony, even water—either in the polymer molecule or in physical additives.

Related problems must be considered in individual products. Bromine, chlorine, and antimony add to the smoke of a fire, while phosphorus and water do not, and some metal oxides can actually reduce it. Toxicity of combustion gases is a major concern; but the main problem is that oxidation of carbon compounds in an enclosed space—indoors—produces carbon monoxide, no matter whether the carbon compounds are wood or plastics. Other problems include the cost of flame-retardants, difficulties in processing, and loss of mechanical or thermal properties.

The designer must balance all of these in each product, and choose the optimum solution to the problem.

Health and Environment

Whenever new chemistry is introduced into the environment, there arises the question of its effect on our health. On the positive side, plastics packaging of food prevents contamination and spoilage, and prolongs its useful life; and use of plastics in medicine has made major contributions to health and longevity. On the negative side, there have been several occasions where plastics chemistry has caused health problems; whenever these have been identified, they have been solved successfully and quickly.

This leaves a large grey area, in which people who do not understand chemistry may combine ignorance and fear of anything new, and try to roll back the material progress of modern science and industry. Unfounded fear of plastics has led to many popular and even political attempts to limit or ban their use. This has certainly had a retarding effect on our ability to develop their new uses to maximum advantage. Such problems have been seen throughout history—steel plows, balanced diets, immunization and medication, all have had to overcome popular fears before they could offer their benefits to mankind; so the industry should not be discouraged when plastics encounter similar difficulties.

Recycling

Modern science and industry have provided a growing supply of material products. When they reach the end of their useful life, they become solid waste, and disposing of it has become a growing problem. Worst of all is over-packaging to stimulate sales, so discarded packaging is the major contributor to this solid waste. Plastics are not the major component of solid waste; but because of their low density, bright colors, and relative weather-resistance, they are the most obvious component. It would be desirable to remove them from solid waste by recycling.

Since 85 percent of the plastics market is thermoplastic, it is theoretically recyclable. In industry, individual thermoplastic materials are almost always recycled immediately for purely economic reasons. Post-consumer waste, on the other hand, presents serious problems. Voluntary recycling of PET soda bottles has been most successful, because they are easily separated, cleaned, and converted into bulk fiber products. Voluntary recycling of HDPE milk and water bottles has been fairly successful, because they are easily separated and cleaned, but development of markets has been slower in coming. Most other plastics occur in smaller amounts and more diversity, so voluntary efforts have been very limited. The greatest problem is that, when several materials are combined in a single product, separation is difficult to impossible. Ultimately, the success of recycling will depend very much on greater cooperative efforts by government, industry, and consumers.

Recycling rarely reproduces virgin plastic materials. With repeated recycling, quality decreases, and potential uses decrease. At some point, the final step should be incineration. This produces useful energy, returns carbon dioxide and water to the ecocycle, and reduces final solid waste to an absolute mini-

mum. Present incineration technology suffers from old inefficient equipment and non optimal operation. Sooner or later, we will have to make the effort to develop, build, and operate incinerators for maximum efficiency and minimal harm to the environment.

PART III. PLASTIC PROCESSING

RHEOLOGY

Fundamental Concepts

Plastic processing is primarily the flow and shaping of viscous liquids. The scientific study of this flow is called *rheology*. Assuming laminar shear flow, viscosity is defined as the ratio of shear stress to shear rate.

$$\eta = \frac{\sigma}{\gamma}$$

If this ratio is constant, it indicates a simple Newtonian fluid. For most plastic materials, however, increasing shear rate disentangles polymer molecules and aligns them in the direction of flow, so increasing shear rate decreases resistance to flow (viscosity) (Fig. 15.26), and this non-Newtonian behavior is defined as *pseudoplastic*.

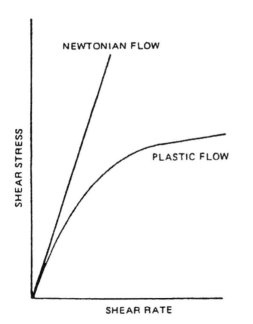
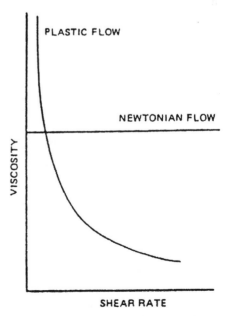

Fig. 15.26. Polymer rheology. (Berins, M. L. (Ed.), *SPI Plastics Handbook*, p. 56, 1991, Copyright © Kluwer Academic/Plenum Publisher, New York. Used by permission of the copyright owner.)

TABLE 15.24   Polymer Activation Energy

Polymer	Activation Energy, kJ/gm·mol
Silicone	16.7
Polyethylene	27.8
Polypropylene	39.6
Polyethylene terephthalate	79.2
Polystyrene	104.2
Polycarbonate	116.7
Poly-a-methyl styrene	133.3

With increasing temperature, viscosity decreases in a manner approximately described by the Arrhenius equation, and the

$$\eta = Ke^{-E/RT}$$

resulting activation energy E correlates with the rigidity of the polymer molecule (Table 15.24). The terms are defined as follows: K is a constant characteristic of the polymer and the test method, e is the natural log base 2.718, E is activation energy, R is the gas constant, and T is absolute temperature in °K.

With increasing pressure, free volume between polymer molecules decreases, flow becomes more difficult, and viscosity increases.

A major factor in polymer viscosity is molecular weight, M, where the experimental

$$\eta = KM^a$$

exponent "a" represents the kinetics of disentangling polymer molecules from each other in the melt (Fig. 15.27) ($a = 1$). Beyond a critical molecular weight M_c (typically 5000–40,000), the difficulty of disentangling molecules multiplies viscosity exponentially ($a = 3.4$).

Instrumental Measurement of Flow Properties

Capillary rheometers measure the effect of pressure on volumetric flow through a cylindrical capillary. They are popular in practical work because shear rate and flow geometry are similar to conditions in extrusion and injection molding. They cover a wide range of shear

rates, and they give practical information on die swell, melt instability, and extrudate defects. Their main disadvantage is that they require a number of mathematical corrections to convert to true viscosity.

Cone and plate rheometers solve one problem, by providing constant shear rate. They can also be designed to measure torque, dynamic properties, normal stresses, and forces in other directions. A disadvantage is that they are limited to low shear rates.

Parallel plate viscometers are used for very high viscosities at low shear rates. Measurement of shear rate is difficult.

Coaxial (concentric) cylinder viscometers provide nearly constant shear rate. A disadvantage is that they are limited to liquids of low viscosity.

Extensional viscometers are useful to measure tensile viscosity in processes such as stretch orientation.

Dynamic or oscillatory rheometers measure viscous and elastic modulus in shear or tension. Energy dissipation produces a phase difference, so stress, strain, and phase angle can be used to characterize complex viscosity behavior.

Practical Effects of Flow Properties

Melt viscosity is the most critical practical property for the process engineer. When stiff molecules give high viscosity and slow flow

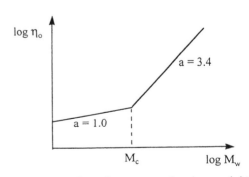

Fig. 15.27. Viscosity vs. molecular weight. (Harper, Charles A., *Modern Plastics Handbook*, p. 5.13, Copyright © 2000 by The McGraw-Hill Companies and used with permission of the copyright owner.)

rates, higher temperature and/or shear-sensitivity (pseudoplasticity) can often be used to decrease the melt viscosity and thus increase processability.

Extrusion defects are primarily due to melt elasticity. When shear rate occurs faster than polymer molecules can disentangle from each other, they simply stretch elastically and store potential energy. When they exit from the die, they release this energy and recover elastically. The resulting extrudate diameter is then greater than the die diameter. This is called *die swell*. The effect is aggravated by high molecular weight, high pressure, low temperature, high shear rate, and short L/D ratio. Two practical ways to compensaste for die swell are to: (1) reduce the die diameter, or (2) pull the extrudate away from the die at a velocity at least as great as it went through the die.

Melt fracture is the occurrence of distorted extrudate coming from the extruder. It is caused by flow disturbance at the point where flow cross-sectional area is rapidly reduced from the large diameter of the melt feed to the much smaller cross-sectional area of the die orifice. Here again it is aggravated by melt elasticity.

Land fracture is a fine surface roughness on the extruded product. It is caused by friction between the melt and the wall (land) of the

TABLE 15.25 Plastic Processes

Process	Use in Industry (%)
Extrusion	34
Injection Molding	31
Blow Molding	13
Calendering	6
Thermoforming	6
Coating	5
Compression Molding	3
Rotomolding	2

die. It is solved by addition of lubricants to reduce melt/die adhesion (see Table 15.25).

EXTRUSION

Extrusion is the process of forming a material continuously through an opening. Most extruders do this by rotating a screw inside a stationary heated cylindrical barrel, to melt the polymer and pump the melt through a suitably shaped orifice (Fig. 15.28). This is used for direct manufacture of finished products such as film or pipe. It may also be used to feed a second process such as injection molding, blow molding, coating, laminating, or thermoforming. It is also important in compounding—blending polymers with additives to improve overall balance of

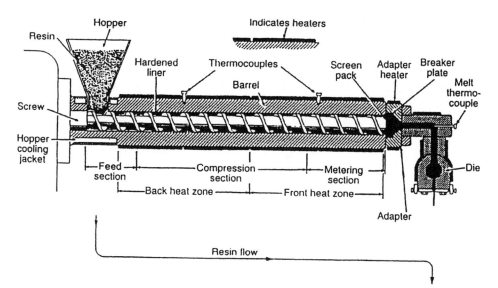

Fig. 15.28. Extruder. (Harper, Charles A., *Modern Plastics Handbook*, p. 5.19, Copyright © 2000 by McGraw-Hill Companies and used with permission of the copyright owner.)

properties—and in recycling plastics out of solid waste.

Basic Functions

The extruder generally performs six successive functions: solids conveying, melting ("plasticating"), melt conveying (pumping), mixing, devolatilization (venting), and forming. While a complete process may involve all six, some processes may omit one or more of these steps.

Solids conveying is carried out in two steps: the feed hopper and the back (entrance) portion of the screw. The feed hopper is an inverted cone or pyramid, in which solid pellets or powder flow downward from the force of gravity. If they flow poorly ("arcing" or "bridging"), the problem may be solved by installing a vibrator or a stirrer ("crammer/feeder") in the hopper, or by coating the particles with a small amount of a chemical that acts as an "external lubricant."

Once the solid particles enter the back end of the screw, they are carried forward and compressed by the rotation of the screw and the friction between the solids and the barrel. Friction can be increased by roughening the surface of the barrel, and particularly by grooving the internal surface of the barrel.

Melting of the solid polymer is the result of heat from two sources: conduction from the heated barrel, and friction between the barrel and the polymer. Most of the heat comes from friction between the barrel, the viscous melt film, and the remaining solid polymer. This frictional heating is most efficient when the melt viscosity is high, and when the melt film is thin. Thinness of the melt film depends on designing and maintaining a very small clearance between the screw flights and the barrel surface.

Melt conveying is the forward motion of the molten polymer through the extruder, due to the pumping action of the rotating screw. This simple "drag flow" M_d is proportional to melt density, down-channel velocity, and cross-sectional area of the screw channel. In most cases, however, there is also a pressure gradient as the melt moves downstream, either

positive pressure approaching the exit die, or negative pressure following the solid/melting zones; so this "pressure flow" M_p must be subtracted from the drag flow to calculate the net output of the extruder. Since pressure flow is inverse to melt viscosity, which is non-Newtonian and temperature sensitive, this complicates the calculation considerably.

Mixing in the extruder is important for homogenization of temperature and pressure, and especially for uniform blending with additives. In a simple single-screw extruder, melt flow is fairly linear and provides little mixing. A variety of ingenious modifications of screw design have been developed to build in mixing elements. Some improve distributive mixing of the liquid melt, to homogenize temperature and pressure fluctuations. Others build in higher shear (dispersive mixing), to break down particles of additives and blend them uniformly into the molten polymer. And still others combine the two types of action.

Twin-screw extruders permit much wider variation in design and performance. The screws may be non-intermeshing, just touching ("tangential"), partially or fully intermeshing. They may be co-rotating or counter-rotating. They are assembled ("programmed") of different sections ("elements"), designed for feed, melting, conveying, distributive melt mixing, dispersive shear for additives mixing, and sealing pressure or vacuum at the vent or the die. The barrels are also programmed of elements to provide functions such as feed ports, venting, and abrasion resistance. Their higher mixing efficiency makes them particularly useful in compounding with additives, processing polyvinyl chloride, reactive extrusion, and devolatilization.

Devolatilization can be used to remove up to 5 percent of volatile impurities from the plastic melt. The first melt conveying (metering) zone builds up melt pressure. Then channel depth is increased abruptly in the vent zone, the melt is decompressed, and volatiles escape through the vent. After this the melt enters a second metering zone, which builds up melt pressure again, and feeds it to the die.

Die forming forces the melt into the shape and dimensions desired in the final product.

Temperature, pressure, viscosity, die design, flow rate, and flow patterns must all be optimized and controlled closely to make a suitable product. The system is so complex that it is managed more by experience and rule-of-thumb than by theoretical design. Typical guidelines include small approach angles, land length = 10 × land clearance, avoid abrupt changes in geometry, no dead spots, generous radii, and thin uniform wall sections.

Calibration is a technique for maintaining the shape and dimensions of the product from the time the melt exits from the die until it can be cooled enough to solidify and stabilize it. Depending on the size and shape of the product, calibrators can use water-cooled plates, internal mandrel, internal air pressure, external vacuum, or a post-extrusion die to change the shape of the molten extrudate before it solidifies.

Major Processes and Products

Blown film is produced from a single-screw extruder by extruding a tube, cooling it with external and/or internal air streams, stretching it in the machine direction by pulling it away from the die ("draw-down"), stretching it in the transverse direction (typically 2–4 × "blow-up" ratio) by internal air pressure up to 34 KPa (5 psi), flattening it by passing through nip (pinch) rolls, and winding it onto a cylindrical roll. Optional post-stretching operations may include flame or corona surface treatment for wettability/adhesion, sealing, slitting, and bag-making.

Flat film, sheet, and coating are produced from a single-screw extruder with a high L/D ratio (27–33/1), which feeds the molten polymer through a flat die. The die opening is adjusted to control the thickness of the film, and the film is solidified in a cold water bath, or preferably, for transparency and gloss, over two or more water-cooled steel chill rolls. Compared to air-cooled blown film, water-cooled flat film generally has higher clarity. The take-off and wind-up line may include automatic feed-back thickness control, surface treatment, and/or slitting.

For extrusion coating, the substrate—paper, plastic, or metal foil—is preheated, and may be pretreated, before the extruder deposits a layer of molten polymer onto its surface. Low molecular weight and high temperature help the polymer to flow into a uniform adhesive coating. The laminated layers pass between pressure and chill rolls, and optionally through surface treatment, printing, and slitting before collecting on the final windup rolls.

Coextrusion produces multilayer laminates in a single process step. Two or more extruders feed different molten polymers into a multi-manifold die which layers them directly, or into a modular feedblock which layers them before feeding them into the die. This is used primarily in the packaging field, to sandwich an impermeable barrier layer between two commodity outer-film layers, and often includes adhesive tie layers to bond the barrier layer to the outer layers.

Pipe, hose, and tubing are extruded through an annular (ring) die. L/D ratio is typically 24/1 or greater. Rigid vinyl is the leading material, and often requires a conical twin-screw extruder. The molten pipe is solidified by water-cooling. Pipe dimensions are controlled initially by the die, but then finally calibrated by pull-off rate, internal mandrel, vacuum, or compressed air. Flexible tubing is collected on a wind-up unit; while rigid pipe is hauled off by a caterpillar puller to a cut-off saw and stacker.

Profile extrusion of siding, window frames, gasketing, and other shapes is complicated by the effects of their asymmetry on heating, viscosity, cooling, and dimensional control. This generally requires modification of the die, and vacuum-driven calibration of the extruded product.

Wire coating extrudes plastic insulation around electrical wire and cable as it passes through a T-shaped crosshead die. The entire process line includes pay-off and capstan to feed and preheat the wire, extrusion-coating, water-cooling, spark testing, diameter and eccentricity controls, and take-up capstan and wind-up. The most common "pressure-coating" die applies the plastic coating inside the die; whereas, for larger wire and cable, the "tubing (tool) die" applies the coating as wire and plastic exit from the die, using internal vacuum to pull the molten coating onto the wire. A fine wire coating line can run at speeds up to 10,000 ft/min.

INJECTION MOLDING

Introduction

Extrusion and injection molding are the two leading methods of converting plastic materials into solid products. Extrusion produces continuous products with a fixed cross-section; injection molding produces discrete products with more complex shapes. Modern injection molding is a very fast, automated process for large-scale manufacture of complex products at minimum cost. An injection molding machine can be used for many different jobs and for many years, so the amortization per job can be very low. On the other hand, it requires a different mold to make each product, so the cost of the mold must be amortized over the life of its individual production run.

Injection Molding Cycle (Fig. 15.29)

The injection molding cycle involves a sequence of events:

- Closing the mold
- Melting the plastic material
- Injecting the melt into the mold
- Filling the mold cavity to form the product
- Cooling the molten product to solidify it in the mold
- Opening the mold and ejecting the solid product

The entire cycle is usually completed within a minute or less.

Closing the Mold. One half of the mold is attached firmly to the melting/feeding end of the injection molding machine. The other half slides horizontally to open or close the mold. The moving half is driven mechanically or hydraulically. When it closes the mold, it is held firmly by clamping action to withstand the injection pressure.

Melting the Plastic Material. The plastic pellets are fed through a hopper into an extruder screw, typically with an L/D ratio of 20/1. The first half of the screw compresses the pellets and squeezes out air. The third quarter of the screw melts the pellets, 70 percent by friction, 30 percent by conduction from heater bands on the extruder barrel. This is called "preplastication." The fourth quarter of the screw pumps the melt forward to the front of the extruder, where it goes through a one-way valve to prevent backflow. As the molten plastic accumulates at

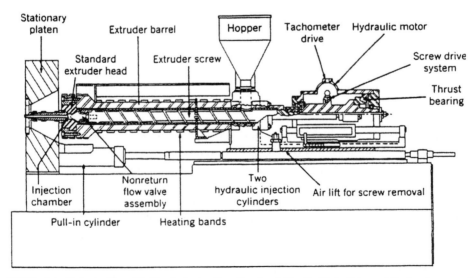

Fig. 15.29. Injection molder. (Rubin, Irvin I., *Handbook of Plastic Materials and Technology*, p. 1235, Copyright © 1990 by John Wiley & Sons, Inc.; used by permission of John Wiley & Sons, Inc.)

the front of the screw, this pushes the screw back to form a reservoir of molten material.

Injecting the Melt into the Mold.   When the reservoir contains the desired (preset) volume of molten material, hydraulic force pushes the screw forward rapidly, injecting the melt into the mold cavity. For this reason it is called a "reciprocating screw." In order to fill the mold completely, and make up for shrinkage during cooling with or without crystallization, the screw and the one-way valve maintain high "boost" ("hold") pressure during this part of the cycle. An important auxiliary function is "venting," tiny openings to allow air to escape from the mold, so that the molten material can fill the mold completely and flawlessly.

Filling the Mold Cavity to Form the Product.   The mold cavity is designed and machined to form the shape of the finished product. This is itself a complete art and science, based partly on experience, and increasingly on computerized engineering principles. Some major considerations are fast uniform flow, avoidance of degradation, minimization of orientation/anisotropy, fast cooling/solidification, shrinkage and dimensional tolerances, and of course final properties of the product.

Cooling the Molten Product to Solidify it in the Mold.   The molten product must be solidified before it can be removed from the mold. This is accomplished by flowing cold water through channels machined into the mold. Computerized design of the cooling channels, and refrigeration of the cooling water to speed the cooling process, are major considerations here. Since heat transfer through organic polymers is slow, the design favors thin-wall products, generally under $\frac{1}{4}$ in. Polymers which can crystallize rapidly offer promise of faster, more economical molding cycles. Polymers which crystallize slowly, and amorphous polymers which stiffen gradually, often require longer molding cycles.

Opening the Mold and Ejecting the Solid Product.   When the plastic product has cooled sufficiently to be solid and retain its shape, the moving half of the mold is opened automatically. With luck, some products will eject from the mold spontaneously; this depends on design of the mold and the product, low polarity of the plastics versus the high polarity of the metal mold, and the use of external lubricants to help release the product from the mold. In most cases, however, the moving half of the mold is fitted with "knockout pins." When the mold is opened, the knockout pins automatically project into the mold cavity, and press gently on the product to force it away from the mold surface. Then, as the mold closes for its next cycle, the knockout pins automatically retract again.

Variations and Details

Drying.   Although most plastics are quite resistant to water, the powder or pellets may still absorb or adsorb small amounts of moisture during storage and handling. In the heat of molding, this moisture evaporates, causing microscopic voids ("blushing") or macroscopic voids (bubbles), or even hydrolysis to lower molecular weight and less desirable properties. Thus many or most plastics require pre-drying before molding. They may be dried on shallow trays in a circulating-air oven, or by passing hot dry air through the hopper as they are fed into the extruder. Or they may be dried by opening a vent midway down the extruder screw to allow the steam to escape.

"Two-Stage" Injection Molding Machine. Instead of using a reciprocating screw to melt the plastic and to push the melt into the mold, the extruder screw may feed the molten plastic into a separate reservoir, and then a separate plunger can feed the melt into the mold.

Electric Drive.   This may replace the hydraulic drive in some injection molding machines. There has been very active discussion of this procedure in recent years.

Multicavity Molds.   While the mold can be machined with a single cavity to produce a single product, in most cases it is more efficient to machine multiple cavities into the mold and fill them all with a single shot of molten material. This requires a system of runners (tunnels) to distribute the melt to all the cavities, which in turn requires much more sophisticated engineering design to balance them all equally.

Hot Runners. When the molten plastic is pumped into the water-cooled mold, the cooling system solidifies both the plastic product in the mold cavities and also the plastic material in the runners. Later the solid runners must be separated, reground, and reused. This is an extra burden on the process. An alternative is to avoid cooling the runners, and actually keep them hot, so that the molten polymer in them remains ready for the next shot into the mold.

Thermosetting Plastics. While injection molding is used primarily for thermoplastic polymers, it is so efficient and economical that processors working with thermosetting plastics may suffer from higher processing costs. This has led some of them to adapt the injection molding process to thermosets. This may involve some modification of the reactivity of the thermosetting plastics and/or the injection molding machinery, gentle heating to melt the reactive materials without curing them prematurely, and use of warm or hot molds to finish the curing process quickly and thus speed the molding cycle.

runners, gates, flash, and rejects of imperfect products. In theory, thermoplastic polymers should be perfectly recyclable. In practice, inplant scrap can be kept clean, reground, and recycled. While 100 percent recycle is theoretically conceivable, most processors simply determine how high a percent of recycle they can blend with their next batch of virgin material, without any harm to their process or product. In most cases they reuse all of their scrap internally, without any contribution to solid waste. This is simply both good economics and good environmental practice.

REACTION INJECTION MOLDING

Overview

Reaction injection molding (RIM) is a fast, low-pressure, low-temperature, low-cost process for one-step conversion of reactive liquids into large finished solid plastic products. Liquid polyol and liquid diisocyanate are mixed by impingement, pumped instantly to fill a large mold cavity, and polymerize/cure rapidly to form a thermoset polyurethane product. The cured polymer may be a stiffly flexible product such as automotive bumper covers, front ends, and trim; or a rigid foamed product such as furniture and housings (cabinets) for computers, business machines, TV, and radio.

Ingredients

The basic reaction to form polyurethanes is

$$HO\text{-}R\text{-}OH + O{=}C{=}N\text{-}R'\text{-}N{=}C{=}O \longrightarrow \overset{\displaystyle O \; H \quad\;\; H \; O}{\underset{\displaystyle \text{Polyurethane}}{\text{-}R\text{-}O\overset{\|}{C}\text{-}\overset{|}{N}\text{-}R'\text{-}\overset{|}{N}\text{-}\overset{\|}{C}\text{-}O\text{-}}}$$

$$\underset{\text{Polyol}}{HO\text{-}R\text{-}OH} + \underset{\text{Diisocyanate}}{O{=}C{=}N\text{-}R'\text{-}N{=}C{=}O}$$

Instant Inplant Recycling. Injection molding produces a certain amount of scrap:

The *polyol* may be a polyether or a polyester:

$$\underset{\text{Polyether}}{HO(CH_2\overset{\overset{\textstyle CH_3}{|}}{CH}O)_nOH} \qquad\qquad \underset{\text{Polyester}}{HO(CH_2CH_2O\overset{\overset{\textstyle O}{\|}}{C}CH_2CH_2CH_2CH_2\overset{\overset{\textstyle O}{\|}}{C}O)_nCH_2CH_2OH}$$

Long polyols (high *n*) give flexible polyurethanes; short polyols give stiff, rigid, and/or crystalline polyurethanes. Branched polyols, with three or more –OH groups, give cross-linked thermoset polyurethanes; short multi-branched polyols give rigid products. Replacing part or all of the polyol by a *polyamine*

$$H_2N-R-NH_2$$

gives faster reaction and therefore shorter cure cycles; and also contributes increased hydrogen-bonding, which produces higher rigidity and strength.

The *diisocyanate* is generally toluene diisocyanate (TDI) or methylene diphenyl isocyanate (MDI) or oligomers based on them.

mixing head. Here they are mixed by high-speed impingement. The mixed liquid system is very reactive, so it must be pumped into the mold cavity to fill it as rapidly as possible. Molding itself is a low-pressure process, rarely more than 0.7 MPa (100 psi), so the mold can be rather light construction; but since molds and products are generally large, the total force needed to clamp the mold is still considerable. The reaction is exothermic, and the temperature rise could over-cure the product, so mold cooling is required. The entire process is rapid, and can be accelerated further by use of polyamines and catalysts to speed the polymerization/cure reactions. It has proved particularly attractive for mass production of large polyurethane parts.

2,6-TDI   2,4-TDI   MDI

Foaming agents are either volatile liquids or a trace of water. For flexible products, only a trace of foaming is needed to optimize the product. For rigid products, a higher degree of foaming is used to produce light-weight products.

Other ingredients include catalysts to control/balance the polymerization/cure reactions, surfactants for foam uniformity, fillers for stiffening, pigments for coloration, flame-retardants where needed, and especially short glass reinforcing fibers to increase strength and dimensional stability in Reinforced RIM (RRIM).

Equipment and Process

Raw material holding tanks are warmed to 30–38°C (85–100°F) to keep them at optimum viscosity. The raw materials are measured and pumped by piston pumps, to deliver exactly equivalent amounts of the two reactants, bring them up to 14–21 MPa (2000–3000 psi), and shoot them into the

Other Polymers

Once RIM had been developed for polyurethane molding, the industry began asking whether it could be applied to other fast polymerization/cure reactions. It was quite readily applied to nylon 6 monomer casting, but the cost of the caprolactam monomer appeared non-competitive. It was also suggested for epoxy cure and possibly other fast reactions, but none of these have yet been developed commercially.

STRUCTURAL FOAM

Definition

When a solid plastic is expanded to moderately lower its density, it is called a "structural foam." Bone and wood are natural structural foams, which benefit from this moderate degree of expansion. Most plastic materials can similarly be expanded during melt processing, and also gain many benefits as a result.

General Description

A blowing agent is added to the plastic material, to compress gas into the melt during processing. It can be compressed nitrogen gas, which is inexpensive but difficult to disperse; or it can be a "chemical blowing agent," usually an organic azo compound $RN=NR$, which gives much more uniform dispersion in the melt, and reacts to produce compressed gases in the hot melt.

In extrusion, when the hot melt exits from the die, the compressed gas expands and foams the extrudate. In injection molding, the hot melt is injected into the mold, but the quantity is reduced ("short shot") so that there is not enough melt to fill the mold; the compressed gas expands, and the melt foams and fills the mold. Either way, the foamed melt cools and solidifies, producing a product with a solid skin and a somewhat expanded internal foam structure. It looks like a solid product, but it is lighter in weight and offers a number of advantages.

Degree of Expansion

Plastics with solid densities of 0.9–1.2 are generally expanded 13–25 percent, giving foamed densities of 0.75–0.9. At this modest degree of expansion, they look and perform like solid products.

Benefits

Use of 13–25 percent less material is an obvious economic advantage. Low pressure in the mold (1.4–3.4 MPa, 200–500 psi) permits construction of a lighter mold and operation at lower pressure, both of which provide further economic advantage. A low degree of foaming is enough to expand the molding against the walls of the mold, eliminating any accidental "sink marks." Higher degree of foaming increases rigidity 3–4× over a solid part of equal weight. Low-pressure molding also eliminates molded-in stresses, which would weaken the product and could also cause post-molding warpage. This in turn permits design and production of more complex parts in a single molding step, eliminating need for later machining and assembling.

Problems in Structural Foaming

Foamed products tend to be thicker, and contain gas bubbles. Both of these factors reduce heat transfer, taking longer to cool and solidify the product, and thus lengthening the manufacturing cycle.

Tiny foam bubbles in the skin of the product tend to form swirl patterns which are unsightly in consumer products. Several modified injection molding processes offer promise for reducing or eliminating this problem: (1) The *Farrel/USM Process* injects the melt into an expandable mold at high pressure, then expands the mold to lower the pressure and permit foaming. (2) The *Coinjection Process* uses two reciprocating screws to feed the mold; the first feeds a nonfoamable short shot which will form the solid skin, and then the second feeds a foamable second short shot which expands, pushing the skin ahead of it and filling the interior with foam. (3) The *Gas Counterpressure Process* seals the mold, compresses nitrogen gas into it, then injects a short shot of melt containing blowing agent, and then uses gradual controlled venting to release the nitrogen gas and allow the foamable melt to expand, forming a solid skin and a foamed core.

LOW-DENSITY FOAMS

When plastics are foamed to low densities, containing more air than polymer, they acquire unique new properties and applications. Major uses are in crash padding and thermal insulation. Closed-cell foams are outstanding for flotation, rigidity, and insulation; while open-cell foams are outstanding for softness, resilience, and comfort.

While foam production and properties may all belong to one unified theoretical basis, commercial practice is different for each of the major polymer families; so it is best to review them individually.

Polyurethane

The largest quantity of low-density foam is polyurethane, made by mixing liquid polyols with liquid diisocyanates, pouring the mixed liquid and allowing it to foam, polymerize, and cure rapidly into its finished form. The auxiliary ingredients—foaming agent, tertiary amine and organotin catalysts, surfactant, and optionally flame-retardant—are mixed with the polyol before it is mixed with the diisocyanate. The two liquid streams are metered by piston or gear pumps, and fed at high pressure into an impingement chamber where they mix by turbulence. This mixture is very reactive, so it is poured rapidly into the desired form or location, and allowed to polymerize, foam, and cure into the finished product. The largest amount is open-cell soft flexible foam; a smaller amount is closed-cell rigid foam for thermal insulation.

Flexible foam is made from long-chain diols with a small amount of triols for cross-linking to give strength and resilience. Foaming is produced by reaction between measured amounts of isocyanate and water to liberate carbon dioxide. Molded products are made by pouring the reactive liquid mixture into a mold cavity at 50°C. They foam and cure in 2–10 min, after which they are temporarily crushed or vacuum-shocked to open the cell walls and insure softness. Then they are allowed to condition for several hours to finish the cure. The major products are auto seating and headrests, and furniture cushions.

Large slabstock is made by pouring the reactive liquid mixture into a moving paper form, up to 8 × 5 × 50 feet long, which takes up to 5 min. Polymerization, foam rise, and cure are exothermic, and the interior of the slab can reach 140–170°C. The rough surface is trimmed off, and the slab allowed to condition 12–24 hr to finish the cure. The finished slab is sliced to the desired thickness for furniture, mattresses, public transportation seating, textile backing (sportswear), carpet backing, and packaging.

Rigid foam is made from short-chain polyols with a typical average functionality of four hydroxyls per molecule, to give high cross-linking for rigidity. Foaming is produced by volatile organic solvents, which are boiled by the exothermic polymerization/cure reaction. Sheet, slab, and block are made by

pouring the reactive liquid mixture onto a moving paper form, or directly onto facing panels for laminated structures. Pour-in-Place technique is used to fill the wall cavities of refrigerators and freezers, holding them at about 40°C to control the reaction. Spray Coating insulation is applied to buildings, tanks, and pipes in the field, using a highly-catalyzed mixture that will foam and gel in less than 10 sec, so that it will stay in place without running down; the coating can be built up to 2 in. thick in a single pass, more by repeated passes.

Rigid polyurethane foam is used primarily as thermal insulation for buildings, trucks, rail cars, shipping containers, tanks, pipelines, cold-storage warehouses, and frozen food display cases.

Polystyrene

Polystyrene is foamed by swelling with pentane, heating to soften the polymer and vaporize the pentane, and allowing it to expand at atmospheric pressure. Extrusion thus produces foamed sheet and board stock, which are used mainly as thermal insulation in commercial refrigerators and freezers, and also in food packaging, roof and wall insulation, and pipe insulation.

Moldable beads are produced by suspension polymerization, swelling with pentane, warming to soften and "pre-expand" the beads, pouring them into a mold, and steaming to expand them fully, soften them, and fuse them together into a finished product. This produces drinking cups, molded packaging, board stock, and display and novelty products.

Loose fill for packaging is produced by extruding and chopping polystyrene into various shapes, swelling with pentane, and steaming to expand them into "peanuts."

Polyvinyl Chloride

Flexible PVC foams are generally laminated to layers of cloth and/or solid vinyl, and used for leatherlike clothing and luggage, upholstery in autos and furniture, and resilient flooring. They are made by polymerizing vinyl chloride in emulsion, spray-drying to a fine powder, mixing into liquid plasticizer at room temperature to form a viscous paste, compounding with barium/zinc stabilizer/catalyst and azodicarbonamide foaming agent, roller-coating onto a moving belt of textile or paper or solid vinyl film, oven-heating to dissolve the resin in the plasticizer and activate the foaming agent, and cooling to room temperature to gel the plasticized vinyl and stabilize the flexible foam. An alternative for large-scale production is to compound general-purpose PVC with plasticizer, stabilizer/catalyst, and foaming agent, calender and/or laminate sheet below the activation temperature of the azodicarbonamide, and then pass through a hotter oven to foam the sheet.

Polyethylene

Polyethylene foam sheet is used mainly in package cushioning, and also in roof insulation; and extruded profile is also used as construction sealants and pipe insulation. It is made mainly by tandem extrusion. The first extruder melts the polyethylene and blends in liquid volatile organic foaming agent. The second extruder cools it to the optimum foaming temperature, and pumps it through a die into room-temperature air, where it expands into foamed sheet.

For higher performance, cross-linked foamed sheet can be produced by radiation or chemical cross-linking. For radiation cross-linking, polyethylene is melted in an extruder and mixed with powdered chemical foaming agent, extruded into sheet, cross-linked by electron beam radiation, and foamed in an oven. For chemical cross-linking, polyethylene is melted in an extruder and mixed with peroxide cross-linking agent and chemical foaming agent, extruded into sheet, and passed through a two-stage oven for cross-linking and foaming.

Molded foam for package cushioning, flotation devices, and sports equipment is made from foamed pellets. Polyethylene pellets are foamed with volatile organic liquid and cross-linked by peroxide or electron beam. The foamed pellets are poured into a mold, the mold is closed, and the pellets are softened and expanded further to fuse into a molded foam product.

Blow Molding

Blow molding uses compressed air to blow and expand a hot plastic tube ("rubbery melt") in a female mold cavity, until the plastic conforms to the walls of the mold. This has developed into a major way to produce plastic bottles, and also more specialized hollow shapes such as fuel tanks, seat backs, tricycles, surfboards, and so on. The leading material is HDPE for milk, water, and household chemicals. Second largest is PET for carbonated beverages. A number of other thermoplastics are blow molded in smaller amounts for more specialized uses.

There are two ways to produce the plastic tube ("parison"): injection molding and extrusion. For injection blow molding, a test-tube shape is first injection molded, then transferred into a bottle mold and blown. For extrusion blow molding, the extruder produces the tube continuously, and a rotating or alternating series of bottle molds take turns clamping around the tube and blowing it. Injection blow molding is used for bottles up to 500 ml in size, while extrusion blow molding is used for bottles 250 ml and larger.

Stretch blow molding holds the parison above its glass transition temperature (T_g) and stretch-orients it to increase modulus, strength, impact resistance, transparency, and impermeability. This is most important for PET, and is also used for PVC, polypropylene, and polyacrylonitrile.

Multilayer blow molding uses a parison containing two or more polymers in concentric layers, and produces a multi-layer laminated bottle or fuel tank. In general this can combine the best properties of each layer. Sandwich structure food packaging, with a mid-layer of ethylene/vinyl alcohol copolymer (EVOH), gains impermeability to oxygen. Sandwich structure fuel tanks of HDPE/nylon/HDPE gain impermeability to gasoline.

To speed the cooling portion of the blow molding cycle, chilled air can help. For more extreme action, liquid carbon dioxide or nitrogen can give even faster cooling.

Surface fluorination has been claimed to make polyethylene less permeable to gasoline and nonpolar solvents in general. Fluorine gas can be used to blow the container, which treats only the inside surface. Or the finished container can be fluorinated both inside and outside in a single post-treatment. This produces a fluorinated layer 20–40 nm thick.

THERMOFORMING

Thermoplastic sheet can be softened by warming, placed in or on a mold, pressed gently but swiftly to conform against the walls of the mold, cooled to solidify, and trimmed to separate the product from the surrounding unused sheet.

Sheet can be produced by extrusion or calendering. For use in thermoforming, the sheet must be extremely uniform. For improved properties, the sheet may be biaxially stretch-oriented up to 300 percent. For large automatic production runs, it is fed continuously off a roll; for short runs, large, and/or specialty jobs, it may be cut into individual sheets and fed one at a time.

The sheet must be preheated to make it soft and pliable ("rubbery melt"). For fastest processing and best final uniformity and properties, it should be as hot as possible, without losing melt strength or beginning thermal degradation. There are three methods of preheating. (1) Convection oven is slow but very uniform. (2) Conduction heating is done with electrically-heated Teflon-coated aluminum plates. (3) Radiant heating is most efficient, especially when the infrared wavelengths are chosen to match the maximum absorption frequencies of the plastic material.

There are three ways to press the warm sheet gently and swiftly against the surface of the mold. (1) Vacuum forming is the most popular, pulling the soft sheet against the mold surface, with atmospheric pressure applying the driving force. The vacuum must be at least 25 in. Hg to give the fastest possible cycle. (2) Pressure forming (compressed air), at pressures of 140–860 kPa (20–125 psi), is faster and gives

better final properties. (3) Mechanical force (tensile, flexural, or compressive) can also drive the warm sheet to conform to the mold surface. In some cases, two or all three of these methods may be combined for optimum performance.

The best molds are made of aluminum with water-cooling channels cut into them. Tooling costs are low, and heat transfer is high. The formed plastic sheet must be held against the mold surface until it is cooled to the solid state. This is often judged by the heat deflection temperature at 455 kPa (66 psi).

Whether the process uses continuous sheet or individual sheets, the product is formed from the center of the sheet, and the edges must be trimmed off to give the final product. Die cutting knives are probably most common, but a variety of other mechanical and thermal techniques are also in use. The trim may be 10–70 percent of the original sheet. It is reground, blended up to 50 percent with virgin resin, and recycled directly into the same process. Considering the variability of recycle and blending, this requires careful control to maintain virgin quality.

Overall, thermoforming is a very useful method of fast low-cost production for a great variety of plastic products, from packaging to building to automotive parts.

ROTATIONAL MOLDING ("ROTOMOLDING")

Hollow products can be made by placing powdered plastic in a closed mold, tumbling and heating it until the plastic has coated the walls of the mold, cooling it to solidify the product, and opening the mold to remove the product.

Polyethylene is by far the most popular material. Others used occasionally include vinyl plastisol (liquid rather than powder), nylon, polypropylene, and polycarbonate. The powder is usually ground to 35 mesh.

The most popular machine is a carousel design with 3–4 arms operating independently, providing separate stations for loading, heating, cooling, and unloading. Molds are generally made of aluminum, two-piece plus clamps.

Slow rotation gives the most uniform products. The pattern of biaxial rotation must be determined by trial and error. Heating is best done in a gas-fired oven with a fan for circulation.

Cooling is a two-stage process, first a fan and then a cold water spray or mist.

Advantages are low capital investment, design freedom, strain-free products, and little or no post-molding secondary operations. For initial costing, multiply the cost of the raw material by 5.

Limitations: blow molding is faster and uses a wider variety of plastics.

Typical products are tanks (20–85,000 L, 5–22,500 gal), containers for packaging and materials handling, portable outhouses, battery cases, light globes, vacuum cleaners, garbage cans, surfboards, toys, traffic barricades, display cases, and ducting.

POWER COATING

A solid plastic may be ground to a solid powder and then used to apply a plastic coating to a metal product, either for decorative reasons (color) or for functional performance such as insulation or corrosion resistance. Plastics most often used in this way include nylon, vinyl, acrylic, polyethylene, polypropylene, and epoxy. Coating without solvents is beneficial both economically and environmentally, and 100 percent utilization of material eliminates waste. After the powder is bonded to the metal surface, it is often reheated to flow into a more uniform coating and, in the case of thermosetting resins, to complete the cure reaction.

There are three techniques for applying the powder coating to the surface of the metal product: fluid bed, electrostatic fluid bed, and electrostatic spray.

Fluid Bed Coating

The equipment is simply a horizontal box with a finely-porous shelf near the bottom. Powdered plastic, ground to 20–200 μm, is poured above the shelf. Compressed air is fed in below the shelf, percolating up through the pores, and percolating the powder so that it rises and flows much like a liquid. Sometimes the box is also vibrated to produce greater uniformity.

The metal product to be coated is preheated to a temperature which will melt the powder. Then the product is dipped into the fluid bed. The powder particles melt and flow onto the metal surface. Coatings up to 2.5 mm (0.1 in.) thick can be applied in a single dip.

Typical products are electric motors, electronics, transformers, valves, pumps, refinery equipment, and appliances.

Electrostatic Fluid Bed

In a modification of the fluid bed technique, the product to be coated, instead of being preheated, is passed over the bed, and the powder is attracted to it by a static charge. Then it is passed through a heating oven to fuse it into a finished coating. This process can be used for either discrete or continuous products.

Electrostatic Spray

Conventional electrostatic spray coating pumps a liquid coating formulation through a spray gun, which puts an electrostatic charge on the liquid, and sprays charged droplets toward a grounded metal product. The droplets are attracted to the product, where they discharge and adhere to the metal surface.

Plastic powders can be applied in a similar way. The solid plastic is ground to a 30–200 μm powder, fluidized, and conveyed by compressed air through a spray gun which uses high voltage to apply an electrostatic charge to the particles, and then sprays them at the grounded metal product. Spraying is normally done in a hood, to protect the worker and to collect and recycle the excess powder. The charged powder is attracted to the metal product, attaching a coating 50–75 μm thick on a cold product, or up to 250 μm on a preheated product. A post-fusion step melts and flows the powder into a uniform coating and, in the case of thermosetting resins, also completes the cure reaction.

CALENDERING

Basically, a viscous bank of excess thermoplastic or rubber is pressure-formed between a pair of parallel co-rotating rolls to form a thin "film" or a thicker "sheet." Most calenders consist of four rolls in L- or Z-shape, plus additional feed rolls and post-calender laminating, patterning, and/or wind-up rolls.

PVC is the most commonly calendered plastic, especially in plasticized flexible form. Other materials commonly calendered are rubber, ABS, polyurethane, and thermoplastic elastomers.

The plastic material delivered to the calender must be molten, homogeneous in composition and temperature, and at optimum viscosity for the calendering operation. For rigid vinyl, optimum temperature may be 180–190°C; for flexible vinyl, 10–20°C lower.

Nip pressures between the rolls are typically 160–1050 kN/m (900–6000 lb/in.) of roll face.

A surface pattern can be applied by the calender roll or by a post-embossing operation. Feeding hot sheet to a cold embossing roll works well. The pattern is measured by a profilometer.

Calendering can also be used to apply a plastic surface to a substrate web such as metal, cloth, or paper.

Capital investment for a calender line may be about $5 million. Operating cost may be about $500 per hr. Calendering is a very high-speed method of producing plastic film and sheet, but high-speed economy must be balanced against quality of the finished product.

VINYL PLASTISOL PROCESSING

Plastisol is a specialized technique for making flexible vinyl products. It is also sometimes called "paste" or "dispersion" technology. Basically, powdered PVC is stirred into liquid plasticizer at room temperature to form a viscous liquid or spreadable paste. This can be poured or spread into the shape of the final product, heated to fuse, and cooled to gel into the final solid product. Simple pouring or gentle spreading require no heavy equipment, and do not damage delicate substrates such as paper and cloth. Consequently, this accounts for about 10 percent of all vinyl production.

About 90 percent of PVC is made by suspension polymerization and processed as viscous melt at high pressure in heavy-duty equipment. Some 10 percent of PVC is made by emulsion polymerization and spray-drying, for use in plastisol processing. The resin is stirred into liquid plasticizer at room temperature, along with stabilizer and other optional additives, to form a viscous liquid or easily spreadable paste. This is poured or spread into the form of the desired final product. It is then heated, passing through several phases. At first the resin particles absorb the liquid plasticizer, swell, and form a gel. On further heating, the resin melts and dissolves in the hot plasticizer, forming a hot solution. On cooling, the resin crystallizes somewhat, turning the hot solution into a hot gel. On cooling to room temperature, the gel solidifies to a firm rubbery plastic.

If the plastisol is too viscous for pouring at room temperature, it may be thinned with volatile solvent; on heating, the solvent evaporates, leaving a firmer final product. This is called "organosol" technology. Alternatively, a polymerizable plasticizer may be added to thin the plastisol; on heating, it polymerizes to a solid ingredient, giving a firmer final product. This is sometimes called a "rigisol."

If molds are required to shape the plastisol into a finished product, they are generally light-weight low-cost molds of aluminum, electroformed nickel, or ceramic. Oven heating is slow but common. The major handicap is slow heat transfer, so overall production cycles may take 4–20 min. Several different methods are described below for converting plastisols into finished products.

Dip Coating. A product is preheated and dipped into the plastisol. The plastisol gels onto the surface of the product. It is withdrawn from the liquid, allowed to drain off excess liquid, and then placed in a 191–204°C oven to heat it in a few seconds to about 177°C and fuse the gel to a homogeneous solution. This is then cooled in air or water to solidify it to the final rubbery coating. Typical products are tool handles, kitchen implements, and electrical insulation.

Dip Molding.   Instead of a heated product, a heated male mold is dip-coated in the same way. In this case, the final flexible vinyl product is stripped off of the mold. A typical product is medical gloves.

Slush Molding.   This is the reverse of dip molding. A female mold is used to give any desired surface finish on the product. The mold is preheated, plastisol is pour into it, and gels onto the surface of the mold. The excess liquid plastisol is poured out, and the internally- coated mold is heated to gel, then cooled to solidify the rubbery product, and the product is stripped out of the mold. Typical products are arm rests, head rests, road safety cones, anatomical models, dolls, toys, and auto parts.

Hot–Melt Molding.   Hot molten plastisol is injected into a cooled mold to make products like fishing baits and novelties.

Rotational Casting.   Like rotomolding, the plastisol is poured into a cold mold, which is rotated and heated to gel the plastisol onto the walls of the mold. It is then heated to fuse, cooled to solidify, opened, and the product removed. Typical products are volley balls, basketballs, dolls, and auto parts.

Open Molding.   Plastisol is poured into an open mold, heated to gel and fuse, cooled to solidify, and stripped out from the mold. Typical products are auto air filters, oil filters, truck flaps, and place mats.

Closed Molding.   Plastisol is filled into a closed mold, heated to gel and fuse, then cooled to solidify. A typical product is switch mats for automatic door openers.

Spray Coating.   Liquid plastisol can be spray coated up to 1.25 mm (50 mils) in a single pass. The plastisol is formulated to be liquid enough to spray, then firm enough to hold on a vertical surface without running down. It is heated to gel and fuse, then cooled to form the final rubbery coating. A typical product is tank linings.

Continuous Coating.   Doctor blade or roll coating applies plastisol continuously to a moving web of metal or cloth, or two successive coatings apply a solid vinyl skin and a foamable vinyl core to a fabric backing. Typical products are house siding, conveyor belting, and resilient flooring.

Silk-Screen Inks.   Plastisol coatings can be applied a few mils thick on cloth to produce T-shirts and athletic uniforms.

Organosol Coatings.   The low viscosity of organosol formulations permits coatings under 10 mils thick, particularly for chemical resistance.

LIQUID CASTING PROCESSES

Reactive liquids may be mixed and poured to form a solid plastic product. They may be two co-reactive monomers or prepolymers, or one monomer/prepolymer plus a catalyst or curing agent. They are primarily thermosetting plastics and elastomers. Rigid plastics are primarily epoxy, polyester, and acrylic. Elastomers are generally room-temperature-vulcanizing (RTV) polymers: polysulfide, polyurethane, and silicone. They are used mostly to provide electrical insulation and mechanical and environmental protection for delicate electrical/electronic products such as resistors, coils, solenoids, capacitors, transformers, printed circuit boards, opti-electronics, and light-emitting diodes. There are a number of such liquid casting processes, generally carried out at room temperature with or without a later heating step for complete cure.

Casting.   The reactive liquid mixture is poured into a mold, and allowed to polymerize and cure. This is used to make solid tires for industrial vehicles, and to make decorative items such as simulated wood trim, furniture, picture frames, and lamp bases.

Potting. Delicate electronic devices are placed inside an empty shell, and the space is filled by pouring the reactive liquid mixture into it and curing it.

Encapsulation/Embedment. After potting is completed, the outer shell is removed, and the solid plastic is the container.

Conformal Coating. Using a thixotropic reactive liquid, the product is dipped into the liquid and removed with a coating of the liquid on it. This is then heat-cured to solidify the coating and make it permanent.

Impregnation. A porous product can be dipped into a fluid reactive liquid, which soaks into the pores and fills them completely. The liquid is then cured to leave the product completely filled with solid polymer in the pores.

Syntactic Foam. Hollow glass, ceramic, or plastic spheres are dispersed in the reactive liquid system before it is cast. When the liquid is polymerized and cured, the hollow spheres make it a unicellular foam. The air "bubbles" in the cells make it low-density, low dielectric constant and loss, and very resistant to compressive forces such as hydrostatic head in deep-sea equipment.

Aside from simply casting at atmospheric pressure, vacuum is often used to remove air bubbles and volatiles which would give an imperfect casting, or to help promote porous impregnation. Small specialty runs can be done manually, with no capital investment but high labor cost; larger production runs can be automated with only modest capital investment. Since many of these liquid systems may be volatile and unpleasant to work with, worker protection often includes ventilation and/or protective clothing for eyes, skin, and lungs.

COMPRESSION MOLDING AND TRANSFER MOLDING

Thermosetting plastics and rubber are heated to soften ("plasticate") them, and then pressed at 14–28 MPa (2000–4000 psi) for about a minute to cross-link ("cure") them. Then the mold is opened and the finished product is removed.

Outstanding Properties. Compared to engineering thermoplastics, thermosets can offer equal or superior properties at lower cost. Major advantages include rigidity, creep resistance, dimensional stability, impact strength, heat resistance, and chemical resistance.

Typical Applications. Common compression-molded thermoset products include electrical equipment, appliance handles and knobs, dinnerware, distributor caps, under-the-hood parts in general, automatic transmissions, brakes, and pumps.

Compression Molding. This press has two horizontal platens facing each other. The upper one is generally stationary; the lower one moves vertically, driven by mechanical, hydraulic, or pneumatic power. The upper mold half is fastened to the upper platen, the lower mold half to the lower platen. Most molds are electrically heated. In semi-automatic operation, the operator puts the granular molding powder in the lower half of the mold, and presses a button which closes the press, holds it till cure is complete, and opens the press again; then the operator removes the molding from the mold, and repeats the cycle. In fully automatic molding, the entire cycle is automatic and does not require an operator.

Transfer Molding. This preheats the molding powder in the upper part of the mold. Then a plunger pushes the fluid material down into the (lower) mold cavity and holds it there until cure is complete. This gives more fluid flow than compression molding, and avoids damage to delicate molds or molded parts.

Additional Considerations. Compared to injection molding of thermoplastics, compression molding is less capital intensive, more labor-intensive, and takes a longer molding cycle. Transfer molding can equal injection molding cycle times. Compression and transfer molding do less damage to reinforcing fibers, so they can give more impact-resistant products.

The molding cycle can be shortened by preforming and preheating. The granular molding powder is cold-pressed into a pill of the desired weight, and then dielectric preheating for 10–20 sec brings it to molding temperature before it is inserted into the mold. This gives better flow and faster cure.

"Venting" and "breathing" must often be included in the molding cycle. Venting allows trapped air to escape through small grooves as the molding powder is compressed and flows into a solid part. Breathing opens the mold about 3 mm ($\frac{1}{8}$ in.) for a second or two, early in the molding cycle, to allow escape of water and other volatiles from the condensation cure reaction. This permits production of bubble-free parts.

Ejector pins help to separate the finished molding from the mold cavity. These are driven mechanically during the mold-opening stage of the cycle.

Post-cure ovens may be used to help complete the cure cycle, especially to control gradual cooling of the molded part.

Deflashing is often required to remove excess trim ("flash") from the edges of the molded product. This may be done by tumbling in a rotating drum, or by an air-blast grit.

REINFORCED PLASTICS PROCESSING

Adding short-fiber reinforcements to thermoplastics can produce major increases in modulus, strength, and dimensional stability; it makes melt processing more difficult, but does not change it qualitatively. Adding fibrous reinforcement to thermosetting plastics produces outstanding modulus, strength, impact strength, dimensional stability, and heat and chemical resistance; but it requires entirely new methods of processing to convert them into finished products.

The polymers most often used are unsaturated polyesters and epoxy resins; for ultra-high performance, polyimides and other specialty resins are also used in small amounts. The reinforcing fibers are primarily glass; for ultra-high performance, aramide, carbon/graphite, and metallic fibers are also used in small amounts.

The polymers are generally polymerized to low-molecular-weight "A-stage" resins, which are still liquids or fusible solids, and potentially very reactive. In pure form they can be stored for months to years. The liquids are easily mixed with reinforcing fibers and other ingredients, and shaped into the form of the ultimate products. Addition of peroxide initiators to polyesters, or of amine or anhydride curing agents to epoxies, liberates their high reactivity to permit fast cure cycles to completely thermoset products. Some processes use heat and/or pressure to complete the shaping and cure cycle, while others are carried out simply at ambient conditions. These variables offer a wide variety of processes which can be used to manufacture reinforced thermoset plastic products.

Matched Die Molding Processes

By analogy with compression molding, thermosetting resins plus fibrous reinforcement can be pressed between matching mold halves, with heat and moderate pressure, and cured directly into finished products. Typical cure cycle is about a minute. Half of all reinforced thermoset products are made in this way. Major applications are in the automotive, electrical/electronics, appliance, and business machine industries.

Bulk Molding Compound (BMC). Liquid resin, $\frac{1}{4}-\frac{1}{2}$ in. long glass fiber, and simple filler (typically calcium carbonate or clay) are mixed in a dough mixer or cold extruder. Group II metal oxide is added to gel the liquid resin (Table 15.26). This doughy mix ("premix") is placed in a hot mold, pressed to flow and fill the mold cavity, and held until cured.

TABLE 15.26   Typical Mix for a BMC

Resin	22%
Fiber	25%
Filler	53%

Sheet Molding Compound (SMC). An SMC machine pours liquid resin and 1-in. glass fiber onto a moving belt, passes through calender rolls to make a good sheet, and then through an oven to begin cross-linking and thus gel the resin. The sheet is placed in a hot mold and pressed to shape and cure it. While the short fiber in BMC is more moldable, the longer fiber in SMC gives greater strength and impact resistance.

Cold Press Molding.   Occasionally the mix may be placed in a cold mold and pressed at about 345 kPa (50 psi). The cure cycle is considerably longer.

Preform.   A metal screen is made in the shape of the final product. Glass fiber is chopped 2 in. long and sprayed uniformly all over the shaped screen, using vacuum on the back side of the screen to assist the process. A small amount of binder, typically 5 percent of polymer in latex form, is sprayed onto the fiber to hold its shape. It is then removed from the screen, placed in the mold, saturated with an equal weight of liquid resin, and the mold is pressed at 1380 kPa (200 psi) and heated until cured, typically 3–15 min. This early process has been largely replaced by SMC.

Prepreg.   Fabric is impregnated with 25 percent of liquid resin and laid in the mold. To insure isotropic properties, or to maximize properties in a specific direction, successive layers of impregnated fabric are carefully oriented in different directions. The mold is closed, pressed, and heated till cured. Products made from such impregnated fabrics have much higher strength than simple random fiber reinforcements.

Resin Transfer Molding (RTM).   Reinforcing fibers are distributed uniformly in the mold and the mold is closed. Liquid resin is injected into the mold until the excess comes out of the vents. The mold is pressed and heated, similarly to preform molding, until cure is complete.

Open Molding

Instead of applying pressure in a closed mold, the mix of liquid resin and reinforcing fiber may be laid into an open mold, and optionally pressed gently at room temperature until cured. To accelerate low-temperature cure, more active catalyst systems are added. Alternatively, the assembly may be UV or oven cured. This requires less capital investment but more skilled labor, so it is useful for prototype and small production runs. It permits unlimited size, so the largest reinforced thermoset products are made in this way, for example, large tanks and whole boat hulls.

Hand Lay-Up (Contact Molding).   A layer of liquid resin is applied to the surface of the mold. A layer of glass fiber mat (low strength) or fabric (high strength) is hand laid over it. Liquid resin is poured over it, and brushed or rolled (squeegeed) into it. The process is repeated to build up the desired thickness of the product. The assembly is allowed to stand until cured.

Spray-Up.   Instead of hand lay-up, continuous glass roving and liquid resin are fed into a gun, which chops the glass fiber, mixes it with the resin, and sprays it into the mold. This can be automated for lower labor cost and greater uniformity. Spraying is often followed by hand-rolling to expel air and densify the assembly. Then it is allowed to stand until cured. Products are similar to those from simple hand lay-up.

Vacuum-Bag Molding.   After hand lay-up, the assembly is covered with an air-tight film, typically polyvinyl alcohol, occasionally nylon or other material. Then a vacuum is pulled on the underside of the film, to let atmospheric pressure squeeze out air and excess resin. Use of a hand-held paddle may help. The assembly is allowed to stand until cured. Products are void-free, and quality is better than simple hand lay-up.

Pressure-Bag Molding.   This is similar to vacuum-bag molding, but 345 kPa (50 psi) air

TABLE 15.27 Typical Properties of Filament-Wound Plastics

Glass fiber	50–85%
Tensile modulus	34–48 GPa (5–7 $\times$ 10^6 psi)
Tensile strength	550–1700 MPa (80–250 $\times$ 10^3 psi)
Flexural strength	700–1400 Mpa 100–200 $\times$ 10^3 psi)

pressure is applied to the outside of the cover film. A rubber bag may be used to facilitate the process.

Autoclave Molding. This is similar to pressure-bag molding, but it is carried out in an autoclave to apply the pressure to the outside of the cover film.

Rubber Plug Molding. Liquid silicone rubber is cast into a female mold cavity and cured. The plug is removed, and layers of mat or fabric are built onto the plug. This assembly is then inserted into the mold cavity, pressed, and heated until cured.

Special Processes

Filament-Winding. This requires a mandrel to shape the desired finished product. Continuous filament or woven tape is fed through a liquid resin bath to impregnate it, and then wound onto the mandrel in a calculated pattern to optimize the final properties (Table 15.27). The assembly is oven-cured. A collapsible mandrel can then be removed from the plastic product; or the mandrel can be left as a part of the finished product. These are the strongest plastic products ever made. Typical products are pipes, tanks, and pressure bottles. Other suggested products include rocket motor cases, railroad hopper cars, automotive springs, drive shafts, ship hulls, housing modules, helicopter rotor blades, and helicopter tail sections.

Pultrusion. Filaments, woven tapes, or fabrics are fed continuously through a liquid resin bath, through a shaping die, and through a curing oven, then cut to any desired length. This makes continuous products of any desired cross-section. Typical products include fishing rods, flag poles, tool handles, ladder rails, tubing, and other electrical, corrosion-resistant, construction, and transportation applications.

Continuous Laminating. Reinforcing fiber and liquid resin are deposited on a moving belt, densified between squeeze rolls, passed through a curing oven, and cut to length. This process is used for mass production of glazing, panelling, and roofing.

Centrifugal Casting. Reinforcing fibers are distributed inside a circular mold. The mold is rotated, and liquid resin is distributed inside it to impregnate the fibers. Rotation of the assembly is continued inside an oven until it is cured. This process is used for making pipes, tanks, and hoops.

Foam Reservoir Molding. Flexible open-cell polyurethane foam is impregnated with liquid resin, faced with glass fiber mat, and gently compression molded. This squeezes the liquid resin into the glass fiber surface mat. Heat curing produces a sandwich structure of low density, high flexural and impact strength.

REFERENCES FOR PART I

1. Sperling, L. H., *Introduction to Polymer Science*, 3rd ed., p. 2, John Wiley & Sons, New York, 2001.
2. Rudin, A., *The Elements of Polymer Science and Engineering*, 2nd ed., pp. 132–141, Academic, San Diego, CA, 1999.
3. Rudin, A., *The Elements of Polymer Science and Engineering*, 2nd ed., p. 150, Academic, San Diego, CA, 1999.
4. Billmeyer, F. W., *Textbook of Polymer Science*, 2nd ed., p. 16, John Wiley & Sons, New York, 1971.
5. Stevens, M. P., *Polymer Chemistry*, 3rd ed., pp. 70–74, Oxford University Press, New York, 1999.
6. Nielsen, L. E., and Landel, R. F., *Mechanical Properties of Polymers and Composites*, pp. 18–23, Marcel Dekker, New York, 1994.

7. Nielsen, L. E., and Landel, R. F., *Mechanical Properties of Polymers and Composites*, p. 21, Marcel Dekker, New York, 1994.

8. Nielsen, L. E., and Landel, R. F., *Mechanical Properties of Polymers and Composites*, pp. 17–18, Marcel Dekker, New York, 1994.

9. Painter, P. C., and Coleman, M. M., *Fundamentals of Polymer Science*, 2nd ed., pp. 284–290, Technomic Publishing Company, Lancaster, PA, 1997.

10. Nielsen, L. E., and Landel, R. F., *Mechanical Properties of Polymers and Composites*, pp. 50–51, Marcel Dekker, New York, 1994.

11. Sperling, L. H., *Introduction to Polymer Science*, 3rd ed., pp. 230–233, John Wiley & Sons, New York, 2001.

12. Billmeyer, F. W., *Textbook of Polymer Science*, 2nd ed., John Wiley & Sons, New York, 1971.

13. Billmeyer, F. W., *Textbook of Polymer Science*, 2nd ed., pp. 270–271, John Wiley & Sons, New York, 1971.

14. Cauaher, C. E., Jr., *Polymer Chemistry*, 4th ed., p. 265, Marcel Dekker, New York, 1996.

15. Steyens, M. P., *Polymer Chemistry*, 3rd ed., p. 206, Oxford University Press, New York, 1999.

16. Odian, G., *Principles of Polymerization*, 2nd ed., p. 182, John Wiley and Sons, New York, 1981.

17. Steyens, M. P., *Polymer Chemistry*, 3rd ed., pp. 208–209, Oxford University Press, New York, 1999.

18. Stevens, M. P., *Polymer Chemistry*, 3rd ed., p. 213, Oxford University Press, New York, 1999.

19. Billmeyer, F. W., *Textbook of Polymer Science*, 2nd ed., pp. 317–318, John Wiley & Sons, New York, 1971.

20. Stevens, M. P., *Polymer Chemistry*, 3rd ed., p. 222, Oxford University Press, New York, 1999.

21. Stevens, M. P., *Polymer Chemistry*, 3rd ed., pp. 223–225, Oxford University Press, New York, 1999.

22. Stevens, M. P., *Polymer Chemistry*, 3rd ed., pp. 234–235, Oxford University Press, New York, 1999.

23. Cauaher, C. E., Jr., *Polymer Chemistry*, 4th ed., p. 277, Marcel Dekker, New York, 1996.

24. Stevens, M. P., *Polymer Chemistry*, 3rd ed., pp. 237–245, Oxford University Press, New York, 1999.

25. Carraher, C. E., Jr., *Polymer Chemistry*, 4th ed., p. 282, Marcel Dekker, New York, 1996.

26. Stevens, M. P., *Polymer Chemistry*, 3rd ed., pp. 246–249, Oxford University Press, New York, 1999.

27. Stevens, M. P., *Polymer Chemistry*, 3rd ed., p. 173, Oxford University Press, New York, 1999.

28. Billmeyer, F. W., *Textbook of Polymer Science*, 2nd ed., pp. 355–357, John Wiley & Sons, New York, 1971.

29. Stevens, M. P., *Polymer Chemistry*, 3rd ed., p. 174, Oxford University Press, New York, 1999.

30. Shaffer, K. A., and DeSimone, J. M., *Trends Polym. Sci.*, **3**, 146 (1995).

31. Stevens, M. P., *Polymer Chemistry*, 3rd ed., p. 174, Oxford University Press, New York, 1999.

32. Billmeyer, F. W., *Textbook of Polymer Science*, 2nd ed., pp. 358–359, John Wiley & Sons, New York, 1971.

33. Billmeyer, F. W., *Textbook of Polymer Science*, 2nd ed., pp. 359–361, John Wiley & Sons, New York, 1971.

34. Stevens, M. P., *Polymer Chemistry* 3rd ed., p. 175, Oxford University Press, New York, 1999.

35. Stevens, M. P., *Polymer Chemistry*, 3rd ed., pp. 194–195, Oxford University Press, New York, 1999.

36. Billmeyer, F. W., *Textbook of Polymer Science*, 2nd ed., p. 231, John Wiley & Sons, New York, 1971.

37. Billmeyer, F. W., *Textbook of Polymer Science*, 2nd ed., p. 227, John Wiley & Sons, New York, 1971.

38. Stevens, M. P., *Polymer Chemistry*, 3rd ed., p. 227, Oxford University Press, New York, 1999.

39. Stevens, M. P., *Polymer Chemistry*, 3rd ed., pp. 273–275, Oxford University Press, New York, 1999.

40. Sperling, L. H., *Introduction to Polymer Science*, 3rd ed., pp. 342–343, John Wiley & Sons, New York, 2001.

41. Birley, A. W., Haworth, B., and Batchelor, J., *Physics of Plastics*, Carl Hanser Verlag, Munich, 1992.

42. Williams, M. L., Landel, R. F., and Ferry, J. D., *J. Am. Chem. Soc.*, **77**, 3701 (1955).

43. Powell, P. C., *Engineering with Polymers*, Chapman and Hall, London, 1983.

44. Birley, A. W., Haworth, B., and Batchelor, J., *Physics of Plastics*, pp. 283–284, Carl Hanser Verlag, Munich, 1992.

45. Nielsen, L. E., and Landel, R. F., *Mechanical Properties of Polymers and Composites*, p. 253, Marcel Dekker, New York, 1994.

46. Nielsen, L. E., and Landel, R. F., *Mechanical Properties of Polymers and Composites*, pp. 342–352, Marcel Dekker, New York, 1994.

REFERENCES FOR PART II

American Plastics Council

Billmeyer, F. W. Jr., *Textbook of Polymer Science*, John Wiley & Sons, New York, 1984.

Brydson, J. A., *Plastics Materials*, Butterworth Scientific, London, 1982.

Craver, C. D., and Carraher, Jr., C. E., *Applied Polymer Science, 21st Century*, Elsevier, Amsterdam, 2000.

Deanin, R. D., *Polymer Structure, Properties, and Applications*, Cahners, Boston, 1972.

Goodman, S. H., *Handbook of Thermoset Plastics*, Noyes Publications, Westwood, NJ, 1998.

International Plastics Selector, *Plastics Digest: Thermoplastics and Thermosets*, D. A. T. A., Englewood, CO, 1995.

Harper, C. A., *Modern Plastics Handbook*, McGraw-Hill, New York, 2000.

Lubin, G., *Handbook of Composites*, Van Nostrand Reinhold, New York, 1982.

Modern Plastics Magazine, *Modern Plastics Encyclopedia Handbook*, McGraw-Hill, New York, 1994.
Plastics News
Rubin, I. I., *Handbook of Plastic Materials and Technology*, John Wiley & Sons, New York, 1990.
Sears, J. K., and Darby, J. R., *The Technology of Plasticizers*, John Wiley & Sons, New York, 1982.

REFERENCES FOR PART III

Harper, C. A., *Modern Plastics Handbook*, McGraw-Hill, New York, 2000.
Klempner, D., and Frisch, K. C., *Handbook of Polymeric Foams and Foam Technology*, Hanser, Munich, 1991.
Plastics News
Rubin, I. I., *Handbook of Plastic Materials and Technology*, John Wiley & Sons, New York, 1990.

16

Rubber

D. F. Graves*

INTRODUCTION

The word "rubber" immediately brings to mind materials that are highly flexible and will snap back to their original shape after being stretched. In this chapter a variety of materials are discussed that possess this odd characteristics. There will also be a discussion on the mechanism of this "elastic retractive force." Originally, rubber meant the gum collected from a tree growing in Brazil. The term "rubber" was coined for this material by the English chemist Joseph Priestley, who noted that it was effective for removing pencil marks from paper. Today, in addition to Priestley's natural product, many synthetic materials are made that possess these characteristics and many other properties. The common features of these materials are that they are made up of long-chain molecules that are amorphous (not crystalline), and the chains are above their glass transition temperature at room temperature.

Rubber products appear everywhere in modern society from tires to biomedical products. The development of synthetic rubber began out of the need for countries to establish independence from natural products that grew only in tropical climates. In times of conflict the natural product might not be available, and its loss would seriously threaten national security. Synthetic rubber, then, became a strategic concern during World Wars I and II.[1] Beyond the security issue, the need for materials with better performance also provided a strong impetus for the development of new rubbery materials. In particular, improvements in oil resistance, high-temperature stability, and oxidation and ozone resistance were needed.[2] Research today is driven to develop materials with even better performance in these areas. In the 1980s and 1990s tires with lower rolling resistance were demanded by car manufacturers to improve fuel economy. This was accomplished, in part, by developing functional tread polymers which chemically bond to fillers resulting in a dramatic reduction in the hysteresis (energy loss) of the tire tread. These new functional polymers will be discussed in this chapter.

*Firestone Polymers, Division of Bridgestone/Firestone, Inc., Akron, OH.

TABLE 16.1   Synthetic Rubber Production by Type, 2003

Type	Description	Capacity (Metric Tons)
SBR-solid	Styrene/butadiene rubber	2,635,000
BR	Butadiene rubber	2,018,000
IR	Isoprene rubber	258,000
CR	Chloroprene (or neoprene)	244,000
EPDM	Ethylene/propylene terpolymer	883,000
NBR	Nitrile or acrylonitrile/butadiene rubber	303,000
Others		1,025,000
Total		7,366,000

Source: *Worldwide Rubber Statistics 2004*, International Institute of Synthetic Rubber Producers, Inc., by permission.

TABLE 16.2   Worldwide Rubber Consumption Forecast, 2008 (Metric Tons)

Total new rubber	19,442,000
Natural	9,042,000
Total synthetic	10,400,000
SBR	4,114,000
Nitrile	424,000
Polybutadiene	2,572,000
EPDM	1,048,000
Other Synthetics	2,242,000
% Natural	46.5
% Synthetic	53.5

Source: *Worldwide Rubber Statistics 2004*, International Institute of Synthetic Rubber Producers, Inc., by permission.

The worldwide demand for rubber was estimated to be 19.4 million metric tons for 2008,[3] excluding latex materials. Of this demand, 54 percent is synthetic rubber of various kinds. Rubber has been classified by use into general purpose and specialty.

The major general purpose rubbers are natural rubber, styrene–butadiene rubber, butadiene rubber, isoprene rubber, and ethylene–propylene rubber. These rubbers are used in tires, mechanical goods, and similar applications. Specialty elastomers provide unique properties such as oil resistance or extreme heat stability. Although this differentiation is rather arbitrary, it tends also to classify the polymers according to volumes used. Styrene–butadiene rubber, butadiene rubber, and ethylene–propylene rubber account for 78 percent of all synthetic rubber consumed.

The 2003 synthetic rubber capacity by type is presented in Table 16.1, and Table 16.2 lists the worldwide rubber consumption forecast by type for 2008.

RUBBER CONCEPTS

Several key principles (outlined below) greatly help in understanding the performance of elastomeric materials. This outline should present these concepts well enough for use in the following discussions of specific polymer types. However, the reader should consult the references for a more complete understanding of these principles. The concepts are classified as those that relate to polymer structure, those that relate directly to physical properties, and those that relate to the use of the material, as listed below.

1. Polymer structure:
   (a) Macrostructure
       • Molecular weight
       • Molecular weight distribution
       • Branching
   (b) Microstructure
   (c) Network structure

2. Rubber properties:
   (a) Elasticity—the retractive force
   (b) Glass transition temperature
   (c) Crystallinity

3. Rubber use:
   (a) Compounding
   (b) Processing

POLYMER STRUCTURE

Macrostructure

Molecular Weight. The single most important property of any polymer is the size or length of the molecule. The polymerization process consists of combining many of the simple monomer molecules into the polymer chain. Most of the monomers used to produce rubbers are either gases or low-viscosity liquids under normal conditions; upon polymerization they form liquids whose viscosity increases to extremely high values as the chain length is increased. At very low chain length, this increase is linear with molecular weight until the chains are long enough to become entangled. Above the entanglement molecular weight, the viscosity increases to the 3.4–3.5th power of molecular weight increase.[4] In addition to viscosity, a great many other physical properties of any polymer depend upon the molecular weight.[5]

Molecular Weight Distribution. A given polymer sample is composed of many polymer chains, which in most cases are not of the same length. This variability can be a result of the synthesis process or of possible random scission and cross-linking that can occur upon processing. For economic reasons, it is not possible to separate the various polymer chains by length prior to use; so it is important to characterize this distribution in order to describe the polymer and understand its performance. As with any distribution, no single number is a totally satisfactory descriptor.

The commonly used molecular weight parameters are the number, weight, and *z* average molecular weight, which are defined, respectively, as:[6]

$$M_n = \frac{\sum_i N_i M_i}{\sum_i N_i}$$

$$M_w = \frac{\sum_i N_i M_i^2}{\sum_i N_i M_i}$$

$$M_z = \frac{\sum_i N_i M_i^3}{\sum_i N_i M_i^2}$$

where N_i is the number of moles of species i, and M_i is the molecular weight of species i.

Although there are many different statistical ways to describe any population, the above parameters have been widely used because they are readily understood in physical terms, and they can be measured directly in the laboratory.[7] A fourth parameter, the dispersion index, frequently is used to characterize the breadth of the distribution. This parameter is simply the ratio of the weight to the number average molecular weight, with 1.0 being the lowest possible number (i.e., all chains of exactly the same length). Typical values for commercial polymers are in the 2–5 range, with those under 2 considered relatively narrow and those over 2.5 considered broad in distribution. The measurement of these molecular weight averages once was a time-consuming task, but with the development of gel permeation chromatography (GPC), also referred to as size exclusion chromatography, the measurement of these distributions has become commonplace.[8] Units are even available that automatically sample polymerization reactors, process the sample, and perform all necessary calculations to provide data for process control. These units can have multiple detectors, thereby providing compositional distribution as a function of molecular weight.

Branching. The concept of a polymer chain implies two ends per chain. However, because of the nature of the process used to form the polymer, the chain may contain one or more branch points, resulting in multiple ends per chain. These chain ends can have an adverse effect on polymer performance. Branching, molecular weight, and molecular weight distribution have been shown to affect processability as well.[9] The optimum macrostructure often represents a compromise between processing and ultimate performance. Branching can also be measured using the GPC technique with special detectors.

Microstructure

In the formation of elastomers from diolefin monomers such as butadiene or isoprene, there are a number of possible structures. Since the control of these structures is critical in obtaining optimum properties, this area has received great attention from the synthesis chemist. The possible polyisoprene structures are:

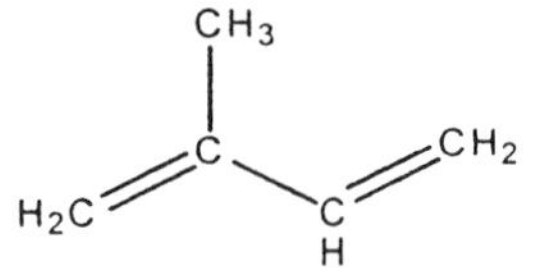

chain mobility, polymer composition also defines the solubility parameter of the polymer, which is a critical property relative to the type of solvents in which the polymer is soluble, the ability of the polymer to accept and hold oil, and the relative compatibility of the polymer with other polymers.[11] Basically, these properties all involve polymer–solvent

isoprene

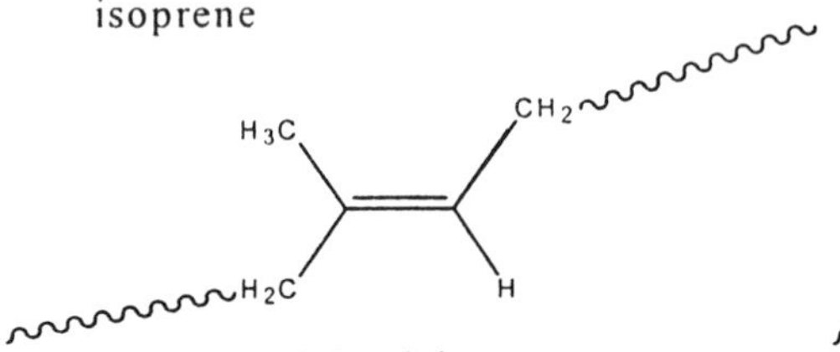

trans-1,4 polyisoprene

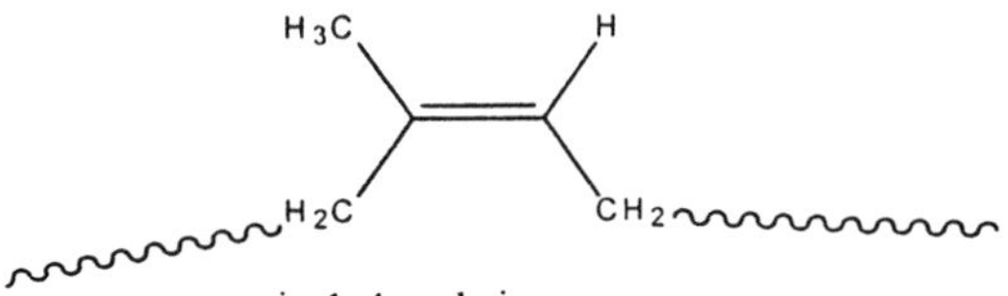

cis-1,4 polyisoprene

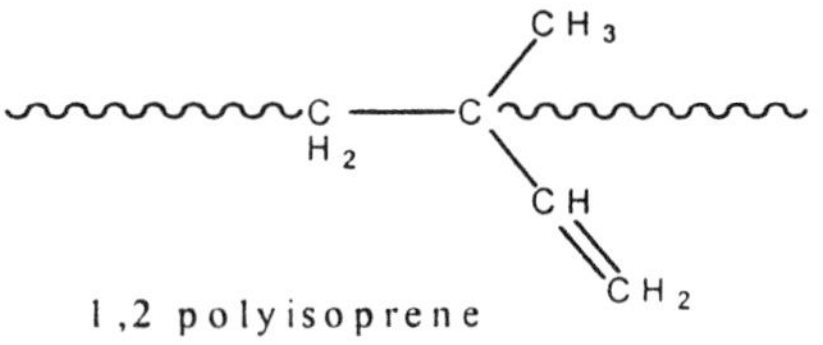

1,2 polyisoprene

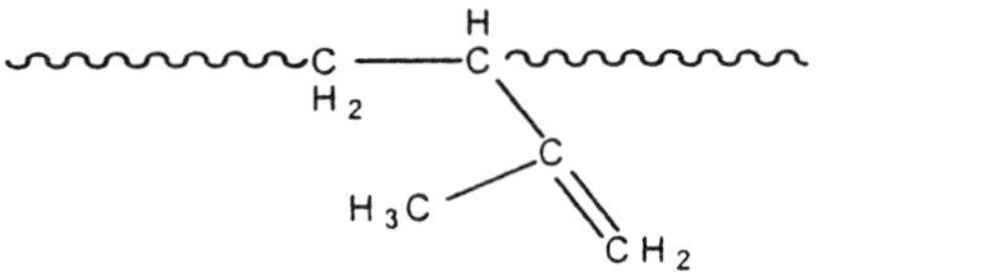

3,4 polyisoprene

For butadiene (no methyl group) the 3,4 form does not exist. The 1,2 addition is referred to as *vinyl addition*.

The polymer in natural rubber (from the *Hevea brasiliensis* tree) is pure *cis* polyisoprene; gutta percha and balata are composed of the *trans* isomer.

Many of the commercial synthetic elastomers are synthesized from more than one monomer, such as styrene–butadiene and ethylene–propylene rubbers. The properties of the resultant polymer depend on the ratio of the two monomers in the polymer and upon the distribution of the monomers within the chain.

If the monomers are uniformly distributed within the polymer chain, the ratio of monomers will define the flexibility of the polymer chain.[10] Because many properties depend on this chain mobility, polymer composition is carefully controlled. In addition to

interaction, with the difference being the increasing molecular weight of the solvent.

In addition to the relative ratio of the monomers, the arrangement of the units in the chain is important. This arrangement is referred to as the copolymer *sequence distribution*. In the previous discussion, the assumption was made that the comonomer units were well mixed in the polymer chain. If this is not the case, parts of the chain can reflect properties of the corresponding homopolymer. It is thus possible to produce polymers that have significantly different properties in different parts of the polymer chain. A most dramatic example of this can be found in styrene–butadiene–styrene or styrene–isoprene–styrene thermoplastic elastomers. The properties of these unique materials will be discussed in the section "Thermoplastic Elastomers."

Network Structure

A critical requirement for obtaining engineering properties from a rubbery material is its existence in a network structure. Charles Goodyear's discovery of vulcanization changed natural rubber from a material that became sticky when hot and brittle when cold into a material that could be used over a wide range of conditions. Basically, he had found a way to chemically connect the individual polymer chains into a three-dimensional network. Chains that previously could flow past one another under stress now had only limited extensibility, which allowed for the support of considerable stress and retraction upon release of the stress. The terms "vulcanization," "rubber cure," and "cross-linking" all refer to the same general phenomenon.

For most rubber systems the network is formed after the polymer is compounded and molded into the desired final shape. Once cross-linked, the material no longer can be processed. If cross-linking occurs prior to compounding or molding, the material is referred to as *gelled*, and it cannot be used. Most rubber is used in a compounded and cured form. There is an optimum *cross-link density* for many failure properties such as tensile strength and tear which will be discussed in the next section.

The one general class of polymers that fall outside this concept is the thermoplastic elastomers, which will also be discussed later.

RUBBER PROPERTIES

Elasticity—The Retractive Force

The fact that cross-linked rubber materials can be extended to several times their original length and return to that original length when released is certainly their most striking feature. This is in contrast to crystalline solids and glasses, which cannot normally be extended to more than a fraction of their original length and also to ductile metals which can be extended to large deformations but do not return to the original length after the stress is removed.

There have been both statistical and thermodynamic approaches to solving the problem of rubber elasticity leading to a phenomenological treatment; however, these methods are beyond the scope of this chapter. The important, and most interesting, result of these theories is that rubber elasticity arises from changes in entropy of the network. Rubber molecules are capable of geometric isomerization, examples of which are *cis* and *trans*, just like any other unsaturated organic molecule. Upon stretching the network of chains, no "configurational" changes take place. That is, *cis* is not changed to *trans* and thus there is no configurational contribution to entropy. Elasticity *does not* arise from stretching or deforming covalent bonds either. What does take place are rotations about single bonds in the chain backbone during stretching. It is these "conformational" changes which give rise to the entropy decrease upon stretching. The chains do not like being in a stretched state because there are many conformational states not available to them until the stress is released. The chain ends are held apart at a statistically unfavorable distance, which gives rise to the entropy decrease. The following expression, for extensions >10%, shows that the elastic force, f, is directly proportional to the absolute temperature, or the elastic response of the rubber is entirely governed by the decrease in entropy (S) which it undergoes upon extension.[12] The term ∂L is the change in length of the sample:

$$f = -T \left[\frac{\partial S}{\partial L} \right]_{T,V}$$

This expression was derived for constant temperature and volume experiments.

The shear modulus of the rubber network is related to the molecular weight between cross-link points or M_c. The lower the molecular weight of chains between cross-links (network chains), the higher the cross-link density and the higher the modulus. This is shown in the following expression:

$$G_0 = \frac{\rho RT}{M_c}$$

where G_0 is the elastic shear modulus, ρ is the density, R is the gas constant, and T is absolute temperature. There is an optimum cross-link density for ultimate strength properties, above which the highly cross-linked network no longer can dissipate strain energy in the form of heat (hysteresis) so all the energy goes into breaking network chains and the material becomes brittle. Below this optimum cross-link level the material simply has too much viscous flow and pulls apart easily. Therefore, the rubber chemist must optimize the state of cure if high strength is desired.

Glass Transition Temperature

In order for a polymer to behave as a rubbery material, it is necessary for the chain to have great mobility. As the temperature is lowered, the ability of the chain segments to move decreases until a temperature is reached where any large-scale motion is prevented. This temperature is referred to as the glass transition temperature (T_g). Below this temperature the rubber becomes a glassy material—hard and brittle. Above this temperature amorphous plastics, such as polystyrene, can exhibit a rubbery character if the molecular weight is sufficiently high. All rubbery materials, then, must have glass transition temperatures below room temperature. For good low-temperature properties, it obviously follows that a low glass transition temperature polymer is required. The control of the glass transition temperature of the polymer is critical for many properties other than low-temperature use. For example, the wet traction and wear of a passenger tire have been shown to greatly depend directly on the polymer glass transition temperature.[13]

Control of this property is possible by controlling the structure of the polymer chain. Monomers with bulky side groups restrict chain mobility and thus raise the glass transition temperature. The composition of copolymers and the ratio of polymer blends often are determined by the desired glass transition temperature of the final product.

The glass transition temperature is usually measured using thermal methods such as differential scanning calorimetry (DSC) which looks at the change in heat capacity when a material goes through its glass transition. Another very useful method is to use dynamic mechanical properties where the polymer is subjected to a temperature sweep in a dynamic mechanical spectrometer from very low temperature $(-120°C)$ to well above room temperature $(+100°C)$. If the modulus is plotted vs. temperature, there will be a sharp decrease as the polymer warms to above its T_g. There will also be a peak in the energy loss property known as tan delta. The tan delta vs. temperature plot is very useful for predicting tire properties such as rolling resistance and wet traction.

Crystallinity

Polymer chains can exist in spatial arrangements that are orderly enough to allow the chains to form crystalline structures. The existence of strong interchain interactions via polar forces, hydrogen bonding, or ionic groups can facilitate crystallization. The existence of crystallization is very important for many plastics and fibers, but crystallinity cannot be appreciably present in rubbery materials, as the corresponding restriction in chain mobility could preclude the very chain mobility needed for rubbery properties. Rubbery materials must have both their melting temperature (if they have a crystalline point) and their glass transition temperature below the use temperature.

Crystallinity can be reduced by disruption of the order in the chain by copolymerization.[14] For example, both polyethylene and polypropylene are crystalline plastics, whereas ethylene–propylene rubber produced at about a 50:50 ratio is an amorphous elastomer. Compositional excursions much outside this range lead to crystalline materials.[15] For some materials, such as natural rubber, that are close to crystallizing, stretching the chains can align them sufficiently for crystallization to occur. Such polymers can exhibit excellent gum properties and improved strength in the uncured state that greatly facilitate processing.

Attempts have been made with some success to produce other polymers that exhibit this property of natural rubber. Although the melting temperature can be matched by appropriately disrupting the crystallizable structure through controlled introduction of another monomer, an exact match is not possible because the extent of crystallinity and the kinetics of crystallization will differ.

Crystallinity can be measured using the same thermal and dynamic mechanical methods described for measuring T_g however, the melting transition is much sharper than the T_g because it is a first-order transition compared with second-order for the T_g.

RUBBER USE

Compounding

The rubber industry began when Charles Goodyear developed the first useful rubber compound: natural rubber plus sulfur. The concept of mixing materials into rubber to improve performance is still of primary importance today. Without compounding, few rubbers would be of any commercial value. Any given rubber application will have a long list of necessary criteria in addition to cost, encompassing appearance, processing, mechanical, electrical, chemical, and thermal properties. Developing such compounds requires a broad knowledge of material science and chemistry combined with experience. The use of designed experiments can greatly facilitate selecting the optimum compound formulation.

The major components in a compound are curatives, reinforcing agents, fillers, plasticizers, and antidegradants.

Curatives. The function of curatives is to cross-link the polymer chains into a network; the most common ones are the sulfur type for unsaturated rubber and peroxides for saturated polymers. Chemicals called accelerators may be added to control the cure rate in the sulfur system; these materials generally are complex organic chemicals containing sulfur and nitrogen

atoms. Stearic acid and zinc oxide usually are added to activate these accelerators. Metal oxides are used to cure halogenated polymers such as polychloroprene or chlorosulfonated polyethylene.

Reinforcing Agents. Carbon black and silica are the most common reinforcing agents. These materials improve properties such as tensile strength and tear strength; also, they increase hardness, stiffness, and density and reduce cost. Almost all rubbers require reinforcement to obtain acceptable use properties. The size of the particles, how they may be interconnected (structure), and the chemical activity of the surface are all critical properties for reinforcing agents. In tire applications, new polymers are currently being developed which contain functional groups that directly interact with carbon black and silica, improving many properties.

Fillers. Fillers are added to reduce cost, increase hardness, and color the compound. Generally they do not provide the dramatic improvement in properties seen with reinforcing agents, but they may have some reinforcing capability. Typical fillers are clays, calcium carbonate, and titanium dioxide.

Plasticizers. These materials are added to reduce the hardness of the compound and can reduce the viscosity of the uncured compound to facilitate processes such as mixing and extruding. The most common materials are petroleum-based oils, esters, and fatty acids. Critical properties of these materials are their compatibility with the rubber and their viscosity. Failure to obtain sufficient compatibility will cause the plasticizer to diffuse out of the compound. The oils are classified as aromatic, naphthenic, or paraffinic according to their components. Aromatic oils will be more compatible with styrene–butadiene rubber than paraffinic oils, whereas the inverse will be true for butyl rubber. The aromatic oils are dark colored and thus cannot be used where color is critical, as in the white sidewall of a tire. The naphthenic and paraffinic oils can be colorless and are referred to as nonstaining.

Antidegradents. This group of chemicals is added to prevent undesirable chemical reactions with the polymer network. The most important are the antioxidants, which trap free radicals and prevent chain scission and cross-linking. Antiozonants are added to prevent ozone attack on the rubber, which can lead to the formation and growth of cracks. Antiozonants function by diffusion of the material to the surface of the rubber, thereby providing a protective film. Certain antioxidants have this characteristic, and waxes also are used for this purpose.

Processing

A wide range of processes are used to convert a bale of rubber into a rubber product such as a tire. The first process generally will be compounding. Typical compounding ingredients were discussed previously. In many compounds more than one rubber may be needed to obtain the performance required. Uncured rubber can be considered as a very high-viscosity liquid; it really is a *viscoelastic* material possessing both liquid and elastic properties. Mixing materials into rubber requires high shear, and the simplest method is a double roll mill in which the rubber is shear-mixed along with the other compounding ingredients in the bite of the mill. Large-scale mixing is most commonly done with a high-shear internal mixer called a Banbury. This mixing is a batch process, although continuous internal mixers also are used.

The compounded rubber stock will be further processed for use. The process could be injection or transfer molding into a hot mold where it is cured. Tire curing bladders are made in this fashion. Extrusion of the rubber stock is used to make hose or tire treads and sidewalls. Another common process is calendaring, in which a fabric is passed through rolls where rubber is squeezed into the fabric to make fabric-reinforced rubber sheets for roofing membranes or body plies for tires. The actual construction of the final product can be quite complex. For example, a tire contains many different rubber components some of which are cord or fabric reinforced. All of the components must be assembled with high precision so that the final cured product can operate smoothly at high speeds and last over 50,000 miles.

NATURAL RUBBER

More than 500 years ago, the people of Central and South America were using a product that they collected from certain trees to make balls and to coat fabric to make it waterproof. This material they called cauchuc, which means "weeping wood." Today we know the tree as the *Hevea brasiliensis* and the material as natural rubber. Although a number of plants produce rubber, the only significant commercial source is the *Hevea* tree. Natural rubber initially had only limited applications because it flowed when hot and had poor strength. In 1839 Charles Goodyear found that when combined with sulfur and heated, the material changed into cured rubber with properties much as we know them today. The development of the pneumatic tire in 1845 combined with the dramatic growth of the automotive industry led to a rapid increase in the demand for natural rubber.

Although the tree was indigenous to Brazil, seeds were taken to England where they were germinated, and the plants were sent to the Far East. Rubber plantations were in existence by the late 1800s, and in the 1920s plantations were begun in West Africa. Because of a leaf blight disease, essentially all natural rubber now comes from plantations in Africa and the Far East rather than Central and South America.[16]

The production process starts with the trees. Over the years considerable biological research has been done to produce trees that grow faster, produce more latex, and are resistant to wind and disease damage. Once such an improved tree has been identified, buds are grafted from the tree onto root stock. All such trees are referred to as clones and will have the same characteristics. It typically takes 6–7 years of growth before a tree is ready for rubber recovery. Peak rubber production is reached at 12–15 years of age. Another major development in improving tree performance has been the use of tree stimulants, which has resulted in an overall yield increase of 30 percent without adverse effects on the trees.

The production process of natural rubber in the tree is not yet fully understood. However, it involves a long series of complex biochemical reactions that do not involve isoprene as a monomer, even though the resulting polymer is 100 percent *cis* 1,4 polyisoprene. Because the tree makes the product, the rubber production process is really one of recovery.

The recovery process starts with tapping of the tree, which involves manually removing thin sections of bark at an angle so that as the latex is exuded from the damaged living latex cells, it will flow down to be collected in a cup. The depth of the cut is critical, as a tool shallow cut will not allow optimum latex yield, and too deep a cut will damage the tree. The damaged latex vessels will seal off after several hours. A preservative is added to the collection cup to prevent coagulation of the latex. The tapper then collects the latex and takes it to a collection station, from which it is shipped to the rubber factory. Additional preservative is added at the collection station to assure stable latex. The latex contains around 30–45 percent rubber. Approximately 3 percent of the solid material is nonrubber materials, consisting primarily of proteins, resins, and sugars.

The latex is processed in one of several ways, depending on the desired final product. If it is going to be used in the final product application in the latex form (such as for dipped goods and adhesives), it will be concentrated to 60 percent or higher solids. The most common process is centrifugation. The latex separates into the high-solids product and a low-solids skim material that contains a much higher percentage of the nonrubbery components; rubber produced from skim is generally of a lower quality because of these impurities. A variety of chemicals can be added to the latex to provide the necessary preservation and mechanical stability. As the final use of the latex will involve destabilizing the rubber particles, care must be taken not to overstabilize the latex concentrate. Being a natural product, the latex tends to change upon aging, a factor that also must be compensated for in the process.

Dry rubber is produced from the latex first by dilution, then by coagulation with organic acids, and finally by formation into sheets or crepe. Rubber smoked sheets are made by working the coagulated sheets between rolls to remove as much of the nonrubbery components as possible, followed by drying for up to a week in a smokehouse. The smoke serves as a fungicide that prevents biological attack of the rubber. To provide a more well-defined uniform product, Technically Specified Rubber (TSR) processes have been developed. These processes involve converting the coagulated rubber into rubber crumb, which is further washed, dried, and baled. Constant-viscosity grades of natural rubber have been developed by chemically reacting the aldehyde groups, which otherwise would lead to cross-linking upon storage. In the pale crepe process the latex is carefully selected for colored bodies (from carotene) and treated with sodium bisulfite to stop enzyme activity. The rubber is extensively washed through rollers to remove serum, as this can lead to yellowing. The wet crepe is dried under carefully controlled temperatures and in the absence of light to assure optimum properties. The previously described processes involve considerable investment and are practical only for plantation operations. Small shareholders may allow their latex to coagulate naturally and sell it to processors. Such a product will vary greatly, depending on the specific history of each rubber slab.[17]

Uses

The largest use of natural rubber is in the manufacture of tires. Over 70 percent of its consumption is in this area. The next largest use is as latex in dipped goods, adhesives, rubber thread, and foam. These uses account for approximately another 10 percent. The remainder is used in a variety of applications such as conveyor belts, hoses, gaskets, footwear, and antivibration devices such as engine mounts.

Because of the high stereoregularity of natural rubber, the units in the polymer chain can form very orderly arrangements, which result in crystallization upon storage at low temperatures or upon stretching. Stored crystallized rubber may be converted to its original amorphous state by heating. Several crystalline forms are reported, with melting points varying from 14°C to 36°C.[18] Although crystallization upon storage can be a problem to users, the ability to crystallize reversibly upon stretching accounts for many of the unique properties of natural rubber. Specifically, the ability of natural rubber to be used as a gum polymer (unfilled vulcanizate) depends on this property. The crystallites that form act both as filler and as temporary cross-links, providing high tensile properties.[19] In tire fabrication this property is reflected in natural rubber stocks possessing high green (uncured) strength and excellent building tack. In the final product, strain-induced crystallization provides tear and cracking (cut growth) resistance.

Although many other polymers can crystallize, only natural rubber has been found to have the necessary combination of rate of crystallization, degree of crystallization, and melting point to provide all the properties discussed above. Natural rubber has good flexibility and resilience. In truck and bus tires, it is used extensively with blends of polybutadiene to give the low heat buildup needed as well as wear and cut growth resistance. In passenger tires, natural rubber is used in the sidewalls and carcass areas. These areas require the building tack, ply adhesion, and hot strength properties that it imparts. Synthetic rubber is used almost exclusively in passenger tire treads.

The largest use of natural rubber latex is in the dipped goods area. Products include balloons, surgical and examination gloves, and prophylactics. The rapid spread of AIDS has led to a dramatic increase in the surgical and examination glove market. Latex also is used to make adhesives, rubber thread, and foams. However, natural rubber is being replaced in these two latex areas by urethanes and synthetic rubber latex. Some of the replacement has been driven by skin allergy problems which are caused by the natural proteins in the latex. Although natural rubber latex is

a mature product, research continues on improving its uniformity, stability, and performance.[20]

POLYISOPRENE

Faraday discovered in 1826 that natural rubber was composed of a hydrocarbon with a ratio of five carbons to eight hydrogen atoms; and in 1860, G. Williams isolated isoprene by collecting the distillate from the heating of natural rubber. By 1887, scientists in France, England, and Germany had converted isoprene back into a rubbery material. Because this offered a potential for manufactured "natural" rubber, research was undertaken to find ways to obtain isoprene from sources other than rubber itself.[21]

Monomer Production

The primary source of isoprene today is as a by-product in the production of ethylene via naphtha cracking. A solvent extraction process is employed. Much less isoprene is produced in the crackers than butadiene, so the availability of isoprene is much more limited. Isoprene also may be produced by the catalytic dehydrogenation of amylenes, which are available in C-5 refinery streams. It also can be produced from propylene by a dimerization process, followed by isomerization and steam cracking. A third route involves the use of acetone and acetylene, produced from coal via calcium carbide. The resulting 3-methyl-butyne-3-ol is hydrogenated to methyl butanol and subsequently dehydrogenated to give isoprene. The plants that were built on these last two processes have been shut down, evidently because of the relatively low cost of the extraction route.

Polymer Production Process

The free-radical catalysts were found to produce a product that did not have the tack, green strength, or gum tensile of natural rubber. Whereas natural rubber is an essentially pure *cis*-1,4 structure, the emulsion product was of mixed microstructure. This precluded the ability of the latter to undergo strain-induced crystallization, which is required to obtain many of the desired natural rubber properties.

In 1955 investigators from the Firestone Tire and Rubber Company and the B. F. Goodrich Company announced the synthesis of polyisoprene with over 90 percent cis-1,4 structure. The work at Firestone was based on lithium metal catalysts, whereas the work at Goodrich was the result of using Ziegler–Natta type coordination catalysts.[22,23]

Use

Although considerable interest was generated by these discoveries, their commercial success has been rather limited. The lithium-based polymers were found to produce up to 94 percent cis, which still was not high enough to provide the properties of natural rubber. Polymers made with the coordination catalysts have cis contents of up to 98 percent, providing products that can more closely serve as replacements for natural rubber than the lithium-based polymers. In comparison with natural rubber, they offer the advantage of a more highly pure rubber (no nonrubber material) and excellent uniformity. For economic reasons, polyisoprene has seen only limited success. Several of the plants built to produce polyisoprene have been either shut down or converted for use to produce other polymers. In terms of synthetic rubber production in 2003, only 258,000 metric tons were produced (excluding centrally planned economy countries, CPEC), which represented only 3 percent of total synthetic rubber production. In comparison, in 2003 natural rubber usage was 7,554,000 metric tons.[24] Evidently because of its strategic importance, the Former Soviet Union (FSU) continues to rely heavily on polyisoprene.

STYRENE–BUTADIENE RUBBER

The largest-volume synthetic rubber consumed is styrene–butadiene rubber (SBR). In 2003, SBR solid rubber accounted for 41 percent of all synthetic rubber. If SBR latex and carboxylated SBR latex are included, its share increases to 55 percent. The major application of solid SBR is in the automotive and tire industry, accounting for approximately 70 percent of the use. Therefore, SBR has been tightly tied to the tire business.[25]

Initially, SBR was developed as a general purpose alternate material to natural rubber. In the United States the thrust came early in World War II when the U.S. supply of natural rubber was cut off. The basic technology was developed in Germany in the late 1920s, and by 1939 Germany had 175,000 metric tons of capacity in place. The first U.S. production was 230 metric tons in 1941, but by 1945 there were more than 850,000 metric tons of capacity. Basically, in a period of five years the emulsion SBR business as we know it today was put in place. By 1973 U.S. capacity had increased to almost 1,400,000 metric tons, but in 1989 it had contracted to 881,000 tons close to the 1945 capacity.[26]

Monomer Production

The production of butadiene monomer is discussed below in polybutadiene section "Polybutadiene." The largest volume of styrene is produced by the alkylation of benzene with ethylene to give ethyl benzene, which is then dehydrogenated to give styrene.[27]

benzene   ethylene   ethyl benzene   styrene

Polymer Production Process

SBR is produced by two different processes: emulsion and solution. The emulsion process involves a free-radical mechanism, whereas solution SBR is based on alkyllithium catalysis.

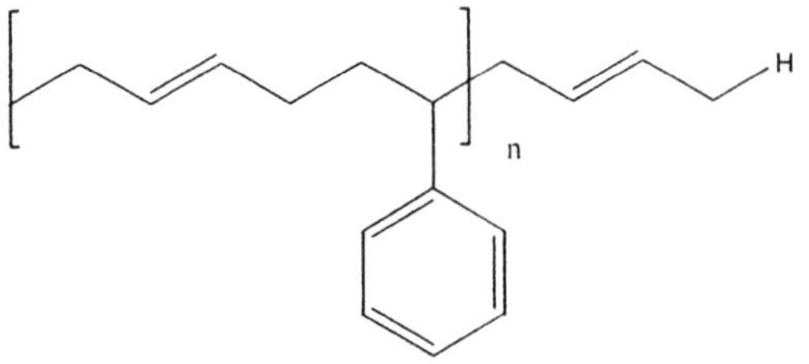

copolymer of butadiene and styrene

Emulsion Process

The formula developed to provide SBR during World War II was standardized, with all rubber plants owned by the U.S. government. The standard recipe is listed below:[28]

Component	Parts by Weight
Butadiene	75
Styrene	25
n-Dodecyl mercaptan	0.5
Potassium peroxydisulfate	0.3
Soap flakes	5.0
Water	180

Initiation occurs through reaction of the persulfate with the mercaptan, as shown below:

$$K_2S_2O_8 + 2RSH \rightarrow 2RS\bullet + KHSO_4$$

$$RS\bullet + M \rightarrow RSM\bullet$$

Chain propagation occurs by the growing chain free radical attacking either the butadiene or styrene monomer. The active radical chain can react with mercaptan to form a new mercaptyl radical and a terminated chain. The mercaptyl radical then can initiate an additional chain. The molecular weight of the chain P can be controlled by the concentration of mercaptan via this chain transfer mechanism.

$$P\bullet + RSH \rightarrow PH + RS\bullet$$

$$RS\bullet + M \rightarrow RSM\bullet$$

Termination also can occur by the reaction of two free radicals, through either combination or disproportionation reactions.

$$P\bullet + P\bullet \rightarrow P\text{-}P \text{ combination (bimolecular coupling)}$$

$$P\bullet + P\bullet \rightarrow P\text{-}CH=CH_2$$
$$+ PH \text{ disproportionation (hydrogen-free radical transfer)}$$

The mercaptyl radical also can react with growing chains, to lead to termination.[29]

Polymerization is initially carried out at 50°C until conversion of 70–75 percent is reached, at which time the polymerization is terminated by the addition of a free-radical scavenger such as hydroquinone. Polymerization beyond this point results in excessive free-radical attack on the polymer chains. Products made under such conditions have poor properties due to excessive branching and gelation. Unreacted butadiene and styrene are removed by flashing and steam stripping. Antioxidant is added to the latex, followed by coagulation with the addition of polyelectrolytes and salt-acid. The coagulated crumb then is washed, dried, baled, wrapped, and packaged for shipment. Because of the soap and other chemicals in the formulation, most emulsion polymers will contain about 7 percent of nonrubber residues. The emulsion process flow sheet is shown in Fig. 16.1.

It was soon discovered that polymers made at lower temperatures had significantly better properties, especially in tire treads. This was mainly because they contained fewer low molecular weight species. A lower temperature process, using "redox" chemistry, was eventually developed. It used peroxides or hydroperoxides with a reducing agent such as a water-soluble transition metal salt which were active even at 0°C.

Mercaptans are also used as chain transfer agents to provide a mechanism for molecular weight control. Commercially these types of polymerization are carried out at 5°C and are referred to as "cold" polymerizations to differentiate them from the previously discussed "hot" systems. A typical formula is listed below.[30]

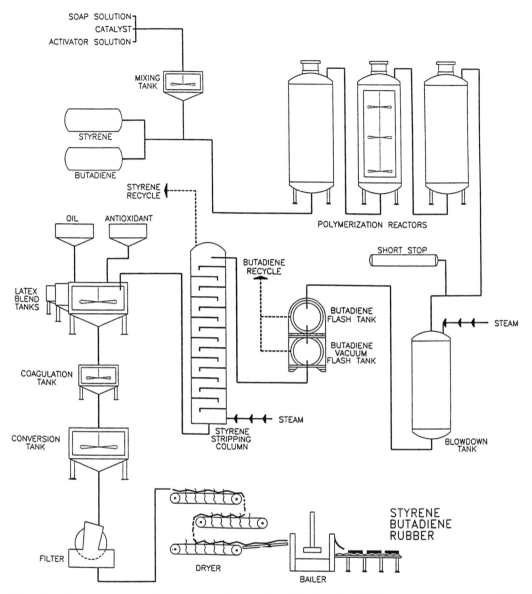

Fig. 16.1. The production of styrene-butadiene rubber. (*Modified by D.F. Graves from a drawing in the Vanderbilt Rubber Handbook, 1990 Edition, Copyright the R.T. Vanderbilt Company, Inc., by permission.*)

Component	Parts by Weight
Butadiene	71
Styrene	29
tert-Dodecyl mercaptan	0.18
p-Methane hydroperoxide	0.08
Ferrous sulfate heptahydrate	0.03
Trisodium phosphate decahydrate	0.50
Tetrasodium ethylenediaminetetraacetate	0.035
Sodium formaldehyde sulfoxylate	0.08
Rosin acid soap	4.5
Water	200

The improved tire wear of cold polymerization SBR led to the very rapid replacement of hot SBR for most applications. This change was relatively easy to make, as all the equipment could be used with the only modification required being the addition of reactor cooling, which is achieved with either the reactor jacket, internal coils, or both.

It later was found that even more improvements could be realized by polymerizing to very high molecular weights and then adding petroleum-based oils to the latex prior to coagulation. The oil is absorbed by the rubber which, upon coagulation, produces oil-extended polymers. For tread applications, oils of higher aromatic content were preferred because of their excellent compatibility with the rubber. Typically 37.5 parts of oil are added, although grades containing up to 50 parts have been produced. Very high molecular weight polymers thus can be processed without requiring excessive energy to mix them. The oil also allows these tough polymers to be processed without excessive degradation. Carbon black masterbatches also are produced. In this process carbon black is added to the latex prior to coagulation, and the black, along with oil, is incorporated into the latex in the coagulation step. These products offer the user the advantage of not having to handle free black in their mixing operation, and can provide additional compounding volume for manufacturers with limited mixing capacity.

Not all emulsion SBR is converted to dry rubber for use. There is a variety of applications where the latex can be used directly in the final fabrication process. This technology logically grew out of the latex technology developed for natural rubber. For latex applications the particle size distribution can be critical because of its effect on viscosity and performance variables, as when used to provide impact strength in plastics such as ABS.[31] Careful control of the mechanical stability of the latex also is critical, as these systems must destabilize under relatively mild conditions such as those in a coating operation. A number of processes have been developed to control particle size via partial destabilization of the latex. Among the commercial methods are: careful control of a freeze–thaw cycle, controlled shear agitators, high-pressure colloid mills, and the addition of chemicals such as hydrocarbons or glycols.

A special variation of SBR latex containing terpolymerized vinyl pyridine is used in the tire industry to provide adhesion of organic fiber tire cords to rubber stock. The vinyl pyridine SBR latex is combined with resins and coated on the fiber by a dipping process. The adhesive is set by a controlled temperature and tension process to control the shrinkage properties of the cord.

Solution Process

The discovery of the ability of lithium-based catalysts to polymerize isoprene to give a high *cis* 1,4 polyisoprene was rapidly followed by the development of alkyllithium-based polybutadiene. The first commercial plant was built by the Firestone Tire and Rubber Company in 1960. Within a few years the technology was expanded to butadiene–styrene copolymers, with commercial production under way toward the end of the 1960s.

The copolymerization with alkyllithium to produce uniformly random copolymers is more complex for the solution process than for emulsion because of the tendency for the styrene to form blocks. Because of the extremely high rate of reaction of the styryllithium anion with butadiene, the polymerization very heavily favors the incorporation of butadiene units as long as reasonable concentrations of butadiene are present. This observation initially was somewhat confusing because the homopolymerization rate of styrene is seven times that for butadiene. However, the cross-propagation rate is orders of magnitude faster than either, and it therefore dominates the system. For a 30 mole percent styrene charge the initial polymer will be almost pure butadiene until most of the butadiene is polymerized. Typically two-thirds of the styrene charged will be found as a block of polystyrene at the tail end of the polymer chain:

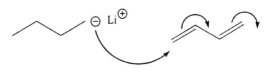

anionic initiation of butadiene by butyllithium

Several methods have been proposed to overcome this problem. In one, the styrene and part of the butadiene are charged initially with butadiene metered at a rate equivalent to its incorporation into the chain. A second approach involves adding both monomers at a relatively slow rate so that the equilibrium monomer concentration reaches a pseudosteady state that will produce polymer at the desired composition.[32] This process can be done in either a batch or a continuous mode.[33]

In addition to these reaction engineering approaches to produce uniformly random copolymer, the chemistry may be changed by the addition of polar agents such as amines and ethers. This action results in bringing the reactivity rates much closer together. The change in chemistry also is reflected in the microstructure of the butadiene portion of the polymer. Whereas in the nonpolar system the vinyl content of the butadiene portion is around 10 percent, in polar systems vinyl contents of 30–40 percent typically are obtained when a 20 percent styrene polymer is randomized. Higher styrene contents require higher modifier levels, resulting in even higher vinyl contents. An added complication with polar modifiers is their ability to react with the growing chain, resulting in undesired termination.

In spite of these complications, all recent U.S. expansions or announced plants for SBR have been for solution polymers. The ability to better design the polymer structure and produce special functional polymers (described below) accounts for most of this shift.

Functional Solution SBR

The driving force toward functional solution SBR is its improved hysteresis properties for passenger tire treads. The fact that anionic SBR has a much more narrow molecular weight distribution compared with emulsion gives it lower hysteresis. However, the big advantage is the relatively stable growing chain ends which can be chemically modified to improve interaction with carbon black and silica in tire compounds.[34] This modification can lead to a dramatic reduction in rolling resistance, which is critical for automotive manufacturers who must meet government-mandated fuel economy targets. The most active functional end-groups contain either organotin or certain amines. Termination with tin

"live SBR"   tin-coupled SBR

carbon black   quinone functionality on CB surface

tetrachloride is the easiest and most popular method which generates a four-armed star polymer. The polymer–tin bonds break down during mixing of the compound and both lower the compound viscosity and create active sites for reaction with carbon black surfaces. The creation of this "carbon-bound rubber" effectively prevents the carbon black from agglomerating on a microscale to form hysteretic, three-dimensional networks. The breaking of these networks during the deformation of a tire tread is a major source of rolling resistance. Polymers with amine end-groups also show good activity with carbon black. Termination with silane esters is usually used to obtain interaction with silica fillers producing the same decrease in hysteresis. There have recently been efforts to make functional anionic initiators, some of which could be used to make low hysteresis rubber.[35] However, the main use for such polymers have been in adhesives and other nontire applications. One study comparing an emulsion polymer, its solution counterpart, and a chemically modified version of the solution polymer showed a 23 percent hysteresis reduction in going from emulsion to solution and an additional 15 percent reduction for the chemically modified polymer, to provide an overall reduction of 38 percent.[36] To date, this type of chemical modification is only possible using anionic techniques.

POLYBUTADIENE (BR)

Next to SBR, polybutadiene is the largest volume synthetic rubber produced. Consumption was approximately 2,018,000 metric tons in 2003.[37]

Monomer Production

Butadiene monomer can be produced by a number of different processes. The dominant method of production is as a by-product from the steam cracking of naphtha to produce ethylene. The butadiene is recovered from the C-4 fractions by extractive distil-

lation.[38] "On-purpose" butadiene is generally produced by dehydrogenation or oxidative dehydrogenation of four-carbon hydrocarbons.[39,40]

Polymer Production Process

Polybutadiene is usually produced by alkali metal, and transition metal coordination solution processes. Most production is based on the solution processes because of the ability to obtain preferred microstructures by these routes.

Alkali-metal-based polymerization (usually organolithium) produces a product with about 36 percent *cis*, 54 percent *trans*, and 10 percent vinyl. The polymerization process is conducted in an aliphatic hydrocarbon under an inert atmosphere in either a batch or a continuous mode. Because of the characteristics of this polymerization system, polymers of extremely narrow molecular weight distribution and low gel can be produced.[41] The narrowest distribution is produced via batch polymerization. Coupled (star branched) and end-functional polybutadienes are possible using organolithium technology due to the living anion on the chain end which is available for further reactions. Upon the addition of polar agents, such as ethers or amines, the organolithium initiators can produce polybutadienes with vinyl contents up to 100 percent.[42] The vinyl content can be controlled by the ratio of modifier to catalyst and the polymerization temperature, with lower temperatures favoring increased vinyl formation. Even with high vinyl contents such polymers do not crystallize because of the atactic nature of the vinyl units. High *cis* polybutadiene is produced via solution processes using Ziegler–Natta type transition metal catalysts. The major commercial catalysts of this type are based on titanium, cobalt, nickel, and neodymium.[43] Typically the transition metal is used in the form of a soluble metal salt, which can react with an organoaluminum or organoaluminum halide as a reducing agent to give the active species. Because of the active nature of transition metals, the polymer solutions are treated to deactivate or remove such materials

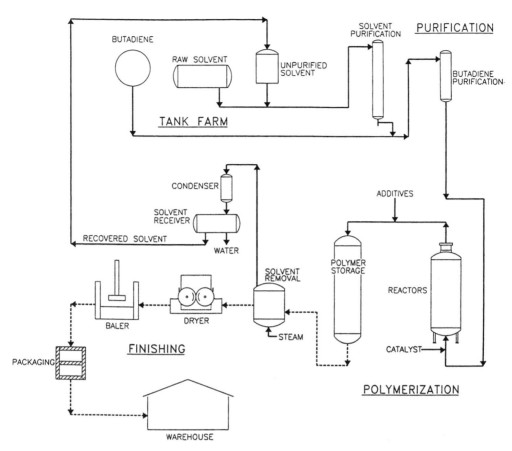

Fig. 16.2. Flow diagram for a typical solution process for the manufacture of polybutadiene. (*Courtesy of Firestone Polymers, Akron, Ohio.*)

from the final product. All of these catalysts produce products with 90 percent or higher *cis* content. The neodymium system is reported to produce the highest *cis* (98–99%) with the most linear chain structure.[44] The highest branched *cis* BR is produced with the Co system with Ni giving intermediate branching.

All the solution processes require high efficiency in recovering the solvent. The most widely used process consists of termination of the polymerization and the addition of antioxidant to the polymer solution. The solution may be treated to remove catalyst residue and then transferred into an agitated steam stripping vessel in which unreacted monomer and solvent are flashed off, leaving the rubber as a crumb slurry in water. The water–crumb slurry then is dewatered and dried. The recovered monomer/solvent is recirculated to a

series of distillation columns to recover monomer and purify the solvent. As both the anionic and the coordination catalyst systems are highly sensitive to impurities such as water, the purification system is very critical for satisfactory process control (Fig. 16.2).

Uses

The major use for polybutadiene is in tires, with over 70 percent of the polymer produced used by the tire industry. Cured polybutadiene has excellent low-temperature properties, high resiliency, and good abrasion resistance due to its low glass transition temperature. However, this same fundamental property also leads to very poor wet skid resistance. For this reason, polybutadiene is blended with other polymers such as natural rubber and

SBR for use in tread compounds. In general, polybutadiene is a poorer-processing polymer than SBR, but this is generally not a problem as it is blended with other polymers in use. The very high *cis* polymers have the potential for strain-induced crystallization, which can lead to improved green strength and increased cut growth resistance in the cured product. High *cis* polybutadiene is reported to have a melting point of 6°C.[45]

The other major use for polybutadiene is as an impact modifier in plastics, in particular high impact polystyrene (HIPS) and acrylonitrile–butadiene–styrene resin (ABS). In the HIPS application the rubber is dissolved in the styrene monomer, which is then polymerized via a free-radical mechanism. A complex series of phase changes occurs, resulting in small rubber particles containing even smaller polystyrene particles being incorporated into a polystyrene matrix. The rubber is added to increase impact strength. Because of the unique morphology that is formed, low levels of rubber (typically around 7%) provide rubbery particles having a volume fraction of 30–40 percent. This morphology leads to high impact at very low rubber levels, providing good stiffness and hardness.[46]

There is also a fairly large market for high *cis* BR in solid core golf balls. In this application, the polymer is compounded with zinc acrylate and the mixture is cured with peroxide.[47] This produces an ionically cross-linked compound that has outstanding resilience. The covers are also ionomers with superior cut resistance. In the last few years the golf ball market has been shifting away from the traditional wound ball to these new solid core balls that use polybutadiene.

ETHYLENE–PROPYLENE RUBBER

There are two general types of polymers based on ethylene and propylene: ethylene–propylene rubber (EPM) and ethylene–propylene terpolymer (EPDM). EPM accounts for approximately 20 percent of the polyolefin rubber produced. Comprising a totally saturated polymer, these materials require free-radical sources to cross-link. EPDM was developed to overcome this cure

limitation. For EPDM a small amount (less than 15%) of a nonconjugated diene is terpolymerized into the polymer. One of the olefinic groups is incorporated into the chain, leaving its other unsaturated site free for vulcanization. This ensures that the polymer backbone remains saturated, with corresponding stability, while still providing the reactive side group necessary for conventional cure systems. The nonconjugated dienes used commercially are ethylidene norbornene, 1,4 hexadiene, and dicyclopentadiene. The selection of the termonomer is made on the basis of the reactivity of the termonomer, both in polymerization and in vulcanization. The estimated 2003 worldwide consumption (excluding CPEQ) was 883,000 metric tons.

Monomer Production

Ethylene and propylene are produced primarily by the cracking of naphtha. They also are available from the fractionation of natural gas. Ethylidene norbornene is produced by reacting butadiene with cyclopentadiene. 1,4 Hexadiene is produced from butadiene and ethylene. Dicyclopentadiene is obtained as a by-product from the cracking of heavy feedstocks to produce ethylene.

Polymer Production

There are two processes used to produce EPM/EPDM: solution and suspension. In either case a Ziegler–Natta type catalyst is used (aluminum alkyl or aluminum alkyl chlorides and a transition metal salt). The most generally used transition metal is vanadium in the form of the tetrachloride or the oxytrichloride.[48] The solution process is similar to that used for other solution polymers. The polymer cement can be finished by stream stripping and drying of the resulting crumb.[49]

In the suspension process, the polymer is suspended in the monomer propylene. This process offers the advantages of being able to operate at higher solids owing to the lower viscosity of a suspension compared with a solution at comparable solids. Other advantages are simple heat removal by the evaporative cooling of the propylene, more uniform reactor temperature profile, and ease of production of high molecular weight or semi-crystalline polymers.[50]

A specially developed titanium-based catalyst has been used in the suspension process for EPM and EPDM where the termonomer is low-boiling. The advantages claimed, in addition to those characteristic of the suspension process, are better structural control and high catalyst efficiency, resulting in a high-purity product without requiring catalyst removal.[51]

The polymer composition for both EPM and EPDM is usually in the 40/60 to 60/40 ethylene/propylene ratio. Outside these ranges, the polymer will start to crystallize because of either polyethylene or polypropylene blocks.

Use

EPM/EPDM polymers exhibit outstanding resistance to heat, ozone, oxidation, weathering, and aging due to the saturated backbone. They have low density, are miscible with aliphatic and naphthenic oils, and maintain acceptable properties at high filler loadings. They are used in single-ply roofing, wire and cable, automotive parts, impact modification of polypropylene, and viscosity index additives for automotive oils. They also can be used in producing thermoplastic olefin elastomers by blending with polypropylene, which may be partially grafted or cross-linked by dynamic vulcanization. These "polymer alloys" will be discussed in the section "Thermoplastic Elastomers." Although at one time EPDM was expected to become the major polymer for tires, this market has not materialized for a variety of processing and performance reasons.[52]

BUTYL RUBBER

Butyl rubber is one of the older synthetic rubbers, having been developed in 1937. Because of the saturated nature of a poly-olefin elastomer, the commercial polymer is actually a copolymer of isobutylene and isoprene. The isoprene is added to provide cure sites. In addition, halogenated (bromo or chloro) derivatives are available.

The halogenated products improve the mixing and cure compatibility with the more common unsaturated rubbers such as natural or styrene–butadiene rubber.

Monomer Production

Isobutylene is obtained as a by-product from petroleum and natural gas plants. The monomer must be highly purified to assure high molecular weight.

Production Process

Butyl rubber is produced at very low temperature (below $-90°C$) to control the rapid exotherm, and to provide high molecular weight. The process consists of charging isobutylene along with isoprene (2–4%) with an inert diluent such as methyl chloride to a reactor to which a Friedel–Crafts catalyst is added. The polymerization is very rapid, and the polymer forms in a crumb or slurry in the diluent. Heat is removed via the reactor jacket. The slurry is steam-stripped to remove all volatiles. The catalyst is neutralized, and antioxidants are added to the slurry prior to drying.[53] The halogenated derivatives are produced by the direct addition of the halogen to a solution of the isobutylene–isoprene polymer.

During the last 10 years another type of butyl rubber was developed which is derived from a copolymer of isobutylene and p-methylstyrene.[54] They are subsequently brominated to varying degrees producing different grades of the elastomer. Bromination occurs selectively on the methyl group of the p-methylstyrene providing reactive benzylic bromine functionality, which can be used for grafting and curing reactions.

brominated poly(-isobutylene-p-methylstyrene)

Properties and Use

The most important characteristics of butyl rubber are its low permeability to air and its thermal stability. These properties account for its major uses in inner tubes, tire inner liners, and tire curing bladders. Because of the poor compatibility of butyl with other rubbers (with respect to both solubility and cure), the halobutyls are preferred. The brominated *p*-methylstyrene-containing butyl rubbers are used in a number of grafting reactions for tire applications and adhesives. Other uses for butyl rubber are automotive mechanical parts (due to the high damping characteristics of butyl), mastics, and sealants.[55]

NITRILE RUBBER

Nitrile rubber was invented at about the same time as SBR in the German program to find substitutes for natural rubber.[56] These rubbers are copolymers of acrylonitrile–butadiene, containing from 15 to 40 percent acrylonitrile. The major applications for this material are in areas requiring oil and solvent resistance. The estimated worldwide consumption in 2003 was 303,000 metric tons.[57]

Monomer Production

The production of butadiene is discussed in the diene section "Polybutadiene." Although several routes have been developed to produce acrylonitrile, almost all now is produced by the catalytic fluidized-bed ammoxidation of propylene.

Polymer Production

The polymerization process parallels the emulsion process used for styrene–butadiene rubber. Either a hot or a cold process can be used, with the cold polymerization providing the same improved processing and vulcanizate properties as seen in SBR. Polymerizations are carried to 70–80 percent conversion and terminated to avoid gel formation. The latex must be stripped to remove unreacted butadiene and acrylonitrile.

butadiene-acrylonitrile copolymer

Properties and Use

As the acrylonitrile content increases in the polymer chain, the properties change predictably. The glass transition temperature increases approximately 1.5°C for each percent increase in acrylonitrile. Properties such as hysteresis loss, resilience, and low-temperature flexibility will correspondingly change. The oil resistance increases with increased acrylonitrile content, as does the compatibility with polar plastics such as PVC. The major market for nitrile rubber is in the automotive area because of its solvent and oil resistance. Major end uses are for hoses, fuel lines, O-rings, gaskets, and seals. In blends with PVC and ABS, nitrile rubber acts as an impact modifier. Some nitrile rubber is sold in latex form for the production of grease-resistant tapes, gasketing material, and abrasive papers. Latex also is used to produce solvent-resistant gloves.[58]

HYDROGENATED NITRILE RUBBER

During the last 15 years several companies have developed hydrogenated grades of nitrile rubber to both improve its thermal stability and solvent resistance. Although the hydrogenation of a polydiene backbone was done as early as the 1920s, real commercial products with acrylonitrile were not introduced until the mid-1980s.[59]

Hydrogenated NBR (HNBR) is produced by first making an emulsion-polymerized NBR using standard techniques. It then must be dissolved in a solvent and hydrogenated using a noble metal catalyst at a precise temperature and pressure.[60] Almost all the butadiene units become saturated to produce an ethylene–butadiene–acrylonitrile terpolymer.

These "post-polymerization" reactions are very expensive so HNBRs usually command a premium price. HNBR is usually cured with peroxides, similar to ethylene–propylene elastomers, because it has no unsaturation for a conventional sulfur cure system.

Uses

HNBR has many uses in the oil-field, including down hole packers and blow-out preventers, because of its outstanding oil resistance and thermal stability. For the same reasons, it has also found uses in various automotive seals, O-rings, timing belts, and gaskets. Resistance to gasoline and aging make HNBR ideal for fuel-line hose, fuel-pump and fuel-injection components, diaphragms, as well as emission-control systems.

CHLOROPRENE RUBBER

Chloroprene rubber (Neoprene—trade name of DuPont) was one of the earliest synthetic rubbers, first commercialized in 1932. It has a wide range of useful properties but has not become a true general purpose synthetic rubber, probably because of its cost. It does possess properties superior to those of a number of general purpose polymers, such as oil, ozone, and heat resistance; but for these properties other specialized polymers excel. Polychloroprene thus is positioned between the general purpose elastomers and the specialty rubbers.

Monomer Production

Chloroprene monomer production starts with the catalytic conversion of acetylene to monovinylacetylene, which is purified and subsequently reacts with aqueous hydrogen chloride solution containing cuprous chloride and ammonium chloride to give chloroprene.[61]

Production Process

Polychloroprene is produced by using an emulsion process. Two general types of processes are used: sulfur modified and unmodified. In the sulfur modified process, sulfur is dissolved in monomer and is incorporated into the polymer chain. Upon the addition of thiuram disulfide-type materials and under alkaline conditions, some of the sulfur bonds are evidently cleaved to give the soluble polymer. In the unmodified process chain transfer agents are used. If neither the sulfur modified nor the chain transfer system is used, the resulting polymer is a gelled tough material. Typical polymerization systems consist of rosin acid soap emulsifier and persulfate catalyst. Conversions of 80–90 percent are obtained. Polymerizations are run at around 40°C. For the modified polymer, the thiuram disulfide is added after polymerization, and the latex is aged to allow the peptization (chain scission) reaction to occur. Acidification stops the peptization reaction. The latex is vacuum-stripped and coagulated using a cold drum dryer process. The coagulated rubber is washed and dried.

The polymerization produces primarily *trans*-1,4-polychloroprene. The *trans* content can be increased somewhat by lowering the polymerization temperature:[62,63]

polychloroprene

Properties and Uses

Polychloroprene is stable to oxidation and ozone. It also is flame resistant, and its oil resistance is better than that of general purpose rubbers. Its major disadvantage, other than cost, is relatively poor low-temperature properties. Because of the high stereoregularity, polychloroprene will strain crystallize, giving good tensile to unfilled stocks. At low temperatures the polymer can crystallize, making processing more difficult. Polymer made at lower temperatures will show higher unfilled tensile properties and more rapid crystallization due to the higher *trans* content (i.e., less disruption of the crystal structure). The major end uses are conveyor belts, V-belts, hoses, and mechanical goods such as wire insulation, O-rings, and gaskets. It also has found use in single-ply roofing and adhesives.[64]

SILICONE ELASTOMERS

Silicone elastomers represent a rather unique group of polymers in that they consist of alternating silicon–oxygen bonds to form the polymer chain backbone. Side groups off the silicon atoms are selected to provide very specific properties that differentiate one type from another. The most common side group is the dimethyl structure. Replacement of small amounts of the methyl group with vinyl provides sites for cross-linking. Phenyl groups are used to improve low-temperature properties. Fluorosilicones are produced by replacing the methyl with trifluoro-propyl units. The addition of bulky phenyl side groups leads to an increase in the glass transition temperature. However, the disruption caused by such groups leads to the desired reduction or elimination of crystallization, which is critical for low-temperature properties. Such polymers have glass transition temperatures around $-110°C$, which is the range of the lowest T_g carbon-backbone polymers.

Monomer Production

The actual polymerization process involves a ring-opening reaction of dimethyl-substituted cyclic siloxanes. The preparation of the cyclic materials starts with the production of pure silicon via the reduction of quartz with coke in an electric arc furnace. The silicon metal then reacts with methyl chloride to give a mixture of silicones, from which dimethyldichlorosilane is removed by distillation.[65] Subsequent hydrolysis gives the cyclic dimethylsiloxane.

Polymer Production

The polymerization process involves an equilibrium ring-opening reaction carried out in the bulk state, which can be catalyzed by acids or bases:

polydimethylsiloxane

Uses

Silicone rubber offers a set of unique properties to the market, which cannot be obtained by other elastomers. The Si–O backbone provides excellent thermal stability and, with no unsaturation in the backbone, outstanding ozone and oxidative stability. The very low glass transition temperature, combined with the absence of low-temperature crystallization, puts silicones among the materials of choice for low-temperature performance. The fluoro-substituted versions provide solvent, fuel, and oil resistance along with the above-mentioned stability advantages inherent with the silicone backbone.

The gum polymer has rather poor tensile properties when cured, but these properties can be greatly improved by the use of silica-reinforcing agents. These systems exhibit some of the greatest improvements in properties by filler addition; and because this improvement is significantly higher for silica than for other reinforcing agents, it is assumed that direct bonding occurs between the silica and the polymer. The silicone materials also may have very low surface energy, which accounts for their nonstick characteristics. Because of the inherent inertness of the materials, they have been widely used for medical purposes within the body. The largest use of silicone is in sealant and adhesive applications.

Several different methods have been developed to cure silicones. Free-radical cures are possible for those polymers containing vinyl groups. The largest-volume process, however, involves room-temperature vulcanizations, which can employ either a one-component or a two-component system. In the one-component system, a cross-linking agent such as methyltriacetoxysilane is used.

With exposure to moisture, hydrolysis of the cross-linking agent leads to the silanol reactive cure site, so such materials must be compounded and stored free of moisture.[66]

POLYURETHANE RUBBER

A wide range of materials is included in this class. The common feature is the use of chain extension reactions to provide products with acceptable commercial properties. The chain extension reaction effectively reduces the actual number of chain ends, thereby eliminating the generally poor properties observed when very low-molecular-weight polymers are cross-linked. The chain extension step involves the reaction of a difunctional polymeric polyol with difunctional organic isocyanates to give the polyurethane:

$$OCN\!-\!\!R\!-\!\!NCO \ + \ HO\!-\!\!P\!-\!\!OH$$

di-isocyanate   polymeric diol

$$\longrightarrow \ \left[\!-OOCHN\!-\!\!R\!-\!\!NHCOO\!-\!\!P\!-\!\!OH\right]_n$$

polyurethane

The most used polyols have a polyester or a polyether backbone. A wide variety of isocyanates are used, with toluene di-isocyanate, *m*-phenylene di-isocyanate, and hexamethylene di-isocyanate the most common.

Raw Materials

The largest-volume polyether used is obtained from propylene oxide polymerized under basic conditions. Polyester polyols are produced from a number of different materials involving diacids and diols to give the ester linkage. Aliphatic polyesters generally are used for elastomers to impart chain flexibility.

The production of isocyanates is based on the reaction of phosgene with primary amines. Toluene di-isocyanate is the most frequently used di-isocyanate.

Uses

The urethane elastomers are complex-segmented or block polymers. Soft, noncrystalline blocks are provided by the polyether or aliphatic polyester long chains, whereas stiff, hard blocks are produced by the reaction of aromatic di-isocyanates with low-molecular-weight materials such as diols or diamines. The hard blocks can phase-separate to provide a physical rather than a chemical cross-link, similar to those to be discussed for the butadiene–styrene thermoplastic elastomers, although the domain size is much smaller for the polyurethanes. Chemical cross-links can be introduced by using tri-functional materials or by adjusting the stoichiometry to allow additional reactions with the urethane or urea structures to give allophanate or biuret linkages.

Polyurethane rubbers can have high tensile strengths, excellent tear strength, and good abrasion and chemical resistance. The greatest disadvantage is the hydrolytic instability of the urethane linkage. A major use is in automotive bumpers and facias. These materials are made in a reaction injection molding process. Castable urethanes are used to produce solid rubber wheels and printing rolls. Millable urethanes can be processed on conventional rubber equipment. Unsaturation can be introduced to allow the use of conventional rubber cure systems. Cures also are possible by reaction of the active hydrogens with materials such as high molecular weight polyfunctional isocyanates that are nonvolatile at cure temperatures.

Thermoplastic polyurethane elastomers are processed by injection molding and other processes used for thermoplastics. Small gears, seals, and even automotive fender extensions can be produced by this means.

By selecting from the large number of possible reactions and stoichiometry, properties can be tailored to meet a very wide range of applications.[67–69]

MODIFIED POLYETHYLENE RUBBERS

Elastomeric polymers can be produced by the chlorination or chlorosulfonation of polyethylene. Both products start with polyethylene, either in solution or in aqueous suspension, which then is reacted to give the specified degree of substitution to obtain the desired properties. Sufficient substitution is necessary to disrupt the regularity of the polymer chain, changing it from the crystalline polyethylene plastic into amorphous elastomers.

Chlorinated Polyethylene

The chlorinated products contain around 40 percent chlorine. These materials must be stabilized with metal salts, like other chlorinated elastomers and plastics. Peroxide cross-linking generally is used. Being saturated, the materials have excellent weather and ozone resistance and can be used over a temperature range of -65–$300°F$. The high chlorine content imparts oil resistance and relatively slow rates of burning. Typical applications where this combination of properties is required include hoses for chemical or oil resistance, tubing, and belting. In comparison with plasticized PVC, these materials have better low-temperature properties and do not suffer a loss of plasticizer because none is required.

Chlorosulfonated Polyethylene

Reaction with sulfur dioxide in addition to chlorine introduces cross-linking sites into the polymer chain. Sulfur contents in the range of 1.0–1.5 percent are used, with chlorine contents of 25–40 percent. Curing is accomplished by using metallic oxides, sulfur- bearing organic compounds, and epoxy resins. These materials have outstanding ozone resistance and show little color change upon light exposure. Good resistance to oils, heat, oxidation, weather, and corrosive materials also is exhibited. Applications include pond and pit liners, coated fabrics, light-colored roofing membranes, wire and cable insulation, chemical hose, and belting.[70,71]

THERMOPLASTIC ELASTOMERS (TPE)

This class of elastomeric materials is called "thermoplastic" because they contain thermally reversible cross-links of various types. The types of crosslinks vary from phase-separated

polystyrene domains, such as in styrene–butadiene–styrene (SBS) elastomers to ionic cluster cross-links in the ionomers. The beauty of these noncovalent interactions to form cross-links is that when the material is heated, the cross-links are broken. This allows the polymer to flow and be processed, and also recycled. When cooled, the cross-links reform and the material becomes strong again. All TPEs are two-phase systems where there is a soft, rubbery "continuous phase" and a hard "dispersed phase" which does not flow at room temperature. The first two polymers to be discussed are "block copolymers" and "ionomers."

Block Copolymers

A very popular and useful TPE is made from blocks of styrene and butadiene monomers using anionic polymerization techniques, which was described in the solution SBR section above. They are made up of short chains of polystyrene (usually 8000–15,000 MW), followed by a much longer chain of polybutadiene (about 60,000 MW), and capped off by another short chain of polystyrene, hence the name SBS. Similar polymers are prepared using isoprene instead of butadiene (SIS). The differences between SBS and SIS will be discussed later in the subsection "Uses."

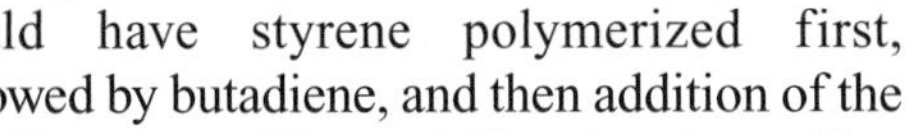

| block PS segment | polyBd segment | block PS segment |

The linear polymers, as shown above, can be built up by the sequential addition of monomer or by coupling the living anionic chains using compounds like dichloro dimethylsilane. Hence, the base polymer would have styrene polymerized first, followed by butadiene, and then addition of the coupling agent. If a multifunctional coupling agent such as silicon tetrachloride is used, a radial block or "star-branched" SBS is formed.

The polystyrene is highly insoluble in the polybutadiene so the PS chains cluster together and phase-separate into domains. Since there is much more polybutadiene (PBD) than polystyrene, the PBD becomes the continuous phase containing dispersed particles of PS which act both as cross-links and reinforcing agents. Every PBD chain is tied to a PS chain on both ends so a very strong cross-linked network is formed. If the PBD is tied to only one PS, then a "diblock" polymer is formed which has very little strength. When this network is heated to above the glass transition of polystyrene (100°C) the PS domains break down and begin to flow, so the polymer can be processed by injection molding or extrusion. Upon cooling to below 100°C, the domains (cross-links) reform and the material becomes strong again. SBS elastomers can have tensile strength as high as conventional thermoset elastomers which may approach 4000 psi.

adhesives because the isoprene segments tend to undergo chain scission during aging instead of cross-linking, which is observed in butadiene polymers. This leads to better retention of adhesion after aging; however, the SIS polymers usually have poorer initial strength compared with SBS. Another high-volume use is in toughening of asphalt compounds for paving, crack sealants, and roofing. The SBS improves rutting and low-temperature performance in paving and crack resistance in roofing applications.[73] SEBS would be the elastomer of choice in many adhesive and asphalt applications because of its superior aging properties; however, its high cost is

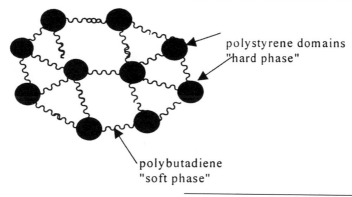

polystyrene domains "hard phase"

polybutadiene "soft phase"

The butadiene blocks can be hydrogenated, as mentioned above with hydrogenated nitrile, to form SEBS polymers having better thermal stability and chemical resistance. The EB stands for ethylene–butylene, which are the structures formed after the butadiene segments have been hydrogenated.

Other block copolymers which are useful are based on polyesters, polyurethanes, and ethylene–propylene. The first two have been discussed in other sections and the ethylene–propylene blocks will be discussed below in the subsection "Metallocene Polymers."

Uses

SBS copolymers are used in a wide variety of applications because of their clarity, toughness, and ease of processing. A major application is hot melt adhesives where they are compounded with hydrocarbon resins and oil.[72] The SIS polymers are very popular in

prohibitive. SBS also is used widely to toughen polystyrene and high-impact polystyrene. This polyblending technique is used to toughen a number of plastics. Various other injection molding and extrusion applications include shoe soles and toys. SBS is limited in use because of its poor high-temperature performance.

Ionomers

Ionomers are copolymers in which a small portion of the repeat units have ionic pendant groups on usually a nonpolar backbone. The ionic groups tend to separate themselves into domains similar to the polystyrene segments in the SBS rubber because they are insoluble in the nonpolar polymer chains. Therefore, these ionic clusters serve as cross-links up to temperatures where they tend to disassociate. Most commercial grades of ionic elastomers are based on ethylene and propylene monomers.

The polymer backbone usually contains sulfonyl or carboxylic acid groups and the metal counterion can be zinc, calcium, sodium, or lithium. The properties are highly dependent on the metal cation because they determine the temperature at which the ion clusters disassociate. The ionic cross-links usually impart outstanding tensile and tear strength properties because these cross-links are very good energy absorbers (hysteretic) due to their mobility. A process known as ion hopping provides one source for absorbing energy.

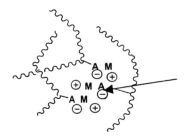

clusters of ionic groups forming a crosslink
M^+ = metal cation, A^- = anion

Uses

Solid golf balls are a good example of ionic cross-linked materials.[74] The covers are usually lithium or sodium cross-linked plastics but the cores are also a type of ionomer. The cores are high *cis* polybutadiene which is compounded with zinc diacrylate (ZDA) and then peroxide cured. During the curing process, the ZDA grafts to the BR backbone creating a material with a very high cross-link density where the cross-links are clusters of zinc carboxylates with very high resilience. Other applications include heat-sealable food packaging, automotive trim, footwear, foamed sheets (mats), and interlayers for bulletproof glass.

Metallocene Elastomers

Metallocene catalysts are the latest innovations to make a big impact in the polymer industry. They have been used mostly to make new polyolefin plastics, such as very-high-molecular-weight, bullet-proof polyethylene, but they have also been used to make elastomers. The catalysts make very regular "stereospecific" polymers similar to the Ziegler–Natta catalysts.

They are based on various metals. Such as zirconium, complexed with cyclopentadienide anions. This type of compound is called a "zirconocene" and is used with organoaluminum to make highly regular polymers. The catalyst has the ability to flip back and forth from making *atactic* to *isotactic* polypropylene in the same polymerization. The alternating tacticity of the polymer breaks up the crystallinity of the chains and yields an elastomer. Metallocene catalysts are currently very expensive and cannot yet polymerize dienes such as butadiene, so they have only enjoyed limited commercial success in elastomers. However, this is one of the most intense fields of polymer research and many new product breakthroughs are expected in the near future.

Rubber–Plastic Alloys

A discussion on thermoplastic elastomers would not be complete without mentioning the elastomers produced by simply blending rubbers and plastics in an internal mixer using a process known as "dynamic vulcanization." The simplest of these elastomers is based on polypropylene and EPDM rubber. Products are made with various rubber contents (hardness values) by simply mixing the rubber and plastic in a Banbury or an extruder at high temperature while cross-linking the EPDM in situ with a phenolic curing resin or other curing agents such as sulfur or peroxides. The resulting blend processes like polypropylene but is actually an elastomer because of the cross-linked rubber phase which it contains. It can be reprocessed and recycled like other TPEs. The process is described in an excellent review article by S. Abdou-Sabet.[75] Other alloys are based on nylon and NBR to obtain better thermal stability and solvent resistance. Constant improvements are being made in polymer alloys by using different combinations of plastics and rubbers and also new grafting and cross-linking chemistry to achieve properties more like conventional thermoset rubbers.

Uses

The applications for this type of elastomer have been limited because of the melting point of the plastic phase and poor compression set. For these reasons, alloys probably will never be successful as tire materials but they have many other automotive applications, including instrument panels, cowl vents, body panels, and bumpers. They also are used in appliances, lawn and garden equipment, and as grips on tools.

PLASTICIZED POLYVINYL CHLORIDE

Generally one thinks of polyvinyl chloride as a rigid plastic, which it is, with a glass transition temperature around 85°C. However, the addition of polar chemicals such as dioctyl phthalate can reduce the glass transition temperature below room temperature, producing a rubbery material.

Monomer Production

Vinyl chloride is produced primarily from ethylene, which is converted to ethylene dichloride either by chlorination or oxychlorination. The ethylene dichloride is cracked to give vinyl chloride and hydrogen chloride.

Production Process

Polyvinyl chloride is produced by the free-radical polymerization of vinyl chloride. Bulk, emulsion, solution, and suspension polymerization processes have been used.

The plasticized product can be produced by mixing the polymer and plasticizers at elevated temperatures, also by dry blending in which the plasticizer is absorbed into the resin and then heated. Solution blending is sometimes used, as well as the plastisol process, in which fine polyvinylchloride powder is dispersed in the plasticizer which is relatively stable until it is heated.

Properties and Use

Plasticized polyvinyl chloride can be regarded as the first thermoplastic elastomer, as it is used in an uncross-linked form. Because of the lack of cross-linking, this material exhibits high rates of creep and stress relaxation. As with other thermoplastic elastomers, these disadvantages worsen as the temperature is increased. Although the polymer is saturated, it must be stabilized for use to prevent dehydrochlorination at processing temperatures. Because of the high chlorine content, polyvinyl chloride has excellent flame resistance as well as good electrical properties. Even at high plasticizer levels, these materials have marginal elasticity when compared with most other elastomers. Despite their shortcomings, they are used in many areas where they compete with other rubbers.

One of the larger uses is for wire and cable insulation. The flame resistance makes this the material of choice for residential wiring, extension cords, and so on. Inexpensive garden hose represents another large-volume use. Sports balls can be produced by rotational molding of plastisols. The low fabrication cost allows such products to dominate the lower-price-range market.[76,77]

FLUOROCARBON ELASTOMERS

The fluoroelastomers were developed as specialty materials for high heat applications and solvent resistance. By 2003 there was over 14,800 metric tons of fluoroelastomer capacity worldwide.[78] The elastomers were by-products of Plunkett's 1938 discovery of polytetrafluoroethylene. Copolymers of olefins with vinylidene fluoride were found to be leathery, whereas elastomers were made if tetrafluoroethylene (TFE) or trifluoropropene were used. The first commercial product was produced as a copolymer of vinylidene fluoride and chlorotrifluoroethylene (CTFE) and was called Kel-F. The later, more stable polymers used hexafluoropropene (HFP) with TFE. They were first commercialized by DuPont under the trade name Viton and then a few other companies added their own fluoroelastomers. These elastomers are usually prepared by radical polymerization in emulsion using catalysts such as ammonium persulfate and sometimes with chain transfer agents such as carbon tetrachloride or halogen salts.[79] Caution must be exercised when using these polymerizations because the fluoromonomers can be explosive. They are usually isolated by coagulation and normal rubber drying techniques and sold in the form of pellets, slabs, or rubber crumb.

As a result of being very unreactive, the fluoroelastomers cure very slowly and usually require a high-temperature post-cure. The curatives are designed to remove hydrogen fluoride to generate a cure site which can react with a diamine or bisphenol. They are also cured with organic peroxides. Most fluoroelastomers can be compounded using normal rubber processing equipment such as rubber mills and internal mixers. Processing aids such as dioctyl phthalate or waxes can be used to obtain smoother extrusions and better mold release.

copolymer of TFE with vinylidene fluoride (VF)

Uses

Fluoroelastomers have outstanding heat resistance when cured. Some vulcanizates have almost indefinite service life at temperatures up to 200°C. The perfluoro polymers, such as DuPont's Kalrez, have short-term useage at up to 316°C and extended service as high as 288°C. This polymer is extremely expensive and is only offered in the form of finished parts, usually O-rings, seals, or gaskets. The fluoroelastomers also have excellent solvent and ozone resistance making them ideal for automotive fuel hoses. Although many of the applications involve small seals, O-rings, gaskets, and hose, the single largest applications is flue-duct expansion joints. The polymer's resistance to high temperature and wet acidic flue gases are critical in this application.

REFERENCES

1. Whitley, G. S., Davis, C. C., and Dunbrook, R. F. (Eds.), *Synthetic Rubber*, John Wiley & Sons, New York, 1954.
2. Blackely, D. C., *Synthetic Rubber: Their Chemistry and Technology*, Chapter 2, Applied Science Publishers, London, 1983.
3. *Worldwide Rubber Statistics 2004*, p. 6, 11, International Inst. of Synthetic Rubber Producers, Inc., Houston, 2000.
4. Fox, T. G., Grateh, S., and Loshaek, S., *Rheology—Theory and Applications*, Vol. 1, Academic Press, New York, 1956.
5. Billmeyer, F. W., Jr., *Textbook of Polymer Science*, John Wiley & Sons, New York, 1984.
6. Odian, G., *Principles of Polymerization*, pp. 19–24, McGraw-Hill, New York, 1970.
7. Flory, P. J., *Principles of Polymer Chemistry*, Chapter 7, Cornell University Press, Ithaca, NY, 1953.
8. Rodriguez, F., *Principles of Polymer Systems*, pp. 138–142, McGraw-Hill, New York, 1982.
9. Rudin, A., *The Elements of Polymer Science and Engineering*, p. 126, Academic, New York, 1982.
10. Rudin, A., *The Elements of Polymer Science and Engineering*, p. 402, Academic, New York, 1982.
11. Krause, S., *J. Macromol. Sci.*, **C7**(2), 251 (1972).
12. Aklonis, J. J., *Introduction to Polymer Viscoelasticity*, John Wiley & Sons, New York, 1972.
13. Oberster, A. E., Bouton, T. C., and Valaitis, J. K., *Die Angewandte Markomoleculare Chemie*, **29/30**, 291 (1973).
14. Flory, P. J., *J. Chem. Phys.*, **17**, 223 (1949).
15. Natta, G., in *Polymer Chemistry of Synthetic Elastomers*, Part 1, J. Kennedy and E. Tornqvist (Eds.), Chapter 7, John Wiley & Sons, New York, 1969.
16. Semegen, S. T., and Cheong, S. F., *Vanderbilt Rubber Handbook*, pp. 18–41, R.T. Vanderbilt Company, Inc., 1978.
17. Webster, C. C., and Baulkwill, W. J., Chapters 1 and 11, *Rubber*, John Wiley & Sons, New York, 1989.
18. Brandrup, J, and Immergut, E., *Polymer Handbook*, Vol. III, p. 54, John Wiley & Sons, New York, 1975.
19. Treloar, L., *The Physics of Rubber Elasticity*, Clarendon, Oxford, 1949.
20. Poh, W. N., *Elastomers*, p. 12 (1989).
21. *Synthetic Rubber: The Story of an Industry*, International Inst. of Synthetic Rubber Producers, New York, 1973.
22. Stavely, F. W., *Ind. Eng. Chem.*, **48**, 778 (1956), presented at Div. Rubber Chem., ACS, Philadelphia, 1955.

23. Eng. Patent 827365, to Goodrich–Gulf Chem., Dec. 1954, priority data.
24. *Worldwide Rubber Statistics 2000*, International Inst. of Synthetic Rubber Producers, Inc., pp. 6, Houston, 2004.
25. *Worldwide Rubber Statistics 2000*, International Inst. of Synthetic Rubber Producers, Inc., p. 19, Houston, 2000.
26. Renninger, T. J., presentation to International Rubber Study Group, Ottawa, Sept. 1990.
27. Faith, W. L. et al., *Industrial Chemicals*, p. 731, John Wiley & Sons, New York, 1957.
28. Dunbrook, R. F., *India Rubber World*, **117**, 203–207 (1947).
29. Gardon, J. L., *Rubber Chem. Technol.*, **43**, 74–94 (1970).
30. Storey, E. B., *Rubber Chem. Technol.*, **34**, 1402 (1961).
31. Kuntz, L, *J. Poly. Sci.*, **54**, 569 (1969).
32. Bouton, T., and Futamura, S., *Rubber Age*, **3**, 33 (1974).
33. Mostert, S., and Van Amergongen, G., British Patent 1,136,189, 1968.
34. Day, G., and Moore, D., paper at 26th Annual Meeting, IISRP, May 1985.
35. Quirk, R. P., *Rubber Chem. Technol.*, **69**(3), 444 (1996).
36. Oshima, N., Salcacobore, M., and Tsutsumi, F., paper at 27th Annual Meeting, IISRP, May 1986.
37. *Worldwide Rubber Statistics 2000*, International Inst. of Synthetic Rubber Producers, Inc., p. 9, Houston, 2000.
38. Ponder, T., *Hydrocarbon Processing*, **55**(10), 119–121 (1976).
39. Womeldroph, D., *Am. Petrol. Inst.* (May 14, 1958).
40. Welch, M., *Hydrocarbon Processing*, **57**(11), 131–136 (1978).
41. Adams, H., Farhat, K., and Johnson, B., *Ind. Eng. Chem. Prod. Dev.*, **5**(2), 127 (1966).
42. Halasa, A., Schulz, D., Tate, D., and Mochel, V., in *Advances in Organometallic Chemistry*, F. Stone and R. West (Eds.), Vol. 18, Academic, New York, 1980.
43. Tate, D., and Bethea, T., *Encyclopedia of Polymer Science and Engineering*, Vol. 2., pp. 568–572, John Wiley & Sons, New York, 1985.
44. Lauretti, F., and Gargani, L., 27th Annual Meeting, IISRP, June 1987.
45. DeChirico, A., Lamzani, P., Eaggi, E., and Bruzzone, M., *Makromol. Che.*, **175**, 2029 (1974).
46. Buckenell, C., *Toughened Plastics*, Applied Science, London, 1977.
47. Maruko, T., US Patent No. 6,071,201 *(2000) Solid Golf Ball.*
48. Natta, G., *J. Poly. Sci.*, **51**, 411 (1961).
49. Lukach, C., and Spurlin, H., *Copolymerization*, G. Ham (Ed.), Interscience, New York, 1964.
50. Crespi, G., and DiDrusco, G., *Hydrocarbon Processing*, **48**, 103–107 (Feb. 1969).
51. Vandenberg, E., and Repka, B., in Ref. 12, Chapter 11.
52. Scagliotti, F., Milani, F., and Galli, P., paper at 26th Annual Meeting, IISRP, May 1985.
53. Miles, D., and Briston, J., *Poly Technol*, p. 299, Chemical Publishing, New York, 1963.
54. Powers, K. W., US Patent No. 5,162,445 (1993) *Para-alkylstyrene/isoolefin copolymers.*
55. "Butyl Rubber Reaches 50 Year Mark," *Elastomerics*, 30–31 (Mar. 1988).
56. Konrad, E., and Tschunkur, E., U.S. Patent 1,973,000 to 1. G. Farbenindustrie, 1934.
57. *Worldwide Rubber Statistics 2000*, International Inst. of Synthetic Rubber Producers, Inc., p. 9, Houston, 2004.
58. McKenzie, G., paper at 27th Annual Meeting, IISRP, May 1986.
59. Hashimoto, K., et al., paper at 26th Annual Meeting, IISRP, May 1985.
60. Buding, H., US Patent No. 4,581,417 (1986), *Production of Hydrogenated Nitrile Rubber.*
61. Whitley, G. S., Davis, C. C., and Dunbrook, R. F. (Eds.), *Synthetic Rubber*, John Wiley & Sons, New York, 1954, p. 770.
62. Miles, D., and Briston, J., *Poly Technol*, p. 305, Chemical Publishing, New York, 1963.
63. Blackely, D. C., *Synthetic Rubber: Their Chemistry and Technology*, Chapter 2, Applied Science, London, 1983, pp. 175–194.
64. Brodrecht, L., *Chemical Economics Handbook*, SRI International, 1989.
65. Polmateer, K., *Rubber Chemistry and Technology*, **16**(3), 470 (1988).
66. Semegen, S. T., and Cheong, S. F., *Vanderbilt Rubber Handbook*, pp. 216–232, R.T. Vanderbilt Company, Inc., 1978.
67. Natta, G., in *Polymer Chemistry of Synthetic Elastomers*, Part 1, J. Kennedy and E. Tornqvist (Eds.), Chapter 8, John Wiley & Sons, New York, 1969.
68. Morton, M, and Fetters, L, *Polymerization Processes*, Chapter 9, John Wiley & Sons, New York, 1977.
69. *Handbook of Elastomers*, pp. 643–659, Marcel Dekker, New York, 1988.
70. Broadrecht, L., Mulach, R., and Tauchiya, K., in *Chemical Economics Handbook—Elastomers*, SRI International, Menlo Park, CA, 1989.

71. Semegen, S. T., and Cheong S., *Vanderbilt Rubber Handbook*, pp. 18–41, R.T. Vanderbilt Company, Inc. 1978.
72. Kraton Polymers for Adhesives & Sealants, Shell online literature, Shell Chemical Web site.
73. Bull, A., and Vonk, W., Shell Chemicals Technical Manual TPE 8: 15 Report, 1988 (Shell Web site).
74. Maruko, T., US Patent No. 6,071,201, *(2000) Solid Golf Ball.*
75. Abdou-Sabet, S., *Rubber Chem. Technol.*, **69**(3), 476 (1996).
76. Penn, W., *PVC Technology*, Applied Science, London, 1971.
77. Miles and Briston, *Polymer Technology*, p. 159, Chemical Publishing, New York, 1963.
78. *Worldwide Rubber Statistics 2000*, International Inst. of Synthetic Rubber Producers, Inc., Houston, 2004.
79. Morton, M., *Rubber Technology*, p. 410, Chapman & Hall, London 1995.

17

The Agrochemical Industry

A. M. Malte* and A. T. Lilani*

INTRODUCTION

Scope of the Chapter

This chapter deals with the chemicals used in agriculture mainly to protect, preserve, and improve crop yields. The term "agrochemical" is used broadly. Much agrochemical research and some advanced development is directed toward the introduction of genes that may provide disease, insect, or viral resistance into plants or other organisms. Further progress is being made in improving the protein, fat, or carbohydrate composition of the plant itself. Microorganisms are being propagated and currently marketed that are insecticidal (e.g., *Bacillus thuringiensis*) fungi that are herbicidal, bacteria that are fungicidal, nematodes that are widely biocidal, and so on, are all products or candidate products for use in agriculture.

Arbitrarily excluded from discussion in this chapter are those substances that serve as fundamental nutrients, which are treated in Chapters 24 and 22 on fertilizers and nitrogen

technology, respectively. Nevertheless, it is the current practice of the farmer, particularly in advanced agriculture, to integrate nutritional and plant-protection application schedules, and even provide single formulations that include both fertilizers and pesticides. Further, plant nutrition at this stage of scientific sophistication is far more complex than the older classical "N-P-K" applications alone.

Many chemicals that accelerate plant growth act as hormonal agents, modifying plant metabolic processes at some stage of development. Because these substances are manufactured and marketed by the agrochemical industry, they are included as subject matter here. Also included in this chapter are chemicals that are significant to public health. Many organisms are vectors in the dissemination of human and animal disease. Because products of the pesticide industry control the insect, the rodent, the mollusk, and so forth (the vectors), they often are the most effective and sometimes the only practical means for controlling some of the most serious health problems of humankind, especially, but not exclusively, in the underdeveloped countries. An historical analog would be the use of rodenticides in the control of plague.

*Gharda Chemicals, Ltd., Dist. Thane, Maharashtra, India.

History

It is probable that farmers' treatment of crops with foreign substances dates back into pre-history. The Bible abounds with references to insect depredations, plant diseases, and some basic agricultural principles such as periodic withholding of land in the fallow state. Homer speaks of "pest-averting sulfur." More recently, in the nineteenth century, there was a great increase in the application of foreign chemicals to agriculture. Discovered or, more precisely, rediscovered was the usefulness of sulfur, lime sulfur (calcium polysulfides), and Bordeaux mixture (basic copper sulfates). With the exception of the organic compound formaldehyde, inorganic chemicals provided farmers with their major weapons.

The earliest of the organic compounds generally were chemicals derived from natural products or crude mixtures of chemicals in states of very elementary refinement. Extracts of ground-up plant tissue were useful in the control of insects. Such extracts were employed in agriculture quite often before the chemist had elucidated the structure or synthesized the molecule responsible for biological activity. These extracts included the pyrethroids, rotenoids, and nicotinoids, which continue to be derived in large part from plant extracts. Crude petroleum fractions were recognized for their effectiveness in the control of mites, scale, and various fungi, as well as for their phytopathological properties.

Although a few synthetic organics were already known, the great revolution in the use of organic chemicals in agriculture roughly coincides with the period of the onset of World War II. The more important of these discoveries were DDT (invented by Othmar Zeidler in late 1800s, insecticidal properties discovered by Mueller in 1939), 2.4-D (Jones patent—1945), benzenehexachloride (ICI and French development—ca. 1940), and the organic phosphate esters (Schrader—begun in the late 1930s, revealed in the 1940s).

These new chemicals were so enormously more potent than their predecessors in their biological activity (frequently by orders of magnitude) that they very rapidly displaced almost all of the chemicals previously employed. The classical chemicals of today, some discovered in the 1950s and 1960s, are predominantly extensions of this almost revolutionary transition from inorganics to synthetic organics that dates from the period of World War II (Table 17.1). It is fair to say that within the United States the agrochemical industry since the late 1950s has been dominated by synthetic organic chemicals.

ROLE OF THE AGROCHEMICAL INDUSTRY

The world's population, which stands as 6 billion today and will continue to grow over the years, needs food. This food needs to be grown on the arable land available today, without further destruction of forest resources, wetlands, or mangroves. Thus, improving farm productivity by reducing losses experienced during crop cycles and post harvest is a major challenge. Agrochemicals, which are selective toxicants, provide an important tool. Available agrochemicals need to be deployed judiciously and newer ones must be developed to minimize impacts on water supplies and food quality. Presently, it is believed that more than a third of global farm output is lost due to ineffective pest control.

The agrochemical industry has the responsibility of using the power of science and technology for providing the world population with not only sufficient food but with food of increased nutritional value. On another, less obvious front, it needs to be noted that some 75 percent of the global population use medicinal plants for primary health care, and these plants also require crop protection. In yet another related area, public health, vector control is more dependable than are curative drugs in combating diseases such as malaria, yellow and dengue fevers.

The agrochemical industry is large, complex, and involves many facets and many players. Listing several of these, from multinational manufacturer to individuals who apply the product in the field, will help in visualizing the breadth of the industry.

TABLE 17.1   New Chemistry

Chemical (Chemical Class) (Year of Reporting)	Company's Name	Structure	Biochemistry	Mode of Action	Known Profile of Use	Environmental Fate
Acetamiprid (Neonicotinoid) (1992)	Nippon soda		Nicotinic acetyl-choline receptor agonist	Systemic insecticide with translaminar activity and with contact and stomach action	Control of Hemiptera, Thysanoptera and Lepidoptera by soil and foliar application on a wide range of crops especially vegetables, fruit, and tea	*Plants.* Slowly degraded on or in plants, forming five identified metabolites. *Soil.* DT_{50} in clay loam 1 day, in light clay 1–2 days. DT_{50} for total residues 15–30 days
Clothianidin* (Nitromethylene neonicotinoid)	Takeda		Nicotinic acetyl-choline receptor agonist	Insecticide affecting the synapses in insects' central nervous system	Control of sucking and chewing insects by soil, foliar and seed application on rice, fruit, and vegetables	—
Dinotefuran* (Nitromethylene neonicotinoid) (1998)	Mitsui		Nicotinic acetyl-choline receptor agonist	Insecticide active by ingestion and contact; also exhibits root systemic activity	For control of a range of Hemipterous and other pests	—
Nitenpyram (Nitromethylene neonicotinoid) (1993)	Takeda		Nicotinic acetyl-choline receptor agonist	Systemic insecticide with translaminar activity with contact and stomach action	Control of aphids, thrips, leafhoppers, whitefly and other sucking pests on rice and glass house crops	*Soil.* DT_{50} in soil 1–15 days, depending on soil type

(continued)

TABLE 17.1   Continued

Chemical (Chemical Class) (Year of Reporting)	Company's Name	Structure	Biochemistry	Mode of Action	Known Profile of Use	Environmental Fate
Thiamethoxam (Neonicotinoid) (1998)	Novartis		Nicotinic acetyl-choline receptor agonist	Insecticide with contact stomach and systemic activity	For the control of aphids, whitefly, thrips, ricehoppers, ricebugs, ealybugs, whitegrubs, olorado potato beetle, flea beetles, wireworms, ground beetles, leaf miners and some lepidopterous species. Foliar and soil applications: cole crops, leafy and fruity vegetables, potatoes, rice, cotton, deciduous fruits, citrus, tobacco, soybeans. For seed treatment: maize, sorghum, cereals, sugarbeet, oil seed rape, cotton, peas, beans, sunflower, rice, potatoes	*Animals.* Quickly and completely absorbed, rapidly distributed in the body and rapidly eliminated. *Plants.* Degradation/metabolism has been studied in six different crops with soil, foliar and seed treatment application. The qualitative metabolic pattern was similar for all types of applications and for all studied crops. *Soil.* Soil DT_{50} (median) 51 days
Propoxycarbazone* (Sulfonylamino-carbonyltriazoli-none) (1999)	Bayer		Amino-acid synthesis inhibitor	Herbicide translocated both acro-petally and basipetally within both xylem and phloems	Post-emergence control of annual and some perennial grasses and some broadleaf weeds in wheat, rye, triticale	*Animals.* Rapid and nearly complete (>88%) within 48 hr, primarily via feces; 75–89% unchanged parent compound in urine and feces. *Plants.* The unchanged parent compound and its 2-hydroxypropoxy metabolite are regarded as the relevant residues for plants. *Soil.* Soil DT_{50} c. 36 days. Field dissipation DT_{50} c. 9 days

Name	Company	Structure	Mode of action	Absorption	Uses	Metabolism and fate
Flucarbazone—sodium (Sulfonylaminocarbonyltriazolinone) (1999)	Bayer	*(chemical structure: OCF$_3$, SO$_2$NCON, N–CH$_3$, OCH$_3$, Na$^+$)*	Amino-acid synthesis inhibitor	Herbicide absorbed through foliage and roots and translocated acropetally and basipetally	Wheat for post-emergence control of grass weeds especially *Avena fatua* and *Setaria viridis* and some broad-leaved weeds	*Animals.* Almost completely excreted via feces and urine within 48 hr. *Plants.* Extensively metabolized. The relevant residues are the parent compound and the *N*-desmethyl metabolite. *Soil.* Average soil DT$_{50}$ 17 days. Not mobile in soil
Metosulam (Trizolopyridine) (1993)	Dow Agro-Sciences	*(chemical structure: CH$_3$O, CH$_3$O, SO$_2$NH, Cl, Cl, CH$_3$)*	Amino-acid synthesis inhibitor	Herbicide readily taken up by roots and foliage	Post-emergence control of many important broadleaf weeds including *Galium aparine, Stellaria media, Chenopodium* spp., *Amaranthus retroflexus* etc. in wheat, barley, rye, maize	*Animals.* Rapidly absorbed. DT$_{50}$ <1 hr, extensively metabolized in rodents, much less in dogs, and excreted with metabolites 3-hydroxy (aliphatic oxidation) and 5-hydroxy (*O*-demethylation) in urine (DT$_{50}$ 54–60 hr in rodents; 73 hr in dogs). *Plants.* Poorly absorbed metabolized by hydroxylation of the ring methyl, to give a 3-hydroxymethyl- metabolite and its glycoside. *Soil.* Field DT$_{50}$ in the 0–10 cm horizon has a mean value of 25 days degradation via the 5- and 7-hydroxy analogue to 5-amino-*N*-(2,6-dichloro-3-methylphenyl)-1*H*-1,2,4-triazole-3-sulfonamide and CO$_2$. Does not have leaching potential
Flumetsulam (Triazolopyrimidine)	Dow Agencies	*(chemical structure: F, F, NHSO$_2$, N–N, N, N, CH$_3$)*	Amino-acid synthesis inhibitor (ALS or AHAS)	Systemic Herbicide absorbed by roots and leaves of plants and translocated to growth points	Used alone and in combination with trifluralin or metolachlor for control of broad-leaved weeds and grasses in Soyabeans, field peas, maize	*Animals.* Rapidly cleared via urine and feces with no metabolites. 5-Hydroxy metabolite found in the hen. *Plants.* DT$_{50}$ in maize 2 hr soya beans 18 hr, *Chenopodium* 131 hr. Metabolites depend on the species; 5-hydroxy or 5-methoxy derivatives are common. *Soil.* DT$_{50}$ in soil (25°C, pH 6–7, o.m. content 2–4%, 1–2 mos DT$_{50}$ in soil (pH 6–7, o.m. content 2–4%) 1–2 months

(continued)

TABLE 17.1 Continued

Chemical (Chemical Class) (Year of Reporting)	Company's Name	Structure	Biochemistry	Mode of Action	Known Profile of Use	Environmental Fate
Florasulam (Triazolo-pyriidine)	Dow Agencies (1999)		Aminoacid synthesis (ALS or AHAS)	Herbicide taken up by inhibitor and shoots, and translocated in both xylem and phloem	Post-emergence control of broad-leaved weeds both roots *aparine, Stellaria media, Polygonum convolvutus, Matricaria* spp., and various cruciferae in cereals and maize	*Soil.* DT$_{50}$ 2–18 days; neither florasulam nor its degradates should leach especially *Galium*
Diclosulam (Triazolopyri-midine)	Dow Agencies		Acetolactate synthase inhibitor	Herbicide taken up by roots and foliage and translocated to new growing points. Lethal amounts in meristems halting cell division and resulting in plant death	Soil applied control of broad-leaved weeds in Peanuts, soyabeans	*Animals.* Metabolized primarily by dealkylation of the ethoxy group and hydrolysis of the sulfonamide linkage *Soil.* Dissipation occurs primarily through microbial degradations oil DT$_{50}$ (in a wide variety of soils) c. 33–65 days
Cloransulam-methyl (Triazolo-pyrimidine) (1997)	Dow Agencies		Acetolactate synthase inhibitor	Herbicide having plant meristems as the primary site of activity	Control of broad-leaved weeds in Soyabeans. Applied to the soil surface or incorporated pre- or post-emergence	*Animals.* In female rats, excreted mainly via the urine; in male rats, excreted in both urine and feces. After 72 hr, <0.1% of the dose was found in any tissue *Soil.* Photolysis on soil surface, DT$_{50}$ 30–70 days (corrected for metabolism) The apparent transformation DT$_{50}$ in aerobic soils 9–13 days (est.). May be mobile

Name (year)	Company	Mode of action	Type/properties	Uses	Structure	Metabolism/fate
Pyrithiobac-sodium (Pyrimidinyl oxybenzoic analogue) (1991)	Ihara/ Kumaio	Amino-acid synthesis inhibitor (ALS or AHAS)	—	Pre- and post-emergence control of wide range of broad-leaved weeds in cotton	(structure: Cl, CO_2Na, OCH_3, N, S, OCH_3)	*Animals.* More than 90% excreted in urine and feces within 48 hr; the major excreted metabolite was the *O*-desmethyl derivative. *Plants.* At 62 dat, no residues were found; major metabolites were the phenol formed by mono-demethylation, and its glucose conjugate. *Soil.* Microbial and photochemical degradation play a major role in degradation t; DT_{50} in silty soil 60 days
Azoxystrobin (Strobilurin) (1992)	Zeneca	Inhibitor of mitochondrial respiration	Fungicide with protectant, curative, radicant, translaminar and systemic properties	Control of a number of pathogens in cereals, rice, vines, cucurbits, potato, tomato, peanuts, peach, turf, banana, pecan, citrus, coffee	(structure: N, CO_2CH_3, CH_3O, O, CN)	*Animals.* Majority of radiolabel is excreted in the faeces. Of a large number of metabolites, only the glucuronide of azoxystrobin acid is present at >10% of the administered dose. *Plants.* Metabolism was extensive, but parent azoxystrobin was the only major (>10%) residue. *Soil.* In soil, in the dark, six identified metabolites were formed; over 1 year, 45% of applied radiolabel is evolved as CO_2. DT_{50} 1–8 weeks. Low to moderate mobility in soil; typical K_{oc} for azoxystrobin c. 500
Diafenthiuron (Strobilurin) (1988)	Novartis	Converted by light into the corresponding carbodiimide, which is an inhibitor of mitochondrial respiration	Insecticide, Acaricide having contact and stomach action. Also shows some ovicidal action	Control of phytophagous mites, Aleyrodidae, Aphididae, and Jassidae and some leaf feeding pests in cotton, various field and fruit crops, ornamentals, vegetables	(structure: $CH(CH_3)_2$, $NHCSNHC(CH_3)_3$, $CH(CH_3)_2$, O)	*Animals.* Major portion excreted with the feces. Degraded to yield its corresponding carbodiimide, which, in turn forms urea and fatty acid derivatives. *Plants.* Shows a complex metabolism pattern. *Soil.* Diafenthiuron and its main metabolites show a strong sorptivity to soil particles. Degradation in soils proceeds rapidly: DT_{50} <1 hr to 1.4 days

(continued)

TABLE 17.1 Continued

Chemical (Chemical Class) (Year of Reporting)	Company's Name	Structure	Biochemistry	Mode of Action	Known Profile of Use	Environmental Fate
Fenamidone* (strobilurin) (1992)	Aventis		Inhibitor of mitochondrial respiration by blocking electron transport at ubihydroquinone; cytochrome oxido reductase	Protectant and aerative fungicide	Under development for control of a range of Oomycete diseases	—
Kresoxim-methyl Strobilurin type: (Strobilurin) (1992)	BASF		Inhibitor of mitochondrial respiration by blocking electron transfer between cytochrome b and cytochrome C_1 control	Fungicide with protective, curative, eradicative and long residual disease control	Control of scab, powdery mildew, mildew, scald, net blotch, glume blotch on apples, pears, vines, cucurbits, Sugarbeet, cereals, vegetables	*Animals.* Widely distributed and quickly eliminated; no bioaccumulation. The major routes of excretion were feces and urine. Thirty two different metabolites were identified *Plants.* Residues in cereals and pome fruit at harvest are <0.05 mg/kg, in grapes and vegetables <1 mg/kg *Soil.* Rapidly degraded. In soil, DT_{90} (lab.) <3 days, the main metabolite is the corresponding acid. Very mobile in soil. However, in lysimeter studies, only low levels of kresoxim-methyl and its metabolite were found in leachates
Tebufenpyrad (Pyrazole) (1993)	Mitsubishi chemical		Mitochondrial respiration Inhibitor. Acts as an inhibitor	Nonsystemic Acaricide active by contact and ingestion.	Control of all stages of *Tetranychus, Panonychus, Oligonychus, Eotetranychus* spp.,	*Animals.* Metabolite is N-[4-(1-hydroxymethyl-1-methyl-ethyl) benzyl]-4-chloro-3-(1-hydroxyethyl)-1-methyl-pyrazole-5-carboxamide *Plants.* As for animals

Name / Type / Company (Year)	Mode of action	Structure	Activity	Uses	Metabolism / Fate
(continued from previous entry)	of electron transport chain at site I		Exhibits translaminar movement following application to leaves	on fruit, vines, citrus, vegetables, hops, ornamentals, melons, cotton	*Soil.* Aerobic degradation occurs in soil, DT_{50} 20–30 days
Trifloxys-trobin Strobilurin type: (Strobilurin) (1998) — Novartis	Inhibits mitochondrial respiration by blocking electron transfer at the Q_0 centre of Cytochrome bc1	*(chemical structure: CH_3, CF_3, $O-N$, CH_3O-N, CO_2CH_3)*	Mesostemic broadspectrum fungicide with preventive and specific curative activity and displaying rain-fastness. Redistributed by superficial vapor movement and also have translaminar activity	For control of powdery mildew, leaf spots, rusts, bunch and fruit rots of cereals, Pome fruit, grapes, Peanuts, bananas, vegetables	*Animals.* Absorbed from the gastrointestinal tract, rapidly metabolized and quickly and completely eliminated from the body. *Plants.* Rapidly degraded. *Soil.* Dissipates rapidly. DT_{50} 4.2–9.5 days. No leaching potential
Famoxadone* Strobilurin type: (Oxazolidinedione) (1996) — DuPont	Inhibits mitochondrial electron transport, by blocking ubiquinolcytochrome C oxido-reductase at complex III	*(chemical structure: oxazolidinedione ring, CH_3, O, $N-H$, phenyl, phenoxyphenyl, O)*	Protectant translaminar and residual Fungicide	Control of mildew, potato and tomato late and early blights, wheat leaf and glume blotch and barley net blotch in grape, potato, tomoto, wheat, barley	*Animals.* Elimination is rapid. Unmetabolised famoxadone was the major component in the feces; mono- (at 4'-phenoxyphenyl) and di- (also at 4-phenylamino) hydroxylated famoxadone were the primary fecal metabolites. In urine, products arising from cleavage of the heterocyclic ring were found. Metabolism was complex, involving hydroxylation, cleavage of the oxazolidinedione-aminophenyl linkage, cleavage of the phenoxyphenyl ether linkage and opening of the oxazolidinedione ring

(continued)

TABLE 17.1 Continued

Chemical (Chemical Class) (Year of Reporting)	Company's Name	Structure	Biochemistry	Mode of Action	Known Profile of Use	Environmental Fate
						Plants. In grapes and potatoes, famoxadone was the main residue; no residues were found in potato tubers. In wheat, famoxadone was extensively metabolised, primarily by hydroxylation, followed by conjugation *Soil.* In laboratory soil, DT_{50} 6 days (aerobic), 28 days (anaerobic). Degradation routes include hydroxylation (at the 4'-phenoxyphenyl position), ring opening (with formation of a glycolic acid derivative), and is primarily microbial
Fenazaquin Strobilurin type: (Oxazolidinedione) (1992)	Dow Agro-sciences	O–CH$_2$CH$_2$— ⋯ —C(CH$_3$)$_3$	Inhibitor of mitochondrial electron transport chain by binding with complex I at co-enzyme site Q	Contact Acaricide with good knockdown activity on motile foms as well as true ovicidal activity. Preventing eclosion of mite eggs	Control of *Euteranychus, Panonychus, Tetranychus, Brevipalpus phoenici* in almonds, apples, citrus, cotton, grapes, ornamentals	—
Fenpyro-ximate Pyrazole (acaricide) (1990)	Nihon Nohyaku	(CH$_3$)$_3$COOC— ⋯ —CH$_2$O—N=C—H ⋯ N–N, CH$_3$, CH$_3$	Inhibitor of mito-chondrial electron transport at complex I	Acaricide having quick knockdown activity against	Effective against some phytophagus mites. Control of Tetranychidae, Tarsonemidae, Penuipalpidae and	*Soil.* DT_{50} 26.3–49.7 days

Name (year)	Company	Structure	Mode / Activity	Use	Target / Metabolism
Quinoxyfen (Quinoline) (1996)	Dow Agrosciences		Growth signal disruptor. Mobile, Protectant fungicide acting through inhibition of appressorial development. Active through systemic acropetal and basipetal movement and by vapor transfer	Control of powdery mildew in cereals	Eriophyidae. In citrus, apple, pear, peach, grapes. larvae, numphs and adults mainly by contact and ingestion. *Plants.* Only slightly metabolized in wheat, with low residues found in the crop. Extensively photodegraded on the wheat leaf surface, giving multiple polar degradation products. On grapes and cucumbers, the main residue was unchanged quinoxyfen. *Soil.* DT$_{50}$ (field) 123–494 days (biphasic); nonleaching. The main metabolite in the soil (also classed as nonleaching) was formed by hydroxylation at the 3-position of the quinoline ring; a minor metabolite (DCHQ), formed by cleavage of the ether bridge, was observed, especially in acidic soil
Cyclanilide Strobilurin type: (Oxazolidinedione) (1994)	Aventis		Inhibits Polar auxin transport. Plant growth regulator	Cotton and other crops	*Animals.* Rapidly excreted, primarily as unchanged cyclanilide. *Plants.* Little degradation occurs in plants; cyclanilide is the major residue. *Soil.* Low to moderate persistence, DT$_{50}$ c. 16 days under aerobic conditions.
Degrades					Primarily by microbial activity. Medium to low mobility

(continued)

TABLE 17.1   Continued

Chemical (Chemical Class) (Year of Reporting)	Company's Name	Structure	Biochemistry	Mode of Action	Known Profile of Use	Environmental Fate
Diflufenzopyr (Semi-carbazone) (1999)	BASF		Inhibits auxin transport, apparently by binding with a carrier protein on the plasmalemma	Systemic, post-emergence herbicide	Control of annual broad leaved and perennial weeds in maize	*Animals.* Partially absorbed and rapidly eliminated; 20–44% of the dose was eliminated in urine and 49–79% in feces. Total radioactive residues in tissues <3% of the administered dose. Eliminated primarily as unchanged parent compound. *Soil.* Average DT_{50} in field soil 4.5 days. Very mobile Metabolities also very mobile. However, based upon proposed use, US EPA does not expect diflufenzopyr to reach drinking water
Flumioxazin (N-phenyl-phthalimide)	Sumitonio		Protoporphyrinogen oxidase inhibitor	Herbicide absorbed by foliage and germinating seedlings tissue	Control of many annual grasses, pre- and post-emergence in soyabeans, peanuts, orchards and other crops	—
Flumiclorac-pentyl N-phenyl phthalimide (1998)	Sumitono valent		Protoporphyrinogen oxidase inhibitor	Fast acting, contact herbicide. When applied to foliage of susceptible plants. It is readily absorbed into plant	Control of problem broad leaved weeds including *Xanthium strumarium, Chenopodium album, Ambrosia artemisifolia, Datura stramonium, Amarnthus sp., Sida spinosa, Euphorbia maculata, Abutilon theophrasti,* pre- and post-emergence in soyabeans and maize	*Plants.* In soya beans and maize, the major metabolite is 2-chloro-4-fluoro-5-(4-hydroxy-1, 2-cyclohexane-dicarboximido) phenoxyacetic acid formed by reduction of the tetrahydrophthaloyl double bond and hydroxylation; other metabolic pathways include cleavage of the ester, and cleavage of the imide linkage. *Soil.* Rapidly degraded in soil: DT_{50} 0.48–4.4 days in loamy-sand soil (pH 7); degradates have DT_{50} c. 2–30 days. The a.i. is immobile in soil; degradates have low to medium mobility

Common name (type) (year)	Manufacturer	Mode of action	Uses	Fate
Cinidon-ethyl N-phenyl-phthalimide (1998)	BASF	Protoporphyrinogen IX oxidase inhitor —	Post-emergence control of annual broad-leaved weeds especially *Galium aparine*, *Lamium* sp. and *Veronica* sp. in winter and spring small grain cereals	*Animals.* Following limited, but rapid, absorption, and widespread distribution in organs and tissues, the a.i. is extensively metabolised and rapidly excreted. *Plants.* The a.i. is extensively metabolised. *Soil.* Readily biodegradable. Soil DT$_{50}$ 0.6–2 days (lab., aerobic conditions, 20°C); rapidly mineralized
Azafenidin (Triazolinone) (1998)	Dupont	Protoporphyrinogen oxidase inhibitor and shoots.	Herbicide absorbed through roots citrus, grapes, olives, Improves the efficacy of other post-emergence herbicides and increases the speed of action of contact herbicides	*Soil.* Degrades in soil by microbial and photolytic process. In the field, in a range of 4 soils, mean DT$_{50}$ c. 25 days mean DT$_{90}$ c. 169 days. There was minimal movement in soil column leaching studies
Carfentrazone-ethyl (Triazolinone) (1993)	FMC	Protoporphyrinogen oxidase inhibitor	Post-emergence control of a wide range of broad leaved weeds especially *Galium aparine*, *Abutilon theophrasti*, *Ipomoea hederacea*, *Chenopodium album* and several mustard species in cereals	*Animals.* c. 80% is rapidly absorbed and excreted in the urine within 24 hrs. The major metabolite was the corresponding acid. Further metabolism appears to involve oxidative hydroxylation of the methyl group or dehydrochlorination to form the corresponding cinnamic acid. *Plants.* Rapidly converted to the free acid, which is hydroxylated and then oxidised at the triazolinone methyl to form the dibasic acid; DT$_{50}$ (carfentrazone-ethyl) <7 days, DT$_{50}$ (carfentrazone) <28 days

Herbicide absorbed by foliage with limited translocation

(continued)

TABLE 17.1   Continued

Chemical (Chemical Class) (Year of Reporting)	Company's Name	Structure	Mode of Action	Biochemistry	Known Profile of Use	Environmental Fate
Fluthiacet-methyl (Thiadiazole) (1993)	Ihara/ Kumai (Also reported by Ciba-Geigy AG)		Selective, herbicide requiring light for activity	Protoporphyri-nogen oxidase inhibitor	Post-emergence control of broad-leaved weeds e.g., *Abutilon theophrasti*, *Chenopodium album*, *Amaranthus retroflexus*, *Xanthium strumarium* in *maize* and soyabeans	*Animals.* Within 48 hr 80% is eliminated via the feces, 14% via urine. Metabolism proceeds via hydrolysis of the methyl ester, isomerisation at the thiadiazole ring and hydroxylation of the tetrahydropyridazine moiety *Plants.* Organosoluble metabolites are similar to those in animals *Soil.* DT$_{50}$ (hydrolysis, pH 7) 18 days (photolysis on soil) 21 days (UV light) 2 hr. In loam soil, DT$_{50}$ 1.2 days (25°C, 75% of max. water capacity
Butafenacil (Pyrimidindione) (1998)	Novartis		Non-selective contact herbicide, rapidly absorbed by the foliage. Translocation occurs only within leaves	Protoporphyri-nogen oxidase inhibitor	Control of a wide range of annual and perennial broadleaved weeds in fruits, orchards, vineyards, citrus, non-crop land	*Soil.* Rapidly degraded in soil: DT$_{50}$ 1–2 days

Name	Company	Structure	Mode of action	Type	Uses	Fate
Carpropamid (MBI: dehydrase) (1994)	Bayer		Inhibitor of melanin biosynthesis, by inhibiting the dehydration reactions from Scytalone to 1,3,8-trihydroxy naphthalene and from vermelone to 1,8-dihydroxy naphthalene	Systemic, protective fungicide	Control of *Pyricularia oryzae* as protective treatment or seed treatment	*Animals.* Readily excreted via feces and urine. Metabolized oxidatively, mainly in the liver. *Plants.* Absorbed by the roots and translocated to the shoots. The major residue in rice was carpropamid. *Soil.* Metabolized oxidatively under paddy soil conditions; CO_2 was the major metabolite. The calculated half-lives ranged from several weeks to several months, resp. Low mobility
Fenoxanil* (MBI: dehydrase)	BASF, Nihon Nohyaku		Melanin biosynthesis inhibitor Inhibits dehydratase enzymes, which dehydrate scytalone to trihydroxy naphthalene and vermelone to dihydroxy naphthalene	Systemic, protective fungicide with residual effects	Under development for control of rice blast by foliar or into-water application in rice	—
Benzofenap (Pyrazole)	Mitsubishi/ Acentis		*p*-Hydroxyphenyl pyruvate dioxygenase inhibitor	Systemic Herbicide, absorbed principally through root and bases of target weeds	Used in combination with pyributicarb and bromo-butide, controls annual and perennial broadleaved weeds in rice	*Plants.* No detectable residues in rice crops (detection limit 0.005 ppm). *Soil.* DT_{50} 38 days. Nonmobile

(continued)

TABLE 17.1   Continued

Chemical (Chemical Class) (Year of Reporting)	Company's Name	Structure	Biochemistry	Mode of Action	Known Profile of Use	Environmental Fate
Isoxaflutole (Isoxazole) (1995)	Aventis		p-Hydroxy-phenyl pyruvate dioxygenase inhibitor	Herbicide, Systemic by either root or foliar uptake.	For pre-emergence or pre-plant broad spectrum broad broad-leaved weeds control in maize	*Animals.* Rapidly excreted *Plants.* Residue levels at harvest are very low, and comprise mainly a nontoxic metabolite *Soil.* Degradation proceeds via hydrolysis and microbial degradation, with final mineralization to CO_2. Isoxaflutole and its major metabolites are nonmobile under field conditions
Mesotrione (Triketone) (1999)	Zeneca		p-Hydroxy-phenyl pyruvate dioxygenase inhibitor	Herbicide, Uptake is foliar and via the root, with both acropetal and basipetal translocation	Pre- and post-emergence control of broad leaved weeds such as *Xanthium strumarium, Ambrosia trifida, Abutilon theophrasti, chenopodium, Amaranthus* and *polygonum* spp. and some grass weeds in maize	*Soil.* Stable to hydrolysis under sterile conditions at pH 5–9, with <10% degradation after 30 days (25°C). Degradation is influenced by soil pH; DT_{50} 31.5 days (pH 5.0% o.c. 2.0) to 4.0 d (pH 7.7% o.c. 0.9).
Sulcotrione (Triketone) (1991)	Zeneca		p-Hydroxy-phenyl pyruvate dioxygenase inhibitor	Herbicide Absorbed predominantly by leaves but also by roots	Post-emergence control of broad-leaved weeds and grasses in maize and sugarcane	*Animals.* Rapidly excreted in the urine, the major metabolite being 4-hydroxysulcotrione *Plants.* Deactivated by the formation of 2-chloro-4-methylsulfonylbenzoic acid *Soil.* DT_{50} 1–11 days. The major metabolite is 2-chloro-4-methylsulfonylbenzoic acid

Name (year)	Company	Structure	Mode of action	Uses	Fate
Halofenozide (Diacylhydrazine) (1997)	Rohm & Hass	Cl—C6H4—CONHN(C(CH3)3)—CO—C6H5	Ecdysone agonist	Systemic, ingested insecticide active by root application. Interferes with moulting affecting larval stages of insects. Also reduces fecundity in treated adults and have some ovicidal properties	*Soil.* Soil dissipation DT$_{50}$ (field) 42–267 days (five sites); turf DT$_{50}$ 3–77 days
Methoxyfenozide (Diacylhydraxine) (1997)	Rohm & Hass	[structure: 2-methyl-3-methoxybenzoyl–N(C(CH3)3)–CO–3,5-dimethylphenyl]	Ecdysone agonist	Insecticide active primarily by ingestion, also with contact, ovicidal and root systemic activity	Control of lepidopterous larvae in vines tree fruits, vegetables, row crops	*Animals.* Rapidly absorbed, metabolized via phase II conjugation and eliminated. *Soil.* Aerobic soil metabolism DT$_{50}$ 336–1100 days; field DT$_{50}$ 23–268 days
Tebufenozide (Diacylhydrazine) (1996)	Rohm & Hass	CH3CH2—C6H4—CONHNCO(C(CH3)3)–3,5-dimethylphenyl	Ecdysone agonist	Insecticide Lethally accelerates moulting process	Control of lepidopterous larvae in rice, fruit, row crops, nut-crops, vegetables, vines and forestry	*Animals.* 16 whole-molecule metabolites are formed as a result of oxidation of the alkyl substituents of the aromatic rings, primarily at the benzylic positions. *Plants.* In apples, grapes, rice, and sugar beet, the major component is unchanged tebufenozide. Small amounts of metabolites result from oxidation of the alkyl substituents of the aromatic ring, primarily at the benzylic positions. *Soil.* Metabolic DT$_{50}$ in soil 7–66 days; DT$_{50}$ for field dissipation 4–53 days. No mobility below 30 cm

(continued)

TABLE 17.1   Continued

Chemical (Chemical Class) (Year of Reporting)	Company's Name	Structure	Biochemistry	Mode of Action	Known Profile of Use	Environmental Fate
Chromafenozide (Diacylhydrazine) (1996)	Nippon Kayaku; Sankyo		Ecdysone agonist	Insecticide Initiating a precocious incomplete lethal moult	Control of lepidopteran larvae in rice, fruit, vegetables, tea, cotton, beans, and forestry	Animals. Rapidly excreted with 48 hr and is not persistent in tissues and organs. The major component excreted is unchanged chromafenozide Plants. Many minor metabolites are detected in small amounts, but the major component is unchanged chromafenozide. Soil. DT_{50} for field dissipation 44–113 days (upland soil), 22–136 days (paddy soil)
Pyrimethanil (Aminopyrimidine) (1992)	Aventis		Inhibitor of methionine biosynthesis leading to inhibition of the secretion of enzymes necessary for fungal infection	Fungicide Protectant in Botrytis and both protective and curative action in Venturia	For control of grey mould on vines, fruits, vegetables and ornamentals and of leaf scab on pome fruit	Animals. Rapidly absorbed, extensively metabolised and rapidly excreted. No evidence of accumulation, even on repeated dosing. Metabolism proceeds by oxidation to phenolic derivatives which are excreted as glucuronide or sulfate conjugates Plants. Little metabolism occurs in fruit Soil. Rapid degradation, DT_{50} 7–54 days. Low potential for leaching
Cyprodinil (Anilino pyrimidine) (1994)	Novartis		Inhibitor of methionine biosynthesis and secretion of fungal hydrolytic enzymes	Systemic Fungicide with uptake into plants after foliar application and transport throughout the tissue and acropetally in the xylem. Inhibits penetration	Control wide range of pathogens like Tapesia yallundae, T. acuformis, Erysiphe spp., Pyrenophora teres, Rhynchosporium secalis, Botrytis spp., in cereals, grapes, pome fruit, stone fruit, strawberries, vegetables, field crops and ornamentals, barley Alternaria spp., Venturia spp. and	Animals. Rapidly absorbed and almost completely eliminated with urine and faeces. Metabolism proceeds by 4-hydroxylation of the phenyl and 5-hydroxylation of the pyrimidine rings, followed by mono- or di-sulfation. No evidence for accumulation or retention of cyprodinil or its metabolites Plants. Metabolism mainly via hydroxylation of the 6-methyl group of the pyrimidine ring, as well as hydroxylation of the phenyl and pyrimidine rings Soil. DT_{50} 20–60 days Formation of bound residues the major route for dissipation. Immobile in soil

Common name (year)	Company	Biochemical mode of action	Type of action	Uses	Structure	Metabolism and fate
			and mycelial growth both inside and on the leaf surface	monilinia spp.		
Fentrazamide (Tetrazolinone) (1997)	Bayer	Cell division inhibitor. Primary target site may be fatty acid metabolism	Herbicide inhibiting cell division in root and meristem	Control of barnyard grass *Echinochloo* spp. and annual sedges in rice for pre-emergence	Cl, N, CH_2CH_3, O, N=N	*Animals.* The main pathway of biotransformation proceeded via hydrolytic cleavage of the parent compound *Plants.* No parent compound was detected in any plant fraction *Soil.* Thoroughly degraded and mineralized Calculated half-lives were in the range of a few days and several weeks, respectively. Immobile
Flufenacet (Oxyacetamide) (1995)	Bayer	Cell division inhibitor. Primary target site may be fatty acid metabolism	Pre- and early Post-emergence herbicide	Selective herbicide with broad spectrum grass control and control of some broad leaved weeds in maize, soybeans, sunflower, wheat, rice	$CH(CH_3)_2$, N, CH_2C, O, N–N, S, F_3C, F	*Animals.* Rapidly excreted. Metabolism takes place via cleavage of the molecule, followed by conjugation of the fluorophenyl moiety with cysteine and formation of a thiadazolone and its various conjugates *Plants.* Rapidly and extensively metabolized; no parent compound was detected, even at early sampling dates *Soil.* Rapidly degraded, immobile
Dithiopyr (Pyridine) (1994)	Rohm & Hass	Inhibits cell division by disrupting spindle microtubule formation	—	Pre-emergence and early post-emergence control of annual grass and broad-leaf weeds in turf	F_3C, N, CHF_2, $COSCH_3$, $CH_2CH(CH_3)_2$, CH_3SOC	*Animals.* Rapidly absorbed, extensively metabolized and rapidly excreted *Soil.* DT_{50} in soil 17–61 days, depending on the formulation type. The major soil metabolites are the di-acid, the normal mono-acid and the reverse mono-acid; these metabolites, themselves, dissipate almost completely within 1 year
Thiazopyr (Pyridine) (1994)	Rohm & Hass	Inhibits cell division by disrupting spindle	Herbicide causing root growth inhibition and	Pre-emergence control of annual grass and some broad-leaved weeds in tree fruit,	S, $CH_2CH(CH_3)_2$, CO_2CH_3, N, F_3C, CF_2H	*Animals.* Rapidly and extensively metabolized and eliminated. Oxidized by rat liver microsomes via sulfur and carbon oxidations and via oxidative de-esterification

(continued)

TABLE 17.1 Continued

Chemical (Chemical Class) (Year of Reporting)	*Company's Name*	*Structure*	*Biochemistry*	*Mode of Action*	*Known Profile of Use*	*Environmental Fate*
			microtubule formation	swelling in meristematic regions	vines, citrus, sugarcane, pineapple alfalfa, forestry	*Plants.* Initially metabolized in the dihydrothiazole ring by plant oxygenases to the sulfoxide, sulfone, hydroxy derivative and thiazole, and is also de-esterified to the carboxylic acid *Soil.* Degraded by both soil microorganisms and hydrolysis. Average DT_{50} 64 days (8–150 days). Minimal mobility: The monoacid metabolite also has limited mobility
Acibenzolar-*S*-methyl Plant activator/ Plant host defense induces (1995)	Novartis	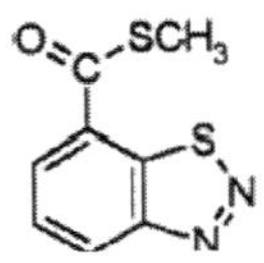	Acts as a functional analogue of the natural signal molecule for systemic activated resistance, salicylic acid	Activates plants' natural defense mechanism (systemic activated resistance [SAR]). Has no intrinsic fungicidal activity	For control of fungal infections in wheat under development against a range of diseases in rice, bananas, vegetables, and tobacco	*Animals.* Rapidly absorbed and also rapidly almost completely eliminated with urine and feces. No evidence of accumulation or retention of acibenzolar-*S*-methyl or its metabolites *Plants.* The metabolism proceeds via hydrolysis with subsequent conjugation with sugars, or by oxidation of the phenyl ring followed by sugar conjugation *Soil.* Dissipates via hydrolysis; DT_{50} 0.3 day. The product further degrades, DT_{50} 20 days; metabolites become completely degraded and mineralized. Strong adsorption to soil, low mobility
Pethoxamid Acetamide (2001)	Tokuyama	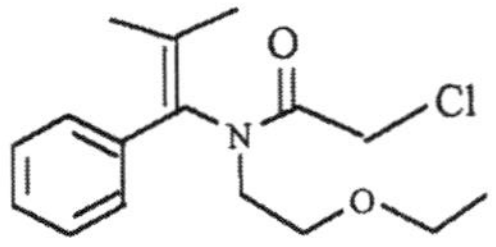	Presumed to be acting by inhibiting fatty acid biosynthesis	Precise mode of action not yet been clarified. Absorbed by roots and young shoots after application to soil surface	Controls grass weeds including *Echinochloa cruss-galli, Digitaria sanguinalis* and *Setaria geniculata* and broad leaf weeds such as *Amaranthus retroflexus Chenopodium album*, *Convolvulus arvensis* and *Polygonum pericaria*	—

The "players" include:

1. Large, multinational companies engaged in discovery, manufacture, and distribution of agrochemicals, seeds, and other products of biotechnology.
2. Large corporate entities engaged in the manufacturer and distribution of off-patent agrochemicals.
3. Companies engaged in the formulation and distribution of agrochemicals purchased from (1) or (2) above.
4. Retailers who make end-use products available to growers.
5. Extension workers from governmental or nongovernmental sources who provide guidance on the proper use of agrochemicals in the field.
6. Professional consultants who perform activities in (5) to individual growers.
7. Pest control operators who are professionally trained to properly apply restricted use of agrochemicals as well as other agrochemicals deployed in disease vector control and termite control.

For the year 2000, it was estimated that the global sales revenue of the agrochemical industry amounted to US$30 billion. This breaks down as follows.

Herbicides	US$14 billion	73% for cereals, maize, soybeans, fruits, vegetables
Insecticides	US$8 billion	71% for fruits, vegetables, cotton, rice
Fungicides	US$6 billion	70% for cereals, fruits, vegetables
Others	US$1 billion	

The revenues were distributed globally approximately as follows: North America—27.2 percent; Far East—26 percent; Europe—25.5 percent; Latin America—14.8 percent; and the rest of the world—6.5 percent.

CHARACTERISTICS OF THE AGROCHEMICAL INDUSTRY

Among the distinguishing characteristics of the agrochemical industry are: (1) the multitude of chemical agents employed, (2) a limited price range (which derives from the limited chemical complexity, in turn driven by by the economics of agricultural production), (3) a fairly rapid obsolescence of the chemicals used, and (4) a high degree of government regulation for the production, application, shipment, and use of agrochemicals.

Government Regulation

In the United States, the first state laws on insecticides were enacted in 1900 to establish standards of purity for the arsenical Paris green (copper acetoarsenite) which is no longer used in agriculture in the United States. Gradually these laws were extended to cover a wide list of inorganic compounds and plant extracts, many of them, like Paris green, extremely toxic to humans. Included in this group are such compounds as arsenic combined with copper, lead, and calcium; phosphorus pastes for ants and roaches; strychnine in rodent baits; thallium in ant and rodent baits; and selenium for plant-feeding mites. Mercury, both as a corrosive sublimate and as calomel, was used as an insect repellent and later as a seed disinfectant. Sodium fluoride was a common ant poison, and sodium cyanide, calcium cyanide, and HCN itself were general fumigants (Table 17.2). Nicotine sulfate was used generally in the garden and on the farm. These compounds, among the most toxic of any known at that time, were widely marketed without supervision under any of the early state laws. There was no provision for public health, either in regulating the amounts applied or regarding the possible danger of minute amounts (residues) remaining on the marketed produce. The need to protect the applicator, farmer, laborer, and the general public against the dangerous qualities of the insecticides, or their residues on crops, provided the motives for all the legislation that followed.

Governmental concern was first related to standardization of the manufactured chemical and protection of the farmer in relation to the product that he or she purchased. This was then extended to the handling of the chemical in interstate commerce, to the protection of the consumer of raw agricultural products

TABLE 17.2 Fumigants

Chemical (Chemical Class)	Structure	Biochemistry	Mode of Action	Environmental Fate
Chloropicrin	CI_3CNO_2	—	Fumigant	—
Dazomet (Methyl isothiocyanate precursor)		Nonselective inhibition of enzymes by degradation products	A pre-planting soil fumigant, acting by decomposition to methyl Eothocyanate	*Plants.* Following application to strawberries, noresidues of dazomet or of its degradation products methyl isothiocyanate, dimethyl- or monomethylthiourea were detected at >0.01 ppm in the fruit *Soil.* In the presence of moisture, undergoes degradation to methyl-(methylaminomethyl) dithiocarbamic acid, which then undergoes further degradation to methyl isothiocyanate, formaldehyde, hydrogen sulfide and methylamine
1,3-dichloropropene (Chloroalkene)	—	—	Soil fumigant nematicide	—
Methyl Bromide	CH_3Br	—	Fumigant insecticide and nematicide	*Animals/Plants.* Metabolism not totally elucidated; inorganic bromide ion is formed

(e.g., apples, corn, and lettuce), and, in other legislation, to the protection of the consumer of finished goods (e.g., canned juice, margarine, cereal food, meat, and milk). Included in this legislation were provisions that protect the shipper of the chemical, the applicator of the chemical, and all personnel proximal to the application of the chemical. Legislation now regulates chemicals applied to crops or foods as protective agents—pesticides, emulsifiers, solvents, packaging materials (wax, container materials, plasticizers, antioxidants, etc.).

Toward the end of the 1960s, a new area of concern arose: the effect of the manufacture and application of pesticide chemicals *on the environment* was recognized. Concern for the environment was the subject matter of Rachel Carson's book, *Silent Spring*, which was published in 1962. In 1970, this new focus led to the establishment of the Environmental Protection Agency (EPA; aka US-EPA), which was given authority to regulate virtually all aspects of agrochemical manufacture and use in the United States. Since its inception, the principal objectives of the EPA's agrochemical activities have been to: (1) establish procedures that ensure that new pesticides will

not pose unreasonable risks to human health and the environment, and (2) terminate the use of those previously registered pesticides that exceed certain risk criteria. Among the requirements called for are studies on mammalian toxicology (including lifetime animal feeding studies), environmental chemistry (persistence, mobility, etc.), and effects on fish and other wildlife. New product registrations are granted only after EPA scientists and administrators are satisfied that use of the product does not pose unreasonable risk to humans or hazard to the environment. All aspects of the environment are considered. Soil, air, and water (streams, lakes, oceans, rivers, marshes, and underground aquifers) are matters of environmental concern, as are the living organisms that reside therein. The protocols employed to ensure safety are complex and not infrequently are at the boundaries of scientific capability. The registration of any new product is a highly complicated and expensive procedure. As to the second objective, much the same criteria are used in judging whether or not to allow continued use of previously registered pesticides.

Although comprehensive regulatory legislation was developed first and most extensively in the United States, all the technically developed nations of the World now regulate the manufacture, sale, and use of agricultural chemicals. The criteria used are not unlike those which were developed over the past several years in the United States. In Europe, the European Economic Community (EEC) in 1991 adopted Directive 91/414/EEC which, included the following goals.*

1. Coordinate the overall arrangements for authorization of plant protection products within the European Union. Whereas it is intended to coordinate the process for considering the safety of particular substances at the Community level, individual Member States have responsibility for product authorization.
2. Establish a list of active substances which have been shown to be without unacceptable risk to humans or the environment.
3. Maintain an up-to-date listing (Annex I of the Directive) of active substances which have been authorized.
4. Member States can authorize the sale and use of plant protection products only if they are listed in Annex I.

Under provisions of directive 91/414/EEC, all existing agrochemicals are being reviewed and new ones approved using common database criteria. The review process started in 1993 and is expected to be completed by 2008. Many agrochemicals will cease to be used in the European Union because of the failure of participating companies to submit full dossiers on them. Additionally, water quality directive 98/83/EC demands detectability for a given agrochemical in water below 1 ppb.

The 29 country member Organization for Economic Cooperation and Development Working Group offers a common platform to the national pesticide regulators for discussion of activities on conventional, biological, and microbial pesticides.

On a broader scale, several years ago the Food and Agriculture Organization of the United Nations adopted the International Code of Conduct on the Distribution and Use of Pesticides.[†] A few of the many provisions of that Code are listed below to give a general idea of its thrust.

1. Governments have the overall responsibility and should take the specific powers to regulate the distribution and use of pesticides in their countries.
2. The pesticide industry should adhere to the provisions of this Code as a standard for the manufacture, distribution, and advertising of pesticides, particularly in countries lacking appropriate legislation and advisory services.
3. Manufacturers and traders should supply only pesticides of adequate quality, packaged, and labeled as appropriate for each specific market.

Manufacture of Agrochemicals

The manufacturing route to an agrochemical can have multiple options. The route chosen may depend on commercial availability of desired reagents, engineering capabilities, byproduct formation, separation techniques, and so on. This is on par with the manufacture of drugs, dyestuffs, or speciality chemicals. One feature unique to agrochemicals is a detailed and precise label preapproved by a regulatory body on the product safety, usage instructions, compatibility statements, and other statements as deemed necessary for the needs of medical professionals and general public information.

Classes of Agrochemicals

Agrochemicals that control insects by growth regulation, or by mortality through contact or stomach action are called *insecticides* (Table 17.3a,b). Those that control competing weeds (grasses, broad leaved plants, or sedges) through preplant incorporation, preemergence, early post- or post-emergence application with respect to the main crop are called *herbicides* (Table 17.4a–d). These

*http://www.pesticides.gov.uk/
[†]http://www.fao.org/ag/agp/agpp/pesticid/Code/References.htm

TABLE 17.3a   Insecticides for Sucking Pests

Chemical (Chemical Class)	Structure	Biochemistry	Mode of Action	Environmental Fate
Acephate (Organophosphorous)		Cholinesterase inhibitor	Systemic insecticide	*Animals.* Metabolized to methamidophos (*q.v.*). *Plants* Residual activity lasts for c. 10–15 days. The major metabolite is methamidophos (*q.v.*). *Soil.* Readily biodegraded and non-persistent; soil DT$_{50}$ 2 days (aerobic) to 7 days (anaerobic). Methamidophos (*q.v.*) has been identified as a soil metabolite
Imidacloprid (Neonicotinoid)		Acts as an antagonist by binding to post-synaptic nicotinic receptors in the insects' central nervous system	Systemic insecticide with translaminar activity and with contact and stomach action. Readily taken up by plant and further distributed acropetally, with good root-systemic action	*Animals.* The radioactivity was quickly and almost completely absorbed from the gastrointestinal tract and quickly eliminated (96% within 48 hr, mainly via the urine). Only c. 15% was eliminated as unchanged parent compound; the most important metabolic steps were hydroxylation at the imidazolidine ring, hydrolysis to 6-chloronicotinic acid, loss of the nitro group with formation of the guanidine and conjugation of the 6-chloronicotinic acid with glycine. All metabolites found in the edible organs and tissues of farm animals contained the 6-chloronicotinic acid moiety *Plants.* Metabolized by loss of the nitro group hydroxylation at the imidazolidine ring, hydrolysis to 6-chloronicotinic acid and formation of conjugates; all metabolites contained the 6-chloropyridinylmethylene moiety *Soil.* The most important metabolic steps were oxidation at the imidazolidine ring, reduction or loss of the nitro group, hydrolysis to 6-chloronicotinic acid and mineralisation. Medium adsorption to soil, imidacloprid and soil metabolites are to be classified as immobile

| Methamidophos (Organophosphorous) | 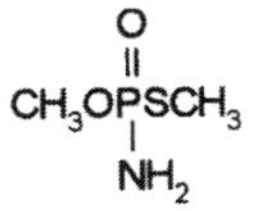 CH_3OPSCH_3 with O (double bond) and NH_2 | Cholinesterase inhibitor | Systemic insecticide with contact and stomach action | *Animals.* Absorbed rapidly and distributed uniformly among all organs and tissues. More than half of the radioactivity was rapidly eliminated from the body, mainly via urine and respiratory air. Radioactivity remaining in the animal was incorporated into endogenous compounds (carbon-1 pool) and eliminated with the natural turnover of these compounds. Metabolism in the rat was by deamination and demethylation.
Plants. Taken up rapidly and translocated into the leaves
Soil. Rapidly degraded in soil; field DT_{50} c. <2 days |
| Fipronil (Phenyl pyrazole) | 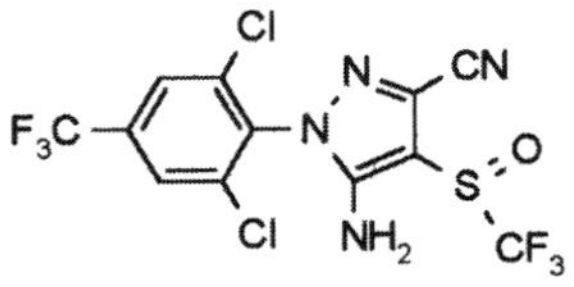 (structure: phenyl pyrazole with Cl, Cl, F_3C, CN, NH_2, $S=O$, CF_3) | Acts as a potent blocker of the GABA-regulated chloride channel | Broad spectrum insecticide, toxic by contact and ingestion. Moderately systemic, Good to excellent residual control following foliar application | In plants, animals and the environment, fipronil is metabolized via reduction to the sulfide, oxidation to the sulfone, and hydrolysis to the amide
Animals. Distribution is rapid. Elimination is mainly via the feces as fipronil and its sulfone. The two major urinary metabolites were identified as conjugates of ring-opened pyrazole products. The distribution of radioactive residues in tissues was extensive after 7 days
Plants. Uptake of fipronil into plants was low (c. 5%). At crop maturity, the major residue components were fipronil, the sulfone, and the amide. Following foliar application to cotton, cabbage, rice and potatoes, at crop maturity, fipronil and the photodegradate were the major residue components
Soil. Readily degraded: major degradates in soil (aerobic) are sulfone and amide, (anaerobic) are sulfide and amide. Present a low risk of downward movement in soil |

(continued)

TABLE 17.3a   Continued

Chemical (Chemical Class)	Structure	Biochemistry	Mode of Action	Environmental Fate
Monocrotophos (Organophosphorous)		Cholinesterase inhibitor	Systemic insecticide with contact and stomach action	*Animals.* In mammals, following oral administration, 60–65% is excreted within 24 hr, predominantly in the urine *Soil.* Rapidly degraded in soil; DT_{50} (lab.) 1–5 days
Endosulfan (Cyclodiene organochlorine)		Antagonist of the GABA receptor-chloride channel complex	Nonsystemic insecticide with contact and stomach action	*Animals.* The principal route of elimination is feces; most of the radioactivity is excreted within the first 48 hr. Metabolized rapidly to less-toxic metabolites and to polar conjugates. *Plants.* The plant metabolites (mainly endosulfan sulfate) were also found in animals 50% of residues are lost in 3–7 days (depending on plant species) *Soil.* DT_{50} 30–70 days. The main metabolite was endosulfan sulfate, which is degraded more slowly. DT_{50} for total endosulfan (alpha- and beta-endosulfan and endosulfan sulfate) in the field is 5–8 months. No leaching tendency
Butocarboxim (Oxime carbamate)		Cholinesterase inhibitor	Systemic insecticide with contact and stomach action	*Animals.* Metabolized to butoxycarboxim, and excreted in the urine as butoxycarboxim and its degradation products *Plants/Soil.* The methylamine moiety is split off, and the sulfur atom is oxidized to sulfoxide and sulfone. DT_{50} in soil 1–8 days DT_{50} for metabolites 16–44 days
Disulfoton (Organophosphorous)		Cholinesterase inhibitor	Systemic insecticide, absorbed by roots, with translocation to all parts of the plant	*Animals.* ^{14}C-disulfoton is rapidly absorbed, metabolized, and the radioactivity excreted in the urine. The main metabolites are disulfoton sulfoxide and sulfone, their corresponding oxygen analogues and diethylthiophosphate

Flucythrinate
(Pyrethroid)

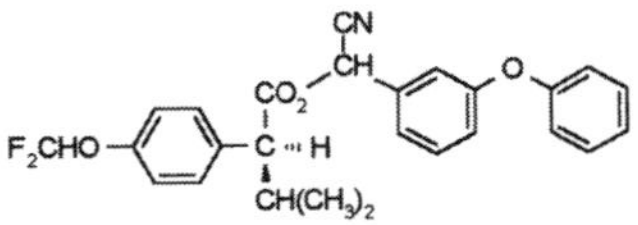

Nonsystemic insecticide
with contact and stomach
action

Plants.   Very rapidly metabolized. The
metabolism is the same as in animals

Soil.   Very rapidly degraded. The metabolism is
similar to that in animals and plants. It exhibits
medium to low mobility in soil

Animals.   60–70% is eliminated within 24 hr,
and >95% within 8 days, in the feces and urine.
In the facets, the parent compound makes up
most of the material excreted, but in the urine and
in tissue, several metabolites are present. The major
route of degradation is through hydrolysis, with
subsequent hydroxylation of the hydrolysis products

Soil.   Immobile DT_{50} c. 2 months

TABLE 17.3b   Insecticides for Chewing Pests

Chemical (Chemical Class)	Structure	Biochemistry	Mode of Action	Environmental Fate
Carbaryl (Carbamate)		Weak cholinesterase inhibitor	Insecticide with contact and stomach action and slight systemic properties	*Animals.* Does not accumulate in body tissues, but is rapidly metabolized to nontoxic substances, particularly 1-naphthol. This, together with the glucuronic acid conjugate, is eliminated predominantly in the urine and feces. *Plants.* Metabolites are 4-hydroxycarbaryl, 5-hydroxycarbaryl and methylol-carbaryl. *Soil.* DT$_{50}$ (aerobic) 7–14 days in a sandy loam and 14–28 days in a clay loam
Fenitrothion (organophosphorous)		Cholinesterase inhibitor	Nonsystemic insecticide with contact and stomach action	*Animals.* Rapidly excreted in the urine and feces. After 3 days c. 90% has been excreted by rats, mice and rabbits. The most important metabolites are dimethylfenitrooxon and 3-methyl-4-nitrophenol. *Plants.* DT$_{50}$ 4 days; 70–85% is degraded within 2 weeks. Major metabolites are 3-methyl-4-nitrophenol, the oxygen analogue and their decomposition products desmethylfenitrothion, dimethylphosphorothionic acid and phosphorothionic acid. *Soil.* DT$_{50}$ 12–28 days under upland conditions, 4–20 days under submerged conditions. The major metabolites under upland conditions are 3-methyl-4-nitrophenol and CO$_2$, whereas, under submerged conditions, the major decomposition product is aminofenitrothion
Methoxychlor (Organochlorine)			Insecticide with contact and stomach action	*Animals.* Degradation in animals is principally by O-dealkylation to the corresponding phenol and diphenol, and by dehydrochlorination to 4,4′-dihydroxybenzophenone
Chlorpyrifos (Organophosphorous)		Cholinesterase inhibitor	Nonsystemic insecticide with contact, stomach and respiratory action	*Animals.* Rapid metabolism occurs, the principal metabolite being 3,5,6-trichloropyridin-2-ol. Excretion is principally in the urine. *Plants.* Residues are metabolized to 3,5,6-trichloropyridin-2-ol which is conjugated and sequestered. *Soil.* Field DT$_{50}$ for soil-incorporated applications 33–56 days for soil–surface applications 7–15 days. Primary route of degradation is transformation to 3,5,6-trichloropyridin-2-ol, which is subsequently degraded to organochlorine compounds and CO$_2$

| Oxamyl (Oxime carbamate) | (CH₃)₂NCOC = NOCONHCH₃ with SCH₃ | Cholinesterase inhibitor | Contact and systemic insecticide, absorbed by foliage and roots | *Animals.* Hydrolyzed to an oximino metabolite (methyl *N*-hydroxy-*N′*,*N′*-dimethyl-1-thiooxamimidate) or converted enzymically via *N*,*N*-dimethyl-1-cyanoformamide to *N*,*N*-dimethyloxamic acid. Conjugates of the oximino compound, the acid, and their monomethyl derivatives constituted over 70% of thermetabolites excreted in the urine and feces |

$(CH_3)_2NCOC = NOCONHCH_3$ with SCH_3

Cholinesterase inhibitor

Contact and systemic insecticide, absorbed by foliage and roots

Animals. Hydrolyzed to an oximino metabolite (methyl *N*-hydroxy-*N′*,*N′*-dimethyl-1-thiooxamimidate) or converted enzymically via *N*,*N*-dimethyl-1-cyanoformamide to *N*,*N*-dimethyloxamic acid. Conjugates of the oximino compound, the acid, and their monomethyl derivatives constituted over 70% of thermetabolites excreted in the urine and feces

Plants. Hydrolyzes to the corresponding oximino compound which, in turn, conjugates with glucose. Total breakdown into natural products has been demonstrated

Soil. Degraded rapidly in soil, DT_{50} c. 7 days

Pyridaphenthion (Organophosphorous)

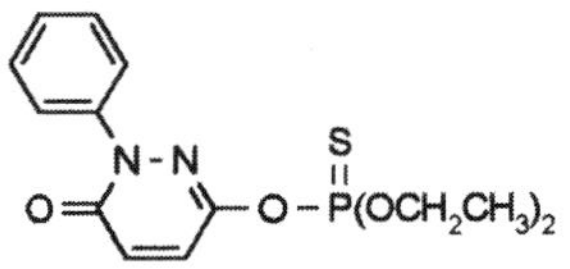

Cholinesterase inhibitor

Insecticide with contact and stomach action

Animals. In mice and rats, the parent compound, *O*-ethyl-*O*-(3-oxo-2-phenyl-2*H*-pyridazine-6-yl) phosphorothioate and the corresponding phosphate are found

Plants. In rice, phenyl maleich ydrazide, *O*,*O*-diethyl thiophosphoric acid, and PMH glycoside are formed

Soil. DT_{50} 11–24 days

TABLE 17.4a   Cerebral Herbicides

Chemical (Chemical Class)	Structure	Biochemistry	Mode of Action	Activity	Environmental Fate
Isoproturon (Urea)		Photosynthetic electron transport inhibitor at the photosystem II receptor site	Selective systemic herbicide, absorbed by roots and leaves, with translocation	Graminicide and broadleaf weeds controller	*Animals.* 50% is eliminated within 8 hr, predominantly in urine *Plants.* Degradation mainly via hydroxylation of the isopropyl group to 1,1-dimethyl-3-[4(2'-hydroxy-2'-propyl)phenyl] urea; *N*-dealkylation also occurs *Soil.* Undergoes enzymic and microbial demethylation at the nitrogen, and hydrolysis of the phenylurea to 4-isopropylamine DT_{50} 6–28 days
Bromoxynil (Hydroxybenzonitrile)		Photosynthetic electron transport inhibitor at the photosystem II receptor site, also uncouples oxidative phosphorylation	Selective contact herbicide with some systemic activity. Absorbed by foliage with limited translocation	Broadleaf weed controller	*Animals/plants.* Metabolism by hydrolysis of the ester and nitrile groups with some debromination occurring *Soil.* DT_{50} *c.* 10 days. Degraded by hydrolysis and debromination to less toxic substances such as hydrobenzoic acid
Tribenuronmethyl (Sulfonyl urea)		Branched chain amino acid synthesis (ALS and AHAS) inhibitor. Acts by inhibiting biosynthesis of the essential amino acids valine and isoleucine, hence stopping cell division and plant growth	Rapidly absorbed by foliage and roots and translocated throughout the plant	Broad leaf weed controller	*Soil.* Degrades by hydrolysis and direct microbial degradation. Hydrolysis is faster in acidic than alkaline soils. DT_{50} 1–7 days

Compound	Structure	Mode of action	Use	Spectrum	Metabolism
Clodinafoppropargyl [(2-(4-aryloxyphenoxy) propionic acid]		Fatty acid synthesis inhibitor, by inhibition of acetyl COA carboxylase (Assase)	Post-emergence, systemic herbicide	Graminicide	*Animals.* Hydrolyzed to the corresponding acid *Plants.* Rapidly degraded to the acid derivative as major metabolite *Soil.* Undergoes rapid degradation to the free acid (DT_{50} <2 hr) and then further to phenyl and pyridine moieties which are bound to the soil and mineralized. The free acid is mobile in soil, but is further degraded with DT_{50} 5–20 days; negligible leaching potential
Carfentrazoneethyl (Triazolinone)		Protoporphyrinogen oxidase inhibtor leading to membrane disruption	Absorbed by foliage with limited translocation	Broad leaf weed controller	*Animals.* About 80% excreted in urine within 24 hr. Major metabolite is corresponding acid. Further metabolism appears to involve oxidative hydroxylation of the methyl group or dehydrochlorination to form corresponding Cinnamic acid *Plants.* Converted to free acid, which is hydroxylated and oxidized at triazolinone methyl to form the dibasic acid. DT_{50} <7 days *Soil.* Degradation by microbial action. Strongly adsorbed to sterile soils. In nonsterile soils, rapidly converted to free acid, which has low soil binding. In laboratory soil DT_{50} is a few hours
Flufenacet (Oxyacetamide)		Inhibits cell division and growth. Primary target site may be fatty acid metabolism	Pre and early post-emergence herbicide	Broad spectrum herbicide controlling grasses and broad leaved weeds	*Animals.* Rapidly excreted by rat, goat, and hen. Metabolism via cleavage of the molecule followed by conjugation of flurophenyl moiety with cysteine and formation of a thiadozolone and its various conjugates *Plants.* Rapidly metabolized, residues accounted based on total amount of N-flurophenyl-N-isopropyl derived residues *Soil.* Rapidly degraded in soil. No threat of leaching

Table 17.4b   Maize Herbicides

Chemical (Chemical Class)	Structure	Biochemistry	Mode of Action	Activity	Environmental Fate
Atrazine (1,3,5-triazine)		Photosynthetic electron transport inhibitor at the photosystem II receptor site	Selective systemic herbicide, absorbed principally through roots, but also through foliage, with translocation acropetally in the xylem and accumulation in the apical meristems and leaves	Cross spectrum weed controller	*Animals.* Rapidly and completely metabolized, primarily by oxidative dealkylation of the amino groups and by reaction of chlorine atom with endogenous thiols. Diaminochlorotriazine is the main primary metabolite, which readily conjugates with glutathione. More than 50% of the dose is eliminated in the urine and around 33% in feces within 24 hr. *Plants.* In tolerant plants, readily metabolized to hydroxyatrazine and amino acid conjugates, with further decomposition of hydroxyatrazine by degradation of side-chains and hydrolysis of resulting amino acids on the ring together with evolution of CO_2. *Soil.* Major metabolites are desethylatrazine and hydroxyatrazine; DT_{50} 16–77 days
Bromoxynil (Hydroxybenzonitrile)		Photosynthetic electron transport inhibitor at the photosystem II receptor site, also uncouples oxidative phosphorylation	Selective contact herbicide with some systemic activity. Absorbed by foliage with limited translocation	Broadleaf weed controller	*Animals/plants.* Metabolism by hydrolysis of the ester and nitrile groups with some debromination occurring. *Soil.* DT_{50} c. 10 days: Degraded by hydrolysis and debromination to less toxic substance such as hydrobenzoic acid
Dicamba (Benzoic acid)		Synthetic auxin (acting like indolylacetic acid)	Selective systemic herbicide, absorbed by the leaves and roots, with ready translocation throughout the plant via both symplastic and apoplastic systems	Broadleaf weed controller	*Animals.* Rapidly eliminated in the urine, partly as a glycine conjugate. *Plants.* Degradation rate varies greatly with species. In wheat, the major metabolite is 5-hydroxy-2-methoxy-3,6-dichlorobenzoic acid, while 3,6-dichlorosalicylic acid is also a metabolite. *Soil.* Microbial degradation occurs, the principal metabolite being 3,6-dichlorosalicylic acid. DT_{50} <14 days

Name (class)	Structure	Mode of action	Activity	Use	Metabolism
Clopyralid (Pyridinecarboxylic acid)		Synthetic auxin (acting like indalyl acetic acid)	Selective systemic herbicide, absorbed by roots and leaves with translocation both acropetally and basipetally and accumulation in meristematic tissue	Broad leaf weed controller	*Animals.* In rats, there is rapid and almost quantitative unchanged elimination in urine. *Plants.* Not metabolized in plants. *Soil.* Microbial degradation occurs. Major product is CO_2
Metolachlor (Chloroacetamide)		Cell division inhibitor	Selective herbicide, absorbed predominantly by the hypocotyls and shoots. Inhibits germination	Grass weed controller	*Animals.* Rapidly oxidized by rat liver microsomal oxygenases via dechlorination, O-demethylation and side-chain oxidation. *Plants.* Metabolism involves natural product conjugation of the chloroacetyl group and hydrolysis and sugar conjugation at the ether group. Final metabolites are polar, water-soluble, and nonvolatile. *Soil.* Major aerobic metabolites are derivatives of oxanilic and sulfonic acids. $DT_{50} = 20$ days
Nicosulfuron (Sulfonylurea)		Branched chain amino-acid (ALS) and (AHAS) synthesis inhibitor. Acts by inhibiting biosynthesis of essential amino acids valine and isoleucine, hence stopping cell division and plant growth	Selective systemic herbicide, absorbed by roots and leaves with rapid translocation in xylem and phloem to the meristematic tissues	Graminicide and broadleaf weed controller	*Animals.* Nicosulfuron and its metabolites do not bioaccumulate. Hydrolysis of the sulfonylurea bridge and hydroxylation were the main metabolic pathways. *Plants.* Degraded rapidly. DT_{50} 1.5–4.5 days main metabolic pathways were hydrolysis of the sulfonyl urea bridge to form the pyridine sulfonamide and pyrimidine amine, and hydroxylation on the pyridine ring. *Soil.* DT_{50} 24–43 days

(continued)

Table 17.4b   Continued

Chemical (Chemical Class)	Structure	Biochemistry	Mode of Action	Activity	Environmental Fate
Diflufenzopyr (Semi-carbazone)		Inhibits auxin transport, apparently by binding with a carrier protein on the plasmalemma	Systemic, post-emergence herbicide	Graminicide and broad leaf weed controller	*Animals.* 20–44% of oral dose eliminated in urine and 49–79% in feces. Intravenous administration showed excretion of 61–89% in urine. Elimination DT_{50} in urine and feces was about 6 hr. Eliminated mainly as the parent compound *Soil.* Average DT_{50} in field: 4.5 days very mobile
Isoxaflutole (Isoxazole)		*p*-hydroxyphenyl pyruvate diooxygenate inhibitor	Systemic by either root or foliar uptake	Graminicide and broadleaf weed controller	*Animals.* Rapidly excreted *Plants.* Residues at harvest comprise mainly of a nontoxic metabolite *Soil.* Degradation via hydrolysis and microbial degradation with final mineralization to CO_2. Mobile under simulated conditions. However, under field conditions, residues remain in the surface horizons

TABLE 17.4c   Rice Herbicide

Chemical (Chemical Class)	Structure	Biochemistry	Mode of Action	Activity	Environmental Fate
Imazosulfuron (Sulfonylurea)	(structure: pyrimidine with two OCH₃ groups, SO₂NHCONH linkage to chloro-imidazopyridine)	Branched chain amino acid synthesis (ALS and AHAS) inhibitor. Acts by inhibiting biosynthesis of the essential amino acids valine and Isoleucine, hence stopping cell division and plant growth	Absorbed by plants mainly through roots and translocated throughout the plant. Inhibits shoot growth and restarts root development	Broad leaf weed and sedge controller	Unknown
Anilofos (Organophosphorous)	(structure: Cl–phenyl–N(CH(CH₃)₂)COCH₂S–P(=S)(OCH₃)₂)	Inhibits cell division	Selective herbicide, absorbed through roots and to some extent, through leaves	Grass weed and sedge controller	*Soil.* Metabolizes into Chloroaniline and CO_2 DT_{50}:30–45 days
Azimsulfuron (Sulfonylurea)	(structure: pyrimidine with two CH₃O groups, NHCONHSO₂ linkage to pyrazole bearing CH₃–N and N=N–N–CH₃ tetrazole)	Branched chain amino acid synthesis (ALC or AHAS) inhibitor. Acts by inhibiting biosynthesis of essential amino acids valine and isoleucine, hence stopping cell division and plant growth	Post-emergence herbicide with mainly foliar uptake, translocated in xylem and phloem	Broad leaf weed and sedge controller and control of *Echinochloa* spp.	*Animals.* >95% was extracted within 2 days 6–73% in unmetabolized form. The major metabolic pathway was O-demethylation followed by pyrimidine ring hydroxylation and subsequent O-conjugation, a pyrimidine ring-cleaved guanidine was also identified *Plants.* Metabolism was rapid; little parent compound was found in any plant tissue at maturity *Soil.* The most significant mechanisms are indirect photolysis and soil metabolism, together with chemical hydrolysis

(continued)

TABLE 17.4c   Continued

Chemical (Chemical Class)	Structure	Biochemistry	Mode of Action	Activity	Environmental Fate
Mefenacet (Oxyacetamide)		Inhibits cell division and growth	Selective herbicide	Graminicide	*Animals.* Degrades to N-methylaniline which is subsequently demethylated, acetylated and hydroxylated to 4-aminophenol and its sulfate and glucuronide conjugates. *Plants.* Besides 4-aminophenol, benzothiazolone and benzothiazoylacetic acid are found, both of which are formed by hydroxylation. *Soil.* Metabolites formed are benzothiazole and benzothiazolyl-acetic acid
Cyhalofop-butyl (Aryloxyphenoxy propionate)		Fatty acid synthesis inhibitor, by inhibition of acetyl CoA Carboxylase (ACCase)	—	Graminicide	*Animals.* Metabolized by hydrolysis to acid which may further break down to other metabolites which in turn are rapidly excreted. *Soil.* Rapidly metabolized to acid. DT_{50} 2–10 hr in field
Oxadiazon (Oxadiazole)		Protoporphyrinogen oxidase inhibitor	Selective contact herbicide	Graminicide and broadleaf weed controller	*Animals.* 93% is eliminated within 72 hr, predominantly in the urine. *Plants.* Rapidly metabolized. Metabolites do not accumulate. *Soil.* Strongly adsorbed by soil colloids and humus with very little migration or leaching. DT_{50} c. 3–6 months

Name (class)	Structure	Mode of action	Use	Fate
Oxadiargyl (Oxadiazole)	(CH₃)₃C — (oxadiazolone ring, N–N), 2,4-dichloro-5-(prop-2-ynyloxy)phenyl; HC≡C–CH₂	Protoporphyrinogen oxidase inhibitor	Selective herbicide active mainly pre-emergence; effects being at germination. It is not absorbed by plants	Broad leaf weed and grass and annual sedge controller	*Animals.* Rapidly excreted with no accumulation *Plants.* Very low levels of residues at harvest in lemons, sunflowers and rice *Soil.* DT_{50} (lab, aerobic) 18–72 days
Propanil (Anilide)	Cl — NHCOCH₂CH₃ ; Cl (3,4-dichlorophenyl)	Photosynthetic electron transport inhibitor at the photosystem II receptor site	Selective contact herbicide with a short duration of activity	Broadleaf weed and grass controller	*Animals.* The major metabolic pathway in microsomal incubations was acylamidase hydrolysis to 3,4-dichloroaniline *Plants.* Hydrolyzed by an aryl acelamidase to 3,4-dichloroaniline and propionic acid as metabolic intermediates *Soil.* Rapid microbial degradation to aniline derivative occurs. Degradation products are proportionate which is rapidly metabolized to CO_2 and 3,4-dichloroaniline which is bound to soil
Quinclorac (quinoline-carboxylic acid)	CO₂H, 3,7-dichloroquinoline-8-carboxylic acid; Cl, N, Cl	Synthetic auxin (acting like indolyacetic acid), also inhibitor of cell wall (cellulose) biosynthesis	Rapidly absorbed through the foliage	Graminicide	*Animals.* More than 90% excreted in the urine within 5 days *Plants.* Systematically translocated to the roots and to the leaves *Soil.* Only slightly adsorbed by the soil. Depending on soil type and organic matter content, the chemical is relatively mobile. Degraded by micro-organisms, 3-chloro-8-quinolinecarboxylic acid being the major metabolite

(continued)

TABLE 17.4c   Continued

Chemical (Chemical Class)	Structure	Biochemistry	Mode of Action	Activity	Environmental Fate
Bensulfuron-methly (Sulfonylurea)		Branched chain amino acid synthesis (ALS and AHAS) inhibitor. Acts by inhibiting biosynthesis of the essential amino acids valine and isoleucine, hence stopping cell division and plant growth	Selective systemic herbicide, rapidly absorbed by root and foliage with rapid translocation to meristematic tissue	Broad leaf weed and sedge controller	*Animals.* Almost completely bio-transformed and rapidly excreted in urine and feces $\quad$ *Plants.* After uptake by rice, converted to a nonherbicidal metabolite $\quad$ *Soil.* DT_{50} 4–20 weeks on Planagan and Keyport silt loam soils. In rice fields, DT_{50} in water averages 4–6 days
Pretilachlor (Chloroace-tamide)		Cell division inhibitor	Selective herbicide, readily taken up by hypocotyls, mesocotyls, and coleoptiles, and to a lesser extent by roots of germinating weeds	Broad leaf weed, grass, and sedge controller	*Animals.* Substitution of the chlorine atom for glutathione to form a conjugate. Cleavage of the ether bond to yield an ethyl alcohol derivative. Both metabolites are susceptible to further degradation $\quad$ *Plants.* Substitution of chlorine atom to form a conjugate. Cleavage of the ether bond to yield an ethyl alcohol derivative. Hydrolytic and reductive removal of the chlorine atom $\quad$ *Soil.* Applied to paddy water, disappeared from the water by adsorption to the soil, where it is rapidly degraded under practical conditions, median DT_{50} (lab) 30 days. Due to strong soil adsorption, unlikely to leach

TABLE 17.4d   Soyabean Herbicides

Chemical (Chemical Class)	Structure	Biochemistry	Mode of Action	Activity	Environmental Fate
Glyphosate (Glycine derivative)	HO₂CCH₂NHCH₂P(OH)₂ with =O	Inhibits 5-enolpyruvyl-shikimate-3-phosphate synthase (EPSPS), an enzyme of the aromatic acid and biosynthesis pathway. This prevents synthesis of essential aromatic amino acids needed for protein biosynthesis	Systemic herbicide, absorbed by rapid translocation throughout the plant	Non-selective herbicide	*Animals.* In mammals, following oral administration, glyphosate is very rapidly excreted unchanged and does not bioaccumulate *Plants.* Slowly metabolized to aminomethylphosphonic acid ([1066-51-9]), which is the major plant metabolite *Soil/environment.* In soil (field), DT_{50} 3–174 days, depending on edaphic and climatic conditions. In water, DT_{50} varies from a few to 91 days. Photodegradation in water occurs under natural conditions, DT_{50} ca 28 days; no substantial photodegradation in soil was recorded over 31 days. In a lab. whole system with water and sediment, DT_{50} ca 14 days (aerobic), 14–22 days (anaerobic). The major metabolite in soil and water is aminomethylphosphonic acid
Pendimethalin (Dinitro-aniline)		Microtubule assembly inhibitor	Selective herbicide absorbed by roots and leaves	Graminicide and broad-leaf weed controller	—
Bentazone (Benzothiadiazinone)		Photosynthetic electron transport inhibitor at photosystem II receptor site	Selective, contact herbicide, absorbed by leaves with very little translocation, but also absorbed by roots, with trans-location acropetally in the xylem	Broad leaf weed controller	*Animals.* In rats, the major metabolic routes for pendimethalin involve hydroxylation of the 4-methyl and N-1-ethyl groups, oxidation of these alkyl groups to carboxylic acids, nitro-reduction, cyclisation and conjugation (J. Zulian, *J. Agric. Food Chem.*, 1990, 38, 1743)

(*continued*)

TABLE 17.4d **Continued**

Chemical (Chemical Class)	Structure	Biochemistry	Mode of Action	Activity	Environmental Fate
					Plants. In plants, the 4-methyl group on the benzene ring is oxidised to the carboxylic acid via the alcohol. The amino nitrogen is also oxidized. At harvest time, residues in crops are below the validated sensitivity of the analytical method (0.05 ppm) *Soil/environment.* In soil, the 4-methyl group on the benzene ring is oxidized to the carboxylic acid via the alcohol; the amino nitrogen is also oxidized. DT_{50} in soil is 3–4 months (A. Walker & W. Bond, *Pestic. Sci.*, 1977, 8, 359). Kd ranges from 2.23 (0.01% o.m., pH 6.6) to 1638 (16.9% o.m., pH 6.8) (H. J. Pedersen et al., *Pestic. Sci.*, 1995, 44, 131)
Fluazifop-*P*-butyl (Aryloxyphenoxypropionate)		Fatty acid synthesis inhibitor, by inhibition of acetyl CoA carboxylase (ACCase)	Quickly absorbed through the leaf surface, hydrolyzed to fluazifop-*P*- and located through the phloem and xylem, accumulating in the rhizomes and stolons of perennial grasses and meristems of annual and perennial grasses	Graminicide	*Animals.* In mammals, fluazifop-*P*-butyl is metabolized to fluazifop-*P*, which is rapidly excreted *Plants.* In plants, fluazifop-*P*-butyl is rapidly hydrolyzed to fluazifop-*P*, which is then partly conjugated. Ether cleavage gives the pyridone and propionic acid metabolites, which may both be further metabolized or conjugated *Soil/Environment.* c. 5800. In moist soils, rapid degradation of fluazifop-*P*-butyl occurs, DT_{50} <24 hr. The major degradation product is fluazifop-*P*, which is hydrolyzed to 5-trifluoromethylpyrid-2-one, and 2-(4-hydroxyphenoxy)propionic acid, both of which are further degraded, ultimately to CO_2

Compound (structure)	Mode of action / use	Soil/environment · Animals · Plants
Imazethapyr (Imidazo-linone) (CH₃)₂CH, CH₃, O, N, H, N, CO₂H, N, CH₃CH₂	Germicide and broad leaved weed controller Systemic herbicide absorbed by roots and foliage with translocation in xylem and phloem and accumulation in meristetic regions Branched chain amino acid synthesis (ALS and AHAS) inhibitor reducing levels of valine, leucine, isoleucine leading to disruption of protein and DNA synthesis	*Soil/environment.* In laboratory soil (40% MHC, pH 5.3–7.7), DT_{50} 2–9 days (20°C). Field DT_{50} <4 week. Koc 39–84. For degradation route, see fluazifop-*P*-butyl *Animals.* In rats, following oral administration, 92% was excreted in the urine and 5% in the feces within 24 hours. Residue levels in blood, liver, kidney, muscle, and fat tissues were <0.01 ppm after 48 hr *Plants.* Rapidly metabolized in non susceptible plants; half-life in soya beans 1.6 days. The primary metabolic route in maize is oxidative hydroxylation at the carbon atom of the ethyl substituent on the pyridine ring *Soil/environment.* Half-life in soil 1–3 months
Sulfentrazone (Triazolinone) Cl, CH₃, N–N, N, CHF₂, Cl, CH₃SO₂NH, O	Graminicide and broadleaf weed controller Herbicide absorbed by roots and foliage, with translocation primarily in the apoplasm and limited movement in pholem Protoporphyrinogen oxidase inhibitor (Chlorophyll biosynthesis pathway)	*Animals.* In rats, nearly all of administered sulfentrazone is excreted in the urine within 72 hr *Plants.* In soya beans, over 95% of the parent sulfentrazone is metabolized to the nonpolar, ring-hydroxymethyl analogue within 12 hr. This analogue is also rapidly converted, over the same time period, to three polar metabolites, two of which are glycosidic derivatives and one a nonglycoside metabolite *Soil/environment.* Stable in soil (DT_{50} 18 months). In water, stable to hydrolysis (pH 5–9), but readily undergoes photolysis (DT_{50} <0.5 days). Low affinity for organic matter (Koc 43), but is mobile only in soils with high sand content. Low potential to bioaccumulate

interfere with chlorophyll formation and/or activity, and various other metabolic processes in the weeds at much higher efficiency than the main crop. Agrochemicals that control fungal diseases are called *fungicides* (Table 17.5a,b), and those with plant hormonelike action are called *plant growth regulators* (Table 17.6). Agencies such as WHO (World Health Organization) and the US EPA separate agrochemicals into different classes as per agreed upon toxicity levels, as a useful guide to consumers. In many countries, field usage of more toxic agrochemicals is restricted to trained personnel only.

DELIVERY SYSTEMS OF AGROCHEMICALS

Agrochemicals are generally formulated with surface-active agents in dry forms (WP, WDG) or wet forms (EC, SC) for efficient delivery at the site of action. This might include sticking to the foliage, translaminar action in the leaves or through uptake by the root system. An efficient delivery system plays an important role in optimizing dosages per hectare of a given agrochemical in order to achieve maximum efficiency of pest control while reducing risk to personnel and livestock. Selective uses of appropriate sprayers, nozzles are other aids for the same purpose. Good farm practices during spray application and clean-up controls wasteful nontarget dissipation.

Obsolescence of Agrochemicals

Resistance to agrochemicals occurs through the natural selection. Fungi with short life cycles exhibit the most pronounced resistance development. Insects are in an intermediate category with respect to life cycle, while weeds (plants) take much longer. Resistance management is brought through rotational or combination uses of agrochemicals with differing modes of action. Sublethal dosages are strictly to be avoided, as are extreme over-applications. These measures improve the functional lifetime of a product. These practices are very pertinent since major breakthroughs with a new class of agrochemicals

usually occur only once in 25 years or so. Availability of safer, more efficacious, cost effective agrochemicals tends to render prevailing agrochemicals obsolete.

Even with all of the advances to date, it would be wrong to state that all issues of plant protection are adequately addressed. There are still issues defying solutions, and these will continue to attract R&D efforts. As higher and higher farm productivities are achieved, coupled with more judicious distribution of food, the growing global population should enjoy higher levels of nutrition. Agrochemicals will continue to play an important role in health care, protection of farm produce, production of medicinal plants of value, and offer recyclable feedstock to produce chemicals currently derived solely from diminishing petroleum feed stocks. Hence, the value of the agrochemical industry should be judged more on its multiple impacts on the improved human condition rather than its modest size.

PRODUCTS OF THE AGROCHEMICAL INDUSTRY

Many of the chemical structures of agrochemicals demonstrating similar target-(enzyme) specific biological activity can be rationalized under the term *BIOISOS-TERISM*. Bioisosterism is a phenomenon where molecules possessing related structure have similar or antagonistic properties. Bioisosterism is the biological analogue of isosterism, which is the close physical similarity of molecules or ions having the same number of atoms and valence electrons, such as CO and N_2. This similarity is thought to explain certain analogies among the physical constants of molecules.

No two substituents are exactly alike. Any substitution impacts size, shape, electronic distribution, lipophilicity, pKa, chemical reactivity, susceptibility to metabolism, and the like. The bioisosteric approach is the total change induced by substituent replacement on the potency, selectivity, duration of action, bioavailability and toxicity, of an agrochemical. The following groups are examples of

TABLE 17.5a Systemic Fungicides

Chemical (Chemical Class)	Structure	Biochemistry	Mode of Action	Environmental Fate
Benalaxyl (Acylalanine)		Nucleic RNA-polymerase inhibitor	Systemic fungicide absorbed by roots, stems, leaves with translocation acropetally to all parts of the plant including subsequent growth	*Animals.* Rapidly metabolized, and eliminated in the urine (23%) and feces (75%) within 2 days *Plants.* Slowly metabolized to glycosides in plants *Soil.* Slowly degraded by soil microorganisms to various acidic metabolites. DT_{50} in silt loam soil 77 days
Carbendazim (Benzimidazole)		Inhibits betatubulin synthesis	Systemic fungicide, absorbed through roots and green tissues, with translocation acropetally. Acts by inhibiting development of the germ tubes, the formation of appressoria and the growth of mycelia	*Animals.* 66% was eliminated in the urine within 6 hr *Plants.* Readily absorbed by plants. One degradation product is 2-aminobenzimidazole *Soil.* 2-Aminobenzimidazole has been found as a minor metabolite. DT_{50} in soil 8–32 days under outdoor conditions. Mainly decomposed by microorganisms
Diethofencarb (N-phenyl carbamate)			Systemic fungicide, readily absorbed through leaves and roots and translocated throughout the plant	*Animals.* 98.5–100% of ^{14}C was excreted within 7 days. The major metabolic routes were deethylation of the 4-ethoxy group, cleavage of the carbamate linkage, acetylation, and finally formation of the glucuronide and sulfate conjugates *Plants.* Readily degraded in plants *Soil.* Readily degraded in soil; DT_{50} <1–6 days under aerobic conditions; only very slightly degraded under anaerobic sterilized conditions

(continued)

TABLE 17.5a Continued

Chemical (Chemical Class)	Structure	Biochemistry	Mode of Action	Environmental Fate
Fenpropidin (Piperidine)	*(chemical structure)*	Ergosterol biosynthesis inhibitor, by inhibition of steroid reduction (sterol—Δ^{14}—reductase) and isomerisation (Δ^8 to Δ^7—isomerase)	Systemic foliar fungicide with translocation acropetally in the xylem	*Animals.* Rapidly absorbed, distributed, metabolised and excreted in the urine and feces. No bioretention potential. *Plants.* Relatively rapid and extensive degradation. Principal metabolic pathway involves hydroxylation of the piperidine ring and oxidation of the tertiary butyl group. DT_{50} in wheat and barley plants c. 4–11 days *Soil.* Strongly adsorbed and extensively degraded, DT_{50} 58 (loam)—95 (sandy loam). Fenpropidin and its metabolites have little or no tendency to leach
Fenarimol (Pyrimidine)	*(chemical structure)*	Ergosterol biosynthesis inhibitor	Systemic fungicide, translocated acropetally within the plant	*Animals.* Rapidly excreted *Plants.* Forms numerous photodegradation products. *Soil.* DT_{50} 14–130 (average 79) days
Furametpyr (Oxathiin)	*(chemical structure)*	Inhibitor of mitochondrial succinate oxidation	Fungicide with systemic and translaminar action	—
Imazalil (Imidazole)	*(chemical structure)*	Steroid demethylation inhibitor	Systemic fungicide	*Animals.* 90% is eliminated in the metabolized form within 4 days *Plants.* Transformed into (-2,4-dichlorophenyl)-1*H*-imidazole-1-ethanol *Soil.* DT_{50} (field) 4–5 d; DT_{90} (field) 54–68 days

Metalaxyl-*M*
(Phenylamide:
acylalanine)

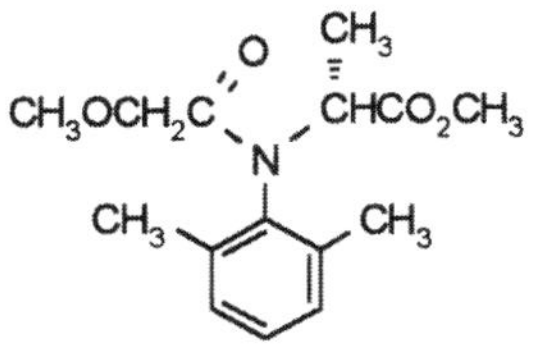

Inhibits protein synthesis in fungi, by interference with the synthesis of ribosomal RNA

Systemic fungicide, absorbed through leaves, stems and roots

Animals. Rapidly absorbed and also rapidly and almost completely eliminated in urine and feces. Metabolism proceeds via hydrolysis of the ester bond, oxidation of the 2-(6)-methyl group and of the phenyl ring and *N*-dealkylation. Residues in tissues were generally low and there was no evidence for accumulation or retention of metalaxyl-*M* or its metabolites

Plants. Metabolized by more than four types of phase I reaction (oxidation of the phenyl ring, oxidation of the methyl group, cleavage of the methyl ester and *N*-dealkylation) to form eight metabolites; at phase II, most of the metabolites are sugar conjugated

Soil. DT_{50} in soil 21 days (realistic range 5–30 days).

Propiconazole
(Triazole)

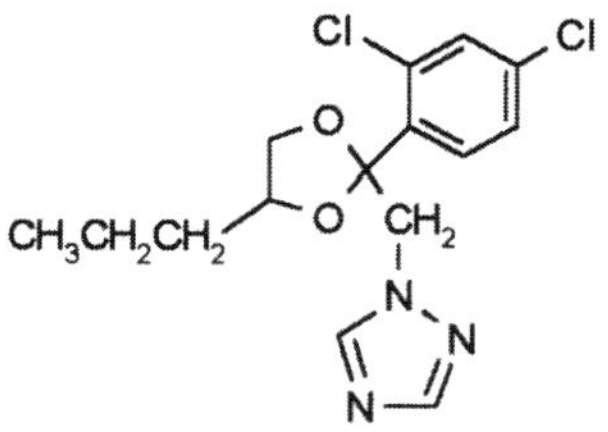

Steroid demethylation inhibitor

Systemic foliar fungicide, translocated acropetaly in the xylem

Animals. Rapidly absorbed and also rapidly and almost completely eliminated with urine and feces. Residues in tissues were generally low and there was no evidence for accumulation or retention of propiconazole or its metabolites. The major sites of enzymic attack are the propyl side-chain and the cleavage of the dioxolane ring, together with some attack at the 2,4-dichlorophenyl and 1,2,4-triazole rings

Plants. Degradation through hydroxylation of the *n*-propyl side-chain and deketalisation of the dioxolan ring. After cleavage of triazole, triazolealanine is formed as the main metabolite. Metabolites are conjugated mostly as glucosides

Soil. DT_{50} in aerobic soils (25°C) 40–70 days. The main degradation pathways are hydroxylation of the propyl side-chain and the dioxolane ring, and finally formation of 1,2,4-triazole. Immobile in soil

(*continued*)

TABLE 17.5a Continued

Chemical (Chemical Class)	*Structure*	*Biochemistry*	*Mode of Action*	*Environmental Fate*
Spiroxamine (Spiroketalamine)		Sterol biosynthesis inhibitor acting mainly by inhibition of Δ^{14}—reductase	Systemic fungicide, which readily penetrates into the leaf tissue followed by acropetal translocation to the leaf tip	*Animals.* Highly absorbed followed by fast elimination from the body (>97% within 48 hr). The radioactivity was readily distributed from the plasma into peripheral compartments. The main metabolite was the compound oxidized to the carboxylic acid in the *t*-butyl moiety. Metabolism proceeds either via oxidation of the *t*-butyl moiety to yield the carboxylic acid compound or via des-alkylation of the amino group resulting in the des-ethyl and des-propyl derivatives of spiroxamine *Plants.* Extensively metabolized by oxidation, desalkylation and cleavage of the ketal structure; the resulting metabolites bearing a hydroxylated *t*-butyl group or an aminodiol were further conjugated *Soil.* Readily degraded, ultimately to CO_2; oxidation on the *t*-butyl moiety and des-alkylation of the amine are the primary reaction steps. The des-alkylated compounds were either further oxidised to the corresponding acids or further degraded to a ketone metabolite. Soil DT_{50} (lab. and field) in the range 35–64 days. Bound rapidly to the sediment
Triadimenol (Triazole)		Inhibits gibberellin and ergosterol biosynthesis and hence the rate of cell division	Systemic fungicide absorbed through roots and leaves with ready translocation in young growing tissues but less ready translocation in older, woody tissues	*Animals.* Metabolized mainly by oxidation of the *tert*-butyl moiety to the corresponding alcohol and then to carboxylic acid. A small fraction of these compounds was conjugated *Plants.* The most important breakdown reactions are conjugation with various sugar compounds (especially hexose) and oxidation at the *tert*-butyl moiety. The resulting primary alcohol is likewise partly conjugated

Name (Group)	Structure	Mode of action	Uptake and translocation	Fate
Triadimefon (Triazole)		Steroid demethylation (ergosterol biosynthesis) inhibitor	Systemic fungicide, absorbed by roots and leaves with ready translocation in young growing tissues but less ready translocation in older, woody tissues	*Soil.* Triadimenol (*q.v.*) is a degradation product of triadimefon (*q.v.*). Degradation involving hydrolytic cleavage leads to the formation of 4-chlorophenol. DT_{50} (sandy loam) 110–375 days; (loam) 240–270 days *Animals.* 83–96% is excreted unchanged in the urine and feces within 2–3 days. However, metabolism occurs in the liver, mostly to triadimenol (*q.v.*) and its glucuronic acid conjugates. Half-life in blood plasma is c. 2.5 hr *Plants.* In plants, the carbonyl group is reduced to a hydroxyl group, with the formation of triadimenol (*q.v.*) *Soil.* In soil, the carbonyl group is reduced to a hydroxyl group, with the formation of triadimenol (*q.v.*). DT_{50} of triadimefon in sandy loam c. 18 days, in loam c. 6 days
Triforine (Piperazine)		Ergosterol biosynthesis inhibitor	Systemic fungicide, absorbed by leaves and roots with translocation acropetally	*Soil.* A range of nonfungitoxic metabolic end-products are formed, presumably including piperazine. DT_{50} in soil c. 3 weeks. Does not accumulate in the environment

TABLE 17.5b Contact Fungicides

Chemical (Chemical Class)	Structure	Biochemistry	Mode of Action	Environmental Fate
Bordeaux mixture (inorganic)	—	Cu^{++} is taken up by the spores during germination and accumulates until a sufficiently high concentration is achieved to kill the spore cell	Foliar fungicide	*Animals.* Copper is an essential element and is under homeostatic control in mammals
Chinomethionat (Quinoxaline)		—	Selective, non-systemic contact fungicide	*Animals.* Rapidly metabolized, and c. 90% is eliminated within 3 days in the feces and urine. The main metabolite is chinomethionat acid (dimethylmercaptoquinoxaline-6-carboxylic acid), which also occurs in the conjugated form *Plants.* No penetration of the a.i. or metabolites in the fruit pulp was observed. The only metabolite detected was dihydromethylquinoxalinedithiol *Soil.* DT_{50} in standard soil 1 and 2: 1–3 days
Chlozolinate (Dicarboximide)		Inhibitor of Lipid Peroxidation in mitochondrial membranes	Contact fungicide	*Animals.* Readily absorbed, metabolized and excreted. Metabolites identified in urine are: 3-(3,5-dichlorophenyl)-5-methyloxazolidin-2,4-dione, *N*-(3,5-dichlorophenyl)-2-hydroxypropionamide, *O*-1-carboxyethyl-*N*-3,5-dichloro-phenyl carbamate, and *N*-(3,5-dichloro-2(or 4)-hydroxyphenyl)-2-hydroxypropionamide and its sulfate and glucuronide conjugates *Plants.* Undergoes hydrolysis and decarboxylation processes, giving the same metabolites as those identified in animals. *Soil.* In silt-loam, sandy loam and clay loam soils, hydrolysis and decarboxylation occur; aerobic DT_{50} <7 hr
Chlorothalonil (Chloronitrile)		Conjugation with, and depletion of thiols (particularly glutathione) from germinating fungal cells, leading to disruption of glycolysis and energy production, fungistasis and fungicidal action	Nonsystemic foliar fungicide	*Animals.* Not well absorbed following oral dosing. It reacts with glutathione in the gut lumen, or immediately on absorption into the body, to give mono-, di- or tri- glutathione conjugates. These may be excreted through urine or feces, or subject to further metabolism resulting in thiol or mercapturic acid derivatives. In ruminants, the 4-hydroxy metabolite may also be present *Plants.* The majority of the residue remains as parent compound. The most abundant metabolite, 4-hydroxy-2,5,6-trichloroisophthalonitrile, is generally <10% of applied parent

Common name (class)	Structure	Mode of action	Type	Degradation
Epoxiconazole (Triazole)		Inhibitor of C-14-demthylase in sterol biosynthesis	Preventive and curative fungicide	*Soil.* Low mobility to immobile. In aerobic and anaerobic soil studies, DT_{50} is 5–36 days. Degradation is faster in biotic aquatic systems, typical DT_{50} (aerobic) <8 hr, (anaerobic) <10 days. *Animals.* Readily excreted via feces. No major metabolites, but a high number of minor metabolites were identified. The important metabolic reactions were cleavage of the oxirane ring, hydroxylation of the phenyl rings and conjugation. *Plants.* There is extensive degradation. *Soil.* Degradation is by microbial activity, DT_{50} c. 2–3 months
Prochloraz (Imidazole)		Steriod demethylation (ergosterol biosynthesis) inhibitor		*Animals.* Rapidly metabolized initially by cleavage of the imidazole ring and quantitatively eliminated from the body. Although absorption following dermal exposure is low, residues in plasma and tissues are rapidly eliminated from the body. *Plants.* The primary metabolite, N-formyl-N'-1-propyl-N-(2-(2,4,6-trichlorophenoxy)ethyl)urea, is formed from cleavage of the imidazole ring. This is degraded to N-propyl-N-(2-(2,4,6-trichlorophenoxy)ethyl)urea, which occurs in both free and conjugated forms. Other metabolites include 2-(2,4,6-trichlorophenoxy)ethanol, 2-(2,4,6-trichlorophenoxy)acetic acid, traces of 2,4,6-trichlorophenol and conjugates of the above. Little unchanged prochloraz is present. *Soil.* Degrades in the soil to a range of mainly volatile metabolites. Well adsorbed onto soil particles, and is not readily leached DT_{50} 5–37 days
Cozeb Alkylenebis (dithiocarbamate)	[-SCSNHCH$_2$CH$_2$NHCSSMn-]$_x$ (Zn)$_y$	Non-specific thiol reactant, inhibiting respiration	Fungicide with protective action	*Plants.* Extensively metabolized, forming ethylenethiourea, ethylenethiuram monosulfide, ethylenethiuram disulfide, and sulfur as transitory intermediates. Terminal metabolites are natural products, especially those derived from glycine. *Soil.* Rapidly degraded by hydrolysis, oxidation, photolysis, and metabolism. DT_{50} in soil c. 6–15 days
Sulfur (inorganic)	—	Nonspecific thiol reactant, inhibiting respiration	Nonsystemic fungicide	*Plants.* Degradation proceeds primarily by microbial reduction. *Soil.* Slight oxidation to the volatile oxides

TABLE 17.5b Continued

Chemical (Chemical Class)	Structure	Biochemistry	Mode of Action	Environmental Fate
Vinclozolin (Dicarboximide)		—	Nonsystemic fungicide	*Animals.* The major metabolic routes are epoxidation of the vinyl group, followed by hydration of the intermediate epoxide, and by hydrolytic cleavage of the heterocyclic ring. Eliminated in approximately equal proportions in the urine and feces, with the principal metabolite being *N*-(3,5-dichlorophenyl)-2-methyl-2,3,4-trihydroxybutanamide *Plants.* The primary metabolites are (1-carboxy-1-methyl) allyl 3,5-dichlorophenylcarbamate and *N*-(3,5-dichlorophenyl)-2-hydroxy-2-methyl-3-butenamide. Alkaline hydrolysis leads to loss of 3,5-dichloroaniline from vinclozolin and its metabolites. The metabolites exist as conjugates *Soil.* Metabolism occurs by loss of the vinyl group, cleavage of the 5-membered ring and eventual formation of 3, 5,-dichloroaniline. Soil degradation takes place with half-lives of several weeks, and mainly leads to the formation of bound residues
Ziram (Dimethyldithio-carbamate)	[(CH$_3$)$_2$NCS$_2$]$_2$Zn		Inhibitor of enzymes containing copper ions or sulfonyl groups	Basic contact, foliar fungicide	*Animals.* Mostly eliminated within 1–2 days leaving 1–2% of the dose in the tissue and carcass after 7 days *Plants.* The major metabolite is dimethylamine salt of dimethyldithiocarbamic acid; tetramethylthiourea, carbon disulfide and sulfur can also be formed. Dimethyldithiocarbamic acid can be present as the free acid or as the metabolic conversion products DDC-β-glucoside, DDC-α-aminobutyric acid and DDC-α-alanine *Soil.* A. aerobic DT$_{50}$ 42 hr. Unlikely to leach

TABLE 17.6   Plant Growth Regulators

Chemical (Chemical Class)	Structure	Biochemistry	Mode of Action	Environmental Fate
Chlormequat chloride (Quaternary ammonium)	$ClCH_2CH_2N(CH_3)_3^+$ Cl^-	Gibberellin biosynthesis inhibitor	Inhibits cell elongation. Also influences developmental cycle, leading to increased flowering and harvest. May also increase chlorophyll formation and root development	*Animals.* 97% is eliminated within 24 hr, principally unchanged *Plants.* Converted to choline chloride *Soil.* Rapidly degraded by microbial activity. DT_{50} in 4 soils averaged 32 days at 10°C; 1–28 days at 22°C. Low to medium mobility
Ethephon (Ethylene generator)	$ClCH_2CH_2\overset{O}{\overset{\|}{P}}(OH)_2$	—	Plant growth regulator with systemic properties. Penetrates into the plant tissue and is decomposed to ethylene, which affects growth processes	*Animals.* Rapidly excreted intact via the urine, and as ethylene via the expired air *Plants.* Rapidly undergoes degradation to ethylene *Soil.* Rapidly degraded and strongly adsorbed; unlikely to leach
Gibberellic acid (Gibberellins)		—	Shows physiological and morphological effects on the plant parts above soil surface at extremely low concentrations. Translocated Plant Growth Regulator	—
Indol-3-ylacetic acid (Auxin)		—	Affects cell division and cell elongation	*Soil.* Rapidly degraded in soil

(continued)

TABLE 17.6   Continued

Chemical (Chemical Class)	Structure	Biochemistry	Mode of Action	Environmental Fate
Isoprothiolane (Phosphorothiolate)		Inhibits penetration and elongation of infection hyphae, by inhibiting formation of infection peg or cellulose secretion	Absorbed by leaves and roots with translocation acropetally and basipetally	—
Maleic hydrazide		Inhibits cell division in the meristematic regions.	Absorbed by leaves and roots with translocation in xylem and phloem	*Animals.* 43–62% of the dose excreted unchanged within 48 hr *Plants.* Various acids, e.g., succinic, fumaric, and maleic, are found as metabolites *Soil.* DT_{50} c. 11 hr
Mepiquat chloride (Quaternary ammonium)		Inhibits biosynthesis of gibberellic acid	Plant Growth regulator, absorbed and translocated throughout the plant	*Animals.* c. 48% is excreted in the urine c. 38% in the feces, with <1% remaining in the tissues. The unmetabolized material constitutes c. 90% in each case *Soil.* DT_{50} 10–97 days at $20 \pm 2°C$ and 40% of maximum water-holding capacity
1-naphthyl acetic acid (Synthetic auxin)		—	Plant Growth Regulator with auxin like activity	—
2,4-Dichlorophenoxy acetic acid (aryloxyalkanoic acid)		Synthetic auxin, acting like indol-lacetic acid	Salts are readily absorbed by the roots while esters are readily absorbed by foliage. Translocation occurs, with accumulation principally at the meristematic regions of shoots and roots. Acts as a growth inhibitor	*Animals.* Elimination is rapid, and mainly as the unchanged substance. Following single doses of up to 10 mg/kg, excretion is almost complete after 24 hr, although, with higher doses, complete elimination takes longer. The maximum concentration in organs is reached after c. 12 hr *Plants.* Metabolism involves hydroxylation, decarboxylation, cleavage of the acid side-chain, and ring opening *Soil.* Microbial degradation involves hydroxylation, decarboxylation, cleavage of the acid side-chain, and ring opening. Half-life in soil <7 days. Rapid degradation in the soil prevents significant downward movement

Name (class)	Structure	Mode of action	Biological activity	Environmental fate
Cyclanilide (Anilide)		Inhibits polar auxin transport	—	*Animals.* Rapidly excreted, primarily as unchanged cyclanilide *Plants.* Little degradation occurs in plants; cyclanilide is the major residue *Soil.* Low to moderate persistence, DT_{50} c. 16 days under aerobic conditions. Degrades primarily by microbial activity. Medium to low mobility
Inabenfide (Pyridine)		Inhibits gibberellin biosynthesis	Plant Growth Regulator which shortens lower internodes and upper leaf blades	*Animals.* The major urinary metabolite is 4-hydroxyinabenfide *Plants.* Metabolized to inabenfide ketone *Soil.* Half-life under Japanese paddy field conditions, c. 4 months
Thidiazuron (Phenylurea)		Cytokinin activity	Plant Growth Regulator absorbed by the leaves which stimulates formation of an abscission layer between the plant stem and leaf petioles, causing dropping of the entire green leaves	*Animals.* Metabolism involves hydroxylation of the phenyl group, followed by formation of water-soluble conjugates. Following oral administration, the compound is excreted in the urine and feces within 96 hr *Plants.* Only small amounts of residue (normally <0.1 mg/kg) are likely in cottonseed *Soil.* Strongly absorbed by soil. DT_{50} in soil c. 26–144 days (aerobic), 28 days (anaerobic)
Paclobutrazol (triazole)		Inhibits gibberellin and sterol biosynthesis and hence the rate of cell division	Plant Growth Regulator taken up into xylem through the leaves, stems or roots and translocated to the growing sub-apical meristems. Produces more compact plants and enhances flowering and fruiting	*Soil.* Soil DT_{50} 0.5–1.0 years in general; in calcareous clay loam (pH 8.8, 14% o.m.), DT_{50} <42 days; in coarse sandy loam (pH 6.8, 4% o.m.), DT_{50} >140 days

(continued)

TABLE 17.6   Continued

Chemical (Chemical Class)	Structure	Biochemistry	Mode of Action	Environmental Fate
Tribufos (Phosphorothioate)		—	Plant Growth Regulator absorbed by leaves. Stimulates foration of an abscission layer between the plant stem and the leaf petioles, causing the dropping of entire green leaves	*Animals.* Rapidly absorbed and metabolized; 96% of the administered radioactivity was excreted within 72 hr. Metabolism proceeds by hydrolysis followed by methylation and successive oxidation of butylmercaptan, yielding the main metabolite (3-hydroxy)-butylmethylsulfone *Plants.* Unmetabolized tribufos is the primary residue in treated cotton *Soil.* Very strongly adsorbed, leaching is extremely unlikely. The half-life under field conditions is 2–7 weeks. The main metabolite is 1-butane sulfonic acid
Flurprimidol (Pyrimidinyl carbanol)		Gibberellin synthesis inhibitor	Plant Growth Regulator absorbed by the leaves and roots with translocation in the xylem and ploem. Reduces internode elongation	*Animals.* In mammals, the skin forms is significant barrier to absorption. Following oral administration, excretion follows in the urine and feces within 48 hr, and more than 30 metabolites have been identified. No accumulation potential *Soil.* Degradation in soil under aerobic conditions leads to more than 30 metabolites
Chlorpropham (Carbamate (mi))		Mitosis inhibitor (Microtubule organization)	Absorbed by roots and coleoptile and readily translocated acropetally	*Animals.* The principal metabolic route is by hydroxylation at the para position and conjugation of the resultant 4-hydroxychloropropham with sulfate. There is also some hydroxylation of the isopropyl residue

6-Benzylaminopurine (Cytokinin)	Stimulates RNA, RuDP-carboxylase, NADP-gleceraldehyde-3-phosphate-dehydrogenase, protein synthesis	Synthetic cytokinin, little translocated	*Plants.* Three major metabolites have been identified, isopropyl N-4-hydroxy-3-chlorophenylcarbamate, isopropyl N-5-chloro-2-hydroxyphenylcarbamate and 1-hydroxy-2-propyl-3'-chlorocarbanilate, These aglycones are found in plants as water-soluble conjugates of glucose or other plant components *Soil.* Microbial degradation leads to the production of 3-chloroaniline by an enzymic hydrolysis reaction, with liberation of CO_2. DT_{50} in soil c. 65 days (15°C), 30 days (29°C) *Animals.* Almost all of administered ^{14}C was excreted in urine and faeces. Three metabolites were identified *Plants.* More than nine metabolites were identified. Urea is an end product *Soil.* 16 Days after application to soil at 22°C, 6-benzylaminopurine had degraded to 5.3% (sandy loam) and 7.85% (clay loam soil) of applied dose. Other studies indicate DT_{50} 7–9 weeks

bioisosterism:

(A) $-CO_2H$, $-SO_2NHR$, $-SO_3H$, $-PO(OH)_2$, $-PO(OH)NH_2$, $-CONHCN$

(B) $-F$, $-Cl$, $-Br$, $-CF_3$, $-CN$, $-SCN$, $-N(CN)_2$, $-C(CN)_3$

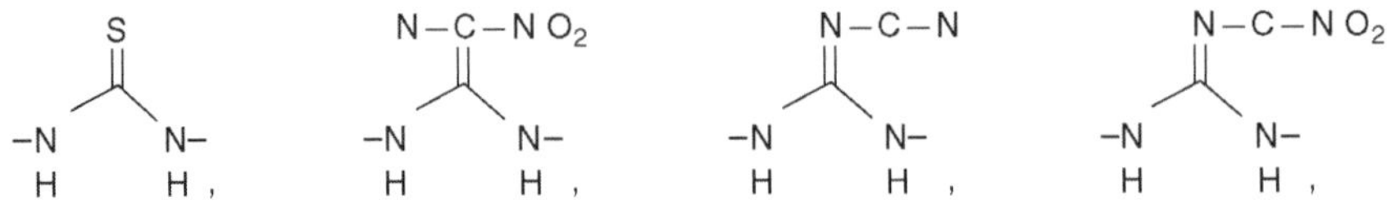

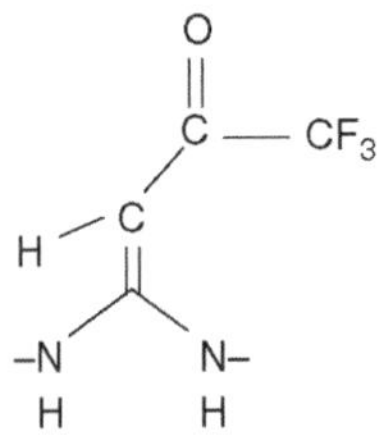

(C) $-OH$, $-NHCOR$, $-NHSO_2R$, $-CH_2OH$, $-NHCONH_2$, $-NHCN$, $-CH(CN)_2$

(D)

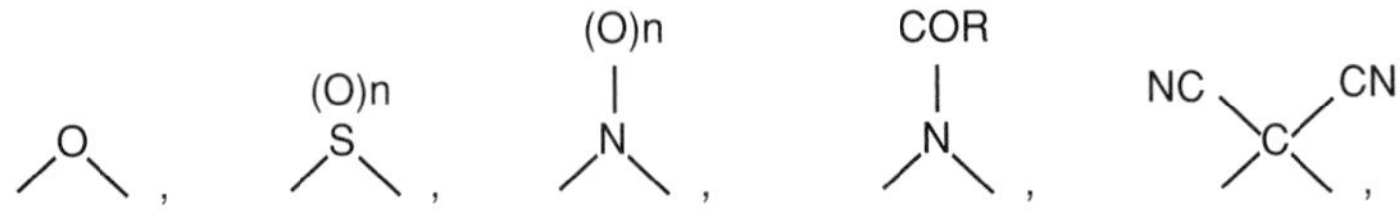

(E) $-(CH=CH)_n$, 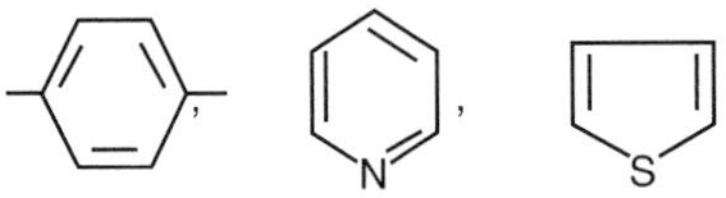 , and so on

(F)

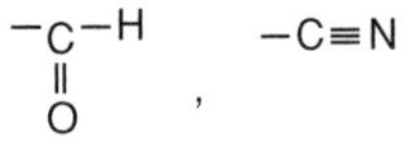

(G)

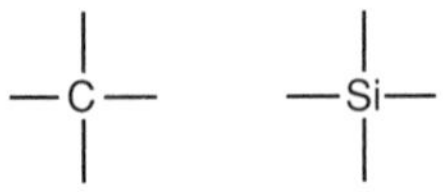

The concept of bioisosterism has been used to theoretically evaluate structural variation within the lead structures of synthetic or natural origin prior to and during the preparation of molecules of specified efficacy, safety, stability, and so on.

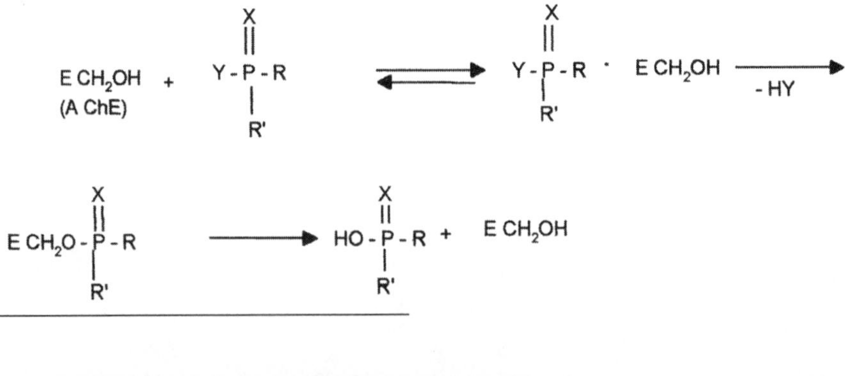

STRUCTURAL BASIS OF AGROCHEMICALS

Organophosphorous Agrochemicals

Insecticides (Tables 17.3a,b). Organophosphorous insecticides bind to Acetylcholinesterase (AchE), that is, where R and R′ are lower alkyl alkoxy, alkylthio, or substituted amino groups; X is oxygen or sulfur; and Y is a good leaving group, for example aryloxy groups substituted with electron with drawing substituents.

Others are: fosamine (herbicide used on noncrop areas, in meadows and pastures), piperofos (used in rice), and anilofos (used on rice).

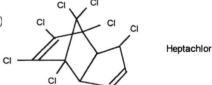

Y can also be a mixed (C, N, S) heterocycle as in the case of the nematicide Fostiazate.

Organochlorine

$R_1 = R_2 = Cl, \quad R_3 = H\text{-}DDT$
$R_1 = R_2 = Cl, \quad R_3 = OH\text{-}Dicofol.$

Organochlorine—Cyclodiene

Nematicide
(Fosthiazate)

Heptachlor

Endosulfan

Herbicides (not numerous) (Tables 30.4a–d).
Organophosphorous herbicides are derived from amino acids, for example,

H₂N-CH₂-CO₂H   (OH)₂P-H₂C-NH-CH₂-CO₂H   Glyphosphate
Glycine derivative   *N*-(phosphonomethyl) glycine

Homoalanine derivate

Ammonium 4[Hydroxy(methyl)phosphonoyl]homoalalinate

Carbamates   *Ureas*

$R_1 - O - C - NH - CH_3$ (with O double bonded to C)

$R_1 = $ (naphthalene with methyl)   Carbaryl

$R_1 = $ (benzofuran with two CH_3)   Carbofuran

$R_1 = $ (benzene ring) $CH(CH_3)_2$ Isoprocarb

Ureas structure: R_1, R_2 on ring, $-N-C-R_3$ with O double bond and H on N

$R_1 = CH(CH_3)_2$, $R_2 = H$, $R_3 = N$ (with two CH_3)   Isoproturon

$R_1 = R_2 = Cl$, $R_3 = N - (CH_2)_3 CH_3$ (with CH_3)   Neburon

$R_1 = R_2 = Cl$, $R_3 = N$ (with two CH_3)   Diuron

Pyrethroids

Structure: R_1, R_2, $C=C$, H, cyclopropane with CH_3, CH_3, $C-O-R_3$ with O double bond

(a) $R_1 = R_2 = CH_3$, $R_3 = CH_2-N$ (tetrahydrophthalimide)   Tetramethrin

(b) $R_1 = R_2 = CH_3$, $R_3 = $ (cyclopentenone with CH_3 and $CH_2-CH=CH_2$)   Bioallethrin

(c) $R_1 = R_2 = Cl$, $R_3 = -CH_2-$ (phenoxyphenyl)   Permethrin

(d) $R_1 = R_2 = Cl$, $R_3 = -C-$ (with OH and CN, phenoxyphenyl)   Cypermethrin

(e)

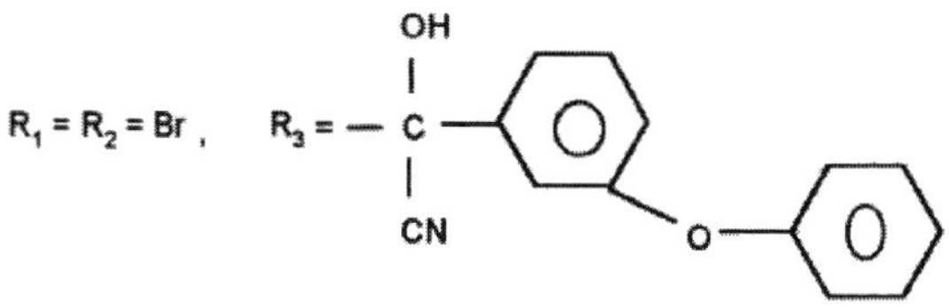

$R_1 = R_2 = Br$, $R_3 = $ Deltamethrin

(f)

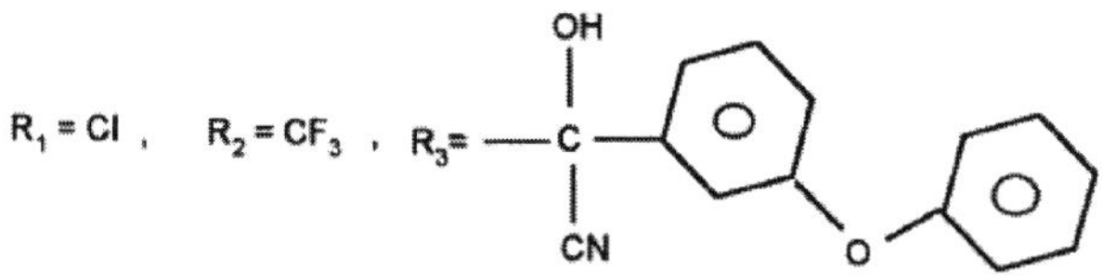

$R_1 = Cl$, $R_2 = CF_3$, $R_3 = $ Lamdacyhalothrin

(g)

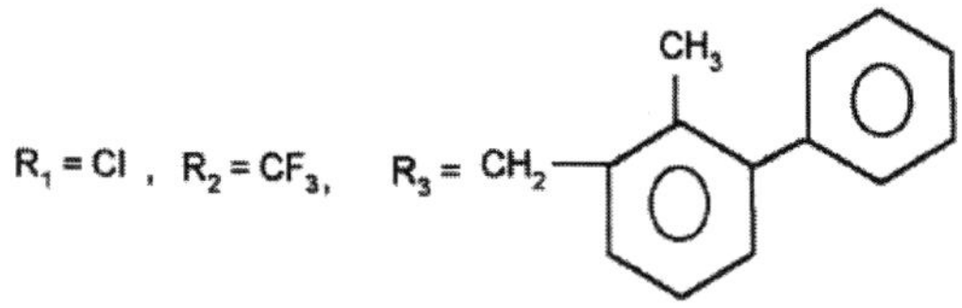

$R_1 = Cl$, $R_2 = CF_3$, $R_3 = CH_2$ Bifenthrin

Nonester pyrethroids/"Si" in lieu of "C"

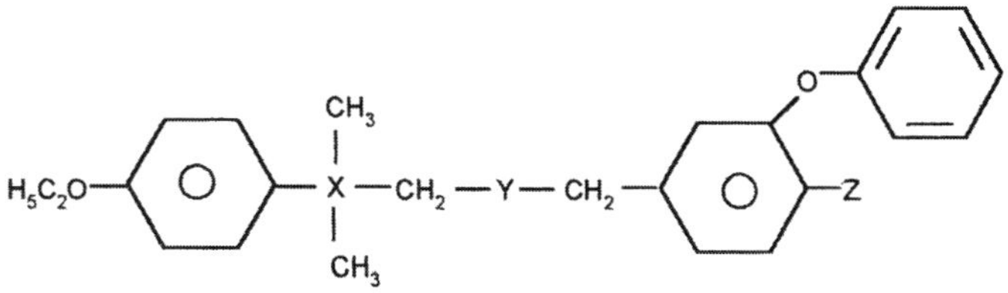

$X = C,$ $Y = O,$ $Z = H$ Ethofenprox
$X = Si,$ $Y = CH_2,$ $Z = F$ Silafluofen

Triazenes

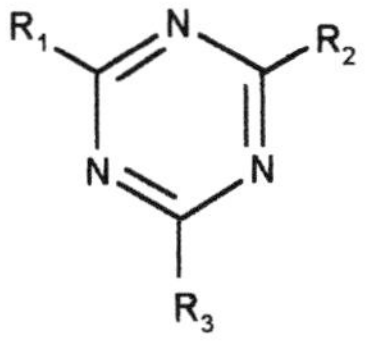

$R_1 = Cl,$ $R_2 = NHCH_2CH_3$, $R_3 = NHCH_2CH_3$ Simazine

$R_1 = Cl,$ $R_2 = NHCH_2CH_3$, $R_3 = NHCH(CH_3)_2$ Atrazine

$R_1 = Cl,$ $R_2 = NHCH_2CH_3$, $R_3 = NH\text{-}C(CH_3)_2$ Cyanazine
$\qquad\qquad\qquad\qquad\qquad\qquad\qquad\quad |$
$\qquad\qquad\qquad\qquad\qquad\qquad\qquad\; CN$

Azoles

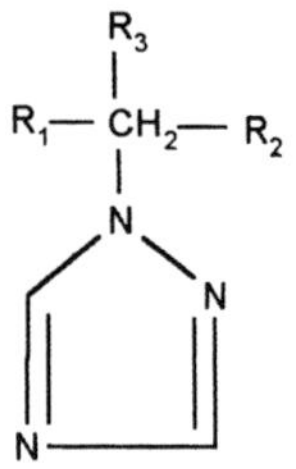

R$_1$ = Cl—⬡—Cl   R$_2$ = (CH$_2$)$_3$CH$_3$, R$_3$ = OH   Hexaconazole

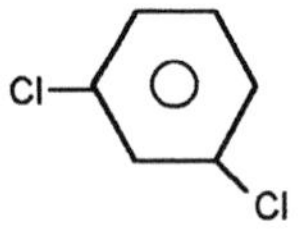

R$_1$ = Cl—⬡—Cl   R$_2$ + R$_3$ = [dioxolane ring]—CH$_3$ Propiconazole

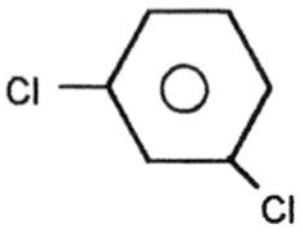

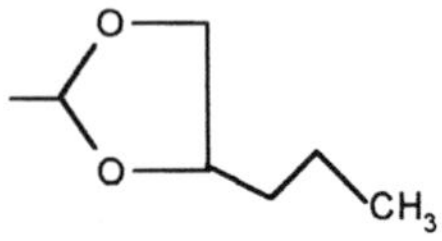

R$_1$ = Cl—⬡—CH$_2$CH$_2$-,   R$_2$ = - C(CH$_3$)$_3$,   R$_3$ = OH   Tebuconazole

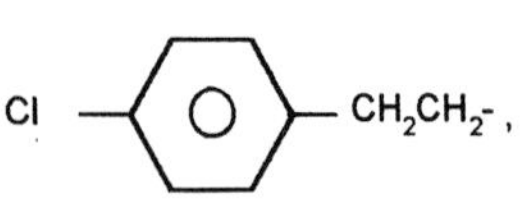

Dinitroanilines

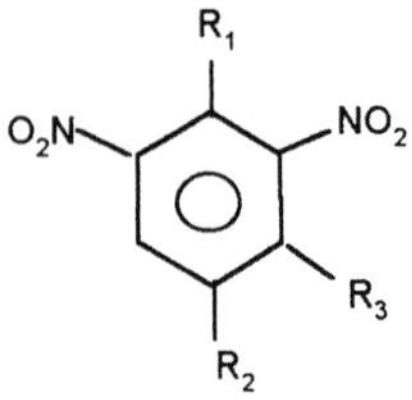

R$_1$ = NHCH(CH$_2$CH$_3$)$_2$,   R$_2$ = CH$_3$,   R$_3$ = CH$_3$ Pendimethalin

R$_1$ = N(CH$_2$CH$_2$CH$_3$)$_2$,   R$_2$ = CF$_3$,   R$_3$ = H   Trifluralin

Chloracetanilides

Sulfonyl Ureas

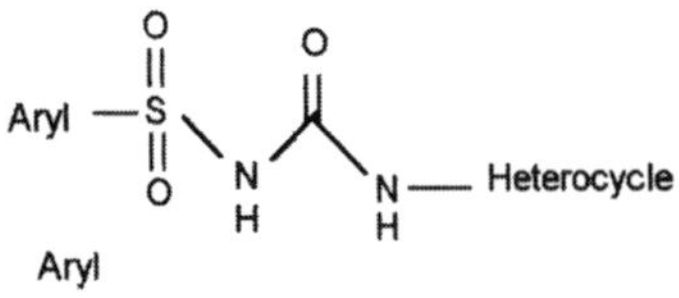

Aryl —S(=O)(=O)— N(H) — C(=O) — N(H) — Heterocycle

Aryl   Heterocycle

Rice Herbicide
Bensulfuron

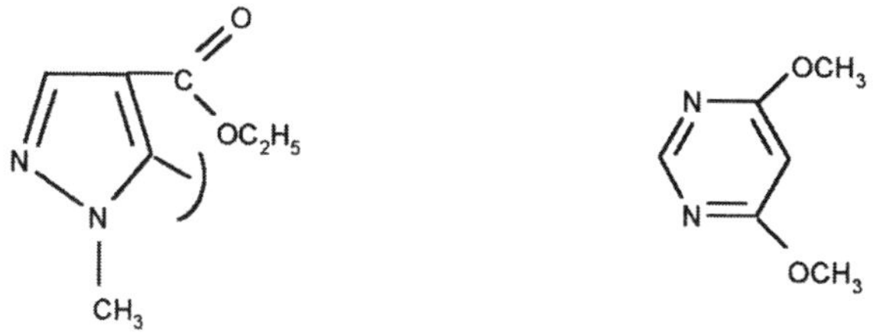

Rice Herbicide
Pyrazosulfuron

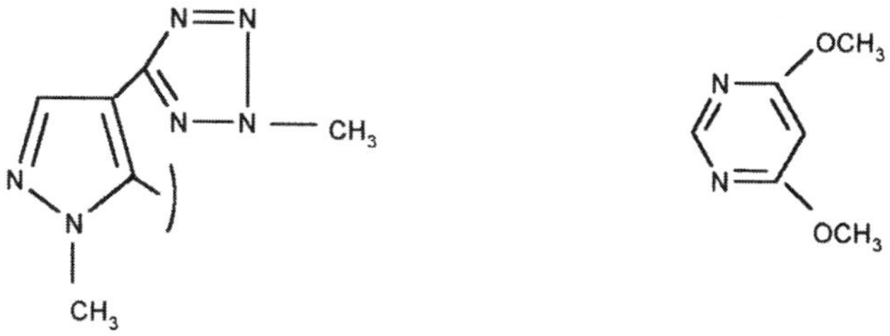

Rice Herbicide
Azimsulfuron

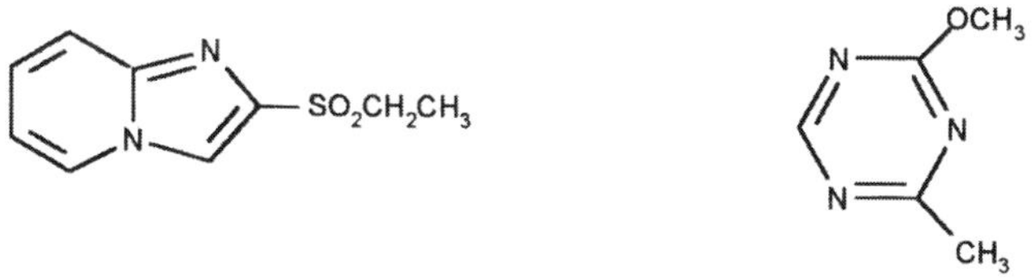

Wheat Herbicide
Sulfosulfuron

Wheat Herbicide
Metsulfuron

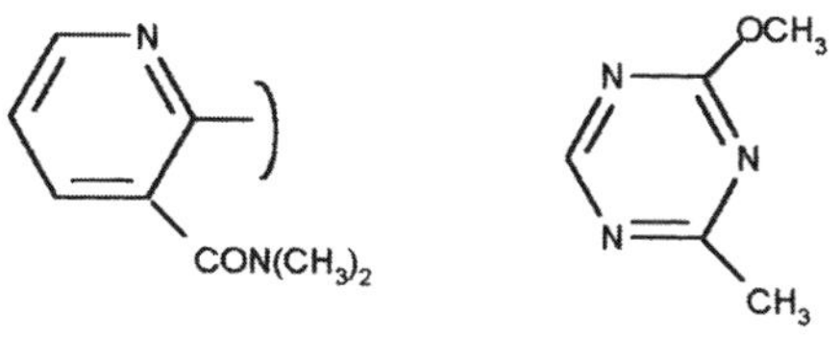

Corn Herbicide
Nicosulfuron

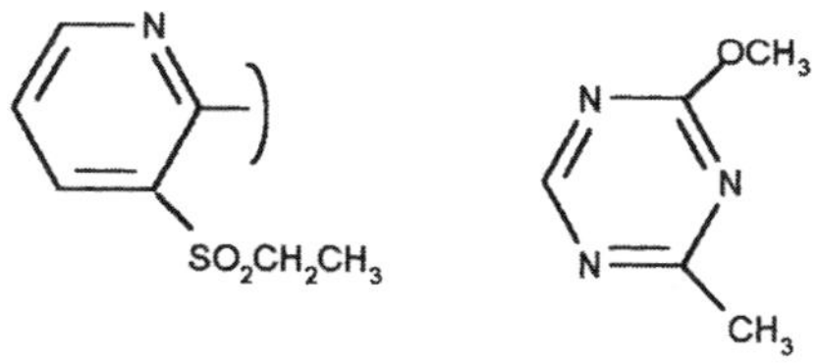

Corn Herbicide
Rimsulfuron

Aryloxyphenoxypropionic Acids
(Cereal Herbicides)

Clodino/cyhalo/diclo/fenoxa/fluazi/
haloxypropaquiza-fops

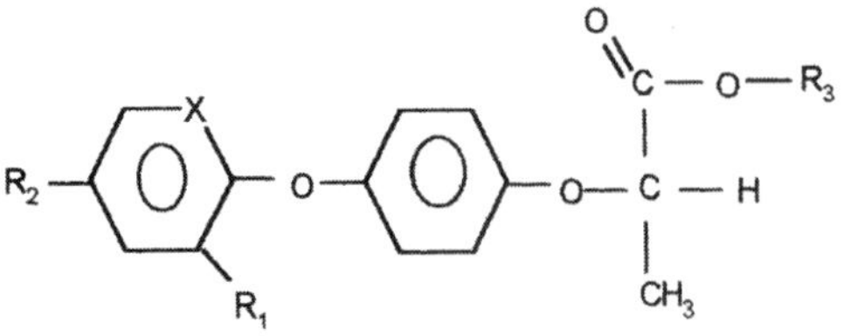

1. X = H, R = H, R_2 = H, R_3 = Methyl–Diclofop
2. X = H, R_1 = F, R_2 = CN, R_3 = n-Butyl–Cyhalofop
3. X = N, R_2 = Cl, R_1 = F, R_3 = CH_2–C≡CH–Clodinofop
4. X = N, R_1 = H, R_2 = CF_3, R_3 = n-Butyl = –Fluazifop 4.5

Diarylether Carboxylic Acids (Cereal Herbicides)

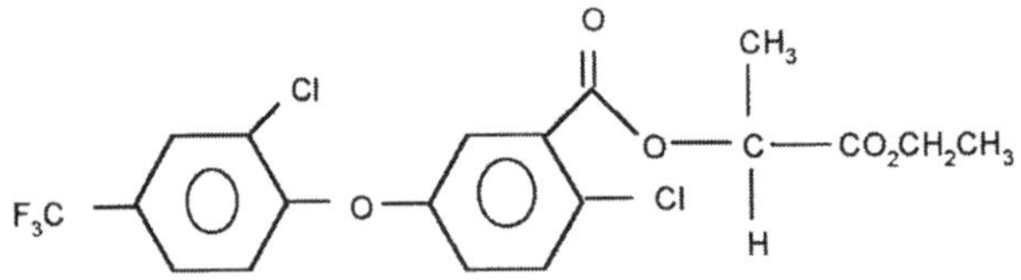

HC-252

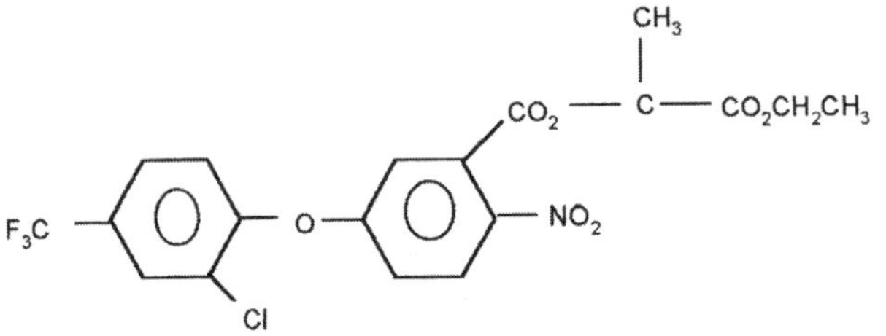

Fluoroglycofen

Benzoyl Ureas (Insect Growth Regulators)
(Table 17.7)

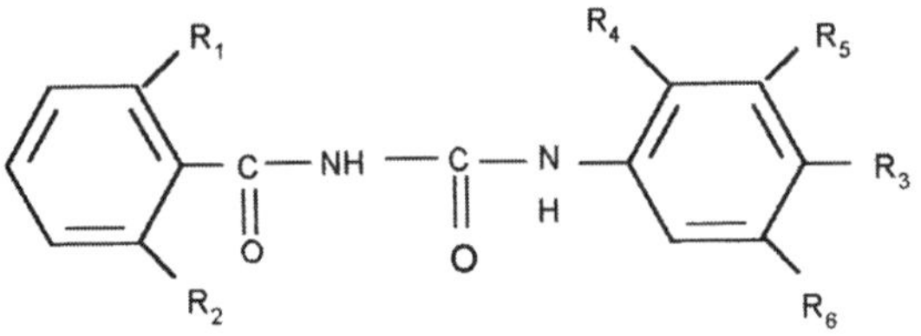

R_1 = R_2 = F, R_3 = Cl , R_4 = R_5 = R_6 = H Diflubenzuron

R_1 = R_2 = F, R_3 = 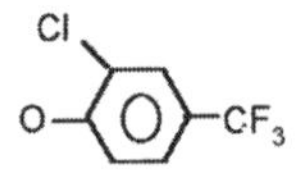 R_4 = F, R_5 = R_6 = H Flufenoxuron

R_1 = R_2 = F, R_3 = OCF_2CHF_2 , R_4 = H , R_5 = R_6 = Cl Hexaflumuron

Pheromones

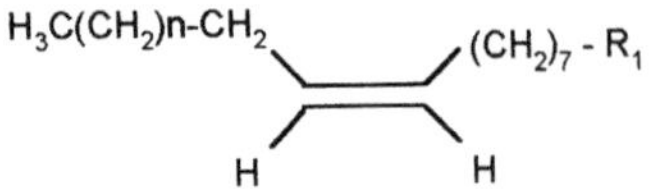

R_1 = CH_3 , n = 11 Mating hormone of housefly

R_1 = $OCOCH_3$, n = 1 Sex pheromone of oriental fruitmoth

TABLE 17.7 Insect Growth Regulators

Chemical (Chemical Class)	Structure	Biochemistry	Mode of Action	Environmental Fate
Diflubenzuron (Benzoylurea)	CONHCONH— with difluorophenyl and 4-chlorophenyl groups	Chitin synthesis inhibitor and so interferes with the formation of insect cuticle	Nonsystemic insect growth regulator with contact and stomach action. Acts at the time of insect moulting or at hatching of eggs	*Animals.* Elimination is partly as the unchanged parent compound in the feces, partly as hydroxylated metabolites (for c. 80%) and as 4-chlorophenylurea plus 2,6-difluorobenzoic acid (for c. 20%). The intestinal absorption is strongly related to the dosage administered—the higher the dosage, the more (relatively) is excreted unchanged in the feces. *Plants.* Nonsystemic. Nonmetabolized on plants *Soil.* Strongly absorbed by soil/humic acid complex and is virtually immobile in soil. Rapidly degraded in soil, with a half-life of <7 days. The principal degradation products are 4-chlorophenylurea and 2,6-difluorobenzoic acid
Cyromazine (Triazine)	H_2N— triazine ring —NH-cyclopropyl, NH_2	Interfere with moulting and pupation	Insect Growth Regulator with contact action. When used on plants action is systemic. If applied to leaves, it exhibits a strong translaminar effect, applied to soil it is translocated acropetally after absorption by roots	*Animals.* Efficiently excreted, mainly as the parent compound *Plants.* Rapidly metabolized. The principal metabolite is melamine *Soil.* Cyromazine and its main metabolite melamine are moderately mobile. Efficiently degraded by biological mechanisms
Fenoxycarb (Carbamate)	phenyl—O— —$OCH_2CH_2NHCO_2CH_2CH_3$	—	Nonneurotoxic insect growth regulator with contact and stomach action. Exhibits a strong juvenile hormone activey, inhibiting metamorphosis to the adult stage and interfere with the moulting of early instar larvae	*Animals.* The major metabolic path is ring hydroxylation to form ethyl [2-[p-(p-hydroxyphenoxy)phenoxy]ethyl]carbamate *Plants.* Rapidly degraded in plants *Soil.* Low mobility in soil, no bioaccumulation. Relatively fast degradation: DT_{50} 1.7–2.5 months (lab.), few to 31 days (field)

TABLE 17.7 Continued

Chemical (Chemical Class)	Structure	Biochemistry	Mode of Action	Environmental Fate
Hydroprene (Juvenile hormone mimic)	$(CH_3)_2CHCH_2CH_2CH_2CH_2$—$C=C$—$C=C$—$CO_2CH_2CH_3$ (with CH_3 and H substituents)	—	Prevents metamorphosis to viable adults when applied to larval stage	*Plants.* Degradation principally involves ester hydrolysis, *O*-demethylation, and oxidative splitting of the double bond. *Soil.* Rapidly decomposed, DT_{50} is only a few days
Buprofezin (Cyclic urea)	cyclic urea structure with S, $=NC(CH_3)_3$, N—$CH(CH_3)_2$, O, and phenyl group	Probable chitin synthesis and prostaglandin inhibitor. Hormone disturbing effect, leading to suppression of ecdysis	Contact and stomach action, not translocated in the plant. Inhibits moulting of nymphs and larvae, leading to death. Also suppresses oviposition by adults, treated insects lay sterile eggs	*Animals.* Low residues were found in nearly all ruminant and poultry tissues. Extensive metabolism was observed, with a large number of minor metabolites being produced. *Plants.* Limited metabolism in most plant species; minor metabolites indicate a pathway involving hydroxylation or oxidative loss of the *tert*-butyl group, followed by opening of the heterocyclic ring. *Soil.* DT_{50} (25°C) 104 days (flooded conditions, silty clay loam, o.c. 3.8%, pH > 6.4), 80 days (upland conditions, sandy loam, o.c. 2.4%, pH 7.0)

Strobilurins (Fungicide)

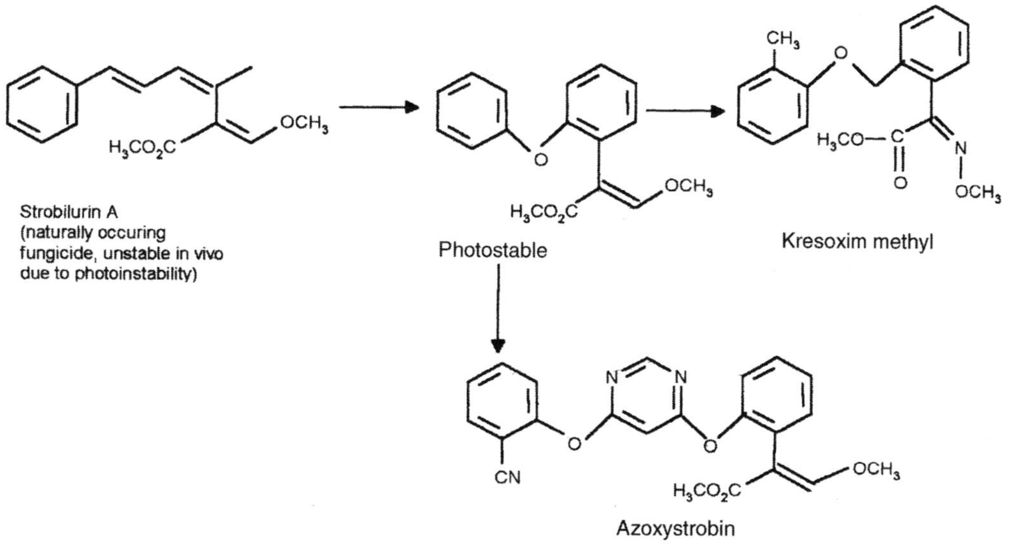

Strobilurin A
(naturally occuring
fungicide, unstable in vivo
due to photoinstability)

Photostable

Kresoxim methyl

Azoxystrobin

Neonicotinoids (Sucking pest control insecticides)

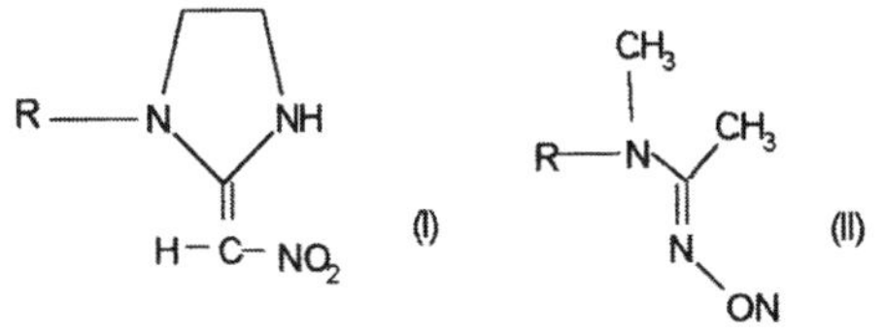

LC_{90} (Against Rice Plant hopper)

R = (a) 4-chlobenzyl—40 ppm

(b) 3-pyridinobenzyl—08 ppm—improved spectrum of activity

(c) 6-chloropyridino-3-benzyl—0.32 ppm (I)—Imidacloprid

(d) 6-chloropyridino-3-benzyl-(II)— Acetamiprid

a......d having isosteric relationships. Distance between "N" in pyridine to acidic proton in (I) crucial to activity.

ROLE OF CHIRALITY

Whenever a chemical structure has one or more asymmetric centers or double bonds, either diastereoisomerism or geometric isomerism is possible. Usually only one of the specific stereoisomers is responsible for all or most of the biological activity. An example is

cypermethrin,

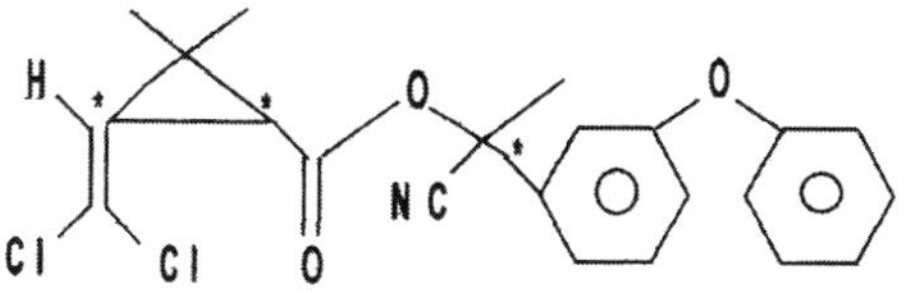

Cypermethrin

which possesses three asymmetric centers and hence eight possible isomers. Of eight possible isomers, the 1R-*cis*-S isomer is the most biologically active isomer, and the bromo analogue is deltamethrin, commercialized in 1984. In the search for low-dose agrochemicals, chiral synthesis plays an important role. For example, one methodology, as used for deltamethrin, is the separation of *cis* and *trans* acids, resolution of the *cis* acid using a chiral amine to prepare the 1R-*cis*-acid. Condensation of the 1R-*cis*-acid chloride with in situ generated cyanohydrin, derived from metaphenoxybenzaldehyde, offers a mixture of "1R *cis* ∝ S" and "1R *cis* ∝ R" deltamethrin. The crystallization of the mixture under conditions of epimerization allows isolation of high purity 1R *cis* ∝ S deltamethrin. Today, stereoisomer separations are also relatively facile using chiral phases in preparing HPLC.

Table 17.8 lists examples of chiral agrochemicals.

TABLE 17.8 Chiral Chemistry

Chemical (Chemical Class)	Structure	Biochemistry	Mode of Action	Environmental Fate
Deltamethrin (Pyrethroid)		Prevents sodium channels from functioning so that no transmission of nerve impulses can take place	Nonsystemic insecticide with contact and stomach action	*Animals.* Eliminates within 2–4 days. The phenyl ring is hydroxylated, the ester bond hydrolyzed, and the acid moiety is eliminated as the glucoside and glycine conjugates. *Soil.* Undergoes microbial degradation within 1–2 weeks DT_{50} in field <23 days. Soil photolysis DT_{50} 9 days. No risk of leaching
S-Metolachlor (Chloroacetamide)		Cell division inhibitor	Selective herbicide, absorbed predominantly by the hypocotyls and shoots; inhibits germination	*Animals.* Rapidly oxidized by rat liver. Oxygenases via dechlorination, O-demethylation and side-chain oxidation, conjugation by glyutathione S-transferases. *Plants.* Metabolism involves dechlorination and conjugation to glutathione-S-transferases, followed by further degradation to polar, water soluble, nonvolatile metabolites. *Soil.* Major aerobic metabolites are derivatives of oxalic and sulfonic acids; DT_{50} (field): 11–30 days. DT_{90} (field): 36–90 days
Epoxiconazole (Azole)		Inhibits C^{14} demethylase in sterol biosynthesis	Preventive and curative fungicide	*Animals.* Readily excreted via feces. Metabolic reactions are cleavage of the oxirane ring, hydroxylation of the phenyl rings and conjugation. No major metabolites; high number of minor metabolites. *Soil.* Degradation by microbial activity. DT_{50} a. 2–3 months

Name	Structure	Mode of action	Absorption	Metabolism
Dimethenamid (Chloroacetamide)	dimethenamid / dimethenamid-P	Cell division inhibitor	Herbicide absorbed by coleoptile	*Animals.* Metabolites in rat, goat, and hen include glutathione, cysteine, and thioglycolic acid. *Plants.* Metabolism in maize leads to thiolactic acid besides above. *Soil.* Rapidly degraded in soil, probably through microbial action, with DT_{50}: 8–43 days, depending upon soil type and weather conditions. Photolysis DT_{50} on soil a. 7.8 days
Mecoprop-*P* (Aryloxyalkanoic acid)		Systemic auxin, acting like indolylacetic acid	Selective herbicide, absorbed by leaves, with translocation to the roots	*Animals.* In mammals predominantly eliminated as conjugates in urine. *Plants.* Hydroxylated at the methyl group with formation of 2-hydroxy methyl-4-chlorophenoxy propionic acid. *Soil.* Degraded predominantly by microorganisms to 4-chloro-2-methyl phenol, followed by ring hydroxylation at the 6-position and ring opening. DT_{50} (aerobic) : 3–13 days

Basis of Chemistry Used in Synthesis of Agrochemicals

Agrochemicals belong to a variety of chemical classes. Each chemical class demands certain key building blocks. Some examples are: cyanuric chloride (triazines), iminodiacetic acid (glyphospate), 2,6-disubstituted anilines (chloracetanilides), dialkoxythiophosphoryl chlorides (organophosphorous group), metaphenoxybenzaldehyde/alcohol (pyrethroid), isocyanates—alkyl/aryl (carbamates / ureas), sulfonyliso-cyanates (sulphonyl ureas), dialkyl/dihalovinyl-dimethyl-cyclopropane carboxylic acid chloride (pyrethyroids), dihalophenols (phenoxyherbicides), chlorocresols (dichlorophen, etc.), ethylenediamine (dithiocarbamates).

During the synthesis of specific building blocks (key intermediates, penultimate intermediates, or target-active ingredients) a whole range of chemical reactions are deployed. Examples obviously include aromatic electrophilic and nucleophilic substitution, Diels Alder cycloadditions, telemerizations, "N," or "0" alkylations, and so on. New and safer reagents are often employed, for example diphosgene/triphosgene/diarylcarbamates, in lieu of phosphene in the manufacture of carbamates and sulfonyl ureas. Toxic reagents are generated for immediate chemical reaction (consumption) avoiding hazards of storage through leakage. As an example, the standard production of alkyl isocyanates for production of carbonates—from alkalimetal cyanates and alkyl halides or sulfates—can be replaced with the in situ generation of arylisocyanates from arylamines and urea to produce target aryl alkyl ureas.

Case Study—Chemistry and Manufacture of Metolachlor

A typical example of agrochemical manufacture is provided by the maize herbicide (Table 17.2b), *metolachlor*.

The chemistry of metolochlor synthesis is shown below, followed by a description of the manufacturing process.

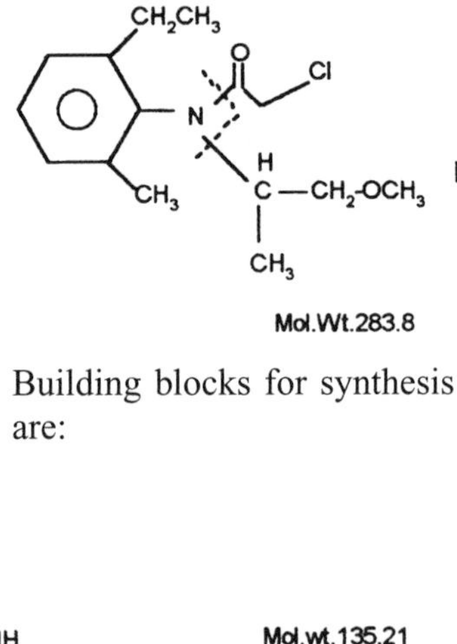

CH2CH3
O
Cl
N
H
CH3
C —CH2-OCH3
CH3

Metolachlor (Maize

Mol.Wt.283.8

Building blocks for synthesis of metolachlor are:

A) 2-ethyl-6-methyl-aniline

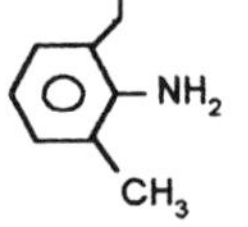

NH2
CH3

Mol.wt.135.21
bp. 226.8°C

B) Chloroacetyl chloride

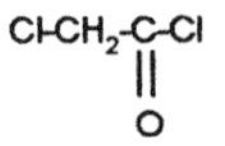

Cl-CH2-C-Cl
O

Mol.Wt.
bp 105-6°C

C) Methoxyacetone

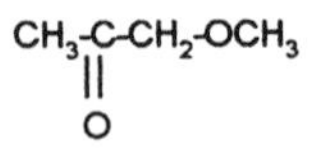

CH3-C-CH2-OCH3
O

Mol.Wt.
bp 118°C

Synthesis

A + B

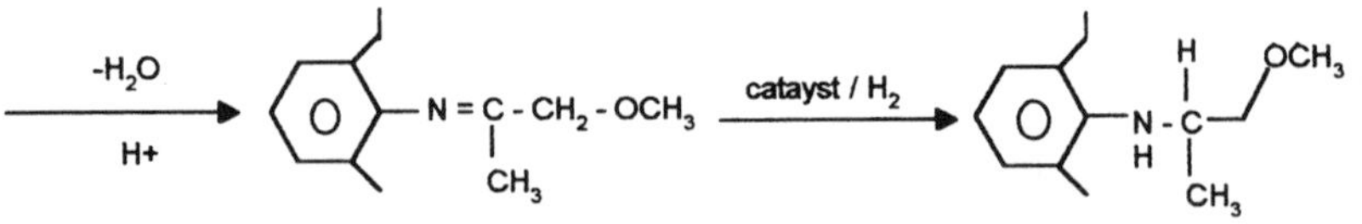

-H2O
H+
N = C - CH2 - OCH3
CH3
catayst / H2
H
OCH3
N - C
H
CH3

I

II

(D)

(E)

$$\text{E} \xrightarrow[\text{III}]{\quad \underset{\text{ClCH}_2\text{-C-Cl}}{\overset{\overset{\displaystyle O}{\|}}{}} \quad} \text{Metolachlor}$$

The nature of the catalyst employed during the hydrogenation of (D) offers (E) as a racemate of 1S and 1R isomers, or enriched in 1S isomer which leads into *racemic (R/S) metolachlor* or *S-metolachlor*. A chiral catalyst allows the 1S isomer enrichment to be favored.

The raw materials used in the commercial synthesis of metolachlor include 2-ethyl 6-methyl-aniline, which can be built up from

ortho-toluidine and ethylene (ex ethanol if required) at high temperature and pressure over metallic aluminum. Methyoxyacetone is built up from

propylene oxide and its reaction with sodium methoxide in methanol, which favors opening from the least hindered side and results in

$$\underset{(F)}{\overset{\displaystyle H}{\underset{\displaystyle OH}{H_3CO\text{-}CH_2\text{-}\overset{|}{\underset{|}{C}}\text{-}CH_3}}}$$

the methylether of 2,3-propanediol. Due operational care is necessary in handling propylene oxide as it is extremely flammable and poisonous.

The methyl ether of 2,3-propanediol can be oxidized to methoxyacetone in the vapor or liquid phase using a suitable catalyst or oxidant, respectively. To achieve production of high purity metolachlor, which is a liquid (mp. 62.1°), it is necessary to control the quality of all input materials, monitor reaction progress, terminate each reaction step effectively, and use solvents that can be effectively eliminated to very low levels without volatilization losses of metolachlor (100°C/0.001 mmHg).

Reaction step I is conveniently carried out in a stainless steel reactor using a hydrocarbon solvent to expel water of the reaction as an azeotrope. Completion of the reaction can be determined by measuring the water of reaction or through disappearance of both the 2-ethyl-6-methyl aniline and methoxyacetone by gas chromatography (GC) or high performance liquid chromatography (HPLC).

In the reaction step II, the energetics of saturation of an olefinic Schiff base linkage are favorable and quantitative over most hydrogenation catalysts. 1S isomer enrichment can be achieved through use of a chiral catalyst and the enrichment may be monitored through chiral HPLC. A stainless steel reactor is considered adequate for this operation.

Reaction step III is "*N*-acylation." It is conveniently carried out in a refluxing chlorinated aliphatic hydrocarbon; the hydrogen chloride gas which is evolved should be scrubbed in aqueous caustic soda. Precautions are necessary when handling the highly corrosive chloroacetyl chloride. This step should be carried out in a glass lined reactor.

Effluents from the reaction are the water of reaction in step I, and hydrogen chloride gas in step III.

Usage of Agrochemicals

Agroclimatic conditions, soil type, irrigation, as opposed to rain as the source of water, and other factors all govern the nature of crop cultivation and also the nature of pest attack. Soil preparation, seed dressing, pre-sowing, pre-plant incorporation, pre-emergence, early-post-, or late post-emergence, right through harvest are all various stages during which agrochemicals find use. Protection of farm produce or seeds from pest attack is dependant on the use of agrochemicals. Ectoparasite control of livestock and poultry is another area of agrochemical usage. Disease vector control from flies, mosquitoes, and cockroaches on one hand and protection of dwellings from termites and other wood borers on the other are both significant consumers of agrochemicals. Different insecticides are used to control insects during the different stages of insect life cycle. The habitat of the insects, whether dwelling above or below the soil surface, and the different feeding habits of individual groups of insects dictate the use of specific insecticides. Use of herbicides depends upon the weed spectrum as per crop and climatic conditions. The competing growth stages of the weeds, as compared to the crop, as well as nature of weeds (e.g., grasses, broadleaf weeds, or sedges) and the efficacy characteristics of individual herbicides need to be synchronized for effective control. Biannual versus annual, or perennial weeds need be treated with different specific herbicides.

The various fungicides have either specific or broad spectrum efficacy. These can be deployed for either prophylactic or curative action. Beyond the farm crops and horticulture, maintenance of turf and lawns free of insects and weeds, and the preservation of natural and agricultural forestry are significant markets for insecticides and herbicides.

All agrochemicals must be used at well-researched recommended dosages. Sublethal dosages as well as single agrochemicals used repeatedly leads to resistance development. It is therefore important to use agrochemicals with an independent mode of action in rotation or in well-conceived mixtures to decrease resistance development. Different formulations of a given agrochemical are deployed when the intention is for soil application for absorption by root system or by foliar uptake. Different sprayers with specific nozzles allow better efficacy. Aerial spraying demands yet different formulations.

Differential selective or broad spectrum agents include nematacicides (Table 17.9), acaracides, ovicides, molluscicides (Table 17.10), algicides (Table 17.11), bactericides (Table 17.12), and rodenticides (Table 17.13), which control damage from nematodes, mites, eggs, snails/slugs, algae, bacteria, and rodents, respectively.

Insecticides control insects with differential feeding habits, which have a reasonably predictable pattern per crop and agroclimatic condition. Systemic insecticides (Table 17.3a) find use to control insects that feed by sucking plant juices. Insecticides with contact and stomach action control insects feeding (Table 17.3b) on foliage, stems, and fruits. Insect growth regulators (Table 17.7) find a complementary role to make insects more prone to control. Plant growth regulators (Table 17.6) direct nutrition to fruit formation by controlling vegetative growth. Herbicides (Tables 17.4a–d) are used for farm preparation, preplant incorporation, pre-early post, and post-emergence application to control competing weeds. Fungicides (Tables 17.5a,b) are used for seed dressing (control from soil fungi), and for the prophylactic and curative protection from fungal attack. Economic threshold levels, use of pheromone traps (Table 17.14) to measure insect population and the like are tools to use agrochemicals judiciously. Agrochemicals, when used rotationally or in mixtures of two with differential modes of action (e.g., mixture of an insecticide with insect growth regulator, mixture of herbicides or fungicides with different modes of action) help tackle the problem of resistance development.

BIOTECHNOLOGY FOR FARM PRODUCTIVITY

Biotechnology using recombinant DNA has affected farm productivity in some areas and may become more significant over time. A full discussion of this interesting subject is beyond the present scope of this chapter, but mention of some aspects may be appropriate.

The determination of DNA sequences of genes for resistance to pests and pathogens, cloned from a number of crop species, suggests the existence of certain domains in the protein products. These are shared among

TABLE 17.9   Nematicides

Chemical (Chemical Class)	Structure	Mode of Action	Biochemistry	Environmental Fate
Aldicarb (Oxime Carbamate)	CH_3 $\|$ $CH_3S-C-CH=NOCONHCH_3$ $\|$ CH_3	Systemic nematicide with contact and stomach action	Cholinesterase inhibitor. Metabolically activated to Aldicarb sulfoxide	*Animals.* Absorbed rapidly and completely; >80% is excreted in the urine within 24 hr, >96% within 3–4 days. Aldicarb is oxidized to the sulfoxide and sulfone, which undergo further metabolism *Plants.* The sulfur atom is oxidized to sulfoxide and sulfone groups. Further degradation leads to the formation of oximes, nitriles, amides, acids, and alcohols which are present in the plant only in conjugated form *Soil.* Sulfur atom is oxidized to sulfoxide and sulfone groups. Further degradation leads to the formation of oximes, nitriles, amides, acids, and alcohols
Fenamiphos (Organophosphorous)	CH_3 CH_3S —⟨benzene ring⟩— $\overset{\overset{O}{\|\|}}{O}PNHCH(CH_3)_2$ OCH_2CH_3	Systemic nematicide with contact action. Absorbed by roots with translocation to the leaves	Cholinesterase inhibitor	*Animals.* In mammals, following oral administration, there is rapid metabolism involving oxidation to the sulfoxide and sulfone analogues, followed by subsequent hydrolysis, conjugation and excretion via the urine, some *N*-dealkylation also occurs *Plants.* Degradation is by thiooxidation and hydrolysis. The major metabolites are fenamiphos sulphoxide and fenamiphos sulfone *Soil.* Degradable on soil surfaces. Duration of activity in soil is c. 4 months. Compound with low mobility. Soil DT_{50} (aerobic and anaerobic) several weeks, The major degradation products are fenamiphos sulfoxide and fenamiphos sulfone and their phenols
1,3-dichloropropene (Chloroalkenes)	—	Soil fumigant nematicide.	—	—

TABLE 17.10 Molluscicides

Chemical (Chemical Class)	Structure	Biochemistry	Mode of Action	Environmental Fate
Metaldehyde		—	Molluscicide with contact and stomach action. Poisoned slugs secrete large quantities of slime, desiccate and die. Their mucus cells are irreversibly destroyed	*Soil.* Aerobic and anaerobic microorganisms in soil decompose metaldehyde to CO_2 and water
Methiocarb (Carbamate)		Cholinesterase inhibitor	Molluscicide with neurotoxic toxic effect	*Animals.* Rapidly absorbed and excreted, principally in the urine, with only a small proportion in the feces. Metabolism involves hydrolysis, oxidation, and hydroxylation, followed by excretion in free or conjugated form. There is a continuous decrease of activity in all organs. *Plants.* Methylthio group is oxidized to sulfoxide and sulfone, with hydrolysis to the corresponding thiophenol, methylsulfoxide-phenol, and methylsulfonyl-phenol *Soil.* Degradation is rapid. The importance metabolites are methylsulfinylphenol and methylsulfonylpehnol
Niclosamide (Anilide)		—	Molluscicide with respiratory and stomach action	*Animals.* The major metabolite in the urine was the reduced compound 2′,5-dichloro-4′-aminosalicylanilide. Several labile conjugates were also detected. The major constituent in the faeces was unchanged niclosamide, although considerable amounts of 2′,5-dichloro-4′-aminosalicylanilide were also present *Soil.* Degradation followed pseudo-first order kinetics, DT_{50} 0.3 days

TABLE 17.11   Algicides

Chemical (Chemical Class)	Structure	Biochemistry	Mode of Action	Environmental Fate
Nabam (Multi-site alkylenebis (dithiocarbamate)	NaSCSNHCH$_2$CH$_2$NHCSSNa	Non-specific thiol reactant, inhibiting respiration	—	*Plants.* The principal metabolite is ethylenethiourea. Other metabolites include ethylenethiuram monosulfide, ethylenethiuram disulfide, and sulfur
Dichlorophen (Chlorophenol)		—	Contact action	—

TABLE 17.12 Bactericides

Chemical (Chemical Class)	Structure	Biochemistry	Mode of Action	Environmental Fate
Bronopol	$\begin{array}{c} Br \\ | \\ HOCH_2-C-CH_2OH \\ | \\ NO_2 \end{array}$	Oxidation of mercapto group of bacterial enzymes. Inhibition of dehydrogenase activity leads to irreversible membrane damage	—	*Animals.* Rapidly absorbed and rapidly excreted, mainly in the urine. The major metabolite is 2-nitropropane-1,3-diol *Plants.* Biochemical degradation leads to the metabolite 2-nitropropan-1,3-diol
Kasugamycin (Antibiotic)		Protein synthesis inhibitor, inhibits binding of Met-RNA to the mRNA-3OS complex thereby preventing amino acid incorporation	Systemic bactericide	*Animals.* Mostly excreted in the urine within 24 hr. After oral administration to rats at 200 mg/kg, 96% of administered dose remained in the digestive tract 1 hr after administration *Plants/soil.* Degraded to kasugamycinic acid and kasuganobiosamine; finally degraded to ammonia, oxalic acid, CO_2 and water
Oxolinic acid (Pyridone)		—	Systemic bactericide	—
Probenazole		—	Systemic bactericide, absorbed by roots and translocated acropetally	*Soil.* Half-life in soil <24 hr (alluvial or volcanic soil)
Tecloftalam (Benzoic acid)		—	Bacteriostat	*Soil/Environment.* DT_{50} 4–10 days undergoing loss of chlorine from the benzoic acid ring

Streptomycin
(antibiotic)

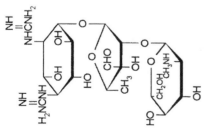

Inhibits protein synthesis by binding
to the 3OS ribosomal subunit,
causing a misreading of the genetic
code

Bactericide with
systemic action

—

TABLE 17.13   Rodenticides

Chemical (Chemical Class)	Structure	Biochemistry	Mode of Action	Environmental Fate
Bromadiolone (Coumarin anticoagulant)		Second generation anticoagulant rodenticide which also blocks prothrombin formation	—	*Soil.* Leaching behavior is inversely related to clay and organic matter content of soils. In soil column and soil layer studies, 97% was remaining in the top soil, with 0.1% in leachate
Phosphine[a]	PH_3	—	Respiratory, metabolic and nerve poison. Evolves nonflammable mixture of phosphine, ammonia and carbon dioxide	*Animals.* Phosphine is probably metabolised to nontoxic phosphates. *Plants.* In stored products, phosphine undergoes oxidation to phosphoric acid
Chlorophacinone (Indandione anticoagulant)		Blocks prothrombin formation and uncouples oxidative phosphorylation	—	*Animals.* 90% is eliminated in the feces within 48 hr in the form of metabolites

[a]Includes aluminium phosphide, zinc phosphide, magnesium phosphide.

TABLE 17.14   Insect Pheromones

Chemical (Chemical Class)	Structure	Biochemistry	Mode of Action	Environmental Fate
Codlemone (Pheromone)		—	Acts by disruption of mating, either in trapping or in disorientation mode	*Plants.* Rapidly oxidized on the surface of leaves *Soil.* Rapidly degraded in soil to CO_2
Disparlure (Pheromone)		—	Acts both as an attractant and as a mating disruptant	—
Dodeca-7,9-dienyl Acetate (Pheromone)	 (*7E*,9*Z*)- isomer	—	Acts as an attractant and by disruption of mating in the disorientation mode	—

(continued)

TABLE 17.14 Continued

Chemical (Chemical Class)	Structure	Biochemistry	Mode of Action	Environmental Fate
Gossyplure (Pheromone)	(Z,Z) - (Z,E) -	—	Acts as an attractant and by disruption of mating	—
Methyl eugenol (Synthetic pheromone)		—	Synthetic attractant	—

plant species and are shared among genes, which can provide resistance to viruses, fungi, bacteria, nematodes, and insects. It may be possible to design durable resistance genes that have a broad spectrum of activity. Modification of genes controlling key steps in signal-transduction pathways and resulting in the generation of defense responses may influence the speed and type of plant response. This may then lead to pest and disease resistance. Herbicide tolerance is already widely used in the United States. Genetic modification of secondary plant metabolism through the synthesis of novel toxins or increased levels of existing toxins and through synthesis of molecules which regulate pest or pathogens colonization, development or morphogenesis may enhance crop resistance to pests and pathogens. Knowledge of plant signaling processes and the role of secondary plant metabolites in regulating insect behavior offer opportunities to manipulate relevant biosynthetic pathways. This, in turn, may increase the ability of plants to deter pest colonization, inhibit pest development and reproduction, and attract natural enemies of its pests through semiochemical production.

Genes of microbial, animal or plant origin, coding for toxins, inhibitors of pathogenicity factors, such as the lecithins, or degradative enzymes, such as chitinases and proteases, may enhance resistance to pests or pathogens when expressed in transgenic plants (e.g., genetically modified (GM) crops expressing *B. thuringiensis* (Bt) toxins) (Table 17.15). However, the effect of GM crops on predator and parasites, as well as on pollinators, need to be monitored. Also, development of resistance to GM plants by target insects requires careful monitoring. Genetically modified predators and parasites for enhanced adaptability or baculoviruses with modified genes for toxin production for enhanced efficacy of viruses as biological control agents are currently being investigated. Virus resistant squash and papaya are already available in U.S. markets. Good levels of antibody expression in plants have not yet been

achieved due to technical problems. Discovery of new essential processes of pests and pathogens and their genetic make-up/gene function could lead to novel chemistries having such processes as target systems. Combination of genomics and biochemical screening procedures may lead to a range of new chemical control methods for agrochemical industry while increasing the screening rates of the molecules. Genetic processes are also used to understand pesticide resistance within pest populations, leading to high-resolution diagnostics for resistant alleles, especially for pests with multiple resistance mechanisms. Molecular techniques, used in ecological research to elucidate the structure of populations and to estimate gene flow between populations occupying discrete habitat patches, help in the formation of ecological modification strategies for maintaining biodiversity as farmland promoting sustainable pest management. The risk assessment of the release of GM organisms in the environment may be aided by assessing the gene flow among populations in agricultural systems.

Regulatory Scene Regarding Products of Biotechnology

The health effects of eating genetically modified foodstuffs, especially the allergenicity, are being investigated by the FAO/WHO Codex Alimentarius Commission, and its relevant subsidiary bodies have reflected the result of these studies. An Intergovernmental Task Force of Foods Derived from Biotechnology is being established to develop standards, guidelines, and recommendations as deemed to be appropriate for foods developed through biotechnology.

In the United States, the U.S. Department of Agriculture (USDA), EPA, and FDA are involved in the regulation of products of biotechnology. State and federal statutes and standards must be satisfied prior to registration. At present, no requirements are in place for the varietal registration of new crops. The

TABLE 17.15   Biological Agents

Biological Agents	*Usage*
Amblyseius spp. (a number of predatory mite species)	Control of thrips
Anagrus otomus	A wasp, parasite of leafhoppers
Aphelinus abdominatis	A wasp, parasite of aphids used in protected crops
Aphidius colemani	A wasp, parasite of aphids used in protected crops
Aphidoletes aphidimyza	Predatory midge, consumer of aphids in protected crops
Chrysoperla carnea	Entomophagus lacewing larva
Cryptolaemus montrouzieri	Beetle, consumer of mealybugs used in orchards, vines, and protected crops
Dacnusa sibirica	A wasp, parasite of leaf miners used in protected crops
Diglyphus isaea	A wasp, parasite of leaf miners
Encarsia formosa	A wasp, parasite of glasshouse whitefly—most widely used
Hippodamia convergens	Ladybird—consumer of aphids and other pests
Leptomastix dactylopii	A wasp, parasite of mealybugs used in horticultural and fruit crops
Metaphycus holvodus	A wasp, parasite of soft scales used in orchards and in protected crops
Orius spp.	Predatory bug. A number of species are used for control of thrips
Phytoseiulus persimillis	Spider mite consuming mite, used in protected crops
Trichogramma spp.	A number of species of wasp, parasitic of Lepidotera in protected crops
Bacillus sphaericus	Used against mosquito larvae
Beauveria bassiama	Entomopathogenic fungused under development for use on control of a wide range of coleopteran, homopteran, and heteropteran pests
Helicoverpa zea NPV	Nuclear polyhedrosis virus used for control of heliothis and Helicovenpa in cotton and tobacco
Heterorhabditis bacteriophora and *H. megidis*	Insect parasitic nematodes used for control of Japanese beetles, black vine weevils, etc.
Mamestra brassicae NPV	NPV used for control of lepidoptera
Metarhizium anisopliae	Entomopathogenic fungus under development for control of locusts
Spodoptera exigua NPV	For control of beef armyworm in various crops
Steinernema spp.	Parasitic nematode that searches for enters and kills target pests
Ampelomyces quisqualis	Hyperparasite of the Erysiphaceae genus, which causes powdery mildew diseases
Bacillus subtilis	Seed treatment for control of *Rizoctonia solani, Fusarium* spp., *Alternaria* spp., *Aspergillus* spp., etc. in cotton, legumes of other crops
Candida oleophila	Selective fungal antagonist for control of post harvest diease in citrus and pome fruit
Eiliocladium spp.	Used for the control of foliar pathogens in seedlings and as a post-harvest treatment. *G. vorens* is used for the control of soil diseases
Streptomyces griseoviridis	Bacterium used for control of *Fasaurium* and other pathogens
Trichoderma spp.	Mitosporic fungi used for control of range of soil/foliar pathogens
Microlarinus lareynii	Attacks seedheads of puncture vine
Urophora sirunasova	Gallfly attacks seedhead of yellow star thistle
Bangastemus orientalis	Weevil—attacks seedhead of yellow star thistle
Coleophora parthenica	Attacks stem of Russian thistle

following table gives a picture of the present regulatory review process.

the structural features of such secondary metabolites will act as important leads in the

New Trait/Organism	Regulatory Review Conducted by	Reviewed for
Viral resistance in food crop	USDA	Safe to grow
	EPA	Safe for the environment
	FDA	Safe to eat
Herbicide tolerance in food crop	USDA	Safe to grow
	EPA	New use of companion herbicide
	FDA	Safe to eat
Herbicide tolerance in ornamental crop	USDA	Safe to grow
	EPA	New use of companion herbicide
Modified oil content in food crop	USDA	Safe to grow
	FDA	Safe to eat
Modified flower color ornamental crop	USDA	Safe to grow
Modified soil bacteria Degrades pollutants	EPA	Safe for environment

FUTURE DIRECTIONS

Allelopathy

Herbicides comprise 60–70 percent of total usage of pesticides in developed countries. Plants develop resistance to herbicides which in turn demands altered management strategies through improved herbicides and/or herbicide mixtures. This may involve different modes of action, for short-term efficacy and long-term control to be effective.

Allelopathy is the action of secondary metabolites in plants, algae, bacteria or fungi, which influences the growth and development of other species. This may help in overcoming such problems of resistance through development of crop varieties having greater ability to smother weeds, use of natural phytotoxins from plants or microbes as herbicides, and use of synthetic derivatives of natural products as herbicides. Another way allelopathy may be used in agriculture is through isolation, identification, and synthesis of the active compounds from an allelopathic plant or a microorganisms species. Knowledge of

development of future herbicides to tackle resistance phenomena.

Biorational Approach to Chemical Synthesis

Knowledge of pest biochemistry will open multiple target sites to which agrochemicals of appropriate structure could be directed. This is to say that new modes of action will emerge and may overcome some of the presently unmet needs. These include such items as:

- Need for insecticides with *multiple modes of action*
- Need for *higher selectivity* herbicides to control weeds that resemble the main crop
- Fungicides and insecticides of *high safety* to preserve agricultural produce
- Control agents for plant intake to *minimize damage* from soil fungi, soil insects, and nematodes
- *Safer and more effective* rodenticides and bird repellants
- Antivirals and antibacterials

REFERENCES

1. *Future Research, Development and Technology Transfer Needs for UK Crop Protection*, A report prepared by the British Crop Protection Council, 1997, August.
2. *Evaluation of Allergenicity of Genetically Modified Foods*. Report of a joint FAO/WHO expert consultation on Allergenicity of Foods derived from Biotechnology, 2001, January 22–25.
3. *Biotechnology in Crop Protection: A BCPC Appraisal of Progress and Prospects*. A report prepared by British Crop Protection Council, 1999, November.
4. *Acreage*, Released by National Agricultural Statistics service (NASS), Agricultural Statistics Board, US Department of Agriculture, 2001, June.
5. Regulatory information downloaded from various US, EEC, and Australian sites.
6. *Pesticide Manual*, 12th ed., CDs Tomlin (Ed.), 2001.
7. Synthesis and Chemistry of Agrochemicals IV, *ACS Symposium Series* 584, D. R. Baker, J. G. Fenyes, and G. S. Basarab (Eds.), 1995.
8. *Kirk—Othmer Encyclopedia of Chemical Technology*, 4th ed., J. I. Kroschwitz and M. Howe-Grant (Eds.), 1992.

18

Petroleum and Its Products

S. Romanow-Garcia* and H. L. Hoffman**

THE NATURE OF PETROLEUM

Petroleum is a diverse mixture of hydrocarbons—chemical combinations of primarily hydrogen and carbon. Complete combustion of hydrocarbons yields the end products of carbon dioxide (CO_2) and water (H_2O). However, incomplete combustion results in a composite mixture of other products such as CO_2, H_2O, carbon monoxide (CO), and various oxygenated hydrocarbons. Since burning petroleum consumes air, nitrogen compounds are also formed. In addition, other elements are associated with hydrocarbon compounds such as sulfur, nickel, and vanadium.

Petroleum is found at great depths underground or below seabeds. It can exist as a gas, liquid, solid, or a combination of these three states, which is very common. Drilling efforts are used to reach and extract gaseous and liquid deposits. These products are brought to the surface via piping. Once found in a reservoir, gas usually flows under its own pressure. Conversely, discovered liquid

hydrocarbons may flow on their own due to pressure from the reservoir or may be forced to the surface by submerged pumps. Also injection of fluids and gases provides a driving force to push liquid hydrocarbon through rock strata. Solid or semisolid petroleum is brought to the surface though several methods: by digging with conventional mining techniques, by gasifying or liquefying with high-temperature steam, or by burning a portion of the material in the ground so that the remainder can flow to the surface.

Hydrocarbon Forms

As mentioned earlier, petroleum is any product that is primarily composed of hydrogen and carbon bonded compounds. These compounds can be further categorized by their characteristics.

Natural gas is the gaseous form of petroleum. It is mostly the single-carbon molecule—methane (CH_4). When natural gas is associated with liquid petroleum underground, the methane will come to the surface in admixture with some heavier hydrocarbons. The gas is considered a wet gas; the heavier hydrocarbons

---

*Hydrocarbon Processing, Houston, Texas.
**Hydrocarbon Processing, Houston, Texas—Retired.

are isolated and purified in natural-gas processing plants. Gas processing yields ethane (an important petrochemical feedstock), propane [liquefied petroleum gas (LPG)], butane (refinery blending stock), and hydrocarbon liquids (natural gas condensate). When the underground natural gas is associated with solid hydrocarbons such as tar or coal, the methane will have few other hydrocarbons and is considered a dry gas.

Crude oil is the common name for liquid petroleum. In some literature, one will see reference to "petroleum and natural gas," suggesting petroleum and crude oil are used as synonymous terms. Some crude oils have such great density that they are referred to as heavy oils and tars.

Tar sands are small particles of sandstone surrounded by an organic material called bitumen. The bitumen is a highly viscous hydrocarbon that clings tenaciously to the sandstone; thus, it is easy to think of the mixture as a solid form of petroleum. Yet, it is a mixture of high-density liquid on a supporting solid.

Oil sands are true petroleum solids. Curiously, oil sands do not contain petroleum crude oil; it is an organic material called kerogen. The kerogen can be heated to yield a liquid called shale oil, which can be refined into conventional petroleum products.

Largest Energy Supplier

In Chapter 19 the point is made that coal offers an abundant primary energy source. Yet, present and proposed environmental legislation deters future coal usage. Due to stringent stack-emission restrictions for power generation, utilities are seeking "cleaner fuel" options to replace coal. Utility companies view natural gas as the "cleaner fuel" option. Petroleum feedstocks contain sulfur, which is strictly regulated on emission permits. Natural gas usage is growing within the power/utility industry. Yet, petroleum remains the major fuel source used in transportation, manufacturing, and home heating.

Primary energy sources are defined as those sourced to natural raw materials. Electricity is not included because it is a secondary energy source; it is generated by consuming one or more of the other natural energy sources. To put petroleum consumption into perspective, the primary energy sources considered here are: petroleum crude oil, natural gas, coal, hydropower (water to generate electricity), and nuclear energy. The quantities reported here will exclude energy from wood, peat, animal waste, and other sources, despite their importance to some localities.

The common practice is to relate energy units to a common product, in this case, to petroleum liquid. For example, world consumption of crude oil and liquids (condensates) from natural gas in 1999 reached 149.72 Quadrillion (10^{15} BTUs)—Quad. If the amount of energy from other sources were converted to equivalent barrels of oil, the total world energy consumption in 1999 would be 380 Quads.[1] The relative distribution of these sources is shown in Fig. 18.1. More energy comes from oil than from any other single source.

Another view to consumption is that the world consumption of crude oil and liquids from natural gas in 2001 reached 76 million barrels per day (MMbpd). North America is the largest energy consumer at 24.1 MMbpd. The Asia–Pacific regional demand is steadily increasing. In 2001, Asia–Pacific comprised 27.2 percent of the world's oil demand, up from 20 percent in 1990.[2] Crude oil's share of primary energy consumption was 39.4 percent in 2001. Thus, petroleum oil and natural gas remain the steadfast energy sources globally, as shown in Fig. 18.1.[2]

From Well to Refinery

Crude oil production for various countries is shown in Fig. 18.2.[2] The Middle Eastern countries produce more oil than they consume; the extra production is gated for export. Conversely, the United States and Western Europe consume much more crude oil than they produce (Fig. 18.3).[2] This condition demonstrates the great importance of worldwide petroleum movements. The difference between production and consumption for any

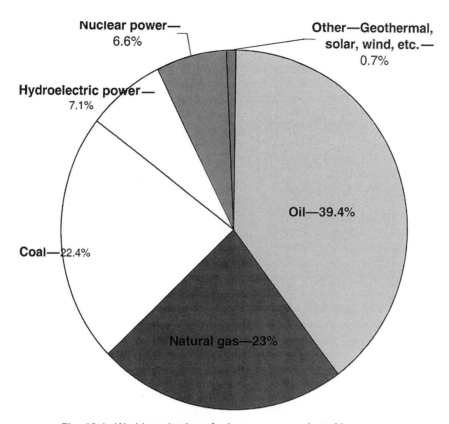

Fig. 18.1. World production of primary energy selected by groups.

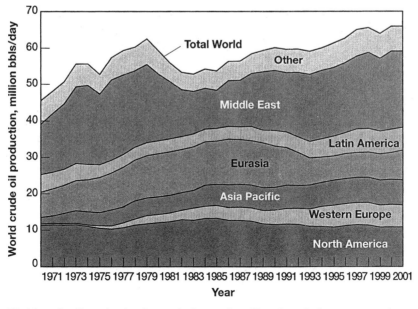

Fig. 18.2. World crude oil production by producing region. (Data from industry sources.)

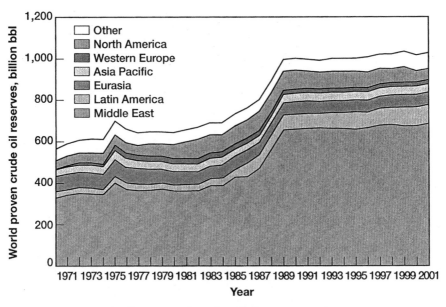

Fig. 18.3. World proven crude oil reserves. (Data from industry sources.)

one period is balanced by oil being added to or removed from extensive areas around the world.

The growth of world refining capacity attempts to keep pace with rising demand for petroleum-based products.[2] Curiously, refining capacity has surged ahead in some regions, notably the Middle East and the Asia–Pacific region, over the last 30 years. However, in developed markets, refinery throughput was almost flat during the 1980s and 1990s. New and larger state-of-the-art refining facilities were more energy efficient and had lower operating costs per barrel of refined products. Consequently, smaller refiners could not compete against new facilities and subsequently had to shut down operations. Another factor in refining growth is the time to construct processing units. In highly industrialized countries such as the United States, Japan, and Western Europe, mounting environmental regulations and stiff emission and performance laws have all but stifled the construction of new grassroots refineries. Construction of new refining capacity must overcome a long list of federal, state, and local governmental requirements. In the United States, the last grassroots refin-

ery was constructed in 1974, before the onslaught of the Clean Air Act of 1970. New governmental regulations are focused on strict reductions on emissions (air, water, and solids) to improve air quality for high-density populated areas and high-density industrial regions. Consequently, construction of new facilities in developed markets is negligible.

In industrialized nations, new refining capacity will be realized by the expansion of existing facilities that are permitted by the local regulatory agencies. Even expanding existing facilities is exempted from environmental constraints. With new construction and capacity expansion, operating companies must cut emissions below present permitted levels. Thus, operating companies must install more intrusive emission reduction/control technologies and equipment to eliminate release from new and existing plant equipment.

Refiners have become particularly adept in using technology to find incremental capacity from existing processing equipment. Thus, the refining industry can process more crude oil with present equipment. In spite of this, the number of refineries is decreasing; yet, capacity increases incrementally.

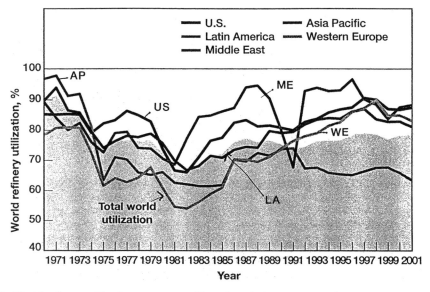

Fig. 18.4. World refinery utilization as percent. (Data from industry sources.)

Advancements and innovations in processing methods and catalyst systems have enabled construction of large, highly integrated, and complex refining complexes. New materials of construction have aided in building larger facilities; thus, smaller, older refineries cannot compete with newer, larger facilities. The number of operating facilities continues to decrease; yet capacity rises. Less efficient facilities shut down their operations since they are handicapped in producing refined products (Fig. 18.4).

In 1984, the number of refineries operating in the United States peaked at 318 facilities with a refining capacity of 18.62 MMbpd.[3] Technological advancements in processing methods and catalytic systems have enabled refiners to increase the capacity of existing units incrementally. Newer processing units are larger than the earlier versions. Refiners are applying economies of scale to disperse the product costs of refineries. Since 1981, the number of United States operating refineries has decreased to 155 in 2000 with a total operating capacity of 16.52 MMbpd, as shown in Fig. 18.5. Notably, smaller and less competitive refineries were shut down. Equally important, the utilization of operating United States

refineries rose over this same period, from a low of 68.6 percent to 92.6 percent.[3]

Technology helps refiners to push the boundaries of manufacturing, especially in the average-size refinery. In 1975, the averaged United States refinery had an operating capacity of 60,000 barrels per day (bpd). Innovations in catalyst technology and equipment design enabled the construction of larger vessels and reactors, and the introduction of ancillary equipment to support processing operations. Thus in 2000 the average refining capacity for a United States refinery exceeded 100,000 bpd, nearly double the capacity from 1975. In the United States, refiners have avoided constructing grassroots facilities to meet rising demand for products.

As the wave of environmental regulations continues to be levied against the refining industry, more consolidation is anticipated. More companies will leave segments of fuel manufacturing due to capital investments with diminishing returns?

Distribution of Crude Oil and Refined Products. Crude oil and its refined products are viewed as commodity products; thus, they are easily traded and transported to market. Many methods can be used to deliver crude to

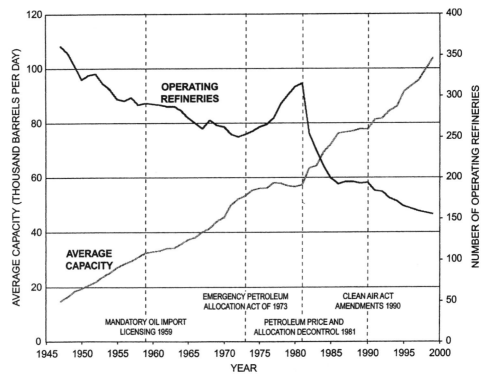

Fig. 18.5. Average capacity and number of U.S. operating refineries. (Data from industry sources.)

oil refineries. For example, United States refinery can receive feedstock crude oil via pipelines, tank trucks, barges, and ocean-going vessels—supertankers. Most refineries are located near port facilities to receive feedstocks from barges and supertankers.

PRODUCT

Refining crude oil involves breaking carbon-to-carbon (C–C) bonds of the complex hydrocarbon compounds and adding hydrogen. Such efforts are done thermally and catalytically. The distinction between refined products and petrochemicals often is subtle. In general, when the product fraction is from crude oil that includes a fairly large group of hydrocarbons, the fraction is classified as a refined product. Examples of refined products are: gasoline, diesel fuel, heating oil, lubricants, waxes, asphalts, and petroleum coke.

By contrast, when the product from crude oil is limited to only one or two specific hydrocarbons of fairly high purity, the fraction is called a petrochemical. Examples of petrochemicals are: ethylene, propylene, benzene, toluene, and styrene. Refined products are defined by the fraction's boiling point and may be composed of various hydrocarbons. Multiple compounds compose refined-product fractions. In contrast, petrochemicals are single-compound fractions, which are required for feedstocks for other petrochemicals and polymers. More processing and separation (distillation, extraction, etc.) operations are used to extract petrochemical products from processing streams. Thus, more identifiable petrochemical products are processed than refined products. Many specific hydrocarbon compounds can be derived from crude oil. However, these hydrocarbons lose their individual identity when they are grouped together as a refined product.

Refined Products

Most refined products at the consumer level are blends of several refinery streams. Product specifications determine which streams are suitable for a specific blend. Part of the difficulty in learning about refining lies in the industry's use of stream names that are different from the names of consumer products.

Refining is considered a "dirty" processing effort. Product separation of refined product streams is not as clean as efforts to process petrochemicals. Refiners have the flexibility to blend final fuel products. Thus, there is no exact recipe used by all refiners to produce consumer products. Multiple crude oils are processed and then blended to meet consumer fuel product specifications. Table 18.1 lists the refining streams that are blended to produce consumer products. The consumer products are familiar. However, within a refinery these products are blended from portions of crude oil fractions from the listed reforming process units. To complicate the situation further, not all refineries are configured identically. Many different processing operations can be used to refine and separate product streams to blend the products listed in Table 18.1.

For example, gasoline at the consumer level may be called benzol or petrol, depending upon the country where it is sold. In the early stages of crude oil processing, most gasoline components are called naphthas. Kerosene is another

example. It may be called coal oil to denote that it replaces stove oil (or range oil) once derived from coal. Historically, Kerosene gained significant importance as a replacement for whale oil for lamps. In the early 1900s, refining efforts were directed to supplying Kerosene–lamp oil. However, in the 1920s, a new energy form—electricity—began to displace Kerosene usage. Thus, early refiners sought other products to compensate for this market loss.[4] Today, Kerosene fractions are directed to jet fuel and high-quality No. 1 heating oil.

Product Specifications

Product applications and customer acceptance set detailed specifications for various products properties. In the United States, the American Society for Testing Materials (ASTM) and the American Petroleum Institute (API) are recognized for establishing specifications on both products and methods for testing. Other countries have similar referee organizations. For example, in the United Kingdom, it is the Institute of Petroleum (IP). In Germany, it is the Deutsches Institute suer Normung (DIN). In Japan, it is the Ministry of International Trade and Industry (MITI).

A boiling range is the major distinction among refined products, and many other product properties are related to the products in these boiling ranges. A summary of ASTM

TABLE 18.1   Several Names for the Same Material

Crude Oil Cuts	*Refinery Blends*	*Consumer Products*
Gases	Still gases	Fuel gas
	Propane/Butane	Liquefied petroleum gas (LPG)
Light/heavy naphtha	Motor Fuel	Gasoline
	Aviation turbine, Jet-B	Jet fuel (naphtha type)
Kerosene	Aviation turbine, Jet-A	Jet fuel (kerosene type)
	No. 1 Fuel oil	Kerosene (range oil)
Light gas oil	Diesel	Auto and tractor diesel
	No. 2 fuel oil	Home heating oil
Heavy gas oil	No. 4 fuel oil	Commercial heating oil
	No. 5 fuel oil	Industrial heating oil
	Bright stock	Lubricants
Residuals	No. 6 fuel oil	Bunker C oil
	Heavy residual	Asphalt
	Coke	Coke

specifications for fuel boiling ranges is given in Table 18.2.[5] Boiling range is also used to identify individual refinery streams; in a later section we use the example of crude oil distillation. The temperature that separates one fraction from an adjacent fraction will differ from refinery to refinery. Factors influencing the choice of cut-point temperature include: crude oil feedstocks, type and size of downstream processes, and market demand for products.

Other specifications can involve either physical or chemical properties. Generally, these specifications are stated as minimum or maximum quantities. Once a product qualifies to be in a certain group, it may receive a premium price by virtue of exceeding minimum specifications or by being below maximum specifications. The only advantage of being better than the specifications is an increase in the volume of sales in a competitive market.

TABLE 18.2 Major Petroleum Products and Their Specified Boiling Range[5]

Product Designation	ASTM Designation	Specified Temp. for Vol. % Distilled at 1 atm °F		
		10%	50%	90%
Liquefied petroleum gas (LPG)	D 1835			
Commercial propane		—[a]		—[b]
Commercial butane		—[a]		—[c]
Aviation gasoline (Avgas)	D 910	167 max	221 max	275 max[d]
Automotive gasoline	D 439			
Volatility class A		158 max	170–250	374 max[e]
Volatility class B		149 max	170–245	374 max[e]
Volatility class C		140 max	170–240	365 max[e]
Volatility class D		131 max	170–235	365 max[e]
Volatility class E		122 max	170–230	365 max[e]
Aviation turbine fuel	D 1655			
Jet A or A-1		400 max		—[f]
Jet B		—[g]	374 max	473 max
Diesel fuel oil	D 975			
Grade 1-D				550 max
Grade 2-D				540–640
Grade 4-D		—	not specified	—
Gas turbine fuel oil	D 2880			
No. 0-GT		—[h]	not specified	—
No. 1-GT				550 max
No. 2-GT				540–640
No. 3-GT		—	not specified	—
No. 4-GT		—	not specified	—
Fuel oil	D 396			
Grade No. 1		419 max		550 max
Grade No. 2		—[h]		540–640
Grade No. 4		—	not specified	—
Grade No. 5		—	not specified	—
Grade No. 6		—	not specified	—

[a]Vapor pressure specified instead of front end distillation.
[b]95% point, −37°F max.
[c]95% point, 36°F max.
[d]Final point, 338°F max.
[e]Final point, all classes, 437°F max.
[f]Final point, 572°F max.
[g]20% point, 293°F max.
[h]Flash point specified instead of front end distillation.

The evolution of product specifications will, at times, appear sadly behind recent developments in more sophisticated analytical techniques. Certainly, the ultimate specification should be based on how well the product performs. Yet, the industry has grown comfortable with certain comparisons, and these standards are retained for easier comparison with earlier products. Thus, it is not uncommon to find petroleum products sold under an array of tests and specifications, some seemingly measuring similar properties.

It is behind the scenes that sophisticated analytical techniques prove their worth. These techniques can identify the specific hydrocarbons responsible for one property or another. Suitable refining processes are devised to accomplish a desired chemical reaction that will increase production of specific hydrocarbon products.

When discussing refining schemes, major specifications will be identified for each product category. It should be kept in mind that a wide variety of specifications must be met for each product.

Product Yields

As changes occur in product demand and specifications, refiners continuously adjust the configuration of internal processing streams. The challenge remains that increasing the volume of one fraction of crude oil processing will lower volumes of other product fractions. Thus, adjustments of one processing stream, especially major processing units, affect downstream processing streams and end-product volumes.

Refined product demand is seasonal. Demand for heating oil is higher during winter than during mid-summer. Equally important, gasoline demand fluctuates from summer highs, known as the driving season, and then declines in fall and winter. Refiners begin ramping up gasoline production over heating oil in early spring to meet anticipated demand and have sufficient gasoline supply in the system for distribution. Refiners try to avoid storing products.

Notably, fuel specifications for industrialized countries mandate blending winter- and summer-grade gasolines. Regions that do not meet air-quality specifications—known as nonattainment areas—are strictly regulated on the sale and distribution of the proper gasoline types. Refiners constantly estimate how much of a particular gasoline type to blend without overprocessing. Linear program (LP) models are extensively used to evaluate how best to use a crude stock to process designated products with the available processing capabilities. LPs are models that refiners can use to predict product yields with the resources available. They are gaining increased importance in assisting refiners in optimizing resources to avoid waste and maximize yields. These models are extensively used to estimate how changes in operating conditions, feeds, and new processing units/equipment will affect facility operations.

A barrel of crude oil has limited value, if any, to consumers. Its true value is the number of value-added products that can be extracted from the crude oil using various chemical reactions and separation methods. Thus, the refining operation is the first step in the transformation of crude petroleum oil into consumer products. So what are the possible products from a barrel of crude oil? Figure 18.6 lists the average breakdown of a barrel of oil by a United States refinery. As shown in Fig. 18.6, over 75 percent of the product yield from a refined barrel of oil is fuel based. In this example, United States refineries are focused on gasoline production, whereas European refineries focus on diesel product. Yet, refineries can also produce value-added petrochemicals for adjacent facilities.

Petrochemicals

The portion of crude oil going to petrochemicals may appear small compared with the volume of fuels yielded by refining operations; however, the variety of petrochemicals is large. Table 18.3 lists the many products derived for petrochemical applications. Many of these products are described in Chapter 10. A few are included here in as much as they compete with the manufacturing of fuels. Despite their variety, all commercially manufactured petrochemicals account for the

Product	Gallons per Barrel	%
gasoline	19.5	44.1
distillate fuel oil	9.2	20.8
(includes both home heating oil and diesel fuel)		
kerosene-type jet fuel	4.1	9.3
residual fuel oil	2.3	5.2
(Heavy oils used as fuels in industry, marine transportation, and for electric power generation)		
still gas	1.9	4.3
coke	1.8	4.0
asphalt and road oil	1.3	3.0
petrochemical feed stocks	1.2	2.7
lubricants	0.5	0.1
kerosene	0.2	---
Other	0.3	---

Figures are based on 1995 average yields for U.S. refineries. One barrel contains 42 gallons of crude oil. The total volume of products made is 2.2 gallons greater than the original 42 gallons gallons of crude oil. This represents "processing gain".
Source: API

Fig. 18.6. Product breakdown from one barrel of crude. (Source: API.)

TABLE 18.3 Petrochemical Applications

Absorbents	De-emulsifiers	Hair conditioners	Pipe
Activators	Desiccants	Heat transfer fluids	Plasticizers
Adhesives	Detergents	Herbicides	Preservatives
Adsorbents	Drugs	Hoses	Refrigerants
Analgesics	Drying oils	Humectants	Resins
Anesthetics	Dyes	Inks	Rigid foams
Antifreezes	Elastomers	Insecticides	Rust inhib.
Antiknocks	Emulsifiers	Insulations	Safety glass
Beltings	Explosives	Lacquers	Scavengers
Biocides	Fertilizers	Laxatives	Stabilizers
Bleaches	Fibers	Odorants	Soldering flux
Catalysts	Films	Oxidation inhib.	Solvents
Chelating agents	Finish removers	Packagings	Surfactants
Cleaners	Fire-proofers	Paints	Sweeteners
Coatings	Flavors	Paper sizings	Synthetic rubber
Containers	Food supplements	Perfumes	Textile sizings
Corrosion inhib.	Fumigants	Pesticides	Tire cord
Cosmetics	Fungicides	Pharmaceuticals	
Cushions	Gaskets	Photographic chem.	

consumption of only a small part of the total crude oil processed.

REFINING SCHEMES

A refinery is a complex processing methodology involving a massive network of vessels, reactors, distillation columns, rotating/compression equipment, heat exchangers, and piping. The total scheme can be subdivided into a number of unit processes. In what follows, only the major flow streams will be shown, and each unit will be depicted by a simple block in the flow diagram.

Refined products establish the order in which each refining unit will be introduced. Only one or two key product(s) specifications are used to explain the purpose of each unit. Nevertheless, the reader is reminded that the choices among several types of units and sizes of these units are complicated economic decisions. The trade-offs among product types, quantity, and quality will be mentioned only to the extent that they influence the choice of one type of processing technology over another.

Feedstock Identification

Each refinery has its own range of preferred crude oil feedstocks from which a desired product portfolio can be obtained. Crude oil typically is identified by its source country, underground reservoir, or some distinguishing physical or chemical property. The three most frequently specified properties are density, chemical characterization, and sulfur content.

API *gravity* is a contrived measure of density:

$$\text{API} = \frac{141.5}{sp\ gr} - 131.5$$

where *sp gr* is the specific gravity, or the ratio of the weight of a given volume of oil to the weight of the same volume of water at a standard temperature, usually 60°F. An oil with a density the same as that of water, or with a specific gravity of 1.0, would then be a 10°API oil. Oils with a higher than 10°API are lighter than water. Because the lighter crude oil fractions are usually more valuable, a crude oil with a higher °API gravity will bring a premium market price.

Heavier crude oils are receiving renewed interest as supplies of lighter crude oil dwindle and increase in price. Heavy crudes are those with an 20°API or less. Generally, heavier crudes fetch a lower price on the market. However, heavier crudes will require more processing to convert the high-boiling-point fractions into desired lighter products. Thus, refiners balance the cost of more expensive light, sweet feedstocks against capital investment to refine cheaper, heavy, sour crude oils.

A *characterization factor* was introduced by Watson and Nelson to use an index of the chemical character of crude oil or its fractions.[6] The Watson characterization factor is defined as

$$\text{Watson } K = \frac{(T_B)^{1/3}}{(sp\ gr)}$$

where T_B is the absolute boiling point in degrees Rankine (°R), and *sp gr* is the specific gravity compared with water at 60°F. For a wide boiling point range of material such as crude oil, the boiling point is taken as an average of five temperatures at which 10, 30, 50, 70, and 90 percent are vaporized.

A highly paraffinic crude oil might have a characterization factor as high as 13, whereas a highly naphthenic crude oil could be as low as 10.5. Highly paraffinic crude oils may also contain heavy waxes, which make the oil viscous and difficult to flow. Thus another test for paraffin content is used to measure how cold a crude oil can be before it fails to flow under specific test conditions. The higher the pour-point temperature, the greater the paraffin content for a given boiling range.

Sweet and *sour* are terms that refer to the sulfur content of the crude oil. In the early days, those terms designated the smell of the oil. A crude oil with a high sulfur content usually contains hydrogen sulfide, the gas associated with rotten eggs. Such crudes with high sulfur levels were called sour. Without this disagreeable odor, the crude was judged as sweet. Today, the distinction between sour and sweet is based on analytical assessment of sulfur content. A sour crude oil is one with more than 0.5 weight percent (wt.%) sulfur, whereas a sweet crude has less than 0.5 wt.% sulfur. It is estimated that 81 percent of the world's crude oil reserves are sour.[7]

ASTM distillation is a test prescribed by the American Society for Testing and Materials to measure the volume percent distilled at various temperatures.[5] The results often are reported the other way around: the temperatures at which given volume percents vaporize.[8] These data indicate the quantity of conventional boiling range products occurring naturally in the crude oil. Analytical tests on each fraction indicate the kind of processing

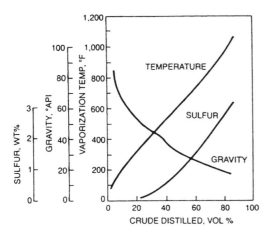

Fig. 18.7. Analysis of light Arabian crude.

that may be needed to make specific products. A plot of boiling point, sulfur content, and API gravity for fractions of Light Arabian crude oil is shown in Fig. 18.7. This crude oil is among those most traded in international crude oil markets.

From Fig. 18.7 it can be seen that the mid-volume range of Light Arabian crude oil has a boiling point of approximately 600°F, a liquid density of nearly 30°API, and an approximate sulfur content of 1.0 wt.% . These data are an average of eight samples. More precise values would be obtained on a specific crude oil if the data were to be used in design work.

Because a refinery stream spans a wide boiling range, the crude oil analysis data would be accumulated throughout that range to provide fraction properties. The intent here is to demonstrate the relationship among volume distilled, boiling point, liquid density, and sulfur content.

Crude Oil Pretreatment

Crude oil comes from the ground admixed with a variety of substances: gases, water, and dirt (minerals). The technical literature devoted to petroleum refining often omits crude oil cleanup steps. It is often assumed that the oil has been previously pretreated before entering the refining process. However, cleanup is important if the crude is to be transported effectively and processed without causing fouling and corrosion. Cleanup occurs in two ways: field separation and crude desalting.

Field separation is the first attempt to remove gases, water, and dirt that accompany crude oil extracted from the ground. As the term implies, field separation is done onsite at the production operation. The field separator is often no more than a large vessel that gives a quieting zone to permit gravity separation of the three phases: gases, crude oil, and water (with entrained dirt).

The crude oil is lighter than water, but heavier than the gases. Therefore, the crude oil appears within the field separator as a middle layer. The water is withdrawn from the bottom and is disposed of at the well site. Gases are withdrawn from the top and piped to a natural-gas processing plant or reinjected back into the reservoir to maintain well pressure. Crude oil from the middle layer is pumped to the refinery or to storage to await transportation by other methods.

Crude desalting is a water-washing operation done at the refinery to further clean up the crude oil before processing. The crude oil pretreated by field separators will still contain water and entrained dirt. Water-washing removes much of the water-soluble minerals and entrained solids.

If these crude-oil contaminants were not removed, they could cause operating problems during the refining process. The solids (dirt and silt) can clog equipment and deposit on heat-transfer surfaces, thereby reducing processing heat-transfer efficiency. Some solids, being minerals, can dissociate at high process temperatures and corrode major equipment. Other solids and minerals can deactivate catalysts used in refining processes.

Crude Oil Fractions

The importance of boiling range for petroleum products has already been discussed in Table 18.2. The simplest form of refining would isolate crude oil into fractions having boiling ranges that would coincide with the temperature ranges for consumer products. Some treatment steps might be added to

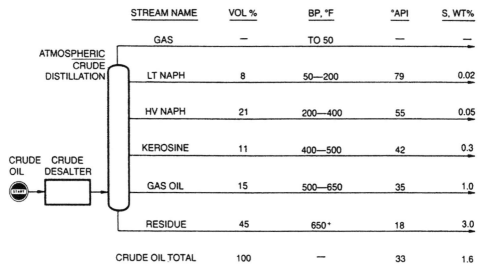

STREAM NAME	VOL %	BP, °F	°API	S, WT%
GAS	—	TO 50	—	—
LT NAPH	8	50—200	79	0.02
HV NAPH	21	200—400	55	0.05
KEROSINE	11	400—500	42	0.3
GAS OIL	15	500—650	35	1.0
RESIDUE	45	650⁺	18	3.0
CRUDE OIL TOTAL	100	—	33	1.6

Fig. 18.8. Separating desalted crude oil into fractions.

remove or alter undesirable compounds, and a very small quantity of various chemical additives would be included to enhance final properties.

Crude oil distillation separates the desalted crude oil into fractions of different boiling ranges. Instead of trying to match final product boiling ranges, the fractions are defined by the number and type of downstream processes.

The desalting and distillation units are shown in Figs. 18.8–18.10 along with the crude fractions from the crude distillation column. The relationships between some finished products and downstream processing steps will be expanded upon later in the chapter.

GASOLINE

Gasoline is blended from several refining processes, as shown in Fig. 18.8. Depending on the individual refinery configuration, gasoline blending streams are separated and refined. Figure 18.9 depicts a light-fraction processing scheme.[9] A straight-run gasoline stream is separated from the top portion of the atmospheric crude distillation column, which has a boiling range of 90–200°F and is very paraffinic.

The next cut gasoline stream from the crude distillation column is the naphtha cut. This stream has a boiling range of 200–365°F and contains a significant portion of naphthenes, aromatics, and paraffins. Thus, the naphtha cut is hydrotreated and reformed to upgrade this stream into a gasoline blending stock.[9]

In present-day refineries, the fluid catalytic cracking (FCC) unit has become the major gasoline-producing unit. The FCC's major purpose is to upgrade heavy fractions, that is, gas oil from the atmospheric and vacuum distillation columns and delayed coker, into light products. Atmospheric gas oil has a boiling range of between 650–725°F.[9]

The crude oil feedstock heavily influences the product slate for the refinery and the downstream processing required to meet the refinery's product goals. Fuels are blended to meet product specifications of volatility, sulfur content, and octane number. Most important, refiners constantly seek to optimize their blending programs to meet product goals without giving up product.

The automobile engine's drive train sets the specifications for gasoline. Notably, as automobile manufacturers design more sophisticated engines, in response refiners must adjust their operation to refine and blend fuels that are compatible with newer engines.

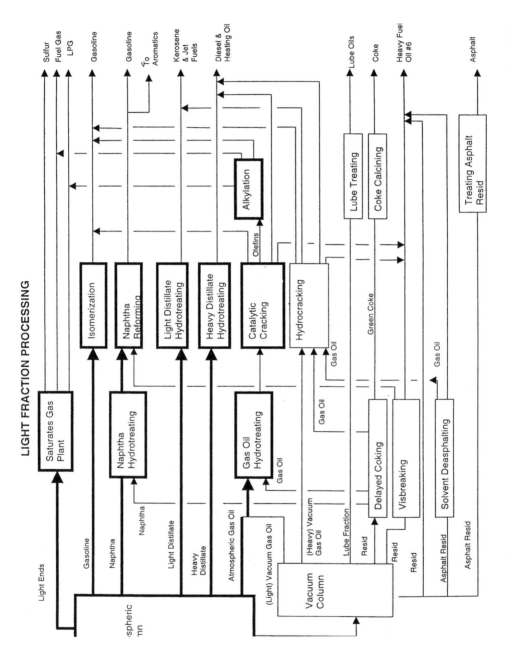

Fig. 18.9. Process flow diagram for a modern integrated refinery. (Reprinted with permission of Ed Ekholm, Copyright 1999 by Ed Ekholm; all rights reserved.)

TABLE 18.4   Anticipated Fuel Specifications Under Auto Oil Program and U.S. Requirements[10]

Period	Up to 1999	2000	2005	2008 (Expected Values)	2000 (U.S. only)	2006
Gasoline						
Sulfur content	<500 ppm	<150 ppm	<50 ppm[b]		<170 ppm	30 ppm
Benzene content	<5% v/v	<1% v/v	<1% v/v[a]		<1% v/v	—
Aromatics content	—	<42% v/v	<35% v/v		<25 v/v	
Diesel fuel						
Sulfur content, ppm	<500	<350	<50	<30	<500	<10
Cetane number, min	49	51	53[a]	54–58	40	—
Polynuclear aromatics	—	<11% w/w	<6% w/w[a]	<4–1% w/w	—	—
Density, kg/l max	<0.86	<0.845	<0.845[a]	<0.830–0.825	—	—
Distillation T95, °C	<370	<360	<360[a]	<350–340	—	—

% v/v: percent by volume; % w/w: percent by weight; [a]Expected values; [b]Germany is promoting 10 ppm sulfur for gasoline by 2003.
Source: Linde Technische Gase GmbH.

Worldwide, new environment legislation has set product specifications for fuels. Table 18.4 lists the quality standards for automotive gasoline and diesel.[10] These mandates are geared to lower tailpipe emission from vehicles. Sulfur content and volatility will be strictly limited in future fuel requirements.

Volatility

A gasoline's boiling point is important during its aspiration into the combustion chamber of a gasoline-powered engine. Vapor pressure is a function of the fuel's boiling point. Boiling range and vapor pressures are combined as the concept, *volatility*.[11]

The lighter components in gasoline are used as a compromise between two extremes: enough light components so that adequate vaporization of the fuel–air mixture provides an easy engine start in cold weather, but too many light components can cause the fuel to vaporize within the fuel pump and cause vapor lock.

Environmental studies suggest that light gasoline components are detectable in the atmospheres of large metropolitan areas. New environmental laws limit the volatility of gasoline, so refiners must use other processing streams to meet volatility requirements. However, the fuel must provide performance to consumers, for example, by minimizing chamber deposits and spark-plug fouling in the engine.

Sulfur Content

Sulfur compounds are corrosive and foul-smelling. When burned in an engine these compounds form sulfur dioxide and other oxides referred to as SO_x in engine exhaust. These compounds recombine and form sulfur trioxide and sulfuric acid mist, which is released as engine exhaust. Efforts to improve air quality are targeted at reducing vehicle engine exhaust of toxins and SO_x compounds. Thus, many new environmental regulations (Table 18.4) are focused on reducing the sulfur content of fuels. All crude oils contain some sulfur concentration. How much desulfurization is needed is dependent on the feedstock and product slate.

Caustic wash or other enhanced solvent-washing methods are a sufficient pretreatment to remove sulfur compounds from light naphtha. The sulfur compounds in light naphtha are mercaptans and organic sulfides that are readily removed by these washing processes. Heavy naphtha is harder to desulfurize. This stream has a higher sulfur content, and, equally important, the sulfur is embedded in complex hydrocarbon compounds and rings. Washing efforts are more effective on mercaptans, which are not usually present in heavy

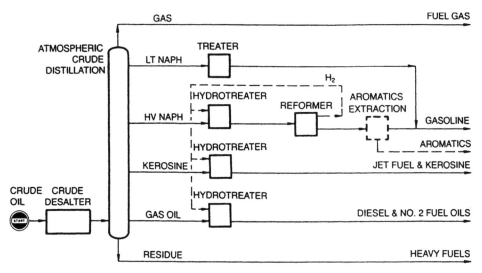

Fig. 18.10. Low-investment route to modern products.

naphtha. More aggressive methods are needed to break the compound structures and release the sulfur. Hydrotreating is one effective method to reduce sulfur content in hydrocarbon streams.[12]

Hydrotreating is a catalytic process that converts sulfur-containing hydrocarbons, that is, sulfides, disulfides, and ring compounds such as thiopenes, into low-sulfur liquids and hydrogen sulfide. This technology is widely used throughout the refinery, as shown in Figs. 18.9 and 18.10.[9] This process is operated under a hydrogen-rich blanket at elevated temperatures and pressure. The process is a hydrogen-consuming step, because the hydrogen replaces sulfur on the hydrocarbon molecule.[9]

Nitrogen and oxygen compounds are also dissociated by hydrotreating. Consequently, hydrotreating provides additional benefits of hydrodenitrification, which improves downstream operations. For nitrogen and oxygen compounds, the products from hydrotreating are ammonia and water, respectively. Thus, these contaminants will be separated in the off-gas and are easily removed by conventional gas-treating processes.

Octane Number

Another condition to keep gasoline engines running smoothly is that the fuel–air mixture starts burning at a precise time in the combustion cycle. An electrical spark starts the ignition. The remainder of the fuel–air mix should be consumed by a flame front moving out from the initial spark.

Under some conditions, a portion of the fuel–air mix will ignite spontaneously instead of waiting for the flame front from the spark. The extra pressure pulses that occur from spontaneous combustion are usually audible above the background sounds of the engine running and give rise to a condition know as "engine knock." The engine pings and rumbles when under "knock conditions." This condition is undesirable; it is a waste of available power.

The *octane number* is a measure of a fuel's ability to avoid knocking. The octane number of gasoline is determined in a special single-cylinder engine where various combustion conditions can be controlled.[5] The test engine is adjusted to trace the knock from the fuel being rated. Various mixtures of iso-octane and normal heptane (*n*-heptane) are used to find the ratio of the two reference fuels that will give the same intensity of knock as that from an unknown fuel. Defining iso-octane as 100 octane number and *n*-heptane as 0 octane number, the volumetric percentage of iso-octane in heptane that matches knock from the unknown fuel is reported as the octane number of the fuel. For example, 90 vol.% of iso-octane and 10 vol.% *n*-heptane establishes a 90 octane number reference fuel.

Two types of octane number ratings are specified, although other methods are often used for engine and fuel development. Both methods use the same reference fuels for essentially the same test engine. Engine operating conditions are the difference. In the *research method*, the spark advance is fixed, the air inlet temperature is 125°F, and the engine speed is 600 rpm. The other method is called the *motor method*; it uses variable spark timing, a higher mixture temperature (300°F), and a faster engine speed (900 rpm).

The more severe conditions of the motor method have a greater influence on commercial blends than they do on the reference fuels. Thus, a motor octane number (MON) of a commercial blend often has a lower research octane number (RON). Consequently, blended fuels use an arithmetic average of both ratings—MON and RON—and can be abbreviated as (R + M)/2.

Catalytic reforming is the principal process used to upgrade the octane number of naphtha for gasoline blending.[13] Reforming uses catalysts to reshape the molecular structure of hydrocarbons to raise the octane number of the process stream. Naphthenes are converted to aromatics; paraffins are isomerized to isomeric forms.[9] Reforming efforts are most effective when used on heavier molecules; a greater increase in octane number can be attained by reforming heavy naphtha cuts.

Reforming catalysts typically contain platinum or a mixture of platinum and other metal promoters on a silica–alumina support. Only a concentration of platinum is used, averaging about 0.4 wt.%. The reforming process is a highly endothermic process. Desulfurized feeds are preheated to 900°F, and the reactions are done at various pressures (50–300 psig), which are dependent on the licensed process used.[12] At elevated temperatures and pressures, the catalyst is susceptible to coking, which decreases catalyst efficiency. Thus, refiners must regenerate the catalyst to maintain process efficiency. Reforming catalyst can be regenerated in situ by burning off the coke from the catalyst. Newer developments now use continuous regeneration of the reforming catalyst in which three reforming reactors are stacked one on top of the other. Gravity flow moves the catalyst from the top to the bottom and sends it to a regeneration step in which a dry burn removes the coke. The regenerated catalyst is then returned to process. Also, reforming feeds are pretreated to remove poisons that can kill precious-metal catalysts.

Hydrotreating is an effective method to pretreat reforming feedstocks (Fig. 18.10). Combining hydrotreating with reforming is most effective. Due to cyclization and dehydrogenation of hydrocarbon molecules in the reformer, hydrogen is a by-product of this operation.[14] Notably, by-product hydrogen from the reform can be directed to the hydrotreating operations. Thus, the reformer can provide the refinery with the hydrogen supply for hydrotreating. A rule of thumb is that the catalytic reformer produces 800–1200 scf/bbl = standard cubic feet per barrel (scf/bbl) for naphtha. The excess hydrogen is available for hydrotreating other fractions in separate hydrotreaters.

DISTILLATES

Jet fuel, kerosene (range oil), No. 1 fuel oil, No. 2 fuel oil, and diesel fuel are all popular distillate products from the 365–650°F fractions of crude oil.[9] Distillates are further classified as light distillates with a true boiling point range of 365–525°F, and heavy distillate cuts have a true boiling point range from 525–650°F. Light distillates are blended into kerosene and jet fuels. Heavy distillate cuts are used to blend diesel fuels and home heating oils.[9]

Some heating oil (generally No. 2 heating oil) and diesel fuel are very similar and are sometimes substituted for each other. Home heating oil is intended to be burned in a furnace for space heating. Diesel fuel is intended for compression–ignition engines.

Distillates are lower cuts from the atmospheric crude distillation column (Fig. 18.10); thus, these refinery streams may have high sulfur concentrations due to the feedstock that is processed. Newer product specifications limit sulfur concentrations in consumer products, especially diesel. Consequently, distillate streams must be upgraded. Hydrotreating improves the product properties of distillate products; notably it reduces sulfur content. More important, hydrotreating hydrogenates unsaturated hydrocarbons so that they will not contribute to smoke and particulate emissions, whether the fuel is burned in a furnace or used in an engine.

Residuals

Crude oil is seldom distilled at temperatures above 650°F. At higher temperatures, coke will form and plug the lower section of the crude oil distillation tower. This bottom fraction from the atmospheric crude column has a true boiling point range of 650–725°F and is often referred to as *atmospheric gas oil* or *residuals*. This fraction is traditionally not vaporized. Atmospheric gas oil must be upgraded extensively; it can be severely hydrotreated to break apart the complex ring compounds and saturate them into lighter products. This stream can be sent to a catalytic cracker to further upgrade this heavy fraction into gasoline, diesel, and home heating fuel.[9] The heaviest cut of the atmospheric crude distillation is often referred to as the *long residuum*. This fraction is further processed via a vacuum distillation column.

PRODUCING MORE LIGHT PRODUCTS

The refining scheme shown in Fig. 18.9 is a simplified view of an integrated refinery. More processing steps can be added and are dependent upon the product slate of the refinery and the cracking slate anticipated for its design. If the refinery is a gasoline refinery, the cracking process will be directed toward producing light products, gasoline at the expense of diesel and heating oil. Conversely, if diesel is the desired end-product, gasoline product is sacrificed to produce more distillate streams. United States refineries are predominately gasoline-oriented. Conversely, Western European refineries are diesel-fuel-oriented. In Western Europe, the demand for gasoline is projected to decline over the next 10 years, and demand for middle distillates increase.[15] Therefore, Western European refiners must adapt operations not only to

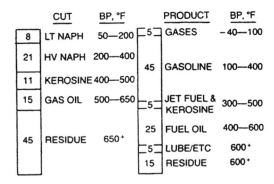

Fig. 18.11. Light Arabian crude oil compared to U.S. product deliveries.

produce more diesel but also to minimize gasoline production.

For industrial areas, where the principal demand is for transportation fuels or high-quality heating oils, the simplified refining scheme is shown in Fig. 18.9. In the case of a United States refinery, the emphasis is directed into converting more of the crude oil into lighter products, namely gasoline. If this refinery processed a Light Arabian crude, nearly 80–85 vol.% of the products would be lighter than the boiling temperature of 650°F compared with the 55 vol.% existing in the feed crude oil (Fig. 18.11). More than half of all United States products are gasoline and lighter distillates (boiling temperatures below 400°F) compared with 29 vol.% in the crude oil. This comparison is unique for these circumstances. Refining technologies and catalysts have advanced from the earlier days. Notably, refiners can process a variety of crudes into valued end-products. Crudes that require more intense processing, that is, high-severity hydrocracking and hydrotreating, fetch a lower price on the market and are often referred to as *opportunity crudes*. Refiners processing lower-cost, sour, heavy crude must make more capital investment in processing capability. However, these refiners pay less for their feedstocks and risk attaining payback on capital investments through volume processing of lower-cost feedstocks. The design of an integrated refinery constantly balances the future price of oil against the return on investment for refining capacity. Thus, the decision on what types of crude to process will affect the design and operation of a refinery.

Cracking

As mentioned earlier, the refining process involves adding hydrogen to carbon molecules. Notably, the desired hydrocarbons are much lighter products; thus, the refining process strives to break the large, complex hydrocarbon molecules into smaller molecules and add hydrogen to the open bonds. Cracking processes typically break hydrocarbon molecules into two or more smaller molecules. Thermal cracking uses high-temperature (above 650°F) and a long residence time to sever hydrocarbon bonds. Higher pressure facilitates the cracking process; however, the capital investment for a high-pressure reactor is greater than the expense for low- or medium-pressure reactors.

In thermal cracking, the formation and deposition of coke on piping and equipment walls is an unwanted side reaction. Another option to crack crude oil into desired products at lower pressures and temperatures is *catalytic cracking*. Catalytic cracking splits the molecules quicker and at lower temperatures. Catalysts are used to promote the desired reaction rate for the process.

Catalytic cracking involves large reactors with large fluidized catalyst beds. As in the case of thermal cracking, coke is also formed during the catalytic process, which can deposit on the catalyst and hinder its activity. Thus, this process uses reactor–separators and regenerators to remove the catalyst and regenerate it and return the catalyst to the process. The fluidized bed mixes the feed with the catalyst to optimize contact time. The catalyst is separated from the hydrocarbon products. A portion is regenerated; the remainder is returned to the catalytic cracker reactor.

Catalytic cracking is very effective in upgrading heavy refining streams, such as gas oils, into motor gasoline stocks thereby increasing the octane number for product streams. This process produces less gas and coke as compared with thermal cracking operations. Catalytic cracking also yields more liquid products, which can be tailored toward gasoline or diesel fuel and home heating oil products. Different operating conditions and catalysts will define the product mix from a catalytic cracker.

Several factors determine the best feeds for catalytic crackers. Heavy feeds are preferred; thus, the lower boiling point is about 650°F. The feed should not be so heavy that it contains an undue amount of metal-bearing compounds or carbon-forming material. Deposition of metals and coke can quickly deactivate the catalyst.

Visbreaking is a mild, once-through thermal cracking process. It is used to crack resid products into fuel-oil specifications. Although some light products such as naphtha and gasoline are produced, this is not the purpose of the visbreaker.

Coking is another matter. It is a severe form of thermal cracking in which coke formation is tolerated to attain additional lighter liquids from the heavier, dirtier fractions of crude oil. In this process the metals that would foul catalysts are laid down with the coke. The coke settles out in large coke drums that are removed from service frequently (about once a day) to have the coke removed by hydraulic methods. Several coke drums are used to make the process continuous; thus, one drum is online while the other is being emptied and readied for the next cycle.[12]

Hydrocracking converts a wide variety of heavy refining product streams into light products; fuels and distillates. A robust catalyst system is used to desulfurize, denitrify, and hydrocrack the feed.[9,12,14] The process combines hydrotreating and catalytic cracking goals. However, hydrocracking is a more capital-intensive and operating-intensive step. The operating pressure is higher (up to 3000 psi); consequently, thick-wall vessels are used as reactors (up to 9 in. thick). Products from a hydrocracker are very clean (desulfurized, denitrified, and demetalized) and will contain isomerized hydrocarbons in greater quantity than from conventional catalytic cracking. This process consumes a large quantity of hydrogen, which adds considerably to its operating costs.

Vacuum Distillation

As mentioned earlier, most consumer products are light products: those with boiling points less than 400°F. However, in the refining of a crude oil, a significant portion of the products has a true boiling point above 650°F.

Atmospheric distillation is least effective in converting heavier products into lighter components. A second distillation column under vacuum is needed to further separate the heavier parts of crude oil into lighter fractions. Some fractions from the vacuum units have better quality than atmospheric distillation cuts because the metal-bearing compounds and carbon-forming materials are concentrated in the vacuum residue.

Reconstituting Gases

Cracking processes to convert heavy liquids into lighter products also create gases. Another option to make more liquid products is to combine the gaseous hydrocarbons. A gas separation unit may be added to a refinery to isolate individual types of gases. When catalytic cracking is part of the refining scheme, a large quantity of olefins (ethylene, propylene, and butylene) is co-produced. Two routes are available to reconstitute these gaseous olefins into gasoline blending stocks, as outlined below.

Polymerization uses a catalytic process to combine two or more olefins to make polymer gasoline. The double bond in only one olefin is changed to a single bond during each link between two olefins. Thus, the product will still have some double bonds. This process was developed in the 1940s to produce high-octane aviation fuel. However, the olefinic nature of polymer gasoline does have a drawback.[4] The gum-forming tendencies of the polymer gasoline are problematic especially during long storage in warm climates. The olefins continue to link up and form larger molecules—gum or sludge—which are undesirable. Some refiners still use catalytic polymerization of light ends. It is a lower-cost process, both in terms of operating and investment costs. However, due to new environmental specifications, hydrogenation of polymer gasoline may be necessary to meet emission standards for Bromine Number.[9] Hydrogenation can reduce the octane number of the polymer gasoline.

Alkylation catalytically combines light olefins—propylene and butylenes—with isobutane to produce a branched-chain paraffinic fuel.[14] Alkylate is a great blending component for the gasoline-blending pool. It has a high octane number (usually above 94), low vapor pressure, and is almost sulfur free.[16] Present-day alkylation processes are carried out in the presence of sulfuric or hydrofluoric acids. New health and safety issues are promoting research on solid-acid technologies. Some successes have been demonstrated in the laboratory and in pilot studies; however, no commercial units have been built.

The *ether* process combines an alcohol with an iso-olefin. In the United States, a weight percentage of oxygenate (2 wt.%) content is mandated for reformulated gasoline (RFG). The most common oxygenate currently used is methyl tertiary butyl ether (MTBE). Methanol and the iso-olefin form of iso-butylene are reacted to form MTBE. Other alcohols, such as ethanol, may be reacted with iso-butylene to form ethyl tertiary butyl ether (ETBE). Methanol can be reacted with iso-amylene, another iso-olefin, to form tertiary amyl-methyl ether (TAME). Of all the mentioned ethers, MTBE is the one most widely used as a gasoline-blending component.[16]

A MODERN REFINERY

A refining scheme incorporating the processes discussed above is shown in Fig. 18.9. The variations in this flow diagram are numerous. Types of crudes processed, product slate, and competitive quality goals of products are just a few factors that influence the processing needs for a refining complex. Many other processes play an important role in the final scheme. A partial list of these processes would be: dewaxing lubricating oils, deoiling waxes, deasphalting heavy fractions, manufacturing specific compounds for gasoline blending (alcohols, ethers, etc.), and isolating specific fractions for petrochemical applications. See Fig. 18.12.

Petrochemicals

Refining crude provides many products, depending on the types of products sought.

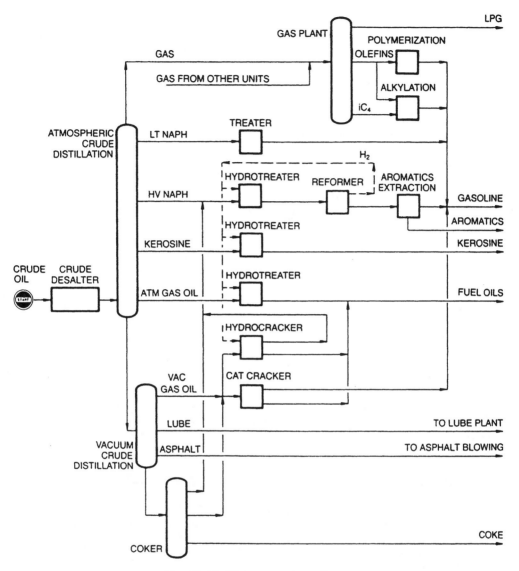

Fig. 18.12. High conversion refinery.

Lighter products from the refining of crude have higher value as petrochemical feeds than as gasoline-blending components. In particular, olefins and aromatic compounds possess higher values as petrochemicals than as gasoline components.

Ethylene is a key building block for the petrochemical industry. It is usually made by thermally cracking gases—ethane, propane, butane, or a mixture of these—as they exist in refinery off-gases. When gas feedstocks are scarce or expensive, naphthas and even whole crude oils have been used in specially designed ethylene crackers. The heavier feeds provide significant quantities of higher-molecular-weight olefins and aromatics.

Aromatics are typically concentrated in product streams from the catalytic reformer. When aromatics are sought for petrochemical applications, they typically are extracted from the reformer product stream by solvent extraction or distillation extraction. A common solvent used is sulfolane; new processes now use *n*-formylmorpholin as the extractive solvent.[12]

The mixture of aromatics is typically referred to as BTX and is an abbreviation for benzene, toluene, and xylene. The first two components, benzene and toluene, usually are separated by distillation, and the isomers of the third component, xylene, are separated by partial crystallization.[17] Benzene is the starting chemical for materials such as styrene, phenol, and many fibers and plastics. Toluene is used to make a number of chemicals, but most is blended into gasoline. Xylene usage is dependent on its isomer. *Para-xylene (p-xylene)* is a precursor compound for polyester. *Ortho-xylene (o-xylene)* is the building block for phthalic anhydride. Both compounds are widely used to manufacture consumer products.

PROCESS DETAILS

Thus far, the refining units have been described as they relate to other units and to the final product specifications. At this point,

typical flow diagrams of some major processes will be presented to highlight individual features. In many cases the specific design shown is an arbitrary choice from the work of several qualified designers.

Crude Desalting

Salts such as sodium, calcium, and magnesium chloride are generally contained in water suspended in the oil phase of hydrocarbon feedstocks.[9] Other impurities are also present in crude oils as mechanical suspensions of silt (dirt), iron oxides, sand, and crystalline salt.[14] These contaminants must be removed before processing the crude oil feeds; thus, the best method is mixing the crude oil with water and creating an emulsion.[12]

A typical flow diagram is shown in Fig. 18.13 The desalter operation is incorporated into the preheat train of the crude distillation unit to conserve energy. Depending on the characteristics of the hydrocarbon feedstock,

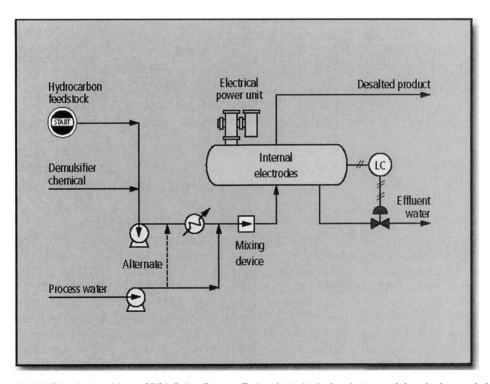

Fig. 18.13. Electric desalting—CB&I. Baker Process Technology. Includes: heater, mixing device, and electrostatic settler. (Source: *Hydrocarbon Processing*, 2004 Refining Process Handbook, CD-ROM. Sept. 2004: Copyright 2002 by Gulf Publishing Co., all rights reserved.)

the crude oil is preheated to 150–300°F. The crude oil is preheated with crude unit products and pumparound reflux to the desired temperature. The operating pressure is 40 psig or more. The elevated temperature reduces oil viscosity to improve mixing, and the elevated pressure suppresses vaporization. The washwater, 3–6 vol.%, can be added upstream and/or downstream of the heat exchanger(s). The combined streams pass through a mixing device—a throttling valve or emulsion orifice—and create a stable water-in-oil emulsion. The properties of the emulsion are controlled by adjusting the pressure drop across the mixing device. Trace quantities of caustic, acid, polymers, other chemicals are sometimes added to promote treatment.[12]

The emulsion enters the desalter vessel where a high-voltage electrostatic field is applied. The electrostatic field causes the dispersed water droplets to coalesce, agglomerate, and settle to the lower portion of the vessel. The various contaminants from the crude oil concentrate in the water phase. The salts, minerals, and other water-soluble impurities are discharged from the settler to the effluent system. Clean, desalted hydrocarbon product flows from the top of the settler and is ready for the next processing step.

Additional stages can be used in series to gain additional reductions in the salt content of the crude oil. Two stages are typical, but some installations use three stages. About 90 percent of the emulsified water can be recovered in one step, whereas 99 percent recovery is possible with a two-step process.[9] The additional investment for multiple stages is offset by reduced corrosion, plugging, and catalyst poisoning of downstream equipment with the cleaner crude feed.

Crude Distillation

Single or multiple distillation columns are used to separate the crude oil into fractions determined by their boiling range. Common identification of these fractions was discussed using Fig. 18.12, but should only be considered as a guide. Many refining schemes can be used to alter the type of separation made at this point.

A typical flow diagram of a two-stage crude oil distillation system is shown in Fig. 18.14. The crude oil is preheated with hot products from the system and desalted before entering the fired heater. The typical feed to the crude-fired heater has an inlet temperature of 550°F, whereas the outlet temperature may reach 657–725°F. Heater effluent enters the crude distillation (CD) column, where light naphtha is drawn off the overhead tower. Heavy naphtha, kerosene, diesel, and cracking streams are sidestream drawoffs from the distillation column. External reflux for the tower is provided by several pumparound streams.[12]

The bottoms of the CD, also known as atmospheric residue, are charged to a second fired heater where the typical outlet temperature is about 750–775°F. From the second heater, the atmospheric residue is sent to a vacuum tower. Steam ejectors are used to create the vacuum so that the absolute pressure can be as low as 30–40 mm Hg (about 7.0 psia). The vacuum permits hydrocarbons to be vaporized at temperatures below their normal boiling point. Thus, the fractions with normal boiling points above 650°F can be separated by vacuum distillation without causing thermal cracking. In this example (Fig. 18.14), the distillate is condensed into two sections and withdrawn as two sidestreams. The two sidestreams are combined to form cracking feedstocks: vacuum gas oil (VGO) and asphalt base stock.

Atmospheric distillation is an energy-intensive process. With pressure to reduce operating costs, new design efforts are investigating energy conservation on the CD column. A new atmospheric distillation process by TECHNIP FINAELF uses a progressive distillation strategy to minimize total energy consumption, as shown in Fig. 18.15. In this processing scheme, two pre-flash towers separate the light products—LPG, naphtha (light, medium, and heavy), and kerosene—from the crude feed to the main atmospheric distillation column. The light products are fractionated as required in a gas plant and rectification towers.

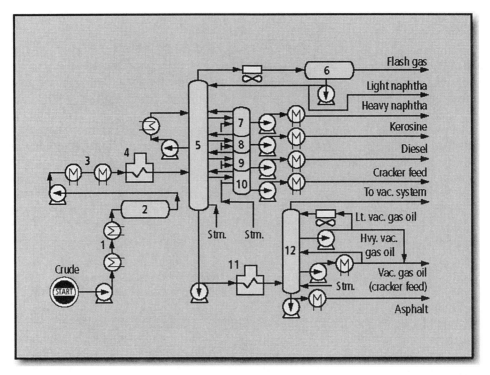

Fig. 18.14. Crude distillation—Foster Wheeler. Includes: exchanger heaters (1,3), desalter process (2), fired heater (4), main fractionator (5), overhead gas accumulator (6), sidestream strippers (7,8,9,10), second fired heater (11), and vacuum fractionator (12). (Source: *Hydrocarbon Processing*, 2004 Refining Process Handbook. CD-ROM. September 2004: copyright 2004 by Gulf Publishing Co., all rights reserved.)

The topped crude is typically reduced by two-thirds of the total naphtha cut. The bottoms from the second pre-flash tower are sent to the charge heater and directed to the main distillation column and produce four product streams: heavy naphtha, several kerosene cuts, and bottoms residue. The residue is further processed in a vacuum column and produces VGO and several distillate streams.

Incidentally, the total refining capacity of a facility is reported in terms of its crude-oil handling capacity. Thus, the size of the first distillation column, whether a pre-flash or an atmospheric distillation column, sets the reported size of the entire refinery. Ratings in barrels per stream day (bpsd) will be greater than barrels per calendar day (bpcd). Processing units must be shut down on occasion for maintenance, repairs, and equipment replacement. The ratio of operating days to total days (or bpcd divided by bpsd) is called the "onstream" factor or "operating factor."

The ratio can be expressed either as a percent or a decimal. For example, if a refinery unit undergoes one shutdown period for one month during a three-year duration, its operating factor is $(36 - 1)/36$, or 0.972, or 97.2%.

Outside the United States, refining capacity is cited in metric tons per year. Precise conversion from one unit of measure to the other depends upon the specific gravity of the crude oil, but an approximate relation is 1 barrel per day equals 50 tons per year (tpy).

Hydrotreating

Hydrotreating is one of the more mature refining processes still practiced today. Refiners began using catalytic hydrotreating in the 1950s to remove undesirable materials from refining product streams, as shown in Fig. 18.9. This process effectively removes contaminants such as sulfur, nitrogen, olefins, metals, and aromatics.[9,12,14] The chemistry of

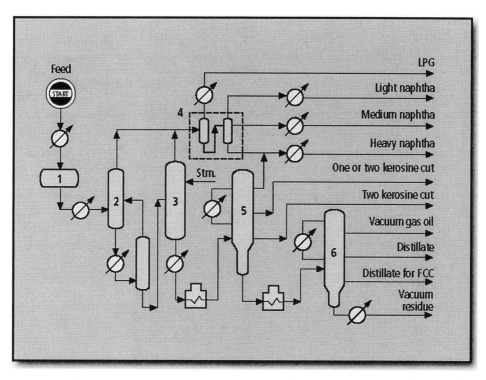

Fig. 18.15. Crude distillation—TOTALFINAELF and Technip. Includes: desalter and preheater (1), preflash towers (wet and dry) (3,2), gas plant and rectifier tower (4), main fractionation tower (5), and vacuum fractionation tower (6). (Source: *Hydrocarbon Processing, 2004 Refining Process Handbook*. CD-ROM. September 2004: copyright 2004 by Gulf Publishing Co., all rights reserved.)

hydrotreating can be further divided into three categories: hydrodesulfurization, hydrodenitrification, and hydrodearomatization (saturation of olefins and saturation of aromatics).[12,14]

The utility of most hydrotreating efforts is desulfurization. Sulfur-containing hydrocarbons are present in crude oil with many varying forms and boiling points. New product specifications limit the amount of sulfur present in finished fuels; thus, greater effort will be needed to remove more sulfur-containing compounds throughout the fuel-blending pool. Notably, more complex, high-boiling-point sulfur compounds must be extracted from the blending pool to meet lower fuel specifications, as shown in Table 18.5.[18]

Depending on the severity of the operation, hydrotreating is done at elevated temperatures and pressures. Higher temperatures and pressures are needed to open the complex ring compounds and remove the sulfur molecules. A typical flow diagram is shown in

Fig. 18.16.[12] This process converts atmospheric and vacuum residue into lighter products. The oil feed is preheated with product streams and a charge heater, and mixed with a hydrogen-rich gas. This mixture is charged to the main reactor and passed over a fixed-bed catalyst system where exothermic hydrogenation reactions occur. Proper internals are needed in the reactor to evenly distribute the feed throughout the catalyst bed and prevent channeling. Product separation is done by a hot high-pressure separator, cold high-pressure separator, and fractionator. In the first high-pressure separator, unreacted hydrogen is taken as the overhead; it is scrubbed to remove hydrogen sulfide (H_2S). The cleaned hydrogen is recycled back to process. In the second high-pressure separator, the remaining gases and light products are removed from the liquid product. If the feed is a wide-boiling range material from which several blending stocks are to be made, the hot,

TABLE 18.5   Major Gasoline Sulfur Compounds Normal Boiling Point (NBP) and Hydrocarbon Boiling Range

Component	NBP*	Boiling Range °F
Ethyl mercaptan	95	70–90
Dimethyl sulfide	99	75–80
Iso-propyl mercaptan	126	110–130
Tert-butyl mercaptan	147	120–150
Methyl ethyl sulfide	151	130–140
n-Propyl mercaptan	154	115–130
Thiophene	183	140–200
Iso-Butyl mercaptan	191	180–200
n-Butyl mercaptan	204	185–200
Dimethyl disulfide	230	190–200
2-Methyl thiophene	234	200–250
3-Methyl thiophene	239	210–270
Tetrahydrothiophene	250	220–260
1-Pentyl mercaptan	259	245–255
C_2 Thiophene	278	250–310
C_1 Tetrahydrothiophene	306	260–320
Hexyl mercaptan	307	290–340
C_3 Thiophene	317	300–340
C_2 Tetrahydrothiophene	318	300–340
C_3 Tetrahydrothiophene	329	320–340
C_4 Tetrahydrothiophene	340	320–360
C_4 Thiophene	361	340–380
C_5 Thiophene	411	390–420
Benzothiophenes and others	427+	400+

*Reprinted with permission of Gulf Publishing Co., 2002. Copyright, all rights reserved.

high-pressure separator is followed by a fractionation column. The fractionator separates the treated feed into several liquid product streams; naphtha, middle distillate, VGO, and a very clean hydrotreated resid product.

The feed for hydrotreating can be a variety of different boiling-range materials from naphtha to vacuum residues. Generally, each fraction is treated separately to permit optimum operating conditions, the higher boiling-point materials require more severe treatment conditions. For example, naphtha hydrotreating can be done at 200–500 psia and at 500–650°F with a hydrogen consumption of 10–50 scf/bbl of feed. Conversely, a residue-hydrotreating process can operate at 1000–2000 psia and at 650–800°F, with a hydrogen consumption of 600–1200 scf/bbl.[19]

Hydrotreating is a versatile cleanup step; however, it is a large hydrogen-consuming process. Most refineries are able to meet their hydrogen-processing demands with hydrogen recovered from the catalytic reforming process. However, as refiners intensify hydrotreating efforts to meet tighter specifications for products, hydrogen demand will increase. Consequently, one option to balance hydrogen consumption is to construct onsite hydrogen plants to meet present and future hydrogen needs. Purchasing hydrogen from over-the-fence suppliers is another option.[20]

Catalyst formation constitutes a significant difference among hydrotreating processes. Refiners must address reducing sulfur concentrations to lower levels: 15 ppm for diesel and 30 ppm for gasoline. Consequently, the activity and efficiency of the hydrotreating catalysts become even more vital. Presently, cobalt–molybdenum (CoMo) and nickel–molybdenum (NiMo) catalysts are the preferred hydrotreating systems. CoMo catalysts are very effective at breaking carbon–sulfur (C–S) bonds, and NiMo catalysts are more effective at hydrogenation. Just using both systems—CoMo and NiMo—will not guarantee optimum results.[14,21] To obtain the very low ppm levels, desulfurization efforts become more specified at the compound that must be reacted to remove the sulfur from the product stream.

Catalytic Reforming

This process upgrades naphtha (light distillates) into aromatic-rich streams that can be used for octane enhancers for gasoline blending or as a petrochemical feedstock. Originally the process was developed in the 1950s to upgrade low-octane, straight-run gasoline to high-octane liquids, as shown in Table 18.6.[14] This process converts naphthenes into corresponding aromatics and isomerizes paraffinic structures to isomeric forms.[9] The naphtha charge is a varying mixture of C_6–C_{11} paraffins, naphthenes, and aromatics. In a catalytic reformer, aromatic compounds pass through the system unchanged, whereas naphthalenes react selectively to form aromatics.[14]

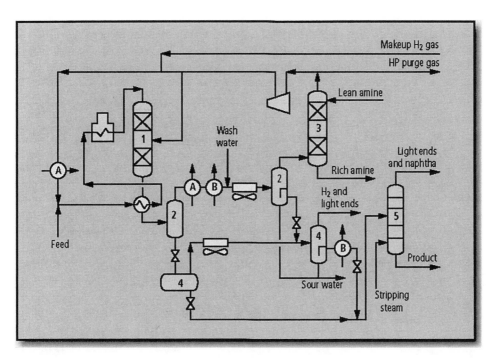

Fig. 18.16. Hydrotreating—Chevron Lummus Global LLC. Includes reactor (1), hot high pressure separator (2), hot low pressure separator (3), cold high pressure separator and product fractionator (4). (Source: *Hydrocarbon Processing, 2004 Refining Process Handbook.* CD-ROM. September 2004: copyright 2004 by Gulf Publishing Co., all rights reserved.)

TABLE 18.6   Aromatics have Higher Octane Numbers[6]

		Octane Number, Clear	
Hydrocarbon Homologs		*Motor*	*Research*
C$_7$ hydrocarbons			
n-paraffin	C$_7$H$_{16}$ (*n*-heptane)	0.0	0.0
naphthene	C$_7$H$_{14}$ (cycloheptane)	40.2	38.8
	C$_7$H$_{14}$ (methylcyclohexane)	71.1	74.8
Aromatic	C$_7$H$_8$ (toluene)	103.5	120.1
C$_8$ hydrocarbons			
n-paraffin	C$_8$H$_{18}$ (*n*-octane)	−15[a]	−19[a]
naphthene	C$_8$H$_{16}$ (cyclooctane)	58.2	71.0
	C$_8$H$_{16}$ (ethylcyclohexane)	40.8	45.6
Aromatic	C$_8$H$_{10}$ (ethylbenzene)	97.9	10.4
	C$_8$H$_{10}$ (*o*-xylene)	100.0	120[a]
	C$_8$H$_{10}$ (*m*-xylene)	115.0	117.5
	C$_8$H$_{10}$ (*p*-xylene)	109.6	116.4

[a]Blending value at 20 vol.% in 60 octane number reference fuel.

In the reformer, multiple reactions occur simultaneously. This process is endothermic and is subject to carbon laydown; thus, refiners must regenerate reforming catalysts. Several catalyst-regenerating approaches are possible. Semi-regenerative processes use moving-bed catalyst reactors. The catalyst bed reactors are placed side-by-side, and hydrogen is used to lift and convey the catalyst to the next bed, except for the last bed where it is regenerated, as shown in Fig. 18.17.[12] Other reforming designs use a continuous moving

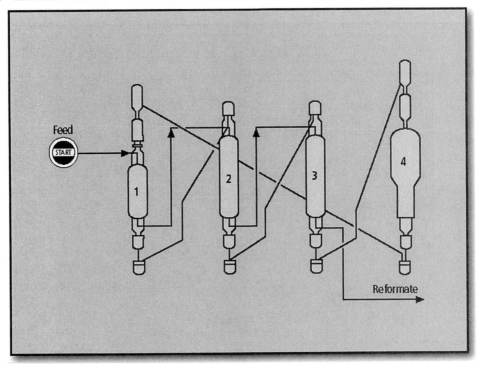

Fig. 18.17. Catalytic reforming—Axens. Includes moving-bed reactors (1,2,3) and regenerator (4). (Source: *Hydrocarbon Processing, 2004 Refining Process Handbook.* CD-ROM. September 2004: copyright Gulf Publishing Co., all rights reserved.)

bed to continuously regenerate a portion of the catalyst. The reactors are stacked on top of each other, and gravity moves the catalyst through the bed. From the last reactor, the catalyst is lifted by nitrogen or hydrogen to a catalyst collection vessel. The catalyst is regenerated in a regeneration tower and returned to process as shown in Fig. 18.18.[12]

In the catalytic reforming process, the feed is pumped to operating pressure and mixed with a hydrogen-rich gas before heating to reaction temperatures. The net hydrogen produced is a by-product of the dehydrogenation and cyclization reactions. Several reactions occur:

- Dehydrogenation of naphthene
- Isomerization of paraffins and naphthenes
- Dehydrocyclization of paraffins
- Hydrocracking and dealkylation of paraffins[14]

Reforming catalysts promote these reforming reactions. Isomerization is a desired reaction, especially to raise the octane value of the product. However, hydrocracking is an undesired side-reaction that produces light gases. Higher operating pressures are used to suppress hydrocracking. Unfortunately, higher operating pressures suppress reforming reactions also. Generally, a compromise is made between the desired reforming and undesired hydrocracking. The effects of operating conditions on competing reactions are shown in Table 18.7.[23]

In the late 1960s, it was discovered that adding certain promoters such as rhenium, germanium, or tin to the platinum-containing catalyst would reduce cracking and coke formation. The resulting bi-metallic and tri-metallic catalysts facilitate a lower operating pressure without fostering hydrocracking conditions. Earlier reforming pressures ranged around 500 psig; with improved catalyst systems, such operations now use operating pressures of 170–370 psig.[12] Advances in continuous catalyst design permit using operating pressures as low as 50 psig.[12]

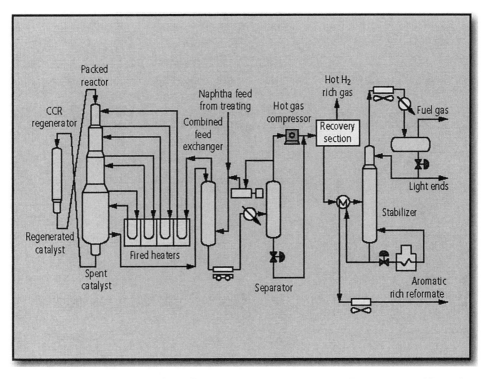

Fig. 18.18. Catalytic reforming—UOP LLC. Includes stacked reactor(s) and regenerator with product separation (Source: *Hydrocarbon Processing, 2004 Refining Process Handbook.* CD-ROM. September 2004: copyright 2004 by Gulf Publishing Co., all rights reserved.)

TABLE 18.7 Favored Operating Conditions for Desired Reaction Rates[23]

Feed	Reaction	Product	Desired rate	To Get Desired Rate Press.	Temp.
Paraffins	Isomerization	Iso-paraffins	Inc.	Inc.	Inc.
	Dehydrocyclization	Naphthenes	Inc.	Dec.	Inc.
	Hydrocracking	Lower mol. wt.	Dec.	Dec.	Dec.
Naphthenes	Dehydrogenation	Aromatics	Inc.	Dec.	Inc.
	Isomerization	Iso-paraffins	Inc.	Inc.	Inc.
	Hydrocracking	Lower mol. wt.	Dec.	Dec.	Inc.
Aromatics	Hydrodealkylation	Lower mol. wt.	Dec.	Dec.	Dec.

Operating temperatures are also critical. The listed reactions are endothermic. The best yields occur along isothermal reaction zones, but are difficult to achieve. Instead, the reaction beds are separated into a number of adiabatic zones operating at 500–1000°F with heaters between stages to supply the necessary energy to promote reaction of heat and hold the overall train near or at a constant temperature. Three or four zones are commonly used to achieve high-octane products.

Catalytic Cracking

This process upgrades heavier products into lighter products, as shown in Fig. 18.19. Catalyst systems are used to catalytically crack the large, heavy hydrocarbons into

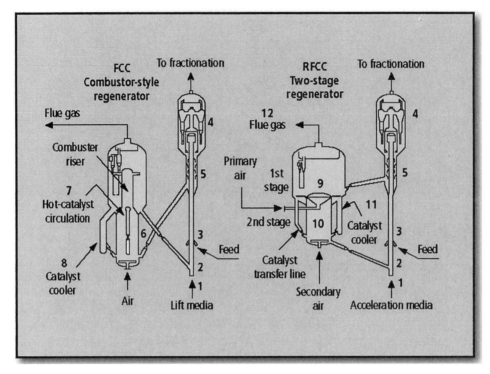

Fig. 18.19. Fluid catalytic cracking—UOP LLC. Combustor style is used to crack process gas oils and moderately contaminated resids, while the two-stage unit is used for more contaminated oils. (Source: *Hydrocarbon Processing, 2004 Refining Process Handbook*. CD-ROM. September 2004: copyright 2004 by Gulf Publishing Co., all rights reserved.)

smaller compounds. The products contain a higher hydrogen-to-carbon ratio than the feed. Consequently, excess carbon is produced, which can deposit on the catalyst and piping and equipment walls; both effects are undesirable.[9]

A typical catalytic cracking unit is shown in Fig. 18.19. The unit comprises two large vessels: one to react feed with hot catalyst and the other to regenerate the catalyst by burning off the carbon deposits with air.[12] The key feature of the catalytic cracking operation is the short contact time between the catalyst and feed to produce the desired product. Extended contact between the feed and catalyst promotes overcracking and the creation of gases. The activity of newer catalysts selectively promotes primary cracking reactions. New catalysts use a distributed matrix structure; zeolites are applied to the surface and pores of the catalyst, thus increasing the number of active sites.[24] With more active sites, the

activity of the catalysts increases and less contact time is necessary.

The short contact time is accomplished by using a transfer line between the regenerator and the reactor vessels. Most of the reaction occurs within the riser section.[9,12,14] A termination device can be used to separate the catalyst from the products that are taken quickly as overhead. The main reactor vessels contain cyclone separators to remove the catalyst from the products and provide additional space for cracking the heavier fraction of the feed.

In the fluidized catalytic cracking (FCC) process, the feed is injected into the reactor through a feed-nozzle system and mixed with the catalyst. The atomized oil mixes with the catalyst and ascends the riser. The cracking process—*riser cracking* or short-time contacting—has several advantages. This system can operate at high temperatures, thus promoting the conversion of feed into gasoline

TABLE 18.8 Typical Gasoline Pool Composition of a Refinery[25]

Gasoline Blendstocks	Percent of Pool Volume	Percent of Pool Sulfur
Alkylate	12	—
Coker naphtha	1	1
Hydrocracked naphtha	2	—
FCC naphtha	36	98
Isomerate	5	—
Light straight-run naphtha	3	1
Butanes	5	—
MTBE	2	—
Reformate	34	—
Total	100	100

Reprinted with permission of Gulf Publishing Co., 1999. Copyright, all rights reserved.

and olefins. It minimizes the destruction of any aromatics formed during cracking. The net effect is gasoline production with two to three higher octane.

The catalyst is regenerated at high temperatures (1300–1400°F). Coke that is deposited on the catalyst is quickly burned off with high-temperature air. Newer catalysts are rugged and can withstand the rigors of extreme heat and fluidizing.

The catalytic cracking unit is often referred to as the gasoline workhorse of a refining unit. As shown in Fig. 18.9, feeds to the catalytic cracking unit are gas oils from the atmospheric and vacuum distillation columns and delayed coker. These heavier fractions also carry metals such as nickel, vanadium, and iron. More important, sulfur compounds concentrate in the heavier product fractions. Table 18.8 lists a typical mass balance for sulfur.[25] FCC blendstocks comprise 36 percent of the volume of the gasoline pool. However, this stream also contributes 98 percent of the sulfur concentration to blended procucts.[25] As specifications on sulfur concentrations in diesel and gasoline tighten, more efforts are focused on how feeds and product streams from the FCC are pre- and posttreated for sulfur concentrations.

Coking

Coking is an extreme form of thermal cracking. This process converts residue materials—

products that are pumpable but not easily changed into lighter products through catalytic cracking. Coking is a less expensive method to convert these residual fractions into lighter products. In the coking process, the coke is considered a by-product; its creation is tolerated in the interest of converting the bulk of the residuals into lighter products. The by-product coke can be sold as feedstock to power-generating utility companies. Electricity utility operations burn clean coke to generate high-pressure steam and power. Thus, refiners can sell coke to their over-the-fence power companies and, in return, purchase steam and electricity.[26]

A typical flow diagram for a delayed coker is shown in Fig. 18.20. Several processing configurations are possible. In this example, the feed is sent directly to the product fractionator to pick up heavier products to be recycled to the cracking operation. The term "delayed coker" indicates that the furnace adds the heat of cracking, and the cracking occurs during the long residence time in the coking drums. The feed and recycled products are heated by the coker heater to the desired operating temperature (900–950°F) and then sent to the coking drum, where partial vaporization occurs in addition to mild cracking at pressures ranging between 15 and 90 psig. Overhead vapors from the coke drum are sent to the fractionator and separated into lighter products such as refinery fuel gas, LPG, (coker) naphtha, and light and heavy gas oils.[12] The by-product coke accumulates in the coke drum.

The coking process uses several parallel coking drums. One drum is online while the other is being emptied. At the end of the coking cycle, the coke is steamed to remove any residual oil-liquid.[12,14] This mixture of steam and hydrocarbons is sent to the fractionator to recover hydrocarbons. The drum is cooled with water and then drained. After the coke drum is unloaded, high-pressure water jets are used to cut away the coke from the drum. Conveyors are used to move the coke to storage.

Fluid coking is a proprietary name given to a different type of coking process. In this process the coke is suspended as particles in

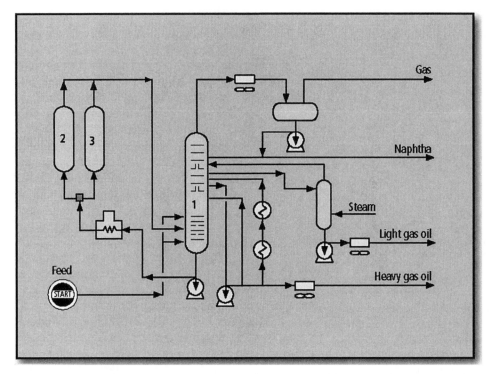

Fig. 18.20. Coking—Foster Wheeler and UOP LLC. Includes feed/product fractionator (1), coke drums (2,3), and vapor recovery. (Source: *Hydrocarbon Processing, 2004 Refining Process Handbook*. CD-ROM. September 2004: copyright by Gulf Publishing Co., all rights reserved.)

fluids that flow from a reactor to a heater and back again. When the coke is gasified, the process is called *flexicoking*. Both fluid coking and flexicoking are proprietary processes of ExxonMobil Research & Engineering Co.

A flow diagram for the flexicoking process is shown in Fig. 18.21. The first two vessels are typical of fluid coking, in which part of the coke is burned in the heater to provide hot coke nuclei to contact the feed in the reactor vessel. The cracked products are quenched in an overhead scrubber where entrained coke is returned to the reactor. Coke from the reactor circulates to the heater where it is devolatilized to yield light hydrocarbon gas and residual coke. A sidestream of coke is circulated to the gasifier, where, for most feedstocks, 95 percent or more of the gross coke is gasified at elevated temperature with steam and air. Sulfur entering the system is converted to hydrogen sulfide, exits the gasifier, and is recovered by a sulfur-removal step.

Hydrocracking

Before the late 1960s, most hydrogen used in processing crude oil was for pretreating catalytic reformer feed naphtha and for desulfurizing middle-distillate products. Later, sulfur requirements for fuels were lowered and became an important consideration. The heavier fractions of crude oil are the fractions with the highest sulfur concentrations and are more difficult to treat. With a constant decline in demand for heavy fuel oils, refiners needed to convert heavier fractions into lighter products. Thus, hydrocracking became a possible solution to the problem.

Figure 18.22 is a typical flow diagram of a hydrocracking process. The process is similar to hydrotreating. The feed is pumped to operating pressure, mixed with hydrogen-rich gas, heated, passed through a catalytic reactor, and distributed among various fractions. Yet this process significantly differs from hydrotreating. In the hydrocracking process, operating pressures are very high: 1500–3500 psia.

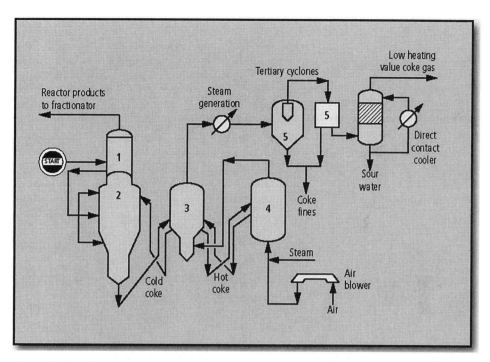

Fig. 18.21. Fluid coking (Flexicoking)—ExxonMobil Research and Engineering Co. Includes: reactor (1), scrubber (2), heater (3), gasifier (4), and coke fines (5). (Source: *Hydrocarbon Processing 2004 Refining Process Handbook.* CD-ROM. September 2004: copyright 2004 by Gulf Publishing Co., all rights reserved.)

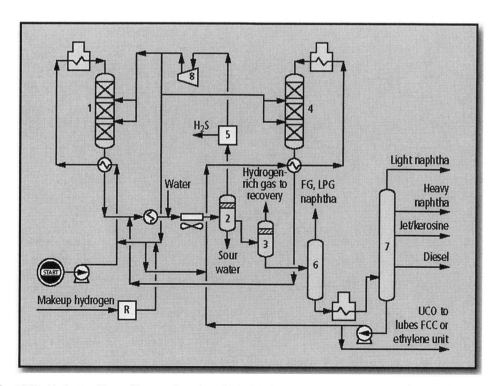

Fig. 18.22. Hydrocracking—Chevron Lummus Global LLC. Includes: staged reactors (1,4), HP separators (2,5), recycle scrubber (3), LP separator (6), and fractionation tower (7). (Source: *Hydrocarbon Processing 2004 Refining Process Handbook.* CD-ROM. September 2004: copyright 2004 by Gulf Publishing Co., all rights reserved.)

Hydrogen consumption is also greater: 1200–1600 scf of hydrogen per barrel of feed, depending on the extent of cracking.[19] If the refinery has a high hydrogen demand due to hydrocracking needs, construction of an onsite hydrogen plant may be necessary.

Hydrocracking catalysts perform a dual function. They drive both hydrogenation and dehydrogenation reactions and have a highly acidic support to foster cracking reactions. The hydrogenation–dehydrogenation components of the catalyst are metals such as cobalt, nickel, tungsten, vanadium, molybdenum, platinum, palladium, or a combination of these metals. The acidic support can be silica–alumina, silica–zirconia, silica–magnesia, alumina–boria, silica–titania, acid-treated clays, acidic-metals phosphates, or alumina, to name a few.[27]

Greater flexibility is attributed to most hydrocracking processes. Under mild conditions, the process can function as a hydrotreater. Under severe conditions (high pressure and temperatures) this process can produce a variety of motor fuels and middle distillates, depending on the feedstock and operating variables. Even greater flexibility is possible if the process is tailored to convert naphthas into liquefied petroleum gases or convert heavy residues into lighter products.

Hydrocracking is a swing process; it is a treater and a cracker. Thus, this process function can be incorporated into a number of different places within a refining scheme. As a cracker, it can convert feeds that are too heavy or too contaminant-laden to go to catalytic cracking. As a treater, it can handle high boiling-point fractions such as heating oil and saturate this fraction to provide good burning quality.

With pending low-sulfur fuel specifications, hydrocracking efforts will be increased to break complex hydrocarbon compounds and expose embedded sulfur molecules. Notably hydrocracking significantly upgrades feeds to downstream processes and fuel products, especially diesel products.

Alkylation

Another method to convert light olefins into gasoline-blending stocks is alkylation. In this process, light olefins—propylene, butylenes, and amylenes with isobutane—are reacted in the presence of strong acids to form branched chain hydrocarbons. These branched hydrocarbons, often referred to as alkylate, have a high-octane value; thus, it is an excellent contributor to the octane pool.[9,12,14,16]

A flow diagram of an alkylation unit using sulfuric acid is shown in Fig. 18.23. Alkylation traditionally combines isobutane with propylene and butylene using an acid catalyst, either hydrofluoric (HF) acid or sulfuric acid. The reaction is favored by high temperatures, but competing reactions among the olefins to give polymers prevent high-quality yields. Thus, alkylation is usually done at low temperatures to deter polymerization reactions. Temperatures for HF acid-catalyzed reactions are approximately 100°F, and for sulfuric acid they are approximately 50°F.[14] Notably, some acid loss occurs with this process. Approximately 1–1.2 lb of HF acid/bbl of alkylate is consumed, and 25–30 lb of sulfuric acid/bbl of alkylate is consumed. The alkylation feed should be dried and desulfurized to minimize acid loss. Because the sulfuric-acid-catalyzed reactions are carried out below normal atmospheric temperatures, refrigeration facilities are needed.

As shown in Fig. 18.23, dry liquid feed containing olefins and isobutane is charged to a combined reactor–settler. In this example, the reactor uses the principle of a differential gravity head to circulate through a cooler before contacting a highly dispersed hydrocarbon feed in the reactor pipe. The hydrocarbon phase, generated in the settler, is sent to a fractionator, which separates LPG-quality propane, isobutane recycle, n-butane, and alkylate products. A small amount of dissolved catalyst is also removed from the propane product by a small stripper tower.

Environmental and safety concerns regarding acid-based processes are promoting research and development efforts on solid-acid alkylation processes. Liquid catalysts pose possible risks to the environment, employees, and the general public from accidental atmospheric releases. Also, these acid

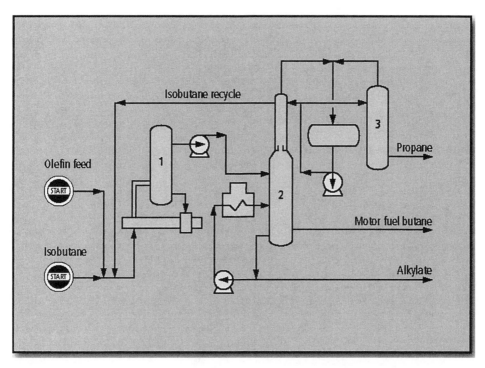

Fig. 18.23. Alkylation—Technology Solutions Division of ConocoPhillips. Include combination reactor/settler (1), main fractionator (2), and small propane stripper (3). (Source: *Hydrocarbon Processing, 2004 Refining Process Handbook.* CD-ROM. September 2004: copyright 2004 by Gulf Publishing Co., all rights reserved.)

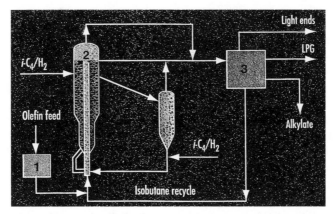

Fig. 18.24. Alkylation—UOP LLC. Solid Catalyst System removes impurities from the feed (1), clean feed, recycled isobutene, and catalyst and feed to main reactor (2), and final product are separated in fractionation section (3). (*Hydrocarbon Processing*, 79, No. 11. Nov. 2000: copyright 2000 by Gulf Publishing Co., all rights reserved.)

catalysts must be regenerated, another reliability and safety issue. Thus, research efforts are directed at investigating other methodologies to produce high-octane alkylation gasoline component streams.

UOP LLC has developed two alternate processes for liquid-acid alkylation. The direct alkylation method, Alkylene, uses a packed moving catalyst bed. The feed is pretreated to remove impurities such as diolefins, sulfur, oxygen, and nitrogen compounds. These components suppress catalyst activity and can also permanently deactivate the catalyst. As shown in Fig. 18.24, the olefinic feed

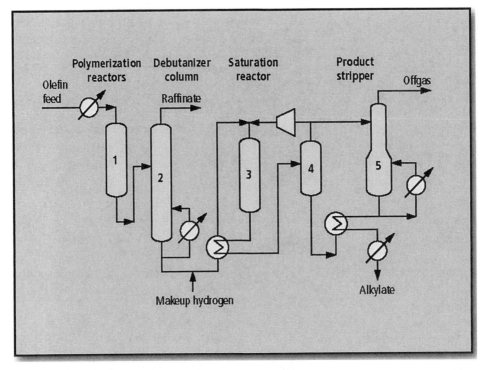

Fig. 18.25. Alkylation (indirect)—UOP LLC. Process uses solid catalyst; major processing steps include: polymerization (1), hydrogenation reactor (2), hydrogen recovery (3), and product stabilization (4). (Source: *Hydrocarbon Processing, 2004 Refining Process Handbook.* CD-ROM. September 2004: copyright 2004 by Gulf Publishing Co., all rights reserved.)

and isobutane are combined and injected at the bottom of the riser, where the alkylation reaction occurs. At the exit of the riser the catalyst is separated from the hydrocarbons and flows by gravity to the reactivation zone.[12,28] The hydrocarbon stream is sent to the fractionation (distillation) section, where alkylate product is separated from the light paraffins and LPG product. Isobutane is recycled back to the main reactor. For this process, the feed is partially dehydrogenated to remove diolefins.

A second indirect alkylation process, InAlk, is also a solid catalyst process (Fig. 18.25).[12,29] InAlk combines two commercially proven technologies: polymerization and olefin saturation. Isobutylene is reacted with light olefins (C_3-C_5) in a polymerization reactor. The resulting mixture of iso-olefins is saturated in the hydrogenation reactor. Excess hydrogen is recycled and the product is stabilized to produce a paraffinic gasoline blending stream. Yet, new solid-acid alkylation processes face tech-

nical challenges. Solid-acid catalysts remain more difficult to regenerate and have a shorter service life. Research to overcome these operating problems is continuing.

Ether Processes

Refiners have always incorporated ethers into the gasoline pool when needing to increase octane. Ethers provide a high-octane stream with low vapor pressure. Beginning in 1995, United States reformulated gasoline (RFG) was required to have 2 wt.% oxygenate content. The choice of oxygenate was left to the refiners' discretion. Initially, the ethanol industry had hoped that ethanol would be selected as the primary oxygenate for RFG. However, refiners searched for other options. Methyl tertiary butyl ether (MTBE) became the oxygenate of choice for blending RFG. It is produced by reacting methanol with isobutylene, as shown in Fig. 18.26.[38] Other ether compounds can be made by a similar

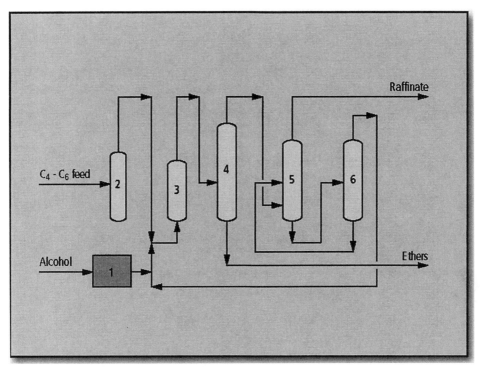

Fig. 18.26. Ethers—Axens. Includes: alcohol purification (1), hydrocarbon purification (2), main reactor (3), reactive distillation system (4), wastewater system to remove water from the raffinate (5), and product fractionation column (6). (Source: *Hydrocarbon Processing, 2004 Refining Process Handbook.* CD-ROM. September 2004: copyright 2004 by Gulf Publishing Co., all rights reserved.)

process. Ethyl tertiary butyl ether (ETBE) is produced from ethanol and isobutylene; tertiary amyl methyl ether (TAME) is made from methanol and isoamylene. MTBE holds the lion's share of the oxygenate market.

Presently, MTBE's usage is at the center of a highly emotional, political struggle. As refiners began using MTBE at the 2 wt.% concentration, this ether was detected at trace levels in drinking water supplies in areas required to use RFG. Leaking underground storage tanks (USTs) were identified as the primary source for MTBE found in drinking water. The ether is found in water due its high affinity for water. Legislation was (and is) in place that required UST owners to install safeguards to prevent leakage; however, leakage did occur.

Because drinking water was affected by leakage from USTs, a political effort is underway to ban the use of MTBE in gasoline. At the time of this writing, much debate is ongoing. The State of California and several other states have banned using MTBE in RFG. However, the consequence of shortages in RFG supplies has raised some doubts and caused second thoughts on the MTBE ban.

Oxygenate usage is also required for gasolines supplied to nonattainment areas. The sister ethers—ETBE and TAME—are under the same disfavor as MTBE and are not logical replacements. Ethanol can be a substitute for MTBE; however, it also brings other problems. Ethanol usage raises the Reid vapor pressure of the fuel, which is also limited by fuel specifications. In the United States, ethanol receives an agricultural subsidy that elevates its price. In the United States market, ethanol capacity is located far from refining centers and fuel terminals; thus, transportation, distribution, and storage issues are challenges for ethanol.[30] Of course, the ultimate fate of MTBE capacity constructed in the mid-1990s to meet RFG blending demand is yet to be determined.

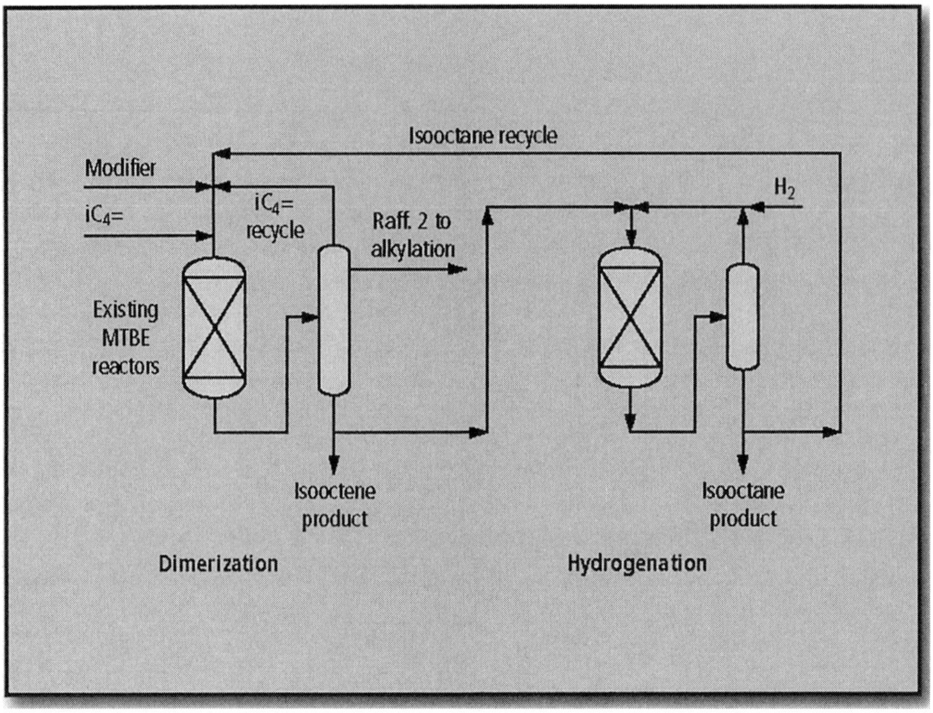

Fig. 18.27. Isooctane—Lyondell Chemical and Aker Kvaerner. Includes dimerization section and hydrogenation section. (Source: *Hydrocarbon Processing, 2004 Refining Process Handbook*. CD-ROM. September 2004: copyright 2004 by Gulf Publishing Co., all rights reserved.)

Should MTBE be banned, what would be the logical replacement(s)? There are several options available. Several refiners opted to build MTBE capacity and avoid purchasing the ether on the open market. MTBE units were an option to use the facility's isobutylenes. Several licensed processes can be used to convert existing MTBE units. Kvaerner and Lyondell Chemical Co. offer technologies to convert an MTBE unit to produce iso-octane, as shown in Fig. 18.27.[12] Snamprogetti SpA and CDTECH also have an iso-octene/iso-octane process. These processes can use various feedstocks such as "pure" iso-butane, steam-cracked C_4 raffinate, 50/50 iso-butane/isobutene feeds, and FCC butane–butane streams. The process selectively dimerizes C_4 olefins to iso-octene and then hydrogenates the iso-octene (di-iso-butene) into iso-octane. The processes were developed to provide an alternative to MTBE. The dimerization reactor uses a catalyst similar to that for MTBE processes; thus, the MTBE reactor can easily be converted to

iso-octane service. The product is low sulfur with 100 octane number, great for gasoline blending.[31]

FUTURE TRENDS

The refining industry is constantly required to meet cleaner fuel specifications. The only certainty is that, globally, fuels are becoming cleaner. Yet, clean fuels comprise a very broad spectrum that is totally dependent on the market served. In Asia and some parts of Europe, lead removal still remains a key issue. These refiners strive to maintain octane while phasing out lead. In other markets, more complex issues remain to be solved.

In developed markets, cleaner-fuel issues revolve around several product specifications. The fuels market is at various stages of removing sulfur and aromatics from gasoline and diesel. For global organizations, formulating a clean-fuels agenda is a very market-dependent issue, as shown in Table 18.4.

Costs and Drivers

So how much investment will be made to process low-sulfur fuels? A recent Energy Information Administration (EIA) report estimates that U.S. refiners will invest $6.3–9.3 billion to reach full compliance with the ultra-low sulfur diesel (ULSD) rule through 2011.[33] For Europe, refiners made investments of nearly $22.9 billion from 1997 to 2005 to meet gasoline and diesel specifications. From 2005 to 2015, an additional $9.7–14 billion may be spent on improving the quantity and quality of middle distillates in Europe.[15] Thus, considerable investment will be necessary to bring cleaner fuels to market.

A true unknown, beside how much a clean-fuels program will cost individual organizations, is how many refiners will make such an investment. Developed markets suffer from diminished demand growth for products. Mediocre returns from earlier environmental projects taint possible returns on future spending. With such a backdrop, hesitancy on expected spending is anticipated. Yet, the deadlines draw nearer, and plans must be formulated.

Drivers for investment are directly linked to market demand. In the European Union, gasoline demand is declining, whereas demand for diesel is increasing with modest increases for treating gas oil. Heavy fuel oil demand is also declining. Under such market forces, the E.U. product market is shifting and demand for middle distillates is increasing from 310 million tons (310 MMt) in 1997 to 390 MMt in 2015.[15] Production of LPG, naphtha, and other products is expected to increase.[15]

Under such conditions, the focus of retrofitting and revamping existing units will be optimizing middle-distillate production. Consequently, to meet higher quality requirements for middle distillates, E.U. refiners are projected to invest in hydrocracking capacity. Such investments may include standalone units and moderate conversion of units upstream of the FCC.[15] Refiners will also raise hydrotreating capability.

Technology Options

Due to the individuality of each refinery, multiple solutions are available. Selection will be directed toward final product slates for each facility. For gasoline-oriented facilities, several options are available. These are discussed below.

Sweetening. This application is most effective in treating straight-run (SR) gasoline streams. Amines are used to remove mercaptan species from the hydrocarbon stream. Caustic converts the mercaptans into disulfides, which are extracted by gravity separation or an extractive solvent.[12]

Hydrotreating. This treatment is used widely throughout the refinery. Hydrogen is reacted with the processing stream with a catalyst to remove sulfur compounds. Several licensed technologies are available at varying temperature and pressure ranges (Table 18.9).[34–36] Hydrotreating, depending on the severity of process conditions, effectively removes sulfur, nitrogen, metals, carbon residue, and asphaltene from the hydrocarbon stream. Mild hydrotreating—low pressure and temperature—removes sulfur and trapped metals and is done as a cleanup/upgrade step for downstream processes such as hydrocrackers, FCCs, resid catalytic crackers, and cokers. Processing streams typically treated include: naphthas, kerosenes, distillates, and gas oil (Fig. 18.9). This process consumes hydrogen.

For a conventional hydrotreating process, the process stream is heated and mixed with hydrogen. This stream is charged to a reactor filled with a high-activity catalyst. Single and multiple reactors can be used. Excess hydrogen is used. The reactor effluent is cooled and separated, and the hydrogen-enriched gas is recycled. Depending on sulfur levels of the feed, the hydrogen recycle may be amine scrubbed to remove hydrogen sulfide (H_2S). The liquid product is steam-stripped to remove lighter components and residual H_2S, and/or fractionated into multiple products.[12]

TABLE 18.9   Available Licensed Post-treating Technologies[34-36]

Approach	*Process Name*	*Technology Provider*
Conventional hydrotreating	Hydrotreating	Many
Selective hydrotreating	Scanfining	ExxonMobil
	Prime G	IFP
	Octgain 125	ExxonMobil
	Octgain 220	ExxonMobil
	ISAL	UOP LLC
Catalytic distillation	CDHydro/CD HDS	CD Tech
Adsorption	S Zorb SRT	Phillips Petroleum
Olefinic alkylation	Olefin alkylation of thiophenic sulfur	BP
Extractive mass	Exomer	ExxonMobil and Merrichem

As sulfur levels for products continue to decrease, refiners strive to systematically remove sulfur from the blending pool. Notably, back integration to strategically and systematically remove sulfur compounds throughout the processing scheme is needed. Lower sulfur specs now mandate removing complex sulfur species from various blending streams.

Hydrotreating is an effective method to desulfurize products. However, as desulfurization requirements continue to increase, newer, high-activity catalyst systems will be required to meet desulfurization targets. Existing capacity can be retrofitted with better catalyst systems. Improved reactor internals can more efficiently distribute feed throughout the reactor, optimize contact of reactants with the catalyst, and increase desulfurization efficiency. Refiners may elect to raise operating severity by increasing operating temperatures. However, higher operating temperatures can reduce the service life of the catalyst. Raising operating pressures is not viable and will depend on the pressure rating of the existing reactor. Yet, the new fuel specifications may warrant installing new reactor capacity.

As processing conditions increase, that is pressure and temperature, the hydrotreating process emerges more as a hydrocracking/ hydrodesulfurization/hydrodearomatizing event. Deeper desulfurization of processing/ product streams is inevitable.

Sulfur Segregation.   Another option is to concentrate sulfur compounds into various streams and selectively treat them. Refiners can undercut* product; however, such tactics will reduce yields. For diesel, undercutting will lower diesel yield and increase gas-oil products, an undesirable consequence.

Desulfurization.   As refiners strive to meet tighter restrictions on sulfur in product and blending streams, desulfurization technologies are attracting more interest. For gasoline, licensed posttreatment processes that are targeted at specific streams have high interest. Half the battle for clean fuels is identifying the highest sulfur contributors to the blending pool. For gasoline-oriented facilities, the FCC unit is the highest contributor. Nearly 95–98 percent of the sulfur present in blended gasoline is linked to FCC product cuts. Notably, FCC technology is instrumental in upgrading heavier refining streams into desired products. The school of thought for FCC is split between pretreating FCC feed and posttreating. Pretreating offers some great benefits. In particular, hydrotreating feed streams to the FCC upgrades the feed by removing sulfur and nitrogen compounds and saturates some

---

*Undercutting is using distillation temperatures that are below the specification for the product. The refiner loses some light product to higher-temperature products as insurance to meet light-product specifications. It is product "give-away" and not a good practice.

aromatics. It can increase LPG and gasoline production, while reducing regenerator-SO_x emissions, light-cycle oil (LCO), and clarified-slurry oil (CSO) yields, and minimizing coke formation.[25,37] More important, it is reported that hydrotreating does not affect naphtha octane values.[25] However, hydrogen consumption becomes a limiting issue. The available hydrogen balance will affect how much hydrotreating can be done economically.

Another pretreat option is to replace hydrotreating the FCC feed with partial conversion hydrocracking operations. There are several variations to the process. However, partial conversion hydrocracking operation splits the FCC feed. Difficult-to-convert materials such as LCO and coker gas oils can be upgraded with hydrotreating and hydrocracking before being sent to the FCC unit. UOP LLC's Unicracking process uses two reactors and separates the hydrotreating and hydrocracking processes into distinct zones, as shown in Fig. 18.28. The sulfur content determines the severity of the hydrotreating reactor. This process is estimated to yield more naphtha and distillates than other desulfurization methods.[25] It also produces a higher quality diesel (cetane index of 50).

The Axen's mild hydrocracking process combines an ebullated-bed, mild hydrocracking process with an inline fixed-bed to pretreat FCC feed. The T-Star process can treat vacuum gas oil (VGO) and deasphalted oils, and offers high selectivity toward diesel products. The efficient catalyst system enables more flexibility for processing various FCC feeds. The feed streams are cracked first and then hydrotreated.

Post-treating processes hold keen interest also. Table 18.9 lists several post-treating methods. Many of the newer processes recently have been commercially demonstrated and several commercial units have been completed and are now in operation. How effective these new processes will prove to be remains to be seen.

Tomorrow's fuels will contain less sulfur to be compatible with the more sophisticated engine designs of new automobiles. Fuels specifications will further reduce the concentrations of gasoline compounds that are listed and/or considered to be toxic. Refiners will continue to upgrade heavier components to lighter products and refineries will continue to use innovative catalysts and processing equipment to cost-effectively manufacture fuels and petrochemical feedstocks. We have barely touched on the sophisticated engineering needed to transform a barrel of crude oil into consumable products.

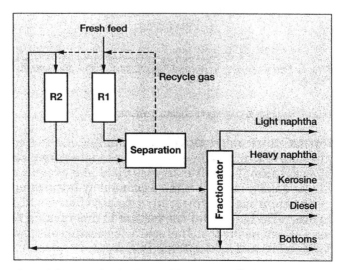

Fig. 18.28. Two-staged, partial-conversion hydro cracking process.[25] (Copyright 1999 by Gulf Publishing Co. and reprinted by permission of the copyright owner. All rights reserved.)

REFERENCES

1. U.S. Energy Information Administration, International Energy Annual 2002, www.eia.doe.gov/oiaf/ieo.
2. Cambridge Energy Research Associates (CERA), *Global Oil Trends 2002*, 2001.
3. National Petroleum Council, "U.S. Petroleum Refining: Assuring the Adequacy and Affordability of Cleaner Fuels," June 20, 2000.
4. *Hydrocarbon Processing*, 34-D–35-D (Feb. 2002).
5. American Society for Testing and Materials, *1990 Annual Book of ASTM Standards*, Vols. 05.01 and 05.02, 1990.
6. Watson, K. M., and Nelson, E. F., "Improved Methods for Approximating Critical and Thermal Properties of Petroleum Fractions," *Ind. Eng. Chem.*, 25, 880 (1933).
7. Hoffman, H. L., "Sour Crudes Limits Refining Output," *Hydrocarbon Processing*, 107–110 (Sept. 1973).
8. Ferrero, E. P., and Nichols, D. T., "Analyses of 169 Crude Oils from 122 Foreign Oil Fields," U.S. Department of Interior, Bureau of Mines, Information Circular 8542, 1972.
9. Eckholm, E., *Refining Fundamentals: Part 2 Light Fraction Processing*, February 1999.
10. *Hydrocarbon Processing*, 13–16 (Sept. 2000).
11. Unzelman, G. H., and Forster, E. J., "How to Blend for Volatility," *Petroleum Refiner*, 109–140 (Sept. 1960).
12. "Refining Processes 2002," *Hydrocarbon Processing*, CD-ROM, September 2004.
13. Huges, T. R., et al., "To Save Energy When Reforming," *Hydrocarbon Processing*, 75–80 (May 1976).
14. Meyers, R. A. (Ed.), *Handbook of Petroleum Refining Processes*, 2nd ed., McGraw-Hill, New York, 1997.
15. Birch, C. H., and Ulivier, R., "ULSG Diesel Refining Study," Purvin & Gertz, Nov. 17, 2000.
16. Eastman, A., et al., "Consider Online Monitoring of HF Acid When Optimizing Alkylation Operations," *Hydrocarbon Processing* (Aug. 2001).
17. "Petrochemical Processes 2001," *Hydrocarbon Processing*, 71–146 (Mar. 2001).
18. Golden, S. W., et al., "Use Better Fractionation to Manage Gasoline Sulfur Concentration," *Hydrocarbon Processing*, 67–72 (Feb. 2002).
19. Corneil, H. G., and Forster E. J., "Hydrogen For Future Refining," *Hydrocarbon Processing*, 85–90, (Aug. 1990).
20. Ratan, S., and Vales, C. F., "Improve your Hydrogen Potential," *Hydrocarbon Processing*, 57–64 (Mar. 2002).
21. Skiflett, W. K., and Krenzke, L. D. "Consider Improved Catalyst Technologies to Remove Sulfur," *Hydrocarbon Processing*, 41–43 (Feb. 2002).
22. American Petroleum Institute, *Technical Data Book*, 4th ed., Publ. No. 999, 1983.
23. Jenkins, J. H., and Stephens, J. W., "Kinetics of Cat Reforming," *Hydrocarbon Processing*, 163–167 (Nov. 1980).
24. "HP Innovations," *Hydrocarbon Processing*, p. 33 (Nov. 2000).
25. Shorey, S. W., et al., "Use FCC Feed Pretreating Methods to Remove Sulfur," *Hydrocarbon Processing*, 43–51 (Nov. 1999).
26. *Hydrocarbon Processing*, 15–16 (Dec. 1999).
27. Sullivan, R. F., and Meyer, J. A., "Catalysts Effects on Yields and Product Properties in Hydrocracking," American Chemical Society, Philadelphia, April 6–11, 1975.
28. UOP, Product sheet—Alkylene, 1999.
29. UOP, Product sheet—Indirect Alkylation (InAlk), 1999.
30 "HP Insight," *Hydrocarbon Processing*, p. 13 (Apr. 2002).
31. Tsai, M. J., et al., "Consider New Technologies to Replace MTBE," *Hydrocarbon Processing*, 81–88 (Feb. 2002).
32. Heck, R. M., et al., "Better Use of Butenes for High-octane Gasoline," *Hydrocarbon Processing*, 185–191 (Apr. 1980).
33. EIA, "The Transition to Ultra-low Sulfur Diesel: Effects on Process and Supply," May 2001.
34. Avidan, A., et al., "Improved Planning can Optimize Solutions to Produce Clean Fuels," *Hydrocarbon Processing*, 47–53 (Feb. 2001).
35. "2001 Clean Fuels Challenge," transcript-Question 2, NPRA Clean Fuels Conference, Houston, Aug. 28–29, 2001.
36. Fredrick, C., "Sulfur Reductions: What Are the Options?," *Hydrocarbon Processing*, 45–50 (Feb. 2002).
37. Nocca, J. L., et al., "The Domino Interaction of Refinery Processes for Gasoline Quality Attainment," NPRA 2000 Annual Meeting, San Antonio, March 26–28, 2000.
38. "Petrochemical Processes 1995," *Hydrocarbon Processing*, 109–111 (Mar. 1995).

Index